HANDBOOK OF APPLIED INSTRUMENTATION

OTHER McGRAW-HILL HANDBOOKS OF INTEREST

AMERICAN INSTITUTE OF PHYSICS · American Institute of Physics Handbook
AMERICAN SOCIETY OF MECHANICAL ENGINEERS · ASME Handbooks:
   Engineering Tables   Metals Engineering—Processes
   Metals Engineering—Design   Metals Properties
AMERICAN SOCIETY OF TOOL AND MANUFACTURING ENGINEERS:
   Die Design Handbook   Handbook of Fixture Design
   Manufacturing Planning and   Tool Engineers Handbook
     Estimating Handbook
ARCHITECTURAL RECORD · Time-saver Standards
BEEMAN · Industrial Power Systems Handbook
BRADY · Materials Handbook
CARROLL · Industrial Instrument Servicing Handbook
CONDON AND ODISHAW · Handbook of Physics
CONSIDINE · Process Instruments and Controls Handbook
CROCKER · Piping Handbook
DAVIS · Handbook of Applied Hydraulics
DUDLEY · Gear Handbook
FACTORY MUTUAL ENGINEERING DIVISION · Handbook of Industrial Loss
   Prevention
FLÜGGE · Handbook of Engineering Mechanics
HARRIS · Handbook of Noise Control
HARRIS AND CREDE · Shock and Vibration Handbook
HEYEL · The Foreman's Handbook
HUSKEY AND KORN · Computer Handbook
JURAN · Quality Control Handbook
KALLEN · Handbook of Instrumentation and Controls
KING AND BRATER · Handbook of Hydraulics
KNOWLTON · Standard Handbook for Electrical Engineers
KOELLE · Handbook of Astronautical Engineering
KORN AND KORN · Mathematical Handbook for Scientists and Engineers
LASSER · Business Management Handbook
LAUGHNER AND HARGAN · Handbook of Fastening and Joining of Metal Parts
LeGRAND · The New American Machinists' Handbook
MAGILL, HOLDEN, AND ACKLEY · Air Pollution Handbook
MANAS · National Plumbing Code Handbook
MANTELL · Engineering Materials Handbook
MARKS AND BAUMEISTER · Mechanical Engineers' Handbook
MAYNARD · Industrial Engineering Handbook
MAYNARD · Top Management Handbook
MORROW · Maintenance Engineering Handbook
PERRY · Chemical Business Handbook
PERRY · Chemical Engineers' Handbook
PERRY · Engineering Manual
ROSSNAGEL · Handbook of Rigging
ROTHBART · Mechanical Design and Systems Handbook
SHAND · Glass Engineering Handbook
STANIAR · Plant Engineering Handbook
STREETER · Handbook of Fluid Dynamics
TRUXAL · Control Engineers' Handbook

HANDBOOK OF APPLIED INSTRUMENTATION

DOUGLAS M. CONSIDINE, *Editor-in-Chief*

Hughes Aircraft Company
Culver City, Calif.

S. D. ROSS, *Associate Editor*

The Michener Company
Philadelphia, Pa.

McGRAW-HILL BOOK COMPANY

New York   San Francisco   Toronto   London

HANDBOOK OF APPLIED INSTRUMENTATION

12426

CONTRIBUTORS

R. K. ABELE, B.S. (E.E.), *Supervising Engineer, Radiation Detection Group, Instrumentation and Controls Division, Oak Ridge National Laboratory, Oak Ridge, Tenn.* (*Nuclear-radiation Detectors*)

C. M. ALBRIGHT, JR., M.S., *Director, Pioneering Research and New Product Development, Remington Arms Co., Inc., Bridgeport, Conn.* (*Analysis Instruments*)

W. FAY ALLER, *Vice President, Engineering, The Sheffield Corporation, a subsidiary of The Bendix Corporation, Dayton, Ohio.* (*Parts Dimension Measurement and Control*)

VIGGO O. ANDERSEN, B.S. (E.E.), *Staff Engineer, Non-Linear Systems, Inc., Del Mar, Calif.* (*Basic Electricity and Electronics for Instrumentation Engineers*)

ROBERT F. BARBER, B.A. (M.E.), M.S. (Ind. Eng.), *Manager, Pulp and Paper Division, The Foxboro Company, Foxboro, Mass.* (*Evaporator Control Systems*)

ROBERT BARCLAY BEAHM, B.S. (Agr. Eng.), M.S. (Agr. Eng.), *Systems Engineer, Taylor Instrument Companies, Rochester, N.Y.* (*Food Industry Instrumentation*)

PAUL F. BEACHBERGER, B.E.E., M.Sc., *Manager, Advanced Systems and Development Department, Eclipse-Pioneer Division, The Bendix Corporation, Teterboro, N.J.* (*Aircraft and Aerospace Vehicle Instrumentation*)

CLYDE BERG, Ph.D. (Ch.E.) *Director, Clyde Berg Associates, Long Beach, Calif.* (*Solids Level*) *and* (*Pilot-plant Instrumentation*)

L. BERTRAND, B.S. (Ch.E.), *Engineering Service Division, Engineering Department, E. I. du Pont de Nemours & Company, Inc., Wilmington, Del.* (*Distillation-column Control*)

DON G. BLODGETT, B.S. (E.F.), *Superintendent of Electrical System Performance, The Detroit Edison Company, Detroit, Mich.* (*Electric Power Generation and Distribution Control*)

DONALD C. BRUNTON, Ph.D. (Physics), *Director of Research, Industrial Nucleonics Corp., Columbus, Ohio.* (*Nuclear-radiation Detectors*)

ALLAN L. BURTON, B.A., M.A. (Physics), *Director of Research, Veeder-Root, Inc., Hartford, Conn.* (*Counters and Digital Indicating Devices*)

NATHAN COHN, B.S. (E.E.), *Vice President—Technical Affairs, Leeds & Northrup Company, Philadelphia, Pa.* (*Electric Power Generation and Distribution Control*)

D. M. CONSIDINE, B.S., *Manager, Marketing Planning, Hughes Aircraft Company, Culver City, Calif.* (*Definition and Classification of Variables*)

N. S. COURTRIGHT, M.S. (M.E.), *Manager of Education, Bailey Meter Company, Cleveland, Ohio.* (*Steam Power Plant Instrumentation*)

L. E. CUCKLER, *Product Manager, Industrial Instruments, Aeronautical and Instrument Division, Robertshaw-Fulton Controls Company, Anaheim, Calif.* (*Moisture Content of Materials*)

STEVEN DANATOS, B.S., M.S. (M.E.), *Associate Editor, Engineering Practice, Chemical Engineering, McGraw-Hill Publishing Company, New York, N.Y.* (*Measuring and Transmission Methods*) *and* (*Summary of Controller Types and Final Control Elements*)

W. J. DARMODY, M.E., *Executive Assistant and Liaison Metrologist to National Standards Laboratories, The Sheffield Corporation, a subsidiary of The Bendix Corporation, Dayton, Ohio.* (*Parts Dimension Measurement and Control*)

RICHARD C. DEHMEL, Ph.D., *Vice President—Engineering, Curtiss-Wright Corporation, Wood-Ridge, N.J.* (*Aircraft Flight-simulation Instrumentation*)

J. H. DENNIS, B.S., *Vice President, Manufacturing, Bailey Meter Company, Cleveland, Ohio.* (*Steam Power Plant Instrumentation*)

HENRY S. DRINKER, B.S. (M.E. and E.E.), *Manager, Systems Development Division, The Foxboro Company, Foxboro, Mass.* (*Evaporator Control Systems*)

W. T. DUMSER, B.S. (E.E.), *Supervisory Engineer, Republic Flow Meters Company, a division of Rockwell Manufacturing Company, Chicago, Ill.* (*Instrument Panelboard Design and Construction*)

GERALD L. EBERLY, B.S. (M.E.), *Technical Staff, Harris D. McKinney, Inc., Philadelphia, Pa.* (*Calorific Value*), (*Fluid-density and Specific-gravity Measurement*) *and* (*Conditioning the Instrument Air Supply*)

H. O. EHRISMAN, B.S. (E.E.), *General Sales Manager, The Foxboro Company, Foxboro, Mass.* (*Pulp and Paper Production Instrumentation*)

P. A. ELFERS, B.S., *Director, Fisher Governor Company, Marshalltown, Iowa.* (*Liquid-level Measurement*)

A. K. FALK, B.S. (E.F.), *Interconnection Coordinator, The Detroit Edison Company, Detroit, Mich.* (*Electric Power Generation and Distribution Control*)

W. HARRISON FAULKNER, Jr., B.S. (E.E.), *formerly Vice President, Engineering and Development, Tracerlab Division of Laboratory for Electronics, Inc., Waltham, Mass.* (*Thickness Measurement of Sheet and Web Materials*)

D. B. FISK, *Manager, Timer Engineering, Clock and Timer Department, General Electric Company, Ashland, Mass.* (*Electrical Variables*)

E. ROSS FORMAN, B.S. (M.E.), M.S. (Bus. Adm.), *Senior Engineer, Catalytic Construction Company, Philadelphia, Pa.* (*Fundamentals of Automatic Control Engineering*)

J. W. FRIEDMAN, B.S. (E.E.), *Supervisory Engineer, Panel Division, Minneapolis-Honeywell Regulator Co., Philadelphia, Pa.* (*Preparation of Wiring Diagrams for Instrument Applications*)

G. D. GOODRICH, A.B. (Physics), *Supervisor, Technical Sales, Statham Instruments, Inc., Los Angeles, Calif.* (*Acceleration Measurement*)

JOHN R. GREEN, B.S., *Consultant in Industrial Instrumentation for Metal Producing and Ceramic Industries, Villa de Verde en Tubac, Tumacacori, Ariz.* (*Steel Production Instrumentation*)

F. K. HARRIS, B.A., M.S., Ph.D. (Physics), *Electrical Instruments Section, Electricity Division, National Bureau of Standards, Washington, D.C.* (*Measurement Errors*)

W. F. HICKES, B.S.E. (M.E.), *Assistant to Chief Engineer, The Foxboro Company, Foxboro, Mass.* (*Humidity and Dew Point*)

T. W. HISSEY, B.S. (E.E.), *Assistant Sales Manager, Electric Power, Leeds & Northrup Company, North Wales, Pa.* (*Electric Power Generation and Distribution Control*)

JOSEPH F. HORNOR, B.S. (Ch.E.), *Planning Adviser, Control Systems, General Products Division Development Laboratory, International Business Machines Corporation, San Jose, Calif.* (*Glass and Ceramics Industries Instrumentation*)

W. H. HOWE, B.S. (E.E.), *Chief Engineer, The Foxboro Company, Foxboro, Mass.* (*Definition and Classification of Variables*) *and* (*Factors in Selection of Measurement Methods*)

R. S. HUNTER, A.B., *President, Hunter Associates Laboratory, Inc., McLean, Va.* (*Photometric Variables*)

JOHN JOHNSTON, JR., B.S., *Engineering Manager, Instrument Products Division, E. I. du Pont de Nemours & Company, Inc., Wilmington, Del.* (*Instrumentation Practices in the Process Industries*)

HAROLD C. JONES, B.S.E.E., M.S.E.E., *Electrical Engineering Department, University of Maryland.* (*Environmental Test Instrumentation*)

J. B. JONES, B.S. (**Ch.E.**), *Engineering Service Division, Engineering Department, E. I. du Pont de Nemours & Co., Inc., Wilmington, Del.* (*Distillation-column Control*)

PAUL KAUFMANN, B.S. (**Ch.E.**), *Technical Representative, Engineering Sales, Electronic Associates, Inc., Long Branch, N.J.* (*Applications of Analog Computers*)

DONALD B. KENDALL, B.S., *Manager, Product Engineering Department, Toledo Scale Company, Division of Toledo Scale Corporation, Toledo, Ohio.* (*Force Measurement*) *and* (*Weight and Weight Rate of Flow*)

JEROME KOHL, B.S. (**Ch.E.**), *Coordinator of Special Products, General Atomic Division of General Dynamics Corporation, San Diego, Calif.* (*Radioisotopes in Instrumentation*) *and* (*Basic Electricity and Electronics for Instrumentation Engineers*)

ROBERT C. LANGFORD, Ph.D. (**E.E.**), *Director of Research, Aerospace Group, General Precision, Inc., Little Falls, N.J.* (*Speed Measurement*)

S. A. LAURICH, *Manager, Process Plants Division, Struthers Scientific & International Corporation, Warren, Pa.* (*Crystallizer Instrumentation*)

J. RUSSELL LeROY, M.S. (**Ch.E.**), *Control Systems Coordinator, International Business Machines Corporation, Dallas, Texas.* (*Applications of Digital Computers*)

ALFRED H. McKINNEY, B.S. (**Chem. Eng.**), *Consultant, Instrument Section, Engineering Service Division, Engineering Department, E. I. du Pont de Nemours & Company, Inc., Wilmington, Del.* (*Control of Solids-drying Operations*) *and* (*Process Laboratory Instrumentation*)

JEROME B. McMAHON, B.S. (**M.E.**), *Consulting Engineer, Chicago, Ill.* (*Flow of Fluids*)

H. V. MILES, B.S. (**Ch.E.**), *Manager, Pulp and Paper Development Division, Dorr-Oliver Incorporated, Stamford, Conn.* (*Filtration Instrumentation*)

J. H. MILLER, B.S., *Retired. Formerly Vice President for Research and Engineering, Weston Instruments, a subsidiary of Daystrom, Inc., Newark, N.J.* (*Summary of Temperature-measurement Methods*)

S. B. MOREHOUSE, B.S. (**E.E.**), *Assistant to the Vice-President, Systems Department, Leeds & Northrup Company, North Wales, Pa.* (*Electric Power Generation and Distribution Control*)

JOHN H. MORRISON, B.S. (**C.E.**), *Product Manager, Helicoid Gage Division, American Chain & Cable Company, Inc., New York, N.Y.* (*Pressure and Vacuum Measurement*)

RICHARD MUMMA, B.S. (**M.E.**), *Product Manager, Fischer & Porter Company, Warminster, Pa.* (*Viscosity and Consistency*)

W. H. PEAKE, *Assistant Professor of Electrical Engineering, Ohio State University, Columbus, Ohio.* (*Acoustical Measurements*)

LOUIS F. POLK, JR., B.S., M.B., *formerly Assistant Vice President and Manager, Instruments and Systems Division, The Sheffield Corporation, a subsidiary of The Bendix Corporation, Dayton, Ohio.* (*Parts Dimension Measurement and Control*)

HENRY REINECKE, JR., M.S. (**E.E.**), *Systems Manager, Non-Linear Systems, Inc., Del Mar., Calif.* (*Basic Electricity and Electronics for Instrumentation Engineers*)

CARL W. SANDERS, B.S. (**Ch.E.**), *Consulting Engineer in Instrumentation, Engineering Services Division, Engineering Department, E. I. du Pont de Nemours & Company, Inc., Wilmington, Del.* (*Automatic Control of Heat Exchangers*)

W. B. SCHULTZ, B.S. (E.E.), *Principal Engineer, Digital Systems, Leeds & Northrup Company, North Wales, Pa.* (*Electric Power Generation and Distribution Control*)

L. M. SILVA, B.S., Engr. Phys., *Formerly Associate Director of Research and Engineering, Beckman Systems Division, Beckman Instruments, Inc., Fullerton, Calif.* (*Instrumentation Data Processing*)

HANS SVANOE, *Consultant, Struthers Scientific & International Corporation, Warren, Pa.* (*Crystallizer Instrumentation*)

R. A. TERRY, *Manager, Marketing Services, Packard Bell Computer Corporation, Los Angeles, Calif.* (*Electrical Variables*)

E. G. THURSTON, *Senior Research Engineer, Hallicrafters Company, Chicago, Ill.* (*Acoustical Measurements*)

J. D. TRIMMER, Ph.D., *Head of Physics Department, University of Massachusetts, Amherst, Mass.* (*Radiation Fundamentals*)

V. S. UNDERKOFFLER, (M.A., Physics), *Head, Systems Equipment Section, Research and Development Department, Leeds & Northrup Company, North Wales, Pa.* (*Nuclear Reactor Instrumentation*)

JAMES R. WALKER, B.S., M.S., *Research Director, Gemco Electric Company, Detroit, Mich.* (*Position Measurement and Control*)

R. E. WILSON, B.A., Ph.D. (Physics), *Research Physicist, Hughes Company, Tucson, Ariz.* (*Temperature*)

ROBERT J. WILSON, *Sales Manager, Photoswitch Division, Electronics Corporation of America, Cambridge, Mass.* (*Applications of Photoelectric Controls*)

JOHN G. WOOD, B.S. (Physics), *Manager of Engineering Tracerlab Division of Laboratory for Electronics, Inc., Waltham, Mass.* (*Thickness Measurement of Sheet and Web Materials*)

C. W. WORK, B.S. (M.E.), *formerly Mechanical Engineer, Hagan Chemicals & Controls, Inc., Pittsburgh, Pa.* (*Combustion Instrumentation*)

PREFACE

The following kinds of people will find this handbook of particular value:

1. The *manufacturing, process,* and *instrumentation engineer*—the person who analyzes the needs and benefits of instrumentation and control and who selects and in some cases designs the instrumental equipment best fitted for a given industrial process, scientific application, or military and aerospace use.
2. The *instrument and control user* who relies on measuring, controlling, and data-processing devices to provide greater efficiency, safety, and quality of the program or process with which he is concerned. The user may be a process engineer, a plant superintendent, a laboratory scientist, or an astronaut.
3. The *technically inclined or technically trained top-management man* who desires to comprehend instrumentation so that he may take full advantage of applied instrumentation in achieving greater profit yields and higher quality in a competitive market or in obtaining the utmost in accuracy and in the saving of time in research and development programs.
4. The *technical student* who may plan on specializing in instrumentation and control; and the *technical student* who may not plan on such specialization, but who realizes that, regardless of what technical field of endeavor he may pursue, he soon will encounter a need to know the fundamentals of applied instrumentation.
5. The *business student* who realizes that instrumentation and control represent one of the most certain means in today's tight profit squeeze of getting more from his investment in manpower and facilities.
6. The *educator*, particularly in technical schools and universities, who can base one or more complete instrumentation and control courses on the crisp and logically outlined format of this handbook.
7. The *self-starting type of instrument and controls craftsman* who knows that a broadened knowledge of how instruments and controls are applied, of what their economic end values are, of how they are used in industries and fields beyond his own, will place him in a good position for career advancement.

The authors and editors of this handbook have placed much stress on the objective of bringing about a greater interchange of applicational data from one field to the next—in other words, a merging of the disciplines of instrumentation. That is why this handbook includes detailed descriptions of instrumentation applications from nearly all of the major industries and scientific fields. The

ix

editors carefully surveyed and screened the many areas of instrumentation in a process of selecting those fields which portray a representative cross section of the subject.

The user of this handbook is urged to canvass the descriptions of those fields which may be somewhat foreign to him for ideas which may spark an instrumental solution to his particular problem. The editors have found in their own instrumentation experience that such crossbreeding of applied technology can save hundreds and even thousands of man-hours and the accompanying costs for development facilities and materials. Careful use of this handbook may save you from reinventing a device or solution that already exists for your problem.

Much stress has been given in the development of this handbook on the "how to" or application engineering aspects of instrumentation. Although many of the early sections of this handbook provide concise descriptions of the "what" of instrumentation, this latter area is covered in more detail in the "Process Instruments and Controls Handbook" (McGraw-Hill). These two handbooks together form a complete compendium on the subject.

Following are thumbnail descriptions of selected subsections of this handbook:

Sec. 1-1—Definition and Classification of Variables

A complex subject is understood best when properly structured. Instrumentation and control are based upon the measurement of literally scores of variables. Some of these variables are thermal in nature; others involve radiation; some are concerned with mechanical forces; still others relate to numerical quantities and rates; some are geometric in character; some involve physical properties, or chemical properties, or electrical characteristics of matter. The classification of variables in this subsection represents a *first in the permanent literature* on the subject.

Sec. 1-2—Measurement Errors

No matter how well designed or how perfectly made, there always are exacting requirements for instruments and controls that go beyond the current state of the art. Hence it is extremely important to the designer and user of precise instruments to appreciate the meaning of errors, how to analyze them, how to predict them, how to compensate for them. This subject is thoroughly covered by an experienced National Bureau of Standards scientist and world authority on this subject.

Sec. 1-3—Factors in Selection of Measuring Methods

Few fields of technology rival instrumentation in the almost endless variety of methods and choices available. For example, there are at least ten basically different methods for measuring temperature, a dozen ways to measure pressure, over a score of ways to measure liquid level, and so on. But astute analysis usually points to the one best measurement for a given application. The big question always is—"How do you go about finding this one best way?" This handbook section is devoted to providing the answers to such complex and provocative questions.

Sec. 2-1—Temperature

What is the basic, physical nature of temperature? What significance do temperature measurements have in the industrial plant? In the laboratory? In nuclear work? In aerospace research? How are temperature standards established? How are they used in the daily calibration of detectors? These basic considerations, often overlooked in the permanent literature, are covered in this subsection.

Sec. 2-2—Calorific Value

The ability of substances to produce heat is an important variable to industry and research. But, to date, the descriptions of this topic have been confined to very specialized books and journals. This subsection is another first in the permanent literature of instrumentation.

Sec. 3-1—Radiation Fundamentals

Principally within the past decade or so, radiation has become important as a tool of instrumentation. Much material on radiation instrumentation has appeared in the literature, but very little of a summary nature on the fundamentals of radiation as interpreted for the instrument designer and user. This subsection, prepared by an outstanding radiation physicist, meets this objective.

Sec. 3-2—Nuclear-radiation Detectors

Radiation detectors no longer are confined to nuclear-reactor research but are required in such fields as medicine, biological research, aerospace exploration, and food processing. This subsection provides a comprehensive review of radiation detectors written to be understood by the men of many fields who may be called upon to use them. The authors have pioneered their design, development, and application for many years.

Sec. 3-3—Radioisotopes in Instrumentation

Not so many years ago the use of radioisotopes for tracing the flow of complex substances in complex systems was new and novel. Today this practice is commonplace. But, again, the literature has lagged behind the art. This subsection, representing a comprehensive review of the subject, is another first in the permanent literature on instrumentation.

Sec. 3-4—Photometric Variables

A few years ago the human eye was the primary instrument for assessing such properties of materials as gloss, sheen, brightness, and close gradations of color. This subsection, prepared by a world authority on the subject, reviews a new generation of photometric instruments.

Sec. 3-5—Acoustic Measurements

The effects of noise on office and factory efficiency and the many applications of ultrasonics in industry and medicine have required the development of a new line of measuring instruments. Two of the nation's pioneers in this field describe what instruments are available, how they can be used, and what instruments still are needed.

Sec. 4-1—Force Measurement

With the tremendous forces encountered in aerospace, nuclear, chemical, and other technologies today and with the invention of numerous new methods for the measurement of such forces, this subsection is of much importance.

Sec. 4-2—Pressure and Vacuum Measurement

This subsection provides a comprehensive tabular and pictorial summary of mechanical, electrical, pneumatic, and other methods for the measurement of pressure and vacuum.

Sec. 5-1—Flow

In addition to supplying a comprehensive but concise review of available flow-measurement devices, this subsection analyzes the importance of flow as a variable and reviews the basic physical and hydraulic theories which underlie the design and application of flow-measuring-and-controlling devices. The author has been associated with the flow-measurement field for nearly thirty years.

Sec. 5-2—Acceleration Measurement

Acceleration measurement has taken on increasing importance in industry and particularly in the fields of missile technology and aerospace exploration. In this subsection the characteristics of accelerometers and their application engineering are described.

Sec. 5-3—Speed Measurement

With the trend toward greater and greater automation of industrial and other physical processes, the engineer is called upon more frequently to specify the most suitable speed-measurement method. These factors are covered in detail in this subsection.

Sec. 5-4—Weight and Weight Rate of Flow

Sophisticated instrumental methods for application to bulk solids came much later than in the case of the fluid systems. But progress in recent years has been rapid. This progress, together with a summary description of available hardware and weight control systems for both bulk and discrete solids, is detailed in this subsection. The author has devoted over twenty years in this area of technology and is responsible for numerous scientific advancements.

Sec. 5-5—Liquid Level

Probably no variable represents so many instrumental methods and possibilities for its measurement. Over thirty years of experience in this field have enabled the author of this subsection to pass along detailed recommendations concerning the best liquid-level measurement system for a given application.

Sec. 5-6—Solids Level

This is another variable involved in the instrumentation and automation of bulk-solid handling systems. Very slow in its initial development, numerous schemes of measurement have become available recently. These are described by an author of many years of experience and are passed along to the user of this handbook in an effort to assist the designer and user in selecting the best measurement method for his application.

Sec. 6-1—Parts Dimension Measurement and Control

Dimension measurement is one of the cornerstones of automation in all industries which handle discrete units, such as the metalworking, mechanical fabrication, and assembly industries. The authors of this subsection are recognized world authorities in this field.

Sec. 6-2—Thickness Measurement of Sheet and Web Materials

The advent of plastic kinds of materials gave much impetus to the production of materials in sheet and web form and also gave cause for the modernization of the earlier metallic films and foils. Improvements in X-ray and other radiation measurement principles greatly aided the development of new thickness gages. The authors of this subsection describe an entire cafeteria of thickness-measurement devices and provide detailed application guidance.

Sec. 6-3—Position Measurement and Control

Position is another key measurand in many automation systems. Unfortunately, the past literature on this subject has been spotty and incomplete. The authors of this subsection targeted on compiling the first really comprehensive review of an applicational nature to be found in the permanent literature.

Sec. 7-1—Fluid-density and Specific-gravity Measurement

The summary presented in this subsection is designed to assist the instrumentation engineer in his selection of the right system for a given application. This is another subject that heretofore has not been covered adequately in the permanent literature.

Sec. 7-2—Humidity and Dew Point

These important variables are encountered in almost all industrial processes and scientific research. What are the instrumental methods for their measurement? What are the basic mathematics and theory concerning them? How can you go about selecting the best measurement method for your need? The author of this subsection proceeds to answer these questions in a logical fashion.

Sec. 7-3—Moisture Content of Materials

For those handbook users who have attempted heretofore to find descriptions of moisture-measurement methodology in one reference, the work of the author of this subsection will be deeply appreciated. Another first in the permanent literature.

Sec. 7-4—Viscosity and Consistency

Few variables are more difficult to comprehend. The author of this subsection carefully explains the basics of rheology, after which he documents the units and scales of viscosity. This is followed by a tabular and pictorial "how to select" guide.

Sec. 8-1—Analysis Instruments

This subsection provides an insight to the selection and application of analytical instrumentation, especially for on-line measurements. As more and more computers enter the process-control loop, analytical instrumentation for measuring the chemical composition of raw materials, of intermediate and sidestream materials, and of final products will become mandatory. The author describes numerous analytical methods and provides an excellent structuring of this extremely complex subject.

Sec. 9-1—Electrical Variables

The basic instrumentation for measurement of current, emf, and other electrical characteristics is described in terse, application engineering style.

Sec. 10-1—Fundamentals of Automatic Control Engineering

Scores of books and hundreds of articles have appeared on this subject, but the author of this subsection faced the challenge of organizing his material in such form that it would be meaningful not only to the student who may be learning of automatic control theory for the first time but also to the technical manager, plant superintendent, scientist, and aerospace engineer who may not desire to become completely expert in the field, but who does require a day-to-day working knowledge of the subject. Because of years of experience as a systems engineer in applying instruments and controls, the author is eminently well qualified to appreciate the many viewpoints of users of this handbook.

Sec. 10-2—Application of Photoelectric Controllers

The author of this subsection, in describing available photoelectric controls and citing numerous examples of their practical application, provokes the imagination and creativity of the instrumentation and processing or manufacturing engineer. Often the ingenious application of photoelectric controls can greatly simplify and reduce the cost of applications that might initially appear to require a rather exotic control system.

Sec. 10-3—Controller Types and Final Control Elements

This subsection provides a terse, tabular, and pictorial summary of electric, pneumatic, mechanical, and hydraulic types of controllers and final control elements, including valves and dampers. The stress is on "how to select"; the design details of the hardware will be found in the "Process Instruments and Controls Handbook" (McGraw-Hill).

Sec. 11-1—Applications of Analog Computers

The practical considerations of applying analog computers to the supervision and control of manufacturing and processing are the objectives of this subsection. Unlike the run-of-mill coverage of this relatively complex subject in the periodical literature, where too often the material is highly mathematical in content or too nebulous in reducing the subject to a practical application science, this subsection will assist the handbook user in determining where analog computers can be used and where some other solution to the problem may be best.

Sec. 11-2—Application of Digital Computers

Unless one follows this rapidly progressing subject on a day-to-day basis, he will find it difficult to amass information quickly that will enable him to make a determination of the practical applicability of digital computers to his problems. The author of this subsection,has structured the subject for quick and convenient comprehension and presents applicational details in a realistic fashion.

Sec. 11-3—Instrumentation Data Processing

This subsection presents a practical, immediately usable review of available data-processing equipments with just enough discussion of principles and theory to assist the handbook user in selecting the right equipment for his job. The author, a veteran data-processing scientist, brings to the permanent instrumentation literature for the first time, in the editors' opinions, a well-organized evaluation and assessment of this subject. Suddenly what appeared to be an extremely complex subject takes on an air of simplicity.

Sec. 11-4—Counters and Digital Indicating Devices

This provides a terse review of these workhorses of instrumentation that are found in practically every systems engineering and electronics laboratory—with growing application for process and machine control.

Sec. 12-1—Steel Production Instrumentation

This subsection represents the most thorough review of this kind attempted to date in the permanent instrumentation literature. The author brings to this section over forty years of concentrated attention to the instrumentation needs of this industry. Handbook users in other industries where temperature measurements and control and where complex, interlocked fuel-firing controls are encountered will find much to consider for their own needs. The steel industry not only has brought earlier instrumentation applications to a high degree of refinement but has also developed many new instrumental methods for controlling new processes. These are covered in this subsection.

Sec. 12-2—Glass and Ceramics Industries Instrumentation

The most thorough review of this kind attempted to date in the permanent instrumentation literature. Many concepts for instrument applications in other high-temperature industries can be found in this subsection. For the executive of a glass or ceramics plant, this subsection sets a high standard against which to measure his own application of instruments and controls for the upping of product quality simultaneously with profit.

Sec. 13-1—Instrumentation Practices in the Process Industries

In this subsection one of the newer generation instrumentation and systems-engineering executives and recent President of the Instrument Society of America reviews the subject of instrumentation from the standpoint of technical management. This subsection brings out the way a company should organize for the engineering and maintenance of instruments and controls and provides numerous ideas on how to sell instrumentation upstairs to top management.

Secs. 13-2 through 13-7—Instrumentation of Heat Exchangers, Filters, Crystallizers, Distillation Columns, Dryers, and Evaporators

The following observations apply to all of these subsections. About thirty years ago, the chemical engineer came up with a conceptual, technological breakthrough by recognizing that various operations, such as those mentioned above, appear in repeated fashion in numerous processes. By zeroing in on the study of these so-called unit operations, rather than in attempting to study each process entirely on its own, the chemical engineer literally saved years in advancing the science of chemical processing.

Surprisingly, and difficult to explain, the literature has included to date only spotty references to the instrumentation and control of these well-defined unit operations. Possibly the most important *first* in this handbook is this series of comprehensive analyses of the needs for and the "how to" of instrumenting the chemical unit operations. It is fully expected that this pioneering editorial attack will be copied many times over in the years to come. The eminence of the authors of each of these subsections in their field is beyond question.

Sec. 13-8—Pulp and Paper Production Instrumentation

This is an industry which has had to face the profit squeeze for many years. Striving for efficiency through automation is not a new concept in this industry. Consequently, users of this handbook from other fields can gain much by way of ideas from the manner in which the advantages of instrumentation have been squeezed out and refined for pulp and paper production. The author is one of the nation's leading authorities on this subject.

Sec. 13-9—Food-industry Instrumentation

Revolution is a rather mild word when applied to the progress and changes which have taken place in the food industry over the past few years. For example, the sanitary requirements have placed a hardship on instrumentation engineers for years, but these problems largely have been licked by new instrument designs which afford "in process" cleaning on an automatic cycle. Older applications, such as pasteurization, have been modernized to include electronic instrumental approaches. All of these applications, the conventional and the new, are described in this subsection by an engineer who has specialized in the instrumentation for this particular industry for a number of years. To the technical managers of food plants, this section should provide considerable "food for thought."

Sec. 14-1—Automatic Combustion Control Systems for Boilers

Despite strides in other energy processes, the boiler plant will remain on the industrial and commercial scene for many years. The instrumentation and control of boilers, from the small portable type to the largest installations, has gone through years of refinement. The handbook user will find in this section an account of instrumental means which reflect the mingling of new concepts with well-established methods.

Sec. 14-2—Steam Power Plant Instrumentation

Any handbook user with a steam power plant, large or small, will find described in this subsection the "how to" of selecting and applying instruments and controls.

Sec. 14-3—Electric Power Generation and Distribution Control

This subsection puts together in one place a summary of instrumental and control methodology and thus represents a breakthrough in the permanent instrumentation literature. Before this, the person interested in the subject had to pore through scores of periodicals and actually depend heavily on manufacturers' literature. All of the authors of this subsection are recognized authorities in their field. One of the first industries to demonstrate realizable economic savings through the application of digital control computers.

Sec. 14-4—Nuclear Reactor Instrumentation

As the most recent energy process, nuclear fission instrumentation coverage in the literature has been skimpy and scattered. The eminent author of this subsection has carefully organized and structured this complex subject and, for the first time, has brought it into focus in one place in the permanent instrumentation literature.

Sec. 15-1—Process Laboratory Instrumentation

Instrumentation in the process laboratory differs quite drastically from the needs of the large-scale production process. Usually the laboratory scientist does not have the funds available and often does not have the knowledge to apply instruments and controls that literally are larger than his laboratory apparatus. Often he does not have the room or the time. Fortunately, specialized instruments have been made available commercially for such use and these are described in this subsection. But, more important, the author describes scores of ingenious homemade devices that the laboratory scientist can put together in a relatively short time—methods that are adaptable to the measurement of variables in a very small space or of very minute quantities. This subsection represents the collection of these ideas by an expert in the field over a period of years.

Sec. 15-2—Pilot-plant Instrumentation

Steadily the concept of piloting instrumentation techniques in the pilot plant is taking hold as well as the piloting of chemical and physical principles and of processing equipment. A part of this slow recognition of applying adequate instrumentation to pilot plants stems from the sad lack of literature on this subject. A thorough literature search may uncover a few articles. The author of this subsection not only describes available hardware for these applications but also develops justification for instrumentation expenditures in pilot plants and a philosophy for guiding these applications.

Sec. 15-3—Environmental Test Instrumentation

Greatly accelerated by the developmental needs of defense and military products, the means to test these products under severe conditions prior to use now comprises the relatively new field of environmental test instrumentation. The author of this subsection, a national expert in this area of specialized instrumentation, describes basic types of vibration-measuring instruments. Shock and acceleration testing, measurements of simulated altitudes, temperature and humidity measurement problems peculiar to environmental simulation, testing of components and products in acoustical environments, and explosion testing are among the many subjects covered. This subsection will be of particular value to those handbook users in the aerospace and associated industries and to all other users who have a growing concern for the pretesting of components, materials, subsystems, and entire systems prior to their actual intended use.

Sec. 15-4—Electronic Laboratory and Research Instrumentation

Surprisingly little attention has been given in the professional literature to the assemblage of data on the many kinds of instruments used in the typical electronics-type laboratory. In this subsection an outstanding staff of engineers has developed this "big picture" for the first time—with emphasis on a review of the kinds of instruments available, their characteristics, and guidance toward selection for specific laboratory applications.

Sec. 16-1—Aircraft and Aerospace Vehicle Instrumentation

A comprehensive review of flight and attitude instruments, navigational instruments, engine and motive power instrumentation as applied to modern air and spacecraft. The author of this subsection is a pioneer and outstanding authority in his field.

Sec. 16-2—Aircraft Flight Simulation Instrumentation

As air and spacecraft become more and more complex and sophisticated in design and use, the need to simulate flight conditions for crew training has grown. The author, who pioneered and invented the first flight simulator, shows how instrumentation is used to assist crew members in determining and rectifying such conditions as troubles with power plant, fuel system, flight controls and autopilots, electrical and hydraulic systems, landing gears, cabin conditions, radio, and communications. The author also develops the economic advantages of instrumented flight simulation where large crews must be trained.

Sec. 17-1—Basic Electricity and Electronics for Instrumentation Engineers

Although a wealth of material can be found in the literature on basic electricity and electronics, seldom has this subject been treated exclusively from the viewpoint of the instrumentation engineer and designer. In addition to reviewing rather exhaustively the fundamentals of the subject, the authors bring the subject right up to date with inclusion of information on solid state and microwave principles. The authors are particularly well qualified to do this because of their many years of experience in applying these basic principles to the design of instrumental systems.

Sec. 17-2—Preparation of Wiring Diagrams for Instrument Applications

This is usually a "take it for granted" kind of subject, which undoubtedly explains why it has been covered so poorly in previous literature. The application of instruments and controls to any use requires first the preparation of an electrical systems diagram which shows how all of the electrical interlocking shall be accomplished. Preparation of such diagrams can be speeded up and made more accurate when specialized techniques such as those described in this subsection are applied. The author has specialized in this area of engineering for over twenty years.

Sec. 17-3—Instrument Panelboard Design and Construction

The design and application of centralized instrument and control panelboards now comprise a specialized art. The selection of instruments often is dictated by the manner in which measurements will be displayed. The distance from the point of measurement to signal display, whether electrical or pneumatic transmission is used, the needs of the operator, the space and funds available—all of these factors determine what form an instrument panelboard or control console shall take. The author of this subsection considers all of these varying conditions and provides an insight into their effect on panelboard design. Numerous examples are cited to show the effect of end results on cost.

Sec. 17-4—Instrument Air Supplies

Pneumatic instrumentation has maintained the pace of progress and still represents a substantial investment where new installations are concerned. Many installations require long runs of piping and complex pneumatic relays and actuators. The very best pneumatic system can fail quickly because of a poor, dirty air supply. The author of this subsection brings to the handbook user a fund of engineering information collected over the years to assure the design of a good instrument air supply system.

Indexes

The editors of this handbook have made every effort to keep all possible users in mind. Hence the ratio of index pages to text pages is much higher than is found in the average technical text or handbook. Further, three special classified indexes are included to assist the handbook user in locating reference information in a minimum of time.

Douglas M. Considine
Editor-in-Chief

S. D. Ross
Associate Editor

CONTENTS

xix

XX   CONTENTS

CONTENTS

CONTENTS

HANDBOOK OF APPLIED INSTRUMENTATION

Section 1

MEASUREMENT VARIABLES

By

D. M. CONSIDINE, B.S., *Manager, Marketing Planning, Hughes Aircraft Company, Culver City, Calif.; Member, Instrument Society of America, American Institute of Chemical Engineers, American Marketing Association, American Ordnance Association, National Sales Executives, American Society of Military Engineers, American Management Association. (Definition and Classification of Variables)*

STEVEN DANATOS, B.S., M.S.(M.E.), *Associate Editor, Engineering Practice, Chemical Engineering, McGraw-Hill Publishing Company, New York, N.Y.; Member, American Chemical Society; Registered Professional Engineer (N.J.). (Measuring and Transmission Methods)*

F. K. HARRIS, B.A., M.S., Ph.D. (Physics), *Electrical Instruments Section, Electricity Division, National Bureau of Standards, Washington, D.C.; Member, Institute of Electrical and Electronics Engineers, Washington Academy of Sciences. (Measurement Errors)*

W. H. HOWE, B.S.(E.E.), *Chief Engineer, The Foxboro Company, Foxboro, Mass.; Member, American Society of Mechanical Engineers, Institute of Electrical and Electronics Engineers, Society of Sigma Xi; Registered Professional Engineer (Mass.). (Definition and Classification of Variables) and (Factors in Selection of Measurement Methods)*

DEFINITION AND CLASSIFICATION OF VARIABLES

By W. H. Howe* and D. M. Considine†

Measurement and control systems exist and are important because the machines, processes, equipments, and systems which comprise industrial, commercial, and military endeavor involve phenomena which are *not* steady state. Rather, such equipments and systems involve conditions which are continuously changing.

The quantities or characteristics measured (which serve as the basis of control) logically are termed *variables*. These are often termed measurement variables, instrumentation variables, or process variables. The last term is confined to the process industries. Because this handbook covers applications of instrumentation beyond the process field, the simple term *variable* is used here.

CLASSIFICATION OF VARIABLES

Variables may be classified in several ways, each with its advantages and limitations. Two basic classifications will be outlined here. The first is in accordance with the character of the variable itself, that is, thermal variables, radiation variables, electrical variables, etc. This follows closely the pattern of Secs. 2 through 9 of this handbook. This classification is most useful in considering the characteristics of the variables themselves.

In considering the measurement of variables, classification by characteristic leads to considerable redundance. Typically, temperature is often measured by converting temperature variation to pressure variation and then measuring the pressure variation; equally common, temperature may produce a millivoltage in a thermocouple with millivoltage-measuring equipment used for record, control, and other responses; temperature is converted into other variables for convenience in measurement. Conversely, the self-balancing potentiometer recorder, originally developed primarily as a temperature-measuring device for use with thermocouples, is now commonly used to record the value of such variables as speed, density, chemical composition, and many others.

This leads to a second classification by types of measurement signal. Specific transducers convert practically all variables to a few common types of measurement signals, with a correspondingly small number of indicating, recording, and similar responsive devices operative from these measurement signals. A classification in accordance with measurement signals directly derived from the variables furnishes a useful guide to available methods for measurement.

The two classifications will be outlined separately. A tabular presentation coordinates the variables with the commonly available measurement signals used for their measurement. The following classification by variables is arranged, in general, in the same order as the material in Secs. 2 through 9 of this volume. This list is not exhaustive. It does cover the variables discussed in this handbook, and some additional items.

* Chief Engineer, The Foxboro Co., Foxboro, Mass.
† Manager, Marketing Planning, Hughes Aircraft Co, Culver City, Calif.

CLASSIFICATION BY VARIABLES

A. Thermal Variables

Thermal variables relate to the condition or character of a material dependent upon its thermal energy.

1. Temperature: The condition of a body which determines the transfer of heat (thermal energy) to or from other bodies.

2. Specific heat: The property of a body defining the relation between change of temperature and change of thermal energy level.

3. Thermal energy variables: Enthalpy and entropy related to total and available thermal energy of a body.

4. Calorific value: The characteristic of a material which determines the quantity of heat (thermal energy) which will be produced or absorbed by the body under specific conditions.

B. Radiation Variables

Radiation variables relate to the emission, propagation, and absorption of energy through space or through a material in the form of waves; and, by extension, corpuscular emission, propagation, and absorption.

5. Nuclear radiation: Nuclear radiation is the radiation associated with the alteration of a nucleus of the atom.

6. Electromagnetic radiation: The electromagnetic radiation spectrum includes radiant energy from the emission at power frequencies through the radio transmission bands; radiant heat; infrared, visible, and ultraviolet light; and the X and cosmic rays. Gamma radiation from nuclear sources is a form of electromagnetic radiation.

7. Photometric variables are variables such as color, gloss, reflectance, etc., concerned with visible light.

8. Acoustic variables include audible sounds and similar inaudible waves in gases, liquids, and solids, especially those at ultrasonic frequency.

C. Force Variables

9. Total force: A force is any physical cause tending to modify the motion of a body.

10. Moment or torque (rotational force) is the distance along the shortest line between the point of application of a force and a reference axis multiplied by the component of the force acting at right angles to this shortest line.

11. Pressure and vacuum, unit stress, are the force per unit area in a fluid or solid.

D. Rate Variables

These variables are concerned with the rate at which a body is moving toward or away from a fixed point. Time is always a component of a rate variable.

12. Flow is the number of unit volumes of material passing a given point within a fixed time.

13. Speed or velocity is rate of motion measured by the distance moved by a body in a unit time. Speed or velocity may be linear or angular. Speed usually refers to motion of solids.

14. Acceleration is the time rate of change of velocity. Acceleration may be either linear or angular.

E. Quantity Variables

These variables relate to the total quantity of material which exists within specific boundaries.

15. Mass is the total quantity of matter within specified boundaries.

16. Weight is a measure of mass, based on the gravitational attraction.

F. Time Variables

17. Elapsed time.
18. Frequency: Frequency of a periodic quantity is the number of periods occurring in a unit time.

G. Geometric Variables

These relate to the position or dimension of a body. Geometric variables are related to the fundamental standard of length.
19. Position is the measurement of the location of a body with respect to a fixed coordinate system.
20. Dimension is the distance between two fixed points.
21. Contour or shape is the relative location of a group of points representative of the surface being measured.
22. Level (liquid or solid) is the height or distance of the surface of a material referred to a base reference level.

H. Physical Property Variables

Physical property variables are concerned with the physical properties of substances, with the exception of those related to mass and chemical composition.
23. Density and specific gravity: Density is the concentration of matter; that is, the quantity of matter in a given volume.   Specific gravity is the ratio of the density of a material to the density of water at specified conditions or, less commonly, the ratio of the density of a gas to the density of air at specified conditions.
24. Humidity: Humidity is the quantity of water vapor in an atmosphere. Absolute humidity is the weight of moisture in a unit volume or referred to a unit total mass of gas.   It is occasionally expressed in terms of pressure of water vapor.   Relative humidity is the relation between the existing pressure of water vapor in an atmosphere and the saturated vapor pressure of water at the same temperature.
25. Moisture content is the quantity of free water contained in a substance.
26. Viscosity is the resistance of a fluid to deformation under shear.
27. Structural characteristics are the crystalline, mechanical, or metallurgical properties of substances such as hardness, ductility, lattice structure, etc.

J. Chemical Composition Variables

28. These variables, too numerous to list here, are concerned with the properties of substances related to composition.

K. Electrical Variables

29. Voltage or electromotive force: Electromotive force is that property of an electrical system which tends to produce an electric current in a circuit.   Voltage is expressed as the difference of potential between any two points of a circuit.
30. Electric current is the rate of transfer of electricity.
31. Resistance and conductance: Resistance is the property which determines for a given current the rate at which electric energy is converted into heat.   In d-c circuits, resistance is the numerical ratio of electromotive force divided by current. Electrical conductivity is the reciprocal of electrical resistance.
32. Inductance is the property of an electric circuit or of two neighboring circuits which determine the electromotive force induced in one of the circuits by a change of current in either of them.
33. Capacitance is the property of an electrical system comprising electrical conductance which determines displacement currents in the system for a given time rate of change of potential.

34. Impedance of an electric circuit subjected to an a-c potential is the numerical ratio of applied potential divided by the resultant current flow.

CLASSIFICATION BY MEASUREMENT SIGNALS

"Measurement signals" are defined for this classification as a small group of variables (ten) employed in measuring systems for conversion, transmission, and utilization of the responses of sensors operating directly from the measured variables. As is apparent from perusal of Secs. 2 through 9 of this volume, for most measurements, a change in the measured variable is converted to a change in some other variable (a measurement signal), which in turn operates the display, initiates control action, or may be converted to another measurement signal. Flow measurement using an orifice or venturi tube is typical. The orifice plate or other primary device develops a differential pressure (measurement signal). This differential pressure may operate an indicating or recording meter directly, or may be converted to a pneumatic (or electric) transmission signal (a second measurement signal), which in turn operates a display, initiates control action, or may feed into a computer whose output is another measurement signal.

The use of measurement signals provides for the measurement of all the different variables by a combination of specialized primary transducers, together with a small number of responsive devices associated with the small number of measurement signals.

Measurement signals fall into ten simple divisions.

A. Motion

All displays of the value of a measured variable are based on some form of motion. Motion is also a common input to transducers, controllers, computers, and other devices responsive to measurement.

101. *Mechanical motion* of an indicating pointer, recording pen, or other solid element is the most usual form of measurement display. Mechanical motion, either linear or angular, is an equally common input to other responsive devices.

102. *Liquid displacement* is used for display in glass-stem thermometers, glass-tube manometers, and the like. It is also used as a transmission signal in metal-tube liquid-filled thermometer systems and in some other devices.

103. *Motion of a light or electron beam* is used as a display in oscilloscopes, oscillographs, light-beam galvanometers, and the like. It is also used as a position sensor in some applications where negligible reaction force from the sensor is essential.

B. Force

This is a common type of signal used in conversion, transmission, and utilization of measurements.

104. *Total mechanical force* is often used as a control input, as a conversion element in force-balance devices, and for signal transmission over distances measured in inches. It is readily derived from and converted into motion, or to or from static or differential pressure.

105. *Pressure (force per unit area)* in fluids is a measurement signal very commonly used for transmission of measurement. Both static and differential pressure are used, ranging from inches of water differential pressure, developed by an orifice plate, to pressures up to 1,000 psi or above, developed in sealed, gas-filled thermometer systems. A 3- to 15-psi pneumatic pressure (or, less commonly, a 3- to 27-psi) is accepted as a standard for pneumatic transmission of measurement and control signals.

C. Electrical Signals

Transducers are available for converting practically all variables to corresponding electrical measurement signals. Many variables, particularly radiation variables, chemical composition variables, and, of course, electrical variables, produce an electrical measurement signal.

106. *Voltage or current signals* have a fixed relationship between the measured variable and the voltage or current signal.

107. *Voltage-and-current relation signals* are those in which the relationship between voltage and current is the significant characteristic of the measurement signal. Where the change in the measured variable produces a change in impedance in the measuring circuit, the relationship between voltage and current or between input and output voltages or currents defines the measured value.

D. Time-modulated Measurement Signals

A number of "on-off" time-modulated signals are used for the transmission of measurements, particularly over considerable distances.

108. *Pulse-duration signals* usually operate with a constant-cycle duration between 1 and 15 sec, the relation between the time the circuit is closed and the time the circuit is open, during each cycle, representing the value of the variable. Pulse-duration signals are also used for integration, without regard to the transmission distance.

109. *Frequency signals* with variation of frequency representing variation of the value of the measured variable are often used for transmission of measurement, particularly over carrier and radio transmission circuits. Rotational speed is sometimes converted to frequency as a measurement signal, without regard to transmission distance.

110. *Pulse-code modulation* in various forms is used for measurement signal. This is basically a digital technique. The measurement signal may be simply a count of the number of pulses within a given time interval, or may be a fully coded binary or binary-decimal signal. Pulse code is commonly used in connection with digital computers, digital data loggers, and like apparatus, but has as yet found little application to transmission where both input and output signals are of the analog type.

MEASUREMENT SIGNALS FOR VARIABLES

Measurement signals are so closely related to the measuring systems as to be practically inseparable. In considering measurement signals and systems, four factors are of major importance.

1. Types and characteristics of transducers available for converting the variables to the measurement signals. Table 1 outlines the types in common usage.

2. The transmission characteristics of the measurement signals.

3. Output devices responsive to the measuring signals for producing indications, records, control and computer inputs, etc.

4. Transducers available for converting from one measurement signal to other measurement signals.

Careful consideration of these factors in relation to the complete instrumentation for a plant or process unit can be most useful in selection of an optimum system.

Table 1 lists the variables and the measurement signals. In this tabulation, an X indicates that a standard transducer is commercially available for converting the variable to the signal. An O indicates that the variable is already in measurement signal form.

Only direct conversions are included. For example, the tabulation indicates a transducer for converting pressure to electric voltage and current relationship; the strain-gage pressure cell is a typical device for this purpose. In this transducer, pressure is converted to mechanical motion of the structure supporting the strain gage, and the gage measures this motion. This is considered as a direct conversion, in spite of the two-stage operation. On the other hand, no transducer is indicated for flow-to-liquid displacement, in spite of the fact that in some cases an orifice unit, which converts flow to differential pressure, is mounted integrally with a mercury manometer, which converts this differential pressure to liquid displacement. This is

Table 1. Variables vs. Measurement Signals

Variable	Mechanical motion 101	Liquid displacement 102	Motion of light or electron beam 103	Total force 104	Pressure 105	Voltage or current 106	Voltage-and-current relationships 107	Pulse duration 108	Frequency 109	Pulse code 110
1 Temperature..........Note 1	X	X	...	X	X	X	X	...	X	
2 Specific heat...................	Note 4									
3 Thermal energy variables.......	Note 4									
4 Calorific value.................	Note 4									
5 Nuclear radiation...............	...	...	...	...	...	X	X	...	...	X
6 Electromagnetic radiation.......	X	...	...	...	...	X	X			
7 Photometric variables..........	Note 5									
8 Acoustic variables.............	Note 5									
9 Total force.....................	X	X	X	O	X	X	X			
10 Moment or torque..............	X	X	X	X	...	X	X			
11 Pressure or vacuum............	X	X	X	X	O	X	X			
12 Flow..........................	X	...	...	X	X	X	X	...	X	
13 Speed or velocity.............	X	...	...	X	X	X	X	...	X	
14 Acceleration..................	X	...	X	X	X	X	X			
15 Mass..................Note 2										
16 Weight........................	X	...	...	X	X	...	X			
17 Elapsed time..................	X	...	...	...	...	X	X	...	...	X
18 Frequency....................	X	...	...	...	...	X	X	...	O	X
19 Position......................	X	X	X	X	X	X	X	X	X	X
20 Dimension....................	Note 6									
21 Contour......................	Note 6									
22 Level.........................	X	X	...	X	X	...	X	...	X	
23 Density and specific gravity...	X	...	...	X	X	X	X	...	...	X
24 Humidity..............Note 3	X	...	...	...	X	X	X			
25 Moisture content.............	X	...	...	X	...	...	X			
26 Viscosity.....................	X	...	...	X	X					
27 Structural characteristics.......	Note 7									
28 Chemical composition..........	Note 8									
29 Electrical voltage.............	X	...	X	X	...	O	X	X	X	X
30 Electrical current.............	X	...	X	X	...	O	X	X	X	X
31 Electrical resistance...........	...	...	...	...	...	X	X	X	X	
32 Electrical inductance..........	...	...	...	...	...	X	X	X	X	
33 Electrical capacitance..........	...	...	...	...	...	X	X	X	X	
34 Electrical impedance..........	...	...	...	...	...	X	X	X	X	

NOTES: 1. Temperature is measured by radiation. Since all bodies radiate and absorb electromagnetic wave energy, dependent upon their temperature, no transducer is involved in the relationship between temperature and radiation. However, the radiation is used as a measure of temperature, especially at high temperatures.

2. Mass is almost invariably measured by the gravitational effect, that is, by weight.

3. Humidity is measured by dew-point temperature. This applies to direct measurement of temperature of pure water whose vapor pressure is equal to the vapor pressure to be measured, and also to saturated lithium chloride vapor pressure in saturated lithium chloride elements.

4. No simple transducers exist for the measurement of specific heat, calorific value, entropy, enthalpy, and the like. Either measurements are derived from computations based on measurements of other variables, or specialized equipment is used to maintain a number of conditions constant, so that one variable, usually temperature, varies in a predetermined relation to changes in the measured variable.

5. Photometric and acoustical variables each include several distinct variables. The sensor for most photometric variables is some type of photocell; the sensor for most acoustic variables is some type of microphone. Both have electrical outputs. The relationship among the variable, the sensor, the associated equipment, and the measurement signal varies with the particular measurement. In almost all measurements except acoustic frequency, considerable peripheral equipment beyond the sensor is involved.

6. Dimension and contour, defined as the relative position between points, are almost invariably measured by a positional measurement, with a selected point of the dimension or contour maintained in a predetermined relationship to a reference point on the position-measuring system.

7. Structural characteristic variables include so wide and varied a group of characteristics that there appears to be no useful common basis for discussion of their measurements.

8. Chemical characteristic variables include so wide and varied a group of characteristics that there appears to be little useful common basis for discussion of their measurements. For most measurements no simple direct transducers exist; measuring equipment is, in general, elaborate; in most cases, the measurement signal is an electrical voltage or current, or an electrical voltage and current relationship.

considered to be two separate transducers. The line of demarcation between one transducer operating in two stages and two transducers in a single mounting is, of necessity, somewhat arbitrary. However, since there are transducers to convert every measurement signal to every other measurement signal, in one or more stages, and since it is possible to assemble any number of transducers in a single package, the necessity of some division is obvious.

The transducers indicated in the tabulation may operate over the full range of the variable, or may operate over a limited span or under limited conditions. For example, transducers which convert temperature to liquid displacement operate between -100 and $+1000°F$; similarly, transducers (magnetic flow meters) for converting flow to voltage-and-current relationship measure only conducting liquid flows.

The variables are arranged in the same order as in the classification of variables. This is basically the order in which these variables are discussed in Secs. 2 through 9 of this handbook.

REFERENCES

1. Linebrink, O. L.: The Measurement Gap, *ISA J.*, vol. 8, no. 2, pp. 38–39, February, 1961.
2. Glennan, T. Keith, A. T. Waterman, A. B. Kinzel, H. F. York, H. W. Russell, and A. V. Astin: Why Are Better Standards Needed? *ISA J.*, vol. 8, no. 2, pp. 39–41, February, 1961.
3. Hall, G A., Jr.: Measurement Standards in Science and Industry, *ISA J.*, vol. 8, no. 2, pp. 42–44, February, 1961.
4. Wildhack, W. A.: NBS—Source of American Standards, *ISA J.*, vol. 8, no. 2, pp. 45–50, February, 1961.
5. Axman, B.: Evaluating Measurement Standards, *ISA J.*, vol. 8, no. 2, pp. 54–57, February, 1961.
6. Staff: Bibliography on Measurement Standards, *ISA J.*, vol. 8, no. 2, pp. 71–74, February, 1961.
7. Staff: A New ISA Guide to Transducers, *ISA J.*, vol. 9, no. 8, pp. 39–42, August, 1962.
8. Keller, E. A.: A Classification System for Measurement and Control, *Trans. IRE*, vol. I-9, pp. 38–42, May, 1958.
9. Churchman, C. W., and P. Ratoosh: "Measurement: Definitions and Theories," John Wiley & Sons, Inc., New York, 1959.

MEASUREMENT ERRORS

By Forest K. Harris*

The universal presence of uncertainty in physical measurements must be recognized as a starting point in a discussion of errors in measuring systems. These errors arise in (1) the measuring system itself, and (2) the standards used for calibration of the system. This statement applies equally to control systems.

This subsection describes (1) sources of error, (2) evaluation of errors, including systematic and random errors, and (3) rules regarding the expression of data, especially significant figures.

DEFINITION OF ERROR

In making any physical measurement, the essential purpose is to assign a value, consisting of an appropriately chosen unit and an associated numeric, which will express the magnitude of the physical quantity being measured. For example, in the measurement of a temperature, the chosen unit may be degrees Fahrenheit and the associated numeric may be 110. Thus, 110°F. The extent of failure in exactly specifying this magnitude, and hence the departure of the stated value from the true value of the quantity, constitutes the error of measurement.

TYPES OF MEASUREMENT

When considering and evaluating measurement errors, it is helpful to keep in mind the scheme of measurement employed. Some of the more common types of measurement are described as follows.

Direct Comparison

A measurement may consist of comparing the quantity being measured with a standard of the same physical nature. In such cases, the ratio between, or the difference of the standard from the unknown magnitude, is determined. The use of a Wheatstone bridge to determine a resistance in terms of a known resistance and a ratio is an example of this technique.

Adjustment to Equality. A calibrated magnitude is adjusted by known amounts until it is equal to the unknown. This is done in the determination of mass with a chemical balance, or in the measurement of voltage with a potentiometer. This type of measurement may be considered as a special case of direct comparison.

Direct Actuation of a Physical System

Some property of the measurand is used to operate a suitable indicating device, and its magnitude is read from an appropriate scale. For example, the magnitude of an electric current may be measured by the torque which it produces on the moving system of an ammeter, and the value is read from the angular deflection of the instrument pointer. Or, a temperature may be measured by the expansion of liquid in a thermometer, and its value read from the height of the liquid column in the capillary.

* Electrical Instruments Section, Electricity Division, National Bureau of Standards, Washington, D.C.

UNIVERSALITY OF ERROR

Whatever scheme of measurement is used, the value of the numeric assigned as a result of the measurement to describe the magnitude of the measurand will be in error to a greater or lesser extent, i.e., will depart somewhat from the true value of the quantity. No measurement, however elaborate or precise, or how often repeated, can ever completely eliminate this uncertainty. Thus, the true value of a measured physical quantity can never be stated with complete exactness. One of the most important phases of the art of measurement consists in the reduction of measurement errors to limits that can be tolerated for the purpose at hand.

Errors are unavoidable in the comparison of an unknown with a reference standard, or in the calibration and use of a measuring system. The value of the reference standard itself is also uncertain by an amount that depends on the whole chain of measurements extending back to the primary standards designed for maintaining all measurements on a common basis.

The established system of electrical units may be taken as an example. By virtue of certain relations between electrical quantities, values of all of them can be established in terms of the ohm and the volt. Values for these latter units are physically maintained by means of a group of wire-wound resistors and a group of saturated cadmium cells, which serve as primary standards at the National Bureau of Standards, Washington, D.C.

The values assigned to the primary standards are the experimental realization, through an elaborate series of measurements, of their legal definition, in terms of the fundamental mechanical units of length, mass, and time. Uncertainties are present at every step of this measurement procedure, extending back to the basic standards—the prototype meter bar and kilogram preserved in the vaults of the International Bureau of Weights and Measures in Sèvres, France.*

The skill of the various scientists who carried out these measurements was such that the uncertainties in values assigned to the primary standards, accumulated in the process of transferring from the mechanical prototypes, is probably less than ten parts in a million for the ohm and slightly more than this amount for the volt. Values assigned to laboratory reference standards, by comparison with these primary standards and with other standards derived from them, are somewhat more uncertain, because of the further measurement procedures involved. Also, the laboratory reference standards themselves are of a lower quality, and are inherently less stable than the primary standards.

Reference standards for laboratory use are readily available for which uncertainties of value are well within a hundredth of a per cent. Some of the best laboratory standard deflecting instruments (ammeters, voltmeters, wattmeters) have errors of not more than a tenth of a per cent, while the errors of the usual grade of working laboratory instruments may amount to a quarter or half per cent, and the better grades of switchboard instruments are in the one per cent accuracy class. Thus, in the case of indicating instruments, the uncertainty of value which must be accepted may be much more a result of inherent limitations in their design, construction, operation, and stability than of uncertainty in the working standards used for their calibration.

SOURCES OF ERROR

In addition to the errors which necessarily result from faulty calibration of a measuring system, there are a number of sources of error that should be examined. These include: (1) noise, (2) response time, (3) design limitations, (4) energy gained or lost by interaction, (5) transmission, (6) deterioration of the measuring system, (7) ambient influences on the system, and (8) incorrect interpretation by the observer.

---

* In addition to the physical standards of length and mass, the prototype meter and kilogram, an absolute measurement of the electrical units requires that a standard of time measurement be set up at the time of the absolute measurement, and that a value of permeability be assigned at the place of the measurement.

Noise in Measuring Systems

Noise may be defined broadly as any signal that does not convey useful information. Extraneous disturbances, generated in the measuring system itself or coming from the outside, frequently constitute a background against which the desired signal must be read. The background noise in a radio receiver is a familiar example; likewise, vibration or sudden displacements of the moving system of an indicating instrument may be classed as noise when they result from vibration pickup or shock excitation.

Sources of Noise. Noise may originate (1) at the primary sensing device, (2) in a communication channel, or other intermediate link, or (3) in the indicating element of the system. In amplifying systems, the signal-to-noise ratio sets an upper limit to the useful amplification and, therefore, a lower limit to the magnitude of the wanted signal that can be observed against the background noise. Such disturbances are particularly a problem in electronic amplifiers, where they may arise from fluctuations in the resistance of a circuit element or contact point, or from microphonics in a vacuum tube. Disturbing signals are, of course, more serious at the input end where they undergo full amplification, than at an intermediate or final stage where their amplification is less, relative to the input signal.

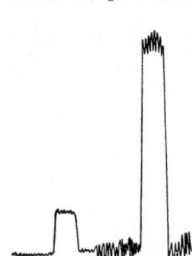

Noise signals may also be picked up by an electrical or mechanical coupling between an external source and an element or communication channel of the system. Another type of noise is the electrical signal produced by static charges generated between the conducting and insulating structure of a cable, as a result of flexing or other mechanical motion of the cable. Wherever signal disturbances are present, they contribute to the uncertainty of the measurement.

FIG. 1. Reproduced from a tracing of voltage record, showing a 0.02-μv signal impressed on a noise background of 0.001 μv. The amplification was increased by a factor of five for the second record. In each instance, the height of the signal is 20 times the height of the random background.

Fundamental Character of Noise. Under the most favorable circumstances, where noise signals are reduced to a minimum through filtering, careful selection of components, shielding, and isolation of the entire measuring system, there still exist certain noise levels that are always present. These random disturbances result from the fact that, in the final analysis, energy is always transferred in discrete, finite amounts (by molecular action in a gas, electrons as carriers of electric charge, and so on). The structure of natural phenomena is not infinitely fine grained. Although the magnitude of fluctuations in the rate of energy transfer is usually small compared with the total transfer involved in a measurement, nevertheless these fluctuations supply a noise background, and limit the ultimate sensitivity to which a measurement can be carried. Examples of sensitivity-limiting mechanisms are Brownian motion in a mechanical system, the shot effect in the heated cathode of a vacuum tube, the Johnson effect in conductors or resistance elements, and the Barkhausen effect in magnetic elements.

Figure 1 shows the voltage background resulting from Johnson noise in a low-resistance network. The figure was traced originally from an actual voltage record. The voltage from the network was fed through an amplifier to a recorder. The background voltage is approximately 10^{-3} μv, and the signal voltage is 20 times as great (2×10^{-2} μv) in each instance. The gain of the amplifier is different in the two records, but the amplitude of the background voltage (noise) is about 5 per cent of the deflection produced by the signal in each record. With either low or high gain in the amplifier, the minimum signal that could be determined against the existing background would be about the same.

Response Time

The time of response of a measuring system to an impressed signal may also contribute to the uncertainty of the measurement. If the signal is not constant in value, lag or delay in the system response results in an indication whose value depends on a sequence of values of the stimulus over an interval of time.

As an illustration of the effect of response time on indication, consider a simple system such as a mercury thermometer in air; or a pressure gage consisting of a bellows connected to a source of pressure by a tube of small diameter. Heat must flow through the wall of the thermometer bulb and be absorbed by the mercury, raising its temperature and expanding it; or gas must flow through the tube to increase the pressure in the bellows and expand it. In either instance, the measuring system has capacitance and is connected through a resistance to the system under measurement. Such a measuring system is known as a first-order system, since its dynamic response can be expressed by a first-order differential equation.

This response is shown in Fig. 2a for a step change in the measurand, and in Fig. 2b for a linear change. It will be observed that, for both types of change, the response equation has a transient term that describes the initial response, and a steady-state

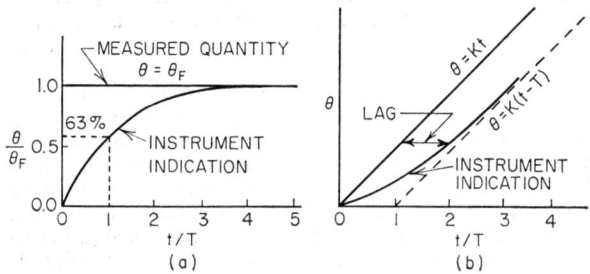

FIG. 2. Response of a measuring system: (a) to a step change; (b) to a linear change.

(a) Response to step change:

$$T \frac{d\theta}{dt} + \theta = \theta_F$$

$$\theta = \theta_F (1 - e^{-t/T})$$

(b) Response to linear change:

$$T \frac{d\theta}{dt} + \theta = Kt$$

$$\theta = K(t - T) + KTe^{-t/T}$$

term (the asymptotic value of the response curve) that describes the behavior of the system after an interval which is long compared with the response time T. It will be seen that, in these simple examples, the indication of the measuring system is a function of its response time as well as of changes in the measurand.

More complicated systems or situations can, in many instances, be analyzed only approximately. But it can be generally stated that, for any system having a finite response time, the indication at any instant is the result of the events which happened over a previous time interval; that the magnitude of the indication not only depends on the variation of the signal over an interval of time previous to the observation, but may also depend in a more or less complicated way on the response characteristics of the system itself.

Design Limitations

Limitations and defects in the design and construction of measuring systems are also factors in the uncertainty of measurements.

Friction. In moving parts, friction not only contributes an uncertain amount to the damping of the system, but, because a certain minimum force is required to overcome the friction and to initiate motion, there results an uncertainty in the rest position of the indicator. This uncertainty can frequently be reduced by gentle tapping of the instrument, such that the forces transmitted to the moving system are just enough to overcome the frictional forces present that prevent the system from assuming its natural rest position. However, vigorous tapping may defeat its purpose by supplying enough force to produce deflection, thus moving the indicator to a new position which is still a function of friction. Also, if the bearings are delicate, they may be injured by vigorous tapping and, thus, increase the friction even more.

Resolving Power. Broadly speaking, this is the ability of the observer to distinguish between nearly equal quantities. In an optical system, such as that of a microscope, resolving power may be stated in terms of the smallest angle at which points in the field of view can be distinguished as separate. If the lenses in the optical train were perfect, this angle would be fixed by the effective aperture of the objective lens and the wavelength of the light used.

For a measuring system in which a scale reading is used to determine magnitudes, resolution would be limited by the smallest fraction of a scale division that could be read with certainty. Since, without a vernier, observers may attempt to read to a tenth of a scale division, but will not generally agree on a reading to better than perhaps two-tenths division, this latter figure probably represents about the limiting resolution of the scale.

In these instances, the limiting (or theoretical) resolving power is an ideal that is never attained because of imperfections or defects in the measuring system. For example, the scale of a good electrical instrument is usually constructed by accurately locating a series of cardinal points (or major scale divisions) by electrical measurements. The scale subdivision between cardinal points is then done by some sort of dividing mechanism. Thus, the actual location of a particular division is uncertain, both by the error of the measurements made to locate the adjacent cardinal points, and by the imperfections of the mechanical device used to subdivide the scale.

Energy Exchanged by Interaction

Wherever the energy required for operating the measuring system is extracted from the measurand, the value of the latter is altered to a greater or lesser extent.

Low Energy Levels. The preceding statement is always true in some degree, but is particularly significant where the available energy is limited in amount. The heat capacity of the thermometer used in a calorimetric experiment must be taken into account. A voltmeter, used to measure the voltage of a high-resistance source, draws current so that its reading is in error by the amount of the internal impedance drop in the source.

Coupling and Feedback Errors. In instances where energy is supplied to the measuring system from an auxiliary source, the value of the measurand may be altered by coupling to the measuring system and consequent feedback of energy. This may occur in electric circuits, for example, through resistive, inductive, or capacitive coupling, and in mechanical circuits through elastic or viscous coupling. Actually, of course, the flow of energy may be in either direction under appropriate circumstances.

Interference. A third type of interaction is interference of an element of the measuring system with the action of the measurand. A simple example is the alteration of flow resulting from the introduction of an orifice plate into a pipe. Another example—a milliammeter, in measuring current in a low-resistance circuit, introduces additional resistance, and may alter the current by a significant amount.

Transmission

In the transmission of information from sensing element to indicator, any (or all) of three types of error may arise: (1) The signal may be attenuated by being absorbed or otherwise consumed in the communication channel; (2) it may be distorted by attenuation, resonance, or delay phenomena whose actions are selective on various signal components; or (3) it may suffer loss through leakage. In any of these circumstances, the signal reaching the indicator will differ in some respects from that at the primary sensing element.

Deterioration of Measuring System

Changes in the measuring system itself constitute a source of error in measurement. Physical or chemical deterioration or other alterations in characteristics of measuring elements may change their response and their indication. The oxidation of weights used in a laboratory balance, the change in resistance of a circuit element through strain relief, or the weakening of a permanent magnet, or a bellows, through aging, are examples. The alteration of thermocouple characteristics in an oxidizing or a reducing atmosphere is another example.

Ambient Influences on Measuring Systems

Of the various ambient conditions that may alter the calibration of an instrument, temperature influence almost always affects the measurement in one way or another. As examples, a change in temperature may alter the elastic constant of a spring, change the dimensions of a measuring element or linkage in the system, alter the resistance of an electric circuit element, or the flux density in a magnetic element.

Other influences that are not so universally active, but that are often important are: humidity, barometric pressure; the presence of smoke or other foreign constituents in the air; the effect of the departure of an unbalanced mechanical system from its proper operating position; and several other influences which come from outside the instrument, but which tend to affect its calibration.

Errors of Observation and Interpretation

Personal errors in the observation, interpretation, and recording of data must also be considered among the sources of uncertainty in measurements.

Parallax Errors. If the observer's line of sight to the pointer of an indicating instrument is not perpendicular to the scale face of the instrument, the reading will be high or low, depending on whether the observer's eye is to the left or right of this perpendicular. The amount of the parallax error will depend on the height of the pointer above the scale, and on the sine of the angle between the line of sight and the perpendicular.

Linear Interpolation of Scales. In estimating the correction to be applied to readings of an indicating instrument, it is common practice to determine by measurement the corrections to be applied at the cardinal points of the scale, and then to determine by linear interpolation the corrections to be applied at intermediate scale points. Such a procedure is simple and convenient, but can be completely justified only if the scale is uniform and if the subdivisions between the cardinal points are correctly located. If the scale is not uniform, or if errors are present in its subdivision, the determination of corrections by linear interpolation can result in error.

Personal Bias of Observer. Another reader error results from conscious or unconscious bias on the part of the observer. Some individuals will tend to favor even (or odd) tenths in estimating indicator position between marked scale divisions, or perhaps will tend to read high in the lower half of the interval and low in the upper half. These bias patterns, of which the observer is not usually aware, vary considerably from one individual to another. They are consistently followed over long periods of time and generally tend to symmetry about the midpoint of the interval. The prevalence of observer bias patterns is such that readings made by a single observer are always questionable by a tenth of a division and are frequently in error by two tenths, even when the observer is consistent in his readings.

Mistakes. Faulty logging, either through writing down an incorrect digit or through transposition of digits, obviously must be classed as an observer error.

CLASSIFICATION OF ERRORS

In estimating the magnitude of the uncertainty or error in the value assigned to a quantity as the result of measurement, a distinction must be made between two general classes of error: (1) systematic and (2) random. Apart from calibration errors which result in the consistent use of incorrectly assigned values and which are, therefore, systematic, the various sources of error give rise in some cases to systematic errors and in others to random errors.

Systematic Errors

Systematic errors are those which are consistently repeated with repetition of the experiment. Faulty calibration of the measuring system or a change in the system that causes its indication to depart consistently from the value assigned in calibration are of this type. Examples might include changes with age of the elastic properties of a spring or diaphragm, or the reduction in strength of a magnet through shock or aging. Failure to take into account the energy extracted from a low-level source to operate the measuring system would also result in a systematic error.

As the observer is usually unaware of the presence or magnitude of systematic errors (since otherwise he could make appropriate corrections), this type of error is difficult to evaluate. It cannot be demonstrated by merely repeating the measurement under identical conditions, as this would lead to results that were consistently wrong. The skillful experimenter may be distinguished from the unskilled by his ability to plan and perform measurements in such a way as to minimize or avoid systematic errors.

In searching for systematic errors and in evaluating them, it is generally helpful to make definite known changes in those parameters of the measurement that are under the operator's control; and to use different instruments or, if possible, a different method of measurement. In this way, errors that are functions of one of the controlled parameters are changed in magnitude; or those that arise from incorrect instrument calibration or that are inherent in a particular method may be altered. At times, it is possible to measure something whose magnitude is accurately and independently known and which is similar to the measurand. Such a measurement constitutes a check on the calibration of the measuring system and should help in evaluating systematic errors that may be present. Checking the accuracy of a micrometer with calibrated gage blocks or the accuracy of a thermometer at one or more fixed points on the temperature scale are of this nature. Another example is the checking of a potentiometer by measuring with it the emf of the standard cell that was used to standardize its current, or alternatively, the measurement of the emf of another standard cell whose value is independently known.

Random Errors

Random errors are those which are accidental; whose magnitude (and sign) fluctuates in a manner that cannot be predicted from a knowledge of the measuring system and the conditions of measurement.

In the measurement of any physical quantity, the observations are influenced by a multitude of contributing factors. These are the parameters of measurement. In an ideal measurement, all the parameters are fixed in value, so that the magnitude of the measurand is completely defined and may be exactly determined. Repeated observations of the magnitude of a quantity differ as a result of the operator's failure to (1) control the parameters closely enough, or (2) to apply proper corrections for their influence. These uncontrolled, or unknown, or uncorrected influences disturb the observations and cause the result to depart from the value it would have if all parameters were completely defined and fixed.

If it is assumed that the various parameters beyond the observer's control (and the uncontrollable residue of those he tries to hold fixed) act in a completely random fashion, probability theory can be used to deduce certain helpful results. Now, combinations of circumstances that produce large departures of the observed from the true value of a quantity will occur less frequently than those producing small deviations, and each influence may be presumed equally likely to cause either a positive or a negative departure. Hence, the general tendency of the effects of all influences will be to cancel one another rather than to be additive. If all measurement errors followed the laws of chance, it could be expected that the true value of a quantity would be the average of an infinite number of observations.

There is an important weakness in the foregoing argument. It may well be that some of the parameters influencing the result of a measurement do not act in a completely random fashion. There is never assurance that all systematic errors have been eliminated from a measurement, or that proper corrections have been made for those errors present. Under such conditions, the average of a large number of observations would not approach the true value of the quantity, but would differ from the truth by the algebraic sum of the uncorrected systematic errors. The laws of probability take no account of an unknown but constant error superposed on deviations that are truly random.

In Table 1, a classification is suggested of the various sources of error with respect to the type of error produced. It will be seen that errors from a particular source cannot, in general, be uniquely classified as either systematic or random. For example, noise is inherently random in nature; but if the response of the indicating system is always in the same direction independently of the signal polarity, then, for a weak signal, the noise present may produce a positive indication that is larger on the average than would be produced, in the absence of noise, by the signal being measured. Thus, for some response systems, noise may give rise to a spurious signal that is always added to the measured value, and its effect is systematic. To illustrate this point further, it is suggested that the reader attempt his own classification of errors, and then look for the exceptions in the various categories.

Table 1. Types and Sources of Measurement Errors

Source of error	Type of error	
	Random	Systematic
Noise..........................	Generally	May be
Response time................	Seldom	Almost always
Design limitations.............	Usually	Sometimes
Energy of interaction..........	May be	Usually
Transmission.................	Sometimes	Usually
Deterioration.................	Seldom	Usually
Ambient influences............	Usually	May be

EVALUATION OF DATA

Statistical procedures are available that make it possible to state from a limited group of data the most probable value of a quantity, the probable uncertainty of a single observation, and the probable limits of uncertainty of the best value that can be derived from the data. It must be borne in mind here that the objective is precision (or consistency) of values rather than their accuracy (or approach to the truth). That this is necessarily so has already been pointed out; the laws of chance operate only on random errors, not on systematic errors. Some of the important results of the theory of probability, as applied to the treatment of data, are given in the following paragraphs.

Arithmetic Average or Mean

The best value that can be obtained from a group of similar measurements of a quantity is usually the arithmetic average or mean.

Suppose, for example, that a series of measurements are made (all with the same care) of the length of a rod, wherein a scale divided into millimeters is used. Estimations are made to tenths of a millimeter. To avoid a systematic error that might result from an attempt to align the end of the rod with the end of the scale, the rod is laid along the scale. The scale position corresponding to each end of the rod is read

The rod is shifted along the scale for each new pair of readings, and the difference between the members of the pair is taken as the observed value of the rod length. As an example, the results of ten such measurements are as follows:

Reading A, mm	Reading B, mm	Difference, observed length, mm
78.8	37.1	41.7
67.4	25.4	42.0
92.1	50.3	41.8
56.8	14.8	42.0
88.1	46.0	42.1
50.5	8.6	41.9
74.7	32.7	42.0
82.4	40.5	41.9
61.6	19.1	42.5
65.4	23.6	41.8

The arithmetic average of the observed values is

$$X = \frac{\Sigma x}{n} = \frac{419.7}{10} = 41.97$$

Standard Deviation

One of the best measures of the dispersion of a set of observations is the root mean square of the deviations of individual observations from the arithmetic average of the set. This is the standard deviation, and can be expressed by

$$\sigma = \sqrt{\frac{\Sigma d_m^2}{n}} \tag{1}$$

where d_m is the deviation of the individual observation from the group mean, and n is the number of observations in the set. Statisticians customarily give as an estimate of σ the quantity

$$s = \sqrt{\frac{\Sigma d_m^2}{n-1}} \tag{2}$$

because it more accurately describes the dispersion of the set when the number of observations is small. It will be noted that the value of s approaches that of σ as n is increased, and differs from it by a significant amount only when n is quite small. If the difference is taken between each of the individual observations of rod length and their arithmetic average in the example given in the preceding paragraph, the standard deviation can be computed as follows:

d_m	d_m^2
-0.27	7.29×10^{-2}
0.03	0.09×10^{-2}
-0.17	2.89×10^{-2}
0.03	0.09×10^{-2}
0.13	1.69×10^{-2}
0.07	0.49×10^{-2}
0.03	0.09×10^{-2}
-0.07	0.49×10^{-2}
0.53	28.09×10^{-2}
-0.17	2.89×10^{-2}

$$\sigma = \sqrt{\frac{\Sigma d_m^2}{n}} = \sqrt{\frac{44.10 \times 10^{-2}}{10}} = 0.21 \text{ mm}$$

$$s = \sqrt{\frac{\Sigma d_m^2}{n-1}} = \sqrt{\frac{44.10 \times 10^{-2}}{9}} = 0.22 \text{ mm}$$

It will be apparent that, even with as small a number as ten observations, there is little or no significant difference between σ and s.

Probable Error of a Single Observation

As an index of the consistency of a set of observations, the probable error of a single observation may be used. It is defined as that deviation from the group mean for which the probability that it will be exceeded is equal to the probability that it will not be exceeded. If the number of observations is large, the probable deviation of a single observation from the mean is given by

$$r = \pm 0.6745\sigma \tag{3}$$

where σ is the standard deviation defined previously. If the number of observations is small, this formula should be modified to

$$r = \pm 0.6745 \sqrt{\frac{\Sigma d_m{}^2}{n-1}} \tag{4}$$

While probable error expresses correctly the range in which the chances are equally good that the numerical value will or will not be found, provided a sufficiently large group of data is available to make this estimate meaningful, it has actually no more significance as a precision index than the standard deviation from which it is derived. Hence, it has largely fallen into disuse, in favor of a statement of standard deviation. It may be found, however, in older literature as an index of the precision of data.

In the foregoing example,

$$r = \pm 0.6745 \sqrt{\frac{\Sigma d_m{}^2}{n-1}} = \pm 0.6745 \sqrt{\frac{44.10 \times 10^{-2}}{9}} = \pm 0.15 \text{ mm}$$

Probable Error of the Mean

This is defined as the amount R by which the mean of a group of observations may be expected to differ (with a probability of 50 per cent) from the mean of an infinite set taken under the same conditions of measurement. It can be calculated by the formula

$$R = \frac{r}{\sqrt{n}} \tag{5}$$

or, if the number of observations is large, by its equivalent

$$R = \pm 0.6745 \frac{\sigma^2}{\sqrt{n}} \tag{6}$$

the symbols having the same meaning as in the foregoing description. Only if the number of observations is large will the value of R, calculated from Eq. (5), accurately represent the departure of the group mean from the mean of an infinite set with a 50 per cent probability. If the number of observations is small, R becomes an increasingly inaccurate index of precision. The present trend in the statistical treatment of data is toward the abandonment of probable error in favor of a much broader and more useful concept, that of "confidence intervals," which are described in the following paragraphs.

In the previous example, the probable error of the mean of ten observations is given by

$$R = \frac{r}{\sqrt{n}} = \frac{\pm 0.15}{\sqrt{10}} = \pm 0.05 \text{ mm}$$

Confidence Intervals

Probable error, previously defined, is a special case of a much broader concept. It is possible by the statistical analysis of data to state a range of deviation from the mean value within which a certain fraction of all values may be expected to lie. Such a range is called a "confidence interval," and the probability that the value of a randomly selected observation will lie within this range is called the "confidence level." If the number of observations is large and their errors are random (the normal distribution of errors), various confidence intervals about the mean value μ are as stated in Table 2.

Table 2. Confidence Level and Confidence Interval Values Where Number of Observations Is Large

Confidence level	Confidence interval	Values lying outside confidence interval
0.50	$\mu \pm 0.674\sigma$	1 in 2
0.80	$\mu \pm 1.282\sigma$	1 in 5
0.90	$\mu \pm 1.645\sigma$	1 in 10
0.95	$\mu \pm 1.960\sigma$	1 in 20
0.99	$\mu \pm 2.576\sigma$	1 in 100
0.999	$\mu \pm 3.291\sigma$	1 in 1,000

If the number of observations is small and the standard deviation σ is, therefore, not accurately known, these intervals must be broadened. Here one would compute $S = d_m^2/(n-1)$, and multiply it by an appropriate factor, as shown in Table 3, to establish the confidence interval, i.e., the interval within which one would expect to find a randomly selected observation, with a particular level of confidence.

It will be noted that the factors in the table corresponding to an infinite number of observations are the same as the factors multiplying σ in Table 3.

Table 3. Factors for Establishing Confidence Interval Where Number of Observations Is Small

Number of degrees of freedom	Number of observations	Confidence level			
		0.5	0.9	0.95	0.99
1	2	$\mu \pm 1.00s$	$\mu \pm 6.31s$	$\mu \pm 12.71s$	$\mu \pm 63.66s$
2	3	$\mu \pm 0.82s$	$\mu \pm 2.92s$	$\mu \pm 4.30s$	$\mu \pm 9.92s$
3	4	$\mu \pm 0.77s$	$\mu \pm 2.35s$	$\mu \pm 3.18s$	$\mu \pm 5.84s$
4	5	$\mu \pm 0.74s$	$\mu \pm 2.13s$	$\mu \pm 2.78s$	$\mu \pm 4.60s$
5	6	$\mu \pm 0.73s$	$\mu \pm 2.02s$	$\mu \pm 2.57s$	$\mu \pm 4.03s$
6	7	$\mu \pm 0.72s$	$\mu \pm 1.94s$	$\mu \pm 2.45s$	$\mu \pm 3.71s$
7	8	$\mu \pm 0.71s$	$\mu \pm 1.90s$	$\mu \pm 2.37s$	$\mu \pm 3.50s$
8	9	$\mu \pm 0.71s$	$\mu \pm 1.86s$	$\mu \pm 2.31s$	$\mu \pm 3.36s$
9	10	$\mu \pm 0.70s$	$\mu \pm 1.83s$	$\mu \pm 2.26s$	$\mu \pm 3.25s$
10	11	$\mu \pm 0.70s$	$\mu \pm 1.81s$	$\mu \pm 2.23s$	$\mu \pm 3.17s$
15	16	$\mu \pm 0.69s$	$\mu \pm 1.75s$	$\mu \pm 2.13s$	$\mu \pm 2.95s$
∞	∞	$\mu \pm 0.67s$	$\mu \pm 1.64s$	$\mu \pm 1.96s$	$\mu \pm 2.58s$

NOTE: This table is a modification and abridgment of Table IV in Fisher and Yates, "Statistical Tables for Biological, Agricultural, and Medical Research," Oliver & Boyd, Ltd., Edinburgh and London.

To obtain the confidence intervals for the group mean from the corresponding intervals for an individual observation, the latter is divided by $\sqrt{n}$. Thus, the expectation that the mean of a group of observations will not differ by more than a certain amount from the theoretical mean of an infinite set can also be expressed in terms of a confidence interval and a confidence level.

In the previous example for the measurement of rod length, $s = 0.22$ mm, so that the confidence intervals for an individual observation corresponding to various confidence levels are:

Confidence level	0.50	0.90	0.95	0.99
Confidence interval	±0.15	±0.40	±0.50	±0.71

For the group average (41.97), the corresponding confidence intervals are:

$$±0.05 \qquad ±0.13 \qquad ±0.16 \qquad ±0.23$$

Thus, the average value (41.97) found from the ten observations made would be expected, with a probability of 50 per cent, to be within 0.05 mm of the mean of an infinite number of measurements. At the 99 per cent level of confidence, it may be stated that the observed and theoretical means differ by not more than 0.23 mm.

Rejection of Data. Occasionally, one individual value in a set of observations is noticeably different from the others, and a decision must be made to either use or discard it. If the observer knows at the time the data are taken that the system is not behaving properly or that a blunder has been made, the observer should immediately discard the data. If the observer is not able to assign a valid reason for discarding it at the time the observation is made, it is questionable whether it should be eliminated at all. The cause should be sought when a single observation is outstandingly different from the others of the set. It may be that the quantity being measured has changed temporarily, or that there has been some other significant change in the conditions of measurement. To eliminate an observation simply because it differs from the others by more than normally would be expected is not sound practice.

A criterion sometimes used for discarding an observation is that its deviation from the mean is greater than four times the probable error of a single observation. This corresponds to discarding data that lie outside a confidence interval for a single observation at a level of 0.993. A better criterion, which would avoid the difficulty of estimating probable error when the set is small and the standard deviation σ is not accurately known, would be to discard data that lie outside the interval corresponding to a confidence level of 0.99 for a single observation. On this basis, not more than 1 in 100 observations would lie outside this range if only random influences were operating to produce dispersion in the data. However, rather than to use such a criterion arbitrarily as a reason for discarding an observation, it is perhaps better to use it as a criterion for thoroughly examining the conditions of the measurement in order to find its cause. Better yet, the interval corresponding to a confidence level of 0.95 may be used as a criterion for the need to scrutinize the measurement procedure and the quantity measured.

In the example of the measurement of rod length, the ninth observation in the set differed from the mean by 0.53 mm. This is outside the interval $(\mu 0.50)$ corresponding to a confidence level of 0.95, but inside the interval $(\mu 0.71)$ corresponding to the 0.99 level. On the basis of the rejection criterion just stated, the observation should be retained.

Comparison of Averages. If two sets of measurements, taken under different conditions, yield different average values (A and B) of a quantity, with different standard deviations σ_A and σ_B, the question of their consistency arises. Is the difference $(A - B)$ consistent with the assumption that random errors alone operate? Or does the difference result from the presence of systematic errors that operate differently in sets A and B? The sets are inconsistent if the difference of their averages is more than twice the sum of their standard deviations. This is a rough but simple test to apply. Thus, A and B are inconsistent (show the operation of systematic errors) if $(A - B) > 2(\sigma_A + \sigma_B)$.

A much more sensitive test is based on the confidence intervals previously described. Furthermore, this test (the t test) is equally applicable to large and small sets. Let the individual values in the first set be $x_1, x_2, \ldots, x_m$ (m observations) with individual differences from the average $d_i = x_i - A$. Let the values in the second set be $y_1, y_2, \ldots, y_n$ (n observations) with the individual differences from the average $f_i = y_i - B$. Then

$$s^2 = \frac{\Sigma d_i{}^2 + \Sigma f_i{}^2}{(m-1)+(n-1)} \tag{7}$$

The estimated standard deviation of an individual measurement is s, and that of the difference between the average is

$$\frac{s}{\sqrt{mn/(m+n)}} \tag{8}$$

The ratio of the difference of the averages to the standard deviation of the difference is

$$t = \frac{A-B}{s} \sqrt{\frac{mn}{m+n}} \tag{9}$$

This computed t ratio is a measure of the consistency of the two sets of measurements, and should be compared with the factor previously used to multiply s in establishing confidence intervals. The critical values of t are given in Table 4 for confidence levels of 0.99 and 0.95, corresponding to 1 chance in 100 or 1 in 20, respectively, that random errors alone cause a larger difference.

Table 4. Values of t Used in Establishing Confidence Levels

Total number of observations $(m+n)$...	4	6	8	10	12	17	32	∞
Degrees of freedom $(m-1)+(n-1)$...	2	4	6	8	10	15	30	∞
Critical value of $t_{.99}$....................	9.92	4.60	3.71	3.36	3.17	2.95	2.75	2.58
Critical value of $t_{.95}$....................	4.30	2.78	2.45	2.31	2.23	2.13	2.04	1.96

NOTE: More elaborate tables of t will be found in most modern textbooks on statistical methods. For example, table 5 in the appendix of W. J. Dixon and F. J. Massey, Jr., "Introduction to Statistical Analysis," 2d ed., McGraw-Hill Book Company, Inc., New York, 1957.

If the computed ratio is significantly larger than the critical value of t, it may be concluded, with a corresponding level of confidence, that there are systematic errors which operate differently in A and B. Of course, the presence of systematic errors that affect both sets of measurements in the same way will not be disclosed by either test.

Propagation of Errors. It frequently occurs that independent measurements $x_1, \ldots, x_n$ are obtained, and that some function of them $f(x_1, \ldots, x_n)$ is of interest. If $f(x_1, \ldots, x_n)$ has continuous first and second derivatives, and if each x_i has approximately a normal distribution with a mean μ_i and standard deviation σ_i, where each σ_i is small, then $f(x_1, \ldots, x_n)$ has approximately a normal distribution with mean $f(\mu_i, \ldots, \mu_n)$ and standard deviation

$$\sigma_f = \sqrt{\sum_{i=1}^{n} \left(\frac{\partial f}{\partial x_i}\right)^2 \Bigg|_{x_i = \mu_i} \cdot \sigma_i{}^2} \tag{10}$$

For practical purposes, since the mean values μ_i are usually not known, the observed x_i may be used in evaluating the partial derivatives. Subject to the stated restrictions, the standard deviation of $f(x,y,z, \ldots)$, where $x, y, z, \ldots$ are independent, is

$$\sigma_f = \sqrt{\left(\frac{\partial f}{\partial x}\right)^2 \sigma_x{}^2 + \left(\frac{\partial f}{\partial y}\right)^2 \sigma_y{}^2 + \left(\frac{\partial f}{\partial z}\right)^2 \sigma_z{}^2 + \cdots} \tag{11}$$

This can be written in the following simple instances as follows:

Let $\qquad f = ax + by \qquad$ then $\qquad \sigma_f \approx \sqrt{a^2\sigma_x^2 + b^2\sigma_y^2}$ (12)

$\qquad\qquad f = kxy \qquad\qquad\qquad\quad \sigma_f \approx k\sqrt{y^2\sigma_x^2 + x^2\sigma_y^2}$ (13)

$\qquad\qquad f = k\dfrac{x}{y} \qquad\qquad\qquad \sigma_f \approx k\sqrt{\sigma_x^2 + \left(\dfrac{x}{y}\right)^2 \sigma_y^2}$ (14)

$\qquad\qquad f = kx^n \qquad\qquad\qquad\ \sigma_f \approx nkx^{n-1}\sigma_x$ (15)

$\qquad\qquad f = ke^x \qquad\qquad\qquad\ \sigma_f \approx ke^x - \sigma_x$ (16)

$\qquad\qquad f = k\log_e x \qquad\qquad\ \sigma_f \approx \dfrac{k}{x}\sigma_x$ (17)

Normal Distribution Law. The normal pattern of distribution about a mean value is shown in Fig. 3. Starting with the assumption that the most probable value of a quantity is the arithmetic mean of a large number of determinations, each made with equal care, the normal law of distribution

$$y = \frac{h}{\sqrt{\pi}} e^{-h^2 x^2}$$ (18)

can be deduced. Here x is the amount that a particular observation deviates from the group mean, and y is the frequency of occurrence of such a deviation in the group.

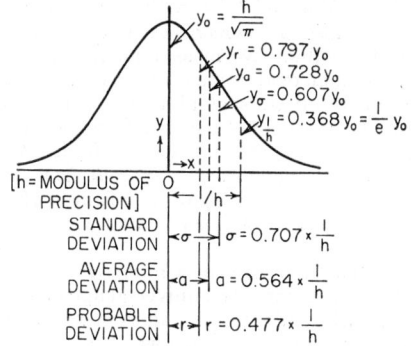

FIG. 3. Indices of precision.

Normal distribution law: $\qquad\qquad\qquad\qquad y = \dfrac{h}{\sqrt{\pi}} e^{-h^2 x^2}$

Probability of deviation between $-x$ and x: $\quad P_x = \dfrac{h}{\sqrt{\pi}} \displaystyle\int_{-x}^{x} e^{-h^2 x^2}\, dx$

The precision (or repeatability) of the observations in the group is indicated by the maximum height $h/\sqrt{\pi}$, and by the deviation $x = 1/h$ at which the height is $1/e$ times the maximum. Thus, the parameter h is an indication of the narrowness or "spread" of the curve and has been called the "modulus of precision."

Certain indices of precision sometimes used (namely, the standard, average, and probable deviation of observations from the mean) are indicated in Fig. 3. The frequency of occurrence of each in a normal distribution is also shown. Ordinates of the normal distribution curve can be considered as probability coefficients; the probability of a deviation x from the mean is given by the area under the curve between x and y. Thus

$$P_x = \frac{h}{\sqrt{\pi}} \int_{-x}^{x} e^{-h^2 x^2}\, dx$$ (19)

The law of normal distribution can also be written in terms of standard deviation as

$$y = \frac{1}{\sigma \sqrt{2\pi}} \exp \left[-\frac{1}{2} \left(\frac{x - \mu}{\sigma} \right)^2 \right]$$ (20)

where σ is the standard deviation and μ is the mean value.

Significant Figures

In computing or stating the results of a measurement, it is important that the recorded figures include all that convey usable information. Frequently, it is also important that figures be excluded that do not convey information of value in the determination. In recording a numerical result, if only those digits that are meaningful are set down, the manipulation of surplus digits is avoided in subsequent computations, and opportunities for arithmetical errors are lessened.

Because data are more easily tested in terms of precision (or repeatability) than in terms of accuracy (or approach to the true value of the quantity), the number of digits making up the significant figures is frequently the recorded statement of the experimenter's precision of measurement. Wherever statistical tests of precision are made, or if there is any possibility that they may be made in the future, the recorded data should include the numerical figures which contain information on the variation between determinations. This applies not only to the worker's notebook, but also to published data that may be subjected to statistical analysis by an interested reader. Usually, the experimenter is also the person best qualified to form a judgment concerning the accuracy of his results. No measurement can ever be considered really complete without some kind of an evaluation of the accuracy of results. If the worker distrusts the accuracy of his results to such an extent that, in rounding off numerical values, he suppresses the information needed to determine their precision, his work is thereby decreased in value. When this must be done, it should be set forth in an accompanying statement if it is not clearly apparent from the results themselves. There are certain conventions regarding significant figures which have attained wide acceptance, but which should be used judiciously and with the foregoing argument in mind.

Retention of Digits. The last digit in a numerical result should represent the point of uncertainty. Although there is no universal agreement on a rule for deciding how many digits to record when tabulating values of known precision, it is usually considered acceptable to retain the last figure which is uncertain by not more than 10 units. Thus, the value 24.3 would at best lie between 24.2 and 24.4, and at worst between 23.3 and 25.3.

Rounded Numbers. In rounding off a number, the last retained digit should be increased by one unit when the first dropped digit is greater than 5, or is 5 followed by digits other than zero. The last digit should not be changed when the dropped digit is less than 5. Thus, if 24.352 were rounded off to three significant figures, it would be stated as 24.4, whereas 24.349 would be stated as 24.3. When the dropped digit is 5 and no further figures follow it, a practice which is frequently followed is to round off to the even number. Thus, 24.35 would round off to 24.4, whereas 24.25 would be rounded off to 24.2. To a limited extent, such a practice improves a value obtained by averaging, since the rounding will cause an increase to the higher digit about as often as a decrease to the lower digit.

Significant Zeros. To avoid misunderstanding and for convenience in multiplication and division, zeros which are not significant, but which serve only to indicate the location of the decimal point, should not be used in the stated number. It is better to indicate the location of the decimal point exponentially, using an appropriate power of 10. Thus, 12,500 is ambiguous as stated. One does not know whether three or five significant figures are intended. If there are only three significant figures, the ambiguity is eliminated if it is written 125×10^2 or 1.25×10^4. If five-place significance is intended, there is no ambiguity if it is written 125.00×10^2 or 1.2500×10^4.

When numbers are to be added or subtracted, they must, of course, first be reduced to the same units and expressed in terms of a common power of 10.

Multiplication and Division. In multiplication and division, we need retain in each factor only the number of digits which will produce in that factor a percentage uncertainty that is no greater than the uncertainty in the factor having the fewest significant figures. Thus, the product of 103.24 and 8.1 would be written $103 \times 8.1 = 83$. The factor 8.1 being known only to about 1 per cent, the factor 103.24 need be considered only to 1 per cent (or 103).

Addition and Subtraction. In these operations, no digit need be retained in the result whose position with respect to the decimal point is to the right of the last significant figure in any number entering the computation. Thus, 24.3 and 2.102 would be added as 24.3 plus 2.1 = 26.4. No digit in the result, farther to the right of the decimal point, would have a meaning.

Slide-rule Computation. If the accuracy of the result need not be better than ¼ per cent, a 10-in. slide rule is adequate for the computation. If accuracy requirements are better than ¼ per cent, machine methods, longhand, or logarithms should be used.

Computation with Logarithms. In computing with logarithms, no more digits need be retained in the mantissa of the logarithm than are significant in the corresponding numerical factor. Thus, the Briggsian logarithm of 103.2 may be written as 2.0137 rather than 2.013679.

Averaging. In taking the average of four or more numbers, an additional digit beyond those of the individual values may be retained as having possible significance.

Precision Index. A number representing a precision index need never be stated to more than two significant figures.

REFERENCES

1. ASTM Manual on Presentation of Data, 1945.
2. Curtis, H. L.: "Electrical Measurements," McGraw-Hill Book Company, Inc., New York, 1937.
3. Harris, F. K.: "Electrical Measurements," John Wiley & Sons, Inc., New York, 1952.
4. "Handbuch der experimental Physik," vol. 1, Messmethoden, Akademische Verlagsgesellschaft Geest & Portig KG, Leipzig, 1926.
5. Eckman, D. P.: "Industrial Instrumentation," John Wiley & Sons, Inc., New York, 1950.
6. Whitehead, T. N.: "Instruments and Accurate Mechanism," Dover Publications, Inc., New York, 1954.
7. Wilson, E. B., Jr.: "An Introduction to Scientific Research," McGraw-Hill Book Company, Inc., New York, 1952.
8. Dixon, W. J., and F. J. Massey, Jr.: "Introduction to Statistical Analysis," 2d ed., McGraw-Hill Book Company, Inc., New York, 1957.
9. Youden, W. J.: Statistical Methods for Chemists," John Wiley & Sons, Inc., New York, 1951.
10. Worthing, A. G., and J. Geffner: "Treatment of Experimental Data," John Wiley & Sons, Inc., New York, 1943.
11. Bäckström, H. E.: *Z. Instrumentenk.*, vol. 50, pp. 561, 609, 665, 1930; vol. 52, pp. 105, 260, 1932.
12. Cerni, R.: Sources of Instrument Error, *ISA J.*, vol. 9, no. 6, pp. 29–32, June, 1962.
13. Youden, W. J.: Realistic Estimates of Error, *ISA J.*, vol. 9, no. 10. pp. 57–58, October, 1962.
14. Topping, J.: "Errors of Observation and Their Treatment," Reinhold Publishing Corporation, New York, 1957.
15. Chatterton, J. B.: The Uncertainty of Measuring Systems, *Trans. IRE*, vol. I-7, pp. 90–94, March, 1958.
16. Entin, L. P.: Is Instrument Zero Output Really Zero? *Control Eng.*, vol. 6, no. 12, pp. 95–96, December, 1959.
17. Entin, L. P.: What about Scale Factor and Resolution? *Control Eng.*, vol. 7, no. 2, pp. 75–77, February, 1960.

FACTORS IN SELECTION OF
MEASUREMENT METHODS*

By Wilfred H. Howe[†]

   Methods for measuring variables provide varying degrees of performance in applied instrumentation. It is the purpose of this subsection to review the factors in selection of the most suitable method for the application.

   Measurements are often made primarily as the basis for automatic control of a process or operation. The continuous-process industries—such as petroleum refining, chemical processing, electric power generation, paper making, food processing and nuclear energy applications—are excellent examples of highly developed automatic control and of the application of measurement as an integral part of the production process. In this subsection the examples of measurement and its application will be drawn from this field of continuous-flow processing. However, the salient points are applicable to the whole field of automatic control.

WHAT IS THE TRUE SIGNIFICANCE OF THE MEASUREMENT?

   Two factors of major importance affect the significance of a measurement. First, how well does the measurement represent the characteristic or condition supposedly being measured? Second, what does this measurement mean with regard to the actual process operation? The measurement obtained can be significant only to the extent that the primary element is exposed to a condition or characteristic which is (1) truly representative and accurately measured, and (2) of significance to the process.

   For example, in a simple temperature measurement, the value being measured is the temperature of the sensitive portion of the primary measuring element—the bimetal strip or the fluid filling or the resistance wire or the thermocouple junction, depending on the type of instrument used.

   How well the measurement represents the true temperature of the process is affected by a number of factors. In the first place, there is the question as to whether there exists one measurable temperature which is truly representative. There is no problem in a well-stirred water bath, but in many operations, stratification, dead pockets, hot spots, and the like may result in important temperature differences between different points of measurement. The desirable procedure is to eliminate these differences. When this is not practical, two things are essential. First, it is necessary to select the best available measurement point or points. Second, and often neglected, it is necessary when using the measurement to recognize that it is less than truly representative.

Instrument Accuracy

   With the primary measuring element suitably mounted with respect to the process, the next point is instrument accuracy. How well does the instrument output correspond to the primary element? In our thermometer example, how close is the indicated temperature to the actual temperature of the primary element? This question of required accuracy is often a major factor in selection.

---

* From a paper, Effective Selection of Measurements for Process Control by W. H. Howe, published in *Instruments & Control Systems*, August 1959, by permission of the copyright owner, Instruments Publishing Co., Pittsburgh, Pa.

   † Chief Engineer, The Foxboro Co., Foxboro, Mass.

High accuracy is generally available at added expense. Statements of accuracy and performance furnished by instrument manufacturers are a reliable guide. It is usually good judgment to pick out the better rather than the cheaper measurement. On the other hand, unrealistic specifications calling for improbable and often unnecessary performance not only can lead to excessive first cost, but also may result in the selection of an unnecessarily complicated measurement which requires a large amount of maintenance effort.

Accuracy is almost invariably specified on a steady-state basis. Dynamic performance—that is, how the measurement output responds on a time basis to changes in the value being measured—is only beginning to receive the attention it deserves. For example, in industrial practice the thermometer primary element is almost invariably mounted in a thermal well to protect it from the process fluid and to make possible withdrawing, checking, and replacing the element without interrupting process operation.   The thermal well, at best, introduces a considerable time lag in the response of the measurement.   When the measurement is used for control, this lag can have rather startling effects, as illustrated in Figs. 1 to 4.

Effect of Instrument Lag on Dynamic Response

These illustrations refer to a commercial heat exchanger, set up under laboratory conditions. Figure 1 shows the general setup. Temperature of the fluid flowing from the heat exchanger is the measured variable.   Two separate measuring systems are used—one a conventional resistance thermometer mounted in a standard socket immersed in the line; the other a special high-speed resistance thermometer directly immersed in the stream to give faster response to stream temperature.

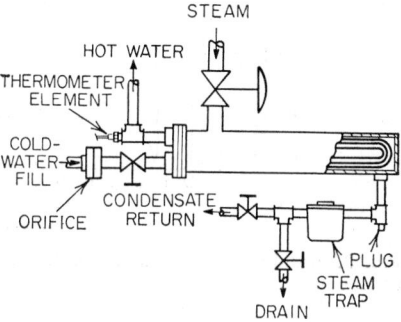

Fig. 1. Temperature of fluid leaving heat exchanger is measured by resistance thermometer in a well, and by a second high-speed resistance thermometer directly in steam.

The curves in Fig. 2 show the results when control was operated from the high-speed measurement.   In the upper graph, the solid line is the temperature recorded

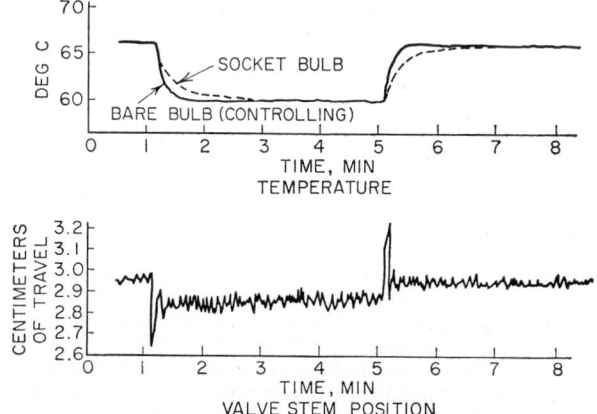

Fig. 2. Response of system to 6° change in set point, using high-speed measurement.   Top shows bulb in socket, compared with high-response element.   Bottom shows valve response.

from the high-speed thermal system; the dotted line shows the measurement from the conventional socket-mounted thermal system.   All curves show the effect when the set point was lowered and raised 6°C.   The lower graph shows the valve response, with almost immediate response and only slight overshoot.

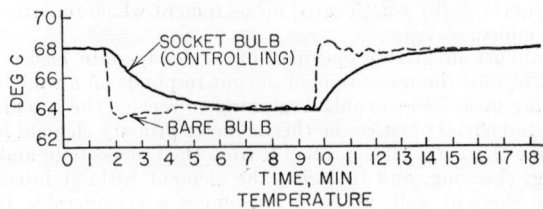

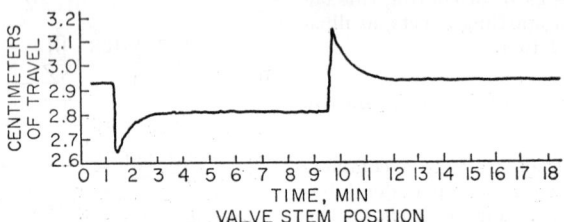

Fig. 3. Results obtained when controller is adjusted properly for use with low-speed thermometer.

Figure 3 shows the results obtained with the controller suitably adjusted for operation from the measurement provided by the low-speed socket-mounted bulb.   The time scale is double that of the previous curve.   Comparing Figs. 2 and 3, note that, with the control operated from the slow measurement and with suitable controller settings, the control of water temperature (as measured by the fast thermometer) was almost as good as the control obtained from the high-speed thermal system.

Measurement Lag Is Dangerous

Figure 4 shows what can happen with the controller operating from a low-speed thermal measurement when the lag of the measuring system is disregarded.   The middle curve shows the response to the control action as measured by the low-speed system.   This measurement shows only a 6° change in temperature.   However, taking the measurement from the fast system as representing the true temperature of the process fluid, actual temperature fluctuated widely, as shown in the upper curve. Note that the temperature scale in this upper curve has to be compressed to get the curve on scale.   A change in temperature set point of about 6° caused an initial swing of more than 30° in true temperature.   The lower curve showing the valve operation indicates that the valve excursions also were excessive.

This example is typical of measurement dynamics and their effects.   If the lags can be eliminated, better control with less critical adjustment and wider dynamic range can be obtained.   However, if lags exist in the measurement, either inherently or resulting from an application practice (such as thermometer wells), it is almost always possible to make allowances for taking care of this without serious effect on process operation.   It is important to recognize and make allowance for the dynamic as well as the static characteristics of measurement if optimum results are to be obtained.

Lag in measurement is particularly dangerous.   If overshoot or instability results from process lag, this shows up normally in the measured record of performance.

However, there is usually no direct indication of instability which results from measurement lag.   Usually there is no high-speed measurement available for comparison, as in the illustrations given.   The measurement can appear smooth and stable—but variations which can adversely affect product quality or processing operation can exist without showing up in the measurement output.

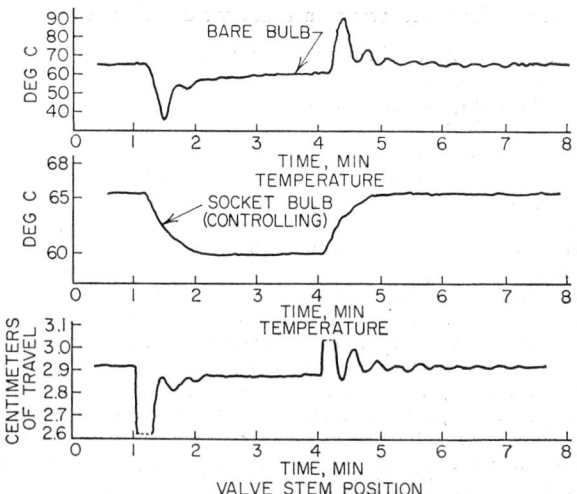

FIG. 4. Control response using low-speed measurement systems and controller improperly adjusted for lag.   Actual temperature (top) varies nearly 70° although bulb in socket shows only 6° change.

Temperature has been used as a simple example.   In more complex measurements the problems are often more difficult.   Typically, analytical devices which measure composition or similar characteristic of a flowing stream usually depend on sampling. This in itself accentuates all the problems discussed.   Regardless of the simplicity or complexity of the measurement, it is essential to understand the problems and limitations, to make every reasonable effort to minimize these limitations, and to recognize the resultant effects on the validity and significance of the measurement.

SIGNIFICANCE OF THE MEASUREMENT ON THE PROCESS

The second phase of significance of measurement is concerned with the actual meaning of the specific measurement with respect to the process operation on which it is applied.   What is to be accomplished as a result of measurement?   It is surprising how often insufficient attention is paid to true significance, resulting in applications which fall short of what is desired.   Three examples will be stated to illustrate the variety of significance which attaches to a specific measurement in differing applications.

Steam Temperature as a Measure of Quality

A simple case—temperature measurement of the steam from a steam boiler is a direct quality measurement.   Knowing the characteristics of the process fluid (dry steam with some superheat), temperature, pressure, and the pressure drop across an orifice or flow nozzle as a primary measurement for flow, the characteristics of the steam output are completely defined.   The temperature is measured directly.   The measurement is commonly recorded and also applied to a control system to maintain the temperature at a desired value for the particular operating conditions. Note that the temperature is of interest in itself. Superheat, which is, in general, of great significance, is derived from a temperature measurement and a pressure measurement. Flow is computed from pressure drop across the orifice, corrected for both temperature and pressure.

Steam Condensing Temperature as a Measure of Paper Moisture Content

A second example of the application of temperature lies in the measurement of condensing temperature of steam used for the drying of paper on a paper machine. Here the objective is to dry the paper so that the moisture content of the final product is at the desired level. Temperature is adjusted in accordance with some independent determination of moisture content. In this operation, steam condensing temperature is not a measure of the paper moisture content itself but, rather, of the processing condition inferentially related to product quality. Constancy and repeatability of the measurement are of importance—but absolute value has relatively little significance. This type of measurement of process condition is common.

Any change in temperature produces a corresponding predictable change in moisture content. Since steam-temperature measurement is simple, reliable, and relatively inexpensive, it is used as a basis for moisture-content control. In usual operation, temperature is adjusted when a change in moisture content is desired. However, moisture content of the sheet depends not only on the temperature but also on a number of other variables, including freeness of the stock supplied to the machine, condition of the press felts, and other variables equally difficult to measure. Hence, measurement and control of temperature of the dryers is useful but not in itself sufficient to produce the desired product quality. This is typical of inferential measurements where a process condition instead of the product characteristic is measured.

The determination of the actual sheet moisture content may take a variety of forms. Typically, the operator may simply judge moisture content by the "feel" of the sheet. On the other hand, moisture content may be continuously monitored by a measurement of dielectric constant, which, in turn, is calibrated against sample gravimetric moisture determinations.

Temperature in the Measurement of Gas Flow

A third type of application of temperature measurement is in the measurement of gas flow. In gas production and distribution, measurement is in terms of standard cubic feet. Considering the fact that major gas-measuring stations can handle hundreds of millions of cubic feet per day, it is obviously impractical to get all this gas to a condition of standard temperature and atmospheric pressure, and then proceed with a volumetric measurement. The American Gas Association Report No. 3, the current standard, expresses total gas flow as the result of a series of computations. There are a set of empirical coefficients, the primary variable being differential pressure across the orifice; then corrections are made for static pressure, temperature, specific gravity, supercompressibility, etc. The American Gas Association has devoted a great deal of work to determining as closely as possible all the variables which affect the measurement. With an installation made in accordance with recommendations, and with suitable measurement of the variables, the total flow of gas can be determined to a high degree of accuracy.

Here the temperature itself is of no interest whatever. However, since a 10°F change in temperature results in a change in density of about 2 per cent, which, in turn, has a 1 per cent effect on the total flow measurement, it becomes important that temperature be accurately known.

The contrast between gas measurement and paper moisture should be noted. In each, a number of variables affect the final result. In gas measurement, the significant variables are measurable; as a result, the flow in terms of standard cubic feet can be determined by computation. In contrast, in papermaking, there are an equal or greater number of variables affecting moisture content, but temperature is about the only one that can be readily measured.   It is also the manipulated variable whereby the final product quality is maintained at a desired value.   Since many of the other variables which affect moisture content are not measurable, we cannot determine the actual value of moisture content.   However, temperature measurement serves a useful purpose in the paper-drying operation.

The same general reasoning applies to other measurements.   The problems involved in some of the more complex analytical determinations require a considerably more detailed study for full appreciation of the significance of some of the measurements.   However, the same general principles apply.   The importance lies in the fact that, regardless of the measurement, significance with regard to the operation should be thoroughly understood as a basis for the selection, the application, and the interpretation of the resulting measurement values.

WHAT IS TO BE ACCOMPLISHED AS A RESULT OF THE MEASUREMENT?

For any specific measurement, the answer to this question is implicit in the basic significance of the measurement.   When the true meaning of measurement is fully understood, there is understanding of what its application can accomplish.   Going back to examples previously cited, the measurement of temperature of steam coming from a boiler can provide a variety of results.   The temperature itself is a matter of considerable interest and can be used directly as the basis of automatic control.   The temperature in conjunction with the pressure and differential pressure provides complete information with regard to quality and quantity of steam produced.   This combined measurement can be used directly both for control purposes and as a basis of plant balance, efficiency, and other determinations.

In contrast, the measurement of the temperature of natural gas in distribution lines has a single purpose: the determination of a correction factor affecting the measurement of the total flow in standard cubic feet per minute.   There is no direct interest in the actual value of the temperature; furthermore, this value is almost never used as a basis of control or other manipulation.

The basic objective of industrial measurement is to contribute to plant operation for maximum profit considered on an over-all basis.   Many measurements obviously are profitable.   In other cases, a balanced judgment is called for as to what can be accomplished as a result of a particular measurement (in terms of improved quality, increased productivity, decreased manufacturing expense, and so on) against what is involved over-all in providing the particular measurement.   The factor of judgment is often particularly important where the choice lies between a simple, easily available measurement of a quantity related only indirectly to the desired value and a much more complicated and expensive measurement with a much more direct correlation to the value desired.

The usual considerations of better quality, greater quantity, and decreased labor are obvious and well recognized.   Safety in many instances is a direct function of measurement—all the way from over-temperature indication of an airplane engine to the automatic "scram" of a nuclear reactor when the period becomes dangerously short.   The use of measurement as the base of cost accounting and material and thermal balances is equally apparent.

Instruments Can Reduce Plant Investment

In most cases, instruments are considered as an addition to the producing plant.   Somewhat less obvious is the case where measurements may be a major factor in reducing rather than increasing the over-all plant investment.   Typically, in petroleum production, a major movement is afoot to substitute automatic custody transfer for the present manual gaging operation on the producing leaseholds.   This involves a major upheaval, including the necessity of new permissive legislation.   A considerable amount of not inexpensive measuring equipment is required for each lease.   However, the saving in the cost of tankage which could be effected by use of automatic custody transfer could result in a considerably lower total cost per leasehold. This is entirely aside from savings from labor and improved accuracy. This whole matter hinges directly on the question of reliability of the automatic measurements. There appears to be a strong presumption that modern measurement devices can be designed to produce more reliable and more accurate measurement than the gage stick and manual sampling and testing procedures now in use.

Process Optimalization

Even more spectacular is the application of measurement to direct optimalization of processing operations. At least one large petroleum corporation is applying an on-line closed-loop digital-computer control to a standard refinery operation. The essential characteristics of the process itself are measured, computed, and stored in the computer memory. Variables in the process are continually measured and fed into the computer. The computer combines this information at a high rate of speed and automatically adjust the varying processing conditions in a manner directly calculated to produce the most profitable operation.

The direct approach to the objective of operation for maximum profit opens up exciting possibilities. There are the usual practical problems hinging, to a considerable extent, on whether the necessary measurements can be made available with sufficient accuracy and reliability. There is the broader question as to the specific approach most suitable for obtaining this optimum operation. The important point is that work is being done and real progress is being made in this direction.

Measurements are presently applied to provide a vast and vital range of information which makes possible modern automation. We can look forward with confidence to measurement applied directly to the objective of maximum profit. In our free world, where the basic force of natural selection operates through the profit motive, maximum profit is the final answer to the question: What is to be accomplished as a result of measurement?

WHAT IS INVOLVED IN OBTAINING THIS MEASUREMENT?

This brings us to the last and final phase—cost of the measurement. We are rapidly approaching the situation where almost any measurement is obtainable at a price. There are exceptions: Quantitative measurements for flavor are still in the future. And there are still many other values which, for a combination of economic and technological reasons, are not measured continuously under plant operating conditions. Usually, however, it is simply a matter of dollars. If the economic value of a particular continuous-plant stream measurement is great enough, nearly anything can be accomplished if the urgency is sufficient and the financing is adequate.

In the mundane field of normal plant operation, the objective always comes back to maximum profit. Over-all cost must be reckoned with. The purchase price of equipment for any particular measurement is easily established. Necessary accessories, installation costs, etc., usually can be estimated with reasonable accuracy. Operation costs are less readily defined in advance. And the factor of reliability as it affects over-all cost of operation must be seriously considered.

The question of what is involved in obtaining the measurement might be well broadened to include the question of what is involved when and if the measurement is not obtained. Interruption, even for a short time, of vital measurements on a critical process may necessitate complete shutdown, with a cost which may be several times the cost of the actual measuring equipment involved. Present techniques of operation on large nuclear reactors are a notable example of this.

To summarize, the first costs are definite. Operating costs, which, in the long run, may far exceed the initial price, are a matter of sound engineering judgment based on experience and on records of previous performance. Complete performance records, essentially measurement on the performance of measurement equipment, are a vital factor in sound engineering selection of measurement.

The three factors which enter into the selection of measurement are significance, application, and cost.   Sound engineering judgment based on these factors is essential to selection of the necessary and sufficient measurements for optimum plant operation.

REFERENCES

1. Churchman, C. W., and P. Ratoosh: "Measurement: Definitions and Theories," John Wiley & Sons, Inc., New York, 1959.
2. Cerni, R. H., and L. E. Foster: "Instrumentation for Engineering Measurement," John Wiley & Sons, Inc., New York, 1962.
3. Considine, D. M. (ed.): "Process Instruments and Controls Handbook," Why Measure and Control?   pp. 1-3 to 1-5, McGraw-Hill Book Company, Inc., New York, 1957.

MEASURING AND TRANSMISSION METHODS

By Steven Danatos*

Individual subsections of this handbook on variables cover a number of the measuring methods employed for converting a variable into a suitable signal for indication, recording, or control. This subsection summarizes common devices which receive electric transducer outputs that can represent a variety of variables, such as temperature, pressure, flow, thermal conductivity, electrical conductivity, pH, redox, and the like. It deals primarily with industrially used devices for measurement of voltage, resistance, capacitance, or inductance.

Covered also in this subsection are methods of *transmitting* or *telemetering* measurements over some distance (from less than 100 ft to many miles). As brought out in the discussion of methods, the distinction between measuring devices per se and transmitters is not always a clear one.

MEASURING METHODS†

Most electrical measurements are made by bridge circuits of one sort or another. These include Wheatstone bridges, inductance and capacitance bridges, potentiometers, and variations thereof. Some operate on direct current, some on alternating, and some may operate on either. Some use both—for example, measure the direct current of a thermocouple, but convert the unbalance of the circuit into alternating current for amplification and operation of the balancing devices. Alternating-current bridges of various kinds have become more common in recent years.

Bridges can, in general, be operated in two different ways: as *deflection* or as *null-balanced* circuits. Deflection bridges are relatively uncommon. The advantages of the null system, wherein the electrical values of the bridge are brought to a balanced condition so that no current flows through the detecting device, are great enough in most cases to outweigh the greater cost. Most balanced bridge circuits used industrially are of the automatic self-balancing type, although manual balancing is common in laboratory instruments, and in some industrial types, such as electrical conductivity meters.

Basic Measuring Circuits

Measuring circuits can be classified as:
1. Direct-voltage measurement by millivoltmeters.
2. Voltage measurement by deflection potentiometers.
3. Voltage measurement by null-balance potentiometers.

Each is reviewed briefly below, after which typical available circuits are described with reference to schematic diagrams.

* Associate Editor, Engineering Practice, *Chemical Engineering*, a publication of McGraw-Hill Publishing Company. This material is based upon a portion of a special report published in the June 12, 1961, issue of *Chemical Engineering*, pp. 231–234.

† See also p. 9-2, Electrical Variables, and p. 15-67, Electronics Laboratory and Research Instrumentation, for related information.

Direct-voltage Measurement by Millivoltmeters. Millivoltmeters are commonly used in the measurement of temperature by thermocouples and of speed with tachometer generators. The meters are usually indicators, but recorders are also made—some in which the millivoltmeter movement operates the pen directly with the use of a chopper bar; others in which the recording mechanism follows the pointer without contact by means of an oscillator-coil pickup system.

Voltage Measurement by Deflection Potentiometers. In the elementary potentiometer circuit of Fig. 1, with battery voltage adjusted against a standard cell, unknown voltages can be read by the position of the slider on the slidewire at zero deflection of a galvanometer. If the slidewire is replaced by a series of tapped resistances, and the galvanometer is calibrated as a millivoltmeter, then the movable contact in the unknown voltage circuit can be set as a point which will give only a small deflection to the galvanometer. Assuming that the instrument were used for temperature measurement by a thermocouple, for example, then temperature can be read as the sum of the calibrated readings of the resistance tap position and the millivoltmeter indication. This method is sometimes used where a low-cost indicator for one or a number of readings is desired and manual operation is satisfactory.

Voltage Measurement by Null-balance Potentiometers. Most instruments measuring voltage as a measure of the variable use a null-type, self-balancing circuit similar to that shown in Fig. 1. This method has the advantage that, because no current is drawn in the balanced condition, the measurement is unaffected by the resistance of the unknown voltage circuit, or by changes in resistance. In Fig. 1, a switch (often operated automatically in self-balancing instruments) is provided for switching a standard cell into the circuit to standardize the battery by means of resistance R. With a thermocouple (TC in Fig. 1) in the circuit (or other unknown voltage to be measured), the instrument is balanced manually or automatically (see below), by moving the slidewire contact until the current through galvanometer G is zero. The position of the contact is a measure of the unknown voltage, and can be suitably calibrated to read directly in terms of the variable or to operate an indicating pointer, recording pen, and/or controller mechanism.

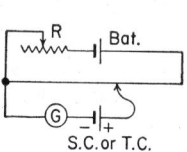

FIG. 1. Elementary potentiometer circuit.

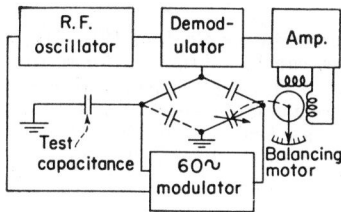

FIG. 2. Capacitance bridge for level.

Typical Bridge Circuits

Numerous commercial measuring devices have been developed since the early, simple potentiometer of Wheatstone bridge circuits. Some of these are described briefly below.

Capacitance Bridge* (Fig. 2). Used mainly for capacitance-level measurements, the bridge has a probe in the tank ("test capacitance" in Fig. 2) which serves as one capacitance in a four-capacitance a-c bridge. In operation, variation in capacitance unbalances the bridge, producing a 60-cycle modulated r-f output that adds to the signal from an r-f oscillator. The signal is then demodulated and amplified to drive a balancing motor in a direction to rebalance the bridge by changing a variable capacitor. The motor also positions an indicating pointer or recording pen.

* Robertshaw, A & I Division.

Capacitance Follower* (Fig. 3). For recording or controlling from millivolt-meters or other low-torque devices, the vane on the pointer is followed without contact by a servo-operated follower. A capacitance bridge circuit similar to that in Fig. 2 detects capacitance between vanes and accordingly drives the follower as well as the recording pen to maintain constant capacitance.

Capacitance Potentiometer† (Fig. 8). Capacitances can be substituted for a pair of resistances in a potentiometer-type circuit (see below).

Resistance Bridge.‡ The common Wheatstone bridge is widely used with resistance thermometers, conductivity meters, thermal conductivity gas analysis, etc. For applications where variations in lead resistance could upset response from a sensitive element, a three-wire lead is used, so that lead-wire resistance is added to both sides of the bridge and hence essentially cancels out. Like potentiometers, Wheatstone bridges used in industrial instruments are commonly self-balanced automatically with some form of detecting mechanism that moves a slidewire contact to a point of zero current flow. Such bridges can use direct or alternating current.

Inductance Bridge (Figs. 21 and 23). Either two or four of the resistances in an a-c Wheatstone bridge can be replaced by inductances (or impedances). Figure 21 shows a type that is self-balancing without the usual self-balancing mechanism. Figure 23 is a type used for precise detection of the position of an instrument element.

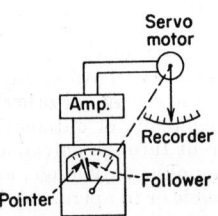

Fig. 3. Capacitance-follower mechanism.

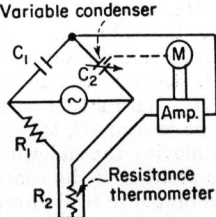

Fig. 4. Capacitance-type resistance-thermometer circuit.

Capacitance Bridge§ (Fig. 4). As noted above for potentiometers, two capacitances can be substituted for resistances in a Wheatstone bridge circuit. Figure 4 shows such a circuit used with a resistance thermometer. Capacitances C_1 and C_2 replace a pair of resistances in the bridge circuit (see Fig. 8, page 1-40). Any unbalance in the a-c bridge is amplified electronically and fed to a reversible motor to alter the value of C_2 until current flow through the detector circuit ceases. Temperature R_2 at the bulb is measured by the position of the balancing capacitor.

Typical Balancing Methods

To balance bridge circuits a number of ingenious and practical industrial methods have been developed. The more common ones are described briefly below.

Microsen System* (Fig. 5). This is an electromechanical balancing system in which an electrical or mechanical input (measurement) is balanced against an accurately proportional d-c output to an indicating or recording instrument. The basis of several variations of the system is a balance beam that is upset by the primary measurement so as to affect an oscillating electronic circuit. Balance is restored to the beam as circuit output reaches a value equivalent to the initial upsetting force. Applied to measurement of a thermocouple potential, the system consists of beam B (Fig. 5) with calibrating spring S and an armature at one end carrying coils C_1 and C_2.

---

* Robertshaw, A & I Division.
† The Foxboro Co.
‡ See p. 15-92, Fig. 44, Wheatstone bridge, and text accompanying it. Also p. 2-15, Fig. 8, Resistance thermometer circuit.
§ The Foxboro Co.

At the other end, the beam is in the field of an oscillating coil O in an electronic oscillating circuit. Voltage output of the thermocouple TC applied to coil C_1 causes rotation of B toward a permanent magnet. But this changes electronic circuit output to the recorder, also changing feedback through C_2, to restore balance. Instead, C_2 and electrical feedback may be omitted, and direct mechanical feedback from the recorder may be used to rebalance the beam. Note that in this application the system acts as a d-c amplifier. The beam is comparable to a millivoltmeter in that the circuit is not "nulled," and lead resistance has to be considered. On temperature, accuracy is said to be $\pm\frac{1}{2}$ per cent of scale. Other applications: pressure, strain gages, pH, and conductivity.

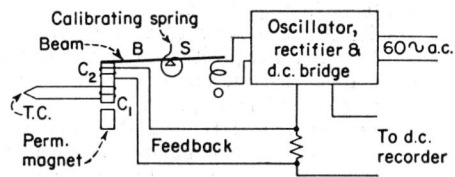

FIG. 5. Microsen balancing system.

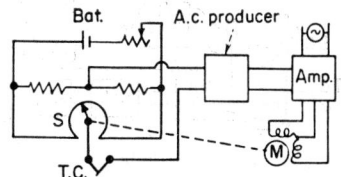

FIG. 6. Dynamaster balancing system.

Dynamaster System [*] (Fig. 6). Thermocouple output is applied to a potentiometer circuit. If not in balance, current in the TC circuit is converted to alternating by a synchronous switch. Alternating-current output is amplified and applied to a balance motor to drive S in the proper direction for restoring balance. This system with modifications can be used to balance circuits where the electrical quantity can be represented by voltage, current, capacitance, or resistance. In resistance thermometers, for example, an a-c bridge is used. Unbalance output, being a-c, can be amplified directly, without synchronous switch.

Brown Electronik [†] (Fig. 7). Thermocouple output is applied to a potentiometer circuit. If the circuit is unbalanced, TC voltage is converted to a-c, the direction of unbalance determining the phase relation between generated alternating current and a-c supply voltage, so as to determine the direction of correction applied by balancing motor M. If slidewire S is not balanced, current flow in the thermocouple circuit is

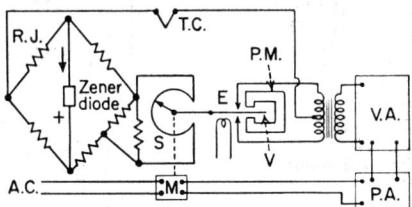

FIG. 7. Brown Electronik balancing system.

converted to alternating by a transformer through the action of a vibrating-reed converter driven by alternating current at 60 cycles by coil E. The transformer output is amplified by voltage and power amplifiers and applied to balancing motor M. If in phase, the motor corrects for overbalance. If 180° out of phase, the motor corrects for underbalance. The Zener diode circuit provides a constant and continuous standard voltage to the measuring circuit from line voltage. Reference-junction resistor RJ compensates for variations in the thermocouple reference-junction temperature.

* Bristol Co.
† Minneapolis-Honeywell Regulator Co., Industrial Division.

Stranducer System. * This is an electromechanical device that balances a d-c millivolt input from a sensing unit against a known voltage. By using four resistance strain-gage elements to form the variable arms of a Wheatstone bridge circuit, infinite resolution of impressed signal is obtained. Change in input drives the motor in one direction or the other, increasing tension on two of the wires, while decreasing it on the other two. Change in stress changes the electrical resistance of the wires, and continues to change it until the bridge is electrically balanced.

Dynalog System † (Fig. 8). This system operates directly from a standard cell without a separate battery and so needs no standardization. The vibrated switch

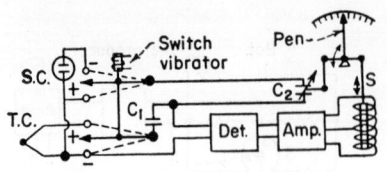

FIG. 8. Dynalog balancing system.

alternately connects the thermocouple and standard cell SC into the capacitor bridge (contacts in upper position), then short-circuits the capacitors through a detector (contacts in lower position), SC charges C_2, TC charges C_1. If charges are unequal, detector output is amplified to drive solenoid bridge S and reposition C_2 and the recorder pen. When balance is reached, short-circuit discharges become equal.

Current drawn from SC is small enough so that the usual battery is not needed. Variations of capacitor bridge are used for other variables expressed as resistance, capacitance, and inductance.   Accuracy is said to be $\pm \frac{1}{4}$ per cent.

Instead of a standard cell, some potentiometers use a regulated d-c source, e.g., Bailey Meter's d-c receiver for thermocouple and other d-c pickups. This has a regulated electronic d-c supply to serve as reference voltage without a standard cell, and with no need for periodic standardization.

Electrosyn System ‡ (Fig. 9). This method uses a magnetic amplifier; requires no vacuum tubes. A primary measurement rotates a rotary differential transformer connected to the amplifier. Alternating-voltage output is amplified, turns a two-phase motor geared to a recorder, other auxiliaries (such as encoders for telemetering or data logging) and, through the cam, turns a second differential transformer, which feeds back an opposing a-c voltage. The motor position, which just balances feedback voltage against primary voltage, represents the current value of the variable.

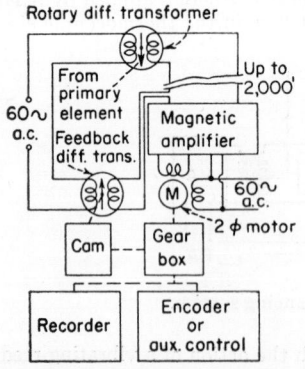

FIG. 9. Electrosyn balancing system.

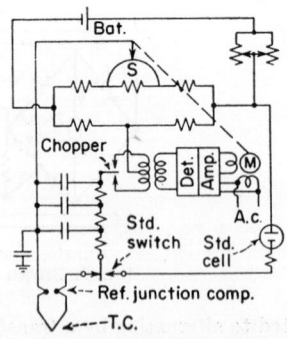

FIG. 10. Speedomax balancing system.

Any change in primary measurement is followed by immediate rebalancing. The primary element and amplifier-recorder can be up to 2,000 ft apart. The system shown is used for mechanical inputs such as pressure and differential pressure. Temperature input from the resistance thermometer is handled similarly, except that a variable resistor is used for feedback balancing.

* Minneapolis-Honeywell Regulator Co., Industrial Division.
† The Foxboro Co.
‡ American Standard.

Nongalvanometer Balancing Systems. Most makers of self-balancing bridge and potentiometer instruments now supply continuous-balance instruments in which the galvanometer formerly used is replaced by some method of electronic detection of unbalance. Characteristically, such instruments balance so rapidly as to be able to carry the pen across the entire chart scale in 2 to 20 sec.

Speedomax System* (Fig. 10). The thermocouple output is applied to the potentiometer circuit. If unbalanced, the excess is chopped to a 60-cycle alternating current, amplified, and used as power supply for one phase of a two-phase motor; the other phase is powered by plant supply. The motor drives a balancing slide-wire in the proper direction to rebalance the thermocouple output. At balance, supply of amplified current ceases. A similar balancing system is used for other bridge circuits.

TRANSMISSION (TELEMETERING) METHODS

The distinction between "transmitters" and "measuring devices" is not always clear. A measuring device is one that converts a primary indication into a position (or into some form of energy that can easily be displayed as position) on a scale, to show the value of the primary variable. Some transmitters do the same thing, though others are primarily relays. If there is a real distinction, it is that transmitters can display the value of the primary variable at a considerable distance from the primary element.

Transmitters for still-longer distances are called "telemeters." In some cases, telemeters are designed to transmit over their own wires; in other cases, over phone wires or by microwave. Transmitters, however, may be pneumatic as well as electrical; but, being shorter-distance senders, they use their own communication connections between sending and receiving devices.

A great many different methods are used for extending the distance that measurements can be sent. In general, these include hydraulic, pneumatic, electrical, and electronic. Short-distance methods of getting away from friction-producing elements, such as stuffing boxes, include mechanical (torque tubes) and magnetic devices.

Hydraulic and Magnetic Types

Hydraulic Transmission† (Fig. 11). One method used for transmitting liquid-level measurements from a tank float employs two transmitting and two receiving bellows, connected by two lines, with the system filled with liquid. The purpose of a

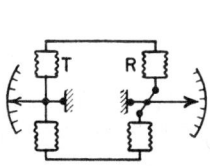

Fig. 11. Hydraulic transmission system.

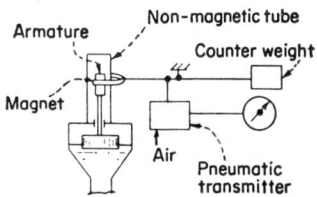

Fig. 12. Magnetic-follower system.

double-bellows system is to compensate for ambient temperature changes. Expansion or contraction, due to temperature, affects each bellows system equally, being canceled out by a link between receiving bellows. However, the movement of the pointer at the transmitting end (representing position of float or other measurement to be transmitted) expands one bellows, contracts the other; hence, it moves the receiving pointer an equal amount.

Magnetic Follower (Fig. 12). Use of magnetism to transfer motion inside a sealed system to outside the system is a principle found in several types of instru-

* Leeds & Northrup Co.
† Liquidometer Corp.

ments. It eliminates a stuffing box, with its chance of leakage and loss of accuracy due to friction. The magnetic-follower method is used in level gages, controllers, flow manometers, and rotameters. A level gage made by Magnetrol, Inc., uses this method to operate level control switches. In Fig. 12, a float in a level gage on a manometer operates an armature in a nonmagnetic tube to position a pneumatic transmitter. In a similar way, the system could be used to operate an electronic transmitter.

Pneumatic Types

Pneumatic transmitters have come into extensive use, especially since development of the pneumatic force-balance type of measuring device. Such transmitters are used for temperature, pressure, differential pressure from flow, level and density measurements, weight, force, and position. With the development of close-coupled control systems with remote set and indication at the panelboard, they are being used to transmit to a controller close to the point of measurement, and to an indicating or recording instrument at a remote panelboard. Similar transmitters, balanced against pressure of a spring, are being used to produce loading pressure for remote setting of controllers in such installations. This method takes the recorder out of the control loop, puts the controller and point of measurement close together, and hence cuts instrument lags to a minimum.

Positioning Nonbalance Transmitter (Fig. 13). Figure 13 illustrates the principle, but shows no particular type. Measurement pointer P positions a flapper F before a nozzle, raising or lowering pressure P_2 by changing the leak at the nozzle. Bellows B moves movable fulcrum M in a direction to oppose change, acts as a feedback so that each output pressure P_2 demands an exact position of F. Thus P_2 is proportional to the position of pointer P. P_2 can be transmitted. For use over distances where a considerable lag in transmission is to be avoided, P_2 is fed to a relay air valve, to increase flow during the time that P_2 is changing.

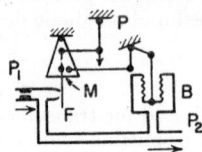

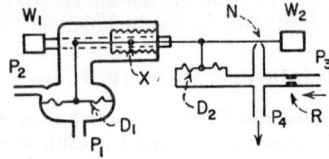

FIG. 13. Positioning nonbalance transmitter.   FIG. 14. Force-balance transmitter, beam type.

Force-balance Transmitter, Beam Type (Fig. 14). Force from any sort of measurement (shown here as that of a differential-pressure diaphragm) is applied to a weight-balanced beam pivoted at point X. In this application, X is inside a bellows used as a frictionless seal for the differential-pressure system. The beam obstructs flow or air from nozzle N, which is supplied with air at pressure P_3 through restriction R. Assuming that P_1 increases, the beam will tilt to raise nozzle pressure P_4, but this increases pressure on diaphragm D_2, which restores the beam to substantially its original position. Hence, P_4 is balanced vs. the differential pressure, either exactly or in some multiple, depending on the relative lever arms. P_4 is transmitted. A similar beam type can be used for ratioing.

Force-balance Transmitter, Null-balance Vector Type* (Fig. 15). This is a newer type than that shown in Fig. 14, in which vectorial forces are combined by linkages. The range is adjusted by changing angle θ by sliding pivot 3 in a calibrated slot. Differential pressure ΔP is measured by a diaphragm, with motion taken out by the lever pivoting at 1, through the seal bellows. The link pivoted at 2 raises or lowers the link pivoted at 3 when ΔP changes; the amount of this movement varying with angle θ. If differential increases; link 2–3 lifts, lowering air pressure in

* Republic Flow Meters Co.

nozzle N.   Reverse air relay R then raises the output pressure, also raising pressure on the feedback diaphragm and moving pivot 2 downward to restore balance.   Output pressure is directly proportional to initial differential pressure.

Force-balance Transmitter, Stack Type* (Fig. 16).   Where Figs. 14 and 15 interposed beams or linkages between measured and output pressures, permitting the lever arm change to change proportionality, several makers supply a force-balance

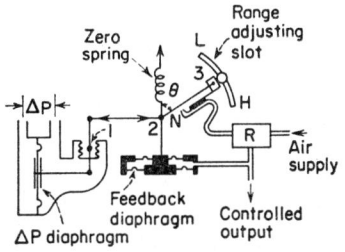

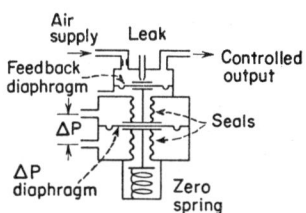

FIG. 15.   Force-balance transmitters, null-balance vector type.

FIG. 16.   Force-balance transmitter, stack type.

instrument in which the balance is direct on opposite sides of a diaphragm or bellows. Figures 16 and 17 show this principle.   In Fig. 16, the resultant of differential pressure on both sides of the lower diaphragm (or a single pressure on the lower side only) is opposed by the controlled air output pressure so as to balance.   Air supply enters through a restriction and flows through the chamber above the upper diaphragm.   If upward pressure from the ΔP diaphragm increases, air loss through the leak decreases, pressure rises in the upper chamber and outlet, meanwhile balancing the increased ΔP. If ΔP falls, leakage increases, and top chamber and transmitted pressure fall.   Transmitted pressure is directly proportional to ΔP.

Air-relay Force-balance Stack-type Transmitter† (Fig. 17).   This is similar in principle to Fig. 16, except that it incorporates an air relay to give rapid changes

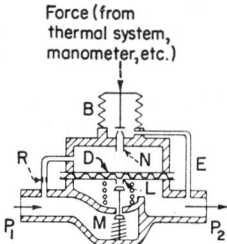

FIG. 17.   Air-relay force-balance stack-type transmitter.

in output pressure.   Air at pressure P_1 enters the transmitter, and is applied through restriction R to nozzle N and top of diaphragm D.   Any force such as that from a liquid-filled thermometer system is applied to top of bellows B, restricting delivery from nozzle N.   Increased back pressure over diaphragm D closes leak L through porous-center diaphragm D, and opens main valve M to increase pressure P_2.   This works through equalizing line E to raise pressure in B, thus almost restoring the initial nozzle opening.   Now output pressure balances force.   Assuming that force then decreases, nozzle delivery increases, pressure over D falls, and leak L bleeds air to lower P_2 until P_2 again equals force.   Response of the instrument depends on force of the calibrating spring under main diaphragm D.   Main valve M acts as relay for quick changes.

* Taylor Instrument Cos.
† Moore Products Co.

Electric and Electronic Types

Several transmitting and telemetering methods are in use that can be classed as electrical rather than electronic in that they do not require vacuum tubes or transistors for amplification. One method now being widely used appears in several guises under different names, although the principles involved are generally similar. Among these are the inductance bridge, the impedance bridge, and the differential transformer. All are a-c bridge circuits in which the degree of coupling between inductances is varied by altering the amount of iron core within a coil.

Resistance Manometer (Fig. 18). This is a device for transmitting the height of mercury in a flow manometer. A spiral of rods in the manometer, arranged to compensate for the square-root relation in flow, is connected to resistances so that as the mercury short-circuits out more or less of the rods, the total remaining resistance is proportional to the flow rate. Flow-rate indication and recording are obtained by passing alternating current from manometer through a bridge circuit that measures the conductance by means of a core that is drawn a certain distance into a coil, the distance depending on the resistance of the manometer. Integration of flow is by means of an instrument similar to a watt-hour meter.

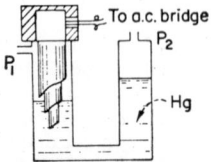
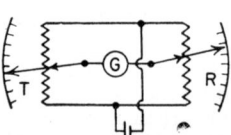

FIG. 18. Resistance manometer.   FIG. 19. Wheatstone bridge transmitter.

Wheatstone Bridge Transmitter (Fig. 19). A null-balance Wheatstone bridge can be used to transmit indications by contact on slidewire T, positioned by the measurement to be transmitted. The unbalance indicated by galvanometer G then operates to move the contact on slidewire R to balance the position, which then reproduces the position of T. Alternatively, transmitting resistance T in Fig. 19 may be varied by temperature (resistance thermometer) or by a force (strain gage), and the resistance change mirrored at R. For example, strain gages are used by Brooks to transmit rotameter readings.

Self-synchronous Motors (Fig. 20). Motors with three-phase stators S_1 and S_2 and two-phase rotors are connected to the same line. The rotors remain stationary unless turned by an external force. If primary rotor R_1 is turned, R_2 will follow closely with a lag of 1 to 3°. Selsyns may be paired to transmit both feet and inches in remote level indication (Shand & Jurs). The "inches" transmitter turns one revolution for each foot. The "feet" transmitter turns one revolution for 60 ft of range. Corresponding indications of two receiver selsyns permit readout in feet and inches.

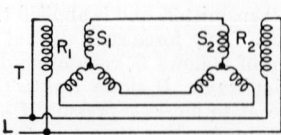
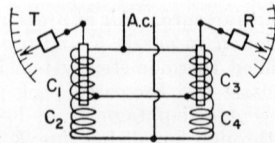

FIG. 20. Self-synchronous motors.   FIG. 21. Inductance bridge.

Inductance Bridge (Fig. 21). This is used for transmitting indications from inside sealed instruments, such as flow-meter manometers, rotameters, etc. Transmitter and receiver coils are connected as shown by three wires in a bridge arrangement and supplied with alternating current. The transmitter soft-iron armature is positioned vertically by the measurement to be transmitted. The receiving armature positions itself similarly. This device is self-balancing without the usual balancing devices. Any unbalance due to armatures being unequally placed in coil pairs results in current flow through the center lead that brings about rebalance by altering flux distribution until balance is reached.

Differential Transformer (Fig. 22). This device is an a-c motion transducer that can be designed to produce an electric output over a full range with any desired range of motion of the armature. The type shown allows long armature travel; other types are designed down to fractions of an inch. Transformers have one or two primaries and two secondaries, generally connected to "buck" each other. Alternating current is applied to primary. Alternating current produced in secondaries depends on position of the armature and amount of coupling so produced. Such devices provide accuracies of $\frac{1}{2}$ to 1 per cent of full range and can be used to transmit forces, pressures, differential pressures, weights, and the like. They can transmit up to 5,000 ft. The type illustrated has linear motion of core. Rotary differential transformers are used in a number of instruments, e.g., Electrosyn transmitter (American Standard, Fig. 9). Several instruments use what is, in effect, a differential-transformer bridge, e.g., Electrosyn transmitter, and a remote level transmitter supplied by Yarnall-Waring Co.

Impedance Bridge (Fig. 23). The transformer with single primary, supplied with alternating current, and double secondary has movable armature positioned by the measurement to be transmitted. Secondaries are in a bridge circuit. Unbalance

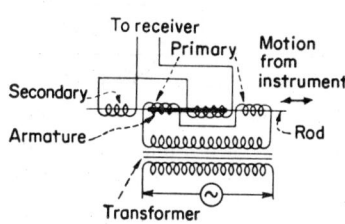

FIG. 22. Differential transformer.

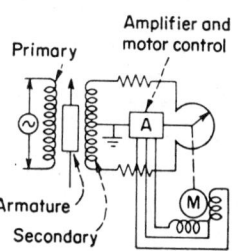

FIG. 23. Impedance bridge.

due to armature not being centered is amplified by A and transmitted to a balancing motor M, which repositions the slidewire to balance. Slidewire position is a measure of the position of armature in the transformer.

D-C Converter. * This variation of Fig. 23 eliminates amplifier and slidewire, uses two diodes to convert output of the secondaries to direct current in a d-c bridge. Unbalance resulting from armature movement is put out as d-c voltage proportional to the armature position.

D-C Transmitter. The Microsen-balance measuring device of Fig. 5 is used to transmit measurements that are initially voltages or forces, converting these to direct current that can be transmitted considerable distances. The receiver employs a second Microsen with feedback to make the pointer position proportional to the signal. The signal positions the rotary solenoid from the output of the receiving Microsen, with a mechanical link from solenoid to balance beam, thus restoring the spring. The solenoid positions the indicating pointer. The mechanically loaded Microsen beam can be used for remote set-point adjustment of electronic controllers.

Telemeters—Impulse and Timed Signal

Impulse and timed-signal methods of telemetering have the advantage of giving accuracy independent of supply-voltage variations. There are several methods:

1. Sending a number of electrical impulses proportional to the value of the variable, and counting pulses at the receiving end.

2. Sending a signal whose frequency is proportional to the variable.

3. Sending a single pulse whose duration, as a fraction of the fixed time interval is the same as the ratio of the variable to its maximum value.

* Bailey Meter Co.

4. Sending a pattern of pulses in code, such as decimal or binary-decimal, corresponding to the digital value of the variable.

Pulse-signal Level Telemeter* (Fig. 24). One of several pulse methods of transmitting continuously varying variables such as liquid level and tank temperature is shown in Fig. 24. The transmitter cycle has three parts. A remote operator signals any individual tank for level and/or temperature. The selected transmitter first sends a pulse group identifying itself, and then sends level, first "feet," then "inches" by eighths. The "inches" disk is driven by perforated steel-float tape, makes one revolution for each foot of float movement, and then advances the "feet" disk one increment through the Geneva gear. When the instrument is asked to transmit, a motor-operated sweep scans each disk for its position, first the "feet" and then the "inches" disk, sending for each a number of pulses corresponding to the disk position. Receiver dials are ratchet-operated by incoming pulses to count exactly.

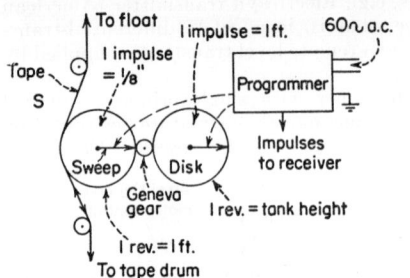

FIG. 24. Pulsed-signal level telemeter.   FIG. 25. Timed-signal telemeter.

Timed-signal Telemeter† (Fig. 25). A rotating cam lifts a follower for a period during each cycle proportional to measured value. This closes a switch and sends a signal over a two-wire line to the receiver. Duration of the signal during each cycle is therefore proportional to the measured variable. The receiver has two clutches—one "increase," one "decrease." The first runs while current is on, the second during part of the cycle when current is off. The transmitted value is the difference. The pen moves only when the difference changes. This works over long distances at high accuracy.

Frequency-code Data Transmission.‡ Physical variables of a process are sensed by instruments compatible with the Bendix Electro-span system to produce an electrical signal that may represent contact closures or openings; or voltage or current. In the Electro-span system, three different elements are used to convert the basic sensor outputs into digital form. Transmission is by means of a series of five binary-bit characters over teletype or telephone lines for remote locations, or by direct wires for nearby sites. If telephone line is used, data are transmitted by a polytonic method of transmission. Transmission of a single five-binary-bit character is performed by simultaneous transmission of five tone signals on the communication line, where the absence or presence of a single tone distinguishes between 1 or 0 for each bit transmitted.

REFERENCES

For descriptive details concerning the circuits and transmission methods briefly outlined here, see the following sections of the "Process Instruments and Controls Handbook," Douglas M. Considine, Editor-in-Chief, McGraw-Hill Book Company, Inc., New York, 1957:

Potentiometers, Sec. 8, pp. 8-89 to 8-106.
Self-balancing Electrical Instruments, Sec. 8, pp. 8-107 to 8-117.
Electrical Bridge Instruments, Sec. 8, pp. 8-118 to 8-136.
Electric Telemetering, Sec. 8, pp. 8-57 to 8-73.
Dynamic Factors in Pneumatic Transmission, Sec. 8, pp. 8-74 to 8-76.

* Shand & Jurs Co.
† B-I-F Industries.
‡ Bendix Corp.

Section 2

THERMAL VARIABLES

By

GERALD L. EBERLY, B.S.(M.E.), *Technical Staff, Harris D. McKinney, Inc., Philadelphia, Pa.; Member, Instrument Society of America.* *(Calorific Value)*

J. H. MILLER, B.S., *Retired. Formerly Vice President for Research and Engineering, Weston Instruments, a subsidiary of Daystrom, Inc., Newark, N.J.; Fellow, Institute of Electrical and Electronics Engineers; formerly, Member, American Society of Mechanical Engineers, and Institution of Electrical Engineers (Great Britain).* *(Summary of Temperature-measurement Methods)*

R. E. WILSON, B.A., Ph.D. (Physics), *Research Physicist, Hughes Company, Tucson, Ariz.; Member, American Association for the Advancement of Science, Instrument Society of America, Washington Academy of Sciences, Washington Philosophical Society; Fellow, American Physical Society.* *(Temperature)*

TEMPERATURE

By R. E. Wilson*

Temperature is probably the most widely measured and frequently controlled of the process variables. Although temperature and its effects are commonly encountered in industry, it does not follow that a thorough understanding of the characteristics of this variable is widespread.

As indicated by the references at the end of this subsection, volumes have been required to exhaustively describe temperature, its theory and many manifestations. In this subsection, an attempt has been made to give a terse summary of the subject, including (1) temperature scales, and (2) establishing and maintaining temperature standards. At the end of this subsection is a tabular summary of the common principles upon which temperature measurements are based and typical examples of temperature-measuring devices.†

There are many industries in which the measurement and control of temperature play a critically important part. In the manufacture of steel, for example, the temperature to which the liquid steel is heated in the open-hearth furnace affects the final product in several ways. Oxygen is an influential constituent of steel, and its solubility in the molten metal rises sharply with increased temperature. The rate of chemical reactions within the liquid itself and between the liquid and its slag increases rapidly with increasing temperature. The segregation of minor constituents on freezing, the crystalline structure of the solid steel, and the nature of the surface of the ingot are all influenced by the temperature of the liquid in the furnace.

The rates of most chemical reactions increase with increasing temperature; this is especially true of organic reactions where the rate in many cases increases by a factor of 2 or 3 for every 10°C rise in temperature. Since life processes are so largely a result of many interdependent organic chemical reactions, it is not surprising that life as we know it is possible only in the relatively narrow temperature range of about 40 Celsius degrees (72 Fahrenheit degrees). The vast differences between life in the tropic, temperate, and arctic regions are largely due to temperature, although light and other factors also contribute.

Table 1 indicates a few of the temperatures that are of interest or importance in various fields.

DEFINITION OF TEMPERATURE

The term temperature is generally used to denote the relative hotness or coldness of a body as determined by its ability to transfer heat to its surroundings. Stated in another manner, we say that there is a temperature difference between two bodies if, when they are placed in thermal contact with each other, the temperature of one body increases and that of the other decreases. Temperature, however, is not a measure of the total quantity of energy in the form of heat which a body possesses; for that depends also upon other factors, including the mass of the body and its specific heat. Thus, a given body contains more heat when it is warm than when it is cool, but two objects at the same temperature may differ greatly in heat content,

* Research Physicist, Hughes Aircraft Co., Tucson, Ariz.
† Details of the various temperature primary elements can be found in D. M. Considine (ed.), "Process Instruments and Controls Handbook," McGraw-Hill Book Company, Inc., New York, 1957.

Table 1. Some Important Points of the Temperature Spectrum*

FAHRENHEIT		CENTIGRADE	Point	KELVIN
		10^{18}	COSMIC RAY PARTICLE [KINETIC TEMPERATURE]	10^{19}
		10^{12}	1 BEV ACCELERATOR PARTICLE [KINETIC TEMPERATURE]	
		10^{9}	GALACTIC COLLISIONS [KINETIC TEMPERATURE]	
3.6×10^{7}	Thermonuclear Reactions	2×10^{7}	H-BOMB IGNITION	
		10^{6}	SOLAR CORONA	
		300,000	A-BOMB FIREBALL AT 45 FT. DIAMETER	10^{18}
40,000	Atoms	25,000	ALMOST ALL HELIUM IONIZED	
		22,000	SURFACE OF HOTTEST STARS [TYPE "O"]	
		20,000	EXPLODING WIRE EXPERIMENTS	
		15,000	SHOCK TUBE EXPERIMENTS	10^{17}
11,700	Molecules	6,500	ALMOST ALL MOLECULES DISSOCIATED	
		6,000	SOLAR SURFACE	
		6,000	TUNGSTEN CARBIDE BOILS	
		5,900	TUNGSTEN BOILS	10^{16}
10,600	Elemental Liquids	5,000	HIGH EXPLOSIVE BOMB	
		4,400	SOLAR FURNACE	
		4,300	PLATINUM BOILS	
		4,160	HAFNIUM CARBIDE MELTS	
		4,000	ARC LIGHT	
6,700	Flames	3,700	DISSOCIATION OF ORGANIC MOLECULES	10^{15}
		3,600	ELECTRIC FURNACE	
6,300	Elemental Solids	3,500	CARBON SUBLIMES	
		3,370	TUNGSTEN MELTS	
		3,300	OXY-ACETYLENE FLAME	10^{14}
		3,100	ROCKET ENGINES [COMBUSTION GASES]	
		3,000	SURFACE OF "COOL" STARS [TYPE "M"]	
		3,000	TUNGSTEN LAMP FILAMENTS	
		2,900	OXY-HYDROGEN FLAME	
		2,800	MAGNESIA MELTS	10^{13}
		2,300	DIRECT NITROGEN FIXATION	
		2,050	ALUMINA MELTS	
		2,000	CERMETS [LOW STRESS USE]	
		2,000	CALCIUM CARBIDE FURNACE	
		1,900	BLAST FURNACE	10^{12}
		1,800	INTERNAL COMBUSTION ENGINE [COMBUSTION GASES]	
		1,700	KITCHEN RANGE FLAMES	
		1,600	OPEN HEARTH FURNACE	
		1,550	IRON MELTS	
		1,400	HOUSEHOLD OIL BURNER FLAMES	10^{11}
		1,375	GLASS FURNACE	
		1,370	PHOSPHORIC ACID BY ELECTRIC FURNACE	
		1,350	IRON "WHITE" HOT	
		1,300	KILNS	
		1,250	CERMETS [HIGH STRESS USE]	10^{10}
		1,200	ACETYLENE BY WULFF THERMAL PROCESS	
		1,100	MOLYBDENUM ALLOY [1,000 HR.—15,000 P.S.I.]	
		1,100	COKE OVEN	
		925	GAS TURBINE [COMBUSTION GASES]	10^{9}
		760	ALLOY STEEL [1,000 HR.—15,000 P.S.I.]	
		845	HYDROCHLORIC ACID FROM SALT	
		810	FUSED METALS FOR HEAT TRANSFER	
		810	GAS TURBINE [METAL]	
		800	IRON "RED" HOT	10^{8}
		860	SUPERALLOY [1,000 HR.—15,000 P.S.I.]	
		750	NITRIC ACID FROM AMMONIA	
		700	THERMAL CRACKING	
		650	STEAM TURBINE [VAPOR, COMING]	
		600	STEAM TURBINE [VAPOR]	10^{7}
		600	METAL SALTS FOR HEAT TRANSFER	
		550	AMMONIA BY HABER-BOSCH PROCESS	
		500	SULFURIC ACID BY CONTACT PROCESS	
		375	CRITICAL POINT OF WATER	
		360	PETROLEUM OILS FOR HEAT TRANSFER	10^{6}
		350	BUTADIENE FROM ETHYL ALCOHOL	
		327	LEAD MELTS	
		315	ALUMINUM ALLOY [HIGH STRESS USE]	
		300	ACETYLENE FROM CALCIUM CARBIDE	10^{5}
		230	INTERNAL COMBUSTION ENGINE [METAL PARTS]	
		100	WATER BOILS	
		95	MOON [SUNNY SIDE]	
		85	CAUSTIC FROM LIME AND SODA ASH	
		82	ATMOSPHERE AT 150,000 FT.	10^{4}
		79	ALCOHOL BOILS	
160	Life Processes	70	HOT SPRING ALGAE	
		58	HEAT POLE OF EARTH [LIBYAN DESERT]	
98.6	Human Life	37	HUMAN BLOOD	
		35	FERMENTATION PROCESSES	
		20	ROOM TEMPERATURE	1,000
		15	EARTH CLIMATIC MEAN	
		0	OCEANIC MEAN	
		0	WATER FREEZES	
−20	Life Processes	−7	PRIMITIVE MOLDS	100
		−40	COLD RUBBER POLYMERIZATION	
		−55	ATMOSPHERE AT 50,000 FT.	
		−70	COLD POLE OF EARTH [VERKHOYANSK, SIBERIA]	
		−117	ALCOHOL FREEZES	
		−130	MOON [DARK SIDE]	10
		−183	OXYGEN BOILS	
−305	Industrial Processes	−187	AIR REDUCTION PLANT	
		−218	OXYGEN FREEZES	
		−230	SURFACE OF PLANET PLUTO	
		−235	AVERAGE OF UNIVERSE	1
		−253	HYDROGEN BOILS	
		−259	HYDROGEN FREEZES	
		−268	HELIUM BOILS	
		−272	HELIUM FREEZES [UNDER PRESSURE]	
		−273	ABSOLUTE "ZERO"	

* Prepared by the James Forrestal Research Center for *Scientific American* and reproduced here by permission of that magazine.

depending on their mass and the material of which they are made. Practically all properties of matter, such as size, color, electrical and magnetic properties, and physical state (i.e., gas, liquid, or solid) change with changing temperature.

TEMPERATURE SCALES

Since accurate temperature measurements are of great importance in industry, in scientific research, and in connection with our physical comfort and health, it is essential that temperatures measured in all laboratories have the same meaning, i.e., that the same temperature scale be used by all. This is accomplished by selecting reproducible fixed points and by dividing the temperature intervals between them into a convenient number of degrees. The most important of these fixed points are the boiling and freezing points of pure substances at specified pressures.

Although the principle of basing temperature scales on reproducible fixed points is now universally recognized, this was not always the case. Galileo is credited with inventing the thermometer about 1595. Galileo's first thermometers were merely glass bulbs on long stems with the stem ends immersed in water. The level of the water in the glass tube forming the stem depended puon the temperature of the air in the bulb above it (as well as upon the barometric pressure and other variables). As this air was warmed or cooled, the water level in the tube became lower or higher. Marks were placed on the tube to serve as "degrees" of convenient size. It can be seen that these early instruments did not permit a high order of precision in the measurement of temperatures.

The first sealed thermometers which represented a considerable advance in the art were made before 1654 and contained alcohol as the thermometric liquid. In 1694 Carlo Renaldini, Professor of Mathematics at Padua, suggested that it would be desirable to adopt a temperature scale based on two fixed points: (1) melting point of ice, and (2) boiling point of water. Unfortunately, his contemporaries were not convinced of the constancy of these points and his suggestion was ignored.

Fahrenheit Scale

The first real step toward reliable thermometers with a reproducible scale was taken by Daniel Fahrenheit. Fahrenheit began working on the problem of thermometers in 1706, and by 1709 he had perfected an alcohol thermometer which he distributed over northern Europe. By 1714, Fahrenheit was making mercury thermometers; and Fahrenheit's thermometers were recognized as being far superior to any that had as yet been produced. Fahrenheit's scale was based on three fixed points. The zero was determined by immersing the thermometer bulb in a mixture of ice, water, and sal ammoniac; the second point, 32°, was obtained by the use of a mixture of ice and water; the third point, 96°, was the temperature determined by placing the bulb of the thermometer in the armpit of a "healthy" man. The present Fahrenheit temperature scale differs somewhat from the original, but on both scales the melting point of ice is 32°. The Fahrenheit scale is in common use in English-speaking countries.

Réaumur Scale

Réaumur, about 1731, was the next to propose a temperature scale, which is still in use in Germany and in some other countries. He chose the freezing point of water for one fixed point. He determined experimentally that alcohol diluted with one-fifth water expanded from 1,000 to 1,080 units of volume between the freezing and boiling points of water; so he called the freezing point 0° and the boiling point 80°.

Celsius (Centigrade) Scale

In 1742, Celsius, Professor of Astronomy at Uppsala, proposed a temperature scale with 0° for the boiling point of water and 100° for the melting point of ice. Two scientists, Christians of Lyons and Stromer of Uppsala, independently inverted the Celsius scale, thereby obtaining essentially the same scale as the centigrade scale of France.

Kelvin Scale

In 1848, Lord Kelvin suggested a method of defining a temperature scale which is not based on the change of some property of a particular substance with temperature. Such a scale is based on the laws of thermodynamics, and is known as a thermodynamic scale. Kelvin's scale is equivalent to a scale based on the change in volume with temperature of a fixed mass of an ideal gas held at constant pressure, or the change in pressure of the gas at constant volume. Fortunately, a gas thermometer using a real gas can be used to determine temperatures on Kelvin's scale. If corrections are made for the nonideality of real gases, the ratio of two temperatures may be determined by measuring the ratio of pressures of a constant volume of a real gas at the two temperatures (or ratio of volumes of a gas at constant pressure). The size of the degree on the Kelvin (thermodynamic centigrade) scale was determined by assigning 100° to the interval between the ice and steam points. On this basis, the zero of this scale, the absolute zero, was approximately 273.16 of these degrees below the ice point. A gas thermometer yielding high precision is a complex instrument and is not adaptable for use in ordinary temperature methods.

In 1954, by international agreement, the Kelvin scale was revised by assigning the temperature of 273.16°K to the triple point of water. On this basis, the ice point is very nearly 273.15°K. The triple point was chosen in preference to the ice point because it is more reproducible.

Rankine Scale

This scale is the equivalent of the absolute thermodynamic scale expressed in terms of Fahrenheit degrees. Thus, the temperature of the triple point on the Rankine scale, corresponding to 273.16°K, is very nearly 459.69°.

International Temperature Scale

To provide a fundamental basis for precise and convenient temperature measurements, a temperature scale, known as the international temperature scale, has been established and adopted, which covers the range from the boiling point of oxygen to

Table 2. Basis for the International Practical Temperature Scale of 1948

Temperature range, °C	Fixed points, °C	Interpolation equation	Standard instrument
−182.97 to 630.5		$R_t = R_0[1 + At + Bt^2$ $+ C(t - 100)t^3]$	
0 to 630.5	Oxygen (b.p. − 182.970) Triple point of water (+0.01) Steam (b.p. 100) Sulfur (b.p. 444.600) (See Notes 1 and 2)	$R_t = R_0(1 + At + Bt^2)$ where R_t = resistance of thermometer resistor at temperature t R_0 = resistance at 0°C A, B, and C are constants	Platinum resistance thermometer
630.5 to 1063	Silver (f.p. 960.8) Gold (f.p. 1063.0)	$E = a + bt + ct^2$ where E = emf of standard thermocouple a, b, and c are constants	Platinum vs. platinum–10% rhodium thermocouple
1063 to ∞	Gold (f.p. 1063.0)	$\dfrac{J_t}{J_{Au}} = \dfrac{\exp\,[c_2/\lambda(t_{Au} + T_0)] - 1}{\exp\,[c_2/\lambda(t + T_0)] - 1}$ where J_t and J_{Au} are the radiant energies per unit wavelength interval at wavelength λ, emitted per unit time by a unit area of a blackbody at the temperature t and at the gold point t_{Au}, respectively. $c_2 = 1.438$ cm deg T_0 = temperature of ice point, °K λ = wavelength	Optical pyrometer

F.p. = freezing point. B.p. = boiling point.
Notes: 1. In the interest of higher reproducibility, it is recommended to use the freezing point of zinc with the value 419.505°C in place of the sulfur point.
2. The temperature 630.5°C to be determined with a standard resistance thermometer.

the highest temperatures of incandescent bodies and flames. This scale is based on six reproducible equilibrium temperatures, or fixed points, to which numerical values have been assigned, and upon specified interpolation formulas relating temperatures between or above these points to the indications of standard temperature-measuring instruments. The international temperature scale was first adopted in 1927 by the Seventh General Conference on Weights and Measures, and was revised in 1948 by the Ninth General Conference. At this conference, the name "Celsius" was recommended to designate the scale having 0° as the ice point and 100° as the steam point. Celsius thus replaces the word centigrade as commonly used in the United States. The abbreviation remains unchanged: °C. In 1960 the Eleventh General Conference changed the name of the scale to International Practical Temperature Scale of 1948 and adopted a new text of the scale. The actions, taken in 1960, do not change the numerical values of temperature as defined by the ITS of 1948. (See Reference 7, vol. 3, part 1, pp. 59–66, "The Text Revision of the International Temperature Scale of 1948" by H. F. Stimson.)

Table 2 lists the six fixed points, the instruments used for interpolating between the fixed points, and the interpolation equations.

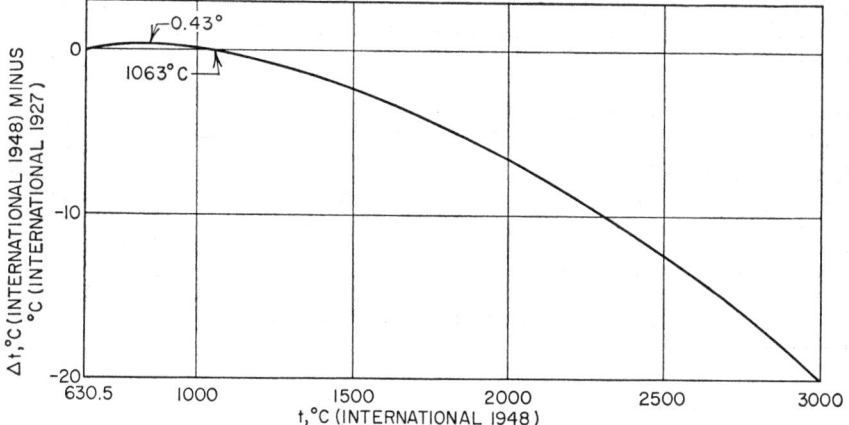

FIG. 1. Difference between the international temperature scales of 1948 and 1927.

In addition to the definition of the international temperature scale of 1948, there are recommendations on the procedures to be used in realizing the fixed points and the requirements which the instruments must fulfill if they are to be standard instruments. The differences between the 1927 and 1948 scales are shown graphically in Fig. 1.

There is no international temperature scale below the oxygen point, but a scale based on the resistance of several capsule-type platinum resistance thermometers which were calibrated with a gas thermometer is maintained at the National Bureau of Standards (Washington, D.C.). This scale is distributed to other laboratories by calibrating similar thermometers by comparison methods.

For the temperature range from approximately 10 to 4°K, there is nothing corresponding to the international scale except the gas thermometer. From 4.2 to 1°K, the vapor pressure of helium is used for the measurement of temperatures in laboratories throughout the world. Below 1°K, the magnetic properties of paramagnetic salts provide both the means of attaining temperatures in this range and the indication of the temperature of the salt.

An ultrasonic thermometer is currently being used at NBS in the range 2 to 20°K. (See Reference 7, vol. 3, part 1, p. 129.)

ESTABLISHING AND MAINTAINING A TEMPERATURE SCALE

The problem of establishing and maintaining a temperature scale is not confined to the National Bureau of Standards, but it is necessary that this be done in industrial and research laboratories as well. In the preceding description of temperature scales, a definition was given of the international temperature scale on which practically all temperature measurements are based. By setting up apparatus suitable for realizing the fixed points, and by calibrating the three standard instruments used for interpolating between these fixed points, the primary temperature scale may be established. To do this requires a considerable investment in equipment and time. Three different methods or combinations of them are commonly used in maintaining temperature scales in laboratories.

Primary Calibration of Three Standard Instruments

The first method involves the setting up of equipment for primary calibrations of the three standard measuring instruments; the resistance thermometer, the platinum vs. platinum-rhodium thermocouple, and the optical pyrometer. This method involves the greatest expenditure of time and money, but it does provide a primary calibration of all three instruments, with the opportunity to repeat these calibrations as required.

National Bureau of Standards Calibration

The second method which is more commonly used is to submit the three standard instruments to the National Bureau of Standards for calibration. In this case the platinum resistance thermometer and the platinum vs. platinum-rhodium thermocouple may be given primary calibrations, and the optical pyrometer is given a comparison calibration using the standard optical pyrometer at the Bureau. The instruments calibrated in this manner and those calibrated in the previous procedure then become the standards of the laboratory and are normally used in turn to calibrate the secondary instruments, such as other thermocouples or liquid-in-glass thermometers.

Secondary Standards

The third procedure which may be used is to rely entirely on what may be called secondary standards, namely, base-metal thermocouples and liquid-in-glass thermometers. Such instruments are calibrated by the National Bureau of Standards if they meet certain specifications, and they may be used to maintain the temperature scale although not with the same accuracy as the primary instruments. In the paragraphs that follow, a brief description of the standard instruments and the calibration procedures will be given.

STANDARD PLATINUM RESISTANCE THERMOMETER

Beginning at the low-temperature end of the international temperature scale, the standard platinum resistance thermometer should be constructed of platinum having a purity such that R_{100}/R_0 is 1.3920 (as required by the international practical temperature scale), where R_{100} is the resistance of 100°C and R_0 is the resistance at 0°C. The platinum wire should be annealed and as nearly strain-free as possible. This can be achieved by winding the thermometer coil in such a manner that it is not constrained either initially or as a result of heating or cooling, and by annealing the thermometer for a sufficiently long period of time at a temperature higher than the highest temperature at which it is to be used. The thermometer should be of the four-lead type, so that the resistance between branch points can be determined either with a precision thermometer bridge or with a high-grade potentiometer. The leads are normally of gold or platinum through the region of the temperature gradient. The constancy of the resistance of the thermometer, as determined by measuring its resistance at a fixed reference point before and after use at high and low temperatures, may be used to determine the adequacy of the annealing procedure and the general reliability of the thermometer.

Triple Point of Water

The temperature assigned to the triple point is $+0.0100°C$. A diagram of a triple-point cell is shown in Fig. 2 where A is the region containing water vapor, B represents liquid water, and C represents ice, thus providing a triple point, as discussed in a later paragraph. A reentrant well permits immersing the thermometer in the cell, as shown in the diagram. The technique for preparing the cell consists of freezing a mantle of ice along the thermometer well by inserting Dry Ice in the well. This method purifies the ice adjacent to the thermometer well, and, prior to use, a thin layer of this pure ice is melted by inserting a warm rod in the thermometer well for a few seconds, to provide an interface between pure water and pure ice to envelop the thermometer resistor.

The reproducibility of the temperature determined with the triple-point cell is of the order of 1 to $2 \times 10^{-4}°C$. To maintain the cell in a usable condition, it is immersed completely in an ice bath which need not be of particularly clean ice. The water surrounding the thermometer in the well serves as the medium for heat transfer so that the thermometer quickly comes to an equilibrium with the temperature of the ice-water interface within the cell.

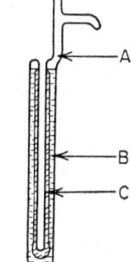

Fig. 2. Triple-point cell.

Boiling Points of Oxygen, Water, and Sulfur

The remaining three fixed points in the standard resistance-thermometer range are boiling points. The ideal situation under which to realize these points is by controlling the vapor pressure of the sulfur, water, or oxygen at the level of the thermometer resistor to the standard value specified in the international temperature scale. This has two advantages: (1) No correction need be made for the difference between the existing pressure and the standard pressure, and (2) there are no fluctuations in temperature resulting from changes in the atmospheric pressure. Such a pressure control requires special provisions to maintain the pressure sufficiently constant.

An alternative method is to allow the boiling-point equipment to be open to the atmosphere, and to correct for the difference between atmospheric pressure and standard pressure by observing the atmospheric pressure with the barometer, and using the relation between the vapor pressure and temperature for the particular substance. With this method, fluctuations in barometric pressure produce temperature fluctuations. The equations to be used for this correction are given below for pressures between approximately 660 and 860 mm of mercury.

Oxygen:

$$t_p = -182.970 + 9.530 \left(\frac{P}{P_0} - 1\right) - 3.72 \left(\frac{P}{P_0} - 1\right)^2 + 2.2 \left(\frac{P}{P_0} - 1\right)^3 \quad (1)$$

Steam:

$$t_p = 100 + 28.012 \left(\frac{P}{P_0} - 1\right) - 11.64 \left(\frac{P}{P_0} - 1\right)^2 + 7.1 \left(\frac{P}{P_0} - 1\right)^3 \quad (2)$$

Sulfur:

$$t_p = 444.6 + 69.010 \left(\frac{P}{P_0} - 1\right) - 27.48 \left(\frac{P}{P_0} - 1\right)^2 + 19.14 \left(\frac{P}{P_0} - 1\right)^3 \quad (3)$$

In these equations, t_p is the temperature at the pressure P, and P_0 is standard atmospheric pressure of 1,013,250 dynes/cm². The design of the boiling-point apparatus must be such that superheating is avoided and that an adequate depth of immersion of the thermometer is possible.

In place of the sulfur point, it is now recommended to use the temperature of equilibrium between solid zinc and liquid zinc (zinc point) with the value 419.505°C (Int.1948). The zinc point is more reproducible than the sulfur point, and the value which is assigned to it has been so chosen that its use leads to the same values of temperature on the International Practical Temperature Scale as does the use of the sulfur point. It is probable that the zinc point will soon replace the sulfur point entirely as a fixed point on the IPTS in the same way that the triple point of water has replaced the ice point.

STANDARD PLATINUM VS. PLATINUM-RHODIUM THERMOCOUPLE

A standard platinum vs. platinum-rhodium thermocouple is calibrated at three freezing points. The temperature of the lowest, the freezing point of antimony, is determined by use of a standard platinum resistance thermometer, and in this way the two portions of the scale are joined. The metals are contained in a crucible, which in turn is heated in the furnace designed so that the metal can be heated to a uniform temperature. The metal is melted and heated a few degrees above the melting point, and is then allowed to cool slowly. The thermocouple is mounted in a porcelain tube with porcelain insulators separating the two wires, and it is immersed in the molten metal through a hole in the center of the crucible cover.

The depth of immersion should be such that the observed electromotive force of the thermocouple is not changed by more than one microvolt when the immersion is increased or decreased by one centimeter. During freezing, the electromotive force should remain constant within one microvolt for a period of at least five minutes. As in the case of materials used for the platinum resistance thermometer, the metals in the thermocouples must be pure in order to provide the correct temperature indications between fixed points.

STANDARD OPTICAL PYROMETER

The method of calibrating the standard optical pyrometer is briefly as follows. The pyrometer lamp current corresponding to the gold point is determined by matching the brightness of a portion of the lamp filament with that of a blackbody immersed in freezing gold. The calibration is extended to lower temperatures by inserting, in turn, various sector disks between the pyrometer and the gold-point blackbody, and observing pyrometer lamp currents required to match the resulting reduced brightnesses. The corresponding temperatures T are then calculated by inserting for J_T/J_{Au} the fractional transmissions of the various sectors used. Optical pyrometers are frequently used for temperature measurements below the gold point although they are not the standard instrument for that region.

The calibration is extended to temperatures above the gold point by using a ribbon-filament lamp and sector disks. The temperature of the lamp is adjusted until its brightness, when viewed through a given rotating sector disk, is equal to that of a blackbody at the gold point. The brightness temperature of the lamp can be calculated as previously indicated. The sector disk is then removed, and the pyrometer lamp current is increased to obtain a brightness match between the pyrometer lamp and the ribbon-filament lamp. This current corresponds to the calculated temperature. By use of sector disks with different angular openings, the pyrometer lamp currents for various temperatures above the gold point may be determined, and a complete calibration is obtained by interpolation.

An alternative (less expensive) method for maintaining the temperature scale in the optical pyrometer range is by use of a calibrated ribbon-filament lamp. If a current-temperature relation for the lamp is known, this may be used to calibrate and check optical pyrometers.

SECONDARY FIXED POINTS

It is desirable, in maintaining a temperature scale in the laboratory, to have apparatus available for occasionally checking the standard instruments. For the resistance thermometer, perhaps the simplest check is to measure its resistance in a triple-point cell or an ice bath. An additional fixed point may be desirable, in order to determine if the ratio of R_{100}/R_0 has changed as a result of use of the thermometer.

The use of these fixed points requires the same precautions concerning purity of material, depth of immersion of the measuring instrument, and care in making measurements as are required for the calibration of the primary instruments at the primary fixed points. The secondary fixed points, however, cannot be used for the calibration of a resistance thermometer as a primary standard.

Benzoic Acid Thermometric Cell

One of the most useful secondary fixed points is that provided by the benzoic acid thermometric cell, which is available from the National Bureau of Standards. The freezing temperature of each individual cell is certified within $\pm0.003°C$, and is approximately 122.358°C. The certified value is determined with standard resistance thermometers.

In addition to checking the calibration of the resistance thermometer by using the ice or triple point and perhaps the benzoic acid point, it is essential for measurements of high precision that ice points or triple points be taken frequently. Since only ratios of resistances are required to determine temperature, the resistance unit may be the absolute ohm or any arbitrary unit, but the resistance bridge must be self-consistent.

Thermocouple and Optical-pyrometer Reference Standards

For checking platinum vs. platinum-rhodium thermocouples, it is perhaps most satisfactory to maintain a thermocouple which is used only as a reference standard, and to intercompare the other standards with this one in a suitable tube furnace. As previously stated, an adequate check of the optical pyrometer may be made, using a calibrated ribbon-filament lamp.

Liquid-in-glass Thermometers as Standards

For those requiring a precision of only 0.1 to 0.01°C, liquid-in-glass thermometers provide satisfactory standards over most of the resistance-thermometer temperature range. As in the case of the platinum vs. platinum-rhodium thermocouple, perhaps the most satisfactory and simplest procedure is to maintain a set of standard liquid-in-glass thermometers which can be compared with the working thermometers as required. It is important that these standard thermometers be provided with ice points so that changes in the volumes of the thermometer bulbs with time, due to a creep of the glass, can be detected. Since the volume of the bulb is very large in comparison with that of the capillary, changes in the bulb volume are the only important ones in such thermometers. In the event that the volume of the thermometer bulb is observed to have changed, one can correct for this by making the corresponding change in all the corrections for the thermometer.

A comparison of the standard thermometers with the working thermometers will require "comparison baths" which may be of the circulating liquid type. A satisfactory design for such baths consists of two vertical cylinders connected at the tops and bottoms. The fluid is circulated from one cylinder to the other and back by means of a stirrer located in the smaller of the two cylinders. The heating and cooling coils are located in this same cylinder. The other, larger cylinder is the working region into which the thermometers are placed. A variety of liquids may be used in such baths to cover the temperature range up to approximately 300°C. If oil is used as the circulating liquid, the range to be covered by any particular oil must be selected so that, at the lowest temperature, the oil is not too viscous to provide satisfactory mixing. In addition, the flash point of the oil must not be so low that the oil becomes a hazard in use. A holder which permits rotating the thermometers under test into the field of view of a reading telescope is desirable for use with such baths.

If large numbers of thermometers are to be tested, and if they are similar in type, it is advantageous to use standards of the same type. This is particularly true if the thermometers are designed for use at partial immersion, so that the determination of the temperature of the emergent mercury column is not required except in the case of the calibration of the standard thermometers.

REFERENCES

1. Stimson, H. F.: International Temperature Scale of 1948, *J. Research Natl. Bur. Standards*, vol. 42, p. 210, 1949.
2. Corruccini, R. J.: Differences between the International Temperature Scales of 1948 and 1927, *ibid.*, vol. 43, p. 133, 1949.
3. Hoge, H. J., and F. G. Brickwedde: Establishment of a Temperature Scale for the Calibration of Thermometers between 14° and 83°K, *ibid.*, vol. 22, p. 351, 1939.
4. Jakob, Max: "Heat Transfer," vol. 1, John Wiley & Sons, Inc., New York, 1949.
5. McAdams, William H.: "Heat Transmission," 3d ed., McGraw-Hill Book Company, Inc., New York, 1954.
6. Zemansky, M. W.: "Heat and Thermodynamics," 4th ed., McGraw-Hill Book Company, Inc., New York, 1957.
7. American Institute of Physics: "Temperature, Its Measurement and Control in Science and Industry," vol. 1, 1941; vol. 2, 1956, vol. 3, 1962, Reinhold Publishing Corporation, New York.
8. Eckman, D. P.: "Industrial Instrumentation," John Wiley & Sons, Inc., New York, 1950.
9. Freeze, P. D.: Bibliography on the Measurement of Gas Temperature, *Natl. Bur. Standards (U.S.) Circ. 513*, 1951.
10. International Symposium on High-temperature Technology, Stanford Research Institute, Menlo Park, Calif., Nevin K. Hiester, General Chairman, October, 1959.
11. Baker, H. Dean, E. A. Ryder, and N. H. Baker: "Temperature Measurement in Engineering," vol. I, John Wiley & Sons, Inc., New York, 1953.
12. Bockris, J. O'M., J. L. White, and J. D. MacKenzie: "Physico-chemical Measurements at High Temperatures," Academic Press Inc., New York, 1959.
13. Hall, J. A.: "Fundamentals of Thermometry," Chapman & Hall, Ltd., London, 1953.
14. Hall, J. A.: "Practical Thermometry," Chapman & Hall, Ltd., London, 1953.
15. Scott, Russell B.: "Cryogenic Engineering," D. Van Nostrand Company, Inc., New York, 1959.
16. Kingery, W. D.: "High Temperature Properties," John Wiley and Sons, Inc., New York, 1959, *Chemical Engineering Progress*, vol. 56, no. 6.
17. Staff: "Thermocouples for Nuclear Plants," *Automatic Control*, vol. 12, no. 3, p. 50, March 1960.
18. Lachman, John C. and F. W. Kuether: Stability of Rhenium/Tungsten Thermocouples in Hydrogen Atmospheres, *ISA J.*, vol. 7, no. 3, p. 67, March 1960.
19. Combes, J. J.: Temperature Measuring Devices, *Automation*, vol. 7, no. 5, p. 87, May 1960.
20. Fourth Symposium on "Temperature—Its Measurement and Control in Science and Industry," sponsored by American Institute of Physics, Instrument Society of America, and the National Bureau of Standards (U.S.), March 27–31, 1961, Columbus, Ohio:

   Lindsay, R. B.: "The Temperature Concept for Systems in Equilibrium."
   Ramsey, N. F.: "Thermodynamics and Statistical Mechanics at Negative Absolute Temperatures."
   Callen, H. B.: "Non-Equilibrium Thermodynamics."
   Shuler, K. E.: "Relaxation of Non-Equilibrium Distributions."
   McNish, A. G.: "The Role of Temperature in Our Measurement System."
   Riddle, J. L.: "Standard Platinum Resistance Thermometry."
   Dauphinee, T. M.: "Potentiometric Methods of Resistance Measurement."
   Evans, J. P.: "An Improved Resistance Thermometer Bridge."
   Droms, G. R.: "Thermistors for Temperature Measurements."
   Kunzler, J. E. and T. H. Geballe: "Germanium Resistance Thermometers."
   Cataland, G., M. H. Edlow, and H. H. Plumb: "Resistance Thermometry in the Liquid Helium Temperature Region."
   Sachse, H. B.: "Low Temperature Measurements with Thermistors."
   Furukawa, G. T.: "Applications of Resistance Thermometry in Calorimetry."
   Berry, R. J.: "The Stability of Platinum Resistance Thermometers at Temperatures up to +630°C."
   Evans, J. P. and G. W. Burns: "The Stability and Reproducibility of High Temperature Platinum Resistance Thermometers."
   Lefkowitz, I.: "The Control Systems Approach to Process Design: Analytical Tools Required for Temperature Control."
   Davis, E. T.: "Considerations in the Design and Selections of an Electrical Control System."
   Courtright, N.: "Considerations in Designing a Pneumatic Temperature Control System."

21. Nalle, D. H.: Accurate Recording of Fast-Changing Temperatures, *ISA J.*, vol. 8, no. 6, pp. 58–59, June, 1961.

22. Slatosky, W. J. and N. Simcic: Computer Controls End-Point Temperature in Oxygen Steelmaking, *ISA J.*, vol. 8, no. 12, pp. 38-41, December, 1961.
23. Bose, B. N.: Thermocouple Well Design, *ISA J.*, vol. 9, no. 9, pp. 89–94, September, 1962.
24. Mandt, R. D.: Fusing Thermocouple Leads, *ISA J.*, vol. 9, no. 11, p. 82, November, 1962.
25. Kingery, W. D.: "Property Measurements at High Temperatures," John Wiley & Sons, Inc., New York, 1959.
26. Din, F. and A. H. Cockett: "Low Temperature Techniques," John Wiley & Sons, Inc., New York, 1960.
27. Schneider, P. J.: "Temperature Response Charts," John Wiley & Sons, Inc., New York, 1963.
28. Fischer, H. and L. C. Mansur: "Conference on Extremely High Temperatures," John Wiley & Sons, Inc., New York, 1958.
29. Payne, H. G.: Temperature Controls—Selection and Application, *Automation*, vol. 7, no. 11, pp. 90–97, November, 1960.

SUMMARY OF TEMPERATURE-MEASUREMENT METHODS

By John H. Miller*

This summary is to assist in making the selection of the appropriate primary element for a given temperature measurement problem. Engineering details of all methods described here can be found in "Process Instruments and Controls Handbook," McGraw-Hill Book Company, Inc., New York, 1957.

Thermocouples

Seebeck discovered, in 1821, that in an electric circuit having two different materials as wires, when the two junctions were at different temperatures, a potential would exist at the terminals in an open circuit; but if the circuit were closed a current would flow. The magnitude of the potential depends on the materials used and the temperature difference between the hot (or measuring) junction and the cold (or reference) junction (see Fig. 3). Since the output is a function of temperature difference, cold-end compensation is usually required. In oxidizing atmospheres at high temperatures, and in some gases and vapors, appropriate protection tubes are necessary to prevent degradation of the couple and consequent errors in temperature indication.

Fundamental characteristics of commonly used thermocouples are given in Table 3.

Table 3. Fundamental Characteristics of Thermocouples

Couple materials	Practical range limits		Inherent accuracy		Applications
	°C	°F	°C	°F	
Platinum vs. platinum-10 % rhodium..	0 to 1450	32 to 2650	0.1	0.2	Where high accuracy is required at high temperatures. Also as standard for calibration
Platinum vs. platinum-13 % rhodium..	0 to 1450	32 to 2650	0.1	0.2	
Chromel-Alumel....................	−200 to 1100	−300 to 2000	0.5	1.0	Commercial furnace temperature indication and control. Also, same as for filled-system thermometers
Chromel-Constantan................	−100 to 1000	−150 to 1800	0.5	1.0	
Iron-Constantan...................	−200 to 750	−300 to 1400	0.5	1.0	
Copper-Constantan................	−200 to 350	−200 to 650	0.2	0.4	Commercial oven temperature indication and control

* Retired; formerly Vice President for Research and Engineering, Weston Instruments, a subsidiary of Daystrom, Inc., Newark, N.J.

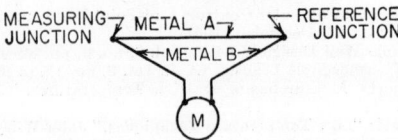

FIG. 3. Simple thermocouple circuit.

Radiation and Optical Pyrometry

These methods of temperature measurement include, in a broad sense, the measurement of heat energy radiated from the hot body in question as a function of its temperature.   In *radiation pyrometry*, the energy is concentrated on a sensitive receiver or transducer which operates, in turn, an indicating or controlling system (see Fig. 4). In *optical pyrometry*, the brightness of a hot filament F is matched by eye at E to the apparent brightness of the hot body.   An ammeter, which responds to the manually

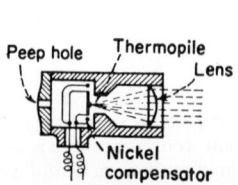

FIG. 4. Typical radiation pyrometer.   FIG. 5. Typical optical pyrometer.

adjusted battery current, is used to heat the filament, and indicates on a scale marked directly in terms of the temperature of the matching hot body (see Fig. 5).   In other systems, the filament current is kept at a fixed temperature, and an optical wedge attenuates the incoming energy to match the filament; the wedge position then indicates temperature.

Calibration generally is based on blackbody radiation, to which a correction is applied when the hot body has an emissivity of less than unity.   Corrections also may be required to take care of energy absorption by intervening windows or atmospheric vapor.   This method is highly convenient for high temperatures; is of limited value below 1000°F.

A summary of the characteristics of radiation and optical pyrometers is given in Table 4.   Figure 6 shows the spectral distribution of radiant energy.

Table 4. Basic Characteristics of Radiation and Optical Pyrometers

Type.....................	Optical system	Receiving device or method
Radiation pyrometer........	Concave mirror or lens system	Thermopile, bolometer, photovoltaic cell
Optical pyrometer.........		Adjustable filament current, optical wedge, fixed filament current

Practical range: 1000°F (540°C) to 10,000°F (5500°C).
Accuracy: ± 10°F (6°C) if proper corrections for target emissivity have been made.
Applications: Molten metal, in bath or being poured; heated billets in furnace or being rolled; furnaces of all types.   (Needs no contact with object measured; can measure temperature of moving objects.)

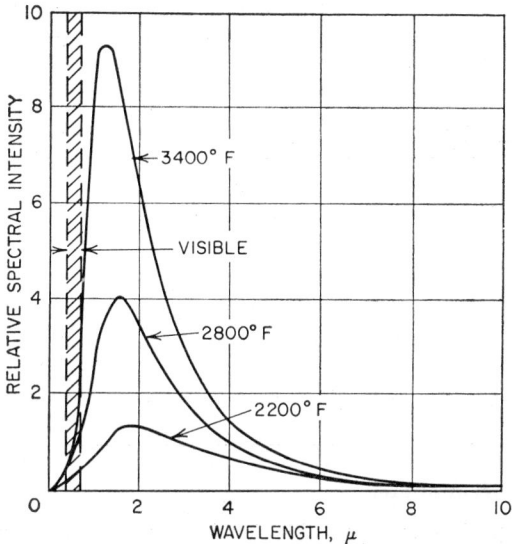

FIG. 6. Spectral distribution of radiant energy.

Resistance Thermometry

All pure metals increase in electrical resistance with rise in temperature. A resistor bulb is simply a length of appropriate wire, suitably insulated, wound on a form, adjusted to a definite resistance at a specified temperature, and contained in a protecting tube. Common resistance values are 10, 25, and 100 ohms. The respond-

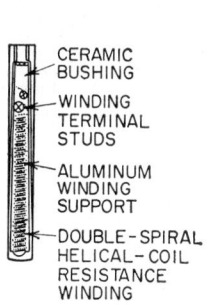

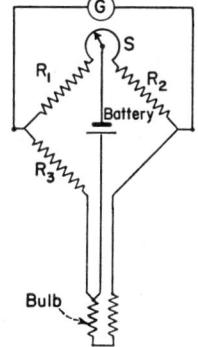

FIG. 7. Cross section of sensitive portion of typical industrial resistance-thermometer bulb.

FIG. 8. Typical resistance-thermometer bridge circuit.

ing device may be an ohmmeter or any instrument which is sensitive to resistance changes. The instrument is adjusted to interpret the resistance value of the bulb in terms of the bulb temperature. A resistor bulb is shown in Fig. 7, and a typical bridge circuit is shown in Fig. 8.

Certain compounds, usually semiconductors such as uranium oxide, show very large changes in resistance with temperature; the resistance *decreases* with temperature rise. These materials are formed into a bead with wire terminals and are termed *thermistors*. They are of small size, are useful for a narrow temperature span near room temperature, and usually require individual calibration after assembly.

Common resistor bulb characteristics are summarized in Table 5.

Table 5. Common Resistor Bulb Characteristics

Material	Resistance increase, (0–100°C), %	Practical range	Inherent accuracy, °C	Remarks
Platinum...	39.2	−258 to 1000°C −432 to 1850°F	0.01	Uniform and high in resistance. Stable in value, expensive
Nickel......	66.0	−150 to 300°C −240 to 575°C	0.50	Large change and high in resistance. Does not run uniform; requires individual adjustment
Copper.....	43.1	−200 to 120°C −330 to 250°F	0.10	Uniform but low in resistance. Stable in value, inexpensive

Filled-system Thermometer

This device comprises an expandable liquid, vapor, or gas contained in a completely sealed system (see Fig. 9). As the fill material in the bulb rises in temperature, the increased volume and/or pressure is transmitted through the fine capillary tube to a receiver (usually a spiral or helix), which is responsive to the volume or

Table 6. Characteristics of Filled-system Thermal Elements

Principle	Expanding material	SAMA* class	Practical range limits†	Inherent accuracy, %	Remarks
Volumetric.........	Liquid	I	−52 to 315°C −125* to 600°F	½–1	Slow in response. Equal divisions
	Mercury	V	−39 to 535°C −38 to 1000°F	½–1	Slow in response. Equal divisions
Pressure...........	Vapor (of a liquid)	II	−40 to 315°C −40 to 600°F	½–1	Medium speed. Unequal divisions
	Gas	III	−240 to 535°C −400 to 1000°F	½–1	Fastest system. Equal divisions

* Scientific Apparatus Makers Association.
† With some liquids, the minimum temperature limit may be as low as −150°C or −300°F.

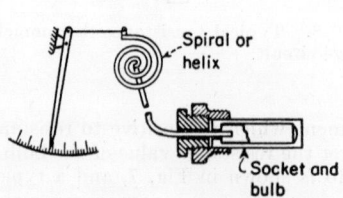

FIG. 9. Filled-system thermometer.

FIG. 10. Industrial-type liquid-in-glass thermometer.

pressure change.   Ambient temperature compensation is required in the receiver; additional compensation also is needed for long capillary tubes.   The scale ranges of individual instruments generally are much narrower than the full ranges given in Table 6.

Liquid-in-glass Thermometers

The common mercury-in-glass thermometer and those filled with colored liquids for low temperatures have been brought to a high state of perfection.   In selection, it should be noted that etched-stem laboratory thermometers are inherently more accurate than industrial types, wherein separate metal scales are attached (see Fig. 10).   In the etched-stem types, attention is required for possible immersion errors which may result if precise instructions of use are not followed.   In all colored-liquid thermometers, possible separation of the column should be checked before the instrument is used.

Characteristics of liquid-in-glass thermometers are summarized in Table 7.

Table 7. Characteristics of Liquid-in-glass Thermometers

Filling liquid	Practical range limits	Best accuracy	Remarks
Mercury..................	−39 to 600°C −38 to 1100°F	0.01°C 0.02°F	Best for general use
Mercury-thallium............	−55 to 0°C −68 to 32°F	0.02°C 0.04°F	For low temperatures in narrow span
"Spirit," alcohol, toluene.....	−100 to 50°C −150 to 120°F	0.1°C 0.2°F	For low temperatures. Must be watched for separation of column

NOTE: Liquid-in-glass thermometers generally are guaranteed to one scale division.   There are large numbers of special thermometers with scale ranges and housings to fit specific applications, e.g., ASTM types for the petroleum industry.

Bimetal Thermometers

When a composite metal strip consisting of two pieces of different metals or alloys welded together is heated, the differential expansion of the metals causes the strip to bend.   Although used for many years in thermostats, an accurate thermometer which utilizes a coil of bimetal was not developed until 1935.   Mechanical and thermal treatments are required for stabilizing the system.   The element usually is contained in a closed tube, with pointer and scale of appropriate size contained in a head.   Included scale angle is approximately 300°.   While no emergent stem correc-

Table 8. Characteristics of Bimetal Thermometers

Type	Practical range		Accuracy
	Limits	Minimum span	
Laboratory..........	−185 to 260°C −300 to 500°F	50°C 100°F	½ % of span
Industrial..........	−185 to 550°C* −300 to 1000°F*	50°C 100°F	1 % of span
Recording..........	−80 to 120°C −100 to 250°F	25°C 50°F	2 % of span

* Should not be used continuously over 425°C or 800°F.   Head temperature should be kept under 300°F; 300 to 600°F with special gaskets.

tion is required, it is necessary that the entire actuating element (1 to 3 in. in length) be immersed.

Characteristics of bimetal thermometers are summarized in Table 8.

Pyrometric Cones

These are used principally in ceramic kilns, and comprise narrow triangular cones made of various ceramic mixtures. They were developed by Dr. Seger in Germany in 1886 and frequently are called *Seger cones*. A series of approximately sixty numbered cones soften and bend over at successively higher temperatures from 585°C (1085°F) to 2015°C (3659°F) in steps of roughly 25°C or 45°C. Tables of temperature equivalents are available for the numbered cones and include data on the *rate* of heating to which the cones are sensitive.

Temperature-sensitive Crayons, Paints, and Pellets

All these items serve to indicate by change in color or appearance when a certain temperature has been *exceeded*. They are available in 12 to 50°F steps and from 113 to 2500°F. They are valuable for indicating excessive temperatures as encountered in electrical equipment, retorts, and piping.

REFERENCES

For descriptive details concerning the various temperature measurement devices briefly outlined here, see the following sections of the "Process Instruments and Controls Handbook," Douglas M. Considine, Editor-in-Chief, McGraw-Hill Book Co., Inc., New York, 1957:

Thermocouples, Section 2, pp. 2-5–20.
Radiation and Optical Pyrometry, Section 2, pp. 2-21–46.
Resistance Thermometry, Section 2, pp. 2-47–59.
Thermistors, Section 2, pp. 2-60–67.
Filled-System Thermometers, Section 2, pp. 2-68–84.
Bimetal Thermometers, Section 2, pp. 2-85–87.
Liquid-in-Glass Thermometers, Section 2, pp. 2-87–93.

CALORIFIC VALUE

By Gerald L. Eberly*

The art of measuring heat of combustion, or calorific value, is known as "calorimetry," and the instruments used in these measurements are called "calorimeters." Calorimeters are available for testing solids, liquids, or gases. They are widely used in industrial laboratories for rapid, accurate fuel tests; in research and development laboratories for calorific measurements; and in college and university laboratories for student instruction.

All commonly used calorimeters for solid and liquid measurement are based on the same principle—a known weight of sample is completely burned in an apparatus in which the heat of combustion is completely absorbed by a known mass of water. By observing with exactness the rise in water temperature, it is possible to calculate the amount of heat liberated. Figure 1 shows a typical temperature rise curve.

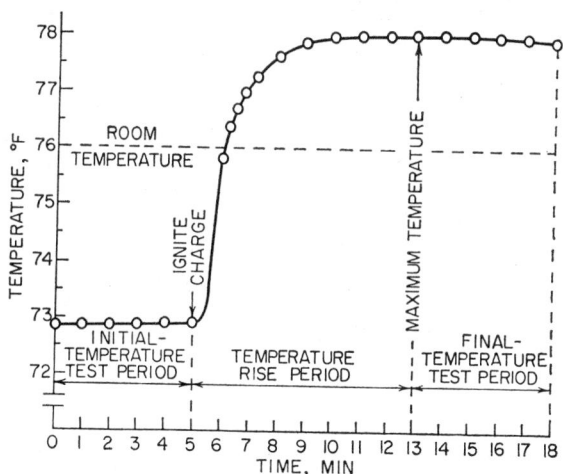

Fig. 1. Typical temperature rise curve for plain calorimeter. (*Parr Instrument Co.*)

Calorimeters differ essentially in the method of oxidation used to burn the sample, as well as in the mechanical construction of the apparatus. The two most commonly used methods of combustion are (1) with oxygen gas under pressure, and (2) with molten sodium peroxide as an oxidizing medium.

Berthelot in 1881 devised the first successful calorimeter, using oxygen under pressure in a closed vessel. Subsequent developments by Mahler (1892), Atwater (1899), and Parr (1912) improved and reduced the cost of the original Berthelot apparatus by variations in design and improvements in the materials for the combustion chamber.

* Technical Staff, Harris D. McKinney, Inc., Philadelphia, Pa.

DEFINITIONS OF THERMAL UNITS

Heat of combustion of a substance is defined broadly as the number of heat units liberated by a unit mass of the substance when combined with oxygen in an enclosure of constant volume. A more exact definition would specify the temperature at which the reaction begins, but the change in heat of combustion with a change in initial temperature is so small that this specification is not necessary for most experimental results. Therefore, the term "heat of combustion" commonly means the heat liberated by the combustion of all hydrogen and carbon with oxygen to form water and carbon dioxide, including the heat liberated by the oxidation of all other elements, such as sulfur and nitrogen, that are present in the sample.

The most commonly used heat units are:

> *British thermal unit* (Btu): The quantity of heat required to raise the temperature of one pound of water one degree Fahrenheit.
> *calorie* (cal): The quantity of heat required to raise the temperature of one gram of water one degree centigrade.
> *Calorie* (Cal): The quantity of heat required to raise the temperature of one kilogram of water one degree centigrade. It is equal to 1000 small calories.

To be precise, these definitions must include the temperature at which the one-degree rise begins, because the heat capacity of water changes slightly as its temperature changes. In the English system, standard measurements are referred to 60°F (equivalent to 15.56°C). This system is generally used for the Btu and the calorie. Where the metric system is used, the results are often expressed in terms of the calorie taken at 20°C instead of 15.56, since the former temperature is closer to that found in the average laboratory. No changes are required in the calculations as long as the metric system is used alone, but if calories per gram on this basis are to be converted to Btu per pound, a multiplication factor of 1.798 is used rather than the 1.8 factor for the English standard.

The heat produced by burning a given weight of material is known by any of several terms, including thermal value, heating value, calorific value, and heat of combustion. The quantity is always expressed as heat units liberated per unit weight of burned sample. The following relationships may be used to convert from one system to another:

$$1 \text{ cal/g} = 1 \text{ Cal/kg}$$
$$= 1.8 \text{ Btu/lb}$$
$$1 \text{ Btu/lb} = 0.556 \text{ Cal/kg}$$

BOMB CALORIMETERS (Fig. 2)

The bomb calorimeter is the most widely used method for determining the heat value of fuels, foods, and other combustibles. It is also used increasingly for the determination of sulfur content in solid fuels, fuel oils, and other combustible materials, since it offers a much faster method. In laboratories, it is used for an increasing number of other difficult oxidation reactions where high pressure is needed to complete the reaction.

With this method, the calorific value of a combustible material is determined by burning an accurately weighed sample in oxygen under high pressure. This is performed in a strong, thick-walled metal vessel, securely closed against leakage from within and contamination from without. Ignition is accomplished by passing a small electric current through a short length of resistance wire in contact with the sample.

Combustion takes place in a few seconds and with almost explosive violence, although there is no external evidence of the reaction. Extremely high shock and static pressures are developed, which decrease rapidly as heat is dissipated to the surrounding medium. All combustion products are retained in the bomb, and distilled water in the bottom absorbs the soluble oxides and acids produced by the reaction. Valves release the residual gases, either to the atmosphere or for collection.

The oxygen-bomb calorimeter consists of three basic parts: (1) the bomb, in which the combustible charge is burned, (2) the calorimeter bucket, or water container, which holds a measured quantity of water in which the bomb, thermometer, and stirring apparatus are immersed, and (3) the jacket, for protecting the calorimeter

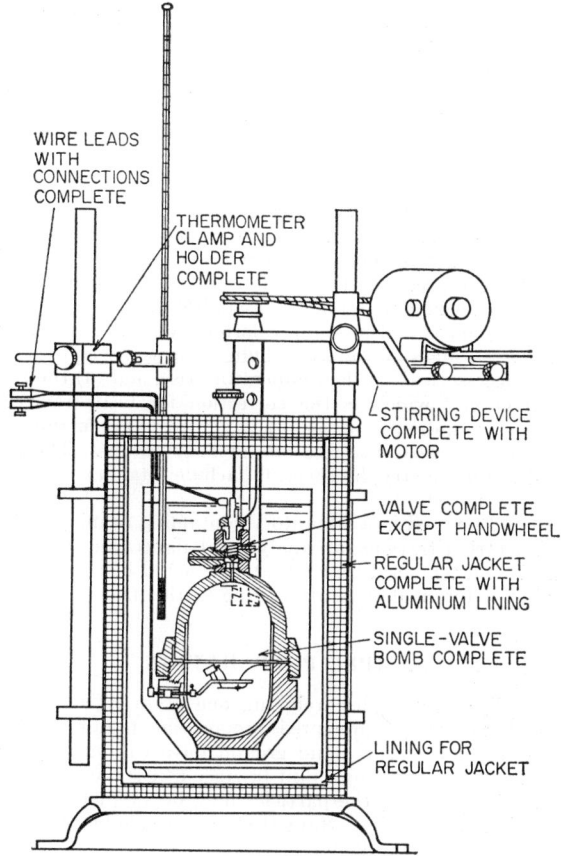

WIRE LEADS WITH CONNECTIONS COMPLETE

THERMOMETER CLAMP AND HOLDER COMPLETE

STIRRING DEVICE COMPLETE WITH MOTOR

VALVE COMPLETE EXCEPT HANDWHEEL

REGULAR JACKET COMPLETE WITH ALUMINUM LINING

SINGLE-VALVE BOMB COMPLETE

LINING FOR REGULAR JACKET

Fig. 2. Cross section of liquid- or solid-fuel calorimeter with stainless steel bomb and plain jacket. (*Emerson Apparatus Co.*)

bucket from the effects of drafts, room-temperature variations, and other thermal disturbances.

Bomb

The bomb is a thick-walled metal vessel with a mechanically sealed cover, which can be removed for inserting the sample and cleaning. Provision is made for supporting the sample, filling with compressed oxygen, igniting, and releasing of residual gas after combustion is completed.

Bucket

The bucket provides for total immersion of the bomb in a measured quantity of water. It requires a stirring device, running at constant speed, to circulate the water for rapid absorption of liberated heat, and to maintain balanced temperature throughout the bucket.

Jacket

The jacket includes two thermal regions; a limited region containing a mass of known heat capacity; and a second region surrounding and enclosing the inner system, having a temperature either uniform or constant, or at least measured during the time of the reaction.

Two jacketing methods are commonly used. In the first method the rise in temperature of the calorimeter bucket is observed while the jacket temperature remains constant. In the second method, the temperature of the jacket is continually adjusted during a test to keep it equal at all times to that of the calorimeter bucket. The second method eliminates the need for a radiation correction, and requires only the observation of the initial and final temperatures, whereas the first method requires the observation of temperatures before, between, and after the initial temperatures. The first method, however, requires only a plain jacketed calorimeter, but the second requires an adiabatic calorimeter to maintain equilibrium.

The quantities to be measured in bomb calorimetry are: (1) the mass of the combustible charge, (2) the mass of water, including the water-equivalent mass of the bomb, bucket, and accessories, (3) the temperature rise, and (4) the thermal leakage, or "radiation correction," which is the correction for heat transfer between the calorimeter and its surroundings. Heat lost or gained in this manner may be lost or gained by conduction, convection, radiation, or evaporation. The term "radiation correction" is used in calorimetry, however, to include total heat transfer between the calorimeter and its surroundings.

Calorimeters should be used in a laboratory containing running water and a drain, apparatus for volumetric titrations, an analytical balance, and other general laboratory equipment. Room-temperature fluctuations should be avoided, and the instrument should not be used immediately after being carried from a cold room to a warm one.

SAMPLE PREPARATION

Calorimeters usually use samples weighing one gram. Care must be taken to avoid overcharging because bomb pressure is proportional to sample weight. Samples of solid materials should be air-dry and ground until the particles pass through a 60-mesh screen. Particle size is important because if it is too large, combustion will be incomplete; and if too small, the particles may be swept out of the combustion capsule and fail to burn. It is important that the sample be truly a representative one of the batch being tested. Pellet presses are available for compacting a fine sample for optimum results.

Some materials of slow-burning characteristics, such as anthracite coal, coke, etc., are difficult to test because they either ignite poorly or else cool below the ignition point before combustion is complete. A small weighed amount of a standard combustible such as powdered benzoic acid can be mixed with these samples. The heat of combustion of the addition is then deducted from the total observed amount, to obtain the net value for the sample.

Highly volatile fuel samples such as gasoline are difficult to weigh accurately, and they often flash when ignited, leaving an unburned carbon residue. A standard open fuel capsule cannot be used for these materials, but good results can be obtained by weighing samples in thin-walled glass bulbs, or gelatin capsules.

ACCURACY

With a reasonable amount of experience, any operator should be able to obtain results within ASTM tolerances. Care must be taken, however, to avoid incomplete combustions and errors in thermometers readings. Under ASTM Standard D 271–46, for example, permissible differences between two or more determinations are to be less than 0.3 per cent in the same laboratory, and 0.5 per cent in different laboratories. These standards are easily met by experienced operators.

STANDARD TESTS

A number of standard tests have been established by various technical societies. Reference to these standards can provide valuable information to the calorimeter operator. Some standard tests are:

Calorific value of coal and coke:
   ASTM Designation D 271–48.
   ASA No. K18.1–1948.
Calorific value of liquid fuels:
   ASTM Designation D 240–50.
   ASA No. Z11.14–1950.
   API No. 517.
   Fed. Spec. VV-L-791e, Methods 2501.2 and 2502.3.
Sulfur in petroleum products and lubricants:
   ASTM Designation D 129–52.
   ASA No. Z11.13–1950.
   API 516–44.
   Fed. Spec. VV-L-791e, Method 5202.7.
Chlorine in oils and greases:
   ASTM Designation D 808-52T.
   Fed. Spec. VV-L-791e, Method 5651.2.
Chlorine in synthetic elastomers:
   ASTM Designation D 833–46T.

GAS CALORIMETRY

Although the instruments used to measure gas heating value are called calorimeters, the techniques differ from those used in solid and liquid calorimetry, as previously discussed in this subsection. Additional terms are also used and are defined below, based on ASTM Standard D 900–55.

Standard cubic foot of gas: The quantity of any gas that at standard temperature and under standard pressure will fill a space 1 cu ft when in equilibrium with water.
Standard temperature: 60°F, based on the international temperature scale.
Standard pressure: The absolute pressure of a column of pure mercury 30 in. in height at 32°F and under standard gravity (32.174 ft/sec²).
Total calorific value (total heating value): The number of Btu evolved by the complete combustion, at constant pressure, of one standard cubic foot of gas with air; the temperature of the gas, air, and products of combustion being 60°F, and all the water formed by the combustion reaction being condensed to the liquid state.
Net calorific value (net heating value): The number of Btu evolved by the complete combustion, at constant pressure, of one standard cubic foot of gas with air; the temperature of the gas, air, and products of combustion being 60°F, and all the water formed by the combustion reaction remaining in the vapor state.
Theoretical air: The volume of air that contains the quantity of oxygen, in addition to that in the gas itself, consumed in the complete combustion of a given quantity of gas.
Excess air: The quantity of air passing through the combustion space in excess of theoretical air.
Combustion air: The air passing into the combustion space of the calorimeter (theoretical air plus excess air).

Gas Calorimeters

Gas calorimeters can be divided into three basic types: (1) total calorific value types; (2) net calorific value types, and (3) inferential types. As pointed out in the definitions above, the net calorific value is less than the total calorific value by an amount equal to the latent heat of vaporization of the water formed during combustion. Type 1 calorimeters make a direct determination of total calorific value by employing principles of operation conforming closely to the conditions outlined in the definition of total calorific value. Type 2 calorimeters employ a number of different operating principles yielding indications related to net heating value, although the results are sometimes presented in terms of total calorific value. Type 3 calorimeters infer calorific values from other characteristics, such as gas analysis, specific gravity, or the appearance of a flame. The most commonly used gas calorimeters are outlined as follows:

 I. Total calorific value types:
    A. Water-flow type:
      1. Manual (Fig. 3).
      2. Automatic recording.
    B. Recording type—Cutler-Hammer.
    C. Thomsen type.
    D. Bomb type.
 II. Net calorific value types:
    A. Thermeter (Cutler-Hammer trade name).
    B. Sigma calorimeter.

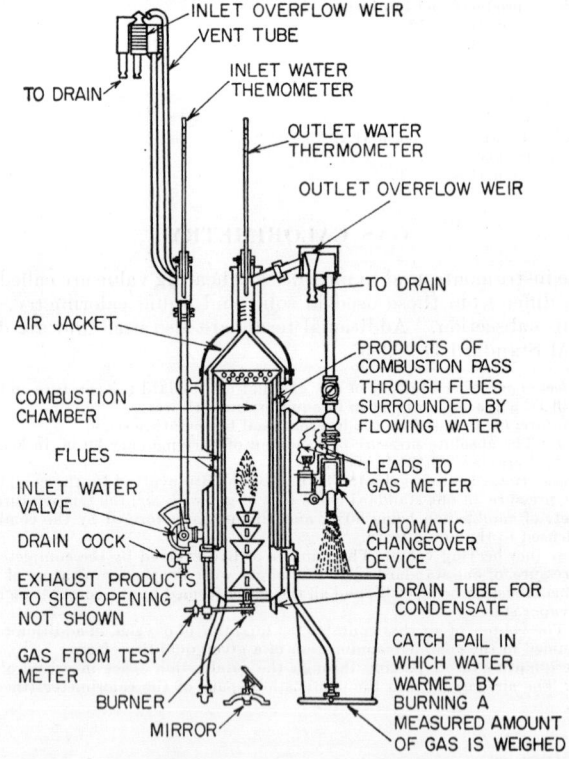

FIG. 3. Manual water-flow gas calorimeter. (*Precision Scientific Co.*)

III. Inferential types:
    A. Parr comparison type.
    B. Caloroptic type.
    C. Gas analyzers:
      1. Absorption.
      2. Mass spectrometer.
    D. Specific gravity units.

Water-flow Calorimeters, Manual

A number of instruments of this type are available, differing in various operating details, but similar in operating principle. In these instruments, the heat to be measured is imparted to a water stream that absorbs the heat of combustion. The mass of water that flows through the calorimeter while a known mass of gas is burned, and the resulting temperature change, provide the data used to calculate the heat absorbed by the water.

$$\text{Observed heating value} = \frac{(\text{temp. rise})(\text{lb of heated water})}{(\text{cu ft of gas burned})(°F)}$$

This equation provides the approximate heating value of the gas. Because combustion takes place at constant pressure, and the temperature of the exiting gases is reduced to nearly that of the inlet water, practically all the water of combustion is condensed. Thus, the instrument permits a direct determination of total calorific value. From the observed heating value, the final heating value is determined by correcting for heat losses, atmospheric humidity, and variation of the specific heat of water.

A modified form of this instrument is available with a temperature and humidity controller which reduces or eliminates the humidity correction factor. Installation, operation, and maintenance of this unit must be given careful attention to achieve optimum accuracy. In routine testing with manual water-flow calorimeters, approximately four determinations of heating value can be obtained in an hour. Figure 3 shows a typical unit.

Water-flow Calorimeters, Automatic Recording

An automatic recording version of the water-flow calorimeter, developed in Great Britain, eliminates the time lag between manual determinations. Known as the Fairweather recording calorimeter, the unit is a water-flow calorimeter with automatic control of water and gas flows and continuous corrections for atmospheric temperature and pressure change.

Reported drift in accuracy from week to week is less than 0.5 per cent, but absolute accuracy depends upon the maintenance of constant ambient conditions. Manual tests can be performed on the unit to check recorded values.

Recording Calorimeter, Cutler-Hammer Type

The most widely used recording calorimeter is the Cutler-Hammer type, which is a flow-type unit using air as the absorbing medium instead of water. The complete calorimeter (Fig. 4) comprises two parts: a recorder; and a tank unit, in which the actual measurement takes place.

Metered quantities of gas and air are delivered to a burner. Liberated heat is then

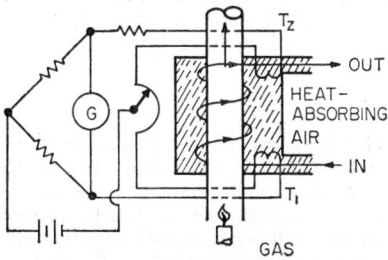

Fig. 4. Continuous gas calorimeter.

imparted to a separate air stream. Temperature rise of the air is measured by resistance thermometers to provide a measure of the heating value. Flue products are cooled to condense water of combustion so that total calorific value is determined.

In the recorder unit, a Wheatstone bridge is used to translate temperature rise into an indication and record of gas calorific value. Scale and chart are graduated in units of Btu per cubic foot and require no correction factors. Wide ranges can be covered, and expanded range instruments are available to provide increased readability over the upper third of the range.

Thomsen Type

This type of calorimeter burns gas in a combustion chamber within the calorimeter. Heat from the products of combustion is given up by passing the gases through a coil immersed in the calorimeter water. Thermometers are used to determine temperature rise, and total calorific values can be determined.

Bomb Calorimeters

Bomb calorimetry differs from the gas-calorimetry methods described heretofore in that volume remains constant, while pressure changes during combustion. These calorimeters are identical in principle with those described for solid- and liquid-fuel measurements. Because of the involved procedure and calculations required, however, their use for gas heating value is restricted almost entirely to university and special research laboratories.

Thermeter

The Thermeter is a modification of the Cutler-Hammer type of gas calorimeter. The operating principle is essentially the same, but the products of combustion are released at a temperature sufficiently high so that none of the water formed in combustion is condensed. Thus, the instrument provides a reading more nearly related to net calorific value than to total calorific value. Although the Thermeter measures net heating value, it is usually calibrated to give a reading in total heat value per standard cubic foot. Suppressed zero ranges are available to cover all of the normally used commercial gases. Accuracy is approximately 1 to 2 per cent.

Sigma Calorimeter

The Sigma calorimeter consists of two parts: a gas-flow regulator, and a gas-burner and recorder unit. Sample gas is burned at the bottom of two concentric steel tubes, setting up a differential expansion between them. Tubes are rigidly connected at the lower end, with the upper end of the outer tube attached to the instrument case, and the upper end of the inner tube fixed to the pen-drive mechanism. Although the measurement is more nearly that of net calorific value, the chart is calibrated in total calorific value. A wide variety of ranges is available, allowing the instrument to be used for all sorts of gases from butane vapor to blast-furnace gas. Accuracy is approximately the same as that of the Thermeter.

Parr Comparison Type

The Parr gas calorimeter is a comparison-type unit in which the rise in temperature in two nearly identical calorimeters is compared. In one, the sample gas is burned; in the other, an equal or multiple volume of a known gas is burned. Pressures and temperatures of the two gases are the same, and the volumes are chosen so that an almost equal temperature rise will occur in each calorimeter.

In this manner, the ratio of the calorific value of sample gas to that of the known gas is obtained. By multiplying the calorific value of the known gas by this ratio, the actual calorific value of the sample is determined.   Speed of testing is about the same as that of the manual water-flow calorimeter.

Caloroptic Type

This type of calorimeter burns the sample gas in a bunsen-type burner.   The size of the air primary opening required for perfect combustion is translated into Btu values.   These values are read directly on the burner tube and adjustable sleeve, which are marked with vernier-type graduations.   A spring-loaded regulator maintains constant gas-sample pressure.   Simplicity and short time lag are the major advantages of this unit.

Gas Analysis

Although not actually a calorimeter, gas-analysis equipment can be used as an inferential way to calculate calorific value.   With an accurate analysis of gas composition, the sample calorific value can be calculated from the percentage by volume of each constituent present, and the heating value of those constituents in Btu per cubic foot.   Both absorption analyzers and mass spectrometers are used in this method, which is limited to laboratory application.

Specific Gravity

Specific gravity instruments, although not designed for calorific value testing, can be used to determine heating values of certain binary mixtures, such as LPG and air. In these cases, a measurement of the specific gravity of the mixture establishes the percentage of fuel gas present.   This assumes that the gas has only one constituent, and that the gravity of that constituent is known.   From the known heating value of the gas and its percentage in the mixture, the calorific value can be calculated.

Sampling

In any analysis of the calorific value of a gas, it is important that the sample be representative of the gas in the main line.   For this reason, the tap should be located on an active line and with adequate mixing where two or more gases are present. The sample line should not be subject to transient changes in temperature, which would cause vaporization or condensation of one or more of the constituents of the sample.   The calorimeter should be located close to the sampling point, and a small size pipe or tube should be used.

REFERENCES

1. Oxygen Bomb Calorimetry, *Parr Instr. Co. Manual* 120.
2. Fuel Calorimeters, Emerson Apparatus Co.
3. Gas Calorimeter Tables, *Natl. Bur. Standards (U.S.) Circ.* C464.
4. Tests for the Calorific Value of Gaseous Fuels by the Water-flow Calorimeter, ASTM Standard D 900–48(55).
5. Mellors, H.: Gas Quality Measurement, *Gas World*, March 26, 1955.
6. Warner, C. W.: Fundamentals of Gas Calorimetry, *Am. Gas Assoc. Paper* CEP-57-9.
7. Standard Methods of Gas Testing, *Natl. Bur. Standards (U.S.) Circ.* 48.
8. Industrial Gas Calorimetry, *Natl. Bur. Standards (U.S.) Technol. Paper* 36.

Section 3

RADIATION VARIABLES

By

R. K. ABELE, B.S. (E.E.), *Supervising Engineer, Radiation Detection Group, Instrumentation and Controls Division, Oak Ridge National Laboratory, Oak Ridge, Tenn.* (*Nuclear-radiation Detectors*)

DONALD C. BRUNTON, Ph.D. (Physics), *Director of Research, Industrial Nucleonics Corp., Columbus, Ohio; Member, American Physical Society, American Nuclear Society, Canadian Association of Physicists; Registered Professional Engineer (Ontario).* (*Nuclear-radiation Detectors*)

R. S. HUNTER, A.B., *President, Hunter Associates Laboratory, Inc., McLean, Va.; Member, Optical Society of America, Inter-Society Color Council, American Society for Testing Materials, Technical Association of the Pulp and Paper Industry, Washington Academy of Sciences, American Association for the Advancement of Science.* (*Photometric Variables*)

JEROME KOHL, B.S. (Ch.E.), *Coordinator of Special Products, General Atomic Division of General Dynamics Corporation, San Diego, Calif.; Member, American Institute of Chemical Engineers, American Nuclear Society; Registered Professional Engineer (Ch.E.) (Calif.).* (*Radioisotopes in Instrumentation*)

W. H. PEAKE, *Assistant Professor of Electrical Engineering, Ohio State University, Columbus, Ohio; Member, Institute of Electrical and Electronic Engineers, Sigma Xi.* (*Acoustical Measurements*)

E. G. THURSTON, *Senior Research Engineer, Hallicrafters Company, Chicago, Ill.; Member, Institute of Electrical and Electronic Engineers, Acoustical Society of America, Audio Engineering Society, American Association for the Advancement of Science, Communications and Control and the Detection and Classification Task Committees of the Anti-Submarine Warfare Committee of the National Security Industrial Association; Registered Professional Engineer (Ohio and Ill.)* (*Acoustical Measurements*)

J. D. TRIMMER, Ph.D., *Head of Physics Department, University of Massachusetts, Amherst, Mass.; Member, American Physical Society, Acoustical Society of America, Instrument Society of America.* (*Radiation Fundamentals*)

RADIATION FUNDAMENTALS

By John D. Trimmer*

Radiation is itself a variable, an object of measurement; but it is also a working part of a great variety of instruments used to measure other variables. Some understanding of radiation is therefore basic to the whole field of instrumentation. Certain general aspects of radiation are treated here. See also Sec. 2, Temperature; subsequent portions of this section; and Sec. 14, Nuclear Reactor Instrumentation.

BASIC CONCEPTS OF RADIATION

Radiation is the spreading out of some form of matter or energy. The moving bits of matter are called particles; the moving forms of energy, waves. The distinction between waves and particles is usually clear-cut. Thus electrons streaming from a radioactive mineral would be recognized as particles, whereas light and sound are examples of waves. Modern physics has shown[1-3] that in some cases (as in

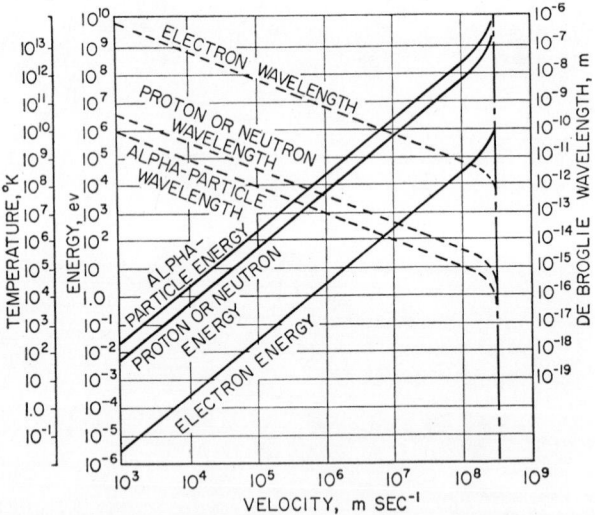

Fig. 1. Energy and wavelength of particles.

passage of electrons through matter, or in light falling on a photoemissive surface) it may be most helpful to view the radiation as a wave-particle combination. The waves associated with particles are called de Broglie waves, and the particles associated with light (and other electromagnetic waves) are called photons.

A beam of light waves all of the same frequency ν may be regarded as a beam of photons, each photon having energy $h\nu$, where h is Planck's constant. The total energy carried by the light beam is the product of $h\nu$ by the number of photons in the

* Head of Physics Department, University of Massachusetts, Amherst, Mass.

beam. Similarly, for a beam of particles of mass m, all having the same velocity v, the total kinetic energy carried by the beam is the product of the number of particles in the beam by the kinetic energy per particle. This kinetic energy is given for ordinary velocities by the formula $(1/2)mv^2$, or by the relativity expression[3,4]

$$mc^2 \left(\frac{1}{\sqrt{1 - v^2/c^2}} - 1 \right) \quad (1)$$

if the velocity approaches c, the velocity of light.

The energy carried by waves may also be expressed in terms of the wave amplitude. For plane electromagnetic waves,[5,6] the intensity (energy per unit area per unit time) is $2.65 \times 10^{-3}E^2$ watts per square meter, where E is the amplitude of electric field strength in volts per meter. For plane sound waves,[7,8] the corresponding formula for intensity in watts per m^2 is $\frac{1}{2}P^2/\rho a$ where P is amplitude of sound pressure, ρ is density of the medium, and a is velocity of sound in the medium—all in mks units.

Particle energies are often expressed in terms of voltage or of temperature. Thus the electron volt (ev, or mev for 10^6 ev) is a unit equal to the energy acquired by a particle carrying one electronic charge e in falling through a potential difference of one volt; it equals 1.6×10^{-19} joule. If a large number of particles interact as in a gas at absolute temperature T, each particle will have an average kinetic energy $\frac{3}{2}kT$, where k is Boltzmann's constant, 1.380×10^{-23} joule per degree. In Fig. 1 the solid lines show the relationship between velocity and kinetic energy, expressed both in ev and in degrees Kelvin.

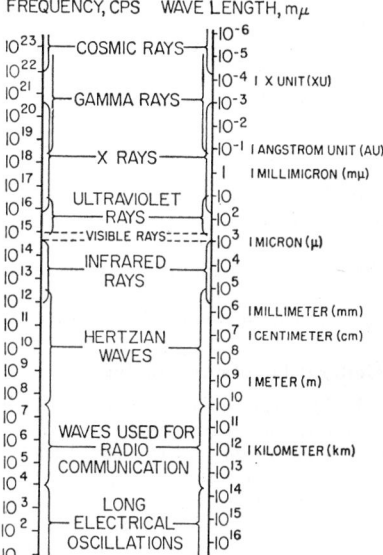

FIG. 2. Gamut of electromagnetic waves.

Table 1. Some Types of Particles

Name or chemical identity	Mass, kg	Electrical charge	
		Sign	Magnitude
Electron (beta particle)	9.11×10^{-31}	−	$e = 1.60 \times 10^{-19}$ coulombs
Positron (positive electron)	9.11×10^{-31}	+	e
Proton (hydrogen nucleus)	1.67×10^{-27}	+	e
Neutron	1.67×10^{-27}		0
Alpha particle (helium nucleus)	6.64×10^{-27}	+	$2e$
Photon (light quantum)	0		0
Atomic or molecular ion	(Atomic or mol wt) $\times 1.66 \times 10^{-27}$	+ or −	ne (usually, $n = 1$ or 2)

The dotted lines in Fig. 1 show the wavelengths of the de Broglie waves for the various particles; this wavelength is h/mv. Thus a beam of so-called "thermal" neutrons (i.e., neutrons in equilibrium with matter at room temperature) would have energy of about $\frac{1}{40}$ ev, average velocity about 2.2×10^3 m/sec, and de Broglie wavelength about 1.8×10^{-10} m. The significance of this wavelength is that such a beam of particles would interact with matter much as would electromagnetic waves (X rays) of the same wavelength. This is the basis of the similar results,[9,10] obtained in X-ray diffraction, electron diffraction, and neutron diffraction. Figure 2 shows frequencies and wavelengths of electromagnetic waves. Table 1 lists some properties of various particle types.

EFFECTS OF RADIATION

When radiation passes into or through matter, interactions occur which in general leave changes both in the radiation and in the material medium. Ordinary optics[11,12] is concerned with the effect on the radiation (light waves)—that is, absorption, reflection, refraction, interference, scattering, diffraction, polarization. Similar subject matter constitutes electron and ion optics.[13,14] Radio propagation, including applications to telemetering, is another problem of following radiation through varying media.[15]

Emphasis is now to be directed to the effect on the medium. The medium may be living or inanimate matter. The effect of light waves on living matter in the human eye is the basis of almost all indicating instruments. Effects of sound waves on the ear account for most of the remainder. Radiation effects on living tissue are also of concern in regard to safety; avoidance of or protection against radiation hazards are the domain of what has come to be called "health physics."[16,17]

General Effects on Matter

Perhaps the simplest effect occurs when radiation falls on an absorbing or reflecting medium; this is the purely mechanical effect of radiation pressure. It is obvious that a stream of particles could exert pressure or torque on an intercepting target. It is less obvious, but no less true, that sound and light waves may also exert pressure or torque. The effects are small, however, and are seldom of practical import.[18,19]

Sound waves are ordinarily detected by their gross mechanical effects, alternating pressure or velocity, on microphone elements.[8] Similarly, radio waves are usually detected by their alternating effects on gross aggregates of matter (antennas and coils).

Another important effect is heating. Thus high-frequency sound or radio waves may be directed into a part of the human body to heat it for therapy.[20] Infrared radiation is commonly used for industrial heating.[21] Wherever radioactive material is present, there is heating due to absorption of the emitted radiation. Presumably, much of the interior heat of the earth is of such origin. The streams of particles in a cyclotron or mass spectrograph create severe heating of the collecting or receiving targets. Heating and other macroscopic effects are secondary consequences of simpler processes.

In Table 2 are listed some of these simpler, individual interactions of radiation with matter. Ionization and photoelectric effect are most important for instrumentation, since they result in readily measured electric currents or voltages. Although from the physics standpoint the number of kinds of interaction is small, there results an enormous complexity of biological[22,23] and chemical[24,25] effects of radiation.

EYE AND EAR AS RADIATION DETECTORS

The eye and the ear are complex structures, whose modes of functioning are not yet entirely understood.[26,27] The instrument engineer is primarily interested, however, in the over-all physiological and psychological powers which permit visual or auditory signals to convey information to the mind. The subject is more complicated for the eye than for the ear, because a single eye perceives a pattern in space as well as in time; whereas a single ear, functioning as a one-dimensional transmission line, responds to time dependence alone. Localization of low-frequency sounds in space requires cooperation of two ears. A single ear permits one to sense direction of a higher-frequency sound by the shadowing effect of the head and by directional properties of the external ear. All aspects of visual and auditory perception[28] may conceivably bear some relevance to instrumentation, but attention is confined here to the general energy range and to the perception of small changes.

Table 2. Interactions of Radiation with Matter

Type of radiation	Name and nature of interaction with matter	Remarks
Charged particles (electron, positron, proton, etc.)	*Excitation.* Raising atomic electron to higher energy level	Atom subsequently emits electromagnetic radiation
	Ionization. Removing one or more electrons from atom	Important physical, chemical, and biological consequences; low to medium energies (e.g., for electrons in lead, below 0.1 mev)
	Bremsstrahlung. Radiation emitted by incident particle as it is changed in direction and energy	Important only for high-energy particles (e.g., for electrons in lead, above 5 mev)
	Elastic Scattering. Change in direction with no change in energy	Caused by nucleus or by atomic electrons
Neutrons	*Absorption.* Neutron "disappears" in a nucleus	Absorbing nucleus may emit a photon, or a particle; or, for a few elements, may undergo fission
	Scattering. Due to magnetic or nuclear forces	Nuclear effect predominates, except in certain magnetic materials
Electromagnetic waves (photons)	*Photoelectric Effect.* Total absorption of photon, with emission of electron from atom	Important for X rays and other photons between 1 ev and 1 mev
	Compton Scattering. Incident photon collides with an electron, both being changed in direction and energy	Here (and in photoelectric effect) disturbed electron may be detected as measure of photon; most important below 5 mev
	Pair Formation. Photon vanishes, replaced by a positron and an electron	Important only at energies above 5 mev
	Nuclear Excitation. Photon absorbed in nucleus	Nucleus emits particle or undergoes fission; important for energies measured in mev
	Atomic excitation (e.g., by visible or ultraviolet light) and *molecular excitation* (e.g., Raman effect, infrared absorption, microwave absorption)	Basis of much spectroscopy; important for energies measured in ev

Radiant energy in the form of light or sound causes a subjective experience (sensation) proportional to some power of the intensity of stimulation.[28] But the exact relation is a function of wavelength. In vision, the most important sensations are luminance (or brightness) and color; in hearing, loudness and pitch. Figure 3 shows

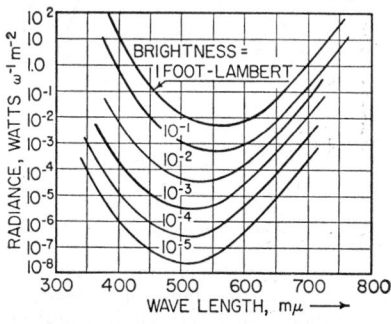

FIG. 3. Equal-brightness contours.

some contours of equal brightness,[29] and Fig. 4 shows equal-loudness contours.[30] The range of intensity of light is from a threshold of about 10^{-9} watts $\omega^{-1}\, m^{-2}$ to a pain threshold (for green light) of about 80 watts $\omega^{-1}\, m^{-2}$. The equal-brightness contours have essentially the same shape for luminance values of 1 foot-lambert or more, and the reciprocal of this curve is called the "standard luminosity function," or "relative luminosity curve."[31] The full intensity range is some 120 decibels for hearing and almost 110 for sight.

In comparing phenomena of hearing and of sight, it is important to keep in mind the different powers of accommodation. The eye and ear are both subject to fatigue and masking, but the ear adapts relatively quickly (1 min or less) compared with the eye (5 to 10 min) from high to low radiation levels.

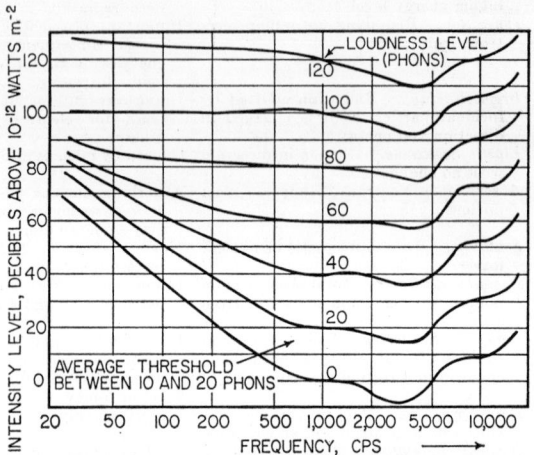

FIG. 4. Equal-loudness contours.

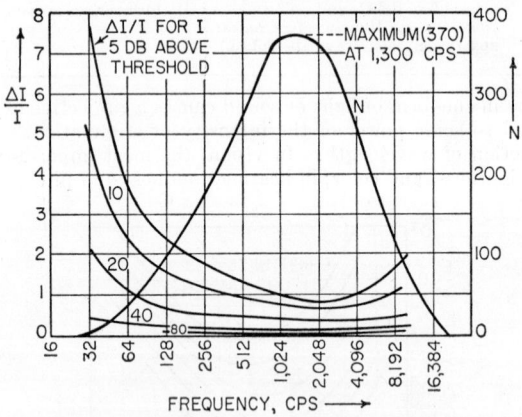

FIG. 5. Auditory intensity discrimination.

Information is conveyed only by a *change* in sensation. So the differential sensitivities, ratios of sensation increment to stimulus increment, are important. Figure 5 shows acoustic intensity discrimination in terms of just-noticeable intensity increment ΔI at various frequencies and at various levels of I.[32] The ear is seen to be most sensitive to intensity increments at the same frequencies (around 2,000) where absolute threshold is minimum. The same figure shows the number N of tones of perceptibly different intensity. For the eye, the value of $\Delta I/I$ depends on the apparent size of the test object; between 6×10^{-4} and 500 millilamberts, $\log (I/\Delta I)$ is proportional to the visual angle subtended by the test object.[33] Figure 6 gives the dependence of $\Delta I/I$ on brightness for two small test objects.[34]

Pitch discrimination in hearing at sensation levels of 30 db or more shows at low frequencies a rather constant *absolute* frequency increment $\Delta \nu$ of 2 or 3 cps required for a just-noticeable pitch change; and above 1,000 cps the *relative* increment $\Delta \nu / \nu$ is roughly constant at 0.003.[35]   The corresponding phenomena in vision would be color discriminations.   Color perception is, however, so complicated as to be outside the scope of this review.[36]

Undoubtedly the most important visual perception, for the subject of instrumentation, is spatial resolution, the apprehension of small spatial separations between portions of the visual field.   This power, often called "visual acuity," is called into play in making readings of pointers on scales and in making coincidence or vernier settings.   Visual acuity is defined as the reciprocal of the minimum effective visual angle, in minutes.   Figure 7 shows the relation among time of exposure, slit brightness, and minimal visual angle subtended by the slit in a test of the observer's ability

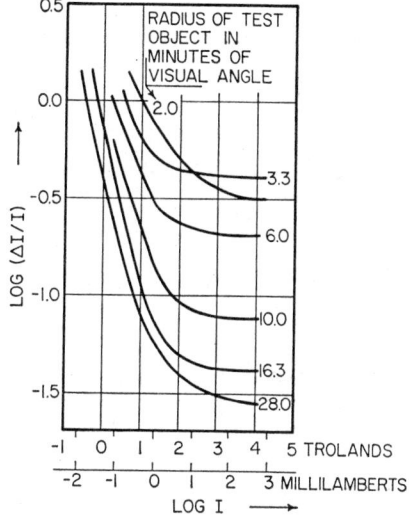

Fig. 6. Visual intensity discrimination.

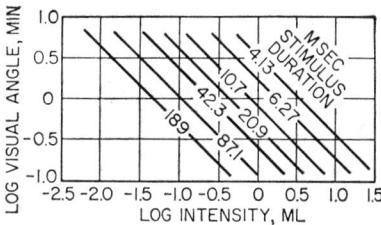

Fig. 7. Visual-resolution test.

to identify the direction of orientation of the slit.[37]   The design of scales for maximum readability has been studied rather fully.[38]   In making coincidence settings of two lines, as on a vernier, an observer is reliable to within an angle of about 12″.   The same limit is a measure of "stereoscopic acuity," the ability to tell which of two identical objects is nearer, if interpreted as the difference in the angles which the interocular distance of the observer subtends at the two objects.[39]

Both in sight and in hearing there are important effects associated with interrupted stimuli.   Critical flicker frequency, the minimum number of flashes per second which produces the appearance of a steady field, depends on illumination and object size; it ranges from less than 10 per second for small, weakly illuminated objects to more than 40 for large, bright objects.   In acoustics, the periodic interruption of a nonperiodic noise can give rise to pitch perception.[40]

ROLE OF RADIATION IN INSTRUMENTATION

As a summary of the part played by radiation in the entire field of instrumentation, Table 3 lists three main items: the human eye and ear, radiation-measuring instruments, and radiation-using instruments.   The classification under the latter two is intended to be representative and not exhaustive.

Table 3. Radiation and Instrumentation

1. Senses of Sight and Hearing

The eye and the ear are radiation receivers by which instrument indications are sensed

2. Instruments for Measuring Radiation

Acoustic Radiation
Sound-level meters
Reverberation meters
Sound-spectrum analyzers
Electromagnetic Radiation
Photometers
Radiometers
Spectrophotometers
Particle Radiation
Ionization chambers
G-M counters
Scintillation counters
Photographic films

3. Instruments in Which Radiation Is Used for Measuring Other Variables

Temperature
Total-radiation pyrometers
Optical pyrometers
Visible Properties of Materials
Colorimeters
Reflection meters
Opacity meters
Turbidimeters
Smoke- and haze-density meters
Chemical Composition and Structure
Refractometers
Polarimeters
Spectrophotometers (emission and absorption)
Diffraction (X-ray, electron, neutron) instruments
Flaw detectors (X-ray, ultrasonic)

Table 4. Quantities Encountered in Radiation Technology

Quantity	Symbol	Units (and magnitudes)	Symbol
Frequency	ν	Cycle per second (kilocycle, megacycle)	cps
Wavelength	λ	Meter	m
		Millimicron (10^{-9} meter)	mμ
Velocity of light	c	Meter per second (2.9979×10^8 m sec^{-1})	m sec^{-1}
Velocity of sound	a	Meter per second (in air at 0°C, 331.7 m sec^{-1})	m sec^{-1}
Velocity	v	Meter per second	m sec^{-1}
Mass	m	Kilogram	kg
Electric field strength	E	Volt per meter	volt m^{-1}
Sound pressure	P	Newton per square meter	newton m^{-2}
Power	ρ	Watt	watt
Electronic charge	e	Coulomb	coulomb
Absolute temperature	T	Degree Kelvin	°K
Planck constant	h	Joule-second (6.625×10^{-34} joule sec)	joule sec
Density	...	Kilogram per cubic meter	kg m^{-3}
Energy	...	Joule	joule
		Electron volt, million electron volt	ev, mev
Boltzmann constant	k	Joule per degree (1.380×10^{-23} joule deg^{-1})	joule deg^{-1}
Solid angle	...	Steradian	ω
Ratio	...	Decibel	db
(Radiation) intensity	I	Watts per square meter [same word and symbol (I) also sometimes used for "intensity of stimulus"—e.g., for brightness]	watts m^{-2}
Loudness level	L	Phon	phon
Radiance	N	Watt per steradian per square meter	watt ω^{-1}m^{-2}
		Lumen per steradian per square meter	lumen ω^{-1}m^{-2}
Luminance (or brightness)	B	Foot-lambert	foot-lambert
		Millilamberts	millilambert
Illuminance	E	Lumen per square meter	lumen m^{-2}
Retinal illuminance	E_h	Troland	Troland

Table 5. Conversion Factors for Luminance (B) Units

y units	x units						
	candle m⁻²	candle cm⁻²	candle ft⁻²	foot-lambert	lambert	milli-lambert	Apostilb
	F multiplying factors						
candle m⁻².............	1	10^{-4}	0.0929	0.292	3.14×10^{-4}	0.3142	
candle cm⁻² (stilb)......	10^4	1	929	2.92×10^3	3.142	3,142	
candle ft⁻²..........	10.76	1.076×10^{-3}	1	3.142	3.38×10^{-3}	3.38	
foot-lambert (equiv. ft-candle)......	3.42	0.342×10^{-3}	0.318	1	1.076×10^{-3}	1.076	
lambert............	3.18×10^3	0.318	296	929	1	1,000	
millilambert........	3.18	0.318×10^{-3}	0.296	0.929	0.001	1	
Apostilb............						0.1	1 (International units)
						0.09	(hefner units)

Table 6. Conversion Factors for Illuminance (E) Units

y units	x units		
	lumen m^{-2} (lux)	lumen cm^{-2} (phot)	lumen ft^{-2} (ft-candle)
	F multiplying factors		
lumen m^{-2}.....................	1	10^{-4}	0.0929
lumen cm^{-2} (phot)..............	10^4	1	929
lumen ft^{-2} (ft-candle)...........	10.76	1.76 × 10^{-3}	1

Table 7. Comparison of Radiometric and Photometric Terms, Symbols, and Units

Physical (Radiometry)	Psychophysical (Photometry)
Radiator (source of radiant energy)	Luminator (source of luminous energy)
Radiation (process)	Lumination (process)
Radiant energy U joule................	Luminous energy Q talbot
Radiant density u joule m^{-3}............	Luminous density q talbot m^{-3}
Radiant flux P watt...................	Luminous flux F lumen
Radiant emittance W watt m^{-2}.........	Luminous emittance L lumen m^{-2}
Radiant intensity J watt ω^{-1}............	Luminous intensity I lumen ω^{-1} (candle)
Radiance N watt ω^{-1} m^{-2}..............	Luminance B lumen ω^{-1} m^{-2}
Irradiance H watt m^{-2}.................	Illuminance E lumen m^{-2} (lux)
Spectral reflectance r..................	Luminous reflectance R
Spectral transmittance t...............	Luminous transmittance T

Adapted from OSA Committee on Colorimetry, The Psychophysics of Color, *J. Opt. Soc. Am.*, vol. 34, p. 245, 1944.

REFERENCES

1. Stranathan, J. D.: "The 'Particles' of Modern Physics," McGraw-Hill Book Company, Inc., Blakiston Division, New York, 1942.
2. Lapp, R. E., and H. L. Andrews: "Nuclear Radiation Physics," 2d ed., Prentice-Hall, Inc., Englewood Cliffs, N.J., 1954.
3. Leighton, R. B.: "Principles of Modern Physics," McGraw-Hill Book Company, Inc., New York, 1959.
4. Einstein, Albert: "Relativity, the Special and General Theory" (translated by R. W. Lawson), Hartsdale House Inc., New York, 1947.
5. Scott, W. T.: "The Physics of Electricity and Magnetism," John Wiley & Sons, Inc., New York, 1959.
6. Stratton, J. A.: "Electromagnetic Theory," McGraw-Hill Book Company, Inc., New York, 1941.
7. Morse, P. M.: "Vibration and Sound," 2d ed., McGraw-Hill Book Company, Inc., New York, 1948.
8. Kinsler, L. E., and A. R. Frey: "Fundamentals of Acoustics," John Wiley & Sons, Inc., New York, 1950.
9. Wollan, E. O., and C. G. Shull: Neutron Diffraction and Associated Studies, *Nucleonics*, vol. 3, pp. 8–21, July, 1948; pp. 17–31, August, 1948.
10. Shull, C. G., and E. O. Wollan: X-ray, Electron, and Neutron Diffraction, *Science*, vol. 108, p. 69, 1948.
11. Jenkins, F. A. and H. E. White: "Fundamentals of Optics," 3d ed., McGraw-Hill Book Company, Inc., New York, 1957.
12. Strong, J.: "Concepts of Classical Optics," W. H. Freeman and Company, San Francisco, 1958.
13. Pierce, J. R.: "Theory and Design of Electron Beams." D. Van Nostrand Company, Inc., Princeton, N.J., 1949.
14. Cosslett, V. E.: "Introduction to Electron Optics, The Production, Propagation and Focussing of Electron Beams," 2d ed., Oxford University Press, New York, 1950.
15. Atwood, S. S., and C. R. Burrows (eds.): "Radio Wave Propagation," Academic Press Inc., New York, 1949.
15a. Desirant, M., and J. L. Michiels: "Electromagnetic Wave Propagation," Academic Press Inc., New York, 1960.

16. *Health Physics* (Journal of Health Physics Society), Pergamon Press, London, June, 1958, et seq.
17. Atomic Energy Commission, Eighth Semiannual Report to Congress, pp. 1–161, Government Printing Office, Washington, D.C., July, 1950.
18. Vigoureux, P.: "Ultrasonics," pp. 73–77, Chapman & Hall, Ltd., London, 1950.
19. Nichols, E. F., and G. F. Hull: The Pressure Due to Radiation, *Phys. Rev.*, vol. 17, pp. 26, 91, 1903. Carrara, N.: Torque and Angular Momentum of Centimeter Electromagnetic Waves, *Nature*, vol. 164, p. 883, 1949.
20. Holzer, W., and E. Weissenberg: "Foundations of Short-wave Therapy," Hutchison's Scientific and Technical Publications, London, 1936.
21. Hall, J. D.: "Industrial Applications of Infrared," McGraw-Hill Book Company, Inc., New York, 1947.
21a. Hackforth, H. L.: "Infrared Radiation," McGraw-Hill Book Company, Inc., New York, 1960.
22. Nickson, J. J. (ed.): "Symposium on Radiobiology: The Basic Aspects of Radiation Effects on Living Systems," John Wiley & Sons, Inc., New York, 1952.
23. Lea, D. E.: "Actions of Radiations on Living Cells," The Macmillan Company, New York, 1947.
24. Second United Nations International Conference on Peaceful Uses of Atomic Energy (Geneva, September, 1958); Papers on radiation effects summarized in *Nucleonics*, vol. 16, pp. 90–91, September, 1958.
25. Dienes, G. J., and G. H. Vineyard: Radiation Effects in Solids, Interscience Publishers, Inc., New York, 1957.
26. Pirenne, M. H.: "Vision and the Eye," Pilot Press Ltd., London, 1948.
27. Wever, E. G.: "Theory of Hearing," John Wiley & Sons, Inc., New York, 1949.
28. Stevens, S. S. (ed.): "Handbook of Experimental Psychology," John Wiley & Sons, Inc., New York, 1951.
28a. Stevens, S. S.: To Honor Fechner and Repeal His Law, *Science*, vol. 133, p. 80, Jan. 13, 1961.
29. Weaver, K. S.: A Provisional Standard Observer for Low Level Photometry, *J. Opt. Soc. Am.*, vol. 39, p. 278, 1949.
30. American Standards Association: American Standard for Noise Measurement, *J. Acoust. Soc. Am.*, vol. 14, p. 102, 1942.
31. OSA Committee on Colorimetry: The Psychophysics of Color, *J. Opt. Soc. Am.*, vol. 34, p. 245, 1944.
32. Riesz, R. R.: Differential Intensity Sensitivity of the Ear for Pure Tones, *Phys. Rev.*, vol. 31, p. 867, 1928.
33. Holway, A. H., and L. M. Hurvich: Visual Differential Sensitivity and Retinal Area, *Am. J. Psychol.* vol. 51, p. 687, 1938.
34. Graham, C. H., and N. R. Bartlett: The Relation of Size of Stimulus and Intensity in the Human Eye, *J. Exptl. Psych.*, vol. 27, p. 149, 1940.
35. Shower, E. G., and R. Biddulph: Differential Pitch Sensitivity of the Ear, *J. Acoust. Soc. Am.*, vol. 3, p. 275, 1931.
36. Judd, D. B.: "Color in Business, Science, and Industry," John Wiley & Sons, Inc., New York, 1952.
36a. "Visual Problems of Color, A symposium held in Teddington, England, in 1957," Chemical Publishing Company, Inc., New York, 1958.
37. Niven, J. I., and R. H. Brown: Visual Resolution as a Function of Intensity and Exposure Time in the Human Fovea, *J. Opt. Soc. Am.*, vol. 34, p. 738, 1944.
38. McCormick, E. J.: "Human Engineering," McGraw-Hill Book Company, Inc., New York, 1957.
39. Jacobs, D. H.: "Fundamentals of Optical Engineering," McGraw-Hill Book Company, Inc., New York, 1943.
40. Miller, G. A., and W. G. Taylor: The Perception of Repeated Bursts of Noise, *J. Acoust. Soc. Am.*, vol. 20, p. 171, 1948.
41. Bethe, Hans A. and Phillip Morrison: "Elementary Nuclear Theory," John Wiley & Sons, Inc., New York, 1956.
42. Blackwood, Oswald H., Thomas H. Osgood, and Arthur E. Ruark: "An Outline of Atomic Physics," John Wiley & Sons, Inc., New York, 1955.
43. Halliday, David: "Introductory Nuclear Physics," John Wiley & Sons, Inc., New York, 1955.
44. Shankland, Robert S.: "Atomic and Nuclear Physics," Macmillan Company, New York, 1955.

NUCLEAR-RADIATION DETECTORS

By R. K. Abele* and D. C. Brunton†

Radiation detectors considered in this section are:

1. Geiger-Mueller counter.
2. Proportional counter.
3. Ionization chamber.
4. Scintillation detector.
5. Semiconductor detector.
6. Electroscope.
7. Photographic film.
8. Chemical dosimeter.
9. Other dosimeters.
10. Activation foil.
11. Calorimeter and thermopile.
12. Cloud chamber.
13. Bubble chamber.
14. Spark chamber.
15. Cerenkov detector.

TYPES OF RADIATION MEASURED

All the above radiation detectors are detectors of ionizing radiation such as alpha particles, beta particles, charged mesons, protons, and heavier charged nuclei. Detection of nonionizing radiation such as gamma rays and neutrons requires a conversion to ionizing radiation before detection.

Gamma rays are absorbed by the material of the detector (walls, gas, or solid volume) and whatever the interaction mechanism (photoelectric effect, Compton scattering, or pair production), electrons are produced which may be measured by the ionizing radiation detectors.

Slow neutron detectors depend on a nuclear reaction for conversion to ionizing radiation. The most common among these is high cross-section reaction (high probability of occurrence) of the boron-10 isotope with a thermal (slow) neutron.

$$B^{10} + n \rightarrow Li^7 + \alpha$$

Both the lithium-7 nucleus and the α particle are released as ionizing radiation. The second common nuclear reaction for detector use is the fission of U^{235}, U^{233}, or Pu^{239}. Again, the reaction has a high cross section, and a thermal neutron will cause the release of the large ionizing energy of the two heavy fission fragments.

Fast neutrons are generally detected through collisions with hydrogen nuclei. The high-energy transfer in such a collision makes the resultant fast proton readily detected as an ionizing particle. This conversion from fast neutrons to protons may be caused to occur exterior to the detector, inside the detector, or integral with the detector material itself; e.g., organic phosphors for scintillation detectors.

Similarly, α and β particles may be detected from a source inside the detector or external to it. In the latter case, a thin window must be provided for entry of the radiation into the detector. The thinness requirement is particularly severe with α particles and limits their usefulness in applied radiation problems.

* Supervising Engineer, Instrumentation and Controls Division, Oak Ridge National Laboratory, Oak Ridge, Tenn.
† Director of Research, Industrial Nucleonics Corp., Columbus, Ohio.

CLASSES OF RADIATION DETECTION REQUIREMENTS

Laboratory Measurements of Radiation

Laboratory radiation measurement may, of course, use every type of detector listed above. The general requirements, however, may be divided into two classes:
Measurement of intensity of radiation.
Measurement of the energy of radiation.
Measurement of intensity of radiation may further be subdivided into rate measurements and integrated radiation-dose measurements.
The common detectors for measurement of the radiation intensity (rate) are:
The Geiger counter.
The ionization chamber.
The proportional counter.
The scintillation detector.
For measurement of a radiation dose integrated over a period of time, the common detectors are:
The electroscope.
The photographic film.
The activation foil.
Less commonly used are chemical dosimeters, calorimeters, and other dosimeters based on change of color, degradation of a material, or mechanical-electric action such as discharge of the static electricity holding glass beads to an insulated surface. The chemical and color dosimeters, in particular, are used generally for integrated doses from high-intensity radiation fields. The calorimeters are used occasionally to determine the total energy output from a radiation source for absolute standards.
The principal detectors for measuring the energy of α, β, and γ radiations are:
Proportional counters (α, β).
Ionization chambers (α, fission fragments).
Scintillation detectors (β, γ).
Photographic film (α, p, mesons, and heavy nuclei).
For special laboratory measurements of high-energy ionizing particles such as those found in cosmic rays or from high-energy machines, there are very special detectors, some costing up to millions of dollars. These include:
The liquid-hydrogen and liquid-helium bubble chambers.
The cloud chambers.
The spark chambers.
The Cerenkov detector.
The cloud chambers are now largely superseded by the much higher efficiency bubble chambers. The latter are often very large, very expensive, and frequently used in conjunction with a high-speed computer. Yet still in the competition in this exotic field is the lowly photographic emulsion. The spark chambers are one of the latest arrivals in the radiation-detection field. They are simply a series of charged plates which can be made to discharge successively by the traverse of an ionizing particle. They are used chiefly in conjunction with coincidence counting techniques of high-energy particles.

Area and Personnel Radiation Monitoring

Radiation monitoring is primarily a measurement of radiation intensity or accumulated radiation dose. Consequently, all the techniques mentioned above are employed.
The Geiger counter, the scintillation counter, and the ion chamber are commonly used for area monitoring to determine the intensity of radiation. The pencil electroscope and the photographic film are trademarks of every radiation area for personnel monitors where the integrated dose by the day, week, month, or throughout a specific operation is desired. For neutron detection, occasionally activation foils may be

employed as area monitors although they are more commonly used to determine the integrated neutron flux associated with a nuclear-reactor experiment.

When area monitoring includes the analysis of radiation to determine the source of contamination, the energy-sensitive detectors must be employed together with some type of β or γ spectrometer. For α-particle determination the proportional counter, the ion chamber, and more recently, the solid-state detector are used. For β-particle determination the proportional counter is normally used unless a magnetic focusing β spectrometer is employed, in which case any β-sensitive intensity detector may be used. Gamma-ray analysis is customarily made with the scintillation detector, although for precision work at higher energies a "pair spectrometer" and a γ-sensitive intensity detector are used. The latter, however, is found chiefly in scientific laboratories and can hardly be called a member of the applied instrumentation class.

Industrial Radiation Measurements

The ion-chamber detector is the cornerstone of industrial radiation measurement with β-ray or γ-ray gages. The inherent stability, high reliability, and the simplicity of the ion chamber make it a first choice for such measurements. The Geiger counter is sometimes used for such purposes where the highest stability is not required and minimum cost is desirable. The Geiger detector is not of itself essentially lower in cost than the ion chamber, but the high internal amplification of the Geiger tube reduces the associated electronic requirement compared with that of the ion chamber. The scintillation detector is employed where the highest detection sensitivity of γ radiation is required even at the expense of highest stability.

The second most common area for industrial application of radiation is in γ-ray industrial radiography where, of course, the detector is the photographic film. In this work, in addition, both area and personnel monitors as described above will be employed.

Other industrial radiation work such as tracer studies commonly employ the Geiger or proportional counters for sample counting.

In the growing area of high-intensity radiation work (radiation fields from 10^3 to 10^8 r/hr) again the ion chamber—although one of very special construction—is the most accepted detector for intensity measurements. For total-dose measurements on individual processed items, colorimetric- and degradation-type detectors are used in which the radiation produces a color change, e.g., cobalt glass or the darkening of a plastic film. For more precision dose measurements the chemical dosimeters are employed or calorimetric techniques are used.

GAS IONIZATION TUBES

One of the simplest devices for radiation detection is a gas-filled diode (Fig. 1). Consider a gas-filled cylinder with a fine axial wire and an ionizing radiation particle traversing the tube volume.[1-3] As the particle traverses the tube, the gas is ionized. This ionization consists of positive ions and electrons. As the fine central wire (anode) is at positive potential with respect to the cylinder (cathode) the electrons move to the wire in a region of increasing field strength and the positive ions move outward to the cathode. The electrons are collected in a short period of time while the positive ions have moved only a short distance out toward the cathode.

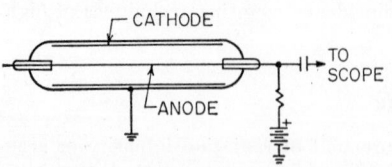

FIG. 1. Gas-filled diode.

At low voltages (ion-chamber region) no additional ions are created by collision and the number arriving at the central wire will be essentially the number produced in the initial ionizing event (neglecting recombination) (see graph, Fig. 2).

The proportional region (Fig. 2) is reached when the counter voltage is raised to that point where secondary electrons are first formed by collisions of the primary electron with gas molecules. As the voltage on the counter is further increased this multiplication of electrons increases until the phenomenon known as the Townsend avalanche occurs. This is a cumulative process in which the initial electrons and their progeny also produce more electrons. The pulse on a proportional counter[1] is given by Eq. (1):

$$V = \frac{Ane}{C} \qquad (1)$$

where V = pulse size, volts
A = gas amplification
n = number of electrons in initial ionizing event
e = charge on electron, coulombs
C = distributed capacity of central wire system, farads

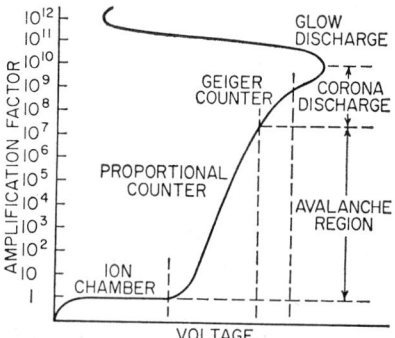

Fig. 2. Relation of voltage and amplification in gas-filled diode. (*Source: Proc. IRE, vol. 37, no. 7, July,* 1949.)

The gas amplification A is defined as the total number of all electrons that are produced (including the original electron) as a result of each initial electron traveling from its origin to the central wire ($A = 1$ for an ion chamber). The proportional region is characterized by amplifications greater than 1, up to values of 10^5 or 10^6. In this region, each discharge is a single avalanche, originates from a primary ion pair, and is limited to a small distance along the wire. The output voltage pulse is proportional to the number of ions formed in the initial ionizing event.

As the voltage is further increased each avalanche produces new avalanches until the discharge spreads along the whole length of the central wire. The amplification increases until it is no longer dependent on the amount of primary ionization and all discharge pulses become equal in amplitude. This is the beginning (threshold) of the Geiger counting region.

Ionization Chambers

An ion chamber may be used as a counting device registering each individual ionizing event or as an integrating device measuring the total current output due to the charge collected on its electrodes from the internal ionization. Counting chambers are used where the number of ionizing events is small. For large amounts of radiation the integrating type is used. Ion chambers may be constructed with parallel plates, coaxial cylinders, or some other particularly desired configuration. Ion chambers have been made of all sorts of materials—metals, plastics, and waxes, to name some.

In the integrating type of ion chamber where construction is often of the cylindrical type, the outer cylinder is usually (but not necessarily) at positive high voltage and the inner signal or collector electrode is insulated from ground by a high-impedance insulator. This insulator must have high surface resistivity as well as volume resistivity, and in most cases this high surface resistance must hold for high humidity conditions.[6] This resistance must be of the order of 10^{15} ohms or higher when measuring small currents. Some of the best materials for this insulator are Teflon, fluorothene, polystyrene, quartz, and ceresin wax.

Surrounding this insulator is a grounded guard electrode. This guard electrode or guard ring serves as an electrostatic shield for the signal insulator, protects it from high-voltage leakage troubles, and may be used to define the collecting volume of the chamber (by defining the electrostatic field). This guard ring must in turn be insulated from the high-voltage electrode of the system. This high voltage to ground insulation need not be so good as the signal insulation.

Saturation. A chamber is said to be saturated when the voltage on its electrodes is sufficient to collect all the ions formed within its volume (Fig. 3). Saturation occurs, in fact, when the electric field strength is sufficient to separate the ion pairs and sweep them to the collecting electrodes before any charge is lost because of recombination of positive and negative charges. The voltage required depends on the geometry used and the gas pressure and to a much lesser extent on the intensity of radiation and the type of gas.

Special consideration must be given to the problem of saturation when an ion chamber is to be used in high-intensity radiation measurements; the chamber should have a high uniform field strength. This is achieved by close spacing and the use of electrodes of such a size (in the case of a cylindrical configuration) that the field between them approaches that of a parallel-plate arrangement.

In practice a saturation curve is run on a chamber with a source of radiation whose magnitude is equal to or greater than the amount of radiation which the chamber may be required to detect. The potential across the electrodes is raised from a few volts to a few hundred volts, if necessary, and the current output is recorded for each voltage used. When no appreciable current increase is evident with increasing voltage the chamber is operating in a saturated condition.

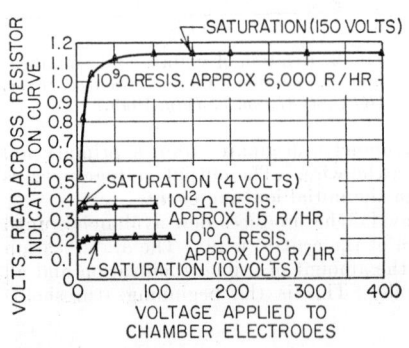

FIG. 3. Saturation curves for ion chamber (volume = 2 cc).

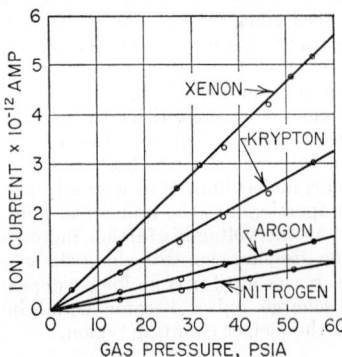

FIG. 4. Effect of gas and gas pressure on ion-chamber current. Ion-chamber volume = 36 cc. Radiation source, 10 mg of Ra-Be with lead shielding.

Current from an Ion Chamber. The saturated current from an integrating-type ion chamber is directly proportional to the amount of radiation present. By definition, one roentgen is that amount of X ray or gamma ray which produces one electrostatic unit of charge per cubic centimeter of air at standard conditions of temperature and pressure.[7] Equation (2) is useful for calculating the approximate current output of this type of chamber.

$$I = \frac{kVPR \times 10^{-13}}{1.08} \qquad (2)$$

where I = output current, amp
k = proportionality constant for type of gas (air = 1)
V = chamber volume, cc
P = gas pressure, atm
R = radiation intensity, r/hr

Ion chambers may be operated with several atmospheres of gas pressure. Gases used for ion-chamber fillings include air, argon, nitrogen, krypton, xenon, and boron trifluoride. Examples are shown in Fig. 4.

Collection Efficiency of an Ion Chamber. The collection efficiency of the chamber, i.e., the ratio of the measured current to the ideal saturation current, is given for a parallel-plate chamber by (as given by Hine and Brownell, "Radiation

Dosimetry," page 165, Academic Press, Inc., New York):

$$f = \frac{2}{1 + \sqrt{1 + \xi^2}} \tag{3}$$

where
$$\xi = \sqrt{\frac{2}{3} \frac{\alpha}{ek_1k_2}} \frac{d}{V}\sqrt{q}$$

where the symbols refer to constants of the gas, the *charge density* q, and the electric field V/d. Normally it is desirable to maintain $\xi \lesssim 0.2$ so that the efficiency of $f \gtrsim 0.99$.

Detection Efficiency of an Ion Chamber. Assuming an ion chamber has been constructed to yield a high collection efficiency of the ionization created in the chamber, how efficient is it in measuring the number of events or the total energy of a given nuclear radiation?

The detection efficiency depends greatly on the nuclear particles being detected. Every beta particle entering the chamber volume will create ionization in the chamber gas and will be detected. The only losses in particle-detection efficiency are the geometric losses because the detector window will subtend some limited solid angle at the source of radiation, and the absorption losses from the material between the source and detector including the detector window. It is extremely important that the ionization chamber be of rigid construction and hermetically sealed whenever high-stability performance is desired since any change of the gas volume or density inside the chamber will change the total mass of gas and hence the ionization created as a beta particle traverses the chamber volume. Where the detector is exposed to a rugged environment, the entrance window may be of sufficient mass per unit area to absorb half the incident radiation or even more. In such cases the particle-detection efficiency is reduced to 50 per cent or lower.

The energy-detection efficiency will in general be considerably lower than the particle-detection efficiency since many of the incident beta particles will traverse the chamber and lose a substantial portion of the incident energy in the chamber wall. It is, of course, possible to design an ion chamber with sufficiently high pressure and electric field strength to absorb totally the energy from a beta-emitting radio-isotope inside the chamber, but such measurement conditions are seldom of interest.

With gamma rays the detection efficiencies are generally much lower. The gamma ray is detected only when it creates a secondary electron in the chamber gas or an electron in the chamber wall whose recoil path is into the chamber gas. The relative importance of the gas filling and the wall effect on gamma chambers depends on the type of gas, the pressure, and the volume of the gas on the one hand; and the area and atomic number of wall material on the other. The specific relationship is also a function of the gamma-ray energy.

At pressures up to 10 atm, the wall effect will predominate over the gas effect for everything but heavy gases such as xenon. For large chambers xenon gas at high pressure is expensive, and it is more practical to increase the chamber detection efficiency by increasing the wall area through the use of multiplate electrodes.[3] The structure then is similar to a multiplate, parallel-plate air capacitor. If the maximum yield in count rate is desired (rather than maximum ion current), the plates may be quite closely spaced ($\frac{1}{4}$ in. or less). They should be of a high-Z material, and a few thousandths of an inch thick. For practical reasons the high-Z material may be used as a coating on a light metal plate (0.010 to 0.020 in. Al) so that the structural material of the chamber absorbs as little of the radiation as is practicable.

The efficiency of a gamma chamber for a given gamma energy is a function of wall thickness. The efficiency increases with wall thickness until the point is reached where few of the secondary electrons produced at the outside surface of the wall can reach the chamber gas. With further increase in thickness the efficiency decreases because of absorption of the gamma rays (Fig. 5).

Compensated Ion Chambers. A special ion-chamber development important in nuclear-reactor control is the gamma-compensation neutron chamber. The neutron flux must be measured accurately in the presence of a high gamma-ray background. This is accomplished by use of a dual chamber, one part sensitive to neutrons plus gammas, and the other sensitive to gammas only. The resultant outputs are subtracted and the sensitivities adjusted so that the gamma effect is canceled.[8]

Proportional Counters

As previously mentioned, the proportional counting region is distinguished by the fact that the voltage pulse produced is proportional to the primary ionization. This makes it possible to differentiate between alpha and gamma or alpha and beta radiation with the electronic equipment that follows the detector. Slow neutrons are thus counted above a background of gamma radiation in a counter tube filled with boron trifluoride gas. Here the pulses produced by alpha particles from the neutron-boron reaction in the gas are much larger than the pulses produced in the gas from gamma radiation. Beta proportional counters are widely used because of their short resolving time (which permits high-speed counting) and their long life.

Counting. When counting beta radiation in the proportional region, the output voltage pulse may vary from a fraction of a millivolt to a fraction of a volt in the same counter because of the energy distribution of the beta spectrum. If the counter is to have a plateau, a region from 100 to 500 volts long where the counting rate rises only 1 to 5 per cent with this increase of voltage, the electronic equipment used must be able to accept pulses over a large range in amplitude. Even though counter construction and filling are good, the counter will not exhibit this plateau characteristic if the electronic equipment is not properly designed and built. For the use of a proportional counter as a beta-ray spectrometer, see, for example, the work of Holloway, Lu, and Zaffarano.[9]

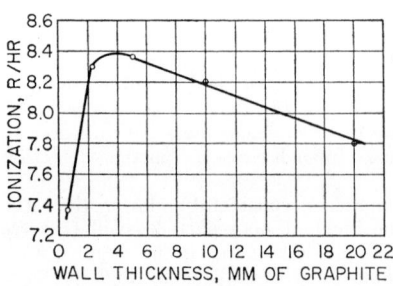

FIG. 5. Variation of ionization, due to a constant source of gamma rays, with wall thickness for a small graphite chamber filled with air. A wall thickness of 4 mm is required to produce equilibrium between the gamma rays and their secondary electrons. For thicker walls, the attenuation of the gamma rays by the walls is noticeable. (*Source: Radioactivity Units and Standards, Nucleonics, October, 1947.*)

Construction of Proportional Counters. Proportional counters are constructed with a variety of shapes and configurations of which the most common is the cylindrical shell with a fine axial wire from 0.001 to 0.004 in. in diameter (Fig. 1). This fine wire is necessary to obtain field strength near the wire sufficient for electron multiplication. The shell is usually operated at ground potential and the central wire at positive high voltage.

Glass envelope counters (with a thin film of metal deposited on the inner surface) may be used; or metal wall counters may be constructed, using glass to metal seals to insulate the central wire. These insulators must be clean, or they will produce high-voltage leakage pulses that are of the same magnitude as the counting pulses.[6] The materials mentioned as good signal insulators for ion chambers are also adequate here. If the counter is evacuated, then filled with gas and sealed, some of the plastics are not acceptable as insulators because of their high vapor pressure.

The counter may be used as a flow counter with the counting gas continuously passing through the counter. This technique is used where solid or liquid radioactive samples are placed in the counter for counting. A radioactive gas may also form part of the fill gas of a counter and thus be counted with no attenuation due to wall thickness.

Mixtures, such as argon plus 10 per cent methane, or argon plus 10 per cent carbon dioxide at atmospheric pressure, operate in the proportional region in the neighborhood of 2,000 volts for a counter whose axial wire is in the order of 0.001 to 0.002 in.

in diameter and whose cathode (shell) is ¾ to 1 in. in diameter. The operating voltage depends on the wire size, cathode size, gas mixture, total gas pressure used, and the electronic counting equipment.

Geiger Counters

The Geiger region for counting begins at that point where all discharge pulses attain the same amplitude. This pulse size is independent of the primary ionization. The Geiger threshold is best located by the use of an oscilloscope to determine the lowest voltage at which all pulses become equal in size. Because of constructional details, the Geiger region is not so clearly defined in some counters as it is in others.

Classification of Geiger Counters. Geiger counters[10,11] are usually constructed with a cylinder and a fine axial wire; the cylinder is operated at ground potential and the wire at positive high voltage. Geiger counters may be classified into two categories as characterized by their filling gases: (1) the non-self-quenching counter filled with the rare gases or air or some mixture of them; and (2) the self-quenching counter usually filled with a mixture consisting of a rare gas and a small percentage of some polyatomic organic quenching vapor (or halogen gas filled—see below).

The start of the discharge is the same in these two types of counters. One ion pair formed in the gas produces an avalanche of electrons which, in turn, produces more avalanches until the discharge spreads the full length of the wire. The electrons are collected on the central wire and the positive ions form a sheath around the central wire. The positive-ion sheath, in both cases, reduces the electric field near the wire so that the counter can no longer discharge. As the positive-ion sheath travels outward to the cylinder it may produce secondary electrons by bombardment of the cylinder. These secondary electrons, in turn, may initiate new discharges.

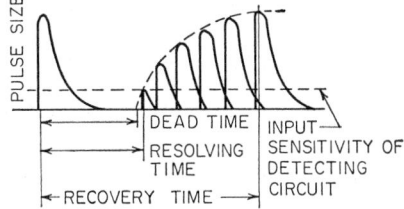

Fig. 6. Dead time and subsequent gradual recovery of pulse size in a Geiger counter.

In the non-self-quenching counter as first used, a resistance of the order of 10^9 ohms was employed across the counter electrodes. Simply stated, the drop in potential across this resistance during discharge lowers the potential across the tube to that point where no new discharge can take place. A large resistor like this necessarily makes the counter inoperative for long periods of time ($\sim$0.01 sec). Electronic circuits can be used to quench this type of counter in a millisecond or less but are not widely used at present because of the development of self-quenching and proportional counters.

In the self-quenching counters, a quench gas such as alcohol may be used with a rare gas such as argon. It has been shown that ultraviolet radiation is emitted during the recombination of ions occurring in the sheath while still close to the wire. It is this ultraviolet radiation which propagates the ionization throughout the length of the wire, since the quantum energies are sufficient in themselves to cause ionization. This radiation can also release electrons from the walls of the tube to produce spurious discharges.

Photons of energies capable of strong ionization and the energy from positive ions are absorbed by alcohol. This absorption process results in the decomposition of the alcohol. Positive-ion liberation of electrons from the wall is thus reduced by this property of the alcohol. Ionization by photons is restricted by the absorption process in alcohol to a region near the wire where they can create no spurious discharge.

After the Geiger counter discharges and produces a pulse it remains inoperative for a period of time called the *dead time*. This was first visually demonstrated by Stever (Fig. 6) with the use of a cathode-ray oscilloscope. The dead time is the time required for the positive-ion sheath to move out from the wire to a position where the electric field can recover so that another avalanche can form.

The *resolving time* of the counter is larger than the dead time and is determined by the point at which the pulse size becomes large enough to again trigger the electronic equipment (see Fig. 6). The *recovery time*, larger still than the resolving time, is that point where the pulse again gains its original amplitude.

All these factors determine the speed at which a counter can operate without losing a large number of counts. The dead time and recovery time are of the order of one to two hundred microseconds for the typical Geiger counter.

Construction of Geiger Counters

A large number of different quenching vapors may be used for filling Geiger counters. Amyl acetate, ether, and alcohol have had wide use. Friedman[10] states that usable self-quenched counters with over thirty different admixtures have been made. A variation of this type of counter has been made by the use of a halogen gas as the quench vapor. The halogen molecule does not dissociate as does the polyatomic molecule, and these counters have an "infinite" life. In actual practice the useful life of an organic quenched counter is of the order of 10^8 counts whereas that of

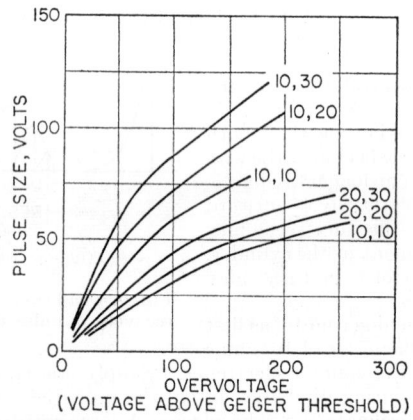

Fig. 7. Pulse size as a function of counter dimensions. (*Source: Proc. IRE, vol. 37, no. 7, July,* 1949.) Note: Tube dimensions are indicated as cathode radius (mm), and wire radius (thousandths of a mm).

the halogen quenched counter may be 10^{10} counts or more. Halogen counters, unlike the organic counters, are not damaged when subjected to voltages above the plateau region.

Geiger counters are made in a variety of shapes other than the most common shape of the cylindrical shell and the axial wire. The information on proportional-counter construction in general applies to Geiger counters. However, because of the difference in amplitude of the output pulses of the two devices and hence the associated electronic equipment, insulation leakage pulses that cannot be tolerated in a proportional counter will not trigger the electronic circuit used with a Geiger counter.

Cleanliness cannot be overstressed in counter construction. The counter should be washed with a detergent or alcohol, followed with several rinses of distilled water and then dried by heating. The fine central wire should have no sharp projections and be free of dust and lint.

A typical counter of 1-in.-diameter shell and 0.004-in. wire, filled with ethyl alcohol to a pressure of 1 cm of Hg and argon to a pressure of 9 cm of Hg, will operate at approximately 1,000 volts. The pulse size varies with the counter dimensions and the voltage above the Geiger threshold (Fig. 7).

A comprehensive review of G-M counters is given by Van Duuren, Jaspers, and Hermsen in *Nucleonics*, June, 1959.[11a] This review includes discussions of end effects, special liquid counters, anticoincidence counter arrangements for reduction of background counting rate, and the use of Geiger tubes as current integrating devices.

SCINTILLATION COUNTERS

In 1949, Coltman[12] and Jordan and Bell[13] reviewed some of the then current work using a photomultiplier tube to count the scintillations from phosphors. Since that time so many new advances have been made that the scintillation detector is no doubt the most important radiation detector in use today.

A scintillation counter consists of a scintillation phosphor optically coupled to a photomultiplier tube and both enclosed in a lighttight shield. The particle to be detected loses energy in exciting and ionizing the molecules of the crystal. Energy in the form of light is radiated from these molecules, and some of this light falls on the photocathode of the photomultiplier tube.

Photoelectrons are ejected from the photocathode and are accelerated to the first dynode where each photoelectron ejects several other electrons by secondary emission. This process of multiplication is repeated at each succeeding dynode by placing it at a higher potential than the preceding dynode. On arrival at the anode or collector a multiplication in the order of 10^6 may thus have been produced.

This charge (initiated by one particle) on arrival at the anode is then used to produce a voltage pulse on the anode circuit capacitance.

Scintillation Phosphors

The prerequisite property of a scintillation phosphor is that it be transparent to its own excited radiation. When an atom of any material is raised to an excited state by an ionizing particle, it will emit light in the process of returning to the ground state. With most materials, the atoms are equally capable of reabsorbing the light thus

Table 1. Inorganic Scintillators[17]

Material	Density, g/cm³, melting point, °C	Wave-length of max emission, A°	β-particle pulse height* (10 μsec)	β-particle light output† (d-c)	Linearity response to heavy particles	Initial decay period‡ (μsec)	Miscellaneous
NaI (Tl)....	3.67 651	4,100	210	210	Linear β, p α/β = 0.5	0.25	Hygroscopic, excellent transparency
KI (Tl).....	3.13 582	4,100	~50	~200	Linear β, p α/β = 1.0	1.0	Excellent transparency
CsI (Tl)....	4.51 621	(Blue)	55	~130	α/β = 0.5	1.1	
LiI (Eu)....	4.06 446	~4,400	75	...	Linear β, p n/β = 0.95§	2.0	Almost colorless, very hygroscopic
ZnS (Ag)....	4.09 1,850	4,500	~100	~400	α/β = 2.0	~10	Crystalline powders
CaWO₄.....	6.1 1,535	4,300	36	...	Linear β α/β = 0.2	6	Small crystals. Chemically inert
CdWO₄.....	7.90 1,325	5,300	21	...	...	8	Small crystals. Pale yellow

* 5819 or 6292 photomultiplier response with ~10 μsec anode time constant, relative to anthracene crystal = 100.

† Photomultiplier anode current, relative to anthracene crystal = 100.

‡ Time for light intensity to fall to $1/e$ of initial value.

§ Thermal neutron pulse height (4.785 mev divided by electron pulse height) energy.

emitted. The peculiar property of phosphors useful for scintillation counting is that they are excited to a particular energy level by ionizing radiation such that the light energies emitted on return to the ground state do not correspond to the resonant absorption energies of that phosphor for light quanta. It is thus transparent to its own excited radiation. This property occurs naturally with certain phosphorescent crystals, particularly those of organic nature. Most of the inorganic crystals require "doping" with another element (notably thallium) to produce this condition.

Table 2. Organic Scintillation Crystals[17]

Material empirical formula	Density, g/cm³, melting point, °C	Wavelength of max emission, A°	β-particle pulse height*	Linearity α/β ratio†	Decay constant‡ × 10⁹, sec	Miscellaneous
Anthracene C₁₄H₁₀.	1.25 217	4,400	100	Linear β > 125 kev α/β = 0.10	32	Large crystals, difficult to grow
Trans-stilbene C₁₄H₁₂	1.16 124	4,100	60	α/β = 0.1	6.4	Good crystal, readily obtained
p,p'-Quarterphenyl C₂₄H₁₈	... 318	4,200	85	...	7	Good crystal, difficult to grow

* 5819 or 6292 photomultiplier response, relative to anthracene crystal = 100.
† Ratio of pulse height to energy for polonium α particles ($E = 5.3$ mev). With that of a high-energy electron ($E_0 = 1.0$ mev).
‡ Time for light intensity to fall to $1/e$ of initial intensity.

Table 3. Liquid and Plastic Scintillators[17]

Solvent	Solute,* g/l	Second solute,* g/l	Wavelength of max emission, A°	β-particle pulse height†	α/β ratio	Decay constant‡ × 10⁹, sec	Miscellaneous
Phenylcyclohexane..	TP, 3	DPH, 0.01	4,500	0.35	...	8.0	Can be used in Lucite containers. Liquid
Toluene...........	TP, 5	aNPO, 0.02	4,150	0.42	...	≦3.2	Photon mean free path ~2 m. Liquid
Toluene...........	PBD, 8	0	3,700	0.49	...	<2.8	Liquid
Polystyrene........	TP, 36	0	3,550	0.28	0.10	≦3.0	Plastic
Polystyrene........	TP, 36	TPB, 0.2	4,450	0.39	0.10	4.0	Plastic
Polyvinyltoluene....	TP, 36	TPB, 0.2	4,450	0.45	0.10	4.0	Photon mean free path ~2 m. Plastic

* Solute abbreviations: TP = p-terphenyl; DPH = 1,6-diphenyl-1,3,5-hexatriene; aNPO = 2-(1-naphthyl)-5-phenyloxazole; PBD = 2-phenyl-5-(4-biphenylyl)-1,3,4-oxadiazole; TPB = 1,1,4,4-tetraphenyl-1-3-butadiene.
† 5819 or 6292 photomultiplier responses, relative to anthracene crystal = 1.00.
‡ Time for light intensity to fall to $1/e$ of initial value.

There are many properties to be considered in the selection of a phosphor for a particular application (Tables 1, 2, 3). Some of these are conversion efficiency of energy in to light out, response of a phosphor to different particles and quanta, linearity of a phosphor to a wide range of energies of a particle or quanta, and decay time of light emission. A multicrystalline ZnS (Ag) phosphor screen ~10 mg/cm² thickness makes a highly efficient alpha detector. Pulses produced in this layer by beta and gamma radiation can easily be biased out electronically because they are many times smaller than the average alpha pulse. Tables 1, 2, and 3 are representative of the materials available and their properties. New phosphors are continually being added to lists and are reported in the current literature.

Photomultipliers

Many different photomultiplier tubes are available today, and new ones are continuously appearing in the market. Table 4 is representative of the properties of these tubes, but for the latest data the standard tube handbooks should be consulted. This gives one a choice of several variables such as size, over-all amplification, spectral response, photocathode and dynode material, and dynode structure (Figs. 8 and 9). In general the focused dynode structure gives smaller electron transit times and transit time spreads than the box structure.[20] Resolution seems to be better on the average in those tubes with the box structure. These characteristics are important in choosing a tube for spectrometry or for timing or coincidence measurements.

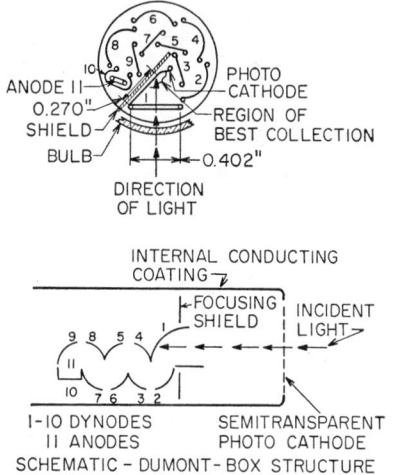

Fig. 8. Two typical photomultiplier tubes.

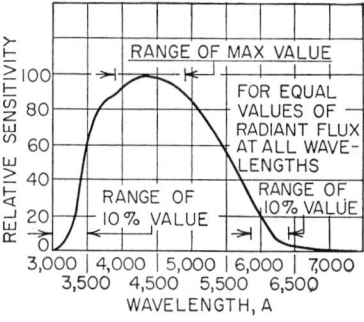

Fig. 9. Spectral characteristics of Dumond photomultipliers with S-11 response.

Radiation-energy Spectrometers. One of the most important applications of the scintillation detector is in making energy measurements. NaI (Tl) is almost universally used with gamma-ray scintillation spectrometers. Energy from the gamma ray may be absorbed in the crystal by one of the three gamma-ray absorption processes: photoelectric effect, Compton scattering, and pair production.

The photopeak, 661 kev, for cesium-137 (Fig. 10) represents the absorption of the total gamma-ray energy in the crystal. Some of these events occur from the photoelectric process, but some also represent the total absorption of the Compton electron and subsequent scattered gamma rays. The relative importance of the photoelectric effect and Compton scattering as contributors to the photopeak depends on the photoelectric absorption coefficient on the one hand and the Compton coefficient and the volume of the crystal (which affects the probability of total absorption of the scattered gammas) on the other. These coefficients are determined by the gamma-ray energy and the atomic number of the detector material. They are available in nuclear handbooks such as Rockwell's "Reactor Shielding Design Manual," D. Van Nostrand Company.

Table 4. Types of Photomultiplier Tubes

Tube type	Nom. diam., in.	Min photocathode diam., in.	Spectral response	No. stages	Avg current amplification at volts*	Dynode structure	Dynode material	Remarks
Dumont								
6365	3/4	1/2	S-11	6	3,000 at 150	Box	Ag Mg	End on flat face
6467	1 1/4	1	S-11	10	215,000 at 105	Box	Ag Mg	End on flat face
6291	1 1/2	1 1/4	S-11	10	215,000 at 105	Box	Ag Mg	End on flat face
6292	2	1 1/2	S-11	10	215,000 at 105	Box	Ag Mg	End on flat face
6363	3	2 1/2	S-11	10	215,000 at 105	Box	Ag Mg	End on flat face
6364	5	4 9/16	S-11	12	215,000 at 105	Box	Ag Mg	End on flat face
K-1328†	16	14	S-11	12	800,000 at 105	Box	Ag Mg	End on curved face
K-1209†	5	4 1/2	S-11	12	200,000 at 95	Box	Ag Mg	End on flat face
K-1213†	3	2 3/4	S-11	12	200,000 at 95	Box	Ag Mg	End on flat face
K-1295†	2	1 1/2	S-11	10	200,000 at 105	Box	Ag Mg	End on flat face
K-1292†	2	1 1/2	Infrared	10	215,000 at 105	Box	Ag Mg	End on flat face
K-1306†	1 3/16	1 3/16 × 9/16	Ultraviolet	9	2×10^6 at 1,000	Circular, focused	Cs₃Sb	Side window
RCA 1P21	1 3/16	1 3/16 × 9/16	S-11	9	2×10^6 at 1,000	Circular, focused	Cs₃Sb	Side window
1P28	1 3/16	1 3/16 × 9/16	S-5	9	1.25×10^6 at 1,000	Circular, focused	Cs₃Sb	Side window
2020	2	1 1/2	S-11	10	1.2×10^5 at 1,250	Circular, focused	Cs₃Sb	End on flat face
5819	2	1 11/16	S-11	10	5×10^5 at 1,000	Circular, focused	Cs₃Sb	End on curved face
6199	1 9/16	1 1/4	S-10	10	6×10^5 at 1,000	Circular, focused	Cs₃Sb	End on flat face
6217	2	1 11/16	S-11	10	6×10^5 at 1,000	Circular, focused	Cs₃Sb	End on curved face
6342	2	1 11/16	S-11	10	1.25×10^5 at 1,250	Circular, focused	Ag Mg	End on flat face
6372	2 9/16	4 3/8 × 2 3/4	S-11	10	6×10^5 at 1,000	Circular, focused	Cs₃Sb	Side window
6655A	2	1 11/16	S-11	10	5×10^5 at 1,000	Circular, focused	Cs₃Sb	End on flat face, curved cathode
6810A	2	1 11/16	S-11	14	12.5×10^6 at 2,000	In line, focused	Ag Mg	End on flat face, curved cathode
7046	5	4 7/16	S-11 (extended)	14	3×10^6 at 2,800	In line, focused	Ag Mg	End on flat face

* Experimental and developmental tubes.

† For Dumont tubes, volts per stage. For RCA tubes, anode to cathode voltage (total).

In the particular case of Cs^{137} and a NaI (Tl) crystal, shown in Fig. 10, only about 25 per cent of the counts fall in the photopeak for a 1- by 1.5-in. crystal whereas about 50 per cent do for a 3- by 3-in. crystal, indicating that at least in the latter case the majority of the photopeak arises from Compton-scattering interaction.

A figure of merit for a NaI (Tl) scintillation detector is its resolution. This is expressed as a percentage and in relation to the photopeak. It is the ratio of the peak width (in energy units) at one-half peak height to the energy at the peak. For cesium-137 a value between 8 and 10 per cent is rather common. For values below

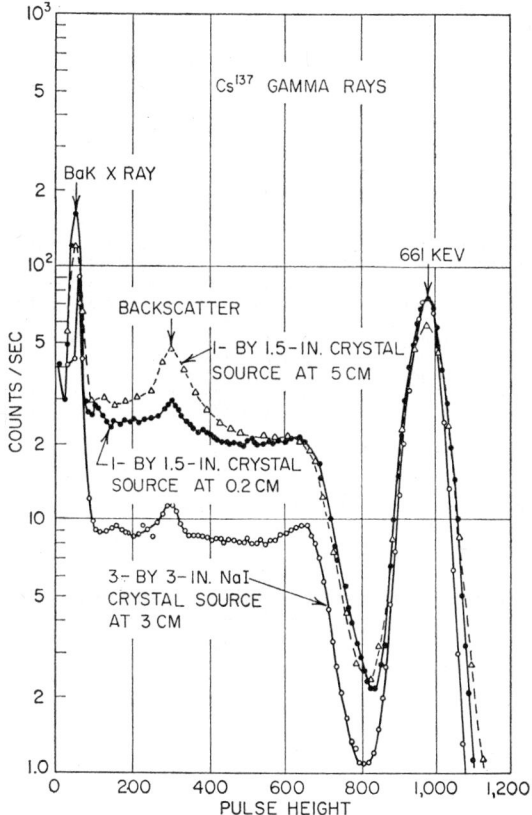

FIG. 10. The response of a 1½- by 1-in. and 3- by 3-in. crystal to gamma rays of Cs^{137} and the response of the 1½- by 1-in. crystal with the source at two distances. (*Source: Ref.* 24.)

8 per cent, much care must be used in the selection of a phototube and the assembly technique of tube to crystal.

Table 5[25] gives calculated photofractions for NaI (Tl) crystals of many sizes.

$$\text{Photofraction} = \frac{\text{area under the photopeak}}{\text{area under the entire pulse height distribution}}$$

Table 6[25] gives the efficiencies for NaI (Tl) crystals. Both these tables are useful in the selection of crystal size for a particular application. Other calculations on efficiencies of NaI (Tl) for a point isotropic gamma-ray source are available.

For a comprehensive discussion of gamma-ray spectroscopy see Bell and Siegbahn.[24]

Table 5. Photofractions for NaI Crystals for a Broad Parallel Beam as Calculated by the Monte Carlo Method*,[25]

Radius, in.	Height, in.	Energy, mev				
		0.279	0.661	1.33	2.62	4.45
1	1	0.816 ± 0.004	0.399 ± 0.007	0.233 ± 0.007	0.147 ± 0.006	0.103 ± 0.006
1	2	0.855 ± 0.003	0.481 ± 0.005	0.286 ± 0.006	0.199 ± 0.005	0.141 ± 0.005
1	4	0.869 ± 0.003	0.530 ± 0.005	0.334 ± 0.005	0.232 ± 0.005	0.172 ± 0.004
1	8	0.869 ± 0.003	0.548 ± 0.005	0.367 ± 0.005	0.258 ± 0.004	0.192 ± 0.004
2	1	0.845 ± 0.004	0.450 ± 0.007	0.286 ± 0.007	0.191 ± 0.007	0.139 ± 0.006
2	2	0.894 ± 0.003	0.576 ± 0.005	0.385 ± 0.006	0.282 ± 0.006	0.213 ± 0.006
2	4	0.913 ± 0.003	0.663 ± 0.005	0.477 ± 0.005	0.361 ± 0.005	0.295 ± 0.005
2	8	0.915 ± 0.003	0.701 ± 0.004	0.543 ± 0.005	0.419 ± 0.005	0.335 ± 0.005
4	1	0.864 ± 0.004	0.485 ± 0.007	0.335 ± 0.007	0.208 ± 0.007	0.184 ± 0.007
4	2	0.919 ± 0.003	0.631 ± 0.005	0.452 ± 0.006	0.346 ± 0.006	0.282 ± 0.006
4	4	0.943 ± 0.002	0.752 ± 0.004	0.592 ± 0.005	0.474 ± 0.005	0.409 ± 0.006
4	8	0.944 ± 0.002	0.811 ± 0.004	0.694 ± 0.005	0.582 ± 0.005	0.519 ± 0.005
8	1	0.869 ± 0.004	0.502 ± 0.007	0.353 ± 0.007	0.222 ± 0.007	0.176 ± 0.007
8	2	0.927 ± 0.003	0.664 ± 0.005	0.490 ± 0.006	0.383 ± 0.007	0.309 ± 0.006
8	4	0.953 ± 0.002	0.802 ± 0.004	0.653 ± 0.005	0.547 ± 0.005	0.479 ± 0.006
8	8	0.954 ± 0.002	0.878 ± 0.003	0.792 ± 0.004	0.693 ± 0.005	0.630 ± 0.005
12	1	0.872 ± 0.004	0.510 ± 0.007	0.358 ± 0.008	0.231 ± 0.007	0.198 ± 0.007
12	2	0.931 ± 0.003	0.674 ± 0.005	0.503 ± 0.006	0.401 ± 0.007	0.319 ± 0.005
12	4	0.958 ± 0.003	0.823 ± 0.004	0.679 ± 0.005	0.569 ± 0.005	0.502 ± 0.006
12	8	0.960 ± 0.002	0.902 ± 0.003	0.828 ± 0.004	0.733 ± 0.004	0.667 ± 0.005
16	1	0.872 ± 0.004	0.511 ± 0.007	0.363 ± 0.008	0.246 ± 0.008	0.202 ± 0.007
16	2	0.934 ± 0.002	0.679 ± 0.005	0.510 ± 0.006	0.407 ± 0.007	0.333 ± 0.007
16	4	0.960 ± 0.002	0.831 ± 0.005	0.692 ± 0.005	0.576 ± 0.005	0.518 ± 0.006
16	8	0.962 ± 0.002	0.912 ± 0.003	0.848 ± 0.004	0.752 ± 0.004	0.687 ± 0.005

* For 5,000 histories.

Table 6. Efficiencies for NaI Crystals for a Broad Parallel Beam [25]

Energy, mev	Height							
	1 in.		2 in.		4 in.		8 in.	
	Monte Carlo*	Analytical	Monte Carlo*	Analytical	Monte Carlo*	Analytical	Monte Carlo*	Analytical
0.279	0.795 ± 0.004	0.798	0.957 ± 0.004	0.959	0.997 ± 0.003	0.998	1.000 ± 0.003	1.000
0.661	0.496 ± 0.005	0.498	0.749 ± 0.005	0.748	0.937 ± 0.004	0.936	0.995 ± 0.004	0.996
1.33	0.374 ± 0.004	0.370	0.613 ± 0.005	0.604	0.847 ± 0.005	0.843	0.976 ± 0.005	0.975
2.62	0.297 ± 0.004	0.298	0.509 ± 0.003	0.507	0.761 ± 0.004	0.757	0.942 ± 0.004	0.941
4.45	0.276 ± 0.001	0.278	0.484 ± 0.002	0.478	0.734 ± 0.001	0.728	0.931 ± 0.002	0.926

* For 5,000 histories.

SOLID-STATE DETECTORS

The solid-state detector is the latest addition to the inventory of applied instrumentation radiation detectors. It is a direct outgrowth of the advances in semiconductor materials in recent years. The solid-state detector may be regarded as a parallel-plate ion chamber in which the plates separate proportionally to the applied voltage.

The solid-state detector is shown diagrammatically in Fig. 11. The sensitive volume of the detector is known as the "depletion layer." This is essentially the extent of the region at the junction of the semiconductor materials where a sharp potential gradient exists and consequently where ionization created by an external influence such as nuclear radiation will be quickly removed with negligible recombination. The thickness of the depletion layer x is given by $x = \frac{1}{3} \sqrt{\rho V}$ approximately, where x is in microns, ρ is the conductivity of the semiconductor material in ohm-cm, and V is the reverse bias voltage applied to the junction.

For the measurement of the energy of ionizing particles, particularly alphas and protons, it is necessary that the particle dissipate negligible energy before entering the depletion layer. Two techniques are used. A very thin layer of p-doped semiconductor may be superimposed on an n-type semiconductor, thus forming a junction very near one outer surface, or the surface-barrier-type diode may be used.[29] In the latter, the discontinuity of space charge within a semiconductor present at a surface is used to provide a high-potential-gradient region. The surface is coated with a thin layer of gold to provide a conductive contact.

The advantages of the solid-state detector are associated with its small size and the high specific ionization for these materials. This specific ionization, the number of ion pairs per electron volt or its reciprocal, the number of electron volts of energy lost by the ionizing particle to create one ion pair, is ten times more favorable for

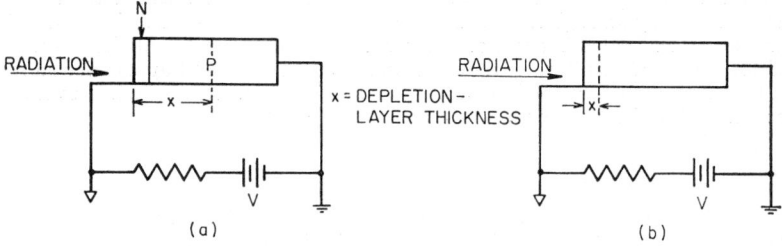

FIG. 11. (a) Junction-diode solid-state detector; (b) surface-barrier solid-state detector.

solid-state detectors ($\sim$3 ev per ion pair) as for gas-chamber detectors ($\sim$30 ev per ion pair). Hence the ionization intensity is an order of magnitude higher in the semiconductor for a given ionizing particle, and as a result, the energy resolution of solid-state detectors is greater than that of gas ion chambers.[29]

However, the solid-state detector does not appear nearly so favorable when compared with the gas ionization chamber for integrated current measurement. The polarizing voltage determines the thickness of the active volume; so this detector cannot show a plateau characteristic similar to that of Fig. 3 for the gas ion chamber. Hence a stable voltage is required for stable operation. More important than this, however, is the fact that the semiconductor is a very low resistance material compared with the gas in a conventional ion chamber. Consequently a background current of the order of 10^{-6} to 10^{-7} amp is passed by the solid-state detector compared with less than 10^{-16} amp by a conventional ion chamber. It is very difficult to stabilize this ohmic current to permit satisfactory measurement of currents of 10^{-9} to 10^{-12} amp such as are associated with common radiation fields.

OTHER DETECTORS

Among the other radiation detectors commonly in use, the photographic emulsion is certainly the most widely employed. The chemical and colorimetric dosimeters are important in high-intensity radiation. The specialized laboratory detectors such as cloud chambers,[30] bubble chambers,[31] spark chambers,[32] and Cerenkov detectors[33,34] are scientific apparatus not normally considered a part of applied instrumentation and will not be discussed further.

Photographic Emulsions

The photographic film is, of course, the standard detector for medical X-ray and industrial radiography. Excellent descriptions of the film properties and their applications are available from the manufacturers (e.g., Ref. 35).

Similar films are used for radiation personnel monitoring in the form of film badges. In addition, photographic emulsions have fulfilled a major detection role in high-energy physics and cosmic-ray research. High-sensitivity thick emulsions have been developed for the recording of individual tracks of ionizing particles, of nuclear events, and even of the sequence of several nuclear events. High-power microscopes are used for scanning and viewing the developed plates, and in many cases detailed analyses are made which identify the nuclear particles and their energies involved in complex nuclear interactions.[36]

The Electroscope

The electroscope, like the film badge, is a common instrument for personnel monitoring.[37] In the common form of the instrument an insulated fiber is supplied with a static charge relative to another electrode. The electrostatic force on the fiber causes it to deflect. The deflection corresponding to a fixed reference voltage is determined as zero dose on the scale. As the electroscope is exposed to radiation, the charge leaks away through the gas and the fiber deflection decreases. The scale of the instrument is calibrated in terms of total radiation dose received. Although primarily used for personnel monitoring, the electroscope still finds useful application in radiation measurement where the integrated dose at a point location is desired over a long exposure such as hours or days.

Activation Detectors

Activation detectors are particularly useful for monitoring reactor neutron fluxes. The method consists simply of determining the induced radioactivity in a sample material when it is exposed to a neutron flux.

The neutron flux is given by

$$f = \frac{A}{N\sigma S} \qquad \text{neutrons}/(\text{cm}^2)(\text{sec}) \qquad (4)$$

where A = activity, disintegrations per second
N = number of atoms of target nuclide
σ = activation cross section for the reaction, cm^2 per atom
S = saturation factor = $(1 - e^{-\lambda t})$ (λ is the decay constant of the nuclide, and t is the exposure time)

A large number of nuclides are available for activation measurements. Suitable nuclides are chosen on the basis of their σ and λ values. The decay constant must be small enough (or the half-life long enough) to facilitate counting of the activated foil and the $N\sigma$ must be large enough for the neutron flux to be measured so that an adequate count may be obtained in a reasonable time, but not so large as to produce a count rate which would produce significant counting losses in the counting system employed.

The counting system, of course, will include a geometrical correction from the foil activation A to the counts per second registered in the counting detector. This factor must be determined experimentally and maintained constant. In the absence of counting losses the accuracy of the method is determined by the statistical variation in N random events and hence is $1/\sqrt{n}$ where n is the number of counts recorded in the counting interval. This accuracy is the standard deviation, or "one sigma" accuracy, in statistical terms.

Counting losses can be very significant with Geiger detectors because of their large dead time, as already discussed, but tend to be negligible with proportional and scintillation counters. If the observed counting rate is n counts/sec and the resolving time of the detector is τ, the actual counting rate N is given by

$$N = \frac{n}{1 - n\tau} \tag{5}$$

More detailed discussion of activation measurements is available in many standard texts. Some of the more recent work is discussed in Ref. 38.

Chemical Dosimeters

Various chemical dosimeters have been developed based on the ability of ionizing radiation to induce chemical reactions. Among these are the ferrous-ferric dosimeters and the cerous-ceric dosimeters. The latter is used for very high radiation doses. The ferrous dosimeter is the most thoroughly studied of these chemical reactions.

Ferrous Sulfate Dosimeter. A method of chemical dosimetry that has been widely studied is that of the conversion of ferrous ions to ferric ions in an 0.8 N sulfuric acid solution by radiation-induced oxidation.[23,39,40] This detector is most useful for high radiation levels, and the data presented pertain to gamma radiation sources. The amount of ferric ion is determined by comparing the transmission of an irradiated sample with that of a nonirradiated sample on a spectrophotometer.

The yield of this dosimeter is expressed in G units, which are the number of molecules produced or converted per 100 ev absorbed. Table 7 gives a listing of these

Table 7. Yields of Ferrous Sulfate Dosimeter[23]

Authority	Radiation	G, molecules/100 ev
Ionization Method		
Hochanadel and Ghormley*............................	Co^{60} γ rays	16.7
Cormack, Hummel, Johns, and Spinks†..............	24.5-mev X rays	15.5 or 16.4
	Co^{60} γ rays	15.8 or 16.0
Calorimetric Method		
Hochanadel and Ghormley*............................	Co^{60} γ rays	15.6 ± 0.3
Lazo, Dewhurst, and Burton‡........................	Co^{60} γ rays	15.8 ± 0.3
Power-input Method		
Hochanadel and Ghormley*............................	...	16.5
Schuler and Allen§..................................	...	15.45 ± 0.11

* From G. J. Hochanadel and J. A. Ghormley, *J. Chem. Phys.*, vol. 21, p. 880, 1953.
† Fom D. V. Cormack, R. W. Hummel, H. E. Johns, and J. W. T. Spinks, *J. Chem. Phys.*, vol. 22, p. 6, 1954.
‡ From R. M. Lazo, H. A. Dewhurst, and B. Burton, *J. Chem. Phys.*, vol. 22, p. 1370, 1954.
§ From R. H. Schuler and A. O. Allen, Paper at Meeting of Radiation Research Society, New York, May 17, 1955.

G values and the methods employed to determine them. These values hold only while oxygen is present in the system. When the oxygen has been depleted, this yield drops. This change in yield is shown by the break in the curve of Fig. 12.

Weiss[39] reports a useful range of 4,000 to 40,000 rep that can be extended to about 200,000 rep by saturating the solution with pure oxygen before irradiation. The yield[40] is independent of quantum energy for 100-kv X rays to 2-mev gamma rays and independent of dose rate from $\frac{1}{60}$ to approximately 200 r/sec.

For dosimetry up to 12×10^6 rep ceric sulfate is used.

Colorimetric Detectors[41]

These detectors are simple, inexpensive, and rugged but rather inaccurate as compared with other detectors. However, their accuracy is probably adequate for civilian-defense purposes against an atomic-bomb blast. Their principle of operation is based on the color change of an aqueous dye solution due to the acid evolved from chloroform when subjected to radiation. They have been found of some use as radiation dosimeters in commercial radiation processing along with the discoloring (degradation) of plastic films.[41a]

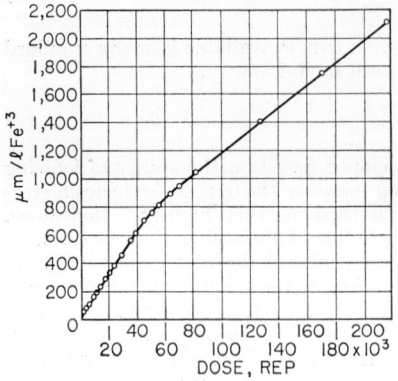

FIG. 12. The oxidation of 0.01 M FeSO$_4$ in 0.8 N H$_2$SO$_4$. (*Source: Ref. 40.*)

Boron Thermopile

An element that has been used for the detection of slow neutrons is the boron thermopile. This detector is made by coating a thermocouple junction with boron. The heating of the junction by the energy given off from the neutron-boron reaction[10] is a measure of the slow neutron flux [neutrons/(cm^2)(sec)]. In actual use alternate junctions of a series of thermocouples are boron-coated. An assembly of chromel-alumel couples joined in series and having 10 coated junctions gives an output of about 5 mv in a neutron flux of 2×10^{12} neutrons/(cm^2)(sec). Fluxes of this magnitude are found only in nuclear reactors.

NUCLEAR-RADIATION DETECTION SYSTEMS

It is possible to tie the various types of nuclear-radiation detectors into complete recording, alarm, and controlling systems, as partially shown in Fig. 13. In many

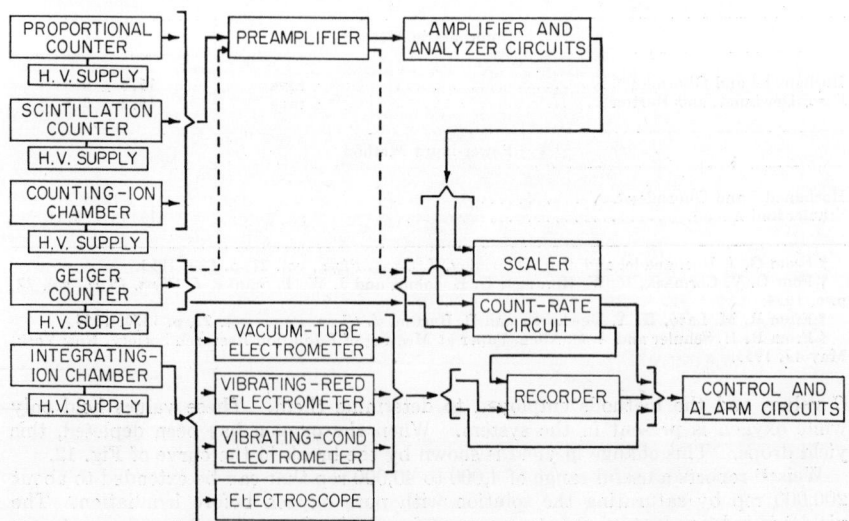

FIG. 13. Various types of nuclear-radiation detectors and how they can be tied into allied equipment to achieve a complete recording, alarm, or controlling system.

Table 8. Selection Guide for Various Types of Nuclear-radiation Detectors

Characteristic	Geiger counter	Proportional counter	Ion chamber		Scintillation counter
			Integrating type	Counting type	
Operating voltage.........	~300 to 2,000 volts. Depends on gas mixture, pressure, and electrode size and configuration	Few hundred to several thousand volts. Depends on gas mixture, pressure, electrode size and configuration	Volts to hundreds of volts. Depends on gas used, electrode size and configuration, and intensity of radiation	Volts to thousands of volts. Depends largely on application of device	Average conditions 600–1,000 volts depending on amplification required. Some 14-stage tubes, 2,400 volts
Gas pressure.........	Few cm Hg to 2 or 3 atm	Few cm Hg to 2 or 3 atm	1 to ~50 atm	Few cm Hg to ~1 atm	No gas filling
Output* pulse size......	0.25 to several volts. Depends on counter length and operating voltage	Millivolts to hundreds of millivolts. Depends on gas amplification, energy, and type of radiation	Fraction of microamperes to milliamperes. Depends on gas used, pressure, chamber volume, energy, and type of radiation	Fraction of millivolts to hundreds of millivolts. Depends on chamber capacitance, energy, and type of radiation	Millivolts to hundreds of millivolts. Depends on operating voltage, type of scintillator, energy, and type of radiation
Type of radiation to be detected†	α or soft β, thin window or windowless (gas flow) counter; β, energy of radiation detected is limited by counter wall thickness; γ, wall thickness is not critical	α or soft β, same as for Geiger counter; β, same as for Geiger counter; γ, same as for Geiger counter; n (thermal), reactions with B^{10} or fissionable isotopes; n (fast), reactions with hydrogen or fissionable isotopes	Same as for proportional counter for α, β, γ, and n. For measuring radiation dose, wall materials and thickness are important as they affect current output	Same as for proportional counter for α, β, γ, and n	α, thin window, zinc sulfide scintillator; β, thin window, anthracene and other scintillators; γ, thin window for energy measurements, sodium iodide, and other scintillators
Useful range‡.........	1. Few cpm to 2 or 3 × 10⁴ cpm. 2. mr/hr to ~100 r/hr. Halogen counters used as d-c output devices	Few cpm to ~6 × 10⁵ cpm. Limited by electronic circuits	mr/hr to 10⁷ r/hr. Upper limit determined by ability to collect all ions formed	Same as for proportional counter	Same as for proportional counter for counting. Can be used in higher radiation fields by measuring integrated current output
Detection efficiency§.....	α, 100%; β, 100%; γ, 1%	α, β, and γ same as for Geiger counter	γ efficiency improved by high gas pressure	α, β, and γ same as for Geiger counter	α, 100%; β, 100%; γ, 10–20%; n depends on scintillator type and size

Table 8. Selection Guide for Various Types of Nuclear-radiation Detectors (*Continued*)

Characteristic	Geiger counter	Proportional counter	Ion chamber		Scintillation counter
			Integrating type	Counting type	
Advantages...........	Large pulse compared with proportional counter, scintillation counter, and counting ion chamber	Pulse size proportional to initial ionization; can discriminate electronically between different types of radiation. Higher counting rates than Geiger counters. Energy measurements	Accurate dosage measurements. Rugged, simple, and stable. Wide range of radiation intensity levels	Same as for proportional counters. However, pulse has no gas amplification as it does in proportional counter	Phosphor light output and hence pulse size is proportional (within limits) to radiation energy. β- and γ-ray spectrometer. High efficiency for γ rays. High counting rates
Limitations...........	Low counting rate limits range of use. Output pulse size is independent of energy of radiation	Smaller pulse size than Geiger counter	Normally does not discriminate between types of radiation; measures total ionization	Very small pulse size	Thin windows for α hard to keep lighttight. Sodium iodide must be sealed from atmosphere, since it is hygroscopic. Very stable high-voltage supply needed. More electronic circuitry required than for Geiger counter or integrating ion chamber

All information in this table is of a general nature. The limits given are approximate. Any of these limits can be extended provided certain conditions are defined as regards detector use and how information from the detector is to be used.

*Pulse height values realized with normally used input circuit constants.

† α = alpha; β = beta; γ = gamma; n = neutron.

‡ cpm = counts per minute.

mr/hr = milliroentgens per hour.

r/hr = roentgens per hour

§ Detection efficiency can be defined as the ratio of the number of particles detected to the number of particles going through the sensitive volume of the detector, expressed as a percentage. Detection efficiency for neutrons is a function of the number of atoms of material available for neutron reaction and cross section of the material for the neutron. Neutron cross section is a measure of the probability of the occurrence of the neutron reaction.

instances, components that are shown on separate blocks may be combined into one complete unit. The equipment shown in Fig. 13 is available from many commercial sources, but not all of it is necessarily available from any one manufacturer.

THE DETECTOR SELECTION GUIDE

Table 8 is a compilation of information that may be valuable in the selection of a particular detector element for a definite application. The radiation-detection elements listed in this table are also available commercially in a large variety of forms with the exception of fission counters and integrating ion chambers. Few ion chambers and almost no fission counters are available as a commercial catalog item.

REFERENCES

1. Korff, S. A.: "Electron and Nuclear Counters," D. Van Nostrand Company, Inc., Princeton, N.J., 1946.
2. Wilkinson, D. H.: "Ionization Chambers and Counters," Cambridge University Press, New York, 1950.
3. Rossi, B. B., and H. H. Staub: "Ionization Chambers and Counters," National Nuclear Energy Series, Div. V, vol. 2, McGraw-Hill Book Company, Inc., New York, 1949.
4. Kohl, Jerome: "Radioisotope Applications Engineering," D. Van Nostrand Company, Inc., Princeton, N.J.
5. Herwig, L. O., G. H. Muller, and N. G. Utterback: *Rev. Sci. Instr.*, vol. 26, pp. 929–936, October, 1955.
6. Glass, F. M.: Methods for Reducing Insulator Noise and Leakage, *Rev. Sci. Instr.*, vol. 20, no. 4, pp. 238–243, April, 1949.
7. Evans, R. D.: Radioactivity Units and Standards, *Nucleonics*, vol. 1, no. 2, pp. 32–43, October, 1947.
8. McCreary, H. S., Jr., and L. T. Baynard: Neutron Sensitive Ionization Chamber with Electrically Adjusted Gamma Compensation, *Rev. Sci. Instr.*, vol. 25, pp. 161–164, February, 1954.
9. Holloway, J. T., D. C. Lu, and D. J. Zaffarano: Large Proportional Counter Spectrometer for the Study of Radioactive Samples with Low Specific Activity, *Rev. Sci. Instr.*, vol. 31, pp. 91–95, February, 1960.
10. Friedman, H., Geiger Counter Tubes, *Proc. IRE*, vol. 37, no. 7, pp. 791–808, July, 1949.
11. Curtiss, L. F.: The Geiger Mueller Counter, *Natl. Bur. Standards (U.S.) Circ.* 490, January, 1950.
11a. Van Duuren, K., A. J. M. Jaspers, and J. Hermsen: G-M Counters, *Nucleonics*, vol. 17, no. 6, pp. 86–94, June, 1959.
12. Coltman, J. W.: The Scintillation Counter, *Proc. IRE*, vol. 37, no. 6, June, 1949.
13. Jordan, W. H., and P. R. Bell: Scintillation Counters, *Nucleonics*, vol. 5, no. 4, p. 30, October, 1949.
14. Birks, J. B.: "Scintillation Counters," Pergamon Press, New York, 1953.
15. Curran, S. C.: "Luminescence and the Scintillation Counter," Butterworth Scientific Publications, London, 1953.
16. Sharpe, J.: "Nuclear Radiation Detectors," John Wiley & Sons, Inc., New York, 1955.
17. Swank, R. K.: Characteristics of Scintillators, *Ann. Rev. Nucl. Sci.*, 1954.
18. Scintillation Counting Today, *Nucleonics*, vol. 12, no. 3, p. 13, March, 1954.
19. Scintillation Counting, 1956, *Nucleonics*, vol. 14, no. 4, p. 33, April, 1956.
20. Scintillation Counter Symposium, *IRE Trans. Nucl. Sci.*, vol. NS-3, November, 1956.
21. Sixth Scintillation Counter Symposium, *IRE Trans. Nucl. Sci.*, vol. NS-5, no. 3, December, 1958.
22. Scintillators and Semiconductors—Symposium Report, *Nucleonics*, vol. 18, pp. 85–100, May, 1960.
23. Price, W. J.: "Nuclear Radiation Detection," McGraw-Hill Book Company, Inc., New York, 1958.
24. Bell, P. R., and K. Siegbahn (eds.): "Beta and Gamma-ray Spectroscopy," Interscience Publishers, Inc., New York, 1955.
25. Miller, W. F., John Reynolds, and William J. Snow: Efficiencies and Photofractions for Sodium Iodide Crystals, *Rev. Sci. Instr.*, vol. 28, no. 9, September, 1957.
26. Wolicki, E. A., R. Jastrow, and F. Brooks: Calculated Efficiencies of NaI Crystals, *NRL Rept.* 4833, Naval Research Lab., Washington, D.C.
27. Friedland, S. S., J. W. Mayer, and J. S. Wiggins: The Solid State Ionization Chamber, *Tech. Mem.* 626, Hughes Aircraft Company, Culver City, Calif., January, 1960.
28. Seventh Scintillation Counter Symposium, *IRE Trans. Nucl. Sci.*, vol. NS-7, nos. 2–3, pp. 178–201, February, 1960.
29. Waller, F. J., J. W. T. Dobbs, and L. D. Roberts: Large Area Surface Barrier Counters, *Rev. Sci. Instr.*, vol. 31, pp. 756–762, July, 1960.
30. Rutherford, E., J. Chadwick, and C. D. Ellis: "Radiations from Radioactive Substances," Cambridge University Press, New York, 1930.
31. Adelson, H. E., H. A. Bustick, B. J. Meyer, and C. N. Waddell: Use of the Four-inch Liquid Hydrogen Bubble Chamber as a Fast Neutron Spectrometer, *Rev. Sci. Instr.*, vol. 31, pp. 1–10, January, 1960.
32. Spark Chamber Symposium, *Rev. Sci. Instr.*, vol. 32, pp. 482–531, May, 1961.
33. Evans, R. D.: "The Atomic Nucleus," McGraw-Hill Book Company, Inc., New York, 1955.
34. Roberts, Arthur: Cerenkov Detector Accurately Measuring Velocity and Direction over a Wide Range, *Rev. Sci. Instr.*, vol. 31, pp. 579–580, May, 1960.

35. "Radiography in Modern Industry," Eastman Kodak Company, X-ray Division, Rochester, N.Y.
36. Yagoda, H.: "Radioactive Measurements with Nuclear Emulsions," John Wiley & Sons, Inc., New York, 1949.
37. Garner, C. S.: Lauritsen Electroscope, *J. Chem. Educ.*, vol. 26, pp. 542–546, February, 1954.
38. Measuring in Core Neutron Fluxes, *Nucleonics*, vol. 20, pp. 41–46, February, 1962.
39. Weiss, J., A. O. Allen, and H. A. Schwartz: *Proc. Intern. Conf. Peaceful Uses At. Energy*, vol. 14, p. 179.
40. Weiss, J.: Chemical Dosimetry Using Ferrous and Ceric Sulfates, *Nucleonics*, vol. 10, no. 7, p. 28, July, 1952.
41. Colorimetric Dosimeter for Penetrating Radiation, *Nucleonics*, vol. 6, no. 6, pp. 66–70, June, 1950.
41a. Van Winkle, Walter, Jr., and C. Artandi: Electron Beam Sterilization of Surgical Sutures, *Nucleonics*, vol. 17, pp. 86–90, March, 1959.
42. Goldsmith, H. H.: Bibliography on Radiation Detection, *Nucleonics*, vol. 4, no. 5, pp. 142–150, May, 1949.
43. Nokes, M. C.: "Radioactivity Measuring Instruments: A Guide to their Construction and Use," Philosophical Library, Inc., New York, 1959.
44. Green, Alex E. S.: "Nuclear Physics," McGraw-Hill Book Company, New York, 1955.
45. Price, William J.: "Nuclear Radiation Detection," McGraw-Hill Book Company, New York, 1958.
46. National Standards for Nuclear Instruments and Controls, American Standards Association, New York, 1959.
47. Anton, N. and M. Youdin: "Some New Aspects of Nuclear Instrumentation in Industrial Electronics," *Transactions of the IRE*, PGIE-5, pp. 51, April, 1958.

RADIOISOTOPES IN INSTRUMENTATION

By Jerome Kohl*

Instrumentation applications of radioisotopes are considered the most important single use of radioisotopes based on yearly dollar savings to industry.[1] These applications include measurement of level, density, thickness, pressure, flow, concentration, and coating thickness. Many of these applications are discussed in the sections of this handbook devoted to measurement variables. In this section basic radiation physics is presented in enough detail to permit understanding the phenomena that make these instrumentation applications possible, and instrumentation applications of radioisotopes not covered elsewhere are briefly discussed.

DEFINITIONS AND DISCUSSION

To clarify the radioisotope applications discussed in this handbook a number of important terms are defined.

Isotope. The word isotope is derived from the Greek words *iso*, meaning the same, and *topos*, meaning place. Atomic species of the same atomic number, i.e., belonging to the same element but having different mass numbers, are called isotopes.

Radioisotope. Those isotopes, distinguishable from other species of atoms with the same atomic number by radioactive transformation, are known as radioactive isotopes or radioisotopes.

Radioisotopes are useful as sources of radiation and as tracers. Their tracer use arises because they exhibit substantially the same chemical behavior as the stable species of isotopes while emitting radiation which permits determining their identity and location.

Table 1 summarizes the principal types of radiation emitted by radioisotopes and gives important properties of these radiations.

Table 1. Principal Types of Radiation Emitted by Radioisotopes

Type	Symbol	Description	Rest mass (O = 16)	Charge (electron = 1)	Range in air	Ion pairs (per cm in air)
Alpha particles	α	Nuclei of helium atoms	4	+2	2–9 cm (for 3–10 mev)	30,000–70,000 (varies with distance from source)
Beta particles	β	Electrons ejected from a nucleus	1/1,840	−1	160–2,000 cm (for 0.5–5 mev)	150–40 (for 0.1–50 mev)
Gamma rays	γ	Electromagnetic radiations produced only in nuclear processes	None	None	15,000 cm ½ value thickness (for 1.5 mev)	$\frac{1}{100}$ of number of pairs produced by same energy β

X rays resulting from the filling of nuclear shells and bremsstrahlung generated by the deceleration of beta particles are two additional sources of electromagnetic radiation produced by radioisotopes that find use in instrumentation.

mev stands for million electron volts; kev stands for thousand electron volts.

None of the above radiations produce detectable amounts of radioisotopes.

* Coordinator of Special Products, General Atomic Division of General Dynamics Corp., San Diego, Calif.

Neutrons. While neutrons are not emitted in the disintegration of radioactive isotopes (except in rare side reactions), they are important in the production of radioisotopes, in the measurement of water content, and in certain level and concentration gages. Neutrons are particles with zero charge and with a rest mass (as compared with relativistic mass) approximately equal to that of the hydrogen nucleus. Neutrons are unstable and decay to a proton and beta particle with a half-life of approximately 12.8 min. In nuclear reactors, neutrons produce radioisotopes by causing fission of fissionable materials, such as uranium-235 or plutonium-239, and by activating substances placed in the reactor through a neutron-absorption reaction. Fission products and activated materials are now available from the United States[2] and many other atomic energy groups and from some private companies.

Neutrons for instrumentation sources are usually obtained from the reactions

Alpha + Be⁹ → neutron + carbon
Gamma + Be⁹ → neutron + 2 alphas

Neutron sources are prepared by mixing together an alpha emitter such as radium (Ra), plutonium (Pu), or polonium (Po) with beryllium (Be) or a strong gamma emitter such as antimony-124 (Sb¹²⁴) (which emits 2.1-mev gammas and decays with a half-life of 60 days) with the beryllium. The gamma-neutron reaction is called a photoneutron reaction. The photoneutron reaction Be⁹(γ,n)2α requires a gamma of at least 1.66 mev. The Be⁹(γ,n)2α reaction is the lowest-threshold photoneutron reaction known. Thus beryllium is commonly used in neutron sources. The reaction $d(\gamma,n)p$, which is the reaction of a gamma with deuterium (d) to produce a neutron (n) and a proton (p), requires gammas of 2.22 mev. Thus some neutrons are always observed from ordinary water, which contains 0.0156 per cent heavy water, when water is subjected to gammas over 2.22 mev. The gamma-neutron sources require shielding for both the gammas and the neutrons, whereas the alpha-neutron sources require only neutron shielding. It is only recently that Pu²³⁹ with its 24,360-year half-life has been made available as an alpha emitter for industrial neutron sources. Because of its long life and no gammas, the Pu²³⁹ is preferred to the 138-day half-life Po²¹⁰ or to 1,622-year half-life Ra²²⁶. The Po²¹⁰ is a pure α emitter while Ra²²⁶ and its decay chain give off alphas, betas, and gammas.

Spontaneous fission of isotopes may someday provide neutron sources. The isotope californium-252 (Cf²⁵²) decays by spontaneous fission with a yield of 3.5 neutrons/fission and a half-life of 60 years.

Bremsstrahlung. The deceleration of electrons (beta rays) in matter produces X rays just like those produced in X-ray tubes. When produced by betas these X rays are called bremsstrahlung. The rate of energy loss by bremsstrahlung is given by the relation

$$\left(\frac{dE}{dx}\right)_{BS} \sim Z^2 EN \tag{1}$$

where $(dE/dx)_{BS}$ = rate of energy loss due to bremsstrahlung
Z = atomic number of absorber
E = beta-ray energy, mev
N = atomic density of absorber

To produce a useful bremsstrahlung source one uses a high Z target, such as lead, bombarded with high-energy betas, such as are obtained from a Sr⁹⁰-Y⁹⁰ source. A 2-mev beta emitter with a lead target will lose 15 per cent of its energy as bremsstrahlung. To avoid the production of unwanted bremsstrahlung, shields for beta emitters are usually made of a low-Z material, such as Lucite.

Half-life. Radioisotopes are unstable; they decay in accordance with the relation

$$N = N_0 e^{-\lambda t} \tag{2}$$

where N = number of unchanged atoms at time t
N_0 = number of atoms present when $t = 0$
λ = constant characteristic of the particular radioactive species, called the decay constant
Lambda is related to the half-life $t_{1/2}$ of the radioactive isotope by the relation

$$t_{1/2} = \frac{0.693}{\lambda} \tag{3}$$

The constant λ relates the rate of disintegration A and the number of radioactive atoms N by the relation $A = N\lambda$, where A usually has the units of disintegration per second. To date scientists have been unable to change the value of λ by varying such factors as temperature and pressure. Half-lives of the known radioisotopes cover the range from a fraction of a second to many years. For example, carbon-10 is a radioisotope with half-life of 20 sec, while carbon-14 has a half-life of 5,100 years.

Radioactivity Units and Source Size Considerations. In radioactivity work, the **curie** is used as a measure of the number of disintegrating radioactive atoms per unit time. The curie (c) is defined as the quantity of any radioactive material having 3.7×10^{10} dps, the millicurie (mc) 3.7×10^7 dps, and the microcurie (μc) 3.7×10^4 dps. For example, 1 mc of phosphorus-32 emits 3.7×10^7 beta particles per sec, while 1 mc of cobalt-60 emits 3.7×10^7 of each of the following radiations per second: 0.3 mev* maximum energy beta ray, 1.17 mev gamma, 1.33 mev gamma. To know the radiations emitted by the decay of a radioisotope requires a knowledge of its decay scheme, which can be found in isotope tables such as Ref. 3. Decay schemes for P³² and Co⁶⁰ are as follows:

* 1 mev = 10⁶ electron volts = 1.60×10^{-6} erg.

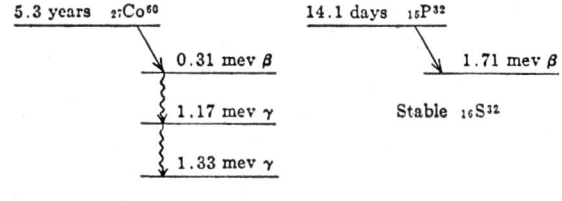

5.3 years $_{27}\text{Co}^{60}$

0.31 mev β

1.17 mev γ

1.33 mev γ

Stable $_{28}\text{Ni}^{60}$

14.1 days $_{15}\text{P}^{32}$

1.71 mev β

Stable $_{16}\text{S}^{32}$

(One can readily measure 4 dps/sec of P^{32}, which represents only 4×10^{-15} g, indicating the sensitivity of the radioactivity technique.)

The amount of a given radioisotope needed for an instrument such as a thickness, level, or concentration gage depends on the following factors:*

1. Statistical precision needed. Because radioisotopes decay in a "random" manner, the laws of probability apply to their decay and to the background radiation due to cosmic rays and naturally occurring radioisotopes such as radium and potassium. The laws of statistics require that the standard deviation of the reading be dependent on the square root of the number of disintegrations measured per measurement interval. A smaller per cent "error" or a higher statistical precision requires either a larger source or longer measurement time or both. (See Zumwalt's work[4] for the effect of source size on beta gage accuracy.)

2. Size and efficiency of detector.

3. Absorption losses between source and detector.

4. Distance between source and detector. (See inverse-square-law discussion.)

Roentgen (r) and Radiological Safety. The roentgen is a measure of the intensity of ionizing radiation in air. One roentgen corresponds to the creation of 1 esu of charge in 1 ml of standard air. The roentgen can be considered as a measure of energy dissipation in air, and its definition has been extended to cover ionization or energy dissipation in other media. One roentgen corresponds to a radiation field dissipating 83.8 ergs/g of air, which dissipates approximately 93.8 ergs/g of body tissue. Ionization in tissue is a measure of physical damage; thus allowable radiation exposures for human tissue are expressed in roentgens. Since the same number of roentgens from various types of radiation produces different amounts of body damage, a term called roentgen equivalent man (rem) is used in stating allowable radiation exposure values. The rem = r $\times$ rbe where rbe, known as relative biological effectiveness, has the values given in Table 2.

Table 2. Values of rbe

Type of radiation	rbe
X and γ radiation	1
β rays	1
α rays	20
Fast neutrons*	10
Thermal or slow neutrons	5

* Neutrons having energies in the range 0.1 to 10 mev. Above 10 mev, the rbe increases rapidly.

The radiation dosage permissible under various situations is given in detail in Ref. 5; it is summarized in Table 3.

Table 3. Permissible Exposure of Individuals to Radiation in Restricted Areas

Locus of radiation exposure	Rems per calendar quarter
Whole body; head and trunk; active blood-forming organs; lens of eyes; or gonads...	1.25
Hands and forearms; feet and ankles	18.75
Skin of whole body	7.5

The dose to the whole body, when added to the accumulated occupational dose, must not exceed

$$\text{MPD} = 5(N - 18) \text{ rems}$$

where MPD is the maximum permissible accumulated dose in rems and N is the individual's age in years at his last birthday.

The radiation dose rate R in r/hr at any distance x in feet from a point gamma-emitting source is approximated by the formula

$$R = \frac{6CE}{x^2} \qquad (4)$$

where C is the number of curies and E is the total gamma energy in mev per disintegration. (It should be noted that, where a radioisotope emits more than one gamma per disintegration, E is the sum of the energies of those radiations. For example, Co^{60} emits two gamma rays of energy 1.17 and 1.33 mev; thus $E = 2.5$.)

* For a full discussion of this subject see J. Kohl, R. D. Zentner, and H. R. Lukens, "Radioisotope Applications Engineering," chap. 4, Statistical Considerations in Radioactivity Measurements, and chap. 6, Calculations for a Tracer Experiment, D. Van Nostrand Company, Inc., Princeton, N.J., 1961.

In addition to external radiation hazards one must also consider radiation exposure from ingested material. Sealed sources and "wipe" tests are used to ensure that radioactive materials do not get into the air or drinking water in harmful quantities. The allowable concentrations in water and air are given in Ref. 5 and for common isotopes are given in Table 4. Note that these values are particularly low for alpha emitters such as Po^{210}, Pu^{239}, Ra^{226} and for the beta emitter Sr^{90} as compared with Co^{60} or P^{32}. Since Sr^{90} is a relatively hazardous radioisotope compared with Kr^{85} and since Kr^{85} will disperse as a harmless gas in case of source breakage, Kr^{85} is gaining favor over Sr^{90} where it can be used as a beta-gage source. The disposal of radioactive materials is governed by Federal regulation,[5] as is their shipment.[6]

Absorption. When alpha, beta, or gamma radiation passes through matter, the radiation interacts with the electrons and the nuclei present in the matter and transfers some of its energy to them. Thus a beta-ray beam leaving a paper sheet possesses less energy (fewer betas and some of these are at a lower energy) than the entering beam. This loss of energy is a function of number of nuclei encountered and the charge and mass of these nuclei. A determination of relative beta absorption is used in beta gages for measuring thickness and of gamma-radiation absorption in density-, level-, and thickness-measuring devices.

Alpha-ray Absorption. Essentially all the alpha particles emitted by a given radioisotope have the same energy and thus the same range in matter. The range-energy relation for alphas is shown in Fig. 1. Because the maximum range for alphas corresponds to a sheet of paper, alphas are useful only for vacuum gages or very thin sheet gaging. However, the low penetrating ability of alphas makes it extremely simple to shield against them. The only hazard in using alpha emitters is that of ingestion.

Beta and Gamma Absorption. Beta particles emitted by any given radioisotope have varying energies up to a maximum which characterizes the particular radioisotope. The difference in energy between the maximum and that of any particular beta is taken up by a neutrino. This varying energy results in beta-absorption curves approximating a logarithmic

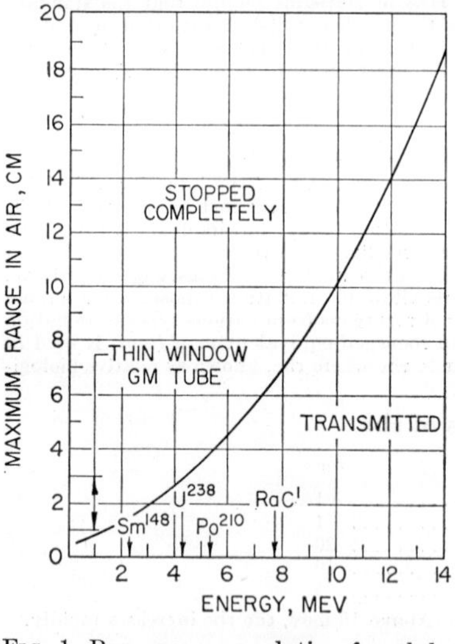

Fig. 1. Range-energy relation for alpha particles. (*Source: Keleket X-Ray Corp.*)

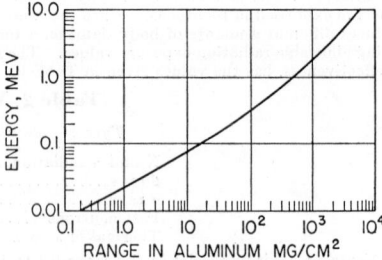

Fig. 2. Range-energy relation for beta particles. (*Source: L. E. Geledenin, "Determination of Beta Particles and Photons by Absorption," Nucleonics, January,* 1948, *pp.* 16.)

relation. The maximum range of betas in aluminum as a function of energy is shown in Fig. 2. The interaction of beta or gamma radiation with matter is approximated by the equation

$$I = BI_0e^{-\mu t} \tag{5}$$

where I_0 = initial beam intensity
I = final beam intensity
μ = absorption coefficient = $0.693/\frac{1}{2}$ thickness
t = thickness of absorber
B = buildup factor = 1 for betas

For beta radiations, μ may be obtained from the relation $\frac{1}{2}$ thickness approximately equals range/7 to 10($\frac{1}{2}$ thickness is the value of t such that $I/I_0 = \frac{1}{2}$). For a beta of known energy, the range in aluminum can be obtained from Fig. 2.

The beta mass absorption coefficient for materials other than aluminum can be determined from the aluminum value using the fact that the coefficient varies about as Z/A, where Z is the atomic number and A the mass number of the absorber.

If t in the absorption equation has dimensions of length, cm or inches, then μ must have units of cm^{-1} or in.$^{-1}$ and is called the *linear absorption coefficient*. If t is expressed in g/cm^2, lb/ft^2, or other units of mass per unit area, then μ must have units of cm^2/g, ft^2/lb, and is called the *mass absorption coefficient*. The two forms of μ are related by the density ρ of the absorber by the relation

$$\mu(\text{linear}) = \mu(\text{mass}) \times \rho \tag{6}$$

in units, for example, $(cm^{-1}) = (cm^2/g) \times g/cm^3$.

For gamma rays, μ is a function of gamma energy and absorber atomic number. Figure 3a gives values of μ for various gamma energies for steel, water, aluminum, and lead. Figures 3b and c gives the total mass absorption coefficients for X and gamma radiation for these and similar materials.

The gamma-ray absorption coefficient is primarily due to three phenomena:

1. *Photoelectric absorption* in which all the energy of the gamma is given to an electron and which is predominant for high-Z absorbers and for gamma energies below 1 mev. This coefficient varies as Z^4.

2. *Compton absorption* and *scattering*, which is the predominant process in the energy region 0.6 to 2.5 mev. This coefficient varies as Z.

3. *Pair production*, which can take place only where the gamma energy exceeds 1.02 mev (the equivalent annihilation energy for the positron-electron pair formed in pair production). The pair-production coefficient varies as Z^2.

Broad-beam or noncollimated-beam attenuation calculations, in which scattered radiation can reach the detector or exposed person, require consideration of a *buildup coefficient B*, values of which can be obtained from Ref. 7. Figures 4 and 5 are useful for broad-beam shielding calculation. For narrow or highly collimated beams where scattered radiation cannot enter the detector, $B = 1$.

X-ray absorption follows the same exponential attenuation law as does gamma-ray absorption. However, for the relatively long wavelength X rays (energies below 100 kev) there are discontinuities in the absorption coefficient at photon energies corre-

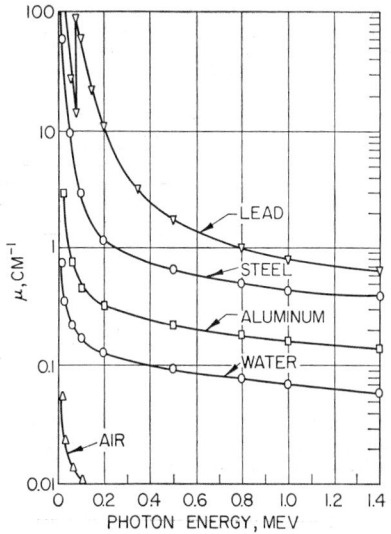

FIG. 3a. Linear absorption coefficients for X and gamma radiations. (*Source: Ref. 7.*)

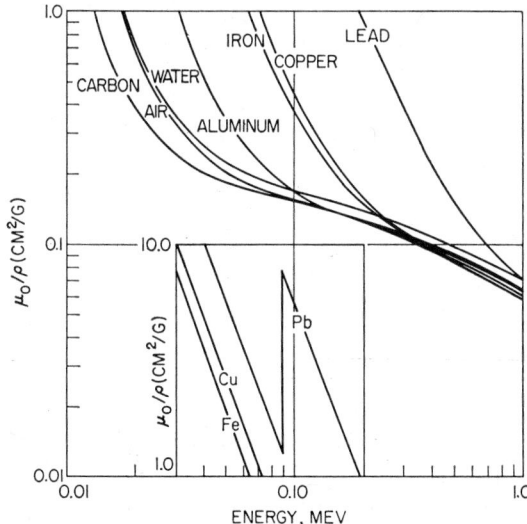

FIG. 3b. Total mass absorption coefficients for X and gamma radiation. The coefficients for air in the range 0.3 to 3.0 mev may be considered identical to those of carbon. Insert extends the range for lead, copper, and iron. (*Source: Ref. 7.*)

sponding to the electron binding energy of the absorber atoms. For example $K\alpha$ X rays* of an element Z are strongly absorbed by the element $Z - 1$ while little absorbed by $Z + 1$. Thus X rays from element Z could be used in a sensitive concentration measuring gage to detect $Z - 1$.

Neutron Absorption. When working with γ, n sources, such as Sb-Be or Ra-Be, it has been found that shielding sufficient to reduce the gamma level to tolerance levels normally suffices for the neutrons.

* $K\alpha$ X rays are emitted by an element when an empty position in the K shell is filled by an electron from the L shell.

For α, n sources, such as Po-Be or Pu-Be, it is necessary to provide a neutron shield which normally comprises a combination of a moderator and absorber, such as borax in water or a boron-loaded plastic. The water or plastic provides the hydrogen nuclei with which the neutrons collide and to which they lose energy. The neutrons are thus "moderated" or "thermalized." The slow or thermal neutrons are

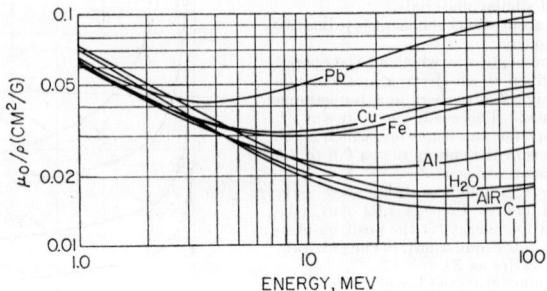

FIG. 3c. Total mass absorption coefficients for X and gamma radiation. The total absorption coefficients for air in the range 0.3 to 3.0 mev may be considered identical to those of carbon. (*Source: Ref. 7.*)

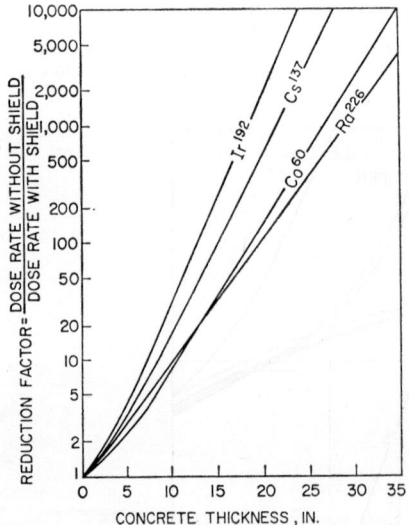

FIG. 4. Broad-beam shielding for absorption of Ir^{192}, Cs^{137}, Co^{60}, and Ra^{226} gamma rays in concrete. (*Sources: Cs^{137}, Co^{60}, and Ra^{226} data: National Bureau of Standards Handbook 54; Ir^{192} data: private communication, Division of Physics, National Research Council, Canada, taken from Ref. 4 with permission of D. Van Nostrand Company.*)

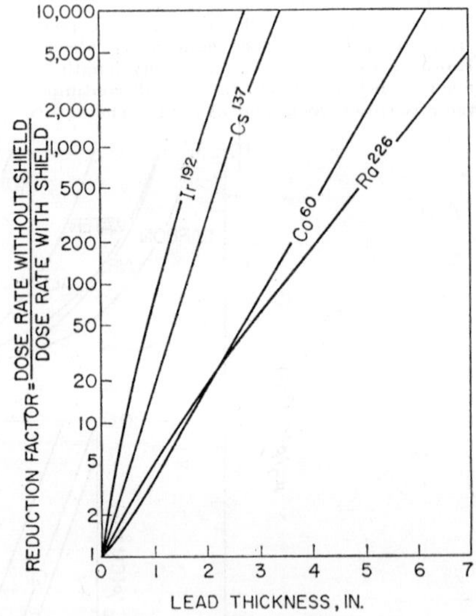

FIG. 5. Broad-beam shielding for absorption of Ir^{192}, Cs^{137}, Co^{60}, and Ra^{226} gamma rays in lead. (*Sources: Co^{50}, Cs^{137}, and Ra^{226} data: National Bureau of Standards Handbook 54; Ir^{192} data: ASTM Bull. 145, taken from Ref. 4 with permission of D. Van Nostrand Company.*)

readily absorbed by the boron. The absorption coefficient of neutrons in an absorber such as boron varies inversely with the neutron energy. Reference 7 covers neutron shielding.

Scattering. Radiation passing near a nucleus or one of its planetary electrons is changed in direction by an interaction between its charge or field and that of the nucleus or electron. Radiation that has been changed in direction is termed scattered radiation. Radiation that has been scattered in the backward direction is termed backscattered radiation. In the case of beta radiation, the amount of

Table 4. Properties of Common Radioisotope Radiation Sources

Radioisotope	Principal type of radiation	Half-life	Energy of radiation	Permissible concentration in		Principal use
				Air	Water	
				7×10^{-12} $\mu c/ml$	3×10^{-5} $\mu c/ml$	
Po210............	α	138 days	5.3 mev	7×10^{-12}	3×10^{-5}	α, n neutron sources
Pu239...........	α	24,360 years	5.15 mev	6×10^{-14}	5×10^{-6}	α, n neutron sources
Ra226...........	α, β, γ	1.622 years	4.77 mev	1×10^{-12}	1×10^{-8}	As γ source for level, as α source for α, n
C14.............	β	5,700 years	0.155 mev	1×10^{-7}	8×10^{-4}	Thin plastic thickness gages
Kr85............	β	10.3 years	0.695 mev	3×10^{-7}	...	Light paper and plastic gages
Sr90 and daughter.....	β	25 years	0.61 mev	1×10^{-11}	1×10^{-7}	Heavy paper, thin metal, rubber gages
Y90.............	β	61 hr	2.18 mev	3×10^{-9}	2×10^{-6}	Heavy paper, thin metal, rubber gages
Ru106 and daughter....	β	1 year	0.04 mev	3×10^{-9}	1×10^{-6}	Heavy rubber, light metals gages
Rh106...........	β	30 seconds	3.53 mev	...	...	Heavy rubber, light metals gages
Fe55............	X ray	2.94 years	6.4 kev	3×10^{-8}	2×10^{-4}	Heavy rubber, light metals gages, sulfur conc. gage
H3 BS...........	Bremsstrahlung (X rays)	12.46 years	17.6 kev and down	2×10^{-7}	3×10^{-3}	Sulfur and lead gage
Sr90 BS........	Bremsstrahlung (X rays)	25 years	2.18 mev and down	2×10^{-7}	3×10^{-3}	Steel, copper, aluminum sheet gages
Cs137 and daughter.....	β	33 years	0.51 mev	2×10^{-9}	2×10^{-5}	Level and density gages
Ba137 m.........	γ	2.60 minutes	0.662 mev	...	...	Level and density gages
Co60............	γ	5.27 years	1.17 mev and 1.33 mev	1×10^{-8}	3×10^{-5}	Level, density, thickness gages
Sb124...........	γ	60 days	2.1, 1.7, and lower mev	7×10^{-10}	2×10^{-5}	γ, n neutron source

The permissible concentration values are from Ref. 5. They are values above natural background permitted in areas not controlled by the radioisotope user; slightly higher values are permitted in areas controlled by the user. In all cases the lowest value given in Ref. 5 is used in this table.

scattering is a function of (1) the thickness of the scattering medium, (2) the energy of the beta rays, (3) the number of nuclei per unit volume, and (4) the charges on the nuclei in the scattering medium.   It is this sensitivity of scattering to Z that enables a backscattering beta gage to measure the thickness of a coating when the coating Z is different from the Z of the base material, or to detect dissolved materials of one Z in a solvent of a different Z.

Gamma rays undergo scattering in passing through matter as a result of an interaction with the electrons in the matter.   Measurement of the scattering of gamma rays is the technique used in certain level and thickness gages.

Inverse-square Law.   Radiation emitted by radioisotopes is uniformly distributed in all directions in space; thus the number of particles or quanta passing through a unit volume at any point distant from the source varies inversely as the square of the distance from the source, or

$$I = \frac{I_0 r_0^2}{r_1^2} \tag{7}$$

where I_0 = radiation intensity at a distance r_0 from a source
  I = radiation intensity at a distance r_1 from the same source

Equation (7) neglects absorption effects and assumes a point source; it is most useful for gamma radiation where the source-to-detector distance is usually much larger than any source or detector dimension.

Radioisotope Radiation Sources.   Important properties of the more commonly used radioisotope radiation sources are tabulated in Table 4.

APPLICATIONS OF RADIOISOTOPES TO INSTRUMENTATION*

Pressure Measurement

The interaction of alphas with a gas (or solid) results in the formation of positive and negative ions, which can be collected and measured as an electric current.   Since the number of ions produced in a gas by alpha particles from a source of fixed size is dependent on the density and composition of the gas, measurement of ionization can be used as a method of measuring pressure or composition.

In the Alphatron, alpha particles emitted from approximately 0.2 mg (for Ra this equals 0.2 mc) of radium in equilibrium with its decay products pass into the gas whose pressure is to be measured. Since radium has a half-life of 1,620 years, the number of alpha particles emitted per unit time is relatively constant; thus the number of ions formed is proportional to the gas density in the chamber.   Ions collected in the pressure chamber provide a current on the order of 2×10^{-10} amp/mm of air pressure.   The Alphatron operates in the pressure range 1 to 10 mm of mercury.[8] Beta rays can also be used for gas density or pressure measurement.   Schumacher describes a beta-particle gas-density gage.[9]

Level Measurement†

The ability of gamma radiation to penetrate steel walls of pressure vessels and brick linings of cupolas and the absorption or scattering of this gamma radiation by material present inside these vessels provides phenomena which permit measuring the level of solids or liquids without the use of floats or other internal devices.   In the radiation-absorption technique, the source of gamma rays, such as radium or Co[60], is placed on one side of the vessel.   The radiation-detecting device is placed on the opposite side, or a number of sources are placed in wells and the detector is located opposite the sources, above the highest fluid level.   The presence of a fluid or solid between the source and detector reduces the amount of gamma radiation that reaches the detector. The change in response of the detector can be caused to operate a relay system, an indicator, or a recorder.   This technique is used in measuring the height of molten iron in cupolas,[10] the height of catalyst in catalyst chambers,[11] and the height of coke in a coking vessel.[12]   (For a 42 in. inside diameter, 61 in. outside diameter cupola, 50 mc of Co[60] is needed.)

Where the vessel is large in diameter and an excessively large source would be required for the radiation-absorption technique, or where the use of this technique is prevented by lack of space or inaccessibility of equipment, radiation scattering can be used to indicate the location of the solid or liquid level.   In this technique the

* For a more complete discussion of radioisotope applications, see General Ref. 4.
† Refer also to Sec. 5 for summaries of level-measuring methods employing radioisotopes.

source and receiver are placed on the same side of the vessel, but the receiver is shielded from direct radiation. When liquid or solid is at the level of the source, gamma radiation is backscattered from this material into the detector which provides an increased signal that indicates the level. The source and detector can be motorized to "find" the level as the point between minimum and maximum signal.

In another version of a backscattering level gage the height of charge in a blast furnace is measured by directing gamma rays from a collimated well-shielded 150-curie Co^{60} source down from the top of the furnace. The rays "bounce" or are scattered from the charge and are picked up by a scintillation detector which is shielded from observing any direct radiation from the source. The scintillation detector is automatically controlled to seek and track the position of maximum signal. The angular position of the detector provides a voltage indicative of the charge level.[13]

The absorption and scattering phenomena are best used for detecting a level that varies by only a few feet, although the absorption technique has been used with a number of sources at different heights and with a number of long detectors to provide a level gage useful over 20 ft of height. For locating the level in pressurized spheres used for storing volatile liquids, two vertical tubes are placed close to each other for the height of the sphere. A source rides in one tube and the detector in the second. The source and detector are connected to a motorized winch and a circuit such that they "seek" the level, which is then indicated by the winch position.

Where it is possible to place a float containing the source in the vessel, the inverse-square-law relationship can be used to provide a continuous measure of level over 20 to 30 ft of height. In this technique the source floats on the surface in a container. The detector is placed at the top of the vessel. As the level drops, the source moves away from the detector and the radiation level at the detector decreases.

Barnartt[14] designed a level gage in which he used a radioisotope neutron source and detector which was sensitive to backscattered neutrons. He found that he could locate the level of a hydrogenous liquid to ± 2 mm with this gage.

Thickness and Coverage Measurement*

The absorption of radiation by matter is a phenomenon that permits measurement of the mass of material between a source and a receiver. For very thin materials (under 5 mg/cm^2, 0.0009 in Al) the absorption of alpha radiation is used to provide a measure of weight per unit area. In the range 0.5 to 1,000 mg/cm^2 (0.00009 to 0.16 in. of Al) beta radiation such as is emitted by Sr^{90}, Kr^{85}, C^{14}, and other radioisotopes is used for measuring weight per unit area. For thicknesses above 1,000 mg/cm^2, bremsstrahlung or gamma sources are used.

Beta Absorption Gages. Beta absorption gages[15,16] use a 5- to 50-mc beta source which is placed beneath the paper, rubber, plastic, or metal sheet. An ionization chamber above the sheet provides a measure of the radiation that penetrates the material.

One recent valuable application of beta gages is in the control of the amount of tobacco in cigarettes.[16]

Beta Backscattering Gages. Beta backscattering gages are used to measure the weight per unit area of a coating material over a base material of different atomic number. They differ from absorption gages in that the source and detector are on the same side of the sheet being measured. Backscattering is used to measure the thickness of sheets, coatings, and films—such as rubber sheet, tin plate on a steel base, and lacquer on a tin base. Sheet measurements utilize a steel roll as the "infinitely thick" base material.

Bremsstrahlung and X-ray Gages. For materials of mass per unit area between the ranges covered by beta and gamma gages, bremsstrahlung and X-ray sources can be used. A bremsstrahlung gage has been used for measuring silt density[16] in lakes and streams.

Leveque et al.[17] describe experiments using X-ray fluorescence to determine chromium coatings on copper in thicknesses under 20μ to better than $\pm 0.05\mu$. Reiffel[18]

* Refer also to Sec. 6 for thickness-measuring methods employing radioisotopes.

used a beam of radioisotope-produced X rays with energy near the critical absorption edge of the element of interest to measure photoemulsion thickness deposited on a cellulose acetate base.

Gamma Thickness Gages.   In the thickness range from 0.050 to 5 or 6 in. of steel, beta radiation is not powerful enough to penetrate the material.   For this range gamma rays, such as are emitted by many radioisotopes including Se^{75}, Ta^{182}, and Co^{60}, are used.   The behavior of gamma gages is similar to that of the X-ray gages commonly used in steel mills for measuring and controlling the thickness of cold-rolled steel.   Scintillation detectors as well as ion chambers are used with gamma gages. Two unusual applications of gamma gages are the measurement of the water content of a snow pack and the continuous weighing of crushed sugar cane moving in free fall from a conveyor belt.[16]

Backscattering Gamma Thickness Gages.   The backscattering of gamma rays can be used to measure thickness when only one side of the gaged material is accessible. This technique has proved particularly useful in measuring the wall thickness of tubes and tanks (up to about 1 in. of steel) to approximately ± 4 per cent and in locating localized thinning due to corrosion.   The detector can be shielded from primary radiation by heavy shielding, or a detector sensitive only to scattered radiation, as described by Putman[19] and Kerry,[20] can be used.   Tolan and McIntosh[21] discuss the use of backscattering to locate defects in tubing.

Neutron Gages.   Reiffel[18] describes a neutron activation method used to determine the thickness of silver plating in the range of 0.0001 to 0.001 in.

Incorporation of Radioisotopes in Material Proper.   Radioisotopes can be directly incorporated into the material whose thickness or coverage is to be measured in order to provide a very sensitive measuring technique.   In a study of the distribution of lubricants on textiles, a rayon manufacturer investigated the uniformity of distribution of the lubricating finish.   Because the finish contained sodium oleate he used radioactive Na^{24} and introduced this into his finish.   Treatment of the yarn was carried out on a laboratory scale and the Na^{24} was measured on samples as small as 1 cm long.   These samples weighed only about 0.2 mg, and each sample contained approximately 0.04 microgram of Na^{24}.

In an experiment to study the distribution of ink on paper, P^{32} was converted to PCl_5, which was added to the varnish used in ink at the rate of 60 $\mu c/cc$ of varnish.[22] The varnish was then added to the ink.   Distribution of ink was determined by measuring its radioactivity.   From such measurements variations in the quality in printing can be related to the quantity of ink, type of paper, and printing temperature.

In the measurement of the thickness of enamel coatings, 18 g of sodium uranylacetate were incorporated in 100 cc of pigment for mixing with a large batch of enamel.[23] A G-M count of the sprayed surface checked micrometer measurements of the baked enamel and permitted determinations of coating thickness on the order of 0.0035 in. over the entire coated area.   It has been suggested that this technique be used for determining the distribution and quantity of such materials as flameproofing compounds and water repellants on cloth, or for determining the distribution and thickness of the coating of wax or other protective coatings on fruit.

Deposit-thickness Gage.   Hull[24] and Sigworth and Fries[25] describe a beta-ray absorption gage used to measure engine deposits without need for disassembling the engine.   A removable plug is made radioactive by coating it with a suitable radioisotope such as Ni^{63}, S^{35}, or Co^{60}; the counting rate of the plug is determined before and after the test; and by calibration with known absorbers the dropoff in measured counting rate can be readily correlated with the thickness of the engine deposit.   The plug is removed for each count.

Tachometers

A patent has been granted for a device whereby a mechanical register is actuated as radioactive material mounted at a point on the circumference of a rotating wheel passes near a detecting device.[26]   It has been suggested that this tachometer be used for the testing and calibration of electric watthour meters.

Position- and Displacement-measuring Devices

A gamma-emitting radioisotope, such as Co^{60}, has been placed in pipeline scrapers or "go-devils" to permit finding them when they get stuck in a pipeline.[27] The technique requires a radioactive source which is mounted in or on the scraper and a portable G-M or scintillation detector carried by the line walker to locate the scraper by its radiation.

An experimental punch-press safety device has been developed in which a small amount of radioactive material is placed on the wrist of a punch-press operator so that the presence of his hands in the dangerous area operates a G-M detector located nearby and prevents operation of the press (see also Sec. 6).

Wear Measurement

The British have used small pellets of Co^{60} placed at different depths in the lining of blast furnaces[28] to measure the thickness of the lining at any time by noting when the various sources disappear.

Gears, bearings, piston rings, and distributor points have been irradiated in the reactors of the USAEC to produce radioisotopes directly in the irradiated material.* The radioactive part is then placed in the equipment whose wear rate is to be measured. Piston rings have been placed in test engines which have been operated with varying cylinder temperature, type of lubricant, and quantity of sulfur in the gasoline.[29] The amount of material worn from the ring is determined by counting radiations from the radioactive metal in the lubricating oil. This technique permits determining quantities of iron worn from a surface as small as a fraction of 1 mg and has the further advantage that it provides a continuous method for measuring wear without requiring dismantling the engine and weighing the parts. Similar experiments have been conducted in France (Ezran, unpublished results) on plastic parts, using as tracer a radioactive filler in the plastic, and allowing measurement after the parts being tested had been in motion for only a few minutes.

A tungsten carbide die has been irradiated and its wear, when used for wire drawing, measured by counting the wire.[28] Autoradiographs† of the wire provided information on the distribution of the die-wear debris and on the amount of wear for various die lubricants.

P^{32} has been added to rubber which has been compounded into tires in a technique for measuring tire rate of wear. These tests by the B. F. Goodrich Research Center (1951) permitted continuous measurement of wear as a function of tire air pressure, temperature, rpm of wheel, and power application without having to wait for fleet tests of large numbers of tires. As used, the technique involved measuring the amount of radioactive rubber left on the road or blown into the air behind the tire.

Wear tests with radioactive isotopes have also been carried out on floor waxes and automotive protective coatings.[30]

Area Measurement

The areas of small irregular sheets of material thick enough to stop all alpha particles can be measured by placing the sheet between a uniform flat source of the alpha radiation and an alpha-sensitive detector and measuring the change in signal resulting from the absorption of the radiation by the sheet. The ionization has been measured using a chamber the same size as the flat source.[28]

* The quantity of radioactivity produced by irradiation depends on the neutron flux, the time of irradiation, quantity of target material and its neutron-absorption cross section. For details see F. E. Senftle and W. Z. Leavitt, Activities Produced by Thermal Neutrons, *Nucleonics*, vol. 6, no. 5, pp. 54-63, May, 1950; or sections on irradiation and activation in General Ref. 4.

† Autoradiography is a technique by which radioactive material is located by permitting it to expose photographic film. For details of this technique see H. Yagoda, "Radioactive Measurements with Nuclear Emulsions," John Wiley & Sons, Inc., New York, 1949.

Uniformity of Mixing

It is possible to use radioisotopes to determine the uniformity of mixing of liquids or dry solids. The British[31] studied mixing of solids consisting of powdered salts, vitamins, oats, and other materials. They used radioactive Na^{24} in $NaCl$ as the tagged compound and determined the quantity of it present in samples taken from the mixer at different times. The amount of Na^{24} in samples taken from 2 tons of poultry mix was determined when 1 mc of Na^{24} had been used. The isotope cost for an experiment of this type is under $10.

Catalysts tagged with Cr^{51}, Sc^{46}, and Ce^{144} have been used to determine mixing patterns in fluid catalytic cracking plants.[32] One mc of a 1-mev gamma emitter uniformly mixed in 500 tons of catalyst will produce a net counting rate of 275 cpm/kg above a background of 200 cpm when 1-liter catalyst samples are counted with a 2-in.-thick, 1¾-in.-diameter NaI scintillation counter.

Oil-, grease-, and gasoline-soluble tagged compounds have been used to determine the quality of mixing in refinery operation.[33–35] Between 200 and 300 mc of P^{32} as an oil-soluble phosphate compound will permit determining the quality of mixing in a 55,000-bbl tank.

Archibald[36] studied the quality of mixing and degree of short circuiting in model sewage-settling tanks using radioactive tracers. Field tests were also made to determine retention times in a 60- by 35-ft primary sedimentation tank, in a 100-ft-diameter circular clarifier, and in a 4.9-acre pond. For the pond test 5 ml (20 mc) of I^{131} tracer were used. A similar test with salt as tracer would require 1,000 lb of salt.

Bogdonova[37] reported on experiments to determine the mixing of molten steel in open-hearth furnaces using Ir^{192}, Fe^{59}, Cr^{51}, and Co^{60}.

Handlos et al.[38] used Kr^{85} to study gas mixing in a catalytic cracking plant. They measured the Kr^{85} concentration using an ionization chamber filled with the gas sample.

Viscosity Determination

Gueron[39] describes an experimental falling-ball viscosimeter for use with nontransparent, highly viscous liquids. With this instrument, viscosity is determined by measuring the time taken for a radioactive ball to travel a fixed distance through the liquid. A clock operated by a G-M counter is started when the ball passes one slit in a lead block which is placed next to the liquid and is stopped when the ball passes a second lower slit. See also viscosity measurement in Sec. 7.

Flow Measurement

The three principal techniques used for measuring flow rate are peak timing, dilution, and total count. These three techniques are illustrated in Fig. 6.

Peak Timing. In the peak-timing technique described by Hull[40] a gamma emitter such as Co^{60} or Sb^{124} is injected quickly at a point close to the section of the pipe in which the velocity is to be found. The time of passage of the peak of the tracer wave is determined using two detectors, such as G-M counters or scintillation detectors, located at a known distance apart and external to the pipe; the detectors are usually connected to a rate meter and a recorder. The time interval for the liquid to travel between the two detectors is calculated from the distance on the recorder chart between two radiation-level peaks, or it is determined by having the arrival of the activity at the first detector start a timer which is stopped by the arrival of the radioactive material at the second detector. The measured time interval is divided by the distance between the observation points to calculate the linear flow rate. If the volume of the pipe between the two points can be determined or calculated, the volume flow rate can also be calculated.

If the width of the tracer wave is very small with respect to the time of its passage, no problem arises concerning just what part of the tracer wave represents the point of injection. If, however, the tracer wave spreads out so that it has an appreciable width, then it is necessary to measure the time between passages of the center of gravity of the curve. This is because under turbulent-flow conditions a Gaussian distribution of the tracer in the line will persist even though the concentration curve will grow broader and flatter with mixing in the line. The peak-timing system cannot readily be used for viscous flow which produces a marked spreading out of the tracer and which provides very different fluid velocities in the center of the pipe compared with the average velocity. Accurate results can be obtained with the peak-timing technique only for highly turbulent flow with effectively a flat velocity profile.

The pulse-timing method can also be used for determining the velocity of bulk solids by a technique discussed by Hull and Bowles.[41]   In their technique, which was used in the TCC catalytic cracking process, the circulation rate of catalyst beads was determined by tagging a few of the beads with Zr^{95}; about $\frac{1}{2}$ mc was added to each of 10

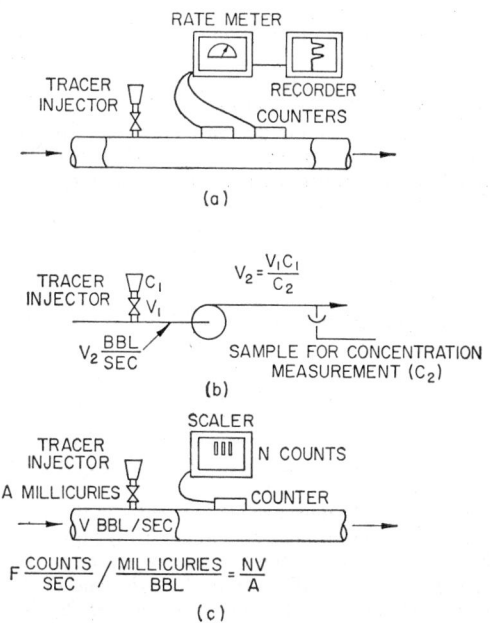

FIG. 6. Basic techniques for measuring flow.   (a) Peak timing; (b) dilution; (c) total count, (*Reprinted with permission from K. A. Kobe and J. J. McKetta, Jr. (eds.), "Advances in Petroleum Chemistry and Refining," vol. 2, John Wiley & Sons, Inc.*)

beads.   The tagged beads were then admitted to the system and one of the catalyst circulating pipes was surrounded at two points with rings of G-M detectors.   When a bead passed the first ring of detectors, it started a timer; as it passed the second ring, it stopped the timer.

The information on the time of passage was combined with data on the density of the flowing catalyst to provide the catalyst circulation rate in tons per hour.   Several of the catalyst flow meters have been used in commercial TCC catalytic cracking plants.   For fluid catalytic crackers, one cannot fix enough tracer on a single particle of the fluidized catalyst to permit following the particle.   Instead, a gamma milker[42] is used to inject a shot of tracer (several millicuries) into the fluid catalyst in a standpipe, and the two-detector technique is then used to determine the flow rate.

Flow by Dilution.   A second method of determining flow rate is the dilution method.   It utilizes the fact that the concentration of mixed tracer in a line resulting from the continuous bleeding of a tracer at a known rate into the line is inversely proportional to the relative flow rates of the line and the bleeder, if the tracer uniformly mixes with the flowing liquid. The concentration of the tracer in the line can be determined by measuring either the beta or gamma activity in samples taken from the line or by using a detector outside the line. The dilution method was illustrated by Karrer, Cowie, and Betz[43] in a measurement of condenser water flow (33,000 gpm). One disadvantage of the dilution technique is that it requires uniform mixing which involves injecting across the total surface area of the flowing fluid and usually mixing of the fluid in a pump or heat exchanger.

Flow by Total Count. Hull[44] developed a third technique for measuring flow rates based on the measurement of the total number of counts from a radioactive tracer which has been added to a flowing stream. The total count bears a simple inverse relation to the flow rate. It can be measured with a detector in or out of the stream and is useful for measuring flow rates in pipes, open channels, rivers, or any type of flowing system for which a calibration constant can be measured or calculated. The number of counts measured is independent of the distribution of the tracer *along* the stream, although the tracer must be uniformly distributed *across* the stream so that the final count can be converted to the flow rate by use of a determined constant. As the flow increases, the time the radioactive atoms are in the vicinity of the counter decreases. Thus the number of counts decreases in direct proportion to the increase in average fluid velocity; i.e., the total count is inversely proportional to the flow rate. The total count and flow rate are related by a constant F in accordance with the equation

$$N = \frac{A}{V} \times F \tag{8}$$

where N = total counts
A = measured amount of radioisotope added
V = flow rate, barrels per second

The constant F is characteristic of the isotope, the counter, and the geometric relationship between the counter and the stream. To determine F, a piece of pipe (similar to the one used in the flow measurement) can be filled with a solution of the tracer of known concentration. The counting rate is measured using the same geometric configuration that is used in the actual flowing conditions. Under certain conditions F can be calculated.

Hull has used this technique to measure (1) flow rates in pipes, where no flow meters exist; (2) leaks in cross streams in heat exchangers; (3) liquid entrainment in distillation towers; and (4) water flow in open canals and rivers.*

In plant tests carried out by Hull, standard deviations of 2 to 5 per cent were obtained resulting from (1) statistical errors in counting; (2) differences in detector placement for various trials or between the calibration and actual runs; and (3) inaccuracies in determining the amount of tracer injected. Other potential sources of error include contributions to the counting rate for sections of pipe not included in the calibration check; loss of tracer due to absorption or precipitation; and corrosion or deposits producing a tube wall of different thickness than that used for calibration.

Where F is known, the equation can be rewritten

$$A = \frac{NV}{F} \tag{9}$$

and A as a fraction of the total amount of tracer injected can be determined; this permits determination of leakage rates. Hull, using this technique, injected 2.3 mc of Cs^{134} into a cooling-water line flowing at 75 gpm to find the leakage rate of water from the line into a gasoline stream. He observed a net count on the gasoline stream of 300 ± 85 counts, corresponding to 0.04 per cent of the 2.3 mc—indicating a leakage rate of 0.03 gpm. Hull also injected 10 mc of Co^{60}-tagged naphthenate into the feed

* For a modification of the total-count technique which improves sensitivity, see D. E. Hull et al. *Isotope Engineering at Large Flow Rates, Trans. Am. Nucl. Soc.,* vol. 3, no. 2, p. 453, December, 1960 In this paper Hull describes a total-count method which uses sampling to decrease the tracer requirement

stream of a distillation tower, permitting detection of entrainment which represented as little as 0.021 per cent of the overhead stream.

Mass-flow Measurement. By combining measurements of fluid density and velocity, it is possible to measure mass-flow rates. Either or both of these variables can be measured with radioisotope techniques. Crompton[45] describes a mass-flow gage for measuring the weight flow of crushed and pressed sugar cane on a conveyor belt. Radiation absorption provides a density reading, and belt velocity establishes the flow information. The over-all error in the system was estimated by Crompton as ±5 per cent. Ohmart[46] describes fluid and slurry mass-flow gages using area meters or orifice meters for measuring flow, and radiation absorption gages for measuring the density of the fluid. It is likely that combinations of radiation devices with other instruments to provide mass-flow-rate information will provide an important measurement technique.

Furnace Flow. The British Iron and Steel Research Association used radon to determine gas-transmission time in a blast furnace. Eight mc of radon was injected into the air stream by means of explosives to give sharp pulses of the radioactive material. Samples of the radioactive gas were then taken every second at the top of the burden. By finding the amount of radioactivity in each sample, it was possible to estimate the mixing and flow velocity in the blast furnace.[47]

Underground Gas Flow. Radioactive A[41] has been used to determine the absence of flow leaks in gas-injection well-casing shoes,[48] and tritium has been used to determine gas-flow patterns and velocities in underground formations.[49]

Sand Flow or Drift. Radioactive sands have been used for following the flow and deposition properties of sand which drifts along beaches and blocks harbor inlets.[50] In these experiments a batch of sand is tagged with an isotope, such as Zn^{65}, and its movement after addition to the sand mass is followed under water or on the beach using portable detectors.

Ionization Anemometer. Since a radioactive source will produce ions in air and these ions can be transported to a measuring device, an ionization anemometer can be constructed consisting of an open ionization chamber and a radioactive source. Recombination of ions will provide a measure of transit time, or ions can be pulsed into the stream and the pulse transit time detected.[51] The British have constructed such an instrument which will measure air velocities in the range 1 to 300 fpm.

Diffusion Measurement

Radioisotopes are used for measuring extremely slow velocities such as are obtained in diffusion in metals, porous bodies, liquids, or gases.[52-54] For measuring self-diffusion, such as cobalt in cobalt or silver in silver, radioisotopes make possible measurements that cannot be obtained by any other technique. Radioactive cobalt has been plated on a cobalt bar which was then heated for the desired time and the surface counted for determination of depth of diffusion as a function of time and temperature.

Permeability

Tritiated water is being used to study the permeability of thin, flexible plastic sheets[28] by clamping a septum of the wrapping over a dish of tritiated water solidified by gelatin. Methane is passed over the upper surface of the septum and into a counter. The counter pulses resulting from the presence of diffused tritium in the methane are measured. The time required to obtain a steady reading is only 3 min in the case of thin thermoplastic materials and 10 min with hygroscopic films. Previously used gravimetric methods took at least 24 hr.

Underground Water Tracing

Radioisotopes have proved to be useful tools for studying flow patterns of underground water supplies or for tracing cross flow between oil wells.

The British have traced subterranean water seepage in the Nubian Desert with radioactive materials.[28]   They traced the source of the water in an effort to determine whether its origin was a natural spring or seepage from a lake.   The basin in which they were working was also tested with tracers to find any flaws that would allow the escape of water.   However, this seems to be an outstanding success which one should not expect to meet in all cases.   Halevy and Nir[55] also have reported on the successful uses of radioisotopes in desert water studies.

Water tagged with I^{131} and other radioisotopes has been extensively used in the United States[56,57] to study the subsurface flow of water used in secondary oil recovery to determine the path, velocity, and carrying strata of the water.

Interface Location

In oil-product pipelines, a number of different products are frequently handled in one line.   To minimize loss of different products at the terminus, Standard Oil Company of California injects an oil-soluble Sb^{124} tagged compound into the line at the time a change in product is made,[58] as illustrated in Fig. 7.   The pumping stations along the route have radiation detectors located ahead of the station which pick up the coming of the interface and automatically operate controls or forewarn the operators so that they may properly handle the new product when it arrives at their particular station.

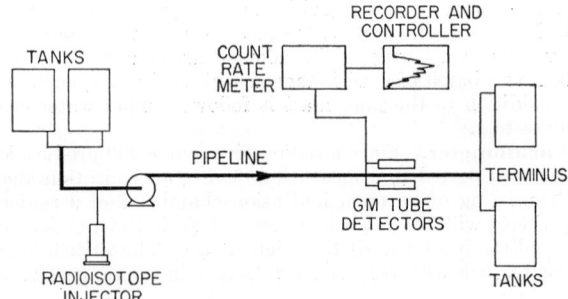

Fig. 7. Arrangement of equipment used in pipeline interface marking.

In an oil-field application of interface location, I^{131} has been added to the acidizing solution.   Since the acid is normally pumped down the center pipe and rises in the annulus around this pipe, attainment of a maximum counting rate by a detector lowered into the well indicates that the acid has passed the counter and risen around it in the annulus.   A counter is placed at the elevation of the formation which the acid is to attack.   The observed counting rate provides information for control of the acid at the desired level.   Figure 8 shows this technique.

To prevent dissolving away of the "roof" of a brine cover with subsequent damage to a pumphouse located on the surface, and to ensure the presence of an oil layer on top of the brine, an oil-soluble radioactive compound has been pumped down a brine well.   A "log" of the well with a radiation logging tool reveals the location of the tagged oil level and thus the elevation of the brine-oil interface and of the top surface of the brine.[59]

Chemical Composition

For additional information regarding determination of chemical composition see Sec. 8.

Using Alpha Rays.   Reiffel[18] describes an analytical instrument for determining a fraction of a microgram of beryllium on a filter paper.   The air sample is drawn through a filter paper on which any beryllium present is collected.   The filter with the sample is next placed in proximity to a sealed 2- to 5-curie Po^{210} alpha-ray source. The resulting α, (n,γ) reaction, which is unique to beryllium, produces neutrons and 4.5-mev gamma rays, either of which can be measured to determine the beryllium content.   Reiffel found that measurement of the 4.5-mev gammas using a scintillation detector provided the best sensitivity.

Using Beta or Gamma Backscattering.   Because beta backscattering is sensitive to the atomic number of the backscattering material, high-atomic-number ions in a low-atomic-number medium can be detected by a backscattering technique.[60] For example, it is possible to detect lead, silver, copper, nickel, or vanadium in water or in a solid, such as paraffin.   For the measuring device, a modified beta backscattering gage, such as is shown in Fig. 9, is satisfactory.   A Sr^{90} source can be used for the

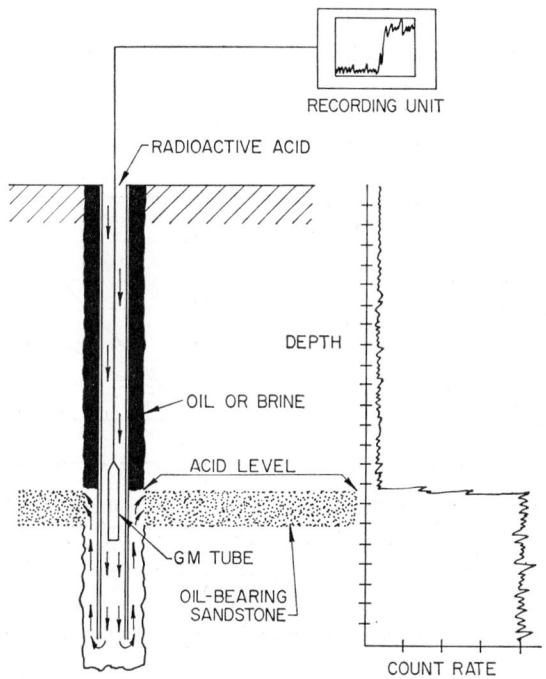

FIG. 8. Arrangement of equipment for acid-interface location with radioisotopes.

beta rays.   Figure 10 indicates the type of measurement obtainable by this technique. Since gamma-ray scattering is also a function of the atomic number of the scattering medium, gamma rays can be used for obtaining compositions.

Using Beta Absorption.   The hydrogen content of liquids is now routinely measured to ±0.03 per cent hydrogen in under 20 min by measuring the absorption of beta rays by the sample.[61]

Smith et al.[62] developed a gas analyzer which makes use of differences in the amount of ionization produced in various gases as a result of the passage of beta rays through them to differentiate between the gases.   The gage is used to measure the argon and ammonia content in the circulating gas stream of an ammonia synthesis plant.

Using X-ray Absorption.   The sulfur content of liquid and gaseous petroleum products in the laboratory and in plant streams is determined by measuring the absorption of X rays emitted by Fe^{55}.[63]   The technique permits determining the sulfur content in the concentration range of 0.05 to 3 $\pm$ 0.02 per cent sulfur.   Measurement of absorption of tritium bremsstrahlung has been used to determine lead and sulfur content.[64]

Using Neutrons. Fast neutrons, such as are obtained from a radium-beryllium source, are slowed down by collision with hydrogen atoms. These slowed-down neutrons can be readily detected. Since the number of fast neutrons slowed down is a function of the hydrogen-atom concentration of the scattering media, the technique can be used for measurement of hydrogen concentrations and thus for water content of soil, fertilizer, cement, etc.[65,66]

The neutron-capture cross sections of various elements differ by measurable amounts. Slow neutrons (such as can be obtained from a neutron source in a water- or paraffin-filled container) are captured preferentially by atoms with a high slow neutron-capture cross section, such as boron or lithium; thus a sample containing boron or lithium can be subjected to irradiation by slow neutrons and the concentration determined by measuring the number of slow neutrons penetrating the sample.[67] The presence of other elements with appreciable cross section can interfere with this technique.

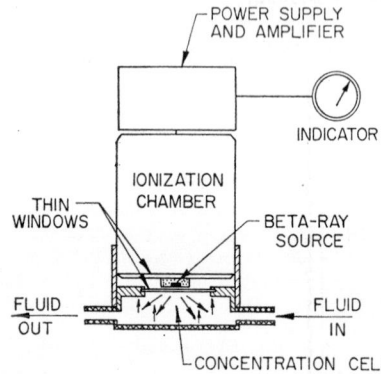

FIG. 9. Backscattering concentration gage. (*Source: Reproduced with permission from Chem. Eng. Progr., vol. 48, no. 12, p. 611, 1952.*)

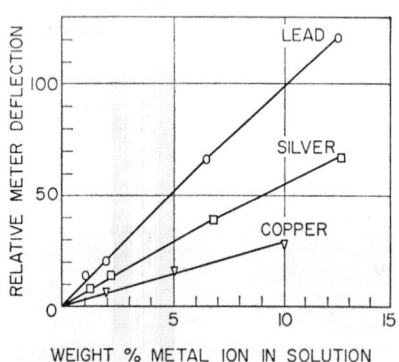

FIG. 10. Type of data obtainable with beta-ray backscattering concentration gage. (*Source: Reproduced with permission from Chem. Eng. Progr., vol. 48, no. 12, p. 611, 1952.*)

In activation analysis[68] a high sensitivity is obtained for elements with high neutron-absorption cross sections by counting the irradiated material for induced radioactivity. A measure of the half-life and energy of the emitted radiation permits identification of the radioisotopes that were formed by the neutron bombardment. From this knowledge, the composition can often be deduced.[69]

Schlumberger Well Surveying Company[70] has developed an instrument for laboratory or plant analysis for nitrogen which they call a "Nuclear Nitrometer." The instrument utilizes (1) a source of neutrons (polonium-beryllium or plutonium-beryllium); (2) a BF_3 detector to monitor the slow neutron flux, and (3) a NaI scintillation detector coupled to a gamma-ray spectrometer to measure the neutron-capture–gamma-ray emission spectra. Capture of neutrons by the nitrogen in the sample results in the production of 10.8-mev gamma rays in accordance with the equation

$$N^{14} + n \rightarrow N^{15} \text{ (unstable)} \rightarrow N^{15} + 10.8 \text{ mev } \gamma$$

This high-energy gamma is readily identified even in the presence of weaker gammas from other neutron-capture reactions (primarily the 2.23-mev hydrogen gamma) by means of the spectrometer. The "Nuclear Nitrometer" was developed for determining protein in organic matter such as corn feed and meal. Within the range 3 to 6 per cent nitrogen content (corresponding to about 20 to 40 per cent protein in corn derivatives) a relative standard deviation of 1.25 per cent of the nitrogen concentration is determined in a 15-min measuring cycle. Recalibration with a standard sample is done once every 4 to 8 hr. The instrument analyzes a 0.5 cu ft granular solid sample.

Determination of Particle Size. Three radioisotope techniques, which measure rate of sedimentation, have been developed for determining particle size. The first utilizes radiation absorption, the second activation, and the third backscattering.

Furnam[71] used radiation absorption to determine the size distribution of uranium oxide which was present at 5 to 30 per cent by weight in a particulate form in liquid NaK.* The low-energy gamma emitter Se^{75} was used as the source of radiation. The equipment automatically vertical scanned, using a drive screw to raise and lower the scintillation detector and the 15-mc Se^{75} source in its 4-in.-thick lead collimator. The collimating system decreased the contribution from scattered gammas and permitted looking at a $\frac{1}{8}$-in.-thick layer of particles in the liquid. The equipment provided the desired settling data.

Abraham et al.[72] activated the powder whose particle-size distribution was to be determined and then measured the radiation from a thin lamina of the suspension as an indication of the weight of the activated powder in the lamina. They compared their radioactivation method and the conventional fractional extraction and weighing method and demonstrated the reliability of the radioactivation technique.

It is also possible to determine the particle distribution of different radioactive components in a mixture of solids as they settle by using a scintillation detector and a pulse-height analyzer to determine the number of gammas emitted by each of the different components. For this type of determination to be feasible, it is necessary that the different radioisotopes have gammas that can be separated by the analyzer.

Connor et al.[73] utilized the beta-backscattering thickness-gaging technique to follow the increase in thickness with time of the deposit that is produced by settling. Where one is working with a water suspension the method is applicable only to compounds having an effective atomic number greater than 16, since one must be able to differentiate between the settled solid and the water. Connor used this method for measuring settling rates of thoria, urania, zirconia, and lithopone powders. It should also be possible to use the beta-backscattering method with the source and detector near the top surface of a suspension to follow the decrease in solid content of the upper layer as the solids settle out of it.

Radiochemical Exchange. Chleck and Ziegler[74,75] have developed a method for the continuous quantitative analysis of certain nonradioactive gases by using a radiochemical reactor bed in which a gas-exchange reaction takes place. The gas sample, containing as little as a few ppm of compounds such as SO_2, NH_3, etc., is passed first through the radiochemical reactor bed where the radioactive material in the bed exchanges with the nonradioactive material in the gas. The sample is then sent to a detector where the amount of radioactivity in the effluent gas is determined. This provides a measure of the concentration of the unknown material in the inlet stream. The effluent radioactive gas after measurement can be passed through an absorber to remove the radioactive constituent.

For example, to measure sulfur dioxide in air, the air is passed over radioactive chlorine dioxide providing an exchange of the radioactive Cl^{36} for the sulfur. By proper design of the reactor this exchange can be made 97 per cent effective. The betas from the Cl^{36} are measured, and the number of betas detected as a function of the sulfur content can be determined by calibration. The chemical reaction can be written

$$SO_2(gas) + Cl^{36}O_3(solid) \rightarrow Cl^{36}O_2(gas)\dagger$$

The 4×10^5 years half-life of Cl^{36} makes the useful shelf life of the reaction cell, in effect, unlimited.

The exchange reaction is affected by gases other than SO_2, for example, by acids

---

* NaK is a sodium-potassium eutectic mixture used as a reactor coolant.
† It is also possible that there is a reaction rather than an exchange, and that the proper equation is

$$Cl^{36}O_2(solid) + 2SO_2 \rightarrow \frac{1}{2}Cl_2^{36} + 2SO_2$$

such as HCl. Such acids if present, however, can be absorbed preferentially out of the sample stream before it reaches the reaction cell. By this exchange technique, it is possible to obtain a probable error of under ±5 per cent for concentrations of SO_2 as low as 0.03 ppm. Other gases, for which concentration determinations by the exchange principle appear possible, include ammonia, phosphorus, and certain arsenic compounds.

Radioisotope Release. Chleck and Hommel[76,77] have also found that certain gases in the parts-per-billion range will release a radioactive gas, such as Kr^{85}, from a three-dimensional organic crystalline "cage" called a "clathrate." The process involves oxidation of the clathrate. The released Kr^{85} can readily be detected by sensitive beta detectors. Since in many cases the amount of Kr^{85} released is proportional to the concentration of the material under analysis, the technique is useful for quantitative gas analysis for materials such as SO_2, O_3, and ClO_2.

Density and Specific Gravity

Since absorption of beta or gamma radiation is a function of the mass of the material between the source and detector, if a cell or pipe of fixed thickness containing a suspension or solution of unknown density is placed between source and detector, the response of the detector will be a function of this density. This type of system has been used for measuring powdered soaps, synthetic detergents, polymers in reactor outlets, dredge suspensions, and a variety of liquids.[16,78,79] The gamma method should be used with caution if the chemical composition of the material being measured varies independently of solids-content variations.

Using gamma-ray scattering, the density of undisturbed soil can be measured on the surface or in a hole to ±1 pound per cubic foot.[65]

Section 7 presents additional information on measurement of density and specific gravity.

REFERENCES

1. National Industrial Conference Board: Radioisotopes in Industry, Studies in Business Policy no. 93, 1959.
2. Catalog and Price List—Radioisotopes, Oak Ridge National Laboratory, Union Carbide Nuclear Co., Oak Ridge, Tenn., May, 1960.
3. Strominger, D., J. M. Hollander, and G. T. Seaborg: Table of Isotopes, *Rev. Mod. Phys.*, vol. 30, no. 2, p. 591, 1958.
4. Zumwalt, L. R.: The Best Performance from Beta Gages, *Nucleonics*, vol. 12, no. 1, p. 55, January, 1954.
5. Part 20, Title 10 of Code of Federal Regulations, Standards for Protection against Radiation, Jan. 1, 1961, and *Natl. Bur. Standards (U.S.) Handbooks* 59 and 69.
6. "Handbook of Federal Regulations Applying to Transportation of Radioactive Materials," Government Printing Office, Washington, D.C.
7. Rockwell, T., III: "Reactor Shielding Design Manual," T.I.D. 7004 Department of Commerce, Washington, D.C., March, 1956; and D. Van Nostrand Company, Inc., Princeton, N.J. Also Kinsman, S.: "Radiological Health Handbook," PB 121784, Office of Technical Services.
8. Gimenez, G., and J. L. Labeyne: *J. Phys., Radium*, vol. 12, 64A, 1951.
9. Schumacher, B. W.: "Beta Ray Gages for Gas Density Measurements," Ontario Research Foundation, Toronto, 1960.
10. Trost, A.: Continuous Control and Measurement of Levels and Densities by Gamma Rays, *Proc. Second Intern. Conf. Peaceful Uses At. Energy*, vol. 19, p. 320, U.N. Press, N.Y 1958.
11. Thornton, D. P.: Atomic Energy Applied to Measurement of Catalyst Level in Cracking Units, *Petrol. Process.*, vol. 5, p. 941, September, 1950.
12. Werstler, C. E., R. J. Niederstadt, and H. A. Lutz: Inside a Coke Drum, *Oil Gas J.*, Aug. 9, 1956.
13. Spooner, R. B.: Scanning Detector Makes Blast-furnace Gamma Gage, *Nucleonics*, vol. 19, no. 5, p. 56, May, 1961.
14. Barnartt S., and K. H. Sun: Using Neutrons for Remote Liquid-level Gaging, *Nucleonics*, vol. 13, no. 5, p. 47, May, 1955.
15. Dixon, W. R.: Beta-ray Thickness Gages, *Natl. Res. Council Can., NRC Bull.* 2358, Ottawa (Division of Physics *Rept.* PR-90), March, 1951.
16. Crompton, C. E.: The Versatility of Radiation Applications Involving Penetration or Reflection, *Proc. First Inter. Conf. Peaceful Uses At. Energy*, vol. 19, p. 124, 1955.
17. Leveque, P., et al.: Some New Applications of Radioelements in France, *Proc. Second Intern. Conf. Peaceful Uses At. Energy*, vol. 19, p. 34, 1958.
18. Reiffel, L.: Measurement and Control Methods Using Radiation, *Proc. Second Intern. Conf. Peaceful Uses At. Energy*, vol. 19, p. 278, 1958.

19. Putman, J. L., S. Jefferson, and J. F. Cameron: Tube Wall Thickness Gage with Selection of Backscattered Gamma Radiations, *AERE Rept.* I/R 1369, Harwell, Berks, England 1954.
20. Kerry, J. P., and E. W. Pulsford: The Circuits of the Portable Gamma Backscatter Thickness Gage, *AERE Rept.* EL/R 1560.
21. Tolan, J. H., and W. T. McIntosh: Investigation of Applications of Compton Backscatter, *NYO* 2779, July, 1960.
22. Buchdahl, R., and M. Polglasse: Measuring Coverage and Film Thickness of Printing Ink and Paint Films, *Ind. Eng. Chem., Anal. Ed.,* vol. 18, no. 2, p. 115, February, 1946.
23. Miller, E. P., and A. R. Cohee: The Use of Radioactive Materials in Measuring the Thickness of Enamel Coatings, *Phys. Rev.,* vol. 59, no. 5, p. 468, 1941.
24. Hull, D. E., and B. A. Fries: Radioisotopes in Petroleum Refining, Research, and Analysis, *Proc. First Intern. Conf. Peaceful Uses At. Energy,* vol. 15, p. 199, 1955.
25. Sigworth, H. W., and B. A. Fries: Measurement of Films and Deposits, U.S. Patent 2,660,678.
26. Allia, D. J.: Revolution Counting Device with Radioactive Control, U.S. Patent 2,566,868, Sept. 4, 1951.
27. Scott, D. B.: Detection of Scrapers in Pipe Lines, *Nucleonics,* vol. 9, no. 3, p. 68, September, 1951.
28. Oxford International Conference on Isotope Techniques, reported in Isotopes in Industry, *Atomics,* vol. 2, no. 9, p. 245, September, 1951.
29. Pinotti, P. L., E. D. Hull, and E. J. McLaughlin: Application of Radioactive Tracers to Improvement of Automotive Fuels, Lubricants, and Engines, *SAE J.,* vol. 57, no. 6, p. 52, June, 1949.
30. Anon.: Floor Waxes—Six Brands Tested Using Radioisotopes, *Consumers' Res. Bull.,* March, 1950.
31. Determination of Quality of Mixing of Dry Granular Solids, *Atomics,* vol. 2, no. 10, p. 285, October, 1951.
32. Singer, E., D. G. Todd, and V. P. Guinn: Catalyst Mixing Patterns in Commercial Catalytic Cracking Units, *Ind. Eng. Chem.,* vol. 49, no. 1, p. 11, January, 1957.
33. Hull, D. E., B. A. Fries, J. G. Tewksbury, and G. H. Keirns: Mixing in Surge Tanks and Stills, *Nucleonics,* vol. 14, no. 5, p. 51, May, 1956.
34. King, W. H., Jr.: Radioisotopes in Petroleum Refining, *Ind. Eng. Chem.,* vol. 50, no. 2, p. 201, February, 1958.
35. Beerbower, A., E. O. Forster, J. J. Kolfenbach, and H. G. Vesterdal: Evaluation of Mixing Efficiency of Processing Equipment, *Ind. Eng. Chem.,* vol. 49, no. 7, p. 1075, July, 1957.
36. Archibald, R. S.: Radioactive Tracers in Flow Tests, *J. Boston Soc. Civil Engrs.,* vol. 37, pp. 49–116, 1950.
37. Bogdonava, N. G., et al.: The Uses of Radioactive Isotopes in Metallurgical Research, *Proc. Second Intern. Conf. Peaceful Uses At. Energy,* vol. 19, p. 180, 1958.
38. Handlos, A. E., R. W. Kunstman, and D. O. Schissler: Gas Mixing Characteristics of a Fluid Bed Regenerator, *Ind. Eng. Chem.,* vol. 49, no. 1, p. 25, January, 1957, and private communication.
39. Gueron, J.: Some Industrial Applications of Radioelements Made at the Commissariat de L'Energie Atomique, includes Determination of Pipe Leakage and of Liquid Viscosity, *Nucleonics,* vol. 9, no. 5, p. 53, November, 1951.
40. Hull, D. E.: Using Tracers in Refinery Control, *Nucleonics,* vol. 13, no. 4, p. 18, April, 1955.
41. Hull, D. E., and R. R. Bowles: Measuring Catalyst Flow Rates in Cat Crackers, *Oil Gas J.,* vol. 51, no. 46, p. 295, Mar. 23, 1953.
42. Newacheck, R., et al.: Isotope Milker Supplies Ba[137] from Parent Cs[137], *Nucleonics,* vol. 15, no. 5 p. 122, May, 1957.
43. Karrer, S., D. B. Cowie, and P. L. Betz: Use of Radioactive Tracer in Measuring Condenser Water Flow, *Power Plant Eng.,* vol. 50, no. 12, p. 118, December, 1946.
44. Hull, D. E.: Total-count Technique in the Refinery, *Ind. Eng. Chem.,* vol. 50, no. 2, p. 199, February, 1958. See also Hull, D. E., and M. Macomber: Flow Measurements by the Total-count Method, *Proc. Second Intern. Conf. Peaceful Uses At. Energy,* vol. 19, p. 324, 1958.
45. See Ref. 16.
46. Mass Flow Meter, Ohmart Company, Cincinnati, Ohio, 1960.
47. *Atomics,* vol. 2, no. 2, p. 41, February, 1951.
48. Kohl, J., R. L. Newacheck, and E. E. Anderson: Locating Casing Shoe Leaks with Radioactive Argon, T. P. 4182, *Petrol. Trans. AIME,* vol. 204, 1955.
49. Welge, H. G.: New Use for Radioactive Tracers, Super Sleuths Trace Flow of Injected Gas, *Oil Gas J.,* vol. 54, no. 17, p. 77, Aug. 29, 1955.
50. Inose, S., M. Kato, S. Sato, and N. Shiraishi: The Field Experiment of Littoral Drift Using Radioactive Glass Sand, *Proc. First Intern. Conf. Peaceful Uses At. Energy,* vol. 15, p. 211, 1955.
51. Cooley, W. C.: Determination of Air Velocity by Ion Transit-time Measurements, *Rev. Sci. Instr.,* vol. 23, no. 4, p. 151, April, 1952.
52. Kurdiumov, G. V.: Investigations of Diffusion and Atomic Interaction in Alloys with the Aid of Radioactive Isotopes, *Proc. First Intern. Conf. Peaceful Uses At. Energy,* vol. 15, p. 81, 1955.
53. Kuczynski, G. C.: New Methods of Measuring Diffusion Coefficient in Solids with Radioactive Tracers, *J. Appl. Phys.,* vol. 19, no. 3, p. 308, March, 1948.
54. Gemant, A.: Measurement by Radioactive Tracers of Diffusion in Liquids, *J. Appl. Phys.,* vol. 19, no. 12, p. 1160, December, 1948.
55. Halevy, E., and A. Nir: "Use of Radioisotopes in Studies of Groundwater Flow," NP-8745 Israel, Water Planning for Israel Ltd., Tel Aviv, April, 1960.
56. Watkins, J. W., and H. N. Dunning: Radioactive Isotopes in Petroleum-Production Research, *Proc. First Intern. Conf. Peaceful Uses At. Energy,* vol. 15, p. 32, 1955.
57. Watkins, J. W., and E. S. Mardock: Use of Radioactive Iodine as a Tracer in Water-Flooding Operations, *J. Petrol. Technol.,* vol. 6, p. 117, September, 1954.
58. Hull, D. E., and J. W. Kent: Radioactive Tracers to Mark Interfaces and Measure Intermixing in Pipelines, *Ind. Eng. Chem.,* vol. 44, no. 11, p. 2745, November, 1952.

59. Kohl, J.: Tracers Find Brine Well Oil Pad, *Chem. Eng.*, vol. 63, no. 9, p. 222, September, 1956.
60. Kohl, J.: Radioisotopes in Process Instrumentation, *Chem. Eng. Progr.*, vol. 48, no. 12, p. 611, December, 1952.
61. Smith, V. N., and J. W. Otvos: Hydrogen Determination and Liquid Analysis with a Beta-particle Absorption Apparatus, *Anal. Chem.*, vol. 26, no. 2, p. 359, February, 1954.
62. Smith, V. N., J. W. Otvos, and D. J. Pompeo: Radiological Gas Analyzer for Ammonia Plant Stream, Paper 56-11-1 Presented at the 11th ISA Conference, New York, September, 1956.
63. Hughes, H. K., and J. W. Wilczewski: K-capture Spectroscopy: Iron-55 X-ray Absorption Determination of Sulfur in Hydrocarbons, *Anal. Chem.*, vol. 26, no. 12, p. 1889, December, 1954.
64. Kannuna, M. M.: Investigations of Tritium Bremsstrahlung as a Means of Determining Sulphur and TEL in Hydrocarbons, *J. Inst. Petrol.*, vol. 43, no. 403, p. 198, July, 1957.
65. Belcher, D. J., T. R. Cuykendall, and H. W. Sack: The Measurements of Soil Moisture and Density by Neutron and Gamma Ray Scattering, *CAA Rept.* 127, October, 1950.
66. See bulletin on Measurement and Control of Per cent Moisture from Industrial Div., Nuclear-Chicago Corp., Des Plaines, Ill., 1961.
67. DeFord, D. D., and R. S. Branam: A Study of Neutron Absorptiometry and Its Applications to the Determination of Boron, *Anal. Chem.*, vol. 30, no. 11, p. 1765, November, 1958.
68. Meinke, W. W.; Nucleonics, *Anal. Chem.*, vol. 28, no. 4, p. 736, April, 1956.
69. Taylor, T. I., R. H. Anderson, and W. W. Havens, Jr.: Chemical Analysis by Neutron Spectroscopy, *Science*, vol. 114, no. 2962, p. 341, Oct. 5, 1951.
70. Anon.: *Chem. Eng. News*, vol. 37, p. 82, 1959; and see *Special Product Bulletin*, "Nuclear Nitrometer" of Schlumberger Well Surveying Corp., Ridgefield Inst. Group, Ridgefield, Conn., 1960.
71. Furnam, S. E.: Gamma Absorptiometric Sedimentation Analysis in Liquid Metal Solid Oxide Systems, KAPL-1648, Jan. 8, 1957, USAEC.
72. Abraham, B. M., H. E. Flotow, and R. D. Carlson: Particle Size Determination by Radioactivations, *Anal. Chem.*, vol. 29, no. 7, p. 1058, July, 1957.
73. Connor, P., et al.: Determination of Particle Size by Beta Backscattering, *J. Appl. Chem.*, vol. 9, no. 10, p. 525, October, 1959.
74. Chleck, D. J., and C. A. Ziegler: A SO₂ Monitor for Air Pollution and Process Control, presented at the Symposium of Applications of Radioactivity in Petroleum Exploration and Production, Tracerlab, Inc., Houston, Tex., 1957; and *Chem. Eng. News*, vol. 35, no. 50, p. 57, Dec. 16, 1957.
75. Chleck, D. J., and C. A. Ziegler: Radio Assay of Non-active Gases, *Chem. Process Eng.*, vol. 40, no. 8, p. 287, August, 1959.
76. Chleck, D. J., and C. A. Ziegler: The Preparation and Some Properties of Radioactive-Quinol Krypton Clathrate Compounds, *Intern. J. Appl. Radiation and Isotopes*, vol. 7, no. 2, p. 141, December, 1959.
77. Hommel, C. O., et al.: Ozone Analyzer Uses Radioactive Clathrate, *Nucleonics*, vol. 19, no. 5, p. 94, May, 1961.
78. Jordan, G. G., V. B. Brodsky, and B. S. Sotskov: Application of Radioisotopes to Control Technological Processes, *Proc. First Intern. Conf. Peaceful Uses At. Energy*, vol. 15, p. 135, 1955.
79. Garrison, W. E., and C. A. Nagel: A Density Meter to Control Sludge Pumping, *Sewage Ind. Wastes*, vol. 3, no. 11, p. 1327, November, 1959.

General References

1. Shumilovskii, N. N., et al.: Radioactive Methods of Control and Regulation of Industrial Processes, AEC-tr-4139, USAEC, Office of Technical Services, Department of Commerce, Washington. D.C., 1959.
2. McKinney, A. H.: Radiation Techniques for Process Measurements, *Chem. Eng. Progr.*, vol. 56, no. 9, p. 37, September, 1960.
3. The Use of Isotopes: Industrial Use, *Proc. Second Intern. Conf. Peaceful Uses At. Energy*, vol. 19, 1958.
4. Kohl, J., R. D. Zentner, and H. R. Lukens: "Radioisotope Applications Engineering," D. Van Nostrand Company, Inc., Princeton, N.J., 1961.

PHOTOMETRIC VARIABLES

By Richard S. Hunter*

This subsection presents the basic optical principles and materials, an understanding of which is important to the instrumentation engineer in the design and application of photometric measuring devices. Photometric variables include light, reflectance, color, gloss, transmittance, and absorptance. Also covered at the end of this subsection is a summary of the various types of photometric instruments and their applications. For further and more specific details, the reader is referred particularly to pages 7-111 through 7-149 of the "Process Instruments and Controls Handbook," McGraw-Hill Book Company, Inc., New York, 1957, and to the other texts listed at the end of this subsection.

Although optics is concerned with radiant energy of the whole electromagnetic spectrum, this section deals chiefly with energy in and adjacent to the narrow band of the spectrum to which the eye is sensitive. Although this band is a small part of the total spectrum, it coincides with the energy center of the solar spectrum.

The sun approximates a blackbody at 5000°K, the spectral curve for which is shown in Fig. 1. About 55 per cent of the sun's energy reaching the earth lies in the 380- to 750-mμ visible range, 37 per cent in the infrared, and 8 per cent in the ultraviolet. Some of the applications, materials, and instruments used in these regions of the spectrum are given in Table 1.

From the instrument engineer's point of view, light is ideal for energy transmission. Except for small effects of little practical importance, one may say that light travels at infinite speed in straight lines until intercepted by objects; it has no mass and is therefore unaffected by gravity and exerts no mechanical force on the objects it strikes; further, when one beam of light crosses another, they have no effect on each other.

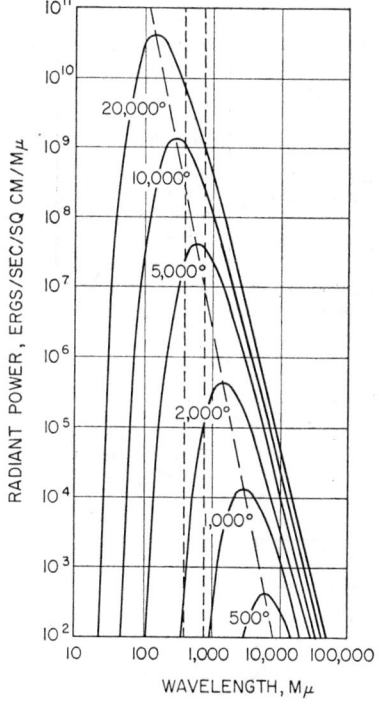

FIG. 1. Blackbody radiation curves for various absolute temperatures plotted on a logarithmic scale. Dotted lines define visible region.

BASIC OPTICAL PROCESSES

The basic optical processes of importance to the instrumentation engineer include (1) emission, (2) reflection, (3) refraction, (4) diffusion, (5) scattering, (6) absorption, and (7) polarization.

* President, Hunter Associates Laboratory, Inc., McLean, Va.

Table 1. Spectral Regions Used for Optical Instrumentation and Applications, Optical Materials, and Typical Instruments

Spectral region	Ultraviolet	Visible	Near infrared	Far infrared
Wavelength range, microns	0.2–0.4	0.4–0.7	0.7–1.5	1.5–20 (and beyond)
Instrument applications	Chemical analyses by emission and absorption spectroscopy and spectrophotometry	Appearance measurements of color, gloss, etc. Chemical colorimetry, turbidimetry, spectroscopy, etc.	Chemical colorimetry, spectroscopy, and spectrophotometry Camouflage	Chemical absorption, spectroscopy and spectrophotometry (especially for complex organic compounds)
Instrument sources	Fluoroscopy Mercury, hydrogen, and other arcs	Incandescent lamp Mercury and other arcs Daylight	Incandescent lamp	Globar and other incandescent sources
Light-path optics	Quartz lenses and prisms Aluminum and other metallized mirrors (glass from 0.3 to 0.4)	Glass, quartz, and clear plastic lenses and prisms Aluminum and other mirrors	Same as for visible	Rock salt, calcium fluoride, and other synthetic crystal lenses and prisms Aluminum and gold mirrors
Wavelength-isolation devices	Quartz prisms Aluminized gratings Liquid and glass filters	Glass and quartz prisms Gratings Glass, gelatin, and interference filters	Same as for visible	Rock salt and other synthetic crystal prisms Aluminized gratings
Light receptors used	Photocells, chiefly S-4 and S-5, electron-emission types	Human eye Barrier-layer photocells All types of emission photocells	S-1 type emission photocells Photoconductive cells Barrier-layer photocells	Thermopiles Bolometers Golay detector and other selective pneumatic detectors
Typical instruments	Beckman spectrophotometer, model DU, and Cary spectrophotometer	G-E, Cary, and Beckman spectrophotometers Most chemical colorimeters, reflectometers, turbidimeters, etc.	G-E, Cary and Beckman spectrophotometers Infrared spectrophotometers	Perkin-Elmer and Baird spectrophotometers; and automatic industrial gas analyzers

Emission

Objects that emit light are of five types:

1. *Incandescent light sources* are solids so hot that they radiate continuous spectra. As temperature increases, the total energy output increases and the wavelength of maximum energy becomes shorter (bluer). This is shown in Fig. 1.

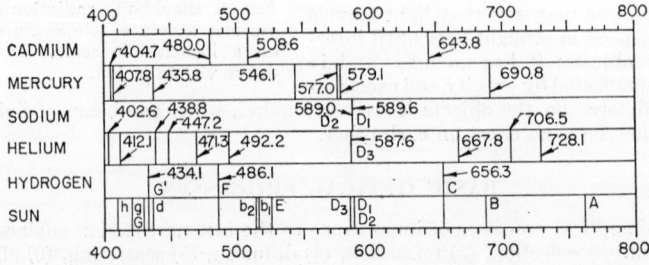

FIG. 2. Spectra of certain gas discharges. The principal Fraunhofer lines in the solar spectrum are shown for comparison.

2. *Gas discharges* which emit characteristic spectral lines when electrically excited. The line spectra of several elements are given in Fig. 2.

3. *Sparks and arcs* are usually combinations of incandescent solid objects emitting continuous spectra and surrounding gases or vapors emitting line spectra.

4. *Fluorescent materials* emit light in spectral regions characteristic of each when each is excited by energy to which it is responsive. In every case, the exciting energy is of wavelengths shorter than the emitted energy.

5. *Luminescent sources* may be phosphorescent, chemoluminescent, radioluminescent, electroluminescent, and thermoluminescent.

Reflection

When a ray of light in air strikes the smooth surface of an object, part is reflected. The reflected ray will leave the surface at an angle equal and opposite to the angle of incidence (see Fig. 3a). The fraction R of the incident ray which is reflected in

(a) *Reflection* causes shiny appearance of smooth surfaces.

$$I = \frac{I_0}{2}\left[\frac{\sin^2(i-r)}{\sin^2(i+r)} + \frac{\tan^2(i-r)}{\tan^2(i+r)}\right]$$

Refraction causes bending of rays entering objects.

$$n = \frac{\sin i}{\sin r} \qquad I_r = I_0 - I$$

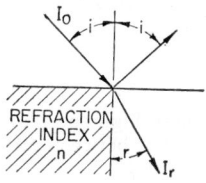

(b) *Diffusion* results from granulation within objects. R_d increases as granule size diminishes to $\lambda/4$.

$$I = I_0 \frac{\omega \cos v}{\pi} R_d(i,v)$$

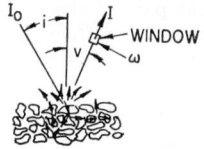

(c) *Absorption* causes objects to be dark and—when it varies with wavelength—to have color.

$$I = I_0 \, 10^{-\beta b} \qquad \text{(at each wavelength)}$$

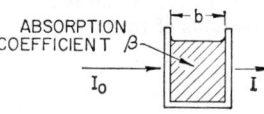

(d) *Combinations* of the above phenomena are responsible for most object appearance.

Symbols used above:

I_0 = incident flux
I = emergent flux
I_r = refracted flux
i = angle of incidence
v = angle of view

r = angle of refraction
ω = solid field angle (of view)
b = thickness
β = absorption coefficient
λ = wavelength

FIG. 3. Physical phenomena responsible for object appearance.

this manner is computed from the Fresnel laws of reflection which show that it increases with:

1. Angle of incidence i (always measured from the perpendicular).
2. Refractive index n.
3. Extinction coefficient k (nearly zero for nonmetal).

The Fresnel laws also involve state of polarization of the incident ray. The complete Fresnel formulas may be found in texts on physical optics. The simplified form applicable only to nonmetals is

$$R = \frac{1}{2} \left[\frac{\sin^2 (i - r)}{\sin^2 (i + r)} + \frac{\tan^2 (i - r)}{\tan^2 (i + r)} \right] \tag{1}$$

where r is the angle of refraction whose sine is $n \sin i$.

Reflection also takes place when a ray of light in an object encounters air. If the object has parallel faces, the *fraction* of the ray reflected back into the medium on leaving is the same as that lost by reflection on entering.

Figure 4 gives the fraction of light reflected by surfaces of water, black glass or black enamel-type paints, gold, and silver. These curves show the manner in which reflectance R increases rapidly as i approaches 90°. It will be noted that the curves for metals are much higher than those for nonmetals. Metals are distinguished from nonmetals by finite values of k and by surface-reflection factors generally much higher than those of nonmetals. The reflection factor of a polished metal, when $i = 0$, is

$$R = \frac{(n - 1)^2 + n^2 k^2}{(n + 1)^2 + n^2 k^2} \tag{2}$$

These high specular reflectances are responsible for the characteristic appearance difference between metals and nonmetals.

Refraction and Refractive Index

That part of the ray of light shown in Fig. 3a which is not reflected enters the object and is refracted. Its direction of travel in the object is represented by angle r, which is related to the direction in air i and the refractive index n by Snell's law

$$n = \frac{\sin i}{\sin r} \tag{3}$$

If the entrant ray is not in air but travels directly from one object into another, the effective refractive index of the interface for both reflectance computations and directions of travel is related to the separate refractive indices by

$$n_{12} = \frac{n_2}{n_1} = \frac{\sin i}{\sin r} \tag{4}$$

Diffusion

Diffusion of light by objects results in part from surface roughness but in the main from granular structure within the object, as indicated in Fig. 3b. When surfaces have roughnesses of the order of 4 μin. or greater, light beams which intercept them are diffused in a manner and to an extent which is determined by the surface structure. However, diffusion results chiefly from refractions and reflections of light rays at the interfaces of media of different refractive index.

These interfaces may be those of fibers and granules in air, such as are found in papers, textiles, powders, and films of wall paint, or they may be pigment granules of high refractive index such as TiO_2, and ZnS in vehicles of lower refractive index, such as linseed oil, glass, and plastic resins. Porcelain enamel, glossy paint, and plastics owe their diffusion to differential refractive indices.

Much of the light which is diffused within an object returns to the surface and leaves in all directions. The reflectance in these directions is described by reference to ideal diffusors. Such an *ideal* diffusor is one for which the light flux reflected from a unit area of surface through a window of solid angle ω in any direction v would be

$$I\omega = I_0 \frac{\omega \cos v}{\pi} R \tag{5}$$

Given an actual, rather than an ideal diffusor, the expression becomes

$$I\omega = I_0 \frac{\omega \cos v}{\pi} R_d(i,v) \tag{6}$$

where v is the angle of view and R_d is the directional reflectance for the surface for the specific i and v directions.

Scattering

When a beam of light traverses a cloud of suspended matter, light is scattered from the beam, by both reflection and diffraction. The intensity, angular distribution, and polarization of scattered light are determined by the size, shape, optical constants, and other properties of the scattering material and surrounding fluid. Thus, from the knowledge of the light scattering by a given system, information can be derived concerning concentration, size, and shape of particles, or molecular weight of large molecules.

In 1871, Lord Rayleigh made the first quantitative study of the laws of scattering by small particles (for particle diameters less than one-tenth the wavelength of light), and in 1908, G. Mie published a more generalized theory. A general physical law for Rayleigh scattering predicts the intensity of scattered light by particles for indices of refraction different from that of the surrounding medium.

One important fact concerning this type of scattering is that the intensity of the scattered light is proportional to the inverse fourth power of the wavelength. From this law, Rayleigh showed that the blue color of the sky is caused by molecular scattering and that light in the region of the sunset is reddish because of the greater loss of blue light, or shorter wavelengths, through scattering.

Measurements with nephelometers and turbidimeters are used to specify the degree of cloudiness or the concentration of suspended material in a fluid.

Absorption

As a ray of light passes through a homogeneous object (see Fig. 3c), it ordinarily loses energy as a consequence of absorption. The capacity of an object medium to absorb light is designated by its absorption coefficient, which when the object is colored varies with wavelength in the visible spectrum. For any wavelength, the same fraction of flux will be absorbed for each unit of distance traveled.

Thus, if a beam of light is passing through amber-colored oil, so that its intensity at 450 mμ in the blue falls to 50 per cent after 1 cm, it will fall to 25 per cent at 2 cm, 12.5 per cent at 3 cm, etc. If T_i (for internal transmittance) is the ratio of emergent to entrant light intensity and b is thickness, then β, the absorption coefficient, is defined by Bouguer's law

$$I = I_0 10^{-\beta b} \tag{7}$$

If the absorbing material is a colorant such as a dye in solution then Beer's (Lambert-Beer) law usually applies

$$I = I_0 10^{-c\beta b} \tag{8}$$

where β is the absorption coefficient of the dye for unit concentration and c its concentration in the solution. Some solutions do not obey Beer's law for all ranges of concentration.

The so-called *failure* of the Lambert-Beer law is described, in various books, particularly, "Analytical Absorption Spectroscopy" (M. G. Mellon, John Wiley & Sons, Inc., New York, 1950). The observed failure to obey the Lambert-Beer law is not a failure in a strict sense. It occurs usually because the composition of the mixture under investigation varies, or in the case of suspensions at high concentrations, where the suspended particles overlap and contribute less absorption than expected on the basis of the sum of their projected areas. Thus the gas, nitrogen peroxide, an equilibrium mixture of NO_2 and N_2O_4, is composed mostly of N_2O_4 at high pressure or low temperature and is nearly colorless; while at low pressure or high temperature, NO_2 is the prevalent form, providing a characteristic yellow to orangeish-red color.

Apparent failure of the Lambert-Beer law also occurs because the wavelengths of absorption are shifted by molecular interactions. For example, an absorption peak for water occurs at 1.953 μ, while the water-absorption peak is shifted to 1.976 μ in 5.5 N HCl. Other types of apparent failure of the law may be ascribed to measurement technique, as discussed in Mellon's book.

Polarization

Polarization of light rays occurs by double refraction of rays entering certain crystals and by surface reflection at angles intermediate between 0 and 90°. Polaroid is a doubly refracting material in which one ray is almost completely absorbed by dye.

The extent to which specularly reflected rays are polarized at different angles of reflection by a surface of $n = 1.5$ is given by the first equation in Fig. 3a. The separate polarized components correspond to the $\sin^2$ and $\tan^2$ components of this equation. It will be seen (in Fig. 4) and can be computed by the aforementioned equation that, at about 60°, the light reflected by such a surface is completely polarized.

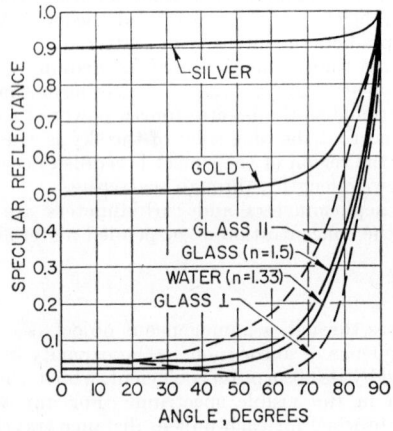

FIG. 4. Specular reflectances by Fresnel equations for smooth surfaces of silver, gold, glass, and water. Separate polarized components for glass are shown by dotted lines.

Combinations of Factors

With most materials met in everyday life, one encounters a combination (see Fig. 3d) rather than simply one of the foregoing optical processes. Diffusion is, of course, a combination of reflection and refraction with some scattering. It is normally accompanied by absorption, which, when spectrally selective, produces color.

Although combinations of structure and the foregoing optical processes are responsible for appearance properties of objects, the combinations are usually so involved that it is impractical to describe quantitatively the processes occurring in most objects.

OPTICAL COMPONENTS AND MATERIALS

In optical instrumentation, general use is made of (1) light sources, (2) mirrors, (3) prisms and gratings, (4) lenses, and (5) spectral filters.

Light Sources

Several forms of light sources are used in instruments.

Incandescent lamps are used to obtain visible and near infrared radiation. Of the great variety of electric lamps available on the market, those with clear envelopes, concentrated filament areas, and highest ratios of light to power are generally used for instrument applications (see Table 2).

Approximate relationships of some of the operating characteristics of one incandescent lamp are shown in Fig. 5. Note that the life of a lamp can be doubled by

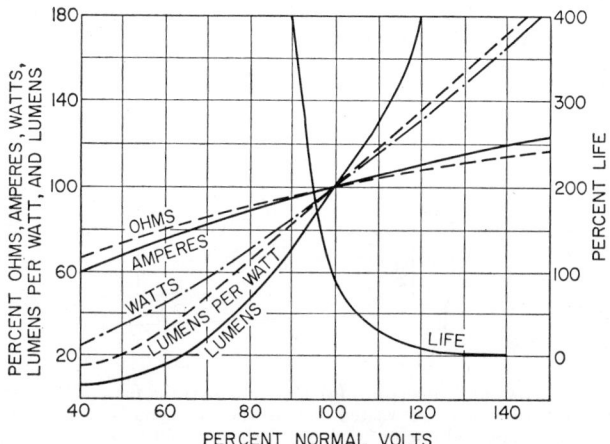

FIG. 5. Effect of voltage variation on operating characteristics of one type of incandescent lamp. (*Source: "IES Lighting Handbook,"* 2d ed., 1952.)

a mere drop of about 5 per cent in voltage. This voltage drop will, however, decrease light output almost 20 per cent and will, of course, make the color of the lamp redder, as was shown in Fig. 1.

Table 2. Typical Lamps Used in Instrument Applications

Designation	Volts	Amp	Approximate lumens	Rated life, hr
No. 1021, hand lantern	4.5	1.25	60	75
No. 1209, auto spotlight	6.1	4.3	400	125
6.5A/T8, sound reproducer	5.0	6.5	800	50
500W, T-20, medium prefocus projection	120	4.2	13,200	50

Mercury lamps are used as sources of light when the specific wavelength characteristics of the mercury spectrum are suitable. The mercury wavelengths of the visible are shown in Fig. 2. Mercury arcs are very efficient sources of blue and ultraviolet radiation and are stable after a warm-up period.

A *Globar* is an element which can be heated to reddish incandescence in air. It is used as a source of energy in the far-infrared region of the spectrum where glass absorbs and incandescent lamps are therefore unsatisfactory.

Mirrors

Mirrors make use of reflection to turn and direct beams of light. They have smooth surfaces and are usually of metals which have high reflection factors. Some of these and their reflection factors are:

Type of mirror	*Rs, per cent*
Silver (slightly yellowish in color, must be protected by glass or lacquer to avoid tarnish)...	90–95
Evaporated aluminum (cleans better when coated with evaporated quartz)...............	85–90
Rhodium (corrosion resistant, and very hard)...	70–75
Chromium (also hard)...	60–65
Stainless steel..	50–60
Semitransparent evaporated (usually chromium).......................................	10–50
Black glass at 45° (see Fig. 4)...	5–6

Common mirror shapes and uses are suggested in Fig. 6.

Aluminum mirrors are used in place of lenses in both the short-wave ultraviolet and long-wave infrared where glass and other lens materials absorb.

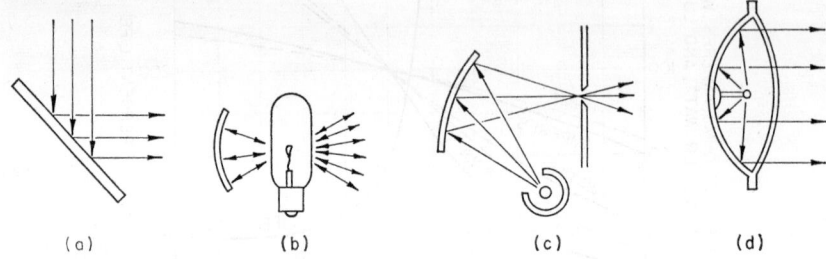

(a)   (b)   (c)   (d)

FIG. 6. Types of mirrors and reflectors: (*a*) Flat mirror turns rays of light striking it; (*b*) spherical mirror returns otherwise wasted light to the small source of a projection lamp to make it brighter; (*c*) spherical mirror condenses light on a slit in an infrared spectrophotometer where a glass lens would absorb; (*d*) parabolic reflector in sealed-beam spotlight directs rays spreading from a small filament into a strong beam of nearly parallel rays.

Prisms

Prisms have flat surfaces and make use of refraction or reflection to turn beams of light in the manners shown in Fig. 7. In many optical instruments, prisms are used in place of mirrors to turn beams of light. Because different wavelengths are refracted differently, prisms are used (as are also diffraction gratings) to separate the wavelengths of the spectrum.

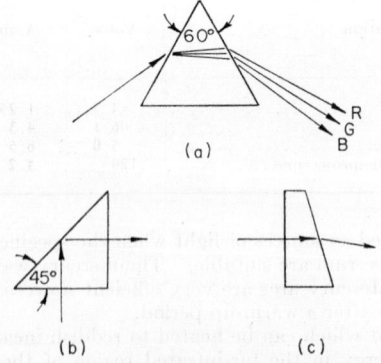

FIG. 7. Three simple prisms: (*a*) 60° prism used to disperse spectrum; (*b*) 45° right-angle prism used to turn beam 90° (*c*) wedge used to divert beam by small angle.

Lenses

Lenses have curved surfaces and make use of refraction to direct and concentrate (or spread) beams of light. They are of glass, clear plastic, or quartz, and have one or both surfaces curved.

Principal lens types are identified in Fig. 8. Lenses with concave surfaces (Fig. 8*f*) spread rays of light and are called negative. Lenses with convex surfaces converge or condense rays of light and are called positive.

Positive lenses with convex surfaces may be planoconvex (Fig. 8*a*), double-convex (Fig. 8*c*), or fluted (Fig. 8*g*). Because different wavelengths are refracted by different amounts, it takes a two-component corrected lens (Fig. 8*b*) to form images with minimum chromatic aberration. Smooth lens surfaces are formed by grinding and polishing against spherical laps.

A lens having the shape of a segment of a sphere converges rays striking near the rim more strongly than those near the center. This is called spherical aberration and can be eliminated by making the shape of at least one surface of the lens aspherical. Aspherical surfaces are readily obtained as in Fig. 8*d*, and *g* by molding. Unfortunately, it is not possible in molding glass to obtain the excellent surface smoothness

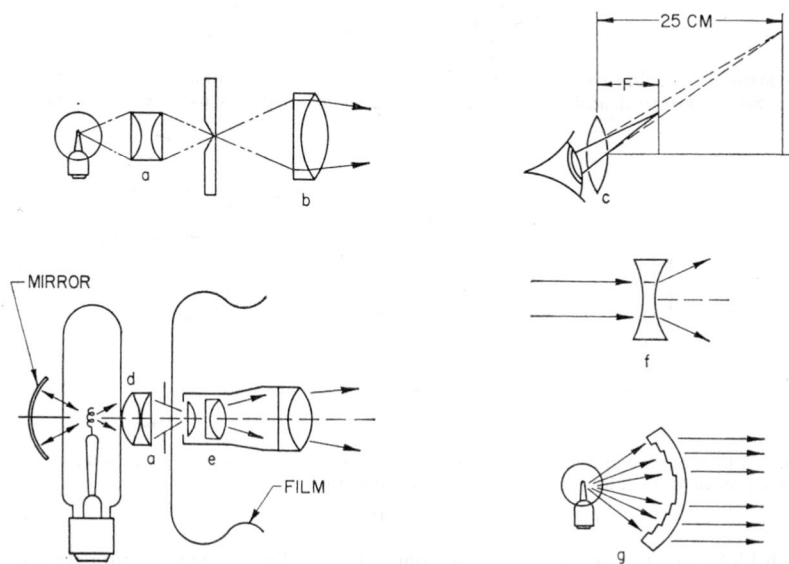

Fig. 8. Some lens types and their uses.

of ground and polished surfaces. Plastic aspherical lenses which can be cast with smooth surfaces are increasing in use.

Focal Distances (Thin Lenses). The ray-converging power of a lens in diopters is $1/F$, where F, the focal length, is given in meters. The focal length is the distance at which light from a distant source is brought to a focus. If one has a doublet of two simple lenses, the combined focal length $F_{1,2}$ is related to the separate focal lengths F_1 and F_2 as follows:

$$\frac{1}{F_{1,2}} = \frac{1}{F_1} + \frac{1}{F_2} \tag{9}$$

Angular Magnification of a Reading Lens. By assuming that an object seen near to with the aid of a magnifying lens is actually thought by the eye to be 25 cm distant (Fig. 8*c*), the angular magnification of a reading lens is said to be $25/F$ where F is in centimeters. Thus, a 10-diopter lens, which would have a focal length of 10 cm, is said to have an angular magnification of 2.5.

The Location of Lenses; Image Magnification. As is shown in Fig. 9, when a positive lens is moved away from a concentrated light source to the focal distance F, the rays it intercepts are bent to form a beam of essentially parallel light. At a greater distance, the rays spreading from the source are converged by the lens to form an image of the source. The distances from lens to source f_1 and from lens to image f_2 are related to the focal length F as

$$\frac{1}{f_1} + \frac{1}{f_2} = \frac{1}{F} \tag{10}$$

When the image is thus formed, the heights of object h_1 and image h_2 are related

$$\frac{h_1}{h_2} = \frac{f_1}{f_2} \tag{11}$$

Light-gathering Power of Lenses. The power of a lens to gather light from a small source and project it as a beam is obtained by squaring the reciprocal of

$$f \text{ number} = \frac{F}{d} \text{ or } \frac{\text{focal length}}{\text{diameter}} \tag{12}$$

The strongest aspherical lenses have f numbers less than 1 and are used as condensers (Fig. 8d). Ground and polished lenses free of aberrations usually have f numbers

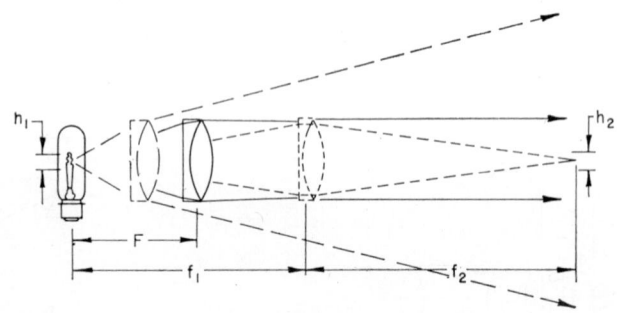

Fig. 9. Effect of changing the position of a lens. Solid lines: distance F; dotted lines: more than F; dashed lines: less than F. Equations applying:

$$1/F = 1/f_1 + 1/f_2 \quad \text{and} \quad h_1/h_2 = f_1/f_2$$

much higher, although some multicomponent lenses (Fig. 8e) have f numbers smaller than 2.

Lens Applications. A lens always should be located so that its faces are perpendicular to the axis of the beam passing through it. Where there is a choice, a lens should always be turned so that the refractions (bending of rays) at its two surfaces are as nearly the same as possible. That is, if the lenses in Fig. 8a were turned around, all refraction would be at the curved surfaces and there would be none at the flat surfaces. This would be undesirable; thus the lenses are properly oriented as shown.

Spectral Filters

These filters make use of spectral absorption or reflection to diminish rays of light passed through them. The ability of a filter to transmit light of different wavelengths is its most important characteristic and is represented by its spectrophotometric curve. Curves for typical filters are shown in Fig. 10. Filters used technically are usually of the following types:

1. Colored glass filters (stable and reproducible)—available from Corning Glass Works, Bausch & Lomb Optical Co., Fish-Shurman Corp., and others.

2. Dyed gelatin (made in variety of colors of varying stability)—made by Eastman Kodak Co. (trade name: Wratten).

3. Interference films passing narrow spectral bands—available from Bausch & Lomb Optical Co., Baird Associates, Farrand Optical Co., Fish-Shurman Corp., and Photovolt Corp.

4. Thin metal films of neutral color and varying transmittance and reflection factors—made by Bausch & Lomb Optical Co. and Evaporated Metal Films Corp.

Light Detectors

Photoelectric detectors generally are used in preference to the human eye because these detectors can operate continuously. There are four basic types of photoelectric detectors:

1. *Barrier-layer* or *photovoltaic cells* which consist of an iron plate with a layer of a semiconductor (usually selenium) and a thin light-transmitting film of a metal such

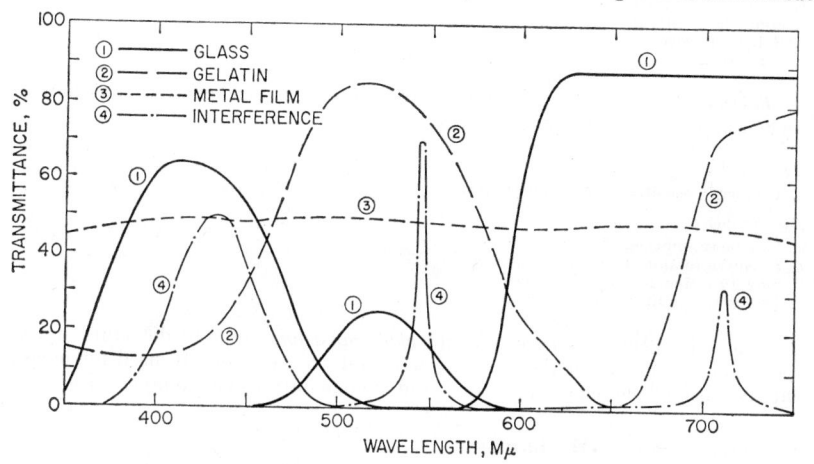

FIG. 10. Spectrophotometric curves of typical filters for four different types. For curves applying to specific filters, reference should be made to the manufacturers of the materials.

as cadmium on top. Photocurrents are generated near the selenium-metal interface without the aid of external voltages:

Advantages:

a. No power supply required.
b. High current output.
c. Closest to eye in spectral sensitivity.
d. Best in manufacturing uniformity (spectral and total).
e. Low dark current.

Disadvantages:

a. Sensitive to temperature.
b. Nonlinear, especially with high external resistances.
c. Subject to fatigue, especially at high illuminations.
d. Sluggish in a-c circuits.
e. Output not readily amplified.
f. Nonsealed cells are subject to attack by moisture and heat.

2. *Vacuum* and *gas-filled photoemission tubes* which have a specially treated metal cathode and a wire anode in a glass envelope. When there is sufficient positive potential, the wire anode collects electrons released by the light striking the cathode.

Advantages:

a. Free from fatigue.
b. Instantaneous response.
c. Signals readily amplified.

Disadvantages:

a. Poor manufacturing uniformity.
b. Different from eye in spectral sensitivity.
c. Some dark current.
d. Currents are small.

3. *Multiplier phototubes* are essentially vacuum-type emission phototubes with built-in amplifiers.

Advantages:

a. Strong signals produced so that light levels as low as those of starlight can be measured.
b. Instantaneous response so that flicker and high-frequency techniques may be used.

Disadvantages:

a. Sensitive to ambient temperature.
b. Poor manufacturing uniformity.
c. Some dark current.
d. Subject to severe fatigue.
e. Sensitive to supply voltage.

4. *Thermopiles* are simply cascades of thermocouples blackened so that they heat up when light falls on them.

Advantage:

a. Uniformly sensitive to all wavelengths.

Disadvantages:

a. Low power efficiency.
b. Sensitive to ambient temperature.
c. Slow to respond to light changes.
d. Delicate—requires careful wiring.

Additionally, *photoconductive cells* and *phototransistors* are light detectors, but they respond most strongly to the near infrared and are infrequently used for measurements because of tendencies to fatigue and change output with temperature.

Light-measurement Mechanisms

Microammeters and galvanometers may be used to measure the currents in photometer circuits. With both microammeter and galvanometer, coil resistances are necessarily high (100 to 5,000 ohms) in order to detect the small current available.

Bridge, push-pull, and feedback circuits are preferred to unaided meters because the circuits can provide more favorable electrical conditions for the photodetectors. Resistance-bridge measurements of the relative outputs of two photocells are widely used in instruments for reflectance and transmittance because they may be made insensitive to variations in light-source intensity and other factors affecting the two photocells identically.

Optical compensators for adjusting two light beams by known amounts make possible photoelectric instruments of high stability and are the only light-measuring devices that may be used with the eye as detector. The two beams of light which are equated may be directed to separate detectors. There are six basic types of optical compensation:

1. *Inverse-square law and modifications.* If light is spreading in all directions from a source or diffusing specimen, the light collected by a receiver can be varied by moving the receiver toward or away from the specimen.

2. *Polarization.* The amount of light in a beam, passing through two parallel polarizers, varies as the cosine squared of the angle between their two axes of polarization.

3. *Cosine-law control of light,* wherein light control is based on the geometric fact that the solid angle subtended by a flat area seen from a distant point varies with the cosine of the angle of view.

4. *Mechanical shutter,* which varies the amount of light in a beam by intercepting part of the beam.

5. *Optical wedge* made from absorbing gelatin, glass, or photographic emulsion.

Mechanically, these wedges are used to control light by moving the wedges across the light beams.

6. *Sectored disk* gives passage of light through an accurately calibrated, rapidly moving, intermittent opening.

Standards for Photometric Measurements

The primary and working standards for different photometric scales are given in Table 3.

Table 3. Standards for Photometric Measurements

Quantities	Primary standards	Secondary or working standards
Light intensity.................	Special blackbody furnace of freezing platinum	Incandescent lamps calibrated against primary standard
Directional reflectance and transmittance with goniophotometer	Imaginary perfect diffusors, reflecting and transmitting	Ceramic panels calibrated for 45°0° directional reflectance against MgO
Directional reflectance for diffuse reflectance	Surface of freshly smoked magnesium oxide	Porcelain-enamel and opaque glass panels calibrated against MgO
Specular reflectance for gloss...	Imaginary perfect mirror	Surfaces of black glass or liquid of known refractive index, ceramic wall tile calibrated against the foregoing
Directional (apparent-diffuse) transmittance	Imaginary, perfectly transmitting and diffusing film	Opal glass pieces calibrated with goniophotometer against directional reflectance standards
Rectilinear (direct) transmittance	Clear air or clean solvent	Calibrated filters may be used, but here the primary standards are readily available

CLASSIFICATION OF PHOTOMETRIC INSTRUMENTS AND USES

The basic components of a photometric instrument (Fig. 11) are (1) a light source E, which directs a beam of light onto a specimen; (2) a receptor mechanism (photocell s and filter t) which accepts a beam of light from the specimen and measures it;

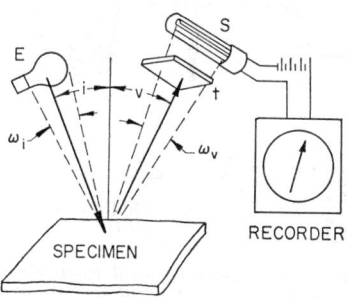

Fig. 11. Elements of a photometric instrument. E, light source; t, filter; s, photocell. Spectral specifications: i, axis of incident beam; v, axis of viewing beam; ω_i, incidence field angle; ω_v, receptor field angle. Solid angles measure extent of nonparallelism in rays of incident and viewing beams, respectively.

and (3) some means of positioning the specimen with respect to the beam of light (not shown in Fig. 11).

To identify the photometric quantity measured by an instrument, it is necessary to give (1) the *spectral* combination of wavelengths involved, which is the spectral product of source energy, filter transmission, and receptor response: (2) the *geometric* conditions of measurement, which are the directions in which light is projected onto, and then taken from, the specimen for measurement; and (3) the *numerical* character of the light-indicating mechanism responding to the photocell.

The conditions of measurement represented in Fig. 11 are:

	Spectral	Geometric
Source..........	E	$i,\ \omega_i$
Receptor........	$t \times s$	$v,\ \omega_v$

No data are given on the numerical response of the indicating mechanism.

Classification of Instruments

Photometric instruments are classified by the spectral and geometric attributes of the light distributions they measure. Table 4 shows four spectral and six geometric types which combine altogether to make 24 classes of instruments. Some, however, have little practical importance. *Spectrally,* four types of instruments are in general use:

1. *Spectrophotometric,* in which a narrow and continuously variable waveband is isolated by prism or grating. Spectrophotometric curves are used for defining the properties of specimens responsible for color, and for spectrochemical analyses.

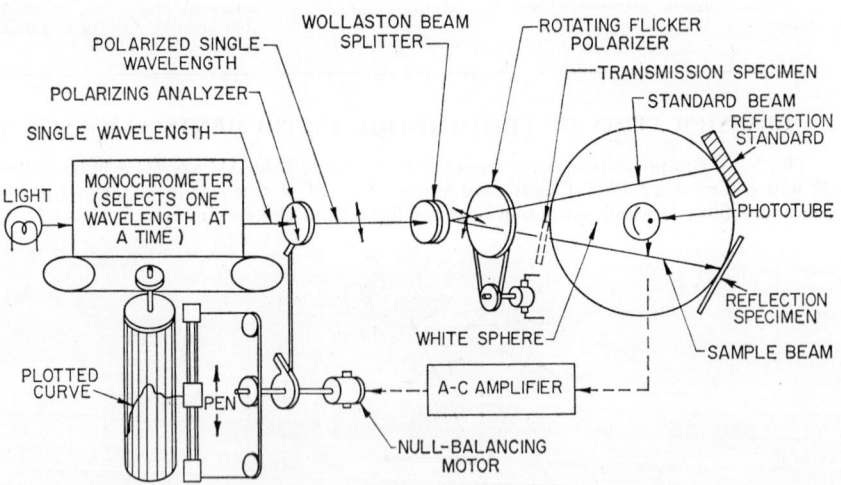

Fig. 12. The recording spectrophotometer (*General Electric Co.*) has a spectrometer-type monochromator in which monochromatic light of any desired wavelength from 400 to 700 mμ is produced by dispersion with prisms and a slit. This light is passed alternately to a sample and a standard, chopped to produce an alternating-current (if unbalanced), and then is balanced by a light-polarizing prism, position of which is recorded as per cent transmission vs. wavelength.

2. *Abridged spectrophotometric,* in which fixed wavebands are isolated by filters Narrow-band filter measurements are used for quantitative chemical analyses by light absorption using standardized techniques.

3. *Luminous,* in which a special filter is used to simulate the response of the **human**

eye to light intensity. Luminous measurements are used for lighting levels, reflecting efficiencies, and gloss determinations.

4. *Tristimulus*, in which three or four special filters are used to simulate the response of the human eye to color. Tristimulus measurements are used for approximate visual color and for small color differences. They are not sufficiently accurate for exact color.

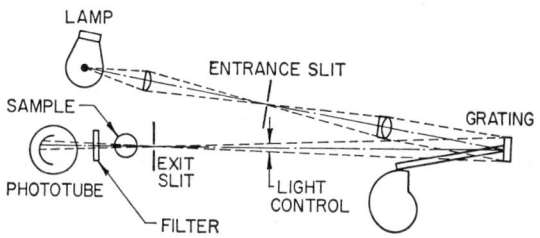

FIG. 13. Diffraction-grating spectrophotometer (*Bausch & Lomb Optical Co.*) produces monochromatic light by using 600 grooves per mm grating. Dispersion of grating is linear and output is constant. Grating is adjustable to cover part of range (375 to 650 mμ).

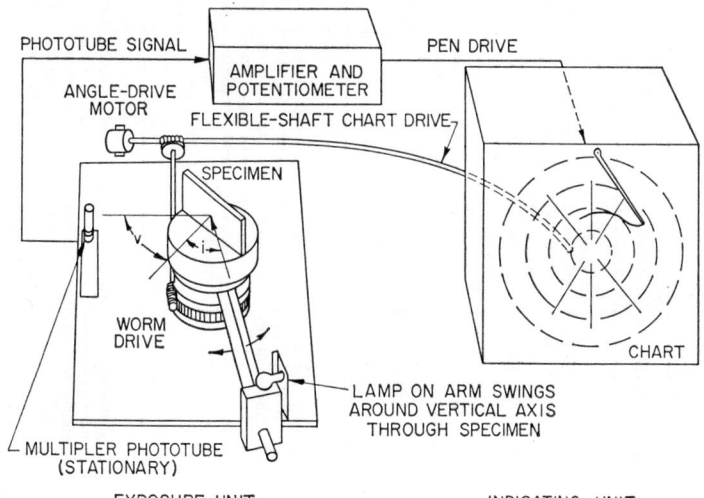

FIG. 14. Simplified diagram of Hunterlab recording goniophotometer in which i is variable and v is fixed according to operator's choice. (*Hunter Associates Laboratory, Inc.*)

Geometrically, six different types of instruments may be identified:

1. *Goniophotometric*, in which directions of view (or sometimes incidence) are varied continuously for studies of light distribution by illuminating units, reflecting surfaces (for research on gloss), and transmitting films and solutions (for size and density of suspended particles).

2. *Light flow in space*, for studies of illumination, brightness, and photographic exposures.

3. *Diffuse reflectance;* two sets of geometric conditions, 45°0° and 0° diffuse, are used for studies of reflecting properties of pigmented, granular, and fibrous materials.

4. *Specular reflectance;* several specular angles, 60°, 75°, 85°, 45°, are used for measurements of gloss, mirror reflectance, coating continuity, and surface condition.

Table 4. Classification and Applications of Photometric Instruments According to Spectral and Geometric Classes of Measurement

Geometric classes of measurement	Spectral classes of measurement			
	Physical measurements		Psychophysical measurements	
	Spectral (fλ) (narrow wavelength band, continuously variable)	Abridged spectral (Δλ) (fixed wavelength bands isolated by filters)	Luminous (Y) (visual intensity)	Tristimulus (Y,x,y) (visual color)
Light flow, light sources F	Spectroradiometers for (F(fλ)). Used for spectral composition of light sources, chemical identification by light emission	Filter flame photometers and other devices for F(Δλ). Used for chemical identification by light emission	Illumination meters, photographic exposure meters, and brightness meters for F(Y) (see Fig. 20)	Tristimulus photometers F(Y,x,y). Used for colors of light sources
Goniophotometric reflectance or transmittance with continuously variable angles R or T(u,v)	Spectrogoniophotometers for R or T(u,v)(fλ). Uses: Not generally used (would be universal instrument)	Filter goniophotometers and light-scattering apparatus for R or T(u,v)(Δλ). Used to measure large molecules in liquid suspensions by light scattering (see Fig. 14)	Goniophotometers for R or T(u,v)(Y). Used for research on gloss, transparency, turbidity, haze, and like appearance properties (see Fig. 14)	Tristimulus goniophotometers for R or T(u,v)(Y,x,y). Used for studies of color associated with gloss, transparency, luster, and like properties (see Fig. 14)
Diffuse reflectance Rd (measured by directional reflectance) 0° diffuse 45°0° geometry	Reflection spectrophotometers. Used for exact color properties of diffusing surfaces and materials (see Fig. 12)	Filter reflectometers and reflection densitometers for Rd(Δλ). Used to measure textiles, reflection photographs, and printing for dye or ink-film density (see Fig. 15)	Reflectometers for Rd(Y). Used for reflecting efficiency of paints and structural materials, contrast-ratio opacity of papers, paints, detergency studies, etc. (see Figs. 15, 24)	Tristimulus reflection colorimeters for Rd(Y,x,y). Used for measurements of near-white colors and color differences between surfaces of similar color (see Figs. 15, 24)

Specular reflectance R_s (nonmetal surfaces are measured for gloss)	*Mirror spectrophotometers* for metals, etc. (see Fig. 12) $R_{s(\lambda)}$. Used for color, etc., of metals and mirrors (see Fig. 12)	*Mirror filter reflectometers* for $R_{s(\Delta\lambda)}$. Not generally used	*Glossmeters.* Uses: $R_{s(Y)60°}$ for paints, medium gloss; $R_{s(Y)75°}$ for paper and printing; $R_{s(Y)85°}$ for paints, low-gloss metals; $R_{s(Y)20°}$ for paints, high-gloss metals; $R_{s(Y)45°}$ etc. for misc. metals (see Fig. 21)	*Mirror colorimeters* for $R_{s(x,z,y)}$. Used for mirror and bronze color, etc.. (see Fig. 15)
Diffuse transmittance T_d Film / Volume	*Diffuse-transmittance spectrophotometers for* $T_{d(\lambda)}$. Used for color properties of translucent films and volumes (see Fig. 12)	*Fluorescence meters* have separate source and receptor filters. Used for chemical identification by fluorescence (see Fig. 16)	*Hazemeters and turbidimeters.* Used for optical quality of plastics, suspended matter in liquids, smoke density, etc. *Diffuse transmittance meters* used for efficiency of lighting materials (see Figs. 22, 24)	*Translucence colorimeters* $T_{d(x,z,y)}$. Use: Color by diffuse transmission (see Fig. 24)
Rectilinear transmittance T_r Film / Volume	*Direct-beam spectrophotometers* $T_{r(\lambda)}$. Used for chemical analyses by spectral absorption and for color studies (see Figs. 12, 13)	*Filter (chemical) colorimeters and transmission densitometers* $T_{r(\Delta\lambda)}$. Used for chemical analyses by absorption, and for photographic densitometry (Figs. 16, 17, 18, 19)	*Transmissometers* for $T_{r(Y)}$. Used for atmospheric transmission and smoke absorption (see Fig. 23)	*Direct-beam tristimulus transmittance meters* $T_{r(x,z,y)}$. Used for color differences between clear films, filters, and liquids of similar color

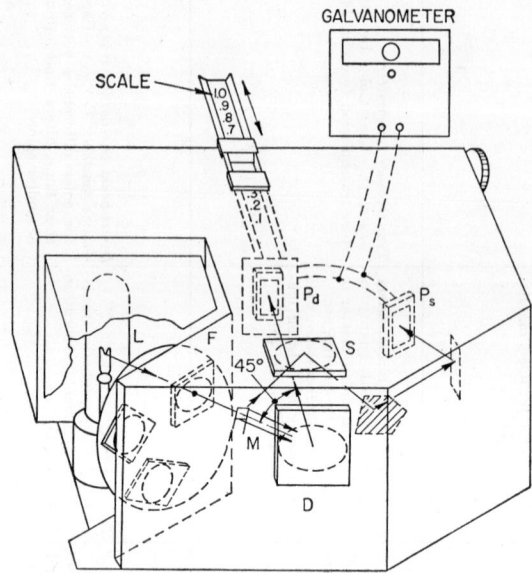

Fig. 15. Hunter multipurpose reflectometer in which beam from lamp L is directed through one of three filters F and split into two parts by mirror M. Light reflected diffusely (45°0°) from the beam striking panel D is picked up by photocell P_d; light reflected specularly (45°, −45°) from the beam striking panel S eventually reaches photocell P_s. P_d is moved until the galvanometer shows its signal to be equal that of P_s; then the scale is read. (*Gardner Laboratory, Inc.*)

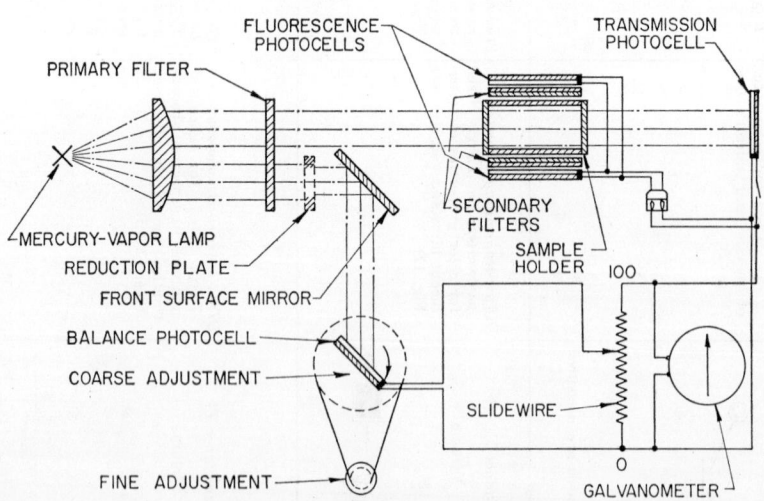

Fig. 16. Photovolt Lumetron transmission colorimeter which can be modified to serve as a fluorescence meter. Only ultraviolet light passes the primary filter. The secondary filters in front of the two fluorescence photocells absorb ultraviolet light; so that only light with wavelengths altered by fluorescence can reach these cells. The signal generated by these two cells is measured in terms of that from a balance photocell. (*Photovolt Corp.*)

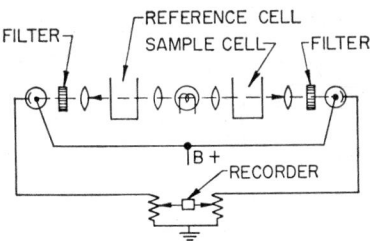

FIG. 17. Photoelectric comparator (*Milton Roy Co.*) uses two similar photocells. Light is passed through the sample to one cell, while some of the same light is sent unobstructed through the reference cell. Cell outputs are then compared in the recorder circuit. Outputs are balanced and relative transmission recorded. Instrument can be used with auxiliary equipment involving colorimetric titrations for the detection of ppm range of many compounds such as silica or total hardness in water.

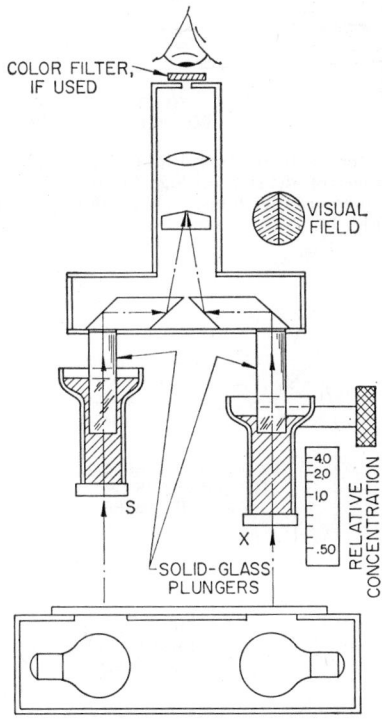

FIG. 18. Dubosq colorimeter with which observer adjusts length of light path through solution *X* for match with density-standard solution *S*.

5. *Diffuse transmittance,* used for measurements of turbidity, transparency, and similar characteristics.

6. *Rectilinear transmittance,* in which the beam passing directly through a specimen is measured for diminution by absorption, reflection, and scattering.

Table 4 shows in a separate block the name, applications, and some of the commercial suppliers of each of the 24 classes of instruments which result from the breakdown into four spectral and six geometric types. It will be seen that some types enjoy little use and others have a wide variety of applications.

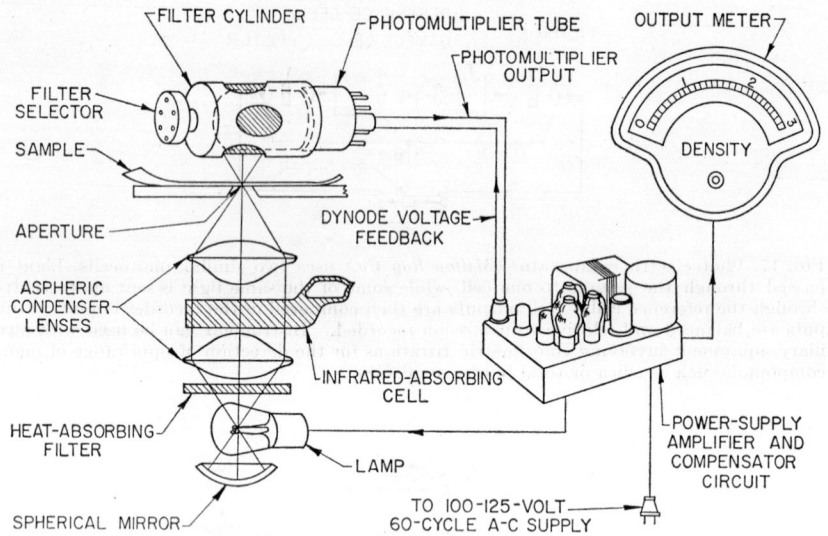

FIG. 19. Color densitometer using photomultiplier tube and stable feedback circuit to indicate directly the values of density ($-\log T$) for narrow wavebands. Specimens usually are color photographs or other films. (*Macbeth Corp.*) For different applications of this same circuit, see Fig. 14.

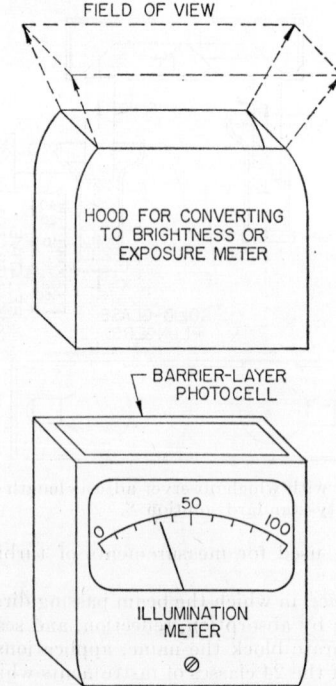

FIG. 20. Photoelectric illumination meter (without hood, it accepts light from all directions) or brightness and meter (with hood, its field of view is limited). (*General Electric Co.*)

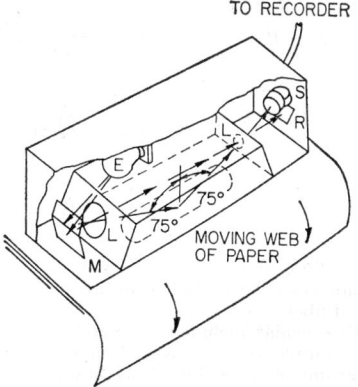

Fig. 21. Exposure unit of 75° (*Hunter-Gardner*) gloss recorder showing how gloss beam and reference beam pass through lenses together so that dust collecting on these lenses will affect both beams identically. Cell outputs are compared by a null-balance potentiometer recorder that records in per cent reflectance.

R = mirror
E = light source
M = 2 mirrors $7\frac{1}{2}°$ out of parallel
L = lenses
S = comparison and test photocells

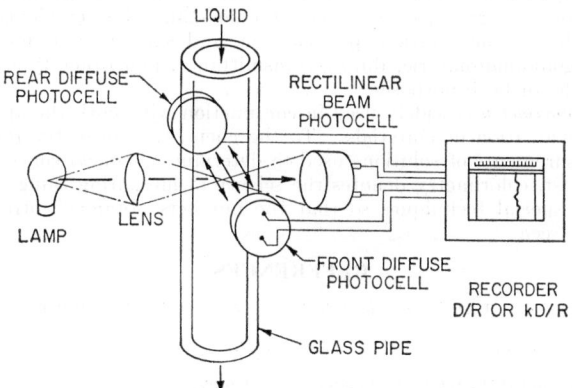

Fig. 22. Photoelectric turbidimeter (*Baird-Atomic, Inc.*) has light transmitted by the liquid in a sample cell. Light is continuously compared with the light scattered by suspended particles in the stream. Phototube output, measured by recording potentimeter, varies as the light is more or less cut off by the suspended particles. The unit is used for monitoring turbidity, color, or other light-absorption properties of process solutions.

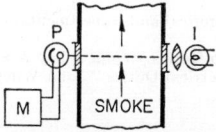

Fig. 23. Smoke and fume-density meter (*Photomation, Inc.*) uses photocell to measure light absorption by smoke in a stack. Another type (*Bailey meter*) uses a hermetically sealed bolometer (resistance thermometer) to measure reduction in the amount of heat received from a sealed-beam spotlight. Null-balance recorder is compensated for temperature.

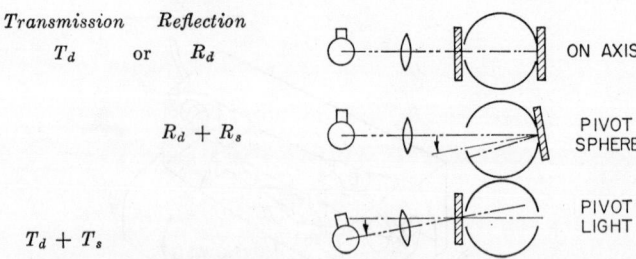

Transmission   Reflection

T_d or R_d ON AXIS

$R_d + R_s$ PIVOT SPHERE

$T_d + T_s$ PIVOT LIGHT

T_d	Turbidity of films and liquids. Color by scattering
R_d	Diffuse color (specular excluded) of opaque surfaces such as paint, paper, textiles, and fibers
$R_d + R_s$	Color, with specular included, of metals, metallic paints, plastics, shiny textile fibers, directionally textured materials
$T_d + T_s$	Diffuse transmission. Color of plastics and translucent films. Color of liquids. Color of transparent films
$T_d/(T_d + T_s)$	Haze of films

(NOTE: d = diffuse, s = specular)

FIG. 24. Types of measurement with the Hunterlab D25P sphere or color difference meter. (*Hunter Associates Laboratory, Inc.*)

Specific instruments in a number of these categories are diagramed in Figs. 12 to 23. Brief descriptions of these instruments and what they do are contained in the legends with the figures.

A distinction should be made between photometric measurements used to evaluate attributes of appearance, such as color and gloss, and those employed for physical attributes such as concentration, particle size, and surface smoothness. Spectrophotometric, goniophotometric, diffuse-transmittance, and other types of measurements are made for both purposes.

The term *colorimetry* is widely used in conjunction with both the appearance and the physical evaluation of materials. To the chemist, colorimetry refers to transmittance measurements of solutions used to determine chemical concentration. To the photometrist, colorimetry denotes the science of measuring things which people see. He uses special techniques so that the numbers obtained correspond to the colors that are seen.

REFERENCES

1. Hardy, A. C., and F. H. Perrin: "The Principles of Optics," McGraw-Hill Book Company, Inc. New York, 1932.
2. Jacobs, D. H.: "Fundamentals of Optical Engineering," McGraw-Hill Book Company, Inc., New York, 1943.
3. Jenkins, F. A., and H. E. White: "Fundamentals of Optics," 3d ed., McGraw-Hill Book Company, Inc., New York, 1957.
4. "Lighting Handbook," 2d ed., Illuminating Engineering Society, New York, 1952.
5. Hunter, R. S.: Photometric Variables, in Considine, D.M. (ed.): "Process Instruments and Controls Handbook," pp. 7-111–7-128, McGraw-Hill Book Company, Inc., New York, 1957.
6. Judd, D. B.: "Color in Business, Science, and Industry," John Wiley & Sons, Inc., New York, 1952.
7. Harrison, V. G. W.: "Definition and Measurement of Gloss," Printing and Allied Trade Associations, London, 1945.
8. Optical Society of America Committee on Colorimetry: "The Science of Color," Thomas Y. Crowell Company, New York, 1953.
9. Valasek, J.: "Introduction to Theoretical and Experimental Optics," John Wiley & Sons, Inc., New York, 1949.
10. Robertson, J. K.: "Introduction to Optics," John Wiley & Sons, Inc., New York, 1954.
11. Herzberger, Max: "Modern Geometrical Optics," John Wiley & Sons, Inc., New York, 1957.

ACOUSTICAL MEASUREMENTS

By E. G. Thurston* and W. H. Peake†

Acoustics, or the science of sound, rigorously defined as the study of compressional oscillations about an equilibrium position of continuous media, is generally divided into two broad areas. These areas depend upon whether or not the sounds under study can be heard by a so-called normal human ear. Subjective acoustics refers to audible sounds, while objective acoustics refers to those outside the audible range.

FUNDAMENTAL THEORIES

Three common subdivisions exist: (1) *subsonic*, including frequencies below 20 cps; (2) *sonic*, frequencies between 20 and 20,000 cps; and (3) *ultrasonic*, frequencies greater than 20,000 cps.

For simplicity, particle motions are considered to have a sinusoidal time dependence. Then, the displacement ξ of a particle (or infinitesimal volume element) of the medium may be written in terms of two fundamental kinematic parameters of the acoustical field

$$\xi = A \sin \omega t = A \sin 2\pi f t \tag{1}$$

where f = frequency and A = amplitude. It follows that the velocity of the particle is

$$v = \frac{d\xi}{dt} = A\omega \cos \omega t \tag{2}$$

and the acceleration is

$$a = \frac{d^2\xi}{dt^2} = -A\omega^2 \sin \omega t \tag{3}$$

These oscillations take place in a medium which can be characterized by two additional acoustical parameters: acoustical impedance Z, and propagation constant k.

If attention is extended beyond consideration of a single point, and the entire medium is considered, it is found that the acoustical disturbance propagates in a wavelike manner. Thus it is necessary to characterize any acoustical variable as a function of space as well as time.

The most important type of acoustical motion is the plane wave, in which acoustical pressures and particle displacements have common phases and amplitudes at all points on any plane perpendicular to the direction of wave propagation (see Fig. 1). The particle displacement at any point may then be written

$$\xi = \xi_{\max} \sin (\omega t - kx) \tag{4}$$

and the medium is said to be excited by a compressional plane wave traveling in the x direction with frequency $f = \omega/2\pi$ and propagation constant $k = 2\pi/\lambda$ where λ is the wavelength. The velocity of wave propagation is $c = f\lambda = \omega/k$ (do not confuse with particle velocity $v = d\xi/dt$).

* Senior Research Engineer, Hallicrafters Co., Chicago, Ill.
† Assistant Professor of Electrical Engineering, Ohio State University, Columbus, Ohio.

The acoustical impedance Z_0 of the medium is defined as the ratio of the pressure variation Δp in the plane wave to the particle velocity v; it is related to the density ρ and the propagation velocity c in the medium as shown in the following.

$$\Delta p = \omega \rho c \xi_{max} \cos (\omega t - kx) \tag{5}$$

$$v = \frac{d\xi}{dt} = \omega \xi_{max} \cos (\omega t - kx) \tag{6}$$

$$Z_0 = \frac{\Delta p}{v} = \rho c \tag{7}$$

Intensity I of a sound wave is the average rate of flow of acoustical energy per unit area normal to the direction of propagation.

$$I = \frac{1}{2} Z_0 v^2 = \frac{1}{2Z_0} (\Delta p)^2 \tag{8}$$

The velocity of propagation of acoustical waves is dependent upon the medium. The general expression is given by $c = \sqrt{dp/d\rho}$, where dp is the change in pressure associated with a change in density.

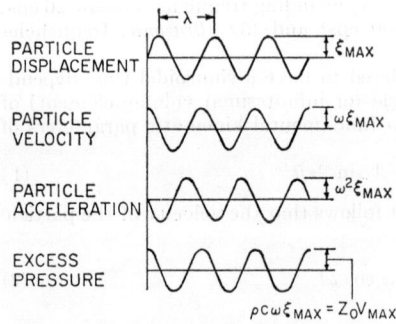

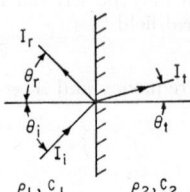

FIG. 1. Phase relations between particle displacement, velocity, acceleration, and excess pressure in a sound wave.

FIG. 2. Reflection and refraction of a plane sound wave at a boundary between two media.

As shown by Fig. 2, acoustical plane waves obey the usual laws of refraction and reflection. When a plane sound wave strikes a plane interface between two media with densities ρ_1 and ρ_2 and propagation velocities c_1 and c_2 and the incident, reflected and transmitted rays, all in the same plane, make angles θ_i, θ_r, θ_t with the normal to the interface, then

$$\theta_i = \theta_r \quad \text{(angle of incidence = angle of reflection)} \tag{9}$$
$$c_2 \sin \theta_i = c_1 \sin \theta_t \quad \text{(Snell's law)} \tag{10}$$

and the intensities are related by

$$\frac{I_r}{I_i} = \left(\frac{\rho_2 c_2 \cos \theta_i - \rho_1 c_1 \cos \theta_t}{\rho_2 c_2 \cos \theta_i + \rho_1 c_1 \cos \theta_t} \right)^2 \tag{11}$$

In the case of normal incidence, this reduces to

$$\frac{I_r}{I_i} = \frac{(\rho_2 c_2 - \rho_1 c_1)^2}{(\rho_2 c_2 + \rho_1 c_1)^2} = \left(\frac{Z_2 - Z_1}{Z_2 + Z_1} \right)^2 \tag{12}$$

An important effect associated with such reflection at normal incidence is the production of a standing wave. The reflected and incident waves interfere to reinforce or cancel each other and the resulting points of minimum and maximum amplitude form a standing wave which is stationary in space. The standing-wave ratio (ratio of maximum to minimum amplitude in the standing wave) depends on the ratio of the amplitude of the reflected to that of the incident wave $\sqrt{I_r/I_t}$.

It is clear from Eq. (12) that, at an interface between two media having widely different acoustical impedances, a large part of the incident energy will be reflected, producing large standing-wave ratios. Sound waves are diffracted and scattered similar to light waves, and like electromagnetic waves they can be guided by tubes or ducts. They can be focused by acoustical reflectors or lenses.

Because of the great range of intensities encountered in acoustical measurements, it is convenient to use a logarithmic scale. Thus an intensity I_i can be specified in terms of decibels (db) vs. some defined reference level I_0, by use of the expression

$$D(\text{db}) = 10 \log \frac{I_i}{I_0} \tag{13}$$

Similarly, a pressure P (or particle velocity or amplitude) is said to have an intensity of D decibels with respect to a reference pressure P_0 if

$$D = 20 \log \frac{P_1}{P_0} \tag{14}$$

It is common practice in liquid-phase acoustics to use the microbar (dyne-cm^{-2}) for the reference pressure P_0. In air acoustics, the reference pressure is usually taken to be 0.0002 dyne-cm^{-2}. This arbitrary value was adopted because it is approximately the sound pressure at the threshold of hearing in human beings at 1,000 cps (approximate frequency of greatest hearing sensitivity).

ACOUSTICAL TRANSDUCERS

Acoustical transducers are divided into two categories: (1) receivers, which convert acoustical into electrical energy; and (2) transmitters, which convert electrical into acoustical energy. There is a large class of transducers which may be used as either receivers or transmitters, but it is usual to consider each function separately since the parameters of interest in the two cases are somewhat different.

Acoustical Receivers

Receivers are divided into three major classes according to their field of application.
1. *Microphones*, which are actuated by sound waves in air.
2. *Hydrophones*, which are actuated by sound waves in a liquid.
3. *Vibration pickups*, which are actuated by the gross motion of a solid.
Microphones. Microphones are classified according to their operating mechanism and their performance or terminal characteristics.
  1. *Carbon-button microphone.* The sound field acts on a packet of carbon granules through a linkage to vary its resistance which, with a source of direct current, produces a d-c voltage directly proportional to the sound pressure.
  2. *Moving coil (dynamic).* The sound field moves a coil in a magnetic field. This induces a voltage in the coil proportional to the sound pressure.
  3. *Ribbon velocity (pressure gradient).* The sound field acts on both sides of a metal ribbon suspended between the poles of a magnet. The resultant motion induces a voltage across the ribbon proportional to the pressure gradient in the acoustical field.
  4. *Piezoelectric.* Stresses in a piezoelectric element result from a sound field and generate an output proportional to pressure.

5. *Electrostatic (capacitor)*. The sound field acts on a stretched metal diaphragm which forms one plate of a capacitor, whose capacity varies according to variations in sound pressure. May be used in a d-c circuit with a source of voltage and a resistor to develop a voltage across the resistor, or in an electronic-tube circuit to modulate the frequency of an oscillator. In both applications the output is proportional to the pressure.

The three characteristics which determine the response of a microphone are sensitivity, directivity, and frequency response. The sensitivity is usually defined as the ratio of the electrical output in volts to the acoustical input in sound-pressure units. Owing to extended development in the field of electronic amplification, this characteristic is of small significance.

The directivity of a microphone is usually presented graphically to represent its relative sensitivity to the angle of arrival of a plane wave perpendicular to a specified axis. The three most common directivity characteristics are (1) omnidirectional, (2) bipolar, and (3) cardioid, as shown by Fig. 3. The omnidirectional, which responds equally to sound in all directions, is usually of the pressure type. Velocity microphones have bipolar response, showing considerable sensitivity in the front and back but little to the sides. The cardioid, or unidirectional microphone, provides even more discrimination against unwanted sound from directions other than the direction of prime sensitivity.

Frequency response of a microphone is expressed for the range through which the output is constant (flat) for constant input pressure. A typical microphone designed for audio frequencies might be classed as flat from 40 to 15,000 cps.

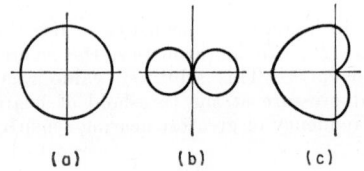

(a)   (b)   (c)

Fɪɢ. 3. Common microphone directionality characteristics: (*a*) omnidirectional; (*b*) bipolar; (*c*) cardioid.

Hydrophones. The three parameters, sensitivity, directivity, and frequency response describe the performance of hydrophones in much the same way as for microphones. A major difference between microphones and hydrophones is that in the latter the sound pressure often acts directly on the element without intermediate mechanical linkages. This is possible because of the much higher acoustical impedance of liquids as compared with air, thus permitting a better acoustical match between the element and the medium.

Vibration Pickups. Vibration pickups are frequently designed so that they may be rigidly attached to the moving body. Hence, the entire pickup unit, including its housing, moves with the vibrating body. The output then is dependent upon the acceleration of the body at the point of contact, and this output can be integrated electrically to produce an output proportional to velocity or displacement.

Other vibration pickups contact the vibrating object through a mechanical linkage which can move relative to the pickup housing. The output of such pickups is proportional to velocity (displacement of the vibrating body).

Performance characteristics are defined in terms of sensitivity (volts per unit displacement, velocity, or acceleration), frequency response (which determines range of usefulness), and maximum acceleration, velocity, or displacement the pickup unit will withstand safely.

Acoustical Transmitters

Acoustical transmitters or projectors convert electrical or other sources of energy into acoustical energy. There are three basic classes: (1) those used to produce sound waves in air, usually called loudspeakers; (2) liquid-phase transmitters or projectors, which are used to produce sound waves in liquids; and (3) whistles, sirens, and all those other sources which are irreversible (cannot be used as receivers).

The parameters of most interest in describing transmitters are: (1) the efficiency (ratio of energy radiated by the transmitter to the energy supplied); (2) frequency response; (3) power limit (maximum amount of power that may be supplied without causing distorted output or damage); and (4) directivity.

Loudspeakers. Almost all loudspeakers embody the moving-coil principle. Alternating current is applied to a coil (usually called the voice coil) which moves in a magnetic field. To the coil is attached a diaphragm or piston, the motion of which generates sound waves proportional to the frequency of the exciting current.

The efficiency of loudspeakers is low, on the order of 5 per cent. They have a frequency response of several octaves, are omnidirectional at low frequencies, but become highly directive (tending toward a narrow beam of projection) as the diaphragm diameter increases beyond a few wavelengths.

Efficiencies of from 30 to 50 per cent may be obtained from loudspeakers by using a horn of gradually increasing cross-sectional area. The directivity will then depend upon the characteristics of the horn.

Liquid-phase Projectors. The chief transducing means used in transmitters for exciting acoustical waves in liquids are piezoelectric, magnetostrictive, and electromagnetic. They are usually designed as resonant devices and are operated at their fundamental frequency. Radiated power decreases rapidly with deviation from the fundamental frequency. The bandwidth ratio, or Q, of such transducers is considered to be equal to $f_r/f_2\text{-}f_1$, where f_r is resonance and f_1 and f_2 are the frequencies above and below the fundamental at which radiated power is one-half the power at resonance.

Piezoelectric liquid-phase projectors may be used from 20 kc to over 100 mc, and magnetostrictive transducers are effective in the range between 10 and 100 kc. Both types have efficiencies on the order of 50 to 70 per cent and bandwidth ratios of from 5 to 20.

Energy-radiating capabilities are limited to the order of $\frac{1}{2}$ to 4 watts/cm². Radiation intensities up to 50 watts/cm² can be obtained by increasing the pressure of the liquid surrounding the transmitter and by providing special cooling means. Radiation limits are determined by the point at which cavitation begins and by physical limitations of the unit.

To increase power-handling capacity and to control directivity, many liquid-phase projectors are made up of arrays of individual transducers.

Nonreversible Projectors. A variety of projectors, used mostly to excite acoustical vibrations in air, are not reversible; i.e., they cannot be used as receivers. Among these are:

1. *Modulated air-flow speaker.* An air stream under pressure is modulated by a valve which is controlled by an electrical signal. Such speakers are used for public-address applications. They have high efficiency and power output, but high distortion and poor frequency response.

2. *Whistles.* A jet of air is impinged against a sharp edge. They have a high efficiency but are unstable with respect to frequency and amplitude, except when coupled to an auxiliary device such as the resonant cavity of an organ pipe.

3. *Sirens.* A stream of compressed air is chopped by a series of blades mounted on a rapidly rotating disk. They have a high efficiency, high intensities, and stable and easily controlled frequencies.

Explosives. Chemical explosions in liquids or solids and spark discharges in air are sometimes used to produce wideband noise energy or for other special purposes, such as sound ranging and seismographic soundings. The frequency characteristics can be controlled by using the explosion to excite a resonant cavity. Such systems have only recently been investigated, especially in the United States. They show promise of producing great output intensities, especially in liquids, with relatively low cost apparatus.

MEASUREMENT OF SOUND LEVEL AND NOISE

Acoustical measurements may be divided into two broad areas: (1) the measurement of sound-field variables and (2) measurements in which the behavior of the sound field is indicative of some other property of interest.

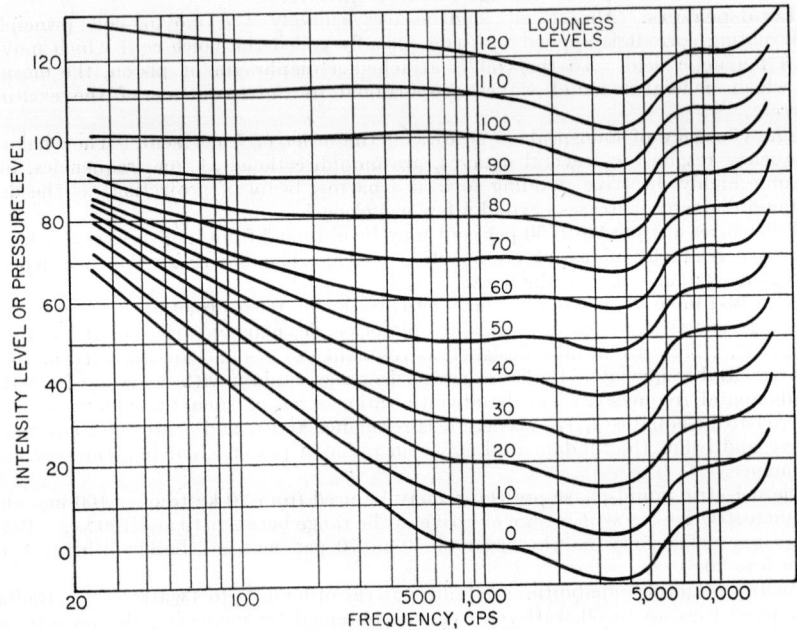

Fig. 4. Equal-loudness contours or Fletcher-Munson curves.

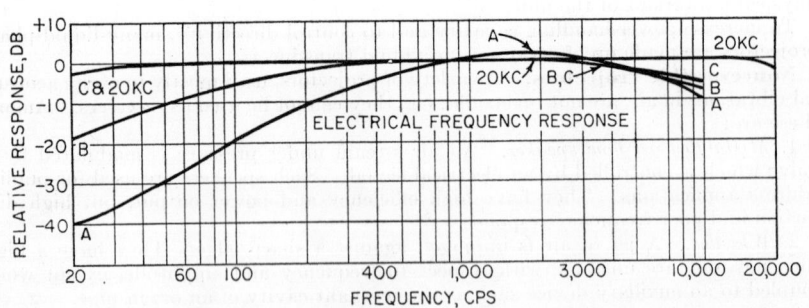

Fig. 5. Frequency response of a typical sound-level meter. Note effect of weighting networks.

Measurements of the first type are important to those specializing in acoustics, but for instrumentation engineers this type of measurement is usually concerned with a problem of noise.

The subject of noise measurement and control is receiving increasing attention. Manufacturers recognize the sales value of proper noise control in their products, but it is in the field of industrial hygiene that most intensive studies are being made.

This results from recognition that objectionable noise is an occupational hazard, and this problem is receiving attention from such divers groups as labor unions, industrial hygienists, otologists, architects, construction experts, and acousticians. The basic instrument for obtaining data on noise conditions is the sound-level meter. This device consists of three functional sections: (1) a calibrated microphone, (2) a frequency-weighing network, and (3) a calibrated amplifier with a meter which indicates sound level directly in decibels.

A difference exists between loudness level and sound-pressure level (related to the difference between subjective and objective acoustics). Sound-pressure level is the actual value of the sound pressure at a point, usually expressed as decibels vs. 0.0002 dyne cm².

Loudness level of a sound is the sound-pressure level of a standard tone (usually 1,000 cps) which sounds equally as loud as the sound under measurement. Loudness levels of pure tones and narrow bands of noise have been investigated by Fletcher and Munson, who have plotted the results. These are commonly referred to as Fletcher-Munson curves or equal-loudness contours (see Fig. 4). The numbers on the contours are the loudness levels of the sound in *phons* (the sound-pressure level of a 1,000-cps tone that is equally loud). Thus, if a certain complex sound wave sounds equally as loud as a 1,000-cps tone having a sound-pressure level of 60 db re 0.0002 dyne/cm², the complex wave is said to have a loudness level of 60 phons regardless of its sound-pressure level.

Sound-level meters usually have settings labeled *A*, *B*, and *C*. These select weighting networks designed on the basis of the equal-loudness contours. Their purpose is to compensate for the frequency characteristics of the ear for pure tones at different levels. The *A* network is 40 db, the *B* network 70 db, and the *C* is approximately flat. Figure 5 shows typical response curves for the three networks.

Loud noise not only interferes with speech intelligibility and causes nervousness and reduced efficiency in personnel but can cause hearing damage. Because of the seriousness and importance of the problem, tentative ratings have been established. These are listed in Table 1.

Table 1. Tentative Ratings of Noise Levels Damaging to Human Hearing for Continued Exposure

Octave band, cps	Limiting level continuous spectrum (white noise), db re 0.0002 dyne/cm²	Pure tones, db re 0.0002 dyne/cm²
20–75	110	
75–150	102	96
150–300	97	88
300–600	95	85
600–1,200	95	84
1,200–2,400	95	83
2,400–4,800	95	82
4,800–9,600	95	81

Most noises contain many frequencies, and hence the values in the second column of Table 1 should be used. In some cases, however, the noise may be concentrated in a very narrow band or be a relatively pure tone, such as a siren, in which case the values of the third column apply.

If only a sound-level meter is available, it is recommended that readings be taken on all three weighing networks, and one suggested rating is based on the reading obtained with the *B* network. Any *B* network reading above 100 db should be regarded as probably unsafe for everyday exposures over a period of months, and steps should be taken for noise reduction or ear protection. Readings below 80 db are probably safe even when the noise is relatively pure tone. For short exposures of perhaps less than 1 hr, these values can be increased by about 20 db.

ACOUSTICS IN INDUSTRY

Low-power Applications

Low-power applications are characterized not only by the low intensity of the sound wave used but also by the fact that the sound wave is not expected to cause any change in the chemical or physical characteristics of the media. Thus such applications are found in testing, inspection, and control, generally utilizing acoustical energy in the ultrasonic range.

Nondestructive Testing. In acoustical inspection procedures, a sound wave of ultrasonic frequency is propagated into the piece under inspection. Defects are located by reflection of the wave from any interface in the object such as a crack. Three techniques in common use are: (1) through testing, (2) pulse testing, and (3) resonant testing.

The through-testing technique utilizes a transducer generating a continuous wave train, and a second transducer at the opposite end to act as a receiver. Both transducers are coupled to the piece with an oil film or the piece is completely immersed in a liquid. If there is no flaw, the transducer output is constant, but any irregularity such as a crack or foreign object will cause some reflection of the transmitted waves and consequent reduced output from the receiver. This method is being used successfully for such applications as testing rocket motors and inspection of tires for internal defects.

In the pulse-testing technique, ultrasonic pulses of short duration (1 to 10 μsec) are projected into the piece under test. The transducer, usually a quartz crystal, acts as both transmitter and receiver. The original pulse is synchronized with the sweep of a cathode-ray tube so that it, and any reflections, can be displayed on the tube. A flaw is indicated by the appearance of a small reflected pulse between the initial pulse and its reflection from the opposite wall of the piece. The position of the flaw in the test object can thereby be determined.

Both these techniques have disadvantages in that complex pieces develop spurious reflections, requiring considerable skill in testing. These methods are superior to X-ray testing in that sound waves are not significantly attenuated in metals, thus permitting inspection of much thicker pieces than with X rays. In addition, discontinuities can be detected which are missed by normal X-ray techniques.

Resonant-testing methods are utilized principally to measure the thickness of materials quickly and accurately with the advantage that the part can be measured from one side only and without any need for drilling holes or otherwise inflicting damage. Resonant testing employs the principle that a vibrating crystal reacts when acoustical resonance exists. This occurs whenever the piece is any integral multiple of a half wavelength.

Standard methods for the use of ultrasonics for nondestructive testing (thickness measurement, flaw detection, etc.) have been dealt with exhaustively in Ref. 12a. More recent techniques, based on the excitation of shear waves, are considered by Worlton.[25]

Flow Measurement. An ultrasonic flow meter developed at the National Bureau of Standards employs two transducers which introduce and receive acoustical waves from the liquid under measurement. These two transducers are switched electrically at a high rate so that each transducer alternately acts as a projector and then as a receiver.

When the sound wave is traveling in the same direction as the liquid, the rate of flow is added to the sound-propagation velocity; in the opposite direction it is subtracted. Thus the flow rate produces a phase difference in the received signal, which is independent of such variables as temperature change. The phase difference is measured electronically and the flow rate indicated directly.

Acoustical flow meters have been described in Ref. 12b, pp. 4–89 and 4–90, and in two articles by Linford.[26,27]

Liquid-level Measurement. Acoustical pulses are used to measure liquid level by two sonic gages, one detecting the surface level and the other sensing the spacing between a series of acoustical reflectors. The measurement is a function of the time taken by a pulse from the surface transducer to be reflected by the surface, and the time for a pulse to be reflected to the second transducer from the acoustical reflectors. The result is converted electrically to indicate digitally in hundredths of a foot and is independent of type of liquid or temperature.

Acoustical or sonic liquid-level indicators have been described in Ref. 12b, p. 5-43.

Consistency Measurements. Techniques for monitoring the consistency of such diverse materials as ice cream and paper-pulp slurries by measuring the attenuation of acoustical waves have also been developed (see, for example, U.S. patents 2,755,662, R. C. Swengel, July 24, 1956; and 2,768,524, R. B. Beard, Oct. 30, 1956).

High-power Applications

Applications for high-power ultrasonic generators utilize violent cavitation to bring about some specific physical or chemical change. Although not thoroughly substantiated it is believed that the effects which take place are produced by shock waves occurring on the implosion phase of the cavitation cycle.

When the maximum value of sound pressure in a liquid becomes greater than the hydrostatic pressure, the liquid is torn apart so that pockets of vapor are formed. This occurs at points where the net pressure is sufficiently negative during the rarefaction phase of the pressure cycle. In the positive phase of the pressure cycle, these "cavitation bubbles" implode violently and generate intense local shock waves. These are useful for cleaning, degreasing, drilling, mixing and dispersing, soldering, and in causing chemical reactions.

REFERENCES

Books

1. Kinsler, L. E., and A. R. Frey: "Fundamentals of Acoustics," John Wiley & Sons, Inc., New York, 1950.
2. Beranek, Leo L.: "Acoustics," McGraw-Hill Book Company, Inc., New York, 1954.
3. Beranek, L. L.: "Acoustic Measurements," John Wiley & Sons, Inc., New York, 1949.
4. Morse, P. M.: "Vibration and Sound," 2d ed., McGraw-Hill Book Company, Inc., New York, 1948.
5. Swenson, G. W., Jr.: "Principles of Modern Acoustics," D. Van Nostrand Company, Inc., Princeton, N.J., 1953.
6. Richardson, E. G.: "Technical Aspects of Sound," Elsevier Publishing Company, Amsterdam, 1953.
7. Huetter, T. F., and R. H. Bolt: "Sonics: Techniques for the Use of Sound and Ultrasound in Engineering and Science," John Wiley & Sons, Inc., New York, 1955.
8. Davis, H.: "Hearing and Deafness—A Guide for Laymen," Murray Hill Books, Inc., New York, 1947.
9. Bruel, P. V.: "Sound Insulation and Room Acoustics," Chapman & Hall, Ltd., London, 1951.
10. Knudsen, V. O., and C. M. Harris: "Acoustical Designing and Architecture," John Wiley & Sons, Inc., New York, 1950.
11. Carlin, B.: "Ultrasonics," McGraw-Hill Book Company, Inc., 2d ed., New York, 1960.
12. McGonnagle, W. J.: "Nondestructive Testing," McGraw-Hill Book Company, Inc., New York, 1961.
12a. McMaster, R. C. (ed.): "Non-destructive Testing Handbook, vol. 2, The Ronald Press Company, New York, 1959.
12b. Considine, D. M. (ed.): "Process Instruments and Controls Handbook," McGraw-Hill Book Company, Inc., New York, 1957.
13. Harris, C. M.: "Handbook of Noise Control," McGraw-Hill Book Company, Inc., New York, 1957.
14. Peterson, A. P. G., and L. L. Beranek: "Handbook of Noise Measurement," General Radio Co., Cambridge, Mass., 1954.

Technical Papers

15. Hardy, H. C.: Noise and Vibration Reduction "Speaks for Itself," *Ind. Power*, May, 1947.
16. Pollack, Irwin: The Effect of White Noise on the Loudness of Speech of Assigned Average Level, *J. Acoust. Soc. Am.* vol. 21, May, 1949.
17. Pollack, I.: On the Measurement of the Loudness of White Noise, *J. Acoust. Soc. Am.*, vol. 23, November, 1951.
18. Beranek, L. L., J. L. Marshall, A. L. Cudworth, and A. P. G. Peterson: Calculation and Measurement of the Loudness of Sounds, *J. Acoust. Soc. Am.*, vol. 23, May, 1951.
19. Kalmus, H. P.: Electronic Flowmeter System, *Rev. Sci. Instr.*, vol. 25, no. 3, March, 1954.

20. Mattiat, Oskar: Ultrasonics and Industry, *Proc. Natl. Electron. Conf.*, vol. 9, February, 1954.
21. Weissler, A.: Sono Chemistry, The Production of Chemical Changes with Sound Waves, *J. Acoust. Soc. Am.*, vol. 25, July, 1953.
22. Kielich, Clayton R.: Sonic Liquid Level Indicator, *Radio T.V. News*, vol. 51, no. 6, June, 1954.
23. Bonvallet, G. L.: The Measurement of Industrial Noise, *Am. Ind. Hygiene Assoc. Quart.*, vol. 13. September, 1952.
24. Beranek, L. L., J. L. Reynolds, and K. E. Wilson: Apparatus and Procedures for Predicting Ventilation System Noise, *J. Acoust. Soc. Am.*, vol. 25, March, 1953.
25. Worlton, D. C.: Ultrasonic Testing with Lamb Waves, *Nondestructive Testing*, vol. 15, no. 4, p. 218, 1957.
26. Linford, A.: Recent Developments in Fluid Flow Metering, *Ind. Chem.*, vol. 36, pp. 107–113, 1960.
27. Linford, A.: Ultrasonic Type Flow Meters, *Fluid Handling*, vol. 87, pp. 99–120, 1957.
28. Hardy, Howard C.: Fundamentals of Noise, Sources of Machine Noise, Analyzing Machine Noise, *Prod. Eng.*, vol. 19, February, 1948; March, 1948; April, 1948, respectively.

Standards—Acoustical Society of America

29. Acoustical Terminology, Z24.1, 1951.
30. Sound Level Meters, Z24.3, 1944.
31. Test Code for Apparatus Noise Measurement, Z24.7, 1950.

Section 4

FORCE VARIABLES

By

DONALD B. KENDALL, B. S., *Manager, Product Engineering Department, Toledo Scale Company, Division of Toledo Scale Corporation, Toledo, Ohio; Member, Institute of Electrical and Electronics Engineers; Registered Professional Engineer (Ohio). (Force Measurement)*

JOHN H. MORRISON, B. S. (C. E.), *Product Manager, Helicoid Gage Division, American Chain & Cable Company, Inc., New York, N.Y.; Senior Member, Instrument Society of America. (Pressure and Vacuum Measurement)*

FORCE MEASUREMENT

By Donald B. Kendall*

Force is defined as any cause that produces, stops, or changes the motion of a body or tends to produce these effects. Change in motion (speed) of a body (mass) is its acceleration, and Newton's second law of motion gives the relationship between force and these factors:

$$F = ma \qquad (1)$$

where F = force, poundals (English system)
     or dynes (metric system)
    m = mass, lb or g
    a = acceleration, ft/sec² or cm/sec²

Although this is the scientific definition of force, in most engineering work the term is used much more loosely—such as the force exerted by a spring (generally measured in pounds). The engineer is not thinking of the ability of the compressed spring to impart a certain acceleration to a certain mass but rather of its ability to close a valve or hold it closed against a certain pressure. Although a rocket engine is designed to impart a definite acceleration to a definite mass, the designer knows that he must provide additional thrust to overcome the force of the friction of the air.

One of the commonest forces to be measured, the force of gravity, is covered in Sec. 5. Other forces for the measurement of which instrumentation is required are:

1. Aerodynamic forces acting on an airplane.
2. Spring forces—compression, tension, or torque.
3. Testing materials—force required to cause deformation or rupture.
4. Thrust developed by jet engines and rockets.
5. Torque (force times moment arm) developed by electric motors or various types of engines.

A force is usually measured by applying it to a calibrated device which resists the force and indicates, or records, its magnitude. Because centuries of development have resulted in highly accurate scales for the measurement of weight, these same scales (see Sec. 5) can be used for measurement of the other forces. In some cases, modifications are required to apply the force properly to the load-receiving element of the scale. For instance, a horizontal force can be measured directly by opposing it with a load cell, but it must be converted to a vertical force by some such means as a bell crank to measure it with an even-arm, beam, or pendulum-type scale—as shown in Figs. 2 and 5. Some forces of a transient or rapidly varying nature require the rapid response of a strain-gage load cell or a piezoelectric system. For some force measurements the high accuracy obtainable from "commercial-type" scales is not required, and less complex equipment is used.

TYPES OF EQUIPMENT FOR FORCE MEASURING

Section 5 describes various types of scales. The even-arm balance and even-arm scale are seldom used for force measurement because of the time consumed in the measuring operation.

The *beam scale* is used for some spring-testing devices because its accuracy permits testing a wide range of springs with high accuracy.

* Manager, Product Engineering, Toledo Scale Division of Toledo Scale Corporation, Toledo 12, Ohio.

Pendulum scales are adaptable to measuring any force which is steady for the two or three seconds necessary for the scale to come to rest.

Springs are frequently used for force measurement. Spring scales generally use special alloy springs and refinements in construction to attain high accuracy over an ambient temperature range from approximately 30 to 120°F.

Many force-testing devices incorporating springs without these refinements are perfectly satisfactory where accuracies of better than $\frac{1}{2}$ or 1 per cent are not required. Such devices are available in capacities from a few grams to 50,000 lb or more.

Strain-gage load cells are excellent force-measuring devices, particularly when the force is not steady. A rapidly varying force can be measured with a strain-gage load cell and an oscilloscope.

Strain gages are sometimes applied directly to the force-developing device and the device is calibrated against strain-gage output. The techniques of bonding and temperature compensation are extremely specialized, however; so caution should be observed if accurate and reliable results are required.

Hydraulic and pneumatic load cells are frequently used where large forces are to be measured.

Piezoelectric effect, where application of pressure to crystals (such as quartz) develops an electrical charge proportional to the pressure, is used for measuring forces, particularly of a transient or rapidly changing variety. Forces from less than 1 lb to 5,000 lb can be measured.

Proving rings, as shown in Fig. 1, determine force by measuring the deflection of a steel ring. Deflection can be measured by a micrometer, using a vibrating reed to

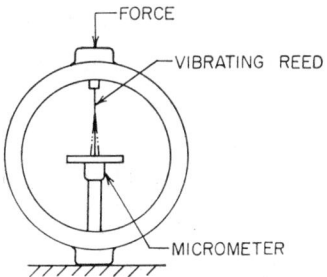

FIG. 1. Morehouse proving ring.

determine when contact is made, or by a linear differential transformer. The rings are very stable and accurate to ±0.1 per cent, but because deflection is not linear the actual force is determined by reference to a calibration curve. The commonest application is for calibrating other measuring devices. Capacity range is 300 to several hundred thousand pounds.

AERODYNAMIC FORCES

To prove a specific design for an airplane, or to test new principles of design, it is common practice to test a model of the airplane or a component incorporating the new principle in a wind tunnel. A wind tunnel is a rectangular or circular tunnel in which propellers circulate the air at the desired velocity for the test. An airplane mounted in the airstream is subjected to the same forces as it would be when flying at the speed of the air flow. Wind tunnels are in operation which will test a moderate-sized full-scale airplane at air speeds up to 500 mph. For supersonic speeds, because of the tremendous power required to circulate a large volume of air, the tunnels are smaller and scaled-down models are tested. In some cases, the "tunnel" is a venturi tube between a pressure tank and a vacuum tank.

Figure 2 shows an airplane in a wind tunnel mounted on a platform. The platform is mounted on four vertical supports, each supported by a scale lever. Three horizontal connections to the platform are connected to scale levers by means of bell cranks. By proper interconnection of levers, six scales measure the forces and moments shown; one scale each for lift, drag, side force, roll moment, pitch moment, and yaw moment. Alternately, load cells with electrical output proportional to the force can be installed in the vertical and horizontal members, and by proper electrical interconnection, the three forces and three moments can be measured directly.

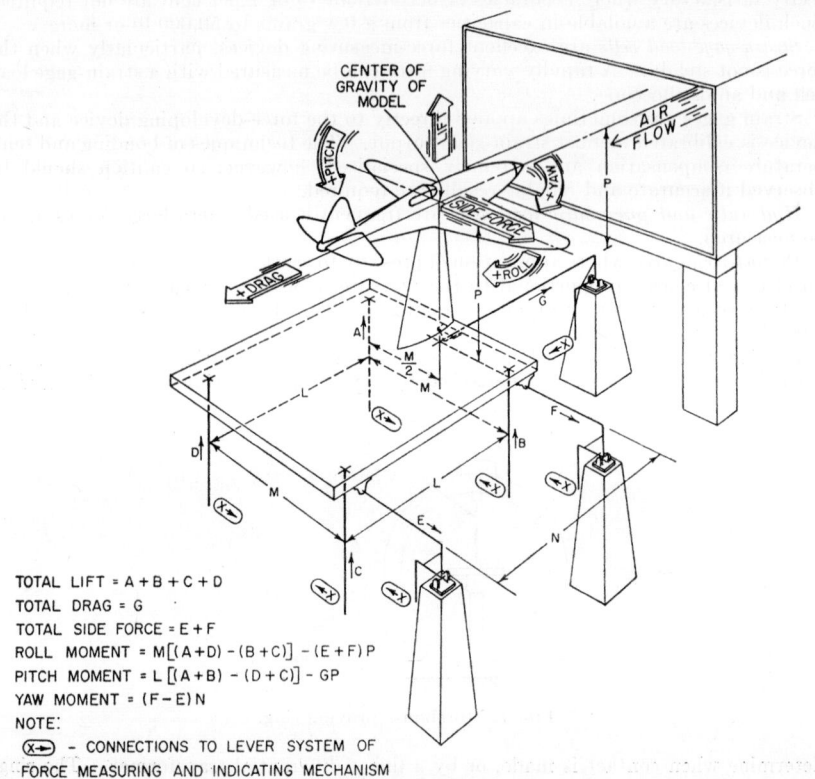

TOTAL LIFT = A + B + C + D
TOTAL DRAG = G
TOTAL SIDE FORCE = E + F
ROLL MOMENT = M[(A+D) − (B+C)] − (E+F)P
PITCH MOMENT = L [(A+B) − (D+C)] − GP
YAW MOMENT = (F − E)N
NOTE:
(X→) − CONNECTIONS TO LEVER SYSTEM OF
FORCE MEASURING AND INDICATING MECHANISM

FIG. 2. System for measuring aerodynamic forces in a wind tunnel.

By testing the airplane or model at different air speeds, level, and rotated to different angles, and by operating the control elements (such as ailerons, rudder, brake flaps, and trim tabs), power requirements and stability of the airplane or model can be determined.

SPRING TESTING

Most springs used by industry are manufactured for a specific application. Design of springs has developed into quite an exact science, but it is quite difficult and tedious to measure the factors—wire diameter, coil mean diameter, number of turns, and modulus of elasticity of the material—which determine a spring's characteristics. It is much easier to measure the force exerted by the spring, preferably at the deflection under which it is to be used.

The force on a spring to cause a definite deflection is

$$F = \frac{LGd^4}{8D^3N} \qquad (2)$$

where F = force, lb
$\quad L$ = deflection, in.
$\quad G$ = torsional modulus of rigidity
$\quad d$ = wire diameter, in.
$\quad D$ = mean coil diameter, in.
$\quad N$ = number of active turns

Because the force at a given deflection varies as the fourth power of the wire diameter, inversely as the cube of the mean coil diameter, inversely as the number of active turns, and directly as the modulus of the material, if the application is critical, 100 per cent inspection may be justified. In other cases, spot checks may be made to assure that the spring-forming equipment remains in adjustment and that the material dimensions and properties remain uniform.

For many applications, measuring the force exerted at a specific dimension is sufficient to determine that the spring meets specifications. This is quite often true of compression- and torque-type springs. Many tension springs are wound with initial tension and the initial tension as well as the rate is important to the application. These characteristics can be determined by measuring the force at two lengths of extension.

For a tension spring without initial tension,

$$R = \frac{F}{L_1 - L_0} \tag{3}$$

where R = rate, lb/in. deflection
$\quad L_1$ = extended length, in.
$\quad L_0$ = free length, in.
$\quad F$ = force, lb, to extend from L_0 to L_1

For a tension spring with initial tension, the rate can be determined by measuring the force at two lengths:

$$R = \frac{F_2 - F_1}{L_2 - L_1} \tag{4}$$

where R = rate, lb/in. deflection
$\quad L_1$ = extended length, in., for all active coils open
$\quad F_1$ = force, lb, at L_1
$\quad L_2$ = any extended length, in., beyond L_1
$\quad F_2$ = force, lb, at L_2

The initial tension can then be determined from

$$F_0 = F_1 - R(L_1 - L_0) \tag{5}$$

where F_0 = initial tension, lb
$\quad L_0$ = free (unextended) length, in.
$\quad L_1$ = an extended length, in. (all active coils must be open)
$\quad F_1$ = force, lb, at length L_1
[F_1 and L_1 can be either F_1 and L_1 or F_2 and L_2 in the data taken for the spring rate in Eq. (4).]

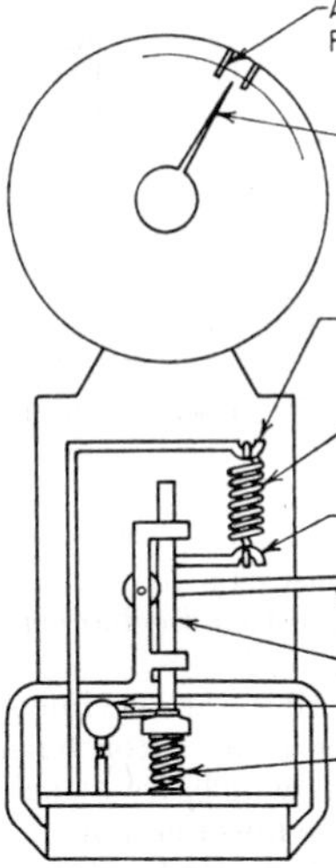

FIG. 3. Spring tester.

For production testing of springs, automatic indicating scales are generally used because of the speed with which the force can be measured. Figure 3 shows a bench scale adapted for manual testing of springs of moderate force. A pinion attached to the operating arm rotates as the operating arm is pulled down. The pinion engages a rack on the power plunger to drive the plunger down. This compresses a compression-type spring against the scale platform until an adjustable-dial indicator, registering with a stop on the scale platform, shows the spring has been compressed to the desired dimension.

A tension spring to be tested would be inserted between the upper tension hook, attached to the scale platform, and the lower tension hook, attached to the power plunger. The scale has a tare poise or other tare-adjusting means for offsetting the dead weight of the spring. Gage blocks of the proper height for specific springs or master springs, whose force at a specific length are accurately known, are generally used to adjust the dial indicator when the scale indicates the desired force. For larger springs the power plunger is generally hydraulically driven.

TESTING OF MATERIALS, JOINTS, AND BONDS

Many properties of materials, such as strength in tension, compression, bending, shear, and torsion, are determined by applying a known force and measuring the resulting yield of the material or applying an increasing force until failure occurs. Strength of welded, riveted, or cemented joints is determined in the same manner.

Equipment for this type of work varies from (1) a simple spring scale with a hook on one end to measure the breaking or yield point of small wire, paper, or plastics to (2) a universal testing machine with capacities to several hundred thousand pounds for measuring strength of concrete and steel. Figure 4 shows a universal testing machine

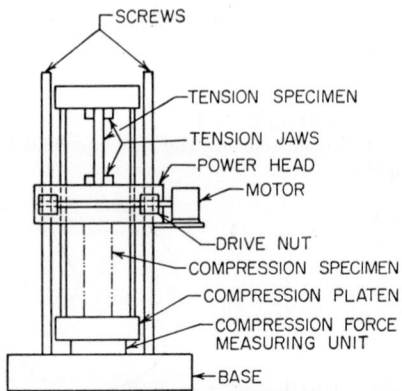

Fig. 4. Universal testing machine.

of one of several designs. Two vertical screws are fixed in the base. A power head is driven toward the base, when force is to be applied, by a motor rotating nuts on the screws. Above the base is a platen supported on the force-measuring device, which may be levers connected to a mechanical scale or one or more hydraulic, pneumatic, or strain-gage load cells.

For compression testing, the specimen is mounted on the platen and force is applied directly by the power head. For tension testing, suitable jaws are attached to a crossbar between posts which are attached to the platen. Similar jaws in the top of the power head grip the lower end of the specimen so that a downward force is exerted on the platen when the power head moves down to apply the force.

THRUST MEASUREMENT

The performance of jet and rocket engines is determined by the thrust force developed vs. the fuel consumed. Figure 5 shows an arrangement for measuring the thrust developed by an airplane-type jet engine. The platform or frame supporting the engine is mounted on three or four vertical posts with antifriction bearings or flexure plates at each end so the platform is free to move in the direction of thrust. The weight of the engine is carried through the posts.   The thrust force is measured by a mechanical scale connected to the platform through a bell crank, or by a load cell mounted horizontally.   Rocket- or missile-propulsion engines are frequently mounted vertically for testing, with strain-gage load cells incorporated in the supporting structure.

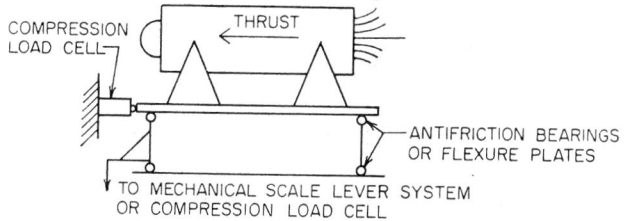

Fig. 5. Method for thrust measurement.

TORQUE MEASUREMENT FOR ENGINE AND MOTOR TESTING

Very small motors can be given a quick torque test by wrapping several turns of string around the shaft, fastening one end of the string to a hanging scale, and pulling on the other end to apply load.   The torque is the spring reading times the arm length, which is the radius of the shaft.

The *prony brake*, shown in Fig. 6, consists of a hollow drum attached to the motor or engine shaft and an arm attached to a band with friction lining which passes around

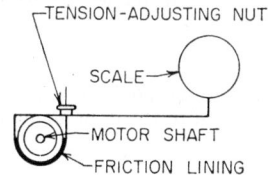

Fig. 6. Arrangement of prony brake.

the drum.   The free end of the arm either is attached to a hanging scale or rests on the platform of a bench scale.   Load is applied by increasing tension on the band. The drum generally contains water to dissipate the heat developed.   The prony brake is inexpensive, but it is difficult to adjust and maintain a specific load.

Most modern engine and motor testing is done with *electric dynamometers*.   One form of dynamometer is essentially an electric generator driven by the engine or motor. The power developed is dissipated into heat or may be channeled into power lines to serve a useful purpose.   Another form is the eddy-current dynamometer, consisting of one or more conducting disks rotating in a magnetic field.   The power developed by the engine or motor is determined by measuring the torque applied to the dynamometer and multiplying by the speed.   Since 1 horsepower = 33,000 ft-lb/min,

$$\text{hp} = \frac{2\pi L \times F \times \text{rpm}}{33,000} \qquad (6)$$

where hp = horsepower
L = radius (length of arm), ft
F = force, lb
rpm = revolutions per minute

Since most dynamometer arms are measured in inches, for L in inches,

$$\text{hp} = \frac{F \times L \times \text{rpm}}{63,025} \qquad (7)$$

To simplify calculations, L is generally established as 63,025 divided by some round number, such as $2,000(L = 31.513$ in.$)$, $3,000(L = 21.008$ in.$)$, or $6,000(L = 10.504$ in.$)$. The round number is then the "dynamometer constant" and the formula becomes

$$\text{hp} = \frac{F \times \text{rpm}}{\text{dynamometer constant}} \tag{8}$$

To determine the rating for a dynamometer scale (or other force-measuring device), the formula may be rewritten as

$$F = \frac{\text{hp} \times \text{dynamometer constant}}{\text{rpm}} \tag{9}$$

Assume a dynamometer with a continuous rating of 200 hp with 150 per cent (300 hp) rating for 15 min, 2,500 to 6,000 rpm, with a torque arm of 21.008 in. (dynamometer constant $= 3,000$), $F = \dfrac{300 \times 3,000}{2,500} = 360$ lb. The scale or other force-measuring device would have to have a rating of at least 360 lb.

A further simplification, common in laboratories with a number of dynamometers of different sizes, is to calibrate the scales in arbitrary units, so that for each dynamometer

$$\text{hp} = \frac{\text{scale reading in arbitrary units} \times \text{rpm}}{1,000} \tag{10}$$

In each case, the pounds pull per arbitrary unit is 63.025 in. divided by the actual length of the torque arm in inches.

In most cases, the dynamometer will be called upon to measure torque in either direction of rotation. When mechanical scales, either spring or pendulum type, are used, a "reversing linkage" (Fig. 7) is required. An actuating bar with two knife-edges in opposite directions is bolted to the dynamometer frame. If the direction of

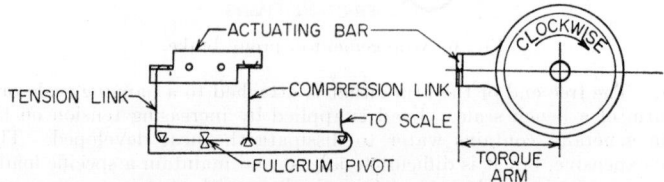

FIG. 7. Dynamometer connection to a scale.

rotation is clockwise, the left linkage, in tension, applies an upward force to its pivot in the lever and a downward force is exerted on the scale. The right linkage is free. If the direction of rotation is counterclockwise, the right linkage, in compression, applies a downward force to its pivot in the lever and a downward force is exerted on the scale. The left linkage is free. The "push-pull" fulcrum pivot is located midway between the tension and compression pivots in the lever, so that the scale reads the same for either direction.

Torque from the dynamometer can also be measured by coupling a "universal" (calibrated in tension and compression) strain-gage load cell to the arm, or by coupling two compression-type hydraulic or pneumatic load cells, one above and one below the arm. The torque can also be measured by mounting two strain gages (see Fig. 8) on a calibrated section of shaft and inserting this unit between the engine or motor and the dynamometer. Slip rings are used to carry the strain-gage leads to the conventional bridge circuit.

To minimize errors in torque measurement, the electrical leads to the dynamometer should be very flexible in the direction of rotation, and the shaft bearings should have the least friction possible. On some large precision dynamometers, the outer races of the shaft bearings are rotated so the bearing friction is always "rolling" and "break-away" friction is never encountered.

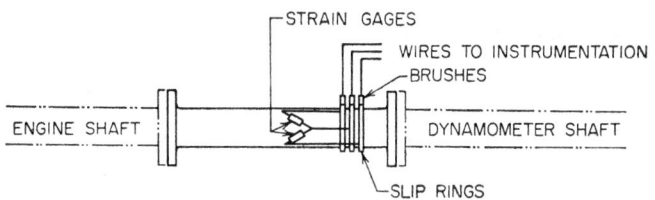

FIG. 8. Strain gage used for torque measurement.

FACTORS AFFECTING ACCURACY OF MEASUREMENTS

The full force must be applied to the measuring device in the proper direction, so that no component of the force will "bypass" the measuring device or affect its accuracy.

Any extraneous forces, such as the weight of the force-producing mechanism, must be allowed for or compensated for before the test starts.

The measuring device must be accurately calibrated under the proper conditions. If it is temperature-sensitive, it must be calibrated at the test temperature, or a correction factor must be used.

Accurate force-measuring devices generally contain relatively delicate parts. They must be properly maintained and protected from abuse.

REFERENCES

1. Ludewig, F. A., Jr.: Digital Transducer: Force Input—Binary Coded Output, *Control Eng.*, vol. 8, no. 6, pp. 107–109, June, 1961.
2. Jones, E.: Piezoelectric and Piezoresistive Strain Gages, *Control Eng.*, vol. 8, no. 9, pp. 134–137, September, 1961.
3. Dickson, T.: Selecting Strain Transducers, *Automation*, vol. 9, no. 11, pp. 85–89, November, 1962.
4. Carleton, R. J.: High-capacity Force Standards, *ISA J.*, vol. 8, no. 6, pp. 38–42, June, 1961.
5. Murray, W. M.: What Are Strain Gages—What Can They Do? *ISA J.*, vol. 9, no. 1, pp. 30–36, January, 1962.
6. Murray, W. M.: Strain Gage Types and Basic Circuits, *ISA J.*, vol. 9, no. 2, pp. 47–51, February, 1962.
7. Catz, J.: Basic Strain Gage Instrumentation, *ISA J.*, vol. 9, no. 4, pp. 50–55, April, 1962.
8. Sanchez, J. C.: Semiconductor Strain Gages—What Can They Do? *ISA J.*, vol. 9, no. 5, pp. 38–40, May, 1962.
9. Starr, J. E.: Strain Gages vs. Temperature, *Auto. Cont.*, vol. 14, no. 3, pp. 19–23, March, 1961.
10. Staff: Electromagnetic and Potentiometer Transducers, *Electromechanical Design*, May, 1959, pp. 35–42.
11. Proceedings of Institute of Physics Conference: "The Measurement of Stress and Strain in Solids," Reinhold Publishing Corporation, New York, 1948.
12. Nalle, D. H.: Fundamentals of Strain Measurement and Recording, *Auto. Cont.*, vol. 14, no. 11, pp. 51–55, November, 1961.
13. Nalle, D. H.: Fundamentals of Strain Measurement and Recording, *Auto. Cont.*, vol. 14, no. 12, pp. 31–35, December, 1961.
14. Considine, D. M. (ed.): "Process Instruments and Controls Handbook," McGraw-Hill Book Company, Inc., New York, 1957.
15. Tatnall, F. G.: Field Testing Techniques Using the Bonded-wire Strain Gage, *ASTM Bull.*, pp. 62–66, July, 1954.
16. Tatnall, F. G., and G. Ellis: Stress-Strain Records at High Straining Rates, *Nondestructive Testing* (a publication of Baldwin-Lima-Hamilton Corp.), November-December, 1953.
17. Stein, P. K.: Advanced Strain Gage Techniques, a mimeographed compendium of some 700 pages prepared by Stein Engineering Services, Inc., Phoenix, Arizona, 1962. (Contains very extensive lists of references on strain gages.)
18. Ruge, A. C.: 250,000-pound Calibration Facility (for load cells), *Instr. Control Systems*, vol. 36 no. 2, pp. 91–92, February, 1963.

PRESSURE AND VACUUM MEASUREMENT

By John H. Morrison*

This subsection briefly reviews some fundamentals and methods for the measurement of pressures ranging from fractions of a mm Hg to 100,000 psig. More details of most devices covered herein are given in Ref. 1 and 2. Instruments employing the mechanical elements vary from the familiar dial-indicating pressure gage to indicating or recording instruments that may include electric or pneumatic transmission with automatic controllers incorporated in the locally mounted or remote receiving instrument. Electrical types generally transmit signals to instruments with bridge circuits as described in Sec. 1.

DEFINITIONS OF PRESSURE TERMS

Pressure may be defined as the action of a force against some opposing force; a force in the nature of a thrust, distributed over a surface; or a force acting against a given surface within a closed container. Usually it is measured as a force per unit area. Figure 1 shows kinds of pressure that gages can measure involving the terms *absolute, atmospheric, gage,* and *differential pressure,* as well as *vacuum.* These and other terms associated with pressure measurements are defined in the following paragraphs.

Absolute Pressure. Fluid pressure measured above a perfect vacuum or above zero absolute pressure (point *A* or *A'* in Fig. 1). Zero absolute pressure is obtained

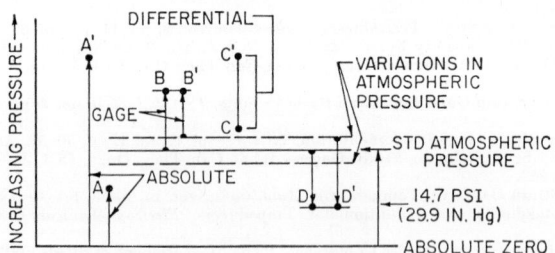

FIG. 1. Relation of pressure types.

only when there is no molecular impingement, which implies a vanishingly small population of gas molecules or of molecular velocity. Simply, the force per unit area exerted on a limiting or bounding surface: the pressure indicated by an ordinary pressure gage added to the atmospheric pressure.

Atmospheric Pressure. The pressure exerted by the earth's atmosphere, as commonly measured by a barometer (hence the synonym "barometric pressure"). At or near sea level the value of this pressure is close to 14.7 psia or 29.9 in. (760 mm) of mercury absolute, decreasing with altitude.

Gage Pressure. Generally, a pressure above† atmospheric, measured by an element that measures the *difference* between the unknown pressure and the *existing*

* Product Manager, Helicoid Gage Division, American Chain & Cable Company, Inc., New York, N.Y.

† See under Vacuum for gage pressures *below* atmospheric.

atmospheric pressure (point B in Fig. 1). Note that, if the *absolute* value of the pressure remains constant and the atmospheric pressure increases, the gage pressure B' decreases; this difference is usually small and in higher-pressure measurements generally insignificant. From Fig. 1, it is obvious that the absolute value of a gage pressure can be obtained by adding the actual atmospheric-pressure value to the gage reading (i.e., if B is 10.3 psig and atmospheric pressure is 14.7, the absolute value of B is 25.0 psia).

Differential Pressure. The difference between two measured pressures, such as on the inlet and outlet process line of a heat exchanger to show pressure drop (C-C' in Fig. 1). Can be indicated by a gage arranged to indicate the pressure difference, often with a zero difference point at mid-scale to permit readings in either direction. Even though both gage elements measure gage pressures, the distinction between "absolute" and "gage" pressures is irrelevant because variations in atmospheric pressure would affect both by practically an identical amount.

Vacuum. A gage pressure below atmospheric, often measured by the same types of elements that measure pressures above atmospheric—i.e., by *difference* between the unknown value and the *existing* atmospheric pressure (point D in Fig. 1). Vacuum values increase toward absolute zero and are usually expressed as "inches of mercury," "inches of water," or "millimeters of mercury" (see under Units of Pressure). Here, too, as with gage pressures, variations in atmospheric pressure have a slight effect on vacuum-gage readings. In the case of vacuums, such variations can be more serious because the entire range down to absolute zero is only 14.7 psi.

From Fig. 1 note that vacuum readings can be converted to absolute-pressure values by subtracting the reading from the actual atmospheric pressure as determined by a barometer.

Static Pressure. The force per unit area acting on a wall by a fluid at rest or flowing parallel to the wall in a pipeline, as at A in Fig. 2. It is also termed *line pressure.*

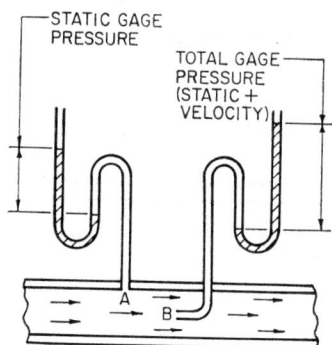

Fig. 2. Relation of static and total pressure.

Total Pressure. Together with the static pressure, which acts in all directions, velocity pressure is acting on the bent tube at point B in Fig. 2, to create a *total pressure* (also known as *kinetic pressure*). Normally, total pressure and static pressure are read as gage pressures. In engineering work, however, especially where computations are involved, absolute values of these pressures are frequently used. Thus total pressures must be specified as gage or absolute.

Velocity Pressure. That due to the speed of flow; it is also known as *velocity head* or *impact pressure.* For an incompressible fluid, or a gas flowing at low velocities, it is equal to $V^2/2g$ where V = rate of fluid flow and g is acceleration due to gravity. Flow is a computed and not an indicated pressure.

Hydrostatic Pressure (or Head). The pressure at a point below a liquid surface due to the height of liquid above it. Its value P at a distance h below the surface of the liquid whose density is d is $h \times d \times g$, where g is the gravity constant and all values are in consistent units. The familiar mercury column or manometer uses such a pressure as a measure of an unknown pressure.

Total Pounds or Tons, or "Tons on Ram." Terms frequently encountered with hydraulic machinery, such as hydraulic presses. They express the force in pounds or tons that acts over a given area, such as the area of a hydraulic ram:

$$P \text{ (lb)} = \text{psig} \times \text{sq in. of area} \tag{1}$$

$$\text{Tons (on ram)} = \frac{\text{psig} \times \text{sq in. of ram area}}{2,000} \tag{2}$$

Conversely, if it is desired to select a pressure-gage range to meet a "tons on ram" need, Eq. (2) is written

$$\text{psi (range)} = \frac{\text{tons on ram} \times 2,000}{\text{sq in. of ram area}} \tag{3}$$

The psi value thus determined is generally doubled to provide a safe working margin and maintain better calibration when the gage is in use.

UNITS OF PRESSURE

Because the original pressure-measuring instruments were U-tube manometers or similar head-type devices, liquid-head units have historically been used in both engineering and scientific work to express values of pressure. Both the English and metric units of length are used. Also, because the type of liquid (usually water or mercury) determines its density, the liquid must be named, such as "inches of water" ("in. H_2O"). Inches, feet, centimeters, and millimeters are commonly used. At high vacuums (near absolute zero pressure), the *micron* (10^{-3} mm Hg) is often used.

For units of liquid head, the density of the liquid varies somewhat with temperature. Hence, to define the unit rigidly, a standard temperature (usually that of maximum density for the liquid) must be named. For water, the standard temperature is 3.9°C or 39°F; for mercury, 0°C or 32°F. However, readings of liquid heads at other than these standard temperatures are seldom seriously in error, and the difference can be neglected except for the most exacting test or scientific work.

Table 1. Pressure Unit Conversion Factors

Pressure units	psi	in. H_2O	ft H_2O	in. Hg	atm	g/cm²	kg/cm²	cm H_2O	mm Hg
1 psi =	1.000	27.68	2.307	2.036	0.06805	70.31	0.07031	70.31	51.72
1 in. water (39°F) =	0.03613	1.000	0.08333	0.07355	0.002458	2.540	0.002540	2.540	1.868
1 ft water (39°F) =	0.4335	12.000	1.000	0.8826	0.02950	30.48	0.03048	30.48	22.42
1 in. mercury (32°F) =	0.4912	13.60	1.133	1.000	0.03342	34.53	0.03453	34.53	25.40
1 normal atmosphere =	14.7	406.79	33.90	29.92	1.000	1,033	1.033	1,033.0	760.0
1 g/sq cm =	0.01422	0.3937	0.03281	0.02896	0.0009678	1.000	0.0010	1.000	0.7356
1 kg/sq cm =	14.22	393.7	32.81	28.96	0.9678	1,000	1.000	1,000	735.6
1 cm water at 4°C =	0.01422	0.3937	0.03281	0.02896	0.0009678	1.000	0.0010	1.000	0.7355
1 mm Hg at 0°C =	0.01934	0.5353	0.04461	0.03937	0.001316	1.360	0.001360	0.001360	1.000

When pressures are expressed as force per unit area, again both English and metric units are used. The most common units of force are the gravitational or weight units:* the pound, ounce, or ton (usually the pound) and the gram or kilogram. The unit of area is almost always consistent with the weight unit—both in English units

* A weight unit of force times g, the acceleration due to gravity, equals an absolute unit of force (seldom used to express pressure). In English units g = approximately 32 ft/sec²; hence 1 lb = roughly 32 *poundals*. In metric units, g = approximately 980 cm/sec²; hence, 1 gram = approximately 980 dynes.

or both in metric units. Common area units are the "square inch" and the "square centimeter."

One other way of expressing pressure is the "atmosphere." This unit simply uses the standard atmospheric-pressure value of 14.7 psi (or its equivalent in metric units) and defines it as "one atmosphere." "Two atmospheres" would be 29.4 psi. The unit is convenient in expressing larger pressures.

Table 1 gives conversion factors for changing pressure values in one unit to their equivalent in another unit. It includes only nine of the more common units (see Ref. 2 for a table covering 16 units).

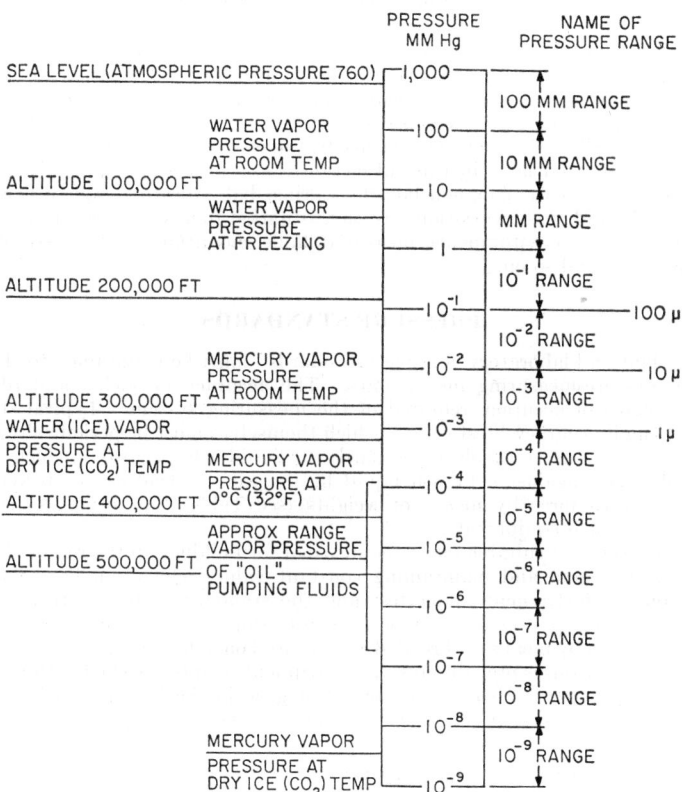

FIG. 3. Subatmospheric pressure ranges and useful pressure values.

Figure 3 charts the subatmospheric-pressure range and some useful values of pressure in this range.

APPLICATIONS OF PRESSURE MEASUREMENTS

In company with temperature and flow, pressure measurements are used extensively in processing or manufacturing industries, in the laboratory, and in other fields such as aerospace, aircraft, and aboard ships. Reasons for measuring pressure are quite varied, and numerous examples are given in later sections dealing with applied instrumentation. Generally, one or more of the following reasons can be found for the measurement (and often automatic control) of pressure:

To Maintain Safe Operating Conditions. A closed vessel or pipe has a maximum safe pressure, varying with its material and construction. Excessive pressure not only can destroy valuable equipment but also, in the process of destroying it, can endanger personnel and adjacent equipment—particularly if inflammable or corrosive fluids are involved. For such applications, absolute and highly accurate readings are often less important than is extreme reliability. Control may only be a high or low limit shutdown type with visible or audible alarm to warn the operator.

To Help Control a Process. Pressure can have a direct or indirect effect on the value of another process variable (such as composition of a mixture in a distillation process). In such cases, its absolute value—consistently and accurately measured—can be more important than minimum maintenance. Simple indication to guide an operator may sometimes suffice; in other cases, recording and automatic control are employed.

To Provide Test Data. In research or quality control, tests on materials and equipment may require very accurate measurement of pressure as a part of other test data. In such applications extreme care in proper selection, installation, and maintenance of the measuring equipment is essential.

Because of this broad variety of user needs, the wide variety of pressure-measuring methods (summarized later in this subsection) have been developed. Even within one method, such as the common bourdon-spring dial-indicating gage, a bewildering array of classes and model variations are on the market to meet user needs for different degrees of accuracy, environments, measured fluids, mountings, and accessories—such as alarms, seals, and so on.

PRESSURE STANDARDS

In the plant and laboratory suitable "standards" must be maintained to check and calibrate pressure-measuring instruments. The accuracy of such standards varies with the degree of accuracy required of the measuring system. Typical standards include (1) high-accuracy "test gages" which themselves must be periodically checked against a more accurate absolute standard; (2) liquid-column testers which compare the tested device against a known liquid head; and (3) dead-weight testers which provide a known force by means of weights (see Table 2 for principles and brief descriptions of such equipment).

In the test setup, a suitable means is provided for building up the required internal pressure in a system and maintaining constant values which can be compared as readings on the instrument under test and the standard or test instrument. One typical source of test pressure is a pump comprising a cylinder in which a plunger compresses a fluid by means of threaded rod (as used on a dead-weight tester). Where steady process pressures prevail, pressure instruments can be checked with reasonably high accuracy by parallel connection of a test gage in the line; plugged fittings are often provided in an installation to permit such checks.

PROTECTION OF MEASURING ELEMENT

As will be noted in the subsequent type descriptions, pressure measurements can cause maintenance problems due to improper selection and installation, because the vital measuring element can be exposed directly to the fluid being measured. In fact, the fluid can be inside the instrument case in dial indicators or field-mounted transmitters employing elastic-member elements such as a bourdon spring or bellows. Hence the prospective user must consider well the properties of the measured fluid and their possible effects on the element.

For example, the fluid may be corrosive to the element's material. In such cases, an element material resistant to this corrosion may be available, or the fluid must be sealed from direct contact with the element. A chemical diaphragm seal, shown schematically in Fig. 4, is one common and effective way to protect bourdon-tube elements from corrosion. The diaphragm is of a material that will withstand the corrosive fluid; the volume above the diaphragm, including the measuring element, is filled with a suitable oil that transmits pressure hydraulically to the element.

Various types of seals are also useful for reasons other than protection of the element

Table 2. Pressure-measuring Devices—Selection Guide

Note. Unless otherwise stated, figures for accuracy given in this table as per cent signify that the average expected error will not be more than plus or minus the stated percentage (which is in terms of full scale). Accuracy figures are difficult to generalize for all types of instruments and in some cases only a qualitative term is used; in other cases an estimated value is given, based upon best available data. Well-designed, installed, and maintained systems may give considerably better accuracy; if the reverse holds, poorer results may be obtained.

Operating principle	Range	Accuracy	Applications and remarks
1. Elastic-member Types			
Bourdon-type Pressure Deflecting Units (see Fig. 1). Tube of flattened cross section, shaped to an incomplete circle (conventional bourdon tube) or spiral or helix, tends to straighten with pressure increase. Motion is multiplied to pointer by cam or gearing	0–40,000 psi (higher with certain types), bourdon to 100,000 psi	Up to ½%	Most indicating gages use conventional bourdon tube. Method is simple, reliable, and usually accurate
Twisted-tube Pressure Element (see Fig. 2). Has hollow metal tube that has been flattened and twisted about a longitudinal axis. One end is sealed; other, open to pressure. Tube responds to pressure or vacuum applied internally or externally, or to pressure difference between internal and external values. Pressure response is converted to electric signal	...	...	Small internal volume, low mass, and high stiffness give the element a maximum natural frequency of 10,000 cps. Small rotational movement of element keeps stresses below static limit. Hence, hysteresis is minimum
Resistance Detector (Fig. 3). Converts gage pressure to electrical resistance by means of capsule (shown) or bourdon tube. Connecting rod positions noble-metal contact wiper over precision potentiometer	0–1½ to 0–5,000 psig	¾–2%	Internal volume of capsule is 0.29 cu in.; of bourdon tube, 0.15 cu in. Response speed 50 msec full scale
Single-bellows Absolute-pressure Gage (Fig. 4). Evacuated bellows installed within vessel whose pressure is to be measured transmits motion outside vessel through seal. Measurement is pressure difference vs. vacuum; hence, absolute pressure	0–200 mm Hg abs	½–1%	Can be supplied in spans as low as 0 to 6 mm Hg abs for vacuum drying, steam condensers, vacuum stills, antibiotic processing

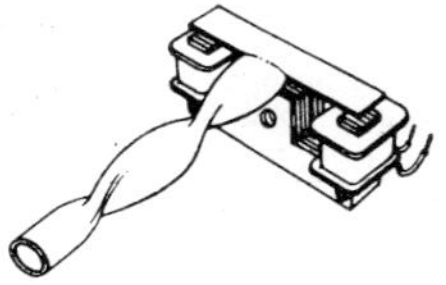

Fig. 1

Fig. 2

Fig. 3

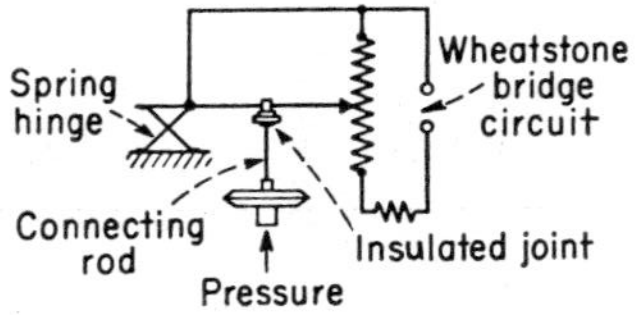

Spring hinge • Connecting rod • Pressure • Wheatstone bridge circuit • Insulated joint

Fig. 4

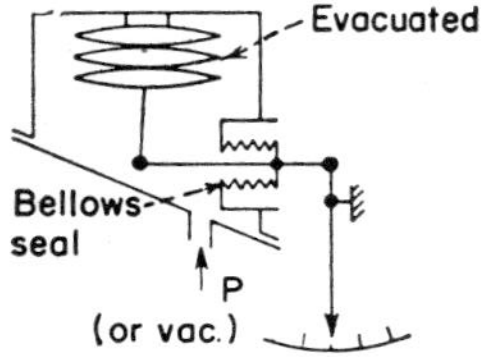

Evacuated • Bellows seal • P (or vac.)

Table 2. Pressure-measuring Devices—Selection Guide (Continued)

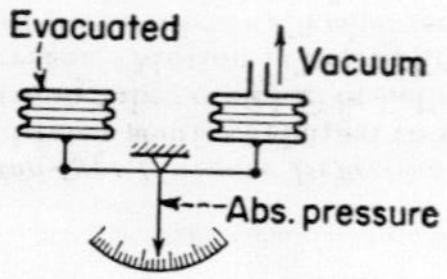

Operating principle	Range	Accuracy	Applications and remarks
Two-bellows Absolute-pressure Gage (Fig. 5). Made with either opposed or beam-balanced bellows; one is evacuated, other connected to unknown pressure	0–25 psia	1% or better	Considered accurate for absolute pressures down to about 5 mm Hg. Uses are similar to those given directly above
Aneroid Manostat (Fig. 6). Evacuated bellows, permanently sealed, provide zero absolute pressure reference. Temperature-compensated spring opposes bellows compression	0.8–60 in. Hg abs	High	Used in dead-end service in static testing systems; also in calibration and testing
Bellows-type Gages (Fig. 7). Include metallic bellows, capsular and multiple diaphragm units, in which pressure is balanced against unit's spring action or against a calibrated spring	Low vacuums to about 100 psi	½–1%	Simple and rugged; widely used in moderate-pressure and low vacuum measuring instruments—especially in recorders
Nested-diaphragm Gage (Fig. 8). A variation of bellows type (Fig. 7) wherein metal diaphragms are nested for compactness with sensitivity	0–30 in. water to 0–30 psi	¼%	Gage withstands high external overpressure without damage or loss of calibration
Stainless-steel Bellows (Fig. 9). Senses pressure in liquid metal. Core of variable transformer is connected to bellows. Variations in pressure are translated into an electrical output, which is indicated by a suitable meter	...	...	Unit directly measures pressure of molten metal. Range is adjusted by linkage in balancing spring. Transformer winding of transmitter and receiver are connected in series, thus providing voltage and temperature compensation

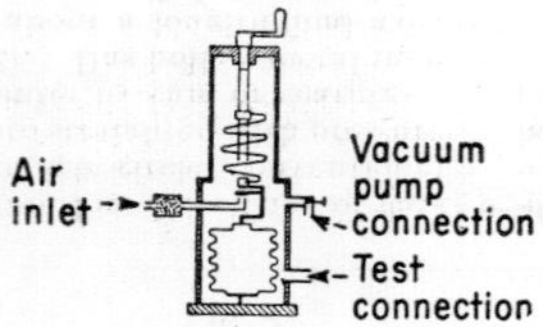

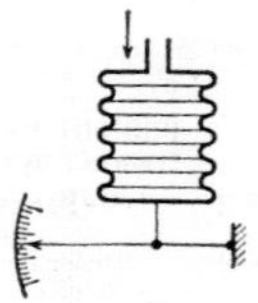

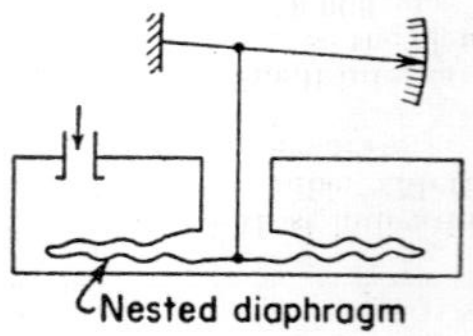

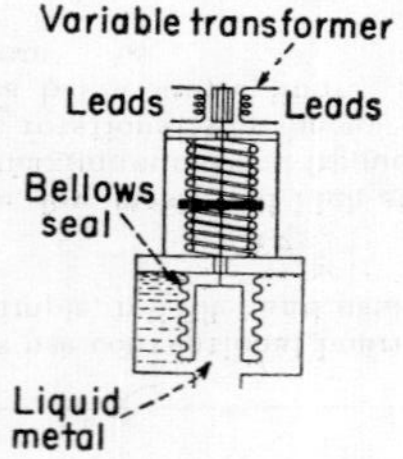

Fig. 5

Fig. 6

Fig. 7

Fig. 8

Fig. 9

Table 2. Pressure-measuring Devices—Selection Guide (Continued)

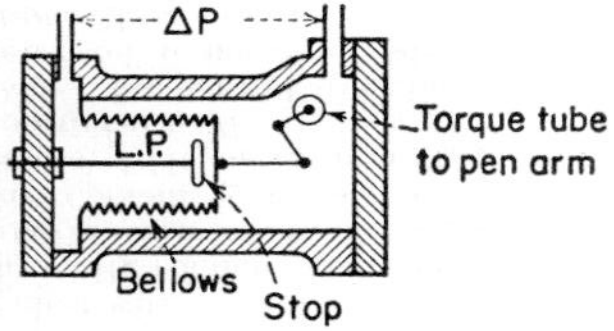

Fig. 10

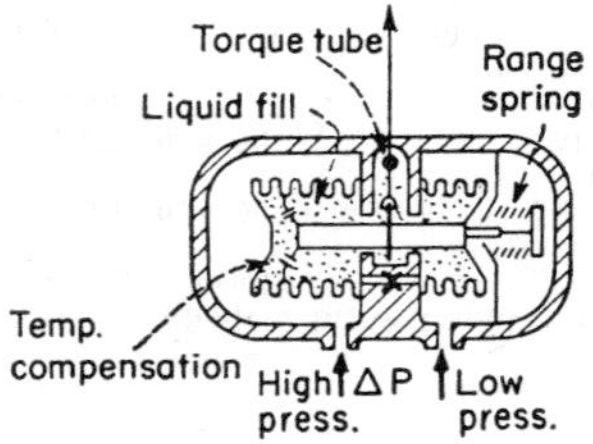

Fig. 11

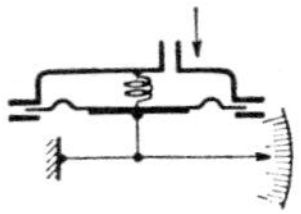

Fig. 12

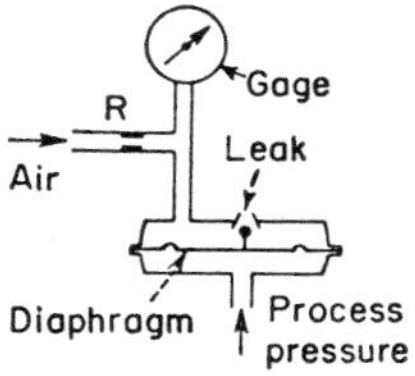

Fig. 13

Operating principle	Range	Accuracy	Applications and remarks
Single-bellows Differential Gage (Fig. 10). Bellows (or diaphragm) in casing is subjected to low pressure on one side, high on other side. Bellows extension measures differential	0–25 to 0–800 in. water differential	Approx 2 %	Simple and rugged. Range easily changed. Requires no mercury. Minimizes maintenance. For general differential-pressure measurements including flow by variable-head methods
Double-bellows Differential Gage (Fig. 11). Opposed bellows in two-part casing are filled with liquid, balanced by spring, and connected together by a shaft passing through a restriction. Arrangement acts like spring-balanced piston with high pressure at one end, low at other	0–10 water to 0–400 psi	½ %	Pulsations are damped by valve controlling the flow of liquid from one bellows to the other. Temperature compensation for liquid expansion is provided by floating-bellows section. Unit is unharmed by overranging. Range is easily changed. Torque tube for transmitting motion eliminates possible leakage
Slack-diaphragm Gage (Fig. 12). Slack diaphragm is balanced by pressure or tension of a calibrated spring	0–120 in. water	About 1 %	Used especially for low pressure and draft such as 0 to 0.5 in. water
Force-balance Gage Seal (Fig. 13). Process pressure is applied to a force-balance pneumatic transmitter; transmitter pressure then activates gage. Restriction *R* in air-supply line, combined with diaphragm-controlled leak vent, exactly duplicates pressure	Any	1 %	Gage segment may either be built into the rest of the unit, or separated from it by as much as 1,000 ft for remote control. Either way, gage is protected against corrosive, viscous, dirty, or radioactive fluids

4-17

Table 2. Pressure-measuring Devices—Selection Guide (*Continued*)

2. Gravitational—Manometers, Bells, etc.

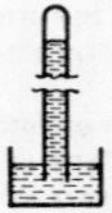

FIG. 14 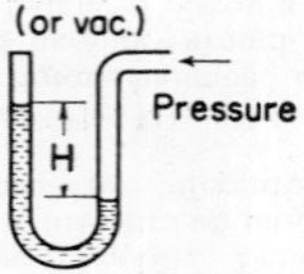   FIG. 15 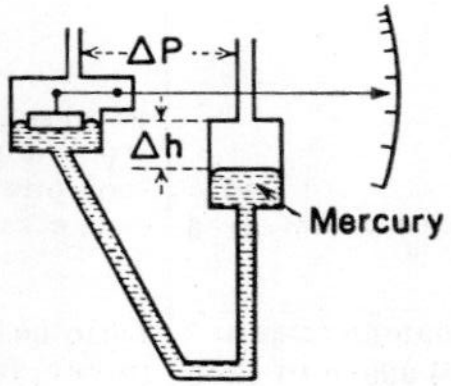   FIG. 16 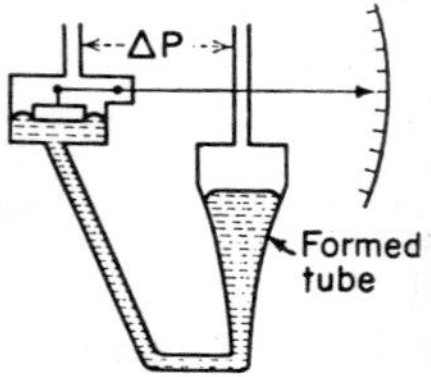   FIG. 17 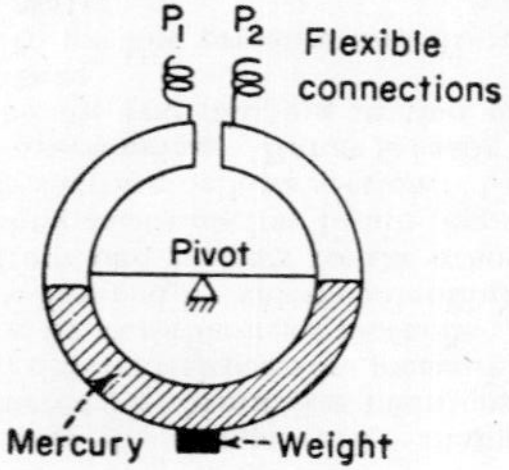   FIG. 18

Operating principle	Range	Accuracy	Applications and remarks
Liquid-column Barometer (Fig. 14). The fundamental instrument for measuring pressure of the atmosphere. Glass tube, closed at one end and filled with mercury, is inverted in a vessel of mercury. Height of Hg column above vessel level is the barometric pressure in mm Hg (corrected to mercury density at 32°F)	Any pressure of atmosphere	Roughly 0.1 mm	Although less convenient than the aneroid type, the Hg barometer is used for precise measurements and for calibration of other instruments
Liquid-filled Glass Columns of Well- or U-type Design (Fig. 15). Measure pressure, draft, or vacuum vs. pressure of atmosphere. Pressure proportional to H	About 2 in. of liquid head	About 0.1 in. of manometer liquid	Simple and accurate, but limited by glass construction and available column height
Inclined-draft Gage. Similar to U-tube but has one inclined leg and a large-diameter well to increase reading accuracy	0.5–50 in. water	Better than 0.01 in. of liquid	Used chiefly for low drafts and pressures where visual indication is sufficient
Glass Liquid-column Manometers. Identical to the type of Fig. 15, except that the two pressures are connected, one to each end of tube	0.5–50 in. Hg	Min division 0.02 in. of liquid	Same as for Fig. 14. Also limited static pressure
Metal Manometers, Plain Tube (Fig. 16). Liquid level, usually mercury, is read by carrying float position outside manometer by pressuretight bearing, torque tube, or magnetic follower	Differentials of 10 in. water min to 400 in. or higher	1%	Most common method of measurement for rate-of-flow meters. Also for differential-pressure, level, and density measurements. Static pressures to 5,000 psi
Metal Manometers, Formed Tube (Fig. 17). Similar to plain-tube type except the range tube is formed to variable parabolic cross section to extract square root for flow	...	1%	Not commonly used, because of expense of accurately forming the range tube. Gives uniform scale for flow
Ring-balance Manometer (Fig. 18). Pressure differential displaces mercury in ring, causing ring to tilt until weight balances displaced Hg. Degree of tilt is measure of differential	About 10 in. water max	1%	Used for general rate-of-flow measurement. Avoids need for pressuretight shaft

Table 2. Pressure-measuring Devices—Selection Guide (Continued)

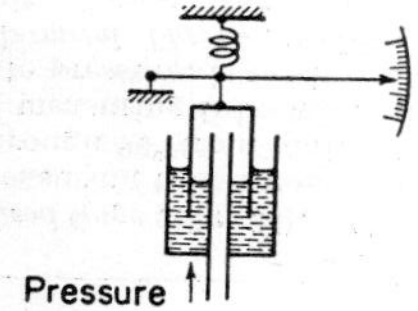

Fig. 19 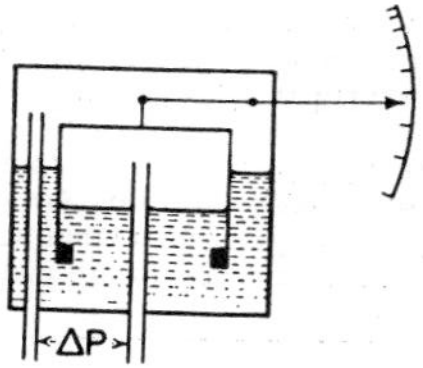 Fig. 20 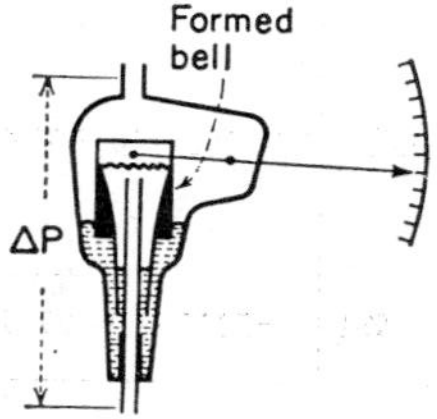 Fig. 21 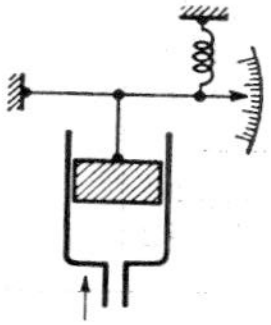 Fig. 22 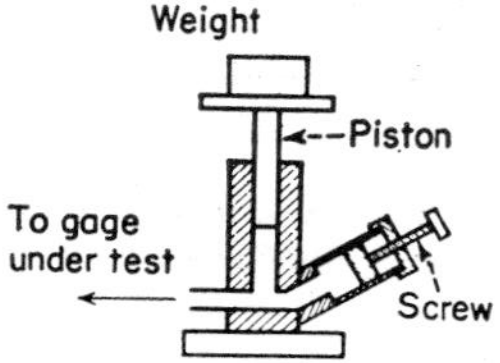 Fig. 23

Operating principle	Range	Accuracy	Applications and remarks
Liquid-sealed Bell (Fig. 19). Inverted bell dips into liquid, which acts merely as a seal. Floating Bell, Single (Fig. 20). Bell is balanced by a calibrated spring. Similar to single-pressure bell, except enclosed for application of pressure to both sides. Bell position proportional to differential, carried by torque tube or pressuretight bearing	Vacuum to about 15 in. water 0.2–12 in. water	1 % or better 1 %	Used for low pressures and vacuums; can be made extremely sensitive Large area of bell gives ample power for operating recording mechanism at very low differentials
Formed Bell (Ledoux Bell) (Fig. 21). Bell shaped internally to parabolic form extracts square root of differential to give uniform scale on flow. Hg seal	212 in. water max differential	2 %	Used for general rate-of-flow metering. Uniform scale facilitates reading and integrating
Piston Gage (Fig. 22). Except for its friction, unit is comparable to slack-diaphragm gage. Pressure is balanced by a calibrated spring. (Variations include weight-loaded and scale-beam type balancing methods)	Vacuum to high, depending on spring	0.2 %	Used mainly for study of pressure cycles, as in the steam-engine indicator. Operates stylus, records
Dead-weight Tester (Fig. 23). Type of piston gage in which test pressure is balanced against known weight when applied to known piston area. Pressure is applied by secondary piston and screw. Weight platform is rotated to avoid sticking	Unlimited	0.1 %	Used as primary standard for calibration of gages

Table 2. Pressure-measuring Devices—Selection Guide (*Continued*)

3. Electrical Types for High Vacuums

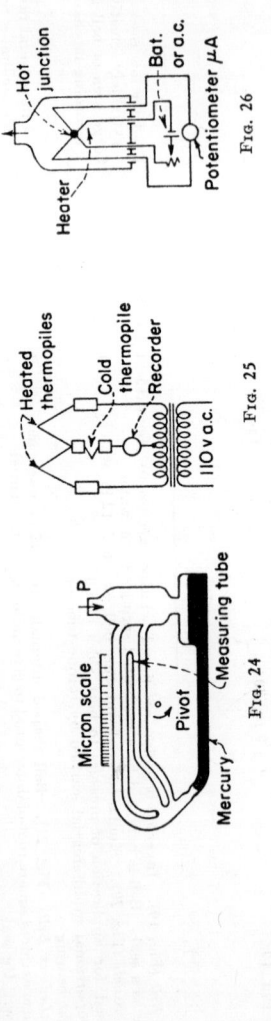

Fig. 24

Fig. 25

Fig. 26

Operating principle	Range	Accuracy	Applications and remarks
McLeod Gage (Fig. 24). Gas at low absolute pressure is trapped in measuring tube by mercury as instrument rotates about a pivot through 90° from horizontal. Mercury compresses gas into top of measuring tube where final volume, expressed as initial absolute pressure, is read	0.05–50 mm; 0.05–5,000μ	1%	Used as primary standard for absolute pressures in microns (0.001 mm). Requires trapping of condensables to avoid error
Noble-metal Thermocouples (Fig. 25). Connected in a bridge circuit. Low-voltage alternating current is supplied to heated thermopiles. Unheated thermopile connected in opposite polarity provides ambient-temperature compensation. The d-c output is measured	0–1,000μ; 0.1–200 mm	High	Rapid response, not damaged by release to atmospheric pressure. Low operating temperature, no oxidation of thermocouples. Measures total pressure as a function of thermal conductivity of the gas molecules remaining
Thermocouple Gage (Fig. 26). Inferential type, which measures thermal conductivity of gases at reduced pressure in terms of pressure. Temperature of heater filament is detected by thermocouple	1–300μ (some to 1,000μ)	. . .	Easy to use but must be calibrated for type of gas to be measured. Extends measurement below field of bellows-type gages and Hg manometers. Fast response
Pirani Gage. Similar in principle to thermocouple gage, except that heated-filament temperature is measured as a resistance thermometer in Wheatstone bridge	1–2,000μ	. . .	Features are similar to those of the thermocouple gage. Uses are also similar, above the operating range of ionization gages

4–20

Table 2. Pressure-measuring Devices—Selection Guide (*Continued*)

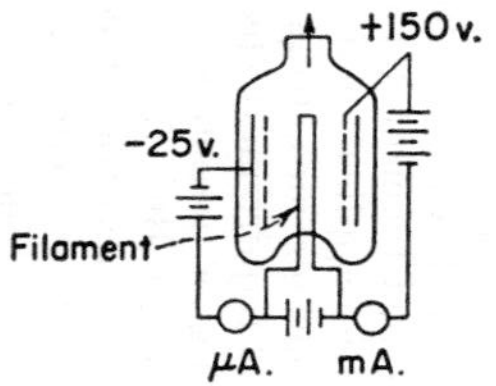

Operating principle	Range	Accuracy	Applications and remarks
Hot-filament Ionization Gage (Fig. 27). Uses three-element vacuum tube containing low-pressure gas to be measured. Molecules are ionized by electrons from filament; they flow to plate, are discharged, yield plate current proportional to pressure	$0.001–1\mu$	...	Suitable for low absolute pressures but must be protected against burnouts if pressure goes above 1μ. Unsuited for gases decomposed by hot filament
Knudsen Gage. A radiometer type in which a mechanical force is produced between one heated surface and one surface at the temperature of the gas by the molecules rebounding from the hotter surface with more kinetic energy than those from the cold surface. This force is proportional to the absolute pressure	$0.001–0.000001$ mm Hg	...	Temperature-measurement difficulties make most radiometer gages unsuitable to plant use

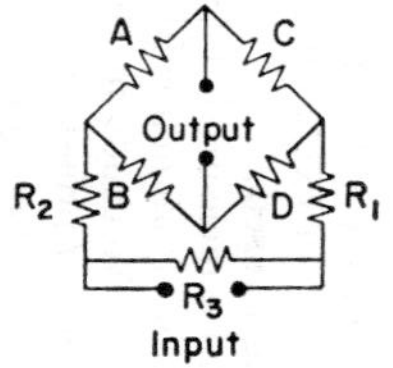

4. Strain Gage

Operating principle	Range	Accuracy	Applications and remarks
Strain-gage Load Cell (Fig. 28). Uses small resistance elements bonded to an element that expands under internal pressure. Stretch increases resistance, which is proportional to strain. Usual method assembles four elements in a Wheatstone bridge so that resistances A and D are strained, B and C are not. Other resistances are provided for temperature compensation and calibration	Unlimited	¼ % or better	Relatively new method, becoming popular owing to high accuracy, ease of transmission to millivolt-meter or potentiometer indicator or recorder, good linearity and reproducibility, negligible hysteresis, quick response. Unbonded types also used.

Fig. 27

Fig. 28

4–21

from corrosiveness of the measured fluid. For example, liquids that contain solids or clog the insides of the element must be sealed. Similarly, measured media that are likely to solidify or freeze at ambient temperatures of the pressure instrument must be sealed.

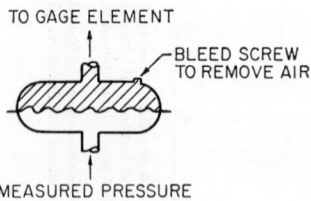

FIG. 4. Schematic of diaphragm seal.

CLASSIFICATION OF PRESSURE-MEASURING METHODS

Pressure-measuring instruments may be classified in several different ways,* each of which has some value in guiding the prospective user or explaining principles. In the selection guide in Table 2 the following grouping is used:

1. *Elastic-member Types.* Where the pressure to be measured deforms a bourdon tube or other elastic member, thus developing a counterbalancing force in a manner that permits calibration to provide a measurement. Methods in this classification can measure absolute, gage, vacuum (low), or differential pressure.

2. *Gravity-balance Types.* Largely those where a liquid column, such as mercury, counterbalances the unknown pressure by the gravitational force of a liquid head. This class includes U-tube manometers and variations thereof. The weight-loaded piston, such as in a "dead-weight tester" may also be included in this class as may liquid-sealed bells—devices which depend on weight to balance the unknown pressure.

3. *High-vacuum-measuring Types.* These are grouped together for convenience in reference because the user undoubtedly knows he has an application of this type when he approaches the selection problem. Most methods for measuring high vacuums (less than 0.1 mm Hg abs) are electrical in nature and require more specialized knowledge of measuring techniques than do methods for higher pressures.

4. *Strain-gage Types.* This is an electrical method (see Ref. 1, pages 3-30 through 3-63, for an extensive discussion of these and other electrical transducers).

In practice, instrument manufacturers and users have shown great ingenuity, adapting physical or electrical principles in a variety of combinations to suit application needs. It is difficult in this summary section to cover all variations, but the methods described give some idea of the possibilities. As one case in point the "null-balance" or "force-balance" principle—either pneumatic or electric—is widely employed in instrumentation for many variables; it may well be adapted, say, in the form of a transmitter with a primary element given in the summary table, even though not specifically covered.

REFERENCES

1. Considine, D. M. (ed.): "Process Instruments and Controls Handbook," Sec. 3, McGraw-Hill Book Company, Inc., New York, 1957.
2. Behar, M. F.: "Handbook of Measurement and Control," 3d printing, Chap. III, The Instruments Publishing Co., Inc., Pittsburgh, Pa., 1959.
3. Feller, Eugene W. F.: "Instrument and Control Manual for Operating Engineers," Chap. II, McGraw-Hill Book Company, Inc., New York, 1947.
4. Carroll, Grady C.: "Industrial Process Measuring Instruments," Chap. 4, McGraw-Hill Book Company, Inc., New York, 1962.
5. Dushman, S.: "Scientific Foundations of Vacuum Technique," John Wiley & Sons, Inc., New York, 1949.
6. Kennard, E. H.: "Kinetic Theory of Gases," McGraw-Hill Book Company, Inc., New York, 1938.

* See Ref. 2, p. 33.

7. Litting, C. N. W., and W. K. Taylor: "An Automatically-controlled Knudsen-type Vacuum Gage," Monograph 36, The Institution of Electrical Engineers, London, 1952.
8. Bayard, R. T., and D. Alpert: Extension of the Low Pressure Range of the Ionization Gage, *Rev. Sci. Instr.* vol. 21, pp. 571–572, 1950.
9. Evans, E. C., and K. E. Burmaster: A Philips-type Ionization Gage for Measuring of Vacuum from 10^{-7} to 10^{-1} mm of Mercury, *Proc. IRE*, vol. 38, p. 651, 1950.
10. Hunt, L. B.: The History of Pressure Responsive Elements, *J. Sci. Instr.*, March, 1944, pp. 37–42.
11. Giacobbe, J. B., and A. M. Bounds: Selecting and Working Bourdon-tube Materials, *Instr. Mfg.*, July–August, 1952.
12. Van der Pyl, L. M.: A Bibliography on Bourdon Tubes and Bourdon Tube Gages, ASME Paper 53-IRD-1.
13. Jones, E.: Piezoelectric and Piezoresistive Strain Gages, *Control Eng.*, vol. 8, no. 9, pp. 134–137, September, 1961.
14. Clarridge, R. E.: Selecting Pressure Transducers, *Automation*, vol. 9, no. 10, pp. 82–87, October, 1962.
15. Dickson, T.: Selecting Strain Transducers, *Automation*, vol. 9, no. 11, pp. 85–89, November, 1962.
16. Kramer, F.: Pressure Switch Selection, *Automation*, vol. 9, no. 12, pp. 77–78, December, 1962.
17. Murray, W. M.: What are Strain Gages—What Can They Do? *ISA J.*, vol. 9, no. 1, pp. 30–36, January, 1962.
18. Murray, W. M.: Strain Gage Types and Basic Circuits, *ISA J.*, vol. 9, no. 2, pp. 47–51, February, 1962.
19. Catz, J.: Basic Strain Gage Instrumentation, *ISA J.*, vol. 9, no. 4, pp. 50–55, April, 1962.
20. Sanchez, J. C.: Semiconductor Strain Gages—What Can They Do? *ISA J.*, vol. 9, no. 5, pp. 38–40, May, 1962.
21. Ortman, G. C.: Photoelectric Pressure Sensing, *ISA J.*, vol. 9, no. 12, pp. 63–64, December, 1962.
22. Aronson, M. H.: "Strain Gage Instrumentation," Instruments Publishing Co., Pittsburgh, Pa., 1958.
23. Leck, J. H.: "Pressure Measurement in Vacuum Systems," Reinhold Publishing Corporation, New York, 1957.
24. Landy, J. J., and G. J. Fiedler: *Automatic Pressure and Flow Controls System Design for Multi-loop Fluid Dynamics Facility with Prominent Distributed Parameters*, technical paper, AIChE Symposium, St. Paul, Minn., September, 1959.
25. Moynihan, G. J.: Circuits for High-ambient, Differential Pressure Measurement, *Control Eng.*, vol. 7, no. 4, pp. 163, April, 1960.
26. Brown, J. O.: Using Pressure Switches to Monitor and Control, *Automation*, vol. 7, no. 5, pp. 97 May, 1960.
27. Buffenmyer, W. L.: Selecting Bourdon-tube Gauges, *Instr. Control Systems*, vol. 34, no. 2, pp. 238–242, February, 1961.
28. Norton, H. N.: Potentiometric Pressure Transducers, *Instr. Control Systems*, vol. 34, no. 2, pp. 244–247, February, 1961.
29. Seegers, H.: Precision Bourdon-tube Gauge, *Instr. Control Systems*, vol. 34, no. 2, pp. 234–236, February, 1961.
30. Newhall, D. H., and L. H. Abott: High-pressure Measurement, *Instr. Control Systems*, vol. 34, no. 2, pp. 232–233, February, 1961.
31. Smith, O. W.: In-system Pressure Calibration, *Instr. Control Systems*, vol. 35, no. 3, p. 114, March, 1962.
32. Staff: Errors in Mercury Barometers and Monometers (abstracts from NBS Monogram 8, "Mercury Barometers and Manometers" by W. G. Brombacher, D. J. Johnson, and J. L. Cross, May 20, 1960, National Bureau of Standards, Wash., D.C.) *Instr. Control Systems*, vol. 35, no. 3, p. 121, March, 1962.
33. Newhall, D. H.: Manganin High-pressure Sensors, *Instr. Control Systems*, vol. 35, no. 11, pp. 103–104, November, 1962.
34. Studier, W.: Quartz Pressure Sensors, *Instr. Control Systems*, vol. 35, no. 12, pp. 94–95, December, 1962.
35. Clark, D. B.: Rare-earth Pressure Transducers, *Instr. Control Systems*, vol. 36, no. 2, pp. 93–94, February, 1963.
36. Damrel, J. B.: Quartz Bourdon Tube, *Instr. Control Systems*, vol. 36, no. 2, pp. 87–89, February, 1963.
37. Norton, H. N.: Piezoelectric Pressure Transducers, *Instr. Control Systems*, vol. 36, no. 2, pp. 83–85, February, 1963.
38. Rutherford, S. L.: Vacuum Techniques, *Instr. Control Systems*, vol. 36, no. 2, pp. 78–81, February, 1963.
39. Benson, J. M.: Thermal Conductivity Vacuum Gauges, *Instr. Control Systems*, vol. 36, no. 3, pp. 98–101, March, 1963.
40. Lafferty, J. M., and T. A. Vanderslice: Vacuum Measurement by Ionization, *Instr. Control Systems*, vol. 36, no. 3, pp. 90–96, March, 1963.
41. Norton, H. N.: Strain-gage pressure Transducers, *Instr. Control Systems*, vol. 36, no. 3, pp. 85–88, March, 1963.
42. Roehrig, J. R., and J. C. Simons, Jr.: Calibrating Vacuum Gauges to 10^{-9} Torr, *Instr. Control Systems*, vol. 36, no. 4, pp. 107–111, April, 1963.
43. Walsh, J. P.: Molecular Vacuum Gages, *Instr. Control Systems*, vol. 36, no. 8, pp. 106–107, August, 1963.

Section 5

QUANTITY AND RATE VARIABLES

By

CLYDE BERG, Ph.D.(Ch.E.) *Director, Clyde Berg Associates, Long Beach, Calif.; Member, American Institute of Chemical Engineers, American Chemical Society, American Petroleum Institute; Registered Professional Engineer (Calif.). (Solids Level)*

P. A. ELFERS, B.S., *Director, Fisher Governor Company, Marshalltown, Iowa; Member, American Society of Mechanical Engineers, Instrument Society of America. (Liquid-level Measurement)*

G. D. GOODRICH, A.B. *(Physics), Supervisor, Technical Sales, Statham Instruments, Inc., Los Angeles, Calif. (Acceleration Measurement)*

D. B. KENDALL, B.S., *Manager, Product Engineering Department, Toledo Scale Company, Division of Toledo Scale Corporation, Toledo, Ohio; Member, Institute of Electrical and Electronics Engineers; Registered Professional Engineer (Ohio). (Weight and Weight Rate of Flow)*

ROBERT C. LANGFORD, Ph.D.(E.E.), *Director of Research, Aerospace Group, General Precision, Inc., Little Falls, N.J.; Member, Institute of Electrical and Electronics Engineers, Institute of Radio Engineers, American Rocket Society, American Nuclear Society. (Speed Measurement)*

JEROME B. McMAHON, B.S.(M.E.), *Consulting Engineer, Chicago, Ill.; Member, American Society of Mechanical Engineers; Fellow, Instrument Society of America; Registered Professional Engineer (Ill.) (Flow of Fluids)*

FLOW OF FLUIDS

By J. B. McMahon*

Flow measurement is used in industry and commerce for two principal reasons: (1) accounting, and (2) providing the basis for controlling processes and operations, especially those of a continuous nature. Transition from batch to continuous operations has stimulated the use of continuous-flow measuring devices.

Flow measurement has been the subject of many years of intense engineering investigation, and many aspects of the subject have arrived at a reasonable engineering maturity and standardization, a situation that is uncommon to many branches of instrumentation. New concepts of flow measurement are being continually evolved, however, and require that an almost constant evaluation of the older, more established methods be made.

It is the purpose of this subsection to tersely outline the broad parameters of flow-measurement applications and methods and to provide a condensed summary of fluid-flow theory.† Table 1 at the end of this subsection provides a limited summary of available flow-measurement devices, including brief remarks regarding operating principles, range and accuracy of measurement, and applications. This table can be useful as a guide to the tentative selection of flow-measuring devices.

RANGE OF FLOW RATES AND OPERATING CONDITIONS

Flow-measurement applications vary from a few ounces per hour, as encountered in research work, to billions of gallons per day, as found in the use of water by large metropolitan centers.

Fluids are metered at temperatures down to minus 300 and up to 2500°F. Meters are in use on fluids flowing at pressures as low as 10^{-1} mm Hg and as high as 15,000 psig. Recent developments promise to increase this latter limit to as much as 200,000 psig.

The flow of all types of gases, liquids, and more recently, of fluidized solids is measured. Flow measurements are made for practically all kinds of industrial and commercial operations—from pure research to the dispensing of final products to the consumer.

Thus, with the wide gamut of applications, flow rates, and operating conditions, it is not surprising to note that there are literally scores of different primary and secondary measuring devices, with an almost unlimited number of ways in which these devices can be combined and applied.

ACCOUNTING APPLICATIONS

Historically, the earliest use of flow measurement was for accounting purposes; and for a determination of the efficiency of water pumps and pipelines. The classical work of Herschel involved measurements of the discharge of water pipelines and of water pumps. The early work of Weymouth and others involved the measurement of natural gas for purchase and sales purposes.

---

* Consulting Engineer, Chicago, Ill.
† Engineering details of the various primary elements and devices used for flow measurement can be found in sec. 4 of Ref. 12.

In industrial plants, flow meters for accounting were used initially for steam and water services. These uses remain common today, and may involve a series of meters which integrate the distribution of steam, water, and other utilities by departments and individual operations. The increasing cost of utilities, such as treated water, steam, and compressed air, have accentuated the need for accurate flow measurement.

When most processing was done by batch operations, it was common to weigh ingredients and final products. Continuous processing required that materials accounting be put on a continuous basis, and thus greatly increased the need for accurate flow meters.

PROCESS CONTROL APPLICATIONS

Metering of fluid flows has become a necessity in most industrial processing plants, largely as the basis for automatic control. Flows must be carefully measured and controlled to avoid upsetting a complex continuous process. Flow measurement is also important in proportioning and ratioing applications.

NATURE OF FLUIDS

Fundamentally, a fluid may be defined as a substance which will flow when subject to shear. In contrast, a solid requires a considerable amount of shear before it will "flow." Fluids can be classified as Newtonian or non-Newtonian.

Characteristics of Flowing Fluids

Many industrial measurements of rate of fluid flow are based upon the effect of velocity change caused by a restriction to the flow in a pipeline. The pattern of flow before, at, and after the restriction is of great significance, because practically all such measurements utilize empirical formulas or constants. To ensure a duplication of test results which establish these constants, the flow patterns of an industrial installation must be the same as those of the test installation. It is important, therefore, that users of such meters understand the factors which influence flow patterns.

Streamline or Laminar Flow. Assume that a thin, rigid sheet is immersed in a body of liquid at rest (see Fig. 1). If the liquid gradually starts to flow past the

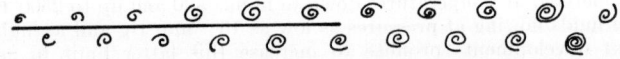

Fig. 1. Whirling eddies formed in turbulent flow along a thin rigid sheet.

sheet at low velocity, a pattern of disturbance forms along the sheet which shows that the individual particles of liquid tend to flow in paths parallel to the main stream flow. The fluid acts as though it is flowing in individual thin sheets or laminae; hence, this type of flow is known as *laminar* or *streamline* flow (also known as *viscous* flow).

If the liquid is constrained between banks, the individual laminae will have differing velocities, from zero at the bank to maximum midway between the banks (due to frictional drag of the bank). The flow, however, still consists of individual sheets which tend to flow parallel to one another at different velocities.

Turbulent Flow. As the velocity is increased, small whirling eddies start to form along the surfaces of the sheet, owing to its frictional drag (see Fig. 1). These eddies roll along the sheet surfaces and maintain their identity for appreciable distances from the end of the sheet. As the velocity is increased further, the eddies tend to intermix, until it becomes evident that the flow is no longer laminar and that there is a pronounced cross flow between the former laminae. All semblance of orderly progression ceases and individual particles assume erratic motions in all directions. Such flow is known as turbulent flow.

Average Velocity in Pipes. In a circular pipe, the same phenomena are observed, modified by the constraining effect of the pipewalls. Experiments have shown that at low velocities the fluid acts as though it flowed in individual tube streams with the stream nearest the pipewall at zero velocity and a gradually increasing velocity of succeeding concentric shells toward the center where maximum velocity exists. With such streamline flow where the shear strain is independent of the rate of shear, the resultant velocity profile is a parabola (see Fig. 2b). The average velocity of the flowing fluid is, therefore, mathematically computed to be one-half the maximum velocity. In this case, experimental results and observations agree very closely with mathematical analyses of the forces involved.

As the velocity is increased in a pipe, the tube streams start to intermix, with resultant cross flows and turbulence. At some rather indeterminate velocity, motions of the individual particles of the fluid become entirely random, and the flow becomes highly turbulent, to an extent that it becomes impossible to analyze the forces mathematically. Experimental results show that the average velocity with turbulent flow is a considerably higher proportion of the maximum velocity, varying from 0.78 to 0.89, depending on (1) the velocity, (2) the pipe size, (3) the fluid density and viscosity, and (4) the roughness of the pipe.[2] Figure 2a shows the velocity profile of such turbulent flow.

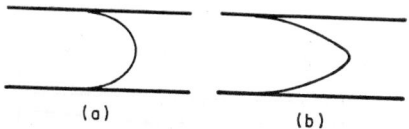

(a)   (b)

Fig. 2. Flow velocity distribution across a pipe: (a) Turbulent flow; (b) laminar flow.

Reynolds Number. In 1883, Sir Osborne Reynolds, an English scientist, proposed a very valuable criterion for the conditions of flow prevailing in a smooth pipe, reported in a paper[1] before the Royal Society. This criterion, known as Reynolds number correlates the factors affecting fluid flow as follows:

$$R_d = \frac{DV\rho}{\mu} \tag{1}$$

where R_d = Reynolds number
       D = diameter of pipe
       V = average velocity of flow in pipe ⎫
       ρ = density of fluid   ⎬ all in consistent units
       μ = absolute viscosity of fluid   ⎭

A number of significant points should be noted regarding this number. In the first place, it is seen that in consistent units of measurement, such as the English or metric system, all dimensions cancel out. Hence, the Reynolds number is what is termed a dimensionless number, having the same numerical value regardless of units. For example, in the English system

$$R_d = \frac{(\text{ft})(\text{ft/sec})(\text{lb/ft}^3)}{(\text{lb/ft-sec})} = \cancel{\text{ft}} \times \frac{\cancel{\text{ft}}}{\text{sec}} \times \frac{\cancel{\text{lb}}}{\cancel{\text{ft}^3}} \times \frac{\cancel{\text{ft-sec}}}{\cancel{\text{lb}}}$$

and in cgs units

$$R_d = \frac{(\text{cm})(\text{cm/sec})(\text{g/cm}^3)}{\text{g/cm-sec}} = \cancel{\text{cm}} \times \frac{\cancel{\text{cm}}}{\text{sec}} \times \frac{\cancel{\text{g}}}{\cancel{\text{cm}^3}} \times \frac{\cancel{\text{cm-sec}}}{\cancel{\text{g}}}$$

Previously in this subsection it was shown that for normal flow in a pipe, the ratio of average velocity to maximum velocity varies as the flow changes from streamline to turbulent (from a constant of 0.5 for streamline flow to between 0.78 to 0.89 for turbulent flow, depending on the factors contained in the Reynolds number). If a graph is plotted of the velocity ratio $V/V_{\max}$ vs. Reynolds numbers, based both on average and maximum velocities, as shown in Fig. 3, a sharp break is seen to occur at a Reynolds number of about 4,000 (curve B). Given $V_{\max}$, as measured by a pitot tube located at the center of the pipe, curve B of Fig. 3 can be used to calculate average velocity V. Conversely, curve A is used to predict V_{max} when $DV\rho/\mu$ is known.

The Reynolds number is of basic importance in many flow calculations, as is brought out in Ref. 12, sec. 4, on Head Flow Meters. With head meters, the coefficient of discharge is shown to be directly related to the Reynolds number, and tabulations have been made from test data relating these two factors. The usefulness of the Reynolds expression is best evidenced in such cases where test data under one set of conditions can be extended to provide data for other actual conditions in a given application. For example, tests on a fluid of one viscosity can be extended to a fluid of a higher or lower viscosity, provided other factors which comprise the Reynolds number are changed a proportionate amount so as to give the same Reynolds number. Such possible calculations reduce considerably the amount of laboratory data needed to predict results over a wide range of pipe sizes, velocities, densities, and viscosities.

The Reynolds concept furnished the means by which it was possible to establish absolute values of present-day orifice coefficients for natural gas measurement from data based on weighed water tests.

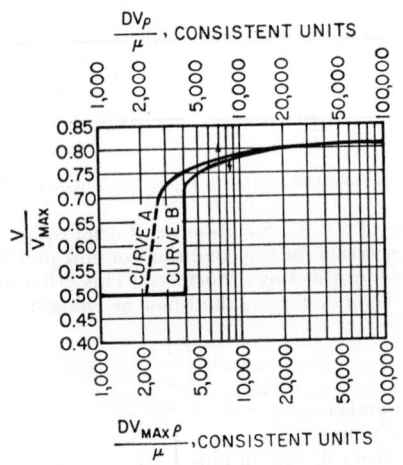

FIG. 3. Ratio of average to maximum velocity vs. Reynolds number. (*Source: W. H. Walker, W. K. Lewis, W. H. McAdams, and E. R. Gilliland, "Principles of Chemical Engineering," 3d ed., McGraw-Hill Book Company, Inc., New York, 1937.*)

Limitations of the Reynolds number, however, include the following:

1. It does not take into account the roughness of pipewall, although it can be usefully incorporated into expressions of velocity distribution which do.[2,3] In the streamline or laminar state, the flow configuration is not affected much by the condition of the pipewall, but in the partially and fully turbulent states, the flow velocity distribution is very definitely affected by the relative roughness of the pipewall.

2. The concept of dynamic similarity which it expresses does not apply universally without modification with respect to pipe size.[2,3]

3. It applies only to fluids in which the viscosity is independent of the rate of shear—the Newtonian fluids.[5] For non-Newtonian fluids, where viscosity varies with rate of shear, the Reynolds number will not adequately express the relation between average velocity and velocity distribution.

However, for Newtonian fluids, the Reynolds number is a very useful tool, since it tells much about the velocity profile in the flowing stream. The velocity profile is extremely important for accurate flow measurement by any of the commonly used inferential methods using orifice plates, venturi tubes, or flow nozzles, or by current meters of the propeller type (turbine meters). The velocity profile also may have an appreciable effect on the accuracy of volume-type meters.

The Reynolds number will tell whether the flow exists in laminar, partially turbu-lent, or turbulent states and, therefore, what the approximate velocity distribution is. This is modified by the effect of pipewall roughness. In the laminar state, the con-dition of the pipewall or channel has very little effect on the flow configuration, but in the partially and fully turbulent states, the relative roughness of the pipewall pro-duces quite definite effects on velocity distribution (see Fig. 2a and b).

It is evident that, because of the greater average angle through which the turbulent flow must turn to get through a sharp-edge orifice, the size of the minimum contrac-tion on the outlet side will be smaller than that for laminar flow. It is the size of this minimum contraction, known as the vena contracta, that determines the capacity of the plate for a given head across it.

Since a sharp-edge orifice operates as though it were virtually frictionless, the viscous drag of the fluid has very little effect. With the rounded-entrance flow nozzle and the venturi tube, there is no vena contracta effect, and the main effect of laminar flow is from the viscous drag. As a result, the correction factors for Reynolds number are the reverse for sharp-edge orifice plates from those for nozzles and venturi tubes. For devices such as pitot tubes and anemometers, which make a spot velocity measurement in a flowing stream and from which flow quantity is interpreted by means of the relation of average velocity to the measured spot velocity, knowledge of the velocity profile is even more important.

BERNOULLI'S THEOREM

Operating principles of several types of flow meters are based upon the law of con-servation of energy, which is expressed in several forms for fluid motion, any one of

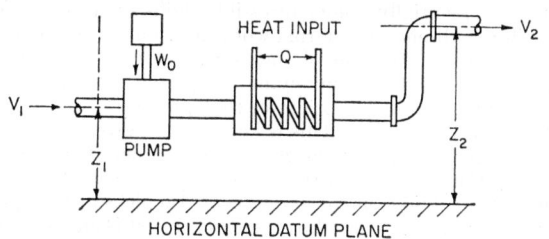

Fig. 4. Theoretical apparatus for energy-balance equations.

which may be termed Bernoulli's theorem. Practical equations applying to operation of certain flow-meter types are derived from this theorem and developed below in terms of a total energy balance and mechanical energy balance.

Total Energy Balance[7]

Consider a unit weight (one-pound mass) of fluid, and let

G = mass velocity, lb/(sec)(sq ft of cross section)
g = local acceleration due to gravity, ft/sec²
J = mechanical equivalent of heat, 778 ft-lb/Btu
p = absolute static pressure, psf
u = internal energy, Btu/lb
V = linear velocity, fps
v = specific volume, cu ft/lb
Z = height above arbitrary horizontal datum plane, ft

Then the potential energy of the fluid relative to the chosen reference level is Z ft-lb/lb of fluid, and the kinetic energy of the fluid is $V^2/2g$ ft-lb/lb. The total energy of the lb of fluid is therefore $Ju + Z + V^2/2g$ ft-lb/lb.

Now, suppose that the fluid is flowing steadily through an apparatus (see **Fig. 4**), with no accumulation or depletion of either energy or matter, and that all conditions at the entrance and exit of the apparatus are steady. Then, the total energy of a sample of the fluid can be altered only by having external work done on, or by, the sample, or by permitting heat to flow into or out of the sample. Therefore, if subscripts 1 and 2 indicate conditions at inlet and outlet, respectively, we obtain

$$\left(Ju_2 + Z_2 + \frac{V_2{}^2}{2g}\right) - \left(Ju_1 + Z_1 + \frac{V_1{}^2}{2g}\right) = JQ + W \tag{2}$$

where Q = heat received from sources external to apparatus, Btu/lb
$\quad W$ = net external work done on the pound of fluid while in the apparatus, ft-lb/lb
A part of the work W is done on the pound of fluid as it is being pushed past the entrance by the fluid behind it; this amounts to p_1v_1. Similarly, on passing the outlet, an amount of work p_2v_2 is done by the fluid on the fluid just ahead of it. Hence,

$$W = p_1v_1 - p_2v_2 + W_o \tag{3}$$

where W_o = work delivered by outside machinery to the average pound of fluid during passage through the apparatus, ft-lb/lb.
If Eq. (3) is used to substitute for W in Eq. (2) the resulting expression is often known as the over-all energy-balance form of Bernoulli's theorem, namely,

$$Ju_1 + Z_1 + p_1v_1 + \frac{V_1{}^2}{2g} + JQ + W_o = Ju_2 + Z_2 + p_2v_2 + \frac{V_2{}^2}{2g} \tag{4}$$

Mechanical Energy Balance[3]

For steady flow of essentially noncompressible fluids under conditions where friction is negligible, in the absence of external work, it is found experimentally that there is little error in an energy balance involving only the terms for the mechanical forms of energy. In corresponding cases involving compressible fluids, however, the total mechanical energy appearing at the outlet of the apparatus (see Fig. 4) often greatly exceeds that at the earlier section. This is because any given element of the moving fluid is undergoing expansion, in the course of which it does mechanical work.

The work done by the fluid element in expanding is expended on the fluid immediately ahead of it, but the fluid element in question also picks up an equivalent amount of mechanical energy from that behind it. The net result is an increase in mechanical energy at the expense of the internal energy of the fluid, or of externally derived heat. This self-expansion work is $\int_1^2 p\,dv$ which should be included in a mechanical energy balance.

When correction is made for self-expansion work, the mechanical energy output is found to be less than the input. In order to balance the equation, a term F is introduced to represent total friction loss due to fluid flow.

Finally, if between sections 1 and 2 a pump actually delivers W_o ft-lb of mechanical work to the fluid, this term must appear on the left-hand side of the equation. The resulting expression is the so-called mechanical energy-balance form of Bernoulli's theorem:

$$Z_1 + p_1v_2 + \frac{V_1{}^2}{2g} + \int_1^2 p\,dv + W_o = Z_2 + p_2v_2 + \frac{V_2{}^2}{2g} + F \tag{5}$$

Together with the total energy-balance form, the above equation serves as the fundamental basis for the solution of problems in fluid flow. As will be seen in Ref. 12, sec. 4, on the various types of flow meters, many assumptions and empirical constants must be derived to provide working equations for the handling of practical problems.

MEASUREMENT METHODS

Methods of measuring fluid flows fall into two broad categories, namely: (1) discrete quantity methods, and (2) inferential methods. These further break down into classes depending on types of meters employed, and are described briefly below, together with a short discussion of solids flow measurement.

Discrete Quantity Methods

Discrete quantity methods employ instruments generally known as positive-displacement meters; in the ASME report, Fluid Meters: Their Theory and Application, they are referred as quantity meters. In general, they may be divided into two classes, namely, meters which count (1) the number of times a definite weight of liquid has passed, or (2) the number of discrete quantities that have passed.

Inferential Methods

Under inferential methods, four types of measurement are used, namely:
1. Spot measurement of velocity.
2. Measurement of total velocity.
3. Measurement of energy change due to a velocity change.
4. Measurement of potential energy.
In any of these, velocity, or rate of flow, is measured and total quantity is secured by integration.

Flow Measurement of Solids

Meters for the measurement of the flow of solids generally operate on a weighing basis. In some cases, the material is discharged into a hopper, which fills until a definite weight is reached, when the inflow is shut off and the hopper discharged. This is identical with the use of weighing tanks for the measurement of liquids.

For continuous measurement of solids flowing in pipes, a variation of the propeller meter is used. On conveyors, a section of the conveyor is isolated, and the variation of weight due to the weighed material is measured (see Sec. 5 on Weight). In many cases, the isolated conveyor section consists of a complete unit, with its own drive operating at a constant speed, and with the weighing instrument included. Integrators can be used for determining total flow. Similar installations have been used to determine variations in density of the solids, by measuring variations in weight of a constant cross section of the material.

Table 1. Flow-measuring Devices—Selection Guide

NOTE: Unless otherwise stated, figures for accuracy given in this table as per cent signify that the average expected error will not be more than plus or minus the stated percentage of full instrument scale. Accuracy figures are difficult to obtain for many kinds of instruments and, in a few cases, no figures are given; in other cases, the figures have been estimated from the best available data. Well-designed, installed, and maintained instruments may give considerably better accuracy; if the reverse situation holds, poorer results may be experienced.

Operating principle	Range	Accuracy	Applications and remarks
Volumetric Meters			
Weighing Methods: Scale-balanced tank automatically fills, cuts off at predetermined weight, records on counter, and discharges—cycle repeats	Moderate	High	For batching and for recording on semicontinuous basis. Temperature correction not required
Piston Meter: Piston driven by fluid pressure shuttles back and forth, recording number of strokes in terms of volume passing. Variations include multiple pistons driving wobble plate; double piston in two-part cylinder; reciprocating-bellows type (used for gas). Piston motion operates poppet or surface-type valve	Up to 1,500 gpm	1 % or better depending on type	Gives volumetric measure of flow. In some types can correct automatically for temperature variations. Generally can handle wide range of viscosities. Most commonly used for liquids (except fuel-gas meters)
Metering Pumps: Many types of positive-displacement pumps can be used for metering of liquids when driven at constant speed. Includes piston-type controlled-volume pumps; piston-type rayon spinning pumps; gear pumps	Low to moderate	High	Depending on type, can be used to provide accurate rate of liquid delivery either in discrete increments, or in pulsationless flow. Can be provided with remote recording or indicating counter
Lobed Rotor Meter (Fig. 1): Two lobed impellers in a close-fitting chamber rotate without contact with chamber or each other, by action of timing gears. Action is similar to a gear pump, liquid or gas carrying around outer ends of chamber	Up to about 1,000,000 cfh	Variable, depending on slip	Used chiefly for gases, but can be used for clean liquid materials. Have low pressure drop. Slip increases at low flow rates; best used at high rates

Rotors

FIG. 1

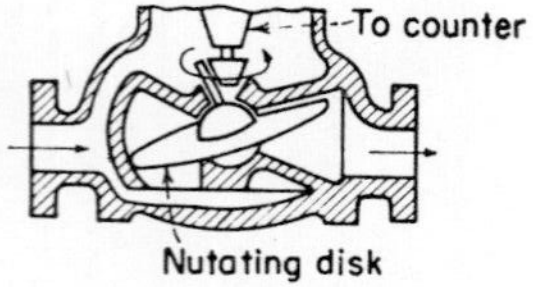

To counter

Nutating disk

FIG. 2

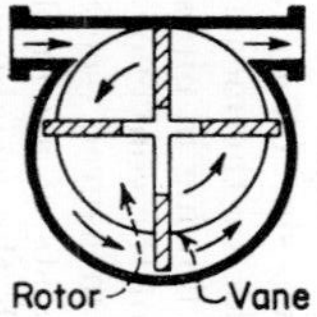

Rotor Vane

FIG. 3

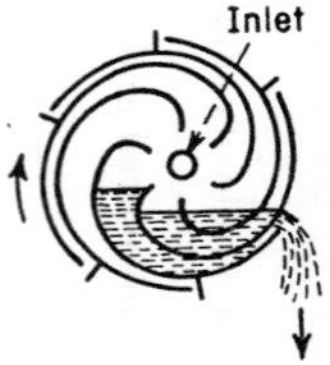

Inlet

FIG. 4

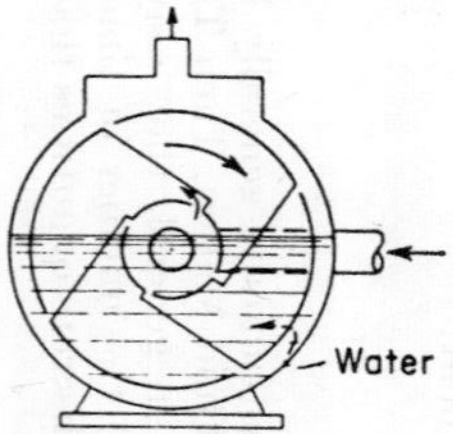

Water

FIG. 5

Device	Range	Accuracy	Remarks
Nutating Disk Meter (Fig. 2): Circular disk in spherical chamber bounded by cones is cut by a vertical partition near discharge point. Spaces above and below disk form three or more separated chambers which progress from suction to discharge side as disk nutates under liquid pressure. End of axis rotates, drives counter	1–1,000 gpm	1% or better	Commonly used for domestic water metering as well as industrial water metering and totalizing of industrial fluids. Can be used for automatic batching with shutoff actuated by meter. Available for high viscosities or pressures, can handle corrosive fluids
Rotary Vane Meter (Fig. 3): Eccentric drum carries radial vanes forced outward by springs against meter casing to form sealed pockets. Fluid pressure rotates drum	Moderate	½% or better	Can be used for liquids or gases. Has virtually no clearance except at ends; so slip is slight
Rotary Bucket Meter (Fig. 4): Curved vanes closed at ends to form buckets are rotated proportional to flow as incoming liquid overbalances drum at right	...	High	Used mainly to measure condensate flow. Is suitable for other liquids
Liquid-sealed Gas Meter (Fig. 5): Cylindrical casing, more than half full of water, contains drum with four "spiral" vanes and gas inlet at center. Gas enters upper compartment, forcing rotation as compartment seeks largest possible gas volume	Wide range from few cfm to large size	0.5% or better	Used for metering fuel gas and for calibrating other meters. Pressure drop as low as 0.1 in. w.c.
Cup Feeders: Shaft carrying scoops, or wheel carrying cups, dips liquid from tank, elevates it, and discharges to a trough. Delivery adjusted by rotation speed	Moderate (low to medium viscosities)	Moderate to high	Used where discontinuous feed is satisfactory, as in feeding small quantity of flotation reagents

Table 1. Flow-measuring Devices—Selection Guide (*Continued*)

Variable-head Meters—Closed Channel

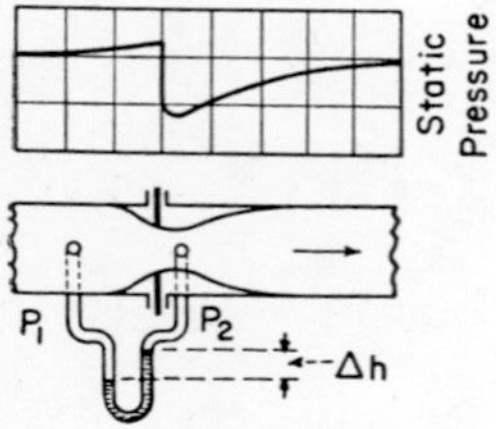

Operating principle	Range	Accuracy	Applications and remarks
Orifices (Fig. 6): A constriction in a pipeline flowing full gives a reproducible and predictable pressure drop across the constriction for various rates of flow. Common type is the square-edged orifice in a thin plate. Such orifices are usually concentric, but may be segmental or eccentric. Pressure connections can be made close to the orifice, but usually are at a point well upstream and at a point of maximum contraction for maximum differential and accurate predictability	Very broad	High, but variable	Orifice characteristics are well known and predictable. Orifices are relatively inexpensive, but have rather high permanent pressure losses. Accuracy depends on care in construction and installation. Used with a variety of types of differential pressure-measuring devices. (See Sec. 4, Pressure Measurement)
Flow Nozzle (Fig. 7): Special streamlined form of orifice designed for installation inside pipe. Lower pressure loss than regular orifice	Moderate sizes	Good	More expensive than orifice plates, but has lower permanent pressure loss
Foster Flow Tube (Fig. 8): Streamlined restriction tube, flanged for installation in pipeline, has ports almost parallel to pipe opening in both up- and downstream slopes to piezometer rings used for pressure connections	All commercial sizes	High	Gives negligible permanent pressure loss. Shorter and easier to install than most venturi-type differential producers. Turbulence has little effect
Venturi Tube (Fig. 9): Streamlined differential producer with practically unity flow coefficient; has 20° entrance cone and 5–6° discharge cone for maximum pressure recovery. Pressure openings connect to piezometer rings at throat and entrance	2 in. diam. and up	1 %	Used mainly for large flow of water, process fluids, wastes, gases. Gives very small permanent pressure loss. Handles suspended solids

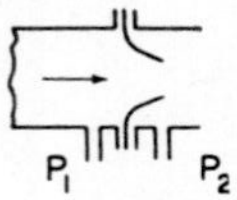

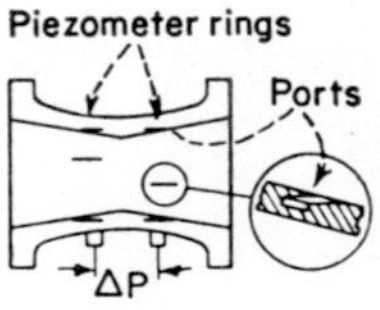

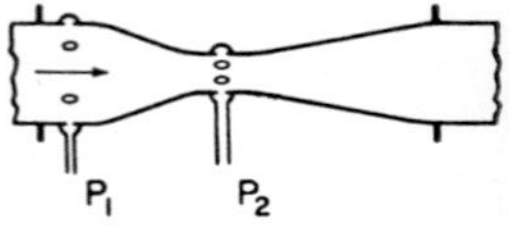

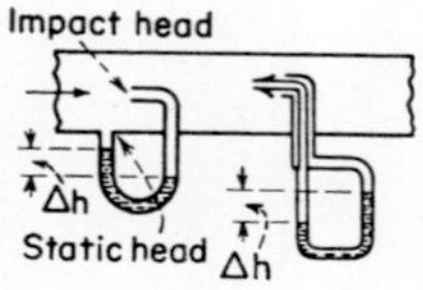

FIG. 6

FIG. 7

FIG. 8

FIG. 9

FIG. 10

Device	Range	Accuracy	Remarks
Pitot Tube (Fig. 10): Measures difference between impact and static pressures in fluid flowing in closed duct. Static-pressure difference is proportional to velocity head, $V^2/2g$. Hence, gives fluid velocity at a point. With velocity traverse, flow rate can be approximated by using average velocity across duct. Must be calibrated	Unlimited	Variable	Simple device; easily installed and capable of use in very large ducts. Since its flow coefficient cannot be predicted accurately, it must be calibrated for specific application

Variable-head Meters—Open Channels

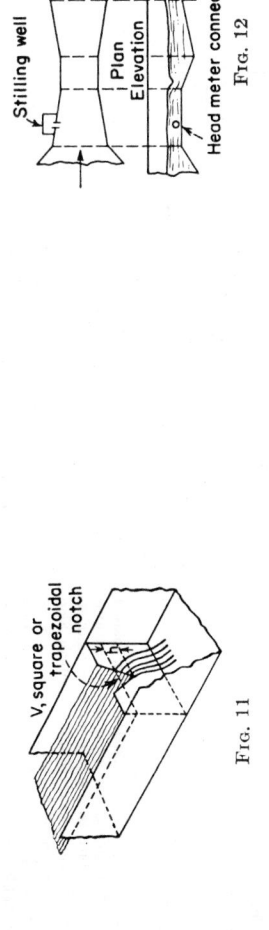

FIG. 11

Device	Range	Accuracy	Remarks
Weirs (Fig. 11): Any obstruction in an open channel causes a backing up of flowing liquid, which varies with flow rate an can be calibrated in terms of head vs. flow rate. Several predictable forms of obstruction (weirs) have been developed—such as the V, square, trapezoidal, "notch" (V illustrated)	Very large	Fair. Best results require calibration	Used mainly for measuring large flows of water (as in irrigation), sewage, or industrial wastes. Usually constructed on the job
Flumes (Fig. 12): Flow channels with gradual rather than sharp restrictions, as in weirs, are called flumes. Flow rate varies with head of quiet fluid behind the restriction, compared with bottom height at restriction. Type shown is Parshall flume. Discharge is determined from empirical formula	Larger than weirs	Variable	Used mainly for measuring large flows of water, sewage, or industrial wastes. Construction usually on the job from published dimensions. (See Parshall, *Trans. ASCE*, 1926)

Stilling well

Plan

Elevation

Step

Head meter connection

FIG. 12

Table 1. Flow-measuring Devices—Selection Guide (*Continued*)

Operating principle	Range	Accuracy	Applications and remarks
Variable-area Meters			

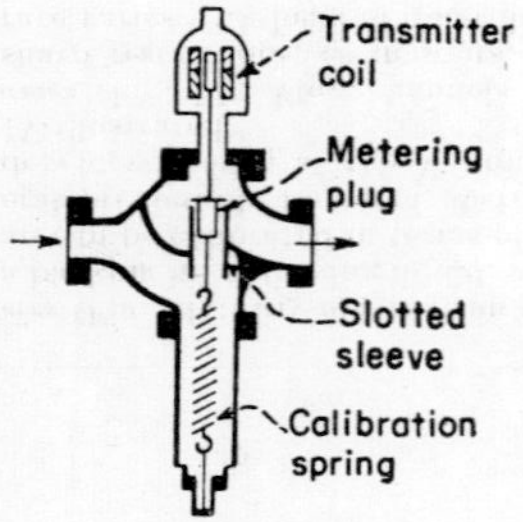

Fig. 13

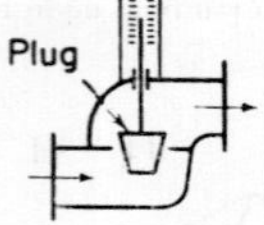

Fig. 14

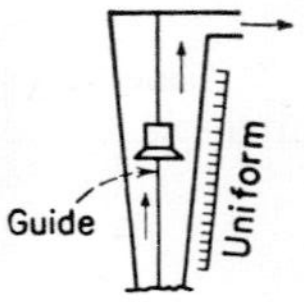

Fig. 15

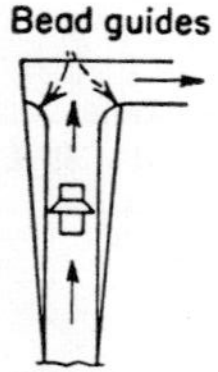

Fig. 16

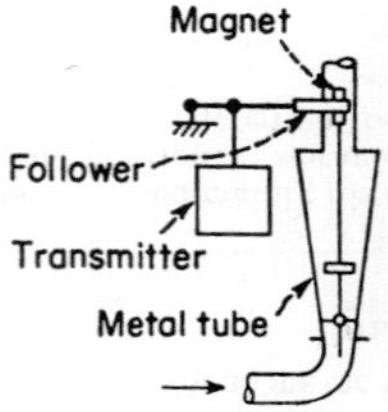

Fig. 16a

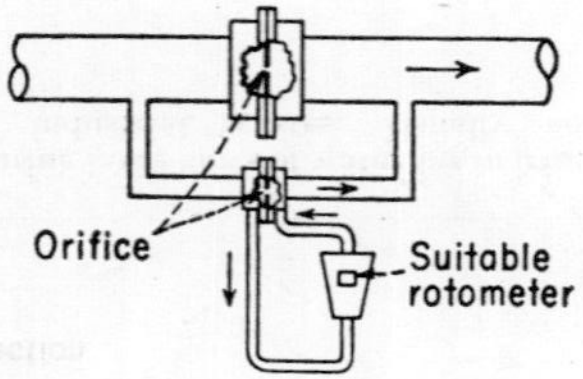

Fig. 17

Operating principle	Range	Accuracy	Applications and remarks
Plug in Ported Cylinder (Fig. 13): Cylindrical plug, loaded by weight or calibrated spring, moves in sleeve containing rectangular slots. Lift is directly proportional to flow rate. Similar design uses rectangular orifice, also makes lift directly proportional to flow	1–4 in.		Used for oils and clear liquids, especially fluids of relatively high viscosity
Tapered-plug Orifice (Fig. 14): With flow against the smaller diameter of the plug, plug rises until upward force due to fluid balances downward force of gravity	Moderate	Sensitivity 0.25%	Sensitive to changes in viscosity
Plug Flow Indicator: Gems Co. makes a flow indicator similar to a lift-plug check valve, but in which the flow against the plug's bottom lifts plug, actuates a magnetically operated switch	½–3 in.	Flow over 1.5 gpm	Does not meter; indicates flow or lack of it. Switch operates warning light
Tapered-tube Rotameter, Free Float: Earliest type of rotameter, using a top-shaped float with sloped flutes to induce rotation, thus centering float in tube. Tube is of glass, tapered uniformly and placed vertically. Upward flow causes float to seek equilibrium height proportional to flow	½–4 in. diam.	0.5–2%	This type is mostly now superseded by types with guided floats or floats insensitive to viscosity

Tapered-tube Rotameter with Center Guide (Fig. 15): Common type where position of float is to be indicated on an external scale by direct movement of a pointer or by electrical transmission. Inductance bridge often used for transmission	½–4 in. diam.	0.5–2 %	Used where visual reading of float level is impractical (opaque liquids) or where transmission is required. Permits use of viscosity-immune floats
Tapered-tube Rotameter with Beaded Tube (Fig. 16): Commonly used for visual-reading rotameters. Tapered glass tube contains three internal beads of constant internal diameter which guide float, taper between beads serving as variable area section	Up to 4 in.	0.5–2 %	Used for stability of floats in visual-read meters. Enables viscosity-immune floats to be used without center guide
Tapered-tube Rotameter, Metal (Fig. 16a): Use of magnetic follower and pneumatic or electric transmission enables use of short metal tapered tube	1–4 in. (high-capacity type: 2–12 in.)	1–2 %	Can handle fluids at high pressure, as well as high temperature, with consistently high accuracy
Kinetic Manometer-Rotameter (Fischer & Porter) (Fig. 17): Orifice or other differential producer is bypassed with a smaller pipeline containing a second orifice, across which connecting piping contains a rotameter to give about half the pressure drop in each. Rotameter flow is proportional to main pipeline flow	Limited only by size of main line from which bypass flow is taken	2 % of max. bypass flow rate	Enables a rotameter to be used for much larger flow rates

Table 1. Flow-measuring Devices—Selection Guide (Continued)

Operating principle	Range	Accuracy	Applications and remarks
Velocity and Current Meters			

FIG. 18    FIG. 19 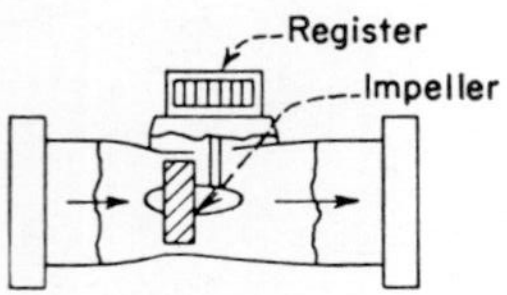   FIG. 20 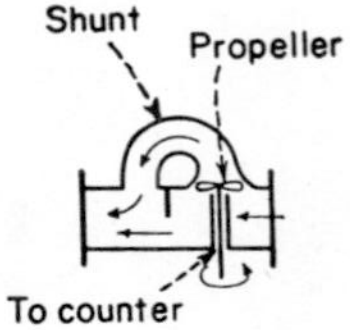   FIG. 21 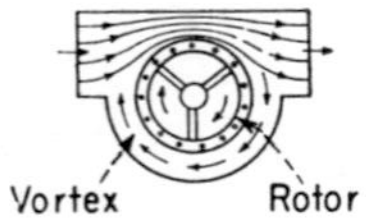   FIG. 22 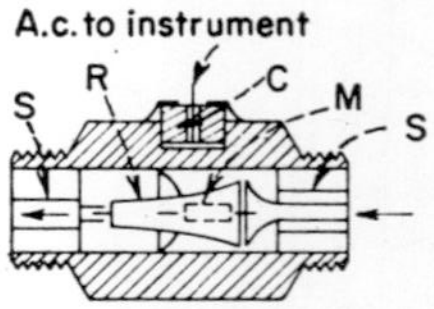   FIG. 23 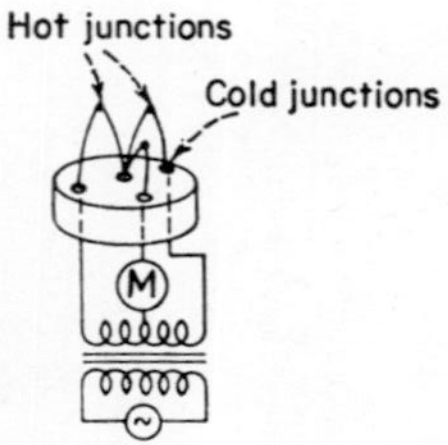   FIG. 24

Shunt   Propeller   Rotor   Vortex

Register   Impeller   To counter

A.c. to instrument   S   R   C   M   S   S   M   S

Hot junctions   Cold junctions

Operating principle	Range	Accuracy	Applications and remarks
Cup Anemometer (Fig. 18): Hollow hemispheres, mounted radially to form a wheel, make a simple current meter. Air current catches open side of cups, producing a rotation speed proportional to air velocity	...	...	Used mainly for wind velocity rather than flow rate
Vane-type Velocity Meter (Fig. 19): Impact of moving air or gas moves a vane against a hairspring. Used in open air or with special jet tubes in ducts or restricted locations	20–24,000 fpm	3 %	Used mainly in heating, ventilating, air conditioning, and exhaust systems
Current Meter (Fig. 20): Turbine wheel in a venturi-shaped tube installed in pipeline rotates at rate proportional to flow rate, driving integrating counter	2–36 in. pipe size	2 % of actual flow	Gives low pressure drop. Used for liquids only. Can stand high temporary over-loads. Subtracts reverse flow. Installed as unit
Shunt-type Current Meter (Fig. 21): Tube with segmental orifice and bypass loop installed in flow line. Turbine wheel at entrance of shunt measures fixed proportion of main flow, records by magnetic-driven counter	1–4 in. main pipe size	2 %	Used for steam, gas, or air only (no liquids). Easy to install, or change orifice for capacity change. Magnetic counter eliminates stuffing box
Vortex Meter (Fig. 22): Cylindrical flow path with semi-cylindrical bulge in side produces a swirling vortex. Light "vortex cage," placed in the position shown, operates the counter by rotating at a velocity proportional to flow rate	10–21,000 gpm; 10–3,100 million cu ft/day	0.5 %	Virtually no mechanical friction. Unit can be used for nonlubricating as well as lubricating liquids. Also suitable as a gas flow meter and, with density compensation, as a gas mass flow meter

5–18

Description	Range	Accuracy	Remarks
Electric Current (Turbine) Meter (Fig. 23): Turbine wheel R, mounted between straightening vanes S in cylindrical passage, is core-shaped to give venturi-like flow, causing end-thrust balance to reduce friction. Rotor R contains magnet M so that rotating field generates a-c in coil C for flow rate measurement	⅛–24 in. and larger	0.5% or better	Capable of high accuracies with low pressure loss. Can have pulse output signal to operate digital-type readouts. Good for pressures to 20,000 psi; temperatures up to 1200°F and down to below zero. Readings can be combined with density compensation for measuring mass flow rates
Hot-wire Anemometer (Fig. 24): Various types; type shown uses thermopile heated by a-c. Cooling by flow is proportional to flow rate; is measured as d-c output of thermopile by millivoltmeter	0–20,000 fpm	2–5%	Used mainly as a directional or nondirectional air velocity meter, but can be enclosed for use as a mass flow meter

Electromagnetic Meters

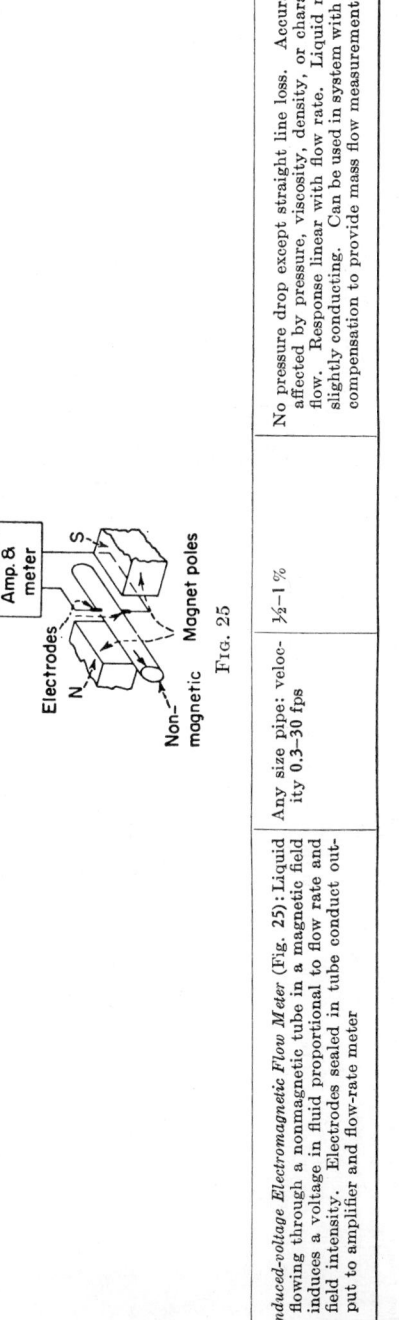

Amp. & meter

Electrodes

N

S

Non-magnetic

Magnet poles

Fig. 25

Description	Range	Accuracy	Remarks
Induced-voltage Electromagnetic Flow Meter (Fig. 25): Liquid flowing through a nonmagnetic tube in a magnetic field induces a voltage in fluid proportional to flow rate and field intensity. Electrodes sealed in tube conduct output to amplifier and flow-rate meter	Any size pipe: velocity 0.3–30 fps	½–1%	No pressure drop except straight line loss. Accuracy unaffected by pressure, viscosity, density, or character of flow. Response linear with flow rate. Liquid must be slightly conducting. Can be used in system with density compensation to provide mass flow measurement

Table 1. Flow-measuring Devices—Selection Guide (Continued)

Operating principle	Range	Accuracy	Applications and remarks
Meters Designed for or Adapted to Mass Flow Measurement			

NOTE: Mass flow meters are designed to measure mass rate of flow (such as lb/hr, g/min) rather than volume flow rate or velocity. They are of two basic types: (1) true mass meters which respond to rate of mass flow, and (2) volume flow-rate meters with automatic density correction. The equation $M = k_2 V$ (where M = mass flow rate, k = a constant, ρ = density of fluid measured, and V = volume rate of flow) shows the relationship between mass and volume flow rate. It is this equation that mass flow systems using volume meters must continuously solve.

Many types of volume rate meters have been adapted with density-measuring devices of different types to measure mass flow rates. These include piston meters, head meters (especially for gases), turbine meters, and electromagnetic meters. One system uses a radiation density detector in the flow line, along with an electromagnetic flow meter, to provide a continuous record of both mass flow rate and density, with optional readout of volume flow rate.

True mass flow meters are of two general types: (1) those in which either heat (Fig. 26) or a substance whose concentration can be detected is added to the flow (e.g. dyes, salts, or radioactive tracers); (2) those that impart angular momentum to the flowing fluid and measure the torque required to impart this momentum (Fig. 27).

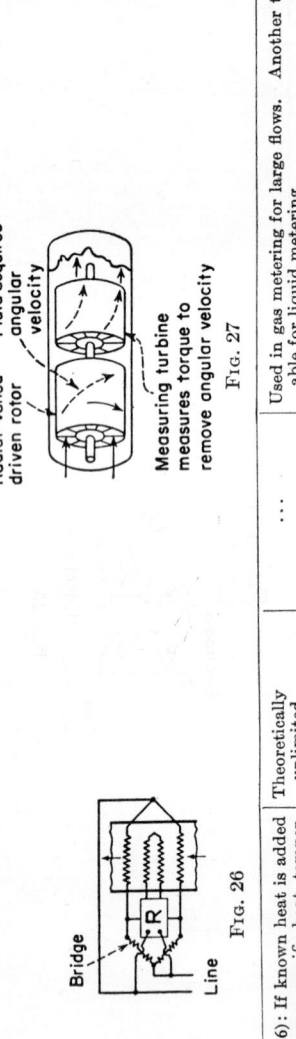

Bridge

Line

FIG. 26

Radial-vaned driven rotor   Fluid acquires angular velocity

Measuring turbine measures torque to remove angular velocity

FIG. 27

Operating principle	Range	Accuracy	Applications and remarks
Heat-addition Flow Meter (Fig. 26): If known heat is added to unknown flow of fluid of known specific heat, temperature rise is inversely proportional to mass flow rate. Type shown varies electric current flow to heater for fixed temperature rise, using resistance thermometers in bridge to regulate heat addition. Current flow is proportional to fluid flow	Theoretically unlimited	...	Used in gas metering for large flows. Another type available for liquid metering
Angular-momentum Type (Fig. 27): Cylindrical body contains two radially vaned rotors. First is rotated at constant speed to impart angular acceleration to the fluid. Second is restrained either electronically or by a spring; measurement is made of torque required to remove the angular momentum imparted by the first rotor. Torque is proportional to mass flow rate	1–15 lb/sec	1% of reading with 5:1 low range	One type can be used on fluids with specific gravity 0.2–1.5; viscosity 0.1–25 centipoises; temperature −67 to 250°F—weighs 6 lb. Can be used for liquids or gases. Can have either analog or digital output

REFERENCES

1. An Experimental Investigation of the Circumstances Which Determine Whether the Motion of Water Shall Be Direct or Sinuous, and of the Law of Resistance in Parallel Channels, *Phil. Trans. Roy. Soc. London*, vol. 174, part III, p. 935, 1883.
2. Folsom, R. G., and H. W. Iversen: Pipe Factors for Quantity Rate Flow Measurements with Pitot Tubes, ASME Annual Meeting, 1948.
3. Harris, Charles W.: An Engineering Concept of Flow in Pipes, *Trans. ASCE*, vol. 115, p. 909, 1950.
4. Pigott, R. J. S.: Flow of Fluids in Closed Conduits, *Mech. Eng.*, vol. 55, p. 497, 1933.
5. Alves, G. E., D. F. Boucher, and R. L. Pigford: Pipe-line Design for Non-Newtonian Solutions and Suspensions, *Chem. Eng. Progr.*, vol. 48, no. 8, pp. 385-393, August, 1952.
6. ASME-API Petroleum Positive Displacement Meter Code.
7. Perry, R. H. (ed.): "Chemical Engineers' Handbook," 4th ed., sec. 5, Fluid and Particle Mechanics, McGraw-Hill Book Company, Inc., New York, 1963.
8. Walker, W. H., W. K. Lewis, W. H. McAdams, and E. R. Gilliland: "Principles of Chemical Engineering," 3d ed., chap. III, Flow of Fluids, McGraw-Hill Book Company, Inc., New York, 1937.
9. Binder, R. C.: "Fluid Mechanics," Prentice-Hall, Inc., Englewood Cliffs, N.J., 1943.
10. Rouse, Hunter: "Elementary Mechanics of Fluids," John Wiley & Sons, Inc., New York, 1946.
11. Ziebolz, H.: Basic Solutions for Flow Measurement, *Rev. Sci. Instr.*, vol. 15, no. 4, pp. 80-87, April, 1944.
12. Considine, D. M. (ed.): "Process Instruments and Controls Handbook," McGraw-Hill Book Company, Inc., New York, 1957.
13. Fluid Meters, Their Theory and Application, 5th ed., Report of ASME Research Committee on Fluid Meters, 1959.
14. History of Orifice Meters and the Calibration, Construction, and Operation of Orifices for Metering, Report of Joint AGA-ASME Committee on Orifice Coefficients, reprinted by ASME, 1935.
15. Caddell, J. R.: "Fluid Flow in Practice," Reinhold Publishing Corporation, New York, 1956.
16. Stearns, R. F.: "Flow Measurement with Orifice Meters," D. Van Nostrand Company, Inc., Princeton, N.J., 1951.
17. Addison, Herbert: "Hydraulic Measurements," Chapman & Hall, Ltd., London, 1946.
18. Ower, E.: "The Measurement of Air Flow," Chapman & Hall, Ltd., London, 1949.
19. Landy, J. J., and G. J. Fiedler: Automatic Pressure and Flow Control System Design for Multi-loop Fluid Dynamics Facility with Prominent Distributed Parameters, *Tech. Paper, A.I.Ch.E. Symp.*, St. Paul, Minn., September, 1959.
20. Staff: Automatic Custody Transfer of Crude Oil at the Lease, *Oil Gas J.*, June 11, 1956, pp. 110-124.
21. Wingo, H. E.: Thermistors Measure Low Liquid Velocities, *Control Eng.*, vol. 6, no. 10, October, 1959.
22. Smith, H. R.: Metering and Proportioning Flow, *Automation*, vol. 6, no. 11, p. 82, November, 1959.
23. Lynch, D. R.: A Low-conductivity Magnetic Flowmeter, *Control Eng.*, vol. 6, no. 12, p. 122, December, 1959.
24. Walk, D. P.: Electromagnetic Instrument Meters Starch Slurry, *Automation*, vol. 6, no. 12, p. 70, December, 1959.
25. Isobe, Takashi, and Hiroo Hattori: A New Flowmeter for Pulsating Gas Flow, *ISA J.*, vol. 6, no. 12, p. 38, December, 1959.
26. Staff: Flowmeter Measures Directly in Pounds, *Autom. Control*, vol. 10, no. 6, p. 32, June, 1959.
27. Decker, M. M.: The Gyroscopic Mass Flowmeter, *Control Eng.*, vol. 7, no. 5, p. 139, May, 1960.
28. Staff: American Petroleum Institute—Installing Rotameters, *Control Eng.*, vol. 7, no. 3, p. 115, March, 1960.
29. Evans, E. C.: Variable orifice Flowmeters Go Linear, *ISA J.*, vol. 7, no. 5, p. 91, May, 1960.
30. Staff: Basic Types of Mass Flowmeters, *ISA J.*, vol. 7, no. 6, p. 52, June, 1960.
31. Staff: Basic Types of Inferential Mass Flowmeters, *ISA J.*, vol. 7, no. 6, p. 54, June, 1960.
32. Staff: A Look at Mass Flowmetering Hardware, *ISA J.*, vol. 7, no. 6, p. 57, June, 1960.
33. Haffner, J. and C. Stone: Novel Mass Flowmeter Measures Angular Momentum and Density of Fluid to Derive a Mass Flow Analog, *Control Eng.*, vol. 9, no. 10, pp. 69-70, October, 1962.
34. Goodrich, J. D.: Measuring Fluid Flow, *Control Eng.*, vol. 9, no. 9, pp. 111-115, September, 1962.
35. Beerbower, A.: Predicting Performance of Metering Pumps, *Control Eng.*, vol. 8, no. 1, pp. 95-97, January, 1961.
36. Goldstein, D. J., R. H. Dick, and D. Harvey Smith: A Viscosity Compensated Flowmeter, *ISA J.*, vol. 8, no. 1, pp. 60-61, January, 1961.
37. Mitchell, W. C.: Flow Ratio Control and In-line Blending, *Autom. Control*, vol. 14, no. 1, pp. 35-38, January, 1961.
38. Staff: Blending System Remembers Component Flows, *Autom. Control*, vol. 14, no. 4, pp. 57-58, April, 1961.
39. Staff: Trends in Flowmeters. *Autom. Control*, vol. 14, no. 12, pp. 44-50, December, 1961.
40. Staff: Computer Monitors Flow Variables, *Automation*, vol. 7, no. 6, pp. 87-89, June, 1960.
41. Miller, E. W.: Turbine Gas-flow Sensor, *Instr. Control Systems*, vol. 35, no. 1, pp. 105-108, January, 1962.
42. Meriam, J. B.: Manometers, *Instr. Control Systems*, vol. 35, no. 2, pp. 114-118, February, 1962.
43. Nolte, C. B.: Wide-range Orifice Meter, *Instr. Control Systems*, vol. 35, no. 7, pp. 174-177, July, 1962.
44. Wood, R. D.: Steam Measurement by Orifice Meter, *Instr. Control Systems*, vol. 36, no. 4, pp. 135-137, April, 1963.

ACCELERATION MEASUREMENT

By G. D. Goodrich*

DEFINITIONS

Acceleration is defined as the time rate of change of velocity, either linear or angular. Linear acceleration is commonly expressed in units of feet per second per second, or meters per second per second. A convenient unit of measure that is often used is the acceleration due to gravity, or g. The standard value of g is 32.174 ft/sec². Angular acceleration may be expressed as radians per second per second—a radian is the angle subtended at the center of a circle by an arc of the circle equal in length to its radius, about 57.29°—or as revolutions per minute per second.

BASIC PRINCIPLES AND MATHEMATICAL RELATIONSHIPS

Nearly all accelerometers consist of a mass and a spring mounted in a case and damped, either electrically or mechanically. Some type of transducing device (which transforms a mechanical motion into an analogous electrical signal, in this case) is incorporated, to provide an electric output which is proportional to the relative motion between the mass and the case. Within the case the mass-and-spring system is constrained to move in one mode of motion, and thus is referred to as a single-degree-of-freedom system. The motion of the mass can consequently be described by the second-order differential equation which reads

$$\frac{d^2y}{dt^2} + \frac{c\,dy}{m\,dt} + \frac{k}{m}y = -\frac{d^2x}{dt^2} \tag{1}$$

where x = displacement of case relative to a fixed point in space
$\quad\ y$ = displacement of mass relative to case
$\quad\ m$ = movable mass
$\quad\ k$ = spring constant
$\quad\ c$ = viscous drag
The type of motion of most interest is sinusoidal motion, as in those cases where the forcing or driving function is sinusoidal in nature. For this case, the motion is assumed to be of the form $x = x_d \cos \omega t$. Using this assumption, the final solution of the differential equation is

$$\frac{y}{x_d} = \frac{\omega^2}{\sqrt{(k/m - \omega^2)^2 + (\omega\,c/m)^2}} \cos(\omega t - \phi) \tag{2}$$

where $\phi = \tan^{-1} \dfrac{\omega\,c/m}{k/m - \omega^2}$

APPLICATIONAL PARAMETERS

The first parameter which is important in the application of an accelerometer is the *natural frequency*. For an instrument described by the above equations, the natural frequency will occur at the 90° phase-shift point. In order to satisfy this condition,

* Supervisor, Technical Sales, Statham Instruments, Inc., Los Angeles, Calif.

the following equation must apply:

$$m\omega_n = \frac{k}{\omega_n} \tag{3}$$

or

$$\omega_n = \sqrt{\frac{k}{m}} \tag{4}$$

and

$$f_n = \frac{\omega_n}{2\pi} = \frac{1}{2\pi} \sqrt{\frac{k}{m}} \tag{5}$$

where f_n is the natural frequency of the system.

Other Frequency Terms

Two other terms which are frequently used in the discussion of an accelerometer are the *undamped natural frequency* and the *resonant frequency*. These terms are often erroneously used interchangeably with the term *natural frequency*. For this reason, it is important to define their meaning.

The *undamped natural frequency* is the frequency at which free oscillations will occur. This is related to the natural frequency by the following equation:

$$f_0 = f_n \sqrt{1 - h^2} \tag{6a}$$

where f_0 is the undamped natural frequency, and h is the damping ratio.

The *resonant frequency* is the frequency at which the maximum amplitude ratio occurs. The resonant frequency is related to the natural frequency by the following equation:

$$f_r = f_n \sqrt{1 - 2h^2} \tag{6b}$$

where f_r is the resonant frequency.

Damping

The second important parameter in the discussion of an accelerometer is the *damping*. The role played by damping in the response characteristics of an accelerometer may be clarified by studying the transient response of the system. For this analysis, it is assumed that the transient response is of the form e^{pt}. If the resulting equation is solved for the smallest value of c (the *damping coefficient* which will just prevent a damped transient oscillation), the solution will yield the equation $c_c = 2\sqrt{km}$, where c_c is called the *critical damping coefficient*. Another useful symbol in a discussion of an accelerometer is the *damping ratio h*, which is given by the equation

$$h = \frac{c}{c_c} \tag{7}$$

The response equations may be simplified and nondimensionalized by the substitution of h and β, where $\beta = \omega/\omega_n$ (the ratio of the variable frequency to the natural frequency). For the case of sinusoidal motion and a constant acceleration of amplitude x_a, the solution for the equation of motion becomes

$$\frac{y}{x_a} = \mu \cos(\omega t + \phi) \tag{8}$$

where $\mu = \dfrac{1}{\sqrt{(1 - \beta^2)^2 + (2h\beta)^2}}$ and $\phi = \tan^{-1} \dfrac{2h\beta}{1 - \beta^2}$

A perfect accelerometer would have $\mu = 1$, independent of frequency, so that its response to a sine wave of displacement would rise as the square of the frequency. The widest possible frequency range over which an accelerometer may be used without change in waveform is obtained in the case where $\mu = 1$ and the phase-shift curve is linear with frequency. The response of an accelerometer to a constant acceleration

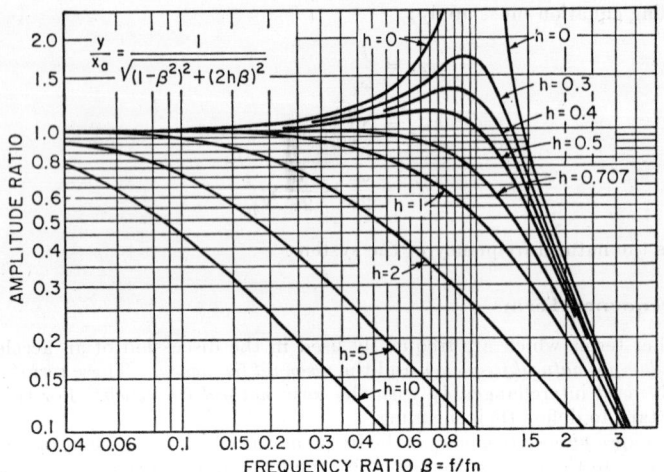

Fig. 1. Response of a seismographic system to a sinusoidal acceleration.

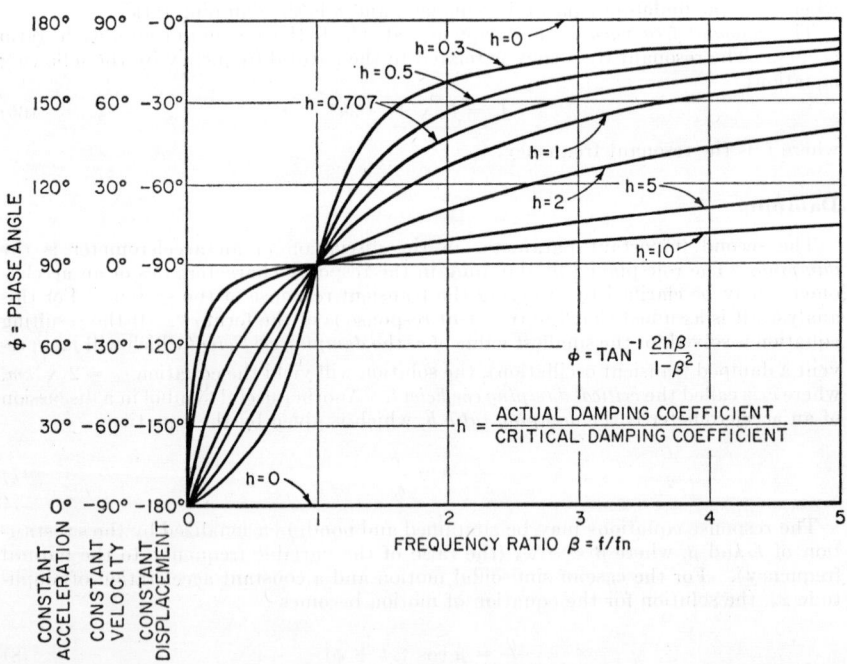

Fig. 2. Phase angle as a function of frequency ratio for a seismographic system.

is shown in Fig. 1, while Fig. 2 shows the phase-shift characteristics as a function of frequency.

Study of the curves in Figs. 1 and 2 indicates that a damping ratio of 0.707 gives a frequency response (±5 per cent) flat out to approximately 0.4 of the natural frequency, and also a phase-shift curve which is very nearly linear with frequency. For these reasons, accelerometers are ordinarily damped at 0.7 critical damping.

If the highest frequency to be measured is a small fraction of the natural frequency of the accelerometer, it may be more desirable to have a lower damping coefficient. Figure 1 shows that the response will be reasonably flat to $\beta = 0.2$ for any low value of h, while Fig. 2 demonstrates that low values of h result in low phase shift or low signal delay.

Recording galvanometers are frequently used in the measurement of dynamic values. If a resistive circuit is used to drive the galvanometer, the response curves in Fig. 1 apply. The over-all response of a resistive gage and galvanometer is obtained by multiplying the respective amplitude ratio curves and adding their phase-shift curves. In many applications, it is necessary to use a galvanometer with a lower natural frequency than the instrument associated with it, in which case the system characteristics are completely dependent upon the galvanometer.

FACTORS AFFECTING INSTRUMENT RESPONSE

The response characteristics previously described are based on the analysis of linear differential equations with constant coefficients. The second-order equation is a reasonable approach, since the design is usually made as simple as possible. From a practical standpoint, however, it is not always feasible to avoid certain factors which cause the instrument to deviate from exact conformance to the ideal solution of the second-order differential equation.

The natural frequency is a parameter of primary importance, since this establishes the frequency range within which the measurements can be made. By shocking the undamped system, the natural frequency can be determined from the observation of the resulting oscillations. Damping is applied either mechanically, as by viscous damping, or electrically, and it can be assumed that the natural frequency will remain the same, provided that the mass and stiffness are not changed. The natural frequency and damping can then be verified by the application of a sinusoidal driving function.

As a rule, oil damping is employed in instruments designed for high natural frequency, since magnetic damping is difficult to apply at these higher frequencies. The effect of damping on the natural frequency is most likely to be noticeable in oil damping, since the real mass of the accelerometer must move oil when it is displaced, and the moving oil has kinetic energy.

In certain instances here, an effect known as *oil pumping* is observed. Oil has an appreciable modulus of elasticity in shear, and if the viscosity of the oil is high, a shear movement of the mass causes elastic deformation of the oil, as well as viscous flow. In such a case, the response is of the form of a fourth-order differential equation. This effect is normally minimized in the design of an oil-damped accelerometer, by proper choice of the mass and stiffness parameters, to keep their product small. The accelerometer may be damped with an oil of relatively low viscosity, thus reducing the coupling to the shear elasticity of the oil. As a consequence, the terms higher than second order become extremely small.

In some instances, the oil pumping effect is very useful, and is employed to reduce the natural frequency of the accelerometer. An example of an application in which a low natural frequency is helpful is the study of acceleration at low frequencies, where high-frequency effects are not desired. Consider an accelerometer which is being used to detect changes in orientation of a vehicle with respect to the earth. The accelerometer is naturally subjected to the structural vibrations of the vehicle, and the effects of this vibration must be eliminated. As a result, if the accelerometer is designed with a sufficiently low natural frequency, there will be no need to filter the signal electrically.

The behavior of the response characteristics under ambient conditions is an important consideration. Figure 3 gives the variations in damping ratios of devices that use magnetic damping, silicone oil and petroleum oil damping. The characteristics here of silicone oil are not the most satisfactory, but are considerably better than those of petroleum oil. Magnetic damping offers more favorable characteristics; but, as noted earlier, this type of damping is usually not practical in high-frequency devices.

An apparent solution to the problem of viscosity variation is control of the ambient temperature. By surrounding the accelerometer with a thermostatically controlled heater jacket, the temperature can be maintained within reasonable tolerances.   However, this method adds size and weight to the accelerometer, and it is difficult to find a thermostat that will operate within the desired tolerances under a high acceleration field.

Gas damping offers an excellent solution of this difficulty, fortunately, since a temperature change which doubles the viscosity of silicone oil changes the viscosity of a gas by only a few per cent.   Figure 4 illustrates the difference between a conventional oil-damped accelerometer and a gas-damped accelerometer.   The components shown include (1) a diaphragm, (2) coupling chamber, (3) porous plug, and (4) case.   The

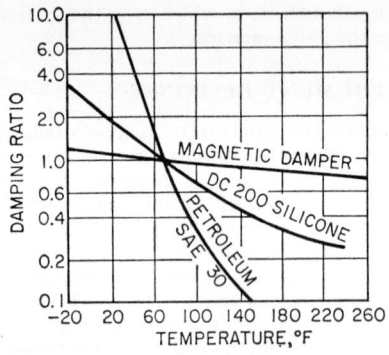

FIG. 3. Variation of typical damping devices with temperature.

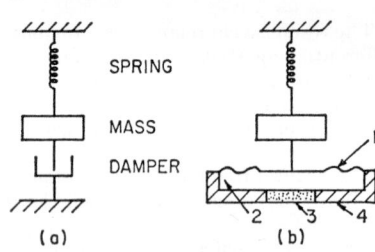

FIG. 4. (a) Conventional oil-damped accelerometer; (b) gas-damped accelerometer.

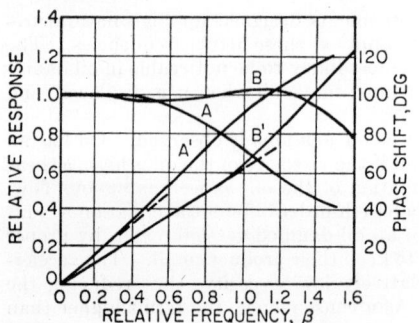

FIG. 5. Relative response, relative frequency, and phase shift: (a) Oil-damped accelerometer (damped 0.7 of critical); (b) gas-damped accelerometer of comparable range.

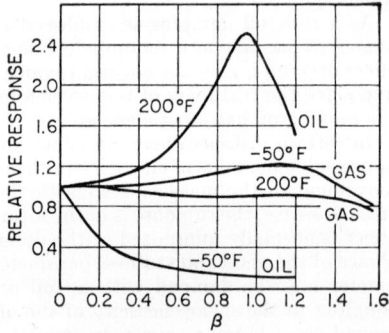

FIG. 6. Comparison of the effects of temperature on the frequency response of an oil-damped instrument and a gas-damped instrument.

diaphragm, which is connected to the sensing element, forces gas through the porous plug and causes energy to be dissipated.

In addition to the advantage that the damping is less affected by temperature changes, the gas-damped accelerometer also offers a much flatter frequency response. Figure 5 shows a comparison between an oil-damped accelerometer (damped 0.7 of critical) and a gas-damped accelerometer of comparable range.   The oil-damped instrument is flat within 5 per cent to $\beta = 0.59$, while the gas-damped accelerometer is flat within 5 per cent to $\beta = 1.4$.   Curves A' and B' represent the phase-shift characteristics of the two types of accelerometers.   It will be noted that A' is linear within $2°$ to $\beta = 0.5$, whereas B' is linear within $2°$ to $\beta = 1.0$.

The gas-damped accelerometer is a multiple-degree-of-freedom system, and the response curve obviously is not defined by calling out the damping and the natural frequency. The most significant parameters for this accelerometer are the flat response and the peak amplitude. As long as these are specified, the other details of the dynamical system are of little importance. Figure 6 shows a comparison of the effect of temperature on the frequency response of an oil-damped instrument and a gas-damped instrument. Throughout the temperature range of −50 to +200°F, the response is flat within ±5 per cent to the conventional β = 0.4, and within ±15 per cent to β = 1.4.

The use of gas damping also makes possible the measurement of acceleration at temperature extremes where oil would be impossible, because of freezing with the cold or evaporating during excessive heat.

Still another factor which affects the response of an accelerometer is the presence of transverse acceleration. Accelerometers are generally thought of as responding only to the component of acceleration along the principal axis, but this is not necessarily true. The presence of high transverse acceleration may actually cause the sensing element to deflect in the transverse axis, thereby producing an output.

A more usual effect of transverse acceleration is inherent in the physical construction of the accelerometer. Because of the manufacturing tolerances required, the sensitive axis may not be precisely perpendicular (or parallel, as the case may be) to the mounting surface. However, the true axis of sensitivity is determinable in practice.

CALIBRATION OF ACCELEROMETERS

Static Calibration

The static calibration of an accelerometer provides information on three major parameters: range, calibration factor, linearity. With low-range accelerometers, the calibration may be very easily and accurately performed by using the so-called $2g$ turnover method. The instrument should be mounted on a leveled platform with its sensitive axis aligned with the earth's gravitational field. It may then be rotated through the $+1g$, $0g$, and $-1g$ positions, thus accomplishing a comparison with a standard. A very accurate calibration can be performed in this way in fractional g steps, since the acceleration applied along the measuring axis will vary as the sine of the angle of rotation.

Static calibration at higher acceleration levels is accomplished by the use of a centrifuge. The axis of rotation should be vertical, so that a $±1g$ ripple is not superimposed on the static calibration. The sensitive axis of the accelerometer must be carefully aligned with a radius of rotation, to avoid shortening of the effective radius. It is also important to determine the exact center of mass, since the accuracy of calibration is directly affected by the length of the radius arm. The center of mass may be approximated by comparison of data from several tests run at different radii and at the same speed. Close control of the speed of rotation is also imperative, since the acceleration produced varies as the square of the angular velocity. Accurate control of speed can be accomplished through the use of a synchronous motor to drive the centrifuge, or through stroboscopic observation of the rotating body.

The linearity of the accelerometer may be determined from a centrifuge calibration by imposing changes of acceleration on the instrument. The static acceleration generated by the centrifuge is given by the expression

$$a = \frac{4\pi^2 N^2 r}{32.174 \times 43,200} = 2.840 \times 10^{-5} N^2 r \tag{9}$$

where r = radius of rotation of the center of gravity of the active mass, in.
N = speed, rpm
a = standard g units

From this expression, it is seen that the acceleration may be varied by changing either the speed of rotation or the radius.

Electrical connections may be brought out from the accelerometer by brushes on slip rings or by thin electrodes which dip in concentric pools of mercury. Worn-in graphite brushes on slip rings have been found to be highly satisfactory, since they display only a negligible contact resistance throughout widely varying ranges of speed. The output voltages are measured with a precision readout device.

Dynamic Calibration

In measuring the dynamic response of an accelerometer whose damping is less than critical, observation of the accelerometer's response to a step function of acceleration is a convenient and frequently used method. After such a procedure, the damping ratio can then be determined from the height of the first overshoot.

The curve shown in Fig. 7 may be used to determine the damping ratio from the amplitude of the first overshoot.

The interval between successive peaks of the oscillatory response determines the period of the free oscillations, but does not give the true natural period. The damping affects the period in which free oscillations will occur, but has nothing to do with the natural frequency.

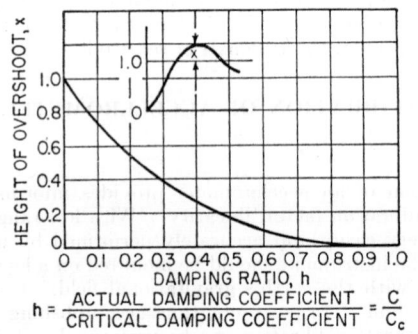

$$h = \frac{\text{ACTUAL DAMPING COEFFICIENT}}{\text{CRITICAL DAMPING COEFFICIENT}} = \frac{C}{C_c}$$

FIG. 7. Height of overshoot of oscillatory transients as a function of damping ratio for a system with one degree of freedom.

The following equation may be used for the determination of the natural frequency— from the period of free oscillations. This method is recommended for accelerometers with a low damping coefficient.

$$f_n = \frac{1}{T_0 \sqrt{1 - h^2}} \qquad (10)$$

where T_0 = period of free oscillations, and h = damping ratio

A more precise analysis of the instrument's characteristics may be obtained by calibrating the instrument on a shake table capable of producing an essentially pure sinusoidal force over a large frequency range.

One of the commonest types of shake table used for this purpose is the electromagnetic vibration exciter. In this type of device, the force is generated by an alternating current flowing in a movable coil, which is positioned in a region of high magnetic-flux density. The magnetic field is derived from a stationary field coil connected to a source of direct current or to a permanent magnet.

The physical characteristics of the magnetic structure ensure that the force generated is dependent only on the magnitude of the current in the moving coil. Since this generated force is as pure as the current supplied to the moving coil, an oscillator and power amplifier with low distortion characteristics must be used. This is a mechanical system, which will resonate at certain frequencies. Accordingly, the system must be calibrated, to ensure that the amplitude produced remains constant while the frequency is varied.

A useful method that is often adopted for monitoring the amplitude of vibration is to attach a velocity-type signal generator coil to the shake table, which moves in the magnetic field of a permanent magnet. Since the velocity coil exhibits phase shift with frequency, this type of monitor may be employed up to frequencies of about 100 cps without phase-shift difficulties being experienced, and roughly up to 500 cps if the phase-shift characteristics of the velocity coil are taken into consideration. Another method frequently used is to mount an accelerometer with known dynamic-response characteristics on the shake table with the test specimen.

Figure 8 illustrates a typical test setup. The accelerometer and signal coil must be isolated, to ensure that there will be no pickup from the strong alternating magnetic field which exists about the driver coil of the exciter. The output of the accelerometer and monitor should be observed on an oscilloscope, to ensure satisfactory waveform.

The natural frequency of the accelerometer may be readily determined by using the setup illustrated above. To accomplish this, the output of the accelerometer and the output of the signal coil are connected to the horizontal and vertical amplifiers of the oscilloscope. The natural frequency of the accelerometer is the frequency at which the two outputs are exactly in phase. In this case, the Lissajous pattern will appear as a straight line. If an accelerometer is used as a monitor, in place of a

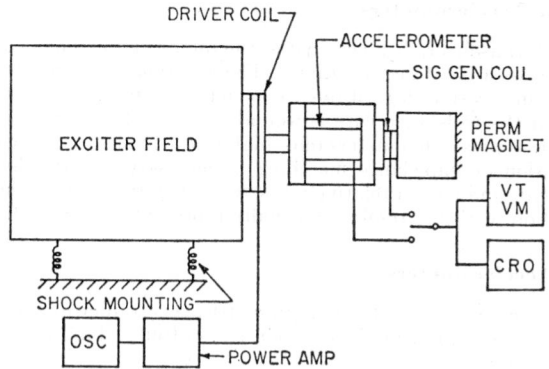

Fig. 8. A typical test setup for dynamic calibration of an accelerometer.

velocity coil, the natural frequency of the accelerometer is the frequency at which the two outputs are 90° out of phase, in which case the Lissajous will appear as a circle. The oscilloscope amplifiers should be checked for phase-shift error, and corrected if necessary.

TRANSDUCING ELEMENTS

Several types of accelerometers which are widely used at present are described in this subsection. The various advantages and disadvantages of each type are discussed briefly.

Strain-gage Accelerometers

In a strain-gage accelerometer, strain-sensitive wires are attached to a fixed frame and a force-summing member. As a rule, the wires are arranged in the form of a Wheatstone bridge. When the force-summing member is displaced, the balance of the bridge is changed, thereby producing an electric output which is proportional to the magnitude of the applied force. Figure 9 illustrates this principle.

Strain-gage accelerometers may be excited by either a-c or d-c voltage. This type of transducer offers continuous resolution, and can be used to measure either static or dynamic phenomena. The bridge can be externally balanced by a simple resistive circuit. The effects of temperature change on the static characteristics of the instrument can be easily compensated. These instruments offer excellent frequency response.

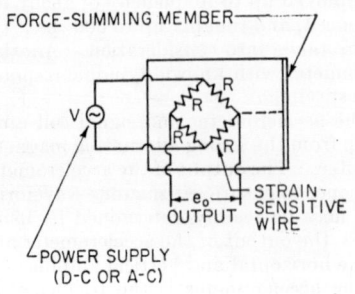

Fig. 9. Strain-gage transducer.

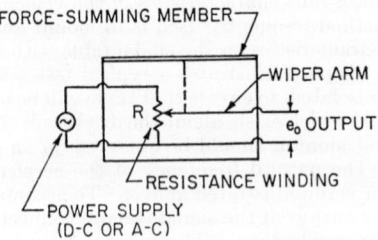

Fig. 10. Potentiometric transducer.

Potentiometric Accelerometers

In this type of transducer, displacement of the force-summing device causes a wiper arm to sweep across a resistance winding. The resistance change which is produced may be linear, sine, cosine, logarithmic, exponential, and so on, depending on the manner in which the resistance wire is wound. Figure 10 illustrates this principle.

Potentiometric-type accelerometers offer high output voltage, and consequently no signal amplification or impedance matching is necessary. These devices require a large displacement, and have a relatively low frequency response. The unit has high mechanical friction, with the resolution usually represented by a finite number.

Piezoelectric Accelerometers

A piezoelectric accelerometer utilizes asymmetrical crystalline materials which, when subjected to stress or strain, produce an electric output. Figure 11 represents this type of accelerometer.

Piezoelectric accelerometers have high output and high frequency response. In addition, they are small in size, ruggedly constructed, self-generating, and they show negligible phase shift. Accelerometers of this type are not capable of measuring static conditions, and are sensitive to temperature changes.

A piezoelectric accelerometer has a high output impedance, and ordinarily it requires impedance matching. This instrument is widely used for the measurement of high-frequency vibration and shock. Basically the piezoelectric accelerometer is an undamped device.

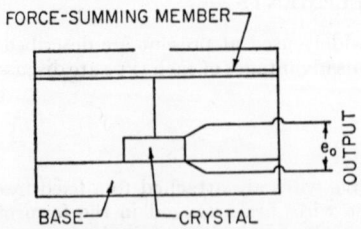

Fig. 11. Piezoelectric transducer.

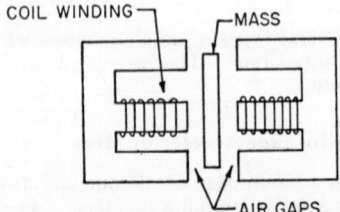

Fig. 12. Inductive transducer.

Inductive Accelerometers

Inductive accelerometers employ a pair of coils and a magnetically coupled mass. When an acceleration is applied, the mass is displaced, and it alters the magnetic coupling path, thereby changing the inductive ratio of the coils.   Figure 12 demonstrates the principle of this accelerometer.

Inductive accelerometers are used to measure static or dynamic accelerations, and they offer high output with essentially continuous resolution.   They also possess a high signal-to-noise ratio, and as a rule they have a very low hysteresis.   Alternating-current excitation is required for these instruments, which must be reactively and resistively balanced.   Inductive accelerometers have a low frequency response, and they are susceptible to magnetic fields.

REFERENCES

1. Sabin, H. B.: Seventeen Ways to Measure Acceleration, *Control Eng.*, vol. 8, no. 2, pp. 106–109, February, 1961.
2. Wolferz, A. H.: Tachometers, Governors, Air-velocity Meters, and Accelerometers, pp. 7-40 to 7-53, in D. M. Considine (ed.), Process Instruments and Controls Handbook," McGraw Hill Book Company, Inc., New York, 1957.
3. Moskowwitz, L.: Accelerometer Calibration, part 1, Measurement of Applied Acceleration, *Instr. Control Systems*, vol. 34, no. 2, pp. 257–260, February, 1961.
4. Moskowwitz, L.: Accelerometer Calibration, part 2, Measurement of Accelerometer Output, *Instr, Control Systems*, vol. 34, no. 3, pp. 467–470, March, 1961.
5. Pinsky, H.: Linear Dead-weight Accelerometer Calibrator, *Instr. Control Systems*, vol. 34, no. 7, pp. 1262–1263, July, 1961.
6. Statham, L.: Bonded vs. Unbonded Strain-gage Transducers, *Instr. Control Systems*, vol. 35, no. 9, pp. 123–124, September, 1962.
7. Kaufman, A. B.: Natural Frequency and Damping, *Instr. Control Systems*, vol. 35, no. 12, pp. 73–78, December, 1962.
8. Pennington, D.: Accelerometer Resonant Frequency, *Instr. Control Systems*, vol. 35, no. 12, pp. 80–81, December, 1962.

SPEED MEASUREMENT

By Dr. Robert C. Langford*

This subsection discusses the measurement of speed, linear or rotational, which is primarily associated with mechanical objects. Common forms and units of measurement include: *frequency* (cycles per second), *rotational speed* (revolutions per minute), *linear speed* (feet per minute), and *differential speed* or speed ratio. Acceleration, or the rate of change of speed, is covered on pages 5-22 to 5-31.

As explained in the following paragraphs, devices for such speed measurements may all be broadly classified as tachometers. Descriptions of such devices are covered in a number of references—see particularly Refs. 4 and 5. Essential principles and general characteristics only of this equipment are contained in this subsection.

CLASSIFICATIONS OF TACHOMETERS

The old "New Standard Dictionary of the English Language," Funk & Wagnall, of 1935 defines *tachometer* as "a contrivance for measuring velocity" and *tachometry* as "the art or science of using a tachometer." Their "New Desk Standard Dictionary," 1953, calls *tachometer* "an instrument for measuring linear and angular velocity, as of machine, the flow of current, etc. (Greek: tachos—swiftness, and meter)."

A more technically strict definition for tachometer in Van Nostrand's "International Dictionary of Physics and Electronics" is "an instrument used to measure angular velocity, as of shaft, either by registering the number of rotations during the period of contact, or by indicating directly the number of rotations per minute."

The "Proposed Recommended Guide for the Measurement of Rotary Speed" issued by the American Institute of Electrical Engineers in April, 1959, narrows the term tachometer to include "an instrument which either continuously indicates the value of rotary speed, or continuously displays a reading of average speed, over rapidly operated short intervals of time."

All these definitions imply that the term of *tachometer* describes any device, whether in one piece or several, sometimes remotely located with respect to one another, which measures rotational speed (velocity) in revolutions per minute. Therefore, the word "tachometer" is used in this meaning here, and the particular parts of the tachometer are called, for example, "tachometer generator" or "tachometer indicator" where necessary.

Speed-measuring devices can be classified in the following ways:

1. *Measurance*, i.e., whether the measured speed is (a) linear, so that the device is a speedometer, measuring distance per unit time, or (b) rotational, so that the device is a tachometer, measuring angular distance per unit time.

2. *Base of measurement*, i.e., whether the measurement is: (a) instantaneous speed, or (b) average speed.

3. *Type of measurement*, i.e., whether the measure of speed is: (a) a basic principle of counting revolutions in a unit of time, or (b) in terms of a secondary effect of speed.

4. *Purpose of measurement:* one or several of the following: (a) to check, or test speed momentarily, (b) to measure speed continuously, (c) to record the speed and speed changes with time, (d) to integrate speed over length of time, or (e) to control speed.

5. *Type of function:* (a) simple tachometer or speed indicator, (b) speed recorder, (c) speed-ratio indicator, (d) speed-difference indicator, (e) speed-deviation indicator, (f) speed-limit signaler.

* Director of Research, Kearfott Division of General Precision, Inc., Little Falls, N.J.

6. *Final use:* For: (*a*) industrial machines, (*b*) aircraft purposes, (*c*) car engines, (*d*) diesel engines for buses, tractors, etc., (*e*) outboard motors, (*f*) high-speed turbines, or (*g*) others.

7. *Construction:* (*a*) hand-type, (*b*) portable, (*c*) attached permanently to the rotating shaft or part, or (*d*) designed as a part of machinery.

8. *Principle of operation:* (*a*) mechanical tachometers, or (*b*) electrical tachometers.

9. *Indication:* (*a*) local, inherent to most mechanical tachometers, (*b*) remote, inherent to electrical tachometers, (*c*) analog—most often d-c indicator, or (*d*) digital—counter, mechanical, or electronic.

In the following portions of this subsection, speed-measuring devices are discussed under the two major headings of Mechanical Tachometers and Electric Tachometers.

MECHANICAL TACHOMETERS

Mechanical tachometers may be classified into three groups:
1. Revolution counter with timed period.
2. Tachometer, measuring instantaneous speed by means of centrifugal force.
3. Resonance tachometer.

Revolution Counter

This type is used with a timing device of some form to determine the number of revolutions in a measured length of time. It therefore measures an average rotational speed rather than an instantaneous rotational speed.

Figure 1 illustrates a simple manually held revolution counter. A worm gear on a spindle is rotated by the shaft whose rotational speed is to be measured. The worm

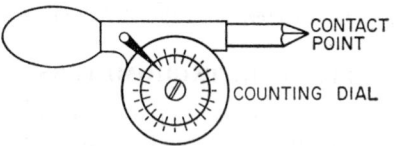

Fɪɢ. 1. Revolution counter.

gear meshes with a spur gear, which in turn moves a calibrated dial, indicating total revolutions. Divisions on the outer dial each represent one revolution of the spindle, while those on the inner dial represent one revolution of the outer dial. A stopwatch, started and stopped simultaneously with the counter, permits calculation of the average speed in revolutions per unit time.

In a second form, sometimes known as a *tachoscope,* a counter is combined with a timepiece. The simplest type (manual) has the two components integrally mounted and simultaneously started when the contact point is pressed against the rotating shaft. Both are stopped when the point is removed. A speed-indicator type of tachoscope, has a revolution counter controlled by an automatic timer. A ratchet arrangement on a measuring wheel frees the wheel for a definite period of time, and a pointer on the wheel indicates on a dial calibrated to read directly in rpm.

A third form of mechanical speed-measuring device is known as a *chronometric tachometer.* It is similar in principle to the speed-indicator tachoscope, but repeats *automatically* its timed period to adjust indication to the average speed for the last timed interval. The interval may be ½ to 30 sec, with shorter periods giving a reading more near the instantaneous speed. Attachments can be mounted on the end of the spindle for measurement of linear speed.

Centrifugal-force Tachometers

An arrangement shown schematically in Fig. 2 employs flyballs, restrained by a spring. Centrifugal force developed by the rotating balls works to compress the spring as a function of speed and accordingly position a pointer. Mechanisms of this type can measure speeds up to 40,000 rpm, and are usually used to make or break circuits for speed control.   Accuracy is about ± 1 per cent.   Variations of this type employ gravity or atmospheric pressure to counterbalance the centrifugal force. Indication can be a pointer on a calibrated scale, level of the meniscus of mercury in a manometer, or reading of a pressure gage.

Resonance Tachometers

A series of consecutively tuned steel reeds (Fig. 3) can be used to determine engine speed on the basis of vibrations created by the machine (e.g., due to reciprocating movement of the piston).   The reed tuned to resonance with the machine vibration frequency will respond visibly, and can be arranged to indicate speed on a scale.   The range of these tachometers is wide (600 to 100,000 rpm), and their accuracy is ± 0.5 per cent or better.   They need contact only with a nonmoving part of the machine.

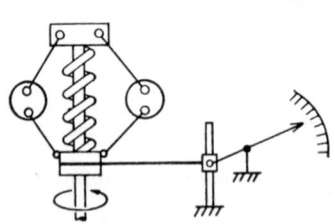

FIG. 2. Centrifugal-force tachometer.   FIG. 3. Vibrating-reed-type speed indicator.

ELECTRIC TACHOMETERS

An electric tachometer consists of a transducer which converts rotational speed into an electrical signal coupled to an indicator.   In proportion to speed, the transducer produces either (1) an analog signal which can be used for analog indication, or (2) pulses which can be digitally counted in terms of revolutions in a unit time.

Eddy-current or Drag-type Tachometers

In the drag-cup type of tachometer (Fig. 4) the transducer produces an analog signal in the form of a continuous drag due to eddy currents induced in the cup.   The current is proportional to the speed.   An analog indicator with linear scale is used.

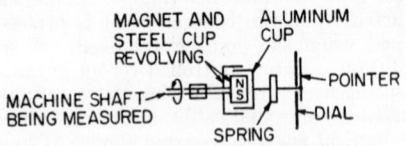

FIG. 4. Drag-type tachometer.

Varieties of this type of tachometer have the following characteristics:
1. The magnetic field of the mechanically coupled, rotating permanent magnet sweeps over a conducting disk or cup and induces in it eddy currents, which in turn produce a torque or drag acting upon the cup against a restraining spring.   Automobile speedometers are of this type, actually measuring angular speed of the wheels. However, the diameter of the car wheel is assumed, and thus the measurement is converted into a linear measurement in miles per hour.
2. Mechanical coupling of the magnet is replaced by an electric drive, consisting of a three-phase synchronous generator, driven by the machine under test, and connected to a three-phase synchronous motor, which in turn drives the magnet of the tachometer.   This is a popular type of aircraft tachometer (see Sec. 16).

3. The permanent magnet can be stationary and a soft-iron rotor produces a revolving magnetic field. This type is very rugged and is used to measure locomotive speed.

4. A conducting cup or disk is located in the field of two coils, electrically 90° out of phase. Only one of these is energized by an a-c source. A signal of the source frequency will appear in the other coil, in proportion to speed of the cup. This type is also used in aircraft tachometry.

5. In one design, developed for industrial instrumentation, which employs pneumatic transmission to remote panel-mounted indicators or controllers, the magnetic drag-disk torque is converted to a transmission signal by a pneumatic force-balance system. Output air pressure, 3 to 15 psi, is directly proportional to speed of rotation. Ranges available are up to 12,000 rpm.

Electric Generator Types

The most common transducers of speed into signals are d-c or a-c generators which produce d-c or a-c voltages measurable by voltmeters of a great variety. Such systems measure speed continuously with accuracies to ±½ per cent over ranges up to about 5,000 rpm. They can easily be geared for speeds beyond this range.

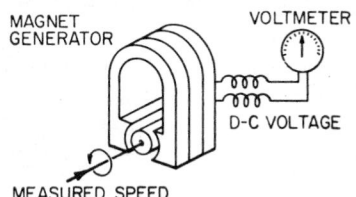

FIG. 5. A-c tachometer.

D-C Tachometers. Figure 5 shows schematically the arrangement of a d-c tachometer which produces a pulsating d-c voltage proportional to the speed. Special attention must be paid to the bearings in the rotor, to maintain the air gap of the magnetic path as uniform as possible (true also of most a-c generators). For greater accuracy, the output voltage can be applied to a high resistance with taps and measured by a null-balance type of potentiometer.

Polarity of the d-c voltage depends on the direction of rotation and can, therefore, be used to indicate this direction by the use of an indicator with its zero point at midscale. Moreover, by using a ratio-indicating instrument, the ratio between two or more speeds can be measured; a generator is used for each speed. Figure 6 shows a typical application of such a system which can be calibrated in terms of per cent of "stretch" or "elongation," as required in textile, steel, and paper mills.

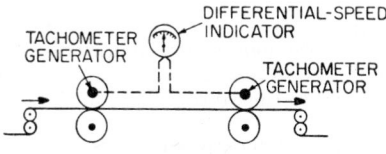

FIG. 6. Differential speed indication.

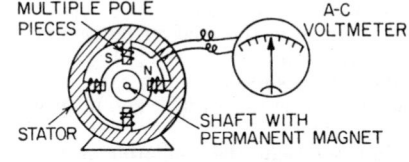

FIG. 7. Basic elements of an a-c tachometer.

A-C Tachometers. Figure 7 shows schematically the elements of an a-c tachometer generator. This type has a stator coil with multiple pole pieces; a magnet on the shaft induces voltage in the stator coil as the magnet passes the pole pieces. The a-c indicator is generally in the form of a permanent-magnet movable-coil device with rectifier. This type can also be used to measure difference in speeds of two sources.

A-C Bearingless Generator. A simpler and less expensive construction of an a-c generator is possible when bearings are eliminated. One method employs a rotor magnet fixed on the machine shaft and the stator on the casing of the machine. With this arrangement, however, the air gap cannot be maintained with high stability, and the a-c voltage effect cannot be so usefully employed as in a generator with bearings. Hence, measurement is made on the basis of signal frequency rather than voltage.

Inductor-type A-C Generator.   This type is found in a great variety of constructions.   A rotating member of soft iron causes flux from a permanent magnet to reverse completely in a coil, inducing in it an alternating current.   With proper shape of stator and rotor, such a generator can produce a frequency output as high as 60 pulses per revolution (particularly useful for measuring low shaft speeds).   The voltage is generally lower than that of the conventional generator, and has a distorted waveshape.   Formerly, the inductor type was used exclusively with analog indicators of the a-c moving-vane type (or d-c instruments with copper oxide rectifiers).   It lends itself better, however, for use with digital-type indicators.

Pulse-shaping Devices.   If the output power of an a-c generator is sufficiently high, use of a pulse-shaping device between the generator and indicator can improve accuracy of indication.   The two most useful devices of this nature are (1) the saturable transformer and (2) the capacitor.

The *saturable transformer* acts as a gate during each cycle, allowing only a part of the energy produced to pass to the indicator, and confining the balance to the generator itself (Fig. 8).   The amount of energy appearing on the secondary side of the saturable

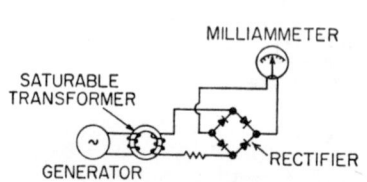

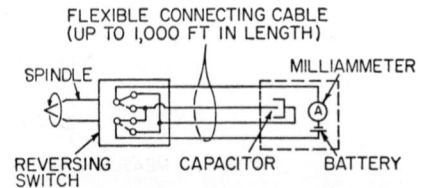

Fig. 8. Schematic circuit of an a-c tachometer with saturable transformers.

Fig. 9. Capacitor-type a-c tachometer.

transformer is constant for each cycle; its value depends on the physical dimensions of the transformer.   Thus, readings of the analog indicator can be made proportional to the *frequency* of the generator pulse output rather than voltage.   This method can be used with all a-c generators, especially the bearingless type, to make the indication very accurate in responding to frequency.

A *capacitor* is often used in a-c applications as a pulse-forming network.   When connected to a source of constant voltage, the capacitor assumes a constant charge. Hence, if the capacitor is allowed to discharge through an indicator, it deflects the indicator proportionally to the number of such charges in each unit of time.

One typical system employing a capacitor (also known as an *impulse*-type tachometer) is shown schematically in Fig. 9.   The pickup head contains a reversing switch mechanically operated from the spindle to reverse twice each revolution.   Thus, battery voltage in the receiver portion of the instrument is applied to the capacitor in each direction.   With each pulse, a current is passed through the milliammeter.   The indicator responds to the average value of such pulses, so that indications are proportional to pulse rates which are, in turn, proportional to the rates of the spindle revolutions (rpm).

In this design, the oscillating switch can be connected directly for speeds of 200 to 10,000 rpm.   With suitable gears, speeds below or above these values can be measured. Properly standardized for battery voltage, such a system is unaffected by temperature, humidity, vibration, or magnetic fields.   In a similar design, where capacitor and switch are connected in one leg of a bridge circuit, higher accuracy of measurement is possible.   Multiple ranges can be obtained by the use of different capacitor values.

Contactless-type Tachometers

Transducers for tachometers come in several forms, to produce pulses from a rotating shaft without being mechanically connected to it.   As a rule, these devices do not produce enough energy to actuate an indicator directly. They generally require an amplifier of sufficient sensitivity. Three types of transducers or "pickups" are briefly described in the following paragraphs.

Electromagnetic Pickup. This is essentially a bearingless generator, as previously described, with a much smaller output voltage and a distorted waveform. It consists of a coil, permanent magnet, and soft-iron piece, forming an open magnetic circuit, which is periodically closed by the insertion of a gear tooth or key to produce pulses.

Capacitive Pickup. A variable capacitor is formed by vanes attached to the revolving shaft under measurement (see Fig. 10). Rotation of the shaft alters the capacitance to ground. The capacitor forms a part of an oscillator tank so that the

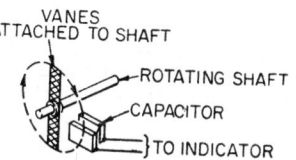

FIG. 10. Principle of capacitive-type contactless tachometer.

number of frequency changes per unit of time is a measure of shaft speed. The pulses thus produced go to amplifying, pulse-forming, triggering circuits where they are converted into constant-amplitude signals. In this way an analog d-c output is produced to provide indicator deflection proportional to speed.

Optical Pickup. A variety of designs use the principle of shaft rotation to interrupt a beam of light falling on a photoelectric or photoconductive cell. The pulses thus obtained are amplified and either (1) counted by an electric counter or (2) shaped to an analog signal before connection to the indicator. In one typical design, a beam of light originating inside the tachometer case hits a white spot on the rotating shaft (or on a spoke of a wheel, etc.). Reflected light falls on a photoconductive cell inside the case, producing pulses in a transistorized amplifier, which in turn causes the indicator to deflect. Such types can measure speeds up to 3 million rpm.

Frequency-type Tachometers

Frequency of an a-c signal from a suitable pickup can be measured and calibrated in terms of speed. For high resolution and accuracy, multipole generators with a large number of poles are preferable (see the previously described pulse-producing transducers). These tachometers can be classified according to the type of frequency meter used as a receiver.

Electronic Counter. This type is the most accurate and has the widest range for speed measurement. It is based on the fundamental principle—revolutions per unit of time.

Frequency Meter. As used for frequency measurements. These types of meters can be calibrated in rpm.

Resonance Frequency Meter. This type can use, for example, a polarized bridge as a null detector and is useful in monitoring speed on an expanded scale, within a very narrow range.

Vibrating Reed. In this type the output from an a-c generator energizes a coil which vibrates reeds at the natural vibration frequency in sympathy with the generator output.

Ignition-type Tachometers

The primary winding of the ignition coil in an internal combustion engine can be connected in series with a battery or magnets and interrupter contacts. The contacts are opened periodically by the revolving shaft, and produce pulses. These pulses are often distorted by high-frequency trains of damped oscillation.

Ignition tachometers are further classified by the way use is made of the developed pulses: (1) systems using low-frequency pulses and (2) systems using high-frequency oscillations.

Systems Using Low-frequency Pulses. Tachometers of this kind are usually connected directly to the high voltage of the ignition primary winding and to ground having a low-pass filter or other means of preventing transients from affecting tachometer indications. For instance, they can have a chopper-type switch that would respond only to low-frequency pulses. The emitted pulses are carefully shaped, by either a saturable transformer or a capacitor, before being conveyed to the indicator.

This kind of tachometer bleeds but little power from the ignition-coil supply and does not upset operation of the ignition system of the engine. Zero-loading the system can be achieved by transistorized amplifiers.

The pulse-shaping arrangements often include the following:
1. A saturable transformer in series with the ignition-coil primary winding.
2. A Zener diode for clipping pulses and for charging the capacitor.
3. A chopper type switch, driven by the pulses, that charges the capacitor from a local miniature battery.
4. A neon bulb across the secondary of a small linear transformer for stabilizing voltage, for charging the capacitor, and for providing a path (when ionized) for discharge through the indicator.
5. A transistorized amplifier, filtering off transients, and shaping pulses more efficiently and providing higher accuracy of indication.

Systems Using High-frequency Oscillations. Instead of trying to filter out the high-frequency transients, one can use them for measuring speed. Tachometers based on this principle of operation are more universal in use because they do not need to be tailored for the particular battery, voltage, or magneto. Two systems operate as follows:
1. A low-value capacitor connected to the high-voltage end of the ignition-coil primary winding picks up the high-frequency oscillation and deletes the low-frequency pulses. Each group of these oscillations, when clipped and rectified by a Zener diode, changes into a unidirectional pulse. The frequency of these pulses is identical with low-frequency pulses produced by the interrupter. When connected to the d-c indicator, these molded pulses produce deflection proportional to the speed of the engine.
2. A radio-type receiver picks up the electromagnetic waves associated with these high-frequency transients and demodulates them into the low frequency of an analog signal.

Stroboscopic Tachometers

For measurement of rotational speed, as well as other cyclic movement, stroboscopes can be used to synchronize a flashing light with rotation of shaft, making it appear to stand still. Usually, a reference mark is put on the shaft or a disk is mounted on the shaft. The speed of the stroboscope itself is measured as a measure of rpm, without requiring contact with the rotating part.

By use of harmonics, the range of this method is practically unlimited, although it is difficult to discern between fundamental and harmonic speeds. This type is most often used in experimental work.

REFERENCES

Handbooks, Encyclopedias, Codes

1. Griffiths, E. A.: "Engineering Instruments and Meters," chap. V, Measurement of Velocity, London, 1920. Over 60 pages devoted to tachometers, speedometers, and miscellaneous speed-indicating devices. Very good and exhaustive treatment of the basic approach to mechanical tachometers; still retains its value.
2. Glazebrook, R. (ed.): "A Dictionary of Applied Physics," reprint of 1920 ed., chap. III, Meters, Peter Smith Publisher, Gloucester, Mass., 1950. Section on tachometers, speedometers, and miscellaneous speed-indicating devices. Describes in detail a number of speed-measuring devices now obsolete but having historical significance.

3. ASME Power Test Code Supplement on Instrument and Apparatus, part 13, Speed Measurement, 1939. Now obsolete classification of the speed-measuring devices with their characteristics and short analysis of factors affecting their accuracy.
4. "Handbook of Measurement and Control," chap. VI, Speed Measurement and Control, by M. F. Behar, The Instruments Publishing Co., Inc., Pittsburgh, Pa., 1959. Perhaps the best and most comprehensive of the earlier presentations of speed-measuring instruments, systematically classified and analyzed.
5. Considine, D. M. (ed): "Process Instruments and Controls Handbook," pp. 7–40 to 7–48 by A. H. Wolferz, McGraw-Hill Book Company, Inc., New York, 1957. Tachometer Section lists several types of up-to-date tachometers and their predecessors. Does not mention digital electronic counter as applied to revolution counting.
6. "The McGraw-Hill Encyclopedia of Science and Technology," vol. 13, p. 382, McGraw-Hill Book Company, Inc., New York, 1959. Article on tachometer is an abbreviation of the A. H. Wolferz section of Ref. 5, supplemented by paragraph on electronic revolution counter.
7. Recommended Guide for Measurement of Rotary Speed, AIEE Reports, April, 1959. Very valuable publication replacing the ASME Power Test Code, part 13. Pays great attention to the classification of speed-measuring methods and their analysis.

Technical Papers

The introduction of digital electronic counters as a speed indicator has resulted in many technical papers on tachometry in recent years. In Great Britain counters used cold-cathode tube (ex decatron); these were particularly adept at using the low frequencies produced mechanically in tachometry. Meanwhile in the United States the development of the very sensitive hot-cathode tube counters permitted measurement of the highest possible frequencies.

8. Akeley, L. T.: Aircraft Tachometer Indicator, Trans. AIEE, vol. 73, part II, pp. 398–402, 1954. An analysis of design factors affecting starting performance. Construction with both permanent-magnet and hysteresis rotor elements free on shaft give best results in frequency (low) starting.
9. Akeley, L. T.: Temperature Errors in Dragmagnet, and J. J. Frazier: Eddy-current Disc Type of Tachometer Indicator, Trans. AIEE, vol. 74, part I, pp. 418–422, Sept. 20, 1955. Analysis of factors producing temperature error in indicator; new design for tachometer compensation.
10. Bland, W. R., and B. J. Cooper: High Speed Precision Tachometer, Electron. Eng., vol. 26, January, 1954. Electronic tachometer of integrating type with ranges 0–8,000 and 10–8,000 rpm for measurement of vibration on rotating machines; reading accuracy 0.01%.
11. Berry, T. M., and C. L. Beattie: A New High Accuracy Counter Type Tachometer, Trans. AIEE, vol. 69, part II, pp. 868–691, 1950. One of the first descriptions of a converter of revolution into pulses and of an electronic counter for speed measurement with accuracy better than ±0.03%.
12. Davis, S. A.: Using Two-phase Servomotor as Induction Tachometer, Control Eng., vol. 2, no. 11, pp. 75–76, November, 1955. How basic two-phase machine can be used as servomotor if power is supplied to both phases, or as induction tachometer if rotor is driven from external source and power is supplied to one phase.
13. Ballard, R. G.: The Magnetic-drag Tachometer, Trans. AIEE, vol. 61, pp. 366–368, 1942.
14. Davidson, G. M., and M. Papalow: Potentiometer Tachometer Has High Sensitivity, Electronics, vol. 29, no. 9, pp. 158–161, September, 1956. Speeds as low as ½₀₀ rpm can be measured by electronic chemical device that is at least 100 times as sensitive as existing tachometers; consisting of linear potentiometer and operational-type differentiating amplifier; tachometer's operating range can be varied over wide limits by changing time constant of feedback network or excitation on potentiometer.
15. Eckels, A. R., and W. R. Peck: Commutatorless D-C Tachometer, Trans. AIEE, vol. 72, part I, no. 9, pp. 625–629, November, 1953. Tachometer based on current-balance principle.
16. Frazier, J. J.: Measuring Aircraft Engine Speed, Gen. Elec. Rev., vol. 60, no. 3, pp. 13–15, May, 1957. Future tachometer design may need radically new approaches as ½₀% accuracies are needed to prevent serious damage to jet engines.
17. Frazier, R. H.: An Analysis of the Drag-up A-C Tachometer by Means of Two-phase Symmetrical Components, Trans. AIEE, vol. 70, part II, pp. 1894–1906, 1951.
18. Frazier, R. H.: Drag-up A-C Tachometer with Constant Current Excitation, ibid., vol. 72, part II, pp. 150–152, 1953.
19. Fearnside, K., and P. A. N. Briggs: Mathematical Theory of Vibratory Angular Tachometer, Proc. IEE London, vol. 105, part 6, no. 7 (monograph 264), pp. 156–66, March, 1958.
20. Feng, T. N.: How to Calculate Performance of A. C. Drag Cup Tachometers, Control Eng., vol. 5, no. 6, pp. 90–92, June, 1958. Simple technique for calculation of output-voltage gradient and its linearity over stated speed range for any drag-cup tachometer, curves for feedback system requirement of zero phase angle between tachometer input and output.
21. Fisher, H. M.: Position versus Tachometer, Speed Matching System, Control Eng., vol. 3, no. 3, pp. 65–71, March, 1956.
22. Girdner, W.: New 60 Cycles per Revolution Generator for Precision Tachometry Measurements, Hewlett-Packard J., vol. 5, no. 3–4, 1953. Description of a converter of revolutions into signals, counting of which gives direct reading of speed in rpm.
23. Goodwin, W. N.: The Weston Magneto-generator, Weston Eng. Notes, vol. 4, no. 1, 1949. Historical notes on the first d-c magnetogenerator for speed measurement, constructed by Dr. Weston in 1896.
24. Goodwin, J. K.: Very High Speed Precision Tachometer, Electron. Eng., vol. 30, no. 359, pp. 18–24, January, 1958. Measuring speed up to 400,000 rpm by high-speed counting tubes, also used as frequency-measuring instrument; description of circuit with diagram.

25. Goodwin, J. K.: Digital Tachometer Aids in Turbine Design, *Electronics*, vol. 32, no. 15, pp. 58–61, Apr. 10, 1959. Counts up to 400,000 rpm in two scales, at accuracy of 0.005 % pulses generated by capacitor unit connected directly to rotating machine, are passed through accurately controlled gate; by generating 30 pulses for each revolution and holding gate open for 3 min, direct indication is provided on counting tubes.

26. Harrington, E. L.: High Speed Revolution Counter, *Electron. Eng.*, vol. 27, no. 326, pp. 142–146, April, 1955. Decatron tubes counter for measuring gas-turbine speed to within 1 rpm at 200,000 rpm.

27. Hilton, E. A.: A New 100 KC Counter for Use in Electronics and Industry, *Hewlett-Packard J.*, vol. 4, no. 3, November, 1952. Description of electronic counter.

28. Tucker, H.: Differenz-Tachometer, *Masler Mitteilunger*, vol. 14, no. 3, pp. 49–53, December, 1955.

29. Knox, L. A.: Approximate Analysis of Drag Cup A-C Tachometer, *Trans. AIEE*, vol. 77, pp. 202–207, Sept. 2, 1958. Improvements achieved in temperature coefficient, output voltage, and speed linearity error of output voltage of precision tachometer by using drag cup of higher resistivity and lower temperature coefficient of resistance; simplified basic tachometer-performance equation; frequency compensation is achieved by *RLC* network in tachometer excitation circuit.

30. Kwast, V. B.: Electrical Tachometer, U.S. Patent 2,958,038, 1960. Differential Tachometer with reference current derived from the same signal tachometer generator.

31. Kwast, V. B.: High Resolution Differential Tachometer with Suppressed Lower Speeds, *AIEE Conf. Paper* CP 61–355. Three differential methods analyzed. Differential bridge allows comparison of signals without interference between sources.

32. Leslie, W. H. P.: Precision Control of Shaft Speed, *Elec. Energy*, vol. 1, no. 1, pp. 2–5, September, 1956.

33. Maddock, A. J.: Tachometers, *Elec. Transducers*, part 2, Control, October, 1958. Short review of five tachometer systems: magnetic drag, d-c generator, a-c generator, contact-making devices, electronic.

34. Mueller, F. A.: Recent Speed and Delay Instruments, *Traffic Engr.*, vol. 25, no. 3, pp. 100–103, 105, December, 1954. Review of devices such as speedometer recording, electric counter boxes, etc.

35. Pollard, P. J.: Speed Indicator Has Expanded Scale, *Electronics*, vol. 30, pp. 188–190, May 1, 1957. Multivibrator triggered by shaped signal from a-c tachometer provides speed indication within 1 % over range from 500 to 5,000 rpm; regulated bucking voltage provides scale expansion to indicate deviations from operating speed with accuracy of 0.1 % on 500-rpm full-scale indication.

36. Reed, C.: Tachometer with High Short-term Accuracy, *Electron. Eng.* vol. 32, no. 384, pp. 103–105, February, 1960. Alternator-speed measuring instrument that indicates speed within range of approx. 10 % of nominal speed; linear scale; accuracy maintained by built-in crystal-controlled oscillator, and calibration is easily checked and adjusted.

37. Strassman, A. J.: RPM Indicator Provides Expanded Scale, *Electronics*, vol. 27, no. 8, pp. 145–148, August, 1954. Electronic tachometer indicates increment of rpm over operating portion of speed range (scale expansion).

38. Schulman, J. M.: Accuracy Tachometry Methods, with Electronic Counters, *AIEE Trans.*, vol. 73, part I, no. 15, pp. 425–426 (Paper 54-290), November, 1954. Three methods of using counters in tachometry.

39. Utyamysheve, R. I.: Meterodyne Instrument for Precise Measurement of Rotational Velocities, *Instr. Exptl. Tech. USSR English Transl.*, **1958**, no. 4, pp. 528–530, July-August. Based on hetero-dyne comparison made of frequency quartz-crystal oscillators; device is used either with photo-electric or induction transducers; working range is 110 to 11,100 rpm and error is plus or minus 0.25 rpm.

40. Wright, M. J.: Electronic Tachometer for Automobiles, *Electron. Eng.*, vol. 31, no. 380, pp. 599–602, October, 1959. Measures fundamental frequency of electric waveform obtained from automobile ignition circuit; waveform is applied to a circuit which combines filtering and limiting functions and feeds fixed-amplitude square wave to frequency-measuring circuit; milliammeter indicates engine speed directly; circuit contains no active components, and is powered by automobile battery.

41. Wolferz, A. H.: Electric Tachometers, *Weston Eng. Notes*, vol. 4, no. 2, April, 1949. Description and characteristics of d-c and a-c tachometer generators. Both use permanent magnet. In d-c generator, 12-coil armature revolves within two-pole ring magnet. Metal (palladium-silver alloy) brushes. In a-c generator, rotor magnet with 8 poles (4 pairs) revolves within stator with 8 coils.

42. Wolferz, A. H.: Differential Speed Indicator, *AIEE Conf. Paper* CP 56-102. Description of the 150–0–150-rpm differential speed indicator over the range of speeds up to 3,000 rpm.

43. Zoha, S. M.: Multipurpose Electronic Tachometer, *J. Technol.*, vol. 3, no. 1, pp. 33–42, June, 1958. Based on electronic pulse technique; measuring speed over wide range with accuracy better than 1 per cent; provides for extra facilities for recording instantaneous speed and for measuring drift of speed of 0.5 per cent of full scale; relay used to cut off power from rotating machine at present speed.

44. Grenfell, K. P.: Operational Amplifier Improves Speed Regulation, *Control Eng.*, vol. 8, no. 12, pp. 71–74, December, 1961.

45. Wilson, R. G.: Digital Systems for Speed Control, *Automation*, vol. 9, no. 9, pp. 62–65, September, 1962.

46. Shenfeld, S., H. R. Manke, and E. F. Soderberg: An Analog-to-digital System for Recording Angular Rotation, *Auto. Control*, vol. 14, no. 10, pp. 40–44, October, 1961.

WEIGHT AND WEIGHT RATE OF FLOW

By Donald B. Kendall*

Weight, which is actually the force with which a body is attracted to the earth, is often confused with mass, which is the quantity of matter in a body. When either buying goods or setting up a process formula, we are more concerned with the quantity of matter than with the force which attracts a body to the earth. The relation between weight (in a vacuum) and mass is

$$W = mg \tag{1}$$

where W = weight
m = mass
g = acceleration due to gravity

UNITS OF WEIGHT

Technically speaking, the metric unit of weight is the *dyne* (the force which will impart to a mass of 1 gram an acceleration of 1 cm/sec²) and the English unit of weight is the *poundal* (the force which will impart to a mass of 1 lb an acceleration of 1 ft/sec²). However, we will use the looser definition of weight as being synonymous with mass. (Webster gives one definition of weight as the quantity of matter in a body ascertained by the balance.)

The unit of weight in the metric system, used commercially in Europe and for scientific work in the United States, is the *gram*. The gram is 1/1,000 the weight of the international prototype kilogram, maintained in Paris. The metric ton is 1,000 kg.

Most commercial weighing in the United States uses the avoirdupois system of weights. The standard for this system is the *pound*, defined as 1/2.20462 international kg. The pound is subdivided into 16 oz or 7,000 grains. The standard or short ton is 2,000 lb. The long or gross ton is 2,240 lb. The troy or apothecaries' system is used for some weighing in the United States. The lb (t) or lb (ap) = 0.823 lb avdp. This pound is subdivided into 12 oz or 5,760 grains. The grain is the same in both systems.

FACTORS AFFECTING WEIGHT

The accepted standard for g is 980.665 cm/sec² or 32.174 ft/sec². Because the earth is not a perfect sphere, however, g is actually 978.039 at the equator and 983.217 at the poles. The value of g also varies with the distance above sea level, because the gravitational force between two bodies varies as the square of the distance between them. Helmert's equation gives the value of g for any location and altitude.

$$g = 980.616 - 2.5928 \cos 2\phi + 0.0069 \cos^2 \phi - 3.086 \times 10^{-6} H \tag{2}$$

where ϕ = latitude and H = elevation, cm.

This variation of g with location means, for instance, that using a spring scale which was calibrated at the location where g = 980.665, a kilogram weight (1,000 g mass)

* Manager, Product Engineering Department, Toledo Scale Division of Toledo Scale Corp., Toledo, Ohio.

would show a weight of 997.3 g at sea level at the equator and a weight of 1002.6 g at sea level at the poles. However, if the mass were weighed on an even balance or any type of scale where the unknown is counterbalanced by known weights, the 1,000-g mass would show a weight of 1,000 g at any location because the change in gravitational force would have the same effect on the known weights as on the mass being weighed. Also, since it is common practice to adjust any scale against master weights at the time of installation, this variation of g with location and altitude has no practical effect on the determination of the amount of matter in a body by weighing it on a scale.

The centrifugal force due to the rotation of the earth also affects the relation between mass and weight at any location except at the poles. However, since the force is the same on weights as on the mass being weighed, this force has no practical effect.

Another factor which affects the relationship between weight and mass is that normal weighing operations take place in the air rather than in a vacuum. Archimedes' principle states that a body immersed in a fluid is buoyed up by a force equal to the weight of the fluid displaced. Of course, if an object being weighed has the same density as the weights used to calibrate the scale, the buoyant effect will be the same and can be disregarded. If the density of the material being weighed is different than that of the weights, this effect can be determined by using a formula for reduction of weighing to vacuo.*

$$M = m + md_a \left(\frac{1}{d_m} - \frac{1}{d_w} \right) \tag{3}$$

where M = true mass of body in vacuo
$\quad m$ = apparent mass of body
$\quad d_m$ = density of body
$\quad d_w$ = density of weights
$\quad d_a$ = density of air

To determine the magnitude of this effect in an extreme case, assume that a scale has been calibrated using cast-iron weights having a density of 7.08, and that a quantity of gasoline having a density of 0.68 is to be weighed at sea level with the density of the air 0.001223. Using Eq. (3), it is found that the actual mass of an amount of gasoline which reads 1,000 kg on the scale is 1,001.63 kg.

Equation (3) can be rearranged and combined to facilitate calculation of the difference in mass for different air densities, as might be encountered at different elevations:

$$M = m \left[1 + (d_{a1} - d_{a2}) \frac{d_w - d_m}{d_w \times d_m} \right] \tag{4}$$

where d_{a1} = density of air at location 1, and d_{a2} = density of air at location 2.

An extreme example would be a comparison between the weight of gasoline at sea level and at a location, such as Las Vegas, N.Mex., where the altitude is $1,960m$ and the air density about 0.000990. Using these figures in Eq. (4), it is found that a quantity of gasoline which weighs 1,000 kg at sea level would weigh 1,000.31 kg at Las Vegas.† However, for materials with densities closer to the density of weights ordinarily used, and for normal variations in elevation, the effect of the buoyancy of the air causes errors which are less than the allowable weighing error in commercial weighing operations.

Thus, although calculations will show that an airplane which weighs 100,000 lb at sea level at the North Pole, will weigh about 100,900 lb when flying at 40,000 ft altitude over the equator, and though the gasoline in an airplane which weighs 1,000 lb at sea level, will weigh about 1001.2 lb at 40,000 ft, the effects of location, altitude, and buoyancy of the air on the relation between mass and weight will be disregarded in the remainder of this subsection.

---

* See "Handbook of Chemistry and Physics," 26th ed., p. 1291.
† A table showing the density of air at various altitudes is given in the "Handbook of Chemistry and Physics," 26th ed., p. 2462.

REASONS FOR MEASURING WEIGHT

The importance of measuring weight was realized early, as this was necessary for even the simpler bartering systems. In modern industry, weight and its measurement are important for many reasons, such as:

Determining the Amount of Raw Material Received

For inventory information and also to confirm invoice figures. Quantities involved vary from a few milligrams of a rare dye to the tons of products carried in a railroad car, or the thousands of tons unloaded from the hull of a ship. Since the actual weight of material received is the figure desired, it is important to distinguish between actual or *net* weight and *gross* weight, which includes the *tare*, or weight of the container.

Also, many materials such as sand, ore, paper, fiber, etc., are subject to considerable variation in the amount of moisture contained. Such material should be purchased on a *dry weight* basis or on the basis of a standard percentage of moisture. If analysis of a sample reveals a certain percentage of moisture *based on wet weight*, then

$$\text{Dry weight} = \frac{100 - \% \text{ moisture}}{100} \times \text{wet weight} \tag{5}$$

If analysis of a sample reveals a certain percentage of moisture *based on dry weight*, then

$$\text{Dry weight} = \frac{100}{100 + \% \text{ moisture}} \times \text{wet weight} \tag{6}$$

Controlling Materials to Processes

In chemical processes, formulas are frequently based on proportioning materials according to functions of their molecular weights. Thus, it is important to weigh the ingredients for proper blending. When a specific solution is to be made, the material to be dissolved and the solvent are both weighed, in order to obtain the proper proportions.

In many physical processes, it is found that equipment will operate most efficiently at a definite load. For instance, a certain pulverizing mill will operate most efficiently at a load of a certain number of tons per hour. For best results, then, the flow of material to the mill is weighed and controlled. Frequently materials to be molded are weighed to assure proper filling of the mold without excess flash.

The range of weights to be measured will vary, from a few milligrams of dye to tint a batch of plastic or paint, to tons of material to charge a blast furnace. The same precautions for obtaining the actual and dry weight are necessary as previously mentioned. Also, in some processes, it may be necessary to make allowance for the buoyant effect of the air, in order to determine the proper mass of a material.

Measuring Shrinkage

In some cases, the loss or waste in a process or operation is determined by comparing the weight of materials going in with the weight coming out. In other cases, waste can be weighed directly.

Inspection

Many products which are molded or cast are weighed, to make sure that the proper density of material is maintained. Containers are weighed to make certain that they contain the prescribed amount of material, and cases of containers are weighed to assure that each contains the proper number of containers and that none of the containers are broken or have leaked. For some products it is possible to check certain dimensions by weighing the product.

Forces

Forces such as torque, the stiffness of springs, etc., can be determined by weighing equipment.

Counting

Objects which are uniform in weight can be counted rapidly by using scales adapted for this use. The counting may be done for inventory or for packaging the product for sale.

Measuring Product for Selling and Shipping

Probably the most common application of weight is in the measurement of the product for selling. Some products are weighed in bulk and some are preweighed into packages or bags. Shipping charges are also determined by weight.

PRINCIPLES USED TO MEASURE WEIGHT

One of five fundamental principles is used in practically all methods of measuring weight:

1. Comparison with known weights.
2. Measuring deflection of a body, using Hooke's law, which states that strain is proportional to stress.
3. Measuring the hydraulic or pneumatic pressure required to support the unknown weight.
4. Measuring the electric current in a coil whose magnetic field supports the unknown weight.
5. Measuring displacement of a liquid, using Archimedes' principle, that a floating body displaces its own weight of a liquid.

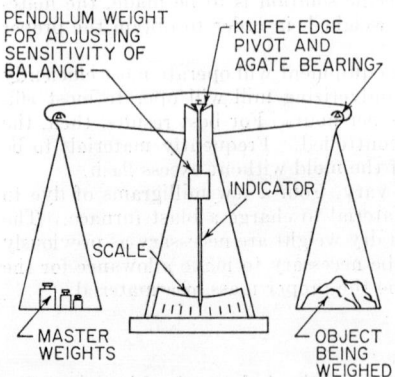

PENDULUM WEIGHT FOR ADJUSTING SENSITIVITY OF BALANCE

KNIFE-EDGE PIVOT AND AGATE BEARING

INDICATOR

SCALE

MASTER WEIGHTS

OBJECT BEING WEIGHED

FIG. 1. Even-arm balance.

There are many ways of using these principles. Fundamental ways are discussed in the following paragraphs. Detailed descriptions of actual devices are found in several references.*

Comparison with Known Weights

Use of this principle may be further classified as:

Direct Comparison with Master Weights. The even-arm balance, shown in Fig. 1, is the oldest known weighing device and also can be made the most accurate. High accuracy is possible because of the simplicity of construction and the high sensitivity this makes available. The operation of weighing takes considerable time, however, and the use of this balance is at present generally restricted to such specialized operations as weighing very small objects or quantities, weighing very valuable commodities such as gold, or the weighing or "sealing" of master weights. Ingenious methods of quickly adding selected counterweights improve the speed of this method of weighing considerably. The even-arm balance is made in capacities of a few grams to 1,000 lb or more.

* Notably in pp. 7-8 to 7-39 of "Process Instruments and Controls Handbook," McGraw-Hill Book Company, Inc., New York, 1957; and "Industrial Weighing," by Douglas M. Considine, Reinhold Publishing Corporation, New York, 1948.

Even-arm Scale. A modification of the even-arm balance, more adapted to industrial weighing, is shown in simplified form in Fig. 2. Speed of weighing is increased by: (1) addition of a dashpot to dampen the oscillations of the system; (2) use of a heavier pendulum (reducing sensitivity) or, in some cases, a spring, to provide more powerful balance restoring forces; and (3) location of the platters above the lever (for convenience in use), and prevention of their swinging by means of check links. Frequently, in actual construction, the positions of the lever and check links are reversed.

It should be noted that in a properly constructed even-arm scale of the type shown in Fig. 2, it is not necessary to center the weights nor the object being weighed in the center of the platters. This principle of construction was discovered by Roberval, a French mathematician, in 1670. The model he made to illustrate his principle is shown in Fig. 3. In this figure, $ABCD$ is a true parallelogram; arms AF, BF, CF', and DF' are all equal. FF', AC, and BD are equal and vertical, and the system is in balance without the weights P and Q. Roberval then discovered that with equal weights P and Q at any position on arms EH and KL, the system will remain in balance, even though the members AB and CD are moved from the horizontal position.*

Briefly, the proof rests in resolving the forces exerted by weight P into a vertical force at B and a couple whose forces are parallel to AB and CD, and the forces exerted

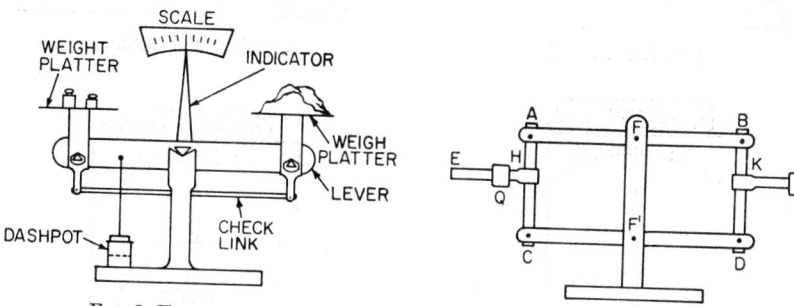

FIG. 2. Even-arm scale.

FIG. 3. Roberval balance.

by weight Q into a vertical force at A and a couple whose forces are parallel to AB and CD. The vertical forces at A and B are equal to weights Q and P. If P is farther out on its arm than Q is on its arm, there will be resultant forces to the right at F and to the left at F'. These forces are equal and opposite, so that even when the unit is displaced from the horizontal position, the only resultant vertical forces on the system are those equal to Q and P at A and B, respectively. Therefore, the system remains in balance.

Even-arm scales, as used in industry, are generally built in capacities from 1 to 100 lb. They are most commonly used for the trimming part of a filling operation or for checking the weight of objects, such as containers of material, which are supposed to be of uniform weight.

Counterbalancing with One or More Calibrated Weights. The steelyard, shown in Fig. 4, is one of the oldest weighing devices, but its principle is used in modern beam scales. Here the weight W is determined by moving the poise until its weight P times its arm B equals weight W times its arm A. The distance the poise is moved is calibrated in terms of weight.

Figure 5 shows the application of this principle to a beam scale. A small poise P' is frequently used to provide a vernier effect for finer weighing, and loose weights can be added at C to increase the capacity of the scale. An indicator at I shows when the poises are properly located to balance the load W.

* Proof of this principle is given in "A Treatise on Weighing Machines," by G. A. Owen, published by Charles Griffin & Co., Ltd., London, 1922.

Automatic Indicating Scale. Figure 6 shows an automatic indicating scale which eliminates the manipulating of the poise of a beam scale. Here the calibrated weight P counterbalances the unknown weight W by the bent lever rotating until $P \times B = W \times A$. Since B increases as the sine of the angle of rotation, the graduations on the chart cannot be uniform when W is connected to a fixed arm A.

Figure 7 shows W connected by means of a ribbon and contoured cam to the bent lever. By proper cam contour, the relationship will result in uniform increments of indicator travel along the chart for uniform increments of weight W. Automatic indicating scales permit much more rapid determination of weight than beam scales where the poise is moved manually.

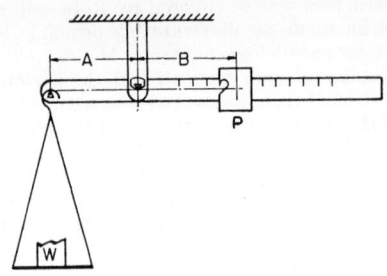

FIG. 4. Steelyard.

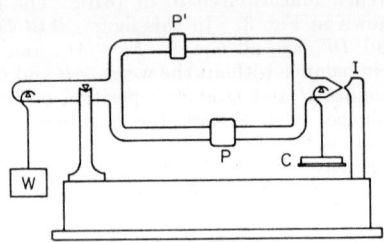

FIG. 5. Weigh beam.

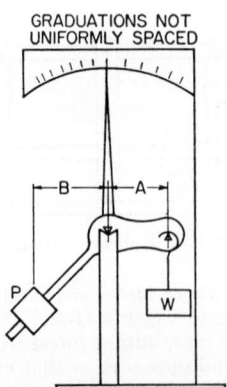

FIG. 6. Automatic indicating scale, hook pull.

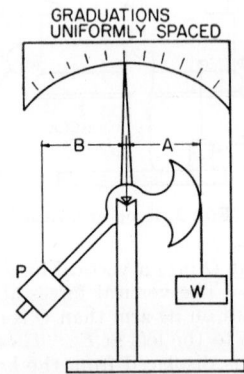

FIG. 7. Automatic indicating scale, cam type.

Both automatic indicating pendulum-type scales and beam-type scales are made in capacities of a fraction to several pounds and, when combined with simple and compound lever systems, are built in capacities from a few pounds to hundreds of tons.

Measuring Deflection of a Body

Most common application of this principle is found in what are called *spring scales*. Coil springs or beams of spring material can be designed to stretch or deflect a definite distance for a definite load. Up to the proportional limit, the deflection is proportional to the load. Calibrating this deflection in terms of weight makes a spring scale.

When the deflection is magnified in some such manner as rotating a pinion by means of a rack attached to one end of the spring, with an indicator attached to the pinion to read a magnification of the deflection on a circular dial, it is found that the rate of deflection for ordinary spring steel varies with temperature. This is caused partially by a lengthening of the spring, but a greater change is caused by a reduction in the modulus of elasticity as the temperature is increased. A change of temperature of 57°F will cause a change in a steel spring deflection of 1 per cent for a given weight. By the use of special alloys and proper spring design, this change with temperature can be greatly reduced, so that it is now possible to build spring scales which are practically unaffected by normal temperature changes. There is some hysteresis and creep however, in any spring, and its effect may be noticeable in the most precise weighing operations.

Spring scales are made in capacities from ounces to several thousand pounds, and, when combined with simple or compound lever systems, are built in capacities from a few pounds to hundreds of tons. In most spring scales, the deflection of the spring is measured directly, and the spring is in the form of a helix or beam so the deflection will be comparatively great.

In recent years, methods such as the *strain gage*, which is capable of measuring extremely small strains, have been applied to weighing. Here the deflection of a piece of steel, of size and proportions to support the load, is measured directly. Since, in the strain gage, compensation must be incorporated in the gage for temperature effects on the resistance and length of the wire, ordinary steel is usually employed and additional compensation incorporated to overcome the effects of temperature on the deflection of the steel. Weight is determined by measuring the change in resistance of the strain gage, usually with a servo-balanced potentiometer. These "strain-gage load cells" are built in capacities from a few pounds to a hundred tons or more.

Measuring Hydraulic or Pneumatic Pressure

Figure 8 shows a hydraulic capsule or load cell constructed so that the load is supported by oil confined by a diaphragm. The pressure developed in the oil is a measure of the weight, and may be determined by a bourdon or other type of pressure gage. These units are built in capacities of 100 to 50,000 lb. If an electrical transducer,

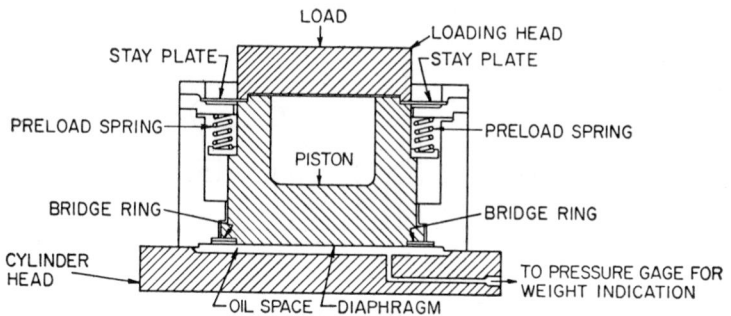

Fig. 8. Hydraulic load cell.

such as a linear differential transformer, is connected to the bourdon tube, the electrical signals from several cells can be added so that a weight supported by several cells can be measured.

Figure 9 shows a pneumatic capsule or load cell constructed so the load is supported by air pressure controlled and measured by conventional pneumatic control equipment. Such units are built in capacities from 25 to 25,000 lb.

Measuring Electric Current

Current flowing in a coil of wire develops a magnetic field, which will attract a magnetic material or another coil of wire with a current flowing in it. A measure of the current flow is a measure of the weight supported.

Because of the problems of varying permeability and permanent magnetism in a magnetic material, as well as field distortion and accurate measurement of current, however, this principle is seldom used except for such applications as determining small changes in weight or following rapid changes in weight or force.

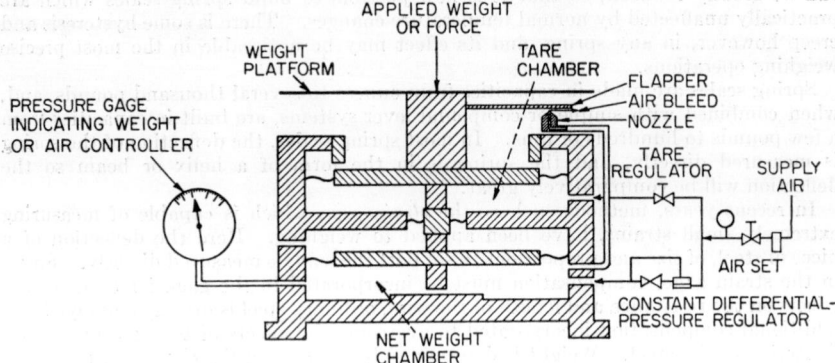

Fig. 9. Pneumatic load cell.

Measuring Displacement of a Liquid

Figure 10 shows a method of determining weight by measuring the depth to which a counterbalance sinks into a liquid. The scale comes to rest when the force due to the difference between the weight of the counterbalance and the weight of liquid displaced balances the weight W. Since evaporation of the liquid would change the reading of the scale for any given weight, mercury is often used to reduce this effect.

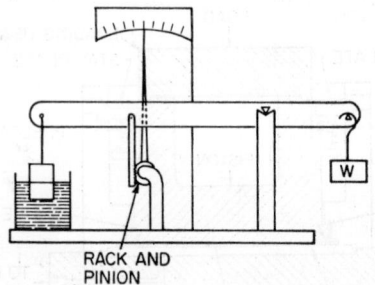

Fig. 10. Hydrostatic scale.

Change of density of the liquid with temperature changes results in a different calibration for the scale. This principle of weighing is seldom used in commercial scales.

It is possible to estimate the cargo in a ship or barge by measuring the difference in displacement loaded and unloaded and computing the weight of water displaced.

BASIC TYPES OF EQUIPMENT

Summarized in Table 1 are the common types of scales and their characteristics. The types are grouped under three classifications:

1. *Basic weighing:* Measuring the weight of an object or a definite amount of a material; also the weighing out of a predetermined amount of material, such as for filling containers.

2. *Continuous weighing:* Weighing material in motion, whether it is a strip of paper or rubber, or bulk material conveyed on a belt.

Table 1. Summary of Scale Types and Characteristics

Equipment	Capacity	Accuracy, %	Comments
Basic Weighing			
Even-arm balance..............	Fractions of gram to 1,000 lb	0.002	Most accurate, also slowest. Requires laboratory conditions for best results
Even-arm scale—may be equipped with beam and poise to reduce number of weights to be handled	Grams to 100 lb	0.05	Generally used for repeated weighings of a definite amount or checking
Beam—generally connected to a lever system. May have separate beam and poise for tare and additional weights to increase capacity	Few pounds to hundreds of tons	0.1	Comparatively inexpensive. Requires operator to adjust poises to weigh unknown. May be preset for repeated controlled weighing
Pendulum—generally connected to a lever system. May have beam for tare and unit weight to increase capacity	Few pounds to hundreds of tons	0.1	Automatic indicating. Easily adapted for printing and data handling and, with cutoffs, to variable controlled weighing
Spring—tare may be offset by prestressing	1–20,000 lb	0.2	High speed due to low inertia. Particularly adaptable to crane scales
Strain-gage load cell—may have tare compensation and range steps	50–400,000 lb or more	0.2	Simple installation. Easily protected against corrosion. "Remote indication" inherent. Easily adapted for printing, data handling, and variable controlled weighing. Relatively expensive
Hydraulic load cell—may use bourdon gage with transducer for electric output	100–50,000 lb	0.2	Simple installation. Requires electrical output to summarize load on two or more cells
Pneumatic load cell—may have tare compensation	25–25,000 lb	0.2	Simple installation. Best adapted for single point of load application. Easily adapted for controls with pneumatic controllers. Source of compressed air required
Continuous Weighing			
Strip—passing over weigh roller between two idler rollers. May be supported on balanced belt	1 oz–100 lb/ft	1	Generally indicates and/or controls deviation from desired weight per foot
Belt conveyor for bulk material—one or more rollers supported on mechanical or load-cell scale. Belt speed and weight per foot are integrated to totalize	500 lb–1,000 tons/hr	1	Simple installation. May be equipped with controls to regulate flow or to stop flow after predetermined amount
Bulk weigher—repeated discharges of hopper after controlled load is accumulated	500 lb–1,000 tons/hr	0.1	May be equipped with controls to stop after predetermined amount or with readout to data-handling equipment. Requires head room for weigh hopper and surge hopper above. Interrupts flow
Proportioning			
Batch weighing—definite amount of each ingredient weighed in one or more scales and delivered to process on demand	500 lb–500 tons/hr	0.1	Easily adjusted for various formulas. Several ingredients may be weighed in one scale. Weights can be recorded for accurate inventory control. Flow is intermittent
Loss-in-weight feeder-supply hopper—with controllable discharge feeder is supported from scale. Rate of discharge is controlled	60 lb–4,000 lb/hr	0.5	Flow continuous for amount in hopper—requires two units per material for fully continuous flow
Continuous feeder—belt feeder mounted on scale equipped with controls to regulate feed from supply hopper	60 lb–750 tons/hr	1	Flow continuous. Separate unit required for each material. Limited to solid materials

Accuracy is expressed as the per cent of error which may normally be expected. Depending on equipment and its installation and care, actual accuracy may be better or poorer than that shown. Accuracy shown assumes operation near capacity of equipment.

3. *Proportioning:* Delivery of two or more materials in a definite ratio, such as for process or blend.

Methods for basic weighing are described, with illustrations, in the previous paragraphs. Some of these same methods have been adapted for use in continuous weighing and proportioning, as outlined in the following paragraphs.

Semicontinuous Feed-belt Weigher (Fig. 11). A scale-balanced weigh belt, running continuously, controls operation of a feed belt. Sections of the belt are fed intermittently with material, so that definite increments of load can be weighed. These increments are small to approach continuous weighing. The method can be used to record the receipt and withdrawal of bulk materials and to batch solids.

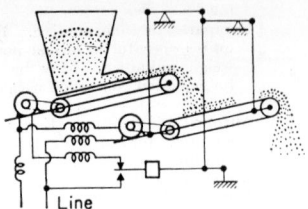 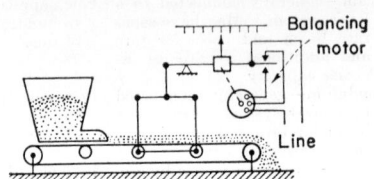

FIG. 11. Semicontinuous feed-belt weigher.   FIG. 12. Continuous conveyor scale.

Continuous Conveyor Scale (Fig. 12). Many equipment designs are based on the concept of totalizing continuous loads on a scale-balanced section of a conveyor belt. Various forms of automatic weighers employed include spring-balanced beams, strain-gage load cells, and motor-driven poise self-balancing beams. A totalizer integrates instantaneous loads with belt speed to correct for any belt-speed variations. This method is commonly used on bulk materials where a continuous record is needed for inventory purposes.

Radioactive Transmission Gage, Absorption Type* (Fig. 13). Beta or gamma radiation from a radioactive source passes to an ionization chamber through a moving sheet or web of the material to be weighed. Attenuation of signal is a measure of the material weight per unit area (or thickness). The electrical signal actuates an electronic recorder. The method is unaffected by water or surface finish, and can be used for such sheet materials as paper, rubber, plastics, coated fabrics, or metals.

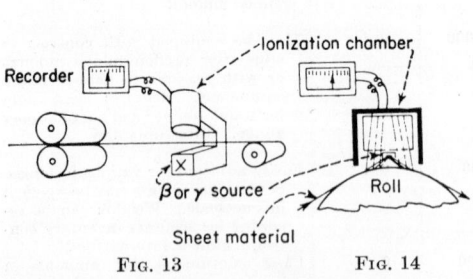 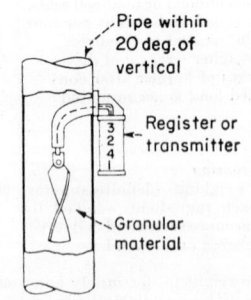

FIG. 13   FIG. 14

FIG. 13. Radioactive transmission gage, absorption type.   FIG. 15. Volumetric solids meter.
FIG. 14. Radioactive transmission gage, backscattering type.

Radioactive Transmission Gage, Backscattering Type (Fig. 14). This method is a variation of the absorption type, where the source and ionization chamber are on the same side of the measured material. Radiation is reflected from a backing plate under the material whose weight is being measured. This type also can be cali-

* See also p. 3-39, Radioisotopes in Instrumentation.

brated in terms of thickness, and is suitable for the same kinds of continuous weight measurement.

Volumetric Solids Meter (Fig. 15). A spiral vane is mounted in an essentially vertical duct ($\pm20°$). Moving solid material in the duct rotates the vane which actuates a digital recorder. Ducts must be 20 in. or larger, and accuracy is ±3 per cent of actual material weight. Being volumetric, the device must be calibrated in terms of material handled to provide *weight* flow.

Weight-rate Meter (Fig. 16). Dry, free-flowing material falls on a constant-speed rotating impeller. The drive motor is mounted on flexure pivots, and is free to move about the axis of its shaft. When particles of material are accelerated to the velocity of the rotating impeller, a torque is created on the motor shaft. This torque, which is proportional to the weight rate of flow, is converted to an air signal by a pneumatic force-balance transmitter. The range of this method is 40 to 200 lb/min, and its accuracy is ±0.2 per cent. A pneumatic signal can be used to control a feeder, or to operate other devices for ratio control.

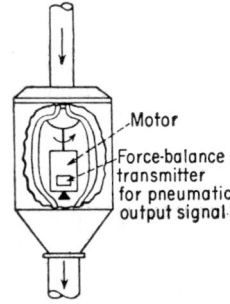

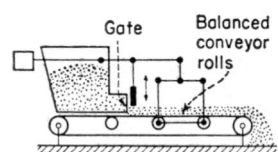

FIG. 16. Weight-rate meter.   FIG. 17. Scale-balanced belt feeder.

Scale-balanced Belt Feeder (Fig. 17). In this method, a scale-balanced belt is mechanically linked to control the gate setting of a feed hopper. Thus, solid materials on the belt are maintained at a constant weight for feeding to a process at a constant rate.

Balanced Weigh Belt, Mechanical Feeder (Fig. 18). A short feed belt, carried on a scale, is fed by a mechanically vibrated tray. Feed rate is regulated by the tray's oscillation which, in turn, is governed by a wedge that is raised or lowered by the scale beam. This method permits use of a variable-speed star feeder in place of the vibrator, and can be used to provide controlled weight rates of feed up to 10,000 lb/hr. It can also be used for proportioning.

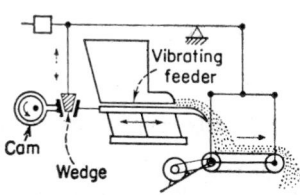

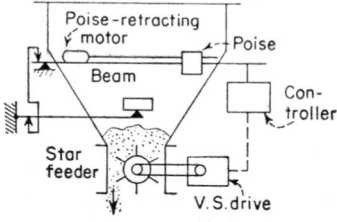

FIG. 18. Balanced weigh belt, mechanical feeder.   FIG. 19. Loss-in-weight star feeder.

Loss-in-Weight Star Feeder (Fig. 19). Material is discharged from a scale-balanced hopper by a star feeder with variable-speed (V.S.) drive. The scale beam counterpoise is retracted continuously by a constant-speed motor. A pneumatic controller holds the beam in balance by matching discharge rate with the feed rate as set by the retracting poise.

Balanced Belt, Electric Vibrating Feeder (Fig. 20). The weight of material on a short, scale-supported feed belt is controlled continuously by a vibrating feeder, which is automatically adjusted by the scale. One type is electric with a magnet used to adjust vibration amplitude.

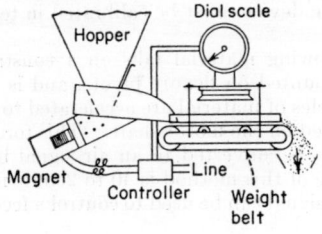
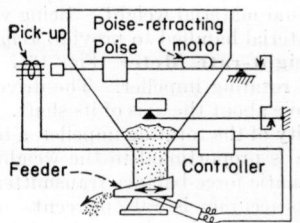

FIG. 20. Balanced belt, electric vibrating feeder.

FIG. 21. Loss-in-weight vibrating feeder.

Loss-in-Weight Vibrating Feeder (Fig. 21). Similar to the method of Fig. 19, except that scale beam balance is picked up electronically, and the material is fed at controlled rate by a vibrating feeder.

AUXILIARY EQUIPMENT

In specific designs of scales, a variety of auxiliary equipment is required to provide the functions of: (1) indicating, (2) recording, (3) computing, (4) totalizing, and (5) controlling. General types of equipment for obtaining these functions are described in the following paragraphs.

Indicating

A mechanical pointer registering with graduations on a chart or dial. Either the pointer or the chart can be driven, with the other held stationary. Drive can be (1) direct in the case of even arm, or pendulum or spring scales, attachments for beam scales; or (2) from bourdon tubes or other spring elements in the case of hydraulic or pneumatic scales. Drive can be by servo system from strain-gage scales or from linear differential transformer or synchro for "remote indication" from pendulum, spring, hydraulic, or pneumatic scales.

Optical projection can be used with a beam of light projected to a chart or with a miniature chart projected to a screen. Optical indication generally reduces the mass of moving elements, thus increasing the speed with which a scale comes to rest.

Digital indication can be provided by driving a mechanical counter with a servo system, by scanning the uncovered portion of a chart with a light beam and counting pulses, or by mechanical or electrical readout of a position encoder.

Recording

Continuous recording may be accomplished by directly coupling to the pen of a recorder or by the use of a suitable transducer.

Instantaneous recording of the weight on a scale may be by (1) direct impression from an etched plate attached to the scale beam or indicating mechanism; (2) printing from wheels or sectors positioned by an encoder attached to the scale; or (3) electric typewriter or adding machine operated by signals from the scale.

Computing

Counting may be accomplished directly by fixed or variable ratio attachments or by feeding unit and total weight information into a computer.

Density may be measured by weighing a definite volume and incorporating the conversion formula into the chart graduations.

Specific gravity may be determined by weighing an object in air and in water.

Percentage formulas, such as moisture content, per cent of material to be added, etc., can be incorporated in charts or dials or by the use of variable ratio scales.

Totalizing

Instantaneous weight from one or more scales may be totalized by (1) feeding signals to an adding machine, (2) counting pulses, or (3) driving a mechanical counter. Tare weight may be subtracted from gross to totalize net.

Continuous flow may be totalized by integrating weight and speed of flow.

Controlling

A **limit switch** may be operated by the motion of a beam or lever or by a change in hydraulic or pneumatic pressure. In a mechanical scale, unless the beam or lever is free from contact with the switch after its operation, a true reading of weight will not be obtained.

A **magnetically operated mercury switch** may be used with the magnet attached to a moving part of the scale.

A **photoelectric relay** is ideal, since the cutting of a light beam applies no force to the scale.

Inductance or capacitance changes, caused by motion of the scale, can tune or detune an oscillator to operate a relay.

A **low-torque potentiometer** coupled to the scale can be connected in a bridge circuit with a remote-set potentiometer to cause cutoff action when balance is achieved.

A **predetermined counter**, operated by pulses from the scale, can be used.

Pneumatic controllers, operating directly from a pneumatic scale or from suitable transducers coupled to other types, can easily provide proportional as well as fixed-point control.

FACTORS AFFECTING SCALE PERFORMANCE

In summary, the following factors should be recognized as affecting the performance of scales:

1. **Selection of equipment** most suitable for the operation. Capacity should not greatly exceed expected maximum requirements.

2. **Installation** should be properly made with adequate supports to avoid excessive deflection or vibration.

3. **Maintenance** of the type accorded precision instruments is required to assure continued accuracy.

4. **Design** of equipment associated with a scale must consider the fact that a scale measures all vertical components of forces applied to the load-receiving element. Connections must be flexible in the vertical plane. If a scale hopper is totally enclosed, pressure inside and outside the hopper must be equalized, or precautions taken to make sure that a difference in these pressures does not apply a vertical force to the scale. If only one end of a tank is supported by a scale, care must be taken to avoid unequal deflection of the supports, which could shift the center of gravity of the load.

5. **Feeders** delivering materials to scales must be controllable within the limits of accuracy expected.

REFERENCES

1. Barnett, L.: Electric Strain-cell Weighing, *ASME Instr. and Regulators Conf. Paper* 53-IRD-12, 1953.
2. Berg, R. H.: Solids Flow and Level Measurements by Continuous Weighing, *ASME Instr. and Regulators Conf. Paper* 53-IRD-14, 1953.

3. Brauer, K.: "The Construction of the Balance," Incorporated Society of Inspectors of Weights and Measures, London, 1909.
4. Briggs, P. R. A.: Strain Gauge Load Cells as Industrial Weighing Elements, *Automation Progr.;* vol. 4; pp. 286–289, 298–299, September, 1959.
5. Carleton, R. J., Jr.: Electric Weighing in Steel-mill Processes, *ASME Instr. and Regulators Conf. Paper* 53-IRD-13, 1953.
6. Considine, D. M.: "Industrial Weighing," Reinhold Publishing Corporation, New York, 1948.
7. Dahle, O.: Heavy Industry Gets New Load Cell, *ISA J.,* vol. 6, pp. 32–37, August, 1959.
8. Evans, G. E.: Maintenance of Weighing Scales, *Iron Steel Eng.,* vol. 34, pp. 86–91; March, 1957.
9. Fischer, A. A., and R. Springer: Air Circuitry Powers Automatic Weighing, *Appl. Hydraulics,* vol. 11, pp. 104–106, May, 1958.
10. Green, M.: Using Load Cells in High-accuracy Weighing Systems, *Control Eng.,* vol. 6; pp. 121–124, October, 1959.
11. Hicks, J. J.: Remote Weight Measurement by Means of Hydraulic Load Cells with Electrical Transmitters, *ASME Instr. and Regulators Conf. Paper* 53-IRD-4, 1953.
12. Kennedy, V. C.: Electronic Weight Determination as a Tool for Control and Measurement Procedures, *ASME Instr. and Regulators Conf. Paper* 53-IRD-8, 1953.
13. Kirwan, J. O., and L. E. Demler: Continuous Weighing Meters and Feeders, *ASME Instr. and Regulators Conf. Paper* 53-IRD-9, 1953.
14. Klein, E.: Processing and Proportioning Materials by Weight, *ASME Instr. and Regulators Conf. Paper* 53-IRD-2, 1953.
15. Knill, B. I.: Weighing While Handling, *Flow,* vol. 13, pp. 58–60, August, 1958.
16. Laycock, G. H.: Load Cell Using Unbonded Resistance-wire Strain Gauges, *Soc. Instr. Technol. Trans.,* vol. 9, pp. 112–122, December, 1957.
17. Lowe, R. P.: Continuous Gravimetric Proportioning, *Instr. and Automation,* vol. 30, pp. 482–485, March, 1957.
18. Lowe, R. P.: Continuous Gravimetric Proportioning Systems, *ASME Instr. and Regulators Conf. Paper* 53-IRD-7, 1953.
19. Maloney, N. G.: High Speed Weighing, *ASME Instr. and Regulators Conf. Paper* 53-IRD-10, 1953.
20. McCullough, F. S.: Problems of Continuous Weighing, *Automation Progr.,* vol. 2, pp. 116–117, 136, March, 1957.
21. McKenna, G. L.: Weighing Enters New Fields via Digital-data Scales, *Iron Age,* vol. 182, pp. 81–83, July 24, 1958.
22. Milroy, J. W., and G. C. Mayer: Characteristics of Components in Pneumatic Weighing Systems, *ASME Instr. and Regulators Conf. Paper* 53-IRD-3, 1953.
23. Mulligan, W. E.: Pneumatic Weigh Cells Improve Automatic Weighing and Blending, *Automation,* vol. 5, March, 1958.
24. Mylting, L. E.: Proportional Batch-type Mixer, *ASME Instr. and Regulators Conf. Paper* 53-IRD-11, 1953.
25. Owen, G. A.: "A Treatise on Weighing Machines," Charles Griffin & Company, Ltd., London, 1922.
26. Richardson, I. H.: Trend toward Automation in Automatic Weighing and Bulk Materials Handling, *ASME Instr. and Regulators Conf. Paper,* 1953.
27. Sanders, A.: Scale Care, *Concrete,* vol. 65; pp. 26–28, October, 1957.
28. Sanders, L.: Weighing and Automation, *Automation Progr.,* voi. 3, pp. 364–367, October. 1958.
29. Smith, R. W.: Testing of Weighing Equipment, *Natl. Bur. Standards (U.S.) Handbook* H37, 1945.
30. Specifications, Tolerances and Regulations for Commercial Weighing and Measuring Devices, *Natl. Bur. Standards (U.S.) Handbook* 44, 1955.
31. Staff: Choosing Scales for Mine Operations, *Eng. Mining J.,* vol. 159, pp. 62–64, Mid-June, 1958.
32. Staff: Digital Printout Techniques for Automatic Proportioning Control, *Elec. Mfg.,* vol. 58, p. 85, December, 1956.
33. Staff: New Ways to Use Electronic Scales, *Steel,* vol. 142, pp. 124–125, Feb. 17, 1958.
34. Staff: Process Control by Weighing, *Process Control & Automation,* vol. 4, pp. 462–464, December, 1957.
35. Staff: Remote Weight Transmission, *Automation,* vol. 3; pp. 60–62, December, 1956.
36. Staff: Scales for Continuous Weighing-control, *Flow,* vol. 12, pp. 61–65, December, 1956.
37. Staff: What Is Load Cell: How Does it Work?, *Design Eng.,* vol. 4, pp. 48–49, 69, May, 1958.
38. Wade, H. T.: "Scales and Weighing," The Ronald Press Company, New York, 1924.
39. Williams, J. C., Jr.: Automatic Weighing Techniques, *Automation,* vol. 6, pp. 34–40, July, 1959.
40. Ziemba, J. F.: Weighing Makes Big Strides, *Food Eng.,* vol. 32, March, 1960.
41. Zilius, R. B.: In-motion Weighing, *Automation,* vol. 3, pp. 44–47, November, 1956.
42. Kennedy, D. W., and C. W. Hibscher: Weighing Scales Couple to Computer, *Control Eng.,* vol. 9, no. 7, pp. 83–85, July, 1962.
43. Ryan, F. M.: Automatic Weighing of Solids, *Control Eng.,* vol. 9, no. 9, pp. 103–108, September, 1962.
44. Williams, J. C., and R. MacLeod: Automatic Batch Weigher, *Automation,* vol. 8, no. 9, pp. 67–68, September, 1961.
45. Staff: High Speed Checkweighers, *Automation,* vol. 7, no. 4, p. 86, April, 1960.
46. Staff: Fast Weighing Increases I-beam Accuracy, *Automation,* vol. 7, no. 5, p. 96, May, 1960.
47. Staff: Centralized Quality Control Monitoring, *Autom. Control,* vol. 14, no. 2, pp. 17–18, February, 1961.
48. Staff: Classifying Glass Blanks by Weight, *Automation,* vol. 8, no. 8, p. 87, August, 1961.
49. Staff: Conveyor Scales Control Bauxite Refining Process, *Automation,* vol. 8, no. 5, pp. 66–67, May, 1961.
50. Cahn, L.: Electromagnetic Weighing, *Instr. Control Systems,* vol. 35, no. 9, pp. 107–109, September, 1962.

LIQUID LEVEL

By P. A. Elfers*

Liquid level is a process variable of key importance in industrial instrumentation for: (1) proper process operation, and (2) cost accounting. Devices for level indication, recording, and/or automatic control range from simple visual gages to local or remote-reading instruments of varying complexity to suit the application.

In this subsection, the importance of liquid-level measurements is reviewed. The fundamental mathematics and physical principles related to these measurements also are given as a background for their application. Table 2 at the end of the subsection is a reference and selection guide designed to give an over-all picture of common methods of liquid-level measurement. Engineering details of such devices are given in Ref. 1.

IMPORTANCE OF LIQUID LEVEL TO PROCESS OPERATION

In many processes involving liquids contained in vessels, such as distillation columns, reboilers, evaporators, crystallizers, or mixing tanks, the particular level of liquid in each vessel can be extremely significant to process operation. A level which is too high, for example, may upset reaction equilibria, cause damage to equipment, or spillage of valuable material. A level that is too low may have equally bad consequences. Combined with such basic considerations, there is a trend in continuous processing toward smaller storage capacity. This reduces the initial cost of equipment, but also accentuates the need for accurate and sensitive control of liquid level.

In the final analysis, effective measurement and control of liquid level in process operations can usually be justified for reasons of economy and safety. To the operator, this variable provides vital information as to: (1) quantity of raw material available for his process; (2) available storage capacity for the products being manufactured; and (3) satisfactory or unsatisfactory operation of his process. A few specific examples will illustrate these important uses of liquid-level measurement and control.

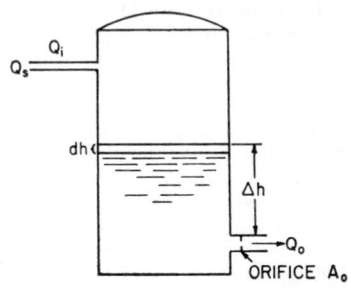

Fig. 1. Basic flow vs. level factors.

Constant Head for Steady Process Flow

Often, steady process flows, such as the introduction of a raw material to a process, are maintained by holding a constant head pressure on the feed line. This can be achieved by control of liquid level in the feed tank whose feed line exits from the bottom of the tank, as shown in Fig. 1.

As regards the theory of this arrangement, in the free flow of liquid through an opening or orifice, or across a weir plate, the quantity of liquid discharged is a function of the level height above the orifice or weir (see Fig. 1). Briefly, the flow is

* Director, Fisher Governor Co., Marshalltown, Iowa.

governed by the equation

$$Q = CA \sqrt{2gH} \tag{1}$$

where Q = quantity, cfs
  C = orifice constant
  A = area of flow, sq ft
  g = acceleration of gravity, 32.2 ft/sec^2
  H = head of liquid, ft

Thus, with the area of flow A constant and the other factors being constants, steady process flow Q will be obtained if the head of liquid H above the orifice or weir is accurately measured and controlled.

Protection of Centrifugal Pumps

Where it is desired to maintain a head pressure against the suction of a centrifugal pump, level of the liquid in the storage tank must be maintained at an optimum value.   If the level drops too low, flashing and cavitation are caused in the pump suction, with resultant erratic pump discharge and extreme wear on pump impellers. If the level rises too high, there may be a loss of accumulator volume in the vessel, thereby affecting the process from an operating viewpoint, as described in the following paragraphs.

Product Quality Control

Warp sizing in the textile industry is a good example of how close control of liquid level directly affects product quality.   The warp yarn is run through a bath of sizing solution which imparts a protective coating on the yarn.   The amount of size absorbed by the yarn is a function of the time during which the yarn is in contact with the size solution.   As the yarn passes through the size solution in a prescribed path (usually around a large cylinder rotating at a fixed rpm), the time of contact is a function of the level height of size solution.   Variation in solution level, therefore, will change this contact time and thus destroy warp uniformity, later causing breakage of threads on the looms.

Efficient Operation of Equipment

Control at One Height.   In many process applications, level must be maintained accurately at a predetermined height, irrespective of load conditions on the process.   Several examples will serve to illustrate this point.

In a steam or vapor generator, such as a boiler, it is desired to maintain level at a predetermined value, in order that two sets of conditions are present at all times, regardless of output from the generator.   The first condition requires that the quantity of liquid inventory in the vessel be maintained in order to provide feed for the evaporation process.   The second condition requires that a vapor volume space be maintained in order to have available storage capacity for vapor, plus a volume space which will prevent carry-over of entrained liquids in the vapor.

In continuous processes, a correct level head in certain equipment is of large importance.   In evaporators, for example, the heating medium may be inside a tube bundle, which must at all times be covered to an optimum depth, thereby requiring precise level control.   Too low a level will uncover heating surface, lowering efficiency of the process.   Too high a level will require greater heat input as head pressure increases, which may result in damage to throughput material or unsatisfactory evaporation action.

Proportional (Wideband) and Averaging Level Control.   In continuous processes, accumulators or storage vessels are introduced between various stages of the process in order to provide storage (inventory).   Process upsets or disturbances are absorbed in such accumulators, and only a minimum of them are passed on to the next phase of the process.

In such process applications, control of level at a constant height is not always desirable. It is more important that the outflow of the storage vessel does not change suddenly and cause an upset in the subsequent process stage. Any sudden increase in input to the storage vessel should be absorbed in the vessel. To accomplish this, "averaging" liquid-level control is used, wherein wideband proportional-plus-reset mode of control is incorporated in the level control instrument.

With this type of control, if the uncontrolled input suddenly increases, the wide proportional band permits the level to rise temporarily, with little change in the controller output regulating outflow from the vessel. If the input remains at its higher value for a period of time, the automatic reset functions to return the level to the set point, and gradually changes the outflow rate. Conversely, if the uncontrolled input suddenly decreases, the level is permitted to drop, but outflow is not permitted to be similarly affected, but is changed at a gradual rate if the decreased input continues.

The size of the vessel required becomes a function of: (1) maximum process upset to be expected in the system and to be absorbed by the accumulator; (2) duration of upset; and (3) elapsed time allowed before full value of continued upset is to be passed along to the next stage of the process.

"Holding time" is often used to describe the function just explained. The size of the vessel also may be limited by physical height considerations, by change of level height allowable, and by the economics of vessel cost. All these become variables for the design and instrument engineers to evaluate, in determining the size of vessels and accumulators.

The actual performance of the liquid-level controller in smoothing out minor upsets and in absorbing them in the accumulator can be further adjusted by proportional-band settings and reset-rate adjustments.*

Level Control Permits Smaller Vessels. In simple single-capacity systems, the utilization of level measurement and control can be advantageously applied to keep the capacity of processing vessels within practical limits. If level measurement and control are used, the size of a mixing or reaction vessel may be small. It is not necessary that large-size vessels be used to handle all available liquid to be processed, since the liquid-level control device will feed only the fluid required to keep the liquid concentration or its height at a predetermined value. Large or bulky vessels thus can be eliminated, with accompanying economy. This also means that a small amount of process material is under reaction or in process, thereby reducing attendant hazards, potential losses, or spoilage.

Importance of Liquid Level to Cost Accounting

The flow meter, weighing scale, and liquid-level gage are the process cost accountant's principal tools for obtaining facts concerning quantities of liquid raw materials and finished products in storage; and of liquids in process. Thousands of liquid-level meters of all types, from the simplest gage stick, often still used for taking inventories of tank farms, to the most sophisticated remote-level-indicating gages are in daily use, principally for cost accounting needs. See paragraphs later in this section on Volume and Weight Measurement from Level.

MATHEMATIC AND PHYSICAL PRINCIPLES

Application of level measurement and control requires an understanding of certain principles which are highlighted in the following paragraphs. These include: (1) relation of flow to level in vessels, (2) capacity vs. level height in variously shaped vessels, (3) theory of buoyancy, and (4) fluid flow.

Relation of Flow to Level in Process Vessels

In designing process equipment and determining the requirements of liquid-level controlling equipment, the engineer often must calculate in advance how a change of

---

* Automatic controllers are fully described in sec. 9 of "Process Instruments and Controls Handbook," McGraw-Hill Book Company, Inc., New York, 1957.

liquid inflow will affect the level in the vessel. In any given vessel system, a stable quantity of inflow to the vessel will equal the outflow. With a stable inflow, height of the level H above the drawoff will increase until the head pressure developed will cause flow through orifice A_o by an amount Q_o which will be equal to the inflow Q_i.

From Eq. (1) and Fig. 1, this may be expressed as

$$Q_i = Q_o = 0.897\ CA_o\ \sqrt{2gH} \tag{2}$$

where Q_i = flow rate into vessel, gpm
$\quad Q_o$ = flow rate out of vessel, gpm
$\quad C$ = orifice coefficient
$\quad A_o$ = orifice area, sq in.
$\quad g$ = acceleration of gravity, 32.2 ft/sec^2
$\quad H$ = height of liquid level above top of outlet orifice, in.

Variation of the inflow or of the orifice area, with the other variables remaining constant, causes a level change. A control valve substituted for the fixed orifice provides an easy means of varying the orifice area. If the orifice area is increased, the level falls until, at the new area, the level stabilizes at a point where its head effect on the orifice causes outflow to equal inflow.

If the orifice area is decreased, the level rises until the product of a smaller orifice value and larger head effect causes outflow to equal inflow. The rate at which the level will rise or fall can be expressed by Eqs. (3) to (5) which follows:

$$Q_o = 0.897\ CA_o\ \sqrt{2gH} \qquad \text{(standard flow formula)} \tag{3}$$

$$\frac{dH}{dt} = (Q_i - Q_o)\frac{231}{A} = \text{height change, in./min}$$

$$= (Q_i - Q_o)\frac{231}{0.785d^2}\ \dagger$$

$$= \frac{294}{d^2}\,(Q_i - Q_o)$$

$$= \frac{294}{d^2}\,(Q_i - CA_o\ \sqrt{2gH})$$

$$= \frac{294}{d^2}\,(Q_i - 7.20CA_o\ \sqrt{H}) \tag{4}$$

where dH = change in liquid-level height, in.
$\quad dH/dt$ = rate of change in liquid-level height, in./min.
$\quad A$ = transverse tank area, sq in.
$\quad d$ = diameter of tank, in.

To find an intervening height H_2 at a time t_2, when Q_i is initially equal to Q_o and is instantaneously changed at time t_1:

$$\frac{dH}{dt} = \frac{294}{d^2}\,(Q_i - 7.20CA_o\ \sqrt{H})$$

$$dt = \frac{dH}{(294/d^2)Q_i - (294/d^2) \times 7.\,0CA_o\ \sqrt{H}}$$

For easy manipulation, let

$$B = \frac{294}{d^2} \times 7.20CA_o \qquad \text{and} \qquad D = \frac{294}{d^2}\,Q_i$$

Then $\qquad dt = \dfrac{dH}{D - B\,\sqrt{H}} \qquad$ and $\qquad dt\displaystyle\int_{t_1}^{t_2} = \int_{H_1}^{H_2}\dfrac{dH}{D - B\,\sqrt{H}}$

† One gallon equals 231 cu in.; one pound per square inch equals 27.7 in. of water.

From table of integrals after transformation into standard form:

$$2x \left[-\frac{\sqrt{H}}{B} - \frac{D}{B^2} \log_e (D - B\sqrt{H}) \right]_{H_1}^{H_2} = t_2 - t_1$$

or

$$\frac{2}{B^2} \left[\sqrt{H}\, B + D \log_e (D - B\sqrt{H}) \right]_{H_2}^{H_1} = t_2 - t_1 \tag{5}$$

It should be noted that if there is a change of Q_i, inflow, the resultant liquid-level height H_2 may be calculated when outflow Q_o again equals inflow, by use of Eq. (3). By use of Eq. (5), the amount of time required for the liquid level to reach any intervening height between H_1 and H_2 can be calculated. Also, by substituting a t_2 figure in Eq. (5), the change in height for any given change in time can be calculated.

If an uncontrolled flow is introduced into a vessel and a level controller is placed on the outflow, the rate of flow and level changes can be calculated. For this detailed discussion, reference should be made to Quantitative Analysis of a Single Capacity Process, by Spitzglass, *ASME Transactions*, November, 1938, and "Automatic Control Engineering," by E. S. Smith, McGraw-Hill Book Company, Inc., 1944.

Capacity vs. Level Height in Various Vessels

Many processing and storage vessels where liquid-level measurement is a factor are cylindrically shaped and mounted vertically on end. Thus, the content for any level height can be simply calculated by

$$\begin{array}{c} \text{Content} = \text{cross section area} \times \text{level height} \\ \text{cu ft} = \pi r^2 h \end{array} \tag{6}$$

where r = inside radius of tanks, ft, and h = level height, ft.

This relation becomes more complex where cylindrical tanks are mounted horizontally. This is best illustrated by calculating a typical problem (dimensions used in the example are given in Fig. 2).*

Example. Determine the volume in a cylindrical tank (flat ends) mounted horizontally, with the following factors known:

$$\begin{array}{ll} \text{Diameter of tank} & = 48 \text{ in.} \\ \text{Depth of liquid} & = 10 \text{ in.} \\ \text{Length of tank} & = 120 \text{ in.} \end{array}$$

Area ACE (shaded portion) = area $ABCE$ − area ABC
Area $ABCE$ = $(2\angle ABD/360) \times$ area of circle

and $\angle ABD$ is found from its cosine which is $14\!/\!24$.

$$\therefore \quad \angle ABD = 54.25°$$
$$\text{Area } ABC = 14 \times 24 \times \sin ABD$$

or

$$14 \times 24 \times \sin 54.25 = 14 \times 24 \times 0.8116 = 272.7$$

$$\text{Area } ABCE = \frac{108.50}{360} \times \pi \times (24)^2 = 545.4$$

$$\therefore \quad \text{Area } ACE = 545.4 - 272.7 = 272.7 \text{ sq in.}$$

$$\text{Volume (U.S. gallons) per ft of length} = \frac{272.7 \times 12}{231} = 14.17 \text{ gal/ft}$$

$$\therefore \quad \text{Total volume} = \frac{14.17 \text{ gal} \times 120}{12} = 141.7 \text{ gal}$$

---

* An alternative method of calculating the capacity of a horizontal cylindrical tank employs the formula

$$A = R^2 \left[\text{arc vers } \frac{H}{R} - \frac{\sqrt{2RH - H^2}}{R^2} (R - H) \right]$$

$$\text{Volume} = A \times \frac{L}{231} \quad \text{gal}$$

where
A = area, sq in.
H = height of liquid in tank, in.
R = radius, in.
L = length, in.
arc vers H/R = angle in radians whose versine is H/R

Values of A are given in a nomograph (Fig. 41, p. 41, of "Design of Diagrams for Engineering Formulas," by Hewes and Seward, McGraw-Hill Book Company, Inc., New York, 1923).

This method of calculation has been used for preparing tables giving capacities of horizontal cylindrical tanks for one foot of length, with the diameter of the tank tabulated against the depth of liquid.

For detailed information concerning capacities of horizontal cylindrical tanks with bulged or dished ends, refer to Perry "Chemical Engineers' Handbook," 4th edition, pp. 6-66–6-69.

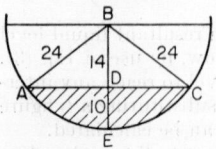

FIG. 2. Dimensions used in calculations of capacities of horizontal cylindrical tanks. [*Source: R. H. Perry (ed.), "Chemical Engineers' Handbook," 4th ed., McGraw-Hill Book Company, Inc., New York, 1963.*]

Regardless of the liquid-level detecting device used, the shape of the vessel has a basic effect on the accuracy of measurement if level measurement is converted to units of weight or volume, but not if the device is calibrated in inches or feet of height. For example, in a horizontal cylindrical vessel (Fig. 2), it is obvious that the same percentage change in level near the bottom of the vessel represents much less volume or weight change than does the same change in level when the liquid height is midway up. In a vertical cylindrical vessel, equal increments of level change represent equal changes in volume or weight, and usually provide the same accuracy of volume or weight measurement.

In considerations of controllability also, the shape of the vessel must be taken into account. With a horizontal vessel of a given diameter half filled with liquid, it is easier to control the liquid level than in a vertical vessel of the same diameter. Unit changes in liquid level represent a much smaller change in volume in the horizontal vessel; thus the same volume addition or withdrawal of liquid changes the level much less in the horizontal vessel.*

Hydrostatic Head

Many liquid-level measuring devices operate on the principle of measuring the *hydrostatic head.* This head may be defined as the weight of liquid existing above a reference or datum line and is expressed in various units, such as pounds per square inch, grams per square centimeter, or feet of liquid measured. The head is a real force, due to liquid weight, and, as shown in Fig. 3, is exerted equally in all directions. It is independent of the volume of liquid involved or the vessel shape. Measurement of pressure due to the liquid head can be translated to level height above the datum line as follows:

$$H = \frac{P}{D} = \frac{P}{D_wG} = \frac{P \times 27.7}{G} \qquad (7)$$

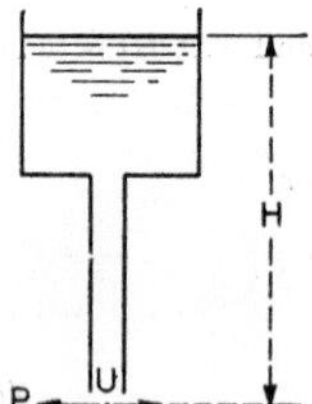

where H = height of liquid above datum line, in.
    P = pressure due to liquid head, psi.
    D = density of liquid at operating temperature, lb/cu in.
    D_w = density of water at 60°F
    G = specific gravity of liquid at operating temperature
(*Note:* Density of water = 62.4 pcf or 1/27.7 lb/cu in.)

FIG. 3. Fluid head.

From this relation it is seen that a measurement of pressure P at the datum or reference point in a vessel provides a measure of the height of liquid above that point, provided the density or specific gravity of the liquid is known. Also, this relation shows that changes in specific gravity of the liquid will affect measurements of liquid level by this method, unless corrections are made for such changes.

If a pressure greater than atmospheric is imposed on the surface of the liquid in a closed vessel, this pressure adds to the pressure due to hydrostatic head, and must be compensated for by the pressure-measuring device used for liquid-level measurement.

* The effects of capacitance in automatic control systems are thoroughly described in subsec. 11-1 of the "Process Instruments and Controls Handbook," McGraw-Hill Book Company, Inc., New York, 1957.

Theory of Buoyancy

Since some liquid-level detectors employ floating, partially submerged, and fully submerged elements, buoyancy must be considered in the design and application of such elements.

Archimedes' principle states that the resultant pressure of a fluid on a body immersed in it acts vertically upward through the center of gravity of the displaced fluid, and is equal to the weight of the fluid displaced. This resultant upward force exerted by the fluid on the body is called the *buoyancy*, or *force of buoyancy*.

Reference is made to Fig. 4. By measuring the difference in weight of a partially submerged element at various degrees of submergence, the level of the liquid in which the displacer element is submerged can be determined. The following equations are useful for determining the forces exerted by the displacer element.

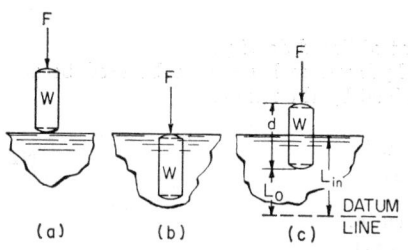

FIG. 4. Relationship between level position and displacer element: (a) Level position at bottom or below displacer; (b) level position at top or above displacer; (c) level position between top and bottom of displacer.

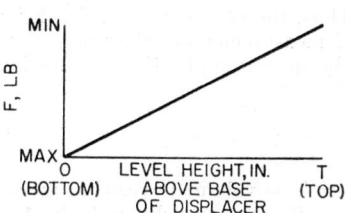

FIG. 5. Force F vs. level height.

Condition 1: Level position at bottom or below displacer (Fig. 4a),

$$F = W \tag{8}$$

where F = force or weight to be supported, lb, and W = weight of displacer, lb.

Condition 2: Level position at top of displacer (Fig. 4b),

$$F = W - \frac{V \times G}{27.7} \tag{9}$$

where V = volume of displacer, cu in., and G = specific gravity of liquid at 60°F.

Condition 3: Level position in intermediate position (Fig. 4c),

$$F = W - \frac{V \times G}{27.7} \times \frac{L_{in} - L_0}{d} \tag{10}$$

where L_{in} = level position in intermediate position, in.
  L_0 = level position at bottom or below displacer, in.
  d = length of displacer, in.

With a cylindrical displacer, F varies as the level position around the displacer varies. The value of F is measured by suitable means, such as a torsion spring, pneumatic force balance, and so on. Thus, this value becomes a function of the level position above the datum line. This relationship is shown in Fig. 5.

Volume and Weight Measurement from Level

In the final essence, liquid-level measurement resolves itself into position measurement, namely, the position (height) of a liquid surface above a datum line. Level measurement, however, need not always be expressed in terms of inches or feet above the datum line, but, with a knowledge of the dimensional and contour characteristics of the containing vessel, can be conveniently interpreted (hence calibrated) in terms of the volume of liquid contained—and further, with information concerning the specific gravity of the liquid, can be expressed in terms of the weight of the liquid in the vessel. The choice of units for a given level measurement varies, of course, with the problem at hand.

Volume Determination. If the purpose of level measurement is to determine the volume of liquid contained in the vessel, then direct measurement of level height is preferable because

$$V = A \times H \tag{11}$$

where V = volume in vessel, cu in.
A = area of vessel, sq in.
H = height of level, in.
Thus, the volume measured is independent of liquid density.
If measurement of the pressure due to hydrostatic head must be used because of the specific application, volume is determined by the relation

$$V = \frac{A \times P}{D} \tag{12}$$

where V = volume in vessel at given level, cu in.
P = pressure due to hydrostatic head, psi
D = density of liquid in vessels, lb/cu in.
In this case, the volume measurement depends on the density of the liquid.

Weight Determination. If the purpose of level measurement is to determine the weight of liquid contained in the vessel, then measurement of the pressure due to hydrostatic head is preferable, because

$$W = H \times D \times A = A \times P \tag{13}*$$

where W = weight of liquid in vessel, lb
H = height of level, in.
D = density of liquid lb/cu in.
A = area of vessel, sq in.
P = pressure due to hydrostatic head, psi
Thus, the measurement is independent of liquid density.
If direct measurement of level height must be used, then

$$W = A \times D \times H \tag{14}$$

and the weight measurement depends upon the density of the liquid.

ERRORS IN THE MEASUREMENT OF QUANTITIES IN STORAGE TANKS†

Where highly accurate measurements of the quantities of liquids stored in tanks are required, it is the usual practice to correct the fluid density for temperature changes, and neglect the other temperature and pressure effects. Generally this is quite justified, but it is at least comforting to be assured of the magnitude of the uncorrected errors. Equations are presented below which allow measurements to be easily and completely corrected to any desired reference conditions.

It has been assumed that the true volume of the container, empty and at its refer-

---

* From Eq. (7), $H = P/D$.
† Material on this subject contributed by A. H. McKinney, E. I. du Pont de Nemours & Co., Inc., Wilmington, Del.

ence temperature, is accurately known, and that the measuring element is without error. Under these ideal conditions, the remaining sources of errors are:
1. The tank may not be perfectly round.
2. The tank wall is stretched, owing to hydrostatic pressure.
3. The fluid is not at its reference temperature.
Corrections for each of these unavoidable conditions are presented as relatively simple, dimensionless expressions in which any consistent data can be substituted. The results are expressed as relative errors e, defined as

$$e = \frac{\text{observed value} - \text{true value}}{\text{true value}}$$

The error e is therefore a decimal, and can be converted to per cent by multiplying by 100. It is positive or negative, depending on whether the observed value is high or low.

Also included is a method by which the temperature error can be completely and exactly compensated for. The correction allows for the expansion of the fluid and the tank at a temperature T_1 below the liquid level, and for the tank and float wire at another temperature T_2 above the liquid level. It allows for an average liquid temperature consisting of one or an infinite number of striations, each at a different temperature. The only requirement is that each horizontal level be uniform.

Necessity for Corrections

The results of these calculations are given below in summary form. For a tank that is not perfectly round, if it appears round to the eye, no corrections are necessary. Similarly, for a tank withstanding hydrostatic pressure; if it is still holding together, the corrections will be almost insignificant.

In the case of temperature errors, however, it is not always safe to neglect corrections. Of course, the greatest errors are involved where the fluid density enters the calculations and is not corrected to the actual liquid temperature. This happens when a pressure-measuring system (Fig. 8a) is used to report volumetric contents, or when a volumetric measuring system (Fig. 8b or c) reads in pounds.

Even using the converse, however, the temperature errors are not always negligible. The worst offender is the use of float and tape, where the tape does not have the same expansion as the tank wall, and the tank is nearly empty.

Summary of Errors in Tank Measurements

1. Error for Tanks "Out-of-Round" (Fig. 6). Assume that the tank is measured by strapping, and that the area is calculated from perimeter P:

$$A' = \frac{P^2}{4\pi} = \text{calculated area}$$
$$A = \pi ab = \text{true area} \tag{15}$$
$$e = 0.298 \left(\frac{a}{b} - 1\right)^{1.952}$$

where a and b = measured dimensions (see Fig. 6).
2. Error Due to Pressure Expansion (Fig. 7):

$$e = \frac{\pi S}{E} \times \frac{H_1{}^2 - H_0{}^2}{H_1(H_1 - H_0)} \tag{16}$$

and
$$S = \frac{6PD}{t}$$

For open tanks, $H_0 = 0$ and

$$e = \frac{\pi S}{E} \qquad (17)$$

where H_0 = pressure above liquid
　　H_1 = pressure at bottom of liquid
　　S = average wall stress ⎫
　　P = average pressure ⎬ psi or any consistent units
　　E = modulus of elasticity ⎭
　　D = tank diameter, ft
　　t = wall thickness, in.

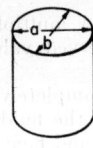
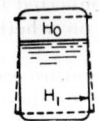
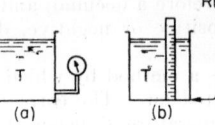
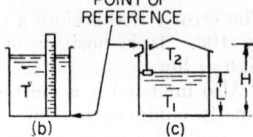

FIG. 6. Error due to "out-of-round" tank.　FIG. 7. Error due to pressure expansion.　FIG. 8. Errors due to temperature (with various measuring systems): (a) Pressure type; (b) rod type; (c) float-and-tape type.

3. Errors Due to Temperature.　Expressions for errors in measurements in volumetric or weight units are given in Table 1 below for three common methods of level measurement.

Table 1. Error Equations for Common Level-measuring Systems

Type of measuring system	Expressions for error			
	Reading, volumetric units	Eq. no.	Reading, weight units	Eq. no.
Pressure type (Fig. 8a)......	$e = -T(f + 2g)$	18a	$e = -2Tg$	18b
Rod (Fig. 8b)...............	$e = -T(2g + c)$	19a	$e = T(f - 2g - c)$	19b
Float and tape (Fig. 8c).....	$e = 3gT_1 - \dfrac{L}{H}T_2(g - c)$	20a	$e = T_1(f - 3g) - \dfrac{L}{H}T_2(g - c)$	20b

T = fluid temperature
f = volume expansion of fluid
g = linear expansion of wall
c = linear expansion of rod or float wire

Temperature Compensation

Figures 9 and 10 indicate electrical and mechanical methods of achieving compensation in equivalent manners. In either case, the wire W is subjected to the same depth of liquid and to the same temperatures as the liquid itself. Its change in length is therefore proportional to rise and fall of the liquid level due to temperature.

To compensate exactly for temperature, add to motion of the float wire, motion of the compensating wire multiplied by a factor J, where

$$J = \frac{3g - f}{g} \qquad (21)$$

In Fig. 9 the change in resistance of the compensating rheostat per inch of wire motion is J times the change in the measuring rheostat per inch of motion of the float wire.

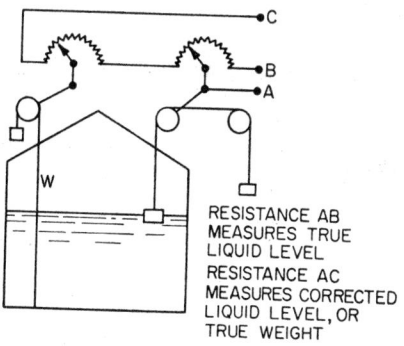

FIG. 9. Electric temperature-compensating system.

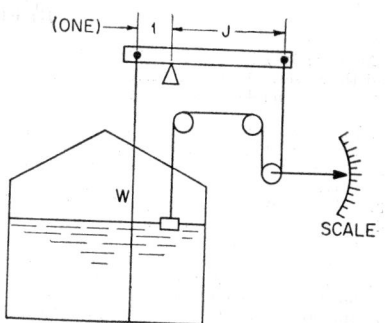

RESISTANCE AB MEASURES TRUE LIQUID LEVEL
RESISTANCE AC MEASURES CORRECTED LIQUID LEVEL, OR TRUE WEIGHT

FIG. 10. Mechanical temperature-compensating system.

Table 2. Typical Values for Thermal Expansion Equation

Material	g, in./in., deg
Aluminum	0.0000255
Concrete	0.0000060
Copper	0.0000141
Steel	0.0000132

Liquids in steel tanks	f, gal/gal, deg	J
Water	0.000207	5.09
Sulfuric acid	0.000576	19.6
Petroleum	0.000955	34.8

Sample Calculations

Example 1. A tank 20 ft in diameter contains 30 ft of liquid of specific gravity 1.7. If the wall is ⅜-in. steel plate, what is the error due to pressure?

$$P = 30 \times 1.7 \times 0.433 = 22 \text{ psi}$$

$$S_{max} = \frac{6PD}{t} = \frac{6 \times 22 \times 20}{\tfrac{3}{8}} = 7,000 \text{ psi}$$

$$e = \frac{\pi S}{E} = \frac{\pi \times 7,000}{30,000,000} = 0.00072 = 0.072$$

Example 2. A manometer reads the pressure at the bottom of a steel tank. What error is introduced if the water temperature is 80 instead of 25°F?
In weight units of calibration

$$e = -2gT = -2(80 - 25) \times .0000132 = -0.00145 = -0.145\%$$

In volumetric units of calibration

$$e = -T(f + 2g) = -55(0.000207 + 0.000026) = -.0128 = -1.28\%$$

Example 3. Using an Invar tape, a gager finds one foot of water in a 30-ft tank. If the liquid is 50°F and the vapor space 30°F above the reference temperature, what error is introduced?
In volumetric units

$$e = -3gT_1 - \frac{L}{H}T_2(g - c) = -3 \times 0.0000132 \times 50 \frac{-29}{1} 30 \times (.0000132 - 0) = -0.0132 = -1.35\%$$

(*Note:* This error of 1.35 per cent would have been almost entirely eliminated by using a steel tape, to balance out the expansion of the steel well, instead of the Invar tape.)

Example 4. A tank has a maximum diameter of 10 ft 6 in. and a minimum diameter of 10 ft 4 in. What error is introduced by assuming it to be a circle of the same perimeter?

$$\frac{a}{b} = \frac{126 \text{ in.}}{124 \text{ in.}} = 1.0161$$

$$e = 0.298 \times 0.0161^{1.952} = 0.000097 = 0.01\%$$

REFERENCES

Many articles have appeared in various periodicals on the subject of liquid level—devices, systems, and their uses. In Ref. 1 there are 35 such references, plus 5 textbook references. Only the more recent and general ones are listed below, together with additional later ones.

1. Elfers, P. A.: Liquid-level Detectors, pp. 5-3 to 5-54 in D. M. Considine (ed.), "Process Instruments and Controls Handbook," McGraw-Hill Book Company, Inc., New York, 1957.
2. Behar, M. F.: Automatic Level Controllers, *Instruments*, vol. 14, no. 11, pp. 326–380, November, 1941.
3. Feller, E. W.: Measuring and Controlling Liquid Level, *Power*, vol. 89, no. 4, pp. 63–86, April, 1945.
4. Nelson, W. C.: Progress in Metals: Liquid Level in Pressure Vessels, *Oil Gas J.*, vol. 45, no. 129, p. 87, Sept. 7, 1946.
5. Electronic Probe Controls Liquid Level, *Prod. Eng.*, vol. 17, no. 10, p. 93, October, 1946.
6. McCaslin, L. S., Jr.: New Invisible Eyes in Liquid Level Control: Gamma Rays, *Oil Gas J.*, vol. 46, no. 134, pp. 100–121, Oct. 25, 1947.
7. How to Drain a Compressor Scrubber by Liquid Level, *Petrol. Refiner*, vol. 27, no. 2, p. 144, February, 1948.
8. Munch, R. H.: Displacement Type Liquid Level Controllers, *Ind. Eng. Chem.*, vol. 40, no. 4, pp. 95A–96A, April, 1948.
9. Ross, S. D.: Liquid Level Measurement in Chemical Processing, *Chem. Inds.*, vol. 62, pp. 924–928, June, 1948.
10. Jones, G. O.: Automatic Liquid Level Controllers for Low Boiling Point Liquids, *J. Sci. Instr.*, vol. 25, no. 7, pp. 239–242, July, 1948.
11. Jones, S. H.: Liquid Level Control Solves Pumping Problem at Badger Mills, *Paper Ind.*, vol. 30, no. 5, p. 372, August, 1948.
12. Camp, A. C.: Control of Multiple Effect Evaporators, *Chem. Eng.*, vol. 57, no. 9, pp. 108–111, September, 1950.
13. Wexler, A.: Measurement and Control of Level of Low Boiling Liquids, *Rev. Sci. Instr.*, vol. 22, no. 12, pp. 941–945, December, 1951.
14. LaGasse, J., and J. Nougaro: Electrical Measurement of Level, *Compt. Rend.*, vol. 234, no. 1, Jan. 2, 1952.
15. Chauvel, L.: Water Level Control in Reservoirs, *Genie Civil*, vol. 129, no. 5, pp. 85–89, Mar. 1, 1952.
16. Hamm, H. W.: Liquid Level Indication and Controllers, *Prod. Eng.*, no. 3, pp. 140–141, March, 1952.
17. Schelle, P. D., and K. R. Schmayer: Problems in Slurry Service Instruments, *Instruments*, vol. 25, no. 4, p. 468, April, 1952.
18. Phillips, D.: Liquid Sensing Device with No Moving Parts, *Prod. Eng.*, vol. 23, no. 6, pp. 134–135, June, 1952.
19. Measuring Liquid Contents of Tanks by Remote Control, *Oil Gas J.*, vol. 52, no. 5, p. 124, June 8, 1953.
20. Ross, S. D., and W. W. Drake: Liquid Level Measurement and Control, *Petrol. Engr.*, vol. 25, no. 8, pp. C7–C10, August, 1953.
21. Ross, S. D.: Liquid Level Control, *Consulting Engr.*, vol. 3, no. 2, pp. 44–47, February, 1954.
22. Williams, W. E.: Liquid Level Indicator for Condensed Gages, *Rev. Sci. Instr.*, vol. 25, no. 2, pp. 111–114, February, 1954.
23. Rod, R. L.: Ultrasonic Liquid Level Indicator Systems, *Electronics*, vol. 27, no. 4, pp. 156–161, April, 1954.
24. Liquid Level Gauge (for Wells), *U.S. Bur. Mines Rept. Invest.* 5060, p. 3, June, 1954.
25. Kovacic, E.: Apparatus for Measuring Small Changes in Liquid Level, *J. Sci. Instr.*, vol. 31, no. 6, pp. 205–206, June, 1954.
26. Coles, R. V.: Level Indicating Record and Control Instruments, *Tele-Tech & Electron. Inds.*, vol. 13, no. 7, pp. 62–64, 106–107, July, 1954.
27. New Liquid Level Gage for Industry (Buoyancy-electric) *World Oil*, vol. 139, no. 6, pp. 216, 218–219, November, 1954.
28. Remote Automatic Tank Gaging, *Petrol. Process.*, vol. 9, no. 11, pp. 1742–1745, November, 1954.
29. Automatic Control of Glass Level in Melting Tanks (Hartford Empire), *Engineer*, vol. 198, no. 5160, p. 858, Dec. 17, 1954.
30. Telemetering System Reports, Remote Liquid Levels, *Automation*, vol. 2, no. 5, pp. 46–48, May, 1955.
31. Fleming, A. E.: Level Control, *Instr. and Automation*, vol. 28, no. 5, pp. 809–815, May, 1955.
32. Wong, J. P.: Exploiting Displacement Level Principle, *ISA J.*, vol. 2, no. 5, pp. 141–144, May, 1955.
33. Kimmell, G. O.: Floatless Liquid-level Controllers, *Inst. and Automation*, vol. 28, no. 10, pp. 1748–1951, October, 1955.
34. Sanders, C. W.: Improving Level Control by Dynamic Analysis, *ibid.*, no. 11, pp. 1918–1919, November, 1955.
35. Holland, J. W.: Some Aspects of Solution Level Control, *Plating*, vol. 42, no. 11, pp. 1412–1413, November, 1955.
36. Kennedy, W. R.: Low Pressure Liquid Level Control in Power Plants, *Soc. Instr. Technol. Trans.*, vol. 7, no. 4, pp. 168–171, December, 1955.
37. Feinburg, R.: Application of Flying Spot Scan Tubes to Level, *Brit. Commun. Electron.*, vol. 2, no. 12, pp. 48–52, December, 1955.

38. Mattio, A. M.: Tips on Using Float Valves for Level Control, *Power*, vol. 100, no. 11, pp. 126–127, 206, 210, November, 1956.
39. Gamma Rays Measure Coke Level, *Petrol. Process.*, vol. 11, no. 12, p. 79, December, 1956.
40. Maimowi, A.: Hot Wire Liquid Level Indicator, *Rev. Sci. Instr.*, vol. 27, no. 12, pp. 1024–1027, December, 1956.
41. Level Indication and Control, *Petrol. Times*, vol. 60, no. 1549, pp. 1149–1150, Dec. 21, 1956.
42. Stoll, H. W.: Principles and Practices in Art of Level Measurement, *Instr. and Automation*, vol. 4, no. 1, pp. 12–17, January, 1957.
43. Scolding, C. G.: Automatic Control of Water Level in Bodies, *Eng. & Boiler House Rev.*, vol. 72, no. 3, pp. 77–79, March, 1957.
44. Young, J. K.: Water Levels Telemetered to Control Plant by Radio, *Water Works Eng.*, vol. 110, no. 3, pp. 240–241, 277, March, 1957.
45. Rod, R. L.: Ultrasonic Level Sensor, *Instr. and Automation*, vol. 30, no. 5, pp. 886–887, May, 1957.
46. Soble, A. B.: Measurement and Control of Liquid Level by Thermistors, *Chem. Engr. Progr.*, vol. 53, no. 9, September, 1957.
47. Jones, E. W.: Role of Gamma-ray Level Detection Systems in Chemical Plants, *Brit. Inst. Radio Engrs. J.*, vol. 17, no. 9, pp. 473–479, September, 1957.
48. Hannula, F. W.: Use Capacitance for Accurate Level Measurement, *Control Eng.*, vol. 4, no. 11, pp. 104–107, November, 1957.
49. Hackman, J. R.: How to Measure Liquids' Interface Levels, *ISA J.*, vol. 4, no. 12, pp. 554–557, December, 1957; vol. 5, no. 1, pp. 60–64, January, 1958.
50. Gerharz, R.: Fluid Level-sensing Device of Simple Design, *Am. Geophys. Union Trans.*, vol. 39, no. 1, pp. 121–122, February, 1958.
51. Airmee: Froth Level Control by Photo Cell, *Instr. and Automation*, vol. 5, no. 5, p. 204, May, 1958.
52. Revesz, G.: Process Instrumentation for Measurement and Control of Level, *IRE Trans. PGIE*, August, 1958, pp. 11–16.
53. Revesz, G.: How to Select Capacitance Level Gauges, *ISA J.*, vol. 5, no. 10, pp. 40–44, October, 1958.
54. Level Control by A.R.D. Relay, *Instr. and Automation*, vol. 5, no. 10, pp. 400–401, October, 1958.
55. Regas, S.: Nuclear Liquid Level Gauges, *Nondestructive Testing*, vol. 16, no. 6, pp. 493–494, November-December, 1958.
56. Hlucham, S. A.: Differential Pressure Instruments for High Temperature Service, Nuclear Engineering and Science Conference, Cleveland, 1959, published by Engineering Joint Council, New York.
57. Waite, D. P.: System Design for Improved Water Level Control of Steam Generators, Nuclear Engineering and Science Conference, Cleveland, 1959, published by Engineering Joint Council, New York.
58. Greenwood, T. L.: Liquid Level Detectors, *Electronics*, vol. 32, p. 49, Jan. 2, 1959.
59. Andrews, H. S.: Survey of Automatic Tank Level Gages, *Control Eng.*, vol. 6, no. 2, pp. 92–95, February, 1959.
60. Walter, L.: Instrumentation in Glass Industry in Great Britain, *Glass Ind.*, vol. 40, no. 3, pp. 125–130, 158, 160, 162, March, 1959.
61. Arlmann, J. J.: Een Neveau Detector voor de Praktijk, *Ingenioren*, vol. 17, no. 19, pp. Ch8–Ch9, May 8, 1959.
62. Nelson, M. P.: Automatic Tank Gaging for Inventory Control, *Petrol. Engr.*, vol. 31, no. 7, pp. C22–C24, July, 1959.
63. Fahnoe, Frederick: Interface Control Using Thermistors, *Chem. Eng.*, Feb. 22, 1960, pp. 153–154.
64. Hudrlik, Roger E.: Thermistor Level Control Works in Difficult Service, *ibid.*, May 16, 1960, pp. 160–161.
65. Mandt, R. D.: How to Calibrate Level Controllers, *ISA J.*, vol. 8, no. 2, p. 75, February, 1961.
66. No 'Tall Tales' Told by Tank-Gaging System, *Chem. Process.*, June, 1961, pp. 97–98.
67. Brook, C. B.: The Dynamics of Level Control: a Mathematical Study, *Private Printing Bull.* TM-7, Marshalltown, Iowa.
68. Brook, R. F.: X-ray Inspection of Liquid Levels, *Automation*, vol. 9, no. 6, pp. 77–83, June, 1962.
69. Hepp, P. S.: Internal Column Reboilers—Liquid Level Measurement, *Chem. Eng. Progr.*, vol. 59, no. 2, pp. 66–69, February, 1963.
70. Karp, H. R.: Level Gaging, *Control Eng.*, vol. 9, no. 9, p. 116, September, 1962.
71. Kay, G. A.: Control of Interface Liquid Level, *Instr. Control Systems*, vol. 35, no. 11, pp. 101–102, November, 1962.

Table 3. Liquid-level Detectors—Selection Guide

NOTE: Unless otherwise stated, figures for accuracy given in this table as per cent signify that the average expected error will not be more than plus or minus the stated percentage of full instrument scale. Accuracy figures are difficult to obtain for many kinds of instruments and, in a few cases, the figures have been estimated from the best available data. Well-designed, installed, and maintained instruments may give considerably better accuracy; if the reverse situation holds, poorer results may result.

Operating principle	Range	Accuracy	Applications and remarks
Visual Methods			
 (a) Tape reel Plumb bob Front view (b) Side elevation FIG. 1	Limited only by practical length of stick, rod, tape, or gage glass	About 0.02 in. Depends upon range and readability of calibrations on stick, rod, or tape	 FIG. 2 Require relatively still liquid surface. For open vessels only. Require manual positioning

Notched Stick: Indicates level above arbitrary data line

Gage Stick: Inserted vertically downward to bottom of tank, calibrated from zero at bottom end of stick in suitable units

Hook Gage (Fig. 1a): Measures distance from point of hook at surface of liquid to a reference datum point. Graduated marks on rod indicate level. [Variations are *point gage* with sharpened point directed downward to contact liquid surface, and *tape-and-plumb bob gage* (Fig. 1b) which employs a metal tape attached to plumb bob—tape is unreeled until bob contacts surface]

Gage Glass (Fig. 2): Attached to side of vessel with lower tap below level and upper tap above normal level. Many special types including *reflex gage* for high pressures, low temperatures, etc.

Table 3. Liquid-level Detectors—Selection Guide (Continued)

Operating principle	Range	Accuracy	Applications and remarks
Float-type Mechanisms			
Ball Float Mechanically Linked to Indicator or Control Mechanism (Fig. 3): Ball floating in liquid inside vessel is connected by rod to indicator or control mechanism (pilot relay or direct link to control valve). Variation is location of float mechanism in external cage mounted like gage glass on side of vessel. Another variation translates float movement to hydraulic or pneumatic pressure for indicator or control	Internal vessel type limited by length of connecting rod to 3–6 ft max. External cage type about 15 in. max	Internal type ±1–3 %; external type ±1–3 %	As for all float-type mechanisms, errors in measurement and control result if float is subject to corrosion or coating by liquid. Most frequently used as simple method of level control where narrow band or on-off control is permissible. Introduces some sealing problems in higher-pressure vessels
Chain- or Tape-float Gages (Fig. 4): Ball or other shaped float is connected by chain or tape to a rotating member with indicator mechanism above liquid surface. (One form employs magnetic-bond principle for following float movement—see below)	Practically unlimited (normally about 35 ft)	±⅛ in. approx. to ±2 %	Overcomes range limitation of mechanically linked ball-float mechanisms. Also provides greater accuracy. Can be used for interface level measurement. Electrical system used for transmission of multiple measurements to one central location
Magnetic-operated Float Devices: Designs of the external float cage type or chain or tape-float gage type use magnetic effects to relate float motion to level indication or control	Varies with system used, such as chain or tape type above	Varies widely	Overcomes problem of stuffing box seal for pressure applications or corrosion problems

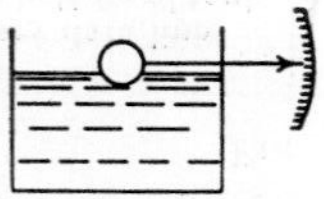

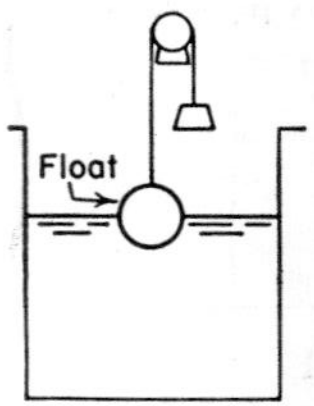

Fig. 3

Fig. 4

Table 3. Liquid-level Detectors—Selection Guide (*Continued*)

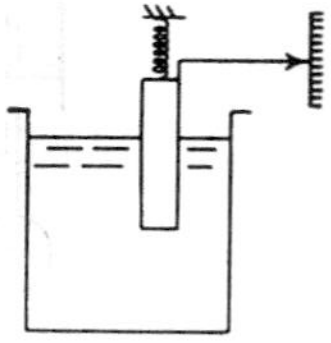

Operating principle	Range	Accuracy	Applications and remarks
Displacer Types			
Weight of displacer (Fig. 5a), held at a substantially fixed position, varies with amount it is submerged in liquid. Amount of force required to balance varying weight is a measure of liquid level. *Torque-tube* or *force-balance* system commonly employed to translate varying displacer weight into level measurement and control action, generally by pneumatic means	14 in.–15 ft for cage-type unit. Wider for internally mounted types	High—torque tube ±0.1%; force balance ±0.3%	Displacer internally mounted, or in cage-type unit. Provides wider range of level measurement with wider proportional control band and higher precision of control. Torque-tube or force-balance system avoids stuffing box problems. Good for interface applications with liquids having relatively same specific gravities
Displacer with Tape (Fig. 5b): Has force-measuring device consisting of a torque tube, an arm, and its contacts, and a solid displacer suitable for the medium. Weight of displaced liquid produces a proportional change in force on the torque tube, causing motor to move tape up and down in response to level. Balance is restored when displacer is accurately positioned in relation to liquid level. Compensation for weight of tape is provided by tape-storage sheave in indicating unit	To 60 ft depth	$\frac{1}{16}$ in.	

Fig. 5a

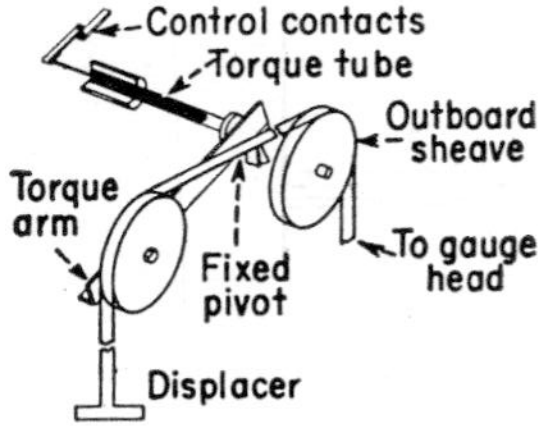

Control contacts
Torque tube
Outboard sheave
To gauge head
Torque arm
Fixed pivot
Displacer

Fig. 5b

Table 3. Liquid-level Detectors—Selection Guide (Continued)

Operating principle	Range	Accuracy	Applications and remarks
Hydrostatic Methods			
			Diaphragm Fig. 8
			Fig. 11
	Fig. 7		
	Air Restriction Vent Diaphragm Fig. 10		
Fig. 6			
Air pressure Fig. 9			
Direct Hydrostatic Pressure Gage (Fig. 6): Measures height of liquid of known density by direct measurement of head above connection point. Differential manometer also can be used with one leg open to atmosphere	Limited only by tank height	±1%	Limited to open or vented vessels. Gage can be pneumatic transmitter for remote indication, recording, or control
Direct Hydrostatic Differential Manometer (Fig. 7): Compensates for pressure over liquid and measures hydrostatic head	Limited by available manometer differential ranges	About ±1%	Suitable for vessels under pressure. Manometers readily available for relatively high operating pressures. Good for interface level measurement

	Range	Accuracy	Remarks
Diaphragm Box with Gage (Fig. 8): Air trapped in closed system of box, tubing, and gage element is compressed by head as measure of level. Differential manometer also can be used for open or closed vessels; in case of latter, one leg of manometer is connected to box; other leg to tank vapor space	Usually low pressure heads—up to about 100 ft	About ±1%	Used in various ways inside and outside tank where direct connection to gage is undesirable. Gage type limited to open tanks
Air-trap System: Similar to diaphragm-box system except that inverted bell is used to trap air in system	Same as above	About ±1%	Diaphragm box limited to liquids below 150°F; gage located up to 500 ft from tank
Air-bubbler System (Fig. 9): Measures hydrostatic head by back pressure created in tube with air bubbling out bottom. Gage used for open vessels (as shown). Differential manometer for closed vessels uses second connection in vapor space. (Essentially: same system can use liquid purge or combination of liquid and air)	Limited only by tank height	About ±1%	Particularly suited for liquids with suspended solids, or corrosive in nature. Gage can be located up to 1,000 ft from tank
Force-balance Diaphragm System (Fig. 10): Balances hydrostatic head acting on metal diaphragm by air pressure in a nozzle-bleed system. For closed tanks, a second compensating unit is connected to tank vapor space and differential manometer is used	Limited only by tank height, but generally not exceeding 300-ft head	About ±1%	Avoids use of air purge where liquid is corrosive or contains suspended solids
Self-operated Control Valve (Fig. 11): Hydrostatic head acts on underside of valve diaphragm, and is opposed by force of spring such that valve will open when head exceeds set valve. (Certain types termed *altitude valves*; one variation uses liquid head on valve diaphragm to balance vessel head)	Limited usually to open tank vessels. Range up to 150 ft	±2-5%	Normally used for open vessels. Advantage is simple and self-operated automatic control of level at one point. Altitude valves are used to hold constant heads in elevated tanks

Table 3. Liquid-level Detectors—Selection Guide (*Continued*)

Operating principle	Range	Accuracy	Applications and remarks
Thermal Methods			
Expansion-tube Unit (Fig. 12): Utilizes inclined tube connected to liquid and vapor space of boiler. Average temperature of tube varies with liquid level, causing proportional change in tube length, which is used to move indicator or to directly position boiler feed valve. (A similar arrangement can be used with thermocouples for measuring the average tube temperature)	Limited. Usually not exceeding 30 in.	About ± 1½ %	Use confined to steam boilers and vapor generators. Simple, rugged, and develops high power output. Must be located close to boiler, and requires relatively constant ambient temperatures. (With thermal-hydraulic unit, connection to indicator or control valve can be up to 600 ft)
Thermal-hydraulic Unit (Fig. 13): Similar to above except that sealed jacket around inclined tube contains water which creates steam pressure in accordance with temperature in tube. Pressure can be indicated on gage or used in hydraulic system to operate boiler feed valve	Limited. Usually 12 in. Range up to 30 in.	About ± 2%	See above

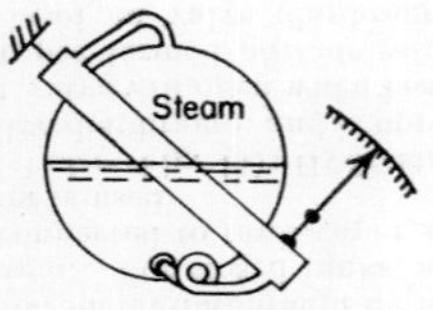

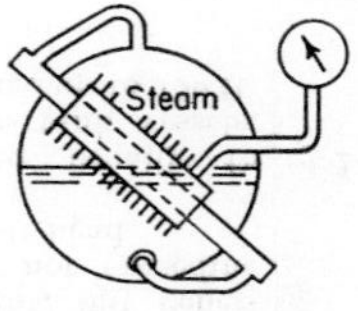

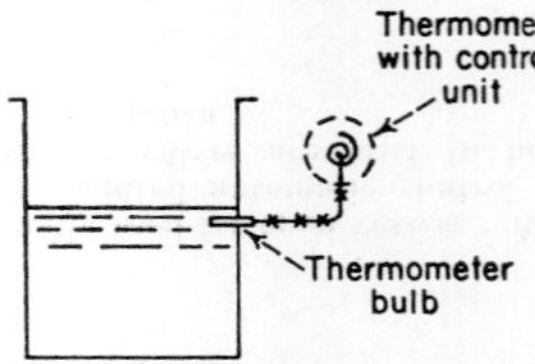

Fig. 12

Fig. 13

Fig. 14

Thermometer-bulb Immersion Method (Fig. 14): Uses conventional pressure-type thermal system with bulb installed at level control or alarm point. Temperature of vessel liquid above or below ambient actuates thermal system when bulb is covered or exposed	Spot type. Usually ± 1 in.	Can be used for single-point level control or alarm. Usually used for high- or low-level cutoff. Instrument can be remotely located from bulb

Weighing Methods

Weight of vessel can be used as a measure of liquid level where vessel shape provides a uniform relationship with changes in level. Various types of weighing scales can be used	Limited primarily by space considerations	Very high	Where practical, this is the most accurate method readily available. Variation of the method is to weigh a small auxiliary vessel attached to top and bottom of main vessel through flexible leads. Especially useful for substances which can not touch or foul mechanical and electrical parts. Examples: food products; rubber latex

5-75

Table 3. Liquid-level Detectors—Selection Guide (*Continued*)

Operating principle	Range	Accuracy	Applications and remarks
Electrical and Electronic Methods			

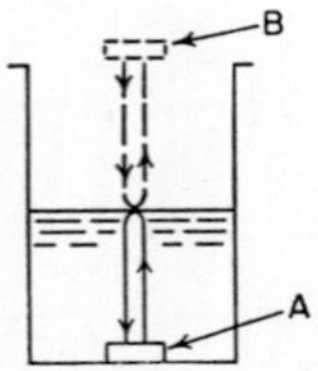

Fig. 15a

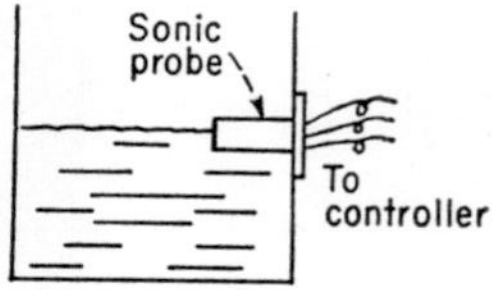

To oscillator and relay

B — A — To controller — Sonic probe

Fig. 15b

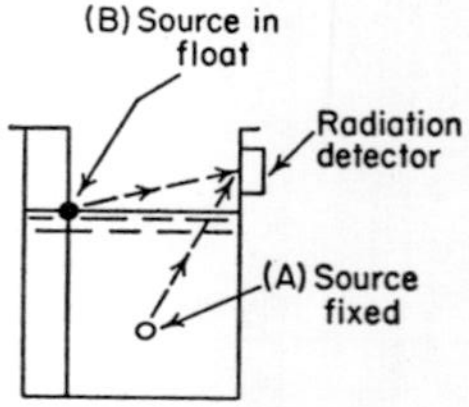

(B) Source in float

Radiation detector

(A) Source fixed

Fig. 16

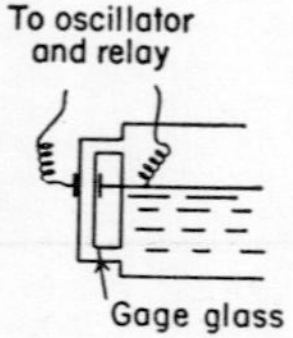

To oscillator and relay — Gage glass

Fig. 17a

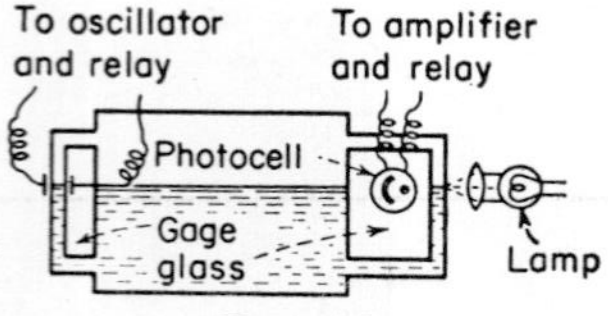

To amplifier and relay — Photocell — Gage glass — Lamp

Fig. 17b

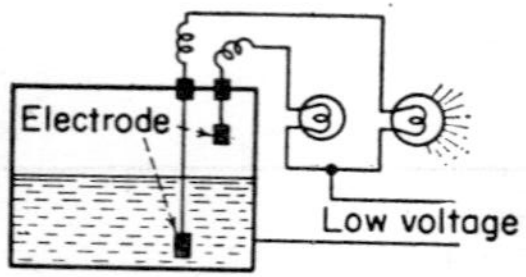

Electrode — Low voltage

Fig. 18

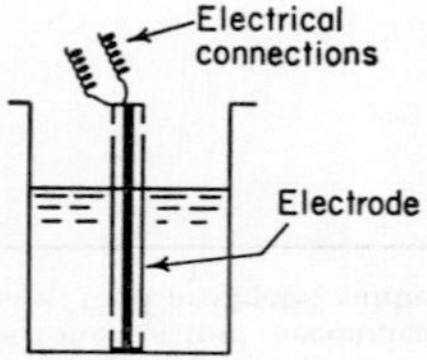

Electrical connections — Electrode

Fig. 19

| | Relatively large | As high as 0.01 in./ft of distance from transmitter to liquid surface | Usually used in deep wells or large storage tanks. Can be used with electrical recorder-controllers |
| | Single point; un-limited; can have series of probes for multiple level measurements | . . . . | Operates independently of density, dielectric constant, temperature, pressure, or conductivity variations. Not affected by foam, froth, or droplets. No moving parts |

Sonic-type Detectors (Fig. 15a): Depend upon transit time of sound waves or ultrasonic energy between transmitter-receiver and surface of liquid either in liquid phase A or in vapor or air phase B. Electric or electronic system converts impulses into measurement of liquid level.

Ultrasonic Probe (Fig. 15b): Operates from probe installed in vessel wall. Surface of probe vibrates at 40 kc when exposed to air; when liquid contacts the probe, oscillations in controller cease; controller actuates alarm or other relay

Method	Range	Sensitivity	Remarks
Capacitance Probe: Glass-covered probe for high or low levels will detect level or level interface by capacitance changes; uses oscillator electronic circuit. Conducting liquid not necessary. Not for continuous measurements	Unlimited	To 1/32 in.	Can distinguish between liquid and foam. Glass-coating resists corrosion (other coatings can be obtained). Available with electric or pneumatic control
Nuclear Radiation Units (Fig. 16): Utilize radiation from radioactive source located in vessel either in fixed location *A* so that vessel liquid varies absorption with level; or in float *B* so that intensity of radiation varies with distance from detector. Geiger counter or gas ionization cell is used in detector circuit, calibrated in terms of level	Limited only by tank height	About 1–3 % is normal. High-range sensitivity up to ±0.1 %	Requires no connections through vessel wall; can be used for very hazardous liquids where most other methods are unsuitable
Gage-glass Automatic Level Detector: Level in gage glass at control point can be detected by (1) condenser plates in oscillator circuit (Fig. 17a) or (2) reduction of light received by photocell (Fig. 17b)	Limited only by available gage-glass ranges	High for spot measurement. ±1/8 in.	Suitable for control or signaling in tanks which can be equipped with gage glasses
Electrode or Probe System (Fig. 18): Probes are installed at desired high and low levels (can be relatively near same height for single-point control with differential). By conduction of low-voltage current through vessel liquid to probe, electric or electronic relays are energized to operate signal lights, alarms, or control elements. (Variation of system is used for measuring liquid level of molten glass)	Unlimited	High	Used for high-low control or signaling. Does not indicate or record level. Electric relays used for relatively good conducting liquids; electronic relays used for liquids with lower conductivities
Electrical Capacitance System (Fig. 19): For dielectric liquids, capacitor is formed by rod with cylindrical metal tube in which capacitance varies with liquid level. For conducting liquids, a single rod with insulated coating is immersed in liquid which itself forms second or ground electrode of capacitor. (One form uses glass coating on electrode)	Practically unlimited	As high as 1/32 in.	Can be used for interface level measurement. Variations include design for use in gage glass for horizontal mounting in vessel. Glass-coated design can distinguish between liquids and foam; others may not. With dielectric liquids, temperature changes may introduce errors. With conducting liquids, scum may affect accuracy

Table 3. Liquid-level Detectors—Selection Guide (*Continued*)

Electrical and Electronic Methods (*Continued*)

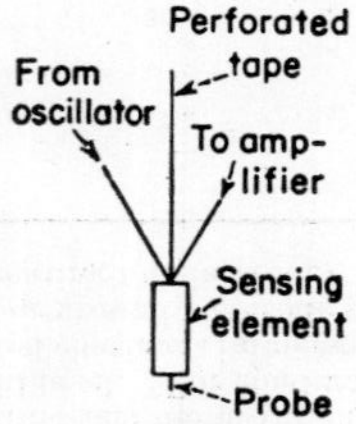

Operating principle	Range	Accuracy	Applications and remarks
Electronic Level Gage (Fig. 20): Level-sensing device is a small radio antenna in an open-end cylinder. Antenna radiates minute r-f signals that are affected by submersion; deeper the submersion, weaker the signal. Return signal, sent up same cable, retains same strength. Electronic unit compares up and down signals, moves sensing unit (as surface follower) by means of servomotor, to hold the radio signal constant. Tape supporting sensor then reports exact level by measuring position of hoisting drum and telemetering value to remote indicator	Unlimited	$\frac{1}{16}$ in.	No possibility of float sticking. Can telemeter long distances. Temperature readings can be telemetered at the same time
Thermistor Type (Fig. 21): Self-heated thermistor installed in vessel changes its resistance when its surrounding medium changes from air to liquid. Thermistor is in a bridge circuit so that this change of resistance operates a relay which can sound alarm shut off a filling system, etc.	Unlimited	0.1 in. around point	Simple system for either level detecting or for the control of filling or emptying

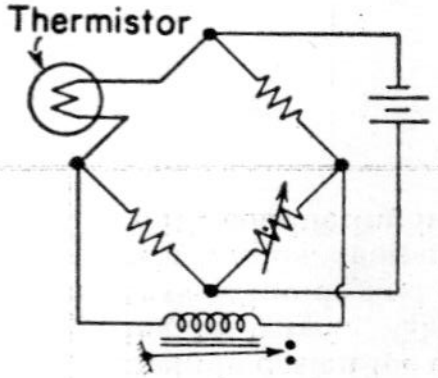

Fig. 20

Fig. 21

5–78

SOLIDS LEVEL

By Dr. Clyde Berg*

With the increasing application of continuous processing, not only to the fluid-handling industries, but also to those industries which handle large quantities of solid materials in bulk, the importance of measuring and controlling methods for solids level is growing.† In this section, the primary elements for solids-level detection are described briefly.

Solids-level detectors fall into two principal categories: (1) *continuous*, which provide a continuity of measurement from the low to the high end of their scales; and (2) *fixed-point*, which provide measurement only at one or several *specific* levels.

Most of the devices in the first category are adapted to pneumatic or electrical proportional control, and hence are particularly advantageous for continuous-processing operations. With these devices, practically all the control techniques long used in the handling of fluids also can be used for solid materials. Many of the fixed-point devices have been available for many years. As contrasted with the continuous type, the fixed-point units largely are used for the actuation of alarm systems in the weighing, storing, and handling of solids. By installing primary devices in multiple, their operation can be made to approach that of the continuous units.

A condensed summary and selection guide of solids-level detectors is given in Table 1.

REFERENCES

1. Wing, Paul Jr.: Exploiting the Displacement Level Principle, *ISA J.*, vol. 2, no. 5, pp. 141–144, May, 1955.
2. Berg, Clyde: *Chem. Eng. Progr.*, vol. 51, no. 7, pp. 326–334, July, 1955.
3. Sittig, Marshall: *Petrol. Process.*, vol. 9, no. 7, pp. 1048–1055, July, 1954.
4. Control of Levels in Hoppers, *Power & Works Engr.*, vol. 47, no. 549, pp. 93–94, 1952.
5. Control Your Bulk Handling with Bin Level Indicators, *Modern Mater. Handling*, vol. 8, no. 2, pp. 64–67, February, 1953.
6. Gamma Rays Measure Coke Level, *Petrol. Process.*, vol. 11, no. 12, p. 79, November, 1956.
7. Eckman, D. P.: "Industrial Instrumentation," p. 262, John Wiley & Sons, Inc., New York, 1951.
8. Berg, Clyde: Solids-level Detectors, pp. 5–55 to 5–70, in D. M. Considine (ed)., "Process Instruments and Controls Handbook," McGraw-Hill Book Company, Inc., New York, 1957.
9. Heacock, W. J.: When Selecting Bin Level Devices . . . , *Plant Eng.*, February, 1962, pp. 133–135.

* Director, Clyde Berg Associates, Long Beach, Calif.
† Engineering details of these devices can be found in subsec. 7-6 of "Process Instruments and Controls Handbook," McGraw-Hill Book Company, Inc., New York, 1957.

Table 1. Solids-level Detectors—Selection Guide

Type of detector	Operating principle	Measurement Fixed	Continuous	Under pressure	Above 600°F	Output signal	Applications and remarks
Grid-response Unit:	Consists of thin metal rings, connected by rods to form a vertical cylindrically shaped grid. Normally only partially immersed in moving solids bed. Grid is connected to torque-tube mechanism which actuates air pilot	...	x	x	x	P & E	Used for continuous control of solids level in process vessels. Has rapid response; also used for indication and recording
Gamma-ray Absorption:	Amount of gamma radiation from a fixed source varies with thickness of solids between source and detector	x	x	x	x	E	Used in solids bins, fluid, and other cat crackers. Requires no opening or connection through vessel wall. Can be used to determine density of fluidized beds with known height of bed
Weighing Device:	Bin containing solids is supported from scale levers or other type of force-balancing system, which can be mechanical, pneumatic, hydraulic, or electric (strain gage or differential transmission)	x	x	x	x	P & E	With known diameter of bin, most accurate method of determining solids level. Relatively expensive, particularly for large bins

Type	Description					Application
Electrical Capacitance:	Insulated vertical rod, permanently installed in bin. Presence of material surrounding rod alters its capacitance in proportion to level of material. Level reading affected by density of material	x	x	x	⋮	E	Used for continuous indicating, recording, and control in both bins and process vessels
Diaphragm:	Flexible diaphragm is exposed to material in bin. As solids level rises, pressure forces diaphragm against counterweighted lever mechanism. Tipping of lever actuates electrical switch	x	⋮	x	⋮	E	One unit can be used to detect either high or low level. Two units detect both high and low levels, and operate signals or conveyors. Intermediate levels detected similarly
Pendant Cone:	Pendant cone is suspended from a universal pivot collar connected to electrical switch. Movement of cone upon contact with solids actuates switch	x	⋮	⋮	⋮	E	Used only for top level detection; shuts off conveyor or operates signal

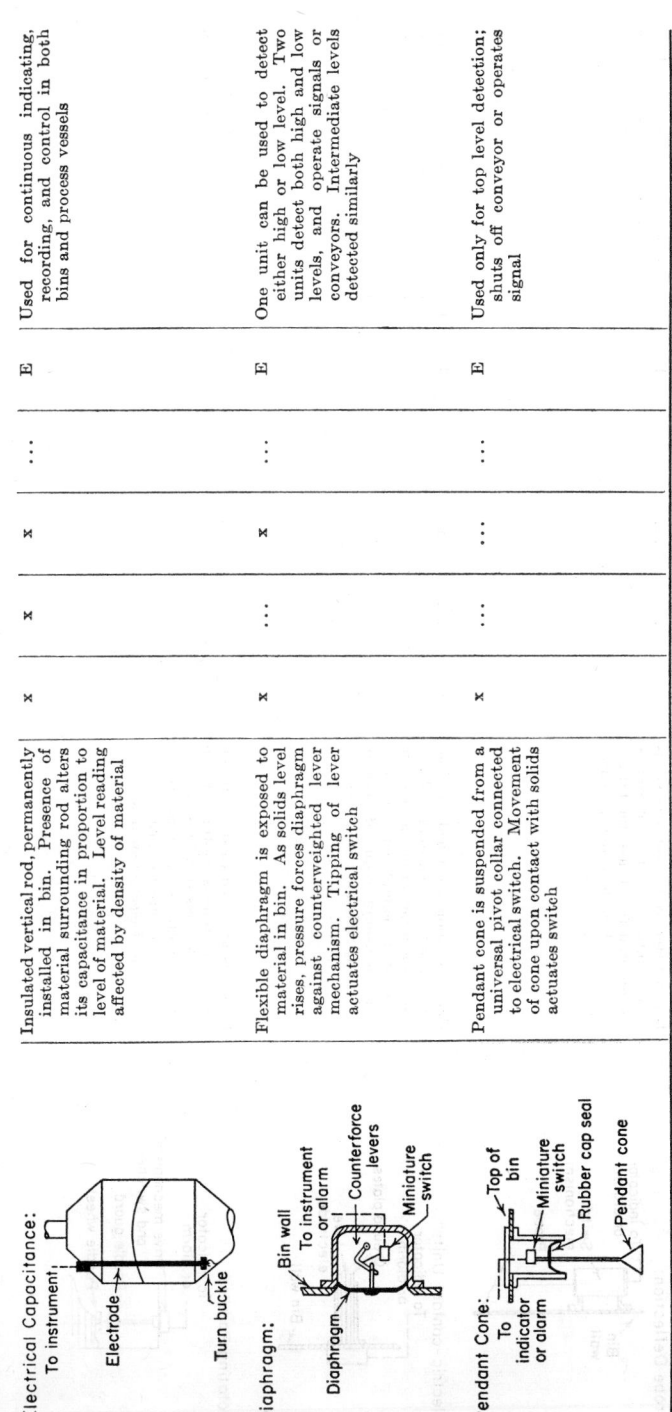

Electrical Capacitance:
To instrument
Electrode
Turn buckle

Diaphragm:
Bin wall
To instrument or alarm
Counterforce levers
Miniature switch
Diaphragm

Pendant Cone:
To indicator or alarm
Top of bin
Miniature switch
Rubber cap seal
Pendant cone

5-81

Table 1. Solids-level Detectors—Selection Guide (Continued)

Type of detector	Operating principle	Measurement				Output signal	Applications and remarks
		Fixed	Continuous	Under pressure	Above 600°F		
Probe Deflection: (To indicator or alarm; Switch mechanism; Probe; Bin wall)	Tapered steel rod penetrates vertically into bin. Top of rod is connected to a brass diaphragm which, when deflected, actuates an electric switch	x	...	...	x	E	Used in solids bins, only for top-level indication. Can operate signal or shut off conveyor
Electric-contact Unit: (To indicator or alarm; Ground plates; Live electrode; Bin wall)	Depends upon electrical conductivity of measured solid. When solids contact probe, an electric circuit is completed. Amplification usually required because of high resistance of probe circuit	x	...	x	x	E	Limited to conductive solids. Can detect high or low levels or operate signals or conveyors. Intermediate levels can be similarly detected
Rotating Paddle: (To indicator or alarm; Drive mechanism and switches; Paddle guard; Paddle wheel)	Paddle is attached to shaft driven by synchronous motor. When rotation is resisted by solid material, motor support rotates in horizontal direction, causing actuation of electric switches	x	...	...	...	E	Can be used only for top-level detection, but operates signals or shuts off conveyor

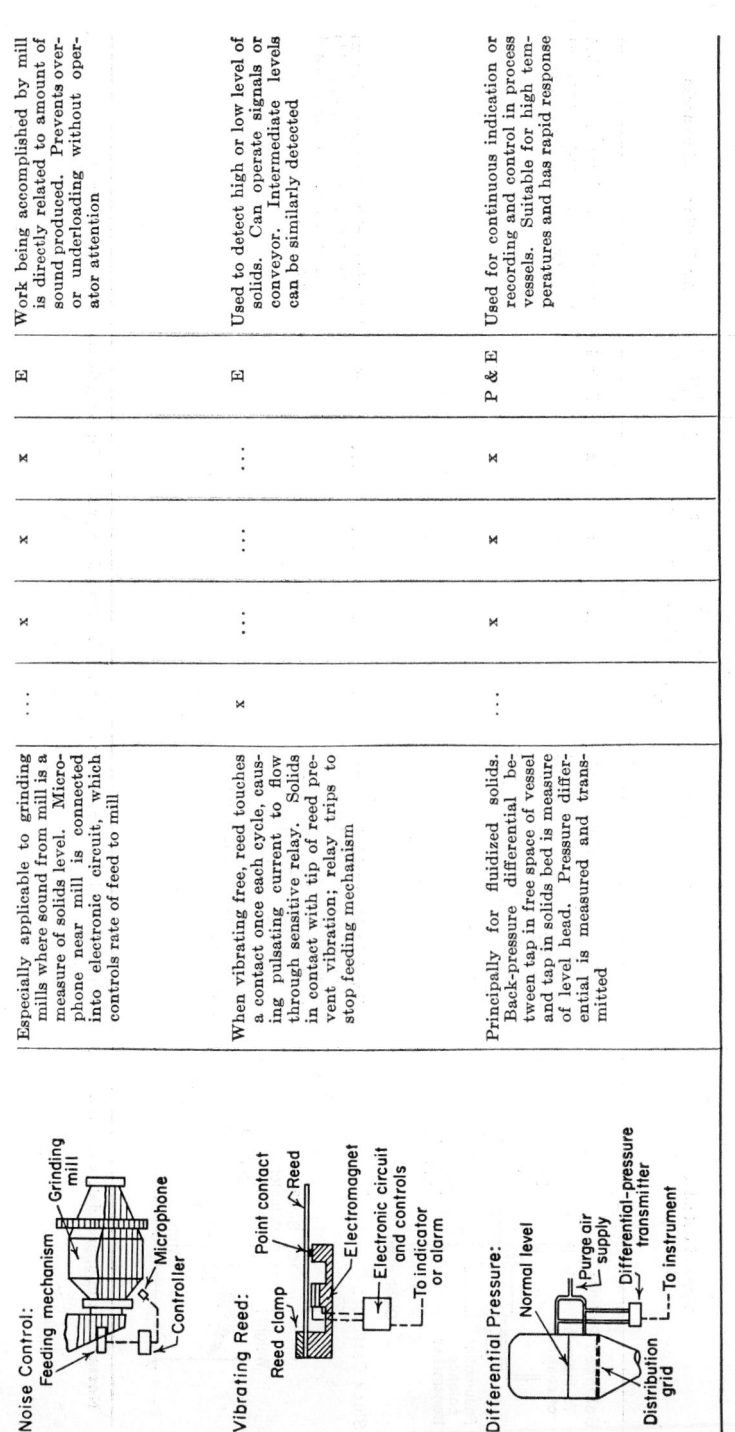

Noise Control:
Feeding mechanism — Grinding mill — Microphone — Controller

Especially applicable to grinding mills where sound from mill is a measure of solids level. Microphone near mill is connected into electronic circuit, which controls rate of feed to mill

Work being accomplished by mill is directly related to amount of sound produced. Prevents over- or underloading without operator attention — E

Vibrating Reed:
Reed clamp — Point contact — Reed — Electromagnet — Electronic circuit and controls — To indicator or alarm

When vibrating free, reed touches a contact once each cycle, causing pulsating current to flow through sensitive relay. Solids in contact with tip of reed prevent vibration; relay trips to stop feeding mechanism

Used to detect high or low level of solids. Can operate signals or conveyor. Intermediate levels can be similarly detected — E

Differential Pressure:
Normal level — Purge air supply — Differential-pressure transmitter — To instrument — Distribution grid

Principally for fluidized solids. Back-pressure differential between tap in free space of vessel and tap in solids bed is measure of level head. Pressure differential is measured and transmitted

Used for continuous indication or recording and control in process vessels. Suitable for high temperatures and has rapid response — P & E

Table 1. Solids-level Detectors—Selection Guide (*Continued*)

Type of detector	Operating principle	Measurement				Output signal	Applications and remarks
		Fixed	Con-tinuous	Under pressure	Above 600°F		
Balanced Paddle:	Balanced paddle on horizontal arm extending into vessel through diaphragm seal is balanced by pneumatic balance transmitter. Contact with solids unbalances arm, transmits to indicator or controller	x	...	x	x	P & E	Designed to measure level of hot catalyst at temperatures to 1100°F. Has rapid response
Slack Detector:	Small weight M on cable is lowered periodically into bin by cycle-controlled winch W. When it strikes solids, slack detector D stops and reverses winch, leaving indicator S at corresponding position cycle; then repeats so that indicated level is corrected each cycle	...	x	...	...	E	Intended primarily for continuous indication or recording of solids level in a bin, but can be used for control

NOTES: An x in appropriate column indicates that device is suitable for that type of service. P signifies pneumatic; E signifies electric output signal.

5-84

Section 6

GEOMETRIC VARIABLES

By

W. FAY ALLER, *Vice President, Engineering, The Sheffield Corporation, a subsidiary of The Bendix Corporation, Dayton, Ohio; Member, American Society of Mechanical Engineers, American Ordnance Association, National Society of Professional Engineers; Registered Professional Engineer (Ohio). (Parts Dimension Measurement and Control)*

W. J. DARMODY, M.E., *Executive Assistant and Liaison Metrologist to National Standards Laboratories, The Sheffield Corporation, a subsidiary of The Bendix Corporation, Dayton, Ohio; Member, American Society for Quality Control, American Ordnance Association. (Parts Dimension Measurement and Control)*

W. HARRISON FAULKNER, Jr., B.S.(E.E.), *formerly Vice President, Engineering and Development, Tracerlab Division of Laboratory for Electronics, Inc., Waltham, Mass.; Member, American Physical Society, Institute of Electrical and Electronics Engineers, American Association for the Advancement of Science; Registered Professional Engineer (Mass.). (Thickness Measurement of Sheet and Web Materials)*

LOUIS F. POLK, JR., B.S., M.B., *formerly Assistant Vice President and Manager, Instruments and Systems Division, The Sheffield Corporation, a subsidiary of The Bendix Corporation, Dayton, Ohio; Member, American Ordnance Association, American Society of Tool Engineers. (Parts Dimension Measurement and Control)*

JAMES R. WALKER, B.S., M.S., *Research Director, Gemco Electric Company, Detroit, Mich.; Member, Institute of Electrical and Electronics Engineers, American Physical Society; Registered Engineer (Mich.). (Position Measurement and Control)*

JOHN G. WOOD, B.S. (Physics), *Manager of Engineering Tracerlab Division of Laboratory for Electronics, Inc., Waltham, Mass.; Member, Society of Plastics Engineers, Technical Association of Pulp and Paper Industry, American Marketing Association. (Thickness Measurement of Sheet and Web Materials)*

6-1

PARTS DIMENSION MEASUREMENT AND CONTROL*

By W. F. Aller,† W. J. Darmody,‡ and L. F. Polk, Jr.§

Important as they are to manufacturing progress, accurate dimensional standards would be but laboratory curiosities without convenient means for applying them on the production floor. For production dimensional gaging, *convenient means* traditionally embraces such factors as versatility, accurate and almost instantaneous readings, ease of setup and operation, and economy of use. Within recent years, convenient means has been expanded to include the concept of automatic measurement, wherein measuring equipment not only senses the magnitudes of variables, but also supplies necessary control signals to guide dependent production equipment.

DIMENSION MEASUREMENT ACCURACY

Achievement of accuracy and precision is not the job of the laboratory, or of the tool and instrument technician, alone. Its accomplishment in the fullest sense requires a constant consideration and reevaluation of objectives, standards, and environment, as well as close attention to human attitudes and elements contributing to industrial skills. A detailed resume of these elements and considerations is given in Table 1.

This table makes it clear that while the accuracy objective may be defined fairly well, the appraisal of the accuracy of measurement actually attained is involved and interwoven with a number of facets of the measuring technique. Generally, engineering studies show that the appraisal of accuracy requires a balanced evaluation of errors or possible errors. These may arise in or be due to the following factors:

1. The measuring instrument.
2. The standard used to set the instrument.
3. The workpiece—its geometrical limitations.
4. The environment, including such factors as temperature.
5. The person performing the measurement—that is, the human element.

Table 1 lists under each of the five elements a group of questions or matters for decision that may be useful in whole or in part for appraisal of accuracy.

A balanced evaluation of the potential or possible errors in these five elements may be defined in whole as skill in measurement. Careful appraisal of these five elements should precede any decision as to an accuracy objective actually attained in a measurement task.

Accuracy Objectives

At all levels of decision and operations, most engineers are familiar with what appear to be paradoxes in accuracy and precision standards demands and methods. These paradoxes, or rather, differences in objectives, invariably determine the nature of accuracy and measurement appraisals.

* The material in this subsection is based on original papers that have appeared in *Automation* and *American Machinist*. See references at the end of this subsection.
† Vice President, Engineering, The Sheffield Corp. (a subsidiary of The Bendix Corp.), Dayton, Ohio.
‡ Executive Assistant and Liaison Metrologist to National Standards Laboratories, The Sheffield Corp., Dayton, Ohio.
§ Formerly Assistant Vice President and Manager, Instruments and Systems Division, The Sheffield Corp., Dayton, Ohio.

Table 1. Appraising Accuracy in Dimension Measurement

(1) Workpiece	(2) Instrument	(3) Standard	(4) Environment	(5) Human element
Clean and free from burrs	Adequate amplification for accuracy objective	Precision proportional to accuracy objective of workpiece measurement	Standard measuring temperature is 68°F	Training
Temperature equalization with instrument and standard	Amplification checked under conditions of use	Calibration by authoritative laboratory	Temperature equalization between standard, workpiece, and instrument. Deficiency of one degree in equalization can introduce error of 6.5 millionths per inch of size	Skill
Essential points of measurement	Effects of friction, backlash, hysteresis, or zero drift	Application of calibration correctness	Thermal expansion effects from heat radiation from lights, heating components, sunlight, and people	Sense of precision appreciation
Accuracy objective proportional to size	Electrical, optical, or pneumatic input to amplifying system functioning within prescribed limits	Effects of wear, damage, burrs, or dimensional instability since last calibration	Effects of cycles in temperature control	Complacent or opinionated attitudes toward personal accuracy
Geometrical truth of supporting features	Contact geometry correct for both workpiece and standard	Thermal expansion effects of materials	Impinging drafts of air may introduce thermal expansion size errors	Open-minded, competent attitudes toward personal accuracy achievements
Consistency of errors in related features, such as pitch diameter to lead-in threads	Contact pressure control functioning within prescribed limits	Effect of modulus of elasticity on contact indentation	Manual handling may introduce thermal expansion errors. Human temperature is 30°F higher than standard measuring temperature. Can affect error in one inch of steel up to 0.0002 in.	Planning measurements techniques for minimum cost, consistent with precision requirements
Geometric truth, such as out-of-round sufficient to assure repetitive readings within accuracy objective	Contacts in correct geometrical relationship and inspected for wear or chipping	Effects of flatness, parallelism, wear, or warpage of gage blocks as they function in measuring contact	Clean surroundings and minimum vibration enhance precision	Appreciation of scope of accuracy evaluation which may be more or less than listed in columns 1, 2, 3, and 4
Deformations from support conditions	Slides, ways, or moving elements not adversely affected by wear or damage	Long end standards supported at points to minimize deflection effects	Adequate lighting	Ability to select high-quality measuring instruments and standards with required geometrical and precision capabilities
Elastic indentation of measuring contacts	Deformation effects in instrument when heavy workpieces are measured	Effects of clamping force on gage blocks used with holders or attachments		
Effects of thermal expansion between measuring temperature and standard 68°F	Auxiliary elements, such as wires, rolls, angles, plates, calibrated and checked for function	Effects of differing geometrical form between workpiece and standard as they function in measuring contacts		
"Hidden geometry," such as lobing that will not be detected by contact conditions	Magnification of diameter error in thread wires when positioned in thread form			

For example, a precise reading taken on a sensitive measuring instrument may not always be an accurate measurement. So here the question of objectives becomes paramount. If inspection is performed with gages, the accuracy objective may be stated as a gage tolerance on a tool or gage drawing or in a standard, such as ASA B1.2 for thread gages. When gages are inspected, the gage checker establishes an accuracy objective for his measurements with respect to gage tolerance.

Where parts are accepted with measuring instruments, rather than gages, the inspector must establish an accuracy objective in order to evaluate conformity of the product with the requirement. How much the accuracy objective should be in terms of the tolerance involved in the acceptance decision is well defined in some cases, such as gage standards.

In other cases, the establishment of the accuracy objective is properly a matter for the measurement laboratory. If the accuracy objective is a large fraction of the part or gage tolerance, it may lead to acceptance of defective material or rejection of good material. Uncertainty of measurement may encroach upon product limits. Recognition of this effect of uncertainty of measurement, i.e., accuracy, usually results in

Table 2. Sizes and Tolerances of Cylindrical Gages

Size range, above–to and including	Tolerances			
	Class XX	Class X	Class V	Class Z
0.030– 0.825	0.00002	0.00004	0.00007	0.00010
0.825– 1.510	0.00003	0.00006	0.00009	0.00012
1.510– 2.510	0.00004	0.00008	0.00012	0.00016
2.510– 4.510	0.00005	0.00010	0.00015	0.00020
4.510– 6.510	0.000065	0.00013	0.00019	0.00025
6.510– 9.010	0.00008	0.00016	0.00024	0.00032
9.010–12.010	0.00010	0.00020	0.00030	0.00040

efforts to limit accuracy of measurement to 10 per cent or less of the tolerance of the feature being measured.

Having established the accuracy requirement for a measurement task, the next problem is to appraise the accuracy actually attained in the measurement techniques. The abundance of high-amplification measuring instruments has emphasized the necessity for differentiation between sensitivity of the measuring device and the accuracy of measurement actually attained. Exhaustive laboratory tests can best make these determinations for industry.

Costs and common sense, of course, enter into the establishment of all accuracy objectives. A precision of measurement not compatible with the tolerance of the feature (part) to be measured can lead to excessive costs and a high incidence of rather futile borderline acceptance problems. An example of such metrological extravagance would be efforts to attain an accuracy of measurement of ±0.00001 in. on a feature having a tolerance of ±0.002 in. On the other hand, an accuracy of measurement of ±0.001 in. on this tolerance would default precision.

A classification of cylindrical-gage tolerances used by American Gage Manufacturers is one reference for establishing precise accuracy objectives (see Table 2). These tolerances represent the permissible variation in the sizes of gages. To ensure that these tolerances are actually fulfilled, the gage laboratory must measure to an accuracy of a fraction of the tolerance. "Tithing" of the XX gage is a precision challenge.

PNEUMATIC GAGING

A basic pneumatic gaging system consists essentially of components that provide a constant-pressure air supply, indicating means, and a metering orifice. The principle of operation is based on the effects of varying the flow of air from the metering orifice. For example, as flow from the orifice is obstructed, pressure in the system will build up to the regulated value, and flow through the system will fall off. Over a significant range of values in such a gaging system, there exists a linear relationship between flow or pressure and the size of the escapement orifice. In many gaging applications, this linear relationship is equated to the clearance that separates a sized metering orifice and an obstruction. The indicating component reflects the change in flow or pressure as a linear measurement.

Basic Applications of Pneumatic Gaging

In practice, the pneumatic gage is used as a comparator; i.e., it measures variables by magnifying the difference in size between a standard and the work being measured. In flow-measuring systems, changes in air flow are indicated by the position of a float in an internally tapered glass tube through which the flow can be directed. Amplification is partially a function of the taper within the tube and the amount of total available flow that passes through the tube. In pressure-measuring systems, changes in pressure are indicated by well-established principles, such as a pointer actuated by a bellows or a bourdon tube. Amplification is obtained by pneumatic and mechanical means.

The orifice instruction can be either the workpiece being gaged or an integral part of the gage spindle. An example of the first is found in the gaging of a bore by inserting a gaging spindle containing two diametrically opposed orifices through which air is flowing. In this kind of application, there is no contact between the gaging spindle and the part being measured; the closeness of fit serves as the obstruction. In the second kind of gage, the gage spindle includes a component that mechanically contacts the work being measured. Movement of this component relative to a sized orifice affects air flow and establishes the basis for measurements.

Basic applications of pneumatic gages are shown in Table 3. Industrial installations range from the simple application of a single gaging spindle, for checking a single dimension, to a multiple sensing application, in which, for example, as many as 40 dimensions may be checked simultaneously. Pneumatic gages are well suited to post-process gaging, in which parts are checked against specifications after machining. Pneumatic units also have been applied to *in-process* gaging.

An example of in-process gaging is a grinding operation in which a gaging head continuously monitors the work or movement of a machine member while work is being processed. The position of the float or dial indicator guides the operator, permitting him to grind to within exact tolerances without frequent stopping to check dimensions.

ELECTROPNEUMATIC GAGING

It is upon the foregoing background that automatic installations involving electropneumatic components have been designed. The basic element in such systems is an air-to-electric transducer.

In some systems, the transducer is a pressure-sensing, limit-type unit which can actuate signal lights or relays by making or breaking either of two independently adjustable contacts. Thus, the transducer is a sensitive pressure switch with two sets of contacts that can provide a signal upon sensing either of two desired pressure values.

One transducer* operates in the pressure range of 5 to 25 psi, responding to pressure changes as small as 0.05 psi. Accuracy and repeatability are within 1 per cent of the normal gage span. A transducer includes two bellows with linear expansion characteristics for the pressure range at which the component normally operates.

In application, contact operating adjustments for the two pairs of contacts are made with the aid of standardized limit gages, or an indicating gage. In effect, the pressure at which a given contact pair is to operate is created in the system: This positions one contact which rests against a bellows. An adjusting knob then is manipulated to

_____
* The Sheffield Corp.

Table 3. Basic Applications of Pneumatic Gaging

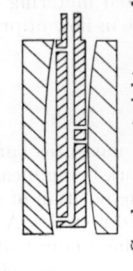

Two diametrically opposed open jets check true diameter of holes having tolerances of 0.005 in. or less

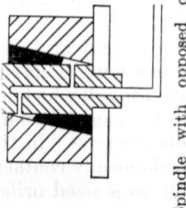

Contact-type gaging head is used in measuring diameters for interrupted bores and bores having a keyway

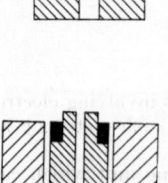

Spindle with opposed open orifices spaced longitudinally checks squareness of a bore axis with a face

Camber or straightness of hole is checked by rotating through 180° a spindle having four jets

Taper is indicated as any spindle is passed through a bore

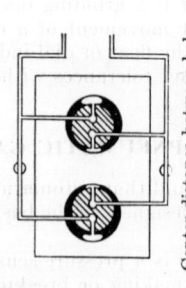

Center distance between holes can be checked by a fixture with two spindles, each having two opposed jets

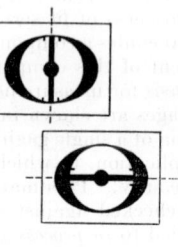

Out-of-round is indicated when any spindle is rotated through 90°

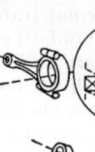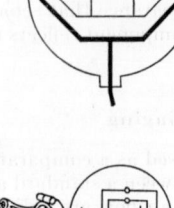

Parallelism of holes can be indicated by combining two spindles of type used in squareness checking

Outside diameter can be checked by two opposed standard jets; tolerance should be 0.002 in. or less

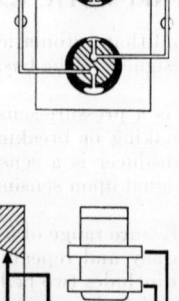

Height, width, or depth can be measured by a suitably fixtured contact-type gaging head

Thickness gage for thin parts and items having close tolerance incorporates two opposed standard jets

Multiple-contact-type gage heads mounted in a suitable fixture can check contour

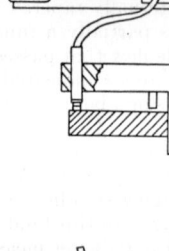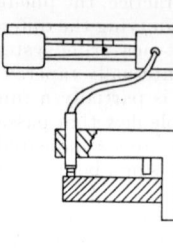

Squareness of surfaces for close-tolerance parts having good surface finish can be checked by a fixtured spindle

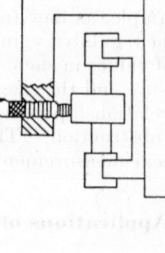

Concentricity can be indicated by one or more contact-type spindles mounted in a suitable gage fixture

positioning the mating contact so that it just opens (for normally closed contacts), or just closes (for normally open contacts). The desired contact action then will be obtained in use when the same pressure is sensed. The other contact pair of the transducer is adjusted in a similar manner to obtain the desired action at a different pressure, enabling the single transducer to respond to either a maximum or a minimum pressure when sensed.

More than one air-to-electric transducer can be used in a single circuit. For example, in addition to one unit that indicates maximum and minimum values, a second unit can be connected in the circuit and adjusted to indicate pressures just below the maximum and above the minimum. This arrangement can be used in warning systems to indicate when critical limits are being approached.

Modular Components

The basic transducer can be incorporated into the design of several modular gaging control components. One component in this class is the differential gaging head. A differential measurement can be made of such variables as taper, taper tolerance, squareness of face to bore, center distance, bend and twist, and parallelism (see Fig. 1).

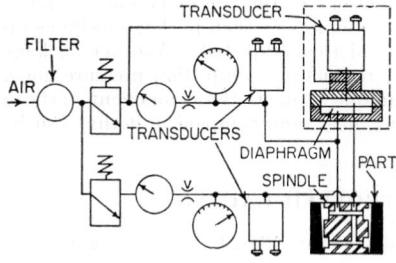

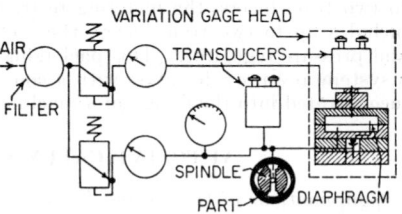

FIG. 1. Incorporation of basic transducer into a modular unit where a differential measurement is desired. Back pressures from two identical circuits are directed to either side of a diaphragm. Deflection of the diaphragm represents a difference between two measurements. The deflection of the diaphragm is sensed by the transducer.

FIG. 2. Modular-variation gage head used to measure the extent to which a value varies in certain types of measurements. In measuring a hole for an out-of-round condition, the part is rotated about a fixed air spindle. Variation in sensed air pressure indicates an asymmetrical hole. Primary operation of modular gage head is to retain maximum pressure sensed and compare the sensing to minimum pressure sensed.

For pneumatic gaging in such applications, two identically balanced air circuits are established as the primary measuring means, each being directed to a separate measuring orifice. Back-pressure circuits connect each measuring circuit to different sides of a diaphragm mounted in a cavity. Significant variations in the back pressure of the two circuits, indicating a defect, cause the diaphragm to deflect. This deflection is sensed by a secondary pneumatic measuring circuit connected to an air-to-electric transducer; and electric contacts are actuated if the back pressure in the secondary circuit reaches preset limits. Separate transducers also can be connected to each primary circuit in a differential gaging application to check the specified dimension being measured at each orifice.

Variation Gaging Head. This is a modular component employing the air-to-electric transducer. The unit measures a variation in size as opposed to identifying the actual size. Typical of such measurements are those to check for an out-of-round condition or concentricity. For example, when a pneumatic gaging spindle is inserted into a workpiece and rotated (or the work can be rotated about the spindle), there will be no significant back-pressure change if the work surface being inspected is truly round. If the work is out-of-round, however, the back pressure will vary over a range with limits determined by the extent that the work is out-of-round (see Fig. 2).

The design of the variation gage head enables it to identify the back-pressure range in such applications. In one unit,* this is accomplished primarily by capturing and retaining the maximum pressure sensed. The back-pressure circuit from the gaging head is directed over two paths to both sides of a diaphragm mounted in a cavity. One of the paths includes a small check valve that opposes flow from one side of the diaphragm cavity. As a gaging spindle is rotated, the maximum back pressure sensed is retained on that side of the diaphragm. With further spindle rotation, the point of minimum back pressure is reached. This pressure is transmitted to only one side of the diaphragm, and the diaphragm is deflected to an extreme point. A secondary pneumatic gaging circuit containing an air-to-electric transducer measures the deflection of the diaphragm, and contacts are actuated if desired limits are exceeded. The primary measuring circuit in a variation gaging application also can contain a separate transducer to check the actual size of the part being gaged.

In addition to the gaging spindles and transducers that have been discussed, a typical electropneumatic system involves such items as an air-line filter, pressure regulator, pressure gage, metering restriction, setup dial, and a slave-relay unit.

Other modular components that include the basic transducer have been designed—primarily to provide various packaged combinations of the necessary and auxiliary devices associated with an installation. For example, one such package includes up to two transducers, the metering restriction, and the setup dial. Another package includes up to two transducers, the metering restriction, setup dial, pressure gage, and pressure regulator. This packaging technique protects the component parts of a system and provides users with a convenient, neat, and compact unit that can be incorporated into the design of a machine.

AUTOMATICALLY CONTROLLED GAGING

Combinations of the various basic gaging devices are used to provide signals and controls for a variety of applications, including:

1. Adjusting movement of cutting or forming tools.
2. Signaling replacement of worn tools.
3. Stopping machines.
4. Warning when part-size trend approaches minimum or maximum limits.
5. Segregating parts at various stages of production.
6. Weighing and checking weight relationships.
7. Connecting or straightening parts automatically.
8. Matching parts for selective assembly.
9. Inspecting finished parts.
10. Classifying and matching parts.
11. Selectively packaging parts.

All the foregoing uses are predicated upon a desired action taking place via electrical signals that are triggered, directly or indirectly, by size measurements obtained from pneumatic gages. And with the accurate creation of a signal, the pneumatic elements in the gaging system have done their job. Thereafter, the capabilities of the systems are determined by electrical techniques and the sophistication of the circuit design and electrical gear in the system. Here it is important to note that the information obtained from the limit-type pneumatic gaging also can serve simultaneously to check the accuracy of an operation, control a machine, segregate parts into size categories, and initiate auxiliary control signals.

Control Circuitry

The electrical and pneumatic circuits illustrated in Fig. 3 provide for the relatively simple postprocess inspection of the bore of a cylindrical part. With this system, good parts can be separated from rejects, and the production machine can be stopped

* The Sheffield Corp.

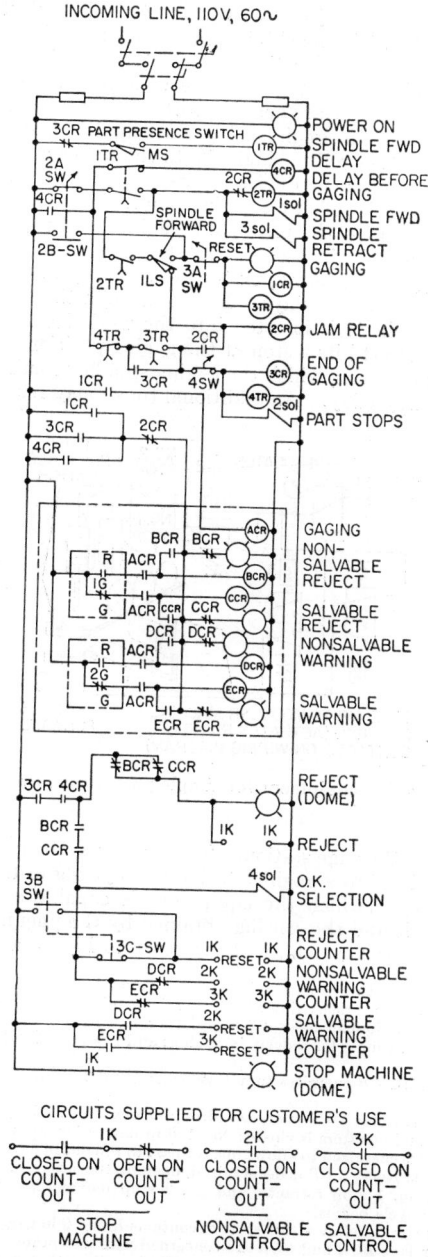

FIG. 3. Circuit for relatively simple postprocess inspection of the bore of a cylindrical part.

if a predetermined number of rejects are gaged. Salvable rejects may be separated from nonsalvable rejects. The machining trend can be followed, by keeping track of the number of parts approaching the reject limits, and the production equipment can be stopped at some predetermined phase of a trend. Inspection rates up to several thousand parts per hour are possible with equipment of this type.

The equipment to which the circuits of Fig. 3 are applied is shown in Fig. 4. Parts from one or several machines roll in single file to the upper end of an inclined chute. A pneumatic-cylinder escapement device allows one part at a time to be separated from the waiting file—and it rolls to a gaging point. Mounted on a slide and actuated by a pneumatic cylinder, the pneumatic gaging head is automatically inserted into the bore of the part to be inspected. As shown in Fig. 3, these straightforward sequencing operations (which follow upon the presence of parts to be gaged) are modified by the action of time-delay relays. For instance, a time-delay relay must time out before the gaging spindle is inserted into the workpiece. This allows for part bounce that occurs as a part hits a stop at the gaging position after rolling down the chute. The time period during which the gaging spindle remains in the workpiece is controlled by a time-delay relay. An additional time-delay relay is used to allow the

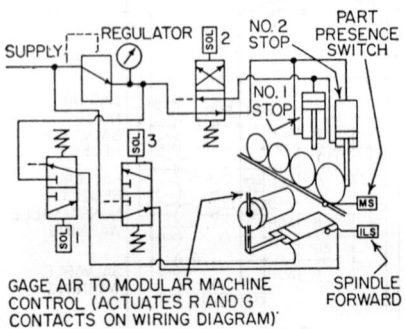

GAGE AIR TO MODULAR MACHINE   SPINDLE
CONTROL (ACTUATES R AND G   FORWARD
CONTACTS ON WIRING DIAGRAM)`

Fig. 4. Application of gaging equipment for simple postprocess inspection of the bore of a cylindrical part.

workpiece to roll out of the gage station. These circuits, then, provide for sequential presentation of workpieces to a gaging station and cycling of the gaging head. The several control actions of which the equipment is capable are determined by the electropneumatic circuit and the sensing obtained by the pneumatic gage.

Detailed operation of the system, with reference to Figs. 3 and 4, is as follows:

1. Gaging cycle commenced when part rolls down escapement, stopping at no. 2 stop and actuating parts-presence switch MS. Time-delay timer 1 TR energizes and starts timing to allow for part bounce.

2. When 1 TR times out: Control relay 4 CR is deenergized and the gage-relay circuit drops out; 2 TR is energized and starts timing for a delay in gaging; and 1 Sol and 3 Sol are energized to bring the gaging spindle forward.

3. When 2 TR times out: 1 CR energizes to initiate gaging and 3 TR is energized to establish gaging time.

4. When 3 TR times out: Relay 3 CR energizes, dropping out 1 TR, 2 TR, 1 CR, and 3 TR in succession; circuit to solenoids and counters is closed; No. 1 stop moves forward; no. 2 stop retracts; gaging spindle retracts; and 4 TR is energized to allow time for part to roll out of gage station.

5. When 4 TR times out: Relay 3 CR is deenergized, opening the circuit to 4 TR; solenoid and counter circuits are deenergized; no. 1 stop retracts; and no. 2 stop moves forward. When following part actuates MS, new gaging cycle begins.

Actions of gaging relay circuit follows from state of contacts R and G in gage heads 1 G and 2 G. Gage head 1 G and associated relays BCR and CCR are concerned with acceptance or rejection. For acceptance, both BCR and CCR must operate each cycle. This requires that normally open contact R be closed and that normally closed contact G remain closed.

Reject counter 1 K is actuated each time a reject part is gaged. Should 1 K count to a predetermined number, the machine is stopped.

Gage head 2 G and associated relays DCR and ECR are concerned with the trend of sensed measurements. For a completely acceptable part, DRC and ECR must operate. This requires that normally open contact R be closed and that normally closed contact G remain closed.

Salvable or nonsalvable warning count will register on counter 2 K or 3 K each time a part in the warning range is gaged. The warning counters will reset whenever an acceptable part is gaged, or a warning part in the opposite range is gaged. Should either counter reach a predetermined count, a control or indicating signal is created.

Role of Counters. It is apparent that in a gaging system based on limit-type gage sensing, counters play an important role in obtaining practical control of a quantity production system. Following statistical theory, the control scheme recognizes that the gaging of any one item can present a distorted picture of true conditions. The out-of-tolerance part must be separated from in-tolerance parts, but it does not follow that an adjustment need be made in the production equipment. Adjustments are required only when dimensional deviations result from controllable causes. The picture of the general trend obtained by examining a consecutive number of parts is one of the effective ways to recognize the occurrence of controllable causes of defects. Thus, the reject counter registers all rejects, but is reset automatically each time an acceptable part is gaged. The machine will be stopped through the reject-counter circuit only when the consecutive number of rejects reaches a predetermined value.

In the circuits for salvable-nonsalvable warning, a count registers on the appropriate counter each time a part is within the warning range. The counters are reset automatically when an acceptable part not in the warning range is gaged, or when a warning signal of the opposite range is obtained. With this arrangement, the warning counters will cause control action only when a predetermined number of consecutive parts are gaged within one or the other warning ranges.

Complex Systems

Automatic gaging machines can be designed for applications much more complex than that just described. As an example, consider a part that is heavy and bulky, where the points to be measured are not readily accessible, where eight measurements must be checked simultaneously, and the part must be classified according to the sensed measurements. The foregoing are the parameters for an automatic machine that inspects and classifies four crankshaft bores and four camshaft bores simultaneously in a six-cylinder automobile engine block. Further, the machine is capable of one hundred cycles per hour.

Up to five blocks can be on the in-line transfer type of machine at one time, occupying the following positions: (1) load, (2) gage, (3) mark, (4) reject, and (5) eject. Blocks are presented to the machine by a transfer mechanism, and enter the load station with the front end of the block first, pan rail up. Moving a block into the load station triggers a limit switch that initiates the transfer through the machine. The blocks are transferred through the machine on flat, serrated rails by a hydraulically actuated, shuttle-type transfer bar with pivoting feed fingers that confine the part. When the gage station is clear, the block is transferred into it from the load station. From above, a shot pin enters a locating hole (also used in the machining operation) at each end of the block.

With the block located, the cam- and crank-bore gaging spindles are pivoted down into the cavities of the block and shifted horizontally approximately one inch so that the spindles enter the bores. The gaging spindles are mounted to a carrier plate which is pivoted on an insert bar. The gaging spindles "float" sufficiently to compensate for bore misalignments up to approximately $\frac{1}{32}$ in. Auxiliary air jets at the leading edge of the spindles help to blast loose dirt and chips out of the bore. The moment the noncontact gaging spindles enter the bores, the gaging circuit is initiated. Each pneumatic gaging circuit for the crank bores contains two air-electric transducers, providing for rejection of out-of-tolerance bores and for classification of in-tolerance bores into one of two size classes of 0.0004 in. each. A single air-electric transducer in each pneumatic gaging circuit for the cam bores provides for either an accept or a reject signal. The signals from the control units are fed into a memory system consisting of electromechanical latching-type relays.

After the gaging operation has timed out, the gaging spindles are retracted and pivoted upward and out of the block. The locating pins are withdrawn, and the block is transferred to the marking station. Here eight spray guns, actuated by signals from the memory, color-code the crank-bore class on the block with either red or blue paint, according to the size group. Blocks that fail to pass inspection are not marked, but are automatically rejected at the reject station.

ENGINEERING CONSIDERATIONS IN GAGING APPLICATIONS

Manufacturing, inspection, and management considerations enter into the development and acceptance of automatic gaging systems. Manufacturing and inspection variables that must be considered include: (1) part size; (2) part configuration; (3) material; (4) finish; (5) production-part changes; (6) tolerance; (7) part handling; (8) type of gage elements; and (9) inspection rate.

For automatic gaging applications, each part must be analyzed individually, and a system devised within the parameters of economic and technical boundaries. Generally, what can be gaged manually can be gaged automatically, keeping in mind that physical dimensions and characteristics of the part to be presented to the gage must be defined; and the manufacturing process must be capable of producing parts within this definition.

Many types of methods and devices are available for transferring simple or intricate and irregular shapes, small or large, into and through the gage. Typical part-transfer methods include gravity-feed, endless-belt, indexing-chain shuttle bar with pivoting feed fingers, cross-feed, and push movements. It is possible, in many instances, for a gage maker to design the gage for adjustable or interchangeable tooling to accommodate changes in part size and shape.

Of greater interest is the method or procedure to follow when a part has a number of check points within an area too small to accommodate all gaging elements. In this case, two solutions are suggested: If gage time permits, a second set of gaging elements is indexed into the part; or the part is transferred to another gaging station to complete the inspection. In any event, close tolerance dimensions should be gaged first; broader tolerances next. With this approach, there is less chance of wasting gaging time on a part destined to be rejected because the finer tolerances are unacceptable.

Part alignment in the gaging station is an important factor in automatic gaging. Wherever possible, locating surfaces for the inspection operation should be the same as those used in machining; or the locating surfaces may be determined by the functional or end use of the part. Gaging elements should "float," to permit slightly misaligned bores or dimensions to be gaged without damaging the tooling or affecting the gage reading. On small, thin-wall parts, the type of chuck or clamping device used to position the part during gaging must not distort the part.

Material and finish of the workpiece dictate the type of gaging elements to use. Nonporous metal and nonmetal parts with surface finishes below 35 rms are generally gaged with pneumatic sensing units, such as open-type air jets, noncontact air-jet spindles, air snaps, air rings, or—if contact-type gaging is desired—with pneumatic gage cartridges.

The gaging of thin-walled parts, narrow lands, and rough or porous surfaces frequently calls for a specific type of gage element. For example, rough or porous surfaces and finishes are gaged with contact-type spindles.

Broad tolerances on the order of 0.005 to 0.080 in. can be gaged directly with contact-type pneumatic-gage cartridges. Tolerances above these also can be gaged with these cartridges by means of deamplifying linkages.

Parts Preparation

As in any precise measuring practice, it is desirable to have clean parts presented to the automatic gage. Where parts are washed in hot or warm liquids, they should be allowed to cool to uniform temperature before being gaged. In gaging holes, auxiliary air jets can be incorporated into the leading edge of the gaging spindle to assist in the removal of loose chips and dirt. Normally, the responsibility for selecting the method for cleaning parts lies with the user of the gage, not with the gage maker.

Controlled environmental temperature is desirable where tolerances on the order of 0.0002 in. or better are encountered. Here, again, part size, wall thickness, dimensions, and tolerances serve to establish the degree of environmental control necessary for achieving the accuracy objective.

How many parts per hour can be gaged with an automatic gage utilizing a pneumatic gaging circuit? Here, too, size, tolerance, material, and method of handling are controlling factors. Quite frequently, the rate of inspection is limited not by the speed of air gaging, but by material handling. Under certain conditions, it is possible to gage at rates up to 4,000 parts per hour. More normal rates are from 1,000 to 1,500 parts per hour.

MANAGEMENT CONSIDERATIONS IN GAGING APPLICATIONS

Unlike manual inspection methods which may allow faulty parts to be accepted because of human errors in the inspection setup, gage reading, or judgement, an automatic gage accepts only parts that are within the prescribed tolerance limits set into the gage. There is no compromise; automatic gaging makes no arbitrary decisions. Thus, the basic reason for automatic gaging—protection of quality. However, unless the inspection problem is one in which the need for accuracy transcends monetary considerations, it will be necessary to show how automatic gaging can cut inspection costs. Experience indicates that the cost analysis should be brief and simple. It should include the number of units inspected and a comparison of present unit inspection costs with the proposed unit cost for automatic gaging.

In addition to direct savings which may be possible with automatic gaging, the analysis should cover those savings accruing through the reduction in floor-space requirements, the smaller numbers of individual gages that will be bought and stocked, and the elimination or reduction of gage records and attendant paper work.

The very essence of automatic gaging is integrated control of the manufacturing process in order to produce uniform high-quality production. As a result, its influence cuts across department lines, and extends into areas and policies beyond those normally ascribed to quality control or inspection departments.

Maximum control over quality and cost in manufacture of the gaging equipment results from the *systems approach* in which the gage maker is responsible for designing, building, calibrating, installing, servicing, and repairing the gaging equipment.

REFERENCES

1. Polk, Louis F., Jr.: Electropneumatic Gaging and Control, *Automation*, vol. 7, no. 3, pp. 78–86, March, 1960.
2. Aller, W. Fay: Measure Your Quality . . . Automatically, *ibid.*, vol. 4, no. 3, pp. 52–58, March, 1957.
3. Darmody, W. J.: How to Appraise Accuracy in Measurement, *Am. Machinist*, vol. 100, no. 22, Oct. 22, 1956.
4. Staff: Electronic Micrometer, *Acft. & Missiles Mfg.*, vol. 2, no. 8, pp. 22–24, August, 1959.
5. Houck, D. R., and W. Murphy: Electronic Gaging and Sorting, *Automation*, vol. 6, no. 12, p. 76, December, 1959.

THICKNESS MEASUREMENT OF SHEET
AND WEB MATERIALS*

By W. H. Faulkner, Jr.† and J. G. Wood‡

Thickness gages for sheet and web materials range in complexity from that of a simple micrometer caliper to that of a beta-absorption gage, the latter of which may be part of a complete, closed-loop thickness control system. In a system of this type (Fig. 1), a beta absorption gage supplies feedback signals to control elements that

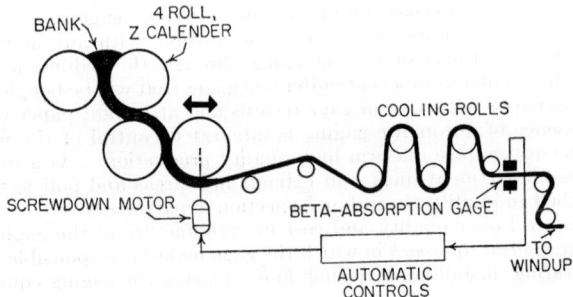

FIG. 1. Application of a beta-absorption gage to measure continuously the thickness of outgoing product in a manufacturing line for plastic sheet stock. Signals from the gage actuate control elements that operate a screwdown motor controlling the spacing between calender rolls. System automatically maintains thickness of final product within preset limits.

continuously adjust the spacing between calender rolls so as to maintain the final thickness of plastic sheet stock within preset limits during high-speed production.

CLASSIFICATION OF THICKNESS GAGES

Because of their variety, thickness gages can be classified in accordance with any of several distinguishing characteristics. For example, they can be divided according to their operating principles, namely, mechanical vs. pneumatic; electrical; or electronic; and so on. On the other hand, such devices can be recognized as either contacting or noncontacting with respect to the materials being inspected. A third possibility is to classify thickness gages according to their principal applications, such as for sampling and spot checking, for setup purposes, or for continuously measuring the thickness of materials moving in process.

* This subsection is adapted from *Automation*, vol. 9, no. 7, 1962.
† *Formerly* Vice President, Engineering and Development, Tracerlab Division of Laboratory for Electronics, Inc., Waltham, Mass.
‡ *Manager of Engineering*, Tracerlab Division of Laboratory for Electronics, Inc., Waltham, Mass.

Contacting vs. Noncontacting Gages

A natural division occurs in the suitability of contacting vs. noncontacting gages for continuous thickness-measuring applications.   For example, an accurate, continuous measurement of sheet thickness or of thickness trend usually is not obtained by a simple dial indicator and contacting-roller assembly—because most materials contain local ridges and depressions.

If a continuous measurement is made by a fixed-anvil-type instrument, only the high spots will be measured, and hence the reading will not be indicative of the average thickness.   When the instrument response is made fast enough to follow all variations of material thickness, the indicator will be in a continuous state of vibration between the highest and lowest thicknesses encountered, and, thus, it will be difficult to determine where the average value may lie, or what the thickness trend may be.

Need for Damping in Continuous Measurements

In practically all industrial processes, it is not possible to eliminate these point-to-point variations in the material being processed, and, generally, such variations do not affect the quality of the final product adversely.   In order to provide a meaningful average-thickness indication, some form of damping is required in the thickness-indicating device.

Single Roller

The contacting member of the instrumental system may take the form of a single roller that rides on top of sheet stock, passing over an accurate surface, or an accurately ground roll.   Large rolls that are ground to an eccentricity or runout of less than 0.0001 in. are difficult, if not impossible, to manufacture and to maintain. However, by introducing sufficient damping or time delay between the contactor and the indicator, variations in the order of millionths of an inch may be accurately gaged, with rolls having a combined runout of several ten-thousandths of an inch.

Differential Roller System

Where sheet materials must be measured on a calender roll because of space limitations—or because of the nature of the material being measured, a more accurate indication can be obtained by means of a differential roller system.   In this system, two rollers are used.   One roller rides on the bare calender roll, and the other rides on top of the stock.   The distance between these two rollers is gaged to give an accurate indication of sheet thickness, independent of the accuracy of the calender roll.

Gage Mounting

The mounting of a contacting-type gaging unit will depend on the type of material being gaged and on what variations are encountered in the pass line of the sheet through the gage.   Where the material is relatively flexible and the pass line is fixed, a double-roller unit may be solidly mounted.   Where more rigid materials are being gaged and where variations in the pass lines are expected, the gaging assembly can be floated on springs which are arranged to maintain the lower reference roll in constant contact with the sheet.

Limit Signals

In some instances, a simple dial indicating thickness gage can be equipped with an electric switching device, arranged to produce pulses whenever preset limits are exceeded.   These pulses are used to actuate limit indicator lights through an electric time-delay circuit.   The indicator lights will be actuated only when the tolerance limits are exceeded for an appreciable time, and they will not flash constantly because of bumps and depressions in the sheet.

Proportional Signal

In another form of contacting gage system, the linear displacement of the contacting rollers is used to actuate a transducer, which will produce an electrical signal proportional to its displacement. The resulting signal may be amplified and measured by metering circuits having any desired response time.

In the magnetic comparator, described later, the movement of a magnetic plunger in the coil system of a differential transformer produces an unbalanced voltage that is proportional to displacement. Displacements of one ten-thousandth of an inch may be measured by this arrangement. The expansion or contraction of the gage rollers with temperature is compensated for in this gage by means of a temperature-sensitive resistor, inserted within one of the bearing spindles and connected into the electric circuit.

SELECTION OF THICKNESS GAGES

Choice of the proper thickness gage for a given application involves the consideration of many factors: (1) some of which are related to the gage itself, (2) some to the process requirements, and (3) some to the environment in which the gage will be used. The following selection criteria should be considered:

1. Will the gage operate satisfactorily under production conditions?
2. Cost—initial, installation, and maintenance.
3. Range of thickness to be measured.
4. Accuracy and reproducibility.
5. Space and support requirements.
6. Frequency of measurement—continuous or intermittent.
7. Safety—personnel considerations, protection required for the measuring element, fail-safe operation.
8. Utilities and services required.
9. Materials of construction.
10. Availability.
11. Accessory equipment required to transmit and transduce the measurement.
12. Compatibility with existing instrumentation.
13. Anticipated life of installation.
14. Experience and skill required of the personnel who will operate the gage.
15. History and reputation of the device.
16. Experience and reputation of the manufacturer of the device.

In addition to knowledge of performance requirements, it is important to have an understanding of the basic design features, advantages, and limitations of available thickness-measuring devices. Descriptions of the major devices, both contacting and noncontacting, are presented in the following pages.

CONTACTING-TYPE THICKNESS GAGES

As implied by the name, gages in this category actually make physical contact with the sheet or web material being measured.

Micrometer Caliper

The familiar precision micrometer caliper is an often-used device for checking the thickness of samples of sheet materials. Normally, the scale graduations are in thousandths of an inch. Units are available with vernier scales that permit readings to ten-thousandths.

Where a micrometer caliper is used to check the thickness of compressible materials, a unit equipped with anvil and spindle ends of large area will provide consistent results.

Reproducibility of readings also will be enhanced if the micrometer caliper incorporates an integral spring ratchet or other arrangement capable of maintaining a constant pressure on the material during measurements.

By checking the thickness of a sample at a number of points according to a regular pattern, a micrometer caliper may be used to provide data for the computation of a reasonably accurate average-thickness figure.

Micrometer calipers also are used to check and set up indirect- and continuous-type thickness gages that may be used on the production line. For this purpose, accurate samples are required. Micrometer calipers should be checked frequently against gage blocks or other standards, to ensure their continued accuracy.

Dial Indicator Gage

The dial indicator gage is used frequently for sampling and spot checking. In this device, the position of a spindle with respect to the body of the gage is indicated on a graduated, circular dial by a needle-type pointer. In order to amplify the motion of the contact point sufficiently to enable a reading of the instrument to thousandths or ten-thousandths of an inch, a precision gear and linkage system couples the spindle to the indicating pointer. High-precision dial indicator gages are available that will indicate increments of 20 millionths of an inch.

Dial indicators are useful for standard sample selection and for production control on a sampling basis. Models are available that are especially adapted to measuring specific sheet materials under constant pressure.

Normally, the ordinary dial indicator is not satisfactory for continuous gaging of moving sheets, but several types are available with rollers which adapt the instrument for intermittent use on moving sheets. These instruments are useful for spot checking of moving materials where physical contact can be maintained between the roller and the material, and where process speeds and sheet flutter are not excessive.

Basis Weight Scale

Occasionally it is desirable to express material thickness in terms of weight per unit area. To determine this property, material having a known area is weighed on a scale. Where many samples must be measured quickly in a production operation, a metal template having the proper area is laid on the torn samples and the samples cut to size. The accurately cut sheets then are hung on a special scale, and a direct weight-per-unit-area reading is obtained. If a sufficient number of samples are obtained, the technique can furnish very accurate information. Basis weight also provides a method for setting up the calibration for and also the direct checking of gages designed for continuous thickness measurement in terms of basis weight.

Magnetic Comparator

In the magnetic comparator, the motion of a movable contact roller is used to move the iron core of a differential transformer. The differential transformer provides an output voltage dependent on the position of the iron core with respect to two secondary windings. This position is a function of the thickness of the material being measured. When used with a suitable electric meter and power source, the magnetic comparator provides a convenient means for measuring material thickness in applications similar to those for dial indicator gages. The magnetic unit is rugged and less likely to get out of adjustment during production measuring operations than a purely mechanical dial indicator gage.

The range of measurable material thickness for a magnetic comparator is about 0.005 to 2.5 in. Accuracy is approximately plus or minus $\frac{1}{2}$ of 1 per cent of full scale. Costs range from \$25 to \$150 (less meter).

Schuster Gage. The thickness of nonmagnetic materials passing over a steel calender roll may be measured continuously by magnetic means. In this type of magnetic gage, commonly referred to as a Schuster gage, two magnetic pole faces in a gage head are arranged to ride on the surface of the material to be gaged, as shown in Fig. 2.

When the material becomes thicker or thinner, the gap between the pole faces and the calender roll varies, causing a change in the magnetic flux between the poles. This change unbalances an a-c bridge circuit, and the amount of thickness variation is indicated on an associated meter and/or recorder (see Fig. 3). In order to maintain accurately the position of the pole faces with respect to the material on the calender roll, the gaging head is mounted on accurately ground rolls which ride on the surface of the material as it passes over the calender.

Gages operating on a similar principle may be used to measure the thickness of nonmagnetic coatings on iron or steel parts, but are not generally adaptable for continuous measurements.

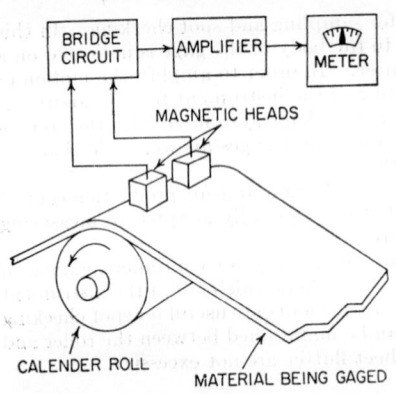

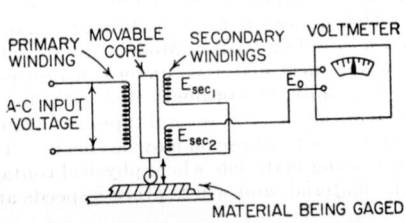

FIG. 2. Application of magnetic gage to thickness measurement of nonmagnetic material passing over a steel calender roll.

FIG. 3. Electric circuitry of magnetic gage. Basic equation:

$$E_o = E_{sec1} - E_{sec2} = kt$$

where E_o = output voltage
E_{sec1} = voltage at secondary winding 1
E_{sec2} = voltage at secondary winding 2
k = calibration constant
t = thickness of material being gaged

Nonmagnetic materials ranging in thickness from 0.0001 to 0.75 in. may be inspected with a magnetic gage. Accuracy depends upon the application, and may range from plus or minus 2 per cent to plus or minus 10 per cent. Costs for magnetic gages range from $200 to $500.

Sonic Gage

The thickness of rigid materials accessible from one side only may be measured by exciting them to vibration by ultrasonic sound waves. A generator of high-frequency sound is coupled to the "wall" to be measured and, by varying its frequency, a value is found where the wall is resonant. From this resonant frequency, the thickness may be calculated or read directly from a calibrated scale (see Fig. 4).

Although refinements of this method allow direct reading of thickness to be obtained quickly, the instrument generally is not suitable for continuous gaging of moving sheets—because accurate gaging requires very close contact between the sound generator and the surface of the sheet to be measured.

Steel sheets from 0.01 to 16.0 in. thick can be measured with sonic gages. Depending on the application, accuracies from plus or minus 0.1 per cent to plus or minus 3 per cent can be obtained. Costs for sonic gages range between $1,000 and $3,000.

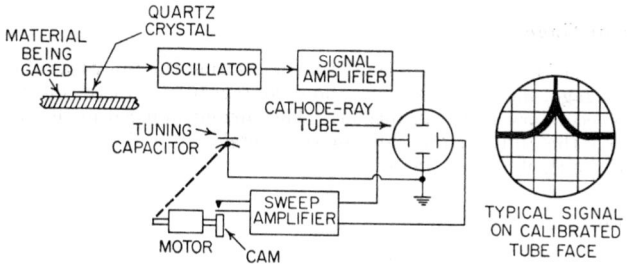

FIG. 4. Application of sonic gage to thickness measurement of a rigid material.   Basic equation:

$$t = 0.5v/f$$

where t = thickness of material being gaged
v = velocity of sound in material
f = resonant frequency

NONCONTACTING-TYPE THICKNESS GAGES

Gages in this category do not require direct contact with the sheet or web material whose thickness is being measured.

Pneumatic Gage

In the pneumatic thickness gage, the material to be measured is inserted between the parallel fixed faces of a specially designed pneumatic member.   The rate of air-flow through a small orifice located in the face of one gaging member is dependent on the thickness of the material placed between the faces (see Fig. 5).

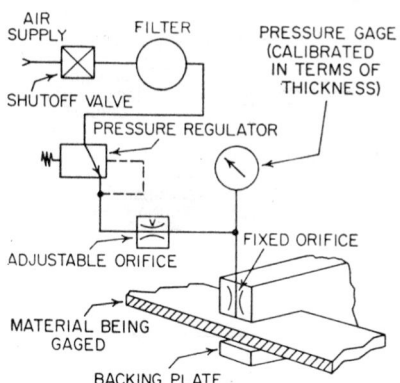

FIG. 5. Application of pneumatic gage to thickness measurement of sheet material.

By means of a regulated air supply and adjustable restrictions, the air pressure existing at the gaging head may be made dependent on the rate of flow of the air through the gaging orifice.   The thickness then may be indicated directly on a bourdon or other suitable pressure gage.   Although usually used for the measurement of samples or for the calipering of machined parts, this type of gage can be adapted for continuous use on small metal-rolling mills.

The range of measurable thickness for a pneumatic gage is about 0.001 to 0.100 in. Maximum full-scale sensitivity may be 0.0006 in. or less.   Costs for pneumatic gages range from $300 up, depending on the application.

Eddy-current Gage

A special type of magnetic gage is used for measuring the thickness of nonmagnetic foils. In this gage, the foil is passed without contact between two coils that form the primary and secondary of a transformer. The current induced in the secondary coil will vary with the eddy-current loss in the metal foil, and hence will provide an indication of the thickness of the foil.

Optical Gages

Although optical techniques may be employed to measure the flatness of surfaces and the accuracy of gage blocks to within a few wavelengths of light, these methods generally are not applicable to production gaging problems. A step gage for measuring the thickness of thin film by comparison of the interference colors produced is available. Films ranging in thickness from 2 to 16 μin. may be measured by color comparison with this step gage, a correction being applied for the refractive index of the material measured. The thickness of transparent materials which can be approached from one side may be measured by the use of a specially adapted microscope. A light-transmission gage has been designed for measuring the weight of thread continuously. The system is subject to large errors where color variations are encountered.

For measurement of transparent films, a type of optical gage has been put on the market. However, no extensive data from industrial installations are available at this time.

Capacitance Gages

In the capacitance-type thickness gage, a continuous measurement is made of the electrostatic capacitance existing between two plates through which the material to be gaged is passed (see Fig. 6). Since the capacitance of a capacitor varies directly

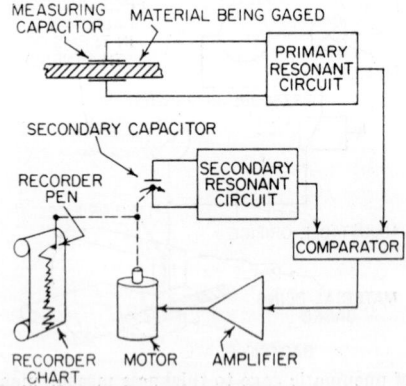

Fig. 6. Application of capacitance-type thickness gage to continuously moving sheet material.

with the thickness and dielectric constant of the material between its plates, it is possible by this means to measure thickness of materials having suitable dielectric properties and possessing dielectric constants different from that of air.

The capacitance thickness gage may be used to gage only those sheet materials which may be classified as electric insulators. This type of gage has been employed successfully to determine the weight per unit area, or thickness, of rubber-coated fabrics, rubber sheets, and plastic materials.

One of the principal limitations of this type of instrument is that it is extremely sensitive to variations in the moisture content of the materials being measured—since the dielectric constant of water may be about ten times higher than that of the material being gaged. In fact, this principle has been applied to moisture-measuring instruments. This limitation restricts use of this gage to materials which do not absorb moisture, or in which the moisture content remains constant.

In one type of capacitance gage, the measuring capacitor is connected in an r-f oscillator circuit as part of a frequency-determining system. Variations in the thickness of the material between the plates of the measuring capacitor cause a corresponding change in the frequency of the oscillator. This frequency then is compared with that of a fixed r-f oscillator. The value of the difference frequency provides an indication of the thickness of the dielectric material under measurement.

In the resonant-circuit-type capacitance gage (shown in Fig. 6), variations in material thickness cause an unbalance between primary and secondary resonant circuits. This unbalance causes a motor to drive a capacitor in the secondary circuit until balance is restored. This balance is attained when the measuring and secondary capacitances are equal or differ only by a constant. A recorder pen, coupled mechanically to the secondary capacitor, indicates the thickness of the material between the capacitor plates.

Resonant-circuit capacitance gages are useful for measuring insulating materials in thicknesses up to about 0.1 in. Accuracies range from plus or minus 1 per cent to plus or minus 10 per cent, depending on application.

Radiation Absorption Gages

The absorption of radiations, such as X rays and those resulting from radioactive decay, is dependent on the thickness of absorbent material interposed between a source of radiation and a radiation detector. Thus, such a system may be used to gage thickness.

As a first approximation, the absorption of radiations in matter may be expressed by the exponential form

$$N = N_o e^{-\mu t}$$

where N = intensity of radiation reaching detector
N_o = intensity of original radiation
e = base of natural or Napierian logarithms and has the approximate value of 2.718
μ = absorption coefficient of material being gaged for type of radiation in use
t = thickness of material being gaged

Radiation-thickness gages all are subject to errors due to the absorption coefficient being dependent upon the composition of the material being gaged. In consequence, such gages should always be calibrated with samples having the same composition as the material to be gaged. The variation in composition which may be tolerated without introducing serious errors will depend on the energy and type of radiation being used, and is best determined empirically for a given process material.

Since the emission of X rays and the occurrence of nuclear disintegrations both are random phenomena, the particle count for radiation arriving at the detector during a given time interval is subject to variations that generally follow statistical laws. The magnitude of these variations will vary inversely with the square root of the radiation source strength and the time constant of the detecting device. Hence, for a given source strength and detector response time, the accuracy of measurement is limited by the magnitude of the statistical variations.

In many applications, the response time of the indicator must be made long enough to provide a representative average value of the sheet thickness, so that statistical variations are not a limitation. In some metal-rolling and other automatic control applications, however, the response time may have to be greater than ideal in order to obtain the required accuracy.

The selection of the type of radiation to be used will depend upon both the absorption coefficient and the range of thickness of material to be gaged. Among the types of radiation gages used for material thickness determinations are: X-ray gages; gamma gages; bremsstrahlung (beta-excited X-ray) gages; beta gages; and radiation back-scattering gages.

X-ray Gages

Two major elements of an X-ray gage are: (1) a source of X radiation; and (2) a radiation detector (see Fig. 7). These elements must be highly stabilized, since most industrial thickness-gaging applications call for the accurate sensing of relatively small percentage changes.

One method of stabilizing a radiation detector is to check the detector against a radioactive source of constant radiation which is presented to the detector during a standardizing period when no X rays are present. The sensitivity of the detector then is adjusted automatically so as to produce a constant output from the standard source. Where photoelectric cells and scintillation detectors are employed, a standard light source may be substituted for the radiation source.

One type of X-ray gage uses a comparison system and two stabilized radiation detectors. In this instrument, the use of a highly stabilized source of X radiation is not required.

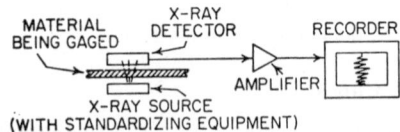

FIG. 7. Application of X-ray gage to continuous measurement of weight per unit area of sheet material.

An advantage of an X-ray gage is that the intensity of radiation from the source can be varied readily by changing the voltage across an X-ray tube. Also, the resulting beam of radiation is collimated easily so as to restrict the field to a small area. This permits the use of relatively high radiation intensities and wide gaging throats. As a result, X-ray gages can be applied to equipment, such as hot-strip steel mills which operate at relatively high ambient temperatures.

X-ray gages provide outputs indicative of weight per unit area of materials being measured. They are suitable for materials where this property ranges from about 0.5 to 100 g/sq cm. Maximum full-scale sensitivities of available gages range from zero up to 40 mg/sq cm. Gage costs fall in the $15,000 to $25,000 range.

Gamma Gages

Gamma radiation obtained from radioactive sources may be substituted for X rays, and is suitable for gaging a similar range of materials. The chief difficulty to be overcome before gamma gages are widely applied is to find a low-cost, reasonably long-lived source of gamma radiation with the proper energy. When suitable artificial radioisotopes, emitting radiation in the desired range, are available at economically feasible prices, gamma gages based on their use should find wide application.

Bremsstrahlung Gages

Bremsstrahlung or beta-excited X-ray gages can be used as a substitute or a replacement for conventional X-ray gages in certain measuring ranges. The bremsstrahlung gage has the advantage of providing a compact, simple, highly stable source of radiation requiring little or no maintenance. The format of this type of gage is similar to that shown for a beta gage (Fig. 8). Data on the maximum gage sensitivities, approximate costs, and normal gaging ranges of bremsstrahlung, beta, and radiation backscattering gages are given in Table 1.

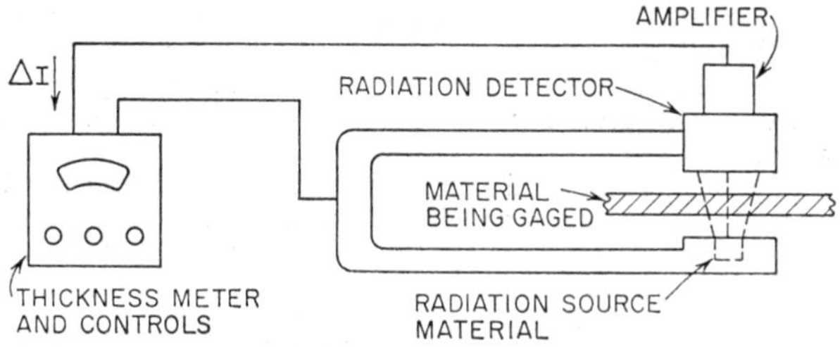

FIG. 8. Application of beta gage to continuous measurement of weight per unit area of sheet material. Basic equation:

$$\Delta I = I_0(1 - e^{-\mu t})$$

where ΔI = change in ionization current
I_0 = ionization current for $t = 0$
e = base of natural or Napierian logarithms (≈ 2.718)
μ = absorption coefficient
t = thickness of material being gaged

Table 1. Characteristics of Radiation-type Thickness Gages

Type of gage	Full-scale sensitivity, mg/cm^2	Approximate cost range	Normal gaging range in mg/cm^2*
BREMSSTRAHLUNG			
Strontium-90	0 – 40	\$5,000 to \$20,000	
BETA ABSORPTION			
Krypton-85	0 – 1.5	\$5,000 to \$15,000	
Cesium-137	0 – 3	\$5,000 to \$15,000	
Strontium-90	0 – 10	\$5,000 to \$15,000	
Ruthenium-106	0 – 10	\$5,000 to \$15,000	
RADIATION BACKSCATTERING			
Strontium-90	0 – 8**	\$5,000 to \$15,000	

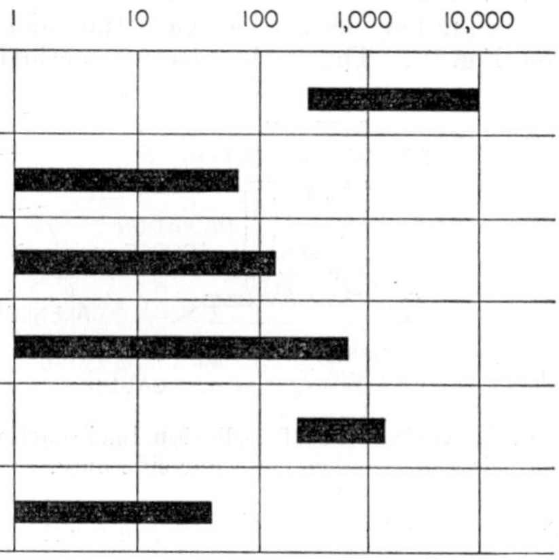

Notes: *If more than one type of source covers the desired range, then the final selection must be based on what other weights per unit area might have to be measured by the same gage or on the desired sensitivity.

**With standard source material. Full scale sensitivity can be increased to 0 – 3 mg/cm^2 with light material.

Beta Gages

The absorption of beta radiation in matter is more exclusively a function of the weight per unit area of the absorbing material than is the absorption of X rays or gamma rays. Thickness gages based on the use of radioactive sources emitting beta radiation have found wide application for the measurement of many sheet materials produced by the process industries.

The beta-absorption thickness gage, or more correctly weight-per-unit-area gage, normally consists of a source emitting beta radiation of the desired energy and an ionization chamber detecting this radiation after it has passed through the material being gaged (see Fig. 8). By the selection of the proper energy of radiation and the refinement of the detecting system for the highest possible stability, almost any desired sensitivity can be obtained. Where very thin materials are to be gaged, the variations of the density of the air surrounding the material between source and detector usually is the limiting factor.

Inaccuracies due to variations in the air column and other factors may be minimized by frequently standardizing the gage, either manually or by automatic means. In the standardizing system usually used, the material being gaged is removed from the gaging throat, and an adjustment is made to give a zero reading on the air column between the source and the ionization chamber. Although the beta gage is less affected by variations in the composition of the material being measured than other gages, the gage calibration should be made using material having the exact composition of that to be gaged. This will ensure maximum accuracy.

In order to minimize the effect of variations in the pass line or variations caused by vibration or flutter of the moving sheet, careful consideration must be given to the design of the source and the detector, to minimize inaccuracies resulting from these causes. Stability in the detecting apparatus usually requires temperature control of the critical circuit elements and a highly stable voltage supply for the gaging circuits. See Table 1 for additional information on beta gages.

Radiation Backscattering Gages

In the radiation backscattering thickness gage, radiation is directed away from the detector unit and the properties of the material to be gaged are evaluated by means of a measurement of the radiation which is scattered back toward the detector. This technique may be used to gage the thickness of sheet materials from one side only (see Fig. 9). The backscattering method usually is applied for the measurement of

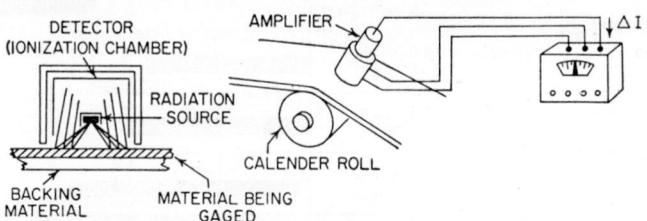

FIG. 9. Application of radiation backscattering gage to thickness measurement of sheet materials or coatings from one side only.

sheet material on calender rolls, or for the measurement of coatings. Since the amount of radiation scattered depends on the atomic number of the reflector, it is required that the sheet material to be gaged have a higher or lower atomic number than the calender roll, or in the case of coatings, than the base material. In order that no inaccuracies will be introduced by variations in the thickness of the calender roll or the base stock, it is necessary that they have a thickness well beyond the range of the radiation being used.

For additional data on radiation backscattering gages, see Table 1.

REFERENCES

For further related information on thickness-measurement principles, it is suggested that the reader turn to the following subsections of this handbook: Radiation Fundamentals, by J. D. Trimmer, p. 3–6 Nuclear-radiation Detectors, by R. K. Abele and D. C. Brunton, p. 3–16; Radioactive Isotopes in Instrumentation, by J. Kohl, p. 3–39; Acoustical Measurements, by E. G. Thurston and W. H. Peake, p. 3–83; and Parts Dimension Measurement and Control, by W. F. Aller, W. J. Darmody, and L. F. Polk, Jr., p. 6–4.

POSITION MEASUREMENT AND CONTROL*

By James R. Walker†

Performance of machine functions involves automatic operation of machine actuators and measurement of resultant machine member *positions* to an accuracy of four decimal places in many cases. The designer of each machine must employ means to appropriately position its movable members and reflect back to the machine logic-control system a quantity proportional to member position with respect to a given reference point. Occasionally, designers have based their choice of position control systems simply on expediency, availability, and simplicity. However, a study of machine characteristics functionally related to those of its position and data control system shows that it is more important for the choice of a system to be based on attainable accuracy, repeatability, reliability, and compatibility with machine components and environment.

MACHINE REQUIREMENTS

Typical requirements for various types of manufacturing equipment are shown in Table 1. Consideration of these data indicates that relatively coarse positioning

Table 1. Typical Machines with Positioning Systems

Machine	Data-storage medium	No. of controlled axes	Repeatability, in.	Total cycling time	Control components
Asphalt batching....	Manual tap switches	6	± 0.1 % of full-scale reading	2 min	Resolvers, transistor amplifiers, relays
Seat-cushion spring welding	Manual tap switches	5	± 0.015	35 sec	Resolvers, transistors, relays, controlled rectifiers, d-c motor
Armature insulator assembly	Card	3	± 0.005	72 sec	Resolvers, transistors, relays, controlled rectifiers, d-c motor
Drilling...........	Tape	2	± 0.001	Depends on number of holes	Potentiometers, tubes, relays, servo valve, hydraulic system
Drilling...........	Tape	2	± 0.001	Depends on number of holes	Potentiometers, transistors, relays, d-c motor, digital contactor
Boring............	Tape	6	± 0.0001	16 sec	Potentiometers, tubes, relays, d-c and a-c motors, magnetic-tooth transducer
Tube bending.......	Card	5	± 0.010	35 sec	Synchros, tubes, relays, hydraulic system
Frame welding......	Tape	2	± 0.010	Variable	Synchros, tubes, relays, hydraulic system
Contour...........	Tape	3	± 0.001	Variable	Synchros, tubes, relays, hydraulic system

* This subsection is based upon material originally published in *Automation*, vol. 8, nos. 4 and 5, April and May, 1961.
† Research Director, Gemco Electric Co., Detroit, Mich.

accuracy and repeatability can be tolerated in equipment, such as cutoff presses, paper trimmers, many handling mechanisms, arc-cutting machines, and tube-bending machines. Many present positioning and data repeat-back systems can satisfactorily meet the needs of such equipment with regard to both accuracy and repeatability. On the other hand, requirements for precision machines indicate a near inadequacy of present systems. Particularly is this true for machining of components for high-speed aircraft and missiles.

Thus the intended purpose for a given machine will be the first factor in determining the type of control system employed, i.e., whether it is to be used for a toolroom machine operated in a controlled environment, a tube-bending machine operated in a service garage, a selective die and cutoff machine for a big production plant, or a blending and mixing system to be operated in an outside location with widely varying environment. After the environmental requirements have been assessed, it is then necessary to consider the degree of required accuracy and repeatability afforded by a projected control system. Then, the final choice of control system for a specific machine application depends upon its reliability characteristics.

These control-system considerations—environmental dependency, accuracy, repeatability, and reliability—are common in varying degree to all machine applications. The principal factors in choosing a control system for one machine or another will lie in the repeatability or accuracy requirements. As the required accuracy becomes greater, it is to be noted that the complexity and extent of equipment necessary to attain this greater accuracy increases many fold.

GENERALIZED CONTROL SYSTEM

The degree of positioning accuracy achieved by positioned machine members is directly related to the characteristics of the principal system components responsible for their control; these are data input, position repeat-back, amplifier, and controlled drive elements. The dependency attached to each one of these components for overall machine accuracy and dependability can be understood from a consideration of the characteristics of each in greater detail.

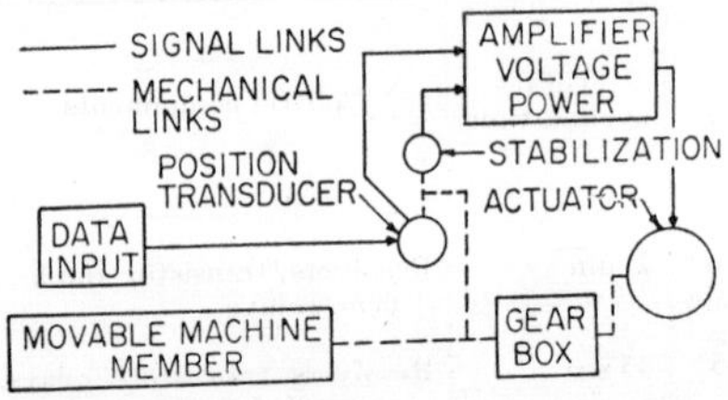

As a basis of common understanding for evaluation of machine control-system requirements, it is possible to represent such a system by a simple block diagram (see Fig. 1). This system contains a data-input device, consisting of either a mechanical or an electrical means, arranged to provide an electrical signal to the input section of an electromechanical transducer. As used in the control scheme shown, a mechanical input to this transducer is provided by the controlled member of the mechanical system. The error signal corresponding to the difference between the data input and the mechanical position of the controlled member exists as an output to the amplifier block.

Fig. 1. Block diagram of generalized control system indicates major elements which may be found in positioning control systems.

Appropriate amplification dictated by system accuracy and repeatability requirements provides excitation for the machine control actuator. Motion of the actuator shaft is converted to appropriate machine position by means of gears, or combinations of clutches and counterrotating shafts, and lead screws. Other systems might use hydraulic or pneumatic transducers, amplifiers, and/or linear actuators. A second or inner control loop is used to achieve dynamic stabilization of positioned machine members, and is necessitated by their inertial characteristics.

POSITION TRANSDUCERS

In order that a machine member can be positioned with an acceptable accuracy for any given set of requirements, it is first necessary to establish the extent of the backlash or dead-band region for the positioning mechanism used with this member. The measuring transducer or repeat-back element and its attendant dead-band characteristics, when employed with the machine member, will determine the amount of dead band or backlash to be included in the control-system loop. Thus, for all types of repeat-back devices whose mechanical input is provided by rotating a shaft, an important consideration is the means of coupling this device to the positioned machine member.

In most cases, a gear train is required to reduce member travel to one revolution or less of the transducer shaft. It is necessary to determine whether any backlash in this train is comparable, when expressed as an arc of the transducer shaft, to the positional accuracy requirement of the machine member. Satisfactory results on the basis of this comparison rest upon the assumption that the electrical and mechanical error factors for the transducer are small ($\frac{1}{10}$ or less) compared with the machine-member positioning tolerances.

If a transducer would have to derive its mechanical input through a gear train used for transmission of power to position a machine member, and this gear train includes a backlash factor large in comparison with required machine tolerances, a type of transducer not employing a mechanical input shaft should be considered. For those machines where it is impossible to utilize such other types of transducers, it is necessary to provide a separate rack and gear train, both exhibiting minimal backlash, to position the mechanical input shaft of the associated transducer.

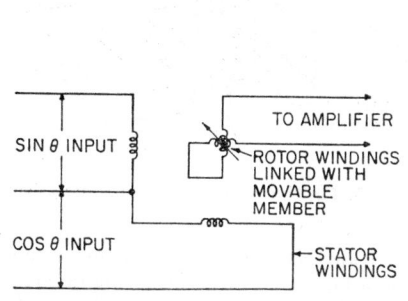

FIG. 2. Schematic circuit for a resolver-type position transducer.

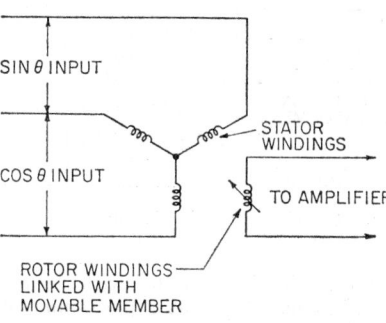

FIG. 3. Schematic circuit for a synchro-type position transducer.

Several common transducers, shown in Figs. 2 through 9, are the resolver, synchro, inductive plate, optical grating, toothed magnetic drum, digital contactor, digital magnetic drum, magnetic pin, magnetic tooth, and multiturn potentiometer. Typical circuits which may be used with these transducers are shown in Figs. 12 through 17.

Effect of Machine Characteristics on Selection of Transducer

Certain other machine characteristics influence greatly the choice of a positional transducer for use on a specific machine. Among these factors are:

1. The controlled member may have undesired movement with respect to the machine base, in a direction transverse to the controlled axis of travel. For example, if the slide and table ways wear nonuniformly, variation in transverse position of a point on the table of a machine may cause a variation in the air gap of a magnetic slot transducer system. This same machine problem results in misalignment of optical transducer systems if the table motion becomes crablike after wear of the slides has progressed.

2. Nondata components of both cyclic and random nature may be superimposed on the true data, because of machine-induced vibrations of the transducer.

3. Contaminants may interfere with transducer performance. For example, iron dust may enter the air-gap region of magnetic-lobe-type transducers; or the optical system for an optical transducer may be partially obscured.

4. Brush position may vary with respect to its segmental disk on a digital-drum-type transducer, owing to machine vibration. The result of this variation is the existence of false pulses of data input which ultimately cause improper machine positioning.

The individual mechanical and electrical characteristics of transducers being employed in present machines can be considered with regard to inherent accuracy, dispersion factors, repeatability, stability, and reliability. A proper evaluation of these factors leads machine designers to employ transducers on the basis of their functional merit.

Resolvers

Among the several shaft-type mechanical-input transducers available, the resolver is perhaps the most versatile because of the many possibilities available for electric input to it. A representative resolver used in machine control systems and a simplified circuit employed for supplying electrical data input to this device are shown in Figs. 2 and 12. A transformer employing a toroidal core configuration is suitably tapped to provide voltages proportional to the sines of various angles and other voltages proportional to the cosines of these angles. The angle is the equivalent of the desired mechanical displacement.

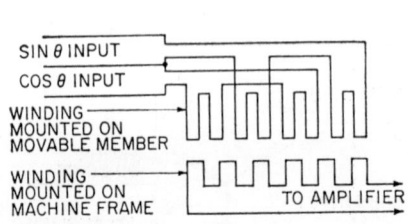

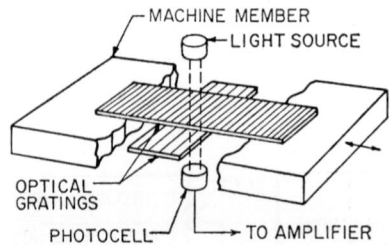

FIG. 4. Schematic circuit for a position transducer utilizing inductive plates.

FIG. 5. Position transducer utilizing optical gratings.

Positioning is controlled with the resolver for any coherent set of sine and cosine data inputs. Mechanical positioning of the data tap switch to a chosen digital position is provided either manually or from a tape-controlled stepping relay. The existence of an error voltage at the resolver output is determined by the difference in commanded position of the resolver shaft and its actual position. Motion of the controlled machine member rotates the resolver shaft until a minimum voltage exists across the rotor output terminals.

As used in the given control system, the resolver shaft is connected to the positioned machine member inside a closed control loop where it will be moved in a direction to reduce its error voltage essentially to zero. This minimal output will, in general, denote the exact machine position given by the input data.

The degree of accuracy attained by use of such a resolver system is influenced principally by inherent manufacturing tolerances and errors derived from harmonics generated within the resolver, due to unequal transformer tap impedance as the data transformer taps are inserted for various data angles. Typical error for a resolver exhibiting the most acceptable positioning accuracy of any commercial type now available is shown in Table 2.

Table 2. Typical Transducer Tolerances

Type	Manufacturer's tolerance
Synchro (3-phase)	4′ of shaft arc
Resolver	4′ of shaft arc
Optical grating	0.00001 in.
Magnetic toothed drum	15′ of shaft arc
Digital drum	35′ of shaft arc
Magnetic pin or tooth	0.00005 in.

False-data Region. The electrical diagram of a resolver in Fig. 12 discloses that the data input or stator winding of the resolver employs a double winding and is disposed in the manner of a two-phase motor stator such as that used with a two-phase induction motor. One point of interest in the electrical characteristic of this device is that the output voltage of the rotor winding varies in a sinusoidal manner with respect to the data readout position. Zero voltage will occur at the readout position corresponding to a chosen data position, and a zero voltage will also occur 180 shaft degrees from the chosen position. Thus, it is seen that the possibility for a false-data region exists, and that consideration must be given to this fact during design of the control system. In order to eliminate the possibility for false positioning of the machine member, the designer chooses a gear ratio for use between the machine member and its transducer which will only permit 180 mechanical degrees of transducer shaft rotation during full member travel.

Synchros

Unlike the resolver, a synchro stator utilizes a three-phase winding configuration, while the rotor either may use a single- or a three-phase winding. Shown in Fig. 3 is a typical synchro similar to that used in machine control systems. The sine and cosine voltages from the digital transformer are supplied to two of the three-phase

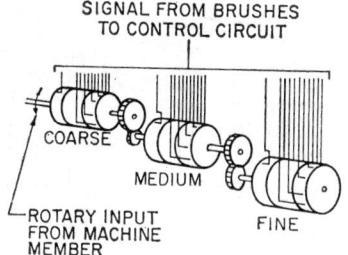

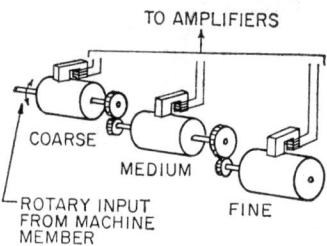

FIG. 6. Digital-contactor-type position transducer.

FIG. 7. Position transducer utilizing digital magnetic drums.

windings of the synchro in a manner similar to that for the two-phase synchro or resolver. In fact, the two types of units often are labeled synchros, with no distinction being made that the two-phase unit is a resolver.

The three-phase synchro and two-phase resolver are constructed in similar fashion, with accuracy and repeatability being determined by the impedance unbalance existing between the sine and cosine sections of the digital transformer. Use of a synchro exhibiting stator winding impedances which are large, in comparison to the internal impedance of the digital transformer, minimizes system error arising from induced harmonic currents due to unbalanced winding impedances in a digital transformer. The smallest deviation to be expected with a synchro exhibiting least possible error is given in Table 2.

Inductive Plates

The general effect of backlash in a transmission link between a transducer and a positioned machine member is to decrease the over-all positioning accuracy of the control loop. Since it is not possible in some machines to reduce the backlash of such transmissions, it is then necessary to utilize transducers not employing a rotatable shaft for mechanical data input.

A transducer using inductive plates is shown in Fig. 4. This device includes an etched stator winding which has been projected upon a dimensionally stable nonconducting surface by a photographic process. The rotor associated with this transducer is constructed in a like manner. Variations in inductor displacement are averaged over a large number of inductors by summing the voltages from a large number of like coil sides contained on the rotor plate. As a consequence, the reproduced rotor and stator inductors need not be printed with a positional accuracy equivalent to that of the final transducer.

A tap switch and double-winding transformer arrangement is used to supply sine and cosine voltages to the two windings contained in the stator plate. This is the same arrangement used with rotating resolvers and synchros. Essentially this transducer is a two-phase synchro or resolver whose windings have been projected onto a linear medium.

The advantage in use of the printed inductive-plate resolver is the elimination of gear backlash between the positioned machine member and the movable element of the transducer. A disadvantage in the use of this unit is the necessity for brushes and conductive bands to feed the output signal from the movable plate to the control system. Some difficulty results from poor contact in the brush-conductive band portion of the transducer, and is evidenced by electrical noise resulting in false-data input to the control-system amplifiers. Further, it is necessary to cover these transducer elements with a shroud which excludes all iron or other ferromagnetic material from proximity to their plate areas.

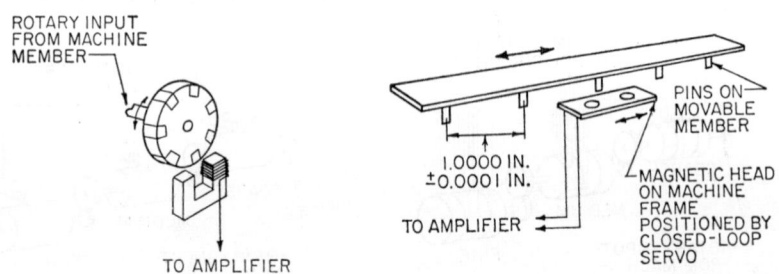

Fig. 8. Principle of position transducer employing a toothed magnetic drum.

Fig. 9. Use of magnetic pins in position transducer.

The accuracy of the printed inductive-plate resolver is equivalent to that of rotary-shaft synchros. Dispersion of data due to generation of harmonics caused by unbalanced impedances in the data transformer also is encountered with the inductive plate resolver. A related type of device is that using bifilar windings on a rod and sensing sleeve.

Pulse Devices

A general class of transducer devices used in machine control systems for measurement and position repeat-back utilizes pulse-producing transducer elements to convert displacement into pulses representing units of displacement. Transducer units included in this class are optical gratings, toothed magnetic drums, digital contactors, and digital magnetic drums.

Typical circuits of commercial units representative of the pulse devices being employed on machine control systems are shown in Figs. 5 to 8. Functioning of the pulse or digital transducers shown can be considered of two types:

1. Transducers which require counting of least-dimension units for machine-member positional measurement. An optical grating with a photocell and a toothed magnetic drum with a reluctance pick-off are examples of pulse-counting systems.

2. Transducers which use logic networks for readout of machine-member position. A digital contactor using a geared sequence of printed-circuit drums contacted by multiple brushes and a sequence of magnetic drum sections with reluctance pick-offs are examples of systems using logic networks to read out positions.

Accuracies for pulse transducers shown in Table 2 are comparable to those for synchros, resolvers, or inductive plates. However, as in the case of the inductive plate devices, optical gratings have an advantage in that they do not require a gear drive between the transducer and the positioned machine member.

Optical grating, toothed magnetic-drum, and digital magnetic-drum systems require use of an amplifier. Amplification is required to raise the power level of the

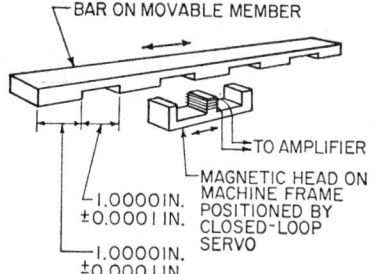

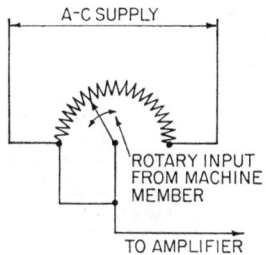

FIG. 10. Principle of the magnetic tooth as used in a position transducer.

FIG. 11. Basic circuit of a multiturn potentiometer as used in a position transducer.

pulses from the associated photocell or magnetic reluctance pick-offs to such a magnitude that they may be used to operate relays in conjunction with other relay logic elements.

Digital contactors display the advantage that they do not require any auxiliary relay or amplifier circuit. However, the digital contactor has the distinct disadvantage that it must employ at least one printed-circuit drum and brush combination,

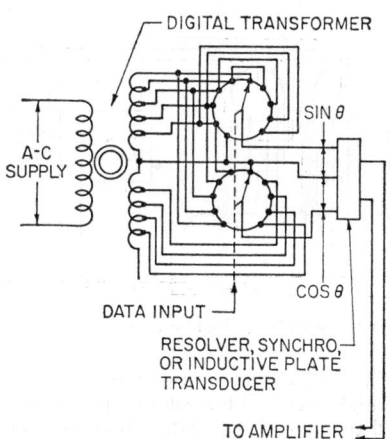

FIG. 12. Digital transformer circuit for resolvers, synchros, or inductive plates.

in which the drum rotates at 1,000 times the input shaft speed of the transducer if the unit is to have positional accuracy comparable with other transducers. In certain machines, this has required the high-speed drum to be rotated at 16,000 rpm or more. As a consequence of this operation, such units have exhibited unacceptably short life. Designers of digital drum-brush transducers have attempted to minimize this difficulty by use of a brush-lifting means designed to remove the brush from the high-speed drum during coarse positioning of the transducer shaft. Generally, unacceptable reliability of operation has attended use of this type of transducer in control systems involving four significant digits or more.

Magnetic Pin or Tooth Transducers

Magnetic-tooth and magnetic-pin transducer systems as presently being used for machine control systems are shown in Figs. 9, 10, and 16. Comparison of positional resolutions for these transducers and other types is given in Table 2 and indicates that greater resolution can be expected from magnetic pin or tooth units.

The magnetic pin or tooth transducer has a useful range of ± 0.150 in. from the readout point. Thus, it is necessary that a transducer of this nature be employed

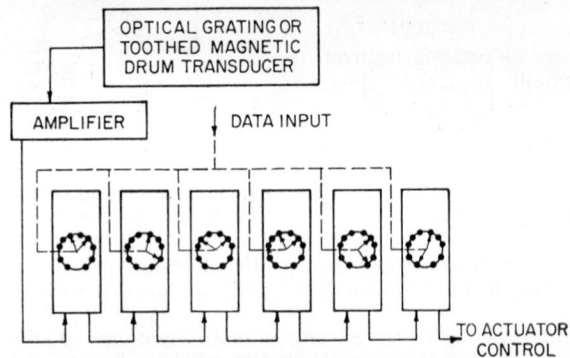

Fig. 13. Tap-switch circuit for optical gratings or toothed magnetic drums.

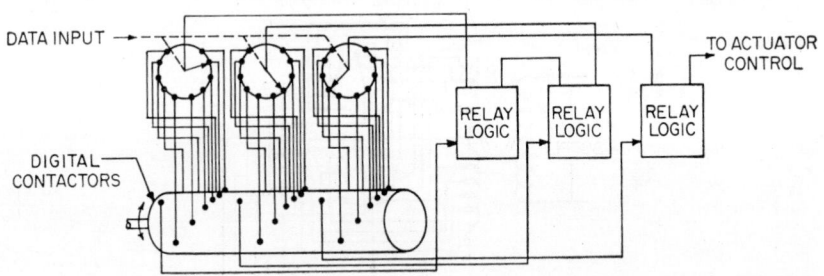

Fig. 14. Switch and relay logic circuit for digital contactors.

with others such as synchros, which can position the machine member within the useful ± 0.150-in. range of the high-speed high-resolution pin or tooth system. A switching means must be provided to successively switch from one transducer to another.

Potentiometers

Multiple-turn potentiometers are being used as position transducers in machine control systems that do not require greater accuracy than the positional equivalent of ± 0.001 in. Circuitry for such a transducer and an electric circuit commonly employed for inserting digital data into the transducer system are shown in Figs. 11 and 17, respectively. An obvious disadvantage of the potentiometer is unreliability related to frictional wear in the slidewire arrangement characteristic of its fundamental design.

TRANSLATION OF TRANSDUCER ERROR SIGNALS

Accuracy of positioning cannot be expected to be better than the accuracy of the error signal obtained from the position transducer. Though this is true, a machine member cannot be positioned accurately, regardless of the transducer used, unless the error signal can be properly translated into movement of the machine.

Transducer Voltage Amplifiers

The generalized control-system block diagram (Fig. 1) shows the use of a voltage amplifier following the transducer in the signal flow direction. The purpose of this unit is to increase the voltage and power level of the transducer electric signal so that it can be used to control a higher-power-level amplifier. Generally speaking, the voltage amplifier should be characterized by the following:
1. High magnitude of voltage gain (10,000 to 100,000).
2. Low distortion for control signals lying in the null region (0.5 to 2.5 mv).
3. Stability of gain and linearity characteristics with changes in ambient temperature.

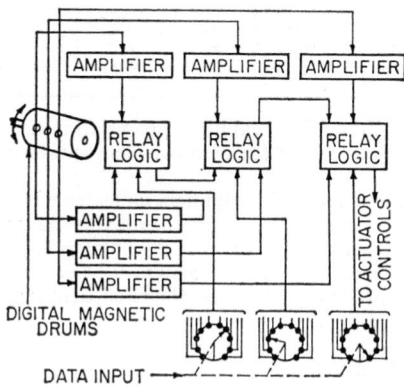

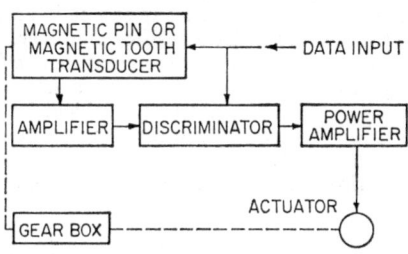

FIG. 15. Switch and relay logic circuit for digital magnetic drums.

FIG. 16. Amplifier and discriminator circuits for magnetic pin and magnetic tooth transducers.

4. High reliability (capable of continuous operation without failure for at least 10,000 hr).
5. Low internal temperature rise, and no requirement for external cooling.

Considering the applicability of these characteristics to all classes of machine control applications leads to the conclusion that each characteristic is important in determining the acceptability of a given machine and its control system. The factors of stability and reliability are especially important for machines which must be maintained in a controlled environment, in order to maintain an accurate relationship of one machine section to another. Although a voltage control amplifier cannot be chosen for use in a machine control system strictly on the basis of its stability and reliability characteristics, these factors are important in determining ultimate application and usefulness of the control systems.

Types of Voltage Amplifiers. Various types of voltage amplifiers used in machine control systems employ control elements exhibiting widely varying characteristics. These voltage amplifiers, differentiated by their principal control elements, are: (1) transistor amplifiers, (2) vacuum-tube amplifiers, (3) magnetic amplifiers, (4) combinations of transistor and magnetic amplifiers, or vacuum-tube and magnetic amplifiers. The data contained in Table 3 indicate fundamental characteristics of these amplifier systems.

Basic differences existing between the systems are in the factors of stability, reliability, and temperature rise. Another difference is in the necessity for external power-supply equipment. In regard to the first three factors, the inherent stability of the vacuum-tube amplifier, when operated under a widely varying environment, is acceptably greater than that of either transistor or magnetic amplifiers. However, a comparison of reliability data for the listed amplifiers would indicate less acceptability for the vacuum-tube amplifier than for the others.

Effects of Temperature Changes. Operation of transistor amplifiers in the environmental range 13 to 43°C results in negligible drift or deviation of control characteristic. In magnitude, the drift is comparable to that for a vacuum-tube amplifier

Table 3. Typical Voltage-amplifier Characteristics

Type	Response time, sec	Voltage gain	Power gain	Auxiliary power supply	Gain change (0–70°C), % drop	Supply frequency, cps
Transistor...................	Carrier rate	10,000	2×10^7	No	32	60
Vacuum tube.................	Carrier rate	22,000	5×10^{10}	Yes	None	60
Magnetic amplifier...........	0.075	10,000	2×10^7	No	4.7	60
Transistor–magnetic amplifier....	0.022	17,200	3.5×10^8	No	6.8	60
Vacuum-tube magnetic amplifier..	0.016	18,900	3.8×10^9	Yes	None	60

operated in a much wider temperature range. To achieve stabilized operation in this restricted thermal environment, it is necessary to utilize temperature-compensating networks within the transistor amplifier.

Consideration of unit temperature rise for the three basic amplifiers indicates minimal rise to be experienced with a transistor or magnetic amplifier, probably never greater than 2°C above room ambient for multielement systems. A rise of 60 to 70°C may be observed for multi-vacuum-tube amplifiers. In certain vacuum-tube control systems employing between two and three hundred vacuum tubes, the cited temperature rise can easily be experienced in a closed cabinet.

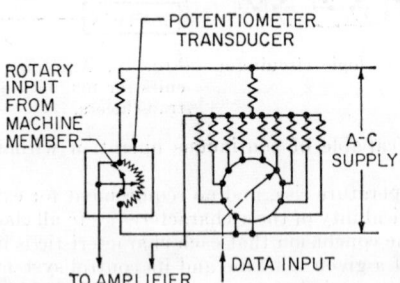

Fig. 17. Decade switch circuit for multiturn potentiometer.

To properly protect control components such as capacitors, resistors, and transformers from this thermal environment in vacuum-tube amplifiers, it is necessary to lower the temperature within the enclosure by means of a filtered air-cooling system. Although such a cooling system appears to solve the related thermal problem when control equipment is initially installed, it usually follows that improper attention is given at a later time to installation of new air filters. Soon maintenance personnel are simply leaving the control enclosure open to obtain natural cooling, and as a result, the equipment becomes covered with residue from associated processes and later fails to operate reliably. From this consideration, it would appear that a transistor or magnetic amplifier, or a combination of the two, offers certain advantages in system operation, providing stable gain is assured in the thermal environment of machine operation.

One further difference indicated by Table 3 is that a full-wave high-gain-type magnetic amplifier exhibits a large time lag when operated at a 60-cps carrier frequency. Its inherent lag can approach 25 to 30 cycles of supply frequency when used in a multistage amplifier capable of a voltage gain of 10,000. As a consequence, this amplifier cannot be used in high-gain control systems to be operated at 60-cycle carrier frequency. If a higher-frequency carrier system can be employed with the machine control system, such as 400 cps, for example, then the response time of 25 to 30 cycles is acceptable.

Power Amplifiers and Actuators

Control-system designers have sometimes questioned the necessity for a division of system amplification means into two sections, i.e., voltage amplifier and power amplifier. However, good reasons for this division rest upon the electrical isolation of circuits for high gain at high impedance from sections of the amplifier controlling larger power. Such isolation of one unit from the other has eliminated the possibility of unwanted or false electrical-signal components combining with control-data components. A division has accordingly been effected, in the 0.1- to 0.2-watt power region, between the output power level of a voltage amplifier and that of a power amplifier. Each of the power-amplifier systems listed in Table 4 utilizes a voltage amplifier of the types found in Table 3. The block diagrams shown in Figs. 18 to 21 indicate the arrangement of several power-amplifier systems.

Table 4. Typical Power-amplifier System Characteristics

Type	Electric	Electric	Electric	Hydraulic	Pneumatic
Components...	Transistor or vacuum-tube amplifier, relay-controlled constant-speed motor	Magnetic or vacuum-tube amplifier, amplidyne generator, d-c motor	Transistor, magnetic, or vacuum-tube amplifier, constant-speed motor, double clutch	Magnetic, transistor, or vacuum-tube amplifier, proportional valve, hydraulic motor or cylinder	Magnetic, transistor, or vacuum-tube amplifier, proportional valve, hydraulic motor or cylinder
Rating........	4 hp	3 hp	3 hp	5 hp	80-lb force
Speed ratio....	2:1	50:1	50:1	400:1	260:1
Time constant, sec..........	0.27	0.2	0.2	0.035	0.080
Drift factor, in./hr	Depends on voltage amplifier	Depends on voltage amplifier	±0.00009	±0.00007	±0.00005
Auxiliary power supply	None	Drive for generator	None	Hydraulic power source	Pneumatic power source
Projected life, hr..........	30,000	10,000	10,000	8,000	8,000
Approach to find position	Controlled direction	Random servo	Random servo	Random servo	Random servo
Approximate cost........	$1,100	$1,900	$1,500	$2,300	$1,200

One of the factors to be considered in the choice of a power actuator and amplifier is the "slewing rate," which is a measure of the ability of a closed control loop to position a member, following input data, with a minimum of lag in position and time. This is one of the power amplifier characteristics known to contribute appreciably to system accuracy and stability, especially in machines continuously positioning movable members according to continuous data command.

As shown in Table 4, the system using a proportional valve with a hydraulic motor provides the greatest dynamic response. For use in a system such as that for controlling contouring machines, the hydraulic power amplifier offers a greater degree of responsiveness than is useful on most machines, because of their inherent inertial properties. Many discrete cycle-positioning systems, as used in table positioning for boring and drilling machines, jaw and anvil positioning in tube-bending machines, and press-head positioning for a multiple-operation press, do not require minimum cycling time for member positioning. Here again, a hydraulic power amplifier offers better response than is useful with such systems.

Successful utilization of high-performance hydraulic power amplifiers in machine control systems requiring high repeatability and reliability is dependent on the initial

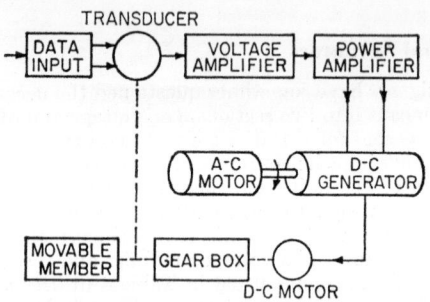

Fig. 18. Electric power-amplifier system, using the position error signal to control output of a d-c generator to a d-c motor.

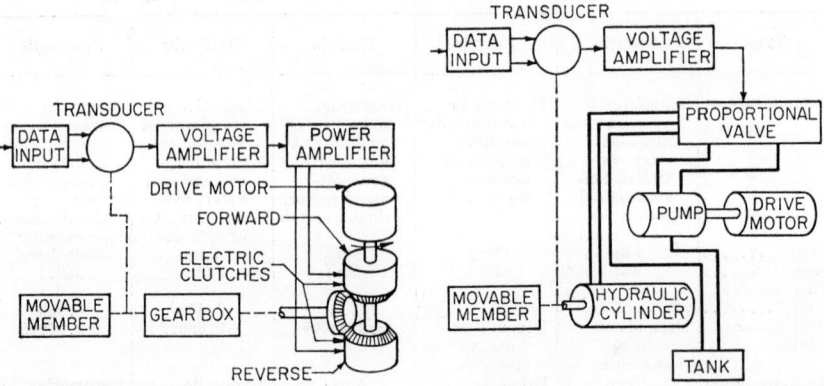

Fig. 19. Electric power-amplifier system, using the error signal to control opposed electric clutches.

Fig. 20. Hydraulic power-amplifier system, using the error signal to control a proportional valve.

conditions of design and adjustment remaining constant during the useful life of the machine. Highly unstable operation, resulting in oscillation of the positioned member and loss of repeatability, often occurs when machine inertial or frictional components change during the life of the machine.

Although accuracy equivalent to that of the hydraulic system is possible with an electric power amplifier, the time response and "settling time" required for the electric motor to finally position its machine member is many times longer. For controlling a machine member characterized by large inertia in a discrete positioning cycle, there is no advantage in using a hydraulic motor due to the low responsiveness of such a machine member. Consequently, an electric-motor control system can be utilized with its characteristically higher reliability and stability at a lower initial cost.

Comparison of Pneumatic, Electric, and Hydraulic Actuators. A comparison of this type reveals the responsiveness of the pneumatic actuator to be much the same as that of electric types. Further consideration of the pneumatic system reveals the necessity for pressure controls and filter systems similar to those needed with hydraulic control systems.

Choice of a power amplifier for use in a machine control system is based on the same factors as cited for the control voltage amplifier, i.e., required stability, reliability, and response time. The choice of a power actuator for a given control system will involve these same factors.

Another factor to be considered in the choice of a power amplifier and actuator motor for machine control is the backlash initially present in the link between the

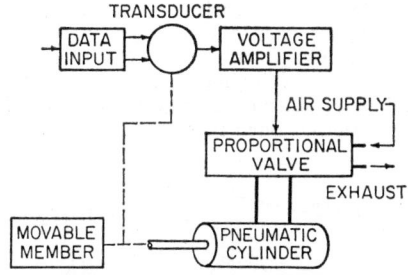

FIG. 21. Pneumatic power-amplifier system, using a proportional valve.

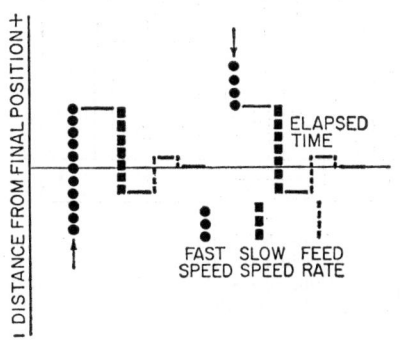

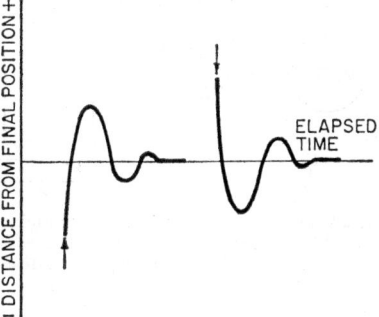

FIG. 22. Typical response for a system using a limited number of fixed positioning speeds and always approaching the final position from a given direction.

FIG. 23. Typical response for a system, using an infinite number of speeds, with the rate of approach proportional to the error and with random direction of approach to the final position.

actuator motor and the positioned member. If the initial design of the machine does not warrant the use of a zero backlash system of gears and lead screw in the drive link, then it is necessary to use a controlled positioning of the machine member so that it always approaches its final position from the same direction at the same speed.

If the necessity exists for minimization of positioning time, a lower settling time will result if a servo approach using variable speed and direction is made to final member position (see Figs. 22 and 23). In the servo approach, the member is positioned at a speed depending on error. Both controlled-direction-approach and random-servo-approach positioning cycles are used in point-to-point systems. However, only a refined random-servo-approach cycle can be used with continuous-positioning systems.

SUPPLEMENTAL CONTROL FACTORS

Positioning-system stability and reliability are also affected by control loops within the principal control loop shown in Fig. 1. These inner loops or inside control blocks are employed for: (1) switching from one transducer to another in accordance with size of control error, (2) anticipating arrival at command point in point-to-point positioning systems, (3) providing constant torque or force within the actuator motor loop to maintain positioning accuracy, and (4) supplying stored data according to the requirements of the control loop.

Switching of Transducers

Values of resolution obtainable for various transducers, as shown in Table 2, are representative of those possible from a single unit operating with a low gear ratio between the positioned machine member and the transducer shaft. Usually an arrangement of transducer and driving member closely geared together will result in ambiguous data transmission from the transducer, by virtue of its rotation through a multiple number of revolutions. Such rotation is permissible if at least one trans-

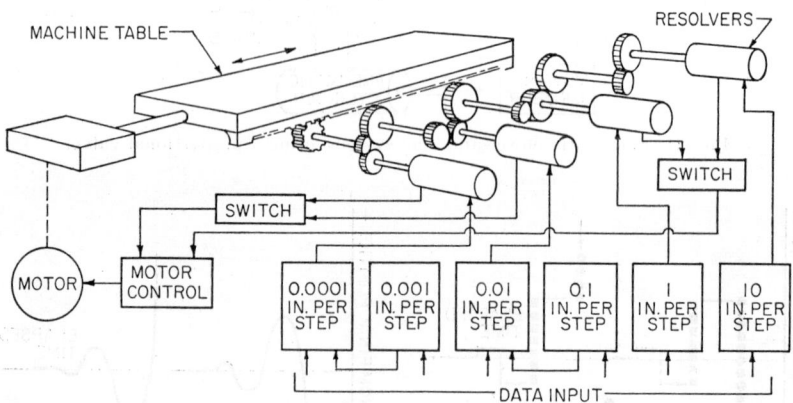

Fig. 24. Four-resolver-position transducing system designed to give a positioning accuracy of ±0.0001 in. without rotating the coarse resolver more than 180°. By keeping that resolver always within its relevant half-cycle of rotation, this system avoids ambiguity. The control system switches to successively finer resolvers as the command position is approached.

ducer of a multiple-transducer system is always kept within its relevant half-cycle of rotation. It is customary for the control-system designer to use a number of transducers and associated decades with a means suitable for switching from one to another. In this manner, it is possible to ensure final selection of the proper half-revolution of the high-speed transducer to achieve desired accuracy.

The block diagram of Fig. 24 shows an arrangement for a multiple-transducer control system to accomplish switching from one transducer control system to another in the basic control loop. This four-resolver system is designed to provide a positioning accuracy of ±0.0001 in. Such a switching system is used in most multiple-transducer systems. Switching from one transducer to another is based on coarse transducer amplitude and is accomplished with relays or diodes.

Two relay switching arrangements are shown in Figs. 25 and 26 and a diode switching circuit in Fig. 27. The circuit in Fig. 25 is for use in a point-to-point positioning system, while the other two circuits are designed for continuous-positioning systems such as those used in contour grinding or milling machines. An advantage of the circuit in Fig. 25 is that it is incapable of oscillating between coarse and fine control during a closed-loop positioning cycle of a machine member. This follows from the fact that, once the amplitude of the coarse transducer has fallen below a preset switch-point value, the coarse transducer is then prevented from commanding the positioning system until another over-all positioning cycle has been initiated.

The switching means of Figs. 26 and 27 allow the coarse transducer to return to command of a positioning system whenever its amplitude exceeds a preset level. For normal functions within a continuous-data system, this type of circuit action is acceptable. The positioned member always will be at or near the command data position, and thus the command will always be given by the fine data transducer. If, however, the positioned member should become temporarily restricted in motion, or if its power is interrupted, then it is possible for the error signal to become large enough so that the command will be taken over by the coarse resolver.

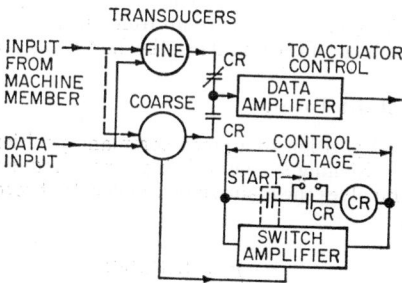

FIG. 25. Typical switching circuits used to transfer control of a positioning action from transducers for coarse increments to transducers for fine increments. Switching from one transducer to another usually is controlled by the coarse transducer.

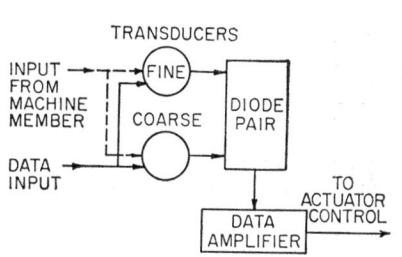

FIG. 26. Relay switching circuit for continuous positioning.

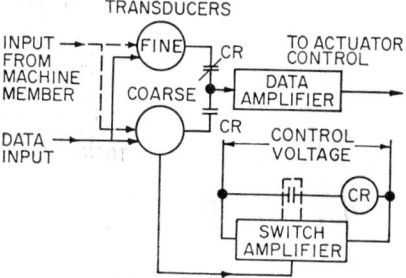

FIG. 27. Diode switching circuit for continuous positioning.

In normal operation, the changeover from the coarse to fine resolver results in no system-position instability. However, it is possible, after some change in machine parameters—friction, backlash, and so on—during the life of the machine, that the originally adjusted gain factors for the coarse and fine data system may be completely unsuitable. As a consequence, an unstable condition characterized by a switching back and forth from one transducer to the other can arise. This condition can be eliminated by a readjustment of the gains in coarse and fine loops to values corresponding to the new parameters of machine operation.

Use of digital-type transducers with multiple-decade preset counters can make it unnecessary to use multiple transducers and their attendant switching means. The elimination of additional components is an advantage; however, the fundamental counting inaccuracy which may be obtained with most electronic counters is a disadvantage.

Command-point Anticipation

Systems for point-to-point positioning of machine members, in drilling, boring, paper cutoff, tube bending, assembly machines, and similar applications, frequently employ two fixed positioning motor speeds and possibly two traverse gear ratios. Such systems require an anticipation signal to reduce traverse rate of the positioned member before the command point is reached.

Variation of the anticipation point with respect to the command point can result in traverse speed reductions being made too near to the command point, and the final accuracy of positioning may be reduced. The amount of such variation in anticipation point can change with input data, and depending on the data elements being used, this effect can be quite troublesome. The principal cause of unplanned variation in anticipation point is the existence of harmonic currents in synchro or resolver transducer windings. Such harmonic currents result from unequal data transformer impedances in the exciting windings of these units as the data transformer taps are switched.

Maintenance of Constant Torque

The closeness of actual position to a given command position, whether a point-to-point or a continuous-positioning system is being considered, depends on still another stability factor. This involves the maintenance of constant torque at the output shaft of the actuator motor.

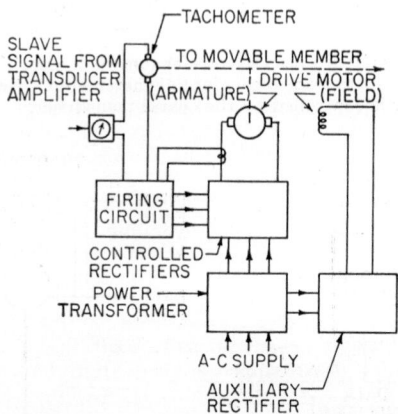

Fig. 28. Block diagram outlining a power-amplifier and motor-actuator circuit to provide constant torque output to the movable member. An arrangement of this type is used to maintain positioning accuracy, as factors like frictional damping change with wear of machine components.

When a machine is new, an electric or hydraulic actuator is adjusted to deliver constant torque to the gear train of the machine member being positioned and for the machine parameters existing with this operation. If the frictional damping component associated with the machine member becomes less as it is subjected to wear, then the loop containing the power amplifier and motor actuator may become unstable and require a reduction of loop gain or an increase of velocity stabilization. The general effect of instability within the power-amplifier–motor-actuator loop is a reduction of accuracy in final positioning of the associated machine member.

The block diagram in Fig. 28 shows a typical power-amplifier–motor-actuator circuit. It effectively provides constant torque at any input or command signal from a selected small minimum value to a chosen maximum. The over-all ratio between maximum and minimum lies between 100:1 and 400:1, depending on the class of control actuator utilized.

Buffer Data Storage

A continuous-positioning system requires an information or data feed rate which is dependent upon its assimilation rate, i.e., the rate of positioning allowed by the process associated with a given machine member. If the process rate varies, an unstable data requirement will be reflected to the data storage system. For example, a contouring machine cutter may not be fed into a workpiece with a uniform force because of an instability within the power-amplifier–actuator loop. This will result in a second instability due to nonuniform data delivery. In turn, this instability will be superimposed upon the first, finally resulting in the machine cutter removing scallops from the workpiece. Such instabilities can be removed by adjustment to compensate for changes in condition of the cutter and changes in frictional losses of associated positioning means.

Multiple-axis control systems, which require correlation of the data and command-point approach for all axes, can be subject to interloop instability resulting from the causes discussed for single-axis loops. Usually, this interaxis instability can be removed by stabilization of each control loop in an individual axis. The process of stabilization must involve rate or amplitude adjustment in all principal and subcontrol loops of the control system, in order to compensate for reduction of frictional components with life of the system.

COST FACTORS

An important consideration, especially for machines to be used in nonmilitary operations, is the cost of the power actuator associated with each controlled axis of the machine. Since the cost of power components for a control system can approach half the cost of the entire system, the control means utilized in this section of the system can serve as a criterion of ultimate cost.

This relative cost of the actuators listed in Table 4 varies greatly and can be compared on the basis of two factors only: (1) ratio of speed variation; and (2) response time. To apply these criteria for cost, Table 4 shows that the hydraulic system can provide a speed ratio of as much as 400:1 with a response time of 20 msec; while a d-c motor-generator system can give a speed ratio of 100:1 and a response time of 350 msec. So it is seen that a highly responsive control capable of positioning a light machine member at a high rate and over a wide variation of rates will, in turn, cost much more than a less responsive control.

Although both controlled direction and servo approaches can be used in point-to-point positioning systems, there is a cost advantage to be gained by using the controlled-direction approach. The initial cost of the servo approach is much larger because of the higher cost of proportional components, such as hydraulic control valves, hydraulic pumps, amplidyne generators, and similar components, as compared to the cost of nonproportional relays. It should be noted, however, that the single-loop approach cycle, which offers a shorter positioning time than that of Fig. 22, requires the same components as the system of Fig. 23.

Another factor which importantly determines control-system cost is the required machine accuracy. It has been shown that the repeat-back accuracy of a control system is proportional to the number of transducers used in the system, and to the stability of their associated switching means and error amplifiers. Investigation of the dependency of machine and control-system cost upon a repeatable accuracy factor indicates that the cost varies approximately as the 2.4 power of accuracy requirements. Hence, it is well to consider this relationship when specifying machine and control requirements, so that the accuracy of the machine does not exceed practical requirements of anticipated production.

A GENERAL POLICY FOR SELECTION OF POSITIONING CONTROLS

The preceding discussion has been devoted to a description of control-system components and their characteristics known to contribute to total machine accuracy, stability, and reliability. An evaluation of the construction and characteristics of a modern programmed machine should be as much concerned with the control system and its components as with the mechanical portion of the machine.

As the individual concerned with final machine and system selection becomes more familiar with the characteristics of various control systems and the advantage of one over another for a specific application, it will be possible for him to choose more exactly that machine and control system which will meet most adequately the conditions of a production requirement.

It is with this in mind that the responsible individual can be advised to choose a control system, including a machine-member positioning means, which will accomplish those multiple machine functions *just necessary* for the production requirement. Choice of a control system including complex functions not necessary to the production program can lead to unwarranted maintenance costs, and maintenance costs increase in more than direct proportion to the complexity of control equipment included with a given machine.

REFERENCES

1. Mallock, R. R. M.: An Electrical Calculating Machine, *Proc. Roy. Soc. London*, Series A, vol. 140, p. 457, 1933.
2. Mynall, D. J.: Electrical Analogue Computing, *Electron. Eng.*, vol. 19, p. 178, June, 1947.
3. Bell, J.: Some Aspects of Electrical Computing, *ibid.*, vol. 23, p. 264, July, 1951.
4. Alexander, W. H.: A Judgment Box, *ibid.*, p. 256.
5. Ridenour, Louis N. (editor-in-chief), and George B. Collins (deputy editor-in-chief): "Massachusetts Institute of Technology Radiation Laboratory Series": Vol. 17: John F. Blackburn (ed.): "Components Handbook," p. 340, 1948; vol. 21: Ivan A. Greenwood, Jr., J. Vance Holdam, Jr., and Duncan MacRae, Jr. (eds.): "Electronic Instruments," p. 372, 1948; vol. 25: Hubert M. James, Nathaniel B. Nichols, and Ralph S. Phillips (eds.): "Theory of Servomechanisms," p. 261, 1947, McGraw-Hill Book Company, Inc., New York.
6. Karplus, W. J., and W. W. Soroka: "Analog Methods," 2d ed., McGraw-Hill Book Company, Inc., New York, 1959.
7. Cooney, J. D., and B. K. Ledgerwood: "Numerically Controlled Machine Tools, *Control Eng.*, vol. 5, nos. 1, 2, 3; January, February, March, 1958.
8. Brewer, R. C.: The Numerical Control of Machine Tools, *Engrs. Dig.*, September, 1958; January, 1959.
9. Numerical Control of Machines, *Am. Machinist*, vol. 104, no. 17, p. 167, 1960.
10. Brewer, R. C.: Recent Developments in the Numerical Control of Machine Tools, *Engrs. Dig.*, August, 1961.
11. "Manual of Digital Techniques," McGraw-Hill Book Company, Inc., New York, 1959.

Section 7

PHYSICAL PROPERTY VARIABLES

By

L. E. CUCKLER, *Product Manager, Industrial Instruments, Aeronautical and Instrument Division, Robertshaw Controls Company, Anaheim, Calif.; Senior Member, Instrument Society of America.* (*Moisture Content of Materials*)

GERALD L. EBERLY, B.S. (M.E.), *Technical Staff, Harris D. McKinney, Inc., Philadelphia, Pa.; Member, Instrument Society of America.* (*Fluid-density and Specific-gravity Measurement*)

W. F. HICKES, B.S.E. (M.E.), *Assistant to Chief Engineer, The Foxboro Company, Foxboro, Mass.; Member, Instrument Society of America; Registered Professional Engineer (Mass.).* (*Humidity and Dew Point*)

RICHARD MUMMA, B.S. (M.E.), *Product Manager, Fischer & Porter Company, Warminster, Pa.; Member, Instrument Society of America, American Institute of Mechanical Engineers.* (*Viscosity and Consistency*)

CONTENTS

7-3

CONTENTS 7–3

Conversion Table and Formulas 7-30
Conversion Table of Moisture Content to Moisture Regain or Air-dry Basis (Table 1). 7-30
Formulas for Moisture Determinations and Conversions 7-30
Formula for Two-stage Drying 7-30
Methods of Moisture Determination 7-31
Thermal (Oven) Drying 7-31
Techniques 7-32
Advantages and Limitations of Oven Drying . . 7-32
Condensed Summary of Drying Techniques for Various Materials (Table 2). 7-33
Distillation Methods 7-34
Techniques 7-34
Advantages 7-34
Limitations 7-34
Chemical-reaction Methods 7-35
Electrical Methods. 7-35
Conductivity Types. 7-35
Capacitance Type 7-36
Humidity-type Measurement. 7-36
Infrared Absorption 7-37
References. 7-37

VISCOSITY AND CONSISTENCY. 7-39

Measurement Theory—Rheology Basics 7-39
Newton's Basic Definition of Viscosity 7-39
Hagen-Poiseuille Law 7-40
Rheology Basics. 7-40
Newtonian Liquids 7-41
Non-Newtonian Liquids 7-41
Bingham Body (True Plastic) 7-41
Pseudoplastic 7-41
Dilatant 7-41
St. Venant Body 7-42
Thixotropic 7-42
Rheopectic 7-42
Identifying Thixotropic and Rheopectic Fluids . 7-42
Apparent Viscosity 7-42
Consistency. 7-42
Units of Viscosity and Viscosity Scales 7-43
Dynamic (Absolute) Viscosity Units 7-43
Kinematic Viscosity Units 7-43
Viscosity Index. 7-43
Viscosity Scales 7-43
Effects of Temperature and Pressure on Viscosity. . 7-44
Temperature 7-44
Pressure 7-45
Viscosity Conversion Factors (Table 1). . . . 7-45
Applications of Viscosity Measurement. 7-45
General Application Guide 7-45
Industrial Applications 7-46
Petroleum 7-46
Chemical 7-46
Food 7-46
Fuel Combustion 7-46
Laboratory Applications 7-47
References. 7-47
Selection Guide for Viscosity and Consistency Instruments (Table 2) 7-48

FLUID-DENSITY AND SPECIFIC-GRAVITY MEASUREMENT

By Gerald L. Eberly*

In industry, the terms *density* and *specific gravity* for fluids characterize the same physical property of the fluid. Hence the two terms are often confused, even though numerical values for the two with a given fluid can be quite different, as defined below. This section gives definitions, common units of specific gravity, uses, and means for measurement and control of this variable.

DEFINITIONS

Density is defined as the mass per unit volume. It is usually expressed, for liquids or gases, in units of grams per cubic centimeter, pounds per cubic foot, or pounds per gallon.

Specific gravity is the ratio of a fluid's density to the density of water, for liquids or solids, or to the density of air for gases or vapors. Ideally, values for specific gravity should always be accompanied by a ratio of two temperatures:

$$1.52^{20°C/10°C} \text{ (often given in tables as } 1.52^{20/10}\text{)}$$

This means that the specific-gravity value represents the ratio of the material's density at 20°C to the density of water at 10°C. Notice that, being a ratio, specific gravity has no units, such as pounds per gallon, as does density.

In practice, specific-gravity figures are often not accompanied by such temperature values. For liquids the figure then means the ratio of the material's density at its operating temperature to the density of water at 4°C. Because the density of water at 4°C is 1.0000 g/cu cm, a value of specific gravity in metric units (cgs) is then numerically equal to the operating density in metric units.

For gases or vapors, the interpretation of specific gravity without given temperature values is quite different. In such cases, a given value means the ratio of the gas density to air density—both at standard conditions of pressure and temperature (generally 60°F and atmospheric pressure). For perfect gases, such a value would also be equivalent to the ratio of densities at any set of identical temperature and pressure conditions, or the ratio of the molecular weight of the gas to the molecular weight of air.

In practice, many gases cannot be assumed to follow the perfect-gas laws to apply corrections for obtaining specific-gravity values at operating temperature from data given for standard conditions. Reference must be made to tables of empirical data.†

COMMON SPECIFIC GRAVITY UNITS

API (American Petroleum Institute). Standard for petroleum products in the United States.

$$\text{Degrees hydrometer scale (at 60°F)} = \frac{141.5}{\text{sp gr}} - 131.5$$

---

* Technical Staff, Harris D. McKinney, Inc., Philadelphia, Pa.
† Such as C. D. Hodgman, "Handbook of Chemistry and Physics," Chemical Rubber Publishing Company, Cleveland, Ohio; Perry's "Chemical Engineers' Handbook," McGraw-Hill Book Company, Inc., New York; or "International Critical Tables," McGraw-Hill Book Company, Inc., New York.

Balling. Used principally to estimate the per cent wort in the brewing industry, but also used to indicate per cent by weight of either dissolved solids or sugar liquors. Graduated in per cent by weight at 60°F or 17.5°C.

Barkometer. Used in the tanning industry. Water equals zero, and each scale degree equals a change of 0.001 in specific gravity.

$$\text{Specific gravity} = 1.000 \pm 0.001n$$

where n equals degrees Barkometer.

Baumé. Widely used to measure acids, light and heavy liquids, and syrups. In 1904, the National Bureau of Standards (U.S.) adopted the following as standard Baumé scales:

Light liquids:

$$\text{Degrees Baumé} = \frac{140}{\text{sp gr}} - 130$$

Heavy liquids:

$$\text{Degrees Baumé} = 145 - \frac{145}{\text{sp gr}}$$

The standard temperature for each of the foregoing formulas is 60°F.

Brix. Used almost exclusively in the sugar industry. The degrees represent per cent sugar (pure sucrose) by weight in solution at 17.5°C. A measurement is expressed as *degree Brix*.

Quevenne. Used for milk testing. Twenty degrees Quevenne signifies a specific gravity of 1.020; 40 degrees Quevenne indicates a specific gravity of 1.040; and so on. One *lactometer* unit is equivalent to approximately 0.29 degrees Quevenne.

Richter, Sikes, and Tralles. Three alcoholometer scales which read directly in per cent alcohol by weight in water.

Twaddle. For industrial liquors heavier than water. The range of specific gravity from 1.000 to 2.000 is divided into 200 equal parts. Thus one degree Twaddle equals 0.005 specific gravity.

REASONS FOR MEASURING SPECIFIC GRAVITY

Density and specific gravity are used almost interchangeably in many industries to express a physical property of a material that may often be an important variable in a process. Other related terms, specialized in use by certain industries, do not change the fundamental reasons for measurement.

As an expression of mass per unit volume, density can be a measure of the amount of a solid dissolved in a solvent, and dilution or concentration to a definite density is a common process-control problem. Density or specific gravity of solutions, such as sulfuric acid, provide a method of determining the concentration. For example, the acid with a specific gravity of 1.5934 (at 60°F, compared with water at 60°F) is a 68.13 per cent solution; if the gravity measures 1.6111, the acid is a 69.65 per cent solution. This illustrates the degree of accuracy required in measurement and the valuable information the measurement can provide. Numerous handbooks give tabulations of such relationships over wide ranges of values.

Particularly with gases or vapors, specific gravity (or density) is markedly affected not only by the composition but also by pressure and temperature. Hence industries involved with the processing or production of gases often find this measurement vital. In some cases, however, pressure or temperature are controlled to prevent variations in specific gravity. Gas-flow measurements are sometimes compensated automatically for pressure or temperatures, or a measure of density is used to provide mass-flow measurements.

In one application,[8] a density-gaging system is used to control the proportioning of four liquid ingredients that go into a final product on a continuous basis. The method is based upon the use of proportioning pumps for basic control of the mixture, with density measurement of the resulting product as a constant quality-control check.

MEASURING METHODS

The common methods for measuring density and specific gravity are described very briefly in this section. For further details on devices, see Refs. 1 and 7.

Hand-type Hydrometer

This device comprises a weighted float with a small-diameter stem (Fig. 1) so proportioned that a portion of its scale will be submerged during the course of measurement. The amount of scale submersion is then an indication of the specific gravity of the fluid in which the device is floated. Hydrometers may be calibrated to any of the scales previously mentioned and usually are accurate to three or four decimal places. These devices are used widely where automatic operation is not required. However, they can be located in a standpipe, equipped with an overflow at reading level, to permit visual readings of a continuously flowing liquid.

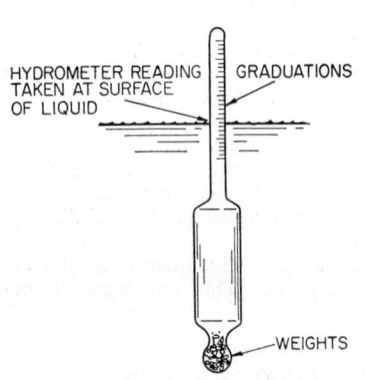

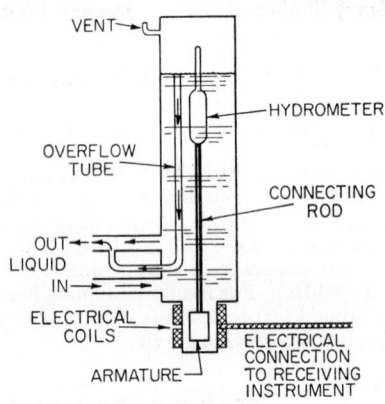

FIG. 1. Simple hand hydrometer.

FIG. 2. Inductance-bridge-type hydrometer transmitter.

Inductance Bridge Hydrometer

In this device, as shown by Fig. 2, the measured liquid is held at a constant level by means of an overflow tube. A glass hydrometer either rises or falls in the liquid as the specific gravity varies. The armature of an inductance coil is attached to the lower end of a rod extending from the hydrometer. This moving armature and the inductance coils constitute the hydrometer transmitter. A similar arrangement, less hydrometer, is duplicated in the receiving instrument. When this system is used, the temperature of the liquid usually is recorded along with the value of the specific gravity so that, if required, temperature corrections can be applied.

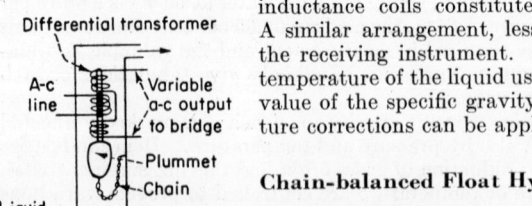

FIG. 3. Chain-balanced float-type hydrometer shown schematically.

Chain-balanced Float Hydrometer

In this instrument, a self-centering hydrometer float, which operates essentially free from friction and which is not affected by surface contamination, is employed. The volume of the "float" is fixed and remains entirely under the liquid surface. As the plummet moves up and down, the effective chain weight which acts upon it varies. For each density within the range of the assembly, the plummet has a definite point of equilibrium (see Fig. 3).

In order to transmit these changes in density, the plummet is attached to a differential transformer whose voltage differential, as a function of plummet displacement, is a measure of specific gravity.   To compensate for a density change due to temperature, a resistance-thermometer bridge senses the temperature change and impresses a voltage across the recorder which is equal and opposite to the voltage transmitted by the pickup coil due to the temperature-induced density change.   Specific-gravity range is 0.6 to 3.5.

Liquid-purged Differential-pressure Method

Two taps are installed in the side of a tank or at different elevations in a vertical pipe.   These taps are led to a pressure-differential-measuring device and are purged with a reference fluid, usually water.   In this way, an automatically suppressed range is obtained, as well as freedom from plugged taps.   In effect, the measured differential pressure is created by two equal columns, one of water and the other of the sample fluid.   The purge rate is small and dilution is negligible.

This system is used frequently in the pulp and paper industry to measure and control the density of green liquor, heavy black liquor, clay or starch slurries, lime slurries, and lime-mud slurries.   Similar materials in other industries are measured by this method.

Air-bubbler Reference-column Method

The reference-column method displaces a known head of the sample liquid and of water from their respective bubbler pipes, as shown in Fig. 4, and compares the two with a differential-pressure-measuring device.   The instrument can be calibrated directly in units of specific gravity.

When it is necessary to correct the specific gravity to a standard temperature, the temperatures of both columns must be known.   However, by arranging the columns so that the sample liquid flows around the reference column, a single temperature measurement will suffice.

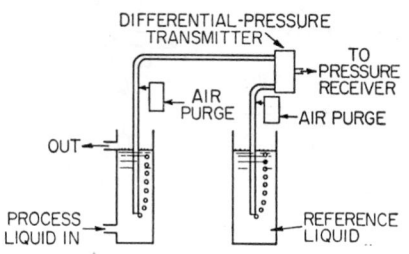

Fig. 4. Specific-gravity measuring system which employs differential bubbler and reference column.

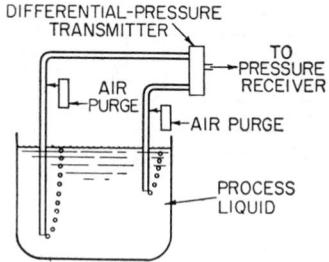

Fig. 5. Specific-gravity measuring system which employs differential bubbler and single vessel.

Air-bubbler Single-vessel Method

One of the simplest and most widely used methods of density measurement is to install two bubbler tubes in the sample fluid so that the end of one tube is lower than the end of the other tube (see Fig. 5).

The pressure required to bubble air into the fluid is equal to the pressure of the fluid at the ends of the bubble tubes.   Since the outlet of one is lower than the outlet of the other, the difference in pressure will be the same as the weight of a constant-height column of the liquid.   Therefore, the differential pressure is directly proportional to the fluid density.

This method is accurate to within approximately 0.3 to 1 per cent specific gravity when used with liquids which do not tend to crystallize in the measuring pipes. Light black liquor, white liquor, and bleach are frequently measured in this manner.

For suppressed range measurements, a constant-pressure-drop range-suppression chamber, as shown in Fig. 6, is connected in series with the low-pressure side.

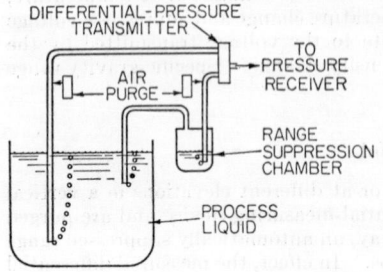

FIG. 6. Specific-gravity measuring system which employs differential bubbler and range-suppression chamber.

Buoyancy Gas Balance

In this instrument, a displacer is mounted on a balance beam in the vessel (Fig. 7). The displacer is then balanced for air and the manometer reading is noted at the exact balance pressure. The air is then displaced by gas and the pressure is adjusted until balance is restored. The ratio of pressure with air to pressure with gas is the density of the gas relative to air. This device is used principally for laboratory measurements and has an accuracy to the fourth decimal place.

Balanced-flow Vessel

Liquid is bypassed through flexible connections into a sampling chamber that is balanced by weights or a pneumatic force-balance transmitter (latter shown in Fig. 8). Weighing the fixed volume of liquid in the vessel provides a method of measuring density with an accuracy of ±1 per cent. Any range of specific gravity or density can be handled on a continuous basis; the method can be used for automatic control by adding a controller in the pneumatic receiver.

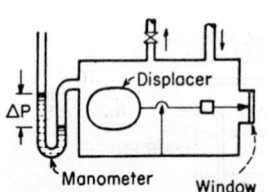

FIG. 7. Buoyancy gas balance.

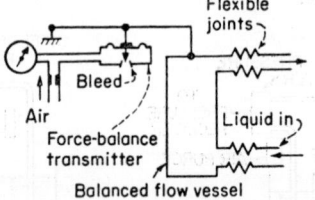

FIG. 8. Balanced-flow vessel.

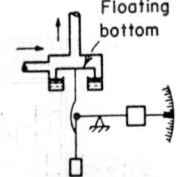

FIG. 9. Gas specific-gravity balance.

Gas Specific-gravity Balance

The weight of a tall column of the gas to be measured is balanced by air pressure acting against the floating bottom of the gas vessel, which is scale-balanced. This method can measure any range of gas specific gravity with a high accuracy. It can be designed to record and control values (see Fig. 9).

Industrial Specific-gravity Displacer

A displacer unit similar to the type used for liquid-level measurement (see page 5-70) can be used to measure liquid gravity. The float is submerged and externally counterbalanced so that it is responsive to changes in gravity. The method can measure any range of specific gravities with an accuracy to the second or third decimal place and is widely used in industrial processes for indication, recording, and if desired, control.

Gas-density Balance

A method using a null-balance electronic potentiometer is illustrated schematically in Fig. 10.   Gas density is measured by the buoyancy of one ball of a rhodium-coated dumbbell which is supported at its center by a quartz fiber.   The ball on the other end of the dumbbell is punctured so that it is not subject to buoyancy effects.   Rotation of the dumbbell due to changes in gas density produces electrostatic force between electrodes E and the quartz-fiber suspension.   Balancing potential is obtained and measured by the amount of light received by phototubes P_1 and P_2.   This potential required to balance the circuit is a measure of gravity relative to air with an accuracy of ½ per cent.   The method is continuous, is compensated for barometric-pressure changes, and has a broad range (0 to 2,000 relative to air).

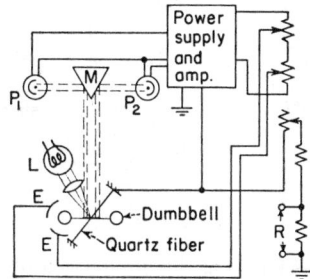

FIG. 10. Electrical gas-density balance.

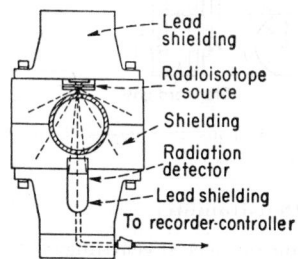

FIG. 11. Gamma-ray density gage.

Gamma-ray Density Gage

Gamma radiation from a radioisotope source passes through the liquid, as shown in the plan view in Fig. 11.   The amount of radiation reaching the detector varies inversely with the density of the stream.   This method can be used to measure the density of solutions, liquids, slurries, and divided solids with an accuracy of some ±2 per cent.   Electrical leads from the detector are connected to an electronic recorder which can be provided with control as required.

Boiling-point Elevation

In this method (Fig. 12) the temperature of a boiling solution is compared with that of water boiling at the same pressure.   For a particular solution, the boiling-point

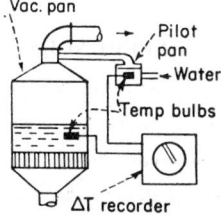

FIG. 12. Boiling-point elevation method.

elevation (difference in temperature) can be calibrated in terms of density at standard temperature.   The method, using resistance thermometers, provides a highly accurate measurement particularly suited to determining the end point of evaporation.   It can be used for solutions involving one dissolved component, or mixtures of fixed composition.

Viscous-drag-type Gas-density Meter

This device incorporates impellers which are driven in standard and test gas chambers, as shown in Fig. 13. Nonrotating impellers on the opposite side of the chambers are coupled together by a linkage which measures the relative drag, as shown by the tendency of the impellers to rotate. The balance point is a function of relative density and the instrument can be calibrated to read directly in density units. Accuracy is about 0.01 specific gravity units.

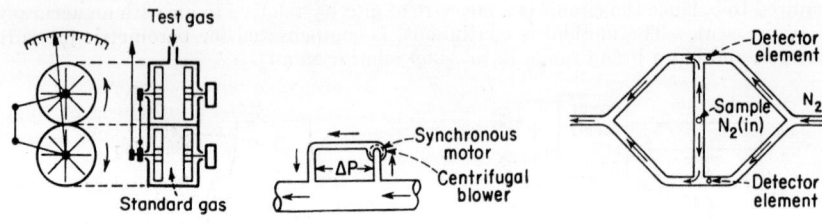

FIG. 13. Viscous-drag gas-density meter.   FIG. 14. Gas densitometer.   FIG. 15. Gas-density detector.

Gas Densitometer

A constant-speed centrifugal blower in a bypass line (Fig. 14) produces a pressure rise that is proportional to the gas density. Differential pressure across the blower is measured by a suitable method, with this value being calibrated in terms of density. The method can be used with a velocity flow meter as a gas mass-flow device. Accuracy for density measurement is ±0.10 per cent.

Gas-density Detector

A pneumatic Wheatstone bridge, mounted in a vertical plane as shown in Fig. 15, contains two thermistor detecting elements. A carrier gas (nitrogen, argon, or other heavy gas) acts as a reference, passing through both arms of the bridge (nitrogen is shown in the illustration). When the flow of this gas is balanced, the detectors are equally cooled and the electric circuit is balanced. The sample of gas to be measured enters a line connecting the two bridge arms (see Fig. 15) and is split in two streams. If the density of the gas is different from that of the reference gas, unbalance of the electrical bridge occurs. Response, in the form of chromatographic peaks, is measured by a recording potentiometer. The gas sample does not come in contact with the detector elements; hence, if it is corrosive, no maintenance problem is created.

REFERENCES

1. Eckman, D. P.: "Industrial Instrumentation," pp. 259–262, John Wiley & Sons, Inc., New York, 1950.
2. Marquard, C. M.: An Accurate Density Meter—How It Works—How It Is Used, *Eng. Mining J.*, vol. 152, no. 1, pp. 78–82, 1951.
3. Wilson, W. A.: Microdetermination of Density, *Metallurgia*, vol. 33, no. 195, pp. 157–160, January, 1946.
4. Anon.: Centrifugal Gas Specific Gravity Meter, *J. Sci. Instr.*, vol. 22, no. 8, pp. 145–146, August, 1945.
5. Fromm, J. L.: Continuous Specific Gravity Recording by the Air Bubble Method, *Chem. Eng.*, vol. 56, no. 6, p. 130, June, 1949.
6. The Instrument Manual, Sec. X, "Measurement and Control of the Specific Gravity of Liquids," United Trade Press, Ltd., London, 1953.
7. Eberly, G. L.: Density and Specific-gravity Measurement, pp. 7-54 to 7-59, "Process Instruments and Controls Handbook," McGraw-Hill Book Company, Inc., New York, 1957.
8. Anon.: Nuclear Gaging System Controls Liquid Density, *Automation*, January, 1962, p. 77.
9. Siggia, Sidney: Continuous Analysis of Chemical Process Systems, Chap. 4, "Density and Specific Gravity," John Wiley & Sons, Inc., New York, 1959.

HUMIDITY AND DEW POINT

By W. F. Hickes*

Humidity refers to the presence of moisture in an atmosphere. When considering the measurement of humidity, the subject often may appear complex because there are two basic ways of quantitatively expressing humidity; that is, on an *absolute* or on a *relative* basis. For example, the U.S. Weather Bureau customarily reports in terms of relative humidity, essentially the degree of saturation of the air. Relative humidity is a term that is descriptive of the ability of air to moisten or to dry materials, and to some extent, relative humidity is a measure of human comfort. However, air which is saturated at 50°F (100 per cent relative humidity) will be quite dry (19 per cent relative humidity) if heated to 100°F. A changing basis of this sort is not convenient for many purposes of computation, as encountered in air-conditioning, combustion, or chemical-process calculations. Thus various absolute units of expression are used, such as dew point; grains of water per pound of dry air; or pounds of water per million standard cubic feet.

IMPORTANCE OF HUMIDITY AS A PROCESS VARIABLE

Humidity affects personal comfort, and a large industry has developed, based on the maintenance of comfortable conditions of humidity and temperature. However, much industrial air conditioning, while it adds to human comfort, is done strictly for process reasons. Following are very abridged descriptions of processes which require humidity control:

Textile Manufacture. This industry was one of the pioneers in humidity control because humidity greatly affects the characteristics of all natural and most synthetic fibers. Under dry conditions, the fibers are brittle and hard to manage. Buildup of charges of static electricity may present a serious fire hazard. Under excessively moist conditions, cotton may mildew and some of the synthetics become weak and are easily overstressed. Humidity is also important because it directly affects the amount of water held by the fibers, a matter of considerable commercial importance in a material sold on a weight basis.

Paper Manufacture. This industry faces problems similar to textile manufacturers. Paper users, specifically printers, find humidity control essential for accurate register in multicolor work.

Drying Operations. The many types of drying processes depend largely on humidity control. If humidity is too high, drying is retarded; if it is too low, the surface of a material may overdry.

Electrical Equipment. Electrical effects also may justify carefully controlled humidity. Low humidities, for example, may be desired to minimize surface leakage or they may be avoided to promote the discharge of static.

Preservation of Materials. Cargoes on ships at sea are protected from water damage (actually condensation damage) by maintaining the dew point of air in the holds lower than cargo temperatures. Many naval vessels are kept in usable condition by maintaining a low humidity within them such that steel will not rust, canvas and leather will not mildew, and leather will not dry out and crack. Food products to be stored, such as apples, may be kept cool and at high humidities to prevent withering,

* Assistant to Chief Engineer, The Foxboro Company, Foxboro, Mass.

or products such as dried eggs or dried milk may be maintained at very low humidities to prevent spoilage.

Combustion Processes. Moisture content of air to a blast furnace materially influences operation. Opinions differ on the relative economies of controlled low humidity for minimum fuel consumption or controlled high humidity to maintain smooth and consistent operation without the expense of dehumidification.

Natural Gas. At the source, natural gas is dehumidified to prevent freezing in transmission lines. Natural gas at the point of distribution is moistened to prevent drying out of packing, drying out of diaphragms, and release of dust and scale in old distribution systems.

DEFINITIONS

Humidity Ratio.* Also termed humidity and absolute humidity, this is the number of pounds of water vapor associated with 1 lb of dry air.

Relative Humidity. The ratio, usually expressed as a percentage, of the partial pressure of water vapor in the actual atmosphere to the vapor pressure of water at the prevailing temperature.

Percentage Humidity. The ratio, expressed as a percentage, of the weight of water vapor in a unit weight of air to the weight of water vapor in the same weight of air if the air were completely saturated at the same temperature. At ordinary temperatures, relative humidity and percentage humidity are, for engineering purposes, the same. Percentage humidity is equivalent to percentage specific humidity.

Dew Point. The temperature of liquid water or ice in vapor-pressure equilibrium with an atmosphere. It is common practice to use dew point as a generic term regardless of whether it is ice or liquid water. More precise usage distinguishes between dew point and frost point, which is the temperature of ice (frost crystals) in vapor-pressure equilibrium with an atmosphere. Below 32°F water can exist as ice or as supercooled water, and the vapor pressures are not the same. Hence, in measuring dew points below 32°F, it is necessary to note which phase is present and to report results accordingly.

Dry-bulb Temperature. The temperature of an atmosphere. The qualification "dry bulb" is used to distinguish the normal temperature measurement from the temperature as measured by the wet bulb.

Wet-bulb Temperature. The dynamic equilibrium temperature reached when the wetted surface of an object of small mass (bulb of a thermometer) is exposed to an air stream. Evaporation of water causes cooling which is counterbalanced by heat absorbed from the air.

BEHAVIOR OF WATER VAPOR

The water vapor in moist air is actually steam, usually superheated steam. Many humidity relationships can be better understood by ignoring the air and considering only the "steam." Consider water in a boiler operating at atmospheric pressure, 14.7 psia. Steam fills the space above the water and steam at 14.7 psia is in equilibrium with water at 212°F. The steam can be superheated above 212°F but cannot be cooled below 212 without causing condensation. Regardless of the amount of superheating of the steam the pressure still remains at 14.7 psia as long as the water temperature is 212°F and vice versa. But if water temperature increases, pressure must increase also; and if water temperature falls, pressure falls also.

The relationship between steam pressure and boiler-water temperature is exactly the same as the relationship between partial pressure of water vapor in an atmosphere and dew-point temperature. The pressure of air tends to slow down evaporation but does not change the basic relationship.

Consider a water surface at 70°F. The steam tables give a saturation pressure of 0.3630 psia for 70°F. Therefore, the water will evaporate into the space above it until the saturation pressure is reached; i.e., the pressure of water vapor (steam) in the

---

* Now termed "mixing ratio" by meteorologists and defined in the Smithsonian "Meteorological Tables" as "the ratio of the mass of water vapor to the mass of dry air (or gas) with which the water vapor is associated."

space is 0.3630 psia. Unless vapor is removed faster than evaporation can replace it, the pressure of 0.3630 psia will be maintained as long as the water remains at 70°F. Also, as long as there is an exposed surface at 70°F no more water vapor can be put into the space. The air above the water is therefore said to be saturated, though strictly speaking it is the space that is saturated. It is in vapor-pressure equilibrium with liquid water at 70°F; therefore, by definition its dew point is 70°F.

Assume that heat is applied to raise the temperature of the air and water-vapor mixture as indicated in Fig. 1. The upper portion is heated to 100°F, but in the lower portion, the temperature at the water surface remains 70°F. Water-vapor pressure remains unchanged at 0.3630 psia. By definition, relative humidity is the ratio of the actual partial pressure of water vapor to the vapor pressure of water at the temperature prevailing. At the top, 100°F corresponds to 0.9496 psia. At the midpoint, 85° corresponds to 0.5960 psia. Hence relative humidities are

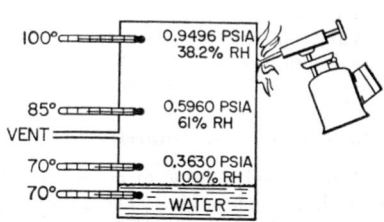

$$\frac{0.3630}{0.9496} = 0.382$$

or 38.2 per cent and $0.3630/0.5960 = 0.610$ or 61 per cent compared with

$$\frac{0.3630}{0.3630} = 1.00 \text{ or } 100 \text{ per cent}$$

FIG. 1. Variation of relative humidity with temperature.

at the water surface. Just as in this example the partial pressure of water vapor, hence dew point, will be substantially uniform throughout any inclosure, even a very large room; but there can be no uniformity of relative humidity unless the temperature is uniform.

It will be noted that in the previous description of relative humidity and dew point the air itself was ignored. Both these terms are functions of water vapor only, and both concepts apply whether or not air or other gases may be present and are independent of the pressure of the air or other gases. However, this does not mean that the pressure can be varied lest the vapor pressure also vary.

Wet-bulb temperature, however, is quite different. The air (or other gas) is included by definition. Hence wet-bulb temperature is affected not only by the amount of moisture present and the dry-bulb temperature but also by the barometric pressure and the composition of the atmosphere.

INSTRUMENTAL METHODS OF HUMIDITY MEASUREMENT

Humidity affects many materials in diverse ways. Thus a wide variety of humidity-measuring instruments is available. They operate on several basic principles and may be classified according to the physical effects on which they are based.

Condensation

A surface in contact with the atmosphere is cooled until condensate (dew) appears. The temperature of the surface when condensation first appears is the dew-point temperature. An instrument based on this principle was invented by Daniell in 1827.

A variation of this method is to cool a sample by adiabatic expansion so that condensation appears as a fog. The expansion ratio to produce a fog and the initial temperature permit calculation of dew point.

Dimensional Change

Most organic materials undergo dimensional change with changing humidity. A typical instrument utilizes human hair arranged so that its expansion with increasing humidity actuates the mechanism. The expansion is, to a close approximation, a linear function of relative humidity. This is the oldest type of humidity instrument and was invented by De Saussure in 1783. Animal membranes, wood, and paper have also been used.   In a variation of this instrument, the humidity-sensitive material is laminated to a thin metal, forming a humidity-sensitive element which operates much like a bimetal thermometer.

Thermodynamic Equilibrium (Wet Bulb)

The bulb of a thermometer is wrapped in a cloth wick maintained continuously wet with water and exposed to an air stream.   The temperature observed is, by definition, the wet-bulb temperature.   This reading, in combination with a reading of the air temperature (dry-bulb temperature), is a measure of moisture content of the air. Dr. James Hutton of Edinburgh is credited with the first use of this principle about 1792, but his instrument had no provision for air velocity.   This is, of course, essential to accurate measurement by this method.

Variations of this basic device utilize ceramic sleeves instead of cloth wicks.   In one design, the wick is eliminated and the temperature of the air stream is measured after cooling by saturation from a water spray.

Absorption—Gravimetric

A measured volume of air is passed through a water-absorbing material, such as phosphorus pentoxide.   The gain in weight of the absorbent is directly the moisture content of the known volume of air.

In a variation of this procedure, the change in pressure when the absorbent is brought in contact with air in a sealed vessel is measured.

Absorption—Conductivity

The amount of moisture absorbed by a quantity of a hygroscopic salt varies with temperature and humidity.   This absorption changes the electrical conductivity between two electrodes in contact with the salt.   The conductivity, when corrected for temperature (usually automatically), can be interpreted as relative humidity.

Electrolytic

The moisture in a measured flow of air is absorbed by phosphorus pentoxide and simultaneously decomposed by electrolysis.   The electrolysis current is, by Faraday's law, directly proportional to the rate of decomposition of water and hence to moisture content of air.

Heat of Absorption

Absorption of water vapor on a solid adsorbent releases heat.   A measurement of temperature change when water vapor is alternately adsorbed and desorbed is interpreted as moisture content.

Vapor Equilibrium

A saturated solution of a hygroscopic salt is maintained at the temperature at which it is in vapor-pressure equilibrium with the atmosphere.   The temperature of the salt converts directly to dew-point temperature.   Conductivity of the solution is utilized to control heating and maintain the equilibrium temperature.

Infrared

Water vapor (and most other compounds) absorb electromagnetic radiation in certain portions of the infrared region.   A measurement of this absorption can be interpreted as moisture content.

The basic methods of humidity measurement just described are listed in Table 1. Ref. 17, pp. 7-60 to 7-72, describes such devices in more detail.

HUMIDITY CONVERSIONS AND CALCULATIONS

Since the different humidity instruments read in different units, relative humidity, wet-bulb temperature, or dew point, conversions from one to the other or to weight units are frequently required. These conversions are not simple addition and multiplication, as in converting °F to °C, but in general require supplementary measurements of temperature and sometimes pressure. Tables are available for these conversions, or they may be made by means of the psychrometric chart. See diagrams H-1, H-2, and H-3 in Sec. 12 of Ref. 17. These charts cover low-, medium-, and high-temperature ranges. Dry-bulb temperatures are given along the abscissa and dew points along the ordinate. The curved line forming the upper left boundary is the saturation line. At saturation, dry-bulb temperature, wet bulb, and dew point are all the same. Parallel to the saturation line are the other lines representing partial pressures of water vapor, 0.8 of saturation, 0.6 of saturation, and so on, corresponding to 80 per cent relative humidity and 60 per cent relative humidity respectively.

Dew point is represented by the horizontal lines. Since each dew-point temperature corresponds to a definite water-vapor pressure these lines also represent vapor pressures and at a fixed barometric pressure convert directly to humidity ratio, pounds water per pound of dry air.

Wet-bulb temperature is represented by lines slanting downward to the right. Dry-bulb temperature, wet-bulb temperature, and dew-point lines all meet at the saturation line; i.e. air saturated at 70° has a dry-bulb temperature of 70°, a wet-bulb temperature of 70°, and a dew point of 70°F. The charts are based on standard barometric pressure of 29.92 in. Hg.

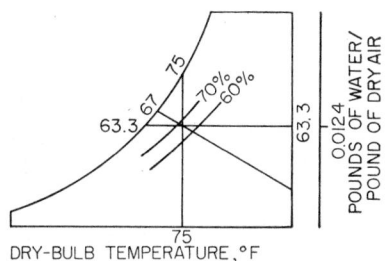

DRY-BULB TEMPERATURE, °F

Fig. 2. Diagram showing use of a psychrometric chart to determine dew point, relative humidity, and humidity ratio.

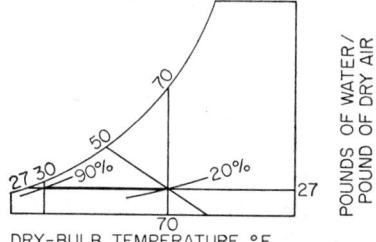

DRY-BULB TEMPERATURE, °F

Fig. 3. Diagram showing use of a psychrometric chart to determine dew point, relative humidity, and wet-bulb temperature.

Examples

Example 1. Find the dew point, relative humidity, and humidity ratio when dry-bulb temperature is 75°F and wet bulb is 67°F.
Solution. Read directly from a psychrometric chart* as indicated in Fig. 2:

        Dew point.................. 63.3°
        Relative humidity.......... 67 per cent
        Humidity ratio............. 0.0124 lb water/lb dry air

Example 2. Air at 30°F dry bulb and 27° dew point is heated to 70°F dry bulb. Find dew point, relative humidity, and wet bulb of the heated air.
Solution. Read directly from a psychrometric chart.* Since addition of heat does not change dew point this is represented by a horizontal line on the chart as sketched in Fig. 3.

        Dew point.................. 27°
        Wet bulb................... 50°
        Relative humidity.......... 20 per cent

For barometric pressures other than 29.92 in. Hg special charts can be drawn but are not generally available. The dry bulb–dew point–relative humidity relationship
* Such as diagram H-2, Ref. 17, which covers medium temperatures.

is independent of pressure and can be read from Ref. 17, diagrams H-1, H-2, and H-3. However, the wet bulb is affected by barometric pressure. Tables WB 235 by Marvin (not included in this handbook) are available for converting wet- and dry-bulb readings to dew point and relative humidity for barometric pressures (in. Hg) of 30, 29, 27, 25, and 23.

Other problems are handled most readily by use of relative-humidity tables, such as H-4 in Sec. 12 of Ref. 17.

Example 3.   Compressed air at 150 psia has a dew point of 85°F.   Calculate dew point after expanding to 40 psia.   A dew point of 85°F corresponds to a pressure of 0.5960 psia.   In expanding, the partial pressure of water vapor will change in same ratio as total pressure.   Therefore, partial pressure after expansion equals $0.5960 \times {}^{4}\%_{150}$ or 0.1589 psia.   From tables, this is found to be equivalent to a dew point of 47.0°F.

Example 4.   Air at 50°F dry bulb has a dew point of 33°F.   Determine grains of water vapor per cubic foot.

Solution.   From Steam Tables, read specific volume of 3,180.5 cu ft/lb for water at 33°.

$$\frac{1}{3,180.5} \text{ (lb/cu ft)} \times 7,000 \text{ (grains/lb)} = 2.201 \text{ grains/cu ft for air saturated at } 32°$$

$$2.201 \times \frac{460 + 33}{460 + 50} = 2.128 \text{ grains/cu ft}$$

Note.   The quantity of water vapor is independent of the pressure and the nature of the air or other gas present.   Exceptions: gases which react with water such as ammonia and sulfur dioxide.

Example 5.   A gas mixture contains 1.5 per cent by volume of water.   Determine dew point at atmospheric pressure.

Solution.   Calculate partial pressure of water vapor

$$14.7 \times 0.015 = 0.2205 \text{ psia}$$

From Steam Tables, the dew point corresponding to 0.2205 psia is 56°F.

SELECTION OF HUMIDITY-MEASURING INSTRUMENTS

The choice of an instrument for humidity measurement is much more complicated than the choice of an instrument for most other variables, such as pressure or flow. Before considering the usual problems of accuracy, cost, and suitability for the range required, it is necessary to decide on the form in which the reading is desired, that is, relative humidity or some form of absolute units, dew point or grains per cubic foot or perhaps wet and dry bulb. Unlike flow or pressure measurement this is not simply a choice of units, readily converted by suitable multipliers, but dictates the whole principle on which the measurement is based.

For some applications, the choice of a humidity-measuring instrument is entirely arbitrary, usually a matter of tradition. For other applications, there are definite advantages in the selection of an instrument reading in the proper units. The basis for choice is the use to be made of the reading (see Table 1).

Applications Involving Hygroscopic Materials

If humidity is measured because of its effect on moisture absorption (regain) of a hygroscopic material such as cotton, wool, paper, leather, rayon, or wood, a reading in terms of relative humidity will be desirable. The moisture content of such materials is, as a first approximation, a function of relative humidity more or less independent of temperature. Hence, in such uses, readings in terms of relative humidity are usually specified. While not true over wide temperature swings, the relationship is sufficiently close for many purposes for the usual variations in indoor temperature of perhaps 5 to 10°F.

It should be noted, however, that the relative humidity which is significant is the relative humidity of the air immediately in contact with the product. If temperatures vary from point to point, relative humidity cannot be uniform and relative-humidity reading at one point is no criterion of results to be expected at another point where temperature is different. It is a very common misdirection of effort to see an

Table 1. Methods of Humidity Measurement—Selection Guide

Device and operation	Range	Applications and remarks
Change in Dimension of Adsorbing Material		

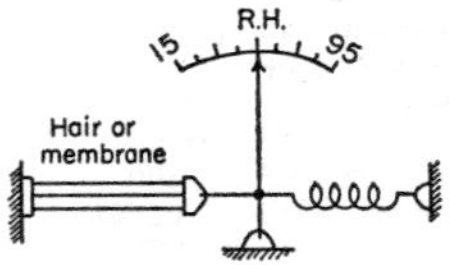

Fig. 1

Device and operation	Range	Applications and remarks
Hair Hygrometer (Fig. 1). Human hair elongates with increasing humidity —is suitably connected to actuate indicating, recording, or controlling instrument	15–95 % RH	Useful at temperatures up to 160°F. Useful, but sluggish at temperatures below 0°F
Membrane Hygrometer (Fig. 1). Same as hair hygrometer except animal membrane is used as sensitive element	20–85 % RH	Useful to 140°F

Cooling by Evaporation of Water

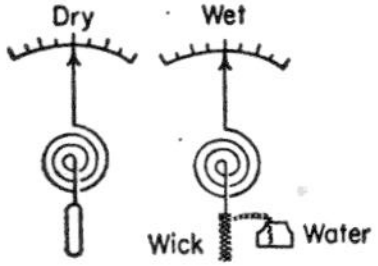

Fig. 2

Fig. 3

Device and operation	Range	Applications and remarks
Wet- and Dry-bulb Psychrometer (Fig. 2). A thermometer bulb in an air stream is cooled by evaporation from a water-saturated wick. This temperature, in combination with air temperature (from a dry bulb) is a measure of humidity	0–100 % RH	Normally useful for wet-bulb temperatures between 32° and 212°F
Sling Psychrometer (Fig. 3). Specific form of wet- and dry-bulb psychrometer using glass thermometers in a frame designed to be swung rapidly to obtain adequate air velocity	0–100 % RH	Commonly used for setting and checking hair hygrometer. Also for routine meteorological observations

7–17

Table 1. Methods of Humidity Measurement—Selection Guide (*Continued*)

Device and operation	Range	Applications and remarks
	Condensation	

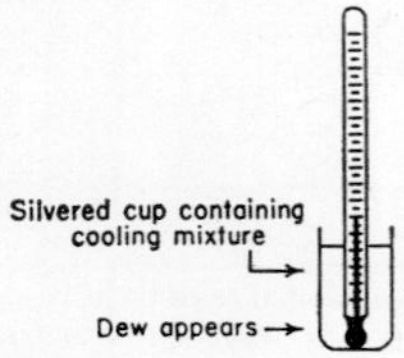

Silvered cup containing cooling mixture

Dew appears →

Fig. 4

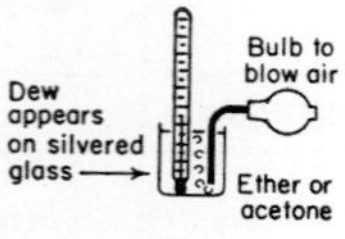

Dew appears on silvered glass →

Bulb to blow air

Ether or acetone

Fig. 5

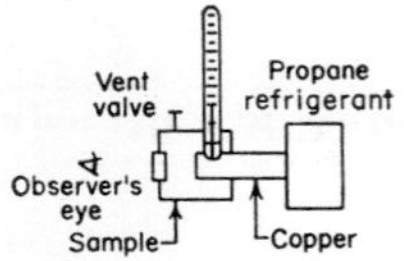

Vent valve

Observer's eye

Sample →

Propane refrigerant

Copper

Fig. 6

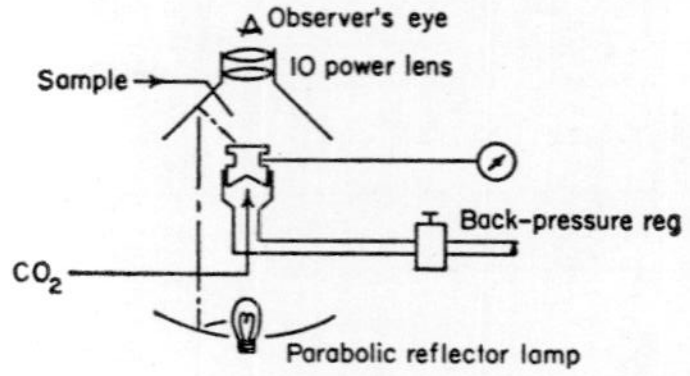

△ Observer's eye

⊙ IO power lens

Sample →

Back-pressure reg

CO₂

Parabolic reflector lamp

Fig. 7

Simple Dew-point Apparatus (Fig. 4). A polished surface is slowly cooled, or heated, until dew (or frost) appears or disappears, respectively. Temperature at first appearance or disappearance is dew point

Laboratory Dew-point Device (Fig. 5). Glass construction; cooled by evaporation of volatile solvent, such as acetone

Industrial Dew-point Device (Fig. 6). Similar in principle to above, but of metal construction. Uses propane or carbon dioxide as refrigerant

Industrial Dew-point Device (Fig. 7). Similar to above (Fig. 7)

Theoretically infinite. Practically used −100 to +212°F dew point

30–70°F of dew point. Useful only at atmospheric pressure

−90 to +70°F dew point at pressures up to 3,000 psi

−76 to +40°F dew point at atmospheric pressure

See specific types below

General laboratory and experimental use. Manually operated indicator

Manually operated indicator. Widely used for natural-gas measurements.

Manually operated indicator. Used to check moisture of cylinder gases

Table 1. Methods of Humidity Measurement—Selection Guide (*Continued*)

Device and operation	Range	Applications and remarks
 Fig. 8		 Fig. 10
 Fig. 9		
Automatic Dew-point Recorder (Fig. 8). Dew or frost is detected by photocell. Controls refrigeration from self-contained compressor unit. Alternately heated and cooled to record on 3-min cycle	− 100 to + 100°F dew point. Atmospheric pressure	Some noncommercial units operate by maintaining a constant thickness and size of dew spot on mirror; i.e., rates of dew deposition and evaporation are equal Continuous indicating and recording
Vapor-pressure-equilibrium Method (Fig. 9). Measures temperature at which saturated solution of lithium chloride is in water-vapor-pressure equilibrium with atmosphere	− 50 to + 142°F dew point. Pressures up to 500 psi	
Adiabatic-expansion Device (Fig. 10). Gas cooled by adiabatic expansion. Pressure ratio to cause condensation is determined. Requires formula and slide rule	Room temperature to − 80°F. Atmospheric pressure only	Manually operated indicator. Widely used for detection of moisture in furnace atmospheres

7-19

Table 1. Methods of Humidity Measurement—Selection Guide (Continued)

Device and operation	Range	Applications and remarks
Absorption		

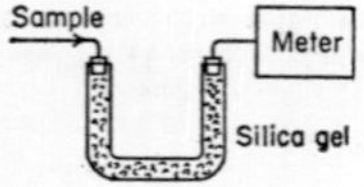

Fig. 11

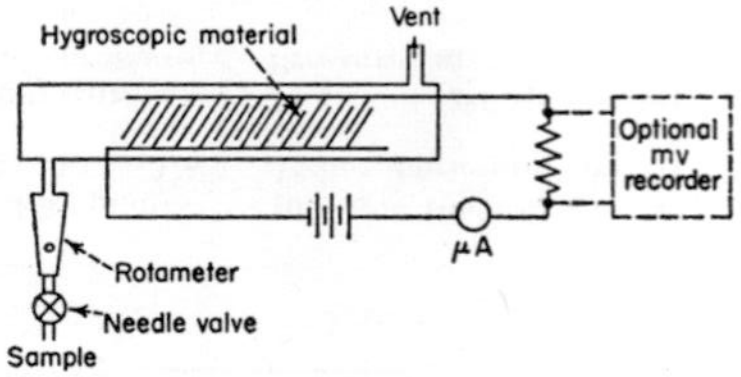

Fig. 12

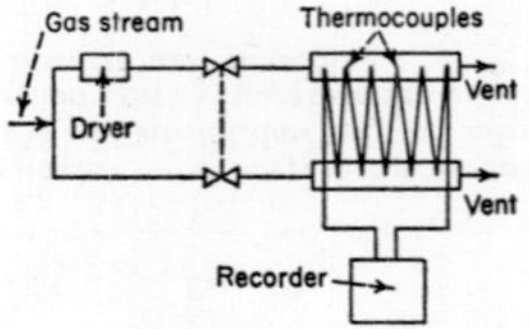

Fig. 13

Device and operation	Range	Applications and remarks
Method Using Desiccant (Fig. 11). Measured quantity of gas is drawn through a desiccant, such as silica gel, phosphorus pentoxide, or calcium chloride. Difference in weight before and after permits calculation of moisture in air sample	Infinite	Used largely as a primary standard for humidity measurement
Electrolytic Water Analyzer (Fig. 12). Measured flow of sample is passed through a cell containing phosphorus pentoxide. Moisture is absorbed quantitatively and continuously decomposed by electrolysis between platinum electrodes. Decomposition current is, by Faraday's law, directly proportional to rate of decomposition of water and, at known flow rate, of moisture content in sample	0–10,000 ppm	General industrial use for low levels of moisture, below 1,000 ppm
Water-vapor Recorder (Fig. 13). Sample stream is divided into two equal parts, one of which is thoroughly dried. Dried and moist sample streams are passed alternately through two adsorbers. Temperature difference beteeen adsorbers is detected by thermopile and measured by peak-to-peak voltmeter which reads directly in parts per million	0–5,000 ppm max	General use for low level of moisture, including hydrocarbon streams as well as air

Table 1. Methods of Humidity Measurement—Selection Guide (*Continued*)

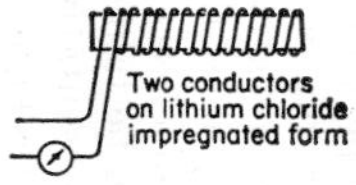

Device and operation	Range	Applications and remarks

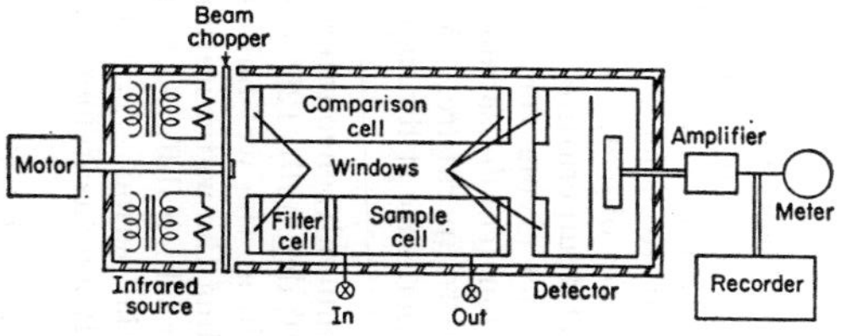

FIG. 14

FIG. 15

Dunmore Cell (Fig. 14). Double-wire winding on insulator is coated with film containing a hygroscopic salt (usually lithium chloride) which becomes more conductive as its equilibrium moisture content increases. Ambient humidity determines conductivity of coating and governs current flow between the wires. Alternating current must be used; measured as direct current by rectifier bridge. Or resistance of element can be measured in Wheatstone bridge. Requires compensator if used at varying temperatures — 10–95 % in steps — Developed largely for testing cabinets and laboratory equipment for simulating industrial conditions. Also used to some extent in process measurements, notably textile

Infrared Analyzer (Fig. 15). Two equal beams of infrared radiation pass through sample cell and comparison cell, respectively. Difference in energy of exit beams is detected and recorded as ppm water vapor — 0–30,000 ppm max — General use for process streams

Notes. The methods in this table are not all-inclusive. Other methods which show good promise for future development include (1) dielectric constant measured by microwave refractometer or r-f capacitance bridge; (2) emission spectroscopy (from spark discharge or low-pressure r-f discharge); (3) diffusion of water vapor through porous walls; (4) use of hot-wire gas analyzer; (5) continuous absorption (or saturation) with flow difference indicated by a pneumatic bridge.

installation designed and operated to maintain nearly perfect control of relative humidity at the point of measurement. Instruments of the highest accuracy must be equipped with the most refined control mechanisms. But the same installation may show temperature differences of 3° between supply air and return and still greater difference from point to point in the space. There can be no uniformity of relative humidity without uniformity of temperature.

Drying Applications

A second major class of humidity applications is in drying, especially the more modern high-humidity dryers. For such applications, the wet bulb has a special significance and a convenience factor. If the product is initially wet, then the temperature it will reach is the wet-bulb temperature, since by definition, the wet-bulb temperature is the temperature reached by a wet object. It will approach this temperature very rapidly because, while product temperature is below dew-point temperature, moisture will condense on the product giving very high heat-transfer rates because of the high latent heat of vaporization (condensation) of water. This continues until the product reaches the dew-point temperature. At this point, condensation ceases and evaporation begins, with temperature still rising, reaching equilibrium at the wet-bulb temperature.

It should be noted that, for conditions commonly encountered in dryers, wet bulb and dew point are very close. Example: 140° dry bulb, 125° wet bulb, and 123° dew point. The product temperature will remain at the wet-bulb temperature until all free water on its surface is evaporated including the water initially present, the water of condensation, and water from the interior which diffused to the surface, diffusion being aided by the high heat input of the initial period. Thus the wet-bulb temperature is significant in drying because it is directly a measurement of product temperature during the initial period of drying.

During the second period of drying (the so-called "falling-rate" period), no form of humidity has any special significance. Wet and dry bulb is the method commonly used.

The final stage of drying as moisture equilibrium is approached is basically the same as the "regain" type of problem first discussed. While relative humidity is, in theory, the significant form of measurement, it is seldom used because temperatures are usually high for most forms of relative-humidity instruments and temperature is controlled so that all forms of humidity measurement convert directly. The wet and dry method is usually used.

In addition to its significance as a measurement the wet bulb has, in certain types of dryers, a factor of convenience. If the dryer operates adiabatically, that is, no heat is added directly to the product by radiation or conduction and no heat is lost so that sensible heat of the circulating air is used to supply latent heat of vaporization of water, the wet bulb will not change as the air passes over the product. While few dryers conform strictly to these conditions, many come close enough to make the relationship usable. When they do, the measurement can be taken at the point most convenient for control, with assurance that it will be representative of conditions generally.

Since accurate wet-bulb measurement requires high air velocity, the measurement is most readily made directly after the fans, frequently in the inlet air duct, a point also conducive to good control. Operating personnel are properly interested in the conditions prevailing at the product, not in some duct. It can be demonstrated, however, in practice as well as in theory that, barring short circuiting of air or pocketing, the wet bulb will be the same at supply duct, at product, and in return duct. This does not mean, however, that the drying action will be the same throughout. The wet-bulb reading must be interpreted in the light of the actual dry-bulb reading at the point in question, that is, at the product.

This effect, of constant wet bulb, is commonly used in lumber dry kilns of the reversing type. Only one wet bulb is used operating alternately on supply air and return air as periodic fan reversals change the function of the duct in which it is installed.

Applications Involving Anticipatory Measurements

In a third class of problems, the humidity is measured at one set of conditions of temperature and pressure and the result is used to predict effects at some other point in the system where conditions of temperature or pressure are different. Typical problems are encountered in the operation of compressed-air systems, especially those supplying control instruments. The moisture content of the air is checked to be sure that outside lines do not freeze on cold nights or perhaps do not fill with condensate. For this problem, it is most convenient to make the measurement in terms of dew point of the air. If it is lower than any probable temperature, no condensation or freezing will occur. The same problem occurs in the operation of natural-gas pipelines though interpretation is not so direct, because hydrates are formed, not ice.

A similar problem arises in warehouses and in the holds of cargo vessels containing any product which would be damaged by condensation. Dew point is the preferred measurement because it can be compared directly with present or anticipated temperatures to determine the possibility of condensation. While a relative humidity of 100 per cent also describes the condensation condition, relative humidity is seldom satisfactory as a measurement in such uses because it is frequently necessary to compare moisture content of the air (dew point) here and now with an anticipated temperature of the air or with the measured or estimated temperature of a product.

Applications Where Moisture Is Considered a Raw Material

In a very similar class of problems, the moisture in the air is regarded simply as one material fed into a furnace or as one ingredient in a chemical reaction. For instance, in controlled-atmosphere heat-treating furnaces, a measurement of moisture is a direct guide to proper control. Dew point is used because it is the most direct measurement. Dew point converts directly to grains of water per standard cubic foot, per cent by volume, or similar units if pressure is constant (or barometric variations are ignored). Dew-point instruments can be calibrated to read directly in such units.

Dew point as a measurement has another advantage for such applications, being independent of temperature. Since it is impractical to operate any type of humidity element directly in a furnace at 1000 to 1800°F, a sample of the atmosphere is drawn off, cooled, and passed through a sample chamber containing the humidity element. The dew point of the gas is not changed by this cooling nor is final temperature important provided only that it is within the safe range for the instrument and not low enough to cause condensation.

Humidity-Temperature Control Applications

When temperature and humidity are both to be controlled, the basis for choice is quite different. At a fixed temperature all forms of humidity measurement are directly convertible. While for reasons of custom or of fundamental significance conditions may be specified as relative humidity, control may be and commonly is based on some other form such as wet and dry bulb. Selection of the measuring instrument is based on its adaptability to the general scheme of control rather than the type of unit in which the result is specified.

HUMIDITY STANDARDS

The primary standard for the testing and calibration of humidity-measuring instruments is the gravimetric method, that is, weighing the quantity of water removed when a carefully measured volume of air is passed through a suitable absorbent. While capable of high accuracy this is not a convenient method for ordinary use, and various alternatives are in practical use. These take two general forms: (1) the production of a precisely controlled atmosphere and (2) measurement of existing conditions by means of a carefully built and carefully operated humidity-measuring instrument.

Divided-flow Humidity Apparatus

For producing a continuous stream of air at a known humidity, the divided-flow principle as illustrated in Fig. 4 has been successful. A stream of air, dried by refrigerating to −80°F, is divided in known proportions into two parallel streams, one of which passes over water or ice at a known temperature, while the other passes over dry surfaces at the same temperature. The streams are then merged and pass over the instrument to be tested. By varying the proportion of air passed over the water (or ice), any desired relative humidity can be obtained. While simple in theory great care is necessary in design and construction if accuracy is desired. The most critical details are (1) absolute uniformity of temperature throughout, including the test chamber (usually obtained by submerging the whole assembly in a temperature-controlled bath), and (2) thorough saturation usually ensured by using two or more saturators in series.

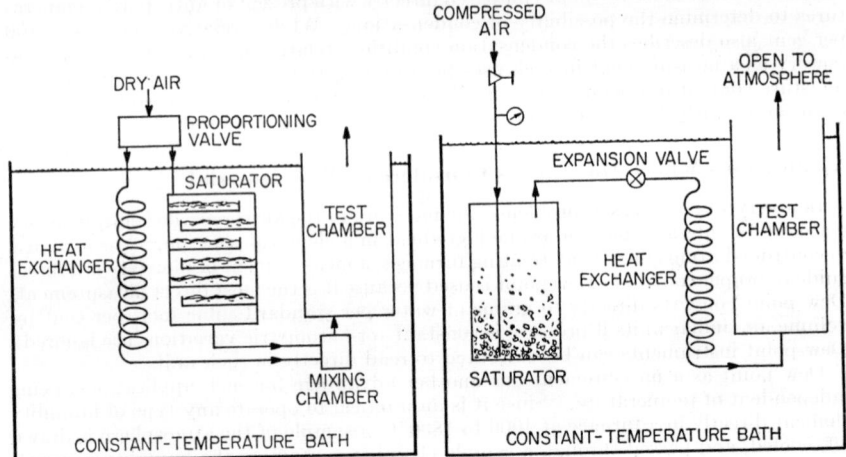

FIG. 4. Divided-flow humidity apparatus.   FIG. 5. Two-pressure humidity apparatus.

Two-pressure Humidity Apparatus

An alternate method saturates air or other gas at an elevated pressure and reduces air pressure to give any desired relative humidity, as shown in Fig. 5. Temperature and pressure in the saturator are selected to give the desired humidity at the desired final temperature and pressure in the test chamber. To ensure uniformity of temperature, the saturator and test chamber may be immersed in a constant-temperature bath. Heating of the air at or before the expansion valve is sometimes required to prevent condensation. If the saturator and test chamber are at the same temperature, the ratio of the two pressures is, very closely, the relative humidity. For precise work, particularly if high pressures are involved, that is, low relative humidity, corrections may be required.

Two-temperature Humidity Apparatus

A third method (Fig. 6), the two-temperature method, saturates air at one controlled temperature and then heats it to a higher controlled temperature. Taking the first temperature as dew point and the second as dry bulb, relative humidity is readily determined from tables or the psychrometric chart.

The preceding methods require fairly elaborate equipment carefully built and operated. For occasional use, convenient methods are available requiring only minimum equipment. These methods are based on the lowering of vapor pressure of water when a chemical is dissolved in it. Accurate data are available on this effect for certain solutions of known strength, and this is therefore the basis of a method for establishing a known humidity in a test chamber. The simplest method of maintaining a known strength of solution is to use a saturated solution. Data on a wide range of such solutions are given in the "International Critical Tables." Selected values from these tables, supplemented by measurements at the National Bureau of Standards, are given in Ref. 5.

Pure salts and distilled water must be used. The solution is made up with enough excess salt to make a slushy mixture. It may be in the bottom of a glass jar used as the test chamber or in a glass, hard-rubber, or enameled-ware tray in the bottom of a metal chamber. The chamber must in any case be tightly closed, as large a surface of solution exposed as possible, and means provided for stirring the air. For small chambers stirring may be omitted if adequate time is given to ensure equilibrium—until successive readings at 15-min intervals agree. The saturated-salt-solution method produces a relative humidity which is only very slightly affected by variations in room temperature; that is, the variation is only about 1 per cent relative humidity for the normal range of room temperature. Note, however, that this is true only if

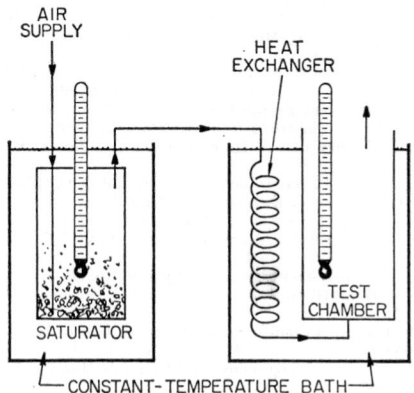

FIG. 6. Two-temperature humidity apparatus.

the entire test chamber and salt solution changes uniformly. If on the contrary the salt solution is at 70°F and the upper part of the chamber where the instrument is located is 72°F, the 75 per cent relative humidity immediately above the salt bath will become 70 per cent at the instrument. In all humidity-test-chamber work it is important to avoid the use of hygroscopic materials such as wood.

The alternate procedure of checking is comparison of the reading of the instrument under test against some instrument accepted as standard. The sling psychrometer is very commonly used for this purpose. While adequate for most industrial purposes it cannot be considered a precision method; agreement within 5 per cent should be considered good unless operated with extreme care. In addition to the sources of error inherent in the use of a sling psychrometer the water evaporated from the wick may change conditions and the temperature at the instrument under test may be different from that at the sling.

A typical case is a room for conditioning and testing samples of paper or textiles. In addition to the air-conditioning controls a hair-element relative-humidity recorder provides a record of conditions in the room. A typical location would be on a shelf above the sample racks, 7 ft above the floor. Periodic checks of the recorder are to be made using a sling psychrometer. Since it takes space to use a sling, the check cannot be made very close to the recorder and will probably be at about 4 ft from the floor in the main aisle. It may safely be assumed that vapor equilibrium will be reached between the two points; but if temperature varies, relative humidity will also vary roughly 2 per cent per °F. The only safe way to do such checking, assuming it must be done in place, is to check temperature at the recorder, reduce readings to dew point, and compare with the dew point reduced from sling readings. Dew points should agree even though circulation is poor provided there are no large additions or losses of water vapor taking place.

A more accurate way of checking is to measure dew point directly using a manually operated dew-point device. If operated with reasonable care, readings should be within 1°, permitting relative humidity calculation to 3 per cent. Better results are possible with good instruments and the best technique.

REFERENCES

1. Marvin, C. F.: "Psychrometric Tables," U.S. Weather Bureau No. 235, Superintendent of Documents, Washington, D.C., 1900.
2. Zimmerman, O. T., and I. Lavine: "Psychrometric Tables and Charts," Industrial Research Service, Dover, N.H., 1945.
3. American Society of Refrigerating Engineers: Psychrometric Theory, "Data Book," 6th ed., Chap. 4, New York, 1949.
4. American Society of Heating, Refrigerating and Air-Conditioning Engineers: Thermodynamics, "Heating, Ventilating, Air Conditioning Guide," Chap. 3, New York (revised annually).
5. Wexler, A., and W. G. Brombacher: Methods of Measuring Humidity and Testing Hygrometers, Natl. Bur. Std. (U.S.) Circ. 512, 1951.
6. Smithsonian Institution: "Smithsonian Meteorological Tables," Washington, D.C., 1951.
7. Ivory, J.: On the Hygrometer by Evaporation, Phil. Mag., vol. 60, p. 81, 1822.
8. Maxwell, J. C.: Diffusion, "Encyclopaedia Britannica," 9th ed., 1877.
9. Carrier, W. H.: Rational Psychrometric Formulae, Trans. ASME, vol. 33, p. 1005, 1911.
10. Whipple, F. W. J.: The Theory of the Hair Hygrometer, Trans. Phys. Soc. (London), vol. 34, 1921–1922.
11. Arnold, J. H.: The Theory of the Psychrometer, Physics, vol. 4, pp. 255, 334, 1933.
12. Carrier, W. H., and C. O. Mackey: A Review of Existing Psychrometric Data in Relation to Practical Engineering Problems, Trans. ASME, vol. 59, pp. 32, 528, 1937.
13. Dunmore, F. D.: An Improved Electrical Hygrometer, J. Res. Natl. Bur. Std., vol. 23, p. 701, 1939, RP-1265.
14. Hickes, W. F.: Humidity Measurement by a New System, Refrig. Eng., vol. 54, p. 351, 1947.
15. Sherwood, T. K.: The Curious History of the Wet Bulb Hygrometer, Chem. Can., June, 1950, pp. 19–22.
16. Jennings, B. H., and A. Torloni: Psychrometric Charts for Use at Altitudes above Sea Level, Refrig. Eng., vol. 62, p. 71, 1954.
17. Considine, D. M. (ed.): "Process Instruments and Controls Handbook," McGraw-Hill Book Company, Inc., New York, 1957.
18. Fishburn, R. E.: Measurement and Control of Humidity, Automation, vol. 10, no. 1, pp. 61–69, January, 1963.
19. Harrison, L. P.: Fundamental Concepts and Definitions Relating to Humidity, 1963 International Symposium on Humidity and Moisture, American Society of Heating, Refrigerating and Air-Conditioning Engineers, New York, 1963.
20. Mathews, D. A.: Review of Lithium Chloride Radiosonde Hygrometer Elements, ibid.
21. Handegord, G. O., C. P. Hedlin, and F. N. Trofimenkoff: A Study of the Accuracy of Dunmore Type Humidity Sensors, ibid.
22. Kobayashi, J., and Y. Toyama: On the Aging Effect of an Electrolytic Hygrometer, ibid.
23. Hedlin, C. P., and F. N. Trofimenkoff: An Investigation of the Accuracy and Response Rate of a Lithium Chloride Heated Electrical Hygrometer, ibid.
24. Handegord, G. O., and C. E. Till: An Application of the Dunmore Electric Hygrometer to Humidity Measurement at Low Temperatures, ibid.
25. Szalai, L.: Experimental Apparatus for the Determination of Equilibrium Relative Humidity in the Temperature Range of 0 to −30°C, ibid.
26. Nelson, D. E., and E. J. Amdur: A Relative Humidity Sensor Based on the Capacitance Variation of a Plastic Film Condenser, ibid.
27. Duggan, S. R., and C. E. Johnson, Jr., Humidity Meter Using Cerium Titanate Elements, ibid.
28. Amdur, E. J., D. E. Nelson, and J. C. Foster: A Ceramic Relative Humidity Sensor, ibid.
29. Smith, P. E., Jr., and Lucien Brouha: Humidity Effects on the Comfort and Well-being of People, ibid.
30. Stine, S. L.: Carbon Humidity Elements; Manufacture, Performance and Theory of Operation, ibid.
31. Marchgraber, R. M.: Transient Behavior of the Carbon Humidity Element, ibid.
32. Nelson, D. E., and E. J. Amdur: The Mode of Operation of Saturation Temperature Hygrometers Based on Electrical Detection of a Salt-solution Phase Transition, ibid.
33. Wylie, R. G.: Accurate Hygrometry with Ionic Single Crystals, ibid.
34. Musa, R. C., and G. L. Schnable: Poly-electrolytic Resistance Humidity Elements, ibid.
35. Benedict, W. S.: Theoretical Bases for Spectroscopic Determination of Humidity, ibid.
36. Staats, W. F., L. K. Foskett, and H. P. Jensen: Infrared Absorption Hygrometer, ibid.

37. Wood, R. C.: The Infrared Hygrometer—Its Application to Difficult Humidity Measurement Problems, *ibid.*
38. Tank, W. G.: A Long-path Infrared Hygrometer, *ibid.*
39. Tillman, J. E.: The Measurement of Water Vapor Density by the Absorption of Vacuum Ultraviolet Radiation, *ibid.*
40. Sivadjian, M. J.: Detection and Measurement of Humidity and Moisture by the Hygrophotographic Technique, *ibid.*
41. Wildhack, W. A., T. A. Perls, and C. W. Kissinger: Continuous Absorption Hygrometry with a Pneumatic Bridge Utilizing Critical Flow, *ibid.*
42. Kennedy, J. E., and L. E. Machattie: The Measurement of Relative Humidity in Confined Spaces *ibid.*
43. Davey, F. K.: Characteristics of Human Hair Affecting the Design and Performance of Humidity Sensing Instruments, *ibid.*
44. Flumerfelt, G. C.: A Sensitive Thermoelectric Water Vapor Recorder, *ibid.*
45. Smith, S. H., Jr.: A Water Vapor Partial Pressure Meter, *ibid.*
46. Czuha, Michael, Jr.: Adaptation of the Electrolytic Moisture Detector to Atmospheric Humidity Measurement, *ibid.*
47. Wexler, Arnold, and R. W. Hyland: The NBS Standard Hygrometer, *ibid.*
48. Greenspan, Lewis: A Pneumatic Bridge Hygrometer for Use as a Working Humidity Standard, *ibid.*
49. Rogers, P. A.: Testing Techniques for the Carbon Humidity Element, *ibid.*
50. Paine, Louis, and H. R. Farrah: Design and Applications of the High Performance Dewpoint Hygrometers, *ibid.*
51. Chamberlin, J. L., J. L. Hartley, and W. D. Huff: Dewpoint Apparatus of High Accuracy, *ibid.*
52. Lene, O. J.: Improvements in Dewpoint Measurements of Gases by the Use of Peltier Devices, *ibid.*
53. Beaubien, D. J., and C. C. Francisco: An Automatic Dewpoint Hygrometer with Thermoelectric Cooling, *ibid.*
54. Brewer, A. W.: The Dew or Frost Point Hygrometer, *ibid.*
55. Cherry, R. H.: A Critical Survey of Thermal Conductivity Gas Analysis in Hygrometric Applications, *ibid.*
56. Wylie, R. G., and D. K. Davies: The Basic Process of the Dewpoint Hygrometer, *ibid.*
57. Greenland, K. M., and S. Martin: Dewpoint Hygrometer and Frost Point Hygrometer, *ibid.*
58. Brendeng, Einar: A Recording Dewpoint Hygrometer, *ibid.*
59. McGavin, R. E., and M. J. Vetter: Radio Refractometry and Its Potential for Humidity Studies *ibid.*
60. Penman, H. L.: "Humidity," Reinhold Publishing Corporation, New York, 1955.
61. Hollander, Lewis E., Jr., David S. Mills, and Thomas A. Perls: Evaluation of Hygrometers for Tele-metering, *ISA J.*, vol. 7, no. 7, p. 50, July, 1960.
62. Fraade, D. J.: Measuring Moisture in Gases, *Inst. Control Systems*, vol. 36, no. 4, pp. 100–105, April, 1963.
63. Amdur, E. J.: Humidity Sensors, *Inst. Control Systems*, vol. 36, no. 6, pp. 93–97, June, 1963.

MOISTURE CONTENT OF MATERIALS

By Lee E. Cuckler*

The quantity of free water in a substance, generally expressed as a percentage of the total weight of the substance, is one of the most common analytical tests which a chemist or engineer is required to make when assessing a raw material or product. This is especially true in those industries where organic materials, such as foods, tobacco, textiles, paper, and biologicals, are handled. Several instrumental methods of moisture testing are available.†

This subsection is designed to provide valuable background information on the character of moisture as a manufacturing variable and to describe fundamental methods of moisture determination. Even when automatic instrumentation is used, the laboratory testing methods described must be employed for instrument calibration and periodic checking. Available commercial methods are reviewed briefly also.

Free moisture (water) is normally present in many materials. The amount present is generally dependent upon the chemical and physical properties of the substance, as well as upon the manner in which the material has been handled. In addition to water being present in the free state, it may be physically or chemically bound to the substance. Thus, when moisture tests are made, the water must be regarded as possibly being present in three distinct phases:

1. *Free or surplus moisture*, similar to the water in a wet sponge.
2. *Physically bound moisture*, as found in many organic substances, particularly colloids.
3. *Chemically combined hydrogen and oxygen*, which under favorable conditions may be chemically released to form either free or physically bound moisture. Water of crystallization is in this category.

NOMENCLATURE

A number of terms are in use to denote the ratio or percentage of moisture in a given material. Some of these terms are peculiar to specific industries. There is no single term universally established which has common acceptance in all industries. Present practice indicates the term *moisture content* as the best descriptive term.

Moisture Content vs. Per Cent Solids

1. *Moisture content*, usually stated as per cent, is the term used when evaluating the water in solids, fibrous, granular, and nonaqueous substances.
2. *Per cent solids* usually is the term applied when analyzing mechanical mixtures of solids in aqueous solutions, such as slurries.

Other terms which are used to denote either the ratio of water to the total product, or the ratio of solids to the total product are:
3. *Consistency*, a term associated with the paper-pulp industry.
4. *Regain*, a designation used in the textile and paper industries.
5. *Air dry*, a term used in the paper pulp and, occasionally, the lumber industry.

* Product Manager, Industrial Instruments, Aeronautical and Instrument Division, Robertshaw-Fulton Controls Company, Anaheim, Calif.
† Many of these methods are described in detail in Ref. 14.

Definitions of Common Terms

Air Dry. The percentage to which drying has been advanced toward a point where the remaining moisture represents 10 per cent of the total weight. This term is applied principally to paper pulp, paper, and occasionally, the lumber industry.

Bone Dry (also Oven Dry and Moisture Free). All the free water has been removed by drying.

Commercial Dry Basis. Moisture content, expressed as pounds of water per pound of solid, as materials leave a dryer or processing machine.

Commercial Moisture Standards. Specific moisture content which has been established by trade practices, such as government specifications and mutual agreement within industries, for various materials.

Consistency. The per cent of weight of air-dry fiber in any combination of fiber and water. This term is used only in the paper-pulp industry.

Dry Basis or Dry-weight Basis. The expression of moisture in a material as a percentage of the weight of the bone-dry material. This term is applied primarily to the lumber and most of the textile industry.

Moisture Content. The quantity of moisture (free water), usually expressed as a percentage, per unit weight or volume of the dry or wet solid. Through common usage, this term is more closely associated with the wet solid (original weight or volume), thus the wet basis.

Moisture Regain. See Dry Basis; also Table 1 and Fig. 2.

Moisture Regain, Commercial. An arbitrary figure formally adopted as the regain to be used in calculating the commercial or legal weight of shipments or deliveries of any specific textile material.

Standard Conditions (Textile Fibers). Standard conditions of a material shall be reached by the material when in moisture equilibrium with a standard atmosphere of 65 per cent (± 2 per cent) relative humidity at 21°C (± 1.1°C). Moisture equilibrium is reached when two successive weighings, no less than 15 min apart, show not more than 0.1 per cent change in weight. Moisture equilibrium shall be approached from the dry side, but not moisture free. Refer to Fig. 1.

Wet-weight Basis. Expresses the moisture in a material as a percentage of the weight of the wet solid. See Moisture Content.

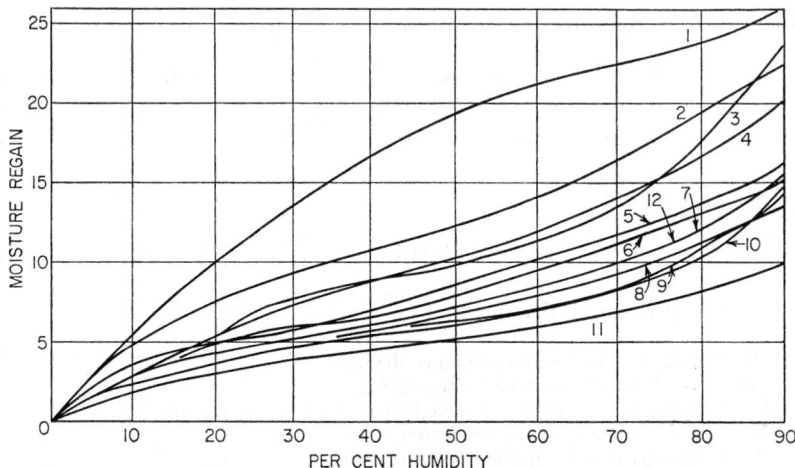

FIG. 1. Hygroscopic moisture at 25°C of natural textile fibers: (1) absorbent cotton; (2) wool, worsted; (3) silk, new yellow; (4) jute; (5) Manila hemp; (6) sisal hemp; (7) Indian cotton; (8) cotton cloth; (9) Egyptian cotton; (10) American cotton; (11) linen; and (12) flax. (*Source: "International Critical Tables," Vol. 2, p. 325.*)

CONVERSION TABLE AND FORMULAS

Regardless of the moisture-measurement method used, it usually is necessary, when tests are being set up or when results are being evaluated, to employ mathematical equations and conversion factors. For example, Table 1 is useful for converting moisture content (wet basis) to moisture regain (dry basis), or to air-dry moisture basis.

Table 1. Conversion Table of Moisture Content to Moisture Regain or Air-dry Basis

Moisture content (wet basis), %	Moisture regain (dry basis), %	Air-dry moisture basis, %
1	1.01	110.0
2	2.04	108.9
3	3.09	107.8
4	4.12	106.7
5	5.26	105.5
6	6.38	104.4
7	7.53	103.3
8	8.69	102.2
9	9.89	101.1
10	11.11	100.0
15	17.65	94.4
20	25.00	88.9
30	42.86	77.8
40	66.66	66.7
50	100.0	55.5
60	150.0	44.4
70	233.3	33.3
80	400.0	22.2
90	900.0	11.1
99	9900.0	1.1

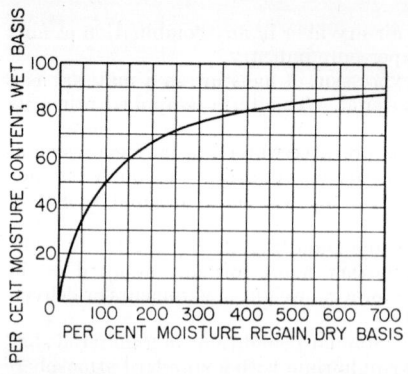

FIG. 2. Relationship between moisture regain and moisture content.

Formulas for Moisture Determinations and Conversions

$$\text{Per cent moisture regain} = \frac{W - D}{D} \times 100 = \%M_R \qquad (1)$$

$$\text{Per cent moisture content} = \frac{W - D}{W} \times 100 = \%M_C \qquad (2)$$

$$\text{Per cent moisture regain (dry basis)} = \frac{100 \times \%M_C}{100 - \%M_C} = \%M_R \qquad (3)$$

$$\text{Per cent moisture content (wet basis)} = \frac{100 \times \%M_R}{100 + \%M_R} = \%M_C \qquad (4)$$

$$\text{Per cent air dry (pulp)} = \frac{100 - \%M_C}{90} \times 100 = \%A_D \qquad (5)$$

where W = weight of wet material before drying
D = weight of dry material after drying
M_R = moisture regain; also referred to as dry basis or regain
M_C = moisture content; also referred to as wet basis
A_D = air dry, usually applied to paper pulp

Formula for Two-stage Drying. Where the moisture is calculated for each stage of drying,

$$\text{Per cent moisture content} = \%M_C = \%M_P + \frac{(100 - \%M_P)(\%M_F)}{100} \qquad (6)$$

where $\%M_P$ = percentage moisture content as calculated on the preliminary dried sample

$\%M_F$ = percentage moisture content as calculated on the final oven-dried sample which may be some fractional portion of the original wet sample

METHODS OF MOISTURE DETERMINATION

A number of techniques exist for the determination of moisture (free water) in a product. The most common methods are:

1. Thermal (oven) drying.
2. Distillation.
3. Chemical reaction.
4. Electrical.

It is generally difficult to select the one of these methods which will give the best accuracy. Each method has its critics and the proper choice depends upon such factors as (1) availability of equipment, (2) type of product, (3) speed required, (4) accuracy, and (5) general recognized practices for measuring the moisture content of such materials.

Thermal (Oven) Drying

Oven drying is probably the oldest, most commonly used laboratory method, and generally the simplest to perform. It is recognized as the standard procedure for checking all other methods. It can be applied to many substances, such as tobacco, textiles, paper, oils, and foodstuffs.

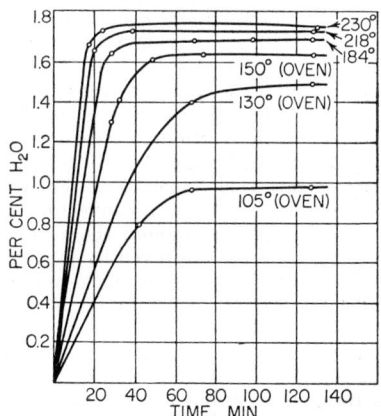

Fig. 3. Removal of water at constant temperature: (*Source; "Analytical Chemistry," Vol. 23, p. 1058, August, 1951.*)

Oven drying signifies the removal of liquids by thermal methods from a liquid, solid, fibrous, or granular material. The application of heat to a moist material will cause volatilization of the water so long as the vapor pressure of the water in the material is greater than that of the surrounding air. Water is liberated rapidly at first and gradually slows up until no more can be liberated without raising the temperature.[2]

A general rule for drying temperature and drying time is difficult to formulate. For reproducible laboratory results, select a time-temperature value on the flattest portion of a drying curve, such as shown in Fig. 3. After a time and temperature have been selected, all samples must be tested at these levels so that representative results will be obtained.   A semiautomatic oven, such as the Brabender moisture tester unit provides a convenient means of making such tests.

Techniques.   *Sampling.*   In grab or spot sample analysis, sample selection is important so that representative sampling of the entire lot is obtained.   Sampling of granular materials is best accomplished by sampling probes or triers.   If a representative sample of a number of containers is desired, each container should be sampled and all the samples blended before testing.

Minimum handling and exposure to ambient surroundings are necessary if errors are to be reduced to a minimum.   The original sample should be large enough that three or more drying samples can be obtained by dividing.   Unless preparation and weighing are to proceed immediately, the material should be placed in an airtight container.

Preparation.   If speed of drying is required, the maximum surface of the sample should be exposed.   This may require grinding, and the careful choice of drying dishes may be an important factor.   The addition of bone-dry aggregate to viscous solutions aids in increasing the drying surface and prevents crusting.

In order to blend the sample and reduce its size to a representative portion of the original sample, dividers or samplers generally are used.   If the material is coarse, it should be sieved before blending or grinding.   If the material has a high moisture content, it should be preliminarily dried to prevent moisture losses during grinding.

Weighing.   Unless the material is approximately at the equilibrium moisture with the air surrounding the scale, the material should be weighed and handled in a covered dish.   Immediately upon removal of a sample from the oven, the container or dish should be loosely covered and placed in a covered desiccant jar for 10 to 20 min, in order to cool the sample before final weighing.

Drying.   High-moisture-content samples are more efficiently handled by predrying in either a water oven or by air drying using a heater and fan.   Often predrying of wet materials is carried out under normal room conditions.   Predrying is particularly important in coarse, granular materials having very high moisture contents which require grinding and other preparation for final drying.   If the moisture is deep-seated, such as in plant and animal tissue, vacuum ovens will prove to be advantageous.

Final drying in multistage drying, or for drying of low-moisture-content samples, may be carried out in either natural-convection, forced-air, or vacuum ovens.   Natural-convection and forced-air ovens may be operated up to approximately 160°C, while vacuum ovens are generally operated from 70 to 110°C.   Forced-air ovens are generally about four times as fast as natural-convection types.   The former also maintain more uniform temperatures throughout the oven.

Summary.   Thermal-drying rates are improved by providing the maximum differential between the vapor pressure of water in or on the substance, and that of the air within the oven.   Drying conditions can generally be improved by use of diminished pressure to minimize diffusion of deep-seated moisture, using desiccanted air, and using oven temperatures as high as possible.

Table 2 is a partial tabulation of drying techniques for various materials from Ref. 4.   Additional moisture specifications can be found in ASTM Standards.[6]

Advantages and Limitations of Oven Drying.   Advantages of oven drying include: (1) not limited to either micro or macro samples, (2) does not require exceptional operator skill, (3) determinations are relatively fast, (4) many tests can be made per day, and (5) equipment is rugged and foolproof, with low cost of upkeep.

It appears that oven drying would offer a high degree of accuracy if the technique were employed carefully, but this is often not true.   One difficulty is that this method does not differentiate between water and other volatile materials.   Thermal methods may lead to either too low or too high indicated moistures, because of thermally unstable substances.   High apparent moisture contents may result in the release of carbon dioxide or volatile oils from the sample.   Low moisture contents may be due to incomplete drying or oxidation of the sample.

Precautionary factors to observe are:

Nonwater Components.   Volatile substances which have an appreciable vapor pressure under the conditions of drying are the most difficult to minimize.   Possible solu-

Table 2. Condensed Summary of Drying Techniques for Various Materials

Material	Sampling	Preparation	Sample weight, g	Oven Type	Oven Temp., °C	Oven Time	Remarks
Soils	Core or auger	Sift and crush	2	Air	105	5 hr	
Fertilizers	Slotted single or double trier	Grind to pass 1-mm-diam sieve	2	Water	95–100	5 hr	Distillation may be used
Insecticides and fungicides	According to requirements	Blend and pulverize		Air	105–110	CW	
Leathers	According to requirements	Wiley mill 4-mm screen	2	Air	100–102	15 hr	Multistage drying for pastes
Plants, feeds, grains, and meats	Slotted trier on grains, others as required	Plants, Wiley mill; feeds, 1-mm sieve	5–10	Vacuum	95–100	5 hr	Distillation may be used
		Grains, 1-mm sieve; meats, grind	2	Air	135 ± 2	2 hr	Multistage drying recommended for moisture contents above 30 %
Cereal, foods, and flour	Slotted tube or core	Grind where necessary	2	Vacuum	98–100	5 hr	Distillation may be used
Dried fruits	As required	Hobart mixer	2	Air	130 ± 2	CW	Distillation may be used
Eggs	As required	Predry if liquid	5–10	Vacuum	70	6 hr	
Coffee, tea, and cocoa	Slotted tube	Grind	2	Vacuum	98–100	5 hr	Dry to constant weight
Oils, fats, and waxes	As required	Shallow dish	5	Air	100–110	5 hr	Dry to constant weight
Malt and wet-spent grains	As required	Grind to No. 30	5–10	Vacuum	80–110	CW	Dry to constant weight
			5	Air	103–105	3 hr	Predry spent grains

CW = constant weight.

tions are collection of the volatilized water as ice, use of an absorbent which is specific to water, or indirect methods.

Chemically Bonded Water. Chemical reaction induced by heat can cause loss of chemically bound water, as in the case of dextrin formations and the hydrolysis of proteins. Reduce by drying at lower temperatures until the bulk of the water is removed and the possibility of the reaction is at a minimum.

Nonwater Solvents. Alcohol or acetone, when used in the preparation of organic substances, cause the moisture analysis to indicate too low a value. The "incomplete drying" may be due to the solvents which have resisted volatilization during drying. Remedy: Humidify the sample in order to replace the solvent with water; then dry in the normal method.

Auto-oxidation. Causes low moisture indication because of the increased weight of the sample resulting from oxidation. More apparent in fats and oils. Can be prevented by drying in an inert gas, such as nitrogen or carbon dioxide.

Thermal Decomposition. Induced by high temperatures if the products of decomposition are volatile. Results in a high apparent moisture content. If possible to determine the critical decomposition temperature, dry below this point. Difficulty is most common with organic substances. Some inorganic materials, such as carbonates, can also decompose.

Distillation Methods

Some researchers believe that distillation methods of moisture determination are more accurate than oven methods and, therefore, often use them for the standardization of oven or electrical methods for control work.

A sample of the material is placed into a flask and is distilled with an excess of a liquid which usually has a boiling point higher than water but is immiscible with it. The water and some of the liquid are then distilled off. The combined vapors are condensed and collected, and the separated water is measured to calculate the moisture content.

Many types of glass distillation apparatus are available as stock items. Generally this apparatus is available for two basic types of distillation.

1. *Distillation of Liquids Lighter Than Water.* Brown-Duvel, Dean-Stark, Bidwell-Sterling, and the Clelland-Fetzer are the best known apparatus. The Brown-Duvel method is described in Ref. 7.

2. *Distillation of Liquid Heavier Than Water.* Hercules, Bailey, and Langland Pratt traps are available for these determinations.

Techniques. The apparatus must be kept clean and dry. Tapered joints must be lubricated before assembling, and preferably treated with graphite to permit easy separation. Dispersing agents are often used with powdered materials, such as cereals, flours, and starches, which have a tendency to lump during distillation.

A distillation rate of approximately 3 drops/sec is about optimum for most materials. A small amount of wetting agent may be used to secure a better meniscus reading of the distilled water.

Choice of the distilling liquid is important. Toluene has a boiling point of 110 to 112°C. This is high enough for most materials, especially if the sample is well dispersed. If the material is heat-sensitive, benzene or mixtures of benzene or toluene are often used. The sample size is chosen so that the amount of collected water falls within a 3.5- to 5.0-ml range.[5]

Advantages. Advantages of distillation methods include:

1. Well suited to materials which contain volatile products other than water.
2. Materials that are heat-sensitive may be reliably analyzed by this method.
3. Considered by some investigators as a "standard" control method for checking other moisture-analysis methods.
4. Oxidation of the sample by air is eliminated because of the sealed system.

Limitations. Limitations of distillation methods are:

1. Not adaptable as a routine procedure as compared with oven drying.
2. Requires more time and space than drying methods.

3. Flammable liquids endanger the operator.

4. Possibility of emulsion formation and difficulty in reading the meniscus. Distillation is not simple. The time required for analysis is 1 to 4 hr plus the time for cleaning and setting up the apparatus. Distillation should not be used if oven methods are applicable.

In analyzing heat-sensitive substances, it may be possible to standardize the oven tests (time and temperature) by means of the distillation method.

Chemical-reaction Methods

The best known and most widely used chemical means is the Karl Fischer reagent titration method. Other methods which are primarily chemical in nature are (1) those utilizing desiccants for the absorption of moisture from a product, (2) evolution of free acidic or basic compounds, (3) evolution of an inert gas, and (4) formation of an insoluble precipitate. In general, these methods fail to distinguish materials containing active hydrogen from water.

A complete description of the stoichiometry of the Karl Fischer reaction can be found in Ref. 10, which covers complete analytical procedures for the use of the Karl Fischer reagent, as well as techniques for moisture determination in many substances.

Automatic burettes of heat-resistant glass are recommended for visual titrations. Suction filling of the burette is preferable over pressure methods with the air inlet protected from moisture contamination by a desiccant. For electrometric titrations, flask adapter units are used.[11]

The titration end point can be determined either visually by the change in color or electrometrically by measuring the oxidation-reduction potential, or by observing the change in current due to polarization of the electrodes placed in the titration flask.

Micro samples can be analyzed as readily as macro samples, provided the proper apparatus and precautions are used. The method is highly recommended for thermally unstable compounds and samples that cannot be readily analyzed by other means.

Electrical Methods

A large number of electrical devices are available for measuring the moisture in a wide variety of products. Primarily the paper and textile industries have had commercial uses for such devices because the economies of over- or underdrying the product are significant.* However, electrical meters have also found a number of applications in measuring the moisture content of such materials as powders, lumber, leather, tobacco, and similar materials. Properly employed, their reproducibility of results can be very high.

Electrical methods can be classified broadly as

1. *Conductivity (or "resistance") types*, generally operating on direct current with a low current impressed across electrodes between which a representative sample of the product is passed.

2. *Capacitance types*, generally operating on alternating current with a pair of electrodes between which the product passes (also known as "dielectric constant" or "a-c impedance" methods).

Conductivity Types. These types employ some form of detector element (see Fig. 5) with electrodes forming part of a basic Wheatstone-bridge circuit with suitable indicator, recorder, and/or controller. Figure 4 shows a typical arrangement of detector for continuous measurement of sheet moisture in a textile mill; it employs a "detector roll" which rotates with movement of the product sheet under it. Models for measuring still objects employ prongs or "needle electrodes" which can penetrate into the material to be measured.

The electrical resistance or conductivity of certain materials varies considerably with moisture content, thus providing a suitable basis for satisfactory measurement.

* For example, see p. 13-76 for use of moisture measurement as part of complete instrumentation for the dry end of a paper machine.

Where high resistance is involved, however, very high quality insulation on the detector electrodes and connecting cable is required. In some materials such as paper, also, the exact relationship between measured resistance and moisture content is affected by such variables as chemical additives, surface condition, electrode contact pressure, and temperature.

Capacitance Type. This type operates on the basis of change in dielectric constant (specified inductive capacitance) of a material when moist and dry. The dielectric constant for paper, for example, is about 2.8 when oven-dried, and that of water is in the order of 80. Hence, a small change in moisture content causes a relatively large change in the dielectric constant.

Capacitance moisture-measuring systems generally employ r-f oscillators whose output is applied to suitable bridge or measuring circuits. The material to be measured is inserted into or between an appropriate electrode (condenser) assembly which is connected to the measuring circuit (see Fig. 5).

One design[16] for measuring paper moisture content employs a condenser assembly composed of about a 4-in.-diameter metal ring within which are other smaller concentric rings all cast into grooves in the surface of a quartz plate. Alternate rings are

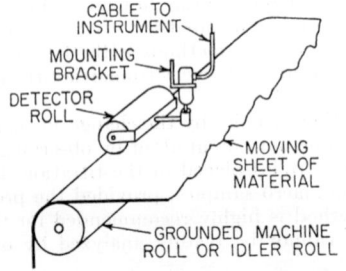

FIG. 4. Typical detector of the conductivity type used in the paper and textile industries.

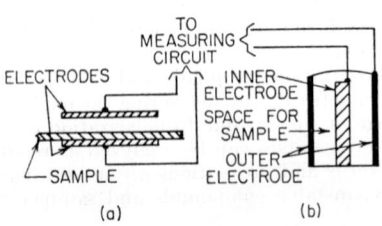

FIG. 5. Two basic forms of capacitance-type electrodes: (a) parallel plates; (b) cylindrical.

connected together to form the two plates of the condenser. The paper forms a dielectric in the fringe electrostatic field established between adjacent rings when the assembly rests on the moving sheet.

By its design the ring-type detector can be used to "scan" across the width of a paper sheet giving instantaneous readings to check the uniformity of moisture content and provide a means for automatic control. The method has been used successfully with a series of individually controlled spray nozzles across the sheet width. As the detector scans the sheet, the resulting measurement is used to adjust the spray so as to maintain the desired moisture content uniformly across the sheet width.

Capacitance-type moisture-measuring instruments are generally used for materials having moisture contents below 20 to 25 per cent. They are not affected by the majority of additive agents; in most designs the sampling electrodes need not contact the measured materials; and they are not sensitive to contact pressure or temperature changes.

Humidity-type Measurement

Humidity-measuring elements described earlier in this handbook (pages 7-11 to 7-26), can be employed to determine moisture content of materials which are not changing their content rapidly. The element is mounted in a small chamber held close and open to the material to be measured. The air in the chamber will have a humidity related to the moisture content of the material; hence measurement of the humidity is a measure of the amount of moisture (see Ref. 14, p. 7-159). Electrical-resistance-type hygrometers employing small measuring elements with relatively high speeds of response are well suited to this type of application. Refer to Fig. 6.

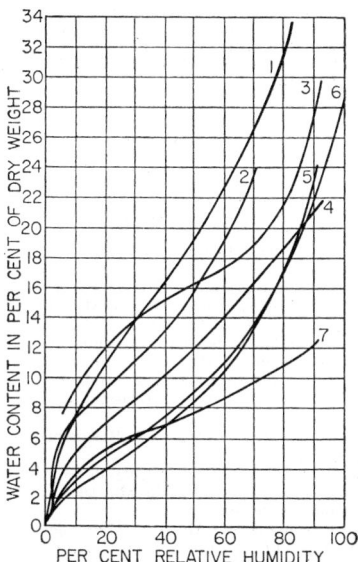

PER CENT RELATIVE HUMIDITY

FIG. 6. Hygroscopic moisture at 25°C of some organic substances: (1) North Carolina leaf tobacco; (2) cigarette tobacco ("Fatima"); (3) sole leather (oak tanned); (4) catgut; (5) soap ("Ivory"); (6) lumber; (7) glue (hide, first grade). (*Source: "International Critical Tables," Vol. 2, p. 325.*) Regarding item (6), all species of lumber have approximately same values at given relative humidities. Rate of absorption or evaporation varies according to the density of the species.

Infrared Absorption

A relatively recent development for measuring paper moisture content utilizes the absorption of infrared rays. A particular wavelength is selected at which the absorption by water is a maximum and that by cellulose is a minimum. Because at some other wavelength just the reverse is true, this technique permits direct measurement of substance (basis weight or the weight per unit area.)[17]

REFERENCES

1. Willits, C. O.: Methods for Determination of Moisture, *Anal. Chem.*, vol. 23, p. 1058, August, 1951.
2. Nelson, O. A., and G. A. Hulett: *Ind. Eng. Chem.*, vol. 12, p. 40, December, 1920.
3. Porter, W. L., and C. O. Willits: *J. Assoc. Official Agr. Chemists*, vol. 27, p. 179, June, 1944.
4. Association of Official Agricultural Chemists: "Official and Tentative Methods of Analysis," 6th ed., Washington, D.C., 1945.
5. Fetzer, W. R.: *Ind. Eng. Chem., Anal. Ed.*, vol. 23, no. 8, p. 1062, August, 1951.
6. American Society for Testing Materials: Various ASTM standards on moisture testing, available from society headquarters, Philadelphia, Pa.
7. "The Brown Duvel Moisture Tester and How to Operate It," *U.S. Dept. Agr. Bull.* 1375, February, 1926.
8. Hoffman, J. F.: *Z. Angew. Chem.*, vol. 21, p. 890, 1908.
9. Fischer, Karl: *Z. Angew. Chem.*, vol. 48, p. 394, 1935.
10. Mitchell, John, Jr., and D. M. Smith: "Aquametry," Interscience Publishers, Inc., New York, 1948.
11. Almy, E. G., W. E. Griffin, and C. S. Wilcot: *Ind. Eng. Chem., Anal. Ed.*, vol. 12, pp. 392–396, December, 1940.
12. Toner, R. K., C. F. Bowen, and J. C. Whitwell: *Textile Res. J.*, part 1, vol. 18, p. 9, 1948; part 2, vol. 19, p. 1, 1949; part 3, vol. 19, p. 11, 1949; part 4, vol. 20, p. 6, 1950.
13. Hlynka, I., B. Martens, and J. A. Anderson: *Can. J. Res.*, vol. F. 27, p. 382, 1949.

14. Cuckler, L. E.: Moisture-measuring Instruments, pp. 7–151 to 7–163, "Process Instruments and Controls Handbook," McGraw-Hill Book Company, Inc., New York, 1957.
15. Lippke, P., and W. H. Rau: The Measurement and Control of Moisture with Hygrotester Instruments, *Pulp Paper Mag. Can.*, vol. 57, no. 3, p. 252, 1956.
16. Gardner, R. C., and F. Church: The Automatic Control of the Transverse Moisture Profile of a Paper Web, *World's Paper Trade Rev.*, Aug. 31, 1961.
17. Poly-Tech Research, Inc.: Continuously Recording Basis Weight without Moisture Errors, *Paper Trade J.*, vol. 145, no. 7, p. 59, 1961.
18. Weise, E. L., and J. K. Taylor: Gas Chromatography in the Determination of Moisture in Grain, 1963, International Symposium on Humidity and Moisture, American Society of Heating, Refrigerating and Air-Conditioning Engineers, New York, 1963.
19. Stanley, E. E.: A Method of Measuring the Moisture Content of Air Directly in Grains per Pound of Dry Air, *ibid.*
20. Watson, Alec: Measurement and Control of Moisture Content by Microwave Absorption, *ibid.*
21. Hughes, F. J., J. L. Vaala, and R. B. Koch: Improvement of Moisture Determination by Dielectric Constant through Density Correction, *ibid.*
22. Outwater, J. O.: A Portable Electronic Moisture Detector for Reinforced Plastics and Its Application, *ibid.*
23. Szuk, Geza: Conductimetric Determination of Surface Moisture of Building Materials, *ibid.*
24. Bouyoucos, G. J.: Plaster of Paris Block as an Electrical Measuring Unit for Making a Continuous Measurement of Soil Moisture Under Field Conditions, *ibid.*
25. Fletcher, J. E.: Use of Capacitance Methods for Determining Quantities of Materials in Mixes, *ibid.*
26. Leroy, Robert: Moisture Measurement by High Frequency Currents, *ibid.*
27. Green, R. M.: Continuous Moisture Measurement in Solids, *ibid.*
28. Rollwitz, W. L.: Nuclear Magnetic Resonance as a Technique for Measuring Moisture in Liquids and Solids, *ibid.*
29. Ladner, W. R.: The Application of Nuclear Magnetic Resonance to the Measurement of Moisture Content in Coals and Cokes, *ibid.*
30. Huet, J.: Field Determination of Moisture and Density in Soils by the Nuclear Method, *ibid.*
31. Miyashita, Yoshio: Measurement and Control of the Moisture in Raw Materials Used in Iron Making Processes by Neutron Moderation, *ibid.*
32. Burn, K. N.: Calibration of a Neutron Soil Moisture Meter, *ibid.*
33. DeBeer, E. E., and E. H. G. Goelen: Measurement of Moisture Content of Soil by Radioisotopes, *ibid.*
34. Hukill, W. V.: Moisture in Grain, *ibid.*
35. Hunt, H. W.: Problems Associated with Moisture Determination in Grain and Related Crops, *ibid.*
36. Young, J. H., J. M. Bunn, and W. H. Henson: Humidity and Moisture Problems Associated with the Handling and Storage of Cured Tobacco, *ibid.*
37. Hughes, F. J., J. L. Vaala, and R. B. Koch: Rapid Moisture Measurement in Flour by Hygrometry, *ibid.*
38. Prentice, J. H.: The Distribution of Moisture in Butter, *ibid.*
39. Feldman, R. F., and P. J. Sereda: Moisture Content, Its Significance and Interaction in a Porous Body, *ibid.*
40. Johansson, G.: Moisture Analysis by Use of Microwaves, *ibid.*
41. Fraade, D. J.: Measuring Moisture in Solids, *Instr. Control Systems*, vol. 36, no. 2, pp. 99–105, February, 1963.
42. Carver, R. L.: Neutrons Measure Moisture Content in Solids, *Instr. Control Systems*, vol. 36, no. 5, pp. 106–107, May, 1963.

VISCOSITY AND CONSISTENCY

By Richard Mumma*

Viscosity and consistency are, in a broad sense, related terms applied to a specific characteristic of fluids, namely, resistance to flow or deformation when subjected to a shearing† force. They can be either a test property or an important process variable, usually of a liquid or liquid-like substance. The distinction between viscosity and consistency, as well as a basic understanding of the multitude of units that have evolved to express measurements, requires some background in rheology—the study of internal mechanics in the flow of liquids and suspensions. This section first gives this background, starting with Newton's early investigations; then the two variables are defined, their units of measurement given, and their applications reviewed.

Viscosity-measuring instruments, of which there are many, are called *viscosimeters*, or, more commonly, *viscometers*. Consistency-measuring instruments are usually called just that, or, where they are used for automatic control, *consistency regulators*. Various measuring methods for the two variables are summarized in Table 2 at the end of this section. Reference 9 (pp. 7-85 to 7-95) describes most of the devices covered n Table 2; several other listed references also cover one or more such instruments.

MEASUREMENT THEORY—RHEOLOGY BASICS

As is true of many variables accepted today as almost commonplace, viscosity as a fluid property evolved from studies of scientists in the laboratory. Hypotheses and theories were developed from experiments and observations by Newton, Poiseuille, and others. Then came the science of rheology.

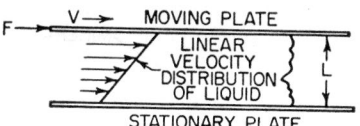

FIG. 1. Newton's definitions of viscosity with two parallel plates.

Newton's Basic Definition of Viscosity

Newton is credited with the first theoretical definition of viscosity, expressing it in terms of a constant force resisting movement of liquid between two flat, parallel planes (Fig. 1). The planes are separated by a unit-thick layer of liquid and move relative to each other at unit speed. The basic relationship stated by Newton is:

$$\frac{F}{A} = \mu \frac{V}{L} \tag{1}$$

* Product Manager, Fischer & Porter Company, Warminster, Pa.
† In fact, the basic definition of a fluid is a substance which undergoes continuous deformation when subjected to a shear stress. (See Ref. 4, section on "Fluids in Motion," for background helpful to an understanding of viscosity- and consistency-measuring problems.)

where F/A (force per unit area) = shear stress
$\quad V/L$ (velocity per unit thickness of layer) = shear rate
$\quad \mu$ = proportionality constant
Whence, viscosity came to be defined as

$$\mu = \frac{\text{shear stress}}{\text{shear rate}} \tag{2}$$

with μ actually being the "viscosity coefficient," later being called simply *absolute* or *dynamic viscosity*. From the basic formula [Eq. (2)], it is obvious that liquids that tend to resist movement more than others (have a higher shear stress) are "more viscous" (have a higher viscosity).

Hagen-Poiseuille Law

A more practical definition of viscosity was developed independently by **Hagen in** Germany and Poiseuille in France. They described it in modern terms as the ratio of shear stress versus shear rate at the wall of a capillary tube:

$$\mu = \frac{PR/2L \text{ (shear stress)}}{4Q/\pi R^3 \text{ (shear rate)}} = \frac{\pi P R^4}{8QL} \tag{3}$$

where P = pressure differential across liquid in tube
$\quad R$ = inside radius of tube
$\quad L$ = length of tube
$\quad Q$ = volume rate of flow of liquid
$\quad \mu$ = absolute (or dynamic) viscosity
The above Eq. (3) is limited to conditions of laminar or viscous (streamline) flow.

Rheology Basics

Out of the early studies of Newton and others grew the science of rheology (from the Greek *rheo*, a flowing) which employs *rheograms* (Fig. 2) to determine and define the characteristics of any fluid.* The force causing deformation (shear stress) is plotted

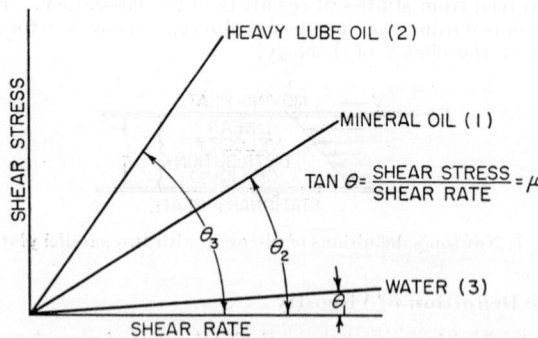

FIG. 2. Rheograms of typical Newtonian materials (constant slope, linear lines with origins at zero).

against the resulting flow (shear rate) under constant conditions of temperature and pressure. Such plots, sometimes called *shear diagrams*, are vital guides to proper understanding of the measurement of viscosity or consistency.

From studies of various viscous materials and their rheograms, the following terms have developed:

* For more information on this science, refer to textbooks listed in the references at the end of this section.

Newtonian Liquids. These are liquids or suspensions in liquids which obey the Newtonian law and thus produce a rheogram consisting of a straight line passing through the graph origin (curve 1, Fig. 2). Generally, pure liquids, most oils (e.g., lubricating, fuel, and hydraulic oils), true solutions, and dilute suspensions are Newtonian, or very nearly so. As shown in Fig. 2, the slope of the straight line is much steeper for more viscous liquids like heavy lube oil (curve 2) than for less viscous liquids like water (curve 3).

Non-Newtonian Liquids. Many everyday fluids do not obey Newton's law— that is, they do not exhibit a straight-line rheogram starting at zero stress and rate— and are therefore said to be *non-Newtonian*. Rheograms of these types of fluids have a variety of curve shapes, as shown in Fig. 3. Note that, by definition, *none of these materials has a constant viscosity at all shear rates.*

From rheological studies, similar-acting non-Newtonian fluids have been given names, as covered briefly below. However, nothing is more misleading in the application of viscosity- or consistency-measuring instruments than to attempt to generalize on the nature of the measured fluid. As discussed below, in order to ensure satisfactory and usable results, it is much wiser to consider every fluid as possibly being unique and

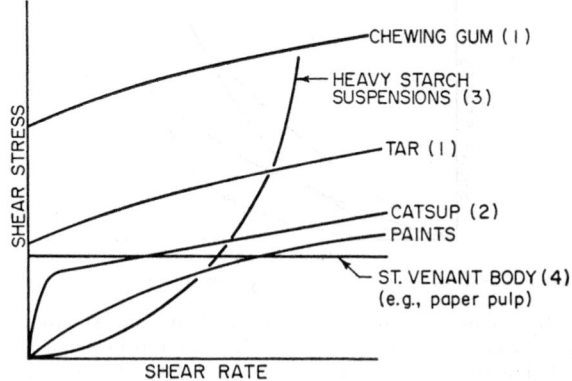

FIG. 3. Rheograms of typical non-Newtonian materials.

be sure of its rheological properties by suitable tests prior to choosing the measuring instrument.

1. Bingham Body (True Plastic). These materials are characterized as having an initial value of shear stress, termed the "yield stress," above which flow starts and after which the relationship is essentially linear. Curves 1 of Fig. 3 for chewing gum and tar are typical. Many rheologists believe that true plastics are rare and that most are really "pseudoplastic."

2. Pseudoplastic. Such materials exhibit an initial slow shear rate with increase in shear stress, until a rather definite "apparent" yield stress is exceeded. Then the curve is nearly linear and the fluid acts like a true plastic or Bingham body. Curve 2 in Fig. 3 for catsup is typical of this behavior.

3. Dilatant. These materials behave just the opposite from pseudoplastics, requiring an ever-increasing applied shear stress per unit increase in shear rate (curve 3, Fig. 3). Although there is no known relationship between this type and pseudoplastics, dilatant materials are sometimes referred to as *inverted plastics*, or *inverted pseudoplastics*, because of their opposite behavior. Heavy starch suspensions and quicksand are examples of dilatant substances; in general, such materials usually contain rather coarse particles in suspension.*

---

* The term "dilatant" comes from the theory that particle suspensions are in a state of minimum voids when the fluid is at rest; any attempt to put it into flow *dilates* the voids and thereby increases resistance to flow.

4. St. Venant Body. This is similar to a Bingham body in that the material has a yield stress, but unlike a Bingham body in that it has practically no slope viscosity (see curve 4 of Fig. 3). Unfilled paper pulp is a typical example. With such materials, "per cent insoluble solids" can be correlated with the yield stress; thus, a single-point measurement completely defines the rheogram (see later discussion of Consistency).

5. Thixotropic. This term is often confused with "pseudoplastic" because a single-plot curve will show decreasing viscosity with increasing shear rate (curve 5 portion with upward arrows in Fig. 4). However, thixotropic materials exhibit a time-dependent property in that the shear rate increases with increasing duration of agitation so that a return plot yields a hysteresis effect (downward part of curve 5 in Fig. 4). Further, upon reagitation, generally they require less stress to create a given shear rate than was required after the first agitation. Hence, rheograms of thixotropic materials depend on the shear history and the elapsed time over which the measurements are made.

Caution: See paragraph below on Identifying Thixotropic and Rheopectic Fluids to avoid misapplication of viscosity- or consistency-measuring instruments.

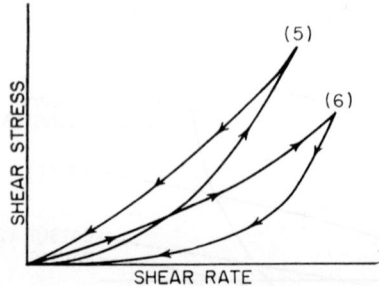

FIG. 4. Rheograms of thixotropic (5) and rheopectic (6) non-Newtonian fluids.

6. Rheopectic. This is another time-dependent non-Newtonian fluid which resembles a dilatant fluid in that the shear rate is higher at lower shear stresses (curve 6, Fig. 4). Such fluids also exhibit a hysteresis effect with time; the apparent viscosity increases, resulting in the downward curve illustrated for curve 6. This is a phenomenon observed with certain thixotropic suspensions under the action of rhythmic shaking or tapping which causes the suspension to set or build up quite rapidly.

Identifying Thixotropic and Rheopectic Fluids. An effective way to determine if a substance is either thixotropic or rheopectic is to plot shear stress versus time with a sample in which shear rate is held at a constant value. This can be done with a viscometer of the disk or paddle (torque) type (see Table 1). If no change in stress is noted, the liquid is not "time-dependent," nor affected by the past history of agitation.

Apparent Viscosity

From the foregoing discussion, it is obvious that the term *viscosity* can be applied in the strict sense of the word only to Newtonian fluids. Because a viscometer may well be used on many non-Newtonian fluids, however, the term *apparent viscosity* has come into use. So long as the user always remembers that this value is a *variable* depending upon the particular rate of shear, and possibly the shear history of the substance, he can often use the concept in process measurement and control in a meaningful way. Also, as will be brought out later, the type of instrument employed will affect the measured value of viscosity.

Consistency

The term *consistency* also relates to the resistance of a liquid or liquid-like material to deformation when subjected to a shear stress. Hence, it, too, is a "flowability" characteristic of the liquid. However, consistency is not related in any known way with viscosity, and different units are used to represent its values. It is a measure of different but similar properties in industry: "palatability" in food, "tackiness" in printing, "thickness" in paint, and "per cent insoluble solids" in paper.

In the paper industry and in parts of the food and mining industries, the expression for consistency denotes the following relationship:

$$\text{Consistency } (\%) = \frac{\text{air-dry weight of solids}}{\text{weight of solids plus water}} \times 100 \qquad (4)$$

Papermakers also use the terms *bone-dry*, *oven-dry*, and *moisture-free* in place of *air-dry* in the above expression. Air-dry pulp usually is considered to contain 10 per cent moisture. Thus, bone-dry consistency (per cent) equals 0.9 times air-dry consistency (per cent).

UNITS OF VISCOSITY AND VISCOSITY SCALES

Viscosity measurement by a variety of means has evolved a corresponding variety of units and scales. Only the more common ones are summarized below.

Dynamic (Absolute) Viscosity Units

From Poiseuille and Hagen [Eq. (3)] came the absolute or dynamic unit of viscosity in cgs units, the *poise* which is equivalent to 1 dyne-sec/cm² (no unit of absolute viscosity exists in the English system). Smaller values of viscosity are commonly expressed in *centipoises* (1 cp = 0.01 poise). The viscosity of water at 68°F is 1 cp.

Kinematic Viscosity Units

Many laboratory viscometers measure *kinematic viscosity*, which is defined as the quotient of dynamic viscosity divided by the density of the fluid (at the same temperature). The unit for kinematic viscosity is the *stoke* (ν). For smaller values there is the *centistoke* (1 cs = 0.01 stoke). The relationship between kinematic and absolute viscosity is thus:

$$\text{Kinematic viscosity (stokes)} = \frac{\text{absolute viscosity (poises)}}{\text{density (g/cc)}}$$

Viscosity Index

This term has been used with reference to petroleum products. It is an empirical number which indicates the effect of change of temperature on the viscosity of an oil. A low index number signifies an oil that changes viscosity by a large amount with a given temperature change. (References 8 and 12 describe this term in detail.)

Viscosity Scales

The majority of viscometers are of the kinematic type, but measurements, particularly in the laboratory, are often reported in *time units* (see summary of devices in Table 2). A few time-based viscosity scales are in common use, namely:
1. Saybolt scales (in the United States).
2. Redwood scales (in Great Britain).
3. Engler scales (in Europe).

All three of these scales are based on the Hagen-Poiseuille law and indicate the time of efflux (under given conditions) for a fixed volume of fluid through a specific capillary or aperture. Kinematic viscosity (and from it, absolute viscosity) can be determined from such scales value by the empirical formula

$$\nu = At - \frac{B}{t} \qquad (5)$$

where A and B = constants applicable to the measuring apparatus (viscometer)
$\qquad\qquad t$ = time of efflux, sec
Commonly accepted values of A and B are as follows:

Viscometer type	Value of constant*	
	A	B
Saybolt Universal	0.0022	1.8
Redwood	0.0026	1.72
Redwood Admiralty	0.027	20
Engler	0.00147	3.74

* These constants do not hold for certain values of t.

The accompanying chart (Fig. 5) provides a convenient means for approximating values of viscosity in the various units. References 4 and 9 include the widely used

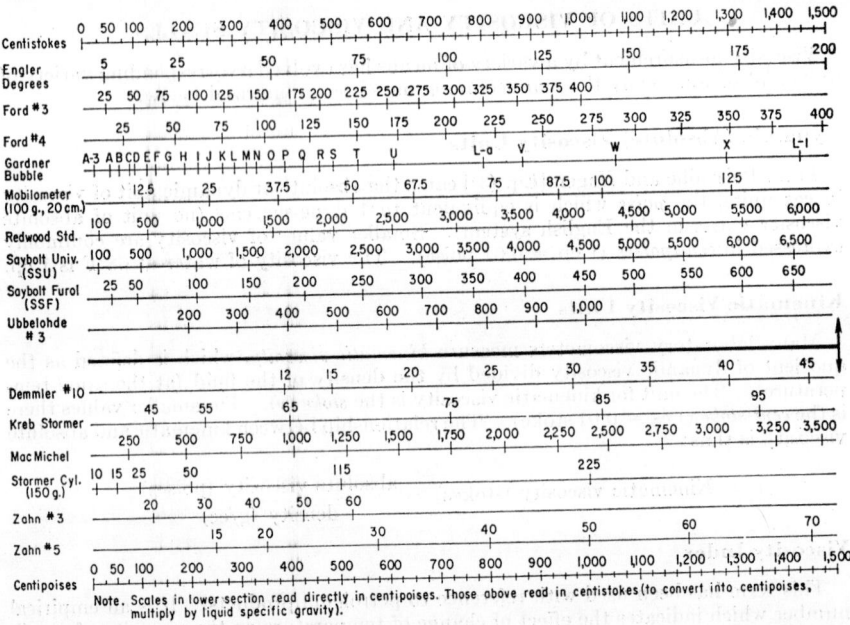

FIG. 5. Chart for converting viscosity units, giving a quick approximation, not absolute accuracy. (*Source: Ref. 13.*)

conversion chart originally copyrighted by the Texas Company in 1921. Table 1 gives useful conversion factors for accurate conversions.

Effects of Temperature and Pressure on Viscosity

Temperature and pressure changes can have marked effects on viscosity. Many basic handbooks as well as the International Critical Tables give tabular data on the viscosities of many fluids at different conditions of temperature and pressure. Following is some general information about these effects.

Temperature. The viscosity of a liquid quite logically decreases with an increase in temperature. For many liquids the plot of viscosity on a log scale versus temperature on a linear scale is essentially a straight line. Empirical formulas for various fluids have been developed to express the effects of temperature. For example, the formula for water is

$$\frac{1}{\mu} = 2.142[(t - 8.435) - \sqrt{8078.4 - (t - 8.435)^2}] - 120 \qquad (6)$$

where μ = absolute viscosity, poises

t = temperature of water, °C

Additional data on temperature effects on viscosity are given in *ASTM Chart D-446*. The absolute viscosity of *gases* generally increases with increase in temperature.

Pressure. Over the pressure range of 0 to 100 atm, with temperature constant, the viscosity of most liquids is essentially constant. Above 100 atm, experiments indicate that viscosity generally increases with pressure, and the rate of increase is greater at higher pressures. At temperatures below about 30°C, however, the viscosity of water decreases with increase in pressure to about 1,000 atm. The absolute viscosity of a gas is independent of pressure, provided the gas is neither very rarified nor very dense.

Table 1. Viscosity Conversion Factors*

1 poise = 1 dyne-second per square centimeter
1 poise = 1 gram per centimeter-second
1 pound (force)-second per square foot = 1 slug per foot-second
1 pound (mass) per foot-second = 1 poundal second per square foot
ν (stokes or cm²/sec) × 0.001076 = ν (ft²/sec)

To convert to—	Multiply the number of—		
	Poises by	$\dfrac{\text{lb (force)-sec}}{\text{sq ft}}$ by	$\dfrac{\text{lb (mass)}}{\text{ft-sec}}$ by
Poises..............	1	478.8	14.88
$\dfrac{\text{lb (force)-sec}}{\text{sq ft}}$	0.002089	1	0.03109
$\dfrac{\text{lb (mass)}}{\text{ft-sec}}$	0.06721	32.174	1

* SOURCE: *Chem. & Met. Eng.*, December, 1945, p. 115.

APPLICATIONS OF VISCOSITY MEASUREMENT

The viscosity-measuring methods summarized in Table 2 may be divided into two broad groups: (1) viscometers capable of continuous measurements in a process, and (2) those suited primarily for laboratory analyses of samples on a batch basis. In either case the general application guide information given below applies. Following this, some typical industrial and laboratory uses are cited.

General Application Guide

The term *viscosity* can be applied only to Newtonian fluids, by its definition. In many industrial applications, however, a viscosity-measuring instrument can be employed to measure the apparent viscosity, as previously defined, provided the application conditions are well understood. An example with reference to Fig. 6 will help explain this.

In Fig. 6 the rheogram is for a non-Newtonian liquid—specifically a pseudoplastic. The substance has an *apparent* yield stress of y_s. If a viscosity instrument is operating at shear rate a (resulting in shear stress a'), the apparent viscosity is reported as the tangent of angle θ_a, or

$$\theta_a = \frac{\text{shear stress } a'}{\text{shear rate } a} \qquad (7)$$

Similarly, a viscosity instrument operating at shear rate b (shear stress b') would measure an apparent viscosity

$$\theta_b = \frac{\text{shear stress } b'}{\text{shear rate } b} \tag{8}$$

From Eqs. (7) and (8) note that different apparent viscosities are reported *for the same substance* because the reported viscosity is a function of both the *substance* and the *instrument*. Thus the term apparent viscosity is meaningless, unless the type of instrument used and other instrumental operating data are also stated.

On the other hand, if the same instrument and operating methods are used, in many cases apparent viscosity is a usable measure of *some* variable—perhaps the variable of prime interest as a basis for process control. Thus, a so-called "one-point viscosity

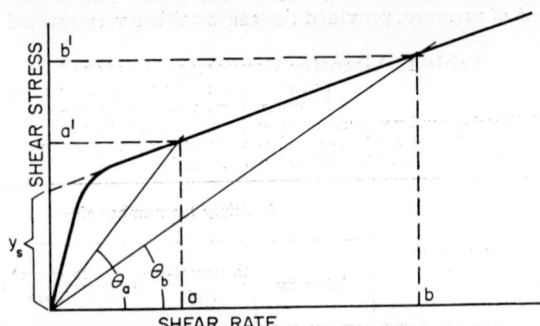

FIG. 6. Rheogram illustrating differences in the apparent viscosities for a non-Newtonian fluid (see text for explanation).

instrument," designed to measure Newtonian viscosity, *can* be applied in a useful manner to measure and control a complex non-Newtonian substance, if the results can be related to the important variable.

Industrial Applications

Following are some typical applications of viscosity (or apparent viscosity) measurement and control in industry.

Petroleum. Products are manufactured to rigid viscosity specifications in blending and fractionation operations. Lubricating, fuel, and hydraulic oils are examples (e.g., in the production of No. 5 fuel oil, Bunker C and No. 2 oil are blended with viscosity measurement of the blend as a basis for automatic flow or ratio control).

Chemical. Viscosity (apparent) is sometimes used to indicate the end point of polymerization reactions which involve non-Newtonian liquids. The variable here can be the key product quality measure in the production of plastics and other synthetic products. Similarly, such a measure can be important in the manufacture of soaps, paints and varnishes, glues, and other chemicals.

Food. Monitoring viscosity (or apparent viscosity) is common in the manufacture of such foods as catsup, mayonnaise, starch solutions and dispersions, and so on. Although the consumer might not be aware of viscosity control, he would certainly notice a lack of it if the product did not satisfy him.

Fuel Combustion. All burners and injection nozzles are designed for optimum performance at a particular viscosity of the fuel oil, to provide the proper mixture of oil and air and hence peak combustion efficiency. Improper viscosity control results in waste of fuel and fouling of nozzles, which then require more frequent cleaning. Control is generally accomplished by measurement of the viscosity of oil leaving a heat exchanger, with regulation on the heat supply to the exchanger.

Laboratory Applications

In research and testing laboratories, viscosity may be one of several variables to be measured as basic data are developed on new products or processes. Quality control of purchased products often includes a viscosity check in the laboratory. A few typical research-type uses of viscosity measurement are jet-engine or missile test and development, high-polymer research, and blood-clotting investigations. In such work, high accuracy is essential in the measurements, and it is important to state the operating conditions of the measurement and the type of instrument employed.

REFERENCES

Textbooks

1. Blair, G. W. S.: "An Introduction to Industrial Rheology," McGraw-Hill Book Company, Inc., Blakiston Division, New York, 1938.
2. Green, H.: "Industrial Rheology and Rheological Structures," John Wiley & Sons, Inc., New York, 1949.
3. Lapple, C. E.: "Fluid and Practical Mechanics," University of Delaware, Newark, Del., 1951.
4. Perry, J. H. (ed.): "Chemical Engineers' Handbook," 4th ed., McGraw-Hill Book Company, Inc., New York, 1963.
5. Barr, G.: "Monograph of Viscometry," Oxford University Press, Fair Lawn, N.J., 1931.
6. Merrington, G. W.: "Viscometry," Longmans, Green & Co., Inc., New York, 1949.
7. Blair, G. W. S.: "Survey of General and Applied Rheology," Pitman Publishing Corporation, New York, 1944.
8. "ASTM Standards on Petroleum Products," American Society for Testing and Materials, Philadelphia.
9. Considine, D. M. (ed.): "Process Instruments and Controls Handbook," McGraw-Hill Book Company, Inc., New York, 1957.

Articles

10. Bingham, E. C.: Some Fundamental Definitions of Rheology, *J. Rheol.*, vol. 1, pp. 507–516, 1930.
11. Merrill, E. W.: Basic Problems in the Viscometry of Non-Newtonian Fluids, *ISA J.*, vol. 2, no. 10. October, 1955.
12. Beerbower, A.: Controlling Fluid Processes with Continuous Viscometers, *Control Eng.*, vol. 5, no. 6, June, 1958.
13. Bates, R. L.: Guide to Industrial Viscometry, *Chem. Eng.*, vol. 67, no. 7, pp. 145–148, April 4, 1960.
14. Hallikainen, K. E.: Viscosity Measurement, *Instr. Control Systems*, vol. 35, no. 2, pp. 137–144, February, 1962.
15. Hallikainen, K. E.: Viscometry, *Instr. Control Systems*, vol. 35, no. 11, pp. 82–84, November, 1962.
16. Nelson, R. C.: Gel-Time Meters, *Instr. Control Systems*, vol. 35, no. 12, pp. 110–111, December, 1962.
17. Nelson, R. C.: Automatic Time Viscometers, *Instr. Control Systems*, vol. 36, no. 2, pp. 110–111, February, 1963.
18. Nelson, R. C.: Automatic Force Viscometers, *Instr. Control Systems*, vol. 36, no. 4, pp. 94–98, April, 1963.
19. Nelson, R. C.: Manual Force Viscometers, *Instr. Control Systems*, vol. 36, no. 5, pp. 115–123, May, 1963.

Table 2. Selection Guide for Viscosity and Consistency Instruments

Schematic diagram	Principle	Application and remarks
	Time to discharge a given volume of fluid through an orifice or nozzle. Principle of viscometers such as Saybolt, Redwood, Engler, Scott, Ubbelohde, Zahn, Parlin cup, and Ford cup. A similar type utilizes the capillary tube in place of the orifice or nozzle. Viscometers of this type include the modified Ostwald, Bingham, and Zeitfuchs.	A laboratory or batch-type instrument commonly used by petroleum and allied industries. Several forms (Zahn, Parlin, and Ford) are commonly used in the paint and varnish industry. With changes in orifice or capillary dimensions, the instrument is suitable for low to moderately high viscosity ranges. Accuracy to $\pm 0.1\%$ is possible if sample and instrument temperatures are closely controlled—usually by immersion in constant-temperature bath. Instrument must be kept clean and should be recalibrated regularly with standard fluids. Viscosity is reported in seconds and can be converted to absolute viscosity units only if fluid is Newtonian.
BUBBLE BALL	Timed fall of ball or rise of bubble. Time of fall of metal ball or rise of bubble through liquid confined in tube is proportional to absolute viscosity because, in both cases, liquid flows in viscous flow through a restriction. Principal of viscometers such as Gardner Laboratory unit and Hoeppler unit (Central Scientific).	A laboratory or batch-type instrument commonly used by the petroleum industry for high-viscosity oils. The ball method can be timed with great accuracy by field coils at the start and finish points. Can be used under high static pressures. Tube can be vertical or at an angle. Suitable for use over wide ranges.
ROTATED CUP	Drag torque on stationary element in rotating cup. Some types use a cylinder, others a paddle, as shown (Brabender). Cup rotates at constant speed; stationary element is restrained by calibrated spring which, by angular displacement, measures torque in terms of viscosity. Principle of viscometers such as McMichael and improved McMichael (Brabender Corp.), Fisher Scientific.	Primarily a laboratory or batch-type instrument, but may be equipped to record. Bowl may be equipped to control sample temperature. Can be used with both Newtonian and non-Newtonian fluids, and, by change of rotational speed, a shear stress versus shear rate diagram can be obtained. Suitable for use over wide ranges.
GUIDE ROD PISTON CYLINDER	Timed fall of piston in cylinder (Norcross Corp.). Submerged plunger is raised automatically, then dropped in cylinder, forming orifice. Time of fall is automatically recorded in terms of absolute viscosity. Similar types, utilizing a holed disk (Mobilometer, Gardner Laboratory, Inc.) or holed cone (Precision Scientific), designed for laboratory use, report results in seconds representing the time for the piston to fall a given distance through the sample.	Used for recording and either manual or automatic control in paints, oil, soaps, plastics, and in similar applications. Suitable for both open or pressurized service. Can be designed to handle viscosity ranges from 0.2 to 200,000 cps with high repeatability.

Table 2. Selection Guide for Viscosity and Consistency Instruments
(*Continued*)

Schematic diagram	Principle	Application and remarks
POSITIVE DISPLACEMENT PUMP, FRICTION TUBE, ΔP, DIFFERENTIAL PRESSURE TRANSMITTER, INDICATOR, AIR SUPPLY	Pressure drop through a friction tube. Sample pumped at constant flow rate through friction tube in viscous flow. Pressure drop across tube is measured by differential-pressure transmitter. Transmitter output is direct solution to Hagen-Poiseuille equation (3) for absolute viscosity. (Westinghouse patent.)	Can provide local indication, recording and automatic control. Can be used for wide range of industrial applications.
MOTOR, A, B, C, CALIBRATED SPRING, FLOW VESSEL, SPINDLE, SAMPLE IN	Torque to rotate a torque element in a liquid (Brookfield Engineering Co.). Synchronous motor drives vertical spindle with disk, paddle, or cylinder submerged in test liquid. Drive is through calibrated spring. Angular lag of spindle behind motor is proportional to viscosity and is measured in various ways. Bristol method shown measures lag of rotating contact C behind A by stationary contact B. Other recorders adapted to this device measure lag angle by change in resistance or capacitance.	Can be used in open or closed vessels under pressure or vacuum. Used for both Newtonian and non-Newtonian liquids or suspensions. Usable over wide viscosity ranges by change of spindle speeds and types. Laboratory or batch-type instruments of similar design with local indication of viscosity are also available. These units are particularly valuable for investigation of non-Newtonian fluids.
RETURN, FLOAT, SAMPLE, FLOW CONTROL	Viscosity-sensitive rotameter (Fischer & Porter Co.). Rotameter floats can be designed to be either viscosity-immune or viscosity-sensitive. If flow rate is held constant, with either a flow regulator or a constant-volume pump, float elevation in a tapered tube is a function of viscosity. Another type of unit employs a viscosity-insensitive float in conjunction with a viscosity-sensitive one to enable manual flow-rate adjustments with a valve prior to reading the instrument.	Suitable for recording or recording and control. Can be used in open or closed systems under pressure or vacuum. Ranges from 3 to 30,000 cps with 10:1 rangeability. Used for both Newtonian and non-Newtonian liquids. Used to control viscosity of fuels to burners and nozzles—also for blending of fuel oils and in other blending operations. Available with temperature compensation.
SONIC PROBE, FLUID, COMPUTER, TEMPERATURE COMPENSATING COMPUTER, RESISTANCE THERMOMETER, RECORDER CONTROLLER, THERMOCOUPLE CONNECTOR	Ultra-viscoson continuous viscometer (Bendix Aviation Corp.). Ultrasonic waves are applied to a thin magnetostrictive alloy-steel blade. Rate at which blade oscillations are damped is a measure of the viscosity of the sample. Relation between damping and viscosity is solved by computer.	Suitable for recording and control. Used for both Newtonian and non-Newtonian liquids. Wide temperature and pressure ranges. Automatic temperature compensation available if required. Ranges to 50,000 cps-g/cc.

Table 2. Selection Guide for Viscosity and Consistency Instruments
(*Continued*)

Schematic diagram	Principle	Application and remarks
	Continuous-consistency instrument (Plastometer—Fischer & Porter Co.). A pump diverts portion of the product stream through a hydraulic "bridge," creating a differential pressure which is a direct measure of sample consistency.	Used to record and control the consistency of baby foods, applesauce, tomato products, and other fibrous or pulpy slurries.
	Graduated-trough consistency-measuring device (Cenco-Bostwick, Central Scientific Co.). This device measures the distance, in a graduated trough, that a material will flow under its own weight in a given time interval.	A laboratory or batch-type instrument used primarily in the food industry by-products of jams, preserves, tomato products, etc.
	Continuous consistency regulator (Fischer & Porter Co.). Paper stock is forced to deform as it flows by sensing element. Resultant force on element is a function of consistency. Pneumatic- or electric-force transmitter measures net force—eliminating static pressure effects. Recorder-controller operates dilution valve to maintain consistency at the set value.	Widely used to record and control the consistency of paper pulp and other fibrous-type slurries. Two units cover consistency ranges of 0.5 to 2.0 % and 2.0 to 8 %. Manufacturer guarantees control to within ± 0.1 %. Suitable for "in-line" or sampling-type applications for line sizes from 6 to 30 in.

Section 8

CHEMICAL COMPOSITION VARIABLES

By

C. M. ALBRIGHT, JR., M.S., *Director, Pioneering Research and New Product Development, Remington Arms Co., Inc., Bridgeport, Conn.; Member, American Society of Mechanical Engineers, Instrument Society of America, Institute of Electrical and Electronics Engineers, New York Academy of Sciences, American Association for the Advancement of Science. (Analysis Instruments)*

ANALYSIS INSTRUMENTS

By C. M. Albright, Jr.*

Editor's Note: The underlying theory and operating details of equipment commercially available for measuring and controlling chemical-composition variables can be found in the "Process Instruments and Controls Handbook," pp. 6-1 to 6-213, D. M. Considine, editor-in-chief, McGraw-Hill Book Company, Inc., New York, 1957.

Many industries are concerned with material processing operations that include changes in physical state or chemical composition. In comparison with the large-scale changes involved in fabrication or assembly operations, these changes in state and composition occur on a microscopic and submicroscopic scale. They involve atomic and molecular interactions and rearrangements which must be measured or characterized and controlled.

APPLICATIONS

Some of the major uses of chemical-composition measurement and control are the following:

Raw Materials

1. Composition analysis to check purchase specifications.
2. Detection of contamination by trace impurities.
3. Analysis check on materials priced on an active-ingredient basis.
4. Continuous analysis of materials delivered by pipeline; water analysis.

Process Control

1. Speedup and improved control by automation of, or replacement of, control laboratory tests on "grab" samples.
2. Improved control by replacement or augmenting of inferential measurements, such as temperature or pressure, with more significant composition data.
3. Use of continuous processes permitted that could not be controlled except by continuous analysis instrumentation.

Process Troubleshooting

1. Temporary use of analysis instruments for process studies aimed at overcoming occasional upsets.

Yield Improvement

1. Continuous analysis of process streams to measure effects of variables influencing yields.
2. Analysis of overflow or purge streams, recirculated material, sumps, and the like, to determine product losses and detect buildup of undesirable by-products that affect yield.

* Director, Pioneering Research and New Product Development, Remington Arms Co., Inc., Bridgeport, Conn.

Inventory Measurements

1. Analytical monitoring of material flowing between process steps and plant areas, to establish consumption and in-process inventory on the basis of active or essential ingredients.

Product Quality

1. Determination of product-composition specifications and adjustment of process to meet those specifications desired.
2. Assessment of structurally dependent attributes, such as color, melting or boiling point, refractive index, etc.

Safety

1. Detection of leaks in equipment.
2. Survey of operating areas for escape of toxic materials from leaks or spills, especially materials not readily detected by human senses.
3. Detection of flammable or explosive mixtures in atmosphere or process lines.

Waste Disposal

1. Monitoring of plant stacks for accidental discharge of toxic or nuisance gases, vapors, or smokes.
2. Analysis of waste streams for toxic or other objectionable materials.
3. Control of waste-treatment or product-recovery facilities.

Research and Development

1. Continuous analysis to speed up research and optimize results.
2. Provison of structural and compositional information not otherwise obtainable.
3. Production of results in a more directly usable form.

CLASSIFICATION OF MEASUREMENT METHODS

Chemical-composition variables are measured by observing the basic *interactions between matter and energy*. All physical and chemical analysis techniques, whether instrumental or not, are fundamentally based on these relationships. This follows directly from the fact that all known matter is made up of a complex but systematic arrangement of particles having mass and electric charge. For all practical purposes, these particles consist of neutrons, having mass but not charge; protons, having essentially the same mass as neutrons with a unit positive charge; and electrons, having negligible mass with a unit negative charge. The neutrons and protons comprise the nuclei of atoms, and each nucleus ordinarily is provided with sufficient orbital electrons, in a progressive shell-like arrangement of different energy levels, to neutralize the net positive charge on the nucleus. The total number of neutrons plus protons determines the atomic weight. The number of protons, which in turn fixes the number of electrons, determines the chemical properties and the physical properties, except mass, of the resulting atom.

Chemical combinations of atoms into molecules involve only the electrons and their energy states. Chemical reactions involving both structure and composition usually occur by loss, gain, or sharing of electrons among the atoms. Every configuration of atoms in a molecule, crystal, solid, liquid, or gas can be represented by a definite system of electron energy states. Moreover, the particular physical state of the molecules, as represented by their mutual arrangement, also is reflected upon these electron energy states. These energy states, which are characteristic of the composition of any particular substance under consideration, can be most readily inferred by observing

the consequences of interaction between the substance and an external source of energy. This external energy source may be in any of the following basic groups:
1. Electromagnetic radiation.
2. Chemical affinity or reactivity.
3. Electric or magnetic fields.
4. Thermal or mechanical energy.
These groups differ fundamentally in their modes of interaction with matter. Moreover, the types of information which these interactions afford may vary considerably in specificity or uniqueness, a situation which sometimes can be controlled by combining techniques. Many properties can be measured or inferred by more than one type of interaction.

Group I Techniques: Interaction with Electromagnetic Radiation

These techniques involve the measurement of the quantity and quality of electromagnetic radiation emitted, reflected, transmitted, or diffracted by the sample.

Electromagnetic radiation varies in energy with radiation frequency, that of the highest frequency or shortest wavelength having the highest energy and penetration into matter. Radiation of the shortest wavelengths (gamma rays) interacts with atomic nuclei, X rays with the inner shell electrons, visible and ultraviolet with valence electrons and strong interatomic bonds, while infrared radiation and microwaves interact with the weaker interatomic bonds and with molecular vibrations and rotation. Most of these interactions are structurally related and completely unique. They may be used to detect and measure the elemental or molecular composition of gas, liquid, and solid substances within the limitations of the available equipment.

The instrumental techniques which are based upon interaction with electromagnetic radiation are outlined as follows:

Emitted radiation:

1. Thermally excited:
   (a) Emission spectroscopy.
   (b) Flame photometry.
2. Electromagnetically excited:
   (a) Fluorescence.
   (b) Raman spectrophotometry.
   (c) Induced radioactivity.
   (d) X-ray fluorescence.

Transmission and reflection measurements:

1. X-ray analysis:
   (a) Absorption.
   (b) Critical-edge absorption.
   (c) Diffraction.
2. Ultraviolet spectrophotometry.
3. Conventional photometry (transmission colorimetry).
4. Colorimetry.
5. Light scattering.
6. Optical rotation (polarimetry).
7. Refractive index.
8. Infrared spectrophotometry.
9. Microwave spectroscopy.
10. Nuclear quadrupole moment.

Group II Techniques: Interaction with Other Chemicals

These techniques involve the measurement of the results of reaction with other chemicals, in terms of amount of sample or reactant consumed, product formed, or thermal energy liberated, or by determination of equilibrium attained.

The selectivity inherent in the chemical affinity of one element or compound for another, together with their known stoichiometric and thermodynamic behavior, permits positive identification and analysis under many circumstances. In a somewhat opposite sense, the apparent dissociation of substances at equilibrium in chemical solution gives rise to electrically measurable valence potentials, called oxidation-reduction potentials, whose magnitude is indicative of the concentration and composition of the substance. While individually all the foregoing effects are unique for each element or compound, many are readily masked by the presence of more reactive substances so that they can be applied only to systems of known composition limits.

The instrumental techniques which are based on interaction with other chemicals are outlined as follows:

Consumption of sample or reactant:

1. Orsat analyzers.
2. Automatic titrators.

Measurement of reaction products:

1. Impregnated-paper-tape devices.
2. Continuous chemical-reaction types.

Thermal energy liberation:

1. Combustion types.
2. Quantitative exothermic reaction of unknown with reactant.

Equilibrium solution potentials (oxidation-reduction):

1. Redox potentiometry.
2. pH.
3. Metal-ion equilibria.

Group III Techniques: Reaction to Electric and Magnetic Fields

These techniques involve the measurement of the current, voltage, or flux changes produced in energized electric and magnetic circuits containing the sample.

The production of net electric charges on atoms or molecules by bombardment with ionizing particles or radiation, or by electrolysis or dissociation in solution or the induction of dipoles by strong fields, establishes measurable relationships between these ionized or polarized substances and electric and magnetic energy. Ionized gases and vapors can be accelerated by applying electric fields, focused or deflected in magnetic fields, and collected and measured as an electric current in mass spectroscopy. Ions in solution can be transported, and deposited if desired, under the influence of various applied potentials for coulometric or polarographic analysis and for electrical conductivity measurements. Inherent and induced magnetic properties give rise to specialized techniques, such as oxygen analysis based on its paramagnetic properties and nuclear magnetic resonance, which is exceedingly precise and selective for the determination of the compounds of many elements.

The instrumental techniques which are based upon reaction to electric and magnetic fields are outlined as follows:

Mass spectroscopy:

1. Nier type.
2. Omegatron.
3. Time of flight.

Electrochemical:

1. Controlled-potential electrolysis.
2. Polarography.
3. Coulometry.

4. Amperometry.
5. "Dead stop" methods.

Electrical properties:

1. Electrical conductivity.
2. Dielectric constant and loss factor.
3. Oscillometry.
4. Gaseous conduction.

Magnetic properties:

1. Paramagnetism.
2. Nuclear magnetic resonance.
3. Electron paramagnetic resonance.

Group IV Techniques: Interaction with Thermal or Mechanical Energy

These techniques involve the measurement of the results of applying thermal or mechanical energy to a system in terms of energy transmission, work done, or changes in physical state.

The thermodynamic relationship involving the physical state and thermal energy content of any substance permits analysis and identification of mixtures of solids, liquids, and gases to be based upon the determination of freezing or boiling points and upon the quantitative measurement of physically separated fractions, as in vapor-phase chromatography. Useful information often can be derived from thermal conductivity and viscosity measurements involving the transmission of thermal and mechanical energy, respectively.

The instrumental techniques which are based on interaction with thermal or mechanical energy are outlined as follows:

Effects of thermal energy:

1. Thermal conductivity.
2. Melting and boiling points.
3. Ice point (crystallization).
4. Dew point.
5. Vapor pressure.
6. Fractionation.
7. Chromatography.
8. Thermal expansion.

Effects of mechanical energy or forces:

1. Viscosity.
2. Sound velocity.
3. Density.

ANALYSIS INSTRUMENTS

Instrumental techniques for the determination of chemical composition, when combined in a single assembly of apparatus, are usually called analysis instruments. Continuous-analysis instruments may be called process stream analyzers or product analyzers.

Analysis is defined as the separation, or resolution, of anything into its component parts so that they are individually distinguishable. In so far as chemical composition is concerned, analysis may involve the determination of the components (1) as to kind, called qualitative analysis, or (2) as to amount, called quantitative analysis.

Where plant measurements are the objective, analysis usually refers to the determination of (1) the amount of one kind of component in the sample stream, less frequently to (2) the ratio of amounts of two kinds of component, and least often to (3) determination of the respective amounts of three or more components.

PLANT AND LABORATORY TECHNIQUES

There are two important distinctions between the procedures of analytical chemical laboratories for qualitative and quantitative determinations and the instrumental techniques employed in plant control laboratories or in automatic plant analysis instruments.

The first distinction lies chiefly in advance knowledge of the probable composition of the sample. Analytical chemistry employs procedures that are useful in resolving into its components a sample about which relatively little is known beforehand. Control laboratory techniques and the continuous-analysis instruments that are in many cases automatized versions of these techniques, on the other hand, require considerable advance knowledge of the sample composition in terms of what kinds of material will be present and their expected ranges of concentration.

The second distinction should be the recognition that some instruments are used by people as tools, both in analytical chemical laboratories and in plant control laboratories, while automatic continuous plant instruments must operate unattended—to determine what is going on while it is happening, without benefit of human interpretation. Such instruments can control the plant process directly if they are properly applied.

PRACTICAL CONSIDERATIONS

Any practical appraisal of the merits of chemical-composition variables for control purposes must recognize certain inherent physical limitations in their measurement. These limitations are:

1. Sample Must Be Representative

Although this requirement may appear obvious, it is a factor that is very frequently overlooked. The sample must be gathered or drawn off in such a fashion that it will consist of the same composition as the body of the processed material. There must be assurance that any changes in conditions, such as temperature or pressure, between the sampling and measuring points cannot influence sample composition. In nearly all cases, the probable composition of the sample must be known ahead of time through some independent method before an analysis technique can be selected.

2. Physical State of Sample

The technique must provide for interaction between the applied measuring energy and the entire sample, as well as for the observation of the total result. This cannot always be accomplished. It is for this reason that a large majority of the techniques are applicable to gases, where the molecules are widely spaced and free to react in a characteristic manner, and that fewer techniques are applicable to liquids, and still fewer can be applied to solids.

3. Uniqueness or Specificity of Method

The selection of method must be tailored to the sample composition and to the information requirements. Some methods or techniques involving atomic and molecular structure are rather universal in that they permit exact identification and measurement of every elemental or molecular constituent present in the sample. These methods are usually the most complex and costly. They are sometimes less sensitive than simpler methods whose only drawback is an inability to distinguish between related substances having similar gross interactions with energy. Where these related substances are known not to be present in the sample, these simpler, less specific methods should always be considered.

PERFORMANCE CONSIDERATIONS

Common standards of acceptable performance, which unfortunately are not met by all instruments, are as follows:

1. Ultimate sensitivity is the smallest change in concentration that can be reliably detected and distinguished from background noise or drift. It will usually be 1 to 2 per cent of the narrowest full-scale range of concentration or less, providing the noise or drift does not exceed this same value. It is very important to recognize that the sensitivity of most analysis instruments varies widely, depending on the particular component being determined.

2. Range is the span of concentrations to which the instrument responds without adjustment. Range adjustments are frequently provided or are feasible to cover most possible concentrations. In connection with range, certain generalized exceptions to linearity should be borne in mind. In the first place, systems that do not follow ideal laws (such as additive volumes upon mixing) may often exhibit anomalies in response to physical measurements. For example, refractive index on binary systems where volumes are not additive may actually show a reversal in value with increasing amounts of the same component.

Another more common situation occurs in measurements of energy absorption, such as X-ray, ultraviolet, visible, or infrared radiation. Energy absorption methods rapidly lose incremental sensitivity or linearity as the absorbing power or concentration of the sample is increased. For example, if a 2-in.-thick sample absorbs 30 per cent of the incident radiation, increasing the cell to 4 in. would increase the absorption to 51 per cent, whereas an increase to 6 in. thick would produce an absorption of 66 per cent, just over twice the initial effect with a cell three times as thick. In the same fashion, increase in concentration of the sought-for component produces less effect at high concentrations.

3. Accuracy is frequently confused with *error*. The two are not synonymous; they are complementary. For example, when an instrument is said to have an accuracy of ½ of 1 per cent, what is usually meant is that the measured quantity can be determined to *within* ½ of 1 per cent; i.e., the instrument reading will be between 99½ and 100½ per cent of the actual value being measured. It would be preferable at least from the standpoint of grammatical correctness, to speak of an *error of ½ of 1 per cent or an accuracy to within ½ of 1 per cent*.

For instruments with uniform scales, the error percentage is usually referred to full scale. This means that for a given range the error is a constant amount; accordingly, the accuracy with which readings can be made for low values on the scale may be rather poor. Moreover, it is important to distinguish between the error attributable to the instrument owing to a failure to be reproducible and the error due to incorrectness in calibration. *Calibration* errors are more frequently given as a per cent of *reading*, since it is usually not a problem to obtain a correct zero point. On the other hand, accurate calibration for upscale readings is far more difficult to achieve than is generally recognized, even with the most exacting laboratory procedures.

Although it is not safe to generalize, the over-all error, including instrument and calibration errors, may frequently amount to as much as 5 per cent of full scale unless unusual precautions are taken. However, for control purposes, relative values are usually much more significant than absolute values, and the reproducibility, or *precision*, of many analysis instruments can be within ⅒ per cent or better. In fact, the analysis instrument is quite often capable of substantially greater precision under plant conditions than can be obtained by a skilled analyst using the finest manual laboratory techniques.

4. Temperature effects frequently are of major importance in analysis instruments. There are two reasons for this:

(a) Chemical systems follow the laws of thermodynamics in many of the energy-matter interactions employed, such as reaction rates, pH, gas or vapor density, and refractive index, with the result that the measured effect may be more responsive to temperature than to the primary variable itself.

(b) The primary elements employed often are primarily responsive to temperature or are affected by ambient temperature variations.

Three methods are employed to overcome the effects of temperature: (1) sample temperature regulation, (2) sample temperature compensation, and (3) thermostating to minimize ambient effects.

SUMMARY OF INSTRUMENTAL TECHNIQUES FOR CHEMICAL ANALYSIS

In Table 1 are listed the most commonly used categories of analysis instruments. This is a severe abridgment of exhaustive tables of instrumental characteristics and principles which appear in the "Process Instruments and Controls Handbook," mentioned earlier. In addition to the factors shown in Table 1, the aforesaid comprehensive tables include much more detailed information on the general and specific objects of measurement (gases, liquids, and solids); degree of uniqueness (specific, nonspecific, and limited); degree of usefulness (nondestructive, destructive, range of concentration, and limitations); availability (laboratory or batch, plant or automatic); relative cost; ultimate sensitivity; whether or not thermostated; error; calibration; response speed; specific sampling requirements; installation and maintenance requirements; and commercial sources of the equipment. In all, over one hundred specific categories of analysis instruments are described in considerable detail.

REFERENCES

1. Chromatograph Optimizes Sulfur Recovery Process, *ISA J.*, vol. 7, no. 3, p. 40, March, 1960.
2. Karasek, F. W., and B. O. Ayers: A Fast Sampling Valve for Gas Chromatography, *ibid.*, p. 70.
3. Bates, Roger G.: "Electrometric pH Determinations," John Wiley & Sons, Inc., New York, 1954.
4. Brode, Wallace R.: "Chemical Spectroscopy," John Wiley & Sons, Inc., New York, 1954.
5. Harley, John H., and Stephen E. Wiberley, "Instrumental Analysis," John Wiley & Sons, Inc., New York, 1954.
6. Kolthoff, I. M., and H. A. Laitinen, "pH and Electro Titrations," John Wiley & Sons, Inc., New York, 1941.
7. Harrison, George R.: "Wavelength Tables with Intensities in Arc, Spark, or Discharge Tube of More than 100,000 Spectrum Lines," John Wiley & Sons, Inc., New York, 1939.
8. Gordy, Walter, William V. Smith, and Ralph Trambarulo: "Microwave Spectroscopy," John Wiley & Sons, Inc., New York, 1953.
9. Koller, Lewis R.: "Ultraviolet Radiation," John Wiley & Sons, Inc., New York, 1952.
10. Milner, G. W. C.: "Polarography and Other Electroanalytical Processes," Longmans, Green & Co., Inc., New York, 1957.
11. Peiser, H. S.: "X-ray Diffraction by Polycrystalline Materials," Reinhold Publishing Corporation, New York, 1955.
12. Dalahay, Paul: "Instrumental Analysis," The Macmillan Company, New York, 1957.
13. Smith, Orsion C.: "Inorganic Chromatography," D. Van Nostrand Co., Inc., Princeton, N.J., 1953.
14. Snell, Foster Dee, and Cornelia T. Snell: "Colorimetric Methods of Analysis," vols. 1–6, 3d ed., D. Van Nostrand Co., Inc., Princeton, N.J., 1948–1961.
15. Willard, Hobart H., Lynne L. Merritt, Jr., and John A. Dean: "Instrumental Methods of Analysis," D. Van Nostrand Co., Inc., Princeton, N.J., 1958.
16. Allport, Noel L., and J. W. Keyser: "Colorimetric Analysis," Chapman & Hall, Ltd., London, 1957.
17. Brimley, R. C., and F. C. Barrett: "Practical Chromatography," Chapman & Hall, Ltd., London, 1958.
18. Blake, G. G.: "Conductimetric Analysis at Radio Frequencies," Chapman & Hall, Ltd., London, 1950.
19. Britton, Hubert T. S.: "Hydrogen Ions," Chapman & Hall, Ltd., London, 1955; D. Van Nostrand Co., Inc., Princeton, N.J., 1956.
20. Zechmeister, L., and L. Cholnoky: "Principles and Practice of Chromatography," Chapman & Hall, Ltd., London, 1951.
21. Zechmeister, L.: "Progress in Chromatography," Chapman & Hall, Ltd., London, 1950.
22. Gaydon, A. G., and H. G. Wolfhard: "Flames, Their Structure, Radiation and Temperature," Chapman & Hall, Ltd., London, 1953.
23. Sawyer, R. A.: "Experimental Spectroscopy," Chapman & Hall, Ltd., London, 1951.
24. Ewing, Galen W.: "Instrumental Methods of Chemical Analysis," 2d ed., McGraw-Hill Book Company, Inc., New York, 1960.
25. Gibb, T. R. P., Jr.: "Optical Methods of Chemical Analysis," McGraw-Hill Book Company, Inc., New York, 1942.
26. Nachtrieb, Norman H.: "Principles and Practice of Spectrochemical Analysis," McGraw-Hill Book Company, Inc., New York, 1950.
27. Biffen, Frank M., and William Seaman: "Modern Instruments in Chemical Analysis," McGraw-Hill Book Company, Inc., New York, 1956.

28. Townes, C. H., and A. L. Schawlow: "Microwave Spectroscopy," McGraw-Hill Book Company, Inc., New York, 1955.
29. Continuous pH Control of Process Solutions, *Automation*, vol. 6, no. 10, pp. 72–75, October, 1959.
30. Fraade, D. J.: Better Refinery Operation with Automatic Stream Analyzers, *Oil Gas J.*, vol. 55, pp. 93–108, Oct. 21, 1957.
31. Glasser, L. G.: Refractometers in Process-stream Analysis, *Control Eng.*, vol. 4, no. 12, pp. 96–101, December, 1957
32. Cost-conscious Chromatographs. *Chem. Eng. Progr.*, vol. 55, no. 6, pp. 108–110, June, 1959.
33. Yanak, J. D., and A. M. Calabrese: Continuous Testers Check Product Performance, part I, Physical-properties Instruments, *Control Eng.*, vol. 6, no. 4, pp. 102–106, April, 1959.
34. Yanak, J. D., and A. M. Calabrese: Continuous Testers Check Product Performance, part II, Chemical-properties Instruments, *ibid.*, no. 6, pp. 99–102, June, 1959.
35. Farrar, G. L.: What's New in Continuous-stream Monitors, *Oil Gas J.*, vol. 57, no. 41, p. 127, Oct. 5, 1959.
36. Barton, P. D.: Sun Is Moving Ahead on Its Product-control Center, *ibid.*, pp. 128–130.
37. Stormont, D. H.: Analyzers Supply Continuous e.p.-i.b.p. Data, *ibid.*, pp. 130–132.
38. Halter, R. C., and L. W. Pohler: Gas Chromatography Monitors Four Refinery Fractionating Columns, *ibid.*, pp. 135–137.
39. Siggia, Sidney: "Continuous Analysis of Chemical Process Streams," John Wiley & Sons, Inc., New York, 1959.
40. Meites, Louis: "Polarographic Techniques," Interscience Publishers, Inc., New York, 1955.
41. Kolthoff, I. M., and James L. Lingane: "Polarography," Interscience Publishers, Inc., New York, 1952.
42. Delahay, Paul: "New Instrumental Methods in Electrochemistry," Interscience Publishers, Inc., New York, 1954.
43. Mullen, Paul W., "Modern Gas Analysis," Interscience Publishers, Inc., New York, 1955.
44. Cassidy, Harold Gomes: "Adsorption and Chromatography," vol. V, "Interscience Publishers, Inc., New York, 1951.
45. West, W.: "Chemical Applications of Spectroscopy," Interscience Publishers, Inc., New York, 1956.
46. Cassidy, Harold Gomes: "Fundamentals of Chromatography," Interscience Publishers, Inc., New York, 1957.
47. Bijvoet, J. M.: "X-ray Analysis of Crystals," Interscience Publishers, Inc., New York, 1951.
48. Richardson, E. G.: "Relaxation Spectrometry," Interscience Publishers, Inc., New York, 1957.
49. Penner, S. S.: "Quantitative Molecular Spectroscopy and Gas Emissivities," Addison-Wesley Publishing Co., Inc., Reading, Mass. 1959.
50. Rogers, Lewis H.: High Sensitivity Continuous Instrumentation for Atmospheric Analyses, *Chem. Eng. Progr.*, vol. 53, no. 8, pp. 381–384, August, 1957.
51. Automatic Measurement of Quality in Process Plants, *Proc. Soc. Instr. Technol.*, Cambridge, 1957.
52. Churchill, J. R.: Emission Spectroscopy Speeds Control of Metals Production, *Control Eng.*, vol. 6, no. 10, pp. 89–94, October, 1959.
53. "Automatic Measurement of Quality in Process Plants," Academic Press, Inc., New York, 1958.
54. Phillips, J. P.: "Automatic Titrators," Academic Press, Inc., New York, 1959.
55. Hecht, G. J., J. A. Edinborgh, and V. N. Smith: A Near-infrared Process Analyzer for Liquid Samples, *ISA J.*, vol. 7, no. 2, p. 40, February, 1960.
56. "Fractionator Control with High-Speed Chromatography," *Autom. Control*, vol. 12, no. 5, pp. 23, May, 1960.
57. Fourroux, M. M., F. W. Karasek, and R. D. Wightman: High-speed Chromatography in Closed-loop Fractionator Control, *ibid.*, no. 5, p. 76, May, 1960.
58. Leisey, Frank A.: Automation Mercaptan Titrator Aids Refinery Processing, *ibid.*, no. 7, p. 67, July, 1960.
59. Fraade, D. J., and E. E. Escher: Chromatography Techniques for Process Control, *Autom. Control*, vol. 14, no. 2, pp. 19–23, February, 1961.
60. Stirling, P. H., and Henry Ho: Successful Sampling: Systems Approach Simplifies Analyzer-sample Handling, *Ind. Eng. Chem.*, vol. 53, no. 3, pp. 57A–62A, March, 1961.
61. Wherry, T. C.: Chromatography for Process Control, *Chem. Eng. Progr.*, vol. 56, no. 9, pp. 49–57, September, 1960.
62. Schall, William C.: Selecting Sensible Sampling Systems, *Chem. Eng.*, May 14, 1962, pp. 157–171.
63. Cassidy, Harold G.: "Fundamentals of Chromatography," Interscience Publishers, Inc., Division of John Wiley & Sons, Inc., New York, 1957.
64. West, W.: "Chemical Applications of Spectroscopy," Interscience Publishers Inc., Division of John Wiley & Sons, Inc., New York, 1956.
65. Cassidy, Harold G.: "Adsorption and Chromatography," Interscience Publishers, Inc., Division of John Wiley & Sons, Inc., New York, 1951.
66. Butz, W. H., and H. J. Noebels: "Instrumental Methods for the Analysis of Food Additives," John Wiley & Sons, Inc., New York, 1961.
67. Reilley, C. N.: "Advances in Analytical Chemistry and Instrumentation," John Wiley & Sons, Inc., New York, 1960.
68. Staff: X-ray Gage Provides Continuous Chemical Analyses, *Automation*, vol. 7, no. 9, pp. 68–9, September, 1960.
69. Tyler, C. M.: Process Analyzers for Control, *Chem. Engr. Progr.*, vol. 58, no. 9, pp. 51–55, September, 1962.
70. Trippeer, W. M., and J. W. Riggle: Optimizing In-process Composition Analysis, *Chem. Engr. Progr.*, vol. 58, no. 9, pp. 56–58, September, 1962.

71. Charlton, K. W., and S. B. Spracklen: Control from Chromatographs, *Control Eng.*, vol. 10, no. 3, pp. 93–96, March, 1963.
72. Willard, Hobard H., Lynne L. Merritt, Jr., and John A. Dean: "Instrumental Methods of Analysis," D. Van Nostrand Company, Inc., Princeton, N.J., 1958.
73. Green, R. M.: Extending Analytical Methods with Process Computers, *ISA J.*, vol. 8, no. 10, pp. 38–40, October, 1961.
74. Neblett, J. B., and F. C. Mears: Linking Computers to Analyzers in Real-time Process Control, *ISA J.*, vol. 9, no. 1, pp. 44–47, January, 1962.
75. Williams, T. J.: How Analyzer Dynamics Affect Control, *ISA J.*, vol. 9, no. 7, pp. 39–42, July, 1962.
76. Maley, L. E.: A Process-stream Refractometer, *ISA J.*, vol. 9, no. 10, pp. 49–50, October, 1962.
77. Kaufman, A., and H. Dodge: Detecting Hazardous Gases, *Autom. Control*, vol. 14, no. 7, pp. 47–51, July, 1961.
78. Devrishian, C.: Analyzing Combustible Gases and Vapors by Catalytic Combustion, *Autom. Control*, vol. 17, no. 5, pp. 41–47, December, 1962.
79. Chaplin, A. L.: Applications of Industrial pH Controls, Instruments Publishing Co., Inc., Pittsburth, Pa., 1950.
80. Lehrer, E.: Integrating Chromatograph Measures Component Volumes, *Control Eng.*, vol. 9, no. 5, pp. 95–97, May, 1962.
81. Karp, H. R., Industrial Process Chromatographs, *Control Eng.*, vol. 8, no. 6, pp. 87–100, June, 1961.
82. Maier, H. J.: Eighteen Streams Analyzed by Chromatographs, *Control Eng.*, vol. 8, no. 8, pp. 88–90, August, 1961.
83. Smith, V. N.: Evaluation of the Performance of Analytical Instruments, *Control Eng.*, vol. 8, no. 10, pp. 93–99, October, 1961.
84. Maley, L. E.: Multiple Stream Sampling Systems Design, *Control Eng.*, vol. 8, no. 11, pp. 91–96, November, 1961.
85. Karasek, F. W.: Stream Analysis and Data Reduction in Pilot Plants, *Control Eng.*, vol. 8, no. 12, pp. 93–95, December, 1961.
86. Weiss, M. D.: Electrochemical Analysis in Process Control, *ISA J.*, vol. 8, no. 1, pp. 62–67, January, 1961.
87. Noble, F. W.: An Ultrasonic Detector for Gas Chromatography, *ISA J.*, vol. 8, no. 6, pp. 54–57, June, 1961.
88. Staff: Review of Instrumental Methods of Analysis Symposium, *ISA J.*, vol. 8, no. 6, pp. 63–64, June, 1961.
89. Crandall, W. A.: A Pilot Plant for Evaluating Analyzers, *ISA J.*, vol. 8, no. 8, pp. 83–87, August, 1961.

Table 1. Analysis Instruments and Techniques

INSTRUMENT OR TECHNIQUE	ANALYSIS OF	ANALYSIS FOR	TO DETERMINE IDENTITY	TO MEASURE QUANTITY	SAMPLING REQUIREMENTS	LIMITATIONS	SPECIFIC OR NONSPECIFIC
COLORIMETRY (reflected or transmitted light)	Gases, vapors, liquids, solutions, slurries, and solids	Dyestuffs, pigments, color indicators, and other substances whose reflectance or transmittance in visible light is characterized by any dispersion vs. wavelength	Not necessarily determined. Color is specified, however, by measurement of relative reflectance or transmittance value vs. wavelength over visible range. Approximations to identity are made by taking measurements at a number of wavelengths, or by filter colorimetry wherein the response to light defined by specially designed filters is measured	Strength of color is determined by measuring amount of light reflected or transmitted at the wavelengths of peak absorption compared with an absolute or standard white	Samples can be liquid in cells, painted panels (wet or dry), and woven fabrics. Amount varies widely. Sample preparation may require considerable study	Sensitivity approximates that of human eye	Nonspecific
DIELECTRIC CONSTANT (dielectric loss)	Liquids, solutions, and suspensions	Substances of high dielectric constant or loss in presence of background having low dielectric constant or loss, such as water in organics, and polar molecules in nonpolar solvents	Not specifically determined except by inference	Change of frequency of tuned circuit (or required change in tuning of circuit or voltage change across circuit) containing sample as dielectric in capacitative element	Sample should be clean, low in conductance, and nominally regulated in temperature. R-f measurements with non-contacting electrodes should be used for corrosive samples	Applicable only to binary and some ternary systems	Nonspecific
EMISSION SPECTROSCOPY	Gases, vapors, liquids, solutions, suspensions, slurries, particulate solutions, and solids	Most elemental constituents of any sample. Especially suited for control analyses of minerals, metals, and alloys	Sample is destructively bombarded by arc or spark to produce radiation of constituent atoms by changes in electron energy levels. Radiation distribution is analyzed by wavelength distribution in spectrometer, producing characteristic patterns of each element	Amount of radiation at characteristic wavelengths can be related to abundance of each element	Sample must be contained in arc or spark source. Electrodes must be of known purity. Special sample-preparation facilities usually required	Can be used to identify 70 elements	Specific
INFRARED ANALYZER (double-beam dispersive type)	Gases, vapors, liquids, solutions, and thin-film solids	Any heteroatomic molecule: specific wavelength, based on separate spectrophotometric data for each analysis. With such data, performance can be predicted	Not determined except by inference. In a particular case, if a pair of infrared wavelengths can be found, one of which is not affected by variation in any constituent and one of which is absorbed only by sought-after component, then analysis can be specific for that component	Measurement of relative amount of infrared radiation transmitted through sample with respect to radiation not affected by measured component	0.1–0.5 cfm of clean, dust-free sample. Must be free of condensate that might deposit in sample cell	Sensitivity affected by presence of interfering substances	Nonspecific
MASS SPECTROMETER (magnetic analyzer)	Gases, vapors, liquids, solutions, and, with proper handling, some solids	Atoms, molecules, and free radicals, and their reaction products. Specialized instruments sensitized to He are widely used as leak detectors	Ionized fragments of sample are separated on basis of momentum by accelerating them with an electric field and deflecting their path with a magnetic field so that a specific mass strikes ion collector. Instrument is qualitatively specific	Ion current, when amplified, is proportional to relative abundance of ions having selected mass, and can be related, for a given instrument, to the quantity of parent molecules in sample	Sample admitted to system at very low pressure, about 10^{-6} mm Hg, usually fractionated through molecular leak; for continuous analysis, however, viscous leak is employed	Mass resolution obtained at expense of sensitivity	Specific

Method	Sample	Measured substance	Principle	Quantitative basis	Sample requirements	Specificity
NUCLEAR MAGNETIC RESONANCE	Gases, vapors, liquids, solutions, suspensions, slurries, particulate solutions, and solids	Any molecule or atom possessing nuclear spin. Characterization and study of interatomic bonds is least suited to O_2 among the commonly encountered elements	Measurement of radio frequency which differs for each nucleus having spin at which nuclei precess in a magnetic field of known strength. Frequency and field strength at which r-f is induced in detector coil owing to precession resonance is characteristic of any particular nucleus	Relative amount of r-f induced or coupled from transmitter coil to detector or receiver coil is proportional to concentration of resonant nuclei per unit volume of sample cell	Any kind of sample is satisfactory, although liquids are generally preferred. Nuclear density is less in gases, giving smaller signal, and resonances are broadened, owing to structural interactions in solids. Temperature must be controlled	Specific
pH (hydrogen-ion)	Liquids, solutions, suspensions, and slurries	H^+ or H_2O^+ content in liquids	(1) Selective potential developed across glass electrode membrane permeable only to H^+ ions; (2) solution potential developed by H^+, such as quinhydrone or antimony electrodes; (3) colorimetric methods employing pH-sensitive color complexes	Magnitude of electrode potential as measured by sensitive voltmeter is related to $-\log H^+$. For each tenfold increase in H^+, the potential of all measuring systems changes 59.1 mv	Sample can be static or flowing. Temperature has inherent effect on pH as well as on electrode response and should be controlled or known	Specific
RADIOLOGICAL GAS ANALYZER (ion chamber analyzer)	Gases and vapors	Detection of gases having high electron density in gases of unknown electron density, or vice versa, such as argon or ammonia in nitrogen	Not specifically determined except in tailored applications where interfering gases are removed	Relative ion current produced in two symmetrical chambers exposed to equal amounts of beta radiation depends on average electron density, or atomic number, of the gases in the two chambers	Sample must be clean, dry, and temperature-controlled. Interfering substances must be removed by absorption	Nonspecific
REFRACTIVE INDEX (differential type)	Liquids, solutions, and sometimes gases and vapors	Optically transparent, nonfouling binary solutions: salts, acids, bases; hydrocarbons; sugars; oils	Not determined. Sample is assumed to be identical in composition with standard when refractive indices coincide	Difference in refractive index or sample from that of reference standard is measured, using photoelectric detectors	Sample must be clean; that is, free of suspended solids, and relatively constant in temperature. Flow can be 100 to 500 ml/min	Nonspecific
SOUND VELOCITY	Gases, vapors	Composition of two-component systems wherein sound velocity differs widely, owing to difference in molecular weight	Absolute sound velocity is physical constant, permitting identification of pure gases, liquids, and solids. Composition of mixtures not specifically determined	Sound velocity through mixture is proportional to mean molecular weight, so that electrical measurement of transmission time of sonic wave or phase shift through sample can be related to proportions of two constituents	Not critical; inlet and outlet for fluid samples must not interfere with sound-wave propagation. Sample temperature is critical and must be closely regulated	Nonspecific
THERMAL CONDUCTIVITY	Gases and vapors	H_2, CO_2, acetone, Cl_2, He, SO_2, and H_2S in air, and sought-for gas wherever it differs widely in thermal conductivity from background gases	Not specifically determined by thermal conductivity alone. Associated techniques, such as selective absorption, may be employed for enhancing selectivity	Proportional to thermal conductivity of mixture as determined by cooling effect on heated wire in chamber into which sample gas diffuses. Effect on heated wire measured electrically with respect to similar wire in enclosed chamber not open to sample	Sample must be clean, dry, and free from suspended particles. Generally should be noncorrosive. Flow should be regulated, as well as absolute pressure of sample	Limited

Table 1. **Analysis Instruments and Techniques** (*Continued*)

INSTRUMENT OR TECHNIQUE	ANALYSIS OF	ANALYSIS FOR	TO DETERMINE IDENTITY	TO MEASURE QUANTITY	SAMPLING REQUIREMENTS	LIMITATIONS	SPECIFIC OR NONSPECIFIC
ULTRAVIOLET SPECTROPHOTOMETER (single- or double-beam)	Gases, vapors, liquids, solutions, and sometimes thin-film solids	Aromatic and other double-bonded organic materials. Cl_2, ozone, NO_2, SO_2. Rare earths, free radicals, and biological materials	Pattern of ultraviolet absorption vs. wavelength is specific for most materials showing strong absorption peaks	Measurement of amount of radiation absorbed in any one of the wavelengths specific to the sample component sought, with respect to the incident radiation at the wavelength of measurement, corresponds to concentration	Laboratory instruments accommodate samples of 10–100 ml, depending on length of sample cell necessary. Not usually employed for continuously flowing samples	....	Specific
VISCOSITY	Liquids, solutions, suspensions, and slurries	Concentration of heavy solutions and suspensions, molecular weight of polymer melts and solutions, and grading of petroleum products	Determined only insofar as viscous properties are characteristic	Techniques vary widely, including rate of fall of objects through sample; discharge rate through orifice; pressure changes across orifice or capillary, retarding torque on rotating or oscillating surface; displacement of bob in tapered tube; and damping of oscillating probe	Vary widely	Varies widely	Nonspecific
X-RAY ABSORPTION	Gases, vapors, liquids, solutions, and solids	Heavy elements in light elements, regardless of chemical composition. Usually not applicable to atoms of atomic weight less than sulfur. Good for lead in gasoline	Not specifically determined except in instruments employing filters or crystal monochromators to isolate X-ray wavelengths specifically located adjacent to a "critical absorption edge" of the element being sought	Ratio of X-ray intensity in absorbing path to that in reference path, as determined by relative counting rate in Geiger-counting-tube detector or similar device	Sample cells usually must be very thin and have windows relatively transparent to X rays, such as Al. Plastics sometimes are useful	....	Limited
CHROMATOGRAPHY	Gases, vapors, liquids, and solutions	All classes of organic compounds that can be vaporized and are thermally stable in the vapor phase. Also such inorganic gases as O_2, N, H_2, CO, CO_2, He, A, H_2S	Thermal conductivity detector (usually) is used to show that a compound has emerged from the chromatographic column. The signal from detector goes to millivolt recorder which shows a peak for each component in sample. Peak height or area can be related to concentration	Thermal conductivity detector, connected to millivoltmeter, records peak for each component in sample. Peak height or area can be related to concentration	0.001–0.5 ml for liquids. 1.0–25.0 ml for gases. Requires electric service and a supply of He	Sensitive to 1 ppm. Empirical calibration. Requires few minutes to run sample	Specific

8–14

Section 9

ELECTRICAL VARIABLES

By

D. B. FISK, *Manager, Timer Engineering, Clock and Timer Department, General Electric Company, Ashland, Mass.; Member, American Institute of Electrical Engineers; Registered Professional Engineer (Mass.). (Electrical Variables)*

R. A. TERRY, *Manager, Marketing Services, Packard Bell Computer Corporation, Los Angeles, Calif. (Electrical Variables)*

ELECTRICAL VARIABLES

By D. B. Fisk* and R. A. Terry†

In the electric power industry, measurement of several variables is essential as an operating guide. The prime variables of interest are direct or alternating voltage and current, power, vars, power factor, frequency, and synchronism. Many such variables also serve as criteria for measurement of speed, temperature, pressure, and other characteristics.

Electrical instruments for such variables usually measure current, regardless of the quantity in which the dial is calibrated. This section reviews briefly the principles and methods generally used in industry to measure the common electrical variables. Electronic Laboratory and Research Instrumentation, page 15-67, also covers some of the same variables and methods, including more fundamental theory and the particular methods applicable to laboratory or research applications.

Instruments used for measurement of electrical variables may be classified as:

1. *Direct-acting instruments:* Those which extract the power for their operation directly from the circuit being measured. These instruments are divided into basic classes according to accuracy (limit of error). The most commonly accepted classes are: 0.1, 0.25, 0.5, 1, and 2 per cent.

2. *Independently powered instruments:* These require little or no power from the circuit being measured. Potentiometers and bridges are in this category.

DIRECT VOLTAGE AND CURRENT MEASUREMENT

Indicators used to measure these variables are usually of the permanent-magnet moving-coil type (d'Arsonval movement).‡ In circuits involving current up to 25 ma,

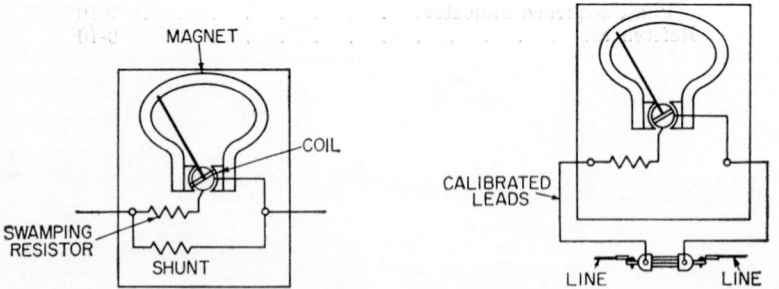

FIG. 1. Milliammeter shunted to measure amperes.   FIG. 2. Millivoltmeter with external shunt.

all the current is passed through the coil. For larger currents, a shunting circuit, as shown by Fig. 1, is used. The method of shunt connection is shown in Fig. 2.

Shunts are usually constructed to produce an IR drop of 50 mv at rated current at

---

* Manager, Timer Engineering, Clock and Timer Department, General Electric Co., Ashland, Mass.
† Manager, Marketing Services, Packard Bell Computer Corp., Los Angeles, Calif.
‡ See p. 15–67 for theory of operation.

the terminals of the shunt leads. This standardization makes any d-c millivoltmeter calibrated for 50 mv full scale suitable for use with any shunt, provided shunt leads having the proper resistance are used with the instrument.

The swamping resistor of Fig. 1, whose value is several times that of the coil, minimizes adverse effects owing to variations in ambient temperature. It is usually made of a material such as manganin, but may be made of a negative-temperature-coefficient material.

For recording, both direct-acting and independently powered instruments are used. Direct-acting recorders are essentially the same as the permanent-magnet moving-coil indicators, except that the system is larger because of the higher power required to move a pen or other recording means.

Although the d'Arsonval is basically a current-measuring device, it can be used to measure d-c voltage by limiting the current through the moving coil to full-scale value for the maximum voltage to be measured. The value of resistance to be inserted is, from Ohm's law

$$R = \frac{E}{I} - R_c \qquad (1)$$

where E = desired full-scale voltage
I = full-scale current rating of instrument, amp
R_c = coil resistance, ohms

The sensitivity of a voltmeter is essentially an expression of the load imposed on the circuit being measured. It is usually expressed in ohms per volt; e.g., a 100-ohms-per-volt instrument imposes a 1-ma load at full scale.

Two other methods for measuring d-c voltage and current are covered thoroughly on pages 15-69 to 15-71. These are (1) the *null-balance potentiometer*, developed to a high degree in a variety of instrument designs for indication, recording, or control; and (2) the *vacuum-tube voltmeter*, used where a high resistance-measuring system is needed.

ALTERNATING VOLTAGE AND CURRENT MEASUREMENT

The effective value of an a-c voltage of any waveform is

$$E = \sqrt{\frac{1}{t}\int_0^t e^2\, dt} \qquad \text{volts} \qquad (2)$$

Since $\frac{1}{t}\int_0^t e^2\, dt$ is the mean value of the square of the voltage, the effective value equals the square root of the mean value of the square of the voltage. This is usually expressed as the *root mean square* or rms value. If the a-c voltage varies sinusoidally, i.e., $E = E_{max} \sin \omega t$, the effective value is $E_{eff} = E_{max}/\sqrt{2} = 0.707 E_{max}$.

Most electrical devices are designed to operate satisfactorily over a given voltage range. Operation below this range (undervoltage) causes loss of performance. If allowed to persist, burned-out equipment may result. Operation above the rated voltage range (overvoltage) shortens the life of the equipment.

If I is substituted for E in Eq. (2), the expression is correct for the effective value of alternating current. Most conductor and other heating effects bear a square-law relationship to the current. Therefore, an accurate knowledge of the value of the current is essential to proper and safe operation of any electrical system or equipment.

The electromagnetic moving-iron meter in several forms is widely used for measuring both alternating voltage and current. Its action is based on reaction to a magnetic field by an iron vane or vanes. As shown by Fig. 3, the pointer assumes a position determined by

$$\frac{1}{T}\int_0^T i^2\, dt \qquad (3)$$

because of the inertia of the moving element.   Because of the exponential function, the scales of such instruments are crowded at the low end.

A longer scale version of instrument is the Thomson inclined-coil mechanism shown by Fig. 4.   The vane, free to move in the magnetic field, tends to line up with the flux, and thereby assumes a position proportional to Eq. (3) because of inertia.   This mechanism is extensively used for alternating current measurement.

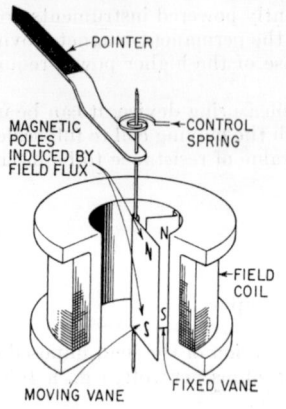

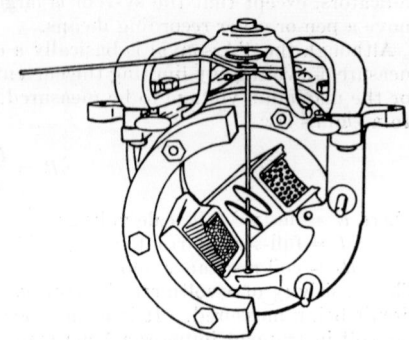

FIG. 3. Essential parts of radial-vane form of repulsion-type element.

FIG. 4. Cutaway view of inclined-coil attraction-type voltmeter.

Another type of attraction mechanism is the solenoid plunger, whereby a soft-iron plunger is pulled into a coil by the induced magnetic field.   This simple mechanism is relatively inexpensive, and is widely used in small instruments for panelboard mounting.

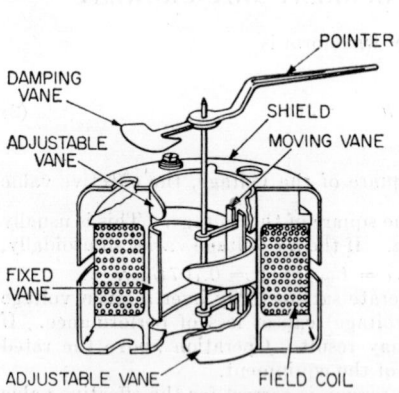

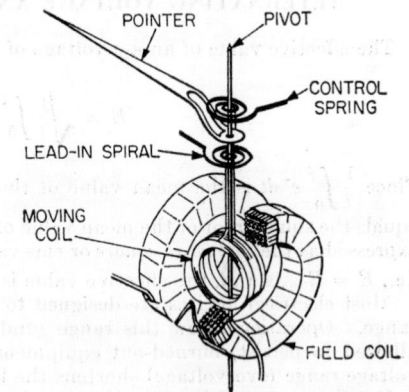

FIG. 5. Cutaway view of repulsion-attraction instrument.

FIG. 6. Dynamometer mechanism.

In the repulsion-attraction mechanism shown by Fig. 5, both principles are utilized to provide relatively high torque and substantial angular movement.   The scales on this type of instrument are more linear than those of attraction or repulsion instruments.

The electrodynamometer, shown in Fig. 6, is widely used for both voltage and current measurements.   It is available with accuracy ratings down to 0.1 per cent.   In the dynamometer, a moving coil is deflected by the induced field of a fixed coil.   The current through a dynamometer type of a-c voltmeter is given by

$$I = \frac{E}{\sqrt{R^2 + \omega^2 L^2}} \tag{4}$$

where R and L are the total resistance and total effective inductance of the instrument; and ω, the frequency in radians per second, equals $2\pi f$ (f being the frequency in cycles per second). R is usually made so high that the reactance can be neglected for ordinary power frequencies.

The instantaneous torque T on the moving coil of a dynamometer is proportional to the product of the instantaneous current i and the instantaneous flux density β of the field set up by the fixed coil. If the coils are connected in series, the current in each is the same. Thus

$$T = K_1 i \beta \qquad \beta = K_2 i \tag{5}$$
$$\therefore \quad T = K_1 K_2 i^2 \tag{6}$$

Portable meters are available for currents up to 60 amp and potentials up to 600 volts. In general, however, instrument transformers are used for currents greater than 10 amp and potentials above 300 volts.

The measurement of a-f and r-f voltages and currents is accomplished by the use of d-c instruments in conjunction with a rectifier or a thermocouple element. In the latter, the thermocouple measuring junction is heated by a wire through which the current passes, or directly by passing the current through a thermocouple bridge.*

Another instrument widely used for a-c voltage measurement is the vacuum-tube voltmeter.* By means of an electronic circuit, a-c potentials are converted to direct current which is passed through a self-contained d-c milliammeter for indication. Scales are direct reading and are frequently available in multiple ranges. Many available instruments can be used both on alternating and direct current. They provide high sensitivity and place negligible load on a circuit because of the high impedance—usually in megohms.

POWER MEASUREMENT

Measurements of voltage and current are necessary as indications of normal operating conditions, but the measurement of power is basic to the control of a process since power is the rate of energy input. In almost all electrically powered processes, proper operation can be assured by maintaining power input at an appropriate level, even though there may be substantial variations in supply voltage and current.

In d-c circuits, power equals EI, but in a-c circuits the relationship between the effective voltage and the effective current determines the power supplied. The total amount of energy supplied in a time t is given by

$$w = \int_0^t P\, dt$$

The average power (total energy supplied divided by the time) or

$$\frac{w}{t} = P_{av} = \frac{1}{t} \int_0^t P\, dt \tag{7}$$

When the term "power" is used, *average power* is meant since rms power has no significance. There is a time or phase difference between E and I by an angle θ:

$$E = E_{max} \sin \omega t \tag{8}$$
$$I = I_{max} \sin (\omega t - \theta) \tag{9}$$

Thus, in a time interval of one period,

$$P = \frac{1}{t} \int_0^t E_{max} I_{max} \sin \omega t \sin (\omega t - \theta)\, dt \tag{10}$$

Hence the average power supplied to a load in a single-phase a-c circuit equals the
* See p. 15–75.

product of the effective voltage and current times the cosine of the angle of lead or lag ($P = E_{\text{eff}}I_{\text{eff}} \cos \theta$).

In a d-c circuit, $P = EI$. So a separate instrument is not required for reading power. In a-c circuits, $P = EI \cos \theta$, where $\cos \theta$ is the load power factor. Since the electrodynamometer evaluates

$$\frac{1}{T}\int_0^T i_F i_M \, dt \tag{11}$$

it will then indicate the effective power because i_F is the effective current to the load, and i_M (the meter current) is proportional to the effective load voltage. The instantaneous torque due to magnetic field interaction between the fixed and movable fields is proportional to the relationship between the load voltage and current, or $EI \cos \theta$. Thus, power is indicated directly with little influence by power factor.

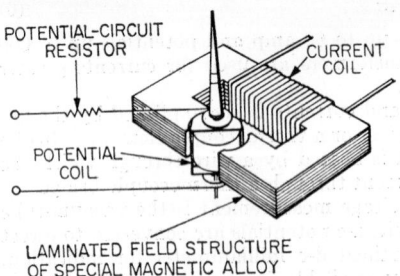

FIG. 7. Iron-core dynamometer mechanism for wattmeter.

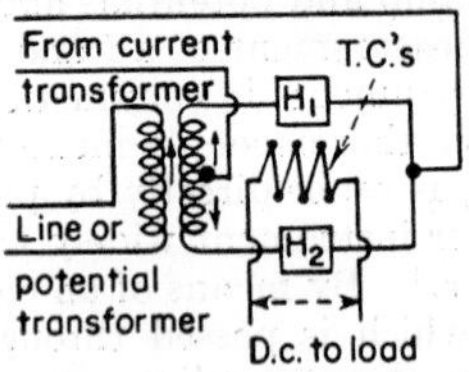

FIG. 8. Thermal converter for power measurement.

Although the classic wattmeter is an air-core dynamometer, iron is often used where the frequency varies as much as ± 10 per cent. Figure 7 shows such a wattmeter having a 90° pointer travel.

Power is also measured by a *thermal converter* which resembles the thermocouple-type a-c meter but has an output proportional to power consumed by a load. This output can be telemetered over long distances. As shown in Fig. 8, the line is connected to the primary of a transformer. Leads from a current are connected into a bridge circuit so that current in one side of the bridge is added to by current induced in the transformer, while current in the other side is opposed by transformer current. Hence, the heat from heaters H_1 and H_2 will be different, depending on the load; and their average, determined by the series thermocouples, will be proportional to the load.

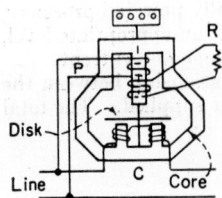

FIG. 9. Induction watt-hour meter

Figure 9 shows a common a-c device for integrating power consumption over a period of time. It consists of a potential winding P on an iron core, with current windings C. The armature is an aluminum disk with shaft behind the core. Interaction of flux produced by the three poles with the induced eddy currents in the disk causes rotation proportional to power. Adjustment of phase relations is obtained with a compensating winding and external resistance R. Revolutions counted by the register represent watthours.

REACTIVE VOLT-AMPERES (VARS) MEASUREMENT

Vars (volt-amperes, reactive) equal the product of the effective voltage and current times the sine of the angle of lead or lag (the component which is 90° out of phase).

$$\text{vars} = EI \sin \theta \tag{12}$$

The measurement of vars not only shows power factor but also provides an indication of what load changes are required to arrive at the desired reactive load. For example, when the vars to a load are excessive owing to large current lag, capacitors can be switched into the circuit to raise the power factor.

In addition to its applicability for the measurement of watts ($EI \cos \theta$), the dynamometer element is widely used to measure vars ($EI \sin \theta$). The wattmeter measures the product of the voltage and the in-phase component of the current of the same frequency. The varmeter measures the product of the voltage and the current, which is 90° out of phase with the voltage. This is done by shifting the phase of the voltage applied to the potential coil. Hence, the flux produced by this coil is in phase with that produced by the quadrature component of the current in the field coil.

In a single-phase circuit, an impedance network is connected in series with the potential circuit, and so the current through the instrument will lag the voltage by 90°. In two-phase circuits, where a 90° difference in phase normally exists, it is only necessary to connect the potential and current coils to opposite phases. In three-phase circuits, autotransformers are sometimes used to obtain the necessary phase shift. Typical connections for a three-phase three-wire circuit are shown by Fig. 10.

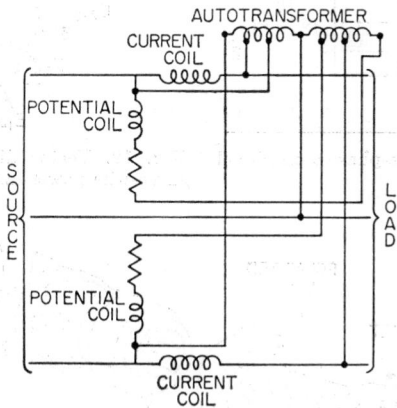

FIG. 10. Three-phase three-wire dynamometer circuit for measuring vars.

POWER-FACTOR MEASUREMENT

The ratio of the power delivered to a load to the product of the effective voltage and current equals the power factor, or

$$\text{Power factor} = \frac{P}{EI} = \cos \theta \qquad (13)$$

Thus, the over-all operation of an a-c system depends to a great extent on the angle of lead or lag between the voltage and the current. For a phase angle of zero, the cosine is 1 and the power factor is unity. This, however, is true only of loads which are purely resistive.

Two basic types of instruments are used for the measurement of this variable, namely, the crossed-coil type and the polarized-vane type. In a single-phase instrument, the crossed coils are coupled to a common shaft and pointer, with the assembly free to rotate. The field-coil flux is in phase with the line voltage; the flux set up by one coil A is essentially in phase with the line voltage; the flux due to the other coil B is substantially in quadrature with the line voltage (90° out of phase).

When the line voltage and current are in phase, coil A aligns with the field coil and coil B has no effect, because its flux is out of phase with the field-coil flux. When line current is 90° out of phase with the line voltage, coil B aligns with the field coil and coil A has no effect. Note that two possible positions exist 180° apart, one for leading current and one for lagging current. Intermediate positions of equilibrium exist for various phase relationships between current and voltage. These instruments are usually calibrated to read from 50 per cent lag to 50 per cent lead, and scales are graduated accordingly to read from 0.5 lag to 1.0 to 0.5 lead power factor.

The three-phase crossed-coil power-factor meter is similar in appearance and basic structure to the single-phase instrument, but the operating principle is somewhat different. In this instrument, the crossed coils of the moving system are connected through resistors across two phases of the circuit, and so the fields of flux of the two coils are opposed (see Fig. 11). Therefore the equilibrium position depends on the ratio of the phase angles between coils A and B with respect to the field. Because this ratio is an index of power factor in a balanced three-phase circuit, the instrument and scale can be calibrated directly in terms of power factor.

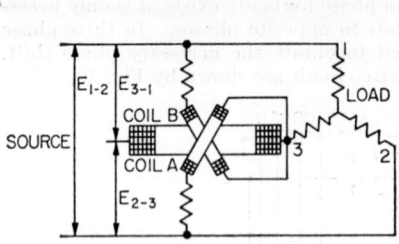

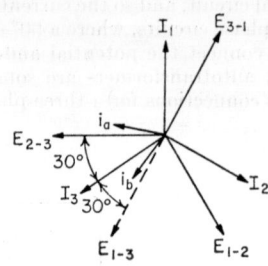

Fig. 11. Three-wire three-phase crossed-coil power-factor meter.

Fig. 12. Vector diagram for three-phase three-wire power-factor meter.

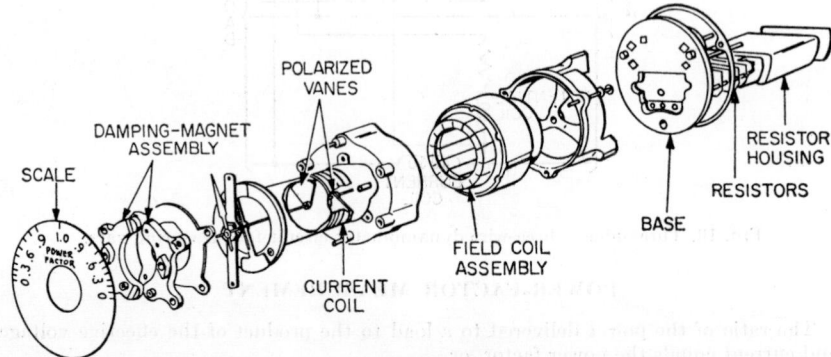

Fig. 13. Exploded view of polarized-vane power-factor meter.

Figure 11 shows the coil connections for a three-phase power-factor meter used on a balanced load. The field current is supplied by one phase, and the coils are connected respectively across this phase and the other two phases. The phase relationship in this type of meter at unity power factor is shown by the vector diagram of Fig. 12. Note that the field current lags one potential and leads the other by 30°. Consequently, the potential coils are fixed at an angle of 60° to each other, to provide a scale distributed over the conventional 90° arc.

With a balanced load, power-factor variation changes the relative phase angle between the field-coil current and the two potential coil currents; hence,

$$T_A = k(\cos 30° + \theta_2) \tag{14}$$
$$T_B = k(\cos 30° - \theta_2) \tag{15}$$
$$T = 2k(2 \cos 30° + \theta_1 + \theta_2) \tag{16}$$

where T_A is coil A torque; T_B is coil B torque; T is resultant torque, which is a measure of the system power factor.

Figure 13 shows a typical polarized-vane power-factor meter, which is also commonly used. The potential is applied to a field to produce a rotating magnetic field.

The line current is passed through the polarizing coil, which induces alternating magnetism in the soft-iron vanes. The vanes align with the angular position of the field at the instant of maximum induced magnetism. Their position of equilibrium is a function of the phase angle between current and voltage.

FREQUENCY MEASUREMENT

The speed of rotating electrical equipment is dependent on the frequency of a-c power, and for this reason measurement and control of frequency is vital to the operation of generation and distribution stations. In most areas of the United States, the frequency is maintained constant. As a result, a-c timing devices remain accurate over long periods of time.

A modified air-core dynamometer will respond to frequency changes without appreciable effect from line-voltage variations. An iron vane is added in the moving-coil assembly, and the field coils are connected in series with resonant circuits. For a 60-cycle instrument, the net torque at mid-scale is essentially zero since the individual torques developed by the two coils are equal and opposite. At frequencies above and below 60 cycles, a net torque results to position the moving-coil assembly away from the midpoint. The iron vane supplies a restoring torque since it tends to align with the axis of the field coils. As both the vane and the coil torques are directly proportional to line voltage, the measurement is essentially independent of this variable.

SYNCHRONISM MEASUREMENT

To interconnect a-c generators and generating stations, they must be alike in voltage and frequency, and in synchronism as to phase. Measurements of these variables must be made whenever interconnection is desired.

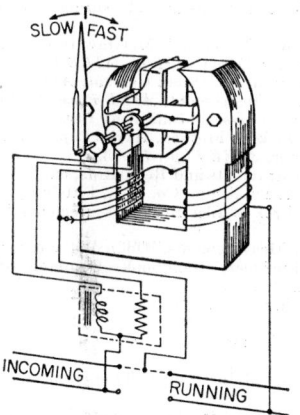

Fɪɢ. 14. Crossed-coil synchroscope.

The basic polarized-vane mechanism shown by Fig. 13 is directly applicable to the measurement of synchronism by connecting the field coil to one system and the current coil (wound for potential measurement) to the other. When the systems are at the same frequency, the position of the vane is dependent on the phase difference between the two systems. If the systems differ in frequency, the vane will rotate in a direction to indicate which system is of higher frequency.

The moving-iron synchroscope depends on the alignment of the iron vanes with the resultant of two magnetic fields rotating in opposite directions. The oppositely rotating fields are obtained by connecting the three-phase winding of the field coil to corresponding phases of the two systems to be compared.

The crossed-coil synchroscope, shown by Fig. 14, utilizes a rotating field produced by the armature coils, and a bipolar field produced by the field coil. One coil of the moving system is supplied with in-phase current through a resistor, and the other coil is supplied with current lagging by 90° owing to the reactor. The field coil is connected to the corresponding phase of the system to be compared.

SPECIALIZED ELECTRICAL MEASURING INSTRUMENTS

Hinged-magnet Voltammeter

This is a device for making quick measurements without direct connection to the circuit. A secondary coil is assembled on a hinged-magnetic circuit, and is connected through rectifiers to a permanent-magnet moving-coil ammeter. Separate potential connections and switching means are sometimes provided for the measurement of voltage, using clip leads. This principle has also been applied to the measurement of power and power factor.

Phase-sequence Indicator

This instrument is used for determining proper connection to a three-phase system of unknown phase sequence. It comprises a simple RC circuit and two neon lamps.

REFERENCES

1. Manual of Electric Instruments, *Gen. Elec. Co. Rept.* GET-1087A, 1949.
2. Giant Size Electric Meters, *Dittmore and Freimuth Co. Form* 632-A.
3. Laws, F. A.: "Electrical Measurements," 2d ed., McGraw-Hill Book Company, Inc., New York, 1938.
4. Drysdale, C. V., and A. C. Jolly: "Electrical Measuring Instruments," Chapman & Hall, Ltd., London, 1952
5. Golding, E. W.: "Electrical Measurements and Measuring Instruments," Sir Isaac Pitman & Sons, Ltd., London, 1942.
6. Knowlton, A. E.: "Electric Power Metering," McGraw-Hill Book Company, Inc., New York.
7. A New Measuring Instrument for Direct Current, *AIEE Paper* 39-92.
8. A New Moving-magnet Instrument for Direct Current, *AIEE Paper* 42-123.
9. A Unique Moving Magnet Ratio Instrument, *AIEE Paper* 44-152.
10. Advancements in the Design of Long-scale Indicating Instruments, *AIEE Paper* 47-22.
11. Power Measurement by the Hook-on Method, *AIEE Paper* 50-111.
12. A Hook-on Power Factor Meter, *AIEE Paper* 51-75.
13. Hayward, H. N.: Electrical Instruments and Regulators, chap. VII in "Handbook of Measurement and Control," The Instruments Publishing Co., Inc., Pittsburgh, Pa., 1959.
14. Aronson, M. H. (ed.): "Handbook of Electrical Measurements," The Instruments Publishing Co., Inc., Pittsburgh, Pa., 1960.
15. Harris, Forest K.: "Electrical Measurements," John Wiley & Sons, Inc., New York, 1952.
16. Pender, Harold: "Electrical Engineers' Handbook—Communication—Electronics," 4th ed., John Wiley & Sons, Inc., New York, 1950.
17. Borden, Perry A., and W. J. Mayo-Wells: "Principles and Methods of Telemetering," 2d ed., Reinhold Publishing Corporation, New York, 1959.
18. Holzbock, Werner G.: "Instruments for Measurement and Control," Reinhold Publishing Corporation, New York 1955.
19. James, S.: "Electricity Meters and Instrument Transformers," Chapman & Hall, Ltd., London, 1952.
20. Banner, E. H. W.: "Electronic Measuring Instruments," Chapman & Hall, Ltd., London, 1958.
21. Smith, Arthur W., and M. L. Wiedenbeck: "Electrical Measurements," 5th ed., McGraw-Hill Book Company, New York, 1959.
22. Laws, Frank A.: "Electrical Measurements," 2d ed., McGraw-Hill Book Company, New York, 1938.
23. Walker, R. M.: Automatic Resistance Measurement System, *Control Eng.*, vol. 7, no. 3, pp. 165, March, 1960.
24. McCaslin, John: "Understanding" Bridework, *ISA J.*, vol. 7, no. 7, pp. 60, July, 1960.
25. Galman, H., and E. Miller: Telemetry Transducers for Electrical Variables, *Control Eng.*, vol. 9, no. 8, pp. 89-91, August, 1962.

Note: Also see references listed in Section 14, "Electric Power Generation and Distribution" and Section 15, "Electronic Laboratory and Research Instrumentation."

Section 10

AUTOMATIC CONTROL

By

STEVEN DANATOS, M.S. (**M.E.**), *Associate Editor, Engineering Practice, Chemical Engineering, McGraw-Hill Publishing Company, Inc., New York, N.Y.; Member, American Chemical Society; Registered Professional Engineer (N.J.). (Summary of Controller Types and Final Control Elements)*

E. ROSS FORMAN, B.S.(**M.E.**), **M.S.** (**Bus. Adm.**), *Senior Engineer, Catalytic Construction Company, Philadelphia, Pa.; Member, Instrument Society of America, American Society of Mechanical Engineers, National Society of Professional Engineers; Registered Professional Engineer (Pa. and N.J.). (Fundamentals of Automatic Control Engineering)*

ROBERT J. WILSON, *Sales Manager, Photoswitch Division, Electronics Corporation of America, Cambridge, Mass. (Applications of Photoelectric Controls)*

FUNDAMENTALS OF AUTOMATIC CONTROL ENGINEERING

By E. Ross Forman*

Techniques of applying automatic control to process operations have evolved rapidly from a relatively crude art to a highly mathematical science. It is not, however, the purpose of this section to develop the advanced aspects of control theory, but rather to summarize certain fundamentals as background for the sections dealing with specific control applications in industry. More detailed and advanced discussions are covered in several of the references cited at the end of this section, notably Section 11 of Ref. 4.

Following a review of control-loop terminology, general application guides are given for the control of frequently encountered process variables. This discussion is followed by descriptions of *unit control systems* which are commonly found or can be applied in a variety of industries; these form handy tools for the solution of new control problems.

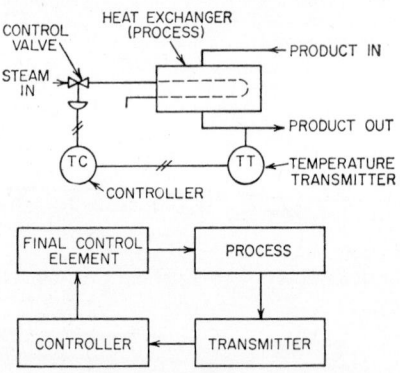

FIG. 1. Basic components of an automatic control system (heat-exchanger circuit).

TERMINOLOGY OF THE CONTROL LOOP

The four basic components of a control system may be considered as (1) the process, (2) transmitter, (3) controller, and (4) final control element. Figure 1 illustrates a heat-exchanger circuit as a typical automatic control system for definition of the basic components.

1. Process Section

Process. The collective functions performed in and by the equipment in which a variable is to be controlled. The equipment does not include any of the control equipment.

Automatic Control System. Any operable combination of one or more automatic controllers connected in closed loops with one or more processes.

Controlled Variable. That quantity or condition which is measured and controlled (flow, temperature, pressure, etc.).

Controlled Medium. Process energy or material in which a variable is controlled.

Manipulated Variable. The quantity or condition which is effected by the automatic controller so as to affect the value of the controlled variable.

Control Agent. The process energy or material of which the manipulated variable is a condition or characteristic.

---

* Senior Engineer, Catalytic Construction Co., Philadelphia, Pa.

2. Transmitter Section

Primary Element. The portion of the measuring means which first either utilizes or transforms energy from the controlled medium to produce an effect which is a function of change in the value of the controlled variable. This effect may be a change of pressure, force, position, electrical potential, or resistance.

Measured Variable. A signal which is a measure of the process variable—generally the output of a measuring instrument or transmitter.

Transmitter. A device for transmitting the signal from the primary measuring element to another point.

3. Controller Section

Controller. A device which produces an output control signal as a function of an input error signal.

Self-operated Controller. A controller in which all the energy to operate the final control element is derived from the control medium through the primary element.

Relay-operated Controller. The energy transmitted from the primary element to the controller is either supplemented or amplified for operating the final control element by employing energy from another source. Most controllers are power-operated because they offer more flexibility. The common power is electricity, air, or hydraulic.

Set Point. The desired value of the process variable.

Error. The difference between the set point and measured variable.

Control Point. The value of the controlled variable which, under any fixed set of conditions, the automatic controller operates to maintain. In some types of automatic controllers (two-position differential gap or floating with neutral controller action) the control point becomes a range of values of the controlled variable rather than a single value. Under ideal and full control, the control point and the set point coincide.

Control Mode. The relationship of the controller output to the input error.

4. Final Control Element Section

Final Control Element. A device which can be manipulated by the controller output signal to regulate the flow of energy or material to a process (*control valves, switches, rheostats, variable-speed pump drives, and dampers*).

Valve Positioner. A device which compares the mechanical position of the valve or final control element with the value called for by the output signal of the controller. Appropriate amplification of the controller signal in proportion to the deviation between the position and desired value is applied to the final control element actuator.

CONTROL EFFICIENCY

The efficiency with which control of a process is obtained is directly related to two effects which must be taken into account in every application. The first of these effects is the load changes. These are changes in the controlled variable which are due to altered conditions in the process. The other is process lag, which is the delay in the time that it takes the controlled variable to reach a new value when a load change occurs.

Process load is the total amount of control agent required by a process at any one time to maintain the process variable at the desired value. Changes in load should result in a signal from the controller to the final control element which holds the controlled variable at a given value. Both the magnitude and direction of the load change are important factors. In a heat exchanger (Fig. 1), for example, the amount of fluid passing through the exchanger and the temperature of the inlet fluid can both cause load changes. A decrease in the flow requires less steam (control agent). A decrease in temperature would require more steam.

Table 1. Summary of Control Modes

The definitions in this table are given in terms related to the output of the controller. In a perfectly linear system the output of the controller becomes the position of the final control element.

It is assumed that the automatic controller operates in a theoretically perfect manner; that is, it detects infinitesimal variations of the controlled variable from the set point and responds instantaneously in accordance with its mode or combination of modes.

On-Off (Two-position). The output of the controller (and consequently the final control element) changes from one of two fixed positions to the other. Normally supplied with differential gap. In this case, the final control element is moved from one of two fixed positions to the other when the controlled variable reaches a predetermined value above the set point. It is returned to the first position only after the variable has passed a predetermined value below the set point.

Control adjustments are made for the set point and the differential gap. Adjustments are commonly expressed in units of the controlled variable or per cent of the controller scale range

ON-OFF CONTROL

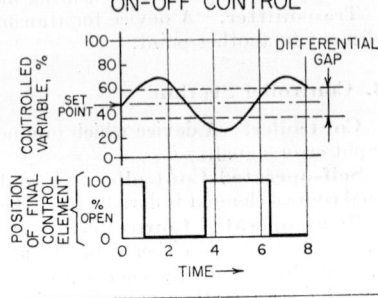

Floating Mode. There is a predetermined relation between the deviation and the rate of motion of the final control element. A neutral zone in which no motion of the final control element occurs is often supplied.

Single-speed floating action is the most popular type of floating action. The final control element is moved at a single rate.

Multispeed floating action moves the final control element at two or more rates, each corresponding to a definite range of value of deviation.

Control adjustments are made for floating speed (applies to single or multispeed controller action). It is the rate of motion of the final control element and is expressed in per cent of full range motion per minute.

FLOATING— SINGLE-SPEED

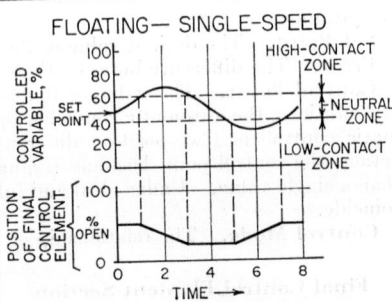

FLOATING — PROPORTIONAL-SPEED

Proportional-speed floating action changes the position of the final control element at a rate proportional to the deviation

Control adjustments are made for floating rate. It is the rate of motion of the final control element corresponding to a specified deviation

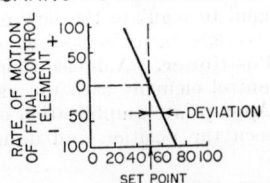

FLOATING AVERAGE-POSITION

Floating average-position action has a predetermined relation between deviation (of the controlled variable) and rate of change of the time-average position of a final control element which is moved periodically from one of two fixed positions to the other

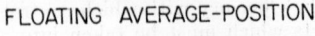
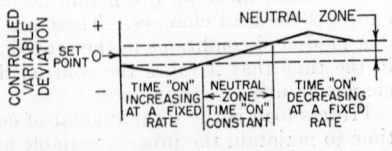
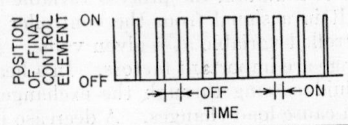

Table 1. Summary of Control Modes (*Continued*)

PROPORTIONAL MODE

Proportional (Throttling, Modulating). There is a continuous linear relation between the value of the controlled variable and the output of the controller (as long as the variable is within the proportional band). Adjustment is made in the proportional band. This is the change in the measured variable which will cause the full change in the output of the controller. It is expressed in units of the controlled variable.

Gain is the reciprocal of proportional band. It expresses the multiplying action of the proportional mode.

Proportional mode is characterized by offset from the control point. Offset is proportional to the proportional-band setting

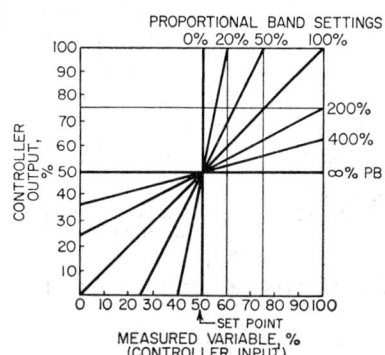

RESET MODE

Reset (Integral). The final control element is positioned in accordance with a time-integral function of the controlled variable. Pure form is floating action without a neutral zone. Often combined with proportional mode.

Adjustments for reset (when combined with proportional mode) are expressed in the terms of period of time that it takes the changing reset signal to duplicate the initial signal created by the proportional-control action alone. Units are repeats per minute or the reciprocal, minutes per repeat.

Reset eliminates offset characteristics of proportional action

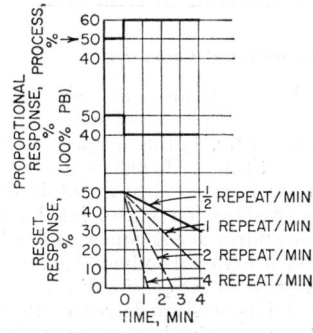

RATE MODE

Rate (Preact, Derivative, Booster, Anticipatory). There is a continuous linear relation between the rate of change of the controlled variable and the output of the controller.

Adjustment is in terms of rate time. It is the time interval by which rate action advances the effect of proportional action on the controller output. It is commonly expressed in minutes

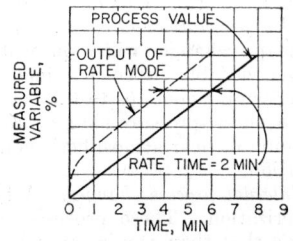

Table 2. Control Mode vs. Application

Mode	Process reaction rate	Load changes		Applications
		Size	Speed	
On-off. Two-position with differential gap	Slow	Any	Any	Large-capacity temperature and level installations. Storage tanks, hot-water supply tanks, room heating
Floating. Single speed with adjust-able neutral zone	Fast	Any	Small	Processes with small dead time. Industrial furnaces, air conditioning
Proportional	Slow to moderate	Small	Moderate	Pressure, temperature and level where offset is not objectionable. Kettle reboiler level, drying-oven temperature, pressure-reducing stations
Proportional plus rate	Moderate	Small	Any	Where increased stability with minimum offset and lack of reset wind-up is required. Compressor discharge pressure, paper-strip edge guiding.
Proportional plus reset	Any	Large	Slow to moderate	Most applications, including flow. Not suitable for batch operations unless overpeaking is allowed
Proportional plus reset plus rate	Any	Large	Fast	Batch control, processes with sudden upsets

Origin of Load Changes

Load changes can come from five major sources:

1. *Changes in demand of the controlled medium for control agent.* In the heat exchanger (Fig. 1) the increase in flow of the fluid or the change in the temperature both constitute a change in demand for the control agent. In both cases, more or less steam is required to maintain the fluid at the proper temperature.

Any combination of associated variables which will cause a change in the demands for the control agent comes under this classification. In a furnace, changes in the amount of material to be heated would constitute a load change.

2. *Change in the quality of the control agent.* If steam is being used as the control agent and its pressure decreases, the temperature of the steam will be lower. Because the steam contains less energy or heating value per pound, more control agent must be added to maintain a constant temperature in the heat exchanger.

Similarly, if the control agent was fuel for a furnace, a lower heating value would mean the necessity of adding more pounds of fuel to get the same heating effect. All corrections for this error must be made by the controller.

3. *Changes in ambient conditions.* Fluctuations in the temperature around an outdoor heat exchanger will require it to need more steam in the winter even though all other variables have remained constant. For this reason, some processes have summer-winter hookups to take care of ambient temperature changes.

Atmospheric-pressure changes can cause problems in processes that are controlled at vacuum or near atmospheric conditions. Such processes include evaporation and distillation.

4. *Internal process changes.* A typical example is the action of a reactor used in polymerization. These processes start out with the reaction being endothermic, meaning that they require energy. Once the reaction is underway, it changes to a exothermic reaction. At one point steam is required to heat the reactor, and later, cooling water is needed to keep the temperature at its required value. In this case, a different control agent is used at different points during the processing in order to maintain temperature control.

5. *Changing control point.* This is an external adjustment to the process. It means new energy requirements in order to meet the new conditions.

Process Lag

Process lag is caused by one or more of three main process characteristics: capacitance, resistance, and dead time.

Capacitance. This characteristic is related to capacity but is not identical. Capacity is a measure of the maximum quantity of energy or material which can be stored within the confines of a stated piece of equipment. It is measured in units of capacity. The volume capacity of an open tank, for example, is the maximum volume of liquid it will hold without overflowing. The weight capacity of a compressed-air tank is the maximum weight of air which it will hold without exceeding safe pressure. Dimensional units for capacity are given below.

Type of process	Unit
Thermal...............	Btu
Pressure..............	Cubic feet
Liquid level..........	Cubic foot, pound

Capacitance is the change in quantity contained per unit of change in a reference variable. It is measured in units of quantity divided by the reference variable.

The two vessels (Fig. 2) have different capacitance even though their capacities are identical. Vessel A has a capacitance of 100 divided by 10 or 10 cubic feet per foot. Vessel B has a capacity of 100 divided by 5 or 20 cubic feet per foot. Other dimensional units for capacitance are shown below.

Type of process	Unit
Thermal...........	Btu/deg
Volume...........	Cu ft/ft
Weight...........	Lb/ft

Large capacitance has a flywheel or inertia effect in a control system. The controlled variable tends to maintain a constant value regardless of load changes. While this characteristic is good for a steady process, it makes it more difficult to change promptly to a new value of the controlled variable.

On liquid-level control, the large capacitance of vessel B would prevent large changes in level due to load. However, it would introduce a time lag in correcting for any load changes which were large enough to affect the controlled variable. Vessel A with small capacitance would

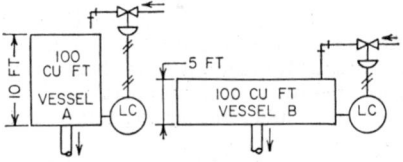

FIG. 2. Examples of capacitance.

respond quickly and be quite sensitive to load changes. In this case, a small load change would cause a change in the controlled variable. If time delay is important, an analysis of the capacitance of the circuit must be considered.

Resistance. This process characteristic is opposition to flow. It is measured in the units of potential change required to produce a unit change in flow. This form of lag is particularly important in temperature control because of the characteristics of heat transfer from the process to the temperature bulb. For example, the film of stagnant gas or liquid on the tubes of heat exchangers often creates a more serious resistance to heat flow than the walls of the tube. Hence thermometers immersed in a fluid stream must have adequate velocity past them to eliminate the film or else a serious delay in temperature transmission is created. Velocities of 60 fpm for liquid and 400 fpm for air are considered necessary for good performance. Dimensional units of resistance are shown below:

Type of process	Unit
Thermal............	Deg/(Btu)(hr)
Pressure............	Psi/(cu ft)(min)

The effects of resistance (transfer lag) are often overcome by the use of the rate mode of control. This mode anticipates the temperature that is in the process by compensating for the lag due to resistance (see Table 1).

Dead Time. This characteristic is any definite delay between two related actions in the process. It is sometimes called *transportation lag* and is measured in units of time.

A good example of dead time is found in a charge heater used in the chemical industry. It takes some time for the oil to go through the long length of heater tubes and finally be discharged into the next process. Temperature control is used on the output in order to assure that the fluid is at the right temperature. If the velocity of the oil is 100 fpm in a heater with 200 ft of tubing, the time required to detect a change of temperature of the oil at the input would be 2 min.

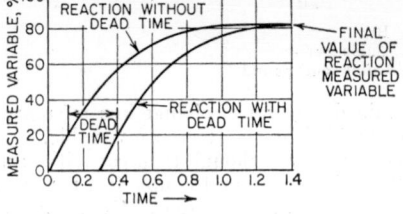

Chemical reactions which take time to complete are another source of dead time. Neutralization control with pH equipment is an example where process dead time is an important consideration.

Dead-time lag causes no change in the process reaction characteristic but rather shifts it in time (Fig. 3). A process change is not sensed until the dead time has passed. This delay is a serious one in many processes and has resulted in several major unit control systems such as cascade and override which correct this condition (see systems described later in this section).

FIG. 3. How dead time shifts reaction curve.

Rate mode is not helpful with dead time since sharp changes in the process variable cause large corrections. Because of the dead time between the change and measurement response, the corrections continue too long, causing instability.

Rate of Load Change

The rate of the change in load is an important criterion in a process-control problem. Abrupt changes in load (step change) are often assumed for purposes of mathematical analysis. However, a slow change is more typical of a process. A slow change is corrected by a narrow proportional band and a high reset rate. This keeps the proportional band shifting constantly as the reset action aids in providing the proper correction.

Different rates of load change cause considerable change in response of the control (Fig. 4). Fast changes cause wide deviations if the proportional band is maintained at the same value as for the small load change. This deviation takes place because the same amount of gain due to proportional action is applied to the large deviation as to the small one. If the load change is great, the multiplying effect of the gain will be great also and reset will still be limited to its original time function.

Proper settings of the reset action can generally overcome such deviations. However, if settings are made for one type of rate change and the rate of the load increases, instability (including resonance) might result. This is particularly true if the "self-regulation" of the system does not limit the offset to a reasonable value.

In start-up of a control system, some anticipation of the rate of load change must be made in order to adjust the controller settings. Setting too "tight" a proportional band and reset rate to give good control with small rates of load change might result in poor control should large rates of load change take place. For single systems, the operator can make appropriate manual corrections at the time of the excessive load change. However, a whole control panel set "tight" can cause unsatisfactory response to spread throughout the process and cause considerable difficulty before the instruments can be adjusted.

The presence of dead time is an additional problem when combined with rate of load change. While small proportional bands will tend to reduce the error as much as pos-

sible, this corrective action takes place after the fact.   This often results in unsatisfactory control, with the correction signal lagging the measured variable.   Use of a
rate mode tends to correct for such delays.

Magnitude of Process Load

It is possible to have a constant rate of process load change and yet have its magnitude vary for different conditions.   Normally, the controller has one setting of its
mode and then no further adjustment is made.   Changes in the magnitude of the load
make it advisable to adjust the controller settings to obtain optimum control.

There are three factors which cause unsuitable control because of the magnitude of
the load change: (1) the performance curve of the final control element, (2) process-
load-reaction-rate relationship, and (3) controller scale characteristic.   Each of these
factors will be considered separately in order to make an analysis simpler.

Effect of Control Element.   Assume that the scale is linear and that there is a
constant reaction rate.   If the valve performance curve is nonlinear (Fig. 5) a correctly operating system at A might cause cycle when the magnitude of the load drops to B.
Figure 5 shows that a 10 per cent valve position change at A causes a smaller change
in the flow than at B.   In effect, the proportional band is being increased when the

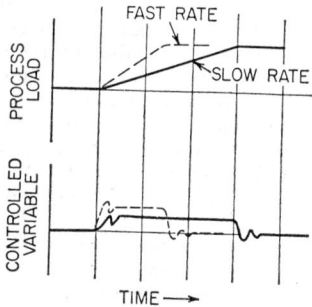

FIG. 4. Curves illustrate larger deviation of
controlled variable for larger rates of proc-
ess-load change (proportional-plus-reset
mode).

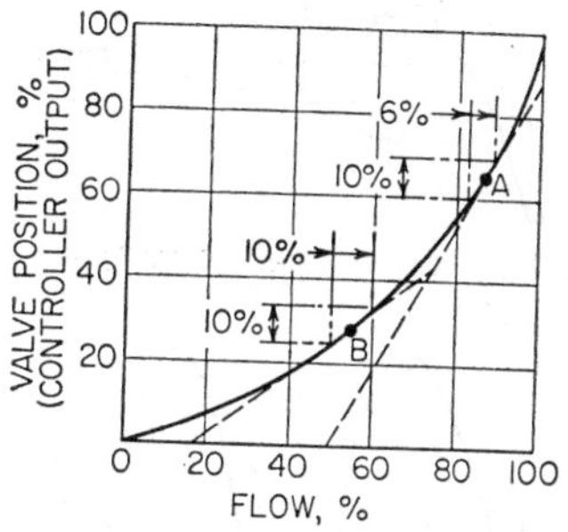

FIG. 5. Curve shows how a nonlinear valve
changes its response to controller output at
different openings.

load moves to B.   An increased gain due to this change in location on the curve will
cause cycling under some circumstances.

On the basis of this analysis, a final control element should generally be linear in
order to maintain a constant proportional band at all points on its curve. However,
from a practical standpoint, a linear valve is limited in its turndown, controlling
satisfactorily down to about one-twelfth its total stroke.   Because of the unknowns
in many new installations, an "equal-percentage" characteristic in the final control
element is usually more desirable. This valve has a constant per cent rate of
load change at constant drop per unit change in valve opening.   Wide load changes
can be absorbed by the valve because of this characteristic, and it is self-compensating
if oversized.

Process Load and Reaction Rate.   The reaction rate of a process varies with the
size of the process load.   Since the reaction curves are not parallel, different controller
settings must be used if stability is to be obtained.   For example, if the proportional
band is kept constant, a small load is likely to cause oscillation at a setting that would
be stable for the larger load.

Because the response of many processes is nonlinear, the use of the equal-percentage
valve characteristic in the final control element aids in canceling out the nonlinearities.
Optimum control setting can be calculated by the Ziegler-Nichols method.[29]

Effect of Controller Scales. Variation in system performance because of non-linearity of the transmitter signal is widely prevalent. The most common examples are flow with its square-root chart; vapor-tension thermometers with a nonlinear chart; and pH with its logarithmic relationship. The performance of these variables is similar to that of changing the process reaction. Again the final control element is used to cancel out the process changes. Another technique is to linearize the signal with a relay or cam device.

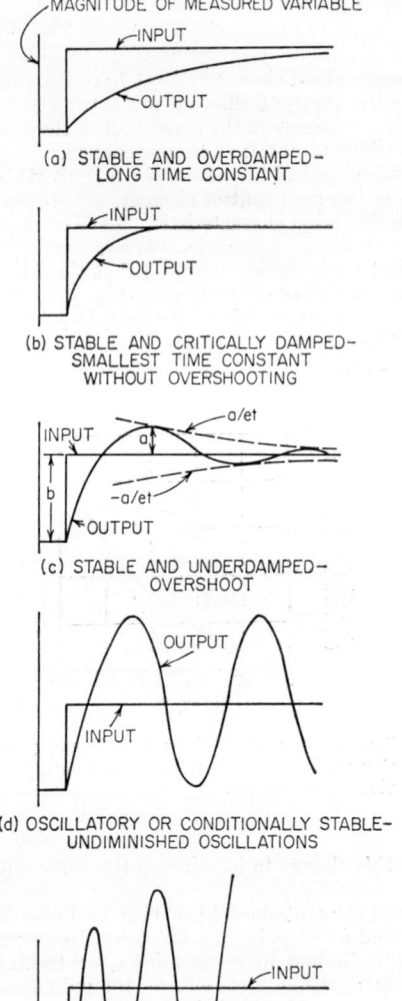

(a) STABLE AND OVERDAMPED–
LONG TIME CONSTANT

(b) STABLE AND CRITICALLY DAMPED–
SMALLEST TIME CONSTANT
WITHOUT OVERSHOOTING

(c) STABLE AND UNDERDAMPED–
OVERSHOOT

(d) OSCILLATORY OR CONDITIONALLY STABLE–
UNDIMINISHED OSCILLATIONS

(e) UNSTABLE–ENLARGED OSCILLATIONS

Fig. 6. Types of response curves. (*Source: Ref.* 28.)

Control Stability

The stability of control is demonstrated by the ability of the measured variable to return to the set point after its disturbance. Stability of control is defined as the properties of the combined effects of control system and process upon one another, whereby the controlled variable is maintained within limits without sustained cycling. Depending on the characteristics of the system, five different types of response in the measured variable can be obtained. Illustrated in Fig. 6, these are: (a) stable and overdamped, (b) stable and critically damped, (c) stable and underdamped, (d) oscillatory or on the threshold of stability, and (e) unstable.[28]

Whenever a closed-loop system is formed, an oscillatory or unstable system is in ever-present danger. Too narrow a proportional-band setting results in a correction whose effect is larger than that of the original disturbance. The effect of this larger correction is then amplified by subsequent corrections so that variations of the output of the controller quickly reach wide proportions.

For a linear system the size of the input step function has no bearing on the stability of the system. For nonlinear systems, however, size of the step may be a very important factor in stability.

The critically damped system is the optimum arrangement. However, rather than chancing an overdamped system with the resulting system sluggishness, a slightly underdamped response is used. In order to get the best response, the first overshoot must be equal to or less than 20 per cent of the input disturbance, and each successive oscillation must fit within an envelope defined by $\pm a/et$ (Fig. 6c).

Different types of processes require distinct criteria as far as the satisfactory nature of the stability of control is concerned. The three criteria which are used are minimum area, minimum deviation, and minimum cycling (see curves of Fig. 7).

Minimum area under the process curve is generally desirable for most applications although it does result in a longer time deviation from the set point. It has the advantage that the deviation is kept on one side of the set point. This is desirable in some applications such as food processing where overshoot would cause burning
of the product.

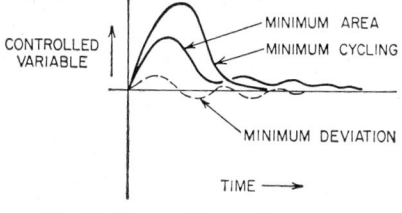

FIG. 7. Types of control stability.

Minimum deviation maintains close control although cycling about the set point results. Keeping the deviations small eliminates unsatisfactory response of the entire system which might take place with a minimum-area criterion. A slightly undamped response results in this curve.

Minimum cycling is used where disturb-
ances of a time duration must be avoided. Many chemical processes have numerous systems in series. A disturbance for any period of time would be reflected throughout the process. Minimum cycling is the result of a stable and critically damped system.

System-Variable Relation

A knowledge of the type of system and the variable being controlled will often enable a prediction to be made about the response of the process. This is helpful in selecting the best instrumentation.

Many processes have a great degree of self-regulation. Processes with little self-
regulation and a combination of lags and loads lead to the real problems in getting satisfactory automatic control. As a generality, the processes which have either an extremely fast or extremely slow response characteristic give the most difficulty. Examples of fast processes are (1) pressure control of pulsating liquid flow; (2) control of the discharge pressure of single positive-displacement pumps, by speed adjustment of the pump, where no accumulation can be tolerated; and (3) control of two series-
connected centrifugal gas blowers.

Examples of slow, long-time-lag processes are (1) distillation control, (2) polymeriza-
tion reactions in vessels containing viscous liquids which must be held for several hours; and (3) liquid or gas analyzers, because of their long sampling time.

There are other processes that provide erratic operating conditions by virtue of their nature. Boiler-drum level control is typical. The difficulty is that an increase in the demand for steam decreases the pressure on the water surface. The water is now superheated and tends to generate a vapor within the liquid, causing the volume of water to increase. This "swell" of the volume of water in the drum produces a higher level, although the actual amount of water in the tank is decreasing because of the increasing evaporation rate. The opposite occurs when the load is decreased, causing a "shrink" in the water level. In both cases the controller receives an erroneous signal. Both two- and three-element control is used to overcome this problem (see Sec. 14).

pH control is usually a problem because the chemical interaction between two liquids is not usually linear and frequently a considerable time lag exists.

Batch Processes

A special control case is that of a batch process involving a relatively slow responding controlled variable. A batch process is when a single quantity of material is treated as a whole sequence of operation in a given period of time and is then discharged as a unit to make room for a new charge. Typical batch processes are the temperature control of heat-treating furnaces, food cookers, or polymerization reactors.

A control problem exists because a large amount of heat is needed to start up the unit but less heat is required when the temperature finally reaches the set point. If the process reaction rate, transfer lag, and dead time are small, on-off control can be used. However, if minimum deviation and lack of cycling are desired, it is necessary to go to proportional control.

Proportional control is adequate where the reaction rate is large or the transfer lag great. However, some degree of offset will exist. If offset is objectionable, it is necessary to add the reset mode. However, the addition of reset will cause overshooting of the process variable (Fig. 8). On start-up, the measured variable is being brought up from zero condition to some higher set point. As long as the process is below the set point, reset action will continue to correct the output signal. Only after the variable has gone above the set point will the correction signal due to reset start to decrease. Overshoot is thus inevitable with the proportional and reset combination. In some processes, an overshoot in the measured variable causes spoilage of the batch.

Overshoot can be eliminated by the addition of rate mode. The rate mode is put ahead of the proportional and reset function of the controller. The effect of this change is to cause the signal output to the controller to be "ahead" of the actual

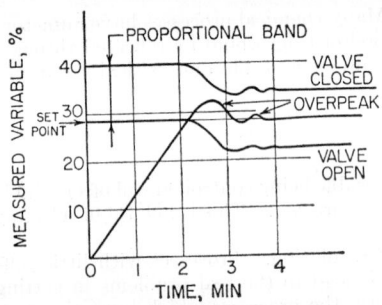 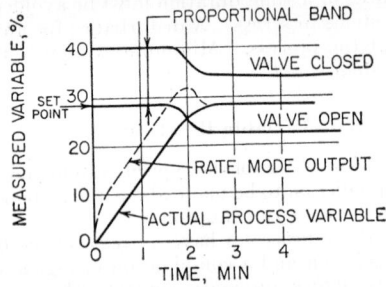

FIG. 8. Overshooting of set point on start-up of batch process.

FIG. 9. Elimination of overshoot with rate mode.

process signal. This leads the controller into believing that the process has already crossed the set point. The reset action starts decreasing at that point which shifts the proportional band downward to a lower position, thus preventing overshoot (Fig. 9).

CONTROL OF BASIC PROCESS VARIABLES

Processes involving the basic variables of flow, pressure, level, and temperature have similar performance characteristics regardless of the application. Such characteristics can be helpful in anticipating instrumentation requirements. A primary criterion of a process is its time constant. This is a function of all the load and lag conditions as well as the physical properties of the variable. A hypothetical curve for a majority of such processes shows flow at one end of the scale with fast response while temperature and chemical analysis are at the other end with a long time constant (Fig. 10).

Flow

Flow is a fast-responding process, having very little capacitance. The largest time lags occur in the measuring and controlling sections of the loop.

Flow signals are measured and transmitted by a variety of equipment. The most common instrumentation is the orifice plate with a differential-pressure transmitter. Most flows take place through the use of pumps or compressors. Consequently, the flow signal has a large quantity of small transient variations due to the pump or compressor. Incompressibility of the fluid and low capacitance of the system mean that the transient signals are not dampened out. Because of this, care must be taken to supply control settings which will ignore the transient pulsations ("noise" or "trash") in the transmitter signal.

The most common final control element is the valve. It is usually placed downstream of the orifice in order to avoid disturbances of the flow stream due to its throttling action which would affect the accuracy of orifice metering. An orifice requires at least 30 diameters of straight run of pipe after a control valve in order to avoid such disturbances. Downstream piping is less critical, five diameters being all that is required to avoid disturbances.

On gases, the location of the control valve depends on whether the flow is to be controlled on the basis of the downstream or upstream pressure. The compressibility of the gas at different pressures results in flow rates that are proportional to the square root of the absolute pressures. One of the differential pressures at the orifice must be absolutely constant if it is desired to maintain a constant flow (Fig. 11).

Where a positive-displacement pump is used, extremely high pressures can be developed. A control valve cannot be put directly in the piping for this reason. Flow

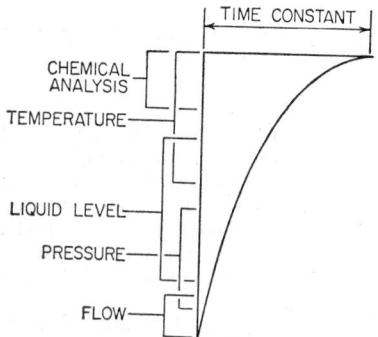

FIG. 10. A comparison of relative time constants for common process variables.

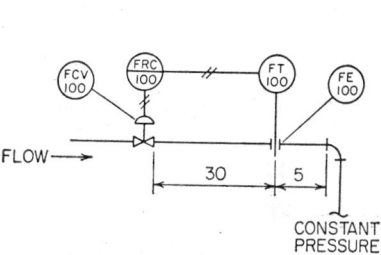

FIG. 11. Orifice installation downstream of valve for flow control of gas with constant downstream pressure.

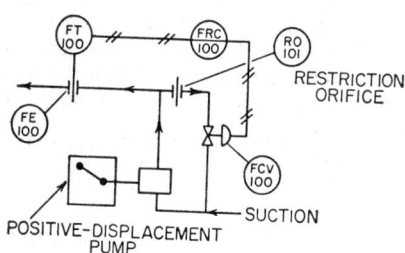

FIG. 12. Typical installation of flow-control system with positive-displacement pump.

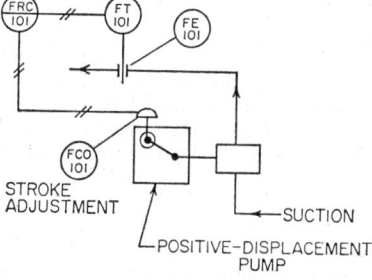

FIG. 13. Flow control by stroke adjustment of positive-displacement pump.

control is achieved by putting the valve in a bypass so that the excess flow is returned to the suction side of the pump (Fig. 12). Pressure-reducing restrictions are often used in the bypass in order that the control valve does not take the entire pressure drop. This avoids excessive wear on the valve.

The valve is never located on the suction side of a fluid-pumping installation. Closure of the valve would result in a low pressure at the suction which could be below the vapor pressure of the fluid. This would generate gases and cause vapor lock to develop on the pump.

Speed control of a positive-displacement pump is often used instead of a valve to control flow. This is done by controlling either the stroke adjustment (Fig. 13) or the input of the driving energy such as steam or electricity.

In centrifugal-pump installations, the valve is located directly in the output line since very little buildup in pressure can occur. The impeller of the pump has enough clearance in the casing so that it can turn without excessive pressure buildup.

Any control of the pump must take into consideration its size and performance. A large pump or compressor installation will have considerable inertia which would prevent fast reaction response. In this case it is better to use a bypass control rather than to control the equipment directly.

Most process control of flow requires a wide proportional band with fast reset. This avoids control action on the "noise" or transient disturbances yet ensures prompt correction of any persistent error signal. The time that the error persists rather than amount of error is more of an indication of its importance as a detection means.

Pressure Control

Pressure control is characterized by large capacity, small transfer lag, and small dead time. The increased capacity is conducive to the self-regulation of the process. This characteristic allows proportional-type controllers to be used in many cases.

Self-operated pressure regulators are often used in pressure control. These devices have a fixed proportional band which is small enough so that the gain is large, and consequently the offset is kept to a minimum. These controllers have low cost, high dependability, and low maintenance. The close coupling of the process directly to the regulator results in quick response to system upsets. Because the device is essentially a spring-opposed system, the accuracy of control is usually held to within 10 per cent of full scale.

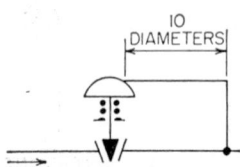

Fig. 14. Installation of self-operated pressure regulator.

The self-operated pressure regulator is installed directly in the line and the control sensing tap is placed about 10 diameters from the unit (Fig. 14). This location eliminates erroneous pressures caused by turbulence, sudden changes in velocity, or line losses.

Shock and vibration difficulties often occur when self-operated regulators are used with liquids. This is due to the sudden opening of the inner valve against the relatively incompressible liquid, which produces hydraulic shock. When the regulator is operating with its plug close to the seat and the line velocity is acting in a direction to close the plug, the Bernoulli effect tends to cause premature closure. When the regulator attempts to correct the problem, hammering takes place. This is usually an indication of too large a regulator for the application.

The self-operated pressure regulator is essentially a tight-shutoff single-seat valve in most cases. Because of this, it fails in the open position owing to either low pressure in the line or diaphragm rupture. If this fail position cannot be tolerated, pilot-operated devices must be used.

Self-operated pressure regulators should be operated at about 50 per cent of full stroke under normal conditions for best results and the least wear on the valve. In applications where there is a wide difference between the maximum and normal flow, two regulators can be used in parallel. The control pressure of one regulator is set for the normal flow condition while the pressure of the second is set approximately 10 per cent lower. When the outlet pressure in the first regulator "droops" because of higher than normal demand, the second regulator cuts in to increase the flow capacity without the loss of pressure control.

Two other types of pressure control are the transmitter-controller-valve arrangement and the pilot-operated pressure control. Pilot-operated pressure controllers use a measuring element which actuates a pneumatic control mechanism that produces an air-pressure signal proportional to the measured pressure. The advantages of these two types of installation are:

1. Greater sensitivity.
2. Adjustable proportional band, reset, and rate.
3. Ease of adjustment.

4. Flexibility (i.e., reversibility, remote location, etc.).
5. Wider spread of pressure settings within a given range.
6. Wider selection of pressure ranges.
7. Can be used on high pressures and pressure drops.
8. Wider flow range.
9. Can be used where flow conditions require large body sizes.
10. Control valves have standard pressure range.
11. Valve can be made to fail in either direction.

When the system has a large capacity, the proportional mode is usually adequate for control. Reset is used where no deviation can be tolerated between load upsets.

An example of control-mode selection is in the pressure control of a distillation column (Fig. 15). Control is accomplished by pressure control of the reflux drum, which allows the noncondensables to be vented. Proportional control could be used in this case because the response is immediate since there are no long transfer lags.

If pressure control of the tower is going to be accomplished by controlling the cooling water on the condenser, there will be an introduction of considerable transfer lags. In this case, reset action is necessary in order to get good control.

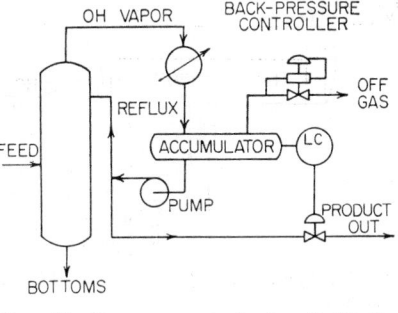

Fig. 15. Pressure control of a distillation column.

Level

Level control is similar to pressure control, usually having a fair amount of capacitance. Some systems have considerable dead time, although transfer lags are small. The capacitance contributes to the self-regulation of the liquid level in atmospheric vessels. Self-regulation is usually small when the vessel is under pressure.

The preciseness of the liquid-level control differs for various processes. This characteristic makes possible three control arrangements: (1) high-low limit, (2) average level, and (3) precise level.

Two-position control with differential gap is used with both electric and pneumatic equipment. This mode of control is satisfactory when the level can cycle between a high and low limit without affecting the process requirements. The level is held within the differential gap without difficulty if the capacitance is large enough. The advantages of this control mode are (1) simplicity of equipment, (2) economy, and (3) lack of wear on valves. This mode requires that the maximum input to the vessel must exceed the maximum discharge in order to ensure that the vessel will not run dry.

Averaging liquid level is used where preciseness of control of level can be sacrificed in order to keep the output of the vessel fairly constant. A typical application is where the process vessel under control is part of a chain of processes where the output of one process is the input of the next. A tank with level control acts as a surge tank to absorb any violent load changes.

The level is permitted to drift gradually up and down. It averages at any point in a given band rather than fluctuating only between the high and low limit as in on-off control. Limit stops are provided to prevent the tank from being either flooded or drained dry before proper control responses can be initiated. A valve positioner is usually required because slow and smooth valve action is required.

Averaging control is obtained with a proportional-plus-reset controller. The proportional band is as much as 400 per cent in order to be very insensitive. The reset time is short, which results in an immediate attempt to correct the level error. However, because the proportional band is large, the amount of correction due to reset is small. Its persistence as time passes causes a small sustained correction, resulting in an overdamped system.

Exact control is necessary in many processes. Proportional-band control often provides close enough level with only slight offset. If the tank is part of a series of processes, this mode prevents severe load upsets at a small sacrifice in the level. Several types of direct-connected float-type controllers are used with proportional mode equal to a small fraction of the travel of the float. The proportional band is as small as $\frac{3}{16}$ in. of level with these devices. However, if there is large capacitance in the vessel, this band is adequate for valve modulation and reasonably close control. The advantages of such equipment are simplicity and economy. Float-operated controllers are often used on tank draw-offs (Fig. 16).

Proportional-plus-reset mode is used when small process capacity exists and there is transfer lag due to multiple capacities. This results in precise and exact control under all conditions.

In many cases, the blanketing pressure in a vessel is considerably higher than the pressure developed by the liquid level alone. A severe fluctuation in the blanketing pressure can cause the discharge flow to vary considerably. As a result, the liquid level is ultimately affected. Self-regulation is absent under these conditions.

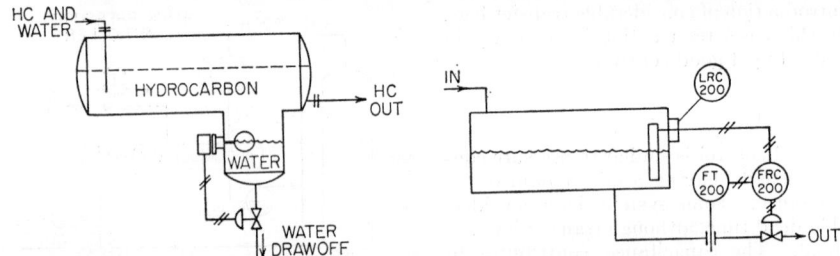

FIG. 16. Typical use of float-operated level controller.

FIG. 17. Cascade control with flow controller reset by level.

In order to compensate for this problem, cascade control is used, with the flow controller being the secondary control (Fig. 17). As flow changes because of pressure swings, the secondary controller immediately corrects the flow before the liquid level can change appreciably. For exact control, both controllers should have proportional-plus-reset mode.

Temperature

Every temperature-control problem is essentially one of regulating heat exchange. Because of the nature of heat transfer, temperature processes are characterized by larger capacities than either flow, pressure, or liquid level. The process reaction rate is also slow. Dead time is often large, particularly in fluid-heating processes such as charge heaters and distillation columns. Some processes such as heat-treating furnaces have small transfer lag and dead time.

The temperature lag involved in the measurement of this variable is an important factor. The thermal element is usually placed in a well to protect it and allow servicing of the element without interrupting the process. The speed of response of the temperature element is dependent on its design, the well material, and the speed of fluid in which it is inserted. The element should not be placed in a stagnant pocket or any place where a low velocity occurs.

Processes involving high capacity and small transfer lag and dead time are suitable for on-off control. Furnaces and bath temperatures fall into this classification. This mode can be used for most electric furnaces, most radiant-tube furnaces, and open-flame furnaces where furnace pressure or fuel-air ratio is not controlled, and where products of combustion are not used as a protective atmosphere. On-off control with differential gap can generally hold control to within about 1 per cent.

Multiposition control is often used on batch furnaces. Such furnaces require high energy input for rapid heat-up but considerably less energy when the load reaches temperature. This refinement of two-position control includes three positions: high, low, and off. Control will cycle between low and off once the process is up to temperature.

Floating control is used where load changes are slow and the process reaction rates are slow to moderate in speed. Transfer and measuring lags must be a minimum. It is often used to control the flow of air through the cooling section of slow-speed continuous furnaces.

Proportional mode is used where closer control is desired. It also provides a continuous flow of power or heat as in gradient heating of continuous strip or webs and nonrecirculating air or gas heating. In these applications, the reaction rate is so high that two-position control might produce large temperature swings. Proportional control is necessary on fuel-fired furnaces in which the fuel-air ratio or furnace pressure are controlled, or where the products of combustion are used as a protective atmosphere.

Proportional control is used where the load changes or measuring time lags are large. For large load changes, the proportional band must be small in order to eliminate offset. When the measuring time lags are great the proportional band can be made wide enough to ensure straight-line control. Reaction rate is not critical with this mode.

On any temperature problem, the capacity on the load side should be higher than on the supply side. A large capacity on the load side is favorable, since it diminishes and smooths out variations. For this reason, equipment such as heat exchangers should be fully loaded in order to avoid cycling when on proportional control.

When the temperature difference between the supply and load is great, the quantity of heat potentially transferable is large and on-off control results in cycling of high amplitude. With proportional control, control is achieved only by wideband control. A small temperature difference aids in obtaining good control.

Proportional-plus-reset mode is used where there is appreciable time lag in sensing the temperature and where there are large and frequent load changes. This condition exists in continuous processes such as heat exchangers and continuous furnaces.

Rate mode is used to overcome transfer lags in temperature processes. Some transmitters have rate built into them in order to overcome the problem of temperature-measuring lag. Rate action combined with proportional-plus-reset mode is very desirable on most temperature controls involving long dynamic lags.

The location of the temperature element often has as much to do with the efficiency of control as other parts of the control loop. The temperature bulb should always be located at a point where the coefficient of heat transfer will be as large as possible. In fractionating columns, vapor leaving a boiling liquid will be at the temperature of the liquid. However, the element should be located in the liquid on the column plate rather than in the vapor space between the plates in order to take advantage of the higher coefficient of the liquid.

In the case of an evaporator in which the liquid has a boiling-point rise, the vapor will be superheated, resulting in a gas film that will surround any bulb located in the vapor space. Here again, the bulb should be located in the liquid.

In order to obtain good temperature control, the flow of the process must be constant even though the temperature varies. Sometimes a pressure or differential-pressure control will stabilize the supply flows. On many distillation and fractionating systems, the temperature control is cascaded onto a flow control. The flow controller immediately corrects all deviations in the flow to or from the unit while the slower temperature control makes the gradual adjustments in the flow rate which are necessary to compensate for change in quality of the feed, or fuel composition.

Humidity Control

Control of moisture in the air is important in the production of foods, textiles, paper, and some chemicals. It is also important in environment chambers and in continuous-curing and drying-process operations.

Humidity control involves the control of both temperature and moisture.* This is

* Refer to p. 7-11 for basics of humidity as a measured variable.

done by a combination of heating, cooling, humidifying, and drying. Usually, refrigeration is used to remove moisture and steam is used to add both heat and moisture. Water sprays can also be used as a source of moisture.

Control of humidity is very close to that of temperature in characteristic. It involves large capacity and moderate transfer lags. The transfer lags exist because the measurement of the humidity is one of temperature. If load changes are not

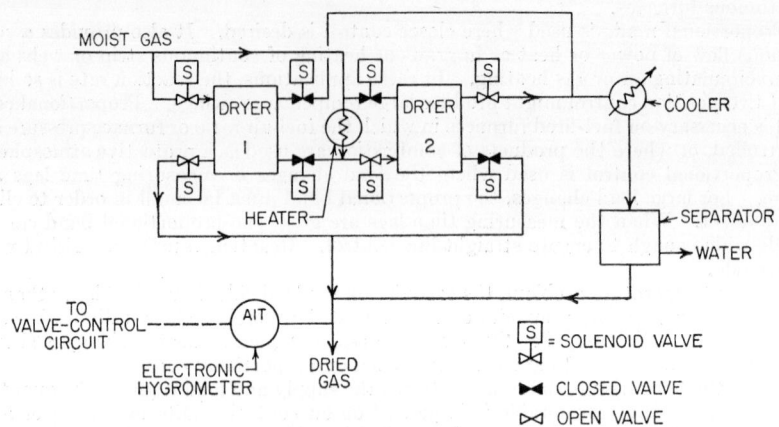

FIG. 18. Dehydrator control using electronic hygrometer.

severe, two-position or proportional control is adequate. However, most processes that require precise control need proportional-plus-reset mode.

Reliable control of the moisture content in gas streams is an important requirement in the operation of natural-gas plants. Excess water reacts with certain hydrocarbons to form hydrates which can seriously impair plant operation. Sales contracts also limit water content in natural gas.

The dehydrating system shown in the Fig. 18 includes twin dehydrating columns so that one can be regenerated while the other is on stream. The electronic hygrometer* monitors the drying cycle and diverts the gas streams from one drying tower to the other whenever the water content increases to a preselected value. The hygrometer also has a chart record so that the outlet-stream humidity record can be kept. Only a small sample flow is required, and this is secured through a bypass system. The sample stream is regulated at a constant pressure, temperature, and flow in order to have consistent readings.

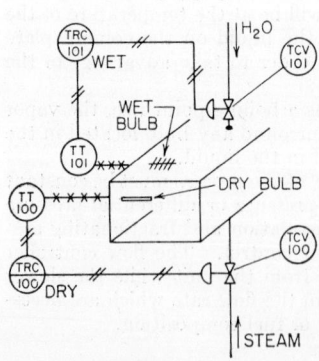

FIG. 19. Humidity-control system for explosive atmospheres.

A typical nonhazardous humidity-control system for explosive atmospheres is shown in Fig. 19. It uses a wet-bulb and dry-bulb system and a recording psychrometer. Each bulb is connected to a separate control circuit. The dry bulb controls the temperature and the wet bulb controls the humidity. Because the system is all pneumatic and the recorder is mechanical, there is no explosion hazard.

The primary elements are located in a duct, with an ejector being used to draw the sample past the bulbs. The wet-bulb control system sends a pneumatic signal to the water-spray control valve. The signal from the dry-bulb control system positions a valve in the steam line which furnishes the heat. Both valves are single-seated for tight shutoff.

* Described under Humidity and Dew Point, p. 7-11.

Chemical Composition

Processes are normally controlled through the chemical and physical variables that produce the product. The settings on the controllers are made as a result of a laboratory analysis of the product. On-stream composition analyzers are available which automatically analyze the product and send a control signal directly to the final control element. This "end-point" control has seen wider application with the development of chromatography, infrared analysis, near-infrared analysis, ultraviolet analysis, mass spectrometry, and refractometry.*

Chromatography is useful in determining the per cent of a given product in a flow stream. It has a great degree of simplicity in operation and is quite sensitive. Its response on a simple analysis (such as isopentane in butanes) requires 4 or 5 min, although high-speed units have been developed to make the analysis in less than 1 min. This characteristic of the instrument results in a certain amount of dead time, which is usually not serious in the type of control problem to which it is applied.

Fractionators and absorbers can be put on end-point control with the use of chromatography equipment. Generally, it can be used on any operation where temperature control is presently being used as a method of composition control.

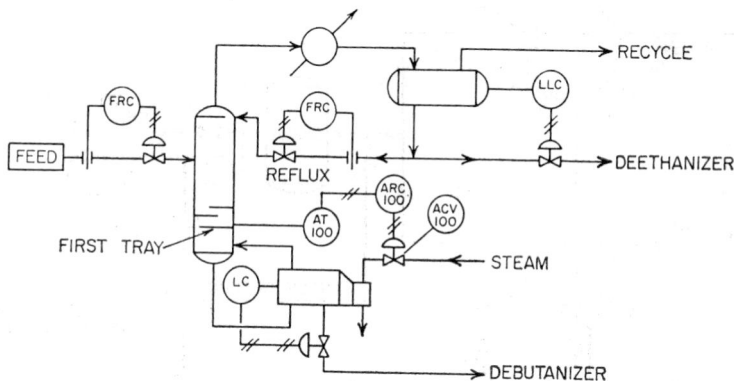

Fig. 20. Chromatography control of depropanizer.

Figure 20 shows automatic control of natural-gas liquids depropanizer with closed-loop control. In this operation, the feed is a composite of raw and natural gasoline from several sources. Feed flow rate and composition vary widely.

The chromatographic analyzer-recorder-controller (ARC) measures propane in the first-tray vapors and manipulates valve ACV on the steam to the reboiler to keep a constant amount of propane in the depropanized natural gasoline. Reports indicate that the analyzer is able to maintain propane in the bottoms product at 0.25 per cent (±0.05).

Besides good quality control, closed-loop chromatographic control saves utilities such as steam by permitting the operator to drive a column only as hard as necessary to maintain minimum product specifications. The technique also allows the less valuable components to be included in the finished product, which makes them available for additional refinement.

Infrared analysis is a technique using the theory that the absorption pattern of various gaseous, vapor, or liquid compounds differ. A wavelength of infrared radiation is transmitted through a sample mixture. Each sample compound in the mixture absorbs specific infrared wavelengths dependent upon its molecular structure. The amount of radiation absorbed is proportional to the concentration of compound.

This technique finds many applications in the chemical and petrochemical field. It is used to measure carbon monoxide on fluid catalytic-cracking units. Other applica-

* Refer to Sec. 8 for summary of chemical-composition measurement.

tions include measurement of ethylene purity during manufacture, measurement of sulfur dioxide loss in stacks of sulfuric acid plants, and isobutane loss on alkylation units.

Near-infrared analysis avoids the requirement of vaporizing a sample. This eliminates the possibility of partial fractionation, decomposition, or even polymerization. It operates on the same principle as the infrared units, but it uses the near-infrared spectrum between the red end of the visible light and extending into the conventional infrared band. It has been used to determine the water concentration in finished solvents such as alcohols or ketones.

Ultraviolet analysis works on the theory of measuring the amount of ultraviolet radiation transmitted through a sample. Its analysis is not so specific as the infrared analysis, and it lacks selectivity. It has been used for ozone monitoring, as well as phosgene, benzene, and toluene composition measurements.

Mass spectrometry is a more elaborate analysis device which operates on the theory that the ions of material are produced and separated according to their masses by magnetic or electrostatic means. This device has been used in the manufacture of

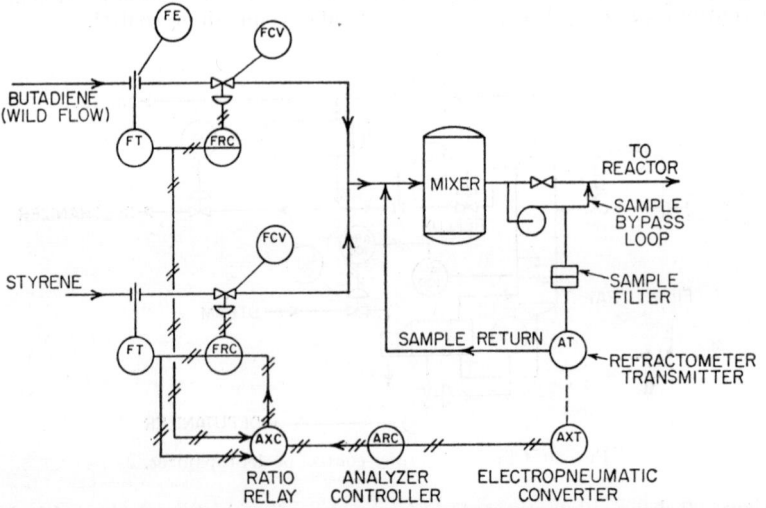

Fig. 21. Styrene-butadiene blending by refractometer.

synthetic rubber and aviation gasoline. It is generally used as a measurement rather than a control device.

Refractometry measures the index of refraction of light of various substances. Since the degree of deflection of light differs with particular substances, it finds wide use in the measurement of both simple and complex mixtures. Figure 21 shows a process refractometer that has been used to control a butadiene-styrene blended feed to a continuous reactor train in a copolymer operation. A ratio of 70:30 (butadiene-styrene) weight per cent is held.

The controlled "wild" flow is butadiene. The styrene is ratioed to the butadiene flow by a ratio controller which receives its command signal from the analyzer controller which in turn receives a transmitted signal from the refractometer.

A standard reference fluid is put on one side of the sample cell. The instrument is adjusted at zero for this sample. The unknown blended-liquid product flowing through the sample side of the cell has its refractive index compared automatically with the sample. The output of the refractometer, which reflects the variation in the refractive index from standard, goes to the controller, which resets the controlled point of the flow ratio on the styrene.

UNIT CONTROL SYSTEMS

In industrial processes, a number of control systems occur which are more advanced than the elementary control loop. Many of these systems are repeated again and again with minor variations. Known as *unit control systems*, they form the basic tool for all instrumentation.

Ratio Control

This is a unit control system where a secondary variable is to be controlled in a direct relationship to the primary variable. For example, an application commonly found in the chemical industry is that of ratioing a gas and liquid flow. Figure 22

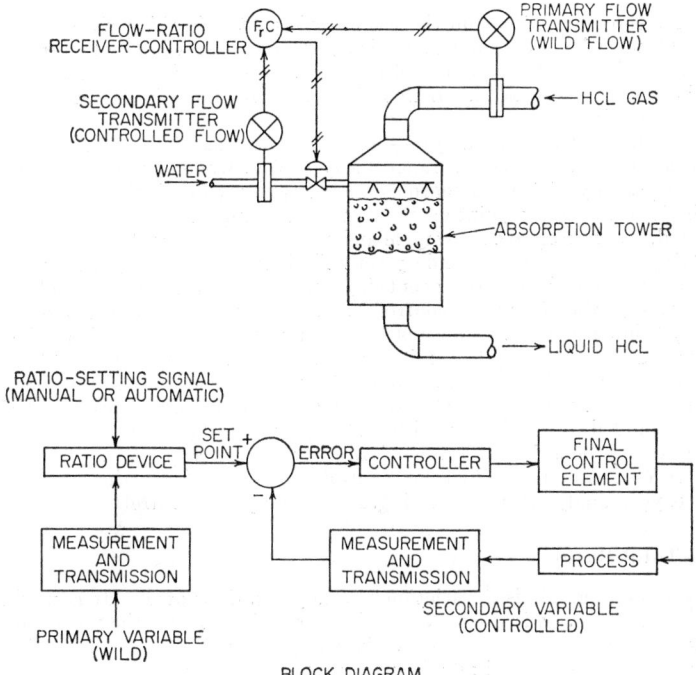

FIG. 22. Ratio control of water to hydrochloric acid gas to form constant concentration acid.

shows a system used in the manufacture of hydrochloric acid of constant concentration by the absorption of anhydrous hydrogen chloride gas in water. The gas is the uncontrolled primary flow from a production unit, with water as the related secondary flow. The water is introduced into the absorption tower as a spray and mixes intimately with gas entering the top of the tower. For all variations in the flow of the gas, there is an exact equivalent amount of water to be added.

The ratio unit is often a separate component, although it may be incorporated in a large case instrument. The signal of the primary transmitter is ratioed (or multiplied) by a factor that is set in the ratio unit either manually or automatically. The output of the ratio unit now becomes the set point on the secondary controller. The secondary controller automatically controls the valve to follow exactly the demands of the ratioed signal from the primary transmitter.

The setting of the ratio relay is a function of the relative ranges of the transmitters.

The formula for calculating the ratio setting is

$$\text{Ratio dial reading} = \frac{(Rc)(Cu)}{(Ru)(Cc)}$$

where Rc = controlled flow rate
Ru = uncontrolled flow rate
Cu = maximum capacity of uncontrolled process transmitter
Cc = maximum capacity of controlled process transmitter

Example. Two pipelines are used to blend fluids. The transmitter on the primary (or wild) flow has twice the range of the secondary (or controlled) flow. It is desired that the controlled flow be exactly 25 per cent of the wild flow at all times.

$$\text{Ratio dial reading} = \frac{(Rc)(Cu)}{(Ru)(Cc)} = \frac{1 \times 2}{4 \times 1} = 0.5$$

There are three limitations on the use of ratio control in this unit control system. Both signals must be in the same units (i.e., gpm, pounds per hour, etc.) and have the same characteristic, either square root or linear. Commercial ratio units are graduated to handle these functions. If there is another relationship, the ratio dial has to have special graduations.

The third limitation is the accuracy turndown or rangeability of the transmitter. Most linear transmitters have a rangeability of 5:1, while square-root transmitters have a rangeability of 4:1. Regardless of the ratio setting, if the secondary flow is below the lowest accurate point on the transmitter range, accuracy of ratioing cannot be expected. Referring to our example and assuming square-root transmitters: If the ratio setting is one-half, the secondary control will lose its accuracy once the primary transducer signal goes below 50 per cent since one-half of value (or 25 per cent) becomes the set point of the secondary controller. The controlled flow could not be held accurately below this point because of the inherent meter-accuracy limitations.

Note that a single primary transmitter may be used for the ratioing of many secondary components by using more than one ratio relay. This enables the solution of a wide variety of ratio problems such as the blending of gasoline where multiple components are added to the single wild flow.

The ratio may be adjusted automatically in many process loops. Typical of this type of process is an air-flow–fuel-flow control in a boiler. Here, the ratio is adjusted from an oxygen analyzer on the stack gas (see End-point Control).

Part-to-Total Ratio

This system (Fig. 23) is a variation of straight ratio control. It is used for two main reasons. First, it is used where it is impossible to meter the uncontrolled flow before the addition of the controlled variable. This problem might occur because of several reasons: (1) the flow line might be inaccessible; (2) high viscosity might make it impossible to measure the flow conveniently (after dilution, the viscosity would be lower and a flow meter could be used); (3) the fluid might be of a corrosive nature. Once it has been added to the main-line flow, the resulting solution might be neutral and more suitable for metering.

The second reason for part-to-total ratio is where it is desired to add the makeup liquid at a ratio to the total flow. This is true in many chemical processes where it is desired to record and meter the total flow and also know that this total flow contains a certain percentage of the desired component. While it is possible to record and control on the basis of a straight ratio control, the part-to-total method makes it much easier to calculate the percentage of a component in the total flow. The exact per cent is directly related to the setting on the ratio device. A two-pen recorder is usually furnished which shows the flow of the added component and the total flow.

Total-quantity Ratio Control

In this system, the quantity of a secondary variable is controlled in a direct ratio to the quantity of an uncontrolled primary variable. The amount of the secondary flow is more precisely controlled by this system than is possible with a ratio system using only instantaneous flow rates. In the straight ratio system, corrections are made to the secondary flow only after changes have occurred without any correction for the error that exists between the detection of the change and the final change of the secondary variable. By totalizing flows and comparing them, this system makes the proper adjustment to ensure that the percentage of the components in the total mixture is precise.

The unit control system finds great use in continuous pipeline blending where long-run operation requires precise flow control of the ingredients. Figure 24 shows a

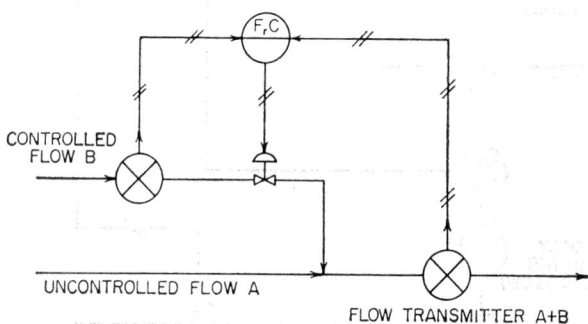

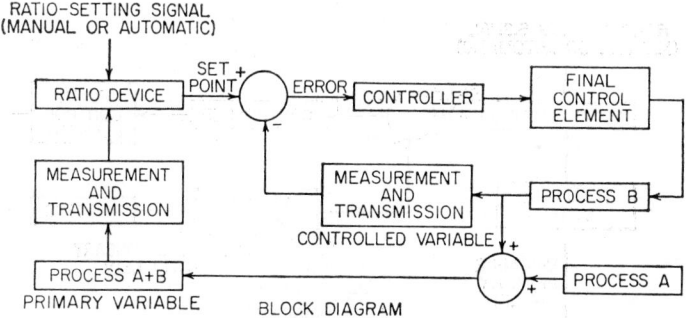

FIG. 23. Part-to-total ratio control of two products to form a third.

typical blending operation consisting of the totalizing devices, ratio adjustment, and a control loop. Numerous lines can be blending by only adding additional ratio devices and the related control loops. Turbine meters or other pulse-generating devices have been employed in such systems, with electrical transmission and control (electric-to-air transducer at valve).

Relation Control

Although relation control is similar to ratio control, it is unique in that it maintains an arbitrary relation between two or more variables such as temperature and pressure or pressure and flow.

There is typically a zone of operation for an axial air compressor where the flow rate is too low or the compression ratio is too high, and the compressor will not operate in a stable manner. This is known as the state of "surge," and instrumentation must be provided to keep the compressor out of this zone by relieving some of its discharge and thereby increasing the flow rate of air through the compressor to avoid surge.

A compressor with atmospheric intake is shown in Fig. 25. Flow rate is measured in the intake and discharge pressure is measured as a direct function of compression ratio since inlet pressure is constant. The discharge pressure signal is adjusted so as to give a set point as desired for the specific relation. Whenever the operating conditions determine a point to the right of the line shown on the curve of Fig. 25, the relief

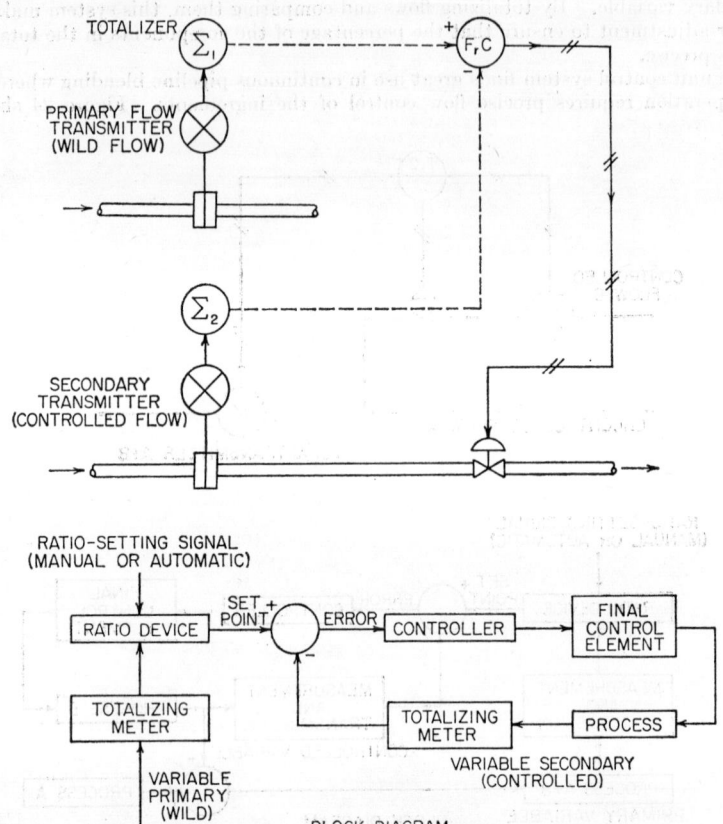

FIG. 24. Total-quantity flow ratio control for pipeline blending.

valve stays closed. If, however, the operating point approaches and tries to cross to the left of the line, the controller opens the valve enough to keep the system on the line, thereby preventing surge. Conceivably, conditions could always call for some relief flow and the system will remain on the line even during "normal" operation.

Another application of relation control is the characterization of a measured temperature to match a suitable vapor-pressure curve. In contrast to the straight-line function of the compressor application, this relation is nonlinear.

Cascade Control

One of the techniques for the increased stability of a complex circuit is the use of cascade control. This type of control accomplishes two important things: (1) it reduces the effect of load changes near their source, and (2) it improves the control circuit by reducing the effect of time lags.

Figure 26 shows a typical installation where cascade control could be used satisfactorily. With straight control as shown, variations in the heating effect of the steam will be sensed only after the change in product temperature has reached the bulb at the kettle output. It is possible that the pressure of the steam is reduced because of demands by other processes on the utility line. A lower head of steam results in a lower flow into the kettle than before. Less heat is being added to the kettle for the

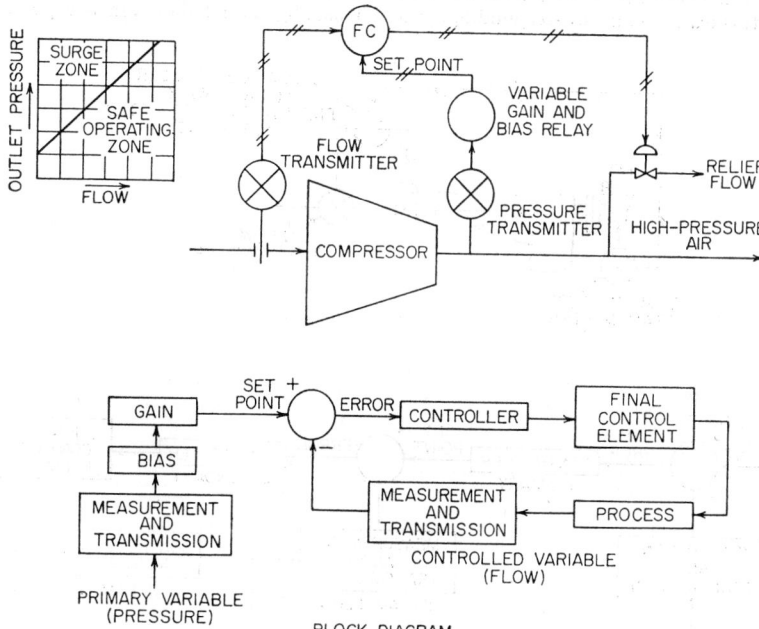

FIG. 25. Relation control of pressure and flow to prevent compressor surge.

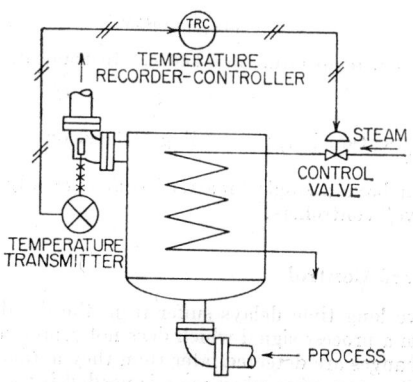

FIG. 26. Direct control of temperature in a kettle.

same valve opening. This decrease in heating effect is not felt until the reduced temperature finally reaches the temperature transmitter at the exit. By this time, however, the flow through the valve might be restored to its original value so that the corrective signal coming from the controller to open the valve further is not required at this time. This results in excess steam heating.

This cycle could be repeated endlessly with the process never catching up to the actual conditions in the kettle. With the master-slave relationship of a cascade controller (Fig. 27) local changes in steam flow are immediately compensated for by the slave flow controller loop.

Cascade control is really a feedback loop (steam flow) within a loop (temperature). As in this example, a slow-responding system is usually cascaded on a fast-responding

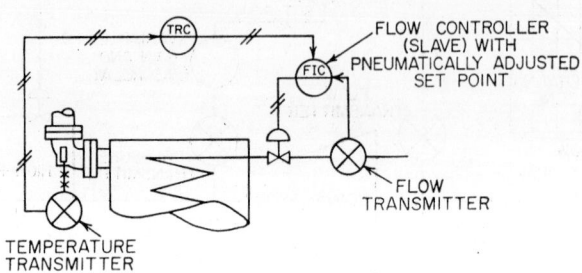

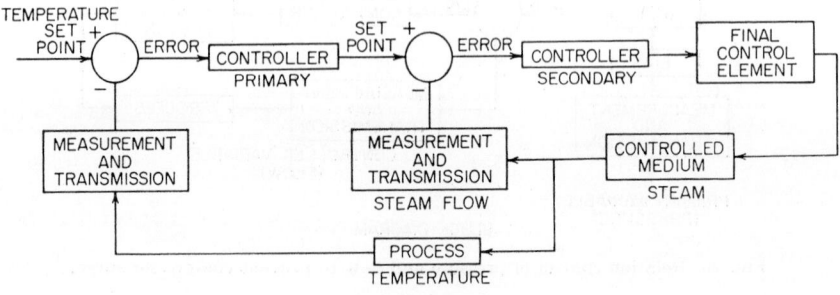

BLOCK DIAGRAM

Fig. 27. Cascade control of temperature in a kettle. Instruments are the same as in Fig. 26, except as noted.

system. Another application is the cascading of level on flow in boiler-drum feed-water control.

Cascade control can have a single "master" controller adjusting the set point of many secondary "slave" controllers.

Impulse Feed-forward Control

Systems which have long time delays suffer from the disadvantage that the controller is working from a process signal which does not represent the true condition of the process. Load changes are detected later than they actually occur, and thus correction is delayed, often occurring when none is needed if the load change had since been eliminated.

A charge heater (Fig. 28) often has the problem where the flow coming from the previous process varies considerably. Large time delays due to the length of the coils in the heater will mean temperature control will never be precise. The temperature being detected at the coil exit does not give a true indication of the possible rise and fall in temperature that will develop in the next instant because of a load change.

In the heater, the product temperature output is controlled by a controller on the charge exhaust line.   If the flow decreases, because of a change in the operation of the previous process, it will be some time before the decreased flow is sensed by increased temperature of the product.   When the temperature controller finally starts to correct by cutting back the fuel, the flow might have been restored and more heat would be needed at that point.   The process would tend to fluctuate and never hold the correct value.

An "impulse relay" (providing derivative or rate action) provides a feed-forward control action that continuously monitors the temperature signal to the control valve.

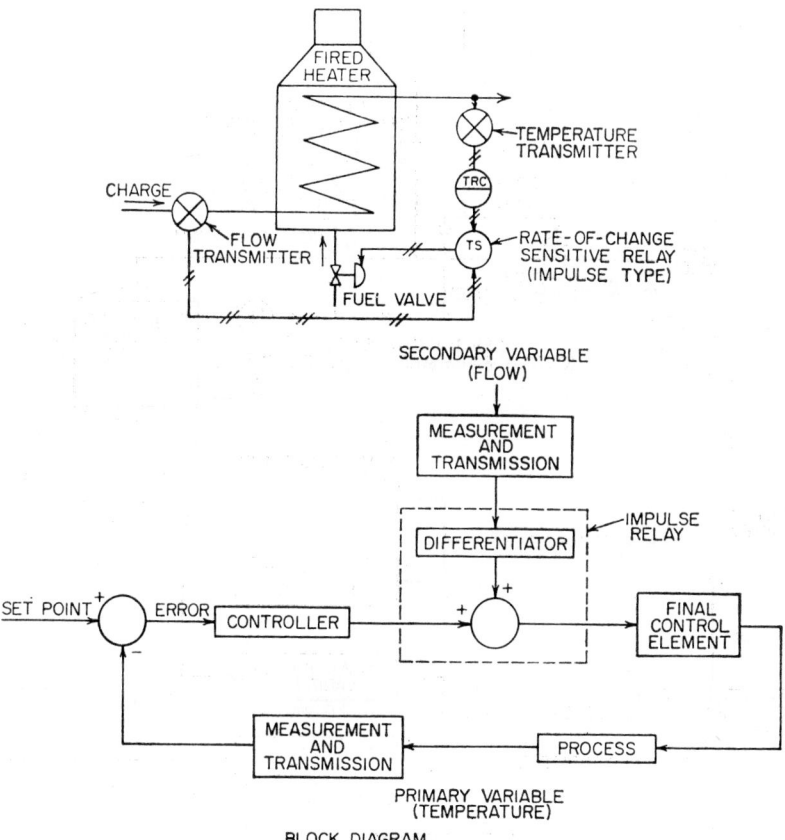

Fig. 28. Impulse feed-forward control of a temperature-controlled charge heater.

If the flow is steady, the temperature signal is passed through the relay without change. Variations in flow rate are detected and added to, or subtracted from, the temperature-control signal.   In this manner the load changes in the flow act immediately on the final control element to compensate for anticipated changes in temperature.

Split-range Control

Split-range control is a unit control system in which a definite sequence of events takes place in order that a certain manipulated variable may have first preference as a means for the process control.

This is illustrated by a hot-water preference system for a deaerating feedwater

heater (Fig. 29).   As the level in the deaerating heater starts to decrease, it is desired
to add hot water from the makeup tank in order to allow for normal level changes.
The use of hot water prevents temperature upsets in the deaerating system which may
cause process difficulties.   However, if the demand for deaerated water persists and
the level continues to fall, the hot-water line in no longer adequate.   At this point cold
water is added from the plant utility line.   In the event of the failure of the control
signal, it is desired to shut off the hot-water valve in order to save its contents.   The

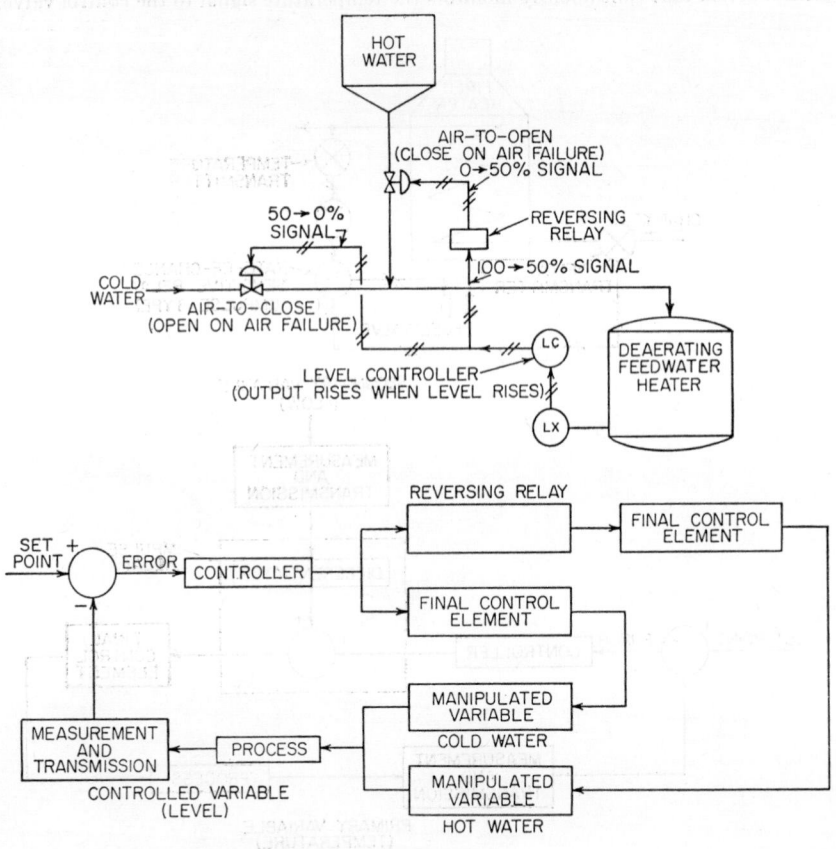

FIG. 29. Split-range control of a deaerating feedwater heater.

cold-water valve is to be opened at the same time to prevent damage to the heater.
In this way, the more valuable contents of the hot-water reserve tank are not wasted.
    This system is accomplished by a half-scale valve range and a reversing relay.
Also, the action of the valve is selected to accomplish the fail-safe requirement upon
failure of the control signal.
    When a level is at the top of the level transmitter in the deaerating heater, 100 per
cent output results.   This signal keeps the cold-water valve closed.   The reversing
relay reverses the signal so that it becomes the minimum output signal, and this keeps
the hot-water valve closed.
    As the level falls to mid-scale (50 per cent output), the cold-water valve still remains
closed.   The hot-water valve throttles to full open since it receives a signal that goes

from zero to 50 per cent because of the reversing relay. Should the amount of hot water be inadequate to hold the level above the mid-point in a tank, the cold-water valve will be throttled open as the level transmitter signal goes from 50 to 0 per cent. At a minimum level in the heater both valves will be wide open.

Failure of the power source (electricity, air, etc.) will result in the hot-water valve being closed to save its contents. The cold-water valve will be open to prevent damage to the heater by overheating.

Override Control

It is often necessary to limit a process variable at some high or low valve in order to avoid damage to the process or to the product. A unit control system of override control can be used to accomplish this need. Applications include temperature-pressure reactor control and pipeline pumping.

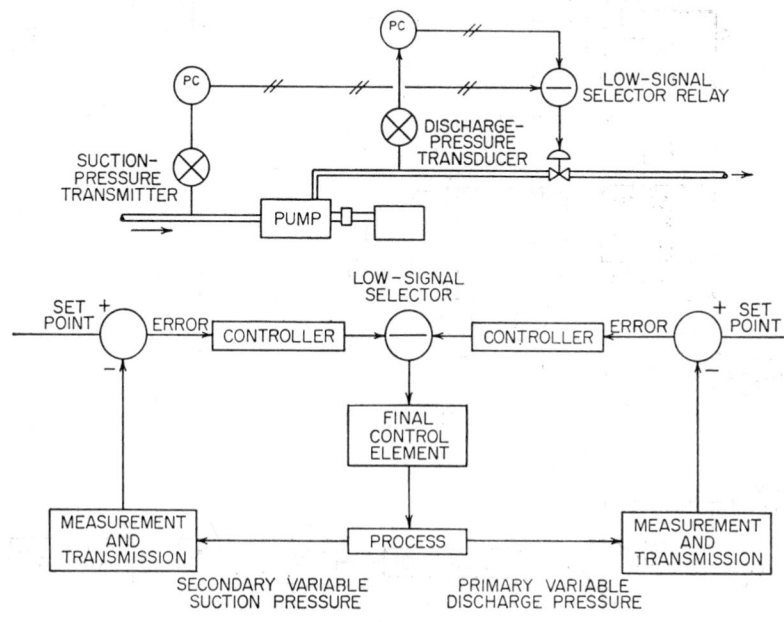

BLOCK DIAGRAM

Fig. 30. Automatic selector control of a pump.

Pipeline pumping stations often have the requirements that both station suction and discharge pressure must be held within safe limits (Fig. 30). In this case there is a normal operating control circuit on one of the possible control variables. This control circuit will continue to function normally and will have its control limited only if the secondary critical variable gets out of hand. At this point, the secondary variable becomes the controlling unit until the process difficulties are remedied.

The two pressure controllers are connected to a low-signal selector. The set point of the suction pressure controller is below the normal operating pressures. Its output signal will be at a maximum because of the positive error above the set-point value.

The discharge controller is a reverse-acting unit and has its set point at the desired pressure. Consequently, its output is normally below that of the suction pressure controller.

The output of both these controllers goes to a low-signal selector relay. Since the relay will select the lower of two input signals for retransmission, the output to the control valve represents discharge pressure control under normal conditions.   When the suction pressure goes below the set point, its pressure-controller output decreases, becoming lower than the discharge signal.   Consequently, the suction pressure controller will take over the operation of the valve and maintain satisfactory operating conditions.

Time-cycle Control

Time cycle is a unit control system which opens and closes one or more circuits during a timed interval.   These circuits can be either pneumatic or electrical.   They

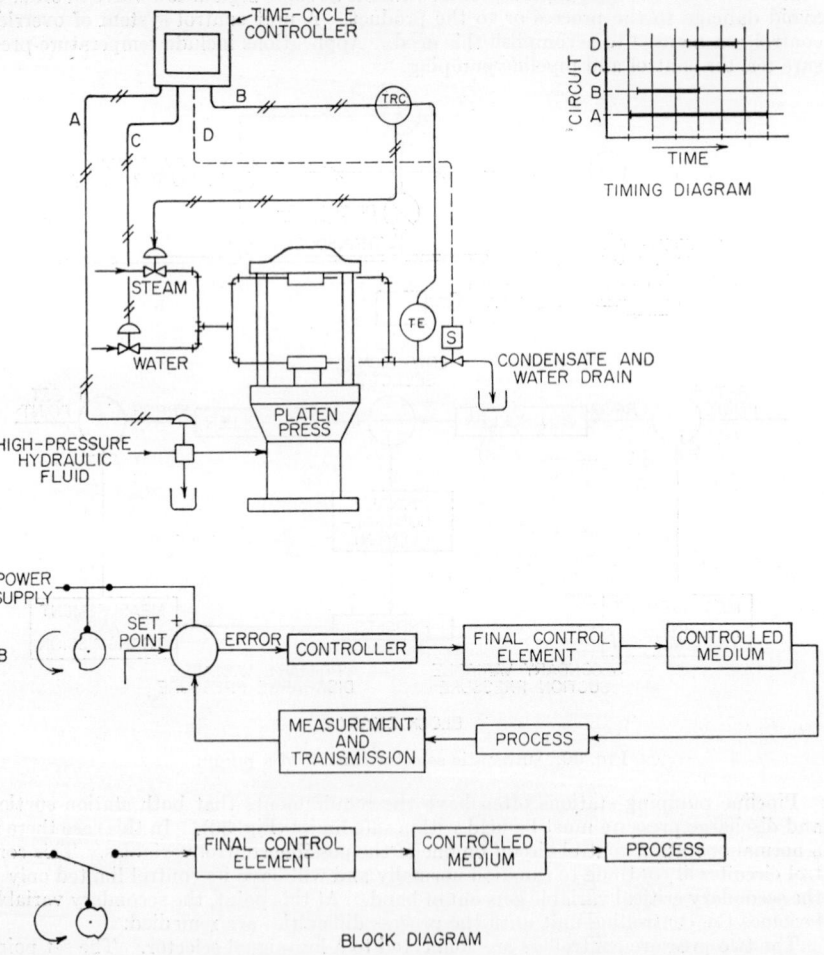

Fig. 31. Time-cycle control of platen press.

can include single function operations such as the activation of an on-off valve or a more complex operation such as a control loop.

Uses of this system include the molding of tires, phonograph records, and rubber goods; sequencing the starting of large motors; sequencing and timing a series of machine operations; temperature sequencing in reactors; regeneration of ion-exchange beds; and fixed-bed catalytic reforming.

A typical platen-press control system is shown in Fig. 31.   Diaphragm motor valves, a solenoid valve, and a temperature controller are operated in correct sequence by the time-cycle controller.

Time-schedule Control

There are many processes where the set point is to be varied automatically over a period of time.   Typical industries include the annealing of steel, batch polymerization, rubber processing, and esterification.

A typical application can involve instrumentation where both rate-of-process change and duration of any one value of the set point are to be accurately controlled.   A popular use of this unit control system is in a temperature process (Fig. 32).   The

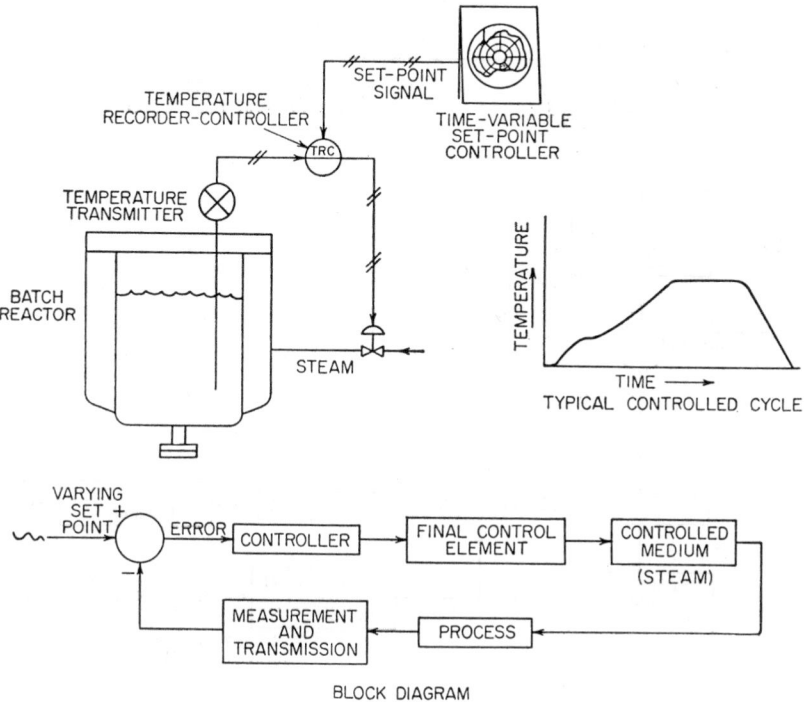

FIG. 32. Time-schedule control of temperature.

instrumentation consists of a complete feedback loop with a set-point feed-forward control acting to guide the process variable.

In such a system only the feedback loop controller contains the control modes.   The set-point control develops a predetermined signal in proportion to the desired value of the process variable for all portions of the time schedule.   In some units, the function generator and the controller can be contained in one case.   Note that a recording controller is supplied in order to have a record of the operation for quality-control purposes.

End-point Control

End-point control is a combination of cascade and ratio unit control systems which enable the ratio of the secondary to the primary variable to be automatically adjusted by the final process variable.

A typical application is shown in Fig. 33 where the end-point analyzer is used to measure and control the two ingredients which form the product upon mixing. A process involving the neutralization of an acidic primary flow stream with a basic fluid by pH control would be a typical example.

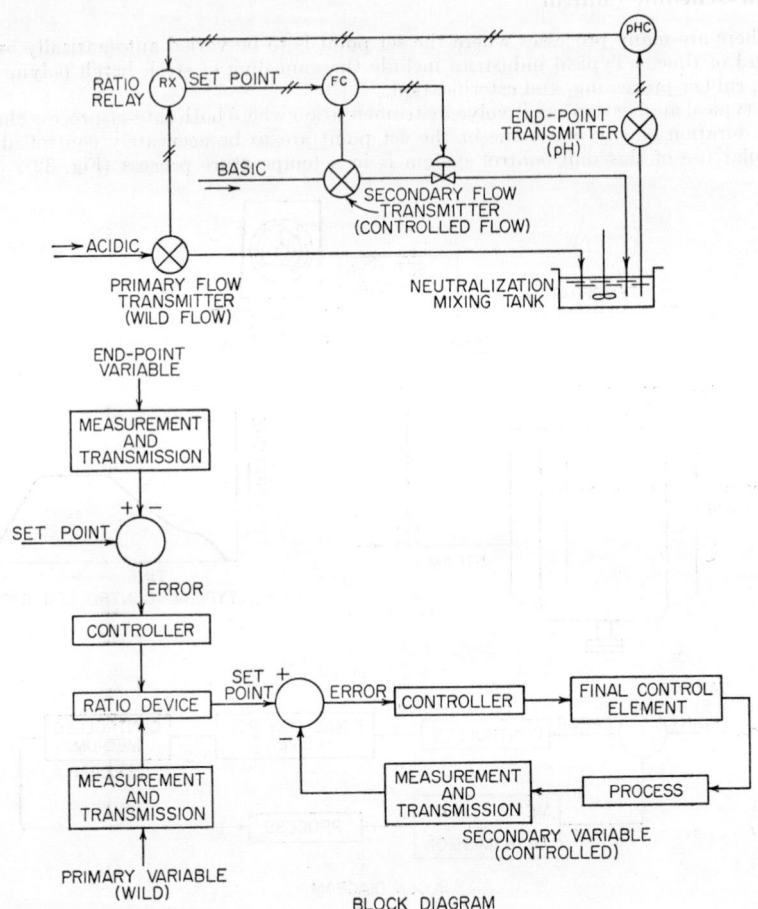

Fig. 33. End-point control.

As the analysis of the final product detects changes from the set point, the controller output adjusts the ratio setting automatically so that the two streams are ratioed correctly. The secondary variable is on control with its set point coming from the ratio unit.

Another common application is the control of the air-fuel ratio in a furnace by measurement of the oxygen content of the exhaust gases.

REFERENCES

1. "Basic Instrumentation Lecture Notes and Study Guide," part I, "Measurement Fundamentals," Instrument Society of America, Pittsburgh, 1960.
2. Cheverton, J. S., K. Losch, et al.: Program Control Applied, *Control Eng.*, September, 1958, pp. 163–182.

3. Chinn, G.: Industrial Control with Pressure Regulators, *Chem. Eng. Progr.*, September, 1960, pp. 58–59.
4. Considine, D. M. (ed.): "Process Instruments and Controls Handbook," McGraw-Hill Book Company, Inc., New York, 1957.
5. Cooney, J. D.: Program Controllers Set the Pulsebeat of Industry, *Control Eng.*, September, 1958, pp. 116–117.
6. Crawford, W. A.: Predicting Process Control Performance with Computers, *ISA J.*, January, 1958, pp. 54–57.
7. Dallimonti, R.: Basic Controller Fundamentals, *Automation*, September, 1960, pp. 54–62.
8. Dutcher, J. L.: How the Application Affects the Choice of Program Controller, *Control Eng.*, September, 1958, pp. 118–121.
9. Eckman, D. P.: "Principles of Industrial Process Control," John Wiley & Sons, Inc., New York, 1945.
10. Forman, E. R.: "The Industrial Instruments Industry," Graduate Thesis, Drexel Institute of Technology, Philadelphia, 1953.
11. Forman, E. R., and W. R. Jensen: Variable Speed Pump Control, *ISA J.*, vol. 5, pp. 36–39, March, 1958.
12. Fourroux, M. D., F. W. Karasek, and R. E. Wightman: High Speed Chromatography in Closed-loop Fractionator Control, *ISA J.*, May, 1960, pp. 76–80.
13. "Fundamentals of Instrumentation for the Industries," Minneapolis-Honeywell Regulator Co., Philadelphia, 1957.
14. Haigler, E. D.: Application of Temperature Controllers, *Trans. ASME*, vol. 60, no. 8, November, 1938.
15. Halfhill, D., and E. Wildanger: Selecting a Magnetic Tape System for Program Control, *Control Eng.*, September, 1958, pp. 134–138.
16. Hecht, G J., A. Edinborgh, and V. N. Smith: A Near-infrared Process Analyzer for Liquid Samples, *ISA J.*, February, 1960, pp. 40–44.
17. Hougen, J. O.: Process Dynamics by Pulse Testing, *Chem. Eng.*, June 12, 1961, pp. 209–212.
18. Kompass, E. J.: New Developments in Program Controllers, *Control Eng.*, September, 1958, pp. 122–127.
19. Koppel, H. H.: A Solid State Electronic Control System for Industrial Process, *Elec. Eng.*, September, 1961.
20. Howe, W. H., P. H. Drinker, and R. M. Green: Progress in Plant Measurements, *Chem. Eng.*, June 12, 1961, pp. 199–204.
21. Mathias, R. A.: Putting Logic to Work in Designing Distributed Program Control Systems, *Control Eng.*, September, 1958, pp. 139–145.
22. Peck, A. R.: Analysis of Fluid-system Dynamics, *Chem. Eng.*, June 12, 1961, pp. 193–198.
23. "Process Control Handbook," Instruments Publishing Co., Pittsburgh, 1960.
24. Rhodes, T. J.: "Industrial Instruments for Measurement and Control," McGraw-Hill Book Company, Inc., New York, 1941.
25. Sills, R. M.: Temperature Control of Heat Treating Furnaces, *Steel*, May 7, 1956.
26. Thackara, A. M.: Fundamentals of Pressure Control, *Instruments Automation*, December, 1955, pp. 2094–2097.
27. Tucker, G. K., and D. M. Wills: "A Simplified Technique of Control System Engineering," Minneapolis-Honeywell Regulator Co., Philadelphia, 1958.
28. Williams, T. J.: Systems Engineering, *Chem. Eng.*, part 8, Aug. 22, 1960, pp. 127–132.
29. Ziegler, J. G., and N. B. Nichols: Dynamic Accuracy in Temperature Measurement, Fourth Annual Instrument Conference and Exhibit, ISA, September, 1949.
30. Wills, D. M.: Cascade Control: When, Why, and How, *Automatic Control*, September, 1961, pp. 36–39.
31. Mamzic, C. L.: Basic Multiloop Control System, *ISA J.*, June, 1960, pp. 63–67.
32. Lloyd, S. G.: Automatic Control Fundamentals, *Automation*, vol. 9, no. 6, pp. 86–90, June, 1962.
33. Lloyd, S. G.: Automatic Control Fundamentals, *Automation*, vol. 9, no. 7, pp. 84–90, July, 1962.
34. Wilson, H. S., and L. M. Zoss: Controller Mode Selection, *ISA J.*, vol. 9, no. 10, pp. 61–62, October, 1962.
35. Young, A. J.: "Process Control," Instruments Publishing Co., Inc., Pittsburgh, Pa., 1957.
36. Lajoy, M. H., and E. A. Baillif: "Process Control Analysis," Instruments Publishing Co. Inc., Pittsburgh, Pa., 1956.
37. Aronson, Milton, and Fred Marton: "Process Control Systems," Instruments Publishing Co., Inc., Pittsburgh, Pa., 1962.
38. Murphy, Gordon J.: "Basic Automatic Control Theory," D. Van Nostrand Company, Inc., Princeton, N.J., 1957.
39. Murphy, Gordon J.: "Control Engineering," D. Van Nostrand Company, Inc., Princeton, N.J., 1957.
40. Clark, R. N.: "Introduction to Automatic Control Systems," John Wiley & Sons, Inc., New York, 1962.
41. Rosenbrock, H. H.: Distinctive Problems of Process Control, *Chem. Eng. Prog.*, vol. 58, no. 9, pp. 43–50, September, 1962.
42. Pavlik, Ernst, and Bruno Machei: "A Combined Control System for the Process Industries," D. Van Nostrand Company, Inc., Princeton, N.J., 1962.
43. Hall, A. D.: "A Methodology for Systems Engineering," D. Van Nostrand Company, Inc., Princeton, N.J., 1962.

Note: See also References on pp. 10-55 and 10-56.

APPLICATIONS OF PHOTOELECTRIC CONTROLS

By Robert J. Wilson*

When light strikes special combinations of materials, a voltage can be generated, a resistance change may take place, or electrons may flow. These three characteristics of the photoelectric effect are employed by industry in various ways for countless photoelectric applications. The field of photoelectrics may be divided into two groups. One is the scientific and industrial field, where measurement of variations in light radiation can be used to control simple and complex industrial equipment, or to detect density, color, temperature, texture, and composition of materials. The second group employs photoelectrics to convert the variations in light radiation into sound and images.

Photovoltaic, photoconductive, or photoemissive cells share one thing in common: their ability to convert light variations into useful electrical changes. Their action can be compared with an ordinary on-off switch. The beam of light directed at the cell is the handle of the switch. By interruption of the light beam, the switch handle is effectively placed in the off position. By restoring the light beam, the switch is returned to the on position. Unlike the switch handle, no energy is consumed in operating the light beam. This permits photoelectric devices to be used on large bulky material, extremely small parts, or freshly painted surfaces, where mechanical actuation would be impractical. The photoelectric instruments and controls used by industry employ three different types of cells.

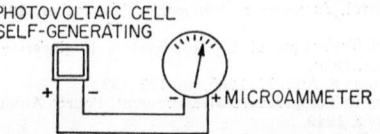

PHOTOVOLTAIC CELL
SELF-GENERATING

FIG. 1. Self-generating photovoltaic cell.

PHOTOVOLTAIC CELLS

When light strikes a junction of certain dissimilar materials, a voltage is produced. This is known as the *photovoltaic effect*. This type of cell is extremely useful because it is self-generating and requires no voltage source to make it operate. Its most common use is in the photographic exposure meter, whereby the cell converts light into electrical energy sufficient to operate a sensitive meter (Fig. 1).

PHOTOCONDUCTIVE CELLS

Special types of semiconductor materials such as selenium, cadmium sulfide, cadmium selenide, lead sulfide, and lead selenide change their resistance value when exposed to light. This is known as the *photoconductive effect*. Cells of this type require a voltage supply and have good infrared sensitivity. They do not have good frequency response, but their high signal-to-noise ratio, small size, and rugged physical characteristics make them ideally suited for certain types of industrial applications.

Photoconductive cells can be used in simple circuts and permit construction of photoelectric controls by anyone who desires to make a reliable control for a special application. Being made of semiconductor materials, they are susceptible to resistance changes due to temperatures. They require cooling and special temperature-compensating circuitry, if they are to be used at high ambient temperatures.

* Sales Manager, Photoswitch Division, Electronics Corp. of America, Cambridge, Mass.

Probably the most spectacular use for the photoconductive cell has been in guided missiles. By detecting heat (infrared radiation) given off by an aircraft's exhaust, the cell permits operation of a guidance system for the missile.

A simple photoelectric control can be made by putting a photoconductor cell in series with a sensitive relay and voltage source (Fig. 2). When the light strikes the photoconductor cell, the cell's resistance decreases and allows sufficient current to flow to energize the relay. A sensitivity adjustment can be added to this circuit by placing a variable resistor in series with the cell (represented by the dotted lines in Fig. 2).

PHOTOEMISSIVE CELLS

Certain types of materials will emit electrons when exposed to light. When such a material is placed within an evacuated glass bulb, it is called a *phototube*. A phototube operates on the *photoemissive effect*. Light strikes a cathode, causing the emission of electrons, which are collected by the plate. This produces a minute current flow.

The phototube is by far the most versatile and useful photoelectric device for industrial applications. Its characteristics can be very linear and extremely fast. The phototube requires a voltage source and because of the minute current flow (usually microamperes) requires an amplifier to make its output usable (Fig. 3).

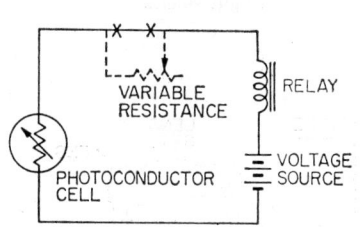

Fig. 2. Simple photoelectric control using a photoconductor cell.

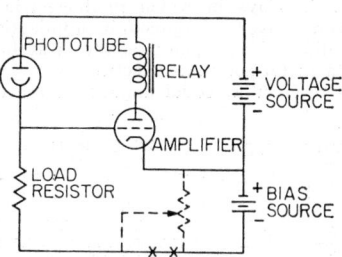

Fig. 3. Simple control circuit employing phototube with an amplifier.

With no light on the phototube, the bias source applies a negative potential to the grid of the amplifier tube through the load resistor. This negative potential is sufficient to keep the tube from conducting. When light is directed at the phototube, current flows and the negative potential is made positive, allowing the amplifier to conduct current and energize the relay. A sensitivity sdjustment can be added to this circuit by placing a variable resistor across the bias source (represented by the dotted lines in Fig. 2).

GENERAL GUIDES TO APPLICATIONS

Photoelectric controls are versatile, inexpensive, easy to maintain, and very reliable. When a photoelectric control is to be selected for a given application, there are certain conditions to consider:

1. Will the control operate from reflected or transmitted light?
2. If reflected light, will it be diffused (Fig. 4a) or specular (Fig. 4b) reflection?
3. Is there sufficient differential between the mark and the background?
4. If transmitted light (Fig. 4c), what is the distance between light source and phototube? Short distances require less light and sensitivity. Long distances require more light and ambient light shields (Fig. 5). Extremely long distances (1,000 ft) require modulated light and receivers tuned to respond only to the frequency of the modulated light.

Another application condition to consider is the size of the object or mark that will interrupt the light beam. Small objects require special lens arrangements for spot focusing (Fig. 6).

Speed of response is also a factor. How long will the object interrupt the light beam? Large objects, moving too slowly, produce chattering and inconsistent relay operation. Special lens arrangements and circuit design must be used. Small objects moving too fast may not interrupt the light beam long enough for operation.

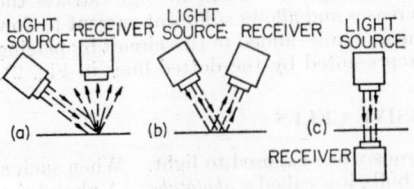

(a) DIFFUSED REFLECTION —
LIGHT SOURCE AT 45° TO SURFACE

(b) SPECULAR REFLECTION —
LIGHT SOURCE AND RECEIVER AT
EQUAL ANGLES

(c) TRANSMITTED LIGHT

FIG. 4. Ways in which photoelectric receivers may be arranged in applications: (a) diffused-reflection light source at 45° to surface; (b) specular-reflection light source and receiver at equal angles; (c) receiver.

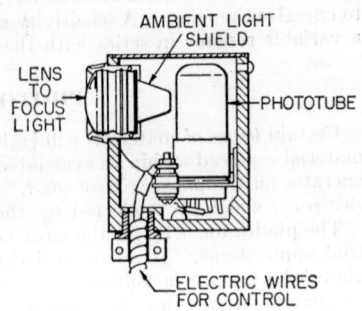

FIG. 5. Typical phototube cell with ambient light shield.

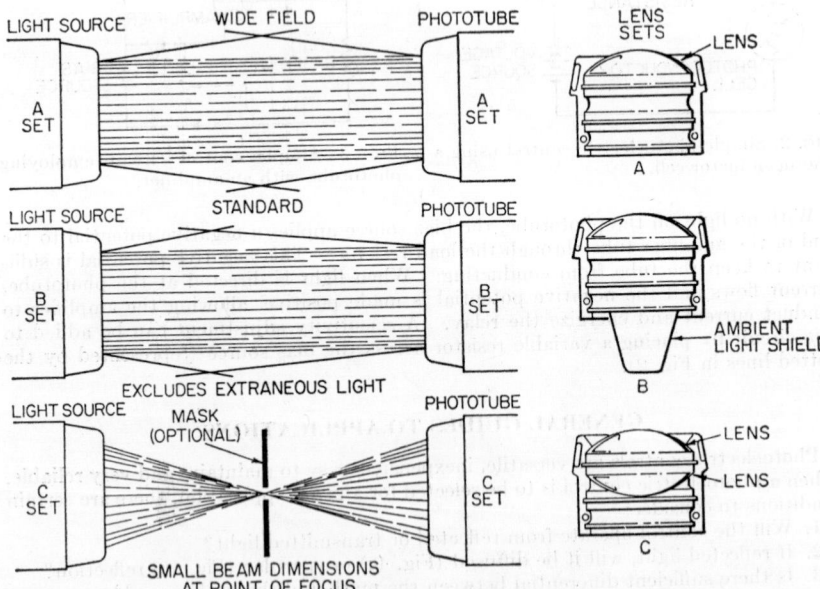

FIG. 6. Various lens arrangements for different application conditions.

Controls with fast response must be used. Areas with high ambient light levels or stray light mean stronger light sources, ambient light shields, and apertures for restricting the field of view.

For most photoelectric applications, normal room ambient light levels will not materially affect the operation of the control. However, care must be taken to position the phototube so it is not directly exposed to outside illumination, such as looking out a large window.

When photoelectric controls are used for measurement, or operating on close differentials, steps should be taken to eliminate as much ambient light as possible. When an application is of such a type that small light changes must be detected and yet operate in a high-ambient-light-level area, it is not desirable to use a percentage-of-light-change type of control, but rather to use an impulse-operated unit. This means that the light level, including the ambient, is always constant and even though they increase or decrease by certain amounts will not materially affect the operation of the unit. The input of the control is designed in such a way that it will respond only to a certain impulse of a definite duration. This greatly minimizes the effect of ambient light and provides reliable operation on smaller light levels with the same ambient conditions.

DETECTION OF CONVEYOR JAMMING

In all industries where conveyors are used to move products or parts, conveyor jamming is a serious problem. Photoelectric delayed-action controls can be used to eliminate this problem (see Fig. 7). The photoelectric control is so located that its control beam is interrupted each time a case passes along the conveyor. As long as the timing circuit receives impulses at regular intervals, because of restoration of the light beam between cases, the load circuit of the photoelectric control will remain closed and will maintain the conveyor motor in operation.

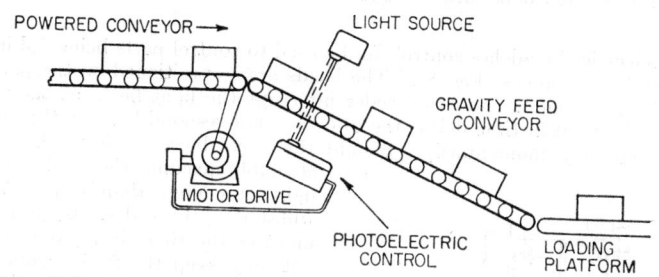

FIG. 7. Use of photoelectric control to prevent jamming of a conveyor.

The timing interval starts each time a carton breaks the light beam, but the timing circuit is reset without relay action each time the beam is restored before the preset timing interval has elapsed. If a jam occurs, causing the cases to butt one against the other, they will no longer permit the light beam to reach the control. The timing circuit will then time out, opening the load circuit and stopping the conveyor motor.

Relay action is reduced to only those occasions when a jam occurs. Since the relay armature motion is the only mechanical motion which takes place, this reduction of its frequency of operation multiplies the life of the equipment and minimizes maintenance requirements.

OTHER TIME-DELAY CONTROL APPLICATIONS

Photoelectric time-delay controls can be applied for conveyor traffic control. This involves filling a multiple-storage conveyor with cases from a single feeder control. Filling of one storage conveyor continues as long as the boxes pass the photoelectric control at regular intervals. When the storage conveyor fills up, the light beam is blocked for a period longer than the time delay, at which time the feeding of cases is automatically switched to another storage conveyor.

Photoelectric time-delay controls can be used for hopper control of both dry and wet materials. When the material builds up to a sufficient height to interrupt the light beam, the control does not immediately respond because of the preset time delay, and the hopper continues to fill. After the preset time delay has elapsed, the control will stop the material flow. The material must now get to a point below the light beam before the control will respond and start the material flowing again. This time-delay feature allows a precise differential to be established so that hopper levels can be maintained.

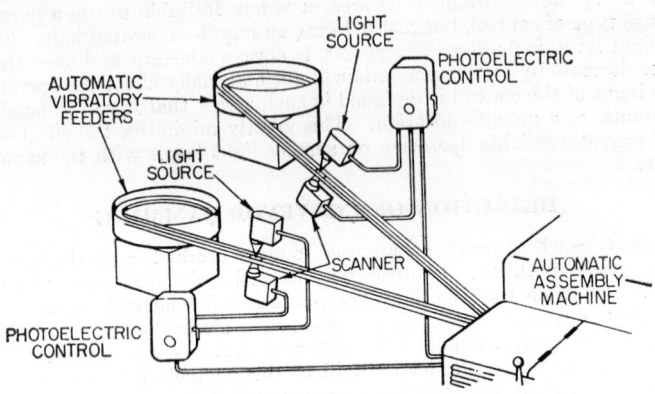

FIG. 8. Control of parts being fed to an automatic assembly machine.

A photoelectric time-delay control can be used to control parts being fed into automatic assembly machines (Fig. 8). The heads are so positioned on the parts feeder track that parts coming from the feeder interrupt the light beam as they pass by. Because of the built-in time delay, the unit does not respond because the light beam is interrupted only momentarily. Should the parts back up from the automatic assembly machine, they will interrupt the light beam longer than the preset time and will shut off the feeder. By proper adjustment of the time delay, the control will not only keep the feeder from jamming but will also make the over-all operation smoother. Parts will not be fed faster than they can be used, and the built-in time delay will keep the feeder from going on and off too often.

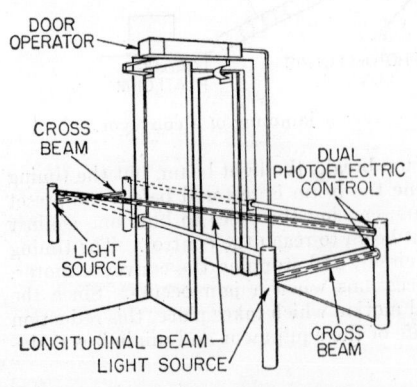

FIG. 9. Photoelectric controls arranged to open doors automatically for short vehicles.

DOOR OPERATOR

Many large industrial plants with automatic mass-production methods are using photoelectric door-operating equipment to assist in material handling. This is especially important in conjunction with the operation of trucks between sections of a plant which are sealed off from the remainder of the factory for such reasons as air conditioning, noise reduction, the presence of hazardous vapors, sanitary regulations, or prevention of heat loss. There are various ways of handling this problem, depending upon the character of the installation.

For example, if the traffic is vehicular only and the vehicles are reasonably long, it is sufficient to use two photoelectric controls, one mounted on either side of the door. Vehicles coming in either direction will break first the light beam on their approach side of the door, which will open the door and keep it open as long as either of the light beams is interrupted. The light beam should be so positioned that the front end of the vehicle will break the second beam on the far side of the door before the trailing end of the vehicle has allowed the approach side light beam to be restored.

If the vehicles are too short to permit this type of operation, a third photoelectric control may be positioned so that it is directed longitudinally along the passage through the doors (Fig. 9). Thus, if the vehicle is anywhere in the passage, the third longitudinal beam will be broken and the doors will remain open. Only when all three beams are restored will the doors close again.

When there is a combination of vehicular and pedestrian traffic, photoelectric controls are used in conjunction with electronic timers to establish time delays sufficient for the vehicle or pedestrian to pass through the door before it closes. In this case, two photoelectric control units and one electronic timer are generally required.

TEMPERATURE-SENSITIVE PHOTOELECTRIC CONTROLS

Photoelectric controls can be used in many ways for indicating, controlling, and recording temperature. For example, a plant had the problem of controlling a conveyor carrying newly forged parts which emerged from a heat-treating furnace at high temperature (2300°F). The conveyor was to be operated only when a part was placed on it, and only for the distance required for the part to reach the next stage of

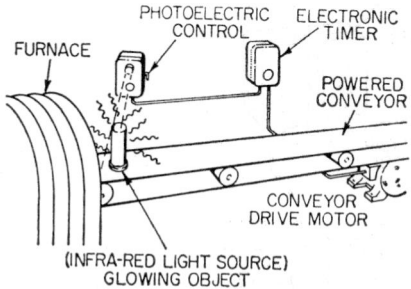

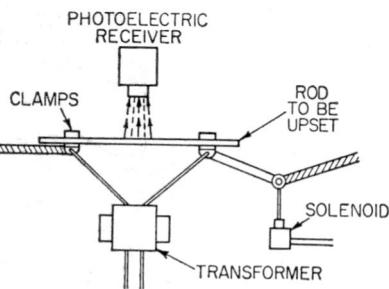

Fig. 10. Arrangement of controls to operate conveyor intermittently in heat-treating plant.

Fig. 11. Automatic rod-upsetting control with heat-sensitive photoelectric receiver.

the process. The parts were ejected from the furnace at varying rates, which precluded the use of a timer alone. Mechanical limit switches contacting the part emerging from the furnace were tried for controlling the conveyor, but contact with the high-temperature parts caused abnormally rapid failure of these switches.

The problem was solved by the use of a highly sensitive photoelectric control and an electronic timer (Fig. 10). The photoelectric control is located where it can view the exit door of the furnace. The glowing white-hot part is an intense source of infrared radiation, which actuates the photoelectric control as soon as the part comes into view. At this point the photoelectric control operates the conveyor which carries the part away from the furnace. At the same time it starts the electronic timer, which keeps the conveyor running for the precise length of time required to position the part for the next operation.

A simple automatic rod-upsetting machine can be made by locating a phototube to detect the temperature increase in the rod to be upset (Fig. 11). When the current passing through the rod has heated it to a certain temperature, the phototube picks up this increase in infrared radiation and actuates a solenoid which upsets the rod.

Chain links are automatically welded by placing a phototube close to the welding tip. When the link comes up to temperature, the phototube detects the increased infrared radiation, shuts off the welder, and automatically indexes the machine to the next chain link. Because photoelectric temperature devices do not have to make physical contact with the heated material, they can be used for applications where it is physically impossible to contact the material such as molten metal, pouring molten metal, checking hot strip steel while it is moving, limiting temperatures of elements in vacuum tubes, and picture tubes by scanning through the glass envelopes. Their temperature range is 1000 to 5000°F.

INSPECTION OF BOTTLE-CAP LINERS

Bottlers need a reliable way to detect the absence of the paperboard liner in bottle caps. Occasionally, the liner may be left out by the cap manufacturer. In other instances the liner may fall out before the cap is applied to the bottle. Reliable detection is important primarily because other bottles in a filled case may be damaged when a "leaker" spills its product in shipment or storage. The cap is usually inspected when it is packed into the shipping carton by the cap manufacturer, and when it is fed into the capping machine. Since the absence of a liner can be detected only by "looking," the installation of a reflection type of photoelectric scanner has been a natural solution.

In one small unit, the scanner incorporates both light source and phototube, with a common lens system (Fig. 12). As the caps move past in a continuous line, the

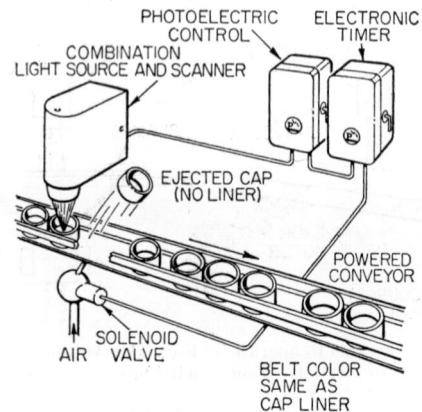

Fig. 12. System for inspection of missing liners in bottle caps.

scanner "looks" into the open ends, instantly recognizing the difference in reflection between the dark liner and the light-colored cap. When it detects a defective cap, the photoelectric relay operates the ejector device, an air blast controlled by a solenoid valve. The duration of the air blast is so accurately controlled by an electronic timer that none of the caps before or after the defective cap are displaced.

INSPECTION OF LABELS

In the food-processing industry, photoelectric controls are used to detect the absence of labels on cans or containers having a reflective surface underneath the labels. In this application one photoelectric control with two light sources and an electronic time-delay relay are used. One light beam is directed onto the photoelectric control across the conveyor line, along which the cans are passing. The other light source is so located on the same side as the photoelectric control that its light is reflected to the phototube from a passing unlabeled can. The photoelectric control recycles the time-delay relay each time it sees a label, and the time-delay interval is set so that the timer will not time out and open its load circuit as long as each can is properly labeled. If, however, an unlabeled can goes by, the photoelectric receiver receives light continuously. As a result, the photoelectric control is not pulsed; no impulse is given to the timer; and the timer load circuit opens, shutting down the conveyor line.

Similar photoelectric inspection devices have been applied to the automotive, dairy, food, and plastics fields.

AUTOMATIC SORTING

In a glassware-manufacturing plant, a main conveyor carries cartons of three types of glassware from the manufacturing area to the warehouse. Once in the warehouse, each type of ware must be sorted and sent to its own storage area. Since the cartons containing any one type of ware differ widely in all dimensions, it is not feasible to sort by size and shape. Before an automatic sorting system was devised, two men were in constant attendance to push cartons from the main conveyor onto the proper distribution conveyors.

Two photoelectric controls solved the problem of automatic sorting (Fig. 13). One is located at one side of the conveyor at the first discharge point in the warehouse.

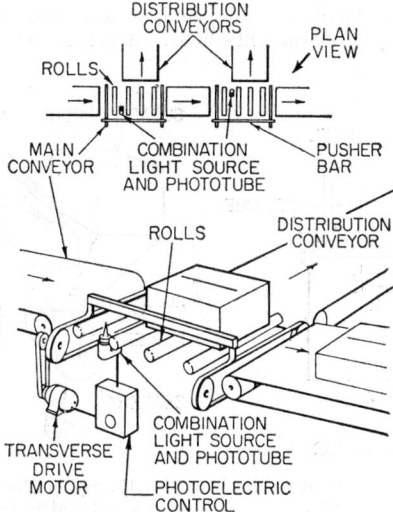

Fig. 13. Automatic sorting system to route different cartons from main to distribution conveyors.

The other is at other side of the conveyor at the second discharge point. The actuating device for the scanners is a small strip of reflecting tape put on by the packer when he assembles a carton.

For one type of ware this strip is located along one edge of the carton bottom and extends almost to the middle of the carton. For the second type of ware the strip is located along the same edge of the carton bottom, but from the middle to the opposite side of the carton. No tape is used for cartons carrying the third type of ware.

The cartons are placed on the conveyors so that the tape is at right angles to the direction of carton travel. The tape on one type of ware is at one side of the conveyor. The tape on the second type of ware is at the opposite side of the conveyor. The scanners "look" at the carton bottoms between the conveyor rollers at each discharging point.

One scanner located at one edge of the conveyor "sees" only one tape location. The other scanner is located at the opposite edge and "sees" only the other tape location. Making observation at the bottoms of the cartons ensures correct focus because all tapes are at a fixed distance from the scanners.

Upon "seeing" a reflecting tape pass, the photoelectric control operates a pusher-bar mechanism which removes the carton from the main conveyor onto the proper distribution conveyor. Cartons without a reflecting tape pass right by the first two distribution conveyors to the third storage area.

The number of items that can be stored is restricted only by the number of reflecting tapes or printed registration marks that can be applied to the carton. Different types of items can be sorted by size, shape, and color. Various fruits are sorted by size and color. Eggs can be candled photoelectrically.

COMPARATOR SYSTEM FOR TURBIDITY, DENSITY, OR COLOR

Turbidity and density of liquids, vapors, and gases, as well as color sorting, can be accomplished by using two phototubes, feeding their output into a photoelectric comparator. Jet and missile fuel is checked for impurities using this system (Fig. 14). One phototube "looks" at the fuel to be checked for impurities; the second phototube "looks" at a sample of fuel known to be pure. Both phototubes are balanced by adjusting potentiometers. Any variation in the transmitted light of fuel cell 2 is indicated on the meter of the photoelectric comparator. Should the transmitted light fall below a preset limit, an alarm will sound. A permanent record is made by the recorder.

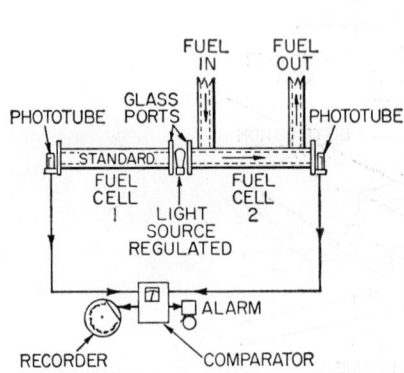

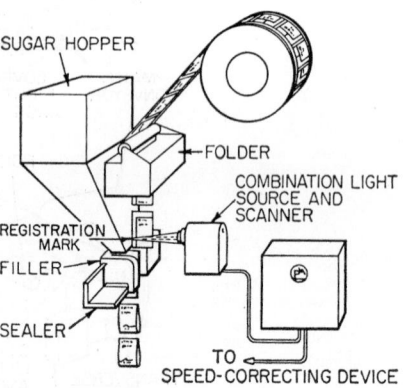

FIG. 14. Comparative system used to check impurities in jet and missile fuels.

FIG. 15. System for controlling registration of printed matter in packaging.

CONTROLLING REGISTRATION OF PRINTED MATTER IN PACKAGING

Photoelectric registration controls are used in the packaging-machine field to keep the advertising matter in register in filling and wrapping machines. If registration is not maintained, the advertising or printed matter quickly moves off center, with the result that the wrapping operation takes place in the center of the design and spoils the appearance of the package. A registration mark is printed at the same time as the design and lettering.

A phototube is positioned to detect this mark either by transmitted light on transparent material (such as cellophane) or reflected light on opaque material (such as paper). When the web is running out of register, the photoelectric control actuates a correcting device which either speed up or slows down the feed rolls.

In packaging sugar, a high-speed high-sensitivity photoelectric control scans the web to locate a registration mark (Fig. 15). The photoelectric equipment operates in conjunction with an articulated linkage in the paper drive in such fashion that, if the registration mark is not viewed by the scanner at the time of the filling, counting, and sealing operations, the web drive is speeded up until the registration mark appears at the time that the operations take place.

Registration controls can also be used to sense the registration marks and stop the web in the correct position for filling and wrapping. This method does not require correcting devices.

SAFETY AND PROTECTIVE USES

Photoelectric controls make ideal safety devices and intrusion alarm systems. All types of industrial machines and punch presses can be protected by ringing the area with a light beam. Intrusion alarms can operate over distances up to 1,000 ft.

Photoelectric flame-failure equipment protects industrial boilers from explosions. A phototube or photoconductor is used to monitor the flame. Should the flame fail, the control circuit immediately stops the flow of fuel to the furnace.

Extremely fast operating fire detectors, utilizing a photoelectric sensing element, are placed in a strategic location in a building or warehouse for the purpose of detecting the presence of fire. A control of this type is constructed in such a way that it responds only to the flicker rate of flame.

Photoelectric explosion suppressors detect and extinguish potential explosions before they have a chance to happen.

Illumination controls turn street lights on and off and regulate the amount of inside light by measuring the amount of outside illumination.

REFERENCES

1. Campbell, N. R., and D. Ritchie: "Photoelectric Cells," Sir Isaac Pitman & Sons, Ltd., London, 1934.
2. Carroll, John M.: "Transistor Circuits and Applications," McGraw-Hill Book Company, Inc., New York, 1957.
3. Henney, K., and J. D. Fahnestock: "Electron Tubes in Industry," 3d ed., McGraw-Hill Book Company, Inc., New York, 1952.
4. Hughes, A. L., and L. A. DuBridge: "Photoelectric Phenomena," McGraw-Hill Book Company, Inc., New York, 1932.
5. Jacobs, Donald H.: "Fundamentals of Optical Engineering," McGraw-Hill Book Company, Inc., New York, 1943.
6. Lange, B.: "Photoelements and Their Application," John F. Rider, Publisher, Inc., New York, 1954.
7. Markus, John, and V. Zeluff: "Handbook of Industrial Electronic Circuits," McGraw-Hill Book Company, Inc., New York, 1948.
8. Walker, R. C., and T. M. Lance: "Photoelectric Cell Applications," Sir Isaac Pitman & Sons, Ltd., London, 1939.
9. Ive, G. A. G.: "Photo-electric Handbook," George Newnes Ltd., London, 1955.
10. Norris, K. H.: Photoelectric Inspector Detects Green Rot in Eggs, *Electronics*, July, 1955, pp. 140-142.

SUMMARY OF CONTROLLER TYPES AND FINAL CONTROL ELEMENTS

By Steven Danatos*

This end section provides a condensed summary of commonly available controller types and the final control elements they so often operate. For a more complete description of available equipment, the reader is referred to Ref. 1, Sec. 9 on Automatic Controllers, and Sec. 10 on Final Control Elements.

CONTROLLER TYPES

Controllers are of many types, although most of them can be classified according to the type of control that they are able to supply. Apart from the control type, there are two general methods of controller operation; self-operating and pilot-operating. *Self-operating controllers* use energy taken either directly or indirectly from the controlled system to operate a valve or other controlled device that regulates the supply of the control medium. *Pilot-operated controllers* control a supply of energy in the form of a fluid under pressure (usually air) or electricity. The controlled pilot fluid then makes the necessary adjustments to the controlled device. An intermediate type of controller uses the energy of the control medium to amplify its responses and so control the control medium.

Self-operated

Self-operated Pressure Regulator (Fig. 1). This controller adjusts its valve opening by the relation between the downstream pressure applied to the bottom of diaphragm D and the pressure of a spring S. If downstream pressure falls, the valve opens wider; while if the pressure rises, the valve closes. Self-operating temperature

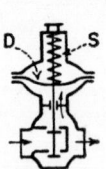

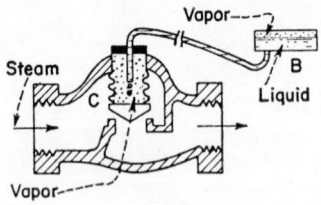

FIG. 1.

FIG. 2.

regulators are similar. Pressure of a filled (usually vapor pressure) thermometer system is applied to a bellows that opens or closes the valve. If the temperature rises, the valve closes to decrease the flow of heating medium (or the valve opens to increase the flow of cooling medium). Since some margin for operation is required, these devices cannot give as close control as pilot-operated controllers. They are used for less critical control problems such as water heating, dry rooms, and jacket-water cooling.

* Associate Editor, *Chemical Engineering*, a publication of McGraw-Hill Publishing Company, Inc. This material is based on a special report published in the June 12, 1961, issue of *Chemical Engineering*.

Hot-chamber Regulator* (Fig 2). This device represents the class of intermediate controllers in which energy from the control medium is used to increase sensitivity of control. Hot chamber C within the steam valve (on the supply side) receives vaporizing liquid from temperature bulb B if the controlled temperature rises. This liquid vaporizes in C and provides energy to close the valve against steam flow. Decrease in temperature at B allows vapor to condense, drawing vapor back from C and increasing the valve opening. Thus valve throttles at opening are determined by demand.

Self-pilot Pressure Regulator (Fig. 3). Except for small sizes, pressure-regulating valves usually have a built-in pilot device operating from the pressure of the fluid being regulated. The valve in Fig. 3 is typical. Where the self-operating type (Fig. 1) gains control only from the downstream side, self-piloted types use fluid that is bled from the high-pressure to the low-pressure side to act as a source of pilot energy. If downstream pressure fails, diaphragm D opens pilot valve P, allowing upstream pressure to force D_2 upward and increase the opening of main valve M. When upstream pressure rises, D_1 closes P, opens L to the low-pressure side, and allows D_2 to fall and the M opening to decrease.

Self-operating Flow Regulator† (Fig. 4). Constant pressure drop is maintained across adjustable and moving orifices. The adjustable orifice is set manually to the desired flow rate. In the device, a compensating sleeve valve, located in the inlet flow, is connected to the moving orifice between the intermediate and outlet chambers.

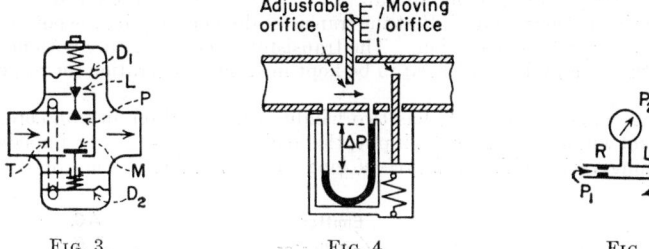

FIG. 3. FIG. 4. FIG. 5.

Pressure difference between chambers operates the moving orifice. Loading weight attached to the sleeve determines the net pressure difference. Increase in inlet pressure causes a rise in intermediate pressure. Sleeve and disk move upward, thus restricting inlet ports until the intermediate pressure returns to the original value above the outlet pressure. Increase in back pressure causes reverse action to take place. Pressure differential within the unit adjusts valve position proportional to desired flow.

Pilots and Relays

Nonrelay Pilots (Figs. 5 to 8). Pneumatic and hydraulic controls use various forms of a pilot device to produce an output pressure that varies in some suitable manner with a primary measurement. Most of them also incorporate a relay to increase the flow of the pilot medium while changes are taking place. For simplicity, relays are omitted in these figures. Figure 5 represents the common flapper or baffle type in which the measurement is translated into the position of the flapper in front of the nozzle. If the flapper approaches, pressure of air P_2 rises and leakage at L decreases. The free-vane flapper (Fig. 6)‡ was developed to overcome the disadvantage of the ordinary flapper (nozzle pressure reacts on the flapper and hence on the measuring system). Vane V moves between opposed nozzles N so that the thrusts of the nozzles on the valve are canceled out. Figure 7 shows the common ball-type pilot in which position of the ball B, controlled by the primary measurement, deter-

* Fulton-Sylphon.
† W. A. Kates.
‡ The Bristol Co.

mines the pressure P_2 by varying the inlet oppositely to the leak L. Figure 8 shows the hydraulic jet pilot used by Askania Regulator Company. A swinging jet pipe positioned by the primary measurement delivers more or less of its impact pressure to the pressure-receiving device, depending on its position. The device moves a piston by swinging back and forth over a partition that separates pipes leading to the two ends of the piston.

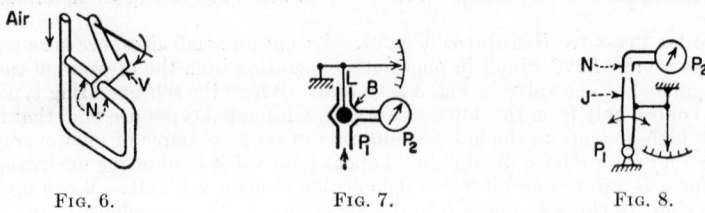

FIG. 6.   FIG. 7.   FIG. 8.

Electrical amplication of minute electric signals gives suitable outputs for recording, control, or telemetering. An elementary vacuum-tube circuit (Fig. 9) permits amplification of a feeble a-c voltage, usually 1 volt or more, and produces an output signal many times stronger. Amplification of signals in the millivolt range with minute power requirements takes place with a transistor (p-n-p type, Fig. 10). The transistor in suitably arranged circuits can produce power gains in the order of 10,000 times. The principal advantage is that the transistor does not require the power supply necessary for vacuum-tube operation. The transistor responds at once with no lag for heating of cathodes and does not need to be kept in standby operation for response to input signals.

The *magnetic amplifier* (Fig. 11) has two windings mounted on an iron core. The signal or control winding carries a d-c input; the load winding, a-c. The impedance of the load winding varies with the degree of magnetic saturation of the iron core. As

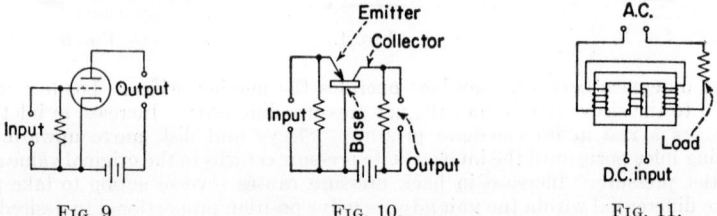

FIG. 9.   FIG. 10.   FIG. 11.

long as the core is not saturated, the impedance of the load winding is high; hence little current flows in the load. When the magnetic flux in the core is increased to saturation by the d-c signal, the impedance of the load winding drops sharply and load current increases in proportion. Load determines maximum current. The magnetic amplifier controls load current between the limits of 3.97 per cent of full value.

Conversion of remote electrical signals to another form of energy large enough to position valve stems or produce shaft displacements requires forms of relays. The relay then operates a pneumatic or hydraulic system to effect necessary motion.

Electrohydraulic Relay* (Fig. 12). Input current causes unbalance of the Microsen beam (see page 1-29, Fig. 5) and operates the pilot valve. Oil supplied by a gear pump is directed to the power cylinder, thus producing movement of the piston proportional to the signal. A feedback spring, which is connected to the output stem through a series of links, expands or contracts in response to stem movement and applies restoring force to the beam. Pilot valve ports are then closed and the output piston is at the required position.

* Robertshaw, A. & I Division.

Electropneumatic Relay * (Fig. 13). A moving coil carrying the input signal is positioned in a magnetic field. The lever arm moves in response to this force. A signal for increased output pressure forces the pilot valve stem down, increasing supply air to the booster-relay diaphragm. The booster valve stem then rises and admits more air to the control valve. When the output pressure balances to a new condition, the booster valve closes on the relay diaphragm, and the feedback diaphragm moves away from the pilot valve stem. The system is then in balance at the required pressure.

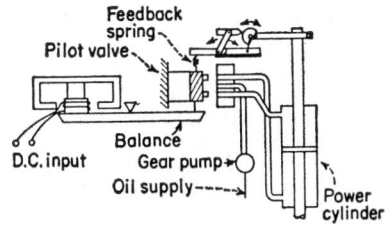

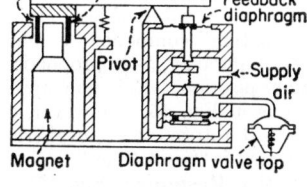

Fig. 12.   Fig. 13.

Pneumatic Controllers

Pneumatic Controller Development (Figs. 14 to 17). This sequence shows four steps in the development of a controller from a relay-pilot-operated on-off controller, through a simple proportional type, to proportional plus reset, and proportional plus reset plus derivative. In Fig. 14, the nozzle-flapper combination of Fig. 5 is combined

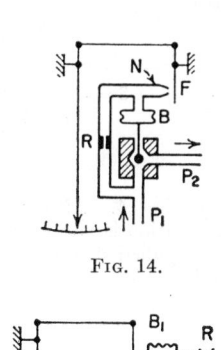

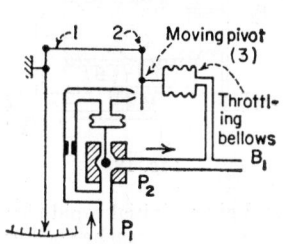

Fig. 14.   Fig. 15.

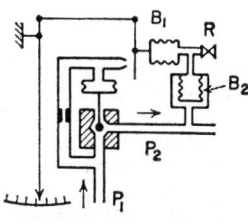

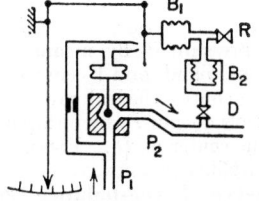

Fig. 16.   Fig. 17.

with the ball pilot of Fig. 7, the latter positioned by a bellows B to act as a relay. Air pressure P_1 flows both to the ball valve and through restriction R to nozzle N. Flapper F is positioned by the measured variable. Changes in position of F through the relay action of B and the ball valve produce such great changes in P_2 that such an arrangement cannot throttle easily and so is used for on-off control.

* Swartwout Div., Crane Co.

To make this controller give proportional action, it is necessary to add feedback as in throttling bellows B_1 in Fig. 15. When a change occurs in the primary measurement, and flapper and relay change P_2 as a consequence, B_1 acts to move the flapper's moving pivot away from the nozzle. Each position of the measurement gives a definite flapper position and definite output pressure P_2. Note that, in the arrangement shown, moving the flapper toward the nozzle decreases P_2. To change the control point of the controller, some way of changing the length of link 1 is used. To change the proportional band (throttling range), the distance from pivot 2 to pivot 3 is changed.

Reset Response. To obtain reset action in the controller of Fig. 15, an additional bellows B_2 and adjustable leak R are added as in Fig. 16. In Fig. 16 the controller has droop, every value of load giving a slightly different control point. To obviate this, a "floating" component is added to the response by allowing the flapper pivot to move temporarily, as in proportional action, but after a time returning the pivot to its original position. P_2 is now applied through a new bellows B_2 to B_1, giving initial response as in Fig. 15. However, after a time determined by the setting of leak R, pressure in B_1 returns to atmospheric and the flapper pivot returns to the initial point. The leak rate is "reset time."

Derivative Response (Fig. 17). Adding derivative or rate response to the controller of Fig. 16 consists merely in adding an adjustable restriction between output line P_2 and reset bellows B_2, thus slowing the down effect of proportional action with a

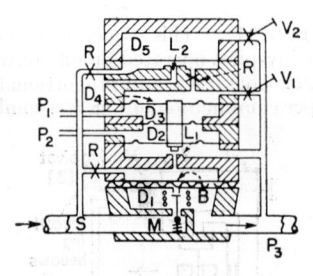

FIG. 18.

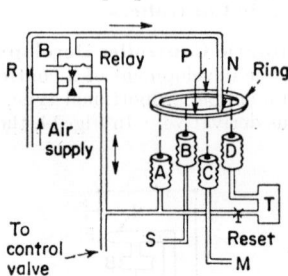

FIG. 19.

change and giving initial rapid change in P_2 almost as in Fig. 14. After a time, this effect wears off as P_2 reaches bellows B_2. Time for equalization is "derivative time." Most actual pneumatic controllers work much as these simplified sketches show, with proper adjustments added. Many actual variations in design detail exist.

Stack-type Force-balance Controller* (Fig. 18). The controller shown has a pneumatic set-point and variable measurement, proportional and reset action, with a relay valve (see page 1-43, Fig. 17). Set pressure P_1 is balanced against measured variable pressure P_2 by means of diaphragms D_2 and D_4. Restriction V_1 controls proportional band, and V_2 controls reset time. Pressure relations on diaphragms D_2, D_3, and D_4 control the leak rate through nozzle L_1 and hence pressure over D_1. In turn, this controls output P_3 by either admitting air at M or exhausting at bleed B through the porous center of D_1. Pressure P_3 under D_2 opposes any change. D_5 controls L_2, thus controlling pressure over D_4 in relation to flow rates through V_1 and R.

Beam-type Force-balance Controller† (Fig. 19). This controller uses a beam in the form of a circular annulus pivoted above by rotatable pivot arm P and pushed upward against pivot arm P by four bellows A, B, C, and D. These are in contact with the annulus. With the pivot point over bellows A and D, the proportional band is zero. With the pivot point over bellows B and C, the proportional band is infinite. At intermediate locations, practical proportional-band percentages are obtained. Bellows B contains set pressure, and bellows C measurement pressure. Bellows A is a throttling bellows, and bellows D a reset bellows.

* Moore Products Co.
† The Foxboro Co

Relations of these four pressures and the pivot points determine the leak at nozzle N. Supply air flows through restriction R to nozzle N, also to the relay valve. Nozzle pressure determines the valve position in the relay, and hence the relation of bleed B and pressure delivered to the control valve and bellows A. Any change in set vs. measurement pressures B and C causes the beam to tilt, depending on the proportional-band setting. Bellows A opposes change while D initially puts in no correction of its own but gradually opposes A (aids change) as reset time runs out.

Electric Controllers

Electric Contact Control, High–Low Type (Fig. 20). Shown is one of several ways in which contacts can be closed mechanically in electric measuring instruments having a pointer. Pointer P swings freely except at intervals when a chopper bar C descends, pushing it against a contact table T. The table tilts if pointer is high or low, making corresponding contact with enough force to complete the necessary circuit either through a relay or by direct action.

FIG. 20.   FIG. 21.

Photoelectric Cutoff. Various types of control can be applied by means of photocell and light beam for example, in scales where a relay is tripped at set weight (see page 10-36 for application data).

Oscillating-coil Control Pickup* (Fig. 21). An instrument pointer carries a vane that swings freely between coils in an oscillating electronic circuit. Coils are on the control-point set arm. If vane V leaves the coils, circuit oscillation changes, operates a relay to increase or decrease, turn on or shut off flow of the control medium. In two-position control, vane movement of 0.006 in. will operate the relay.

Electric-heat Control, Duration-adjusting Type† (Fig. 22). This controls the percentage of time during a short cycle in which electric heat will be on, to give proportional control without current-flow adjustment. Measuring circuits set slidewire S_1,

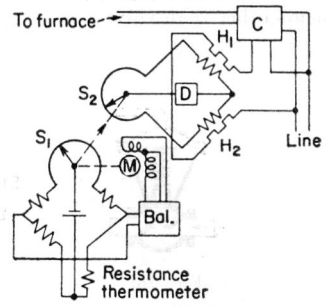

FIG. 22.

which in turn sets S_2 in the control bridge. The relay detector detects unbalance, calling for heat, and closes the circuit to the furnace, at the same time providing heat to a slow heater H_1 and a fast heater H_2, which provide heat to two bridge-circuit resistances R. The effect of heat in the resistances is to rebalance the bridge and cut off contactor C after a period of time proportional to demand.

* Wheelco Instrument Division.
† Leeds & Northrup Co.

A second slower heater adds response also for the time that the temperature has been off the control point so as to reduce the time to return the temperature to the desired value. Condensers and resistances in the output of the second bridge supply reset, derivative, and feedback. Combined outputs of two bridges are fed to the amplifier for control of the final element. **Proportional Electric Control** (Fig. 23). Various ways are used for obtaining proportional control electrically with bridge circuits. A simple circuit with proportional action uses only a bridge in which the measured value sets slidewire S_1. Bridge unbalance is detected by relay R that operates motor valve M in the proper direction, at the same time changing slidewire S_2 until balance is reached, when the valve motor stops.

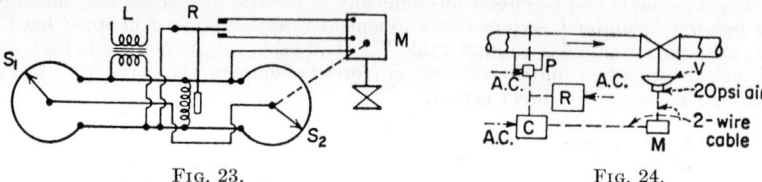

FIG. 23.   FIG. 24.

Electronic-pneumatic Control* (Fig. 24). The primary sensitive element is a resistance thermometer or some element such as a pressure gage, manometer, or other device that provides a potential proportional to the measured value, either direct or with a differential transformer used as a transducer. Measurement is transmitted as alternating current over two wires to the controller, and also to the recorder. The control impulse, which is d-c, goes over two wires to the manual control station and to a power relay at the final control element that controls air supply to the pneumatic control valve. The system eliminates instrument lags and gives a claimed accuracy of $\frac{1}{2}$ per cent.

FINAL CONTROL ELEMENTS†

Most commonly, the final control element in an industrial control system is a valve, since a fluid is likely to be the control medium. Here, however, it may be a positioning device for other purposes, a switch, rheostat, variable-speed drive, damper, etc. Since valves are the most important of these elements, this section will be confined to them, although it should be understood that most of the operators described can be applied to final elements other than valves.

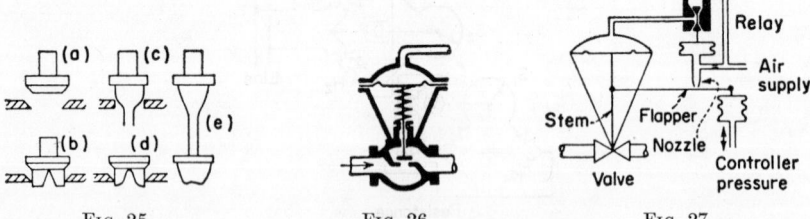

FIG. 25.   FIG. 26.   FIG. 27.

Valve Characteristics. Control valves are of many designs, but in general their characteristics may be described as beveled disk, V port, equal-percentage plug, equal-percentage V port, and linear plug. For throttling service, control valves are usually double-seated so as to be substantially balanced against line pressure. Valves may open on increase in pressure of the control medium, or they may close with medium pressure increase. In most control applications today, the so-called characterized

* Swartwout Div., Crane Co.
† For an excellent survey article on this subject, see C. D. Close, Valve Actuators Tie Precision to Power, *Control Eng.*, September, 1955, pp. 97–104.

valves are used. Figure 25 shows the main inner valve types: *a* is a beveled disk, *b* a V port, *c* and *d* equal-percentage plug and V-port inner valves, and *e* a linear plug. Sketch *e* shows both parts of the double plug as used in balanced valves.

Pneumatic Operators

Valve Design. Many control valves incorporate special features such as finned bonnets for cooling of the stuffing box on service above, about 400°F; special methods of reducing stuffing-box friction or possible leakage, such as bellows-type seals, grease seals, and special packing; extension bonnets to prevent freezing; and special types of closure device, such as pinch valves, flexible-diaphragm valves (Saunders type), and dampers.

Diaphragm-top Valve (Fig. 26). The commonest type of control-valve operator is the diaphragm top, making the valve a "diaphragm motor valve." Diaphragm tops are of two types: air-to-open and air-to-close. In either case, the motion of the diaphragm is opposed by a calibrated spring so as to make the valve-stem position proportional to pressure.

Valve Positioners (Fig. 27). Owing to friction in the stuffing box of a control valve, it is difficult to position the stem accurately in proportion to controller output pressure. Also, there will be hysteresis. A positioner can be used to make the stem position accurately responsive to pressure. Positioners vary in detail but in general are similar in principle to Fig. 27, in which controlled output pressure is applied to one end of a beam supported at the other end by a valve stem. The beam acts as a

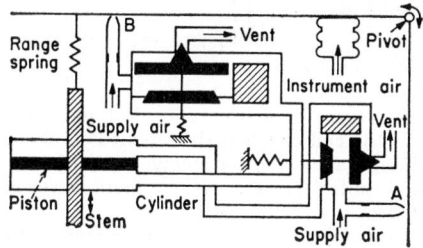

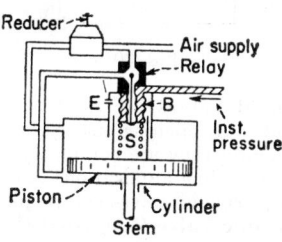

| FIG. 28. | FIG. 29. |

flapper in conjunction with a nozzle and relay air valve. In the air-to-close valve shown, increase in pressure lifts the beam, raises nozzle pressure, and lowers the valve stem (feedback) to bring a definite stem position in response to a pneumatic signal from the controller.

Pneumatically Operated Piston* (Fig. 28). A signal from the control instrument acts on the bellows, causing the two beams at right angles to move about the pivot. Action of each beam is to open the nozzle at *B* and close the nozzle at *A*, or conversely. Supply air is admitted to one side of the piston and removed from the other side by action of pilot relays connected to nozzles at *A* and *B*. A range spring, attached to the piston-rod extension, feeds back piston movement to the upper beam to balance the incoming air signal and thus to position the valve stem. Full plant air pressure can be used on the piston to provide greater power to position the control valve if high fluid forces are present that act upon the valve.

Piston-top Valve† (Fig. 29). Longer stem travel, together with more power for large valves, can be secured with a piston-type valve motor. The design shown has a built-in positioner giving stem position directly proportional to controller output pressure. The piston is loaded on top with constant pressure in *L* from the reducer. Instrument pressure inside double bellows *B* is balanced against calibrating spring *S*, positioning the pilot valve that controls admission of air either under the piston or to exhaust *E*. The position is accurate to 0.001 in.

* Fisher Governor Co., others.
† Annin Co.

Electric and Hydraulic Operators

Combination Valve Operators. For electronic control, a need developed for setting the opening of the final control element from a small electric signal. Combination operators receive an electric signal and produce a controlled output of air or a liquid under pressure. Some of these relay devices have been discussed above under Controller Types (see Fig. 12 for an electrohydraulic operator, and Fig. 13 for an electropneumatic operator).

Electric Valve Operator* (Fig. 30). The d-c signal from the controller is amplified and operates a geared motor driving the valve stem. Valve position is converted to an electrical signal that feeds back into the amplifier circuit where it is compared with the signal. Any difference is amplified to drive the valve to a position exactly proportional to the original d-c signal. Provision is made to operate the valve by handwheel, independently of controller.

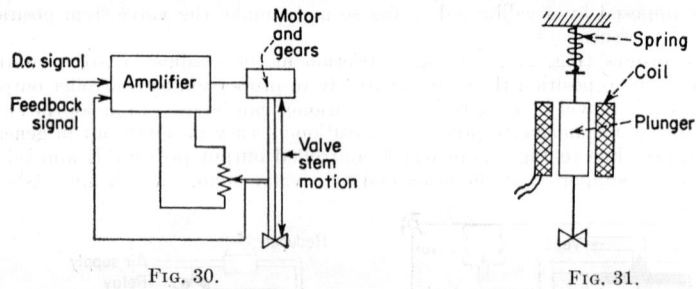

FIG. 30.   FIG. 31.

Solenoid Valves† (Fig. 31). Solenoid valves are of many types, but most are open-and-shut, opening either with current in the coil or when the circuit is broken. Consequently they are commonly used as dump valves and for emergency purposes, leaving throttling applications to conventional control valves. Where more power is required, solenoid valves can be pilot-operated.

Hydraulic Valve Operator‡ (Fig. 32). The hydraulic operator can be applied to valves, dampers, and the like. Primary indication is applied through a spring-balanced beam to a rod controlling the leak at L. Oil (or air) pressure is applied both

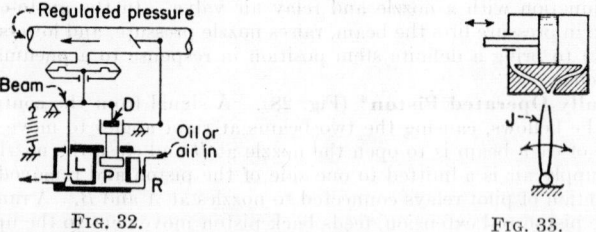

FIG. 32.   FIG. 33.

above and below piston F, the area above being less. Restriction in the line to the cylinder below the piston makes control of under-piston pressure possible with the aid of the leak. Piston movements are converted to the position of the final control element. Through a dashpot they are also transferred to the beam for applying feedback action.

Hydraulic Valve Operator§ (Fig. 33). Primary measurement positions jet pipe J, which projects liquid under 100 lb pressure against openings leading to two sides of

* Conoflow, others.
† Automatic Switch, others.
‡ Republic Flow Meters Co.
§ GPE Controls, Inc.

the piston. When the jet strikes halfway between, pressures are equal. Minute deflection will alter pressure on one side or the other and drive the piston to a new position. For most applications, where the piston should be able to assume intermediate positions, some sort of mechanical feedback would be used to make the piston position accurately proportional to the jet-pipe position. This can be used for driving valves or other final control elements against high pressures, since about 90 per cent of static pressure in the operating liquid is recovered as effective pressure against the piston.

Integral Operator and Valve* (Fig. 34). In the Bendix Fluidal valve, an annular piston varies the flow opening between the valve body and a seat that is supported in the fluid stream by webs. Pneumatic or hydraulic control pressure moves the annular

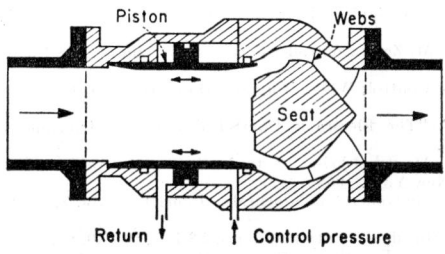

FIG. 34.

piston against the seat in proportion to the incoming signal. The balanced piston with an integral control annulus gives minimum hysteresis and fast response. Pressure range is from 25 to 12,000 psi, temperatures from −65 to 400°F. The valve can respond to frequencies as high as 120 cps in an electrohydraulic control system. Provision is made for remote indication of valve position by electric or pneumatic transmitter.

REFERENCES

1. Considine, D. M. (ed.): "Process Instruments and Controls Handbook," McGraw-Hill Book Company, Inc., New York, 1957.
2. Fitzgerald, A. E., and D. E. Higginbotham: "Basic Electrical Engineering," 2d ed., McGraw-Hill Book Company, Inc., New York, 1957.
3. Lynch, William A., and John C. Truxal: "Principles of Electronic Instrumentation," McGraw-Hill Book Company, Inc., New York, 1962.
4. Cockrell, William D. (ed.): "Industrial Electronics Handbook," McGraw-Hill Book Company, Inc., New York, 1958.
5. Chute, George M.: "Electronics in Industry," 2d ed., McGraw-Hill Book Company, Inc., New York, 1956.
6. Markus, John: "Handbook of Electronic Control Circuits," McGraw-Hill Book Company, Inc., New York, 1959.
7. Truxal, John C. (ed.): "Control Engineers' Handbook," McGraw-Hill Book Company, Inc., New York, 1958.
8. Ahrendt, William R., and C. J. Savant, Jr.: "Servomechanism Practice," 2d ed., McGraw-Hill Book Company, Inc., New York, 1960.
9. Gibson, John E., and Franz B. Tuteur: "Control System Components," McGraw-Hill Book Company, Inc., New York, 1958.
10. Raven, Francis H.: "Automatic Control Engineering," McGraw-Hill Book Company, Inc., New York, 1961.
11. Seifert, William W., and Carl W. Steeg: "Control Systems Engineering," McGraw-Hill Book Company, Inc., New York, 1960.
12. Carroll, Grady C.: "Industrial Instrument Servicing Handbook," McGraw-Hill Book Company, Inc., New York, 1960.
13. Ledgerwood, Byron K.: "Control Engineering Manual," McGraw-Hill Book Company, Inc., New York, 1957.
14. McIntyre, R. L.: "A-C Motor Control Fundamentals," McGraw-Hill Book Company, Inc., New York, 1960.
15. James, Henry Duvall, and Lewis Edwin Markle: "Controllers for Electric Motors," 2d ed., McGraw-Hill Book Company, Inc., New York, 1952.

* Bendix Corp.

16. Humphrey, Watts S., Jr.: "Switching Circuits," McGraw-Hill Book Company, Inc., New York, 1958.
17. Staff: "Uniform Face-to-face Dimensions for Flanged Control Valve Bodies," Instrument Society of America, Pittsburgh, Pa., 1950.
18. Staff: "Standard Control Valve Manifold Designs," Instrument Society of America, Pittsburgh, Pa. 1956.
19. Staff: "Dynamic Response Testing of Process Control Instrumentation," Parts I–IV, Instrument Society of America, Pittsburgh, Pa., 1961.
20. Myles, A. H.: Fast Response Amplifier Controls Large A-C Motors, *Control Eng.*, vol. 9, no. 10, pp. 77–79, October, 1962
21. Wills, D. M.: A Guide to Controller Tuning, *Control Eng.*, vol. 9, no. 8, pp. 93–95, August, 1962.
22. Ryan, Frank M.: Supervisory Control Systems, *Control Eng.*, vol. 10, no. 1, pp. 77–86, January, 1963.
23. Wills, D. M.: Tuning Maps for Three-mode Controllers, *Control Eng.*, vol. 9, no. 4, pp. 104–108, April, 1962.
24. Drinker, P. H.: Choosing Reliable and Practical Electronic Controls, *ISA J.*, vol. 9, no. 7, pp. 43–46, July, 1962.
25. Wilson, H. S., and L. M. Zoss: Controller Settings and Tunings, *ISA J.* vol. 9, no. 11, pp. 78–80, November, 1962.
26. Wills, D. M.: Cascade Control; When, Why and How, *Automatic Control*, vol. 14, no. 9, pp. 36–39, September, 1961.
27. Batcher and Moulic: "The Electronic Control Handbook," Instruments Publishing Co., Inc., Pittsburgh, Pa., 1946.
28. Bower, J. L., and P. M. Schultheiss: "Introduction to the Design of Servomechanisms," John Wiley & Sons, Inc., New York, 1958.
29. Dallimonti, R.: Basic Controller Fundamentals, *Automation*, vol. 7, no. 9, pp. 54–62, September, 1960.
30. Wills, D. M.: Simple Multiloop Control Systems. *ISA J.*, vol. 10, no. 3, pp. 67–70, March, 1963.
31. Hupp, R. E.: Digital Flow Blending, *Instr. Control Systems*, vol. 34, no. 2, p. 252, February, 1961.
32. Kerchner, W. J.: Logical Design of Control Systems, *Instr. Control Systems*, vol. 34, no. 9, pp. 1665–1667, September, 1961.
33. Boonshaft, J. C.: High-power Valve Actuators, *Instr. Control Systems*, vol. 34, no. 10, pp. 1851–1853, October, 1961.
34. Hanssen, A. J.: Valve Sizing for Flashing Liquids, *Instr. Control Systems*, vol. 35, no. 8, pp. 109–110, August, 1962.
35. Harrington, J. and D. J. Lapera: Cryogenic Control Valves, *Instr. Control Systems*, vol. 35, no. 10, pp. 101–103, October, 1962.
36. Casciato, A. C.: Applying Three-way Valves Properly, *Instr. Control Systems*, vol. 35, no. 10, pp. 110–114, October ,1962.
37. Anderson, N. A.: Pneumatic Control Mechanisms, *Instr. Control Systems*, vol. 36, no. 2, pp. 113–118, February, 1963.
38. Wing, P., Jr.: Limitations of Valve Sizing Formulas, *Instr. Control Systems*, vol. 36, no. 3, pp. 131–135, March, 1963.
39. Anderson, N. A.: The Closed-loop Controlled System, *Instr. Control Systems*, vol. 36, no. 5, pp. 126–130, May, 1963.
40. Valstar, J. E.: Step Procedure in Analyzing Control Systems, *Instr. Control Systems*, vol. 36, no. 5, pp. 132–137, May, 1963.
41. Anderson, N. A.: Process Control Simulator, *Instr. Control Systems*, vol. 36, no. 6, pp. 122–124, June, 1963.
42. Edmunds, W.: Practical Dynamic Analysis, *Instr. Control Systems*, vol. 36, no. 6, pp. 127–129, June, 1963.
43. Anderson, N. A.: Control Valves, *Instr. Control Systems*, vol. 36, no. 8, pp. 138–144, August, 1963.
44. Valstar, J. E.: Step Procedure in Adjusting Control Systems, *Instr. Control Systems*, vol. 36, no. 8, pp. 147–150, August, 1963.
45. Anderson, N. A.: Electronic Control Systems, *Instr. Control Systems*, vol. 36, no. 9, pp. 130–135, September, 1963.
46. Laspe, C.: Frequency Response of the Process, *Instr. Control Systems*, vol. 36, no. 9, pp. 138–143, September, 1963.
47. Swartwout, C. J., and F. L. Maltby: Intrinsic Safety, *Instr. Control Systems*, vol. 36, no. 10, pp. 93–97, October, 1963.
48. Lapse, C.: Estimating Controller Actions, *Instr. Control Systems*, vol. 36, no. 10, pp. 109–114, October, 1963.

Section 11

INFORMATION PROCESSING

By

ALLAN L. BURTON, B.A., M.A. (**Physics**), *Director of Research, Veeder-Root, Inc., Hartford, Conn.* (*Counters and Digital Indicating Devices*)

PAUL KAUFMANN, B.S. (**Ch.E.**), *Technical Representative, Engineering Sales, Electronic Associates, Inc., Long Branch, N.J.; Member, Instrument Society of America.* (*Applications of Analog Computers*)

J. RUSSELL LeROY, M.S. (**Ch.E.**), *Control Systems Coordinator, International Business Machines Corporation, Dallas, Texas; Member, American Institute of Chemical Engineers, Instrument Society of America (Senior), Mathematical Society of America.* (*Applications of Digital Computers*)

L. M. SILVA, B.S., Engr. Phys., *Formerly Associate Director of Research and Engineering, Beckman/Systems Division, Beckman Instruments, Inc., Fullerton, Calif.; Member, Institute of Electrical and Electronics Engineers.* (*Instrumentation Data Processing*)

APPLICATIONS OF ANALOG COMPUTERS

By Paul Kaufmann*

This subsection has for its main purpose to provide the reader with some insights into the potential fields of use for analog computers. The examples given are actual not merely proposed, and have been selected as of interest to those concerned with instrumentation. The author does not wish to imply, however, that all actual, applications of this nature are described; those which have come to his attention through appearance in the literature or through personal experience are included.

No attempt is made in this section to cover all the background on operating principles and "hardware" of analog computers. A number of good references covering these are given and the prospective user can obtain a wealth of data from manufacturers. However, some fundamental guide lines to the application of analog computers are included prior to a review of various kinds of applications.

FIELDS OF USE FOR ANALOG COMPUTERS

The analog computer is a device that performs continuous and practically instantaneous computations by means of electrical, electromechanical, or pneumatic components which, when suitably interconnected, behave in the same manner as mathematical or physical processes. It is extensively used in process-control work as follows:

1. As a flexible and economical means of studying physical systems to determine best methods of control.

2. To provide process-operating information by performing off-line computations.

3. For on-line computation as part of a control loop or for operator guidance.

4. As an actual controller.

5. As a means of teaching engineers and operating personnel the dynamic aspects of various processes.

6. For processing data in research and development activities.

The quantity of analog equipment used for simulation, computation, or control is a direct function of problem complexity. It is not necessary to invest in computing capacity beyond the needs of a particular problem. Also, since the analog computer is a parallel device that processes all information simultaneously, the time required to reach a solution is essentially unaffected by problem size. For this reason, simulation studies and computations can be performed in a minimum period of time. Similarly, the analog computer is well suited for incorporation in control circuits where excessive computing times cannot be tolerated.

The accuracy to which an analog computer can solve a problem depends upon the size of the problem. It is not a zero-error device. The digital computer is rightfully rated high in accuracy, but it is error-free only in regard to addition, subtraction, and multiplication. Great accuracy in other mathematical operations is possible but only at the expense of slower solution speeds.

Fortunately, the accuracy requirements of most process-control work are well within the capabilities of the analog computer. As an example, in dynamic-system studies, qualitative rather than quantitative answers are of prime importance. Furthermore, the data available are very often empirical. Physical constants, such as heat capacities or heat-transfer coefficients, are seldom very accurate. In regard to on-line computation, the process variables used as inputs are usually in analog form and, in addition, are measured by means of transducers which are usually not more accurate than $\frac{1}{2}$ per cent.

* Technical Representative, Engineering Sales, Electronic Associates, Inc., Long Branch, N.J.

COMPUTER SYMBOLS FOR MATHEMATICAL OPERATIONS

Many excellent books have been written on computer fundamentals.[1-4]   In addition, technical schools and leading analog-computer manufacturers offer courses in this field.   However, for the convenience of the reader, the following brief review is presented.   Most of the terminology and symbols employed are those which have become standard in the use of electronic analog computers.   Since this type is basically the same whether used for simulation, computation, or control, the same terminology and symbols apply for all these fields.

The basic building block of the electronic analog computer is the operational amplifier, or high-gain amplifier with negative feedback, shown in Fig. 1.   Summing the currents at point S, one has

$$\frac{e_i - e_g}{R_i} + \frac{e_o - e_g}{R_f} = i_g \qquad (1)$$

Also, by definition of the amplification factor A one has

$$e_o = -Ae_g \qquad (2)$$

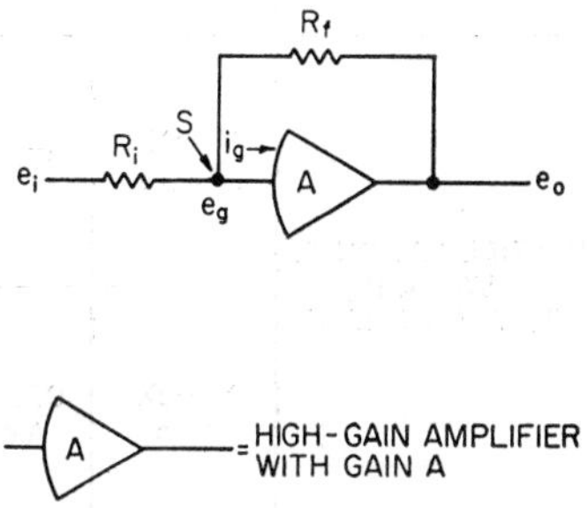

Since A is very large (typically 10^4 to 10^8) and e_o never exceeds 100 volts, e_g and the current i_g resulting from e_g can be neglected.   One then has

$$e_o = -e_i \frac{R_f}{R_i} \qquad (3)$$

Fig. 1. Operational amplifier.

In a similar manner, the mathematical operations listed in Fig. 2 can be derived from the corresponding computer symbols.

Individual operations shown in this figure may be combined.   For example, summation and integration may be performed by one high-gain amplifier with a number of input resistors and a capacitor in the feedback circuit.

For simplicity, some operations are not listed in Fig. 2.   These include multiplication of variables by variables using servomultipliers or quarter-square multipliers and also function generation (arbitrary, logarithmic, X^2, etc.).   These operations are discussed in Refs. 1 to 4.

The equivalent of the high-gain electronic amplifier in pneumatic analog computers is the flapper-nozzle system.   Small movements of the flapper result in large output changes.   In other words, high-gain amplification of an input signal is possible.   Again, by means of negative feedback provided by bellows or diaphragms whose motion tends to nullify the motion caused by input signals, the system gain is reduced with a resulting increase in linearity.   Pneumatic computing devices capable of some or all of the functions listed in Fig. 2 are commercially available.   These devices are adequately described in the literature[5-7] and are recommended for those applications where the accuracy requirements are not stringent and relatively simple calculations are involved.

FUNDAMENTAL APPROACH TO APPLICATIONS

In using an analog computer to solve control problems, the following steps are usually followed:

1. Recognition of a problem that can be best solved by an analog computer.
2. Description of the problem in mathematical terms.
3. Construction of a scaled computer diagram.
4. Fabrication of the computer.

OPERATION	EQUATION	COMPUTER ELEMENT	COMPUTER SYMBOLS
SIGN CHANGING OR INVERSION	$Y=-X$	OPERATIONAL AMPLIFIER	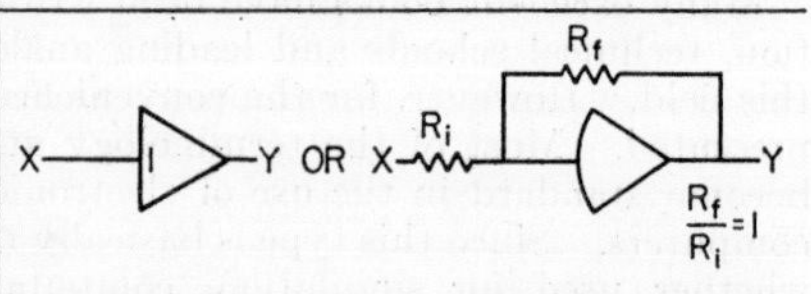
MULTIPLICATION OF A VARIABLE BY A CONSTANT	$Y=aX$ $1 \gtrless a > 0$	POTENTIOMETER	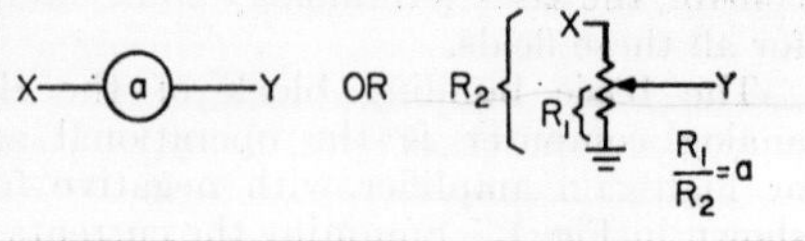
	$Y=-aX$ $a>0$	OPERATIONAL AMPLIFIER	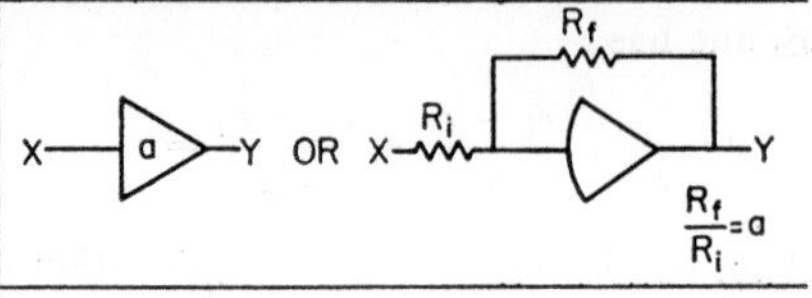
DIVISION OF A VARIABLE BY A CONSTANT	$Y=\frac{X}{a}$	POTENTIOMETER	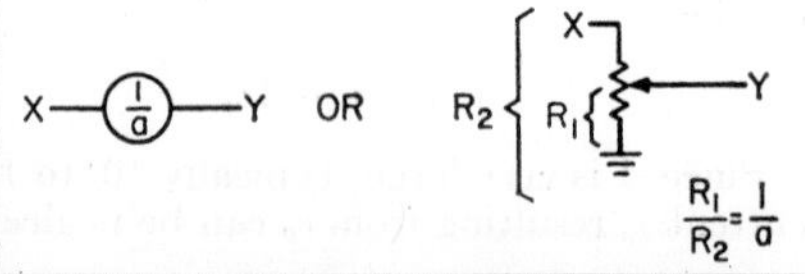

OPERATION	EQUATION	COMPUTER ELEMENT	COMPUTER SYMBOL
DIVISION OF A VARIABLE BY A CONSTANT	$Y=\frac{X}{a}$	OPERATIONAL AMPLIFIER	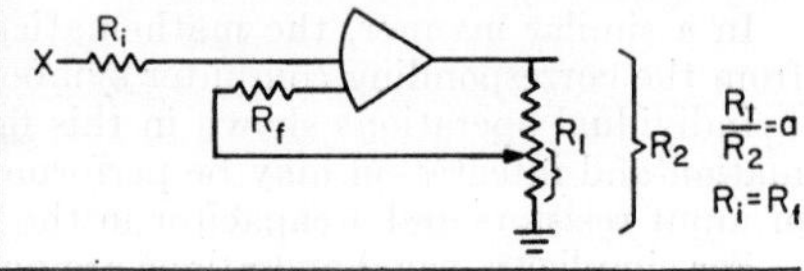
SUMMING	$Y=-(X_1+X_2)$	OPERATIONAL AMPLIFIER	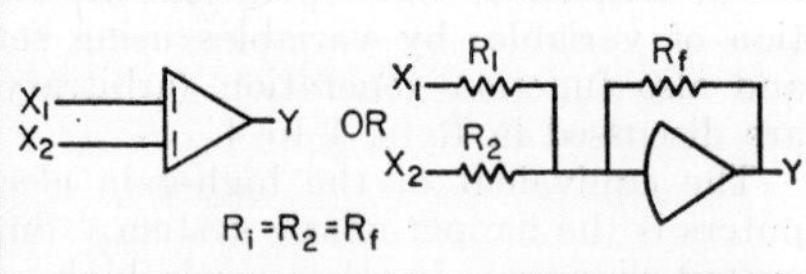
INTEGRATION WITH RESPECT TO TIME	$Y=-\frac{1}{RC}\int_0^t X\,dt$	OPERATIONAL AMPLIFIER	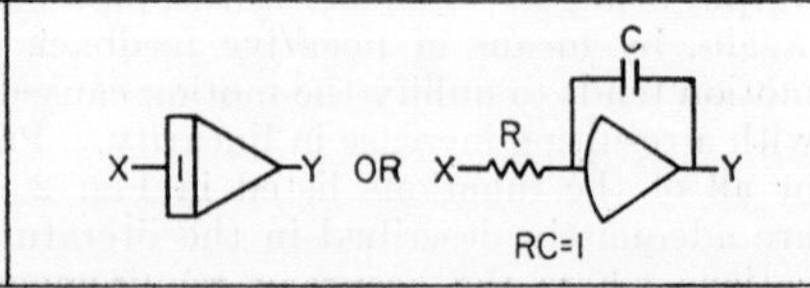

FIG. 2. Mathematical operations and corresponding computer symbols.

Industry abounds with problems which are best solved by analog computers. These problems are often ignored because engineers are not aware of the power of analog-computing techniques or are pressed by other work which seems to be of more immediate concern. As a result, large amounts of money are spent for unnecessarily high maintenance and operating costs or because vital operational guides have not been provided.

As an example, the measurement of the flow of ethylene to a polyethylene reactor has posed almost insurmountable problems to the instrument engineer. Yet, a knowledge of this flow to each of the entry ports of the reactor would be an invaluable aid to the operator.

The extreme pressure (30,000 psi) and the presence of polymer in the ethylene feed stream rule out the use of conventional instruments. If they could be used at all, the cost of maintaining such equipment would be very high.

An attractive answer to this problem would be the ideal primary element—one which does not contact the process stream. Such an approach is feasible through the use of analog-computing techniques. For example, if a heat balance is calculated around the aftercooler which is used to cool the ethylene after compression and prior to entry into the reactor, the ethylene flow can be implicitly measured.

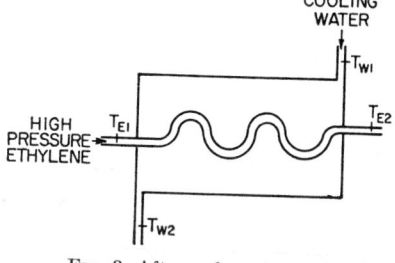

Fig. 3. Aftercooler schematic.

A schematic drawing of an aftercooler is shown in Fig. 3. Since the heat gained by the water is equal to the heat lost by the ethylene, the following equation applies:

$$F_E C_{PE}(T_{E_1} - T_{E_2}) = F_W C_{PW}(T_{W_2} - T_{W_1}) \tag{4}$$

where F = flow rate
C_P = specific heat
T = temperature
E = ethylene
W = water

F_W can be indicated by a rotameter and fixed by a hand valve or an inexpensive flow regulator. Depending upon accuracy requirements, C_{PE} can be assumed to be constant or adjusted according to pressure and temperature measurements by means of additional computer equipment. The differential temperature measurements are accomplished by thermocouples connected to differential amplifiers which raise the small thermocouple signals to the computing voltage level.

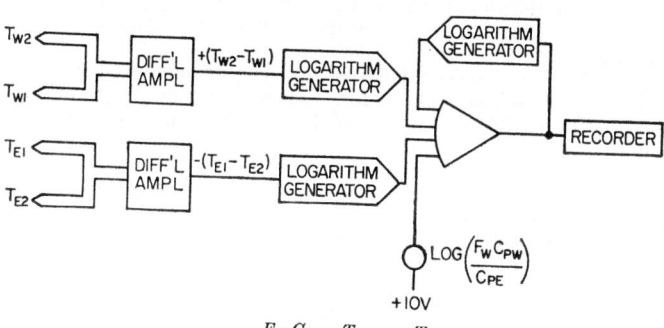

$$F_E = \frac{F_W C_{PW}}{C_{PE}} \frac{T_{W_2} - T_{W_1}}{T_{E_1} - T_{E_2}}$$

Fig. 4. Computer diagram for implicit flow measurement.

The resulting computer diagram is shown in Fig. 4. This is a simplified circuit which remains to be scaled.

SCALING OF COMPUTER CIRCUITS

Scaling is a process in which computer voltages or pressures representing physical variables and their derivatives are made as large as possible without exceeding the computing range of the computer. As an example, consider an electronic analog computer with a computing range of 0 to 10 volts direct current. In this case, the inputs to the operational amplifiers in this computer should never be large enough to cause the output of the amplifiers to exceed 10 volts. Otherwise the stabilizer section of the operational amplifiers becomes ineffective, the summing junctions are no longer essentially at zero potential, and the equations relating inputs and outputs become nonlinear. Computation accuracy is then impaired. Similarly, the standard 3 to 15 psi range of pneumatic computers must not be exceeded.

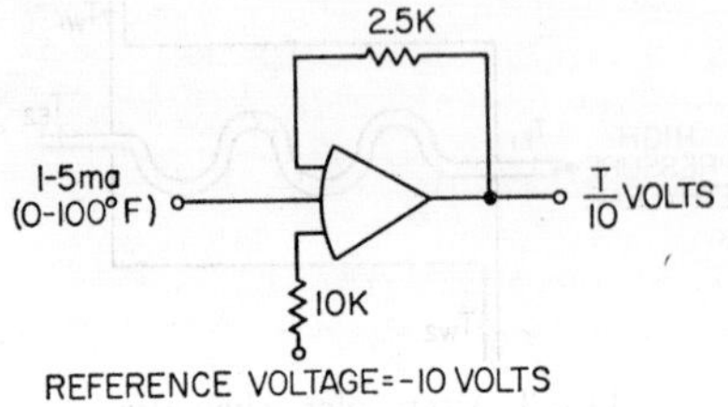

On the other hand, the computer voltages representing the physical variables should be as large as possible in order to achieve maximum computing accuracy. As an example, consider a computer which uses the signal from an electronic temperature transmitter as one of its inputs. Assume that the transmitter has a 1- to 5-ma d-c output corresponding to 0 to 100°F.

FIG. 5. An example of amplitude scaling.

For greatest accuracy, the 0 to 5 ma should be converted to 0 to 10 volts direct current, where 10 volts is the maximum operating voltage of the computer. The scale factor k which relates the voltage in the computer to temperature is calculated as follows:

$$kT_{max} = V_{max} = 10 \text{ volts} \tag{5}$$
$$k = {}^{10}\!/_{100} = \frac{1}{10} \tag{6}$$

A suitable scaled computer diagram is shown in Fig. 5.

The reference voltage impressed across the 10K input resistor produces a 1-ma current. This current is subtracted from the 1- to 5-ma input signal and the amplifier can therefore be viewed as having a 0- to 4-ma input. At maximum temperature (100°F), the adjusted input to the amplifier is 4 ma. Since the high-gain amplifier operates so that the algebraic sum of all currents at the amplifier input is essentially zero, the current flowing through the feedback resistor must be 4 ma. To produce this feedback current, the amplifier output must be −10 volts. Thus the computer circuit has been properly scaled, i.e., the output voltage is as large as possible without exceeding 10 volts.

The output voltage at any particular instant can be converted to the equivalent temperature by means of the scale factor. Thus, if the amplifier output is 5 volts, $T/10$ is set equal to 5 volts; and, upon solving this equation, T is found to be 50°F.

The preceding discussion dealt with amplitude scaling. In simulation studies, time scaling is also of importance. A solution time which is too long is inefficient. On the other hand, a solution time which is too short may result in errors caused by the speed of response of computing or recording equipment. More thorough treatment of amplitude and time scaling may be found in Refs. 1 to 4.

ACCURACY IN COMPUTER APPLICATIONS

A brief discussion of accuracy is in order since, for a given problem, computer cost rises with accuracy. As an example, multiplication of two variables can be accomplished with an accuracy of ±0.1 per cent of full scale using a relatively inexpensive quarter-square multiplier or the same operation can be performed much more accurately by using high-precision logarithm generators which are accurate to ±0.2 per cent of value.

More fundamental to solution accuracy, however, is the basic building block of the analog computer: the high-gain amplifier. Assuming $i_g = 0$ and solving Eqs. (1) and (2) for e_o/e_i we have

$$\frac{e_o}{e_i} = -\frac{A}{A + K + 1} K \qquad \text{where } K = \frac{R_f}{R_i} \tag{7}$$

Ideally, the expression $A/(A + K + 1)$ should be unity, and the only error involved in the use of this simple operational amplifier would depend upon how accurately the ratio R_f/R_i is established. In quality electronic analog computers this ratio is accurate to 0.01 per cent. The extent to which $A/(A + K + 1)$ approaches the value of 1 depends on the values of A and K. For example, with an amplifier gain of 2,000 and $K = 10$, $\frac{1}{2}$ per cent accuracy is the best that can be obtained. With a gain of 2,000 and $K = 100$, a 5 per cent device results. For a 0.1 per cent operation with $K = 100$, a gain of 10^5 is necessary. For a 0.01 per cent operation and $K = 100$, a gain of 10^6 is required. High-grade computing amplifiers have gains several decades higher than 10^6 to allow for possible component deterioration, etc., and thus ensure high-accuracy computations for the life of the computer. By use of amplifiers of this type and accurately ratioed resistors, the contribution of the summing amplifiers in a computer to over-all accuracy is insignificant except for the most complex problems.

In deriving Eq. (7), it was assumed that $i_g = 0$. This is essentially true, at least for short periods of time, if the gain and input impedance of the amplifier are sufficiently high since $e_g = -e_o/A$ and $i_g = e_g/$input impedance. For long-term accuracy (particularly when the amplifier is used as an integrator) stabilized amplifiers, which automatically maintain e_g at essentially zero, are a necessity.

Besides the use of accurate computing components, a number of techniques can be employed to improve accuracy. As an example, where process variables change over reasonably small ranges, perturbation methods can be used. Thus, if X is to be multiplied by Y, the use of even the most accurate quarter-square multiplier will usually result in an error of ± 0.1 per cent of full scale.

On the other hand, we may write

$$XY = (X_A + \Delta X)(Y_A + \Delta Y) \tag{8}$$
$$XY = X_A Y_A + X_A\,\Delta Y + Y_A\,\Delta X + \Delta X\,\Delta Y \tag{9}$$

where X_A is the average value of X and Y_A is the average value of Y.

If the value of ΔX is less than 10 per cent of X_A and ΔY less than 10 per cent of Y_A, the multiplier output $\Delta X \Delta Y$ is only 1 per cent of the product XY. In this way, the error caused by the multiplier has been reduced by a factor of 100. If ΔX and ΔY are larger, the improvement in accuracy is not so great; but in no case is the accuracy worse than for straight multiplication.

The determination of the over-all accuracy of an analog computer can be very difficult, especially if a complex problem is involved. Computer manufacturers will usually supply error analyses with their proposals if so requested. This assures the user that his accuracy specifications will be met. In addition, the computer is usually checked prior to delivery by comparing actual computer outputs with ideal outputs for all possible inputs. If necessary, the passive computing elements are altered so that specified accuracy is obtained.

The errors discussed so far have been static errors, i.e., errors encountered when dealing with stationary or slowly varying computer voltages. The subject of dynamic computing errors is quite complex.[8-11] Such errors are encountered most often in simulation studies where the computer has been time-scaled to increase solutions speed so that the results may be displayed on oscillographs.

SIMULATION USES

As previously pointed out, an electronic analog computer consists of electronic and electromechanical units which can perform specific mathematical operations and therefore, when connected together, can solve mathematical equations. To benefit from the use of an analog computer, the engineer need not be a highly trained mathematician. His job is to describe mathematically the particular problem and program the computer in such a manner that it will solve the equations automatically. The results are then interpreted by the engineer. Of course, the better the mathematical background of the engineer, the more effectively will he be able to take advantage of the capabilities of the analog computer. Certainly a good working knowledge of differential equations and servomechanism theory is highly beneficial. In addition, a knowledge of statistics and linear programming is desirable.

In recent years, servomechanism theory, which was intensively developed during the decade 1940–1950, has been applied to process-control problems.   Unfortunately, the methods of analysis, such as frequency-response and step-response methods, which resulted from this early work, apply only to linear systems or to nonlinear systems that can be considered to be linear over small ranges of the independent variables.

These "paper-and pencil" methods of analysis have had only a limited use in solving process-control problems for several reasons.   One of these is that many processes and pieces of process equipment are very nonlinear over their operating range.   Also, the number of equations involved in even a simple process is so great that a prohibitive time is required for solution.   On the other hand, the analog computer is a very efficient means of solving complex control problems, linear or nonlinear.   The general point of view stated above is illustrated in the following applications.

A Solvent-recovery Process[12]

A block diagram of a simple solvent-recovery system is shown in Fig. 6.   The control loops are not so complex that simulation of this plant is required.   However, it is evident that operating costs can be reduced by recovering as much heat as possible from the several output streams.

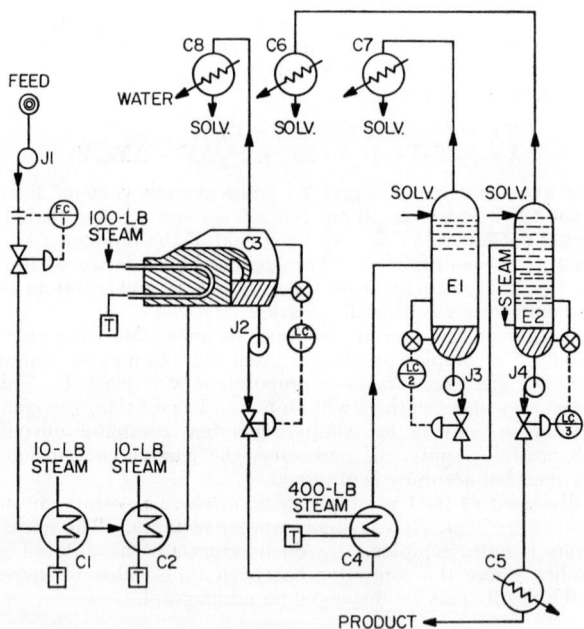

Fɪɢ. 6. A simple solvent-recovery system.

One of several possible schemes for improved operation is illustrated in Fig. 7. Here, the interactions of the various streams are of such complexity as to suggest an analysis by simulation.   Also, frequency-response data around operating points appeared to be of limited value whereas studies of start-up procedures, of operating characteristics at off-normal conditions, and of shutdown characteristics were considered essential.   These studies were conducted on an analog computer.

It is of particular interest that a vigorous simulation completely describing each piece of equipment was considered unnecessary.   Instead, the input-output char-

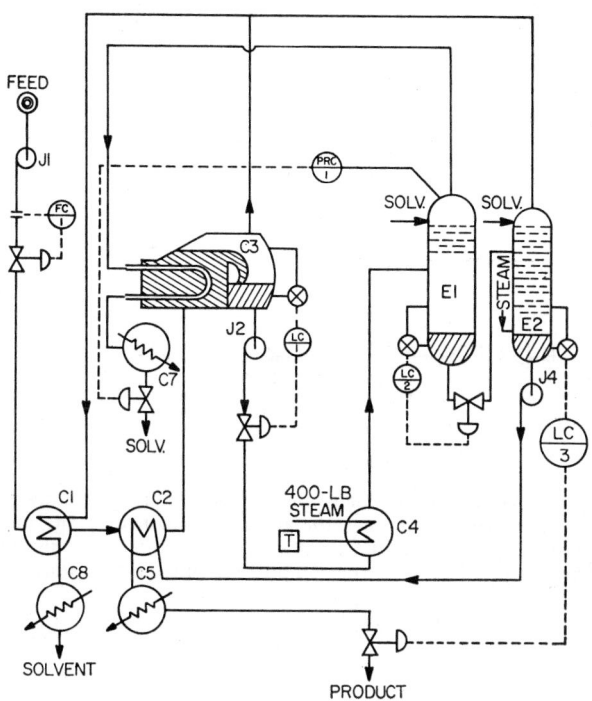

Fig. 7. An improved solvent-recovery system.

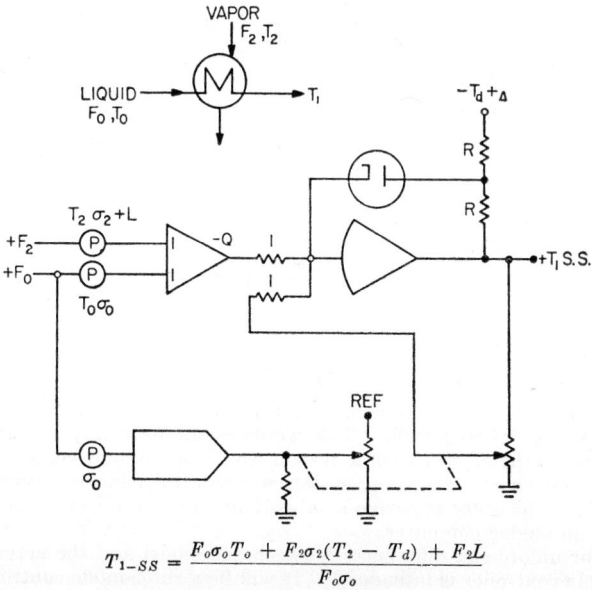

$$T_{1-SS} = \frac{F_o \sigma_o T_o + F_2 \sigma_2 (T_2 - T_d) + F_2 L}{F_o \sigma_o}$$

Fig. 8. Simulation of exchanger $C1$.

acteristics of the plant, element by element, were approximated by analog circuits and then adjusted to duplicate the characteristics exactly.

The steady-state response of exchanger $C1$ was simulated as shown in Fig. 8. This, combined with the time-lag circuit of Fig. 9, resulted in an adequate representation without resorting to the more sophisticated approach which would use a finite difference technique where the heat-exchanger surface is divided into pieces and heat flow through each piece is simulated.

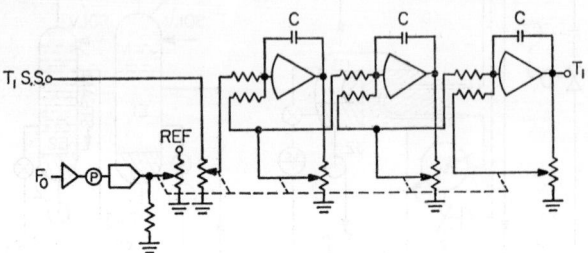

FIG. 9. Variable time-lag circuit.

The validity of this simulation was confirmed by comparing computer results with plant records. Thus the plant operator gained confidence in the analog by comparing the records of the "as-built" plant with the computer results and was quite willing to accept the predictions of the computer with regard to proposed engineering changes.

A pH Control System[13]

Reference 13 provides a thorough discussion of the design of the neutralization phase of a waste-disposal plant. A 12,000-gpm stream with pH varying from 1 to 13 and subject to 2:1 changes in flow is to be maintained between pH 6 and pH 9.

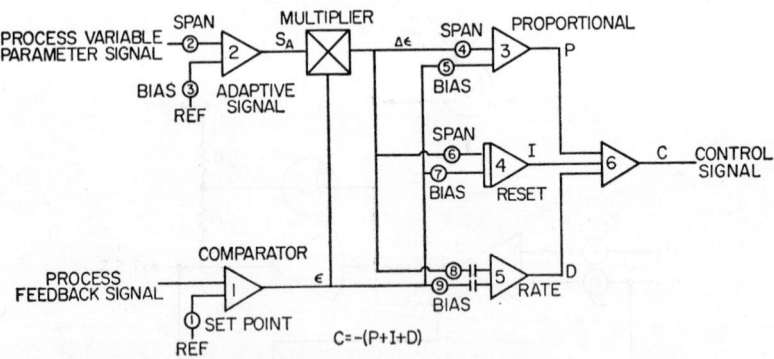

FIG. 10. Analog schematic of three-mode controller for pH control.

A simple approach to this problem would be to use an abundance of capacity in the form of a large stirred tank. This would result in steady operation by self-regulation. Since the cost of such a system would be prohibitive, a more practical solution would be a smaller-capacity system with control achieved primarily through dynamic action. In order to arrive at a workable control scheme, this system was simulated on an analog computer.

To allow for unforeseen differences between the model and the actual plant, the use of a unique controller is intended.[14] It will be a three-mode controller in which the modes are noninteracting and adjusted according to flow rate. This will permit

tight and stable control of the process regardless of changes in process gain and time constants as flow rate varies.
The analog schematic of this controller is shown in Fig. 10.

A Batch Heating Operation[15]

This problem involves a jacketed kettle in which material is to be heated as rapidly as possible to 200°F, held there for a fixed time, and then quickly cooled. By simulation on an analog computer, the optimum control scheme and controller settings were established; the question of whether or not a transient overshoot of less than $\frac{1}{2}$°F could be obtained with all possible batch sizes was resolved; and the effect of changes in steam pressure, ambient temperature, batch reaction, and cooling-water temperature were determined. The procedure followed was "cut and try" and was accomplished in minutes by turning knobs on the computer instead of months by trying various methods of control on the actual process.

Simulation of the jacket-water temperature transmitter is of particular interest in that provisions were made for automatically adjusting the time constants associated with the transmitter as steam and water flow varied. This was accomplished by means of a servomultiplier.

This computing device is essentially a self-balancing potentiometer in which a servomotor drives the arm of a feedback potentiometer and also the arms of other ganged potentiometers mounted on the same shaft. By feeding a voltage to the servomultiplier which is equivalent to a computed turbulence factor, the time constants of the temperature transmitter (determined by the ganged potentiometers) are adjusted in accordance with turbulence.

Nuclear-submarine Steam Generator[16]

The designer of steam-plant control systems for nuclear submarines is faced with problems not encountered in commercial power-plant design. First, space considerations dictate the use of steam drums that are smaller by a factor of perhaps 6. Also, since the steam plant is used for propulsion, considerable swell and shrinkage in drum level occurs.

These dynamic considerations call for a control system more sophisticated than the system conventionally used in such plants. In addition, special instrumentation hardware is required to meet the needs for the reliability and ease of maintenance peculiar to the naval service. Thus this analog-computer study involved not only the simulation of the plant and various control schemes but also the design and the evaluation of control hardware.

Initially, a simple simulation was used to obtain a fundamental "feel" for the operation of the system. Subsequent simulation studies became more sophisticated as the control system and hardware were finalized. Actual control hardware was tested by tying it into the "plant" as simulated on the computer.

An interesting result of this study was that single-element rather than three-element control was found to be suitable.

Vacuum-evaporator Control[17]

Reference 17 contains what is believed to be the first published analog-computer study of the dynamics and control of vacuum evaporators. The risk involved in using rough data for the estimation of process time constants is pointed out. Also, the results of a simplified simulation demonstrated the need for a more complex treatment. Engineers concerned with evaporator control should benefit greatly from this study.

Rolling-mill Control[18]

The availability of radiation gages for thickness measurement has made it possible to design rolling-mill controls that automatically regulate thickness. Because the rolling mill is part of the control loop, it must be understood before the control system can be designed. Because the frequently used tandem mill has a very complex transfer function, an analog simulation was used to solve the derived equations. This study showed the functional requirements of the gage control loop and the optimum adjustments of the winding-reel control.

Other Simulation Studies

The preceding applications were concerned with actual processes. Many other control applications which do not pertain to specific plants have appeared in the literature. Some of these are as follows:

1. Wind tunnel.[19]
2. Chemical reactors.[20,21]
3. Heat exchanger.[22,23]
4. Distillation column.[24]
5. Cement kiln.[25]

Some concluding remarks concerning analog-computer laboratories are in order at this point. First, experience has shown that a team approach is desirable. The most effective combination is two or three systems engineers with a good background in instrumentation and dynamics together with assistance and advice from a number of experts in the various engineering fields.[26]

Such laboratories should have a manual of typical computer circuits for standard pieces of instrumentation and operating equipment. This is a great aid to efficient operation. In addition, control valves and panel-mounted conventional controllers should be available so that valuable computer equipment is not tied up in simulating these devices.

OFF-LINE COMPUTATION

Analog computers used for off-line computation can be considered to be sophisticated calculating machines in which all information required for problem solution is manually inserted, the computer arrives at a solution, and the solution is used as a guide in operating a process.

The benefits which result from the use of analog computers for off-line computation are as follows:

1. Information which is essential to the operation of a process or which will enable a process to be run more efficiently is made available.
2. Operating personnel are relieved of the necessity to perform time-consuming or tedious calculations.
3. Problem solutions are available as needed; computation is essentially instantaneous.
4. Human error is eliminated.
5. Economy. Computer cost is solely a function of problem complexity. It is not necessary to invest in computing capacity that may never be used.

Oxygen-steelmaking Furnace-charge Computer[27]

This process consists of a trunnion-mounted cylindrical furnace which is charged with scrap, hot metal from blast furnaces, and fluxing materials such as burnt lime, mill scale, and fluorspar. Upon completion of the charge, the furnace is turned up and oxygen is blown into the furnace by means of a water-cooled lance. Various thermochemical reactions take place to convert the hot metal to steel. In approximately 20 min, the "blow" is completed, as evidenced by a visual drop in the flame coming from the mouth of the furnace. The batch temperature is measured by means of an immersion thermocouple. If it is within 2880 to 2920°F, the furnace is tapped. If it is not in this range, it must be adjusted to avoid pouring difficulties. These temperature adjustments take time and are therefore costly. They can be avoided if the furnace is charged with the proper proportions of scrap, hot metal, and lime. The hot metal serves as a heat source while scrap and lime act as heat sinks.

Formerly, operating personnel relied on their experience and knowledge of the process to estimate the quantities of charge material required to reach the proper end temperature. By using a special-purpose analog computer for computation of a thermally balanced charge, a substantial improvement in performance was obtained. A block diagram of this computer is shown in Fig. 11.

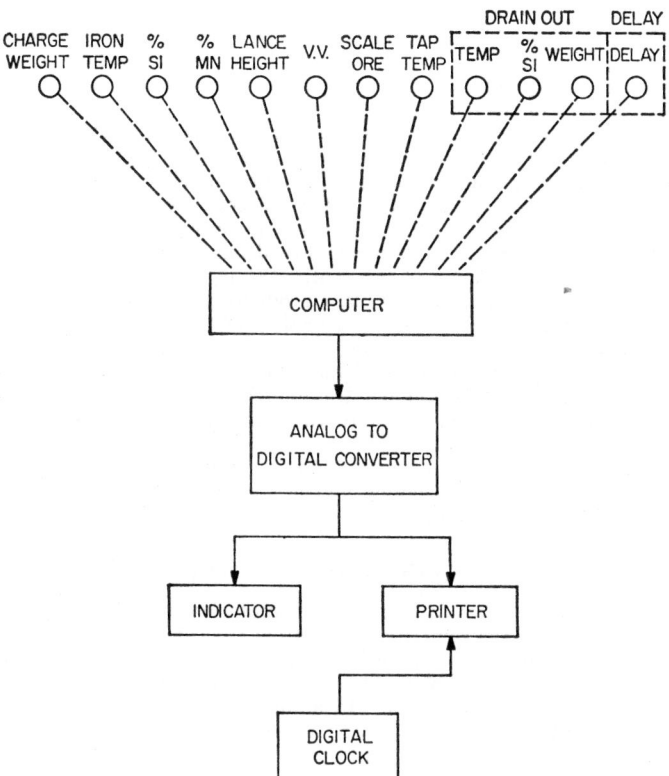

FIG. 11. Furnace-charge computer block diagram.

The operator first dials into the computer the desired tapping temperature along with the charge weight, iron temperature, basicity ratio, lance height, scale and ore, hot-metal silicon and hot-metal manganese content. The computer then solves thermochemical equations and indicates the quantities of scrap, hot metal, and lime required to reach the desired tapping temperature. A digital voltmeter is used to present this information to the operator.

The thermochemical equations which are solved by the computer resulted from an analog-computer study in which differential equations were used to define the transient behavior of the variables. The computer equations are algebraic but, nevertheless, of sufficient complexity to require 18 operational amplifiers for their solution. Multiturn potentiometers with direct-reading dials are used. A printer is included to record all input and output information in addition to time as provided by a digital clock. Computations accurate to ± 0.5 per cent were achieved.

Ice-cream-manufacturing Computer[28]

The preceding application was concerned with the separate solution of three algebraic equations. This application deals with the simultaneous solution of two equations involving two unknowns. The method of steepest ascents is used. A third unknown is then obtained by substituting the first two unknowns in a third equation.

More specifically, the percentage of butterfat, the percentage of serum solids, and the batch total are inserted manually. Three dairy ingredients are then selected. One is a known adjustable quantity, and the remaining two and the water addition are solved for by the computer. The computer does not automatically produce a recipe with a minimum of high-cost ingredients and a maximum of low-cost ingredients. Instead, once the quantity of low-cost ingredients has been specified, the computer automatically determines the proportion of higher-cost ingredients required to minimize the cost of the batch.

An interesting feature of the computer is a manual scaling operation which permits the computing voltages to be made as large as possible and thus provides for maximum solution accuracy. Descaling is accomplished automatically.

Solutions are read out on a digital voltmeter. Results are also recorded on punched cards for the use of the accounting and production-control departments.

The entire program including settings and card punching normally takes about 6 min. Nominal accuracy is ± 0.02 per cent. Maximum analog-computer solution speed is 50 msec. It is claimed that plant production capacity has nearly doubled since the computer and associated equipment were installed.

As mentioned, the previous problem dealt with two equations involving two unknowns. When the number of variables exceeds the number of equations an infinite number of answers is possible. If a unique set of answers is to be obtained, additional constraints must be added to the system. Such is the case in linear programming.

Many linear-programming problems are large enough to justify the use of a digital computer. However, problems having matrices as large as 6×8 are solved most economically by an analog computer.

ON-LINE COMPUTATION

On-line computation differs from off-line computation in that some or all of the information required by the computer comes from transducers that are tied directly into the process. Also, the computer may be an integral part of a control loop. Benefits derived from the use of analog computers for this purpose are similar to those listed for off-line computation. In addition, greater reliability, design simplicity, and economy are possible because analog-to-digital converters are not required. When the computer is part of a control loop, the fast computing speed of the analog computer makes its use mandatory.

Materials-balance Computer*

This computer is used as an operational aid in a polymerization process. It accepts the following signals:

1. The output of a gas chromatograph—weight per cent of monomer and weight per cent of modifier.

2. The signal from a gamma-ray density meter—density of the product stream.

In addition, process-stream temperature and solids density are manually inserted by means of potentiometers with direct-reading dials.

These signals are amplified and combined to give outputs proportional to weight per cent monomer, the ratio of modifier to monomer, and the volume per cent solids in the product stream. Conventional instruments record these computed quantities.

By means of the perturbation technique, which has been previously described, the error in computing volume per cent solids was decreased by a factor of 50. Also, by

* Manufactured under special contract by Electronic Associates, Inc., Long Branch, N.J.

using logarithm generators, accurate to 0.2 per cent of value, the ratio of modifier to monomer is computed with a probable error of 0.25 per cent.

This computer includes a manual calibrator which allows the operator to check computer accuracy. This is accomplished by replacing the normal input signals with two sets of standard signals: one for a 40 per cent output check and the other for a 60 per cent output check.

An interesting design feature in this computer occurs in the amplification of the density signal as shown in Fig. 12. Rather than effecting a gain of 250 in one step, two amplifiers are used for increased accuracy. In addi ion, the use of positive feedback raises the input impedance to several megohms, thus minimizing the effects of loading on the density transmitter.

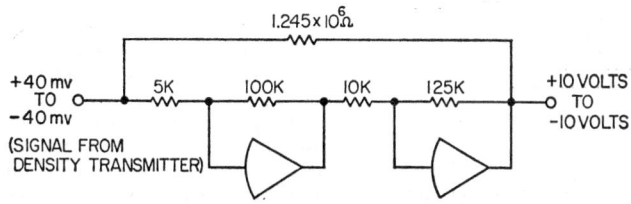

FIG. 12. A simple unloading circuit.

Reactivity Computer[29]

This computer was constructed as a tool for reactor-control studies. However, it can be used as a standard part of reactor instrumentation by operating on electrical signals from standard neutron-measuring instruments and generating a voltage proportional to reactivity.

The author[29] points out that reactivity may have more significance as a basis for scram action during start-up where the transient period is not so important as the alternate steady period, which is a function of reactivity.

Two designs are described. One utilizes a neutron-flux input signal while the other uses a log-rate (of neutron flux) signal. The range and accuracy of the latter computer is essentially independent of the number of neutrons in the reactor. Both designs are based on the fact that there is a unique relationship between the time behavior of the neutron population and the time behavior of reactivity. This approach made it unnecessary to resort to a more involved computation based on measurement of all variables that affect reactivity.

Heat-balance Computer[30]

The process involved consists of a stirred vessel fed by solvent, reactant, and catalyst (refer to Fig. 13). An exothermic reaction takes place in the reactor and product, solvent, catalyst, and unreacted reactant leave as effluent.

The computer calculates production rate by performing a heat balance around the reactor. In addition, concentration of reactor product is computed from the production rate and known reactor-feed information.

Operation of the reactor using these computed values as operating guides proved to have distinct advantages over conventional control systems as evidenced by definite improvements in production rate and product uniformity observed during extensive tests. These dynamic operating guides also provided early detection of erratic operation of the reactor.

The algebraic sum of the heat generated in the reactor and a dynamic correction factor is subtracted from the heat carried out of the reactor to give heat from the reaction. Reactor production rate is then obtained by division of this quantity by an appropriate constant.

The dynamic correction is included to account for heat accumulation or depletion in the reactor as a result of reactor changes. A system based on periodic sampling and storage of present and past voltages proportional to reactor temperatures and a periodic comparison of these voltages by a differential amplifier produces the required dynamic correction.

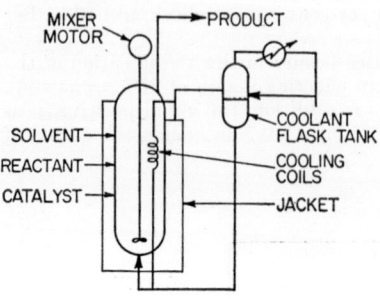

Fig. 13. Stirred-vessel schematic.

The problem of measuring heat loss through the reactor walls was solved by using heat-flow disks.[31] These disks consist of copper silver–telluride differential thermocouples mounted on opposite sides of a disk of known area and thickness. The signal generated is proportional to heat flow through the disk.

The heat-balance computer was eventually used to control production rate by manipulation of catalyst flow.[32] Because of difficulties presented by the long dead time in the process, a unique approach known as "linear predictor" control[33] was used. Simulation studies on an analog computer showed that exact linear predictor control resulted in about 50 per cent improvement in control quality.

Internal-reflux Computer[34]

This device calculates distillation-column internal reflux from measurements of overhead temperature, external-reflux temperature, and external-reflux flow rate. The computed value of internal reflux is then sent to a controller which manipulates the flow of external reflux to maintain internal reflux at a desired value.

This computer serves two functions. It reduces the number of variables which can disturb a column, and it also provides for better column operation since internal reflux is more specific to the separation process than is external reflux.

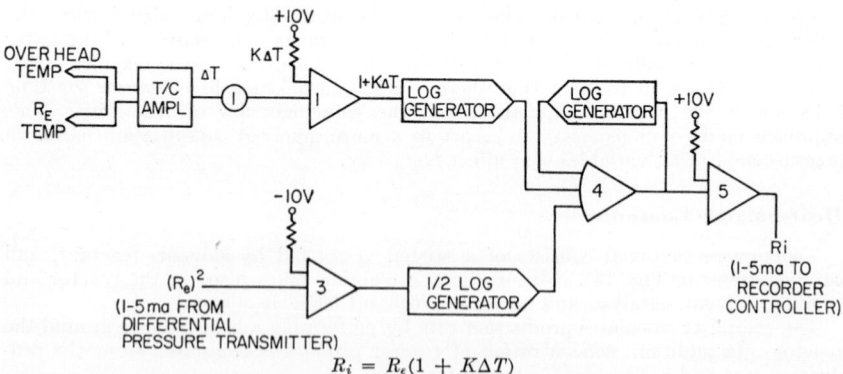

$$R_i = R_e(1 + K\Delta T)$$

Fig. 14. Internal-reflux computer circuit.

Figure 14 shows the applicable computer diagram. The -10-volt reference voltage applied to amplifier 3 provides for zero suppression of the signal from the differential pressure transmitter   The $+10$-volt reference voltage applied to amplifier 5 provides for the required elevated zero output signal.

Because of the simple computation involved, the internal-reflux computer can be constructed using pneumatic computing elements. However, since the computer is an integral part of a flow control loop which cannot tolerate time lags without excessive detuning of the controller, if at all, the electronic version is preferred where long transmission lines are involved.

Feed-enthalpy Computer[35]

This computer calculates the total heat content of the feed to a fractionator by adding the following:

1. Initial feed heat content above some reference temperature.
2. Heat given up by the bottoms-product stream to the feed in the economizer heat exchanger.
3. Heat given up to the feed by the steam in the feed preheater.

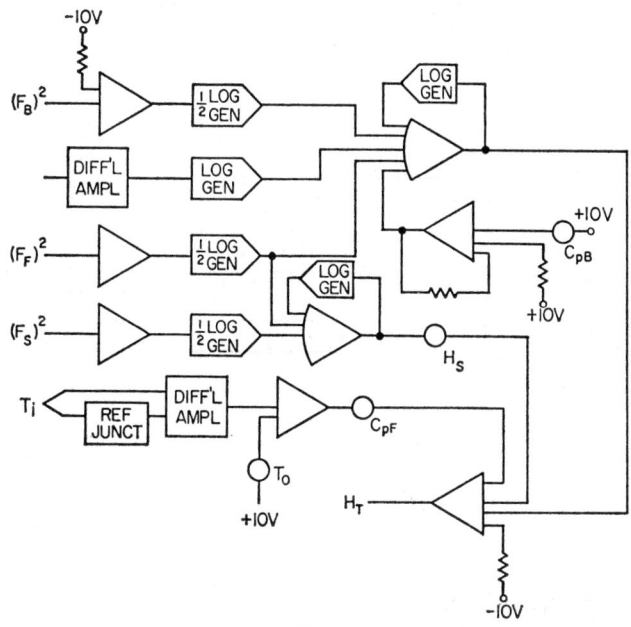

$$H_T = C_{PF}(T_i - T_o) + \frac{F_s}{F_F} h_s + \frac{F_B}{F_F}(T_{B1} - T_{B2})C_{PB}$$

where H_T = total enthalpy of feed referred to T_o
$\quad C_{PF}$ = specific heat of feed
$\quad\quad T_i$ = temperature of feed before entering economizer
$\quad\quad T_o$ = arbitrary reference temperature
$\quad\quad F_s$ = steam flow
$\quad\quad F_F$ = feed flow
$\quad\quad F_B$ = product flow
$\quad\quad h_s$ = difference in enthalpy of steam entering preheater and condensate
$\quad C_{PB}$ = specific heat of bottoms product
$\quad T_{B1}$ = temperature of product entering economizer
$\quad T_{B2}$ = temperature of product leaving economizer

Fig. 15 Feed-enthalpy computer circuit.

The computer output is the heat content of the feed as it enters the column; it is transmitted to a controller which regulates steam flow to the preheater to maintain the desired feed enthalpy.

The simplified computer diagram is shown in Fig. 15. Figure 16 shows how the feed-enthalpy computer is tied into the fractionator. If the feed enters the column at its bubble point or partially vaporized, its temperature will not be a good measure of heat content, particularly if feed composition varies.   Under these circumstances, temperature control of the feed is seldom satisfactory.   Also, because feed flow is not constant, the temperature controller will eventually be detuned in order to achieve stable operation for high feed rates.   Control at low feed rates will be inadequate.

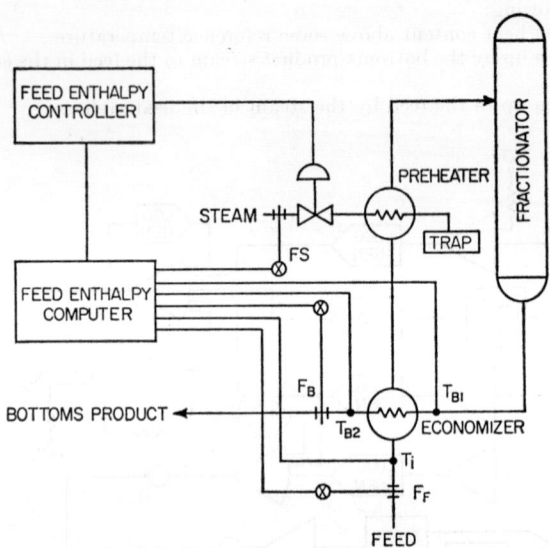

FIG. 16. Feed-enthalpy computer connections to process.

The foregoing effects plus the unstable operation caused by various phase shifts and interactions in conventional control loops serve to justify the use of a feed-enthalpy computer.

Bottoms-product-flow Computer*

This computer is used on fractionating columns to calculate the bottoms-product flow which corresponds to feed rate and composition and the separation which is required in the column.

Use of the foregoing computers, plus another one concerned with reboiler heat input, excludes undesirable disturbances and interactions and compensates for changes in unalterable variables such as feed flow, feed composition, or ambient temperature, which would otherwise affect column operation.   Once "tight" control has been accomplished, a master computer can be used as a means of guiding these individual computers in accordance with over-all requirements.

Depending upon the difficulty of separation and the value of the products, one or all of the preceding three fractionator computers may be justified.   If all are required, they can be consolidated into one computer called, perhaps, the "fractionator computer."

Capacity Control of a Distillation Column[36]

This computer was devised to enable a distillation column to be operated at maximum capacity.   Its proposed usefulness applied principally to production-limited

* Manufactured under license from Philips Petroleum Co. by Electronic Associates, Inc., Long Branch, N.J.

products. For market-limited products, an effort to decrease operating costs would be more pertinent.

Figure 17 is a block diagram which shows how the computer is used. It accepts the signals from the transmitters shown in this figure and computes the vapor capacity

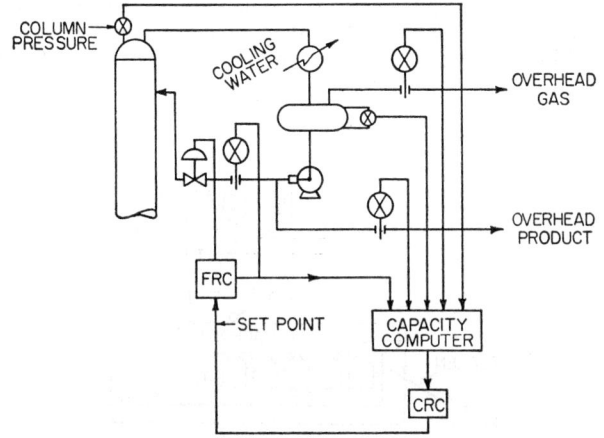

Fig. 17. Capacity-control computer connections to process.

V_2 of the tower and the actual vapor rate V_1. By means of logarithmic techniques, the column efficiency in per cent $(V_1/V_2 \times 100)$ is determined.

Conversion Computer[37]

Investigations which preceded the installation of this computer in a synthetic-rubber plant showed that the polymerization step offered the greatest opportunity for gain. It was in this operation that productivity and product quality were almost completely determined. An analysis also showed that production could be increased and quality improved by proper control of conversion in the polymerization step.

The conversion computer calculates monomer conversion from density measurements and ascertains the reactor operating conditions required to maintain conversion at a given level. More specifically, the computer adjusts the set points of temperature controllers on reactors upstream of the density-measurement points to correct any conversion error. Temperature controllers downstream of the density-measurement point are also adjusted to cancel errors already generated.

An analog computer was chosen because:

1. There was little need for memory or decision-making capabilities in the computer.

2. The number of computations made the analog approach more attractive from an economic point of view.

3. It was felt that plant personnel would require less training than if a digital computer was employed.

The computer also performs certain auxiliary computations and alarm functions. Numerous safeguards are built into the computer to prevent plant operation from being affected by faulty operation of any part of the computer system. It is believed that, if any system malfunction occurs, even the failure of a resistor in a computing element, the operation will automatically be transferred to auxiliary equipment.

Iron-ore Sintering-control Computer[38]

The schematic drawing of a sintering machine is shown in Fig. 18. Iron-ore fines that normally blow through the blast furnace and reduce its yield are converted to usable blast-furnace burden material by the sintering process. As shown, a mixture of fine ores and granular fuel is fed to a continuous grate where the mixture is ignited and burned in a down-draft combustion process. Maximum production results when burn-through occurs at the discharge end of the machine.

Several thermocouples located in the wind boxes under the grate are used as inputs to a computer which calculates the burn-through point. The output of this computer is sent to a second computer which determines the feed rate required to maintain the desired burn-through location. The output of the second computer goes to

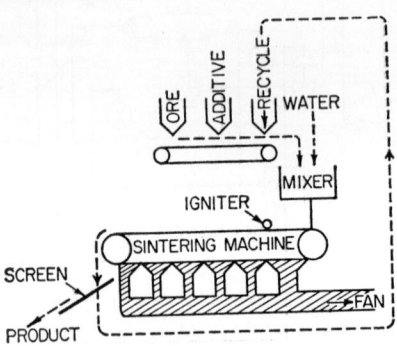

FIG. 18. Iron-ore sintering operation.

an automatic feed-control loop using commercially available weigh feeders to maintain the proper total feed rate and the proper proportions of ore, coke, and recycle.

The control loop is closed by equipment which senses bed level at the feed end of the sintering machine and adjusts grate speed to maintain a preset level.

Other On-line Applications

Other important on-line computing applications of analog computers are the automatic allocation of electric-power-generator loads (Refs. 39, 40, and page 14-45) and airborne gunnery control.[41] At least one chemical company uses analog computers for continuous on-line calculations of mean value and standard deviation. This information is used as a means of controlling product quality. Reference 42 explains how statistical parameters such as the mean, the variance, the autocorrelation function, the Fourier transform, and the power spectrum can be estimated continuously by simple analog techniques.

OTHER APPLICATIONS

The use of an analog computer as a controller was discussed under A pH Control System. A similar application is discussed for wind-tunnel surge control.[43]

An excellent example of the use of the analog computer in medical research is the respiratory computer described in Ref. 45.

A very realistic training device for nuclear-reactor operator training is described in Ref. 45.

Finally, a 300-amplifier computer which is used as a controller, a simulator, and an on-line calculator is discussed in Ref. 46. It is used for ground testing a nuclear reactor capable of propelling a supersonic ramjet missile (Project Pluto).

This reactor is a high-power density, high-temperature device which uses air stored at a high pressure for cooling. Control is effected by three major systems:

1. Reactor control system—maintains desired power level by positioning control rods or vanes.

2. Air flow-rate control system—maintains desired flow rate by positioning a pressure-control valve located upstream of the reactor.

3. Air-temperature control system—controls reactor inlet air temperature by positioning two differentially driven control valves which determine the amount of air which goes through a heater or bypasses it.

An on-line reactivity computer is included to allow the operator to monitor reactivity continuously. Also, analog simulation for system checkout and operator training is used.

Some of the reasons given for choosing a solid-state electronic analog computer for this application were:

1. Proved reliability and performance were established in previous tests.

2. The use of digital techniques would have been impractical in many of the analog subsystems (such as the control-rod-actuation subsystem).

3. The original simulation studies were conducted on an analog computer. The transition to the on-line control and simulation computer was relatively simple.

REFERENCES

1. Johnson, C. L.: "Analog Computer Techniques," 2d ed., McGraw-Hill Book Company, Inc., New York, 1963.
2. Korn, G. A., and T. M. Korn: "Electronic Analog Computers," 2d ed., McGraw-Hill Book Company, Inc., New York, 1956.
3. Karplus, W. J., and W. W. Soroka: "Analog Methods," 2d ed., McGraw-Hill Book Company, Inc., New York, 1959.
4. Rogers, A. E., and T. W. Connolly: "Analog Computation in Engineering Design," McGraw-Hill Book Company, Inc., New York, 1960.
4a. Jackson, A. S.: "Analog Computation," McGraw-Hill Book Company, Inc., New York, 1960.
4b. Smith, G. W., and R. C. Wood: "Principles of Analog Computation," McGraw-Hill Book Company, Inc., New York, 1959.
4c. Hilton, A. M.: Analog "Computation" and Analog Machines, *Electro-Tech.* (*New York*), November, 1960.
5. Mamzic, C. L.: Using Pneumatic Analog Computing Elements for Control, *Control Eng.*, April, 1961, pp. 105–110.
6. Chapin, D. W.: A Pneumatic Computer for Process Control, *ISA J.*, September, 1961, pp. 38–43; October, 1961, pp. 53–55.
7. Stirling, P. H., and H. Ho: Pneumatic Black Boxes . . . Simple Computers for Process Control, *Ind. Eng. Chem.*, August, 1960, pp. 65A–66A.
8. Bode, H. W.: "Network Analysis and Feedback Amplifier Design," D. Van Nostrand, Company, Inc., Princeton, N.J., 1945.
9. MacNee, A. B.: Some Limitations on the Accuracy of Electronic Differential Analyzers, *Proc. IRE*, 1952.
10. Marsocci, V. A.: An Error Analysis of Electronic Analog Computers, *IRE Trans. Electron. Computers*, December, 1956.
11. Murray, F. J.: "Mathematical Error Analysis for Continuous Computers," Project Cyclone, Symposium II, part 2, 1952.
12. Lewis, L. G.: Simulation of a Solvent Recovery Process. *Instr. Automation*, April, 1958, pp. 644–647.
13. Field, W. B.: Design of a pH Control System by Analog Simulation, *ISA J.*, January, 1959.
14. Field, W. B.: "An Adaptive Three-mode Controller for the Process Industries," Digest of Technical Papers, Joint Automatic Control Conference, Boulder, Colo., 1961.
15. Worley, C. W., R. G. E. Franks, and J. F. Pink: Process Control Problems Yield to the Analog Computer, *Control Eng.*, June, 1957, pp. 97–104.
16. Waite, D. P., and E. E. Lynch: Computer Verification of Steam Generator Instrumentation for a Nuclear Power Plant, AIEE Paper, Computers in Control Symposium, Atlantic City, N.J., October, 1957.
17. Johnson, D. E.: Simulation and Analysis Improve Evaporation Control, *ISA J.*, July, 1960, pp. 46–49.
18. Phillips, R. A.: Process Analysis Plus Analog Simulation Yields Better Mill Controls, *Control Eng.*, July, 1958, pp. 100–108.
19. Straight, E., and F. Michaels: Effects of Limiting Error in Feedback Analysis, *ISA J.*, May, 1958, pp. 44–49.
20. Batke, T. L., R. G. E. Franks, and E. W. James: Analog Computer Simulation of a Chemical Reactor, *ISA J.*, January, 1957, pp. 14–18.
21. Williams, T. J.: Chemical Kinetics and the Dynamics of Chemical Reactors, *Control Eng.*, July, 1958, pp. 100–108.
22. Franks, R. G. E.: Maximizing Control Performance and Economy with Analog Simulation, *ISA J.*, September, 1958, pp. 80–84.
23. Hainsworth, B. D., and V. V. Tivy: Dynamic Analysis of Heat Exchanger Control, *ISA J.*, June, 1957, pp. 230–235.
24. Rose. A., and T. J. Williams: Automatic Control in Continuous Distillation, *Ind. Eng. Chem.*, November, 1955, pp. 2284–2289.

25. Min, H. S., P. E. Parisot, J. F. Paul, and J. W. Lyons: Computer Simulation of a Wet-process Cement Kiln Operation, Preprint 202-LA61, ISA Fall Instrument-Automation Conference and Exhibit, Los Angeles, Calif., September, 1961.
26. More, R. L.: The Computer-team Approach to Control System Design, *ISA J.*, December, 1957, pp. 548–553.
27. Slatosky, W. J., and N. F. Simcic: Computer Controls and End-point Temperature, in Oxygen Steelmaking, *ISA J.*, December, 1961, pp. 38–41.
28. Roth, H.: Analog Computer Systems in Blending Processes, Industrial Electronics Symposium (sponsored by IRE, AIEE, and ISA), Boston, Mass., Sept. 20–21, 1961.
29. Stubbs, G. S.: Design and Use of the Reactivity Computer, *IRE Trans.*, PGNS, NS-4(1), 1957.
30. Tolin, E. D., and D. A. Fluegel: An Analog Computer for On-line Reactor Control, *ISA J.*, October, 1959, pp. 32–38.
31. Henley, A.: A New Heat-flow Transducer for Physiological and Medical Research, *Proc. ISA*, vol. 11, part 2, Paper 56-3-5, 1956.
32. Lupfer, D. E., and M. W. Oglesby: The Application of Dead-time Compensation to a Chemical Reactor for Automatic Control of Production Rate, Digest of Technical Papers, Joint Automatic Control Conference, Boulder, Colo., 1961.
33. Smith, O. J. M.: A Controller to Overcome Dead-time, *ISA J.* February, 1959, pp. 28–33.
34. Lupfer, D. E., and D. E. Berger: "Computer Control of Distillation Reflux," *ISA J.*, June, 1959, pp. 34–39.
35. Lupfer, D. E., and W. W. Oglesby: Automatic Control of Distillation Columns, *Ind. Eng. Chem.*, December, 1961, pp. 963–969.
36. Webber, W. C., R. L. Martin, J. F. Pink, and J. T. Hargett: Analog Computer-Controller for Capacity Control of a Distillation Column, paper presented at a Meeting of the American Petroleum Institute's Division of Refining, Statler Hilton Hotel, New York, May 27, 1959.
37. Roquemore, K. G., and E. E. Eddey: Computer Control of Polymerization in a Rubber Plant, *Chem. Eng. Progr.*, September, 1961, pp. 35–39.
38. Schuerger, T. R.: Computing-Control Applied to a Sintering Process, *Control Eng.*, December, 1956, pp. 77–83.
39. Kompass, E. J.: The Early Bird Goes Automatic, *Control Eng.*, December, 1956, pp. 77–83.
40. Travers, R. H., and S. B. Yochelson: First Computer-controlled Power System, *ISA J.*, October, 1957, pp. 454–458.
41. Pfister, R. C., and E. E. Buder: Computing Control Applied to an Airborne Fire-Controller, *Control Eng.*, September, 1957, pp. 138–143.
42. Rubin, A. I.: Continuous Data Analysis with Analog Computers Using Statistical and Regression Techniques, *Princeton Computation Center Rept.* 160, Electronic Associates, Inc., Long Branch, N.J., December, 1960.
43. Russell, D. W., W. K. McGregor, and L. F. Burns: The Analog Computer as a Process Controller, *Control Eng.*, September, 1957, pp. 160–165.
44. Guilliland, M., G. Oliver, and W. Mower: A New Respiratory CO_2 Response Curve Computer, Digest of the 1961 International Conference on Medical Electronics, New York, July, 1961.
45. Schwartzenberg, J. W.: A Simulator for Nuclear Reactor Operator Training, *ISA J.*, September, 1957, pp. 369–373.
46. Finnigan, R. E., and G. G. Nelson: Project Pluto Control System Developments and Test Results, paper presented at the American Rocket Society Space Flight Report to the Nation, New York, October, 1961.
47. Grabbe, Eugene M., Simon Ramo, and Dean E. Wooldridge: "Handbook of Automation, Computation and Control," John Wiley & Sons, Inc., New York, 1958.
48. Susskind, Alfred K.: "Notes on Analog-Digital Conversion Techniques," John Wiley & Sons, Inc., New York, 1958.
49. British Imperial College of Science and Industry: "Modern Computing Methods," Philosophical Library, Inc., New York, 1959.
50. Berkeley, Edmund C., and Lawrence Wainwright: "Computers: Their Operation and Applications," Reinhold Publishing Corporation, New York, 1956.
51. Fifer, Stanley: "Analog Computation: Vols. I and II," McGraw-Hill Book Company, Inc., New York, 1960.
52. Stibitz, George R., and Jules A. Larrivee: "Mathematics and Computers," McGraw-Hill Book Company, Inc., New York, 1956.
53. Svoboda, Antonin: "Computing Mechanisms and Linkages," McGraw-Hill Book Company, Inc., New York, 1948.
54. Staff: Analog Computers Solve Many Control Problems, *Oil Gas J.*, vol. 57, no. 41, pp. 138–140, Oct. 5, 1959.
55. Dunsmore, Chester L.: Computer Analogs for Common Nonlinearities, *Control Eng.*, vol. 6, no. 10, pp. 109–111, October, 1959.
56. Bertran, J. E.: The Role of Computers in Process Control, *AIChE*, Symposium, Technical Paper No. 40, March 16, 1959.
57. West, G. P.: The Role of Computers in Analysis and Design of Control Systems, *Trans. IRE*, PGAC-5, p. 6, July, 1958.
58. Staff: The Design of DAFT: A Digital/Analogue Function Table, *Automatic Control*, vol. 12, no. 2, p. 46, February, 1960.
59. Staff: Factors Influencing the Design of a Solid State Analog Control Line, *Automatic Control*, vol. 12, no. 3, p. 22, March, 1960.
60. Carden, G.: Data File 36—Analog Computer Circuits for Root Locus, *Control Eng.*, vol. 7, no. 4, p. 127, April, 1960.
61. Staff: Computers, *Automation*, vol. 7, no. 3, p. 47, March, 1960.

62. McRainey, J. H.: Role of Computers in Automation, *Automation*, vol. 7, no. 3, p. 48, March, 1960.
63. Harrower, J. A., Jr.: Designing Pneumatic Analog Computing Systems, *ISA J.*, vol. 9, no. 11, pp. 65–69, November, 1962.
64. Staff: Analog Computer's Role in the Processing Industries, *Automatic Control*, vol. 14, no. 6, pp. 45–47, June, 1961.
65. Manton, F. E. S.: Analog Computer Setup Displays Frequency Response, *Control Eng.*, vol. 9, no. 1, pp. 83–85, January, 1962.
66. Stanton, B. D., and A. Bremer: Analog Computer Measures BTU Content, *Control Eng.*, vol. 9, no. 12, pp. 97–99, December, 1962.
67. Wang, H., and C. Shen: Analog Simulation of True Dry Friction, *Control Eng.*, vol. 9, no. 10, pp. 91–92, October, 1962.
68. Shinskey, F. G.: Analog Computing Control for On-line Applications, *Control Eng.*, vol. 9, no. 11, pp. 71–86, November, 1962.
69. Chapin, Ned: "Introduction to Automatic Computers," D. Van Nostrand Company, Inc., Princeton, N. J., 1961.
70. Aiken, W. S.: Don't Measure, Compute!, *ISA J.*, vol. 10, no. 1, pp. 59–60, January, 1963.
71. Mayer, F. K., and E. H. Spencer: Computer Simulation of Reactor Control, *ISA J.*, vol. 8, no. 7, pp. 58–64, July, 1961.
72. Tomovic, R., and W. J. Karplus: "High Speed Analog Computers," John Wiley & Sons, Inc., New York, 1962.
73. Leondes, C. T. (ed): "Computer Control Systems Technology," McGraw-Hill Book Company, Inc., New York, 1961.
74. Korn, J. D., and G. Hannauer: A Practical Approach to Analog Computers, *Instr. Control Systems*, vol. 35, no. 8, pp. 60–69, August, 1962.
75. Clynes, M.: CAT (Computer of Average Transients), *Instr. Control Systems*, vol. 35, no. 8, pp. 87–91, August, 1962.
76. Barclay, J. E.: Process Control Computers, *Instr. Control Systems*, vol. 36, no. 1, pp. 127–129, January, 1963.
77. Staff: Computer Controlled Systems, *Instr. Control Systems*, vol. 36, no. 5, pp. 85–90, May, 1963.
78. Simulation Council Newsletter, published monthly in *Instr. Control Systems*, covers many aspects of analog computer applications.

Editor's Note: A number of references on computers contain information on both analog and digital types. Hence, the reader should also consider references listed at the end of the section "Applications of Digital Computers" for pertinent material on analog computers.

APPLICATIONS OF DIGITAL COMPUTERS

By J. Russell Le Roy*

Digital computers, with their ability to process the masses of data and measurements necessary to gage conditions and evaluate changes in a complex system, are ideal controllers of production operations and processes. Manufacturing, research, medical, communications, military, and governmental groups are realizing the direct benefits of digital-computer control systems:

1. Reduced operating costs—more efficient utilization of material, power, and equipment.
2. Increased production—greater throughput or yield and reduced unit cost.
3. Better product quality control—less quality give-away, less off-specification product.
4. Labor and equipment savings.
5. Safety—immediate proper corrective action can be taken as well as merely activating alarms or printout of special instructions.
6. Research.
    a. Process improvement resulting from detailed analysis of data.
    b. Development of new processing techniques, now impracticable, which become possible if operation is under computer control.
7. Real-time control.

Digital computers are used to improve operations in processes such as catalytic cracking, catalytic reforming, distillation, and polymerization in the petroleum industry; inorganic, petrochemical, and plastic processes in the chemical industry; natural-gas transmission and steam-electric power-generating plants in the public-utilities industry; rolling-mill, blast-furnace, and open-hearth operations in the steel industry; and pulp and paper processing.

Research groups use digital-computer control systems to study high-rate processes such as chemical reactions which require high-speed data-acquisition and processing systems for their evaluation. Analysis of rocket-engine performance is an example. Automotive and air traffic control, television-program control, missile guidance, and medical diagnostics are areas in which digital-computer systems can be justified by the necessity for real-time control.

This subsection is oriented to process industries' application only because other areas are not covered so well in the literature or so well by "off-the-shelf" equipment.

THE CONTROL PROBLEM

In most industrial operations, even with highly efficient automatic control systems, the task of integrating the controllers and of supervising the process becomes the responsibility of the operator. He is confronted with a dynamic operation; 50 or more automatic controllers, each acting independently; and a great number of factors that must be equated and interrelated. These factors include not only those relating to the chemistry or physics of the process but extrinsic economic ones as well.

Equating so many factors involves complicated, long-drawn-out calculations requiring several hours or days for solution if done by desk calculator. When completed, the answers may be useless, for conditions in the process change too rapidly and

* Control Systems Coordinator, International Business Machines Corporation, Dallas, Texas.

frequently. From a practical standpoint, because of the dynamics of the process, the operator relies on his judgment and experience to offset the disturbances, such as feed-composition changes, impurities, and changes in atmospheric conditions. The time lags in detecting the disturbances, the operator's inability to assess analytically the effects of changes in controller set points, and conservatism in making changes which may cause violation of limits preclude maintaining optimal conditions.

The meaning of "optimum" depends on a management decision. It may be maximum profit, maximum yield, minimum cost, or highest quality. As shown in Fig. 1, the justification for a digital-computer control system depends on its ability to compute the optimum point and maintain the system close to it. The frequency

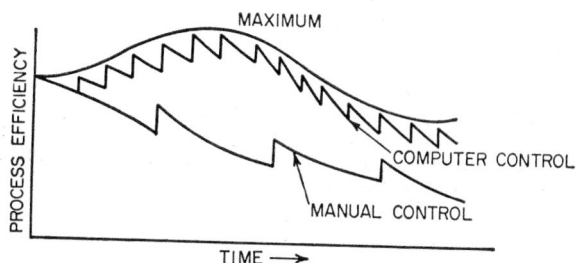

Fig. 1. Effect of computer control.

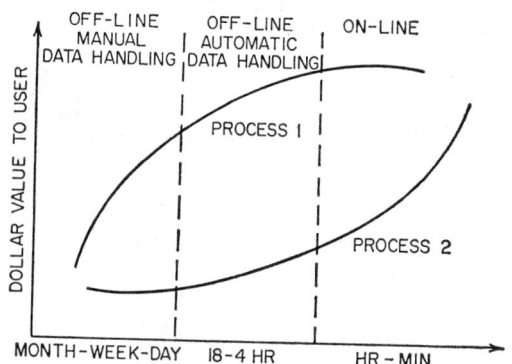

Fig. 2. Frequency of calculation.

of computation must be at least as great as the frequency of disturbance. As illustrated by Fig. 2, this in turn determines the type of control system required.

Characteristics which may be used to identify and to evaluate the applicability of computer control to particular processes are:

1. The annual value of the product is sufficiently large so that even a small percentage improvement is valuable enough to justify computer control.

2. The process is a complex, multivariable one with many production specifications and operating constraints to be observed.

3. The process has a sufficiently long expected lifetime (no foreseeable obsolescence) so that the benefits from computer control will continue for an appreciable time.

4. The process is subject to relatively frequent disturbances which alter the efficiency of operation. Furthermore, it must be possible to adjust the control variables to correct for these changes.

5. The process is technologically advanced, that is, it must be well-understood and well-instrumented so that the necessary quantitative relationships which constitute the model may be obtained and the performance may be measured accurately.

TYPES OF COMPUTER CONTROL

At least three distinct levels exist for applying digital computers to process control. The first level is data reduction, the second is operator guide control, and the third is on-line, closed-loop, process control.

Off-line and On-line

Off-line refers to a computer that is fed data indirectly by means of punched cards, tape, typewriter, or similar equipment. *On-line* refers to a computer that obtains data directly from the process. Degrees between complete off-line and complete on-line and hybrids are possible.

Open-loop and Closed-loop

Open-loop refers to a computer whose output goes to the operator and thence to the process. The system is not, therefore, totally self-correcting. *Closed-loop* refers to a computer that is connected directly to the process. Its output completes the loop to make the system self-correcting.

Data Reduction

As shown in Fig. 3, data reduction involves sampling, analog-to-digital conversion, systematic recording of instrument readings, alarming, data transformation, and possible retransmission of this collected information. Such systems are described in more detail on pages 11-57 to 11-100.

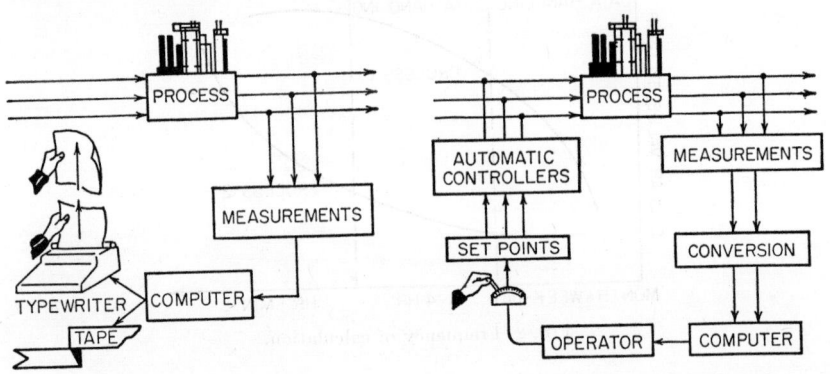

FIG. 3. Data-reduction system.   FIG. 4. Operator-guide control.

A digital computer integrally tied to the data-collection unit provides a powerful system for analyzing process data by mathematical techniques. Empirical methods, correlation techniques, and/or regression analysis of the process data can give better understanding of the process and lead to the construction of a mathematical model.[6,16,18,32,62,82]

Operator-guide Control

This type, shown in Fig. 4, includes data reduction as well as the calculation of guides to assist the operator in controlling the process. The two applications differ principally in the use that is made of the computer. Operator-guide control requires a greater understanding of the interplay between variables.

Typical calculations associated with operator-guide control may be those involved with on-line analyzers such as chromatographs or mass spectrometers, calculation of secondary variables which are functionally related to the measured variable, and calculation of other quantities such as heat and material balances that are of value to the operator in adjusting the process for greater operating efficiency or quality control.

Closed-loop Control

This type, shown in Fig. 5, allows the process to be monitored and controlled by the digital computer on a real-time basis. Immediate action is taken by the computer to maintain the process at its optimum operating point. All elements of operator-guide control are encompassed in closed-loop control, and in addition, electrical outputs to the process from the computer allow the plant to be brought swiftly and safely to its optimum operating point as conditions in the plant change.

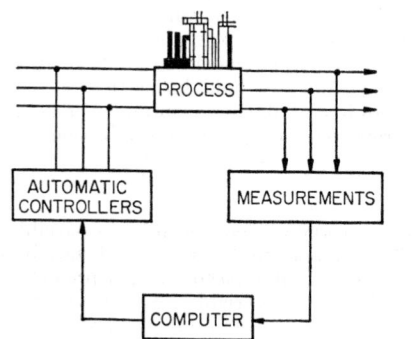

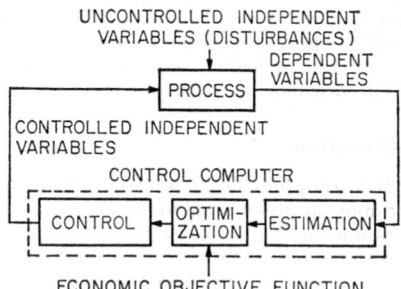

Fig. 5. Closed-loop control.　　Fig. 6. Functional block diagram of a process computer-control system.

Functions performed under computer control may include start-up, shutdown, emergency sequence control, normal process control, and production scheduling— all in an optimal manner.

COMPUTER-CONTROL-SYSTEM FUNCTIONS

As shown in Fig. 6, the essential functions performed by the digital-computer control system are *estimation, optimization,* and *control.* In the estimation step, information from the process is converted to meaningful units, e.g., pounds, pressures, temperatures, and levels, and the performance status of the process is determined. Optimization refers to determination of optimum operating conditions within the region defined by the objective function (a mathematical statement of economic and/or management goals) and constraints. The control function provides for determining adjustments to be made, testing for effects of such changes on systems stability, and making changes in controller set points.

A mathematical model which characterizes the process is an essential element in programming the computer to estimate, to optimize, and to control an operation.

Mathematical Model

The mathematical model describes the interrelationships that are relevant to control of the process within the region of economic, physical, and management objectives and restrictions. The relationships of the independent variables to the dependent variables, expressed as functions, make it possible to evaluate in advance the effects each manipulation of an input will have on the output.

The mathematical model may be developed either analytically or empirically. In the analytical approach, attempts are made to write the relationships between the desired variables, e.g., heat and material balances or reaction equations, without resorting to plant data. This approach presumes an expert knowledge of the technology of the process and the fundamental physical and chemical phenomena that govern it. Such complete knowledge rarely exists for most continuous processes.

The empirical approach is necessary where there is insufficient knowledge about the process to develop a model analytically or where the mathematical difficulties are insurmountable. It uses data obtained from the plant during normal operation or during experimental runs and empirically determines, by regression analysis, the parameters of the chosen functions.

Use of data obtained during normal operation of the plant may seem easiest, but in practice, use of information resulting from carefully preplanned experimental runs with created disturbances has proved far more effective.

In general, a steady-state, or static, model will be required in order to perform the optimization. A static model is based on operations after transients have subsided. In some cases, depending on considerations discussed under the control function, a dynamic model will be required. A dynamic model contains the static model within it but gives, in addition, the relationships that hold during the transient state. In particular, a dynamic model is required if dynamic optimal control is required while a static model is sufficient for repeated steady-state optimization.

Estimation

Before optimizing a process, it is necessary to know what the process variables are doing at present and to predict what is going to happen to the process if all variables continue in the direction they are going. The computer performs this predictive function by reading instruments attached to the process variables, by extrapolating or interpolating other values calculated or obtained from lookup tables stored within it, and by making further calculations.

Quantities like boiler efficiency, catalyst activity, and others, which summarize a considerable amount of information about the state of the plant, are very useful if made available immediately to operators for them to then make appropriate adjustments to the controllers. Providing rapidly computed operating guides of this kind can improve process performance and may be regarded as an early stage in the logical sequence toward on-line, closed-loop digital-computer control.

A further step performed in the estimation function consists of data editing and correction, checking for consistency to eliminate gross errors, giving an alarm when a variable that is measured directly or is calculated exceeds the upper or lower limits established, and periodically printing out other quantities to satisfy accounting requirements.

Optimization

The mathematical model, objective function, and constraints are the keys to process optimization. The mathematical model defines the relationship of variables in the process and allows computations of changes required to effect optimization. The objective function defines management's goals such as maximum conversion, production or yield, or minimum unit cost. In other words, the mathematical function representing the objective function is to be optimized. Invariably constraints or limits will be placed on the operation. Such constraints may represent physical limitations imposed by the equipment used, or limits set by product specifications.

Figure 7 shows part of the model of a hypothetical process with two controllable variables in which the output we are trying to maximize (lb/hr of product) depends on the controllable variables through the relationship shown on the graph. Any fixed point represents a choice of fixed values of flow 1 and flow 2. The corresponding value of the process throughput is indicated by the labeled contour lines, which are curves of constant throughput. There is an upper limit on each flow, represented by straight lines, and a limit on product purity represented by the line along which the impurity is just equal to its specified upper limit.

The region inside the constraints is called the feasible region. Any point within this region corresponds to a choice of values for flow 1 and flow 2 for which the process will operate without violating any of the constraints. Optimization now consists of finding the point within the feasible region which maximizes the objective function, that is, the values of flow 1 and flow 2 not exceeding their maximum values that produce the maximum pounds of product per hour without exceeding the impurity constraint.

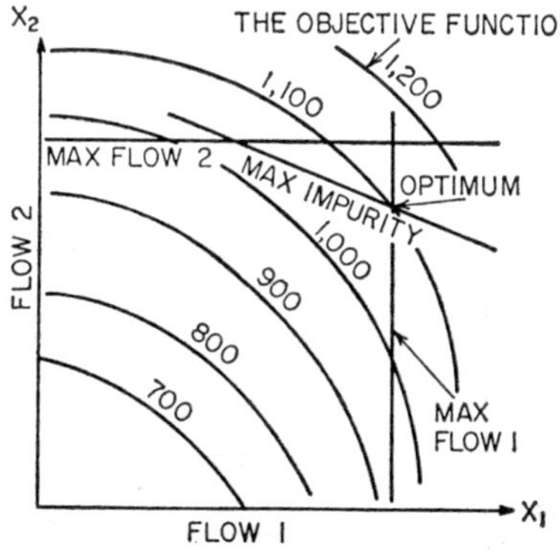

We can find this point readily from our graphic representation. However, when more than a small handful of variables and constraints are involved, the problem can no longer be handled effectively in this way.

If all the constraints and the objective functions are linear, then finding the maximum within the feasible region is a straightforward linear-programming problem.[43,70]

Fig. 7. The constrained optimum.

This problem can be solved by the simplex method of Dantzig for any number of variables and constraints, the only limitation being computing capability.

When the constraints or the objective function are nonlinear, other techniques must be investigated such as gradient methods, Lagrange multipliers, quadratic and nonlinear programming, calculus of variations, and dynamic programming.[5,7,20,30,50,54,70]

Control

The type of control scheme needed depends on:
1. The frequency of disturbances.
2. Dynamic response of the process to changes in the controllable variables.

The frequency of optimization should be rapid enough to ensure that the process is not disturbed away from its optimum position by an economically significant amount during the interval between optimizations.

COMPUTER-CONTROL-SYSTEM CONFIGURATION

Figure 8 is a schematic representation of an industrial process and its on-line closed-loop operation by a computer-control system. The section above the dotted line represents the process without computer control. The process operator takes visual readings from the instruments, manually records and mentally evaluates the readings, and then manually adjusts the set points of the automatic controllers.

The section below the dotted line represents computer control. The instrument readings in the form of electrical signals are automatically sent through the terminal unit, the I/O (input/output) unit, to the computer for evaluation. Analog signals pass through the analog-to-digital converter (ADC). Priority signals (disturbances such as violation of an operational constraint) pass through the interrupt device which causes the computer to evaluate them and initiate corrective action immediately.

Analog signals to the process originate as decimal digits from the computer. They are converted by the digital-to-analog converters (DAC's). The real-time clock can be scheduled, by programming, to interrupt the computer program and initiate control operations at regularly timed intervals. Programs and data are read and punched on paper-tape reader and punch. The process and the computer system are directed either from the process operator's console or from the computer console.

The digital-computer-control system, as in all data-processing systems, can be divided into four types of functional units: *input devices, output devices, storage,* and *central processing unit.*

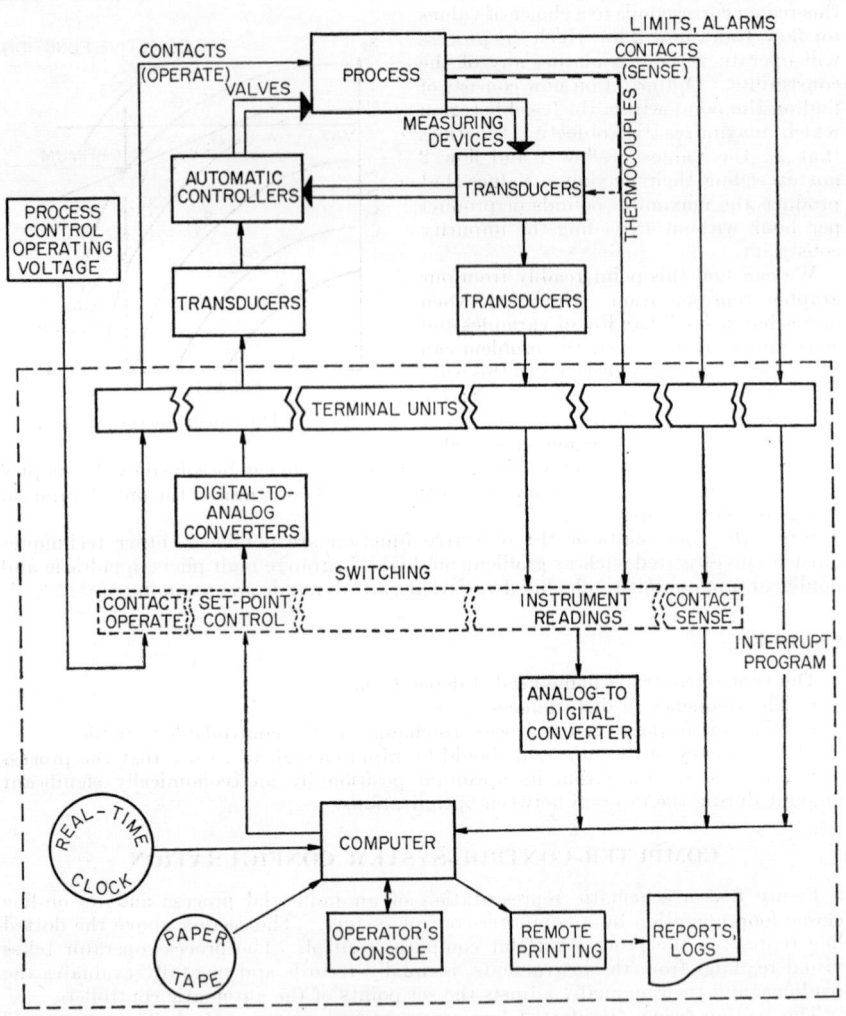

Fig. 8. Elements of a computer-control system.

Input and Output Devices

Input devices read or sense coded data that are recorded on a prescribed medium and make this information available to the computer.   Data for input are recorded in cards and paper tape as punched holes, on magnetic tape as magnetized spots along the length of the tape, and on paper documents as characters printed in magnetic ink.   In the case of the process-control computer, data from an operation can be transmitted through filtering and conversion equipment directly from plant to computer.

Output devices record or write information from the computer on cards, paper tape, magnetic tape, or as printed information on paper. Again in the case of the process-control computer, output signals can be utilized to change conditions in a plant or operation.

Storage

Storage is somewhat like an electronic filing cabinet, completely indexed and almost instantaneously accessible to the computer. All data must be placed in storage before they can be processed by the computer. Information is read into storage by an input unit and is then available for internal processing. Each location, position, or section of storage is numbered so that the stored data can be readily located by the computer as needed.

Central Processing Unit

The central processing unit (CPU) is the controlling center of the entire data-processing system. It can be divided into two parts:
1. The arithmetic-logical unit.
2. The control section.
The arithmetic-logical unit performs such operations as addition, subtraction, multiplication, division, shifting, transferring, comparing, and storing. It also has logical ability—the ability to test various conditions encountered during processing and to take action called for by the result.
The control section directs and coordinates the entire computer system as a single multipurpose machine. These functions involve controlling the input-output units and the arithmetic-logical operation of the CPU, and transferring data to and from storage, within given design limits.

Stored Programs

Each data-processing system is designed to perform only a specific number and type of operations. It is directed to perform each operation by an instruction. The entire series of instructions required to complete a given procedure is known as a program. When placed in the main storage unit in a form that can be interpreted by the computer, it is termed a stored program. The possible variations of a stored program provide the data-processing system with almost unlimited flexibility.

Console

The console provides external control of the data-processing system. Keys turn power on or off, start or stop operation, and control various devices in the system. Data may be entered directly by manually depressing keys. Lights are provided so that data in the system may be visually displayed. The system may also be operated from the console to trace or check out a procedure one step at a time.

Operational Storage Devices

Three types of storage devices are commonly used: core, magnetic drum, and magnetic disk.
Information can be placed into, held in, or removed from computer storage as needed. The information can be:
1. Instructions to direct the CPU.
2. Data (input, in-process, or output).
3. Reference data associated with processing (tables, code charts, constant factors, and so on).

Storage is classified as either main or auxiliary.

When information enters a location, it replaces the previous contents of that location. However, when information is taken from a location, the contents remain unaltered. Thus, once located in storage, the same data may be used may times.

The computer requires some time, referred to as access time, to locate and transfer information to or from storage. In some units access time is so brief that it is measured in millionths of a second.

Magnetic-core Storage. A magnetic core is a tiny ring of ferromagnetic material which can be easily magnetized in a few millionths of a second. Unless deliberately changed, it retains its magnetism indefinitely.

Most modern computing systems utilize magnetic-core storage as the main memory to achieve the high speeds required in complex applications.

Magnetic-drum Storage. A magnetic drum is a steel cylinder enclosed in a copper sleeve. The copper surface is plated with a cobalt and nickel alloy. If an area of this material is placed momentarily in a magnetic field, it becomes magnetized and remains so indefinitely. As the drum rotates at a constant speed, information is written by magnetizing spots or cells as the surface passes a read-write head. When a cell that has been magnetized passes under the read-write head, its magnetic state can be sensed or read by current induced in the read coil.

Magnetic drums were used as the main memory in early computer systems. Because of the inherent inefficiency associated with "recirculating" memories, they are now used primarily as auxiliary memory units.

Magnetic-disk Storage. The magnetic disk is a thin metal disk coated on both sides with a ferrous oxide recording material. Data are stored as magnetized spots located in concentric tracks on each face of the disk.

The advantages of magnetic-disk memory are the relatively low cost for very large capacity and the ability to make large volumes of data randomly accessible when required.

Terminal and Multiplexing Units

These units provide electrical connection of the computer-control system to the process. Signals from the process transmitted by air pressure or some other non-electrical medium are transduced, i.e., converted to electrical signals. Electrical analog or binary signals from instruments or transducers are connected to pairs of terminals in the units. Thermocouples, regardless of type, connect to a special type of terminal block containing a resistance thermometer to provide a reference temperature. In addition to measured values, the status, whether open or closed, of any contact or switch, or presence or absence of a pulse connected to the terminal units, can be "sensed" by the computer.

Output electrical analog signals representing control actions determined by the computer are connected at the terminal units to the devices or controller set points to be operated. Plant pumps, motors, relays, bells, and visual alarms to be controlled by the system are connected to the terminal units for operation by plant voltages switched through computer-operated relay points.

Analog-to-Digital Converter

This unit converts the input analog signals to digital values for use by the computer, while the computer performs other computations. Only one ADC is necessary in most applications because the computer selects only one signal at a time from the process. Conversions can be into either decimal or binary numbers.

The output of ADC's is in the form of four decimal digits plus sign, which provides the precision consistent with current instrumentation and application technologies. If analog-to-digital conversions are in binary form, output is 13 bits plus sign, which is approximately equivalent to four decimal digits. Special applications may, of course, require different degrees of precision.

Digital-to-Analog Converters

The DAC's convert computer-determined set points to electrical analog signals and maintain the signal while the computer performs other computations. One converter is provided for each set point or other analog output to be sent to the process.

Process Operator's Console

The console provides a typewriter for recording console operations and data. It also provides (1) lights for indicating control status, (2) switches for modifying the control procedure and for initiating special procedures, and (3) a decimal keyboard for entering special data.

Additional Typewriters and Printers

These are available to provide additional condition reporting, logging, and data recording.

Interrupt

This feature permits process alarms and certain machine conditions to interrupt computation and initiate special program action immediately. Each condition is treated in order of importance regardless of the sequence of program instructions.

Real-time Clock

The real-time clock provides computer-readable digits for computation of elapsed time between unscheduled events, and furnishes signals to the program at selected intervals to initiate scheduled cycles of computation and data collection.

PROGRAMMING A COMPUTER-CONTROL SYSTEM

Programming the digital computer to perform the control functions may be organized into the following six phases:

Phase Zero: Input, Output, and Program Control

This phase initiates and controls data scanning and recording cycles at specific time intervals. It also controls scanning cycles initiated manually by the operator. Subroutines are included that assist the computer programmers and process operators on procedural matters such as making manual changes to product prices, management objectives, and other data.

Phase I: Data Conversion and Editing

The initial part of the evaluation function (determination of the present position of the process or plant) is performed in this phase. When required by the program, the arbitrary ADC digital values of sampled data (digital readouts, DRO's) are converted to meaningful units such as degrees Fahrenheit, barrels per day, or pounds per square inch. In addition, the data are checked for violation of any imposed limit conditions, for reasonableness, and for validity. Inconsistent data are corrected for subsequent calculations or are used to initiate alarms or special actions. The check of the thermocouple DRO's is an example of the type of limit checking performed on all data conversions.

Phase II: Computation of the Present Operating Condition

Values of unmeasurable but computable variables are computed by means of heat- and material-balance equations and other analytical, empirical, or estimated relationships established during the preliminary study of the process. In addition, the physical-equipment limitations that are currently restricting the values of dependent variables are determined.

Phase III: Optimization

The entire function of optimization is performed in this phase of the program. Using some of the information developed in the evaluation function, the program determines by mathematical techniques whether there is room for improvement without violating limits, computes a move or change in operating conditions that will improve the profits, and determines if this is the "best" move. If it is not a desirable move, that is, if a constraint is violated, recalculation is performed.

Phase IV: Smoothing and Control

To ensure stability of the process, computed control changes are smoothed. Large changes are replaced by several small changes. (In some models, the desired change is dynamically evaluated and applied in increments as large and as rapidly as possible rather than in equal increments.) Changes are translated to set points and are sent to controllers under control of phase zero.

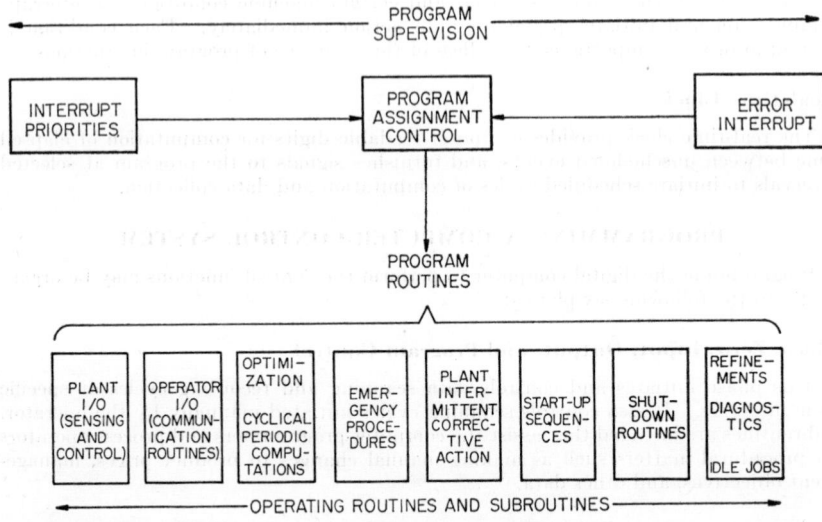

FIG. 9. Computer-control-program routines.

Phase V: Reports and Logs

This phase prepares any intelligible output such as operating logs and condition reports. Actual typing is under control of phase zero. Figure 9 illustrates graphically the magnitude of the programming problem and the need for a master supervisory control or executive program in a sophisticated computer-control application.

The Executive Program

The executive program for a control system should be a highly versatile set of routines. These routines would provide for automatic sequencing of events and consist of a master interrupt control routine which in turn handles such functions as output control, error control, diagnostic control, and main-line interrupt programs.

An executive program, usually provided by the computer vendor, contributes so significantly to the over-all effectiveness of the control system that it should be an integral part of the system specifications.

The main functions of an executive program are described briefly here.

Interrupt Control. The interrupt control routine will decode and sequence interrupt requests. In addition, this section of the executive program provides the communication path to interrupt subroutines.

Input/Output Control. One of the recognizable interrupts is caused by any input or output device. The I/O control program is automatically executed, and from the user's standpoint, he merely specifies the appropriate parameters regarding the message to be transmitted. For output control, an output format generator routine assembles and records the message(s) as specified by the programmer.

Error Control. One of the most important functions of the executive program is the automatic handling of errors. Despite significant design improvements in computers, it is doubtful that 100 per cent error-free systems will exist in the near future. For this reason, it is extremely important that a computer-control system be internally self-checked so that system errors can be automatically detected and corrective action initiated.

Another major function of the error-control routine is to establish an efficient and sound restart procedure for each program or section of the program.

Diagnostic Control. In most process-control applications there is time when the control system is not needed for its primary function. To further contribute to nonstop operation, the executive program includes a diagnostic control routine. The diagnostic control routine allows the programmer or operator to call in automatically and execute various diagnostic routines during the time the control system is not required for process control or monitoring. Upon completion, the diagnostic control program automatically returns control to that section of the main program where it left off.

SPECIFYING A DIGITAL-COMPUTER-CONTROL SYSTEM

Key considerations in specifying a digital-computer-control system follow.

Computer Speed

This is a function of internal logic, instruction type, and average memory access time. Questions such as the following should be asked:

Can input, output and computing be partially or completely overlapped?

Are multiple functions such as reading data, computing, and storing results possible with one instruction or does it take three or more?

What is the average time required to read or fetch data from the computer storage?

Scanning Speed

Control systems use both relay and electronic switching in their input multiplexers. Input signals acceptable to the ADC may be either low-level (millivolt) or high-level (volt). The latter may require amplification (at additional cost) since most instruments provide low-level signals. Present technology and control systems offer switching speeds of from 5 to 20 points or samples/sec on low-level signals. For high-level signals, manufacturers indicate that switching speeds as high as several million samples per record can be attained.

Memory Capacity

A main working memory, usually high-speed magnetic-core, and a random-access auxiliary memory, either magnetic-drums or disk-file, are required by most control systems. The problem, obviously, is to arrive at the optimum ratio of main or relatively high cost to auxiliary or relatively low cost memory requirements. That is, we wish to minimize equipment costs without undue sacrifice in performance of the control system. Factors to be considered in establishing this ratio are flexibility in ability to communicate and make transfers between main and auxiliary memories, access times, internal control and data checks, and ability to optimize file organization.

Word size can also play an important role in system performance. In most optimization routines, linear programming or other iterative techniques are employed. These usually introduce a problem of "round-off" error if the computer word size is too small and may force the use of so-called double-precision arithmetic with a resultant increase in over-all computing time.

Programming Systems

Continuous maintenance and refinement of control-systems programs are usually necessary to obtain most efficient uses of control equipment. This updating is required because of changes in constants, in operating procedures, in economic factors (or objective function), and in process technology. In the extremum, complete rewriting of the control program may be required. The availability of symbolic assembly programs and compilers should be carefully investigated and evaluated in specifying a system. These systems reduce the time and effort of the programmer in going from problem definition to programming, to debugging, to correcting, to operational status.

An additional "software" consideration is access to libraries of general-purpose computer routines available free of charge from several manufacturers.

APPLICATIONS OF DIGITAL-COMPUTER-CONTROL SYSTEMS

The following list of processes and operations indicates the broad base of applications of digital computers to control.[27,59] Several of these are discussed briefly in subsequent paragraphs.

Petroleum Processing.[47]
Crude-oil distillation.[66,31,54]
Catalytic cracking.[21]
Alkylation.[4]
Re-forming.[55,74]
Polymerization.[60]
Gas and liquid products pipeline operation.[52]
Drilling oil wells.
Chemical Industries.[34,40]
Ammonia.[22,37]
Cement.[33,64]
Petrochemicals.[1,58,61,63,69]
Plastics.
Pulp and paper.[24,26,46]
Synthetic fibers.
Synthetic rubber.
Metallurgical Industries.
1. Iron and steel production.
   Blast furnaces.
   Open-hearth operations.[18]
   Rolling mills.[11,13]
   Annealing line.[9]
   Tin-plating line.
2. Nonferrous metals.
   Ore processing.
   Smelting.
   Refining.
Utilities.
Power generation.[48,76,84]
Economic dispatch of power. (See page 14-45.)
Aircraft and Missiles.
Components testing.[77]
Wind-tunnel studies.
Missile-test-stand studies.
Air traffic control.[75]

Other.

Quality control of electronic components.[2,25]
Medical research and diagnostics.[14,15]
Nuclear-energy processes.[3,73]
Television-station program switching.[71]
Highway traffic control.
Railroad classification yard control.[41]
Weather reconnaissance.
Manufacturing industries control and components testing.[28]
Navigational control of ocean-going vessels.
Military and naval systems control.
Communications.[72]

Petroleum-refining Applications

Application of digital-computer-control systems to two of the more common refining processes, crude-oil distillation and fluid catalytic cracking, will be discussed briefly.

Crude-oil Distillation. The compositions of crude petroleums vary widely in composition from almost gaslike materials to semisolid asphaltic materials. One of the first major steps in petroleum processing is separation of the crude oil into various portions according to boiling range.[45,65]

Computer control of American Oil Company's 140,000 bbl/day crude-oil distillation system in Whiting, Ind., is described briefly here. A detailed description of this application of computer control is given in Refs. 31, 54, and 66.

The program leading to installation of this closed-loop digital-computer-control system is an excellent example of the methodical approach required to evaluate a large-scale complex operation. A generalized outline follows:

1. Obtain management decision.
2. Assign technical staff.
3. Make preliminary evaluation.
4. Select operation or process.
5. Analyze process.
6. Develop preliminary mathematical model.
7. Simulate, using off-line computer.
8. Obtain additional performance data.
9. Refine mathematical model.
10. Review economics.
11. Specify and order computer-control system.
12. Control plant on off-line open-loop basis.
13. Install closed-loop system.
14. Test system and update model.
15. Operate, using closed-loop control.

The control problem involves basically:

1. Frequent changes within refinery process units to keep pace with variations in demand for specific products.
2. Ability to meet production goals without waste of potential product or fuel.

Only a part of the information is readily available unless a computer-control system is used. As shown in Fig. 10, the basic functions of the computer are:

1. *Compute Present Position.* This provides a solid base for solving the control problems in the form of precise estimates of variables affected by computer-guided controls. Product qualities from laboratory reports are used when available to update equations in an after-the-fact fashion.

2. *Define Objective.* This provides an economic equation for predicting the over-all economic effects of adjusting computer-guided controls, and the limits that restrict adjustment in a given situation. These are derived from computed variables and the latest information from refinery management regarding productions goals and relative values or products.

3. *Optimize.* This involves finding how to adjust control settings for best results corresponding to maximum economic improvement.

The general sequence of producing information is outlined in Fig. 11.

Two of the methods mentioned earlier for solving the problem of optimization under constraint were applied. The classical Lagrange method of undetermined multipliers was used to adjust flow and enthalpy measurements to minimize errors in heat and material balances. A modified linear-programming technique was used to determine the optimum operating position.

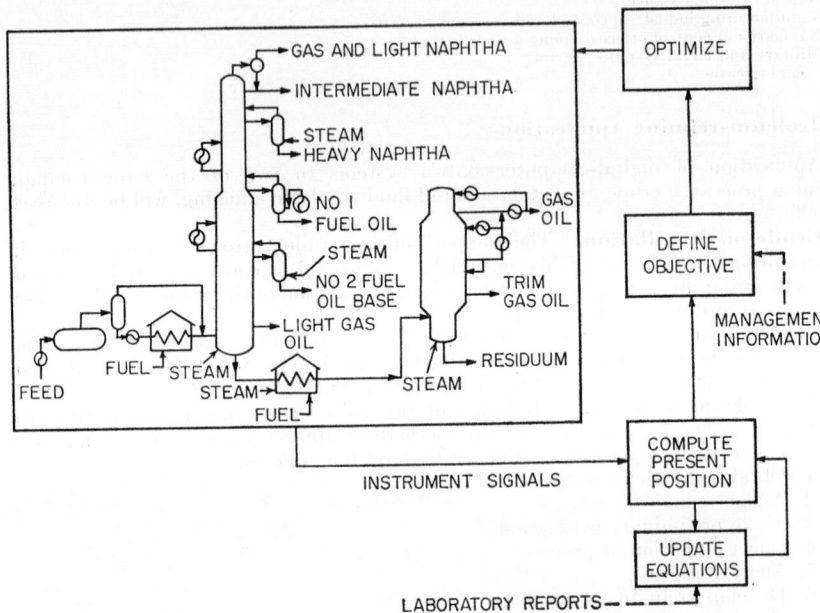

Fig. 10. Basic functions of a computer-control system for crude distillation.

The objective function is derived by considering the economic incentive associated with:
1. Improved yield of higher-value products.
2. Reduced operating costs.
3. Increased service life.
4. Increased capacity.
Constraints include:
1. Physical limits such as furnace tube-skin temperatures.
2. Management limits such as quantity and/or quality of raw materials and products.
3. Move limits or the maximum changes in the variables which can be practically implemented in any one control period.
Direct benefits totaling $\frac{1}{2}$ to 1 cent/bbl of crude oil processed (before taxes or computing costs) are obtained. Main sources are over-all increase in yield of the more valuable primary products and increased capacity.

Fluid-catalytic-cracking Process. The purpose of this process is to convert to gasoline some of the less valuable products emerging from distillation of the crude oil. The heavier hydrocarbon constituents usually add up to more than half, yet their demand and economic value are less than those of gasoline. To make the distillation products more nearly match the products desired, the refiner processes the oil in the catalytic cracker. Most catalytic-cracking processes can convert 50 to 75 per cent of the heavier hydrocarbons into gasoline.

The hydrocarbons under high temperatures vaporize and the molecules break up. The fluid-catalytic-cracking process, using a fine powder as catalyst, speeds up the

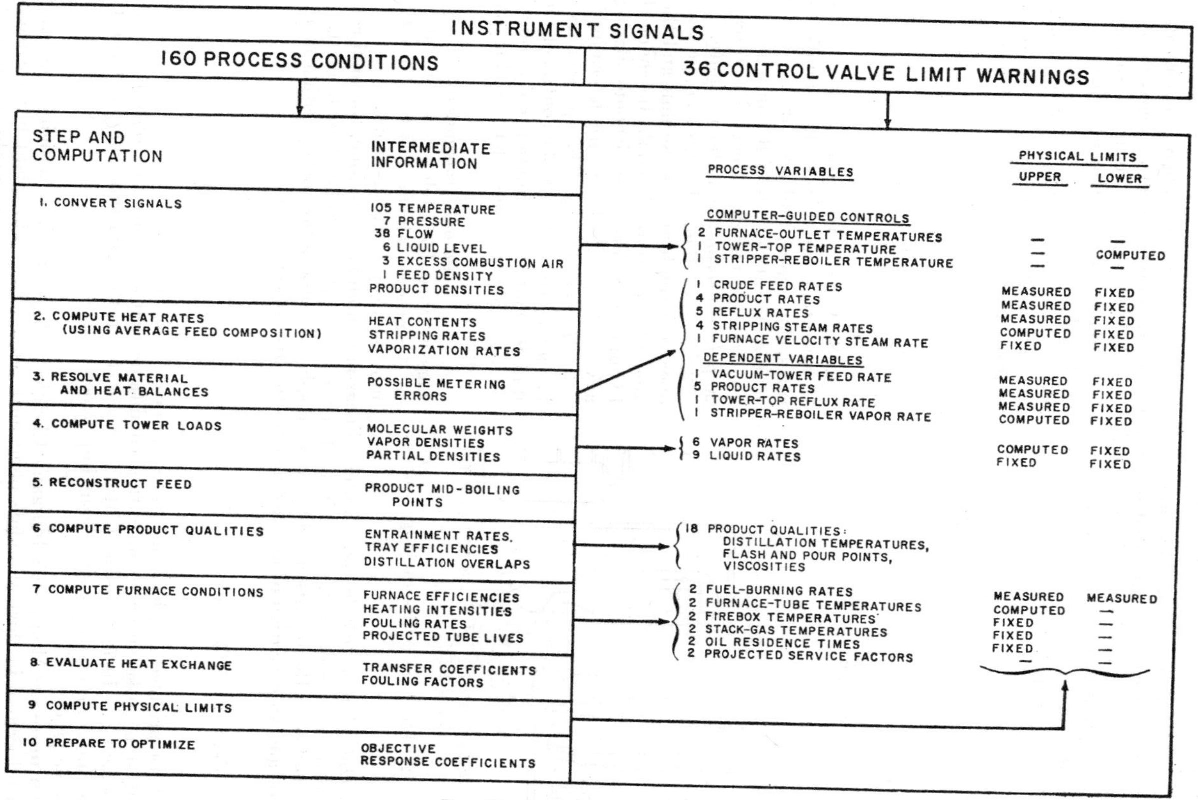

INSTRUMENT SIGNALS

160 PROCESS CONDITIONS	36 CONTROL VALVE LIMIT WARNINGS

STEP AND COMPUTATION | **INTERMEDIATE INFORMATION**

1. CONVERT SIGNALS — 105 TEMPERATURE / 7 PRESSURE / 38 FLOW / 6 LIQUID LEVEL / 3 EXCESS COMBUSTION AIR / 1 FEED DENSITY / PRODUCT DENSITIES

2. COMPUTE HEAT RATES (USING AVERAGE FEED COMPOSITION) — HEAT CONTENTS / STRIPPING RATES / VAPORIZATION RATES

3. RESOLVE MATERIAL AND HEAT BALANCES — POSSIBLE METERING ERRORS

4. COMPUTE TOWER LOADS — MOLECULAR WEIGHTS / VAPOR DENSITIES / PARTIAL DENSITIES

5. RECONSTRUCT FEED — PRODUCT MID-BOILING POINTS

6. COMPUTE PRODUCT QUALITIES — ENTRAINMENT RATES / TRAY EFFICIENCIES / DISTILLATION OVERLAPS

7. COMPUTE FURNACE CONDITIONS — FURNACE EFFICIENCIES / HEATING INTENSITIES / FOULING RATES / PROJECTED TUBE LIVES

8. EVALUATE HEAT EXCHANGE — TRANSFER COEFFICIENTS / FOULING FACTORS

9. COMPUTE PHYSICAL LIMITS

10. PREPARE TO OPTIMIZE — OBJECTIVE RESPONSE COEFFICIENTS

PROCESS VARIABLES	PHYSICAL LIMITS UPPER	LOWER
COMPUTER-GUIDED CONTROLS		
2 FURNACE-OUTLET TEMPERATURES	—	—
1 TOWER-TOP TEMPERATURE	=	=
1 STRIPPER-REBOILER TEMPERATURE	—	COMPUTED —
1 CRUDE FEED RATES	MEASURED	FIXED
4 PRODUCT RATES	MEASURED	FIXED
5 REFLUX RATES	MEASURED	FIXED
4 STRIPPING STEAM RATES	COMPUTED	FIXED
1 FURNACE VELOCITY STEAM RATE	FIXED	FIXED
DEPENDENT VARIABLES		
1 VACUUM-TOWER FEED RATE	MEASURED	FIXED
5 PRODUCT RATES	MEASURED	FIXED
1 TOWER-TOP REFLUX RATE	MEASURED	FIXED
1 STRIPPER-REBOILER VAPOR RATE	COMPUTED	FIXED
6 VAPOR RATES	COMPUTED	FIXED
9 LIQUID RATES	FIXED	FIXED
18 PRODUCT QUALITIES: DISTILLATION TEMPERATURES, FLASH AND POUR POINTS, VISCOSITIES		
2 FUEL-BURNING RATES	MEASURED	MEASURED
2 FURNACE-TUBE TEMPERATURES	COMPUTED	
2 FIREBOX TEMPERATURES	FIXED	
2 STACK-GAS TEMPERATURES	FIXED	
2 OIL RESIDENCE TIMES	FIXED	
2 PROJECTED SERVICE FACTORS		

FIG. 11. Process computations.

reaction selectively to produce more gasoline. The catalyst remains chemically unchanged and can be separated easily and reused continually after regeneration. The heavier hydrocarbon molecules of fuel oil that are cracked result in smaller, lighter, differently arranged molecules, which form some of the constituents of gasoline. The main elements of the process are shown in Fig. 12.

A digital-computer-control system is used at Standard Oil Co. of California's El Segundo, Calif., refinery, to control a 40,000-bbl fluid catalytic cracker in a closed-loop system. A medium-sized solid-state computer with a memory capacity of 1 million digits is used.

Under normal operating conditions the computer goes through six steps every 15 to 20 min.

1. *Scanning.* The computer checks 75 operating variables such as temperatures, flow rates, and pressures. These are scanned at the rate of 40 points a minute, but the unit has a capacity to scan 20 points a second or 1,200 a minute.

2. *Initialization.* In this step, present operating conditions of the unit and the amount each control should be changed to improve performance without violating any of the plant's limits are determined.

3. *Manual entries.* The operator now has a chance to change the control of the unit by adding information not scanned or by making a new demand for the computer.

4. *Accumulations calculation.* Readings at this stage tell where the unit is in the process scheme and how fast it is moving away from this point.

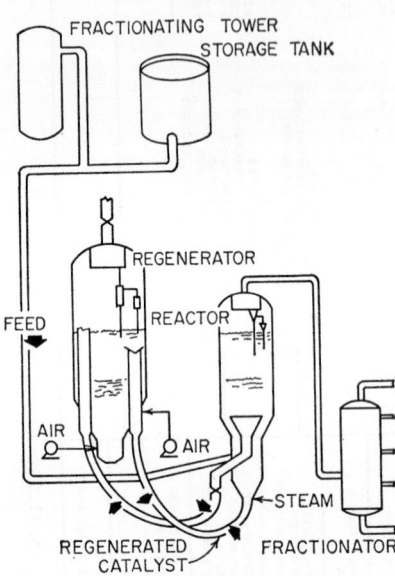

FIG. 12. Fluid-catalytic-cracking process.

5. *Optimization calculation.* Using a linear-programming technique, the final control figures are produced, telling just what the unit ought to do and how to do it.

6. *Output.* Represents a print-out or display of information on operation of the plant, and an execution of process changes to effect optimization.

Chemicals and Petrochemicals

Acrylonitrile. E. I. Du Pont has used a computer to control a large pilot plant built to produce acrylonitrile by a new catalytic process. The efficiency of the catalyst is so sensitive to operating conditions that conventional controls are not satisfactory. In fact, research studies have shown that dynamic computer control is required to achieve a significant improvement in the process and a commercially feasible process.

Ammonia. Monsanto's Inorganic Chemicals Division utilizes a digital-computer-control system in their ammonia plant near Luling, La. This plant has a capacity of about 500 tons of ammonia per day. A typical high-pressure synthesis is used.[22,37] The primary control variables of the Monsanto ammonia process are:

1. Reaction-gas flow to primary reformer.
2. Air flow to secondary reformer.
3. Fuel-gas flow to secondary air blower.
4. High-pressure steam to primary reformer.
5. Low-pressure steam to CO (carbon monoxide) converter.
6. Fuel-gas flow to primary reformer furnace.
7. Fuel-air ratio in primary reformer furnace.

Other important control factors include synthesis-gas feed rate, hydrogen-to-nitrogen ratio at the synthesis reactors and at the CO converter, fuel-gas content at the CO converter, and secondary re-former temperature.

After the important control variables were analyzed, the more important objectives to be met by the computer were found to be: maintain maximum gas flow in spite of changing weather and process conditions; maintain optimum hydrogen-to-nitrogen ratio; immediately and safely reduce raw-materials flow to compensate for any loss of compression; provide fail-safe devices so that instrument or computer malfunctions are detected, and are prevented from affecting the process.

Cement. Portland cement may be described in simple terms as a closely controlled chemical combination of lime, silica, alumina, iron oxide, and small amounts of other ingredients.   A small quantity of gypsum is interground in the final grinding process to regulate the setting time of the cement.   The most common sources of lime are limestone, shells, and marl.   The silica, alumina, and iron oxides are obtained from shale, clay, and sandstones.   Blast-furnace slag, the so-called alkali wastes, and iron ore may also be used as raw materials.

The control problem in cement manufacture is to produce specification product at minimum cost with raw materials having variable compositions.

The most significant economic gain that may be rapidly and directly achieved with the aid of a computer lies in economic blending of the basic raw materials.   The savings potential in this area lie in:

1. Maximizing the use of the least expensive constituents within the chemical requirements (constraints) for the mix.

2. Operating with mix compositions that are consistently close to the required specification limits rather than a safe margin higher than required.

3. Minimizing upsets in the rest of the plant (particularly the kiln) due to composition fluctuations.

4. Eliminating the labor and plant operating time required to upblend selectively off-specification raw mix.

The computer solution of the optimum blend problem can be divided then into three phases:

1. The analyses of the raw materials form the input to a program which calculates the coefficients for a group of linear equations.

2. These linear equations are then solved using a linear-programming technique to determine the optimum weight percentages of each raw material in the optimum blend for the given raw-material analysis.   The optimality criterion used is minimum cost, but others are possible, of course.

3. The weight-percentage values are scaled as a function of the present desired plant feed rate and converted to analog signals which will be transmitted to the closed-loop controllers on the proportioning equipment.

Figure 13 shows a proposed layout for closed-loop computer control of the cement-plant blending operation.

Features of blending operation control system are:

1. Input chemical-analysis information coming directly to the computer from X-ray fluorescence spectrographic equipment which analyzes the material and/or blend on each feed belt.

2. Vibrating feeders and weigh belts are employed in a control system for the blending operation.   Feed rates are held at levels determined by the computer by an automatic controller.   This instrument utilizes continuous feedback of actual weight information to control the vibrating feeders.

The cement kiln represents a complex unit operation in which a series of endothermic and exothermic reactions are simultaneously occurring.   Little is known regarding reaction kinetics, and much remains to be determined about thermodynamic and heat-transfer characteristics.   The size of the equipment involved and the costs associated with its operation indicate that improved control of the unit will be accompanied by very significant cost savings.   Economic benefits should be realized from at least the following sources:

1. Elimination of costs associated with ring removal, by operating the unit without upsets and with a stoichiometrically constant feed.

2. Reducing fuel costs through minimization of process upsets.

3. Improved product quality through steady, consistent operating conditions with fixed temperature zones.
4. Elimination of off-specification clinker.
5. Reduced maintenance costs due to steady-state operation.

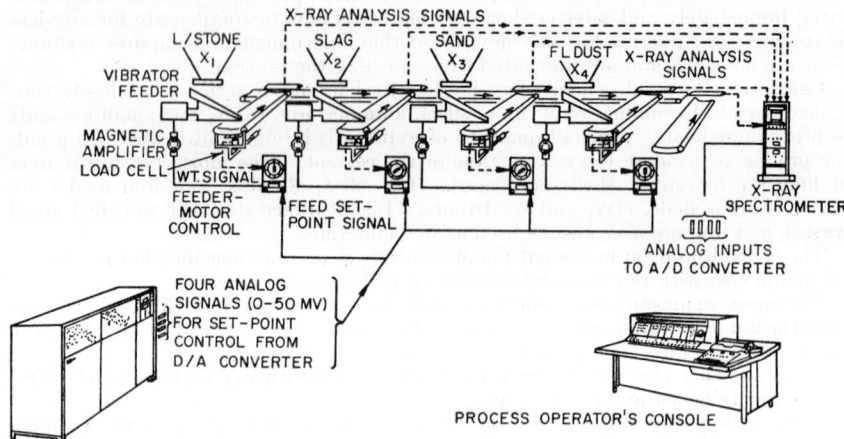

Fig. 13. Proposed layout for closed-loop computer control of the cement-plant blending operation.

Paper. Automatic, on-line computer guidance of a paper machine offers substantial tangible and intangible advantages with systems producing 65 or more tons per day of paper.[24,26,46] This applies especially to those used to produce four to eight different grades during a day.

Among the tangible advantages, particularly in a specialty mill where runs are made to orders on hand rather than to inventory, are reduction of grade change time; reduction of off-grade production; controlled repulping of off-grade paper; reliable moisture control; and increased, uniform production rate. Among the intangible advantages are the competitive advantage of satisfying customers' dependency on uniform quality and quality repeatability when reordering; prompt, reliable production reports for management; and the accumulation of research data.

Potlatch Forests, Inc., Lewiston, Idaho, has a digital-computer-control system currently connected to 46 points on the Fourdrinier or "wet end" of the paper machine. The control system is on-line, open loop and automatically prepares operator-guide instructions for control of the paper machine. Potlatch expects ultimately to tie the computer system to about 200 points along the process line and to close the loop on some points.

Figure 14 is a schematic diagram of the inputs, outputs, and other components of the control system.

Figure 15 shows the over-all logic of the monitor program used to schedule, sequence, and perform the control function. Following the figure counterclockwise from the top, the provisions in the general logic are:

Safe Stop. Stops the program at a safe point for manual entry of miscellaneous data and program modifications from the console typewriter. Otherwise, an interrupted subroutine might be prevented from properly completing its function.

Demand Log. Provides for log entry on demand by the operator. He is able to set an indicator demanding an immediate present-status logging. The program normally provides a log of the machine status every quarter hour during a production run, plus one log entry of machine status immediately after an order change.

New Order. Although the program accumulates a total production figure for any given order, only the operator knows when enough salable paper has been produced to fill the order. When he has reason to terminate production on any order, he enters the order number of the next order to be run (by means of the manual digit entry switches) and presses a button. The button sets an indicator in the computer which is the signal to type the end of run summary, compute the guide and alarms for the new run, type the specifications for the new run (entered along with the rest of the days' orders by means of the console typewriter), type the operating guide, type one log of machine status with red type indicating control changes to be made, and proceed under control of the general logic.

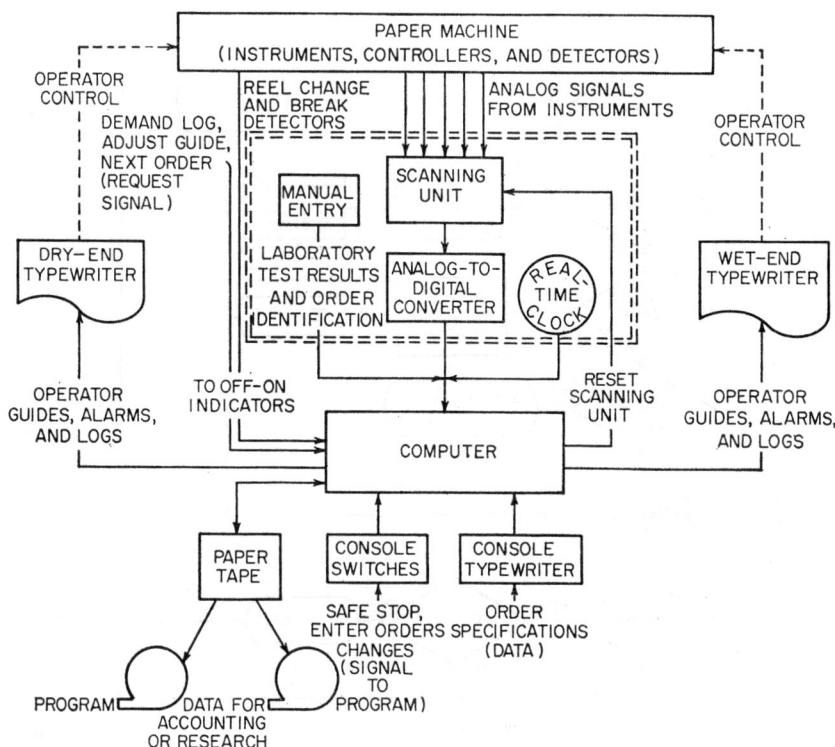

Fig. 14. Computer-control system for a paper machine.

Adjust Guide. To provide for the contingency that the machine cannot be run within the limits provided by the guide and alarms, the operator may signal the control system to adjust the operating guide and alarm limits. This may occur because of conditions in the pulp mill or bleaching plant, or because of wire or felt deterioration. The guide and alarm limits are adjusted to the level at which the machine is actually being operated. They function only until the end of that order run or until another adjustment is signaled.

Enter Day's Orders. Operation of the monitor actually begins when the order clerk signals through the console switches that he is ready to enter the day's orders. When the typewriter is ready, he enters specifications and identifying information for each of the day's orders through the console typewriter.

Periodic Log. The monitor program, by periodically reading the real-time clock, determines quarter-hour intervals. At the occurrence of each logging interval, machine status and alarms are typed on the logging typewriters. If a different interval is desired, it can be provided by a simple program modification.

Scan Cycle. Within the quarter-hour logging cycle, the monitor scans all readings once each minute. Designated readings are accumulated and averaged for the logging cycle. If, during one of the scan cycles, a variable is off limits, the control system prints alarms in red at the right end of the log sheets. The monitor then keeps track of that particular variable. When it comes back on limits, the control

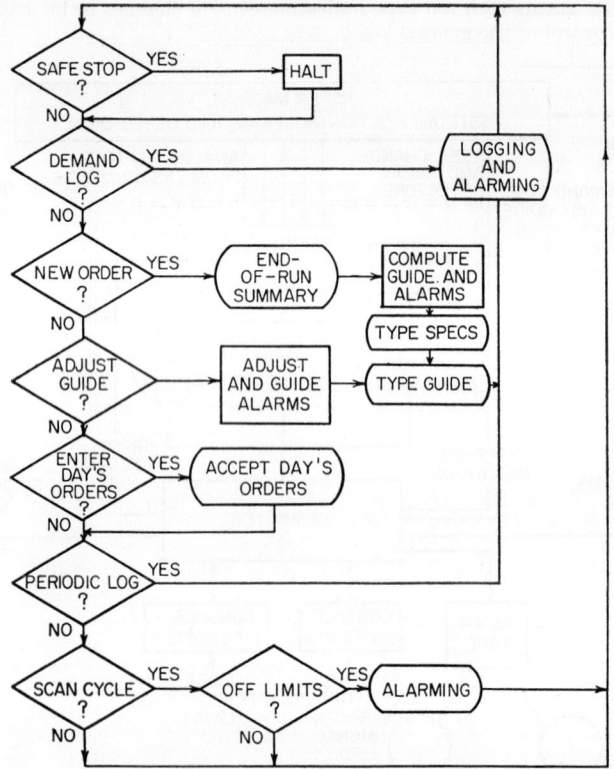

Fig. 15. Diagram of program for computer control of a paper machine.

system types in black in the off-normal summary column. The alarm and return to normal indications of the variable include a designation of the variable and its off- or on-limits value.

Steam-Electric Generating Units

Electric utility companies are constantly endeavoring to produce electric power more economically, while maintaining safe, reliable operation without interruption of service to the customer. These objectives are pursued by improving the operating performance and reliability of individual plant equipment and by reducing the operating costs of the entire system. The upward trend of fuel costs is met by the installation of larger, more efficient generating units. Rising labor costs have been offset by a reduction in plant operating personnel. Such a reduction is achieved by consolidation of widely scattered boiler, turbine, and generator control panels into one central control room.

Increases in unit size and in operating steam temperatures and pressures, coupled with frequent load changes required by new automatic dispatch methods, have increased the complexity of unit operation and control. A highly reliable, fast-acting, supervisory control system is required, to optimize certain manipulated variables properly and to perform calculations vital to improved control.

Digital computers have the potentiality to:

1. Prove a more effective plant equipment design and to allow operation closer to design limits.

2. Improve heat rate by continuous optimization and to relieve operators of tedious logging and other duties.

3. Detect equipment failure at an early stage of fault development and to maintain high standard of safety for men and equipment, thus meeting all the above objectives.

The essential functions in automatic plant operation are collection of plant operational data, correlation and reduction of the data to produce information, evaluation and comparison of the information with references to decision making, and finally, execution of the decisions to achieve the desired operating conditions.

Implementation of these functions in power plants calls for storage of considerable amounts of information, such as norms, past performance, and formulas, which are subject to frequent changes. The digital computer's large memory and flexibility in programming the necessary changes fulfill the desired requirements most economically.

The utilities companies sometimes approach automation in steps which may be described by the following systems:

The first system, which is called operating-guide computer system (OGC), collects, correlates, and computes data to produce information which would guide the operator to possible means of increasing plant efficiency.

The second system involves expansion of the equipment required in the OGC to accomplish automatic start-up and shutdown of the entire generating unit and is identified as OGC + SS.

The third system, which is called computer-control system (CCS), in addition to functions listed under OGC + SS, will actually run the plant. The generating unit is directly under the control of the computer which correlates the information resulting from the OGC functions to produce and execute decisions.

This is an important function requiring a large number of contact sense points for detecting on-off conditions of the plant equipment; contact operate points for on-off switching of the plant equipment; analog inputs for measurements of plant variables; analog outputs which represent the analog equivalent of calculated digital values to adjust the controller's set points; pulse counts to accept plant information in digital form for kilowatthour meters, etc.; and finally a number of priority interrupts which instantaneously interrupt the routine operation of the computer in order to handle high-priority occurrences in real time.

Table 1 compares the number of inputs/outputs and the scanning rates for typical OGC, OGC + SS, and CCS system configurations.

Table 1. Comparison of Computer Systems for Utilities

Computer function	OGC system		OGC + SS system		CCS system	
	No. of points	Rate of points/sec	No. of points	Rate of points/sec	No. of points	Rate of points/sec
Analog input	400	5	700	10	850	35
Digital input	12	...	15	...	20	
Contact sense	25	5	950	300	1,300	1,300
Analog output	5	...	50	...	60	
Contact operate	25	...	480	...	650	
Interrupt	3	...	20	...	25	

In running the plant, the CCS contains preplanned, stored, and automatically available major routine programs for:

1. Starting and stopping the auxiliary equipment.
2. Detecting and alarming off-normal conditions, then taking corrective actions.
3. Starting and stopping subloop control systems.
4. Optimizing plant operation.
5. Taking proper action when a subloop or other equipment fails.
6. Testing the standby, emergency, and running equipment at regular intervals.
7. Performing the normal functions described for OGC.

Application of computers in new steam-electric stations can be justified on the tangible benefits such as labor and fuel savings and on the basis of intangible benefits such as reduction in the possibility of major equipment damage and reduction in generating units outage time.

There are other items, such as reduction in equipment maintenance, reduction in instrumentation, and greater safety for men and equipment, which are difficult to evaluate but can be considered as plus factors.

Fuel and labor savings in existing units are greater than in a new unit because the older units are not automated to the same extent as modern conventional power plants and also because control boards are scattered throughout the plant, thus requiring a greater number of operators.

Steel Industry

Digital-computer-control systems are being used in many steel-industry operations, such as rolling mills, hot-strip mills, continuous-annealing lines, blast furnaces, and basic oxygen furnaces.

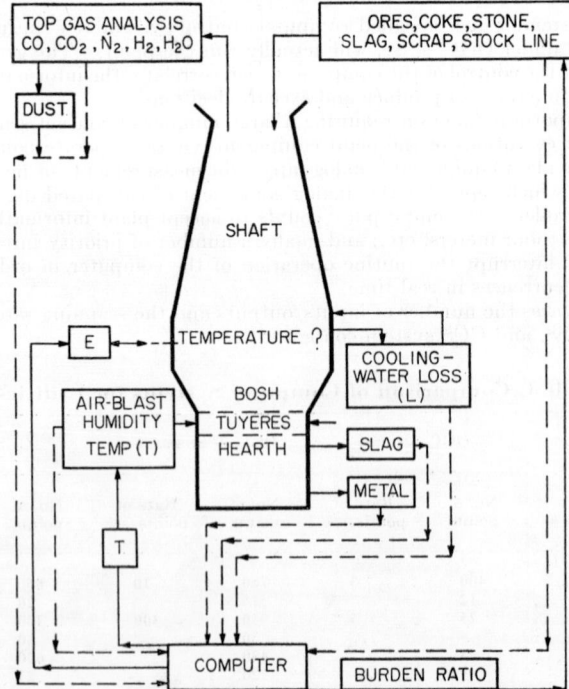

Fig. 16. Primary blast-furnace control system.

The blast furnace is the biggest volume producer and appears to be an attractive area for application of computer control. Benefits include improving the quality of the pig iron, increasing the throughput, and lowering the costs of the operation. Discussion of digital-computer control of a blast furnace follows. Refer to Fig. 16.

The blast furnace is a structure for reducing iron oxide to molten metal. The large amount of heat required for the process is generated by the combustion of coke. This combustion also provides the carbon monoxide necessary for the reduction of the iron oxide.

There are a number of independent control variables which may be manipulated by the operator, to a greater or lesser extent, in order to influence the furnace operation. These variables can be divided into two principal categories, charge variables and operating variables.

Charge variables are C/O (coke-to-ore ratio), St/O (stone-to-ore ratio), and the proportion of each of the many iron-bearing materials to the total ore input. For example, one or more grades of sized ore or of sinter may be charged into the furnace, as well as a variety of different unprocessed ores which, in general, come from different mines and have therefore different analyses and structural properties. It should be noted that the charge variables are all expressed as ratios, rather than absolute flow rates. This follows from the assumption that the furnace level is kept at a given fill mark; charge materials are added in chosen proportions at the necessary rate to maintain the desired level.

The operating variables are blast temperature T_B, blast moisture M, wind rate W, and top pressure P_T. If the furnace is modified to permit oxygen enrichment and natural-gas injection, the two variables O_2 and CH_4 are added to the list of independently controllable operating variables. These quantities can be changed by the operator as desired; their influence is rapid compared with the long delay before adjustments in charge variables take effect.

Smooth operation of the blast furnace is disrupted by several major disturbing factors. These may be viewed as independent variables which affect furnace behavior but are beyond the direct control of the operator, who may institute control action only in response to, or as a reaction to the disturbance. Significant disturbances are:

1. Variations in the analyses of the dry charge materials.
2. Moisture content of charge materials M_c.
3. Shaft efficiency E.
4. Heat loss to bosh and hearth cooling water L_{cw}.
5. Flue-dust losses D.
6. Porosity of the stock column in the furnace r.

The blast furnace may be operating smoothly at optimal conditions, but each time a disturbance occurs, some changes must be made to the process in order to counteract the disturbance or to take advantage of it if the disturbance is a desirable one. The problem is to detect each disturbance and find the new optimum set of conditions under which the process should be operated. The controllable input variables will then be changed accordingly in order to bring the process to the new optimum.

As an example, variations in E, L_{cw}, and M_c drastically affect the heat requirements of the process. When such disturbances occur, the heat input to the system must be changed in such a way as to maintain satisfactory operation. Under present operating practice, the possibility of disturbance is taken into consideration by supplying somewhat more coke than necessary, thereby assuring an adequate heat supply for adverse circumstances. This represents a waste of valuable coke, for the excess energy is removed from the furnace as sensible heat of the top gas. Large savings may be achieved by reducing the coke consumption. This measure would further result in increased production from a given furnace because a greater working volume would be available for processing the ore instead of the coke. Similarly, variations in the analyses of the charge materials and in the flue-dust losses alter the material balances of the process. The latter effects can be countered by adjustments in the coke and stone ratios.

In addition, a significant optimization problem exists with respect to selecting the most economic combination of burden materials which satisfies the desired material relationships. Although this problem in itself is a large off-line linear-programming problem of the "best-mixture" type, some aspects of it may be included in the control scheme with consequent economic advantage.

Wind Tunnels and Missile Test Stands

As regards digital-computer-control systems, wind-tunnel and missile-test-stand requirements are basically the same. Both require a high-speed data acquisition system (DAS) with a computer for data reduction. In many instances the computer is connected directly to the DAS to function as system controller and to provide "quick-look" analysis of test data.

In a typical system, measurements of test unit variables such as pressures, strain-gage readings, angular or linear position, and temperature are changed to proportional analog voltages. These are transmitted to a recording room, digitized, and recorded on magnetic tape, either directly or indirectly from the memory of a digital computer if the computer is on-line. Input channels of 100 or more measurements and analog-to-digital conversions of over 10,000 samples per second are typical.

Additional Considerations

With a little imagination, this list can be extended significantly. But do not limit the "imagineering" process to applications alone. Extend it to utilizing the equipment and techniques already available for controlling complex systems and networks. Consider, for example, adaptive control systems[4,11] in which the control device "learns" with experience, that is, improves its performance on the basis of a combination of observation and analysis. Consider a highly integrated and complex operation such as a petroleum refinery in which major units are controlled by satellite computer systems which receive direction from a centralized digital-control system. This approach can be extended to include use of digital-computer systems for remote control of geographically dispersed operations. In conclusion, there seems to be no limit.

REFERENCES

1. Andrews, A. J., J. R. Parsons, and E. D. Tolin: Controlling a Thermal Cracking Furnace with a Digital Computer, *Control Eng.*, September 1960, pp. 150–153.
2. Arnold, H. H., and J. D. Schiller: Developing a Multi-machine Complex, *Automation*, September, 1961, pp. 52–66.
3. Auricoste, J., R. Chambolle, Y. Panis, and J. Prades: Digital Computers Monitor Nuclear Steam Generator, *Control Eng.*, March, 1961, pp. 127–131.
4. Bellman, R.: "Adaptive Control Processes, A Guided Tour," Princeton University Press, Princeton, N.J., 1961.
5. Bellman, R.: "Dynamic Programming," Princeton University Press, Princeton, N.J., 1957.
6. Bennett, C. A., and N. L. Franklin: "Statistical Analysis in Chemistry and the Chemical Industry," John Wiley & Sons, Inc., New York, 1954.
7. Box, G. E. P.: The Exploration and Exploitation of Response Surface: Some General Considerations and Examples, *Biometrics*, March, 1954.
8. Box, G. E. P.: Some General Considerations in Process Optimization, *Stat. Tech. RES Group, Tech. Rept.* 13, Princeton University, Princeton, N.J., 1958.
9. Bradford, J. T., and R. W. Kirkland: Control Computer System for a Continuous Annealing Line, AIEE Conference Paper 60-976, September, 1960.
10. Brandon, David B.: Let a Digital Computer Run Your Alkylation Plant, *Ind. Eng. Chem.*, October, 1960, pp. 814–820.
11. Mishkin, Eli, and L. Braun, Jr. (eds.): "Adaptive Control Systems," McGraw-Hill Book Company, Inc., New York, 1961.
12. Bristol, Don R.: Computer Application to the Great Lakes 80 Hot Strip Mill, Instrument Society of America, Preprint 125-NY60, September, 1960.
13. Bristol, Donald, and Steven Kraft: Bringing Hot Strip Mills under Automatic Control, *Control Eng.*, September, 1960, pp. 146–149.
14. Buchsbaum, Walter H.: Advances in Medical Electronics, *Electron. World*, November, 1961, pp. 33–36.
15. Bushor, William E.: Medical Electronics, Part I: Diagnostic Measurements, *Electronics*, Jan. 20, 1961.
16. Campbell, D. P.: "Process Dynamics," John Wiley & Sons, Inc., New York, 1958.
17. Ceaglske, Norman H.: "Automatic Process Control for Process Engineers," John Wiley & Sons, Inc., New York, 1956.

18. Cheng, D. K.: "Analysis of Linear Systems," Addison-Wesley Press, Inc., Reading, Mass., 1959.
19. Churchill, R. V.: "Operational Mathematics," 2d ed., McGraw-Hill Book Company, Inc., New York, 1958.
20. Churchman, C. W., R. L. Achoff, and E. L. Arnoff: "Introduction to Operations Research," John Wiley & Sons, Inc., New York, 1957.
21. Computer Added to Big Cat Cracker, *Oil Gas J.*, May 15, 1961, p. 100.
22. Computer Aims to Hike NH₃ Plant's Efficiency, *Chem. Eng. News*, Nov. 7, 1960, pp. 60–61.
23. Computer Automates Army Tank Checkouts, *ISA J.*, October, 1960, pp. 19–22.
24. Computer Controls for Paper Manufacturing, *Western Electronics News*, December, 1961, pp. 34–35.
25. Computer Controls Resistor Production, *Chem. Eng. News*, July 17, 1961, pp. 58–59.
26. Computers Enter Paper Industry, *Chem. Eng. News*, Nov. 20, 1961, pp. 58–59.
27. Computers Start to Run the Plants, *Business Week*, Nov. 5, 1960.
28. Considine, D. M.: "Process Instruments and Controls Handbook," McGraw-Hill Book Company, Inc., New York, 1957.
29. Cosgriff, R. L.: "Nonlinear Control Systems," McGraw-Hill Book Company, Inc., New York, 1958.
30. Courant, R., and D. Hilbert: "Methods of Mathematical Physics," vol. 1, Interscience Publishers, Inc., New York, 1953.
31. Crowther, R. H., J. E. Pitrak, and E. N. Ply: Computer Control at American Oil, I, Application to the Process Unit, *Chem. Eng. Progr.*, June, 1961, p. 39.
32. Davies, O. L.: "Design and Analysis of Industrial Experiments," Hafner Publishing Company, Inc., New York, 1954.
33. Digital Computer to Automate Cement Rock Crushing, Blending, *Pit Quarry*, January, 1959.
34. Dowding, C. W., and F. R. Russell: Process Evaluation in Computer-run Microplant, *Chem. Eng.*, Oct. 30, 1961, pp. 97–103.
35. Dwyer, Paul S.: "Linear Computations," John Wiley & Sons, Inc., New York, 1951.
36. Eckman, Donald P.: "Automatic Process Control," John Wiley & Sons, Inc., New York, 1958.
37. Eisenhardt, R. D., and T. J. Williams: Closed-loop Computer Control at Luling, *Control Eng.*, November, 1960, pp. 103–114.
38. Engineering Staff, GPE Controls, Inc.: Applying the Digital Computer to Open-hearth Operations, *Control Eng.*, vol. 6, pp. 94–100, 1959.
39. Evans, W. R.: Control-system Dynamics, McGraw-Hill Book Company, Inc., New York, 1954.
40. Fluegel, D. A., E. D. Tolin, and J. R. Parsons: Collecting Process Data for an On-line Digital Computer, *Control Eng.*, April, 1961, pp. 147–150.
41. Foulkes, R. M.: The Modern Automatic Classification Yard, *Control Eng.*, March, 1961, pp. 138–141.
42. Gardner, M. F., and J. L. Barnes: "Transients in Linear Systems," vol. 1, John Wiley & Sons, Inc., New York, 1946.
43. Gass, S. I.: "Linear Programming," McGraw-Hill Book Company, Inc., New York, 1958.
44. Grabbe, F., S. Ramo, and D. E. Wooldridge: "Handbook of Automation Computation and Control," vol. 1, "Control Fundamentals"; vol. 2, "Computers and Data Processing"; vol. 3, "Systems and Components," John Wiley & Sons, Inc., New York, 1958.
45. Hengstebeck, R. J.: "Petroleum Processing," McGraw-Hill Book Company, Inc., New York, 1959.
46. "IBM General Information Manual, 1710 Control System for Operator Guide Control of a Paper-making Machine, 1961.
47. "IBM General Information Manual, 1710 Control System for Petroleum Refining," 1961.
48. "IBM General Information Manual, 1710 Control System for Steam-electric Generating Units," 1961.
49. Jury, E. I.: "Sampled-data Control Systems," John Wiley & Sons, Inc., New York, 1958.
50. Kalman, R. E., and R. W. Koepcke: The Role of Digital Computers in The Dynamic Optimization of Chemical Reactions, *IBM Res. Rept. RC-77*, December, 1958.
51. Kalman, R. E., L. Lapidus, and E. Shapiro: Computer Control of Chemical Processes, *Chem. Eng. Progr.*, vol. 56, pp. 55–61, 1960.
52. Karcher, Bill, and Hugh Jacobson: First Computerized Gas Dispatching System, *Am. Gas J.*, February, 1960.
53. Ku, Y. H.: "Analysis and Control of Nonlinear Systems," The Ronald Press Company, New York, 1958.
54. Kuehn, D. R., and H. Davidson: Computer Control, II, Mathematics of Control, *Chem. Eng. Progr.*, June, 1961, p. 44.
55. Lane, James W.: Digital Computer Control of a Catalytic Reforming Unit, ISA Preprint 17-H60, February, 1960.
56. Laning, J. H., Jr., and R. H. Battin, "Random Processes in Automatic Control," McGraw-Hill Book Company, Inc., New York, 1956.
57. Lapidus, L.: On the Dynamics of Chemical Reactors, AIChE Preprint 1, Joint Automatic Control Conference, MIT, Cambridge, Mass., September, 1960.
58. Laspie, C. G., and S. M. Roberts: On-line Computer Control of Thermal Cracking, *Ind. Eng. Chem.*, May, 1961, pp. 343–348.
59. Leondes, C. T. (ed.): "Computer Control Systems Technology," McGraw-Hill Book Company, Inc., New York, 1961.
60. On-line Computer Scores High in Big Test: Control of Refinery Unit, *Chem. Eng.*, Oct. 19, 1959, pp. 102–104.
61. Madigan, J. M.: Computer Controlled Processing, *Chem. Eng. Progr.*, vol. 56, no. 5, pp. 63–67, May, 1960.
62. Mickley, H. S., T. K. Sherwood, and C. E. Reed: "Applied Mathematics in Chemical Engineering," 2d ed., McGraw-Hill Book Company, Inc., New York, 1957.
63. More Computer Control: IBM and Du Pont Study a Tricky New Process, *Control Eng.*, February, 1961, pp. 175–176.

64. Nalle, Peter B., and LeRoy W. Weeks: The Digital Computer—Applications in Mining and Process Control, *Mining Eng.*, September, 1960.
65. Nelson, W. L.: "Petroleum Refinery Engineering," 4th ed., McGraw-Hill Book Company, Inc., New York, 1958.
66. Pendleton, A. D.: Computer Control, III, The Computer Systems, *Chem. Eng. Progr.*, June, 1961, p. 48.
67. Ragazzini, J. R., and G. F. Franklin: "Samp ed-data Control Systems," McGraw-Hill Book Company, Inc., New York, 1958.
68. Ralston, A., and H. S. Wilf: "Mathematical Methods for Digital Computers," John Wiley & Sons, Inc., New York, 1960.
69. RW-300 Computer to Control Oxidation Units at Celanese, *Chem. Eng. News*, Nov. 27, 1961, pp. 51–52.
70. Saaty, T. L.: "Mathematical Methods of Operations Research," McGraw-Hill Book Company, Inc., New York, 1959.
71. Schubert, E. J.: Television Switching Control, *Control Eng.*, May, 1961, pp. 93–96.
72. Shergalis, L. D.: Computer Control of Communications, *Electronics*, Oct. 20, 1961, p. 25.
73. Sherrard, R.: Safe, Reliable Process Monitoring, Part I, What's Needed, *Control Eng.*, December, 1960, pp. 87–91. Part II, Selection, Design, and Specifications, February, 1961, pp. 111–114.
74. Silva, Robert, and R. W. Sonnenfeldt: Computing Control of Catalytic Reforming Process, ISA Preprint 18-H60, February, 1960.
75. Special News Report: New Ideas in Air Traffic Control, *Control Eng.*, April, 1961, pp. 24–32.
76. Summers, William A.: Starting an Electric Generating Station—Automatically, *Control Eng.*, September, 1960, pp. 154–157.
77. Thayer, H. F.: A Centralized Data Requisition Facility for Large Environmental Laboratories, ISA Preprint NY60-120, September, 1960.
78. Tou, J. T.: "Digital and Sampled-data Control Systems," McGraw-Hill Book Company, Inc., New York, 1959.
79. Truxal, J. G., "Automatic Feedback Control System Synthesis," McGraw-Hill Book Company, Inc., New York, 1955.
80. Truxal, J. G. (ed.): "Control Engineers' Handbook," McGraw-Hill Book Company, Inc., New York, 1958.
81. Tsien, H. S.: "Engineering Cybernetics," pp. 63–69, McGraw-Hill Book Company, Inc., New York, 1954.
82. Volk, W.: "Applied Statistics for Engineers," McGraw-Hill Book Company, Inc., New York, 1958.
83. Williams, T. J.: "Systems Engineering for the Process Industries," (reprint of articles appearing in *Chemical Eng.*), McGraw-Hill Book Company, Inc., New York, 1961.
84. Zambotti, Bruno: Data Logging, Scanning, Alarming, Calculating in Power Plants and Substations, *Elec. World*, Nov. 27, 1961, pp. 35–46.
85. Grabbe, E. M., Simon Ramo, and Dean E. Wooldridge: "Handbook of Automation, Computation, and Control," John Wiley & Sons, Inc., New York, 1958.
86. McCracken, D. D.: "Digital Computer Programming," John Wiley & Sons, Inc., New York, 1957.
87. Susskind, Alfred K.: "Notes on Analog-Digital Conversion Techniques," John Wiley & Sons, Inc., New York, 1958.
88. Wilkes, M. V.: "Automatic Digital Computers," John Wiley & Sons, Inc., New York, 1948.
89. British Imperial College of Science and Technology: "Modern Computing Methods," Philosophical Library, Inc., New York, 1959.
90. Berkeley, Edmund C., and Lawrence Wainwright: "Computers: Their Operation and Applications," Reinhold Publishing Corporation, New York, 1956.
91. Engineering Research Associates: "High-speed Computing Devices," McGraw-Hill Book Company, Inc., New York, 1950.
92. McCormick, E. M.: "Digital Computer Primer," McGraw-Hill Book Company, Inc., New York, 1959.
93. Smith, Charles V. L.: "Electronic Digital Computers," McGraw-Hill Book Company, Inc., New York, 1959.
94. Stibitz, George R., and Jules A. Larrivee: "Mathematics and Computers," McGraw-Hill Book Company, Inc., New York, 1956.
95. Wrubel, Marshal H.: "A Primer of Programming for Digital Computers," McGraw-Hill Book Company, Inc., New York, 1959.
96. Long, R. W., R. T. Byerly, and L. J. Rindt: Digital Computer Programs in Electric Utilities, *Elec. Eng.*, pp. 912–916, September, 1959.
97. Staff: Digital Computers—Key to Tomorrow's Pushbutton Refinery, *Oil Gas J.*, vol. 57, no. 41, pp. 140–142, October 5, 1959.
98. Stein, I. M.: The Outlook for Computer Control, *Chem. Eng. Progr.*, vol. 55, no. 4, pp. 86–90, April, 1959.
99. Boycks, E. C., W. Priestley, Jr., and W. F. Taylor: The Performance of a Computer Controlled Pilot Plant, *AIChE*, Symposium, St. Paul, Minn., September, 1959.
100. Staff: Computer Techniques in Chemical Engineering, *AIChE Monograph*, New York, 1959.
101. Muroga, S.: "Elementary Principle of Parametron and Its Application to Digital Computers," *Datamation*, vol. 4, no. 5, pp. 31–34, September/October, 1958.
102. Oshima, S.: The Parametron, *Tsugakkat Shi*, vol. 39, no. 6, p. 56, June, 1956.
103. Yamada, H.: A Parametron Circuit Examined from the Point of Mathematical Logic, *Denshi Kogyo*, special volume, 1956.
104. Martch, H. B., Jr.: The Electronic Computer as Applied to Gas Pipeline Design, *AIChE*, Symposium, May 18, 1959.
105. Giles, B. L.: Design and Analysis of Natural Gas Gathering Systems by Digital Computation, *AIChE*, Symposium, Technical Paper No. 1, May 18, 1959.

106. Kniebes, D. V., and G. G. Wilson: Digital Computer Solution of Gas Distribution System Network Flow Problems, *AIChE*, Symposium, Technical Paper No. 4, May 18, 1959.
107. Bertram, J. E.: The Role of Computers in Process Control, *AIChE*, Symposium, Technical Paper No. 40, March 16, 1959.
108. Lapidus, L., R. E. Kalman, R. W. Keepche, and E. Shapiro: An Investigation of Direct Digital Computer Control of a Simulated Chemical Reactor, *AIChE*, Symposium, Technical Paper No. 42, March 16, 1959.
109. Archer, D. H.: An Optimalizing Process Control, *AIChE*, Symposium, March 16, 1959.
110. Alt, F. L.: "Electronic Digital Computers—Their Use in Science and Engineering," Academic Press Inc., New York, 1958.
111. Ledley, R. S.: Digital Electronic Computers in Biomedical Science, *Science*, vol. 130, no. 3384, Nov. 6, 1959.
112. Braun, E. L.: Digital Computers in Continuous Control Systems, *Trans. IRE*, vol. EC-7, no. 2, p. 123, June, 1958.
113. Gunning, W. F.: Computers in Process Industry Control, *Trans. IRE*, vol. EC-7, no. 2, p. 129, June, 1958.
114. Kalman, R. E., L. Lapidus, and E. Shapiro: "Mathematics is the Key," *Chem. Eng. Progr.*, vol. 56, no. 2, p. 55, February, 1960.
115. Kellett, John W., and Albert S. Perley: Economic Design by Computer, *Chem. Eng. Progr.*, vol. 56, no. 2, p. 67, February, 1960.
116. Ritchings, R. A., and W. A. Summers: Computer Controlled Power Generation, *Automation*, vol. 6, no. 11, p. 90, November, 1959.
117. Braun, E. L.: Comparing Integral and Incremental Process Control Computers, *Control Eng.*, vol. 7, no. 1, p. 113, January, 1960.
118. Burdick, C. W.: Digital Computer Runs Hot Plate Mill, *Control Eng.*, vol. 7, no. 1, p. 126, January, 1960.
119. Staff: The Design of DAFT: A Digital/Analogue Function Table, *Automatic Control*, vol. 12, no. 2, p. 46, February, 1960.
120. Rosen, L.: Characteristics of Digital Codes, *Control Eng.*, vol. 7, no. 2, p. 70, February, 1960.
121. Grohskopf, H.: Cyanamid's Approach to Computer Control, *Control Eng.*, vol. 7, no. 2, p. 70, February, 1960.
122. Madigan, J. M.: Computer Controlled Processing, *Chem. Eng. Progr.*, vol. 56, no. 5. p. 63, May, 1960.
123. Staff: Accelerated Programming for GEVIC in Real Time Applications, *Automatic Control*, vol. 12, no. 3, p. 26, March, 1960.
124. Hendrickson, A. P.: The Story of an Ultra-reliable Computer, *Automatic Control*, vol. 7, no. 7, p. 113, July, 1960.
125. Schuber, E. J.: Statistical Computers Can Really Reduce Data, *Control Eng.*, vol. 7, no. 4, p. 146, April, 1960.
126. Staff: Computers, *Automation*, vol. 7, no. 3, p. 47, March, 1960.
127. McRainey, J. H.: Role of Computers in Automation, *Automation*, vol. 7, no. 3, p. 48, March, 1960.
128. Neblett, J. B., and F. C. Mears: Linking Computers to Analyzers in Real-time Process Control, *ISA J.*, vol. 9, no. 1, pp. 44–47, January, 1962.
129. Eckman, D. P., A. Bublitz, and E. Holben: A Satellite Computer for Control, *ISA J.*, vol. 9, no. 11, pp. 59–64, November, 1962.
130. Hill, C. A., and C. C. Waugh: A Digital In-line Petroleum Blender, *Automatic Control*, vol. 14, no. 6, pp. 21–26, June, 1961.
131. Werme, J. F., M. Maczuzak, and C. H. Terrey: The Building Block Approach to Digital Systems, *Automatic Control*, vol. 17, no. 3, pp. 11–15, October, 1962.
132. Chafets, E. M.: Monitoring and Controlling Computer Time, *Control Eng.*, vol. 9, no. 12, pp. 101–104, December, 1962.
133. Cross, B.: Computers Centralize Inventory Control at Square D, *Control Eng.*, vol. 8, no. 4, pp. 152–157, April, 1961.
134. Dinman, S. B., and R. W. Sonnenfeldt: Priority Interrupt in Control Computers, *Control Eng.*, vol. 8, no. 5, pp. 127–131, May, 1961.
135. Witzel, T. H., and J. L. Wilson: Digital Computers Can Program Analog Computers, *Control Eng.*, vol. 8, no. 6, pp. 135–138, June, 1961.
136. Hagan, T. G.: Hybrid Logic for Special Purpose Computers, *Control Eng.*, vol. 9, no. 10, pp. 73–76, October, 1962.
137. Staff: Cresap, McCormick, and Paget: Central Processors for Medium, Intermediate, and Large Size Computers, *Control Eng.*, vol. 9, no. 10, pp. 103–109, October, 1962.
138. Bhavnani, K. H., and K. Chen: Assigning Confidence Levels to Process Optimization, *Control Eng.*, vol. 9, no. 8, pp. 75–78, August, 1962.
139. Paul, R. J.: Two Ways to Simulate Deadtime Digitally, *Control Eng.*, vol. 9, no. 8, pp. 97–98, August, 1962.
140. McGregor, O., and S. Skoumal: An Economic Comparison of Data Transmission systems, *Control Eng.*, vol. 9, no. 7, pp. 107–111, July, 1962.
141. Eason, T. M.: A Checklist for Designing Digital Systems, *Control Eng.*, vol. 9, no. 7, pp. 113–115, July, 1962.
142. Breedon, D. B., and P. A. Zaphyr: Pros and Cons of Remote Computing, *Control Eng.*, vol. 10, no. 1, pp. 115–117, January, 1963.
143. Staff: Cresap, McCormick, and Paget: Computer Systems Summary—Estimating Time Requirements, *Control Eng.*, vol. 10, no. 1, pp. 119–122, January, 1963.
144. Staff: Cresap, McCormick, and Paget: New Equipment for Medium to Large Computers, *Control Eng.*, vol. 10, no. 2, pp. 99–104, February, 1963.
145. Giusti, A. L., R. E. Otto, and T. J. Williams: Direct Digital Computer Control, *Control Eng.*, vol. 9, no. 6, pp. 104–108, June, 1962.

146. Fluegel, D. A., and L. R. Freeman: Simulating Sampled Date Control, *Control Eng.*, vol. 9, no. 6, pp. 123–125, June, 1962.
147. Massey, R. G.: Three Interlinked Computers to Run New British Steel Works, *Control Eng.*, vol. 9, no. 6, pp. 128–132, June, 1962.
148. Staff: Cresap, McCormick, and Paget: Computer Central Processors, *Control Eng.*, vol. 9, no. 6, pp. 135–140, June, 1962.
149. Wright, R. E.: How to Make Computer Compatible Data Tapes, *Control Eng.*, vol. 9, no. 5, pp. 127–129, May, 1962.
150. Sweeney, J.: Information Control Computers—Messages Pay for Data Links, *Control Eng.*, vol. 9, no. 5, pp. 122–126, May, 1962.
151. Richards, R. K.: "Digital Computer Components and Circuits," D. Van Nostrand Company, Inc., N.J., 1957.
152. Richards, R. K.: "Arithmetic Operations in Digital Computers," D. Van Nostrand Company, Inc., Princeton, N.J., 1955.
153. Siegel, P.: "Understanding Digital Computers," John Wiley & Sons, Inc., New York, 1961.
154. Sangren, W. C.: Digital Computers and Nuclear Reactor Calculations," John Wiley & Sons, Inc., New York, 1960.
155. Coulson, J. E.: "Programmed Learning and Computer Based Instruction," John Wiley & Sons, Inc., New York, 1962.
156. Thurman, C. H.: Existing Plant Gets Computer Control, *ISA J.*, vol. 10, no. 2, pp. 73–75, February, 1963.
157. Aiken, W. S.: Don't Measure, Compute!, *ISA J.*, vol. 10, no. 1, pp. 59–60, January, 1963.
158. Ware, W. E.: Digital Computer Systems Challenge Conventional Instruments, *ISA J.*, vol. 10, no. 1, pp. 57–58, January, 1963.
159. Hendrie, G. C., and R. W. Sonnenfeldt: Computer Reliability, *ISA J.*, vol. 10, no. 1, pp. 51–56, January, 1963.
160. Staff: Computer Monitors Flow Variables, *Automation*, vol. 7, no. 6, pp. 87–89, June, 1960.
161. Staff: Computer-controlled Manufacturing Systems, *Automation*, vol. 8, no. 9, pp. 52–66, September, 1961.
162. Brooks, M. E.: Problems in Programming Control Computers, *Automation*, vol. 10, no. 2, pp. 78–82, February, 1963.
163. Williams, T. J.: Studying the Economics of Process Computer Control, *ISA J.*, vol. 8, no. 1, pp. 50–59, January, 1961.
164. Thomas, R. J.: Justifying Digital Computers for Large Power Consumers, *ISA J.*, vol. 8, no. 5, pp. 43–47, May, 1961.
165. Green, R. M.: Extending Analytical Methods with Process Computers, *ISA J.*, vol. 8, no. 10, pp. 38–40, October, 1961.
166. Donegan, A. J.: Programming Real-time Process-control Computers, *ISA J.*, vol. 8, no. 11, pp. 46–49, November, 1961.
167. McCracken, D. D.: "A Guide to ALGOL Programming," John Wiley & Sons, Inc., New York, 1962.
168. McCracken, D. D.: "A Guide to IBM 1401 Programming," John Wiley & Sons, Inc., New York, 1962.
169. McCracken, D. D.: "A Guide to FORTRAN Programming," John Wiley & Sons, Inc., New York, 1961.
170. Ralson, A., and H. S. Wilf: "Mathematical Methods for Digital Computers," John Wiley & Sons, Inc., New York, 1960.
171. McCracken, D. D., H. Weiss, and T. Lee: "Programming Business Computers," John Wiley & Sons, Inc., New York, 1959.
172. Minto, J. G.: Applying Computer Control to a Production System, *Automation*, vol. 10, no. 3, pp. 77–82, March, 1963.
173. Reed, D. L.: Computers and Programmers in Hybrid Checkout Systems, *Control Eng.*, vol. 10, no. 4, pp. 79–82, April, 1963.
174. Shannon, J. H.: Executive Control Routines for Process Computers, *Control Eng.*, vol. 10, no. 4, pp. 85–88, April, 1963.
175. Leondes, C. T. (ed): "Computer Control Systems Technology," McGraw-Hill Book Company, Inc., New York, 1961.
176. Riordan, H. E.: Pneumatic Digital Computer, *Instr. Control Systems*, vol. 34, no. 7, pp. 1260–1261, July, 1961.
177. Ormord, L. A.: Problems in Digital Recording, *Instr. Control Systems*, vol. 34, no. 8, pp. 1430–1433, August, 1961.
178. Coffin, S. T.: Supplies for Digital Equipment, *Instr. Control Systems*, vol. 34, no. 12, pp. 2256–2259, December, 1961.
179. Azgapetian, V.: Computer Assists Servo Design, *Instr. Control Systems*, vol. 35, no. 8, pp. 81–85, August, 1962.
180. Packert, R. H.: Economic Dispatch by Digital Computer, *Instr. Control Systems*, vol. 35, no. 8, pp. 100–102, August, 1962.
181. Barclay, J. E.: Process Control Computer, *Instr. Control Systems*, vol. 36, no. 1, pp. 127–129, January, 1963.
182. Staff: Computer Controlled Systems, *Instr. Control Systems*, vol. 36, no. 5, pp. 85–90, May, 1963.
183. Simulation Council Newsletter, published monthly in *Instr. Control Systems*, covers many aspects of digital-computer application.

Editors's Note: A number of references on computers contain information on both analog and digital types. Hence, the reader should also consider references listed at the end of the section "Applications of Analog Computers" for pertinent material on digital computers.

INSTRUMENTATION DATA PROCESSING

By Lawrence M. Silva*

INTRODUCTION

The purpose of instrumentation is to obtain information. In the case of performance testing, the information is used to make a decision—for example, whether to accept or not to accept a system, a device, a component, and so on. Instrumentation associated with engineering development, on the other hand, supplies information that is used primarily to monitor the progress of design activities and to plan future action. The latter use also supplies to management control functions, especially in the case of complex engineering projects that require large quantities of information for evaluation in order to make an economic business decision or to plan further activity.

Role of Transducers

To gather information, transducers must be used to convert phenomena of interest into a common language signal acceptable by the instrumentation system. In a typical system, hundreds of transducers are required and an individual test may result in the acquisition of tens of thousands of data points.

Redundant Data

In the instance of engineering test programs, it is typical in many cases that the relevant factors to be considered are not clearly defined prior to the actual execution of physical testing. When this occurs, redundant data are gathered. Such acquisition of redundant data is unfortunate but represents a practical necessity. The result is that large quantities of data must be taken.

It is not unusual to gather hundreds of thousands of bits of raw data during a single test phase in an engineering development program. The task of editing, evaluating, and comprehensive presentation of this volume of data is an integral factor in any instrumentation data-processing program.

Basic Categories of Task to Be Performed

Fortunately, all instrumentation systems have certain factors in common. For convenience, these common factors may be divided into four major categories of tasks to be performed: (1) data acquisition, (2) data evaluation, (3) data reduction, and (4) data presentation.

Data Acquisition. This is largely self-explanatory. The function of the instrumentation system is to gather raw data.

Data Evaluation. This represents one of the most important steps in the process. The acquired raw data are edited. This editing is performed in order to eliminate redundant or useless data.

Data Reduction. After the data are edited, the information must be reduced. Data reduction amounts to organizing the processed data for human consumption. Sometimes this is a relatively difficult task.

* Formerly, Associate Director of Research and Engineering, Beckman/Systems Division, Beckman Instruments, Inc., Fullerton, Calif.

Data Presentation. The final step is that of properly displaying the data in comprehensible format. If this is not done effectively, the value of the foregoing three steps is lost. The data presentation must be in a form that human beings, rather than machines, can understand easily.

BASIC EQUIPMENT CHARACTERISTICS

In a typical instrumentation system, raw information exists in several forms. These may include d-c votages, a-c voltages, varying frequencies, time, or ratios of any two of these quantities. Measurands such as temperature, pressure, force, acceleration, and displacement can be translated into one of the foregoing electrical forms by use of proper transducers.

Each of these sources of raw information must be processed through the four basic steps previously mentioned. If many sources of each type of data signal exist in one system, a basic decision must be made at the outset of the design of the instrumentation system. This decision—how much instrumentation equipment will be provided for simultaneous or parallel gathering of data from the various sources? In order to gather data rapidly from many sources, either duplicate data-processing equipment must be provided or data channels must *time-share* or be *multiplexed* through one piece of processing equipment.

The specific configuration of any given instrumentation system concerning the choice of parallel or multiplexing and time sharing depends in large part upon the following instrumentation characteristics: (1) rate of processing information, (2) accuracy with which operations are performed, and (3) quantity of data to be acquired. The first of these performance specifications involves the number of data words per unit time. The second characteristic can be thought of as the error rate per unit data word. The third factor can be expressed in terms of total data words per run, or per test. Considered in total, these three system characteristics are the key factors in determining the configuration of the data-acquisition system and the overall cost of the system.

DATA SUBSYSTEMS

Most data-processing systems are made up of subsystems. Each of the four major tasks (acquisition, evaluation, reduction, and presentation) may be performed by a different subsystem. In many cases, it is possible to operate each subsystem entirely independent of all other subsystems. An example—the case where data are gathered in real time and recorded. The recording, in turn, is used as an input for editing and evaluation. The data selected for further data reduction are again applied to a computer. The reduced data supplied by the computer are again recorded on tape or cards for use by subsequent equipment. Between each of these steps, the data processing may be interrupted at will, so long as each individual subsystem, that is, each individual step, is allowed to operate as an independent process.

An outstanding exception, of course, is the situation where reduced data are required in real time and at the same time that the raw data are being gathered. Examples include the case of plant operation or dynamic monitoring of a test. In both these cases, all parts of the system must operate at the same pace or rate.

Independent Operation of Subsystems

Since subsystems often operate independently, the output of the data-acquisition subsystem should be of such nature that it is easily recorded in either permanent or semipermanent form. The recorded data usually are transferred to another subsystem that converts them into a form suitable for automatic processing. Quite often, this automatic processing is performed by a high-speed general-purpose computer. Such a high-speed digital computer is justified, particularly in the case of an instrumentation system involving a large amount of data. This situation occurs repeatedly in highly technical testing programs.

Continuously Varying Data (Analog)

If the raw data are of a continuously varying nature or, as it is usually termed, of an analog nature and the data-processing equipment includes a digital computer, then the raw data must be converted from the analog form to the digital form. Even when the acquired data are fed directly to a readout device, such as an electric typewriter, analog data must be converted to digital data inasmuch as the typewriter is a digital instrument.

Converting Transducer Outputs

The outputs of most transducers are typically in electrical form. To convert these voltages and pulses into digital form, equipment such as EPUT meters (events-per-unit-time meters), analog-to-digital converters, and time-interval meters must be included in the instrumentation system. These translators convert the analog input data into digitally coded signals.

Time Sharing of Subsystems

For economic reasons, it is important to time-share translator and recording subsystems over a maximum number of data channels. This requires that the data-acquisition system include apparatus for sequentially scanning the data channels. Further, since typical transducer outputs are incapable of furnishing the necessary power required by the translator, it is necessary to amplify the input analog data signals.

BASIC BUILDING BLOCKS OF GENERAL-PURPOSE DATA-ACQUISITION SUBSYSTEM

The following four elements constitute the basic building blocks of a general-purpose data-acquisition subsystem: (1) recording system, (2) translator, (3) scanner, and (4) amplifier. The configuration of a particular subsystem will depend primarily on the accuracy specifications and information-processing rates required.

Interconnection of Basic Elements

The interconnection of these basic elements for typical systems of various scanning rates is illustrated in Figs. 1, 2, and 3.

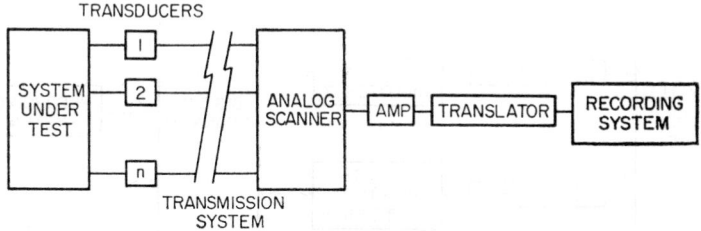

Fig. 1. Data-measuring and -recording system for slow-to-medium speed-scanning rates—1–100 data channels/second.

In these figures, the amplifiers are considered in a generic sense. For d-c signals, d-c amplifiers are required; while for a-c signals, amplifiers and possibly some form of a-c to d-c conversion will be required. Although a-c voltage to digital converters are available, the problems associated with the phase angle of the reference in the converter are a serious limitation.

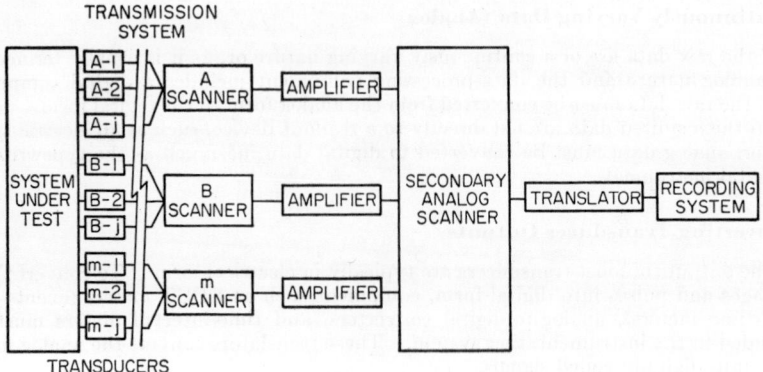

Fig. 2. Data-measuring and -recording system for medium-to-fast scanning rates—20–2,000 data channels/second.

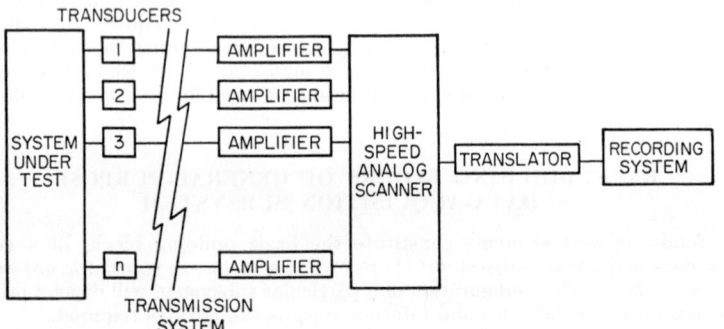

Fig. 3. Data-measuring and -recording system for high-speed scanning rates—1,000–100,000 data channels/second.

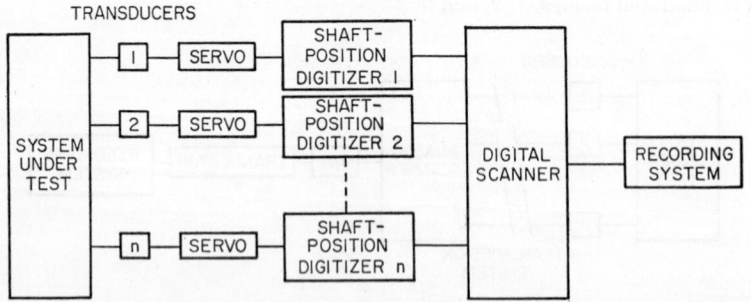

Fig. 4. Data-measuring and -recording system using individual-shaft-position digitizers.

For data channels in which information exists in the form of a frequency or pulse rate and in which medium- to high-speed scanning is required, individual accumulating registers must be provided. The outputs of these registers are scanned and applied directly to the recording system.

A form of data-acquisition system which has found extensive application is shown in Fig. 4. For a given accuracy, this over-all system is considerably more costly than those previously described. This is because of the duplication of reference and feedback elements in each of the servos. Nevertheless, this system has been used extensively in applications in which semiprecision ($\frac{1}{4}$ to 1 per cent) analog recorders are present and in which a digital record is desirable for data-reduction purposes.

Optional Equipment

In addition to the basic measuring and recording system, a data-acquisition system will include one or more of the units or options listed in Table 1. Such options as scalar and offset correction perform the operation $(Y = ax + b)$, where x is the data content and a and b are constants that convert x to engineering units such as °F and psi. These units are used in those systems in which the output data are required in meaningful form for operating personnel. In many cases, this is the extent of the required data reduction.

Table 1. Basic Building Blocks of General-purpose Data-acquisition System

Unit	Function
Programming	Controls system operating modes. Selects signal flow path through system. Furnishes instructions or stored data as required by each unit in system
Clock	Provides real or arbitrary time in digital code to recording system
Manual data insertion	Provides means for operators to insert identification or other data into data recording
Test-vehicle stimulus programmer	Equipment used in conjunction with programming and sequencer control unit to apply known input signals to the vehicle under test
Data identification	Provides digital information to recording system to identify data channel number, frame number, run number—where a frame includes one sample of all data channels and a run is associated with separate distinct test. Channel number typically is omitted if a record of the data-channel sequence is available
Analog signal transmission and termination	System of cables, connectors, and so on required to transmit analog data over extended distances
Raw-data ordering	Patchboard or other similar apparatus for ordering the interconnection of the transducers to the scanner units
Sequence control	Equipment controlled by the programming unit for operating scanners and auxiliary selector units and for controlling the sequence in which data channels and auxiliary equipment are selected. Typical units are classified as being (1) synchronous, (2) asynchronous, or (3) interleaved
Analog filtering	Filtering units either before or after low-level scanners for rejecting noise and power-supply-induced voltages
Sampling	Analog sample and hold equipment for sampling a data channel at a known instant of time or for holding the peak value of a-c inputs for the duration of the sampling period
Visual readout	Equipment for visually displaying the output of the translator or the digital quantity generated or stored in one of the units
Alarm or acceptance limit storage	A memory in which is stored off-normal limits or acceptance limits for individual data channels and, in addition, equipment for comparing high and low limits with the actual data
External control signal	Equipment used in conjunction with sequencer and alarm units to energize on-off control functions associated with individual data channels
Scalar and offset conversion and control	A memory in which is stored the constants required to perform the operation $Y = ax + b$, where Y = digital output data, x = raw input data, a = scalar constant, b = offset constant. In addition, equipment used in conjunction with sequencer and program units to select, for a particular data channel, the appropriate constants and to perform the indicated operation
Output buffer storage	Buffer storage between digital output devices and recording unit
Operator control unit	A console or panel containing all the operator controls
Precision d-c power supply	Power supplies for energizing d-c bridge transducers with 0.005 to 0.1 % long-term accuracy
Thermocouple reference-junction assembly	Equipment for maintaining the reference junction of a pair of thermocouples at a known temperature. Alternately, equipment for generating a corrective emf which is applied to the measuring system
External referencing	Equipment for referencing the translator or analog-to-digital converter to an external source to permit measurement in the form of ratios
Bridge balancing	Equipment for use with bridge-type transducers to adjust the scalar gain constant and to compensate zero offset errors

MEASURING-SYSTEM INPUT CIRCUITS

A measuring system should introduce negligible errors in the raw data. This is a prime objective of system design. Thus, if a system is to measure the voltage generated by a transducer, the system should not, under any circumstances, introduce d-c or a-c voltage or current into the transducer circuit. Furthermore, the amount of power required by the measuring system should be held to a minimum. This latter requirement can be restated simply—*the measuring system should not load the source.*

Measuring-system and Transducer Impedance

If a transducer has an impedance Z_T, the measuring-system input impedance should equal or exceed

$$\frac{Z_T}{\text{Fraction full-scale accuracy}} = \text{minimum measuring-system input impedance} \quad (1)$$

Thus a system which is to measure a 1,000-ohm transducer to an accuracy of 0.01 per cent should have an input impedance greater than 10 megohms.

Voltage Offsets and Currents

The next most obvious source of error associated with a measuring system is the error introduced by voltage offsets and currents appearing at the input terminal of the measuring system. These voltage offsets are introduced by input amplifiers, thermoelectric emf's generated in the input wiring, or improper grounding procedures within the measuring system proper.

Spurious currents I_N appearing at the input terminals of such a measuring system usually are the result of poor insulation, faulty shielding, or improper isolation of input circuits from other elements in the measuring system. In systems in which transistors are used in input amplifiers, a spurious noise current always will exist at the input terminals because of the fluctuations of I_{co} in the input stage.

Errors associated with these voltage offsets and currents should be appreciably less than the fractional full-scale accuracy of the system. In mathematical terms—the transducer full-scale output voltage x fractional full-scale accuracy should be less than or equal to the maximum measuring-system offset, where measuring-system offset equals voltage offset $+ Z_T I_N$.

Common-mode Error

Another source of error, known as the common-mode error, is more difficult to visualize and will predominate in most installations. This error is the result of the common-mode voltages associated with multiple-ground returns, transducer circuits, and measuring-system input circuits. A simplified equivalent circuit of a transducer, transmission line, and measuring-system input circuit is shown in Fig. 5.

In typical test installations, the distance between the transducers and the measuring system will vary from a few feet to a mile. If d-c and a-c voltmeters are connected between grounded objects at each of these locations, d-c and a-c voltages will appear, ranging from a few millivolts to several volts. The d-c voltages are due to electrolytic currents; the a-c voltages are due typically to inductive pickup from power apparatus, radio stations, and so on.

Voltages between the two grounds appear at the three measuring-system terminals (A, B, and ground). These noise signals appear as a voltage common to both transducer leads, on top of which the transducer output signal is superimposed. The term "common-mode" voltage is a carry-over from differential amplifier art in which the signal is contained as the difference of two voltages varying with respect to ground. Later on in this subsection, it will be shown that it is possible to determine the voltage across the terminals A to B very accurately without actually measuring the difference of the voltages from terminals A and B to ground.

Other sources of common-mode voltage are associated with the transducers or the actual measuring system. In bridge-type transducers, the output is the difference of voltage appearing across the two arms of the bridge. Since it is common practice to ground one side of the bridge excitation source, these output terminals are not referenced directly to ground but appear a few volts above ground. Thus the excitation source appears as a common-mode voltage to the measuring system.

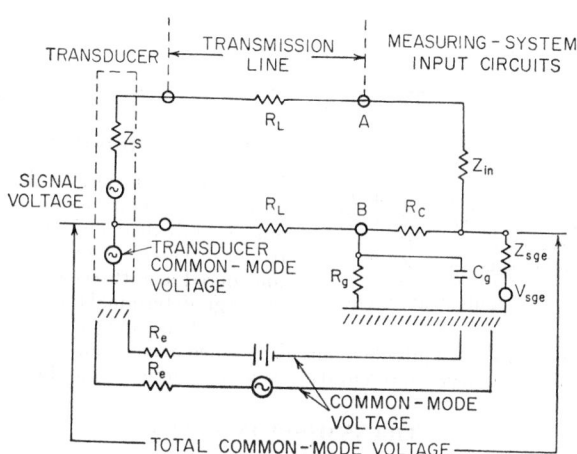

Z_{in} = measuring-system input impedance
R_L = lead resistance
Z_s = source resistance
R_g = measuring-system resistance to ground
C_g = measuring-system capacity to ground

R_e = resistance through earth between ground points
R_c = internal-coupling resistance
Z_{sge} = impedance in series with power-frequency source
V_{sge} = power-frequency voltage appearing between terminals of measuring system and earth ground

Fig. 5. System input circuits.

Differential-transformer-type transducers typically will have a small common-mode voltage. This is due to inadequate shielding of primary and secondary circuits. In general, there will always be some form of common-mode voltage associated with the transducer system. In Fig. 5, this is indicated by a common-mode voltage source appearing between one of the transducer terminals and ground.

GROUNDING

Since the earth's resistance between two distant points is independent of the distance and typically less than 100 ohms, it is obvious that, if the transducer and measuring systems are grounded, large currents will flow through the signal circuits and lead resistances because of the common-mode voltage. These unknown and variable currents will seriously impair the performance of the measuring system.

An immediate solution to this problem is (1) to assume that the transducer terminals will be referenced to ground, either directly or through a power source; and (2) to require that the measuring-system input circuits float with respect to ground. Floating means that the measuring system responds only to voltages appearing across terminals A and B and that the impedance of the measuring system to ground is arbitrarily large. In practice, it is convenient to maintain $R_g > 100$ megohms, while C_g will vary from 100 to tens of thousands of micromicrofarads.

The requirement that the input circuits should float implies complete isolation of these circuits from other grounded circuits in which power-supply voltages and a-c voltages exist. In practice, this isolation never is complete and a voltage V_{sge} will

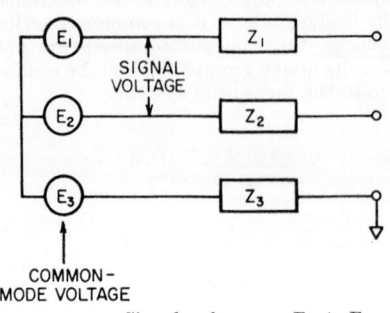

COMMON−
MODE VOLTAGE

Signal voltage $= E_1 + E_2$
Signal impedance $= Z_1 + Z_2$
Common-mode voltage $= E_3$
Common-mode impedance $= Z_3$

FIG. 6. Transducer equivalent circuit.

exist in series with an impedance Z_{sge}. This signal ranges from a few microvolts to a volt. The impedance exists between one of the measuring-system input terminals and ground. Because of the variety of input-circuit configurations, no general rules can be given regarding this potential source of error. The error is dependent on how effectively the voltage V_{sge} can couple into the input circuits.

Transducer Equivalent Circuits

Actual transducer equivalent circuits generally are more complex than the simplified equivalent circuit shown in Fig. 5. In most cases, the actual transducer is considered as a three-terminal device with two signal terminals and one ground terminal. This can be represented accurately by either a wye or delta equivalent circuit. Since most transducers are basically voltage generators, the wye representation is more convenient and is shown in Fig. 6.

TRANSMISSION LINES

A main reason for transducing variables, such as pressure, temperature, and force, to an electrical analog voltage is to facilitate the accurate transmission of data for a considerable distance. The cables used to transmit the transducer signals, however, are themselves a large potential source of error—because of the common-mode voltage that exists in the input circuits. Although transmission lines inherently are distributed-parameter systems, they can be specified adequately at the low frequencies of interest by a small number of lumped constants. Neglecting noise voltages and currents generated by the cables, the transmission lines can be represented by three impedances in a delta configuration.

At the frequencies of interest in practical measuring systems, the impedances in Figs. 7 and 8 consist of parallel combinations of leakage resistance and cable capacitance.

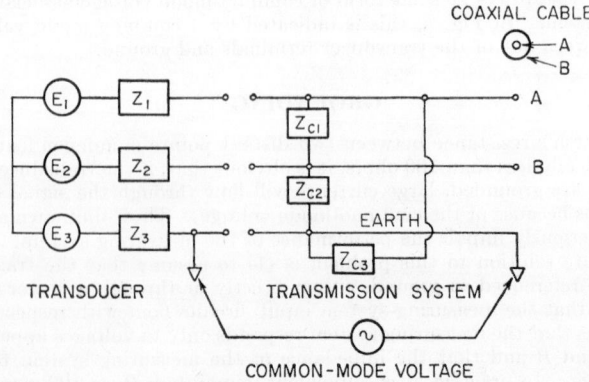

FIG. 7. Lumped-parameter representation of coaxial signal transmission system.

Coaxial System. In the coaxial system shown in Fig. 7, the impedance Z_{C1} will load the transducer and produce errors in proportion to $(Z_1 + Z_2)/Z_{C1}$. Z_{C3} can introduce appreciable errors into the signal circuits since the impedance Z_2 is common with both signal and common-mode circuits.

As an example, consider a 120-ohm strain-gage circuit in which impedances Z_1 and Z_2 will be 60 ohms, Z_3 approximately 50 ohms, and E_3 approximately 5 volts

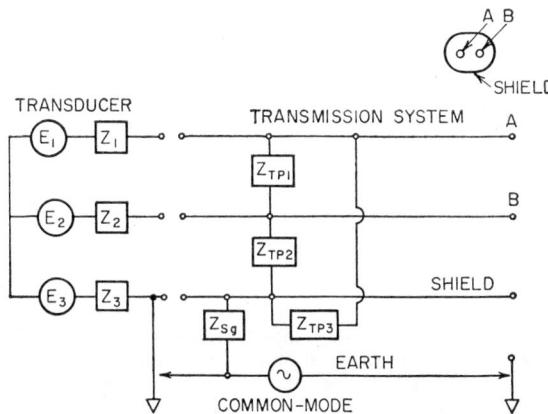

FIG. 8. Lumped-parameter representation of twisted-pair shielded signal-transmission system.

d-c. If the leakage resistance between the outer conductor and ground is 12 megohms, this will cause a voltage drop of 25 volts in impedance Z_2, which will appear as a signal-voltage offset. This order of magnitude of leakage is not uncommon in systems in which moisture is allowed to collect on the outer insulation jacket of cables.

The situation, of course, is much worse when alternating voltages and currents are distributed along the length of the cable and account for the potential difference existing between isolated grounds. If the cable has a capacitance of 10 $\mu\mu$f/ft to ground and is 1,000 ft long, then the total capacitance to ground will be 10,000 $\mu\mu$f; or about 250,000 ohms at 60 cps and 40,000 ohms at 400 cps. Thus, a 1-volt 60-cps common-mode voltage will cause $\frac{1}{4}$ mv of 60-cps alternating current to appear across Z_2 in series with the signal. The situation, of course, is six times worse for 400-cps common-mode voltages.

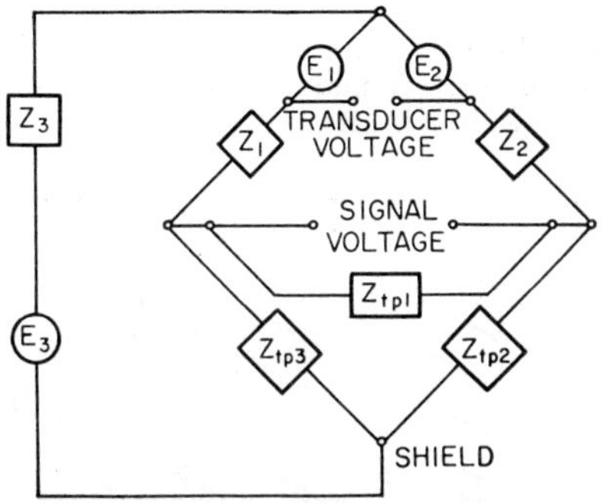

Twisted-pair Shielded System. This situation, as shown in Fig. 8, is much improved as a result of the guard action of the shield. Note that a-c common-mode voltages appearing along the length of the cable are harmlessly short-circuited by the shield. The shield should be grounded at the transducer only. If it

FIG. 9. Bridge representation of twisted-pair shielded transmission system.

were grounded at both the transducer and the measuring system, then the currents in the shield would be increased appreciably because of the conductive path through the shield and the 100-ohm ground return and the common-mode rejection would be degraded.

In most measuring systems, the shield is connected to the measuring system and used to reference the input circuits directly to transducer ground. In this manner.

the system common-mode voltage is effectively equal to the transducer common-mode voltage and the common-mode voltages due to circulating ground currents are eliminated.

If the circuit shown in Fig. 8 is changed, as indicated in Fig. 9, it is evident that the transducer and transmission system form a bridge. Thus, if common-mode voltages are to be eliminated across the signal voltage terminals, then the bridge should be balanced.

$$\frac{Z_1}{Z_{tp3}} = \frac{Z_2}{Z_{tp2}} \tag{2}$$

This is a practical expedient for transducer circuits in which severe common-mode problems exist. Typically, its use will require adding shunt capacitance and resistance across the appropriate cable impedance.

NOISE

It is characteristic of most d-c transducers used in instrumentation systems that the full-scale output voltage is in the range of 1 to 100 mv. If an accuracy of 0.1 to 0.01 per cent is required for the system, it is important to minimize the error introduced by noise which appears in series with the input signal and which can be attributed to components in the system input circuits.

In well-designed d-c vacuum-tube amplifiers, this noise will be a combination of thermal or Johnson noise and flicker-effect noise. The predominating source will depend on the input impedance level. For transistor amplifiers, the input noise will depend critically on the impedance level and will have a $1/f$ spectrum at low frequencies. This $1/f$ spectrum also is characteristic of flicker-effect noise associated with vacuum-tube circuits.

Johnson Noise. Johnson, or thermal, noise is the summation of the effects of the very short current pulses of many electrons as they travel between collisions; or alternately, it is the Brownian movement of electrons in a resistor or conductor. The rms noise voltage appearing in series with the input signal in a circuit of resistance R is

$$V_{rms} = \sqrt{4KTRB} \tag{3}$$
$$V_{rms} = 1.28 \times 10^{-10} \sqrt{RB} \tag{4}$$

for $T = 25°C$ and where R = effective circuit impedance and, $B = 3$ db bandwidth of circuit. A plot of the thermal-noise voltage vs. circuit impedance for various bandwidths is given in Fig. 10.

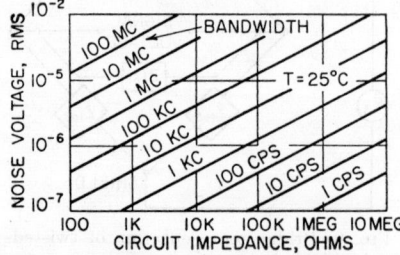

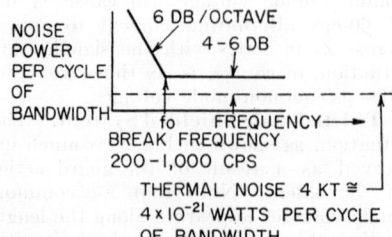

Fig. 10. Plot of thermal-noise voltage vs. circuit impedance for various bandwidths.

Fig. 11. Characteristics of transistor noise.

Percentage-time Noise Peaks. Although the rms noise voltage is a suitable criterion for systems in which long observation times are allowed, it is not a realistic figure for data-acquisition systems in which data are sampled and translated by some form of analog-to-digital converter. In this instance, it is best to interpret the rms noise in terms of the percentage of the time its peak value exceeds the rms value by a factor of 2, 4, and so on. Table 2 lists the percentage-time noise peaks exceed a given peak/rms ratio.

Thus, from Table 2, it is seen that 0.01 per cent of the time the peak noise voltage will be greater than 3.89 times the rms noise voltage. In practice, the relations indicated in the table are observed if the bandwidth is greater than a few hundred cycles. For circuits in which the bandwidth is a few cycles per second, the actual percentage will be found to be slightly greater.

Table 2. Peak to rms Ratio for Johnson Noise

Peak voltage rms voltage	% of time peak voltage is greater than specified peak to rms ratio
1.645	10
2.576	1
3.291	0.1
3.890	0.01
4.417	0.001
4.892	0.0001

At frequencies ranging from 10^{-4} up to 1,000 cps, transistor circuit performance is inherently limited by a noise whose power spectrum falls off approximately in inverse ratio to the frequency and which is not related to thermal noise. The characteristics of this noise are indicated in Fig. 11.

Relationship of Noise Power to Frequency. Note in Fig. 11 that below a frequency f_0 the noise power doubles if the frequency is halved. Thus the rms noise voltage must vary inversely as the square root of frequency. In Fig. 11, the 4- to 6-db value refers to input circuits in which the source resistance is approximately 1,000 ohms. The performance of d-c and audio-frequency amplifiers using vacuum tubes is limited by a $1/f$ type of noise (called flicker noise) which is generated at the cathode of the vacuum tube.

Additive Noise of Components. Since thermal, or Johnson, noise is an inherent property of all passive or active systems or circuits operated at temperatures above absolute zero, it sets a lower bound to the resolution of any system. If amplifiers or other elements in the system generate noise, then this noise must be added to the Johnson-noise component. Assuming each noise source is independent, then

$$\text{rms total noise voltage} = \sqrt{V_j{}^2 + V_a{}^2 + \cdots + V_n{}^2} \tag{5}$$

where V_j = rms thermal noise

V_a = rms magnitude of other noise components

Evaluation of Individual Elements of a System. A ratio, the noise factor or noise figure, commonly is used in such evaluations. The noise factor of any circuit or element of a system is the ratio of the actual noise output power to the noise power which would be delivered if the only source of noise were thermal noise at standard temperatures. In Fig. 11, the ratio of transistor-noise ordinate to the thermal-noise ordinate for a given frequency is the noise factor for a particular frequency and is called the *spot noise factor*. If the noise-factor ratio of each element is expressed in decibels, then the over-all noise factor NF_T of a system or circuit simply is the sum of the individual noise factors.

$$NF = \frac{\text{actual noise power output}}{\text{noise power in output due to thermal noise originating in input circuit}} \tag{6}$$

$$NF = \frac{\text{total mean-square noise voltage}}{\text{mean-square noise voltage due to thermal noise generated by impedance of input circuit}} \tag{7}$$

$$NF_T = N_1 + N_2 + \cdots + N_n \tag{8}$$

Noise Factor and Source Impedance. It is a characteristic of active circuit elements that the noise factor is a function of the source impedance. For transistor amplifiers, the optimum performance is obtained if the source impedance is low, such as 500 to 2,000 ohms.   The opposite situation, however, generally is true for vacuum tubes where the source impedance should be high for best performance at frequencies up to a few kilocycles.   The relationship between source resistance and

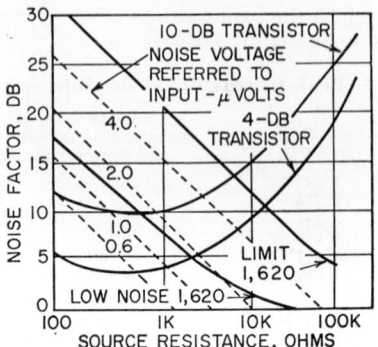

Fig. 12. Relationship between source resistance and noise factor for a low-noise vacuum tube and low-noise transistor circuit.

noise factor for a low-noise vacuum tube and low-noise transistor circuit is given in Fig. 12.

AMPLIFIERS—BASIC CONCEPTS

Although it is possible to build analog-to-digital converters that, without pre-amplification, will convert millivolt and microvolt signals directly to the digital domain, this approach has not been applied extensively because of the difficulty in simultaneously providing filtering or noise rejection and realistic measuring-system input impedances.   With the exception of mechanical-position digitizers or other digitizers requiring one or more seconds to complete a three- or four-digit conversion, it is common practice to provide preamplification and signal conditioning prior to the actual conversion process.

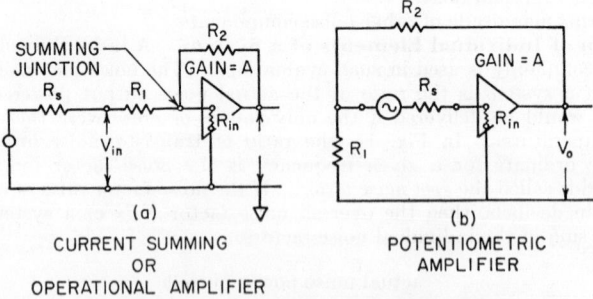

Fig. 13. Amplifier feedback connections.

Feedback Amplifiers.   Preamplifiers used in this connection should have low zero offsets and a high degree of gain stability.   This performance can be realized only with feedback-type amplifiers.   Given an amplifier with an open-loop gain of A, there exist two basic types of feedback networks which can be applied.   All feedback networks are variations of these two basic connections.   The two types are shown in Fig. 13.

Assuming for the moment that A is arbitrarily large, then for the operational amplifier the sum of currents into the summing junction will be zero.

$$\frac{V_0}{R_2} + \frac{V_{in}}{R_1} = 0 \tag{9}$$

$$\frac{V_0}{V_{in}} = -\frac{R_2}{R_1} \tag{10}$$

Since the summing junction is maintained at ground, the input impedance seen by the source is equal to R_1.

Potentiometric Amplifiers. For the potentiometric amplifier,

$$V_0 \frac{R_1}{R_1 + R_2} + V_{in} = 0 \tag{11}$$

$$\frac{V_0}{V_{in}} = -\frac{R_1 - R_2}{R_1} \tag{12}$$

In this configuration, the sum of the voltages in series in the input circuit is arbitrarily small—hence no current will tend to flow from the source and the input impedance that the source V_{in} will see is large. A straightforward analysis of the input and feedback circuits will give the result that the effective input impedance is equal to the actual amplifier input impedance multiplied by the loop gain of the amplifier. Thus,

$$R_{in}\text{ (effective)} = R_{in}A \frac{R_1}{R_1 + R_2} \tag{13}$$

where $A[R_1/(R_1 + R_2)]$ = loop gain. Typical amplifiers have a $A[R_1/(R_1 - R_2)]$ product in excess of 10,000 ohms and R_{in} of 1,000 to 10,000 ohms. Hence the effective input impedance will be from 10 to 100 megohms.

Noise Characteristic of the Input Circuit. For the operational amplifier, the impedance seen at the amplifier input terminals (summing junction to ground) is the parallel combination of R_2, R_{in}, and $(R_1 + R_s)$. Since R_1 must be large to avoid drawing power from the source or affecting the scalar gain constant of the source and further, since R_2 typically will be greater than ten times R_1, the input resistance at the summing junction is approximately given by the parallel combination of R_{in} and $(R_1 + R_s)$, or

$$R_{effective} = \frac{(R_{in})(R_1 + R_s)}{R_1 + R_s + R_{in}} \tag{14}$$

For practical considerations, R_{in} must be at least equal to R_1, letting $R_{in} = R_1$; and $R_1 \gg R_s$.

$$R_{effective} \cong \frac{R_1}{2} \tag{15}$$

For the potentiometric amplifier, the effective open-loop input resistance is, assuming $R_2 > 10R_1$, approximately equal to the sum of the series resistors R_{in}, R_s, and R_1.

$$R_{effective} = R_{in} + R_s + R_1 \tag{16}$$

For low-noise amplifiers $(R_1 + R_s)$ will be equal to R_{in}. R_1 will be small compared with R_s. Hence the effective open-loop input resistance is

$$R_{effective} \cong 2R_{in} \tag{17}$$
$$R_{in} \cong R_s \tag{18}$$

Assuming that an absolute fractional accuracy of Δ is required, then for the operational amplifier $R_1 = R_s/\Delta$ and for the potentiometric amplifier, since the impedance is determined by the loop gain, which is large, no additional restrictions exist on

R_{in}. Recalling that the rms noise voltage is proportional to the square root of the effective circuit resistance, then the ratio of rms noise voltage will be

$$\frac{V_{\text{rms}}(\text{operational})}{V_{\text{rms}}(\text{potentiometric})} = \frac{\sqrt{R_s/\Delta}}{\sqrt{2R_s/\Delta}} \tag{19}$$

$$= \sqrt{\frac{1}{2\Delta}} \tag{20}$$

where Δ = fractional absolute accuracy.

Thus the potentiometric connection has a tremendous advantage in having an arbitrarily large input impedance while the effective noise resistance is of the same order of magnitude as the source resistance. Offsetting this advantage is the fact that a common connection does not exist between the input and output terminals. Since the input circuits of a measuring system must be isolated from ground, it will be shown in the discussion of Fitgo amplifiers that this lack of a common connection is of no consequence.

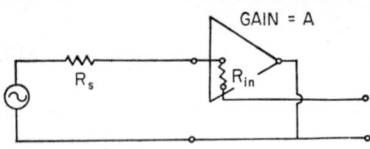

Fig. 14. Unit-gain potentiometric amplifier.

A special form of potentiometric amplifier, indicated in Fig. 14, having unit gain does have a common input and output terminal. Note that the output terminal is grounded and the ground becomes the output. This connection has the advantage of high input impedance AR_{in} and a fractional absolute accuracy of $\Delta = 1/A$ which is independent of resistors. Since A can be greater than 100,000, it is possible to amplify standard cell or other voltage sources very accurately.

INPUT AMPLIFIERS

Most transducers used in instrumentation systems have maximum outputs in the millivolt range. As a result, it usually is necessary to raise the power level of the signal prior to performing analog operations; or the translation to the digital domain with an analog-to-digital converter.

D-C Signals. For d-c signals, an amplifier will be required which has the following characteristics: (1) high input impedance, (2) low noise figure, (3) isolated input and output circuits, (4) high loop gain, and (5) low d-c offset.

Fitgo Amplifiers. The predominating characteristic of amplifiers for this application is the requirement that the input and output circuits be isolated or float. Amplifiers which maintain this high degree of isolation between the input and output circuits are called *Fitgo* amplifiers and are used in systems in which accurate low-level instrumentation is required.

Two basic types of Fitgo amplifiers are shown in Fig. 15. Both these designs depend on transformers to effect isolation between the input and output circuits. Note that, in the feedback connection, the signal and feedback voltage are in series or in the potentiometric connection. Thus the input impedance of the amplifier will be high and equal to the open-loop input impedance of the input modulator multiplied by the loop gain. Further, the noise figure for a given input impedance will be optimum.

Input Modulator. Typically, the input modulator will be a mechanical chopper in order to obtain low d-c offsets. The actual d-c offset referred to the input will be equal to the sum of the chopper offset plus feedback demodulator or d-c transformer offset plus the maximum d-c offset appearing at the output demodulator divided by the d-c to d-c gain of the input modulator (carrier amplifier-output demodulator combination).

Insulation Characteristics of Transformers. In practice, the d-c common-mode performance will be limited only by the insulation characteristics of the transformers. However, the common-mode rejection for low audio-frequency and power-line voltages is critically dependent on the effectiveness of the shielding provided in the two isolating transformers and the transformers used to furnish a-c power to the modulators and demodulators in the signal input circuits.

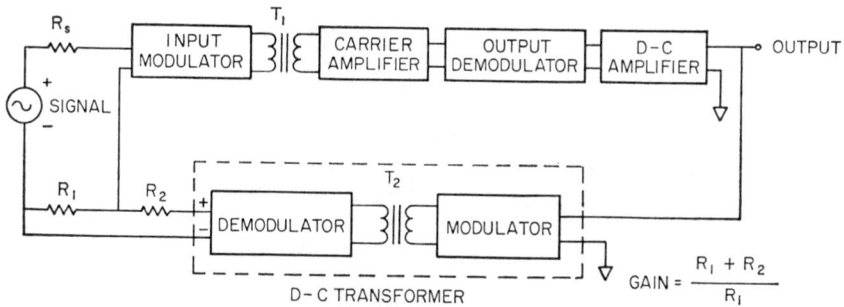

FITGO WITH D-C TRANSFORMER IN FEEDBACK

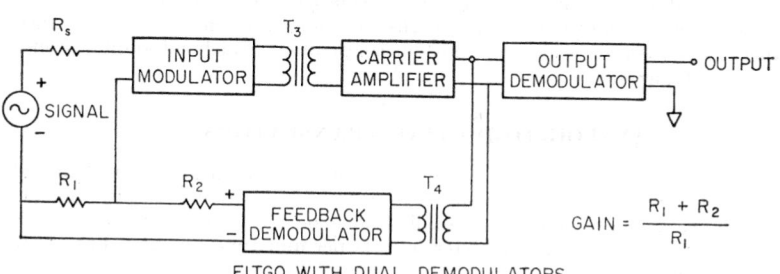

FITGO WITH DUAL DEMODULATORS

FIG. 15. Basic Fitgo amplifiers.

Design Modifications. Modifications of the basic designs just described exist which provide bandwidth in excess of that obtainable with a simple chopper-type amplifier.

The familiar potentiometric recorder (see Fig. 16), which includes either a floating power supply or battery for driving the slidewire, is an illustration of an amplifier which uses a combination of transformer and mechanical motion to effect isolation.

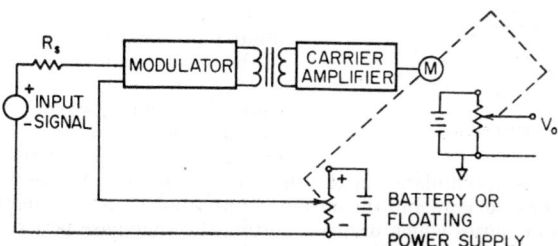

FIG. 16. Isolation of input and output circuits using transformer and mechanical motion.

Although the differential amplifier does not isolate the input and output circuits, it can also be used to amplify low-level signals and provide good common-mode rejection.

The amplifier shown in Fig. 17 amplifies the signal by an amount R_2/R_1 and ideally rejects V common mode. This is accomplished by means of the familiar differential amplifier circuits and two or more chopper amplifiers—one to correct for zero offsets between terminals A and B and the other to correct output voltage offsets.

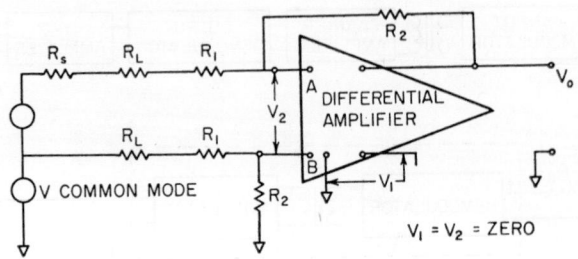

Fɪɢ. 17. Differential amplifier.

The main disadvantages of this design are (1) complexity, (2) low input impedance, (3) high noise figure, and (4) resistance paths to ground are sensitive to unbalanced impedances in the transmission and transducer circuits. Wideband amplification is the principal advantage.

ANALOG-TO-DIGITAL TRANSLATORS

The transducers associated with instrumentation systems in general will provide either an analog voltage output or some form of variable-frequency pulse output. In the latter case, translation to a digitally coded output is effected by a counter which is operated as (1) an EPUT meter (events per unit time) or (2) a TIM meter (time-interval measurement).

Both the EPUT and TIM meters consist of an output counter which counts pulses or cycles for a predetermined period called the *time base*. In the EPUT, the time base is obtained by counting a preset number of cycles of a crystal oscillator. In the TIM, the time base is obtained by counting a preset number of pulses or cycles of the input. In addition to the output counter and time-base units, a gate or switch is required which allows pulses to pass to the output counter and which can be controlled by the time-base unit.

EPUT Meters. A block diagram of an EPUT meter is shown in Fig. 18. The input pulse or frequency signal is applied to the output counter via the gate. When the counting cycle is initiated, the crystal-controlled time-base counter enables the gate to allow input pulses to pass to the output counter. The gate is enabled until the preset count is accumulated in the time base counter. At this time, the time-base counter opens the gate and the output-counter indication is equal to the number of input events or pulses which occurred during the time-base period. If the information in the input pulse train is proportional to frequency.

Fɪɢ. 18. EPUT meter.

$$\text{Information} = K_s f \tag{21}$$

where K_s = scalar constant relating frequency to the actual physical variable.

The number accumulated in the output counter will be equal to the time-base period in seconds multiplied by the input frequency.

$$\text{Digital output count} = \text{time base (sec)} \times f \tag{22}$$

$$= \frac{\text{time base}}{K_s} \times \text{information} \tag{23}$$

Thus, if the time base is 1 sec and the maximum frequency is 10,000 cps, a digital code proportional to the information and accurate to 0.01 per cent ± 1 count can be obtained once per second.

In practice, few transducers supply frequencies as high as 10 kc, and in order to obtain an accurate indication in a *reasonable* period of time, it often is necessary to multiply the input frequency by a fixed constant, using an extremely stable frequency multiplier.

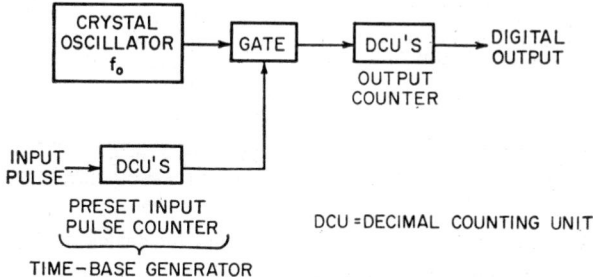

FIG. 19. TIM meter.

TIM Meters. A block diagram of a time-interval measurement meter is shown in Fig. 19. The basic function of a TIM meter is to measure the time interval associated with a preset number of input pulses. In this instance, the gate is controlled by the input or signal counter which counts a preset number of input pulses. During the time that the input pulse counter is accumulating the preset number of input pulses, the input pulse counter enables the gate and allows pulses from the crystal oscillator to pass to the output counter. If the information in the pulse train is inversely proportional to frequency, or proportional to time,

$$\text{Information} = \frac{K_s}{f} \tag{24}$$

where K_s = scalar constant relating frequency to the actual physical variable.

The number accumulated in the output counter will be equal to the ratio of the oscillator and input frequencies multiplied by the preset number n of input pulses which are counted.

$$\text{Digital output count} = f_o(\text{time interval}) = f_o \frac{n}{f} \tag{25}$$

$$= \frac{f_o n}{K_s} \times \text{information} \tag{26}$$

Sampling and Quantizing. Inherent in the operation of EPUT, TIM, or any other form of translator are the concepts of sampling and quantizing. In the EPUT or TIM, the input information is *sampled* for a period determined by the time base. Further, since the variety of different numbers which can be formed by a finite number of digits is limited, the output will have a minimum resolution of 1 count. For example, a 3-digit (999) translator will have resolution of 1 count or an accuracy of ± 0.05 per cent. The operation of generating a digital number which is related to an analog quantity is called *quantization*. Because of the nature of digital or discrete representation, the conversion of an analog signal to a corresponding number can only be an approximation, for the analog signal can take on an infinite number of values.

Quantum Step. Graphically, the process of quantization means that the straight line representing the relationship between input analog information and output information is replaced by a flight of steps as shown in Fig. 20.

The difference between two adjacent discrete values is called the *quantum step*. Clearly, the greatest error inherent in the basic quantizing process amounts to half a quantum step. By reducing the size of the quantum step, the error in the quantizing process may be made arbitrarily small.

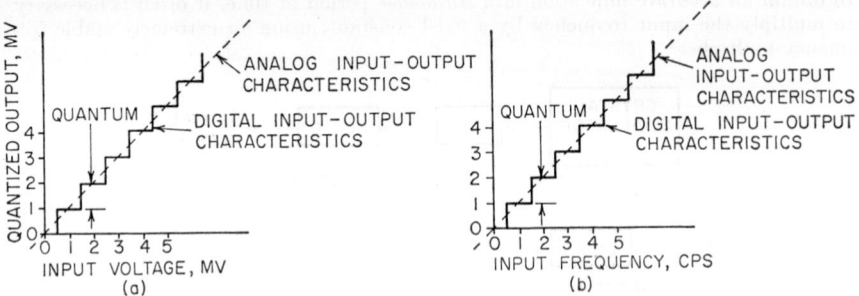

Fig. 20. Relation between input and quantized output for analog-to-digital converters.

In the case of Fig. 20a, the quantum step corresponds to a voltage increment; while in Fig. 20b it corresponds to a frequency increment. To obtain greater resolution in the voltage converter, smaller voltage increments must be detected; while in the frequency case, a longer time is required to permit the measurement of smaller frequency increments.

Later in this subsection, it will be shown that a decimal voltage-to-digital converter requires four voltage decisions or comparisons per digit contained in the output; e.g., a 4-decimal digit converter requires 16 decisions or comparisons. On the other hand, a 4-decimal digit counter must count 10,000 individual pulses to traverse its range.

In the present state of the art, devices exist which can make 0.01 per cent voltage comparisons at a 1- to 5-μsec rate. If it is assumed that 10 mc is a practical limit for decimal counters and that it is possible to operate the frequency translator over a 10-mc range, then the 16-decision converter will obtain samples 10 to 50 times faster than the frequency converter translator. This limitation is characteristic of all translators, frequency, voltage, and so on, which count as part of the quantizing process. The basic difference is the fact that one translator requires $4n$ operations for a complete quantization, while the counting type requires 10^n operations, where n is the number of decimal digits contained in the digital output.

ANALOG VOLTAGE-TO-DIGITAL CONVERTERS

Although a variety of methods exist for translating an analog voltage to a digital code, the majority of these devices fall into two categories:

1. Converters in which there is an intermediate conversion to a time interval.
2. Feedback or comparison converters in which the input is compared with a sequence of voltages controlled by a logical (digital) assembly.

Voltage-to-Time Converters

In the first class of converters, the basic conversion is voltage to time interval. The conversion from time to a digital code is accomplished by a counter as described in connection with the TIM equipment. The conversion from voltage to time can be accomplished by comparing the input voltage with a linear-sweep or sawtooth type of voltage.

Linear-sweep Converter

In Fig. 21, it is assumed that the sweep voltage is linear, that a comparator is available for determining the time t_1 when the sweep voltage passes through zero, and that a second comparator will sense when the sweep voltage equals the output voltage. Then the outputs of the two comparators define a time interval $(t_2 - t_1)$ which is proportional to the input voltage.

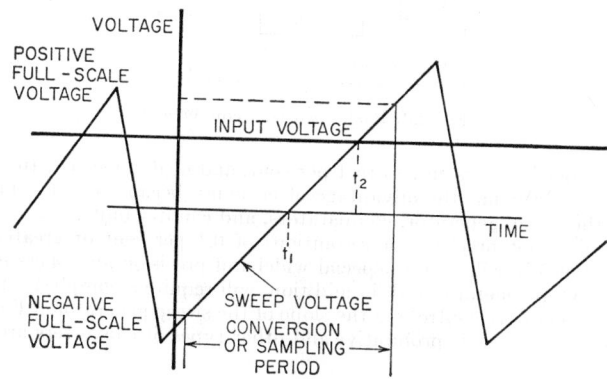

Fig. 21. Voltage-to-time coversion using a linear sweep.

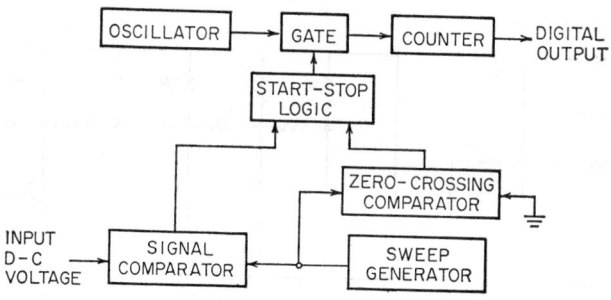

Fig. 22. Linear-sweep ADC.

The basic elements of a linear-sweep converter are shown in Fig. 22. The comparators are devices which provide sharp pulses or step waveforms when one input voltage exceeds the other. If the input voltage is greater than zero, the zero crossing comparator opens the gate and the signal comparator closes the gate. If the input is negative, the operation is reversed by means of a start-stop logic. During the interval that the gate is open, the counter counts pulses from the oscillator.

Accuracy

If an accuracy of 0.1 per cent is required, the oscillator must furnish 2,000 pulses during the conversion period. Further, the comparators must have a voltage sensitivity greater than 0.1 per cent full-scale voltage and a response time less than one pulse period of 1/2,000 of the conversion time.

Typical converters have conversion times of 1 msec to 1 sec for 0.1 per cent accuracy. For a 1-msec converter, the comparators must respond in $\frac{1}{2}$ μsec, since the comparator must be capable of making at least 2,000 decisions during the conversion or sampling

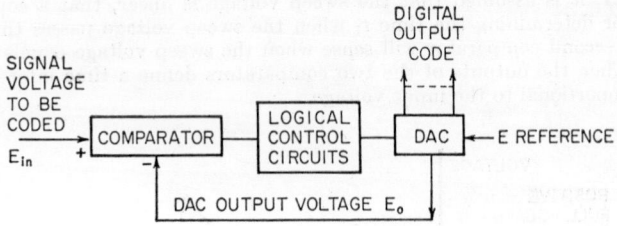

FIG. 23. Basic feedback converter.

period. For modest accuracies, $\frac{1}{2}$ to 1 per cent, and medium speed, 10 to 100 msec, this type of converter has the advantage of economy because of the simple circuits required for the sweep generator, comparators, and counter units.

For accuracies (not precision or resolution) of 0.1 per cent or greater, this type of converter typically will require special wideband precision amplifiers for the comparators and sweep generator and, in addition, will require a complicated referencing apparatus for accurately controlling the slope of the sweep generator. For accuracies of 0.01 per cent, the cost is prohibitive and other types of converters are faster and less costly.

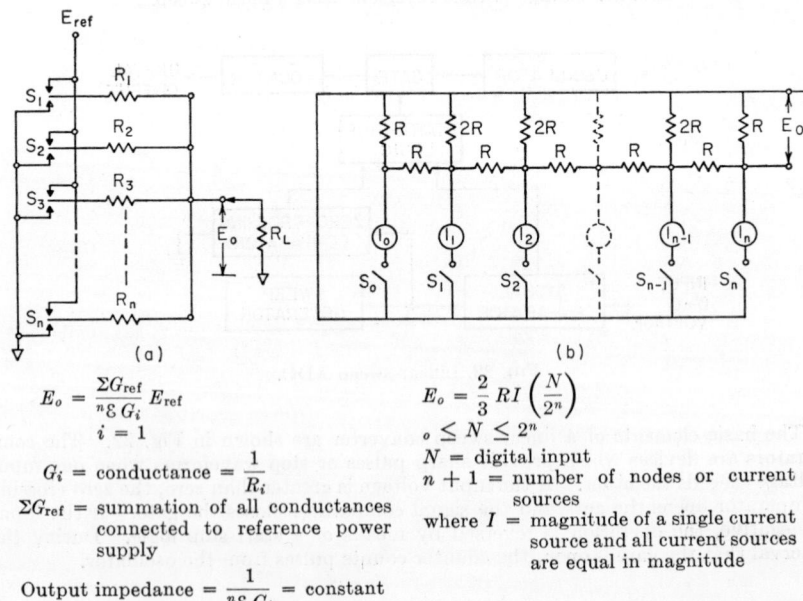

$$E_o = \frac{\Sigma G_{ref}}{{}^{n}\xi\, G_i}\, E_{ref}$$
$$i = 1$$

$$G_i = \text{conductance} = \frac{1}{R_i}$$

ΣG_{ref} = summation of all conductances connected to reference power supply

$$\text{Output impedance} = \frac{1}{{}^{n}\xi\, G_i} = \text{constant}$$
$$i = 1$$

$$E_o = \frac{2}{3}\, RI\left(\frac{N}{2^n}\right)$$
$$o \le N \le 2^n$$

N = digital input

$n + 1$ = number of nodes or current sources

where I = magnitude of a single current source and all current sources are equal in magnitude

FIG. 24. (a) Conductance-adder DAC. (b) Binary-coded digital-to-analog voltage converter using current sources with ladder network.

Feedback Converters

Most precision analog-voltage-to-digital converters are based on the comparison of the signal voltage with a voltage controlled by logical circuits connected to a comparator. In addition to the comparator and logic, these converters include a DAC (digital-to-analog voltage converter). See Fig. 23. The digital output code is obtained from the switch elements associated with the DAC. Typical digital-to-analog converters are shown in Fig. 24a and b.

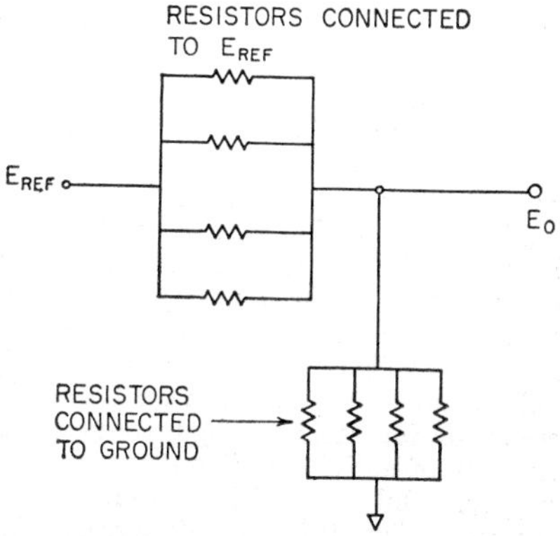

Conductance-adder-type DAC

In the conductance-adder-type DAC, the resistors are connected either to ground or to the reference voltage. Since all resistors are connected to the terminal, the combination forms a simple voltage divider, as shown in Fig. 25. The conductance of the individual resistors is weighted in proportion to the digital value assigned to a particular switch. In Table 3, the values of the resistors are indicated for a 4-decimal-digit DAC using the 2* 4 2 1 decimal code.

FIG. 25. Equivalent circuit of conductance-adder DAC.

Table 3. Four-decimal-digit 2* 4 2 1 Conductance Adder

	Digital value	Conductance	Resistance value for 2,000-ohm output impedance
S_1	4,000	4,000 G_i	5K
S_2	2,000	2,000	10K
S_3	2,000*	2,000	10K
S_4	1,000	2,000	20K
S_5	400	2,000	50K
S_6	200	2,000	100K
S_7	200*	2,000	100K
S_8	100	2,000	200K
S_9	40	2,000	500K
S_{10}	20	2,000	1M
S_{11}	20*	2,000	1M
S_{12}	10	2,000	2 megohms
S_{13}	4	4 G_i	5 megohms
S_{14}	2	2	10 megohms
S_{15}	2*	2	10 megohms
S_{16}	1	G_i	20 megohms

Switch Control

Although many logical techniques exist for controlling the switches in the feedback converter, two methods are favored in practice. In the first, the logic and control circuits contain a forward-backward counter. Each of the counter stages controls a single switch. Thus, if the count is 4,821, the 4,000, 400, 200, 200*, 20, and 1 switches are connected to the reference voltage source. If the output of the comparator indicates the DAC output is greater than the input voltage, the comparator switches the counter into a backward-counting operation and the count is reduced at a fixed rate until the two voltages are equal.

In this servo-type converter, the time to traverse full-scale voltage is equal to the full-scale count (e.g., 10,000) divided by the pulse rate. The pulse rate is limited by the speed of the comparator or the speed of the switches, whichever is the slower. Comparator and switch designs exist which permit decision and switching operations at a megacycle rate. Thus it is possible to traverse full-scale range in 1 msec for a 0.1 per cent converter and 10 msec for a 0.01 per cent converter.

In the second mode of operation, referred to as "successive approximation," the conversion cycle is completed in n steps, where n equals the number of switches. In each step, a single resistor is connected to the reference power supply or a single current source is turned on.

The most significant source (S_1 for conductive adder DAC) is switched in first by the logic circuits. If the error is positive ($E_{in} > E_0$), the resistor is left connected to E_{ref}. If the error is negative, the resistor is returned to ground. The logic circuits then switch in the next most significant source S_2 and the process is repeated. This operation is continued in the order of decreasing significance (digital value) until all switches have been tried. At the completion of the sequence (16 steps for four-decimal-digit converter), the digital value of the input is equal to the sum of individual digital values of the switches connected to the reference power supply. Only 16 decisions are required to obtain a four-decimal-digit (0.01 per cent) answer. Thus this type of converter is inherently fast, and if a 1-μsec decision rate is used, the cycle will require 16 μsec.

Input-signal Variations

In both the linear-sweep and successive-approximation converters, it is obvious that, if the input signal varies during the conversion cycle, the accuracy will be limited to the ratio

$$\frac{\text{Peak-to-peak a-c or noise voltage}}{\text{Full-scale voltage}} = \text{best realizable accuracy} \qquad (27)$$

In case of the successive-approximation converter, the digital indication always will be equal to a voltage which existed during the conversion cycle. However, the exact time of occurrence of the indicated value is uncertain. This time uncertainty introduces additional errors in subsequent data-reduction routine which attempt to recover the input-signal waveform.

SAMPLING

Since most systems which include analog-to-digital converters time-share a converter among a number of inputs, the digital output representation for a particular input is a series of intermittent samples of the input spaced in time. If the inputs are not steady (d-c) but contain a-c components, then the information which can be recovered from the series of samples is critically dependent on the sampling rate and the frequency spectrum of the input signal. The fundamental effect of sampling is independent of the use of the sampled data and is the subject of the famous sampling theorem of Shannon.

Sampling Theorem. According to the sampling theorem, if a signal $f(t)$ is sampled at times $t = -2T, -T, 0, T, 2T$, and so on, at a rate of $1/T = f_s$ samples per second, the frequency components of the signal greater than $f_s/2 = 1/2T$ cps cannot be distinguished from frequencies in the range of 0 to $f_s/2$ cps.

To illustrate the point, assume the sampling rate is 100 samples per second and that the input contains d-c plus a 100-cps component. Then, since 100 cps is greater than $1/2(100) = 50$ cps, according to the sampling theorem, the 100-cps component should appear at some frequency in the range of 0 to 50 cps.

From Fig. 26, it is clear that the 100-cps component appears at the sampling instants as an added d-c component; likewise, the same construction would demonstrate that a 90-cps signal component would introduce a 10-cps variation in the magnitude of the samples. Another example of the effects of frequency folding is given in Fig. 27. In this instance, it is assumed that $\frac{8}{7}$ samples per cycle is taken, or that 8 samples are taken every 7 sec.

To establish the relationship between sampling rate and frequency components in the signal, consider two sine waves, one having a frequency $(f_s/2 + \Delta f)$, the other $(f_s/2 - \Delta f)$. Then, by simple trigonometry, it can be shown that two waveforms

$$\cos\left[2\pi \left(\frac{f_s}{2} + \Delta f \right) t + \zeta \right]$$

$$= \cos\left[2\pi \left(\frac{f_s}{2} - \Delta f \right) - \zeta \right]$$

have equal amplitudes at times which are multiples of the sampling period, or when $t = K1/f_s$. For the example under consideration, $f_s = 100$ cps, then, a frequency of $f_s/2 + \Delta f = 50$ cps $+ \Delta f$ will appear as a component of frequency

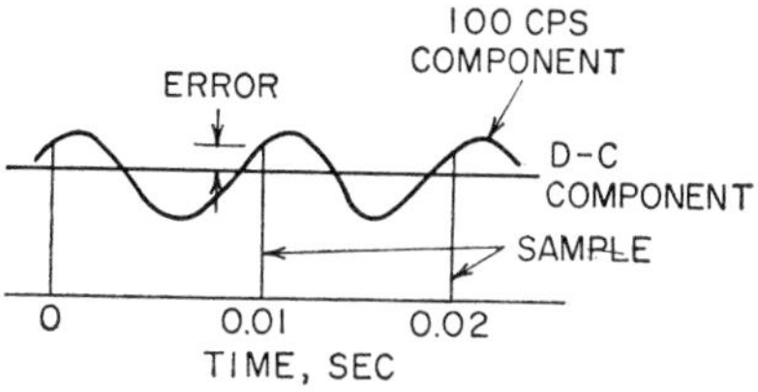

$$f_s - \Delta f = 50 \text{ cps} - \Delta f$$

in the samples. This is illustrated in Fig. 28.

Frequency Folding. The effect of sampling is to fold the original frequency spectrum into a new spectrum contained within one-half of the sampling frequency. These folds occur at every multiple of $1/2T$. Note that the input frequency spectrum includes signal plus all noise sources. In many physical systems, the noise power tends to override the signal power at frequencies above a few cycles per second.

Fig. 26. Sampling errors.

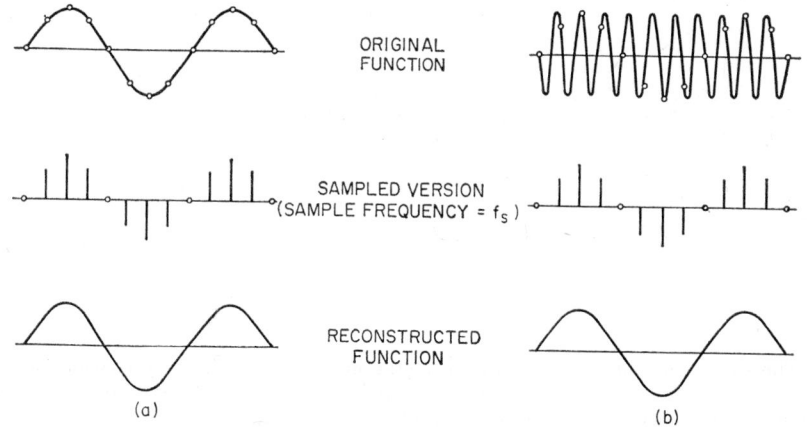

Fig. 27. Sampling theorem. (a) Legitimate application of sampling theorem. More than 2 (actually 8) samples per cycle. Reconstruction identical to original. (b) Disasterous consequences of abuse of sampling theorem. Less than 2 (actually $\frac{8}{7}$) samples per cycle. Reconstruction bears no resemblance to original.

The sampling theorem often is quoted, "It is necessary to take more than two points per cycle of the highest significant frequency component in a signal in order to *recover* that signal." This statement is true. The word "recover," however, is used in the sense that the frequency spectrum of the signal will be preserved by the sampling process. It is not true that high frequencies cannot be passed by low-frequency sampling. The sampling device will fold frequency components greater than $1/2T$ into new low frequencies which are passed by the low sampling rate. The output waveform or frequency spectrum, however, may have little resemblance to the original signal. In dealing with systems that include sampling devices, it is important to remember that sampling is not filtering. Sampling is frequency folding. If high frequencies are present in a signal, either a high sampling rate must be used, or the signal must be filtered to cut out the high frequencies before the sampling operation.

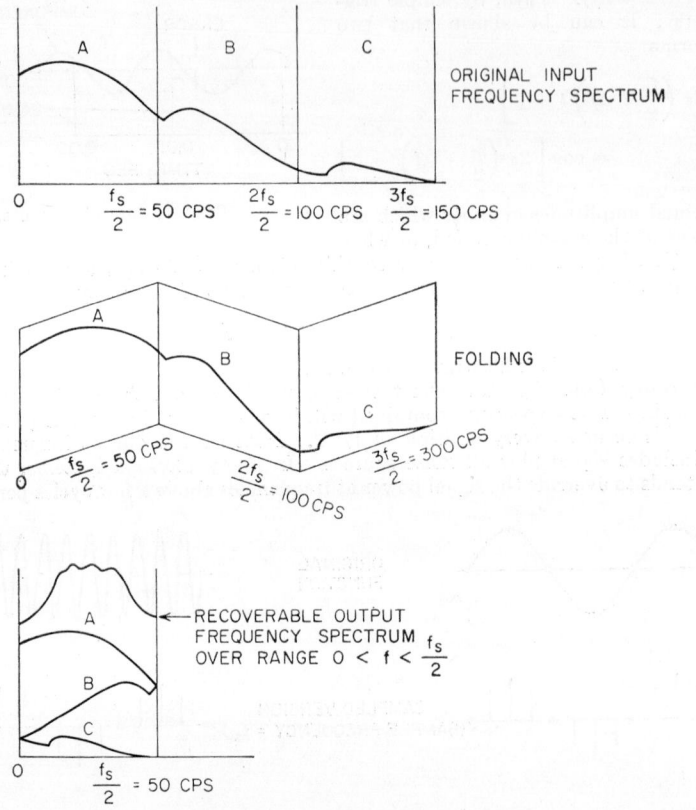

Fɪɢ. 28. Folding of frequency caused by sampling.

The spectrum following a sampling operation can be obtained by considering the Fourier spectrum of the signal, the sampling function, and the spectrum resulting from the modulation of the sampling spectrum by the signal spectrum.

Sampling Operation as a Modulation Process. In this consideration, the signal or information function $f(t)$ is modulated (multiplied) by a sampling function $s(t)$ which is zero everywhere except at the sampling times where it has a unit value. This kind of sampling function is shown in Fig. 29a. The Fourier spectrum of this series of impulses (as shown in Fig. 29b) is a similarly appearing series of lines at frequencies $-2f_s$, $-f_s$, 0, f_s, $2f_s$. The sampling function thus is the summation of continuous-wave (c-w) carriers of equal amplitude at multiples of frequency f_s. Modulation of these c-w carriers by a signal or information function having a frequency spectrum as in Fig. 29c gives rise to a series of reproductions of the signal spectrum around each multiple of f_s, as shown in Fig. 29d. In the illustration, the signal contains frequency components greater than $f_s/2$ and, as a result, the sidebands of the spectrum centered at $-2f_s$, $-f_s$, 0, f_s, $2f_s$, and so on, overlap.

Sidebands. In the ideal case, the signal would contain no frequencies greater than $f_s/2$ and the sideband centered at the carriers $-f_s, 0, f_s$, and so on, would not overlap. Thus the original signal or information could be recovered perfectly by simply passing the sampled signal through a low-pass filter to remove the harmonics of the sampling frequency and the sidebands centered at each harmonic.

In the example of Fig. 28, the sidebands do overlap, and hence the desired spectrum (upper sideband of zero frequency carrier) has added to its components originally from other frequencies. Examination of the spectrum in Fig. 29d over the range 0 to f to $f_s/2$ clearly shows that the effect of sampling is equivalent to folding the spectrum at multiples of $f_s/2$.

Obviously, the error introduced by folding phenomena is a function of the relative magnitude or power content of frequency components greater than $f_s/2$.

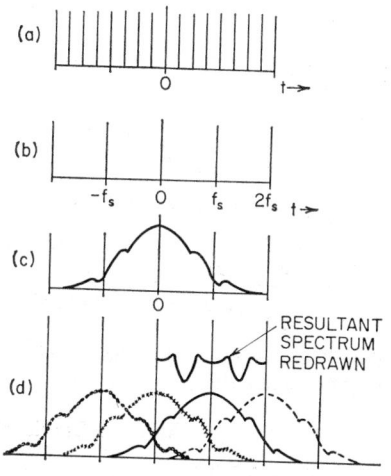

Fig. 29. Sampling operation. (a) Sampling function. (b) Fourier of sampling function. (c) Fourier spectrum of signal. (d) Modulation of (b) by (c).

ERRORS DUE TO THE SAMPLING AND CONVERSION OPERATIONS

In addition to errors introduced by transducers and low-level analog data-processing equipment, the sampling and analog-to-digital conversion operations are a potential source of error peculiar to data-acquisition systems. The sampling error is the result of frequency folding and the converter introduces time and amplitude uncertainties due to a-c components in its input signals. Further, the converter output, when alternating current is present, has no definite relation to a characteristic of the signal, such as an average, mean, and so on. These two sources of error can be controlled by filtering and an operation called "sample and hold." The latter refers to the process of sampling the analog input signal at a known instant of time and holding this voltage until the conversion cycle is complete.

Sampling Errors. It is intuitively evident from the frequency-folding concept that, if the original signal as a function of time is to be recovered subsequent to

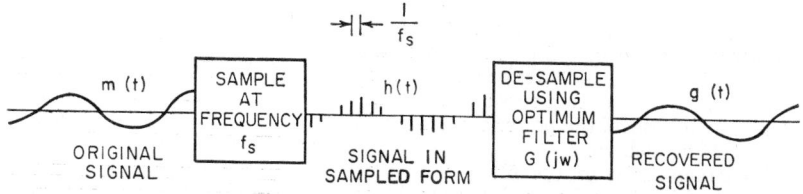

Fig. 30. Data-sampling and -recovery system.

the sampling operation, the signal which is sampled should contain negligible a-c components above one-half the sampling frequency. It is difficult to offer a general rule of thumb that applies to errors introduced by sampling because these errors depend critically on the *sampling frequency* and the *power spectrum* of the signal.

Consider the data-acquisition and reduction process shown in Fig. 30. In the figure, the sampling box includes the amplifiers and ADC. The desampling unit

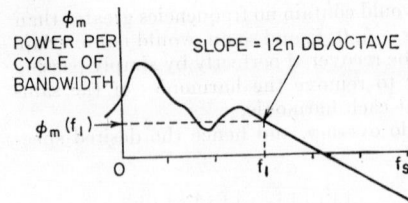

ϕ_m
POWER PER CYCLE OF BANDWIDTH
SLOPE = 12n DB/OCTAVE

$\phi_m(f_1)$

0   f_1 f_s

Fig. 31. Input-signal power spectrum. n = amplitude-spectrum cutoff rate.

includes the digital recorder and either a digital-to-analog converter and filter or a digital computer with a numerical filter program operating on the sample. Assume that the input signal has an arbitrary power spectrum ϕm for low frequencies (see Fig. 31) and that the signal power spectrum m cuts off at a fixed rate of $12n$ db/octave above some frequency f_1. Note that, if the voltage spectrum of the signal falls off at a 12 db/octave ratio $(n = 2)$, the power spectrum (αV^2) will cut off at a 24 db-octave rate. Further, let β equal the ratio of the ordinate at frequency f_1 to the average power per cycle in the band 0 to f_1. Or,

$$\beta = \phi m(f_1) = \text{power contained in signal at frequency } f_1 \tag{28}$$

$$\int_0^{f_1} \frac{\phi_m \, df}{f_1} = \text{average power per cycle in band 0 to } f_1 \tag{29}$$

$$\int_0^{f_1} \phi_m \, df = \text{area under signal power spectrum below frequency } f_1 \tag{30}$$

then it can be shown that the error in the recovered signal, assuming optimum filtering, cannot be less than

$$\text{Min rms fractional error} = \sqrt{I_n \beta \left(\frac{2f_1}{f_s}\right)^{2n-1}} \tag{31}$$

where $I_n = 0.554$ for $n = 2$
$I_n = 0.168$ for $n = 3$
Figure 32 shows the minimum error resulting from sampling for $\beta = 1/10$.

Elimination of Power-line and Other Noise Sources. In practical data-acquisition systems, the low level signals tend to be obscured by power-line and other noise sources, and it usually is necessary to provide a low-pass sharp-cutoff filter to eliminate these sources of noise. It is essential to use a sharp-cutoff filter ahead of the sampling device if the signal power spectrum contains significant energy near or above the folding frequency $f_s/2$. The frequency f_1 conveniently can be the cutoff frequency of the filter and should, of course, be less than $f_s/2$, the folding frequency. The overall spectrum cutoff rate will be the sum of the signal power spectrum and filter power spectrum cutoff rates. In a filter that cuts off amplitude-wise at 12 db/octave, the power spectrum will fall off at 24 db/octave.

The characteristics of the filter, analog or numerical, which is to be used to recover or reconstruct the data, are determined by the sampling frequency and power-spectrum cutoff rate of the signal applied to the sampling device. If the signal power spectrum falls off at a rate of $12n$ db/octave and f_s is the sampling frequency, then a nearly optimum recovery filter should have its 50 per cent amplitude response at $f_s/2$ and the amplitude cutoff rate should be $12n$ db/octave. Since the signal cutoff

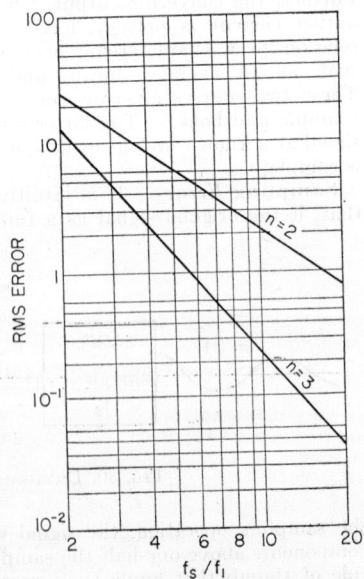

Fig. 32. Minimum rms error vs. sampling frequency for signals with power spectra of the form $1/f^{2n}$ for $f \geqq f_1$.

refers to power and the filter to amplitude, the recovery filter has a cutoff characteristic which is twice the cutoff rate of the signal (see Fig. 33).

If the signal varies during the sampling period of either the linear-sweep or feedback type of analog-to-digital converters, it is characteristic of these devices that

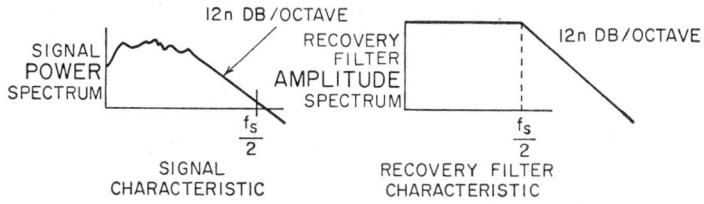

FIG. 33. Recovery-filter characteristics.

the indicated output will have existed during the sampling period. Thus, if the
signal has a peak-to-peak variation of x per cent during the sampling period, the
output indication can be in error up to $x/2$ per cent. This relation is shown in Fig. 34.

If it is assumed that a feedback-type converter is used, then in the situation depicted
in Fig. 34, the voltage is slightly greater than 6.000 during the time the most significant decisions are being made. Hence the
converter will feed back a voltage equal in
magnitude to 6.000 volts. For the remainder of the converter cycle, the input
voltage is less than 6.000 volts; hence all
other voltage increments will be rejected
and the final indication will be 6.000—a
value which existed during the converter
sampling period.

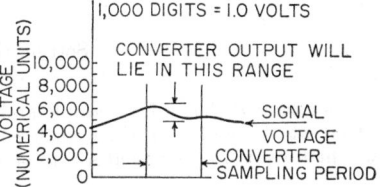

FIG. 34. ADC sampling errors.

In order to obtain a conversion accurate
to x per cent/2, it is necessary to restrict
variations of the input during the sampling
period to x per cent. For example, if it were assumed that the variations were triangular waves of frequency f having a peak-to-peak amplitude of x per cent of full-scale
voltage, then the percentage error would be approximately

$$\% \text{ error} \cong 2t_c f x\% \cong 2\frac{f}{f_c} x\% \tag{32}$$

where t_c = time for ADC to complete a conversion
f_c = sampling rate of converter, e.g., 10,000 samples per second
f = frequency of input variation
x = peak-to-peak amplitude of triangular wave as a percentage of full-scale
voltage

In order to avoid both the time and amplitude errors arising from the fact that
the signal varies during the conversion cycle, a sample-and-hold unit is used to take
a sample of the average value of the signal over a period of time which is small compared with the ADC conversion time. A sample-and-hold unit typically consists of
a condenser to hold the voltage which existed at the instant of sampling and a high-
impedance amplifier. The amplifier is used as a buffer between the condenser and
the ADC input. The actual switching or sampling is done with fast vacuum tubes
or solid-state switches.

SCANNERS

The scanners used in data-acquisition systems function to select or commutate
data sources or signals sequentially to an amplifier, ADC, and so on. This switching
operation can be performed with mechanical relays or, in some instances, with transis-

tors or diodes. The characteristic of these switching devices typically is the determining factor in the amount of equipment required in a data-acquisition system.

Amplifier Switching. If it were assumed that the switches used in scanners or commutators were infinitely fast and introduced negligible errors, the noise associated with preamplifiers would determine how many inputs could be time-shared with a single amplifier. Consider a potentiometric input amplifier with a 1,000-ohm input circuit. The Johnson noise referred to the input of this amplifier would be

$$e_{rms} = 1.28 \times 10^{-10} \sqrt{1,000B} \cong 4 \times 10^{-9} \sqrt{B} \tag{33}$$

Since the signal is being sampled, it is better to consider a peak noise voltage which will not be exceeded for, say, 99.99 per cent of the time. From Table 2, the peak-to-rms ratio for this condition is 4. Hence

$$e_{peak} = 1.6 \times 10^{-8} \sqrt{B} \tag{34}$$

If a feedback amplifier has a bandwidth of B, it will respond to a step input with a time constant T

$$T \cong \frac{1}{W} = \frac{1}{2\pi B} \tag{35}$$

If an accuracy of 0.1 per cent is desired, then approximately seven time constants will be required for the amplifier output to be within 0.1 per cent of the final value

$$\text{Setting time} = 7T = \frac{2B}{7} \tag{36}$$

If it is assumed that a fast sample-and-hold unit is available to sample the amplifier output at the end of 7 time constants, then this amplifier, in principle, could be switched every $7T$ sec; or at a rate of $1/7T$ samples per second.

$$\text{Samples per second} = \frac{1}{7T} = \frac{2\pi B}{7} \tag{37}$$

Now it is only necessary to add the restriction that the peak noise should be less than or equal to some threshold value, e.g., 10 μv. Then, combining the expression for peak and the samples per second

$$\text{Samples per second} = \frac{2\pi B}{7} = \frac{2\pi}{7} \frac{(e_p)^2}{(1.6)^2 \times 10^{-16}}$$

$$= 350,000 \text{ samples per second}$$

$$\text{for } e_p = 10 \ \mu\text{v max} \tag{38}$$

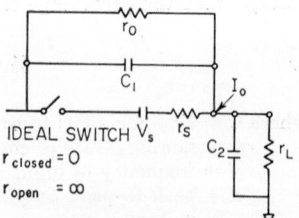

IDEAL SWITCH V_s
$r_{closed} = 0$
$r_{open} = \infty$

r_L = leakage resistance to ground
C_1 = capacity between open terminals
C_2 = stray capacity to ground
r_0 = resistance between open terminals
I_o = current generator appearing at output terminals
V_s = emf in series with closed contacts
r_s = contact resistance

Fig. 36. Analog-switching element equivalent circuit.

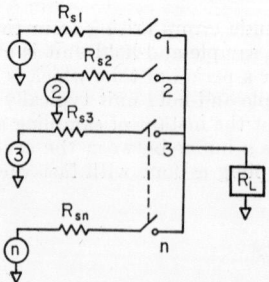

Fig. 35. Analog-switching network.

Thus an ideal amplifier could be switched 350,000 times per second to a 500-ohm source (500 amplifier input impedance) and 99.99 per cent of the time the error would be less than 10 μv. Since practical amplifiers will have a peak-to-peak noise within a factor of 2 of this, a practical figure for the sampling rate would be 350,000/4 = 90,000 samples per second. The factor 4 is due to the fact that the bandwidth must be decreased by a factor of 4 to decrease the noise by a factor of 2.

The figure of 90,000 samples per second is an upper bound on the switching or commutating speed of a low-level analog scanner, assuming an ideal switch which generates no error voltages or currents and in which no coupling exists from the switch drive circuits to the signal circuits.

Table 4 lists the characteristics of a switch which determines its usefulness for switching analog signals. In a typical application for analog signal switches, the individual switches are connected between their respective signal sources and a common input bus, as shown in Fig. 35.

Referring to Fig. 35, assume the individual switches can be described by the equivalent circuit of Fig. 36. The effect of a finite open-circuit resistance R_0 is to cause a

Table 4. Characteristics of a Switch Which Determines Its Usefulness for Switching Analog Signals

Characteristic	Remarks
Operate time	Elapsed time to change state (open or close)
Settling time	Time for contact bounce, switching transients, or triboelectric emf to disappear
Thermal emf's	Thermal emf's generated by switch mechanisms
Open or off impedance	Impedance from output to input terminals when switch is opened. Also, impedance from output terminal to ground, or other leads when in open position
Closed-circuit error voltage or current	Voltage appearing in series with switch or current source at terminals when switch is closed
Open-circuit error currents	Currents appearing at switch terminals when switch is open
Contact resistance or series resistance	Small

small current from each open switch to flow through the source impedance of the closed switch. Assuming all transducers have equal full-scale voltage V_{fs}, then this current will produce an error voltage e_{r_0}

$$e_{r_0} = (n - 1)\frac{V_{fs}}{r_0} R_s \qquad (39)$$

The fractional error X_{r_0} is

$$X_{r_0} = (n - 1)\frac{R_s}{r_0} = \text{fractional error due to shunt resistance across open}$$
$$\text{switch terminals} \qquad (40)$$

Thus, if r_0 were 100 megohms and R_s = 1K, 100 parallel open switches could produce an error of 0.1 per cent.

A similar analysis will show that a loading effect due to the individual leakage resistances R_L also can produce an error

$$X_{RL} = \frac{(n - 1)R_s}{(n - 1)R_s + R_L} = \text{fractional error due to leakage resistance to ground} \qquad (41)$$

The error e_{I_o} due to a current source feeding current into the common bus will be

$$e_{I_o} = (n - 1)I_o R_s = \text{error voltage due to stray current generator} \qquad (42)$$

If n = 100 and R_s = 1K and I_o = 1 mμa (10^{-9}), e_{I_o} will be 100 μv. I_o typically is associated with leakage resistance from the coil circuits to the contacts. If the coil voltage is 10 volts, then an insulation resistance of 10,000 megohms must be main-

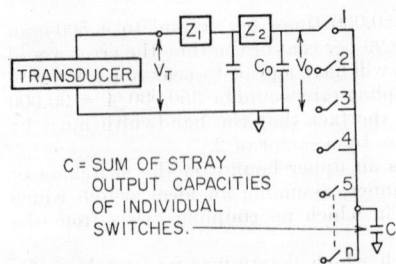

C = SUM OF STRAY
OUTPUT CAPACITIES
OF INDIVIDUAL
SWITCHES.

FIG. 37. Errors due to stray switch capacities.

tained to limit I_o to 1 mμa. In the case of relay units, this error can be minimized by connecting the relay so that the coils are deenergized when the contacts are open or by providing guard circuits.

Effects of Capacity. The effects of capacity C_1 and C_2 in Fig. 36 (equivalent circuit) are more subtle and depend on the equipment associated with the switch. In Fig. 37, a typical measuring-system input circuit is indicated in which low-pass filters are used to reject noise and power-supply components. The series elements may be either inductors or resistors.

When switch 1 is closed, the output capacitor C_o must charge the stray capacity $C = n(C_1 + C_2)$. This will decrease the output voltage V_o and cause a fractional error $X_{C_1C_2}$ if the signal is sampled before the filter has time to return to a steady-state condition.

$$X_{C_1C_2} = \frac{n(C_1 + C_2)}{C_o + n(C_1 + C_2)} = \text{error due to charging of stray switch capacity by output filter condenser} \quad (43)$$

If $n = 100$, $C_1 - C_2 = 10$ $\mu\mu$f, and $C_o = 1$ μf, then $X_{C_1C_2} = 0.1$ per cent.

Sampling Devices. Devices which are used commonly to perform analog signal switching include:

Mechanical	*Electronic*
Stepping switches	Transistors
Mercury relays	Diodes
Chopper-type relays	Photoconductors
Crossbar switches	
Rotary sampling switch	

A summary of their characteristics is given in Table 5.

STEPPING SWITCH

The stepping switch is the most economical switching device available for performing the operations of scanning or selecting where scanning refers to sequential scanning of a number of inputs and selecting refers to the operation of selecting a particular input. Stepping switches can be obtained with from 6 to 50 positions and up to 12 contacts are available per position.

Switch Lifetime. Most stepping switches require periodic maintenance in the form of lubrication and adjustment. The lack of maintenance is the main reason why stepping switches have a reputation for being unreliable. Most commercially available switches will operate in excess of 200,000,000 times if the recommended maintenance procedure is observed.

Critically adjusted stepping switches will operate from 30 to 60 times per second. For reliability and freedom from frequent adjustments, the switch should be operated at much slower rates. Further, circuit design associated with the driving motor should require a minimum reliance on the interrupter contacts. Stepping switches are available which are hermetically sealed in a *special* oil. These switches will provide in excess of 200,000,000 trouble-free operations without maintenance.

MERCURY RELAY

The mercury relay* consists of a hermetically sealed single-pole double-throw mechanism, in which the contacts are wetted with mercury. The mercury is supplied to the contacts by capillary action from a pool of mercury at the bottom of the glass

* A development of Bell Telephone Laboratories.

bulb. The mercury relay, if used within the ratings specified by the manufacturer, is the most reliable mechanical switch available. Life in excess of 5,000,000,000 operations is typical.

Thermal and Spurious Emf's. In low-level multiplexing or scanning service, certain precautions are necessary to assure adequate performance. The structure of a mercury relay is shown in Fig. 38. If low-level signals are switched from the armature A out through pins B, C, D, or E thermal emf's of 50 to 100 μv will exist. If the signal is switched from B through contact arm F and out to C, however, the thermal emf's will be less than 10 μv. The mercury relay operates in from 3 to 5 msec. The contact structure continues to oscillate for 2 to 5 msec following closure. This oscillation or movement of the contacts generates spurious emf's of 1 to 5 mv because of the triboelectric effect and motion in the magnetic field of the coil. These spurious emf's decay rapidly and for all practical purposes are negligible in 8 to 10 msec after energizing the coil.

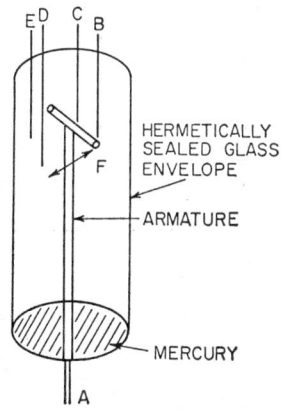

FIG. 38. Mercury relay.

Shielding. Because of the small gap and surface tension of the mercury, the relay makes before breaking. This characteristic is undesirable in many analog switching circuits, and potential applications should be reviewed carefully. By means of proper shielding around the coil and contact structure and shielding of the coil leads, it is possible to use the mercury relay in 10-megohm circuits in which the peak switching transient will be less than 5 mv.

CHOPPER RELAY

In applications in which operating rates in excess of those obtainable with mercury relays are desired, mechanical chopper assemblies can be used. Choppers are relay structures which are used as contact modulators in amplifiers. Typical chopper units have operating times ranging from 3 msec to 200 μsec and a life expectancy in excess of one-half billion operations. Special designs are available that have a double-pole double-throw structure with a stable off position in which none of the contacts are closed. This configuration will permit switching two independent circuits with a single unit and helps to offset the higher cost of a chopper-type relay. Choppers which are designed for low-level applications will have less than 10 μv of thermal emf's and are carefully shielded internally to avoid electrostatic pickup from the coil assembly.

CROSSBAR SWITCH

The crossbar switch is a mechanical matrix in which switch elements are located at the intersections of the horizontal and vertical rows and are called cross points. Switches are actuated by rods attached to magnetic actuators along two sides of the matrix. Each cross point or switch assembly contains from three to six independent single-pole switches. In order to operate a set of switches at a cross point, it is necessary to energize both the vertical and horizontal actuators which are common to that point.

One of the principal advantages of the crossbar is the ability to select a cross point at random. In comparison with a stepping switch, the crossbar is inherently faster when random selection is desired, since the stepping switch must move in a fixed progressive sequence over all intervening positions.

Crossbar switches are available which will operate reliably in 20 msec and which have a life expectancy in excess of 10 million operations/cross point. Typical switches have 100 cross points, 10 vertical columns, and 10 horizontal rows. Thus the crossbar is capable of one billion or more switching operations.

Table 5. Characteristics of Devices Commonly Used to Perform Analog Signal Switching

Switch type	Max operating rate	Closed-circuit series emf, volts	Source of emf	Open-circuit current I_o	Open-circuit shunt resistance R_o	Open-circuit leakage resistance R_L	Closed-circuit series resistance R_s
Stepping	10–30/sec	<10 μv	Thermal	Negligible	100–10,000 megohms	100–100,000 megohms	0.1 ohm
Mercury relay	50–100/sec	<10 μv if used properly	Thermal	Depends on leakage resistance to coil and coil resistance	100–10,000 megohms	100–10,000 megohms	0.1 ohm
Chopper-type relay	100–1,000/sec	5–100 μv	Thermal and magnetic flux from drive coil	Depends on leakage resistance to coil and coil resistance	1,000–100,000 megohms	1,000–1,000,000 megohms	0.1 ohm
Crossbar switch	10–50/sec	<10 μv	Thermal	Negligible	100–100,000 megohms	100–100,000 megohms	0.1 ohm
Single-transistor switch inverse connection	Up to 1,000,000/sec	$\cong 26$ mv/β	β = low-frequency common emitter current gain	Emitter on output side of switch $I_o = I_{eo}\dfrac{1-\alpha_n}{1-\alpha_n\alpha_i}$ Collector on output side of switch $I_o = \dfrac{I_{eo}(1-\alpha_i)}{1-\alpha_n\alpha_i}$	$R_o \cong \dfrac{26 \text{ mv}}{I_{eo}}\dfrac{(1-\alpha_n\alpha_i)}{1-\alpha_n}$ $R_o \cong \dfrac{26 \text{ mv}}{I_{eo}}\dfrac{(1-\alpha_n\alpha_i)}{(1-\alpha_i)}$	1–100 megohms 1–100 megohms	$R_s = \dfrac{26 \text{ mv}}{I_b}\left(\dfrac{1-\alpha_n\alpha_i}{\alpha_o}\right)$ I_b = base current
Two-transistor switch inverse connection	Up to 100,000/sec	X 26 mv/β X = fractional mismatch	Decreased offset due to cancellation of series emf's less than above depending on matching and compensation	…	Same order of magnitude as above	1–100 megohms	$R_s = 2\left(\dfrac{26 \text{ mv}}{I_b}\right)\left(\dfrac{1-\alpha_n\alpha_i}{\alpha_i}\right)$
Silicon diode switch	Up to 1,000,000/sec	(0.5X) volts X = fractional mismatch between diodes in a single switch	Forward voltage drop in diode	10^{-6} to 10^{-9} amp	26 mv/I_o = 1–100 megohms	1–10,000 megohms	$\dfrac{26 \text{ mv}}{I_{oe}V/26\text{mv}}$ V = voltage across diode when conducting. Typical values 40–500 ohms

TRANSISTOR SWITCHES

If a current is passed through the collector-base junction of an alloy junction transistor, it is a characteristic of this type of transistor that (1) the voltage from emitter to collector will be small (0.2 to 2.0 mv), (2) the series resistance appearing between the collector and emitter terminals will be small (2 to 10 ohms), and (3) the foregoing conditions hold for collector-emitter load current flowing in either direction provided the base current is greater than the load current.

If both conditions of the transistor are back-biased, then (1) the impedance between emitter and collector terminals is high (1 to 100 megohms) and (2) a small leakage current flows in both collector and emitter circuits (0.01 to 1 μa). It is obvious that these characteristics describe a switch. An additional characteristic, which classifies this device as a remarkable switch, is the fact that it can be turned on in 1 μsec and off in 1 to 10 μsec.

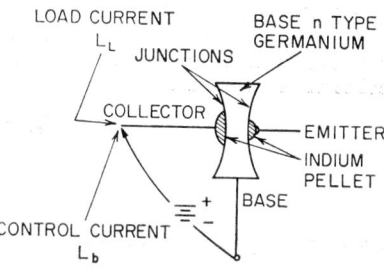

FIG. 39. p-n-p alloy junction transistor.

In normal transistor operation, control current flows through the emitter-base junction. In analog switching applications, the collector-base junction is used. The transistor is said to be in an "inverse connection." Figure 39 illustrates the construction of a typical p-n-p alloy junction transistor.

"Closed" Transistor Switch. The equivalent circuit of the "closed" transistor switch is given in Fig. 40. The voltage V_s is the total series voltage appearing between the emitter and collector terminals with load current flowing. It consists

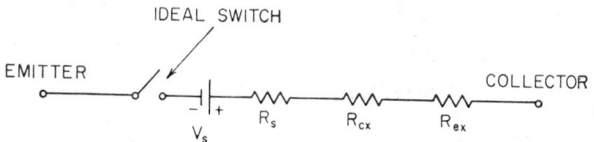

R_s = saturation resistance
R_{cx} = collector-pellet resistance (0.15 to 0.5 ohm)
R_{ex} = emitter-pellet resistance (0.4 to 1 ohm)

FIG. 40. Equivalent-circuit p-n-p transistor switch (closed position).

of two components: (1) a voltage V_f due to transistor action and (2) an IR drop due to the base current flowing through the ohmic resistance of the collector pellet.

For transistors in which β, the grounded emitter current amplification factor, is greater than 30, V_f is approximately equal to

$$V_f = \frac{26 \text{ mv}}{\beta} \tag{44}$$

Thus

$$V_s = \frac{26 \text{ mv}}{\beta} + I_b R_{cx} \tag{45}$$

In a typical analog switching application, V_s will range from $\frac{1}{4}$ to 1 mv.

Saturation Resistance. The saturation resistance is a function of the base drive or control current and the normal and inverted alphas α_n and α_i. The inverted alpha α_i refers to the alpha of the transistor if the actual collector is used as the emitter and the emitter as the collector. If the base drive is three to five times the load current, $\alpha_n > 0.98$ and $\alpha_i > 0.7$. Then

$$R_s = \frac{26\ \text{mv}}{I_b + I_L} \frac{1 - \alpha_i \alpha_n}{\alpha_i} \tag{46}$$

For I_b of 5 ma, R_s typically will be of the order of 1 to 3 ohms.

Reverse Biasing. The equivalent circuit of the transistor when both junctions are reverse-biased is given in Fig. 41. The currents flowing out of the emitter and collector terminals are

$$I_{oe} = I_{eo} \frac{1 - \alpha_n}{1 - \alpha_n \alpha_i} \tag{47}$$

$$I_{oc} = I_{co} \frac{1 - \alpha_i}{1 - \alpha_n \alpha_i} \tag{48}$$

where I_{co} = collector-base current which flows when the collector-base junction is back-biased and zero emitter current flows (open-circuited emitter)

I_{eo} = emitter-base current which flows when the emitter-base junction is back-biased and the collector is open-circuited

Shunt Resistance. This factor R_o is given by

$$R_o = \frac{26\ \text{mv}\ (1 - \alpha_n \alpha_i)}{I_{eo}(1 - \alpha_n)} \tag{49}$$

In the off condition, the true leakage resistances have been neglected. If R_o is large and I_{oe} and I_{oc} are small, then the effects of leakage resistances may not be negligible.

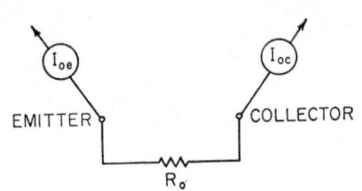

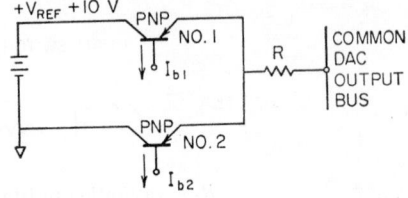

FIG. 41. Equivalent-circuit p-n-p transistor switch (open position).

FIG. 42. Typical conductance-adder element using transistor switches.

In the discussion of conductance-adder DAC's, each resistor was connected by means of a relay to V_{ref} or to ground. The circuit for accomplishing this is given in Fig. 42.

If a current I_{b1} flows as indicated and the base of transistor 2 is placed at $+12$ volts, then transistor 1 is turned on and transistor 2 is turned off. Further, the off current flowing from the emitter of transistor 2 is supplied by the power supply via transistor 1. Likewise the open-circuit resistance of transistor 2 shunts the power supply. Thus the off transistor introduces negligible currents into the DAC resistor network and the accuracy of the switching operation depends only on the characteristics of the on transistor, that is, series resistance and offset voltage. If R is large (20,000 ohms), then the 1- to 3-ohm series resistance is negligible. Further, the 1-mv maximum offset voltage is small compared with 10 volts. Thus it is possible using transistors to switch analog voltages with an absolute accuracy of approximately 0.01 per cent without recourse to compensating techniques. The same considerations apply when current I_{b2} flows and base 1 is back-biased by applying 12 volts.

Use in Low-level Switching Circuits. The fact that transistors are such good switches has prompted their application to low-level switching circuits in which the voltages to be switched are a small fraction of a millivolt. It is obvious that this cannot be accomplished unless some form of compensation or bucking is applied.

A typical circuit in which two transistors are placed back to back in order to buck out the offset voltage is shown in Fig. 43.

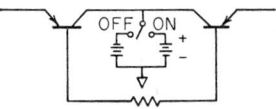

Since the series voltage appearing across a transistor varies with base drive because of the $I_b R_{cx}$ component, it is possible to adjust the potentiometer for a given temperature and set of transistor parameters—so that the two series emf's cancel. This adjustment is sensitive to temperature variations, and it usually is

Fig. 43. Two-transistor low-level switch.

necessary to thermostat the transistors if 0.1-mv stability is desired for short periods.

DIODE SWITCHES

In theory, any nonlinear resistive device can be used as a switch. Its utility in this capacity is fundamentally described by the ratio of the maximum and minimum resistances which can be obtained. In the case of a diode, this is the familiar forward-to-back resistance ratio. Typical silicon diodes have ratios of excess of 1,000,000:1. Thus these devices are potentially capable of excellent performance as switches. Silicon diodes, however, have the characteristic that the forward resistance does not become small until currents of the order of 0.1 to 5 ma are being passed through the diode.

VI **Characteristic.** This characteristic of most junction diodes follows that given by

$$I = I_o(e^{qV/KT} - 1) \tag{50}$$

where I_o = magnitude of current in reverse direction (10^{-9} to 10^{-6} amp).

If the equation is solved for the voltage corresponding to the currents in the range of 0.1 to 5 ma, it will be found that V will range from 0.5 to 0.7 volt. Thus, although the silicon diode is an excellent switch, it has a voltage offset in the on condition which is 500 to 1,000 times greater than that exhibited by transistors. If these diodes are to be used to switch analog voltages ranging from 1 mv to 5 volts, then it will be necessary to match two or more diodes very accurately so that the offsets cancel. As with transistors, these offset voltages vary with temperature, and if reasonable stability is required, thermostating is necessary.

RECORDING SUBSYSTEM

This subsystem records digital inputs from the analog-to-digital converters, time-code generators, and other digital devices in the system. The operation of the recording unit is automatically controlled by the data-acquisition system, and in addition to the recording of data, the system will typically record fixed-parameter data (manual inputs), test-run number, and test-point number.

Subsystem Selection. The choice of recording subsystem for use with a data-acquisition system depends on the characteristics of the equipment used to process the recorded information and the information rate of the data-acquisition system. In most instrumentation data-processing systems, the actual data reduction is accomplished with a digital computer. In the interest of economy and simplicity of the over-all system, it is desirable to record the data in a form which is directly acceptable by the computer or the equipment in the data-processing center which accepts the data. Typical computers are designed to accept data recorded on paper tape, punched cards, or magnetic tape. In most instances, the computer is designed to work with only one of these media, and conversion equipment is required to permit utilization of data recorded in other forms.

Information Rate. Although it is desirable to have the data recorded in a form which is compatible with subsequent machines, often it is impossible because of the rate at which the information is generated by the acquisition system. In these instances, it is necessary to choose the recording medium and equipment based on the information rate of the system and to include necessary conversion equipment to convert the data to a form which is compatible with subsequent equipment. In this situation, the recording system functions in the capacity of a buffer or a device which accepts information at one rate and emits it on command at a rate which the subsequent unit will accept.

Characteristics of Recording Devices. These are listed in Table 6 along with the recording capacity for three-digit data channels. In magnetic-tape units, standard format refers to recording on the tape with the codes and tape format that are used by the computing machine which will process the data.

Table 6. Characteristics of Recording Devices

	Max recording rate, digits/sec	Max reading rate, digits/sec	Remarks
Punched paper tape:			
Mechanical..............	10–240	10–60	
Photoelectric..............	...	100–1,000	
Punch cards:			
Mechanical:			
Serial.................	20		
Parallel................	130–200	...	100–150 cards/min
Photoelectric..............	...	1,200	Up to 900 cards/min
Magnetic tape:			
Standard format...........	10,000	...	Reading rate less than or equal to
Nonstandard format.......	225,000	...	recording rate and determined by
			computer characteristics
Printers:			
Standard typewriter........	7–10		
Line at a time..............	240–1,800	...	120–900 lines/min
Electronic, Xerox..........	10,000		

PUNCHED PAPER TAPE

Standard paper-tape punches operate at 6, $7\frac{1}{2}$, and 10 characters per second. For use with computers, paper-tape punches running at 60, 100, and 240 characters per second are available. The standard density of information on tape is 10 characters per inch of tape length and 5, 6, 7, or 8 channels across the width of the tape.

Tape Reading. This usually is accomplished by putting the tape into the reading position and then advancing a row of pins up to the tape. Where the tape has been punched, the sensing pins move up through the tape and establish an electrical contact. Maximum reading rate is typically 20 to 60 characters per second. Photoelectric readers are available that do not require stopping the tape to read a character. Typical photoelectric readers operate at rates of 100 to 1,000 characters per second.

PUNCHED CARDS

Punched cards have gained universal acceptance for modest speed recording or data processing. Typical cards are 80 or 90 columns wide and 12 rows high. Card-punching speed depends on whether the card is punched a column at a time serially or a row at a time, the latter referred to as "summary punching." In either case, punch operations are performed at the rate of 15 to 25 punching operations (rows or columns) per second. Since there are 12 rows and 80 columns, the summary punch will be 80/12 times faster than the serial punch. Considerable accessory apparatus is required for the summary punch, however, since all 80 digits must be scanned for each row to determine if a hole is to be punched in a particular column.

Scanning. The scanning operation typically involves an 80-digit register and means for gating the appropriate information to the punch magnets as the individual rows pass under the punches. Card-reading machines sense the holes by means of wire brushes or pins which make contact to metal surface through the holes.

MAGNETIC TAPE

Because of the compact nature of the recorded information and the flexibility of magnetic-tape recording systems, magnetic-tape recording of digital information is, at present, the only satisfactory means of recording data from high-speed (5,000 to 30,000 channels per second) data-acquisition systems. In addition to their ability to handle data at higher rates, tape systems have the advantages of a minimum storage volume and the ability to be erased and reused repeatedly. The error rate due to flaws in typical computer-grade magnetic tape will be less than 1 part per billion.

Recording Format. For high-speed data recording, information is recorded on the tape in parallel channels or columns across the tape. Pulse density along the tape varies from 100 to 500 pulses or columns per inch for reliable (error-free) tape systems and from 500 to 2,000 columns per inch for those systems in which a larger error rate is admissible.

Standard tape speeds are $7\frac{1}{2}$, 15, 30, 60, 75, 90, 120, and 150 in./sec and typical machines have start-stop times ranging from 1 to 10 msec. Up to 14 channels per inch of tape width can be provided using parallel adjacent heads. Standard tape widths are $\frac{1}{4}$, $\frac{1}{2}$, and 1 in., providing 3, 7, or 14 channels running the length of the tape. Typical tape formats are illustrated in Fig. 44 for $\frac{1}{2}$- and 1-in. tapes.

Parity Bit. The parity bit serves two functions: (1) a check during the reading operation and (2) a means of generating a clock pulse in non-return-to-zero recording.

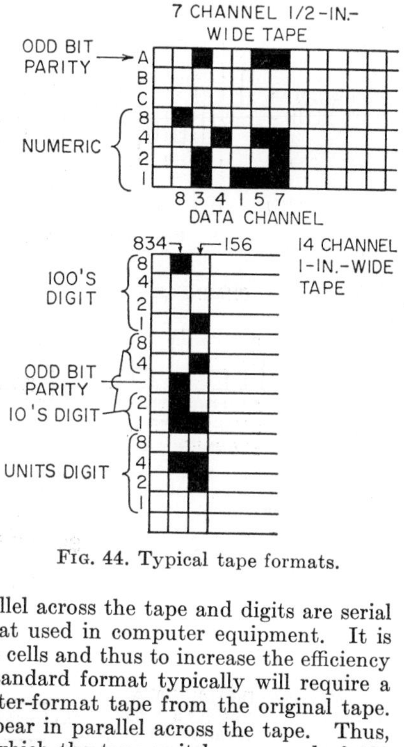

FIG. 44. Typical tape formats.

In the 7-channel $\frac{1}{2}$-in. tape, bits are in parallel across the tape and digits are serial along the tape. This is typical of the format used in computer equipment. It is possible to assign a numerical value to all the cells and thus to increase the efficiency of utilization. However, the use of a nonstandard format typically will require a tape-to-tape converter for generating computer-format tape from the original tape. In the 14-channel 1-in. tape, three digits appear in parallel across the tape. Thus, for a three-digit data-acquisition system in which the tape unit has a speed of 120 in./sec and a density of 500 bits per inch, a recording rate of 60,000 = (120)(500) data channels per second can be accommodated.

Recording Techniques. The methods for recording on magnetic tape fall into two broad categories: (1) those in which the current to the writing heads is allowed to return to zero (RZ) between each pulse and (2) those in which the current is continuously maintained at negative or positive saturation and hence is non-return-to-zero (NRZ). The latter, NRZ, type of operation is more reliable and has found widespread use.

Two forms of "non-return-to-zero" and "return-to-zero" characteristics are illustrated in Fig. 45.

NRZ Recording. In this method of recording, the current or flux changes when the sequence of zeros and ones changes. This is equivalent to assigning a positive current for ones and negative current for zeros.

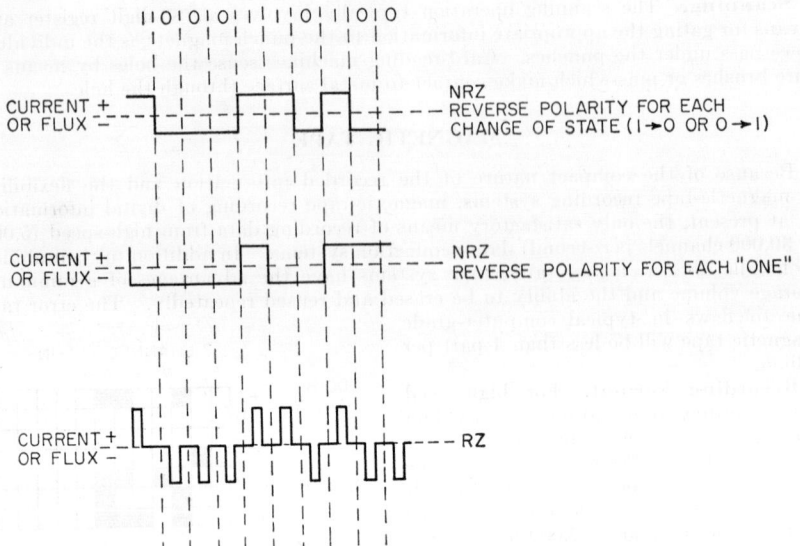

FIG. 45. Magnetic-tape recording techniques.

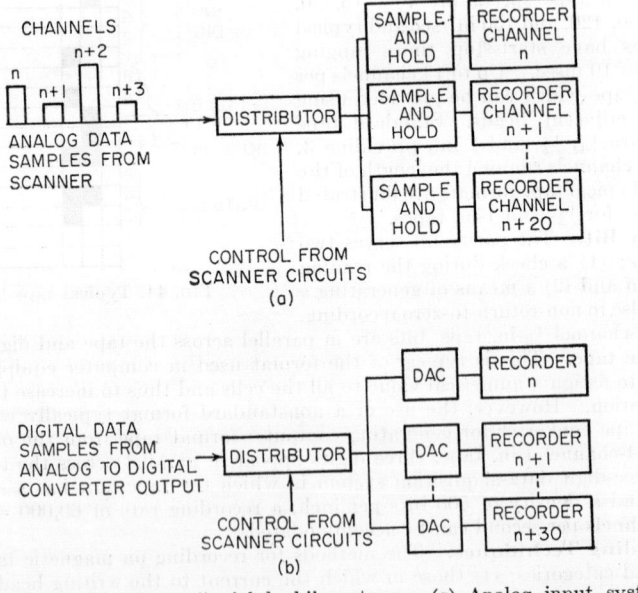

FIG. 46. Real-time elementary "quick-look" systems. (a) Analog input system. (b) Digital input system.

In the second non-return-to-zero recording process, the polarity of current is changed each time a one appears in the sequence. This method has the advantage that, during playback, a voltage will be produced at the read head each time a one is present and no voltage is generated by zeros.

RZ Recording. For comparison, a return-to-zero recording operation is indicated in Fig. 46 in which ones are written with positive pulses and zeros with negative pulses. The return-to-zero operation requires greater bandwidth capabilities for the recording system and is, in a sense, redundant in that the original sequence can be reconstructed knowing only the first digit in addition to the change in the digit sequence.

DATA EDITING AND EVALUATION

Because of the high cost of running tests and setting up information and data-acquisition systems, it is common practice to record large blocks of data within which are contained the specific test results that are sought. The procedure minimizes the chance of missing critical portions of tests that run for a few seconds or minutes. At the conclusion of the series of tests, only certain critical runs are selected for processing through the data-reduction and -presentation equipment. This mode of operation is justified economically on the basis that the incremental cost of recording additional data is negligible, whereas the cost of a test in which insufficient data are accumulated can be great.

Data Costs. In the case of the data-reduction process, the cost is proportional to the amount of data which must be reduced and distributed to plotters, printers, and associated equipment. The cost of the reduction process in terms of computer time, personnel time, and for associated matters is typically large enough to warrant some form of editing or data evaluation during the test prior to the reduction operation—to select that part of the data which is important and should be processed completely.

Forms of Recorders. Since human beings cannot assimilate tabular data effectively, it is necessary to provide editing information in the form of graphs or analog recordings. These analog recordings of the raw or semiprocessed data can be obtained in real time from the data-acquisition system, or subsequent to the completion of the test from the digital record generated by the recording system.

The common forms of recorders include (1) continuous-trace oscillographs, (2) digital-input multistylus analog recorders, and (3) plotters. The bar graph is another form of analog display which has an advantage in applications involving a series of related measurements, e.g., temperature or strain distribution.

In a typical application, only a fraction of the data channels will be displayed in the analog recorders associated with the editing equipment. The emphasis in the design of this equipment is the ability to obtain a "quick look" at the data and not accuracy or quality.

REAL-TIME "QUICK-LOOK" SYSTEMS

Quick-look systems which provide data presentation in real time can obtain their input either from the actual transducer or from the output of components in the acquisition system. The first approach, direct connection of the inputs, is undesirable on the basis that the auxiliary equipment would be unnecessarily expensive because of the floating and restrictive specifications required for input components. If the signals are obtained from sampled analog signals or from digital signals available in the acquisition system, then means must be provided to select the appropriate-signal and to distribute the signal to a designated recorder or position on the bar graph.

Basic "Quick-look" System. The block diagram of elementary quick-look systems is given in Fig. 46. The analog input system requires analog switches in the distributor and some form of sample and hold to provide a steady output to the recorder. In the case of the digital system, the distributor simply is a digital and-gate associated with each DAC. Logical circuits which control the individual gates are slaved to the scanners in the acquisition system. The same control scheme applies to the analog switch distributor used in the analog input quick-look system.

Bar-graph Oscilloscopes. These devices can be used in place of the analog recorders, as indicated in Fig. 47. The bar graph consists of a 10- to 20-in. cathode-ray tube and the additional circuitry required to generate a series of vertical lines originating from a common base line. The individual line height is proportional to the magnitude of the quantity being indicated. In operation, the X-axis position of the beam is controlled by a digital-to-analog converter which decodes a sequence of codes from the scanner to a corresponding d-c voltage. The Y-axis input is a voltage proportional to the magnitude of the signal which is being sampled by the data-acquisition system. The advantages of the bar-graph system include the

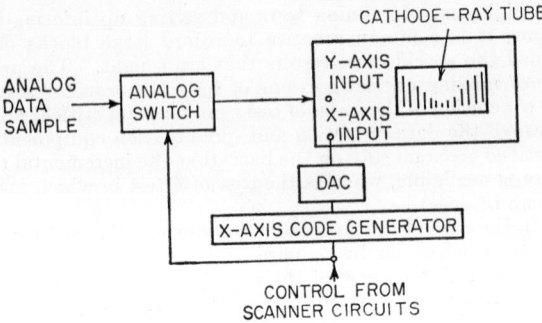

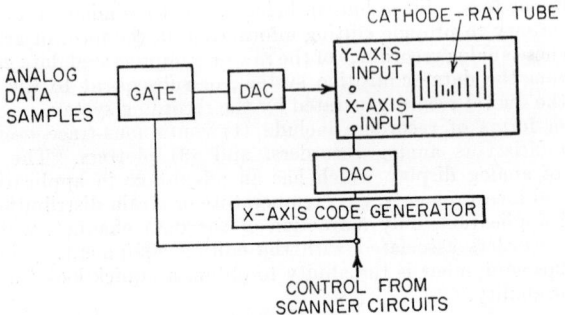

FIG. 47. Bar-graph "quick-look" system.

ability to display from 20 to 50 inputs in a single presentation and the ability to present the simultaneous variation of a group of related variables on a single display. The disadvantages of this device is the lack of a permanent record.

RECORDED-DATA "QUICK-LOOK"

A characteristic of the real-time quick-look system is the requirement of an individual recorder for each channel which must be displayed. If it is assumed that the primary function of a quick-look system is the presentation of selected data channels for evaluation prior to data reduction, then the real-time requirement for quick-look becomes questionable from an economic viewpoint.

The alternate approach is to use a single analog recorder quick-look system and to scan the recorded data once for each data channel which is to be recorded. Thus, in place of a multiple number of analog recording systems (distributor, sample and hold, and recorder), a single analog recording system can be used in conjunction with a tape-search unit. This configuration is indicated in Fig. 48.

Tape-search Unit. This unit controls the operation of the tape recorder and enables the gates when the desired run number, frame number, and channel number are being read out. Since the data rate typically will exceed the capabilities of the analog recorder, it is necessary to provide a core-memory buffer storage for accepting a block from the tape and furnishing these data at a slower rate to the recorder unit. Time typically will be recorded once per frame, and it will be necessary to add or subtract a fixed-time increment to the time code in order to plot the actual time at which the data sample was taken. This is accomplished by an offset translator unit which adds or subtracts the increment digitally at the output of the core memory. The chief advantage of this approach is the ability to plot a function of two variables. It is possible to include amplifiers after the DAC's for scaling the raw data to provide reduced data presentation.

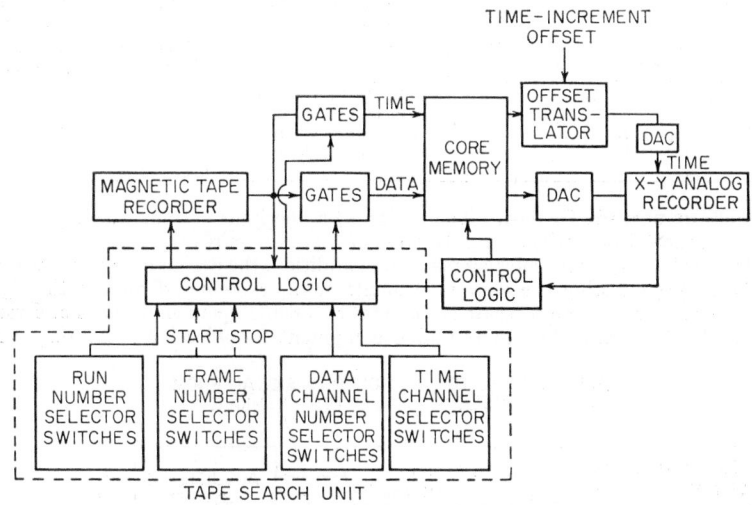

FIG. 48. Recorded-data "quick-look" recorder.

Oscillographic Recorder. An alternate approach that requires substantially less equipment records the data on an oscillographic recorder at the same rate it was recorded originally. This technique eliminates the need for the core memory and X-axis equipment and, as a result, is capable of recording only the variation of an individual data channel vs. time.

COMPUTER-PROGRAMMED CHECKOUT SYSTEMS

The previous discussion has emphasized systems considerations and techniques for data-acquisition systems in which the input signals are predominately d-c voltages from low-level transducers. This situation is typical for test programs in which transducers are used to measure the performance of a test vehicle, engine, or a plant. Another important function of data-acquisition and -reduction systems is the operational testing or checkout of complex electronic and electromechanical systems.

Operational Checkout. This task is different from instrumentation data acquisition in that it is the rule and not the exception that data exist in all conceivable forms, e.g., d-c and a-c voltages, frequency, time, resistance, and so on. Further, the measuring system is required to function over an enormous dynamic range for each of the indicated quantities. For example, a checkout system typically will require the measurement of signals in the 200- to 1,000-volt range for checking power supplies and, in addition, the measurement of millivolt and microvolt signals obtained from transducers or information-bearing signals in the vehicle under test. The same checkout system also may be required to measure the fundamental component of a-c voltages existing in a servosystem in the vehicle to check the stability or performance of a servo. Measuring systems used in checkout or operational testing programs may include measuring subsystems or components for handling one or more of the signals indicated in Table 7.

Table 7. Measuring Subsystems or Components for Handling Input Signals in Checkout Programs

Signal	Range	Remarks
D-C voltage	1 mv to 1,000 volts	Grounded measuring system
Absolute or ratio	10 µv to 1 volt	Fitgo measuring system
A-C voltage, true rms, absolute or ratio	1 mv to 1,000 volts	Grounded measuring system
A-C voltage, peak-to-peak	10 µv to 1 volt	Fitgo measuring system
A-C voltage, average		
Resistance	1 ohm to 10 megohm	
Frequency	1 cps to 1 mc	
Time interval (period)	100 µsec to 1,000 sec	
Impedance (complex)	10 ohms to 100 megohms	

Checkout-system Design. The design of a checkout system which performs a given sequence of operations on a fixed set of inputs requires the specification of the signals and the measuring subsystems and, in addition, the design of the programming unit for interconnecting the various elements of the system and routing the signals through the system. As a result of the many possible operating modes and ranges of a checkout system, the programming equipment for selecting measuring-system

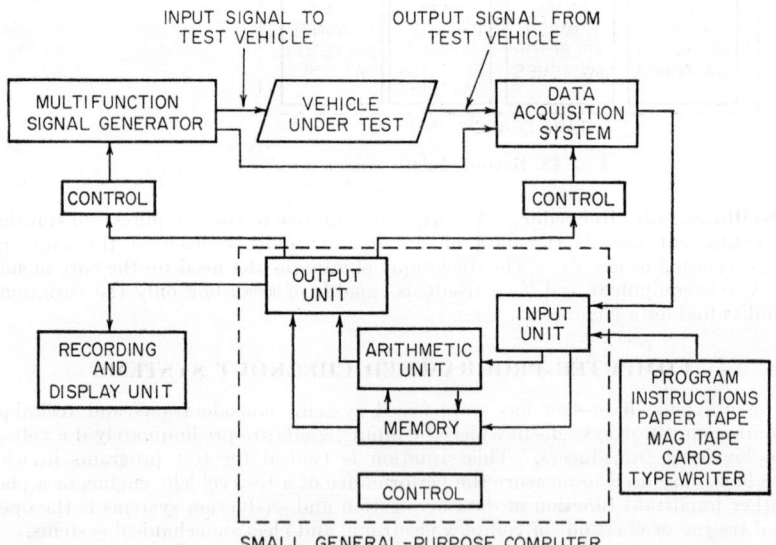

FIG. 49. Computer-programmed checkout system.

components and signal-routing paths will require more equipment than the measuring and recording subsystems. If the programming or operating sequence of the checkout system is fixed, economic justification becomes difficult because of the limited flexibility of the over-all system and the difficulty of changing the program to accommodate modifications of test specifications or new test routines.

Programming. A digital computer with its associated memory is essentially a general-purpose machine that easily can be programmed to perform a series of operations in a prescribed sequence. These series of operations includes instructions to input, output, arithmetic, and memory units. If the data-acquisition system is considered as the input to the computer, then the computer via its output unit is capable of coordinating an endless variety of instructions and operating sequences associated with the data-acquisition system. Further, the computer program can include the instructions for the test-vehicle input-signal unit which provide stimuli or test signals to the vehicles under test. The organization of such a system is shown in Fig. 49.

Computer Instructions. For a particular test program, these instructions will be prepared in advance and stored on cards, paper tape, or magnetic tape. To set up the program, it is only necessary to enter these instructions into the computer

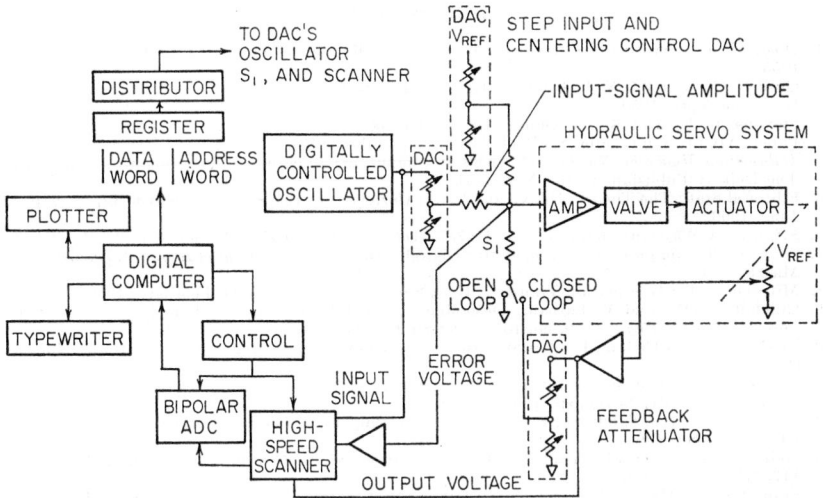

Fig. 50. Automatic servo test system.

and to attach input and output cables to the vehicle. The computer, in conjunction with the stored program, will select the appropriate function generator and adjust the input-signal level. In addition, it will select the appropriate measuring subsystem and set up the interconnections between the various elements of the data-acquisition system. As a final operation, the computer will indicate when data are to be accepted or signals are to be sampled and the disposition of the data.

Typical System. As an illustration of the flexibility and utility of this approach, consider the system of Fig. 50 which is designed to test the performance and stability of hydraulic servosystems. The system included in the figure will automatically perform the following tests and plot or log the data: (1) open-loop frequency response, (2) closed-loop frequency response, (3) gain measurements, (4) stability-gain and phase margins, (5) step response, and (6) automatic centering.

The computer instructions to the checkout system include an address word and a data word. The address selects which digitally controlled element (oscillator, DAC, switch, and so on) is to be adjusted, and the data word contains the magnitude of the quantity (frequency, attenuation, and so on). After the settings are entered on the various system elements, the computer emits a series of pulses indicating sampling times.

Because of the nonlinearity of the servosystem, it is usually the fundamental component which is desired. This component can be obtained by alternately measuring the input and output samples and programming the computer to perform an integration over a sample period to obtain the fundamental component. During the period of the input sine wave, the computer automatically takes 30 samples from both the input and output signals. These samples are integrated and the gain and phase angle of the fundamental is plotted by the plotter. Gain stability is obtained by incrementally increasing the loop gain until the system begins to oscillate. Knowing the open-loop gain and phase shift, the computer can then calculate the gain and phase margins.

The system of Fig. 50, in which the computer can communicate effectively with the measuring subsystem, the input function generator, and the recording subsystem, allows maximum utilization of the investment in a test facility in which test programs and techniques must keep pace with new technology.

REFERENCES

1. Doss, M. P.: "Information Processing Equipment," Reinhold Publishing Corporation, New York, 1955.
2. Gotlieb, C. C. and J. N. P. Hume: "High-speed Data Processing," McGraw-Hill Book Company, Inc., New York, 1959.
3. Stevens, R. F., and W. W. Gaus: "Integrated Data Processing in Process Control," Technical Paper, AIChE Symposium, St. Paul, Minn., September, 1959.
4. *Datamation Magazine:* Various references on computing and data handling, bimonthly publication, The Delreay Publishing Corp., New York.
5. Editorial Staff: The Theory, Design and Use of Correlators, *Automatic Control,* vol. 12, no. 6, p. 17, June, 1960.
6. Sylatins: A Way Out of the System Maze?, *Automatic Control,* vol. 12, no. 4, p. 37, April, 1960.
7. Shaw, R.: Techniques and Equipment for Digital Data Conversion, *Control Eng.,* vol. 7, no. 3, p. 107, March, 1960.
8. Mitchell, L. B.: Information Transfer, *Automation,* vol. 7, no. 2, p. 93, February, 1960.
9. Cundall, C. M., and V. Latham: Designing Sampled Data Systems—Part I. Short Sampling Periods., *Control Eng.,* vol. 9, no. 10, pp. 82–86, October, 1962.
10. Frink, W. J.: In Coding, It's the Structure that Counts, *Control Eng.,* vol. 9, no. 10, pp. 100–101, October, 1962.
11. Kintner, P. M.: Converting Integers: Mathematics Points to the Equipment, *Control Eng.,* vol. 9, no. 11, pp. 99–101, November, 1962.
12. Gelb, A., and H. J. Sandberg: Finding System Settling Time, *Control Eng.,* vol. 9, no. 11, pp. 103–104, November, 1962.
13. Holmes, J. F.: How to Do Your Own Communications Study, *Control Eng.,* vol. 9, no. 11, pp. 114–117, November, 1962.
14. Staff: Cresap, McCormick, and Paget: Other Input-Output Devices for Medium to Large Computers, *Control Eng.,* vol. 9, no. 11, pp. 119–123, November, 1962.
15. Boyd, A. G.: A General Approach to Information Systems Design, *Control Eng.,* vol. 9, no. 8, pp. 100–104, August, 1962.
16. Sheldon, I. R., and T. B. Luzon: Facts and Cautions for Planning Data Communications, *Control Eng.,* vol. 9, no. 8, pp. 105–108, August, 1962.
17. Wilson, M. C.: Encoder Speed Nomograph, *Control Eng.,* vol. 9, no. 7, p. 97, July, 1962.
18. Cundall, C. M., and V. Latham: Designing Sampled Data Systems, Part II. Long Sampling Periods, *Control Eng.,* vol. 10, no. 1, pp. 109–113, January, 1963.
19. Martino, R. L.: Applying Critical Path Method to System Installation, *Control Eng.,* vol. 10, no. 2, pp. 93–98, February, 1963.
20. Wilson, M. C.: Improving Reliability of Shaft Position Encoders, *Control Eng.,* vol. 9, no. 6, pp. 109–112, June, 1962.
21. Holmes, J. F.: Available Facilities for Data Communications, *Control Eng.,* vol. 9, no. 6, pp. 133–134, June, 1962.
22. Staff: Cresap, McCormick, and Paget: Printing Equipment for Digital Computers, *Control Eng.,* vol. 9, no. 1, pp. 91–97, January, 1962.
23. Fitzgibbon, T. P.: Plotter Extracts Data from Computer Curves, *Control Eng.,* vol. 9, no. 4, pp. 95–96, April, 1962.
24. Wolff, C. H.: Nine Criteria for Selecting a Digital Code, *Control Eng.,* vol. 9, no. 4, pp. 128–130, April, 1962.
25. Staff: Cresap, McCormick, and Paget: Random Access Devices for Medium to Large Computers, *Control Eng.,* vol. 9, no. 4, pp. 131–137, April, 1962.
26. Staff: Cresap, McCormick, and Paget: Magnetic Tape Equipment for Larger Digital Computers, *Control Eng.,* vol. 9, no. 3, March, 1962.
27. Helweg, T. W.: When Can You Justify Optical Character Readers? *Control Eng.,* vol. 9, no. 2, pp. 119–121, February, 1962.
28. Staff: Cresap, McCormick, and Paget: Magnetic Tape Equipment for Digital Computers, *Control Eng.,* vol. 9, no. 2, pp. 124–129, February, 1962.
29. Luftig, M.: Punched Tags Let Computer Write the Orders, *Control Eng.,* vol. 9, no. 12, pp. 105–107, December, 1962.
30. Kompass, E. J.: The Whys and Wherefores of Information Systems, *Control Eng.,* vol. 8, no. 1, pp. 103–105, January, 1961.

31. Bain, M.: Automatic Plotting of Digital Computer Results, *Control Eng.*, vol. 8, no. 1, pp. 111–114, January, 1961.
32. Shaw, R. F.: What's Available for Digital Data Transmission, *Control Eng.*, vol. 8, no. 2, pp. 127–133, February, 1961.
33. Ellis, W., G. R. Justus, and W. D. Bell: Systems Talk Through Common-language Pool, *Control Eng.*, vol. 8, no. 2, pp. 135–138, February, 1961.
34. Ruttkay, P.: Eight Ways to Read Punched Tape, *Control Eng.*, vol. 8, no. 3, pp. 152–153, March, 1961.
35. Fluegel, D. A., E. D. Tolin, and J. R. Parsons: Collecting Process Data for an On-line Digital Computer, *Control Eng.*, vol. 8, no. 4, pp. 147–150, April, 1961.
36. Cohen, H. I.: Auxiliary Processors Speed Large Computing System, *Control Eng.*, vol. 8, no. 5, pp. 137–139, May, 1961.
37. Todt, G.: How Dials Direct Pneumatic Messengers, *Control Eng.*, vol. 8, no. 6, pp. 128–130, June, 1961.
38. Bayer, R. H.: Information System Controls Plant over 500-mile Data Link, *Control Eng.*, vol. 8, no. 7, pp. 115–116, July, 1961.
39. Stern, R. K.: The Iterative Analog: Low-cost Speed for Information Processing, *Control Eng.*, vol. 8, no. 7, pp. 117–121, July, 1961.
40. Faulkner, C. E.: New Data Entry Device: The Director Card System, *Control Eng.*, vol. 8, no. 8, pp. 97–100, August, 1961.
41. Cearley, C. R.: The State of Development of Telemetering Components, *Control Eng.*, vol. 8, no. 8, pp. 101–104, August, 1961.
42. Staff: Cresap, McCormick, and Paget: Punched Card Equipment for Medium Size Computers, *Control Eng.*, vol. 8, no. 10, pp. 105–112, October, 1961.
43. Taylor, K.: Get Maximum Reliability from Digital Magnetic Tape, *Control Eng.*, vol. 8, no. 10, pp. 113–115, October, 1961.
44. Staff: Cresap, McCormick, and Paget: Punched Card Equipment for Digital Computers, *Control Eng.*, vol. 8, no. 11, pp. 115–118, November, 1961.
45. Shaw, R. F.: Preserving Accuracy in Digital Data Transmission, *Control Eng.*, vol. 8, no. 11, pp. 119–121, November, 1961.
46. Staff: Cresap, McCormick, and Paget: Punched Paper Tape Equipment for Digital Computers, *Control Eng.*, vol. 8, no. 12, pp. 105–109, December, 1961.
47. Ferguson, W. A.: Establishing Centralized Data Processing, *Automation*, vol. 9, no. 5, pp. 50–56, May, 1962.
48. Staff: Interplant System Collects Production-payroll Data, *Automation*, vol. 9, no. 5, pp. 71–72, May, 1962.
49. Sturgeon, P. B.: Peripheral Equipment Speeds Data Reporting, *Automation*, vol. 9, no. 8, pp. 61–65, August, 1962.
50. Meers, R. E.: Data Collection and Accumulation, *Automation*, vol. 10, no. 1, pp. 70–74, January, 1963.
51. Stanke, W. A.: The Transponder—New Link in Data Digitizing, *ISA J.*, vol. 8, no. 11, pp. 43–45, November, 1961.
52. Claggett, E. H.: Digital Data Processing System for Telemeter-to-computer Linkages, *Automatic Control*, vol. 13, no. 5, pp. 39–43, November, 1960.
53. Malone, J. R.: The 'Missing-Link' in Data Processing Systems, *Automatic Control*, vol. 14, no. 1, pp. 52–55, January, 1961.
54. Kamsler, W. F.: High Speed Data System Solves Low Level Signal Problems, *Automatic Control*, vol. 14, no. 6, pp. 37–44, June, 1961.
55. Canning, R. G.: "Installing Electronic Data Processing Systems," John Wiley & Sons, Inc., New York, 1957.
56. Highland, A. A., and W. F. Williams: Monitoring, Logging, and Control with Data, *Automation*, vol. 8, no. 11, pp. 77–81, November, 1961.
57. McRainey, J. H.: Data—The Crucial Elements, *Automation*, vol. 8, no. 8, pp. 54–70, August, 1961.
58. Scharla-Nielsen, H.: PCM Telemetry Performance, *Instr. Control Systems*, vol. 36, no. 9, pp. 91–95, September, 1963.
59. Leeke, P.: Wideband FM Recording, *Instr. Control Systems*, vol. 36, no. 9, pp. 111–113, September, 1963.
60. Knight, J. P., L. R. Klingler, and D. C. Yoder: Low-level Data Multiplexing, *Instr. Control Systems*, vol. 36, no. 8, pp. 86–89, August, 1963.
61. Staff: Logic Modules, *Instr. Control Systems*, vol. 36, no. 6, pp. 77–80, June, 1963.
62. Baldwin, R. E.: Digital Data Format Converter, *Instr. Control Systems*, vol. 36, no. 6, pp. 82–83, June, 1963.
63. Courtney, E. W.: Telemetry and Supervisory Control, *Instr. Control Systems*, vol. 36, no. 6, pp. 86–87, June, 1963.
64. Long, F. V.: Packaged Turbine Compressor Is Remotely Controlled by Telemetry, *Instr. Control Systems*, vol. 36, no. 6, pp. 89–91, June, 1963.
65. Liberman, R. A.: Synchronous Multibit Pattern Generator, *Instr. Control Systems*, vol. 36, no. 4, pp. 131–134, April, 1963.
66. Lerner, A. M.: TDM Coding Factors, *Instr. Control Systems*, vol. 36, no. 3, pp. 121–123, March, 1963.
67. Bolton, E. A.: No-noise/No-bounce Switch, *Instr. Control Systems*, vol. 36, no. 3, pp. 125–127, March, 1963.
68. Robbins, L.: One-brush Encoder, *Instr. Control Systems*, vol. 36, no. 2, pp. 103–105, February, 1963.
69. Ashman, R. R., and W. E. Chainey: TDM Telemetering, *Instr. Control Systems*, vol. 36, no. 2, pp. 107–108, February, 1963.

70. Bruck, D. B.: Analog Multiplication from DC to AC, *Instr. Control Systems*, vol. 35, no. 8, pp. 93–94, August, 1962.
71. Palevsky, M.: Cost of Data Processing Systems, *Instr. Control Systems*, vol. 35, no. 8, pp. 96–98, August, 1962.
72. Barbour, C. W.: Capcoder A/D Converter, *Instr. Control Systems*, vol. 35, no. 8, pp. 104–105, August, 1962.
73. Georgens, H. H., and L. I. Duthie: Magnetic-logic Telemetric Control System, *Instr. Control Systems*, vol. 35, no. 6, pp. 94–97, June, 1962.
74. Klein, M. L.: Telephonic Transmission of Data, *Instr. Control Systems*, vol. 35, no. 6, pp. 99–102, June, 1962.
75. Johnson, G. N.: Predetection Recording, *Instr. Control Systems*, vol. 35, no. 6, pp. 109–111, June, 1962.
76. Sink, R. L.: PCM (Pulse Code Modulation), *Instr. Control Systems*, vol. 35, no. 6, pp. 114–118, June, 1962.
77. Wilson, M. C.: Gray to Binary Converter, *Instr. Control Systems*, vol. 35, no. 6, pp. 149–150, June, 1962.
78. Pohl, P.: Common-mode Rejection, *Instr. Control Systems*, vol. 35, no. 6, pp. 152–153, June, 1962.
79. Crayford, R. M.: APD Coding and Logic, *Instr. Control Systems*, vol. 35, no. 1, pp. 97–100, January, 1962.
80. McLaughlin, H. J.: Disc File Memories, *Instr. Control Systems*, vol. 34, no. 11, pp. 2063–2068, November, 1961.
81. Whitmer, M.: Magnetic Memory Cores, *Instr. Control Systems*, vol. 34, no. 8, pp. 1427–1429, August, 1961.
82. Ormord, L. A.: Problems in Digital Recording, *Instr. Control Systems*, vol. 34, no. 8, pp. 1430–1433, August, 1961.
83. Verrette, P. A.: Data Recording and Processing in Aircraft-gas-turbine Evaluation, *Instr. Control Systems*, vol. 34, no. 8, pp. 1434–1437, August, 1961.
84. McCabe, P. N.: Twistor Memories, *Instr. Control Systems*, vol. 34, no. 7, pp. 1242–1245, July, 1961.
85. Kodis, R.: Magnetic Core Logic, *Instr. Control Systems*, vol. 34, no. 7, pp. 1246–1249, July, 1961.
86. Engman, G.: Data-bloc and Data-pac Symbolic Logic, *Instr. Control Systems*, vol. 34, no. 6, pp. 1052–1054, June, 1961.
87. Kamsler, W. F.: Basic Telemetry Methods, *Instr. Control Systems*, vol. 34, no. 6, pp. 1068–1070, June, 1961.
88. Even, Arthur D.: "Engineering Data Processing Systems Design," D. Van Nostrand Company, Inc., Princeton, N.J., 1960.
89. Hein, Leonard W.: "An Introduction to Electronic Data Processing for Business," D. Van Nostrand Company, Inc., Princeton, N.J., 1961.

COUNTERS AND DIGITAL INDICATING DEVICES

By Allan L. Burton*

INTRODUCTION

A counter is basically a memory or storage device for totalizing, integrating, or accumulating a series of discrete events. In its most familiar mechanical form, it is used to totalize and indicate numerical values such as the number of machine cycles or movements, the number of items being manufactured or processed, or other numerical information. Counters are used in connection with other devices to indicate linear distance, angular movement or position, the flow of fluids or electricity, or to transform other analog data to digital form. The means by which this is accomplished depends on the type of input device that is used to transfer the data to the counting device and the type of output method that is required for indication.

While counters may be quite complex and often specifically designed to match the characteristics of the input devices, and while the information may be utilized for regulation and/or control, the interpretation and use of the data depend either on an operator or an electrical or mechanical output which, through appropriate circuitry or linkage, is utilized automatically to regulate or control an operation. Different types can (1) integrate with a system for remote indication or printing, so that the data can be correlated with that from other systems and in turn be used for control; (2) be preset so as to shutoff an operation or otherwise adjust it after a given count has been reached; or (3) be connected to feed back the position or location of an input device, such as a potentiometer or synchro, in order to regulate or control an operation.

To be considered in the selection of a counter are the required storage capacity, the amount of driving torque that is available, the method of input available, the method of output that is required, the type of display desired, and the life that is necessary. Other factors, such as the environmental conditions of use, may dictate the materials of construction that are necessary, as well as the general design of the unit.

BASIC CHARACTERISTICS OF COUNTING APPLICATIONS

Since most counting devices involving counting, totalizing, and indicating require interpretation of their data by a human operator, their output is in decimal, or "base 10," form. The characteristics of mechanical and electromechanical counting devices (and many electronic counters—especially those used for industrial applications) are such that both input characteristics and accumulating mechanisms are based on the decimal-counting system. Although some electronic counters may totalize or accumulate counts by means of binary-counting circuitry, their output characteristics are converted back into decimal form for ease of interpretation and compatibility with the other data involved in the application. However, counting devices may be built to have binary, decimal-binary octal, or other types of output. Ordinarily, these types of counters would be coupled with special computing systems and their output characteristics would be of an electrical, rather than a visual

*Director of Research and Development, Veeder-Root, Inc., Hartford,Conn.

type. Combined forms having decimal visual output and some other form of electrical output are not uncommon.

Special forms of nondecimal-base counters such as 12- and 24-hr clocks; foreign nondecimal monetary-system counters; and angle counters indicating degrees, minutes, and seconds are also built. Their deviations from a conventional decimal system are achieved with special mechanical motions and their outputs are ordinarily visual in nature. Figure 1 shows one example—a Longitude Counter.

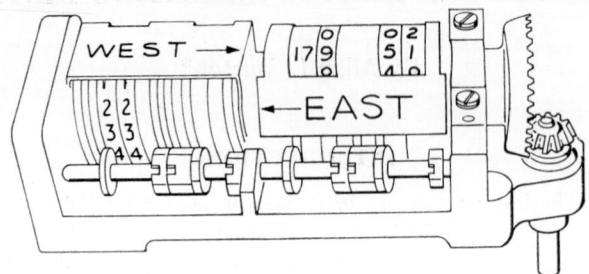

FIG. 1. Longitude counter. Special transfer gearing and figure arrangement permit indication in degrees and minutes. Shutters identify hemisphere and permit unidirectional drive and bidirectional indication when crossing the 0 or 180° meridians.

BASIC FEATURES OF A COUNTER

Because the basic problem of counting involves the totalizing or accumulation of a series of discrete events, a digital rather than an analog type of device is possible. This possibility, in turn, leads to very high storage (memory) capacity and to high accuracy of count. In addition, the mechanical configuration of the conventional type of counter lends itself to a direct-reading, nonambiguous visual presentation of the data stored in the counter.

This latter characteristic leads to the use of counting devices as indicators for quantities and other applications. A typical example is a bearing counter to show course deviation. In such an application, the counter is not being utilized as a storage device to totalize a series of events but is being used to indicate a quantity which may vary bidirectionally around a zero point. The advantages of the counter vs. an analog type of dial-and-needle indicator is that direct-reading presentation eliminates the need for an observer to exercise skill and judgment in deciding on the value of the least significant figures of a reading. The counter-type presentation, also, in cases where a large range of values is required, is without the ambiguity that is inherent in various types of long-scale analog indicators.

The applicability of a counter-type device for use as an indicator depends on the range of quantity to be indicated, its time rate of change of value, and on the conversion means available to couple it to the sensing device being used. In some cases, such as direct-reading indicators for scales, the time rate of change of value of the variable may be so high that a convential counter-type mechanism is not practical to use.

COMMON COUNTER FORMS

The simplest form of counter employs a single dial with numerical values around its periphery and a needle which, as events are counted, moves from zero to higher and higher values. This type of device may have low-inertia moving parts to make it adaptable for high-speed counting, but its finite scale length severely limits its total storage capacity. As the capacity increases, each counting increment becomes smaller and reading the scale more difficult.

The familiar utility-meter dial-type counter is an extension of this type of device where several dials, each coupled to the other with 10:1 direct gearing, are used.

By careful reading of the significant scale indication lower in value than each digit's needle position the total reading of the counter may be obtained. The possibility of making an incorrect reading is very great, and such features as counterrotating dials increase the possibilities of error and introduce serious ambiguity problems.

The commonest type of counter is that having the digits of each significant figure on the periphery of a drum-type wheel. Input is to the lowest-order-digit wheel and a unique type of intermittent gearing accomplishes the transfer, or "carry," function from lower to higher order. The transfer mechanism is such that any possibility of ambiguity in the reading is avoided.

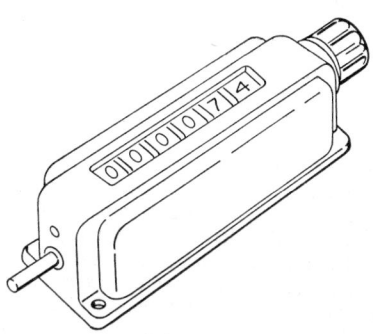

In a decimal-type counter of this type the total storage capacity of the device is a function of the number of wheels [capacity of counter = $(10^n - 1)$ where n is the number of wheels]. The visual indication of accumulated count value is an "in-line" numerical presentation which is read from left to right. The accuracy of reading a

Fig. 2. Conventional mechanical counter, six-digit capacity, rotary reset.

device of this type is independent of the total capacity and is normally within one division of the lowest-order wheel. The odometer counter of an automobile speedometer is a very familiar example of this basic type of counter (see Fig. 2).

In some types of counter the transfer function from lower- to higher-order digit is accomplished by means other than the intermittent gearing mentioned above.

BASIC COUNTER SYSTEM ELEMENTS

The block diagram (Fig. 3) indicates schematically the elements of a variety of types of counting systems. Under Quantity are listed most of the common variables that

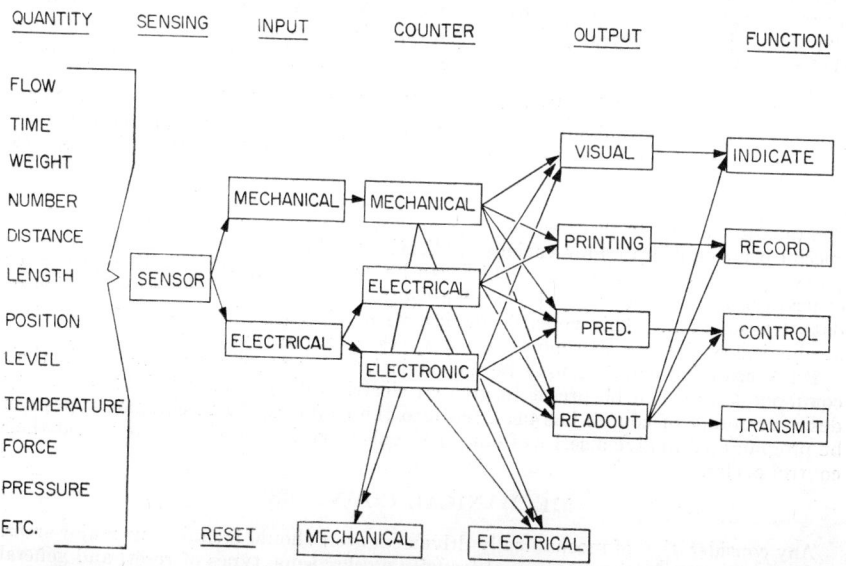

Fig. 3. Basic counter-system elements.

may be counted—or displayed. The first six are thought of as "counted," in that they normally increase serially and the information desired is the total of a number of accumulated incremental units. On the other hand, the remaining variables are more usually indicated in terms of plus or minus differences or displacements from some established reference point.

Increments being counted or indicated must usually be determined by a "sensing" device, and in most cases, a conversion device is required to effect a proper proportional relationship between the units being sensed and the counter mechanism itself. (For example, liquid quantity may be measured—or "sensed"—by a positive-displacement flow meter. The output of such a meter is in the form of output shaft rotation. A particular gear ratio between the output shaft motion and the units indicated on the counter is necessary to make the counter read directly in some conventional unit of fluid measure.)

In general, the senser output may be mechanical or electrical. If mechanical, a mechanically driven counter is ordinarily used. If electrical, the output is usually in the form of voltage pulses and the counter is either magnetically or electronically operated. In the case of differences or displacements to be indicated, the output is more likely to be a mechanical displacement or an electrical change in voltage that is proportional to the displacement from some reference point. An electrical servo-system drive may be used to drive a mechanical counting device where the sensor output is electrical.

Thus sensing (and/or conversion) devices vary considerably according to application, and their characteristics are important factors in choosing a counter. Many of these devices will be found described in other sections of this handbook.

CLASSIFICATION OF COUNTERS

"Counters" themselves may be classified into three types: mechanically driven, electromagnetically driven, and electronic. Both the rate of generation of counting information and the output function characteristics are important factors that help determine what kind of counter is to be used.

With the exception of the input means, the internal construction and the output-function features of the mechanical and the electromechanical types of counters are quite similar. The discussions on transfer mechanisms, types of reset, and output characteristics as covered under Mechanical Counters are also applicable to the electromechanical types. The electronic counter is quite different and is discussed separately.

The functions desired of a counter determine the "output" characteristics that are chosen for it. The visual-output function is by far the most common, but the increasing use of many types of data-recording systems and the trend toward automatic control of machines and equipment have created a constantly growing demand for other types of output. In the data-recording field one of the most common requirements is to be able to print the counter totals. This has given rise to a variety of special printing counters for various applications. The electrical readout counter is a rather special form of counter which can be used in conjunction with a printer or other form of data-recording device as a control counter or for remote indication or data transmission. This versatile device, with relatively simple auxiliary electro-mechanical equipment, can handle many types of output functions.

Predetermining counters may be mechanical, magnetic, or electronic and in their commonest form can be preset to give an electrical or mechanical signal after any desired number of counts has been fed into the counter. More complex types can be programmed to give a series of output signals at various settings or to repeat their control cycles.

MECHANICAL COUNTERS

Any consideration of mechanically driven counters should include four major areas. These are types of drives, principal transfer mechanisms, types of reset, and general specifications.

Drives

Direct drive, probably the most straightforward of any counter drive, operates through a direct connection between the drive shaft and the first (or "unit") wheel of a counter. This type of drive provides 10 counts per revolution of the drive shaft and is often used when indication applications and interpolation between full-figure readings may be required.

Revolution drives are frequently provided through 10:1 gearing between the drive shaft and the first counter wheel. One count per revolution results. Industrial applications where machine shaft revolutions are counted often require this type of input.

Geared drives giving ratios other than 10 counts per revolution or 1 count per revolution can be devised and are common in certain applications.

Ratchet counter drives provide counts of cycles of oscillating or reciprocating motion. The counting stroke is generally composed of "pretravel," the "counting stroke" proper, and "overtravel." The "counting stroke" proper is frequently 36°, with the pretravel and overtravel something less than this amount. Return motions are frequently actuated by internal return springs. Some units count on the return stroke, using the counting (or power) stroke to cock the return spring which drives the wheels when released.

Rotary ratchet drives are very similar to the ratchet drive described above. The principal difference is the elimination of mechanical stops for the counting motion. One count is recorded for each 36° of shaft motion. Oscillating motion greater than 36° but less than approximately 60° will also provide one count.

Revolution ratchet drives provide one count per revolution through the actuation of a ratchet mechanism by a cam mounted on the drive shaft. Additive counting can be achieved by rotating the shaft in either direction. Ratchet counting can also be achieved by oscillating the drive shaft. In this case, stops are frequently provided to control the oscillating motion.

Other forms of special counter drives are available, such as those which provide counting proportional to drive-shaft motion in one direction only or to motion in either direction.

Transfers

Interrupted gear or pinion transfer is most common (Fig. 4). A circular locking disk, usually with two gear teeth extending over 36° of its circumference, is mounted to the lower-order wheel. This meshes with a pinion, with every other tooth partially cut away to obtain a locked sliding action where there are no teeth on the locking disk. The pinion also meshes with a 20-tooth gear attached to the higher-order wheel. This combination results in a 36° motion of the higher-order wheel when the pinion is rotated by the two transfer teeth, and a locked position for the higher-order wheel when the pinion is sliding over the locking disk. Ambiguity of reading at transfer is thus minimized. The number of counting wheels can be chosen to allow any desired storage capacity, but gear-tooth clearances must be precisely held to avoid excessive lag in the position of the higher-order wheels.

Geneva transfer mechanisms are frequently used to achieve higher speed. The transfer action is lengthened from 36° to as much as 72°, thus decreasing the acceleration of the transfer mechanism and of the driven wheel. The lowered impact loading significantly reduces wear and lengthens the life of a high-speed unit.

While external transfer gearing is most common, gearing internal to the wheel can be used where appearance is a major factor and where closer figure spacing is desirable. The mechanical action, however, tends to increase wear and reduce life.

Direct 10:1 gearing is also occasionally used. To avoid ambiguity of reading, lever, cam, or governor mechanisms are frequently provided to bring full figures into the reading window at lower speed or when a counter is stopped to be read.

Ratchet transfer mechanisms are also frequently used (Fig. 5). A staggered com-

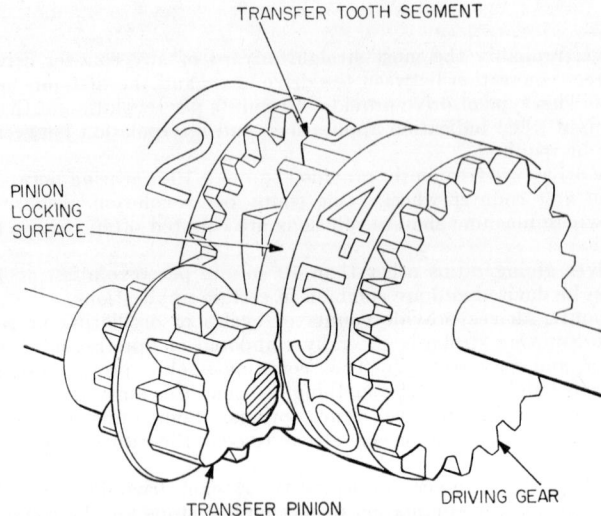

Fɪɢ. 4. Interrupted gear transfer. Between transfers, the pinion and left wheel are held stationary by two of the pinion's long teeth and the right wheel's locking ring. To transfer, the right wheel's transfer element engages the pinion teeth, turning the pinion to the next locking position and the left wheel to the next figure.

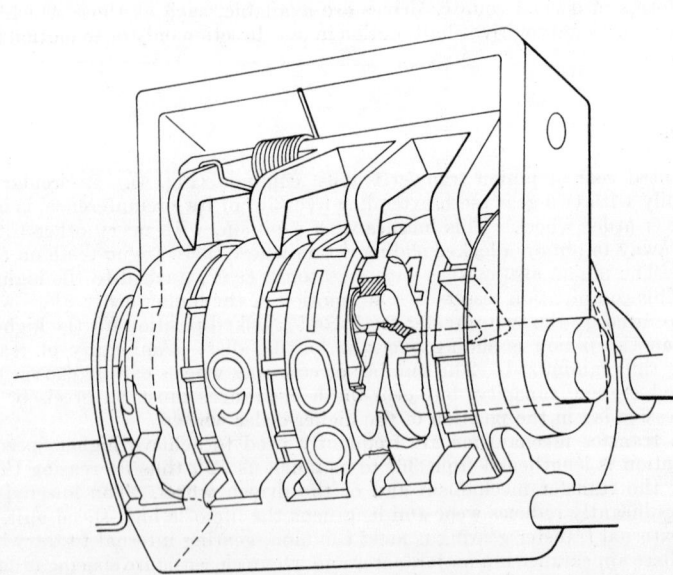

Fɪɢ. 5. Ratchet transfer. Individual pawls of the drive yoke are phased to transfer by allowing engagement with ratchets of higher-order wheels only when the pawls have dropped into the deep teeth of the lower-order ratchets. The cam-actuated lower fork prevents wheel overthrow. The splined shaft engages the wheels' reset pawls for the top coming reset.

posite pawl transfers successive wheels as the pawl teeth drop into deep transfer notches at the "nine position" of the lower-order wheels.

Special varieties of transfer mechanisms employing triggering devices, cams, etc., also have limited applications. These are occasionally used to produce a more even distribution of drive torque by storing energy in spring-loaded transfer mechanisms through a majority of the driving-wheel motion and firing at the transfer point.

Resets

The utility of a counter for almost all applications is greatly increased by having it capable of being reset to all zeros. The basic pinion-type transfer counter described above can be reset only by running it forward or back, sequentially, to the all zero position. However, if the locked gear train composed of the various order wheels and their transfer pinions can be uncoupled between each successive counting wheel, then the individual wheels can be simultaneously rotated back to the zero position, after which the transfer gear train can be reengaged.

One form of reset commonly used in counters manufactured in the United States uses a two-piece counting-wheel construction (Fig. 6). A pawl attached to the shell

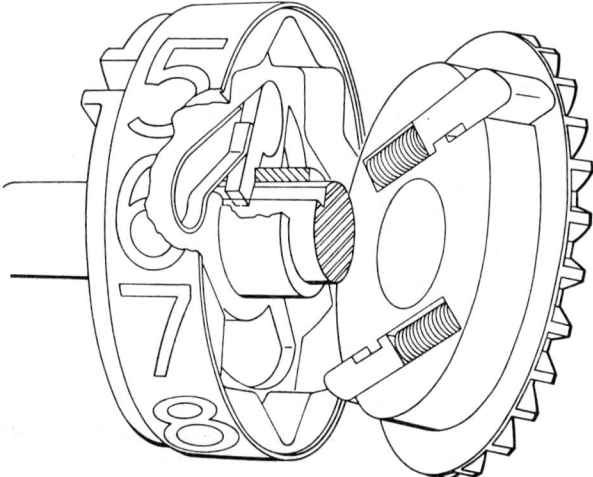

FIG. 6. Figure wheel with pawl and spline reset. While counting, the wheel is driven, top going, on the stationary reset shaft by a drive gear through two drive pawls. Turning the splined shaft top going resets the wheel through the single reset pawl, as the wheel cams past the drive pawls of the gear are held stationary by the preceding pinion.

of the wheel which contains the visible figures and is integral with the locking disk and transfer tooth action is picked up by rotary motion of the main counter shaft (which contains a spline). Sufficient rotational force of the shaft overrides the spring-detented coupling between this section of the wheel and the 20-tooth gear, allowing the wheel shell and its attached locking ring to rotate to zero position without disturbing the rotary position of the pinions. This type of reset is quite inexpensive but is not usually satisfactory where bidirectional drive is required for a counter.

An inexpensive type of reset often found in counters built in Europe utilizes a solid-wheel construction but mounts the transfer pinions on a spring-loaded shaft. This gives enough movement to enable disengagement of the pinions when the wheels are rotated simultaneously and imparts a bouncing action to the pinions which continues until the wheels come to the zero position.

A third method of reset is the so-called "heart cam reset" (Fig. 7), which has the advantage of being operable by means of a short linear motion rather than a full 360°, or greater, rotation of a shaft.   Upon actuating the reset lever (or push button), a mechanical linkage cams the pinion shaft, containing all the transfer pinions, out of mesh with the counting wheels, and a set of fingers, one for each wheel, then enter between the wheels and engage heart-shaped cams.   The cams act to rotate the wheels to the zero position.   Release of the reset mechanism reengages the pinions and retracts the heart-actuating fingers.

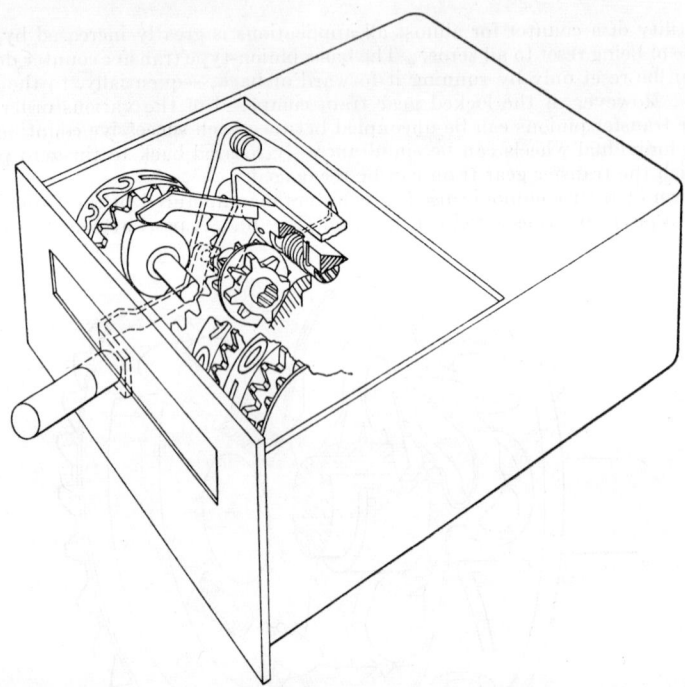

Fig. 7. Heart cam reset.   Pushing the reset button moves the pinion shaft to disengage the pinions and, by forcing followers against the hearts of each wheel, cams wheels to reset position.   Releasing the button returns the pinions to proper engagement with the wheels.

The heart cam reset principle can be easily adapted for solenoid actuation and is capable of much faster resetting than are the other methods   This becomes important when the counters are used as control devices and where delay caused by resetting might result in the loss of input data.

In resetting any counter it is ordinarily necessary either to disengage or in some other way to override the connection from the counter input to the drive mechanism. This can be a problem, particularly in precision counting applications, and the means of accomplishing it must be considered when choosing a counter for use in any particular system.

Specifications

Speed.   Typically, mechanical counters of a commercial variety have operating speeds of approximately 1,000 counts per minute or 100 rpm of the right wheel. Maximum speed for interrupted gear transfer counters is in the area of 2,500 rpm. At this speed, however, a means of reducing the impact of transfer between the first and second wheels, such as a geneva mechanism or controlled direct gearing, should be used to provide increased life. Some forms of controlled direct gearing will enable counting with mechanical counters up to speeds of 6,000 rpm of the first wheel. It should be noted that, generally, the less complicated and lighter wheels of nonreset counters can attain higher speeds than their reset counterparts.

Drive torque is dependent upon the frictional load of bearings and moving components and upon the forces required to provide transfer acceleration. As a result, torque to a significant extent depends upon the speed of the counter. Counters are frequently rated by their static torque. Commercial counters of average size frequently fall within the 1- to 2-oz-in. torque range. Counters of precision quality will have torques as low as 0.02 oz-in. (Torque is, in some respects, proportional to the life of the counter and will increase by approximately 50 per cent throughout most of the counter's life, climbing rapidly by even greater amounts at the end of its life. Torque depends to a great extent upon the materials of the moving parts, the bearings, number of figures, size of moving parts, and manufacturing tolerances. Because of the inertia of the moving parts the dynamic, or operating, torque of a counter can be many times the static torque.)

In considering the moment of inertia, the significance of wheel size, material, and construction can be readily seen.

Figure size is also of considerable importance as it frequently controls the over-all size of the counter as well as the counter wheels. As mentioned above, wheel size has a considerable effect upon the moment of inertia and therefore upon speeds, torque, and life. Mechanical counter figures generally range between $\frac{1}{8}$ and 1 in. high; $\frac{3}{32}$ in. might be considered a practical minimum and 2 in. a maximum. Larger figures can be achieved without tremendous counter wheels through special forms of ingenious displays used for special applications.

Material and structure, particularly of the counter wheels, are also significant. Of particular concern is the effect upon weight and moment of inertia. Materials frequently used for moving counter components are Delrin, nylon, styrene, zinc, and tin die castings, aluminum, steel, and brass.

Life is, of course, one of the most significant counter specifications. As can be seen, it is to a great extent variable, depending upon virtually all the above specifications. Speed possibly has the greatest effect, with life dropping quite rapidly as speed increases, largely because of the greater impact of transfer. Tolerances, with respect to surface smoothness and component balance, also affect life. The life of a commercial counter can be considered to vary between 20 and 100 million counts with an average of approximately 50 million. Maximum life for a mechanical counter of precise construction, and operating under special conditions, could be 500 million counts. Environmental specifications should not be overlooked in any counter application considerations. These include temperature range, vibration, shock, humidity, fungus, salt spray, sand and dust, and explosive atmospheres. Particular attention should be paid to possible natural combinations of these elements.

ELECTROMECHANICAL COUNTERS

The basic difference between an electromechanical counter and the mechanical counters is in the input means. The electromechanical form of counter is ordinarily operated by an electrical pulse rather than a mechanical motion. In cases where events or units are being counted, it is usually possible to substitute an electromechanical counter for a mechanical unit by having the mechanical drive member operate a switch which is used to control on-off voltage pulses which are then applied to the counter.

An important advantage of this type of operation is that the counter need not now be mounted in conjunction with the sensing element but can be remotely located. Electromechanical counters are often mounted in multiple groups and are used to give summarizing information on production statistics for supervisory checking, or for general multiple-channel data storage and indication. Another advantage of remote location is that the counter can be removed from hazardous or difficult-to-read areas.

Most forms of electromechanical counters are unidirectional and, while internally similar to mechanical counters, usually have an input mechanism that is most commonly some form of ratchet or other intermittent drive means to transform a linear reciprocating motion to a rotary motion. Originally, the ratchet-drive mechanical counter was equipped with a solenoid and each operation of the solenoid advanced the wheels one count. The usefulness of this type of counter quickly became apparent and special units were specifically designed for electromechanical drive. The desire for high-speed operation at reasonable power-consumption levels, and for packaging counters in large groups for panel mounting, led to the development of the small compact panel-mounted high-speed electromechanical counter of today.

Most commercially available electromechanical counters can be classified according to counting rate. There are units counting up to about 600 counts/min, those with a capacity of about 1,200 counts/min, and smaller higher-speed units with capabilities of 2,400 to 3,000 counts/min. As the desired counting speed increases, the counter elements usually become smaller because of the need for obtaining minimum inertia of the mechanical parts.

Power consumption customarily varies with the size of figure presented and with the speed capability. Six-watt consumption for a 3,000 counts/min unit with six digits and with $\frac{3}{16}$-in. figures is considered an efficient counter. Lower speed ratings can result in much lower wattage consumption. Some units available with a maximum speed of 500 to 600 counts/min operate on only 1 or 2 watts.

Although some electromechanical counters use iron-core solenoid driving units, it is more common (especially in higher-speed counters) to find clapper-type solenoids such as are used in relay construction. Either single- or double-stroke drives may be used. In the single-stroke mechanism the pull-in of the clapper rotates the lowest-order wheel by 36° through a ratchet and pawl linkage. The return stroke of the clapper is accomplished by spring action and the wheels are at rest during the return stroke. Two-stroke drives move the wheels 18° on the power stroke and the additional 18° on the return stroke.

A typical example of this type of drive is the so-called "inverse escapement" where a forked verge exerts a camming action against a star wheel connected to the lowest-order counting wheel. On the power stroke, one side of the verge cams the wheel through 18°, and on the return stroke, the other side of the verge carries the wheel through the remaining 18°. In general, two-stroke motions are found in the higher-speed electromechanical counters.

Other types of drive are used, one of the commonest being the stepping motor. These units are capable of high speed and, because they have no ratchets and pawls, are usually capable of longer life than conventional counters. However, they are expensive and limited in output torque unless one goes to large-sized units. They also usually require more complex drive circuitry than the more conventional electromechanical counters.

The desire for high speed with low operating power has resulted in the wide use of plastic components in electromechanical counters. Nylon and Delrin are the materials most commonly found in the highest-performance-type units.

Electromechanical counters are basically d-c operated devices, but many models are available capable of operating on alternating current. These units are usually equipped with copper shading coils to increase the residual magnetic storage in the solenoid core and to help eliminate a-c hum and chatter. The a-c units are ordinarily restricted in maximum operating speed to about 1,500 counts/min. D-C counters can usually be equipped with compact full-wave rectifiers, which makes them capable of operation with a-c power input.

Bidirectional magnetic counters, commonly known as add-and-subtract counters, are available for use in certain special applications. They are ordinarily equipped with two power plants, one of which adds counts and one of which subtracts counts from the total reading. These units are particularly useful for applications where a count balance is required, such as inventory or stock-control applications. A source of error exists if it is possible simultaneously to enter add and subtract pulses into the counter. This problem can be overcome by the use of a differential in the drive train which algebraically totalizes simultaneous input pulses.

A special type of electromechanical counter which has many applications is the so-called "single-wheel" unit. In its basic form, this is a counter having a single digital counting wheel and a switch capable of being actuated between the nine and zero position. This transfer switch can be used to operate a higher-order single-wheel counter, and units with any desired number of digits can thus be built up. Each unit would have its own power supply. Ordinarily the counters would also be equipped with switches, opening at zero, which would allow them to be reset by pulsing to zero. Individual power supplies allow counts to be fed into the higher-order digits; thus additions to the counter by factors of 10 are possible.

The familiar stepping switch, or Strowger switch, is an example of a single-wheel counter. These units can be built up into multidigit counters quite easily. Data-processing equipment makes extensive use of single-wheel counters, or "accumulators." These units are not ordinarily equipped for visual readout but do have electrical readout means and can therefore be programmed into a data-handling system. Where simultaneous input to more than one digit at a time is desired, it is necessary to make provision for storage of transfer pulses so that driving the next higher-order digits can be delayed until direct input-counting pulses have ceased.

ELECTRONIC COUNTERS

Electronic counters were initially developed to be used in conjunction with Geiger or Geiger-Mueller counting tubes for the detection and quantitative measurement of ionizing radiation in the radioactivity and nuclear-measurement fields. In these areas, counters are capable of detecting and totalizing individual alpha or beta disintegration particles. Many counting systems of this type are used today in the medical and industrial fields where nuclear-radiation phenomena are widely applied.

Originally, the electronic counters were constructed with binary counting elements. However, during the past several years, gas-filled cold-cathode decimal-type counting tubes have been developed and are now used extensively in industrial electronic counters.

The electronic counter has established a place for itself in a variety of industrial applications where the more conventional mechanical and electromechanical counters have been unable to offer the same features as the electronic counter.

Electronic counters, for industrial use, are capable of rates of count ranging from 1000 to 100,000 counts/sec. This high-speed capability accounts for many of its applications. The inertialess characteristics of an electronic device make resetting of an electronic counter almost instantaneous, and the electrical output characteristics of the individual decade counting tubes make it quite simple to connect manually settable multiple-position switches in parallel to each of the counting decades (Fig. 8). Therefore, when the counter is set to any particular count value, an output pulse (which can be used to operate a control relay) will be generated when the reading matches the value set into the switches. These two features combine very satisfactorily to create a predetermining type of counter which is capable of high-speed cycling or repeating predetermining action. In fact, one of the first applications of the electronic counter for industrial use was for the high-speed counting and batching of small parts or items.

Another advantage of the high-speed counting rates and the resultant high resolution is that the counter, with various electrical sensing means, is capable of detecting and counting small objects which would not normally generate a long enough input pulse to be counted by conventional electromechanical counters.

For conventional counting, where speeds are too high to be met by electromechanical counters, many manufacturers combine an electromechanical unit with one or more electronic decades in the lowest-order positions. Inherently, such a counter with one electronic decade has a count-rate capability ten times that of the electromechanical counter and has, also, ten times the life.

Electronic counters equipped with a time base and electronic gating circuits can be used as high-accuracy tachometers or for the very precise measurement of short time intervals. This class of counter is often called an EPUT (events per unit time) meter.

As a tachometer, the time base (which is usually settable for decimal fractions of seconds and minutes) is used to connect and disconnect (or to gate) the repetitive quantity, whose speed is to be measured, to the counter input. For example, a high-speed rotating shaft can have its rotational speed measured with such equipment. The shaft revolutions are sensed by some suitable means so that they can be counted by the counter, and the time base is used to operate an electronic switch—first to connect and then to disconnect the shaft-revolution-sensing means with the counter. The time between connect and disconnect is some decimal part of a second or minute controlled by a manually settable switch. If a time interval of 1 sec is used, the counter reading after one sampling will read directly in revolutions per second.

Conversely, the time required for some operation (such as closing time of a relay) can be measured by equipping the device whose operating time is to be measured with contacts that operate at the start and finish of the cycle to be timed. These switch functions are used to connect and disconnect a known frequency oscillator to the

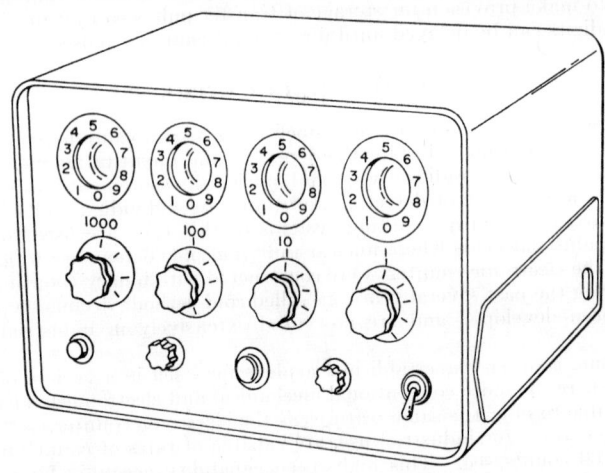

Fig. 8. Electronic repeating predetermining counter. Very high speed counting. Immediate automatic electrical reset. Rotary switches for selecting predetermined quantities. The control relay is actuated at a predetermined number.

counter input circuit. If the oscillator has a 1.0-kc frequency then the counter reading following the measurement of one operating cycle reads the operating time directly in thousandths of a second.

Counters are available equipped with built-in constant-frequency oscillators and gating circuits which allow measurements of this type to be made.

Add-and-subtract electronic counters can, when coupled to high-resolution shaft-position indicators, very accurately indicate the relative position of machine-tool elements. The high count-rate capability of the electronic counter combines with shaft-position encoders having a large number of discrete angular positions and high-sensitivity pickup means to give such counting systems high resolution and make them settable to a high accuracy.

Generally, the input to an electronic counter is an electrical pulse that is generated by a high-speed switch, magnetic pickup, photohead, proximity pickup, or sonic or supersonic sensor. These devices may be used to detect either linear or rotary motion. Most common of this group is the photohead, a device consisting of a light source and lens system to focus a beam of light on a photosensitive receiver in such a manner that the object to be counted will intercept the beam. The receiver may be a special type of vacuum tube, lead or cadmium sulfide cells, built into a transparent plastic capsule or a small glass tube. The more recently developed solid-state devices are similar to transistors and diodes, specially designed to take advantage of their maximum light-sensitive characteristics.

All receivers have one thing in common—an internal change occurs when light strikes them and they revert to the original state when the light is removed. This internal change may be either a change of resistance (as occurs in lead and cadmium sulfide cells) or the actual generation of a voltage which is characteristic of the solid-state devices. Thus, when the beam of light between the source and the photocell is interrupted, the "pulse" is generated which is applied to the input of the electronic counter, thereby registering one count.

Regardless of the input device employed, the electrical pulses generated are not totally suitable for counting devices and often require revision as to their duration and/or amplitude. The counter itself, therefore, must contain appropriate circuitry for pulse amplification, stretching, clipping, or otherwise reshaping its characteristics to make it suitable for counter operation. These functions, as well as the counter itself, may be designed for using either vacuum tubes or transistors depending on the counting rate and economics dictated by the specific application.

If dynamic visual readout is required for the application, each count pulse may be used to trigger a display device such as the numerical in-line Nixie, electroluminescent screens, numerical projection displays, or simple neon lamps which glow and illuminate a sequence of numbers stenciled on a translucent material. One unique device, the glow transfer tube, is employed by many manufacturers as both a counting device and a display providing a small point of "glow" discharge at each of 10 positions around the circumference of the face of the tube. Although not strictly an electronic device, the electromechanical counter is frequently used as an output display for electronically driven counting applications. In addition, electronic counters are frequently being employed for use with high-speed electromechanical printing devices or punched-tape and punched-card mechanisms.

COUNTER OUTPUT CHARACTERISTICS

The most commonly used counters of the mechanical or electromechanical type are those equipped for visual indication of count. All the foregoing material applies primarily to counters having visual-readout means. However, certain features can be built into these basic counters so that they become control devices or offer means of recording their data without the necessity for visual transcription of their reading.

The simplest conversion of a basic counter to a recording device is by replacing the visual-readout wheels with wheels equipped with raised printing characters and incorporating the unit into an assembly (with a printing platen or roller and a suitable frame for holding a card or a mechanism for indexing a tape) past the counter wheels. The visual-readout counter then becomes a printing counter and can be used to produce printed records of the counting operations that are performed.

For electromechanical printing counters, figure alignment of the individual wheels is relatively straightforward since the wheels move incrementally. For mechanical-drive printers it is often necessary to use auxiliary mechanisms which will move the lowest-order wheel into a full-figure position before the printing operation is carried out.

The utility of a straight printing counter has to be evaluated in terms of the application for which it is to be used. A limitation is that a complete printer and counter mechanism is required for each data-input point, and a further limitation exists because information from each counter is normally printed on a separate document. Of course, groups of counters can be arranged to print out on a single document, but the total number of inputs is limited by practical consideration of size of document and mechanical rigidity of the assembly.

A versatile type of counter which can be used for a variety of output functions is the so-called electrical-readout counter. This is basically a standard visual-readout counter with a printed-circuit commutator plate mounted between each wheel. The wheel itself carries a set of brushes, one of which is in contact with a continuous common ring on the circuit plate and the other of which contacts individual segments corresponding to each individual digit on the wheel. The brushes of each wheel are connected together, and each digit is therefore the equivalent of a 10-position rotary switch (in the case of decimal counters). These switching elements can be used to control a solenoid-operated parallel or serial entry printer or other data-recording equipment, for electrical predetermining or presetting action, and for a variety of other possible functions.

One common method of operating a serial-entry, solenoid-controlled printer is to parallel the individual segments of each wheel into a common 10-wire bus system which in turn connects in parallel with the comparable printer solenoids. Then, when print-out is desired, sequentially connecting the common of each wheel to the common of the printer causes a series of input pulses to be fed to the printer, each of which operates the proper solenoid to print the desired digit. Parallel-entry solenoid printers may also be used, but these require individual conductors from each digit of each wheel circuit card to each digit of the printer. They do not require sequential scanning operation from digit to digit.

For predetermining action it is possible to connect the individual digital segments on each circuit card, in parallel, to a multiple-position manually settable switch. Then, with all circuit cards connected in series, it is possible to achieve a closed-circuit path when the counter reading equals the reading set up on the manually set switches. This closed circuit can be used to operate a relay which in turn can be used to control the equipment generating the count. With electrical-reset counters the reset action can also be initiated by this closed circuit, and a repeating predetermining function is thus possible.

The electrical-readout counter can be used to operate any desired type of data-recording equipment such as card punch, paper-tape punch, or teletype. Conversion from decimal electrical-readout counters to devices requiring binary input is most easily achieved by means of a diode matrix.

Electrical-circuit-readout counters are also available with straight binary or binary-coded-decimal readout features. These are particularly useful for input into binary-operated data-recording equipment and are found in some navigational-type counters which are used in conjunction with navigational computers.

A simpler type of mechanism for predetermining control action by a counter is achieved by a mechanical linkage that is ordinarily used to actuate a control switch. The simplest form of such a counter would be either mechanical or electromechanical, would be subtractive, and would have a cam on the left-hand wheel which would operate a switch or other linkage when all the wheels transfer through the zero position to the "all 9's" position. The counter is set to the desired predetermined number and subtractive counting proceeds. At one count more than the preset number the switch operation takes place. In order to make such counters practical, special wheel constructions are used so that the predetermined number may be set up quickly and the counter may be reset to the same point again for repetitive cycles.

If each digit wheel is equipped with a notched cam and a roller or sliding bar contacts each digit's cam surface, then the individual cams can be positioned to the desired predetermined number and the counter can be driven additively from the zero reading up to the predetermined number. At this point all cam notches will be in line and the roller bar can drop down and its linkage will operate the control switch. Mechanically driven counters of this type are widely used on coil winders, textile machinery, etc. Magnetically operated predetermining counters take many forms and are used in many types of production processes and production machines.

REFERENCES

1. Woodson, Wesley E.: "Human Engineering Guide for Equipment Designers," University of California Press, Berkeley, 1954.
2. Jones, Franklin D. (ed).: "Ingenious Mechanisms for Designers and Inventors," vols. I, II, III, The Industrial Press, New York, 1957.
3. Máté, George: Counters Applied to Measurement and Control, *Process Control and Automation*, May, 1961 (published in Great Britain).
4. Temple, L. D.: How to Pick the Right Counter, *Product Eng.*, Apr. 17, 1961.

5. Strong, John: "Procedures in Experimental Physics," chap. VII on Geiger Counters, Prentice-Hall, Inc., Englewood Cliffs, N.J., 1939.
6. Goodeve, T. M.: "The Elements of Mechanisms," Longmans, Green & Co., Inc., New York, 1903.
7. Frank, M. E., and S. T. Schy: Counting on a Magnetic Drum, *Control Eng.*, vol. 8, no. 10, pp. 75–79. October, 1961.
8. Giulie, J. D.: Counting to Control Weight Fill, *Automation*, vol. 9, no. 9, pp. 85–88, September, 1962.
9. Ponstingl, J. C.: Timing Devices for Automatic Control, *Automation*, vol. 7, no. 11, pp. 77–81, November, 1960.
10. Everett, J. E.: Counting for High Speed Control, *Automation*, vol. 7, no. 6, pp. 92–96, June, 1960.
11. Tanzman, H. D.: High Accuracy Time Interval Measurements, *Electron. Industries*, pp. 62–67, January, 1959.
12. McCready, E. A.: Electronics Does the Counting, *Radio-Electronics*, vol. 30, no. 10, pp. 102–104, October, 1959.
13. Russell, R. W.: Applying Electronic Counters, *Automation*, vol. 7, no. 4, p. 80, April, 1960.

Editor's Note: For additional relevant reference material, the reader is referred to the lists of publications which follow the preceding two sections of this Handbook.

Section 12

METAL AND CERAMICS PRODUCTION INSTRUMENTATION

By

JOHN R. GREEN, B.S., *Consultant in Industrial Instrumentation for Metal Producing and Ceramic Industries, Villa de Verde en Tubac, Tumacacori, Ariz. (Retired as Manager of Steel Industry Sales Department, Industrial Division, Minneapolis-Honeywell Regulator Company, Philadelphia, Pa.); Member, Association of Iron and Steel Engineers, American Institute of Mining, Metallurgical, and Petroleum Engineers, American Ceramic Society, Instrument Society of America. (Steel Production Instrumentation)*

JOSEPH F. HORNOR, B.S.(Ch.E.), *Planning Adviser, Control Systems, General Products Division Development Laboratory, International Business Machines Corporation, San Jose, Calif.; Member, Instrument Society of America, Association of Iron and Steel Engineers, American Chemical Society, American Association for the Advancement of Science. (Glass and Ceramics Industries Instrumentation)*

STEEL PRODUCTION INSTRUMENTATION

By John R. Green*

In recent years the application of instrumentation to steel production processes has changed markedly in the techniques used. And this remarkable progress is expected to continue at a rapid rate. However, basic process requirements do not change so fast, and underlying principles that guide instrument selection have a longer life. Hence, the approach in this subsection is along two lines:

1. It gives the principles to guide instrumentation selection for steel production processes, including the philosophy and background that lead to the current "state of the art" and may well direct future trends.

2. The subsection contains a number of specific examples of instrumentation practice in the steel industry, showing the relationship between the process requirements and the applied instrumentation.

Such an approach should prove useful not only to those concerned with the steel industry but also to engineers in other high-temperature industries where similar control and measurement problems exist. One notable example is the manufacture of glass and ceramics (see pages 12-47 to 12-66).

PRINCIPLES TO GUIDE INSTRUMENTATION SELECTION

The steel industry, like many other manufacturing industries, has evolved and developed instrumentation practices, pieces of equipment, and automatic control methods to assist in the volume production of high-quality products at lowest cost. Steel is an outstanding "high temperature" industry, with associated high fuel costs. Thus, the design and development of instruments and control systems to provide rapid and safe heating for both its product and furnace equipment, on a batch or intermittent cycle, has been one of the industry's major contributions to industrial instrumentation.

Many major industries continuously process raw materials into the final salable product, frequently with a variety of coincidental side-stream products. In contrast, the steel industry has production processes which consist fundamentally of a succession of batch-type operations. Intermediate products, complete in themselves, are frequently produced or processed by widely spaced multiples of batch units. They are then transported over considerable distances, and undergo a succession of reheating operations.

Even the "continuous" processes of this industry, with a few notable exceptions, consist essentially of successive or progressive treatment of modular units. These units can vary appreciably in dimension, mass, metallurgy, and physical condition, depending upon both previous processing and final-product requirement.

In its instrumentation requirements, the steel industry has both favorable and unfavorable process conditions. On the unfavorable side are environmental conditions of heat, heavy vibration, and "dirt" in the form of kish, fumes, smoke, flue dust, and mill scale. These conditions under which sensing elements as well as instrument and control mechanisms must operate at the process location are more severe

* Consultant in Industrial Instrumentation for Metal Producing and Ceramic Industries, Villa de Verde en Tubac, Tumacacori, Ariz. Retired Manager, Steel Industry Sales Department, Industrial Division, Minneapolis-Honeywell Regulator Co., Philadelphia, Pa.

than are encountered in any other major industry. This has presented a continuing challenge to instrument manufacturers whose products, to be economically practical, must serve all industries.

The days have long since passed when steel-plant maintenance and service personnel considered a sledge hammer, 36-in. wrench, and brute strength as their most refined tools. Yet there has been some carryover in requests for overheavy construction and "ruggedness." Housing of all possible components of control systems within pressurized and air-conditioned cubicles or rooms at each furnace location has been a practical and economical solution to much of this problem. Completely centralized control rooms can take full advantage and make efficient use of many available instrument designs and computer components. Their adoption, however, must include a realignment or relocation of supervisory and technical organization functions, in addition to solving the purely physical problems of signal transmission over extended distances.

On the favorable side of applying instruments in the steel industry is the fact that requirements for precision measurement and control in most of the processes are not so "tight" as those in many other industries. For example, in temperature measurement, a reproducibility or sensitivity of 1° (or even fractions of a degree)—vital for maximum production or product quality in certain industries—is rarely, if ever, required by the steel industry.

How "Good" Should a Control System Be?

Too frequently the answer to this question is: "As good as we can get." "Good" here means holding the controlled variable as close as possible to the set point, and thus infers the "best" in measuring and controlling mechanisms. Obviously, this would be not only uneconomical in capital and maintenance expenditures, but also in many instances entirely unnecessary as far as steel industry requirements are concerned. A better basic starting point should be: "The simplest, least expensive control system that is capable of producing a final salable product (within its specified tolerances) at highest profit."

As a common example, in the entrance or intermediate zones of continuous furnaces, both product and fuel-supply requirements can be met on a "go or no-go" basis. Yet these zones are frequently equipped with control systems of equal precision to that used in the zone of product discharge, and even the latter may be controlled beyond actual product requirements.

Despite the fact that greater precision in measurement and control of certain physical conditions is as yet unavailable in practical form to the steel industry, much of the instrumentation in current usage has capabilities, complexities, and costs which exceed the actual process requirements.

The problem, of course, lies in identifying and specifying these actual requirements to meet final product specifications, as they exist both today and in the immediate foreseeable future. More "meeting of the minds" between instrument engineer, metallurgist, combustion engineer, operating supervisor, furnace designer, and even product-sales analyst is required to establish practical tolerance requirements before control systems can be most economically specified and selected.

When new furnace or processing equipment is in the design stage, either by steel-plant engineering department or equipment manufacturer, an experienced instrument engineer can frequently point out or suggest minor changes which will greatly improve both the accuracy and simplicity of the required sensing and control system. He can do this without having to resort to complexities or compromises in order to accomplish the required results.

Because "at highest profit" is included in our basic objective for "good" control systems, secondary consideration must be given to the simplest means of reducing direct operating costs. Fuel, in this high-temperature industry, represents a major cost item. Therefore, its most efficient utilization must be provided by any temperature control system, even though its primary objective is to produce a final product within specified tolerances. The system must properly ratio fuel to air in order to maintain the highest available heat head that both the product and refractories can physically withstand. Furnace pressure control is also required to avoid air infiltration or shortened refractory life. In such systems, the simplest equipment that will provide the end results within allowable tolerances should be employed.

For more economical operation, more completely automatic control has become essential because of such factors as (1) continually rising labor costs, combined with a steady decrease in available skilled operators, (2) higher production rates, and (3) faster machine speeds. These same factors have led to advanced computer systems, in spite of the problems inherent in the industry's predominantly batch and semi-continuous processes. Such extension of instrumentation has reduced direct costs, increased production rates, and in many cases made new processes possible. Such advances, however, must be balanced against capital investment, maintenance costs (which are frequently buried), and the training and salaries of skilled technicians in place of operators.

Almost irrespective of its complexity or cost, any system which can increase yield (even by a fraction of a per cent) is worthy of thorough investigation because of the potentially high returns. For this same reason, experimental installations to provide economic proof, can be invaluable. The sources for losses in this intermittent and sequential process industry present some of the greatest possibilities for increasing profits through higher yields that exist today in any industry. Such sources extend from mine tailings, blast-furnace flue dust, slags, skulls and stool stickers in ingot production; through scale in reheating operations; to crop, shear, or side trimming in rolling operations—to say nothing of offgrade or off-dimension finished products.

Selection of Steel-plant Instrumentation

After practical tolerance requirements for any process are established, a selection should obviously be made of instrumentation methods, modes, and media. To express this another way, answers must be found to the three questions of "where?", "how?" and "with what?" These considerations must precede final determination of the specific make and model of the instruments.

Efforts of instrument engineers from both plant and supplier (who must know their product's capabilities and limitations) should be combined with the operator's knowledge based on first-hand and continuous day-to-day contact with the particular process. Even an experienced steel-plant instrument engineer, can always learn something from an operator about the individual plant practice on any piece of equipment or process. This is of particular importance where computer automation is being designed to replace human judgment in the operation and control of a steel-plant manufacturing unit.

The METHOD (or "where") of a control system is an obvious first consideration. Although control from end product is a fundamentally desirable objective, the steel industry's lack of many noncontacting measuring or sensing devices for its moving and intermittent operations greatly limits the use of such control methods.

The application of instrumentation to steel industry processes has been largely dependent on the availability of primary elements to withstand the severe conditions to be measured. The inaccessibility, large physical size, and high temperature of the processing furnaces have presented a difficult task in determining where and how measurements should be made and control applied. The solution of these problems over a period of many years has led to the recognition of certain unifying common denominators which guide the application of instrumentation to this industry.

1. Input and output material and energy flows must be stabilized.
2. An optimum rate of reaction, heating or cooling must be maintained.
3. Product and process equipment may need to be protected from damage, usually by temperature.

Some basic (although frequently overlooked) "method" principles are worthy of mention, as particularly applicable to this industry:

Select as few as possible primary control points, making other measurable control factors either overriding or subservient. This will simplify the over-all control system. It will also lessen the initial and maintenance cost, as well as avoid control systems which "fight each other," in attempting to control too many factors to a fixed value.

Avoid any limitation of "throughput" or production rate, as a method of control, except as a last resort. As has been properly said, for example, "there never has been a continuous reheating furnace built that was big enough." Efforts to get more production should not be hamstrung by a control system based upon "speed."

Avoid any control method (or its only available mode) which can "upset" a process, and will, therefore, cease to be used by operators. For example, the fixed fuel "cutback" system of open-hearth roof-temperature control was perhaps useful in the continuing conflict between the fuel engineer's efforts to get more heat and the refractories engineer's desire to keep furnaces from burning up. The system resulted in such bath reactions and slowdown in "working" of the heat, however, that operators would not use it, and the principle fell into ill repute.

The MODE (or "how") of a control system is a selection which logically follows, and is largely determined by, the method which is to be employed. Definitions and operating characteristics of available control modes—such as an on-off, two-position, proportioning, reset and rate actions—are covered in other sections of this book. Some observations which apply particularly to steel-plant controls are given below.

The development of advanced, refined, and usually more complex control instruments and mechanisms has been motivated principally by the need for (1) greater precision, sensitivity, stability, or speed to meet tighter product specifications, or (2) overcoming unfavorable process characteristics such as longer initial response periods or transportation lags. Where such refinements are required to meet the *actual* product tolerances of steel processes, by all means, select and use them. If they are not *needed* for a "good" control system, however, don't select instruments on the basis of "newest," "most advanced," "fastest," and the like.

Faster speed of response in both measuring and control mechanisms is highly desirable for many high-speed or automated steel processes. The use of designs which are faster than either the process or the sensing element can respond to, however, results only in shortened control-equipment life. This effect has been expressed as "double the speed and quadruple the wear."

An example of variance in product requirement, calling for either an advanced or a simple mode of control, may be found in the temperature control of soaking pits. If the ingots to be reheated are ever sensitive in their metallurgy to "burning" or "scaling" and have a narrow temperature range for fast rolling, the conventional use of automatic reset (and even rate response) as a mode of control is justified. It can provide radically lower fuel rates for "soak out" without any increase in pit temperature. On the other hand, steel composition may always be "tough," track time reasonably consistent, and the subsequent product may have a wide temperature rolling range. Here a simple proportional mode of control with corresponding valve design is adequate. It also costs much less to purchase and maintain.

Because the control of "inputs," rather than control from final products, is so common in the steel industry, interlocked, cascaded, or "piggyback" controls are frequently employed to obtain the end result. In such cascaded systems, the final controller can usually be of a simple mode, because its lags are normally low and its set point is continually being established to satisfy the refined mode of the supervising controller.

In some cases, process-simulation equipment can be employed effectively in determining optimum methods and modes of control for new processes. Such simulators are designed and operated by some instrument manufacturers and research laboratories. They provide the means of feeding process characteristics (lags, masses, transfer rates, sensing times, reaction rates, etc.) into a system by electrical or other means. The system is then controlled by different adjustable modes, and the end result recorded under various conditions of load change or input variations. The obvious problem with such an analytical tool is in the accurate determination, calculation, or selection of the actual process characteristics for input purposes.

Where process equipment is already in operation, and can be purposely upset, an analysis by frequency response or similar techniques can be helpful in determining the most effective mode of control action.

The **MEDIUM** (or "with what") to be selected for control-system operation is again largely dependent on the method and mode requirement of the process. On the basis of their actuating medium, control systems can be divided according to the following classifications:

1. Self-actuated control.
2. Electric control, including electronics.
3. Pneumatic or air-operated control.
4. Hydraulic or liquid-operated control.
5. Mechanical control.

Although the steel industry historically has been electrically minded, pneumatic and hydraulic operation for both signal and control transmission and actuation has become common practice. Each medium has desirable characteristics, which, when

Table 1. Comparative Advantages between Electric and Air-operated Control Systems

	Electric control	Air control
Requirements for actuating medium	Uninterrupted electrical supply at proper voltage and frequency	Constant supply of reasonably clean and dry air or noncorrosive gas
Types of control available......	Two-position Floating Pulsing Narrow-range proportional Proportional with automatic reset Certain forms of rate response	Two-position Narrow-range throttling Wide-range throttling Throttling with automatic reset Rate response
Simplicity of controller.........	May be simpler than air for elementary types Requires auxiliary relays for advanced types of control	Performs all the control functions in the instrument Simpler than electric for advanced types
Simplicity of control mechanism	Incorporates part of balancing or control functions	Extremely simple
Power of control mechanism....	Inherently provides large starting force	May require positioner for high starting force
Number of control mechanism positions	Usually a definite number (less than 100) of available positions	Over 200 positions; with positioner, over 600
Speed of response of control mechanism	Limited by maximum mechanism speed	Practically unlimited. Usually faster than electrical
Resistance to corrosive, contaminating, or explosive atmospheres	Instrument and mechanism both may require special contacts and protective enclosures or explosion-proof construction. Occasionally requires air purging	Self-purging and absence of contacts provide inherent freedom from all corrosion, contamination, or explosive hazards
Practical distance between controller and mechanism.......	Unlimited	Limited
Operation on medium failure...	Usually remains in last position	Usually moves to "safe" position
Initial cost..................	Usually higher than air for complete installation of equivalent type of control	Usually lower than electric for complete system including mechanism and valve
Installation cost..............	May be lower where wiring is not elaborately housed and distance is considerable between controller and mechanism. Usually greater with advanced types of control	If compressed air is not available, usually higher for small installation owing to compressor cost. Does not increase with advanced types of control
Operating cost...............	Energy less costly. Maintaining cost higher	Energy cost usually higher. Less maintenance cost, particularly lower on operator
Ease of maintenance and adjustment	With equivalent basic training usually more difficult for advanced type of control owing to complexity and adjustments	With equivalent basic training, usually simple, particularly with advanced type of control

combined with equipment specifically designed for that medium, provide the simplest and least expensive systems for certain applications.

The relative advantages between different media will continuously shift with the development and availability of new components designed for operation by each medium. A basic comparison of the advantageous and limiting characteristics of electric and pneumatic control operation is given in Table 1.*

A combination of several different operating mediums in the same control system is frequently used to secure the most efficient over-all operating results. Such combinations (either with or without transducers between mediums) are the means of securing the required simplicity, precision, flexibility, ease of maintenance, reliability, and cost, with a minimum of control equipment. As a homely parallel, your mass-produced personal automobile commonly combines mechanical, electrical, hydraulic, and pneumatic (vacuum) actuated control systems within one mechanism. Yet such systems are selected, designed, and manufactured by the most cost-conscious of organizations.

Many steel-plant furnaces are regenerative in operation, or are subject to periodic drawing or casting interruptions. The instrumentation therefore often includes fuel- and time-saving auxiliary systems to regulate fuel readmission rates artificially but automatically during and following each reversal or interruption. Where such systems are employed, the selected control medium must provide the speed and rate flexibility required to trim each individual furnace system for its maximum operating efficiency.

Whenever computer control systems are (or could be) employed, the selection of media for both signal input and control output should be compatible or readily convertible to computer intelligence.

Many steel-plant processes normally under automatic control require occasional transfer to manual operation—either because of process interruption or for servicing purposes. Consideration should be given, therefore, to the selection of mode and medium which can quickly and simply perform "bumpless" transfer between manual and automatic operation.

Instrument Department Organization

The steel-plant instrument department primarily performs a "service" function to both operations and engineering. Hence, whether it is a part of maintenance, electrical, combustion, metallurgical, engineering, or operating departments, is largely a matter of each plant's organizational structure. The important point, however, is that it be centralized in one department with widespread knowledge and skills in the many "trades" which make up modern instrumentation. To send out an "electrician" on a process-unit service call, only to have him find out that a "plumber" is needed (who must then go back for his tools) is too commonly the direct result of a split department.

If subdivision or specialization is required, it is better to do it on the basis of operating processes within the steel plant—rather than by make, type, measurement, or medium employed in any control system. Operating "know-how" can be as important as the skill with instrument "nuts and bolts." Some of the finest instrument servicemen, particularly on crankup jobs, thoroughly study what is going on in the process operations, before they make adjustments (if any) to instruments and controls.

The salary and/or wage classifications applicable to instrument department supervisors and personnel are well worthy of management study, if the combination of experience and diversified skills is to be attracted and retained in this profit-controlling department.

How Should Records Be Used?

The economic value to the steel industry of recording instruments and controllers, in comparison to "blind" or indicating designs, is worthy of consideration. Wherever

* From Instrumentation and Control in Steel Plants, *Iron Steel Engr.*, March, 1956, where further discussion of operating media can be found.

any variable is controlled continuously to a fixed value, a nonindicating or "blind" control system should be given first consideration. Such controllers may be self-actuated, are usually of simplest modes, and find many applications as independent stabilizing or limiting controls in regulating oil temperatures, steam and fuel pressures, or as safety devices. Nonindicating controllers can be supplemented by independent indicating or recording devices, if necessary. Further, if they are designed for audible or visual signals when the variable is beyond a controllable range, they can be used for important functions. Many advanced modes of control are now available in designs which are adaptable to "blind" operation.

Where an indication of any variable is of benefit to the operator, the numerals should be large enough to be easily read by him from his working location. Production has been increased on many furnaces by moving instruments from a catwalk climb to normal working positions.

Recorders pay their way, for operators only, whenever trends, rate of change, or timing have importance in the process, because visual retention cannot be maintained from indicators alone. Slow speed-chart travel emphasizes such trends. A typical example of this use of a recorder is found in recuperative soaking-pit operation. Here the approaching time of soak-out can be forecast from flow records. Furthermore, heating rate on temperature recorders reflects ingot track time and, therefore, predicts approximate drawing availability.

For continuous furnaces, the use of blind controllers, with parallel operation of multiple recorders, which show the entire related operation, can frequently simplify and lower the cost of a control system.

Of course, the prime purpose of records is so that operating supervisory personnel can know what has happened and when. More use of records for this objective, however, could be made in steel plants.

Often, strip-chart records are allowed to disappear on the reroll spindle, while those which must be removed daily (some with flow-integration figures) are looked over and tabulated by the fuel or metallurgical department. Major improvements can result from placing all records with proper identification before *operating* supervisors on a daily, or even turn basis. The information contained in these current records is as important as (and frequently points out the reason for) fluctuations in tonnage, downtime, fuel consumption, or operating rates, as tabulated daily for operating supervision.

After records have made their proper rounds, thus avoiding belated inquiries, for example, as to "what happened on number 10 furnace last Tuesday," there is little value in filling up filing cases (and even vaults) with old records.

In addition to operating records, others which show specified finished-product heat treatment, gage, width, length, coating thickness, surface condition, etc., are finding increased usage as evidence to accompany the customer's shipment. It is entirely possible and probable that such records, in the form of multichannel tape, will be used in computer control of subsequent processing, or even as a basis for establishing the purchase price.

For pilot lines or an experimental process installation, continuous high-speed records of every variable are of great value in analyzing reaction rates, process and transportation lags, as well as controllability of the operation. Unfortunately, budgets for such projects often do not allow enough for instrumentation. Scientists penciling logs from indicators can never provide the means for design analysis of full-scale processing equipment which will have much greater lags, masses, and transportation times Time-coordinated records, at fast speeds, high sensitivity and suppression, are well worth their cost, both in supplying criteria for full-scale design, and in determining the actual control requirements for production units.

INSTRUMENTATION OF STEEL PROCESSES

After considering the underlying philosophy and specialized basic requirements for steel-plant instrumentation, we can now review the principle process applications in steel production. A detailed description of "methods, modes, and media" can be presented as representing current "best" practice. However, an approach from the viewpoint of where and why, rather than with what, is of longer lasting value. New instrument designs, including those for highly desirable end-product measurements, will continue to be marketed and can radically alter good operating practice.

"Standardization" of methods, makes, and means has acknowledged advantages, but for progress rather than stagnation the "best" of new developments must be employed whenever they reach the practical stage.

Some European steel plants employ instrument and control practices which, although theoretically sound, would not be considered at all practical by the United States steel industry. The abundance of skilled technicians and low-maintenance manpower-wage rates, partially accounts for the difference in acceptable and economical practice.

Typical examples of changes in techniques are: the use of oxygen analysis to replace (or at least trim) conventional fuel-air ratioing methods; CO/CO_2 (and H_2) ratios as reduction controls; mass and dimensional measurements by noncontacting means for forming operations. These can all change acceptable control methods and lead to more rapid adoption of completely automatic computer-controlled processes. Therefore, the following process-application descriptions should be viewed as guides to instrument selection, to be seasoned with the salt of experience. New developments should be considered for inclusion in systems, but, above all, the instrumentation should be kept as simple as possible to meet each plant's actual economic requirements.

Blast-furnace Instrumentation

The reduction of ore to molten iron in the blast furnace is an example of a process in which control has been obtained by stabilization of only the input energy and mate-

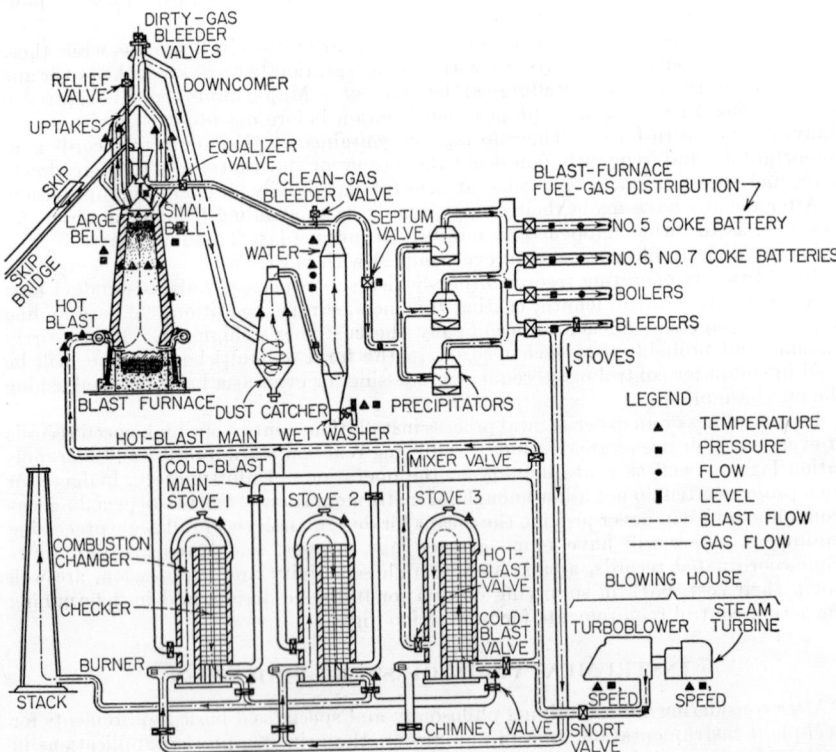

FIG. 1. Principal points of measurement in a blast furnace.

rials. The control of the actual reduction operation is difficult because of the inaccessibility of the furnace interior, its tremendous capacity, and slow response. The measurement of temperature within the furnace and its charge presents severe practical problems.

Input Energy and Material Flows. The principal input energy and material flows which are measured and controlled include:
1. Weight and composition of furnace burden, or charge.
2. Blast volume.
3. Blast moisture content.
4. Blast temperature.

Figure 1 shows a blast furnace and its auxiliaries; symbols indicate the principal locations at which instrumentation has been applied. The trend is toward more extensive instrumentation.

Operation Indicators. It has been found that an indication of conditions within the furnace, such as movement and distribution of material, and the uniformity of flow of gases through the charge can be obtained by observation of certain other variables. These *operation indicators* include: (1) inwall (lining) temperature measurements at various heights above the bosh and around the circumference of the furnace wall; (2) temperature of gas leaving the uptakes of the furnace; (3) top-gas composition, (4) stock-line level (height of material within the furnace); (5) furnace top pressure; (6) pressure drop across the furnace; (7) blast pressure; (8) rate of change of stock-line level.

Stock-line Measurement. This is one of the most useful of the operation indicators. It provides a continuous guide to the movement of material through the furnace and the rate at which the furnace is consuming raw materials and producing products. From knowledge of this measurement, the operators can adjust the intermittent charging rate of the raw materials, level, and other furnace variables to avoid erratic operation. The level can be measured by dropping a sturdy float assembly connected to a winch and cable through the top of the furnace, until it comes to rest on top of the stock (material level). Then, by reference to the amount of cable unreeled from the drum, the position of the float may be determined accurately. When the furnace is charged, the float and cable assembly are withdrawn to avoid burying and damaging the float. Figure 2 shows this equipment.

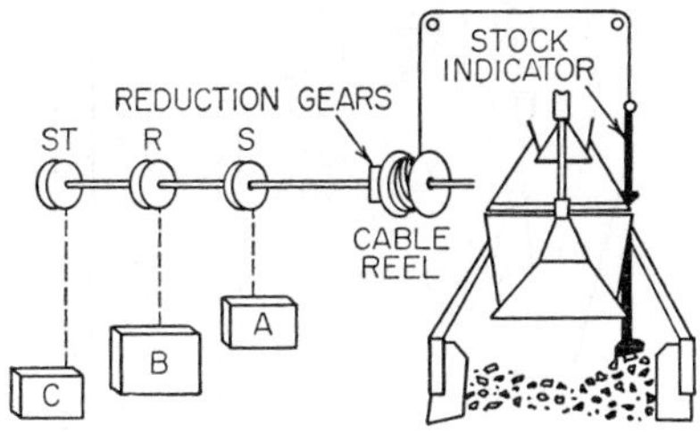

FIG. 2. Blast-furnace stock-line-level measuring system. (S) Slidewires on pinion gear shaft; (R) rotary drum contact switch on pinion gear shaft; (ST) selsyn transmitter on pinion gear shaft; (A) single-pen recorder in stock house; (B) signal light panel on cast house floor; (C) circular indicator in mud gun room.

The large area of the stock line and the irregular movement of the material in the furnace make it desirable to employ two or more stock-level indicators. Stock level is a guide to operators stationed at different locations in the blast-furnace plant. Its position is easily telemetered to these remote locations by several transmitters mechanically coupled to the cable drum. The type of transmitter to be selected is determined by the accuracy required at the receiver. Figure 2 also shows several types of transmitters employed in one installation.

The use of radioactive source and detector systems with staggered placement at normal stock-line level (similar to inwall temperature measurement for detection of "scaffolding") provides by noncontacting means a greatly increased number of stock-level measurements, which are capable of being individually indicated as well as averaged and readily employed in the control of both charging rate and distributor operation. Figure 3 illustrates such equipment (from *Steel*, October 12, 1959, where further description of this method is found).

Furnace Damage Protection. To protect the furnace against damage and obtain efficient operation of its top-gas auxiliaries, records of the following variables have been found useful to show trends: (1) gas washer or scrubber water flow and pressure as well as water inlet and outlet temperatures; (2) temperature of gas leaving washer (with control of water flow, if conservation or thickener capacity is important); (3) exit-gas temperature and dust concentration; (4) pressure drop across auxiliaries in which gas velocity is important such as dust catcher, gas washer, or precipitators; and (5) exit-gas flow and its distribution.

High-top pressure control systems, with suitable interlocks between subsequent gas-cleaning units and charging-bell operations, have materially decreased dust losses and improved production rates and the smoothness of furnace operations. However, the possible effects of high pressure on the equilibrium of gaseous reduction rates in the furnace stack do not appear to have been fully evaluated.

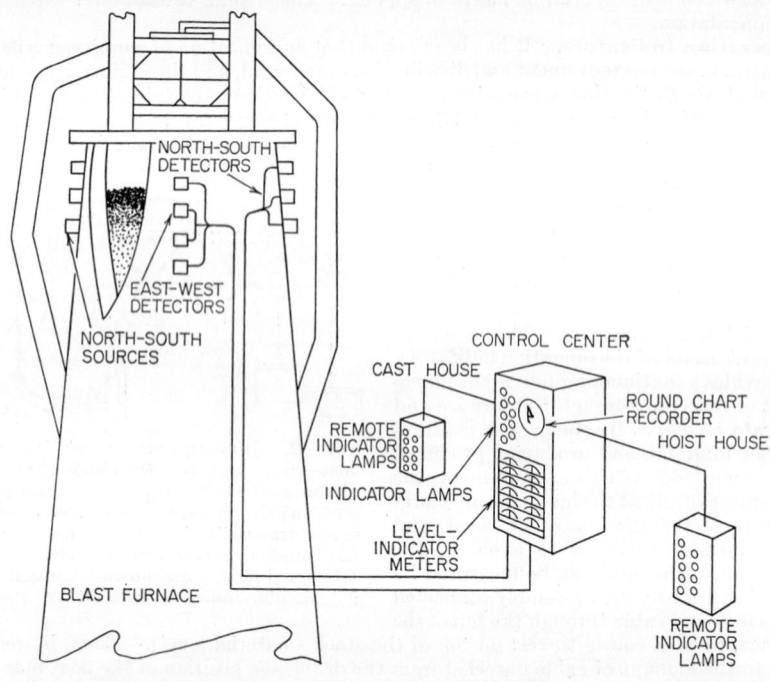

Fig. 3. Nuclear-type gaging system for measuring blast-furnace burden level.

Air-blast Instrumentation. The air blast is an important furnace variable. The blast is preheated in regenerative heat exchangers called *stoves*. These brick-filled, silolike structures are cyclically heated by combustion (within them) of blast-furnace gas. This stored heat is released when the cold blast is passed through the stove. Normally, while one stove is being heated, the second is being held hot, and the third is preheating the blast. To obtain efficient operation of the stoves, the entire mass of brick must be uniformly heated at the end of the heating cycle. In addition, the stove structure must be protected against damage by overheating. A control system which accomplishes these objectives is shown in Fig. 4.

Stove-dome Temperature. In this two-element control system, the primary control is based on a measurement of the stove temperature, because this area attains maximum temperature first and must be protected. Control is obtained by automatically varying the fuel-air ratio-controller set point so that the combustion air flow is increased in proportion to the rise in dome temperature. This action reduces the flame temperature to a value which will permit a safe dome temperature. It also has the decided advantage of maintaining a constant Btu input and of increasing the volume of hot gases flowing through the stove, thereby promoting uniform heating. If, in spite of this control action, the dome temperature continues to rise, a safety interlock is provided to decrease the flow of fuel gas.

In current practice, a radiation pyrometer is used to measure stove-dome temperature. Its construction should permit safe removal of the radiation detector for inspection and preventive maintenance, despite the pressure and temperature under which the stove operates. Furthermore, it should have a positive purge against the stove pressure and flushing air across the pyrometer lens.

Waste-gas Temperature. The secondary element of the system employs the waste-gas temperature leaving the stove as a signal to indicate fuel demand during the heating cycle. In the early stages of heating, the brick in the lower sections of the stove are cold, and the waste-gas temperature is low. Thus, the fuel demand is at maximum. As the heating progresses, the brick throughout the stove begin to

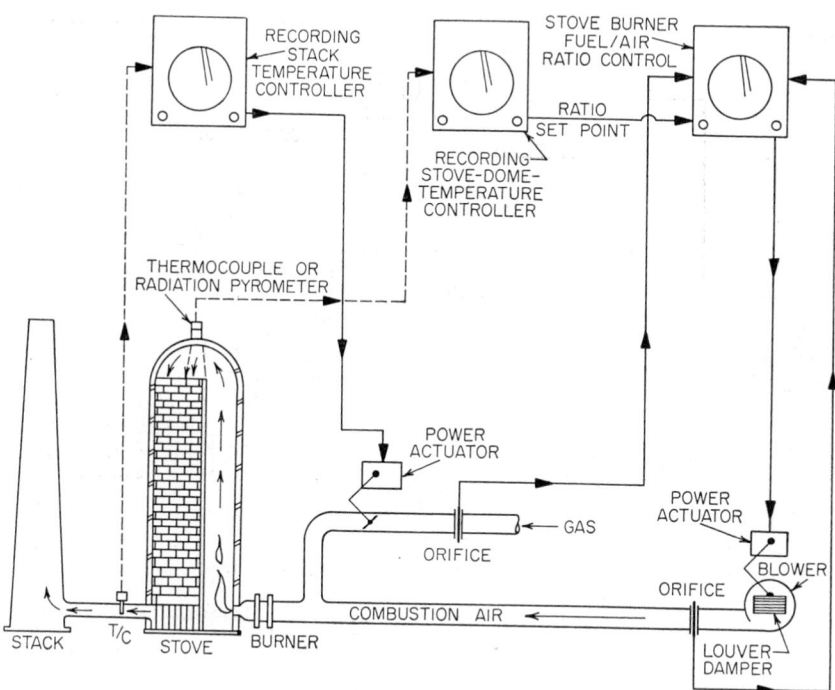

FIG. 4. Blast-furnace stove preheating control system.

equalize in temperature, and the waste-gas temperature rises. The stove-stack temperature can therefore measure uniformity of stove heating, and its control can be used to reduce fuel input as saturation is approached. The ratio control of blast-furnace gas to air flow on stove burners should provide the simplest cascading from such temperature control systems.

Effects of Wind-blown Control. The wind blown will approximate 50 per cent of the total weight of material delivered to the furnace. Thus, a small change in temperature of this large air volume has a pronounced effect on furnace operation. Experience has indicated that the hot-blast temperature affects iron production, iron composition, and refractory life. Because no heat is supplied to the stove while it is on blast, the air temperature leaving the stove is continually decreasing. In addition, when a freshly heated stove is put on blast, the air temperature rises rapidly. The continually varying conditions, which are normal to stove operation, dictate the requirement for a rapid, smoothly responding, and stable control system if a constant hot-blast temperature and smooth-working furnace are to be maintained.

Blast Temperature Control. Figure 5 shows a method of hot-blast temperature control. The temperature of the blast is measured with a sensitive thermocouple located approximately 20 pipe diameters downstream from the mixing valve to assure average conditions. Since the blast main is under 20 to 30 psig pressure, a thermocouple assembly which permits rapid and safe removal is required. An instrument which provides proportional-reset-rate response modes of control has been found necessary because of (1) the widely varying temperature of the air leaving the stove and (2) rapid fluctuations during stove changes and casting.

Control is effected by positioning a valve that admits a portion of the cold blast directly into the hot-blast main or stove-blast valves, to maintain a constant temperature below the minimum outlet temperature of the stoves. Either pneumatic or electric instrumentation may be used. The determining factor usually is preference in regard to the valve actuator. Heavy-duty, high-torque (500 to 1,500 lb-ft) actuators

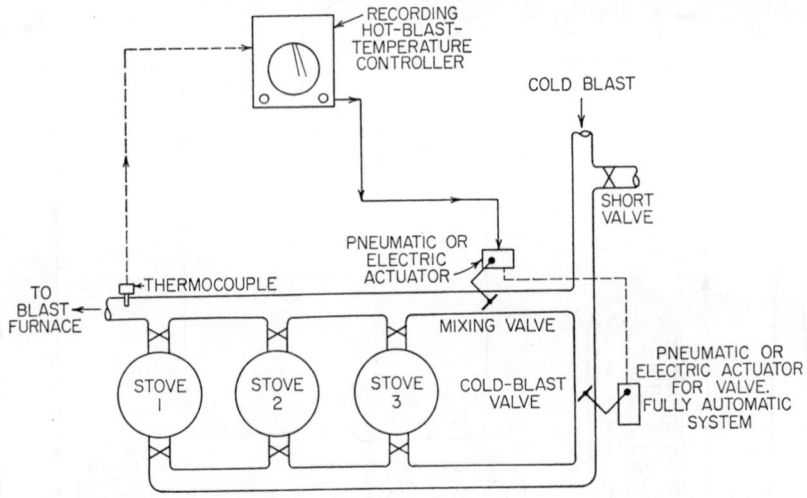

Fig. 5. Hot-blast temperature control system.

are required. A remote, manual-control panel which permits operation of the mixing valve during emergencies usually is provided at the cast house.

Depending on the size of the mixing valve and the main in which it is located, the stove tender may need to manually readjust the amount of air delivered to the mixing valve, especially when a fresh stove is placed on blast. Signal lights which indicate the approach of the mixing valve to its extreme positions inform the stove tender when such readjustment is required. Completely automatic systems which position both the mixing valve and the valve in the cold-blast line to the stoves sometimes are used. In these systems, the motion of the two valves is in sequence, so that whenever the mixing valve has reached its full-open position, the undercut cold-blast valve starts to close, thus forcing more of the cold air through the mixer line, and decreasing the supply to the stove. Figure 5 also shows a dual-valve control system.

Air-blast Moisture. A less obvious raw-material input to the blast furnace is the moisture contained in the air blast. A furnace being blown with 85,000 cfm of wind at a dry-bulb temperature of 80°F and 50 per cent relative humidity will receive about two tons of water per hour from atmospheric moisture alone. Many years of research on the effect of water vapor on furnace operation have resulted in different operating practices, and in several methods of employing instrumentation to correct or compensate for its effect. Initial attempts operated on a "dry blast" basis (with corresponding refrigeration or dehumidifying control systems). These were followed by systems which compensated for variations in atmospheric moisture by automatically adding sufficient steam to the wind to maintain its measured moisture content at a fixed value slightly above the ambient moisture.

The physical chemistry of the processes involved makes clear that the dissociation of water vapor in the furnace will have a cooling effect unless some external source of energy is supplied. An easily controlled source of at least this purely thermodynamic effect is the temperature of the hot blast. The necessary correction has been effected automatically by a cascade control system in which the hot-blast temperature-controller set point is raised as the blast moisture content is increased.

Only recently has the significance of blast moisture begun to be understood. Where the early moisture practice was intended to stabilize moisture variations, additions considerably in excess of atmospheric (up to 24 grains per cubic foot total), are now made with beneficial effects on iron production. As a result, instrumentation for controlling blast moisture content is now in use on many furnaces. The amount of moisture which may be added to the blast is related to other operating variables in the furnace. As further research studies more clearly define this relationship, it can be anticipated that simplified automatic computing systems will be employed to determine the optimum amount of moisture for a particular furnace operating condition—even including top-gas analysis.

In one typical control method, the measuring element employed is an electric hygrometer, located in a sampling chamber adjacent to the cold-blast main. A representative sample is bled from the main to the sampling chamber, where its pressure is dropped and flow rate restricted as illustrated in Fig. 6.

The temperature of the sampling lines and chamber must be kept above the sample dew-point temperature to avoid error due to condensation. Since the cold blast is at a temperature considerably above the ambient, and cools as it leaves the blast main, its temperature is easily regulated by varying the sample flow rate with the atmospheric bleed valve. Then as the ambient temperature drops, an increase in sample flow rate will transfer more heat to the sampling system and

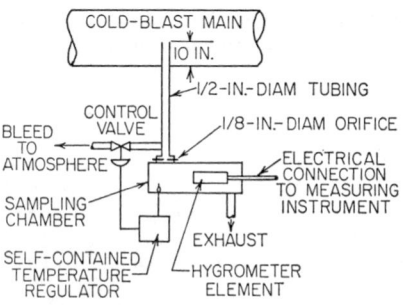

Fig. 6. Blast moisture sampling system.

maintain its temperature. The resultant measurement of grains per cubic foot is used to control steam additions, with interlocks from flow rate or snort-valve operation during casts.

Other Blast-furnace Instrumentation. An accurate control of blowing rate (usually from more than one location) must automatically compensate for changes in temperature, pressure (and moisture) with a stable control of turboblowers. More complex interlocked and multistage control systems are required for blast-furnace turboblowers, illustrating the continuing extension of control instrumentation, in contrast to an indicating pressure gage and centrifugal mercury tachometer on blowing tub, which represented "modern instrumentation" in the blowing room 50 years ago.

Individual tuyere control, for uniform distribution of blast volume—regardless of bustle-pipe connections, wind directions, blast pressure, tuyere clogging, etc.—has become widely employed on large furnaces as a means of reducing "scaffolding" and hot spots. The instrumentation systems vary all the way from multiple recording of differential pressure drops, with manual valve regulation (having an obvious tendency to "choke" the furnace) to complete automatic flow control for each tuyere—including refractory-lined valves, cascaded-control set points from master to assure minimum back pressure, positional signal systems for detection of tuyere plugging, and automatic safety action during casts.

The use of oxygen-enriched blast or hearth introduction requires additional metering and ratio control equipment, with summation in terms of air (or oxygen) equivalents and the process control of oxygen-generating units.   Fuel additions by introduction of natural gas or hydrocarbon liquids require corresponding metering, ratio, and control systems for correlation with coke rate, and may well involve computers for over-all efficiency of the reduction process.

Other instrumentation on the blast furnace proper for operational guides, safety, or accounting purposes may include: (1) Blast-pressure limit signal, indicative of "hanging" furnace; (2) iron temperatures during cast; (3) multiple signal systems on cooler-plate water flows or exit temperatures; (4) pressure and pressure-drop measurements between different elevations in stock; (5) flow measurements from bleeders, with flame detection or reignition systems; (6) rotary distributor operation recorder; (7) temperature-sensing elements around salamander (to indicate potential "breakouts") or recorded temperature rise in underhearth air or water coolants.

The entire stockhouse operation lends itself readily to automatic operation whenever accurate weight measurements and controllable feeder mechanisms are applied. Selectable sequential programming of skip car loading (by magnetic-tape or punch-card systems) from charge-material bins of raw ore grades, beneficiated or pelletized ore, sinter, sized coke, and limestone provide the means of stabilizing and controlling this most important input to the process.

Complete computer control of a blast furnace will receive further attention and application as more "end product" measurements become available and more is known about the complex reactions which account for the operating characteristics of this process.

Blast-furnace Control Center.   A blast-furnace plant has a number of operating sections remote from one another by several hundred feet.   Each of these has a direct

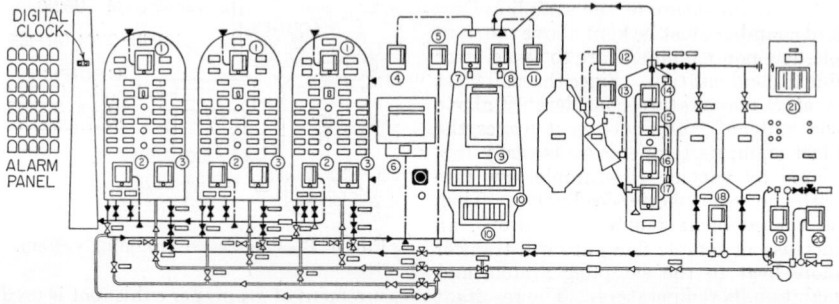

FIG. 7. Blast-furnace graphic panel board.   *Stoves:* (1) dome temperature recorders; (2) stack temperature recorders; (3) gas-flow recorders.   *Furnace:* (4) top pressure recorder; (5) blast pressure recorder; (6) hot-blast temperature controller; (7) stock-level recorder (left); (8) stock-level recorder (right); (9) sequence recorder; (10) tuyere flow indicators; (11) top temperature recorder.   *Venturi scrubber:* (12) water-pressure and flow recorder; (13) differential pressure recorder.   *Cooling tower:* (14) gas out-water in temperature recorder; (15) water-level recorder controller; (16) water-flow recorder; (17) differential pressure recorder.   *Gas flow:* (18) gas to stoves; (19) gas to bleeder; (20) gas to power house. *Temperature:* (21) multipoint dual range temperature indicator.

relationship to other parts of the plant, and all must be integrated to obtain continuous production.   The use of a graphic panel to bring together in one location all of the operating variables on a blast furnace has been shown to exhibit practical advantages. Figure 7 shows the measurements included on one particular graphic panel.*   The panel includes signals, placed at points on the panel corresponding to locations in the plant, and both miniature as well as full-size instruments.   Color and symbol-shaped codings identify types of measurements and permit easy centralized operation of the furnace and its accessory units.

* *Instrumentation*, September–October, 1958.

This instrumentation outline for a blast furnace has been covered in some detail, as an example of analysis which can similarly be applied to many of the steel-plant unit processes.

Open-hearth-process Instrumentation*

Steelmaking by the open-hearth process is a multistage batch operation carried out in a high-temperature regenerative furnace. The raw materials are charged and melted by raising the furnace temperature to approximately 3080°F. The molten charge undergoes chemical reaction, including carbon reduction, and after final adjustments to composition and temperature, is tapped from the furnace.

Major Control Factors. The principal factors which ultimately control the process are (1) charge composition, (2) time, and (3) heat input. The control of heat input, however, requires the manipulation of many secondary variables, such as temperature and its distribution, flow of one or more fuels, combustion air flow, flame distribution, flame radiation, fuel atomizing fluid, combustion-air preheat temperature, length of regenerative firing cycle; furnace pressure; fuel-air ratio, and burner adjustment.

If the open hearth is to be operated at the lowest cost per ton of steel, the following operational factors must be balanced: (1) quantity of steel produced per unit time, (2) fuel consumption, (3) furnace refractory life, (4) melting practice, and (5) raw materials charged.

Instrumentation has been found useful in accomplishing this balance by providing: (1) measurement and control of energy introduced in fuel and air, (2) reproducibility of optimum conditions, (3) measurement of resultant metal temperature and composition, (4) monitoring of critical areas of the furnace structure to prevent damage from excessive temperature, and (5) measurements which aid research in the fundamentals of open-hearth operation.

Charge composition in the form of scrap, pig, hot metal, limestone, ore, alloying elements—as well as weight of resultant metal tapped and its uniform distribution into ingot molds—are important factors in the open-hearth steelmaking process. Measurement of weight, therefore, has received increasing attention. Instrumentation of beam scales, or strain-gage and hydraulic methods, have been widely extended, in spite of the mechanical problems involved and the high ratios of tare to net weights. Development of other practical noncontacting mass measurements will greatly extend this form of instrumentation and computer operation, not only in the production of molten steel but also throughout its subsequent processing into finished products.

The most efficient utilization of fuels, both to attain high furnace temperature, and to promote chemical bath reactions—with a variety of fuels, atomizing agents, oxygen additions, etc.—has led to the development of combustion instrument and control practices (for this batch-type process) which are far beyond those found in any other major industry. As such, and involving regenerative action with different requirements during charging, meltdown, lime boil, and working stages of the heat, they are worthy of some detailed description.

Fuel Measurement and Control. Experience has shown that fuel flow should be indicated and recorded on the instrument panelboard to: (1) display trends in furnace operation, (2) aid in adjustment of controls to optimum levels for each stage of the heat, (3) guide the operator during periods of manual control, and (4) provide records for evaluating furnace operation and cost accounting.

Gas-fuel Flow. This is normally measured with a sharp-edge orifice, frequently constructed of Type 316 steel to minimize corrosion by coke-oven gas. The pressure differential produced by the orifice is transmitted pneumatically or electrically to instruments which are calibrated in flow rate. If the gas pressure and/or temperature at the point of metering vary greatly, the volumetric rate of flow measured by the

---

* Even though the steel industry in the United States has been retiring open-hearth furnaces in favor of oxygen furnaces during the past decade, numbers of open hearths still are in operation and probably will continue to be in operation in other parts of the world. Hence, the inclusion of this material in this first edition of this handbook.

orifice does not accurately indicate the fuel energy being supplied to the furnace. Measuring systems which automatically compensate the indicated flow rate for these pressure and temperature variations assist in obtaining reproducible firing conditions.

Liquid-fuel Flow. The commonly used heavy liquid fuels (Bunker C oil, pitch, and tar) are difficult to meter because of their high viscosity. Although orifice meters may be used, the variable-area and the positive-displacement meters have produced the most satisfactory results. With the latter-type meter, rate of fuel flow is obtained by driving a tachometer generator from the meter shaft. Both these devices produce an electrical signal which is linear with flow.

For accounting purposes, the flow measurement can be automatically integrated and the result indicated on a register-type counter. A separate positive-displacement meter is sometimes used solely for accounting purposes. When this is done, it is essential that means be provided for recalibration of the meter at regular intervals, preferably with the fluid being metered.

The varying flow characteristics of viscous liquid fuels, changes in line pressure of by-product gaseous fuels, plugging of burners, and frequent reversal of firing direction make manual fuel control very difficult. The practice of scheduling the amount of fuel to be used at each stage of a heat cannot be successful unless the specified flow can be continuously maintained. Consequently, the automatic control of fuel flow generally has been accepted as a requirement of stable open-hearth operation. Either pneumatic, hydraulic, or electrical mechanisms may be used to produce a proportional-plus-reset mode of control. Indication of the control point and a convenient, manual means of smooth transfer between automatic and manual are required.

Multifuel Control Systems. The simultaneous firing of two or three gaseous and liquid fuels presents several control problems. Each fuel must be separately metered and controlled.

Calibration in Terms of One Fuel. To simplify the operator's task of coordinating the quantity of each fuel and the total, it has been found desirable to calibrate the instrument scales, or control-point setting knobs, in terms of a single fuel. Thus the gas- or pitch-flow instruments may read in gallons of equivalent oil, or in Btu. Such a calibration assumes that the Btu per unit volume remains constant for each fuel. This assumption is usually permissible, except for mixed gaseous fuels where an additional calorimetric correction is sometimes used. With all meters calibrated in the same units, it is a simple matter for the furnace operator to add or subtract either fuel to obtain the desired total.

Automatic Summarizing Mechanisms. These are used to simplify the operation and to interlock the individual fuel control systems to a master fuel controller. Summarization may be accomplished by mechanical, pneumatic, or electrical methods. Mechanical summarization has limited flexibility for future modification, but has excellent stability in operation. The pneumatic approach presents a good compromise with accuracy, cost, and response. Electrical summarization will produce maximum accuracy and speed of response, but the initial cost may be high to obtain good stability. Any of these methods will yield good results if they are integrated into the original design of the control system. All signals to be summed must be linear with respect to flow. Thus, when an orifice is used, a linearizing device must be provided to extract the square root of the flow signal.

The proportion of each fuel fired will be determined by local conditions, such as availability, stage of the heat, or combustion practice. The control system may be arranged to maintain a predetermined quantity of one fuel (base load), and supply the balance of the total demand from a second fuel. Either fuel may be selected as the base load by switching at the panelboard. In addition, the quantity of the base or secondary fuel may be automatically varied by some other variable, indicative of its availability, such as line pressure or gas-holder height. Figure 8 is a block diagram of a two-fuel system in which gas is assumed to be less available than oil. The amount of gas to be used and the total fuel demand are set manually. The oil control system then automatically supplies the difference between the two.

Such multifuel systems may employ further automatic refinements, such as preset minimums of one fuel; automatic cutoff of a supplementary fuel whenever its volume is reduced below good combustion at the burners; or, in the case of excess fuel "dumping" to the open-hearth shop, the automatic sequencing of such utilization to each furnace. Such a system (as further described in *Iron & Steel Engineer**) must also provide for manual rejection by the operator of such "dumping" whenever the heat stage will not permit it without radical extension of the heat time.

Atomizing Fluid Control. Steam, compressed air, or compressed fuel gas is used to break up liquid fuels into fine droplets, to effect rapid mixing and combustion with high radiating power. Steam is the most widely used.

The application of the atomizing fluid must be carefully controlled to obtain maximum combustion efficiency and the best flame pattern. Whether steam volume or pressure is the critical variable which affects atomization (and this may depend upon burner design), it is essential that the variable be stable, or its value accurately reproduced for a given set of conditions. Conventional pressure and flow control instruments are used for this purpose. When the flow of atomizing steam is controlled, it

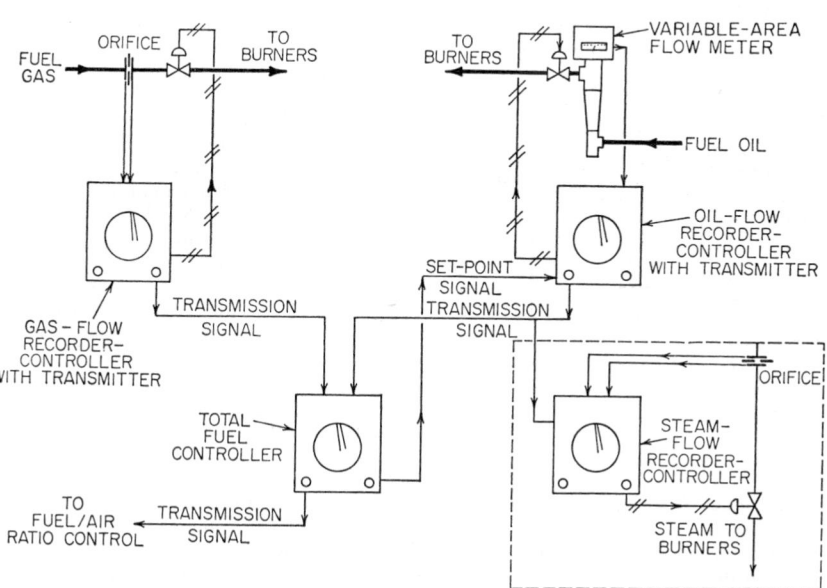

Fig. 8. Multifuel-flow control system with ratio control.

has been found desirable to ratio steam flow to oil flow so as to secure maximum heat release. Such a ratio control system is also shown in Fig. 8. An indication or record of steam temperature is useful in maintaining good atomization.

Fuel Viscosity Control. Good atomization requires not only correct application of the atomizing fluid, but also that the liquid fuel be within a limited optimum viscosity range as measured in seconds Saybolt Universal. This viscosity range is attained by preheating the heavy liquid fuels with steam in a heat exchanger. Because the volume and temperature of oil entering the preheater will vary, the steam flow to the heat exchanger must be adjusted, to obtain a constant exit-oil temperature. A temperature-measuring system with a proportional-plus-reset mode of control may be used, and "blind" systems are frequently employed.

Because viscosity is a function of composition as well as of temperature, reproducibility of a given viscosity cannot be achieved by temperature control alone. Since viscosity changes due to composition will not occur rapidly, it is general practice to make a laboratory determination of the temperature-vs.-viscosity curve and manually change the oil-temperature-controller set point, to maintain a given viscosity.

Indirect control of viscosity through temperature has been supplemented by con-

* March, 1956, issue.

tinuous viscosity measurement of fuels. One method is an adaptation of the variable-area flow meter, where a small constant sample of the fluid is continuously piped through the metering equipment. A second viscosity-measuring system utilizes an acoustic method. The primary element is a small metal probe which is inserted directly in the pipeline.

When viscosity instruments are used directly for control, caution should be exercised in applying the control impulse directly to the steam valve on the fuel-oil preheater. These preheaters usually do not have ideal control characteristics, and require a well-adjusted temperature controller to produce satisfactory results. To avoid an unstable control system, the viscosity control impulse can be applied to vary the set point of the fuel-oil temperature controller, as in Fig. 9. When individual furnace oil heaters are operated from a single fuel-oil main, only one viscosity-measuring system need be employed to control several individual oil-temperature heaters.

Combustion-air-flow Control. Metering and control of the combustion air which is supplied to the open hearth is important, not only to obtain efficient combustion, but also to provide the oxidizing atmosphere needed for steelmaking reactions.

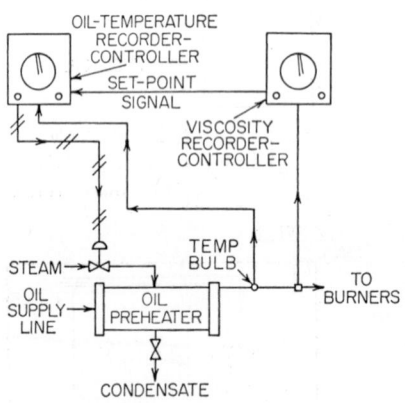

FIG. 9. Fuel-oil–viscosity control system.

Combustion air from forced-draft fans usually is available at static pressures of 1 to 3 in., in ducts of large diameter. Thus, the differential pressure developed for metering must be kept at a minimum. The physical arrangement of the duct-work usually precludes the normal calming runs of straight piping up- and downstream from the primary element. Despite these limitations, a carefully selected and installed flow-measuring system will provide reproducible measurements, even though in variance with actual air delivery to the burner ports after passing through regenerative chambers.

When orifices are used, the orifice-to-pipe-diameter ratio must be large to reduce the permanent pressure loss. Orifices on the inlet side of the fan are frequently used, together with a short entrance section having a bell mouth opening to the atmosphere. The large duct size makes it desirable to use piezometer rings to obtain a representative measurement at the pressure taps. This is particularly necessary on fan-inlet orifices.

Sheet-metal venturis also have been successfully applied to provide maximum pressure recovery, at lower cost than with cast and machined venturis.

For these low-pressure differential measurements, instruments have been developed which produce good accuracy and sensitivity when measuring full-scale differentials as low as 0.5 in. of water. These devices usually provide square-root extraction when required, and sometimes compensation for changes in air density. Both recording and indicating instruments are employed. The flow of air is controlled by electric, pneumatic, or hydraulic actuators which position butterfly dampers in the fan discharge line, or louver dampers at the fan inlet.

Enriching oxygen sometimes is used as a source of part of the combustion air, for a portion of the heat. It is also injected by lance into the bath. Separate oxygen flow, integration, and control systems may be arranged to summarize the quantities of pure and atmospheric oxygen supplied, either as O_2 or air equivalent. A definite relationship between the two quantities then can be automatically maintained.

Fuel-Air Ratio. When the fuel flow to a furnace is varied frequently, it may be advantageous to automatically proportion the air flow so that the ratio between the two is held at a predetermined value. This ratio is usually selected to provide a slight excess of air, above that required for complete combustion. Also, it is changed for different stages of the heat and bath reactions. An economic evaluation between cost and equipment for ratio vs. independent controls should always be made.

When multifuel firing is employed, the problem is more complex. With fuels of similar combustion-air requirements, it is only necessary to totalize the fuel flows in a manner similar to that shown in Fig. 9, and actuate the air-flow control from a signal proportional to this total. However, with fuels of widely different air requirements, the summarizing circuit must be designed to produce a signal proportional to the total air demand rather than to total fuel demand.

Linearization of Flow Signals. Since it is usually necessary to linearize the flow signals to permit their summarization, the signal developed by the summarizing device will also be linear. In applying this linear air-demand signal to an air-flow controller, a device which will square the signal must be employed if the air flow is of the square-root type. Otherwise, the air-flow signal must be linearized.

Basic Considerations of Fuel-Air Ratio Controls. In applying fuel-air ratio controls to an open hearth, several factors must be considered. The air supplied is not solely used for combustion of fuel, but should be considered as one of the steelmaking raw materials. For example, it must be available in sufficient quantity to combine with carbon monoxide when the latter is evolved from the bath. During certain stages of the heat to aid in rapid carbon reduction, it may be necessary to carry a more highly oxidizing atmosphere over the bath than at other times. The air volume may also

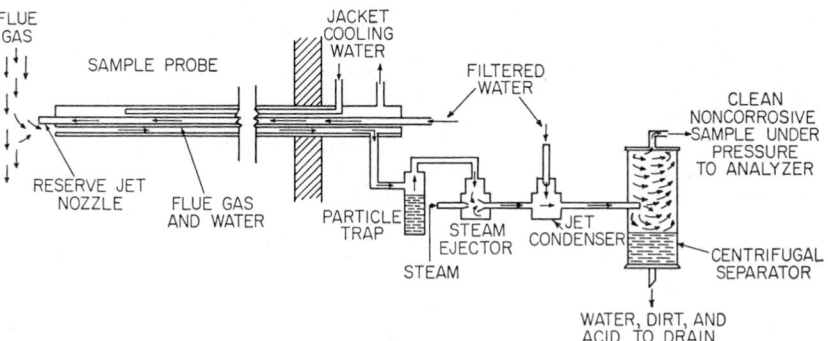

Fig. 10. Waste-gas-sample cleaning system for oxygen measurement.

be varied to aid mixing of fuel and air, or to control the character of the flame. In addition, it is not possible to keep a high-temperature furnace absolutely gastight, and thus varying amounts of unmetered air may be introduced.

These factors make it clear that no one fuel-air ratio will satisfy all stages of an open-hearth heat. They also emphasize the complexity of this control problem and the desirability of stabilizing all possible variables. As greater experience is accumulated, control systems are modified to recognize some of these factors. For example, the fuel-air ratio controller may be adjustably biased by a preset amount to compensate for infiltrated and unmetered air.

Use of Analytical Instrumentation. A direct answer to a varying fuel-air ratio requirement is to employ analytical instrumentation to determine the oxygen demand of the combustion and chemical processes. Although several good types of oxygen analyzers have been available for many years, a satisfactory method of continuously obtaining a representative gas sample at the proper furnace location has been lacking. A satisfactory oxygen-analysis sampling system for the open-hearth process must satisfy a number of requirements: (1) remove solids and corrosive gases from a high-temperature sample; (2) provide a pumping speed which will permit a sample time lag consistent with process response time; (3) permit continuous operation with routine preventive maintenance, rather than emergency maintenance; and (4) prevent contamination of the sample gas with other sources of oxygen.

The extensive development work devoted to this problem has produced several sample cleaning systems which may satisfy these requirements. The system shown in Fig. 10 is representative.

These developments have permitted successful experimental installations of automatic control from oxygen analysis. In applying this measurement to automatic combustion control, it is necessary to consider that, in normal operation of the open hearth, combustion conditions exist during which the oxygen concentration is not controllable. One of these is readily recognized as the period immediately after reversal of gas flow through the regenerators. The existence of this situation leads to the conclusion that a stabilizing control loop should be present to control combustion conditions during this period: the fuel-air ratio or independent air-flow controller. The oxygen analyzer then is employed to adjust or trim, within preset limits, the fuel-air ratio so as to maintain an optimum oxygen concentration in hearth exit gas. During those periods when oxygen concentration is not representative of furnace conditions, this control signal may be locked out of the air and/or oxygen control system without difficulty.

Furnace Pressure Control. Removal of the gaseous products of the steelmaking and combustion reactions is effected by the pressure drop across the furnace. This driving force is created between the forced-air fan, bath reactions, and the stack, or induced-draft fan. The difference between the pressure of the combustion chamber and atmospheric is a measure of the magnitude of the driving force.

Combustion-chamber–Atmospheric-pressure Differential. If this pressure difference is negative with respect to atmospheric, over a large portion of the furnace, cold air will be drawn in through the openings and cracks. This infiltration will reduce the preheated air temperature and increase fuel consumption and melting time. Conversely, if the pressure difference is strongly positive, the high-temperature corrosive furnace gases will be forced into the cracks and joints of the furnace refractories, thereby shortening their life. Furnace pressure is also of importance in controlling the length and shape of the flame. For dollar expended, a well-designed, sensitive, and fast-acting furnace-pressure control system will return greater dividends than any other instrumentation on an open-hearth furnace.

Location of Furnace Pressure Taps. Location and installation of the furnace pressure taps must be made carefully to provide a representative measurement of the average furnace pressure. Areas in which localized transient pressures may develop, or those which may be sensitive to direction of firing, must be avoided. Experience has indicated that a tap made at the center of the furnace roof generally gives the most satisfactory results. To minimize the effect of ambient temperature variation and the difference in elevation along the transmission piping, the atmospheric tap should be made immediately adjacent to the furnace pressure tap, and parallel impulse pipes should be carried back to the measuring instrument.

Because furnace pressure is a relative value, which varies with location in the furnace as well as its temperature, the operating value at the point of measurement is selected to produce approximately atmospheric pressure between the wicket hole and the door-sill level. This area is selected for the region of balanced pressure, since it is the door openings that will provide the greatest infiltration of cold air.

Special-purpose Pressure Measurements. Because the pressure at the point of measurement is very small (0.05 to 0.09 in. of water), specially designed measuring instruments are required. These may be either of the diaphragm or the oil-sealed bell type, having a total measuring span of only 0.2 in. of water, and usually calibrated -0.05 to $+0.15$ in. of water with respect to atmospheric pressure. Their design and narrow span should provide a measuring sensitivity of ± 0.0005 in. of water. Because of the low signal level available for actuating the measuring instrument, these devices may provide only indication and/or control. A record of the pressure is highly desirable, and normally is obtained by a separately actuated recorder. A separate small indicator showing damper position or signals on approaching limits are convenient operating tools.

Speed of Pressure Fluctuations. Controllable pressure fluctuations occur so rapidly and frequently in an open-hearth furnace that manual control is impractical. Control within ± 0.003 in. of water or better is required and obtained in many well-engineered automatic control installations.

Furnace-pressure Control Systems. The final control element is actuated preferably from a proportional-plus-reset mode of control. The actuating medium may be either electrical, hydraulic, or pneumatic. There also has been extensive use of an electrical, two-speed-floating mode of control, for motor positioning of heavy dampers. The damper controlling furnace pressure is usually of the slide type, located in a flue.

Depending on the available equipment, control may also be obtained by positioning the inlet louvers on the induced-draft fan, or bypassing the waste-heat boiler with a fraction of the exhaust gases. A heavy-duty actuator, capable of rapid response, must be used with a slide damper. It is advisable to counterweight the damper, if it weighs more than a few hundred pounds. The physical size of the dampers and their travel are so great that it is frequently necessary to employ reeving. A hand winch or hand-wheel on the automatic operator, to provide a base range of operation and permit operation during failure of the control, is highly desirable.

Electric Pressure Transducers. The availability of sensitive electric pressure transducers designed for industrial conditions has made it possible to effect certain improvements in furnace-pressure instrumentation. If compensated for operation in high and varying ambient temperatures, these transducers can be mounted immediately adjacent to the furnace. When mounted in this manner, the long double run of impulse piping to the panelboard is eliminated, together with its associated time lag and installation cost. The improvement in speed of response of the measuring system is then effective in maintaining furnace pressure during transient upsets, such as following furnace reversal. Because the transducer signal is measured by an electric instrument, recording and controlling functions can be combined in one case.

Any single furnace-pressure measurement is a relative control value, and is normally reset during different heat stages and corresponding firing rates. Hence, the automatic cascading of furnace-pressure set point from fuel rates has some economic possibilities.

Regenerative System Control. The regenerative section of a high-temperature furnace consists of two chambers filled with brick arranged in a pattern of flues. The outgoing waste gases pass through one chamber, and transfer heat to the checker-brick for half of a complete cycle. Simultaneously, the incoming combustion air flows through the other chamber, and by absorbing heat from the brick is preheated. During the second half of the cycle, the flow of gas and air is reversed. Thus, the checker-brick which has cooled from preheating the air, is regenerated by absorbing heat from the waste gases.

Functions of Regenerative Systems. The regenerative system of an open hearth serves two purposes: (1) It preheats the combustion air, so that maximum flame temperature may be developed in the furnace; and (2) it increases the efficiency of the combustion process. To obtain the maximum yield from these objectives, the following variables influencing regenerator operation should be considered: (1) Optimum length of reversal cycle must be determined automatically to obtain the maximum required air preheat with minimum stack losses; (2) uniformity of air preheat, as indicated by the maximum and minimum temperatures attained during each half-cycle must be maintained; (3) checker-brick temperature must not exceed a safe value; and (4) mechanics of the reversal operation must be conducted efficiently.

Regenerative Temperature Measurement. Two methods are in general use for measuring regenerator temperatures. The most desirable location for the primary element is at the furnace end of the regenerator chambers. This area gives an approximation of the final (and most important) air-preheat temperature. At this point, the temperature will approximate 2400 to 2600°F and, if thermocouples are used, they must be of the platinum type, with ceramic protection tubes. The corrosive nature of the waste gases from an open hearth make the frequency and cost of maintenance of thermocouples at this location high. A thermocouple in this location is also handicapped by the local nature of its measurement in a large chamber, radiation and gas velocity errors, as well as protection tube lag.

Use of Radiation Pyrometers. These difficulties have been overcome and some advantages gained by using radiation pyrometers, instead of thermocouples. For example, flame carryover is detected immediately. The radiation devices are sighted at a downward angle, directly on the top course of checker-brick in each chamber. To aid in covering maximum area, the elements should have a large target-to-sighting-distance ratio. Any radiation pyrometer contains an optical system which must be kept clean to obtain reliable measurements. It has been found that a carefully installed unit, with an adequate supply of clean, dry purging air, will operate under severe dust and fume conditions with only an occasional routine inspection. Purging air is introduced in front of the optical system and passes through the tube into the regenerator, with sufficient flow maintained to prevent entry of the open-hearth waste gases into the tube.

Thermocouple Temperature Measurement of Checkers. The second method of regenerator temperature measurement utilizes thermocouples placed in the flue end of the checker chambers or in the stack flue. At these positions, the temperatures normally will not exceed 1400°F but are much less indicative of regenerator operation than those in the hot end. For example, the thermocouple on the ingoing chamber measures a temperature approaching atmospheric rather than air-preheat temperature. The outlet thermocouple does not indicate maximum temperature in the chamber, or flame carryover. Temperature measurements at this point have been shown to be of assistance in detecting short circuiting or dirty checkers, as well as leakage in reversal valves and presence of combustibles when thermocouples are located in the stack flue.

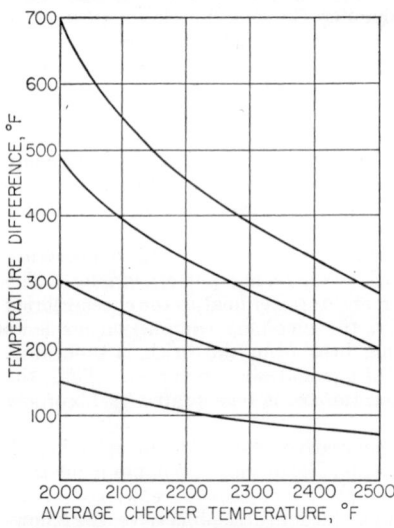

FIG. 11. Temperature difference required to cause reversal vs. checker temperature measured with radiation pyrometer.

Optimum Reversal Cycle. This cycle and the uniformity of air preheat for each cycle are influenced by many variables, but basically are determined by temperature and time. Historically and currently, on many furnaces, the reversal cycle has been controlled from time alone. Timing devices signal or initiate automatic operation of the reversal valves. The manually preset cycle time is determined by the operator, to maintain the desired temperature in the regenerators. The time interval should be changed as the heat progresses, and the time of firing in each direction may need to be varied to compensate gradually for unbalanced furnace conditions, such as during charging. Considerable experience and operator attention are required to make this method of control effective.

To obtain automatic determination of the optimum reversal cycle, reversal must be essentially initiated from some temperature condition. Many special control systems have been designed for this purpose.* One of these, which has been widely employed, actuates reversal from measurement of the difference in temperature between the ingoing and outgoing checker chamber at the hot ends. It is therefore basically independent of temperature level. Reversal occurs when a preset temperature difference is reached.

In order to obtain maximum air preheat when most needed in the finishing stages of the heat, the length of the reversal cycle must be decreased as the average preheat temperature rises. This effect is obtained automatically when the temperature is measured by radiation elements, since the signal developed by these devices is proportional to approximately the fourth power of the target temperature. This relationship is shown in the curves of Fig. 11. If it is preferred to vary the reversal cycle only at the discretion of the operator, the radiation element signal can be modified at the measuring instrument to compensate for its exponential response and provide a linear record. The reversal signal is then given only with attainment of a constant preset temperature difference.

* Further description may be found in *Iron Steel Engr.*, March, 1956.

With any system, equipment may be provided to reverse the furnace if a maximum time interval is exceeded, before reversal from temperature difference. This protective device is useful in the early stages of the heat. In addition, a high limit contact is available to cause reversal if either checker chamber reaches an unsafe temperature at any time during the cycle.

Effecting Reversal in Minimum Time. To achieve maximum regenerator efficiency, it is necessary not only to reverse at the optimum point, but also to effect this reversal in the minimum time. Because of the numerous valves and dampers whose position must be changed, the time required may vary from 10 to 30 sec. Based on a 30-sec interval, the total fuel-off time might be as much as 36 min in a 24-hr period. In addition, of even greater importance is the time required to again establish good combustion conditions and regain the temperature loss during the fuel-off periods. This time amounts to almost a minute for each second saved in actual reversal timing.

Carefully engineered automatic reversal systems markedly reduce this fuel-off time and furnace clearing period after reversal, freeing the operator from a routine duty and improving furnace operation. Systems have resulted in as much as 3 to 8 per cent reduction in fuel rate, as well as in reduced spalling of roof refractory. These advantages may be obtained by automatic equipment whether reversal is actuated from a time or temperature signal.

Automatic-reversal Control Systems. These systems are designed to carry out all reversal operations, following an initiating signal, in a timed or sequence program This program is planned to restore the furnace to previous combustion conditions in the shortest possible time. For example, at the start of reversal, the fuel-air ratio control and the stack damper may be locked in position until reversal is completed. Full steam flow and air flow are introduced into the furnace during reversal, to purge waste gases from the regenerators and clean the oil-burner tip. Fuel is then admitted at a gradually increasing rate so that the furnace is not flooded with partially burned fuels. This artificial but automatic fuel readmission rate is worthy of the closest periodic attention and resetting, because the ideal rate will change with the age and condition of each furnace.

An automatic reversal system should permit manual initiation of the reversal cycle and individual manipulation of each operation, such as of the fuel valves and dampers. It may require establishment of "block" and "blow" conditions. Signal lights which indicate firing direction or completion of remote operations are desirable. Although the specifications of each system must be worked out individually, basic packaged reversal systems with extensive adaptability are commercially available.

Roof-temperature Measurement. The temperature of the roof inside the furnace is an important operating guide, although admittedly requiring more maintenance and service than other open-hearth instrumentation. It is one of the few continuous measurements possible inside the combustion chamber. It is important because maximum production can only be obtained when the roof or other limiting refractory location is operating within some 25 to 50°F of its failure point. Development of a means to measure "fluxing action," rather than surface temperature of refractories, will greatly simplify and extend the usefulness of this high-limit control.

Use of Radiation Pyrometers. Conventional thermopile-type radiation pyrometers usually are employed for this purpose in the United States; successful use has been made of photoelectric-cell pyrometers for roof-temperature measurement in England. In either case, special mounting equipment must be used to protect the pyrometer optical system from flying hot metal and the pyrometer assembly from mechanical damage.

Location of Detecting Element. Target area for measurement has usually been on the roof above the taphole. However, changes in batch-furnace designs and extensive use of all basic roof, as well as modified firing and lancing practices, are continually relocating the region of shortest refractory life—generally toward knuckles, uptakes, fantails, and even checker chambers. Sighting on these limiting areas through open-ended tubes has been the method most commonly employed. This is true in spite of the fact that splashing, dolomite make-up, stringers, etc., all produce measurements in the "unsafe" direction. The front wall is a good location for direct sighting on the roof, if damage to the detecting equipment by charging machines and hot-metal additions can be avoided.   Availability of miniaturized radiation detectors has made it possible to design mounting assemblies (Fig. 12) which can be made small enough to be completely enclosed by the buckstays.

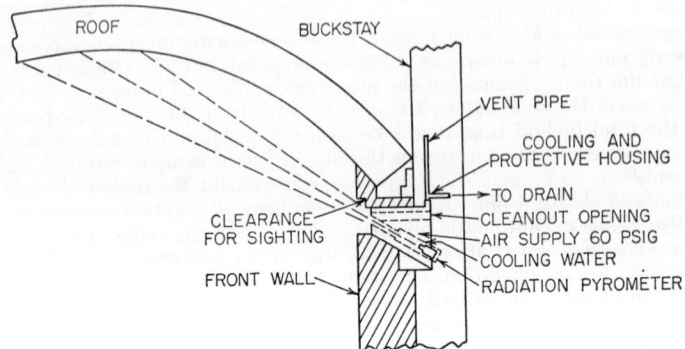

FIG. 12. Front-wall roof-temperature measurement arrangement.

A second method involves a closed-end type of detector.   A specially designed block built directly into the roof (Fig. 13) avoids the problems of hole obstruction and flame interference.   Also, by a progressively lowering thermal gradient, this arrangement provides a temperature measurement in the "safe" direction.   There are, however, serious disadvantages to this method: (1) The equipment cannot be serviced just before roof failure when it is most needed; and (2) maintenance of the pyrometer-detecting element is difficult because of its hazardous location.   On the other hand, the method does have strong proponents who find it useful in spite of these problems.

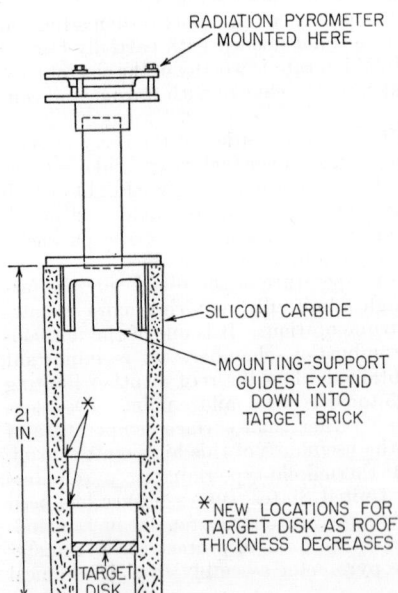

FIG. 13. Refractory-roof block assembly for open-hearth roof-temperature measurement.

With any location or method of using the radiation detector, roof-temperature measurement should not be considered infallible.   Final responsibility for both roof life and heat time should be left to the melter and his helper.   Daily checks with an optical pyrometer, with fuel off, are necessary.   Recording instruments with a calibration adjustment which permit matching the optical measurement and instrument-indicated temperature should be used.

Control of Fuel Valve.   Numerous methods have been employed for controlling fuel input, in order to maintain the desired roof temperature.   Those installations which were unsuccessful failed primarily because of an unreliable measurement.   Reduction of fuel input from a roof-temperature signal will be unpopular with production and operating personnel if attempts are made to operate the furnace well beyond its capacity, or if fuel input is radically reduced with resultant upsets in working the heat.

Several control techniques have been applied including: (1) throttling of air supply to pneumatic fuel-flow controllers when roof temperature approaches its limit; (2)

direct-proportional positioning of a valve in the fuel line; (3) introducing a signal which is proportional to roof temperature as a bias to the fuel demand signal, and (4) even initiating reversal, as a means of "fanning" the furnace. In addition, a number of installations employ several points of temperature measurement on the roof, and use a control system having an auctioneering circuit to select and control from the highest of these temperatures.

A record of the roof temperature is highly desirable to show trends, such as lime boil, and the like. A large indicating pointer is of greatest assistance to the operator.

End-product Measurements. The batch operations of open hearths, electric furnaces, Bessemers, and the numerous oxygen-converter processes for producing molten steel do not lend themselves to "end product" measurements and automatic control. However, instrumentation for many periodic determinations prior to tapping is of great value in meeting composition specifications, as well as affording maximum yield. Speed in taking, transmitting, and evaluating such measurements is of prime importance—particularly in the "fast" converter processes.

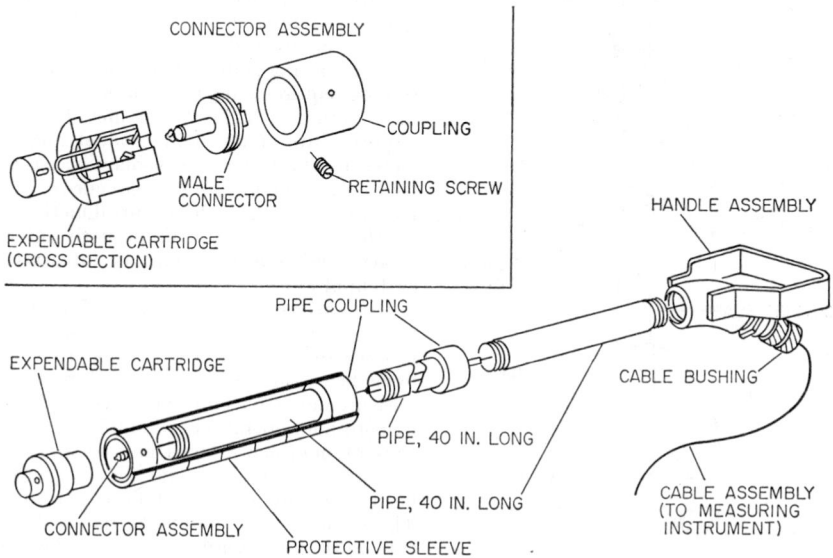

CONNECTOR ASSEMBLY

COUPLING

MALE CONNECTOR

RETAINING SCREW

EXPENDABLE CARTRIDGE (CROSS SECTION)

HANDLE ASSEMBLY

PIPE COUPLING

EXPENDABLE CARTRIDGE

CABLE BUSHING

PIPE, 40 IN. LONG

PIPE, 40 IN. LONG

CABLE ASSEMBLY (TO MEASURING INSTRUMENT)

CONNECTOR ASSEMBLY

PROTECTIVE SLEEVE

Fig. 14. Typical molten-steel thermocouple design.

Molten-metal (or bath) temperature measurements have progressed far beyond optical spoon approximations or "after the fact" readings on the tapping stream. Accurate metal temperature measurements prior to tapping, in addition to being of some benefit in controlling metallurgical composition, have proved to be of great value in increasing yield by reducing skulls in ladles and stool stickers in ingot molds.

Several "blowing tube" designs have been used for measuring steel temperatures below the slag level, employing both optical and radiation principles of operation. Platinum thermocouple designs, as exemplified by the construction illustrated in Fig. 14, have, however, become more universally accepted as most consistently accurate. Costs per reading for the expendable materials and continuous make-up of new elements are relatively high but justified by the value of the measurement to operators.

All sensing-element designs, regardless of whether for open-hearth or blown-metal application, have as an objective the ability to be inserted and removed fast, before they burn up. Ample and continuously available elements, combined with practical operating techniques by first helpers, have been contributing factors to successful operation. High-speed recorders, on rugged and portable dollies, or signal systems from central instrument locations, are part of such systems.

Carbon determinations for open hearths, or other "end point" measurements on converters, have been available for some time. For closest specification control of steel and elimination of ladle additions, spectrographic and X-ray analysis of both slag and metal require high speeds in both the analysis itself and transportation or preparation of samples.

Continuous Casting of Steel. The potentialities for the steel industry of continuous-casting techniques demand extension of modern high-speed and cascaded control methods to rapidly stabilize the number of variables which affect operation of the process. Figure 15 shows the basic components of a continuous-casting machine.

Metal level as maintained in the mold is obviously an important factor. Accurate measurement of this condition has employed X-ray, radioactive materials, and radiation principles. This level depends upon speed of pinch rolls and flow or weight of molten steel delivered from tundish and/or pouring ladle. This flow, as well as cooling requirements, is a function of metal temperature. Physical and metallurgical requirements of the final product are affected by the flow rate of coolant water, as well as its temperature and sequence of distribution. With modern control methods, all these variables can be properly interlocked, and sequenced into an integrated system of automatic regulation.

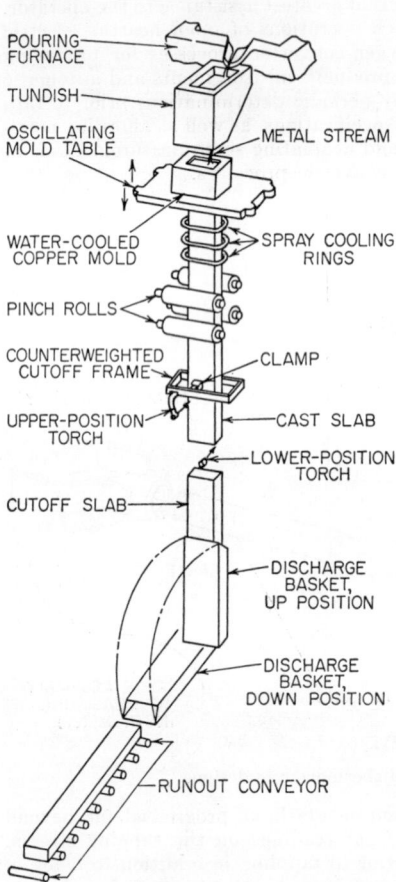

POURING FURNACE

TUNDISH

OSCILLATING MOLD TABLE

METAL STREAM

WATER-COOLED COPPER MOLD

SPRAY COOLING RINGS

PINCH ROLLS

COUNTERWEIGHTED CUTOFF FRAME

CLAMP

UPPER-POSITION TORCH

CAST SLAB

LOWER-POSITION TORCH

CUTOFF SLAB

DISCHARGE BASKET, UP POSITION

DISCHARGE BASKET, DOWN POSITION

RUNOUT CONVEYOR

Fig. 15. Schematic diagram of continuous-steel-casting machine.

Intermittent Open-hearth Measurements. During the course of an open-hearth campaign, many operating problems have been solved by additional periodic measurements beyond those required for hour-to-hour operation. Troubleshooting, evaluation of new designs, and comparison of performance by several furnaces have been aided by specialized measurements. The purpose and equipment required for this work are described in the following paragraphs.

Heating Rate. In heating a new or repaired furnace constructed of silica brick to its operating temperature, the heating rate must be kept slow enough to avoid spalling of the brick. Studies have determined the critical heating rates which may be used for typical open-hearth construction; so the operators then are faced with the problem of heating the furnace at a rate which closely approaches these values. If the actual rate exceeds the critical, the furnace will be damaged, and if it is much slower, the increased heating time will be costly, owing to loss of furnace availability.

To determine the actual heating rate, special thermocouple blocks may be installed at one or more points in the roof. These blocks are drilled to permit insertion of a thermocouple to within $\frac{1}{2}$ to 1 in. of the hot face of the brick. The temperatures thus measured are normally recorded on a portable instrument with a special recorder chart, on which the critical heating rates have been ruled (see Fig. 16). Alternatively, the furnace temperature may be raised by means of a cam-type program controller.

Relative Flame Radiation. The end result which determines the operating value of many open-hearth variables is the relative radiating ability and distribution of the flame. For example, atomizing medium and burner adjustments are evaluated by the intensity of the radiant energy (not optical luminosity) as emitted by the flame. Radiation pyrometers have been widely employed to provide a quantitative measurement of this factor. It is necessary that the optical system of the radiation element be sensitive to wavelengths between approximately 0.5 to 3.8μ or longer. The major portion of the energy emitted by a source at open-hearth flame temperature lies

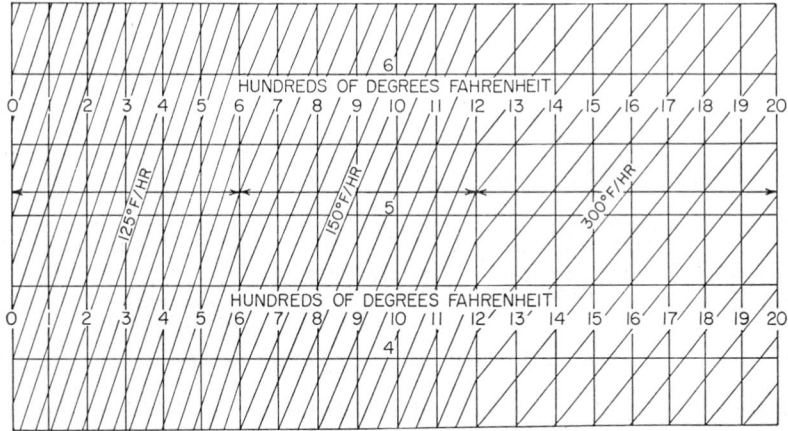

Fig. 16. Temperature record preruled with critical heating rates to guide operators.

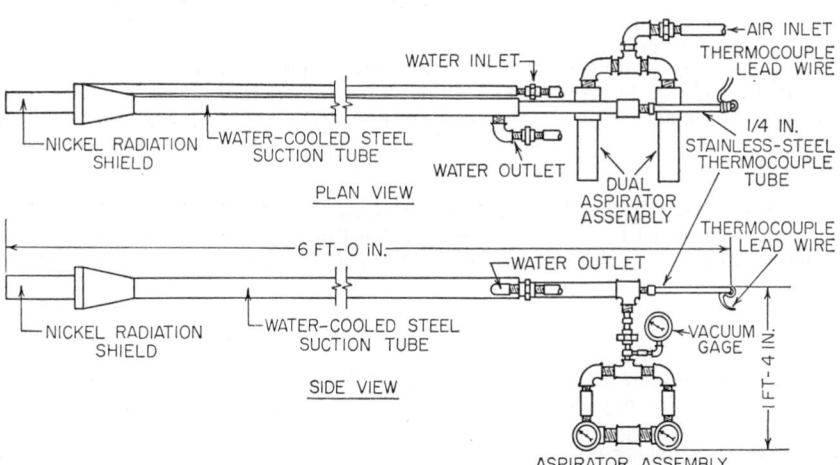

Fig. 17. General design of high-velocity shielded aspirating thermocouple.

between these wavelengths. In use, the radiation element is sighted through the doors of the furnace at the portion of the flame being studied.

The detector usually is held by a helper and may be moved about to obtain a traverse of any region desired. The recording instrument should have a rapid chart speed, in order that all sections of a traverse are defined on the chart.

Combustion-air Preheat Temperature. Under the subsection on regenerative system control, temperature measurements used for controlling the operation of the regenerators are described. Measurements so obtained are subject to a number of well-known errors encountered with high-temperature flowing gases. These measurements represent a compromise between absolute accuracy and a representative guiding temperature which may be measured continuously with easily maintained primary elements. Thermodynamic and combustion efficiency studies require a closer approximation to actual gas temperature than these simplified primary elements can provide. As a result, much work has been done to devise measuring means which reduce the magnitude of these errors.

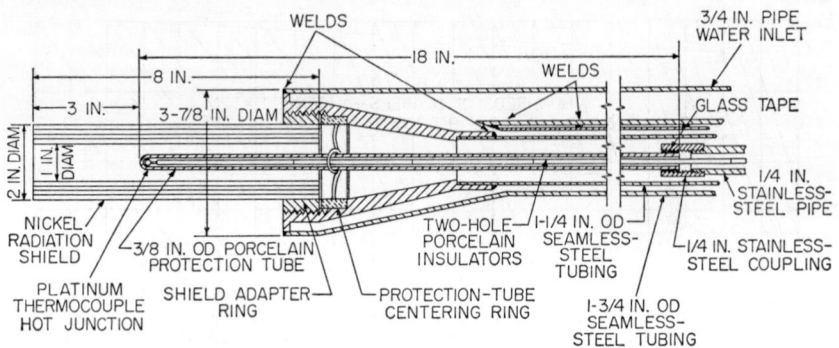

FIG. 18. Details of end of shielded thermocouple.

Figures 17 and 18 show a shielded high-velocity thermocouple which was developed for measuring the air preheat temperature in open-hearth regenerators. Studies with this element have indicated that when it is operated at a gage vacuum of approximately 3 in. Hg, an indication very close to true gas temperature is obtained. As with other shielded thermocouples, it cannot as yet be used for continuous measurement at open-hearth temperatures. Its use on an intermittent basis has provided valuable information concerning the relationship between air-preheat temperature and open-hearth heat time. Infiltration, and its resultant reduction of air preheat, has been more effectively determined by this shielded thermocouple than by any other method.

Steel Reheating

Soaking Pits. The soaking pit is a batch furnace for heating ingots (the product of the open hearth) to a temperature at which they may be mechanically formed. After reaching the desired temperature in the pit, the ingots are permitted to *soak* so that the temperature throughout their cross section is equalized. This heating operation must be controlled to prevent excessive ingot surface temperatures with resultant "washing," and to ensure a pit atmosphere conducive to satisfactory ingot-surface scale conditions. The instrumentation problems are less complex than those of the open hearth, but more precise control must be obtained to secure both highest product quality and maximum yield.

Temperature Measurement and Control. A specific temperature condition is the end result which a soaking pit produces, and not merely the means to it. Accordingly, all other variables may be controlled in relation to temperature. The relationship among temperature of the pit, approximate average temperature of the ingot, time, and fuel are shown in Fig. 19.

Indirect Temperature Measurement of Ingot. It is difficult to obtain a satisfactory direct measurement of ingot temperature which can be used for control. Instead, the temperature of the pit is determined at a point or points which will be representative of the maximum and/or average ingot surface temperature. The relation between this temperature and actual ingot surface temperature may be determined by auxiliary readings with an optical pyrometer during the heating and drawing operations. Both platinum thermocouples and radiation pyrometers sighting in closed-end target tubes are used as primary elements. The temperature-measuring instruments are sometimes supplied with calibrating rheostats so that they can be adjusted to read approximately in terms of actual ingot temperature, as determined by an optical pyrometer.

Location of Primary Element. The most representative location for the primary element must be determined by experience for each pit design, size, and firing method —such as tangential, bottom, corner, or end burner locations. Wherever structurally possible, temperature measurements which are representative of two conditions should be employed for guidance and/or control. The first measurement (most commonly used for automatic control) is representative of the maximum heating condition to which the ingot surface is being subjected. If possible to obtain without undue influence from colder recuperative regenerative areas, the second measurement(s) should be representative of exit-gas temperatures and, therefore, of the "average" ingot temperature at soak-out—when fuel rate has been reduced to merely balance furnace radiation losses.

Requirements of Pit Temperature Controllers. Three requirements of pit operation should be satisfied by the pit temperature controller: (1) Steel should be uniformly heated to within rolling-temperature tolerances; (2) steel must not be overheated; and (3) during removal of ingots for rolling, the last ingots to be removed must not fall below a prescribed minimum temperature.

Wherever a two-element control system is possible, from indications of "maximum" and "average" ingot temperatures, a series-interlocked or switchover-and-recycle method of control can be employed to advantage in securing maximum tonnage at "safe" heating rates.

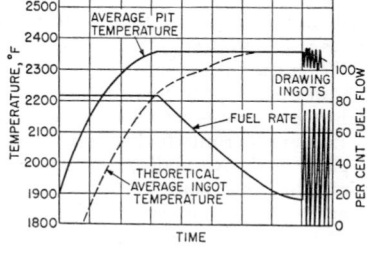

FIG. 19. Soaking-pit temperature and fuel flow for typical cycle.

The temperature measuring and controlling instrument, including the primary element, must be extremely sensitive and have rapid response. During the period when an ingot is being drawn, the pit cover is opened, and all fuel is shut off for approximately 60 seconds. This operation is repeated on a 3- to 4-min cycle until the pit is emptied. To maintain pit temperature, in spite of this heat loss, requires a very rapidly responding control system. Figure 19 shows the effect on fuel flow and temperature during drawing.

When pneumatic control is used, the controlled air pressure may be blocked, and air bled from the pneumatic operator while the cover is open. This condition closes the fuel valve, but causes the automatic reset action of the controller to accumulate. When the controlled air is again placed on the valve, the reset action artificially increases the initial fuel flow to the pit after closing the cover. This action is obtained by means of an electropneumatic valve, which is operated by the craneman, or automatically from cover position.

Temperature control usually is obtained by positioning the fuel or combustion-air valve directly in response to demand, using a proportional-plus-reset mode of control. On some installations, the addition of rate response to the controller has shown some advantage in minimizing temperature loss during the drawing operation.

Heating-rate Control. The lower thermal conductivity of alloy ingots, as compared with carbon steel ingots, reduces the rate at which they may be heated without damaging the surface. This factor may be accounted for by limiting the extent to which the control valve can open during the heating period. Manually adjusted limiting valves sometimes are placed in series with the control valve to accomplish the same result. If more precise and automatic control of the heating rate is required, as with certain grades of alloys, a programmed temperature controller may be used to vary the heating rate as desired.

Fuel Measurement and Control. Soaking pits normally are fired with gaseous fuels although oil is also used. Measurement of the fuel supplied is necessary for accounting purposes, and to develop a signal for the fuel-air ratio controller. The metering problems are similar to those described previously under open hearth.

The fuel valve usually is operated directly from the temperature controller. A number of installations have been made with the combustion-air flow valve positioned by the temperature controller. The fuel valve then follows the air demand with biasing by the air-fuel ratio controller. This arrangement offers greater safety protection, in that it avoids introduction of fuel without combustion air.

When burning fuels, such as blast-furnace and coke-oven gas, it is necessary to provide a pressure regulator to maintain the fuel static pressure. When two fuels are fired simultaneously, it is desirable to employ instrumentation to control the ratio between the two fuels. During the heating period, a high ratio of coke-oven gas to blast-furnace gas may be used. However, during the soaking period when fuel demand is small, a low ratio will provide the high volume of gas desirable for circulation. This ratio may be varied manually or automatically, either from coke-oven-gas availability or progress of soak-out.

Frequently, soaking pits are fired with a mixture of two gases, proportioned to give a definite heating value. An automatic calorimeter may be used to provide control of the heating value by adjusting the ratio of the two gases in the mixture.

As the ingot temperature approaches a uniform value, the fuel rate to the pit will be reduced to the amount necessary to satisfy radiation losses. Therefore, since a low and stable fuel input is indicative of soaked ingot conditions, the position of the fuel valve can be used to operate a signal system, which will tell the operator when the pit is ready to draw.

When the pit cover is removed during drawing, a cover switch automatically shuts off the fuel supply until the cover is replaced. A convenient arrangement during lighting of pits is provided by a manually operated switch that overrides the cover switch and permits fuel supply, even though the cover is open.

Air Flow. Normally, this is metered only to provide a signal for the fuel-air ratio controller. On some recuperative pits, a sheet-metal venturi is used to meter preheated combustion air at the burner. In this position, the metering element accounts for the air which actually infiltrates the recuperator, and thus indicates the volume of air actually available for combustion in the pit. Because the air at the burner is preheated and may vary over a wide range, it is necessary to automatically compensate the flow measurement for these temperature changes, or to vary the fuel-air ratio mechanically from a combustion-air temperature-measuring instrument.

Pit Pressure. The automatic control of furnace pressure at a positive value above atmospheric provides the same benefits previously described for the open hearth. It is particularly necessary during the soaking period, when the volume of waste products is reduced, to prevent air infiltration. Rapidly responding control systems are required to cope with the fluctuating conditions present during the drawing of the pits. When the flue system is arranged so that a single stack serves a battery of pits, and the pits are not equipped with individual pressure controls, it is advisable during drawing to artificially open the stack damper of the battery to prevent the open pit from acting as a stack for the other pits. If each pit is equipped with its own individual pressure control, it is desirable to close the damper when the cover is removed during drawing.

Fuel-Air Ratio. Steel at elevated temperatures is readily oxidized, with formation of scale on the surface. The amount and character of this scale is determined by several factors, but principally by the composition of the atmosphere in which the steel is heated. A slightly oxidizing atmosphere produces a loosely adhering scale which is easily removed in the first pass through the rolling mill. A reducing atmosphere produces a tightly adhering scale that may be rolled into the metal surface and cause a defect. The atmosphere must be controlled very closely to maintain a satisfactory scale condition.

Fuel-air ratio control provides the means of regulating the pit atmosphere. Because of the wide range of fuel flow encountered on soaking pits, manual control of the air flow is unsatisfactory. Also, it is usually necessary to use different fuel-air ratio settings for the heating and soaking periods. However, it has been reported that with compensation for air temperature, little or no manual adjustment of fuel-air ratio is necessary. The most satisfactory solution to this problem is to measure the concentration of oxygen in the waste gas and trim the fuel-air ratio setting to hold a fixed oxygen concentration.

Air-preheat Systems. Soaking pits are constructed for both recuperative and regenerative air preheating. With regenerative systems, automatic reversal is needed to minimize the fuel-off periods and to obtain uniformity. The reversal cycle normally is determined on a time basis, although reversal from temperature difference has been used successfully.

For fastest efficient completions, the reversal operation should follow the same procedures as outlined for the open hearth. Temperature control of regenerative pits may use a method consisting of (1) firing at a maximum fixed fuel rate until a preset pit temperature is reached; (2) blocking the pit until the temperature has dropped to a preset minimum; and (3) resuming firing with interlock of reversal operation to ensure against continued firing in one direction. This method of operation can also provide an adequate means of regulating induced-air volumes, where forced-fan air is not available.

With recuperative firing, a temperature controller may be required to limit the temperature of the metallic recuperators. Control is obtained by varying the amount of air passed through the metal section of the recuperator.

Instrumentation of Older Pits. The addition of automatic controls to older-design soaking pits has resulted in substantial increase in pit heating capacity. Considerable ingenuity is required in some cases to permit adapting the pits to control (either manual or automatic) without complete rebuilding: For example, many of the early pits had no means of controlling air flow. Reference 2-F provides an excellent description of such a problem and its solution.

Relatively simple computer systems based on steel grade, ingot size, pouring temperature, stripping, and track time can assist in the setting of both the pit control systems and rolling schedules.

Continuous Reheating. After heating in soaking pits, ingots are formed into reduced section such as billets, blooms, and slabs, before further mechanical working. These products must be reheated before they can be further rolled. This second reheating usually is done in continuous counterflow furnaces.

In these furnaces, the steel creates the same control problems as when it is heated in a soaking pit. Temperatures must be carefully controlled, and the furnace atmosphere should minimize scaling. Because in a reheating furnace the steel is heated by moving through zones of relatively constant and progressively higher temperatures, certain additional factors must be considered.

The relationship between steel temperature and atmospheric temperature in each zone is a function of the weight rate of steel being heated. Changes in the atmospheric temperature in one zone may affect the temperature of the preceding zone. If the steel travel rate is reduced or halted for rolling-mill reasons, the furnace temperatures must be adjusted to prevent damage to the product. Similarly, after such mill delays, temperature set points must be readjusted gradually to normal operating conditions.

Temperature Measurement and Control. Platinum thermocouples, enclosed in refractory protecting tubes inserted through the roof, may be used for reheating furnace measurements. Since tube breakage and thermocouple replacement may be high, owing to thermal shock, radiation pyrometers sighting in closed-end target tubes have been applied in the same locat'on. With either type of primary element, the temperature measured is that of the furnace interior. This temperature must be empirically related to steel temperature in order to heat the product as desired.

Direct Temperature Measurement. A representative measurement of steel surface temperature furnishes a more satisfactory basis for heating steel than the temperature of the atmosphere surrounding it. It is possible to sight radiation pyrometers directly on the slabs or billets, and obtain a measurement which is close to the actual steel temperature. There are several factors which make the direct measurement of steel temperature rather difficult to apply generally.

The location of the radiation unit must be selected very carefully for each furnace. In making this selection, consideration should be given to (1) avoiding flame interference; and (2) obtaining a sighting path which views steel as much as possible, irrespective of product size. Best results have been obtained by sighting downward at a slight angle from the horizontal so that the target observed is principally the end of the slab. References 1-G and 2-G describe direct measurements of steel temperature in continuous-reheating furnaces.

Methods of controlling continuous furnaces are shown for a three-zone, triple-fired slab-heating furnace in Fig. 20. The instrumentation shown with solid lines is the conventional type. The preheat zone of a furnace of this type is heated only by the hot waste gases flowing toward the exhaust flue at the back, and the temperature in this zone is not controlled.

The heating- and soaking-zone fuel inputs are controlled independently from primary elements located in the roof, or from radiation pyrometers sighting on the steel. The top and bottom section burners of the heating zone may be controlled from the same temperature instrument.

However, because the bottom section will usually require a different fuel input from that of the top section, a device for ratioing bottom-to-top fuel is desirable. Better

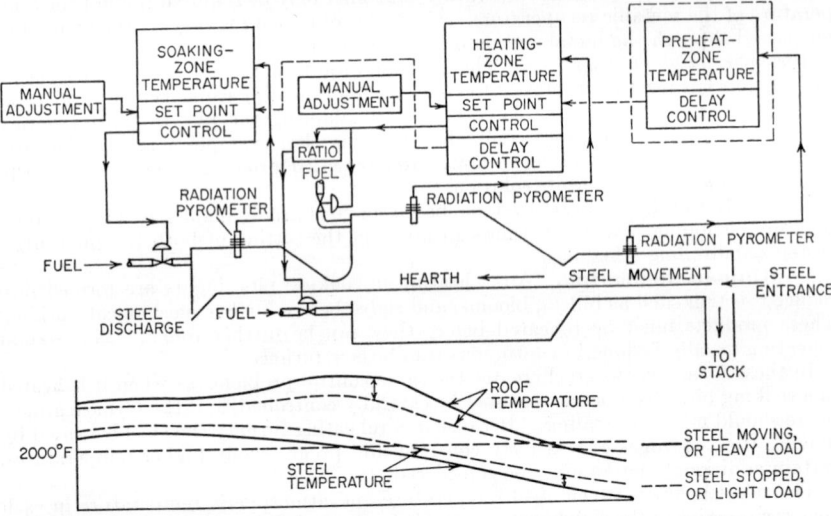

FIG. 20. Temperature control system for three-zone triple-fired reheating furnace.

over-all control can be obtained by operating the bottom section from a temperature measurement of its own. The optimum location for the bottom-section primary element is dependent on furnace design and firing practice.

The temperature relationship between zones, as well as that between atmosphere and steel temperature, must be determined by experience. Using a proportional-plus-reset mode of control, systems of this type have been found to produce more uniform heating than similar furnaces operated manually. As a check on furnace operation, the surface temperature of each slab, immediately after the first roll pass, is scanned by a recording radiation pyrometer. This instrument provides a record readily available to the reheating furnace operator.

Requirements for Manual Control. The control instrumentation just described will correct for the minor and most frequent changes in load or furnace conditions. However, when a major upset in operation occurs, such as a mill delay requiring halting the flow of steel, manual adjustment is necessary to prevent excess steel temperatures. For example, during a mill delay, it is necessary to manually lower the set point of each instrument. The cascade control system shown by dotted lines in Fig. 20 is designed to eliminate the need for these manual adjustments.

The temperature-measuring instrument in the preheat zone varies the set point of the heating-zone instrument in accordance with a predetermined relationship. Thus, when the flow of steel is halted or lighter tonnage is charged, the temperature of the preheat zone begins to rise and the instrument measuring this temperature reduces the set point of the heating-zone instrument. After the steel flow is resumed or tonnage increases, the preheat temperature will drop and the set point will be raised automatically in proportion.

Only the preheat and heating zones are interlocked because it is assumed that in the soaking zone the steel and atmospheric temperatures are nearly the same. When the soaking zone is also used for heating, as it is on most high-production furnaces, it is also necessary to adjust its set point to prevent steel damage.

This double-cascading action operates in the following manner. The preheat-zone measurement of steel temperature transmits a signal at the start of a delay, to reduce the set point of the heating-zone temperature controller. In response to this signal, the heating-zone temperature decreases and, in turn, transmits a signal, generated by a secondary control system, to lower the set point of the soaking-zone temperature control system. The secondary-controller adjustments permit the amount of the reduction in soaking-zone temperature. When steel flow is resumed after a delay

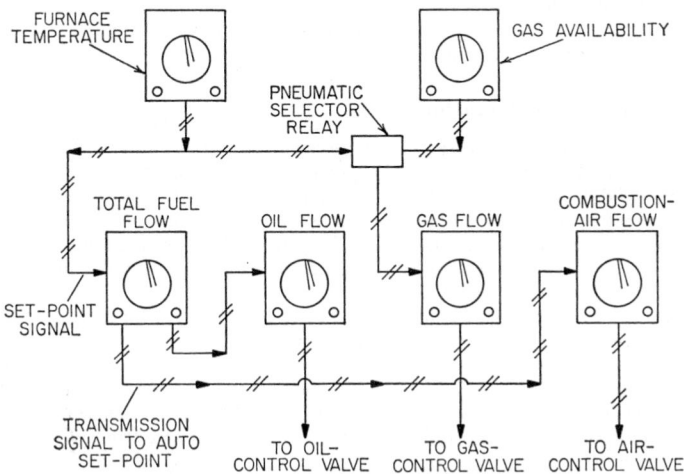

FIG. 21. Two-fuel system with gas-availability limiting control.

the reverse action raises the zone set points to the normal operating level. The cascade control system should have adjustable limiting devices to control the amount and rate of automatic index resetting which can be obtained.

Fuel Control. Reheating furnaces are fired with either gaseous or liquid fuels and, in increasing numbers, with both types simultaneously. With single-fuel firing, the fuel valve is controlled directly by the temperature control instrument. When two fuels are fired together, a flow control system similar to that shown in Fig. 21 is used. The temperature controller positions the index of the total fuel instrument. The latter circuit stabilizes the entire system against deviations which are the result of fuel-supply pressure variations.

The installation shown in Fig. 21 includes a secondary control system which applies gas to the furnace in proportion to both gas availability and furnace fuel demand. Usually, as the supply of gas decreases, its static pressure in the main will drop, and a pressure-measuring instrument can be used to introduce a signal into an availability control system.

In the system shown in Fig. 21, gas is preferentially supplied to the furnace to satisfy the temperature demand, until the gas main pressure drops to a predetermined limiting value. From this point, as the gas pressure further decreases, the gas flow is reduced proportionally until it is cut off. The temperature demand of the furnace during this period is automatically supplied by increasing the flow of fuel oil.   Excellent temperature control has been obtained by this method even though the secondary-fuel availability control rapidly varied the gas flow from 0 to 30 per cent of total fuel. Recording and integrating fuel flow meters furnish accounting information.

Fuel-Air Ratio.   Scale formation must be held to a minimum on continuous reheating furnaces because the steel is rolled into semifinished or finished products.   Accurately maintaining the correct atmosphere requires that combustion air be supplied in proportion to fuel flow.   Because of the widely varying fuel flow required to hold a constant temperature, air flow must be controlled with automatic ratioing equipment.

Furnace Pressure.   This variable must be controlled within narrow limits to obtain good fuel economy and to maintain uniform atmosphere control.   A slightly negative pressure in the vicinity of the exit door will cause excessive infiltration whenever a billet is discharged.   The furnace pressure is usually held equal to atmospheric at the hearth level, and slightly positive at other elevations in the furnace.

Recuperator Instruments.   Because combustion air for continuous reheating furnaces is normally preheated in recuperators, its temperature, as well as that of gas entrance and exit, is measured to assure efficient operation.   High-temperature-limit control systems with simple modes are applied to metallic recuperator sections or to protect economical fan materials.

Mechanical Forming Operations

Temperature Measurements.   The physical properties and metallurgical structure of hot-formed steel, as well as the power required and the life of the forming machinery, are determined partially by the temperature of the metal during working. Portable optical pyrometers, manually operated, have been used for many years to obtain temperature measurements during forming.   The need for more frequent or continuous measurements has made quite general the use of radiation pyrometers sighting directly on the hot steel.

Location of temperature-measuring points depends on the forming operation.   A measurement of steel temperature prior to the first roll pass, but after the scale breaker, protects against roll damage by cold steel.   This location also can serve as an automatic rejection means and can be used to measure temperature distribution across and along steel shape.   When several furnaces are used to feed a single mill, temperature measurement of each slab indicates uniformity of heating from all furnaces.

Hot Strip Mills.   These are operated with the steel at specified temperatures throughout the process.   The slab temperature may be determined after the roughing stads, and again after leaving the holding table before entering the finishing mill. Next, the strip temperature is measured immediately preceding, or following, the last finishing stand.   The strip is cooled by water sprays on the runout table before entering the coilers.   The water cooling is guided by a strip-temperature measurement immediately preceding the coiler.   This complete metal-temperature instrumentation during forming has made it possible to specify accurately the temperature at which products shall be worked in order to control their metallurgical properties.

The response of the measuring system for most applications should be not more than one second for 98 per cent of the final temperature.   This speed is particularly necessary when temperature distribution along the product is to be measured, or when high-speed targets, such as short bars, must be monitored.

Measuring accuracy normally is obtained by calibrating the radiation-pyrometer recorder against an optical pyrometer measurement of the target temperature.   Care should be taken in the making of such a calibration to avoid errors due to the emittance of the steel.   The emittance corrections for optical and radiation pyrometers are not of the same magnitude.   The use of the optical pyrometer should not be extended beyond its sensitive range as a means for calibrating radiation pyrometers for low-temperature targets.   The recorders used with the radiation pyrometers normally provide a measuring-circuit manual adjustment to permit calibration for the actual operating conditions.

The usefulness of temperature measurement will be greatly extended by the development of temperature-sensing elements or sensing systems which are (1) independent of steel emittance; (2) noncontacting, continuous, and high speed; and (3) suitable for recording and automatic control or computer signals.

Other Variables Measurements. In addition to temperature, there are many other physical measurements required in the mechanical forming and rolling of semi-finished or finished steel products. With manual operation of mills, many of these measurements were made with adequate accuracy by visual observation. To apply fully automatic computer and programming operation, however, additional sensing elements are essential. These elements should have the following characteristics: noncontacting, high speed of response, high accuracy, stability, and long reliable life.

Fundamental areas for these additional measurements can be broadly named as follows: (1) weight, (2) dimension (thickness, width, length), (3) force, and (4) location or position. Sections of this handbook on such variables give brief reviews of methods for making these measurements. Below, each is discussed briefly with reference to problems in the steel industry.

Weight Measurement. Measurement of weight, by both strain gage and hydraulic methods, has been mentioned in connection with the steel production processes which precede forming. In such applications, the measurement is a means of regulating the inputs to stabilize these processes. In rolling operations, the actual (and variable) weight of rough or semifinished shapes (when coupled with dimension measurements) provides the only accurate means of setting up the succeeding programmed forming operations; or of computing optimum length cuts for pipe, structural shapes, etc., to increase both production rate and yield. Otherwise these values must be estimated for presetting on punch cards or tape. Improvements in sensing elements, application techniques, or newer methods are required to meet the progress which has already been made in rolling instrumentation.

Dimensional Measurements. Measurements of thickness, length, and width vary greatly in magnitude and accuracy requirements, depending on the individual product to which they are applied. Thickness measurements—by X-ray, beta- or gamma-ray techniques—have provided adequate and stable comparison instrumentation for both hot- and cold-rolled sheets, tubes, plates, and the like. Automatic control is obtained from adjustment of screwdown, tension, and stretch mechanisms. Also measurements are used for control of such operations as sorting or classifying of finished steel products.

Automatic contour control of strip (whenever such rolling practice becomes practical) can be achieved by continuous scanning of sheet width. Measurements of flange and web thickness by similar techniques should reduce roll setup losses on structural and related shapes.

Higher accuracy, greater reliability, and possibly higher speed can be the means of obtaining tighter adherence to specification tolerances with automatic programmed rolling. The result: increases in yield. Again, similar improvements may be required for fully automatic rolling, or meeting other dimensional measuring requirements.

Length measurements employing physical contacting means are limited in both accuracy and speed of product travel. These methods have been supplemented or superseded by noncontacting magnetic-impulse counting and self-balancing optical systems for shearing or cutting operations. Improvements in methods and new techniques can be anticipated to meet the requirements for accurate taped records of deviations from specifications or location of defects on coiled finished products.

For over-all dimensional measuring requirements of rough products immediately prior to forming, a combination or modification of these methods may answer the needs of computer control. Both maximum dimensions (for initial roll spacings) and averaged dimensions (for weight or finished length computation) are highly desirable measurements for development.

Force Measurements. Rolling pressures to be applied, or restricted as high limits, have been measured by frame-mounted strain-gage methods—with compensation for temperature errors. These pressures have also been measured by hydraulic means.

Position or Location Measurements. In order to determine steel entrance and exit positions to roll stands, manipulators, and similar equipment, heat-sensitive and optical devices have been extensively employed. Similar usage in scanning systems may provide some satisfactory length measurements of steel products traveling at high speeds.

Over-all Control from Computers. Coverage of computer operation and design to meet the combined requirements of production, accounting, incentive wage determination, customer sales records, and the like, is found elsewhere as a complete subject.

Batch Annealing

Portable annealing covers are a specialized design of heat-treating furnace for annealing coils of sheet, strip, and wire. They consist of a base on which the steel products are stacked, an inner cover which surrounds the work, and an outer cover which is the source of heat. The inner cover serves to retain a protective atmosphere about the work. The outer cover may provide heat by the direct combustion of fuel between the two covers, or it may be lined with horizontal or vertical radiant-heating tubes.

Although the basic annealing process is simply one of heating, holding at a specified temperature, and cooling, instrumentation is required to effect this cycle economically and to obtain consistent metallurgical results. The major factors of this problem include: (1) maintaining a maximum safe heating rate, (2) providing absolute protection against excessive tube temperatures at all times, and (3) providing means for determining heat saturation and maintenance of steel temperatures within the required limits for a definite period of time.

Temperature Measurements. Temperature measurements are normally made at several locations within an annealing cover by means of thermocouples. The temperature of the cover, or of the radiant tubes, must be obtained to permit high-speed heating without damage. The temperature of the work is required to ensure uniformity and attainment of the desired metallurgical condition. The large size of the charge and its heat capacity necessitate placing several thermocouples throughout the work. Usually top, bottom, and center locations are selected. For critical charges, more points may be measured. Because the work thermocouples must be removed during unloading of the furnace, the insulation must be flexible and must resist abrasion.

The loading, cooling, and unloading operations of these batch furnaces causes only part of the installed thermocouples to be operative at any one time. To obtain maximum utilization of the measuring and controlling instruments, the thermocouple connections at each base are permanently wired to a central panelboard and terminate at plug-in jacks or switches. The wiring then is arranged so that recording and indicating instruments may be connected to thermocouples and corresponding control circuits in any annealing cover as desired.

It is general practice to record both work and cover temperatures during the heating and holding cycle. Additional thermocouples in the work also may be recorded, to provide a complete history of the annealing cycle. After fuel is shut off, the cooling cycle is monitored by manually switching thermocouples to an indicating instrument for intermittent readings.

Recording and controlling instruments have been designed to provide all the control functions in a single instrument in order to conserve panel area. However, work and cover control may be obtained from separate instruments of simpler design and maintenance, to ensure maximum protection of the furnace and its charge.

Automatic Control. The control of cover-annealing cycles involves consideration of several factors. The primary objective is to raise the work to the annealing temperature rapidly and maintain it at this temperature for a specified time. Owing to the mass of steel, the work temperature rises very slowly during the heating period. Therefore, the cover should be operated at a maximum temperature to ensure a high rate of heat transfer. This temperature must be limited to prevent damage to the radiant tubes or cover.

Thus, both the work and the cover temperatures must be controlled. As the work approaches the annealing temperature, the heat head resulting from the cover temperature must be reduced to avoid overheating the work. A fully automatic control system then must interlock work- and cover-temperature signals to produce the optimum control action.

The designs of some direct-and radiant-tube-fired covers lend themselves readily to a two-position mode of control. Therefore, the covers require only simple series or series-parallel circuitry to perform these required functions within the required tolerances. Electric control systems are usually employed for ease of switching, although pneumatic operators with electropneumatic relays have reduced maintainance costs on valve operators.

Switchover-and-Recycle Control. Where either the cover design or product specifications require a proportional mode of control, circuits have been designed specifically to integrate all the foregoing factors. Figure 22 shows the temperatures and valve

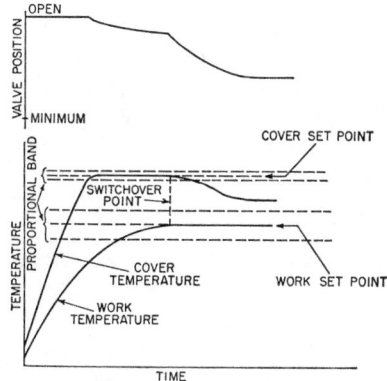

FIG. 22. Annealing-cover cycle with switchover-and-recycle control.

positions during a typical annealing cycle controlled by this method. This control system provides proportional control with manual or automatic reset on both the work and the cover. The control forms are interlocked to provide "switchover and recycle" control in the following manner: Control is effected from the cover temperature until the work temperature reaches its set point; when this occurs, switchover action takes place, and control is placed on the work.

Control is continued from the work as long as the work temperature is above or at its set point, or within an adjustable neutral zone below the set point. When the work temperature drops below the neutral zone, recycle action occurs, and control reverts back to the cover temperature. In addition, the neutral zone is positioned by the work index (bears a fixed relation to this index). Therefore, when the work set point is raised sufficiently, it is possible to effect recycle if the neutral zone is moved above the work temperature. This latter action permits maximum heating rate following a rise in work set point without endangering the cover.

A single-valve control system is operated by the proportional relay. In addition, a high-limit switch may be provided to cut off fuel flow, if the cover temperature should exceed a maximum limit.

For those furnaces in which the relationship between work and cover temperature rarely permits the cover to overheat, the foregoing system is modified to provide proportional control only on the work. The cover temperature is limited by an on-off, or two-position, type of control. With this system, whenever the cover temperature exceeds its set point, the valve is driven slowly toward the closed position until the temperature is again below its set point. After this correction is made, control again is obtained from the work thermocouple.

Programmed Batch Heat Treating

The attainment of many metallurgical structures and physical properties by heat treatment requires that the work be subjected to a precise program of temperatures.

These metallurgical processes include normalizing, malleabilizing, spherodizing, stress relieving, process and isothermal annealing.   The rate of change of temperature during the program also may be critical because: (1) of the variation in size of the work and the resultant necessity of avoiding excessive thermal stresses, and (2) sufficient time must be allotted for the structural changes to occur, or time at a temperature must be minimized to prevent formation of undesirable structures.   Since these operations normally involve a large number of work pieces per cycle, the cost of failure to

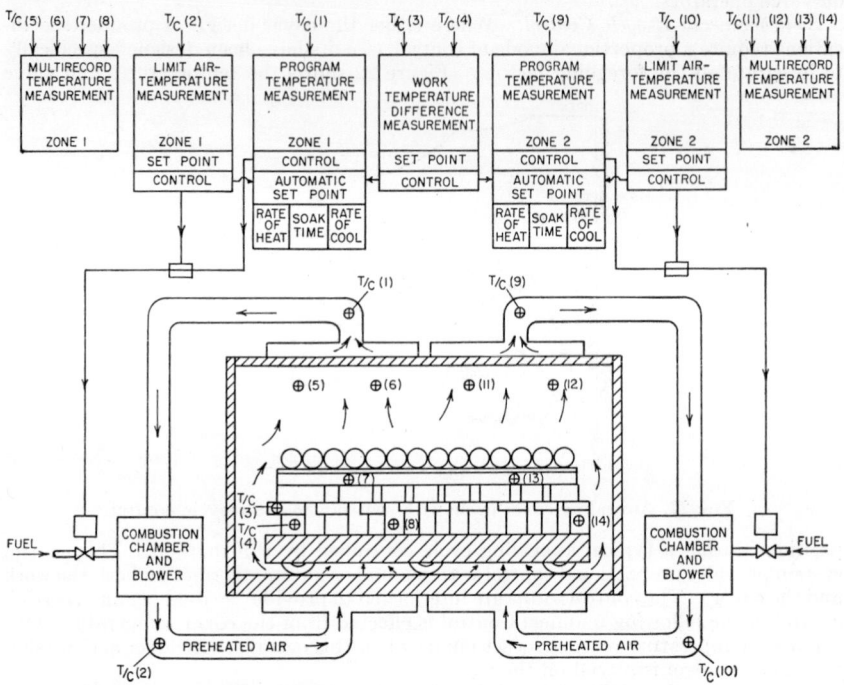

FIG. 23. Automatic program control of stress-relieving car-bottom furnace.

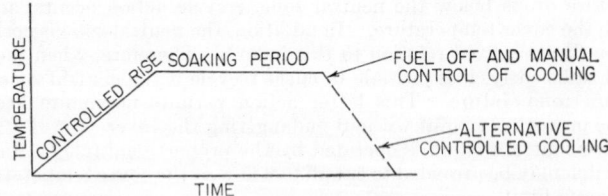

FIG. 24. Typical temperature cycle for stress relieving.

produce the desired heat treatment is very high.   Thus, accurate reproducibility of temperature is needed.

The foregoing requirements have made the use of automatic-programming temperature controllers an economic necessity.   When they are applied to heat-treating furnaces, certain auxiliary equipment has been found desirable.   The specific instrumentation required will vary with the process, and is usually supplied as a complete panel for ease in setting up the program.

Stress Relieving.   Figure 23 shows the application of program controls to a stress-relieving car-bottom furnace of two zones.   The work is heated by circulating externally preheated air through the furnace. The automatic temperature programmers control furnace temperatures in each zone. A typical stress-relieving program might appear as shown in Fig. 24.

The ductwork and blowers are protected from excess temperature by a separate instrument which may interrupt the program control and cut off fuel flow until a safe temperature is obtained. To guard against unequal rates of heating in thick and thin sections of the work, the temperature difference between two cross sections may be controlled. Thermocouples usually are spot-welded to two such cross sections. The differential measuring instrument then is arranged to interrupt the program when the differential exceeds a maximum value. Temperature distribution throughout the furnace and work is determined by strategically located thermocouples, which are recorded on a multirecord instrument.

REFERENCES

General

1. Barlow, D.: Dynamic Quality Control in the Steel Industry, *Cont. Eng.*, vol. 9, no. 9, pp. 139–142, September, 1962.
2. Camp. J. M., and C. B. Francis: "The Making, Shaping, and Treating of Steel," 6th ed., U.S. Steel Co., Pittsburgh, Pa.
3. Hill, R. L.: Steel-Industry Uses of Viscosity Control, *Inst. & Automation*, vol. 28, no. 10, pp. 1732–1734, October, 1955.
4. Massey, R. G.: Three Interlinked Computers to Run New British Steel Works, *Cont. Eng.*, vol. 9, no. 6, pp. 128–132, June, 1962.
5. Perry, J. W.: Status of Recording and Controlling Instruments in the Steel Industry, AIEE Conference on Electrically Operated Recording and Controlling Instruments, Philadelphia, Pa., Nov. 17–18, 1952.
6. Sills, R. M., and G. E. Terwilliger: Steel Opens Three Doors to Automatic Data Processing, *Cont. Eng.*, p. 99, vol. 6, no. 12, December, 1959.
7. Slamar, Frank: Status Report on Computers in Steel Making, *Automation*, vol. 10, no. 6, pp. 52–56, June, 1963.
8. Swain, R. R., and J. A. Hays: Gas Analysis in Steel Mills, *Inst. & Automation*, vol. 28, no. 11, pp. 1940–1944, November, 1955.
9. Thring, M. W.: "The Science of Flames and Furnaces," 2d ed., John Wiley & Sons, Inc., New York, 1962.
10. Webber, J. E.: Distribution Control of Mixed Gas, *Inst. & Automation*, vol. 28, no. 9, pp. 1528–1531, September, 1955.
11. Automatic Control Terminology, *ASME Paper* 52-SA-29.
12. Automation in the Iron & Steel Industry (a collection of papers, from the Fourth & Fifth Annual Conferences on Instrumentation for the Iron & Steel Industry), Instruments Publishing Co., Pittsburgh, Pa., 1956.
13. The Changing Look of Steel Plants, *Steel*, vol. 143, no. 11, pp. 156–165, Sept. 15, 1958.
14. Computers Tackle Bigger Data Job, *Iron Age*, vol. 185, p. 105, May 12, 1960.
15. How Computers Will Help Steelmakers, vol. 143, no. 18, pp. 126–129, *Steel*, Nov. 10, 1958.
16. Spectrometer Speeds Analysis at Up-to-date Steel Mill, *Iron Age*, vol. 186, p. 163, Oct. 20, 1960.

Ore Beneficiation

1-A. Considine, D. M.: "Industrial Weighing," Chaps. 6 and 9, Reinhold Publishing Corporation, New York, 1948.
2-A. Latowski, A. A.: Sintering Plant Instrumentation, *Inst. & Automation*, vol. 31, no. 2, pp. 270–272, February, 1958.
3-A. Marquardt, G. M.: An Accurate Density Meter, *Eng. Mining J.*, vol. 152, no. 1, pp. 78–82, 1951.
4-A. Nilsen, L. A.: Recent and Prospective Sinter Plant Improvement, American Iron & Steel Institute General Meeting, New York, May 23–24, 1951.
5-A. Oram, J. E.: Steelmen Try Automation in the Sintering Process, *Steel*, vol. 144, no. 24, p. 150, June 15, 1959.
6-A. Powers, R. E.: Improving Current Practice in Blast Furnace Sintering, American Iron & Steel Institute General Meeting, New York, May 23–24, 1951.
7-A. Rose, E. H., and D. J. Reed: Sinter Is What You Make It, AIME Blast Furnace, Coke Oven and Raw Materials Conference, Chicago, Ill., Apr. 5–7, 1954.
8-A. Zatkovich, T. R., and B. E. Hamlett: Sinter Plant Control, *Inst. & Automation*, vol. 29, no. 12, pp. 2419–2421, December, 1956.
9-A. Instrumentation of the Heavy Media Separation Process, *Bristol Co., Application Data Publ.* 0.50.1.5-2, Waterbury, Conn.

Blast Furnace

1-B. Agnew, C. E.: Maintenance of Equilibrium in Blast Furnace Operation, Part I, *Steel*, vol. 141, no. 8, pp. 120–134, July 29, 1957.

2-B. Agnew, C. E.: Maintenance of Equilibrium in Blast Furnace Operation, Part II, *Steel*, vol. 141, no. 9, pp. 96–104, Aug. 5, 1957.
3-B. Frost, B. B.: The Performance of Large Hot Blast Stoves, *Proc. AISE*, vol. 22, pp. 80–86, 1945.
4-B. Gillings, D. W.: Blast Furnace Instrumentation, *Inst. & Automation*, vol. 31, no. 2, pp. 256–259, February, 1958.
5-B. Karsten, A. J.: Automatic Stove Changing for Blast Furnaces, *Inst. & Automation*, vol. 31, no. 2, pp. 273–275, February, 1958.
6-B. Mortson, E. T., and S. J. Paisley: Blast Furnace Instrumentation, *Blast Furnace Steel Plant*, vol. 39, pp. 789–799, July, 1951.
7-B. Oram, J. E.: Automation's Taking Over in the Blast Furnace, *Steel*, vol. 144, no. 19, pp. 152–155, May 11, 1959.
8-B. Paisley, S. J.: Inwall Temperatures of Blast Furnaces, *Instruments*, vol. 25, no. 11, p. 1571, 1952.
9-B. Robinson, A. W.: Blast Furnace Automation, *Inst. & Automation*, vol. 28, no. 2, pp. 266–269, February, 1955.
10-B. Zorena, P. J., and T. J. Connors: Blast Furnace Control System, *ISA J.*, vol. 9, no. 12, pp. 43–46, December, 1962.
11-B. Zatkovich, T. R., and B. E. Hamlett: Sinter Plant Control, *Inst. & Automation*, vol. 29, no. 12, pp. 2419–2421, December, 1956.
12-B. Automatic Tuyere Control Cuts Blast Furnace Wear, *Steel*, vol. 144, no. 10, p. 78, Mar. 9, 1959.
13-B. Blast Furnace Is Updated with Nuclear Stock Gages, *Steel*, vol. 145, no. 15, p. 120, Apr. 11, 1960.
14-B. Computer: Key to Predicting Blast Furnace Behavior, *Iron Age*, vol. 185, p. 61, June 2, 1960.
15-B. How to Boost Blast Furnace Efficiency, *Steel*, vol. 145, no. 9, p. 162, Feb. 29, 1960.
16-B. Nuclear Device Keeps Tabs on Blast Furnace Burden, *Iron Age*, vol. 187, p. 134, Apr. 20, 1961.
17-B. Tap Blast Furnaces Automatically, *Iron Age*, vol. 188, p. 96, Oct. 26, 1961.

Bessemer Converter

1-C. Percy, J. W.: Bessemer Converter Bath Measuring, U.S. Patent 2,305,442.
2-C. Percy, J. W.: Bessemer Converter Blow Control Method, U.S. Patent 2,354,400.
3-C. Slatosky, W. J., and N. Simcic: Computer Controls End-point Temperature in Oxygen Steel-making, *ISA J.*, vol. 8, no. 12, pp. 38–41, December, 1961.
4-C. Work, H. K.: Photocell Control for Bessemer Steelmaking, *Trans. AIME*, vol. 145, p. 132, 1941.
5-C. Acme Steel Automates Oxygen Converters, *Steel*, vol. 145, no. 26, p. 80, June 27, 1960.
6-C. Computer Aids in Design of Gale Velocity Furnace, *Steel*, vol. 146, no. 5, p. 84, Jan. 30, 1961.

Open Hearth

1-D. Bayles, A. K., and D. H. Stanton: Electric Control of Open Hearths, *Inst. & Automation*, vol. 29, no. 1, pp. 86–90, January, 1956.
2-D. Churchill, J. R.: Emission Spectroscopy Speeds Control of Metals Production, *Cont. Eng.*, pp. 89–94, vol. 6, no. 10, October, 1959.
3-D. Eisenhart, L. W.: Programming Open Hearth Combustion, *Inst. & Automation*, vol. 31; no. 2, pp. 260–264, February, 1958.
4-D. Ellstrom, C. R.: Two-point Roof Temperature Control of Open Hearth Furnaces, *Inst. & Automation*, vol. 30, no. 7, pp. 1312–1315, July, 1957.
5-D. Francy, R. A.: Getting an Open Hearth Control System to Work, *ISA J.*, vol. 9, no. 5, pp. 32–34, May, 1962.
6-D. Green, J. R., and J. P. Vollrath: Reversing Regenerative Furnaces, *Proc. AISE*, vol. 22, pp. 225–235, 1945.
7-D. Symposium on Firing and Controlling Open Hearth Furnaces, *Proc. AISE*, vol. 29, pp. 531–545, 1952.
8-D. Krouse, Harvey: Viscosity Control of Fuel for Open Hearths, *ibid.*, pp. 398–406.
9-D. Larsen, B. M., and W. E. Shenk: A Completely Automatic Control of Open Hearth Reversal, *Trans. AIME*, vol. 162, pp. 37–48, 1945.
10-D. Marsh, John S.: Significance of Air Temperature in Open Hearth Operation, American Iron & Steel Institute General Meeting, May 23–24, 1951.
11-D. Oram, J. E.: Automation of Open Hearth Expected to Gain Headway, *Steel*, vol. 145, no. 1, pp. 100–102, June 1, 1959.
12-D. Sigl, R. L.: Modern Open Hearth Control, *Inst. & Automation*, vol. 31, no. 2, pp. 265–266, February, 1958.
13-D. Whigham, Jr., W.: Training Program for Open-hearth Fuel Engineers, *Inst. & Automation*, vol. 28, no. 7, p. 1107, July, 1955.
14-D. "Basic Open Hearth Steelmaking," Chaps. 4, 21, 2d ed., 1951, American Institute of Mining, Metallurgical, and Petroleum Engineers, New York.
15-D. British Iron and Steel Research Association: "The Instrumentation of Open Hearth Furnaces," George Allen & Unwin, Ltd., London.

Casting Operations

1-E. Hindson, R. D., and J. P. Orton: Measurement of Steel Bath Temperatures, *Proc. Natl. Open Hearth Comm.*, vol. 35, pp. 143–145, 1952.
2-E. Carney, D. J., and J. M. Oravec: Measurement of Bath Temperatures in the Open Hearth," *ibid.*, pp. 145–151.

3-E. Pearson, Oscar, and F. B. Pearson: Measurement of Steel Bath Temperatures, *ibid.*, pp. 152–161.
4-E. Brady, J. R.: Measurement of Open Hearth Bath Temperatures, *ibid.*, pp. 162–165.
5-E. Faist, C. A.: Production Control of Molten Steel Temperature and Its Effect on Steel Casting Quality, *Proc. Elec. Furnace Steel Conf.*, vol. 8, pp. 210–244, 1950.
6-E. Goller, Geo. N.: The Emissivity of Stainless Steels, *Trans. ASM*, vol. 32, pp. 239–254, 1944.
7-E. Vogt, E. G.: Instrumentation in Vacuum Induction Melting, *Inst. & Automation*, vol. 31, no. 2, pp. 267–269, February, 1958.
8-E. Wensel, H. T., and W. F. Roeser: Temperature Measurements of Molten Cast Iron, *Am. Foundrymen's Assoc. Preprint* 28-13.
9-E. Continuous Casting of Steel Billets, *Metal Progr.*, vol. 63, pp. 87–88, 1953.
10-E. Cupola Blast Control, *Whiting Corp., Bull. FO-7*, Harvey, Ill.

Soaking Pits

1-F. Symposium on Soaking Pits, *Proc. AISE*, vol. 29, pp. 386–393, 1952.
2-F. Cauger, E. H., and J. C. Stamm, Jr.: Operative Results of One-way Fired Recuperative Soaking Pits, *ibid.*, pp. 836–848.
3-F. Isaacs, G. L.: Automatic Control for Regenerative Pits, *Proc. AISE*, vol. 30, pp. 75–83, 1953.
4-F. Pullen, F. R.: Soaking Pit Instrumentation, *Instruments*, vol. 26, p. 724, 1953.
5-F. Robbins, A. F., and H. Meek: Problems of a New Soaking Pit Installation, *Inst. & Automation*, vol. 28, no. 6, pp. 982–985, June, 1955.

Continuous Reheating

1-G. Bloom, Fred S.: Controlling and Measuring Steel Temperatures in Billet Heating Furnaces, *Proc. AISE*, vol. 26, pp. 670–680, 1949.
2-G. Culp, L. D.: Instrumentation for Improved Rolling Temperature Control, *ibid.*, vol. 28, pp. 793–799, 1951.
3-G. Francy, R. A.: Combustion Control of Slab Furnaces, *Inst. & Automation*, vol. 29, no. 8, pp. 1546–1548, August, 1956.
4-G. Yaeger, Carl M.: Instrumentation Factors Affecting Production in Slab Heating Furnaces, *Instruments*, vol. 26, p. 402, 1953.

Mechanical Forming Operations

1-H. Beadle, R. G., and W. E. Miller: Improving Thickness Control in Hot Strip Mills, *Cont. Eng.*, vol. 8, no. 7, pp. 94–99, July, 1961.
2-H. Browning, E. H.: Automatic Card Programmed Control of Reversing Mills, *Trans. IRE*, PGIE-5, p. 64, April, 1958.
3-H. Browning, E. H.: Strip Mill Uses Automatic Programmed Control, *Automation*, vol. 5, no. 6, pp. 64–67, June, 1958.
4-H. Burdick, C. W.: Digital Computer Runs Hot Plate Mill, *Cont. Eng.*, vol. 7, no. 1, p. 126, January, 1960.
5-H. Buza, J. A., Gage Thin Strip Accurately, *Iron Age*, vol. 185, p. 120, Feb. 18, 1960.
6-H. Carlson, G. J., and E. R. Sampson: Transistorized Length Gage Scans Hot Steel, *Cont. Eng.*, vol. 7, no. 4, p. 154, April, 1960.
7-H. Davisson, R. R.: What's Inside the Coil?, *Steel*, vol. 143, no. 10, pp. 106–110, Sept. 8, 1958.
8-H. Dyer, A. C.: Reversing Mill Developments Leading to Punched Card Roll Positioning, *Automation*, vol. 4, no. 3, pp. 75–80, March, 1957.
9-H. Filmer, T., and C. C. Roberts: Feedback Controlled Steel Slab Cutoff, *Automation*, vol. 5, no. 7, pp. 72–74, July, 1958.
10-H. George, William C.: Measuring Strip Steel Without Contact, *ISA J.*, vol. 7, no. 1, p. 80, January, 1960.
11-H. Harris, W. R.: Operating Experience with Regulating Systems for the Steel Industry, *Proc. AISE*, vol. 29, pp. 697–707, 1952.
12-H. Hueschkel, J.: Instrumentation Minimizes the Welding Variable, *Proc. 1st Ann. Conf. on Instrumentation for the Iron and Steel Ind.*, pp. 41–48, The Instruments Publishing Co., Inc., Pittsburgh, Pa., 1951.
13-H. Kuhnel, A. H.: Computer Control of a Rolling Mill Schedule, *Inst. & Automation*, vol. 29, no. 7, pp. 1303–1305, July, 1956.
14-H. Markey, F. J.: Edge Position Control, *Inst. & Automation*, vol. 31, no. 2, pp. 280–281, February, 1958.
15-H. Miller, L. D.: Automated Pressworking, *Automation*, vol. 6, no. 12, p. 40, December, 1959.
16-H. Miller, W. E., and F. S. Rothe: Selection and Analysis of Regulating Systems for Mill Drives, *Proc. AISE*, vol. 29, pp. 819–835.
17-H. Miller, W. E., and R. G. Beadle: Automatic Gage Control of Metal Rolling, *Automation*, vol. 6, no. 10, pp. 76–81, October, 1959.
18-H. Misrahi, J.: Relays and Counters Compute Yard Throughput, *Cont. Eng.*, vol. 8, no. 5, pp. 147–149, May, 1961.
19-H. Mohler, J. B.: Continuous Processing of Steel Strip, *Automation*, vol. 2, no. 12, pp. 37–43, December, 1955.

20-H. Sims, R. B.: Investigating Thickness Control on Small Scale Rolling Mills, *Automatic Control* vol. 7, no. 6, p. 95, June, 1960.
21-H. Sterns, R. B.: Digital Logic Improves Photoelectric Length Measurement, *Cont. Eng.*, vol. 7, no. 12, pp. 114–115, December, 1960.
22-H. Urano, A. S.: Automatic Controls in Pipe Production, *Automation*, vol. 3, no. 7, pp. 46–48, July, 1956.
23-H. Urano, A. S.: Continuous Gaging in Steel Mills Produces Profit, *Automation*, vol. 3, no. 3, pp. 52–57, March, 1956.
24-H. Varlas, G., and A. J. Pasion: Adjustable Voltage Drive System Provides Strip Speed and Tension Regulation, *Automation*, vol. 6, no. 9, pp. 58–62, September, 1959.
25-H. Wiles, A. P.: Keeping Stock on Length, *Cont. Eng.*, vol. 9, no. 9, pp. 117–119, September,1962.
26-H. Automated Conveyors Reduce Unit Costs, *Steel*, vol. 144, no. 3, p. 78, Jan. 19, 1959.
27-H. Card Control Catches on at Automated Hot Mills, *Steel*, vol. 144, no. 18, p. 92, May 4, 1959.
28-H. Die Shear Line is Flexible, Accurate—Electronic Control Allows Sheet Length to be Changed Instantly, *Steel*, vol. 144, no. 17, pp. 114–115, Apr. 27, 1959.
29-H. Electronics Regulates Strip Mill, *Steel*, vol. 143, no. 17, p. 94, Oct. 27, 1958.
30-H. Experts Say Automatic Mill Is in Sight, *Steel*, vol. 144, no. 12, pp. 114–117, 120, Mar. 23, 1959.
31-H. New Hot Strip Mill Gets Computer Control, *Steel*, vol. 146, no. 6, p. 46, Feb. 6, 1961.
32-H. Profile Check at Hot Mill Insures Quality Strip, *Steel*, vol. 146, no. 7, p. 128, Jan. 30, 1961.
33-H. Radiation Gages Control Sendzimir Mill, *Steel*, vol. 144, no. 17, p. 108, Apr. 27, 1959.
34-H. Robot Handles Forging Job, *Iron Age*, p. 96, March, 1961.
35-H. Width Meters Trim Strip Waste, *Steel*, vol. 143, no. 17, pp. 182–185, Oct. 20, 1958.
36-H. X-ray Gains in Process Control, *Steel*, vol. 146, no. 6, p. 92, Feb. 6, 1961.
37-H. Yield Point Control, *Automation*, vol. 6, no. 11, p. 77, November, 1959.

Heat Treatment

1-I. Besselman, W. L.: A Chemo-electrical Method for Controlling Furnace Atmospheres, AIEE Electric Heating Conference, Detroit, Mich., May 26–27, 1953.
2-I. Besselman, W. L., and R. L. Davis, III: Control of Constituent Potentials, U.S. Patent 2,541,857.
3-I. Chamberlin, A. B.: Instrumentation for a Continuous Strip Steel Annealing Line, *Inst. & Automation*, vol. 29, no. 2, pp. 285–287, February, 1956.
4-I. deCoriolis, E. G., O. E. Cullen, and Jack Huebler: Carbon Concentration Control, *Trans. ASM*, vol. 38, pp. 659–685, 1947.
5-I. Eshelman, R. H.: Heat Treating Goes Automatic, *Iron Age*, vol. 187, p. 79, June 8, 1961.
6-I. Gelder, R. H., and W. E. Hand: Instrumentation of a Continuous Batch-Type Annealing Furnace, *Inst. & Automation*, vol. 30, no. 4, pp. 704–706, April, 1957.
7-I. Gier, J. R., Jr.: Method and Apparatus for Determining the Carbon Pressure of Gases, U.S. Patent 2,279,231.
8-I. Hotchkiss, A. G., and H. M. Webber: "Protective Atmospheres," John Wiley & Sons, Inc., New York, 1953.
9-I. Jenkins, Ivor: "Controlled Atmospheres for the Heat Treatment of Metals," Chapman & Hall, Ltd., London, 1946.
10-I. Koebel, N. K.: Dew Point: a Means of Measuring the Carbon Potential of Prepared Atmospheres, *Metal Progr.*, vol. 65, pp. 90–96, 1954.
11-I. Insen, Harold, and Edward J. Rupert: Automatic Carbon Control, *ibid.*, pp. 98–102.
12-I. O'Keefe, T. J.: Continuous Annealing of Specialty Steel, *Inst. & Automation*, vol. 28, no. 10, pp. 1739–1741, October, 1955.
13-I. Webber, J. E.: Continuous-anneal-line Control, *Inst. & Automation*, vol. 29, no. 3, pp. 483–486, March, 1956.
14-I. Furnace Control Adds Precision to Job-shop Gear Production, *Iron Age*, vol. 186, p. 154, Nov. 17, 1960.
15-I. Staff: Furnace Atmosphere Analyzer Increases Heat Treating Accuracy, *Auto. Cont.*, vol. 14, no. 2, pp. 24–25, February, 1961.

Metal Surface Coating

1-J. Camp, J. B.: Classifying Plated Sheet, *Cont. Eng.*, vol. 8, no. 11, pp. 89–90, November, 1961.
2-J. Hand, W. R.: Continuous Galvanizing Instrumentation, *Inst. & Automation*, vol. 31, no. 2, pp. 278–279, February, 1958.
3-J. Miller, W. E.: Computer Enters Hot-Strip Mill to Step Up Yield and Quality, *Iron Age*, vol. 188, p. 113, December, 1961.
4-J. Mohler, J. B.: Parts Handling Key to Automatic Plating Installations, *Automation*, vol. 3, no. 5, pp. 67–70, May, 1956.
5-J. Smith, W. P.: Electrolytic-tinplate Instrumentation, *Instruments*, vol. 26, pp. 1875, 1905, 1906, 1953.
6-J. Webster, R. R.: X-ray Tin Thickness Gage, *Inst. & Automation*, vol. 31, no. 2, pp. 276–277, February, 1958.
7-J. Zimmerman, R. H.: X-rays Check Plating Thickness, *Iron Age*, vol. 186, p. 84, Oct. 13, 1960.
8-J. Flexible Line Automates Hot Dip Galvanizing, *Steel*, vol. 144, no. 17, pp. 104–105, Apr. 27, 1959.
9-J. Non-contacting Lineal System Programs Tinplate Line, *Iron Age*, vol. 185, p. 112, May 26, 1960.
10-J. X-ray Gage Analyzes Materials Without Altering Work Flow, *Iron Age*, vol. 186, p. 110, Aug. 4, 1960.

GLASS AND CERAMICS INDUSTRIES INSTRUMENTATION

By J. F. Hornor*

The material in this subsection is divided into two major parts: (1) glass manufacturing instrumentation and (2) ceramics manufacturing instrumentation. Because furnaces and relatively high temperatures are involved in both industries, the reader is urged to consult both parts of this subsection for information concerning his processing instrumentation problems. To avoid redundancy, instrumentation of certain processes is discussed more thoroughly in part 1 than in part 2.

GLASS MANUFACTURING VARIABLES AND THEIR INSTRUMENTATION

The majority of glass manufacturing, from preparation and charging of raw materials into the furnace, through forming and annealing, is a *continuous process*. The material-handling system has been highly automatic for many years—because the temperature of the glass during processing and the rapid rates of production preclude manual handling.

The glass-melting and forming operation involves many variables, and is perhaps one of the least understood and most complex of all manufacturing operations. Much progress has been made in developing a scientific basis for improvement of this operation. Instrumentation has had a small part in this progress, but it has not been required to make the major contribution of which it is capable. Neither the glass industry nor the instrument industry has invested sufficient research effort to define the specifications of the process control problems. In fact, neither industry has made a thorough engineering study of the application of existing instrumentation technology to the problems of glassmaking and forming.

Present technology offers tools, or can develop others, which are capable of making measurement in glass plants in quantity and with precision not previously obtainable. The industrial digital computer offers a means of studying the technical relationships among these variables and their economic effects, thus aiding management decisions—leading to automatic operation of these plants to maximize profits.

The material in this subsection is based upon current instrumentation practices now carried out in commercial glassmaking—with some emphasis on future requirements and trends.

Molten-glass Depth (Level)

The depth of molten glass within the furnace affects the entire glass-melting and forming operation. Changes in this depth influence the temperature of the glass in the forehearth, the corrosion rate of the refractories adjacent to the glass surface, and the glass quality. But, the most significant effect is observable in the *weight* of glass delivered to the forming machine. Experimental data[1] indicate that in a specific instance, changes in the glass depth as small as ±0.02 in. produced a delivered weight

* Planning Adviser, Control Systems, General Products Division Development Laboratory, International Business Machines Corporation, San Jose, Calif. This subsection was prepared by Mr. Hornor while he was Market Manager, Metals Producing and Ceramic Industries, Minneapolis-Honeywell Regulator Co., Philadelphia, Pa.

change of ±0.75 per cent. Additional unpublished studies by others have shown a similar interdependence between weight and very small glass-depth changes.

Measurement of Molten-glass Level. Glass depth is measured by determining the deviation of the vertical position of the glass surface (level) from a precisely located reference height. An electromechanical system for making this measurement automatically has been applied to most continuous glass furnaces for many years (see Fig. 1).

This system employs a vertically moving probe, which is driven downward until it contacts the glass surface, thereby completing an electric circuit through the glass to the detector of a programmer. The programmer deenergizes the probe-drive motor for a timed period, and then reverses the motor drive to retract the probe from the surface.

During the *delay period*, with the probe touching the surface, the vertical position of the probe (the position of the glass surface) is transmitted by a slidewire geared to

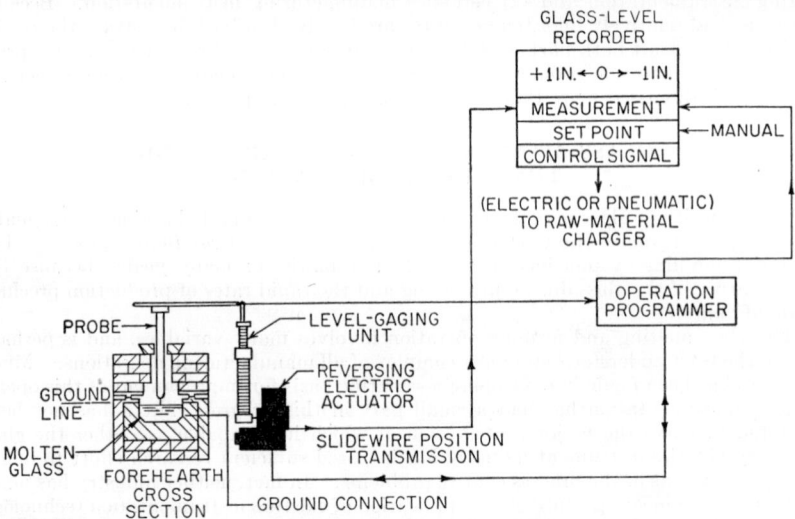

Fig. 1. Electromechanical-type glass-level measurement and control system.

the drive. The slidewire position is measured by a recording potentiometer, and indicated as *deviation in inches* from the reference-glass depth. The recorder pen is operative only while the probe is transmitting the level position, and thus it does not record probe motion.

To complete the cycle, after the delay has expired, the probe is automatically retracted to a position above the glass surface, reversed, and again driven downward until it contacts the glass surface, thus repeating the cycle.

The interval between measurements is approximately 45 sec and is adjustable. This sampling period has been found adequate under favorable circumstances to effect control of the glass level within ±0.01 in., based on measurements made only in the sampling period. The instrumentation has been found capable of detecting a change in level of 0.005 in.

Another automatic method of measuring glass level employs a radioactive isotope,[2,3] and detector mounted on opposite sides of the forehearth adjacent the normal glass level (see Fig. 2). In this position, the "beam" of energy emitted by the isotope is attenuated by absorption in the glass as the level traverses the beam. Thus, the energy reaching the detector and, therefore, its output are a function of the position of the glass level.

This radiation method has been used for many years in Great Britain and other parts of Europe. In the United States, a small number of installations have been made only recently for evaluation. The performance of these installations has not yet been adequately studied and published. The equipment may be expected to require less maintenance than the oscillating probe—since the radiation system requires no moving parts and no components that are exposed to the high-temperature sections of the furnace.

Control of Molten-glass Level. If an error-generating device and controller are added to the level-measuring instrument, the controller output signal may be used to adjust the rate at which raw material is fed to the glass tank—thereby controlling the level. This instrumentation not only stabilizes the level, which is desirable in itself, but also provides means for determining the required material input. If the level remains constant in this continuous process, the input-material flow must equal the output flow. Thus, the difficult problem of measuring the output flow rate of molten glass is avoided.

Because several types of raw-material charging equipments are in wide use, the form of the controller output is varied to match specific requirements. Pneumatic and electric positioning controllers have been employed to operate mechanical speed-changing units coupled to the charging machine. Current-output controllers are used

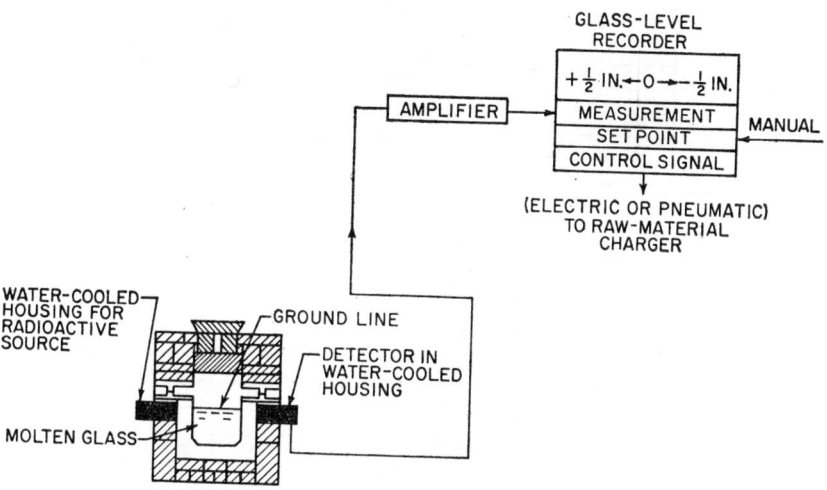

Fig. 2. Radioactive-type glass-level measurement and control system.

to operate magnetic-amplifier-controlled vibrating feeders, or Thyratron-operated motor-speed controls. Because the positioning actuator is located at the charging machine in a high concentration of abrasive dust, completely pneumatic systems have been found useful. To obtain the required regulation of the level, a *two-mode* controller must be applied.

Future Instrumentation of Glass Level. Present instrumentation regulates the level to the limit of the ability of the raw-material charging-unit equipment and of the process to convert the material flow into a level change. Thus, it achieves most of the objectives of glass-level control listed at the beginning of this discussion.

As the tolerances for permissible *gob weight* variation are made smaller, consideration must be given to the effect of level fluctuations between the measurement-sampling periods. A continuous observation obtained with the radiation-measurement system has shown the capability of providing a profile of short-term level changes not previously obtainable.

A record taken from a preliminary study of the application of the radiation system to glass-level measurement is shown in Fig. 3. The measurement here has been greatly amplified to reveal the details of level changes, and would not be used with this sensitivity in connection with a conventional control system. It is interesting to note that, even after allowing for the noise level of the radiation measurement (in this case, equivalent to a level change of 0.01 in.), frequent transient-level changes of significant magnitude are observable. The largest and long-period fluctuations shown in Fig. 3 are those caused by pressure changes during furnace gas-flow reversal. The effects of these changes, of course, have been well known for many years.

A speculative correlation with these level transients may be observed by examining similarly amplified measurements of furnace-pressure changes. A pressure record made on another furnace (unfortunately without a radiation level-measuring device) is shown in Fig. 4. The effect of similar pressure transients to those shown in Fig. 4, on glass level, would depend on the geometry of each furnace and its forehearth.

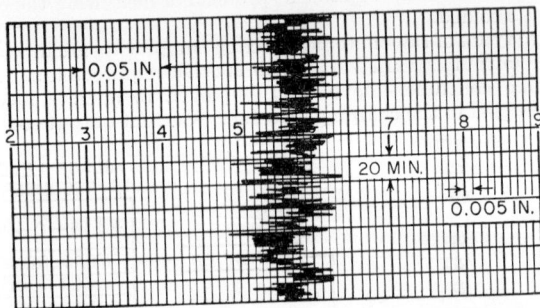

FIG. 3. Section of chart record taken from radioactive-type glass-level recorder.

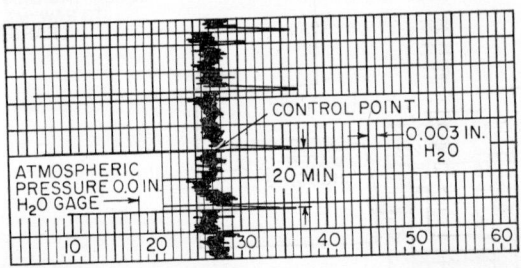

FIG. 4. Section of chart taken from furnace pressure recorder utilizing an electric transducer. Control was achieved by use of a bell-type measuring system with electric multizone control.

Should these transient-level fluctuations prove economically significant to the glass-forming operation, then control methods having a dynamic range similar to the measuring equipment will be required. If it is found that level variations can be held within the required limits by improved stabilization of furnace pressure, the problem is relatively simple. If the level itself must be manipulated, then changes in furnace, or forehearth, design may be required to permit closer regulation. The economic reward, of course, may be large.

The availability of a satisfactory continuous-level measurement now makes possible multiple correlative studies of glass-level action with other variables. These studies will need to consider level movement in each forehearth with respect to that of the main body of glass in a furnace.

Glass-melting Furnace Pressure

A series of pressures, from vacuum at the base of the stack and in the flues to positive pressure in the furnace superstructure, are of interest in operating a continuous glass furnace. These pressures will vary from -1.0 to $+0.15$ in. of water (gage pressure).

Because of its direct effect on the combustion and melting operation, it is the pressure in the combustion chamber above the glass which is automatically measured and controlled. This actually is the *differential pressure* between the inside of the furnace and the atmospheric pressure. Thus, the magnitude of this pressure (whether positive or negative) determines the ease with which gases flow into or out of the furnace.

This pressure difference permits gas flow through every crack and opening in the furnace. When the furnace pressure is negative, cold air flows into the furnace—altering combustion conditions by reducing combustion-air preheat, thus increasing fuel consumption. When the furnace pressure is positive, cold-air infiltration is prevented, but output flow of hot exhaust gases occurs and refractory corrosion may be accelerated.

The furnace-pressure control system must function to stabilize the pressure at a value which tends to optimize these factors. Several variables normal to furnace operation cause large variations in pressure and require that it be automatically regulated. These include: (1) reversal of gas flow on regenerative furnaces, (2) plugging of flues and regenerator chambers, (3) atmospheric conditions, such as wind velocity on natural-draft installations, and (4) changes in the type and volume of fuel fired.

Measurement of Furnace Pressure. The location of the pressure tap(s) in the furnace is somewhat complicated by the large volume of the chamber and the location of the burners. A common location is high in the center of the back wall near the crown. Some installations have used a pressure tap on the furnace front wall. Two pressure taps and a single impulse line to average the pressures—with the taps located on each side of the furnace at the bridge wall—also have been employed.

Experience with a particular furnace design is required to select the location which avoids turbulence and a noisy measurement, but is representative of pressure changes which affect over-all combustion conditions.

Furnace-pressure measuring devices may be expected to have sensitivities of the order of ±0.0005 to ±0.001 in. of water. This performance is necessary to obtain the desired regulation of pressure; but it imposes specific requirements on the construction of the pressure taps and impulse lines—if a useful measurement is to be obtained.

When the measuring system is located at the panelboard, a difference in elevation usually will exist between the pressure tap at the furnace and at the instrument. Any temperature gradient along the vertical sections of the impulse line will create a measurable and varying static pressure due to the difference in air density. Compensation for this effect is obtained readily by running a parallel impulse line to the measuring instrument and connecting it to oppose the furnace pressure. This is why a *differential pressure* instrument is used to measure furnace pressure. The furnace end of the compensating impulse line is terminated at the same elevation as the furnace pressure tap, but is left open to the atmosphere (see Fig. 5). The use of a pressure transmitter reduces this problem, inasmuch as it may be located close to the furnace pressure tap—if it is designed for use in ambient temperatures up to 175°F.

Most furnace-pressure measuring devices contain a considerable enclosed volume, especially the oil-sealed bell type, which must be displaced when large pressure changes occur. The impulse piping must be well designed to avoid excessive measuring lag. The pipe diameter should be large enough to avoid friction effects. A 1¼-in. pipe size has been found optimum for most installations.

Connections at the measuring device normally are ¼-in. pipe size and may be made with flexible tubing if vibration transmitted by the impulse piping is troublesome. Although the effect of this restriction eliminates some noise in the pressure signal, the length of the ¼-in. pipe should be limited to 24 in.—to avoid limiting the dynamic response of the system. Figure 5 shows other refinements which aid in obtaining the best measurement.

The furnace pressure usually is desired to be equal to atmospheric or a few hundredths of an inch of water positive when measured at the glass level. Because of the vertical static-pressure gradient in the furnace, the pressure measured and controlled must be held at a value dependent on the tap location. This is necessary to achieve a balanced furnace pressure at the glass level. A portable inclined-draft gage, with a heat-resisting alloy tube as a pressure probe, is useful for checking that the desired pressure is obtained at the glass level. Considerable care and manual skill are required to make this check meaningful.

Pressures (both positive and negative with respect to the atmosphere) at other locations in the furnace structure may be measured occasionally. These are obtained manually with a portable draft gage, and serve to indicate flue restrictions, stack condition, and plugging in the regenerator chambers. Some installations have been

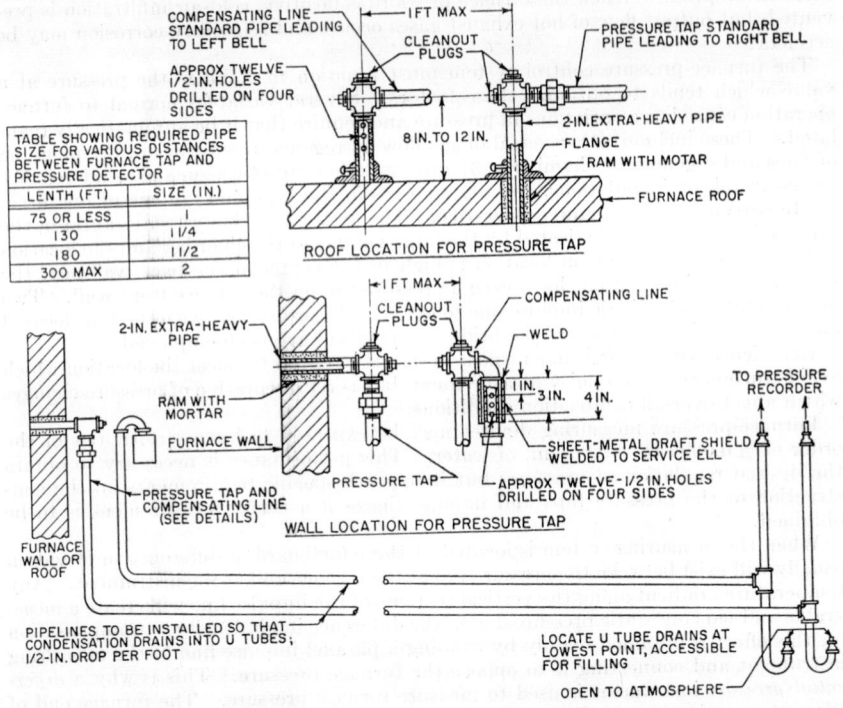

TABLE SHOWING REQUIRED PIPE SIZE FOR VARIOUS DISTANCES BETWEEN FURNACE TAP AND PRESSURE DETECTOR	
LENTH (FT)	SIZE (IN.)
75 OR LESS	1
130	1 1/4
180	1 1/2
300 MAX	2

Fig. 5. Furnace-pressure-tap and impulse-line connections and construction.

made with these pressure taps piped directly to the panelboard and connected through a valve manifold to a permanently mounted inclined-draft gage.

Control of Furnace Pressure. Since most glass combustion systems do not include an induced-draft fan, control of pressure is obtained by positioning a slide damper in the flue system. The low pressure and large volume of exhaust gases require that the flues have a large cross section, and that the dampers be sized in proportion. These slide dampers weigh several hundred pounds and with associated cabling will require high-thrust actuators, together with counterweighting to obtain satisfactory positioning. The dimensions of the flue cause full-scale travel of the damper to be 30 to 48 in., and thus reeving is usually necessary to match actuator motion.

New glass-tank construction has made increasing use of fan-driven draft systems which permit control of furnace pressure by positioning a low-thrust louver damper on the fan intake. These provide better control characteristics than slide dampers, smaller installation costs, and greatly reduced maintenance—because of the absence of cabling and sheaves.

Control of furnace pressure within ±0.003 in. of water is possible. To achieve this performance, the controller must be adjusted to the process characteristics and not (as is often the case) to reduce actuator motion and maintenance. Some slide-damper configurations are uncontrollable. In others, restriction of the flues may place the damper at an extreme position where its motion does not affect pressure. Hydraulic, pneumatic, and electric instrumentation all have been successfully used for furnace-pressure control. Both floating and proportional-plus-automatic reset modes have been employed successfully.

Trends in Furnace-pressure Instrumentation. The development of electric-type furnace-pressure transmitters of high sensitivity and stability makes available a measuring system with dynamics capable of further improving the control of furnace pressure. These transmitters have been employed to study furnace-pressure transients in a glass tank during and following gas-flow reversal; also as an aid to optimum adjustment of the reversal automatic programmer. More important is the fact that these devices provide instrumentation with suitable characteristics for needed research on the variation of furnace pressure at the glass level during normal and abnormal operation; for different types of furnace construction and operating conditions; and to determine the relationship of pressure variations to other operating variables.

The importance of stabilizing furnace pressure generally is recognized by most operators. The limits within which the pressure must be controlled have been rather loosely established by combustion and economic requirements. The possibilities of improving control of glass-output feed rate by manipulation of furnace pressure and glass level need further consideration. Some of the possible relationships have been shown in Figs. 3 and 4.

Glass-tank Combustion Instrumentation

Glass tanks are commonly fired with a single fuel that may be either natural gas, heavy residual fuel oil, or light fuel oil. The choice is determined by the local cost of the available fuel, and usually an alternative to the primary fuel will be piped to the furnace. Gas is measured with an orifice, heavy fuel oil with an area-type meter, and light fuel oil with a low-flow orifice, or a rotameter. Where fuel availability varies seasonally, switching arrangements are provided on a panelboard to permit the operator to rapidly and smoothly change the metering for either gaseous or liquid fuels—without the assistance of an instrument mechanic.

Fuel-input Control. Because of the importance of stability of operating variables on a glass tank, the fuel input is automatically controlled, using a proportional-plus-automatic-reset mode. Automatic control of fuel flow is particularly important when regeneratively fired furnaces are used because of the difference in piping and burner-nozzle restriction for each firing direction. Even with several burners, it is adequate to control only the total fuel to the furnace and to effect fuel distribution between burners manually. The rate of fuel input is established by the operator.

To obtain best results with heavy fuel oil, it is desirable to control the temperature of the oil leaving the preheater, and to regulate the pressure of the oil supply to each furnace. Dual oil filters are an aid to minimizing metering and valve maintenance, but good practice dictates the installation of manual bypass piping around the oil meter and control valve to facilitate cleaning.

Oil and gas flow rates to a glass tank are relatively small, and valves are frequently designed too large. This affects not only over-all performance of the control system, but particularly the uniformity of firing in each direction. The valve employed for flow control cannot be used for shutoff at reversal if best automatic control results are to be maintained.

A fuel-flow recorder is useful for furnace troubleshooting and comparing daily, or shift operating practice. A separate positive-displacement meter with a revolution counter usually is employed for accounting purposes, to obtain total fuel consumption.

Combustion-air-flow Input Measurement and Control. Metering and automatic control of combustion air flow is important on regenerative glass furnaces to quickly reestablish the required flow after reversal. Also, it is necessary to supply uniform air volumes to both sides of the furnace, despite unequal regenerator resistance.

Forced-draft fans usually are employed as a source of combustion air, and adequate static pressure is available to provide a small, but satisfactory differential across the metering orifice. The piping layout associated with the glass tank normally permits sufficient upstream and downstream lengths of piping to provide a good orifice installation, but this should be planned for in the initial layout. Piezometer-ring taps have been found a good investment because of the large diameter of the combustion-air ductwork.

Insufficient attention has been devoted in the past to engineering air-flow metering systems to compensate for air-density variations. When operating a flow control system with a volumetric metering element, such as an orifice, a change of 100°F in air temperature will permit the pounds of oxygen delivered to vary by more than 10 per cent. Seasonal and 24-hr ambient air-temperature variations will affect air flow enough to noticeably alter flame pattern. This troublesome factor in obtaining uniform furnace operation is eliminated easily by employing a flow metering system which compensates for air-temperature variations. It is not necessary to assume the expense of a complete mass-flow system, however, because the only static-pressure variations encountered are barometric, and these are not significant to combustion results.

Control of combustion air flow is effected by positioning a butterfly damper in the ductwork. A preferable arrangement is to use a louver damper on the fan inlet.

Fuel-Air Ratio Control. Supplying the correct amount of air for a given fuel input is particularly important on a glass tank—because the glass-melting process is sensitive to furnace atmosphere composition. The superstructure refractories often are operated near the failure point, and unexpected changes in fuel-air ratio may cause a local rapid temperature rise. The usual reasons for maintaining a suitable fuel-air ratio, viz., economy in fuel consumption, also apply.

Conventional combustion control instrumentation is applied to automatically maintain the ratio as the fuel rate is changed, and particularly to achieve uniformity of firing in spite of furnace gas-flow reversal. Although fuel-air ratio control was not employed widely in the glass industry for many years, most new tank construction now includes it. Its performance in new installations has shown economic justification for adding it to existing furnace instrumentation, and a number of such additions have been made.

Exhaust-gas Composition Measurement. Two significant variables in operating a furnace are not determined by the combustion instrumentation just described. The first is *air volume* which is available for combustion, but is not metered in the forced-draft system. Its sources include burner atomizing air and leaks in the furnace structure. The second is the determination of the fuel-air ratio to produce the desired *heat release*, atmosphere composition, and economy for a specific furnace operating condition.

The needed information is exhaust-gas composition, which usually can be measured by the percentage of oxygen present; but occasionally, when combustion is incomplete, the percentage of combustibles must be measured. These measurements continuously monitor the chemical performance of the entire combustion process, including instrumentation and the furnace structure.

With this measurement, the fuel-air ratio instrumentation may be calibrated over a range of fuel flows and primary air adjustments. The ratio to obtain the desired exhaust-gas analysis then is easily set by the operator. Parasitic air infiltration to the regenerator structure and flues is readily detected by an increase in exhaust-gas oxygen concentration. This analysis not only is an aid to fuel economy, but also is a means of obtaining maximum air-preheat temperature.

Exhaust-gas analysis has been found useful in burner adjustment and primary-air adjustment. Spatz[4] has described its use in adjusting furnace-temperature distribution and providing a "complete picture" of furnace operating conditions. The use of the analyzer to detect and eliminate unburned fuel in the regenerator aids in maximizing refractory life.

Manually operated Orsat apparatus and the more refined, portable, but sampling-type oxygen and combustibles analyzers have been used for many years on glass tanks. But, the full utilization of these measurements have been brought about by the continuous and automatic oxygen-combustibles analysis instrumentation.   This equipment provides more information because it can shorten the time for the measurements and, thus, provide comparable data in several locations in the furnace.   It is more accurate than portable types, when carefully installed and maintained.   Other than maintenance, automatic equipment requires no labor to operate, and thus its information is available at any time to the operator and furnace engineer.

There are undeniable problems in using currently available continuous analyzers. The first and installation costs are quite high.   To maintain the sampling system in reliable operating condition requires much more of the instrument mechanic's labor than is true of most other instrumentation.   If the system components are not selected carefully, and if the installation is not engineered by the purchaser for his particular furnace design, in all probability the results will not be completely satisfactory.

Operators who have persevered to solve these problems have been satisfied that the advantages far exceed the difficulties, and that the costs are quickly paid off.

Automatic Combustion Control.   The question often is raised: Can a continuous oxygen analyzer be used for automatic combustion control and eliminate both

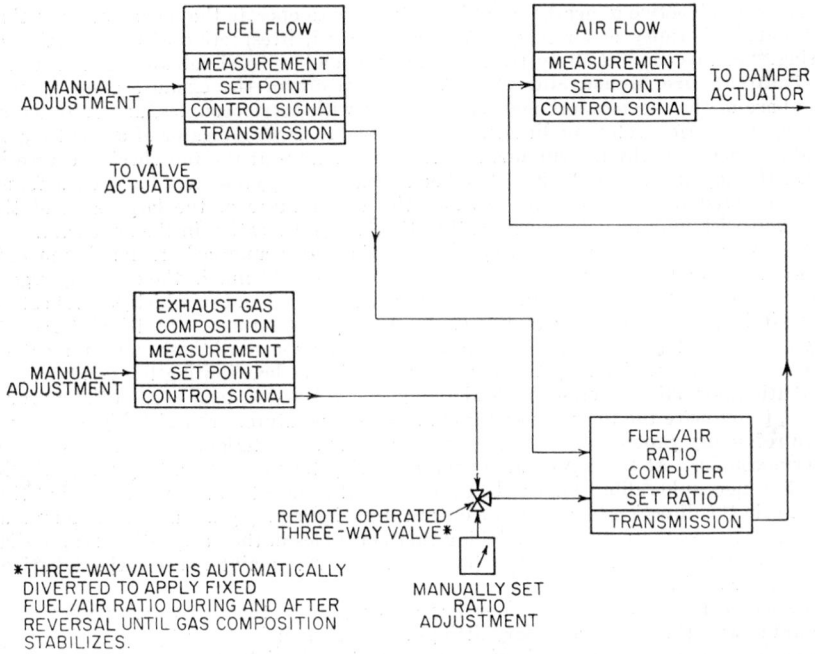

Fig. 6. Automatic-combustion control system employing exhaust-gas analysis.

air-flow measurement and fuel-air ratio control?   Experience has shown that this is possible, but that it is not a very practical control approach on a glass tank.   The factors which cause difficulty include the time lag in the analyzer itself, the frequent downtime for maintenance of the sampling system, and the impracticality of controlling atmosphere composition immediately after reversal.

Automatic control can be obtained best by applying a signal from the exhaust-gas analysis instrument to trim the fuel-air ratio, as shown in Fig. 6.   This arrangement provides the stabilizing action of the independent air-flow control loop, with immediate correction of the air flow, by the fuel-air ratio control—after fuel-input changes.   The analytical instrument control signal then determines the correct fuel-air ratio setting to obtain the desired exhaust-gas composition.   This control configuration does not place complete responsibility for combustion performance on the analytical instrument. Spain[5] provides a brief description and typical records from an exhaust-gas-controlled glass tank. He describes an experience observed in other installations—the furnace-pressure control loop may need to be desensitized, in order to avoid interaction with the exhaust-gas analysis loop.

In an initial installation, the sampling and maintenance problems may not be reduced to a level that encourages automatic control for many months. During this period, the exhaust-gas analysis information applied manually to adjust fuel-air ratio or combustion air flow, will quickly return the initial investment.

Glass-furnace Regenerative-system Control

Maximum combustion air preheat and heat recovery constitute the dual purpose of a regenerative system. An adequate system of instrumentation which will automatically operate the regenerators to achieve this purpose has not yet been developed. The technology and art of the operator and furnace engineer are a prime necessity to achieve good operation.

Another objective of nearly equal importance is to operate the regenerators so that absolute uniformity of air temperature is obtained, irrespective of the direction of firing. In this connection, instrumentation has made several valuable contributions.

Temperature Measurement. The temperature of the exhaust gases leaving the base of the regenerators may be measured with thermocouples, and recorded as an index to regenerator conditions. Of more value as an indicator of current regenerator function is the measurement of the temperature at the top checker course in both the ingoing and outgoing chambers. Radiation pyrometers have been found best adapted for this purpose—because the temperature of the large area of the checker-brick may be determined, rather than a point location in the airstream.

The temperature of the heating chamber indicates approach to maximum safe refractory temperatures and the undesirable presence of flame in the outgoing regenerator. The temperature measured in the incoming chamber which is preheating the air is a function of air temperature entering the furnace. Multipoint recording potentiometers are usually employed to provide a record of these measurements—with provisions for signaling the attainment of high or low limiting temperatures.

Platinum-rhodium thermocouples have been employed instead of radiation pyrometers, to measure the top regenerator chamber temperature. Suitable high-temperature refractory protection tubes are required to avoid deterioration of the platinum thermocouple. At the regenerator peak operating temperature, a long-term drift in the thermocouple calibration can be expected. It is unfortunate and unsafe that this drift is in the direction to reduce the output, thus indicating gradually lower temperatures. The usual reason for employing thermocouples in the past, rather than radiation pyrometers, has been unsatisfactory experience with maintenance of the radiation-pyrometer optical system.

Experience in many installations has shown clearly that a radiation pyrometer mounted with the correct accessory fittings and with an independently clean air supply for each radiation pyrometer will give many months of reliable service without maintenance; or shifts in calibration. A valuable aid in maintenance-free operation is to avoid the installation of shutoff valves in the purge air line. These may accidentally cut off the flow of purge air to the radiation-pyrometer optical system.

When thermocouples are employed, special attention is called to the fact that the use of a double, or inner and outer, protection tube to extend thermocouple life is particularly undesirable in this measurement—since the associated lag is detrimental to the measurement of continuously changing regenerator temperatures.

Automatic-reversal Programming Systems. The regenerative cycle of a glass tank has been most satisfactorily determined on a time basis, with equal cycles in each direction of firing. Twenty-minute reversal cycles are used most frequently. Regenerator checker conditions may require varying cycles, or unequal cycles, for short periods of time as a remedial measure. The possible disturbing effects of the reversal cycle and its frequency make it imperative that reversal occur on a precise schedule and with minimum elapsed time; this is readily seen when it is recalled that with three reversals per hour and a 30-sec elapsed time for switching directions, the fuel is off the furnace 2½ per cent of the time.

Additional firing time also may be lost after restoration of fuel in the new firing direction, because of manual errors, or unsatisfactory but preventable conditions for combustion.

Recognition of these problems has led to the development of the automatically programmed reversal system. These systems execute all of the steps necessary to switch valve and damper positions. They may be adjusted for sequence and timing to optimize conditions for restoring heat release and maximum efficiency in the least possible time.

Automatic-reversal programming systems in the past have been designed in a variety of forms, varying from those which were so simple that they performed no function other than that which could be carried out by a man and a clock, to those which are so complex that they employ secondary protective circuits to guard primary protective circuits.

Experience has shown that a design philosophy may be followed which accomplishes the major objectives of precision repeating of the reversal cycle in the minimum time, together with adequate protection against furnace or reversal equipment failures, and yet employs an instrumentation system which has a low maintenance requirement and a justifiable first cost.

Glass-furnace Temperature Measurement

Except for occasional research investigations, temperatures of the molten glass during the melting process are not measured in commercial furnaces. In part, this results from the difficulty of installing and obtaining reliable performance from thermocouples immersed in the melting section of the furnace. And, in part, it is due to successful technology of operation, based on furnace temperatures measured in the superstructure above the molten-glass bath.

With the increased application of electrical heating directly in the glass bath, the relationship between glass and superstructure temperatures becomes more complex and less effective as an operating guide. New techniques of temperature measurement and the development of more refractory materials than those available in the past offer new tools for developing methods for measurement of melting temperatures. Their application requires extensive research by the glass manufacturers to establish specifications for the primary sensors. This must be followed by a practical effort by the equipment manufacturers to produce practical instrumentation to meet these specifications at a reasonable cost.

Superstructure Temperature Measurements. Control of the glass-melting furnace by superstructure temperature measurements employs a combination of several types of pyrometric devices. Operating temperature levels throughout the furnace are established by intermittent measurements, at a few significant locations, made with an optical pyrometer. To secure reproducible results, it is necessary to obtain the readings with the fuel off. Thus, measurements are made under dynamic cooling conditions. A successful measurement under these conditions requires a standardized manual technique, optical pyrometers that are frequently checked against a secondary standard, and conscientious operators who are well trained. The use of two optical pyrometers which are interchanged for alternate readings provides a continuing check of the equipment without burdening the instrument technician.

Because optical pyrometer readings require the fuel to be shut off, the interval between measurements usually is several hours. As a continuous operating guide, platinum-alloy thermocouples or radiation pyrometers, both employing closed-end protecting tubes, are mounted in the crown. They usually are located on the center line of the furnace to obtain independence of firing direction. One point of crown temperature measurement is located in the refining end of the furnace. One or two others are established in the melting end. For maximum utility, these should be recorded so that the trend and relationships can be observed by the operator.

Another useful guide to furnace operation is obtained by sighting a radiation pyrometer through a hole in the sidewall, between the last port and the bridge wall. In this position, it may be adjusted to read the temperature on the melting-end side of the bridge wall—just above the glass level. The target areas are overlapped to avoid flame interference in each direction of firing. This temperature has been found useful in adjusting flame pattern, and is related to glass temperature in this area. Some installations also have been made with the radiation pyrometer sighting on the refiner side of the bridge wall as a measurement of refiner temperatures.

Operational Problems. Temperature-measuring systems for a glass-melting tank will remain operational and nearly free of short-term maintenance for the complete campaign of a furnace—if they are carefully installed. Thermocouple crown blocks should be made from the most refractory grade of silica brick. They must be installed to fit tightly, and sealed to avoid leaks which will lead to early deterioration. They should be replaced at the end of each campaign, together with the thermocouple.

Nearly as important as care in installation is the need for understanding the basic errors associated with the application—so that the measurements may be interpreted usefully. The operating conditions characteristic of a glass-tank superstructure tend to cause a thermocouple output signal to drift. The corrosive wear on the exposed surface of the protection block reduces its wall thickness, and increases the apparent temperature measured by the thermocouple.

The operating temperature level existing over many months causes a reduction in the thermocouple output signal at a particular temperature. This produces an apparent lowering of the temperature. These errors would not be expected to balance. Finally, it must be recognized that most of these measurements will exhibit dynamic effects, particularly on regenerative furnaces, because of the continuous long-term changes occurring in the furnace as a heat-transfer mechanism. It is for these reasons that glass-tank thermocouples are most usefully employed as short-term trend indicators—with frequent recalibration against optical pyrometer measurements on the furnace.

Use of Direct-sighting Radiation Pyrometers. Although relatively limited use has been made of direct-sighting radiation pyrometers for the melting-end furnace-temperature measurements, excellent results have been obtained. The radiation pyrometer when employed for direct viewing eliminates the lag error of the protection block. It measures a larger surface area than a thermocouple and has greater sensitivity.

These factors become important in protecting the crown from damage under some firing conditions. Radiation pyrometers require no more maintenance than thermocouples in these applications—if certain precautions are observed. Each pyrometer must have a clean purge-air supply, which is independently filtered at each unit. The flow of purging air must be of adequate flow rate (2 to 3 cfm) and uninterrupted. An interruption in air supply for a few minutes will fog the instrument lens. A low-cost differential-pressure regulator on each pyrometer flow line is good insurance. If these recommendations are observed, the pyrometers will operate on a glass-melting furnace for many months without maintenance. It is noteworthy that some coating of the lens by gaseous diffusion is inevitable, but this coating causes no difficulty after an immediate calibration shift, provided that it is not periodically wiped off. Recalibration and overhaul of the radiation pyrometers by the equipment manufacturer, after each furnace campaign, has been found to be low-cost preventive maintenance.

Automatic Control of Furnace Temperatures. Other than on experimental furnaces, little or no attempt has been made to automatically vary fuel rate to control temperature in a regeneratively fired glass tank. The single-chamber construction of the melting end admittedly makes it difficult to achieve control in a specific volume, or zone, without interaction between adjacent control loops. As with nearly all instrument application engineering, the solution to this problem will require changes in process equipment design and integration of suitable instrumentation to achieve automatic temperature control.

Small direct-fired melting furnaces have been in commercial operation for many years with two zones of automatic control.

Forehearth Control in Glass Manufacture

One of the most critical processes in glass manufacture is preparation of the molten glass for delivery to the forming machine. This process consists of a refractory heat exchanger, (forehearth) which increases the viscosity of the glass to the desired value for forming.

Temperature is employed as a measure of viscosity, and control is obtained by regulated cooling of the glass while it is passing through the forehearth. The degree of cooling is regulated by the consumption of fuel above the forehearth glass surface; or by the introduction of electric power, using the glass as a resistive heating element.

It is important to understand that the controllable cooling effect (actually heat input) is only a small fraction of the total cooling load. The remainder of the cooling is obtained by manually adjusted vents and forced-air flow; together with design factors (which are not operating adjustments), such as dimensions and insulation. The design and loading of a forehearth are the result of a highly developed technology which makes it possible to obtain the desired temperature control by automatically regulating a small fraction of the cooling load (heat input).

Glass-temperature–viscosity Relationships. Although viscosity is the major physical property which is adjusted before forming, a suitable, continuous viscometer for molten glass has not been made commercially. The matter is further complicated because glass viscosity is not only temperature- but also composition-dependent.

Glass temperature has been used for many years as a substitute for knowledge of the viscosity. Variations in the temperature-viscosity relationship from day to day, or longer periods, are compensated for by manual changes in the controlled temperature. This compensation cannot be fully effective and always will be after the fact, because of the absence of viscosity measurement.

Additional factors to be considered in forehearth operation are the poor thermal conductivity of molten glass, and with colored glasses, opacity to thermal radiation. These factors constitute system dynamics which must be measured in minutes. The glass enters the forehearth containing viscosity and thermal gradients. It is an objective of forehearth design and operation to minimize or eliminate these gradients. Finally, the ultimate measure of forehearth performance is constancy of *weight per unit* of molten glass delivered to the forming machine. In addition to the effect of glass level on weight control, as discussed earlier, the delivered weight also is dependent on glass viscosity.

These relationships have been presented here in a simplified manner, to provide a background of appreciation for the small tolerances within which forehearth temperatures must be controlled. Glass-container specifications have been met in the past by controlling forehearth temperatures within ±1 to 2°F. As machine speeds increase and as tolerances on weight variation are made smaller, it will be necessary to reduce the permissible temperature variation to less than ±1°F. Unpublished investigations have shown a direct relationship between weight and gob temperature variations down to at least ±0.25°F. Some precision-formed products in large-volume commercial production now employ forehearth temperature control within ±0.1°C.

Use of Direct-sighting Radiation Pyrometers. Most forehearths which are fuel-fired employ direct-sighting radiation pyrometers to measure glass temperature. These pyrometers, when designed to receive radiation not longer than 2.6μ, measure temperature to a depth approximately one inch below the surface. This section of the glass volume has been found sensitive to changes in glass temperatures. Reference 6 contains a discussion of the wavelength sensitivity of radiation pyrometers to radiation from beneath the surface of molten glass. Figure 7 shows sections of a forehearth, and a typical method of installing radiation pyrometers. It is of the greatest importance to the success of this measurement that a clean, dry, and regulated flow of pure air be delivered to each radiation pyrometer. A flow rate of 0.5 cfm is adequate.

The location of the point of temperature measurement is important to operation and is dependent on forehearth design. A radiation pyrometer should not be sighted on the surface beneath a cooling vent.   For control purposes, the burner system of a forehearth is divided longitudinally into independently adjustable zones.   For fore-hearths up to 14 ft in length, two zones are used—the spout and channel.   Longer forehearths may require a third zone at the entrance section.   At least one point of temperature measurement is required in each zone.   For the zone nearest the discharge, the pyrometer is located behind the orifice section, at a point 10 to 15 in. from the center line of the plunger, as shown in Fig. 7.   Measurements in the second or third zones are located in the forward one-third of each zone, near its discharge end. With opaque glasses, the point of measurement should be located about 20 per cent back from the end of the zone.

Both pneumatic and electric systems are used to obtain automatic control.   The spout and channel zones must be controlled from individual instruments.   Adjustable-port valves and valve positioners have been found necessary to obtain the desired control.   A proportional-plus-reset mode of control is required.   Narrow proportional

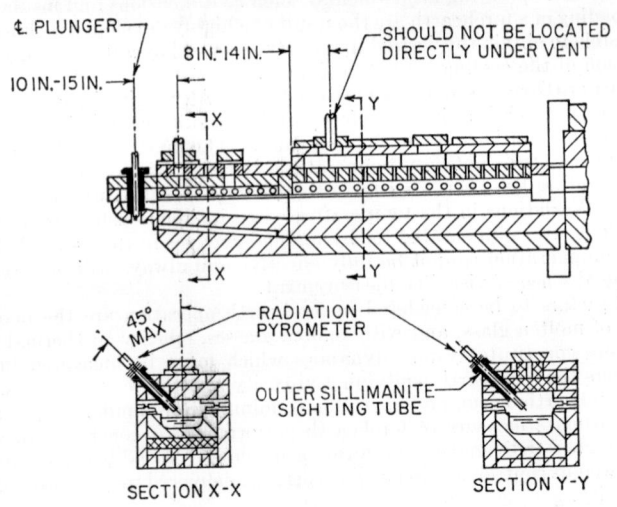

FIG. 7. Cross section of typical forehearth, showing installation of radiation pyrometers.

bands and very slow reset rates are required.   Suppressed instrument ranges with narrow spans, such as 1800 to 2400°F and 1900 to 2300°F give best results.

Trends in Forehearth Operation and Control.   A particularly thorough study of forehearth operation has shown that it is possible to continuously maintain fore-hearth temperatures with ±0.1°C for many days despite normal load variations. Essentially, this has been accomplished by the refinements in instrument design made possible by recent developments in electronic components.

Figure 8 shows the performance of three different control systems on the same elec-trically heated glass feeder.   Note that the record shown is the temperature deviation from the desired set point.   Chart A was made with a conventional electromechanical servo system.   An electronic three-mode controller operated a linearized magnetic amplifier and saturable reactor to control electric heating power.   The temperature was maintained within ±0.5°C.

In the system of chart B, measuring range suppression was obtained by means of a stable electronic reference supply whose adjustable output opposed the temperature signal.   The electronic potentiometer in this section is used to amplify the error signal and to provide a record.   The system, therefore, still contains the electromechanical slidewire rebalanced servo and the error-generating slidewire for the controller.   Chart B indicates that the temperature variation with this system may be limited to ±0.25°C as a result of improving the stability of the measuring circuit through use of external suppression.

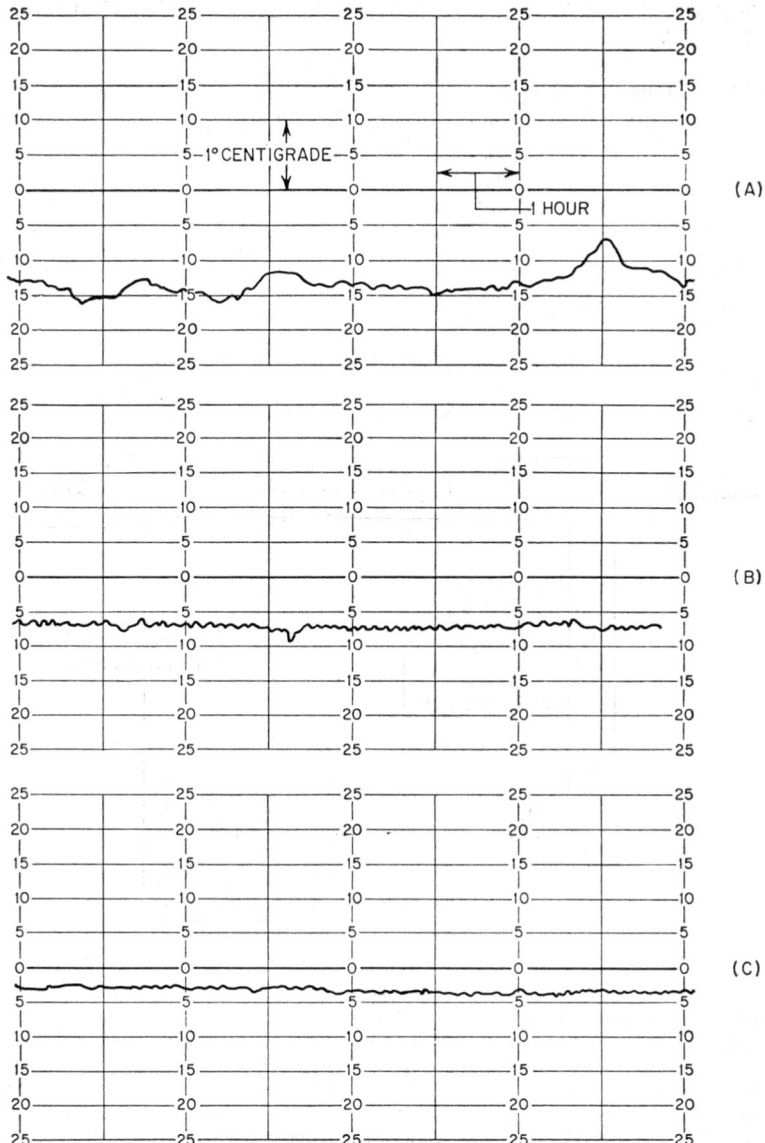

FIG. 8. Performance curves of three different control systems on the same electrically heated glass feeder (see text.)

Chart C was made by a control system which is completely different in concept—except for the voltage reference supply used to obtain suppression. In this system, the electromechanical servo in the potentiometer, the measuring slidewire, the controller error slidewire, and the recording mechanism all have been eliminated. Instead, the system employs a stable d-c amplifier to amplify the difference between the thermocouple signal and the reference signal. The resultant amplified error then is applied directly to a solid-state controller. It will be observed on chart C that the temperature is maintained within $\pm 0.1\,°C$. This performance is obtained over a period of many days in the presence of line-voltage variations of ±10 per cent and ambient variations of 70 to 120°F.

Heat Treatment of Glass

Three forms of heat treatment of glass will be described here: annealing, decorating and bending lehrs, and mechanical forming.

Annealing. All glass products must be annealed after manufacture. The heating and cooling rates during this operation are carefully controlled—not only to relieve permanent stress, but also to avoid breakage by temporary stresses. Most glassware is annealed in continuous conveyor ovens called "lehrs." Their construction and operation are similar to continuous heat-treating furnaces found in the metals industry.

Temperatures are measured with thermocouples or resistance thermometers, whose bulbs are inserted in the atmosphere through the sidewalls. The thermocouples are placed at spaced intervals along the length of the lehr. Alternate thermocouples may be placed on opposite sides of wide lehrs to indicate temperature uniformity. In the fuel-fired zone, a thermocouple is placed on both sides of the lehr, and a manual selector switch permits use of either thermocouple for control.

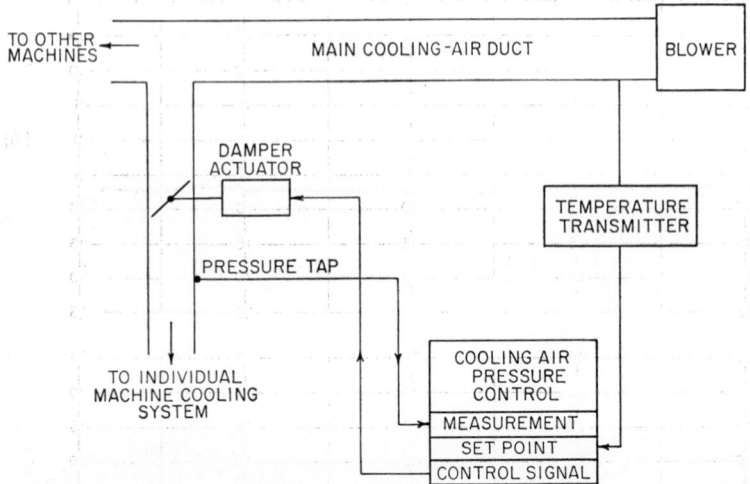

FIG. 9. Mold-cooling air control with temperature compensation.

In general, only the fuel zone is controlled automatically, a single instrument controlling the flow of fuel to all burners. Two-position control has been used widely, especially for uniformly loaded lehrs. For more precise control, a proportional-plus-reset mode of control is used.

The required temperature gradient throughout the lehr is established by experimental runs—with a traveling thermocouple placed in contact with the glass as it passes through a complete cycle. The glass temperatures then may be correlated with the atmospheric temperatures indicated by the permanently installed thermocouples. A set of manual selector switches connected to a single indicating instrument is used to periodically measure the temperature of each thermocouple. Radiation pyrometers have been used to find the actual glass temperature leaving the lehr.

Optical glassware is annealed by a lengthy and carefully controlled cycle in batch kilns. Automatic program controllers are used for this operation. A large number of thermocouples are placed throughout the kiln to permit intermittent determinations of kiln temperature distribution.

Decorating and Bending Lehrs. Most decorative processes used with glassware require that the decoration be fired on a critical temperature schedule and the ware simultaneously annealed. Forming of glass parts by plastic deformation requires carefully controlled temperature processing, as does the laminating of safety glass. Lehrs for these processes will have several zones of automatic control, and may be further divided to provide separate control for each side of the lehr.

Mechanical Forming Operations. Little or no instrumentation is in commercial use at the time of this writing on commercial automatic glass-forming machinery. Instrumentation has been used to study many forming machine variables. These variables have included speed, heat transfer, mold temperature, glass temperature— during and just prior to forming.

An auxiliary to machine operation has been instrumented and should be mentioned because of the successful results that have been obtained. Low-pressure air for mold cooling normally is supplied to each machine from a large duct. Since the blower draws the air from outside the building, the air is subject to daily and seasonal temperature variations. The operator compensates for these variations by adjusting the volume of air delivered to the machine. Figure 9 shows a system which measures the air temperature and adjusts the pressure in the ductwork automatically for compensation. Here, pressure is employed as a measure of air flow. As the air temperature increases, the air flow is increased. These systems have shown increases in production, and make more efficient use of the available cooling air—especially when the system is operating at full capacity.

INSTRUMENTATION OF CERAMICS MANUFACTURING PROCESSES

Instrumentation has been an integral part of ceramic processing equipment for many years. Its economic justification has resulted from improvements in product quality, reduction in rejects, a lowering of fuel costs, and laborsaving. The investment in instrumentation in the large-volume production plants has increased in recent years—as the size and complexity of process equipment has increased. Greater usage of instrumentation also has resulted from extensions to process operations, such as raw-material preparation, which had not previously employed instruments.

Raw-material Preparation

Storing, conveying, weighing, and mixing of raw materials have been completely automated in the large volume plants. Control systems have been designed to monitor and control from a central location the operation of conveyor belts, material feeders, storage-bin contents, size reduction and mixing equipment. The preparation of a specific batch composition is effected by completely automatic means, including the order of addition of specific size materials to obtain maximum mixing uniformity. The engineering of such raw-material preparation plants is a considerable achievement, largely because of the variety of abnormal conditions which must be anticipated. Having recognized them, protective devices must be provided to guard against raw-material-feed shutdown. Reference 7 describes automatic glass batching systems.

Forming Operations

Manual forming is used widely throughout the ceramics industry, and little instrumentation is required. Casting and molding room-temperature control has been found desirable. The latter is particularly useful if the molding area is used as a drying room at night when the room temperature must be increased.

Some instrumentation has been applied to mechanized forming equipment, such as brickmaking equipment. Thus a tachometer connected to the cutting-machine drive records the production rate, calibrated directly in bricks produced per hour.

Drying Operations

The drying of clay products prior to forming must be carried out at a carefully controlled rate to avoid damage to the ware. Moisture reduction is effected by varying the temperature and humidity of the atmosphere surrounding the clay—in either batch or continuous units.   Relative humidity is measured by continuous determinations of the wet- and dry-bulb temperatures with filled thermal systems, thermocouples, or resistance thermometers.   The thermocouple and resistance-thermometer methods permit recording of temperatures at several locations on a single instrument and avoid long runs of capillary tubing.   A small reservoir, having an inexpensive float-operated level control, will ensure an adequate supply of water to saturate the wet-bulb wick.   Relative-humidity electrical-resistance transducers simplify this measuring problem by eliminating the water supply to the wet bulb.   Measuring systems using these transducers may be designed to compensate automatically for variations in the dry-bulb temperature.

The dry-bulb temperature instrument controls the heat input to the drying rooms or oven, usually from steam coils or space heaters.   The wet-bulb instrument controls the addition of moisture to the atmosphere in the case of batch-type drying ovens. Continuous dryers usually have an excess of water vapor.   Control is effected by automatically operated louvers that adjust the ventilation of the oven atmosphere.

Ceramic drying must begin at a slow rate and then may be accelerated gradually. Automatic cam-type program control instruments are used on batch-type ovens to automatically increase the dry-bulb temperature and reduce the atmosphere relative humidity.   This programming is obtained on continuous units by dividing the oven longitudinally into several control zones having individual control of heat input and ventilation.   The set point of the dry-bulb instrument in each zone is set progressively higher than that of the zone preceding it.   Thus, the product temperature is increased as it moves through the oven.   The set point of the wet-bulb controller in each zone is related to that in the previous zone—so that the atmosphere relative humidity decreases in the direction that the work is moving.

Firing Operations

Ceramic products must be fired on a temperature schedule which includes closely controlled rate of heating, positive attainment of firing temperature, without overshoot, and a controlled rate of cooling.   Deviations from this schedule will cause off-specification or damaged products.   Batch kilns for firing ceramics usually are operated from a thermocouple or radiation pyrometer with a closed-end target tube located in the roof.   This measurement is supplemented by intermittent temperature determinations with an optical pyrometer.   These latter measurements indicate temperature distribution throughout the kiln, and provide determinations of product temperature.

The larger batch kilns are controlled manually with the temperature measurements serving as guides.   Automatic programming control of the temperature is increasing in application to batch kilns, because it makes possible unattended operation and reproducibility of optimum firing schedules.

Continuous Tunnel Kilns.   Continuous tunnel kilns require more complete instrumentation for control of firing in order to obtain the precision temperature control required to produce desired product specifications (see Fig. 10).   The heating, firing, and cooling temperature schedule is developed in these kilns by establishing a temperature gradient longitudinally and passing the ware through it on cars.   Fuel is supplied near the center or firing section of the kiln.   The desired temperatures in the preheating zones are obtained by circulating hot air from the cooling sections. Temperature measurement in the preheating and cooling sections is obtained from Chromel-Alumel thermocouples mounted in the roof in several locations.   These thermocouples are connected to a manual selector switch and an indicating instrument—so that a periodic check of the stability of the kiln temperature gradient may be made.   Several of the critical thermocouple signals are recorded on multirecord potentiometers to display operational trends.

Although the temperature gradient through the kiln is determined to a large extent by its design, pressure conditions within the kiln affect the gradient stability   Bell-type furnace-pressure controllers are used to measure and control the kiln draft by adjusting the stack damper.   The furnace-pressure tap location is determined by kiln design, and may be in the firing zone, or in the preheat zone entrance.   Because of the effect of draft on kiln temperature, installations have been made in which the temperature of the preheat zone is automatically controlled by varying the kiln pressure.   Kiln pressure gradient must be controlled at an optimum value to minimize vertical temperature gradients within the kiln.

The temperature in the firing zone of the kiln may be measured with platinum thermocouples located along the roof center lines, and mounted with refractory protecting tubes.   Because of the obvious advantage of direct measurement of the product temperature during firing, radiation pyrometers sighting through the sidewalls on the work also are used.   One radiation detector is used for each side of the kiln, and the detectors are mounted so that the sighting path intersects the cars at an angle.   The angular target path overlaps two cars—so that no discontinuity results from car-

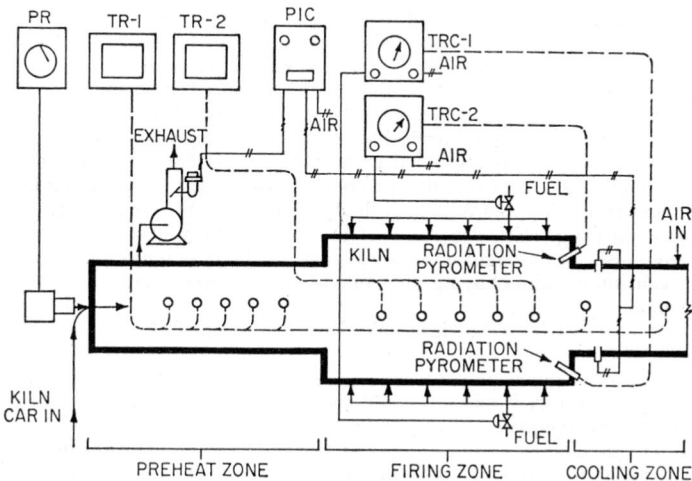

Fig. 10.   Typical ceramic tunnel kiln instrumentation.   (PR) Pressure recorder for car-pusher hydraulic system; (TR-1) multirecord temperature recorder for preheat- and cooling-zone temperature gradients; (TR-2) multirecord temperature recorder for firing-zone temperatures; (PIC) indicating pressure controller for kiln draft; (TRC-1 and TRC-2) recording temperature controllers for firing zone.

to-car movement.   When intermittent pushing of kiln cars is employed, a recessed target section is provided on each car to afford a radiation pyrometer target which will approach blackbody conditions.

Firing-zone Control.   The method of controlling the firing section is determined by the kiln design.   The simplest case is that of a single firing zone with a control thermocouple mounted in the roof.   A single temperature instrument is used to control fuel supply to burners on both sides of the kiln.   Separate control valves should be used for each side.   To improve uniformity of firing across the furnace, separate instruments, primary elements, and valves may be used to control each side independently.   This latter arrangement is employed when radiation pyrometers measure the work temperature.   The kiln also may be designed with several successive firing zones and individual instrumentation for each.   The individual firing zones further may be subdivided for individual control of each side—to gain maximum temperature uniformity.   Although several burners may be used in each zone, a single control valve may be coupled mechanically to one actuator to operate all burners on one side of a zone.

To guide the operator in loading cars in the kiln, a record is made of the pressure in the hydraulic system of the car pusher.   This record indicates the average value of the pressure while a car is being pushed and then decreases sharply at the end of the pusher stroke. The latter drops indicates that another car may be started through the kiln and provides an operation record. In addition, derailment or obstructions in the kiln are indicated by higher-than-average pressures. High- and low-pressure contacts may be mounted in this instrument to signal the operator when either condition occurs.

Increasing evidence of the effect of kiln atmosphere on product quality has made it practical to employ continuous oxygen and combustible analyzers.

Enameling and Glazing Operations. Frit, the raw material for enameling, is prepared by melting refractory and glass-forming materials in a batch or continuous furnace. The molten frit then is tapped from the furnace and granulated by quenching in cold water. The melting temperatures are critical, and must be varied for different types of frit. The corrosive action of the molten material requires the use of radiation pyrometers instead of thermocouples to measure temperatures.

The radiation pyrometers are sighted directly on the molten frit, and provide a measurement which is used to control the fuel supply to the smelter. An air-purged sighting path is used, since during the meltdown period fumes are evolved. Continuous smelters are provided with both a melting and a firing zone. Separate temperature controls are used for each zone. To stabilize combustion conditions, furnace pressure control has been found desirable on both batch and continuous units.

The frit is ground and compounded with other materials to form a solution of the required enamel or glass. Ceramic objects then are coated by spraying or dipping. A ground coat is applied first and then fired, followed by the cover coat. The firing operation is especially critical because of the large variety of defects which it may cause in the finish; or damage, such as warping, to the base material. Firing is conducted in continuous kilns having one zone of control and requiring a proportional-plus-reset mode.

Thermocouples distributed at intervals through the length of the kiln indicate the temperature distribution. Furnace pressure control is required to stabilize operation. Batch kilns also are much used for glaze and enamel firing. Although the furnace atmospheric temperature is measured and controlled, it also has been found necessary to measure the temperature of the enameled objects to obtain consistent results. Radiation pyrometers are used to measure the product temperature in conjunction with an indicating instrument. This instrument operates an audible signal when the desired temperature is reached, warning the operator to remove the glazed material.

Acknowledgments: The author is pleased to acknowledge the significant contributions of the following persons to the material in this section on glass-making processes: Paul M. Spatz, Chief Engineer, Diamond Glass Co., Royersford, Pa.; Richard Post, Principal Application Engineer, and Ernest F. Hucke, Senior Application Engineer, Minneapolis-Honeywell Regulator Co., Philadelphia, Pa.; Harry L. Latham and Louis H. Gauss, Jr., Corning Glass Works, Corning, N.Y.

REFERENCES

1. Griffin, T. W., J. S. Light, and A. W. Russell: Glass Level and Gob Weight Relationship in Automatic Feeder Operation, *Am. Ceram. Soc. Bull.*, vol. 29, no. 5, pp. 183–184, May, 1950.
2. Jones, E. W.: Glass Level Measurement in Furnaces, *J. Soc. Glass Technol.*, vol. 42, no. 205, pp. 62T–69T, 1958.
3. Application of the *AccuRay* Continuous Glass Level Measuring System to Glass Melting Furnace Forehearths, *Ind. Nucleonics Corp. Process Data Sheet* 31-170.01, Columbus, Ohio, 1960.
4. Spatz, Paul M.: Practical Application of Flue Gas Analysis to Glass Melting Furnaces, *Ceram Ind.*, vol. 71, no. 5, p. 72, November, 1958.
5. Spain, Richard W.: "Better Glass Making," Industrial Publications, Inc., Chicago, 1958.
6. Harrison, T. R.: "Radiation Pyrometry and Its Underlying Principles of Radiant Heat Transfer," John Wiley & Sons, Inc., New York, 1960.
7. Covey, Charles W.: Precision Temperature Control for Fiber Glass Production, *ISA Journal*, vol. 7, no. 1, p. 38, January, 1960.
8. Youkers, H. A.: Molding Glass Containers, *Automation*, vol. 7, no. 1, p. 77, January, 1960.
9. Staff: Classifying Glass Blanks by Weight, *Automation*, vol. 8, no. 8, p. 87, August, 1961.
10. Staff: Checking Temperature Gradients in Kilns, *Automation*, vol. 8, no. 6, p. 77, June, 1961.

Section 13

PROCESS INDUSTRIES INSTRUMENTATION

By

ROBERT F. BARBER, B.A.(M.E.), M.S. (Ind. Eng.), *Manager, Pulp and Paper Division, The Foxboro Company, Foxboro, Mass.; Member, Technical Association of the Pulp and Paper Industry.* (*Evaporator Control Systems*)

ROBERT BARCLAY BEAHM, B.S. (Agr. Eng.), M.S. (Agr. Eng.), *Systems Engineer, Taylor Instrument Companies, Rochester, N.Y.; Member, American Society of Agricultural Engineers.* (*Food Industry Instrumentation*)

L. BERTRAND, B.S.(Ch.E.), *Engineering Service Division, Engineering Department, E. I. du Pont de Nemours & Company, Inc., Wilmington, Del.; Member, American Institute of Chemical Engineers, American Society of Mechanical Engineers.* (*Distillation-column Control*)

HENRY S. DRINKER, B.S.(M.E. and E.E.), *Manager, Systems Development Division, The Foxboro Company, Foxboro, Mass.; Member, Technical Association of the Pulp and Paper Industry, Instrument Society of America; Registered Professional Engineer (Mass.).* (*Evaporator Control Systems*)

H. O. EHRISMAN, B.S.(E.E.), *General Sales Manager, The Foxboro Company, Foxboro, Mass.; Member, Massachusetts Society of Professional Engineers, Technical Association of the Pulp and Paper Industry, American Pulp and Paper Mill Superintendents Association, Sales Executives Group of the Recorder-Controller Section of SAMA.* (*Pulp and Paper Production Instrumentation*)

JOHN JOHNSTON, JR., B.S., *Engineering Manager, Instrument Products Division, E. I. du Pont de Nemours & Company, Inc., Wilmington, Del.; Member, American Society of Mechanical Engineers, Instrument Society of America, American Association for the Advancement of Science; on Electrical Code Committee; Registered Professional Engineer (Del.).* (*Instrumentation Practices in the Process Industries*)

J. B. JONES, B.S.(Ch.E.), *Engineering Service Division, Engineering Department, E. I. du Pont de Nemours & Co., Inc., Wilmington, Del.; Member, American Institute of Chemical Engineers.* (*Distillation-column Control*)

S. A. LAURICH, *Manager, Process Plants Division, Struthers Scientific & International Corporation, Warren, Pa.* (*Crystallizer Instrumentation*)

ALFRED H. McKINNEY, B.S.(Ch.E.), *Consultant, Instrument Section, Engineering Service Division. E. I. du Pont de Nemours & Company, Inc., Wilming-*

ton, Del.; Member, Instrument Society of America; American Institute of Chemical Engineers; American Chemical Society; Licensed Professional Engineer (Pa.). (Control of Solids-drying Operations)

H. V. MILES, B.S.(Ch.E.), *Manager, Pulp and Paper Development Division, Dorr-Oliver Incorporated, Stamford, Conn.; Member, American Chemical Society, Technical Association of the Pulp and Paper Industry, American Institute of Chemical Engineers. (Filtration Instrumentation)*

CARL W. SANDERS, B.S.(Ch.E.), *Consulting Engineer in Instrumentation, Engineering Services Division, Engineering Department, E. I. du Pont de Nemours & Company, Inc., Wilmington, Del.; Member, Instrument Society of America. (Automatic Control of Heat Exchangers)*

HANS SVANOE, *Consultant, Struthers Scientific & International Corporation, Warren, Pa.; Member, American Institute of Chemical Engineers; American Chemical Society. (Crystallizer Instrumentation)*

CONTENTS

INSTRUMENTATION PRACTICES IN THE PROCESS INDUSTRIES

By J. Johnston, Jr.*

Manufacturing operations in the process industries depend for their successful execution on a system of instrumentation which has reached such proportions of cost, complexity, and essentiality as to require the most competent engineering judgment in its selection, design installation, and operation. Critical variables must be measured with speed and precision, and their changes must be detected in time for responsible action, which is either automatic or is an educated realignment of processing conditions by human operators. Even when processing steps follow predictable patterns, operating profit in newer processes is in direct relation to the control engineer's understanding of the interdependence of variables and his skill in designing systems to allow for changing requirements or uncontrolled disturbances.

In modern designs, the routine and the obvious are performed automatically, with the more problematical left as a function of operator judgment (helped by preplanning and intelligible presentation of pertinent data on which to base this judgment). This, then, is the scope and purpose of instrumentation in the process industries.

Instrumentation systems represent between 5 per cent (synthetic fibers) and 40 per cent (fine chemicals) of the process-equipment investment. Usually they are installed to provide a centralized information station from which the largest possible number of processing steps can be effectively held under the surveillance of a single operating team. Compact semigraphic arrangements and operator consoles are used which maximize the efficiency of communication without sacrificing adaptability to changing design requirements.

Adaptability has become one of the most important ingredients of control-system design in a processing plant. Process obsolescence, the ever-changing chemical technology, the fickle demands of the market, the forces of competition and the inevitable cost-price squeeze make necessary an ability to *modify* the strategy of operation; to accommodate the desire for higher production at tighter specifications with reduced costs, and to revise operating schedules for faster response to changing customer demands.

Fortunately, almost every branch of instrumentation technology continues to develop faster than the process engineer can exploit it. Concepts of process analysis and optimized operation are ready for use whenever quantitative data on chemical kinetics, heat and mass transfer, and process dynamics have been developed. The instances where this marriage has occurred have produced sufficiently spectacular return on investment to stimulate the desire for its further use.

ORGANIZATION FOR INSTRUMENT ENGINEERING

The enlightened organization has abandoned traditional practices wherein plants are designed with only limited knowledge of the chemical reactions involved. Approximations for heat- and mass-transfer coefficients no longer need be such that the engineer needs excessive safety factors to ensure stable operation and adequate margins of product quality. Excessive capital costs are inherent in such practices. Modern engineers can now take advantage of the lowered cost and tactical capability associated with *quantitative* design for optimized automatic control.

* Engineering Manager, Instrument Products Division, E. I. du Pont de Nemours & Company, Inc., Wilmington, Del.

The newer approach requires very close cooperation and, in fact, a considerable overlap between the functions of process engineer and control engineer. Obviously, advanced mathematical techniques as well as analog and digital computers form important supplements in the hands of such a team. They add speed and confidence to the evaluation of complex relationships which must be investigated where tightly integrated control is desired.

Considerably more attention is now paid to the validity, nature, and precision of the data on which each analysis is based. More time is usually required and higher engineering costs are likely, but time and money are saved in the long run by the avoidance of false starts and by the confidence and speed with which firm decisions are reached.

Additional dividends accrue to those following quantitative engineering practice in the design of process control systems. Traditional process engineering seeks only to optimize operations under steady-state conditions. The comprehensive approach achieves a reduction in the magnitude and duration of upsets and an increase in the speed with which stable production is reached after start-up or change.

Test equipment and specialized procedures are essential to aid the engineer in recognizing achievement of these goals. These include portable instruments that can follow changes in speed, pressure, temperature flow, composition, or specific quality conditions; signal generators and recorders that determine the frequency response of processing units and control-loop components; multichannel recorders and data-reduction systems for correlation studies; and special idealized controllers installed temporarily to evaluate control performance characteristics under actual operating conditions.

Data thus derived from designed experiments and operating history provide the basis for mathematical analysis, process simulation, and plant design. The control engineer must, however, pay particular attention to the circumstances and the conditions of reliability under which these data are obtained. He must consider such factors as the capability of the instruments used, the minimum adequate installation requirements, the intelligence and care with which tests are run, and the ingenuity with which extraneous influences are accommodated and relationships are interpreted.

In all but the smallest processing plants, responsibility for control-systems engineering, operation, and maintenance is vested in a separately identified segment of the organization. This segment is best allied with the operating rather than the service side of the organization. This is because process control is so intimately a part of the operating responsibility. Interpretation of process measurements and adjustment of control set points or other operating parameters are essential to (1) achievement of production goals, (2) maintenance of operating continuity, and (3) adherence to measured standards of product quality.

The instrumentation group must provide a variety of talents and services, including, as a minimum, instrumentation development, process-control-systems analysis, project engineering, and maintenance.

Personnel qualifications somewhat peculiar to the process industries include the need for strength in (1) physical chemistry, fluid dynamics, heat and mass transfer, and reaction systems; (2) analytical techniques required in the definition of the static and dynamic relationships between operating variables in complex chemical reactions; and (3) use of advanced mathematics and statistics in evaluating relationships between variables. These talents are invariably present in some degree within those organizations that are satisfying the responsibilities implicit in process-control-systems engineering. They are, of course, additional to the basic requirements for the practice of instrumentation in all industry.

INSTRUMENT DEPARTMENT FUNCTIONS AND RESPONSIBILITIES

Underlying the functions and responsibilities assigned to the instrumentation department should be a creative interest in helping the company to achieve maximum over-all profit. In addition, there are a variety of specific duties that must be delegated to groups which are best equipped to assume the responsibility. In the interests of efficiency these duties should be clearly defined as to scope and content.

The required talents should be specified and the total organization encouraged to recognize and appreciate the capabilities of its groups.

There is an especial need in the process industries for coordination of the technical responsibilities. This is necessary because chemistry, chemical engineering, and various processing operations involve a somewhat inexact science. So many intangible relationships and crosswise effects are associated, that a tightly integrated and fully informed organization is essential. Inept or uncoordinated actions can produce chaotic and possibly dangerous consequences.

Throughout the department from the recruit mechanic to the senior scientist, knowledge of the *process* is a paramount factor. The strategy of process operation is so much dependent on controller interaction and behavior that to permit an uninformed employee to participate in instrument calibration, repair, adjustment, installation, operation, or selection can be disastrous.

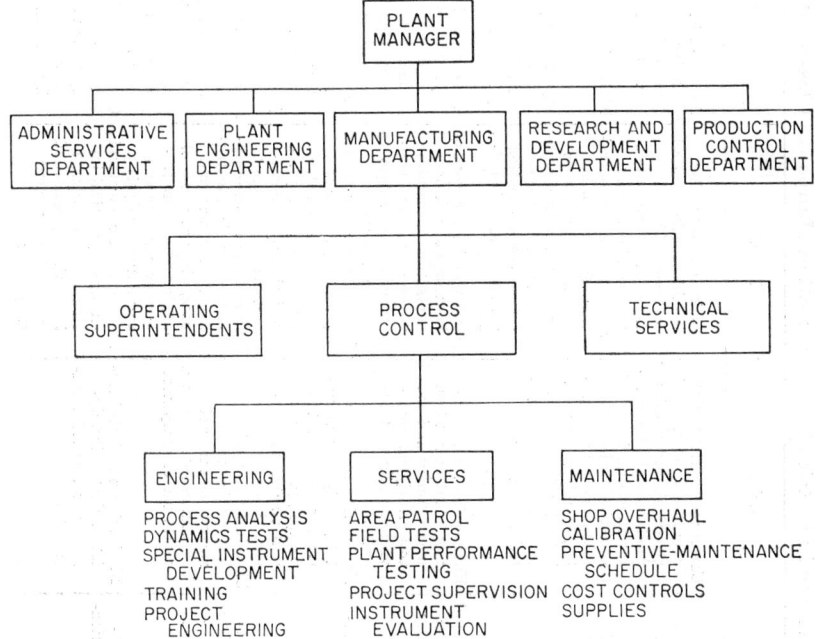

FIG. 1. Typical engineering and maintenance organization for process-plant instrumentation.

Many such factors dictate the need for consolidation of all instrumentation forces within one unifying framework. In addition, there is need for flexibility to meet changing requirements in plant equipment and manpower and for coordinated personnel development programs to meet a rapidly changing technology. The department should retain proprietary responsibility for the operability of these systems, whose design is so intimately tied to incompletely developed process information and whose successful operation is so dependent on sympathetic care and understanding. Personnel must appreciate practical requirements in the development of new and highly specialized instrumentation. They must also know how to allocate funds and effort to the most profitable areas of research and education.

In the organization shown in Fig. 1, the responsibility delegated to each level is relatively obvious. The work performed depends on the nature of the operation, the depth of the dependence on instrumentation, the value of the product, and so forth. In a typical situation, the instrumentation functions, exclusive of supervision, would take the classifications shown on the abbreviated job descriptions in Table 1.

Table 1. Process Industries Instrumentation Manpower Classifications

Position	Qualities	Schooling	Experience	Function	Relationship
Mechanic	Mechanical aptitude, manual dexterity, inquisitive, and learns rapidly	Use of tools, arithmetic operations, mechanical, electrical, and physics fundamentals, interpretation of drawings	Repair and calibration, troubleshooting and adjustment of broad range of instruments on variety of processes	Assure satisfactory availability and performance of installed instrumentation	Cooperates with plant operator and coordinates through technician and supervisor
Technician	Alert, stable, progressive, diagnoses logically and correctly	Use of test equipment, mathematics, electronics, chemistry, process behavior, instrument measurement and control technology	Process equipment operation, diagnosis of instrument and process troubles, repair of complex instrument systems, close association with experienced engineers	Maintain special equipment, diagnose operating troubles, assure peak performance and maximum operating continuity	Cooperates with plant operator, initiates corrective actions
Maintenance engineer	Systematic, careful, sound judge, rapid diagnostician, persuasive leader	Engineering fundamentals, plant operations, measurement and control principles, supervision, equipment operation	Responsibility for work of instrument craft organization of maintenance procedures, supervision of unit start-up and operation	Organize maintenance effort for most economic preservation of operating continuity and adaptation to changing conditions	Coordinates maintenance program for maximum over-all economy, consistent with operating requirements
Project engineer	Meticulous, decisive, cooperative, quick to understand and to use help intelligently	Engineering fundamentals, equipment operation, measurement and control, unit operations, construction	Operation of instrumented processes, design of major process-control installations, close association with manufacturers' engineers	Provide adequate instrumentation for economic operation of proposed plants and processes	Derives information from process engineers; specifies instruments as required
Application engineer	Authoritatively informed on many subjects, analytical, inquisitive, systematic, thoughtful, persuasive, confident	Chemical engineering, automatic control theory, methods of measurement, advanced math, use of computers	Process engineering equipment operation, instrument evaluation and uses, plant test, use of analog computers frequency response and correlation analysis	Investigate and evaluate suitability of instrument systems to process requirements. Specify approach and system most appropriate for needs	Determines process needs in cooperation with process engineering. Defines needs for guidance of project engineer
Development engineer	Patient, inquisitive, ingenious, creative, scientifically inclined	Mechanics, physical chemistry, electronics, dynamics, advanced mathematics	Development of electronic circuits, analytical instruments, automatic inspection systems and special measurement techniques	Conceive and develop workable instrument systems to process requirements not satisfied by commercial units	Sets program based on advice of application engineer and on analysis of operating needs

Table 1. Process Industries Instrumentation Manpower Classifications (*Continued*)

Position	Qualities	Schooling	Experience	Function	Relationship
Specialist—control systems	Analytical, logical, intuitive, realistic, sound judge, able to formulate ideas clearly	Mathematics, statistics, chemical engineering, use of computers, process analysis, advanced control systems analysis	Use of analog and digital computers, dynamics analysis of chemical process, correlation techniques, application engineering	Quantitatively analyze process controls required for optimum operation of existing and proposed units	Investigates process operability and defines control approach in cooperation with research engineers
Specialist—instrumental analysis	Scientifically inclined, ingenious, patient, systematic, deeply interested in physics and chemistry	Physics, chemistry, physical chemistry, nuclear techniques, electronics, mathematics, advanced analytical chemistry	Laboratory analysis, electronic circuit development, fabrication installation, and use of special analytical instruments	Define analytical approach and specify suitable instruments for process stream analysis	Works cooperatively with project engineer or plant's technical group to define analytical needs

Instrument Development

An instrument-development facility in the company is needed to supply the specialized equipment that is required in connection with research work and plant operations. Often no commercial source exists for such equipment. Typical examples include:

1. Custom-designed inspection apparatus.
2. Detection devices based on physical or chemical principles which are unique to the situation involved.
3. Completely new forms of instrumentation whose development represents a potentially superior operating basis.
4. Special combinations of commercially available instruments modified in capability to satisfy more closely the needs of a particular application.
5. Proprietary developments which, if purchased outside the company, would reveal know-how of value to competitors.
6. Some research efforts which may not have obvious relevance to current operations or which may be important to the physics or chemistry of manufacture.

Equipment usually required for an effective instrument development program includes one or more of each of the following items:

Sets of primary and secondary standards plus laboratory calibration and checking apparatus for weight, temperature, voltage, resistance, pressure, spectra, capacitance, gas constants, etc.

Infrared spectrophotometers.

Mass spectrometer.

Emission spectrometer.

Vapor-phase chromatograph.

Radiation sources and detectors.

Precision electrical measuring instruments.

pH and ORP measuring systems.

Electrical and electronic test equipment.

Wet-chemistry analysis instruments.

Plant Test and Process Analysis

Some means is required for collecting and organizing data on laboratory, pilot-plant, and full-scale plant operations. Such data are essential to the development of sound engineering design and the diagnosis of obscure operating anomalies. Many organizations have assembled mobile complements of test instrumentation to obtain experimental data from a number of operating facilities. Others employ the more common practice of using a variety of portable units which can be carried or shipped to the test site. Still others install special instrumentation on a temporary basis at new facilities and pilot plants. For initial operation the instruments record results of experiments during acceptance tests and the period of greatest difficulty of operation. After serving its purpose, the test equipment is disassembled and stored for reuse on subsequent construction.

To support the portable test facility, a data-reduction center and computation machinery are desirable. Also needed are test facilities for checking the performance characteristics of control apparatus and the frequency response of control-loop assemblies. The analog computer is invaluable for exploration of process-control alternatives and for investigating the basis for design of plant equipment. Usually, the following represents an adequate complement of data acquisition and test apparatus:

Infrared analyzer with variety of sources and cells.

Ultraviolet analyzer with variety of sources and cells.

Refractometer.

Gas chromatograph.

pH meters.

Thermal-conductivity analyzer.

Oxygen analyzer (paramagnetic).
Orsat apparatus.
Explosimeter.
Electrical-conductivity bridge and cells.
Set of fast-response flow transmitters.
Set of fast-response pressure transmitters.
Set of fast-response temperature transmitters.
Radiation sources and detectors.
Recording viscosimeter.
Miscellaneous transducers and converters.
Speed recorders.
Power supplies.
Multichannel recorders for electrical voltage and pneumatic pressure.
x-y recorder.
Signal generators.
Scalers.
Oscilloscopes.
Radiation pyrometers.
Load cells.
Densitometers.
Timers and program controllers.

Computation Center

Supplementing the test facility, but available for other than use on instrumentation, the computer center might include the following equipment:
One analog computer with the following components:

100–200	Operational amplifiers.
50–100	Multiplication channels.
25–50	Operational relays.
25–50	Function generators.
25–50	Operational switches.
2–3	Separate consoles.
3–4	x-y recorders.
12–18	Recording channels.

One digital-computer facility of the maximum size and capability that can be afforded. As competence and facility in its use improve, no computer facility ever seems large enough to accommodate the needs that will arise. Any such installation should be made with adequate provision for expansion.

Instrumentation engineering work requiring the computation center ranges from the calculation of flow-orifice diameters to the simulated operation of a complete processing system including its chemical kinetics, energy and material balance, and controller characteristics.

Aside from routine solutions to mathematical equations, the greatest use for the equipment is by process engineers who are searching out the mechanism of a physical or chemical phenomenon in order that an improved understanding will increase their confidence for design and operation. Exploratory work of this nature is gradually developing a body of knowledge which is increasingly needed for quantitative design of controlled processes.

In addition to the development of these mathematical models, the center is an invaluable aid to training. Dynamic models of an actual process can be made to reproduce almost every type of operating difficulty, thus providing the opportunity for realistic trial of alternative ways of overcoming the difficulty. Experience can be gained before the fact in dealing with emergency situations with maximum safety. Alternative choices can be fully evaluated and demonstrated to the satisfaction of all concerned. This is also important in the design phase of a new facility, where unanimous agreement on a course of action results from exploration and comparative evaluation of all suggested alternatives.

After the basic mathematical model is developed, it is expanded to reflect in engineering detail (1) the influence of all critical factors on the various properties of the materials in process and (2) the effect on over-all profit of deviations from each of the influential factors. Thus a basis is derived on which to exercise design judgment and operating strategy for each of the process arrangements deserving serious consideration. To define the process model with the rigor necessary for definition of a control scheme (which, incidentally, requires considerably less rigor than that necessary to specify the equipment design) usually requires that the gaps in the purely analytical synthesis be filled from statistical procedures and by experiments.

The resulting combination represents a working basis for the investigation of process operability. It is an excellent starting point for systems engineering analysis leading to design optimization. Everyone from the research chemist to the operating supervisor is helped to a better understanding of process sensitivities through this powerful tool.

Project Engineering

The instrumentation portion of a new construction project involves details of engineering, specification, and procurement out of all proportion to the purchase value of the instruments. Numerous procedures for streamlining the operation have been developed by groups faced with the problem. Essentially, these take the form of engineering standards for specification and installation. Price agreements are often negotiated for repetitive items. Standard symbols and terminology simplify drafting.

Project requirements are evolved as the design progresses. Basic data become available and decisions on operating strategy are reached in a series of steps. Instrumentation engineering generally must be adapted and changed as the project develops. Controllability of proposed designs should be considered early in their development to minimize changes in the instrumentation. Analog and digital computers can help establish quantitative bases for these control decisions.

An additional useful tool is the scale model, which permits development of the most efficient and utilitarian design layout. At the outset, a very inexpensive representation using blocks and sheets can be used to obtain agreement on general layout. Then, it is profitable to build a model with approximate shapes and to assign pipe, duct, and conduit routes. Questions of process sequence, hydraulic and mechanical efficiency, operator access, ease of maintenance, safety, and expansibility are resolved at this stage.

In the final model, replicas are used; piping, duct, and electrical cabling are placed; measurement and control equipment locations are chosen; possible modifications are visualized and provided for; agreement is reached on the size and location of service facilities, and plans are evolved for the most efficient sequence of construction steps. Special cameras are used to photograph views of the model which will serve as substitutes for drawings and which, when duplicated, will serve each craft in developing bills of material and sketching installation details.

A project engineering section must work with speed and decisiveness in order to minimize engineering costs. This is best achieved by separating all specialized engineering and science functions from the project organization without making the separation so complete as to hinder the use of such specialized talent where needed. Usually, the specialists and their equipment in an operating company are used in programs of research and plant assistance, financed from corporation funds. This tends to produce factual data on which to base design decisions which, when made jointly, minimize misunderstanding and yield superior results at lowest cost.

With nearly 1,000 new instrument items being introduced each year, no project engineering group can operate effectively without keeping abreast of new developments. The instrument-evaluation facility operated by the plant test group can include in its program the comparative evaluation of instruments being considered for project purposes.

There are so many different approaches that can be used in instrumenting a project, and so many varieties of equipment competing for use, that the lessons of experience become an invaluable factor in developing teams able to judge the system configuration best suited to specific needs. For this reason, all specialist engineers should continue their responsibility throughout the life of the project, or until they have witnessed satisfactory operation of their portion of the design.

Handling of Maintenance

Instrument maintenance in a processing plant is generally handled under one of two entirely different philosophies. Consistent with the accepted pattern for care of operating equipment, one philosophy attempts to reduce all work to patterns of systematic jobs and schedules. Success is measured in terms of the reduction in maintenance costs and decrease in shutdowns due to instrument failures. The other philosophy emphasizes continuity of the operating process, even at the expense of maintenance. Success in the latter case is spelled in terms of plant production vs. capacity, consistent with *acceptable* maintenance cost.

Wherever possible, an intelligent combination or balance between the two extremes should be sought. The first, or maintenance-efficiency philosophy, is adopted by those who believe that most problems can be eliminated by establishing a procedure for every possible eventuality. The second, or operational philosophy, is adopted by those who are less interested in the cost of repair than they are in the cost of lost production. Behavior is constantly observed in order to detect the beginnings of imperfect function so that steps can be taken to prevent the need for repair. Naturally, such constant surveillance is costly in terms of maintenance labor, but it is counterbalanced by the prevention of major repair or extended interruption to plant operation.

Manpower and Training

Some organizations consolidate the operating-supervisor and control-technician functions into one, making operating responsibility a highly technical one. Controller adjustment, field calibration, troubleshooting and in-place repair are the function of this area control team who are responsible to operating management but are controlled by the central instrumentation organization.

As a craft, instrumentation maintenance continually faces new problems. New techniques, new instrument designs operating on new principles and set to heretofore impossible tasks, confront the instrument craftsman with increasing frequency. Training programs, factory refresher courses, and shop talks go on constantly. In a never-ending effort to master the changing patterns of plant operations, the instrument man must continuously pursue basic programs in mathematics, physics, and chemistry and learn new flow diagrams and process descriptions.

Where the installation is new, training for instrument maintenance begins 6 months before the plant starts operation. A typical program covers the material outlined in Table 2. It should be conducted by experienced instrumentation engineers who have had additional instruction in the principles of teaching. The course furnishes a refresher in mathematics, physics, chemistry, and electronics; then covers the principles of operation, installation practices, care, and maintenance of each of a number of kinds of instrument suited to each of a number of plant situations. Sources of error are illustrated and evaluated. Methods for circumventing commonly encountered difficulties as shown. Test procedures and diagnostic practices are taught.

During the final phase, agreement should be reached on participation in the contractor's final installation work and checkout of the installed equipment. This and the actual start-up are the most critical, and often the most valuable, training stages.

A training device of value for both the instrumentation technician and the operator is the simulated control center. With it, the instructor can reconstruct almost every conceivable operating situation. The operator can be taught to diagnose the situation and to plan and execute the proper steps for correction. The technician must

Table 2. Process-plant Training in Instrumentation—Program for Advancement from Trainee to Control Mechanic

Subject	Class hours	Shop hours
Math:		
Use of fractions and decimals	2	0
Use of ratio, proportion, and per cent	2	0
Use of square root and squares	2	0
Calculation of area and volume	2	0
Conversions of units	2	0
Use of graphs and calibration curves	2	0
Pressure:		
Fundamentals of pressure	2	
Use of manometers	4	2
Use of elastic-deformation instruments	2	4
Use of pressure calibrators	4	2
Maintenance of pressure gages	6	4
Maintenance of pressure recorders	4	4
Maintenance of pressure transmitters	20	16
Temperature:		
Application of temperature fundamentals	2	
Use of glass-stem and industrial thermometers	4	2
Use of bimetallic thermometers	4	2
Application of filled-system thermometers	2	0
Installation of liquid-filled thermometers	2	2
Installation of gas-filled thermometers	2	2
Use of temperature-calibrating equipment	4	2
Maintenance of temperature transmitters	4	4
Maintenance of temperature recorders	6	4
Flow:		
Flow-measurement fundamentals	2	0
Use of orifice plates	4	2
Use of differential manometers	6	2
Use of flow calibrator	6	4
Maintenance of flow meters (general)	2	4
Maintenance of mercury manometers	6	4
Maintenance of ring-type manometers	6	4
Maintenance of bell-type manometers	6	4
Maintenance of bellows-type manometers	6	4
Use of area-type meters	2	2
Maintenance of glass-tube rotameters	6	4
Maintenance of metal-tube rotameters	6	4
Maintenance of flow transmitters	18	6
Maintenance of displacement meters	6	4
Level:		
Fundamentals of level measurement	2	0
Hydrostatic instrument maintenance	2	2
Bubbler-type instrument practice	6	2
Ball-float instrument practice	4	4
Displacement-float instrument practice	8	6
Mechanized-float instrument practice	6	4
Weight:		
Fundamentals of weighing	2	0
Load-cell practice	8	6
Gravimetric feeder practice	8	6
Automatic control:		
Instrument air supply practice	4	0
Concepts in pneumatic control	2	0
Pneumatic controller operation	2	2
Maintenance of pressure regulators	2	2
Maintenance of temperature regulators	2	2
Controller relay operation	2	4
Controller mechanisms	4	4
Controller circuitry	4	3
Controller test and calibration	4	4
Maintenance of position-balance controllers	6	4
Maintenance of force-balance controllers	6	4
Maintenance of specialized controllers	6	4
Control-valve principles	2	0
Maintenance of diaphragm actuators	8	4
Maintenance of piston actuators	8	4
Maintenance of valve positioners	6	4
Maintenance of control accessories	8	2

13–16

Table 2. Process-plant Training in Instrumentation—Program for Advancement from Trainee to Control Mechanic (Continued)

Subject	Class hours	Shop hours
Process-control practice	6	4
Control-loop diagramming	6	2
Control-loop checking	2	2
Controller adjustment	16	10
Process-control troubleshooting	16	10
Time-cycle controllers	4	2
Interlock systems	4	2
Alarm and signal systems	4	2
Specialized control-loop checking	8	0
Control center check-out and maintenance	16	0
Electricity and electronics:		
D-C electricity	8	0
D-C series circuits	2	2
D-C parallel circuits	2	2
Application of Ohm's law	2	2
Measurement of d-c power	2	2
A-C electricity	4	2
A-C voltage and current relationships	4	2
Inductance, capacitance, impedance	2	2
Troubleshooting a-c circuits	6	4
Electromagnetic theory	8	2
Transformers	2	2
Vacuum-tube voltmeter use	2	2
Oscilloscope use	8	4
Resistor-capacitor coding	8	2
Capacitance circuits	8	4
Diode characteristics and use	8	4
Filter circuits	2	2
Rectifier circuits	2	2
Voltage-divider circuits	2	2
D-C power supplies	2	2
Triode characteristics and use	4	4
R-C amplifier coupling	2	2
Amplifier coupling practice	2	2
Control-grid operation	2	2
Amplifier troubleshooting	6	3
Power-supply troubleshooting	2	2
Feedback effects	2	2
Cathode-follower circuits	2	2
Application of tetrodes and pentodes	2	2
Oscillator troubleshooting	6	3
Use of tube tester	2	4
Thermoelectric principles	2	2
Use of portable potentiometer	6	4
Operation of electronic potentiometers	8	4
Troubleshooting electronic potentiometers	8	8
Troubleshooting resistance bridge	8	4
Troubleshooting capacitance bridge	8	4
Troubleshooting interconnected systems	4	2
Electrical instrument installation practice	8	0
Electronic control-systems theory	4	0
Electronic controller operation	16	4
Controller adjustment	4	2
Controller troubleshooting	12	6
Electrical actuators for control valves	4	2
Electropneumatic transducers	2	2
Control-loop checking	2	2
Control-loop troubleshooting	16	8
Control-center check-out and maintenance	16	0
Analyzers:		
Principles of pH measurement	4	4
Troubleshooting pH measurement	4	4
Troubleshooting electrical conductivity instruments	2	2
Troubleshooting photoelectric instruments	2	2
Fundamental of chemical analysis	16	2
Thermal-conductivity analyzers	8	4
Photoelectric analyzers	8	4
Infrared analyzers	16	8
Gamma- and beta-radiation instrumentation	16	8

be similarly versed, with the added requirement that he cannot assume that all the instruments are operating correctly.

Motion pictures, sound slides, exploded working models, construction kits, and practice instruments are all needed for the instrumentation program. Course materials in the form of lecture guides, texts, films, and demonstrations can be obtained from the Instrument Society of America, from instrument manufacturers, and from numerous technical institutes, vocational schools, and colleges.

Facilities for Maintenance

In field maintenance, process improvement or the diagnosis of process troubles is an integral part of the detection of measurement and control anomalies. It requires essentially the same equipment as that listed under Instrument Development. Additionally, the competent engineer or technician will have available to him a full complement of hand tools, test sets, replacement units, communication gear, and calibration devices.

In the process plant, unrelenting alertness to potential hazards must become second nature. The instrument man must follow safe practices in taking equipment out of service, conducting tests, and placing equipment in service. Vents and test taps must be used. Protective clothing must be worn. Nonsparking tools are nearly always required. Appropriate tests for explosivity, toxicity, corrosion, and high temperature should be made before beginning work. Safety interlocks should be in place and all affected persons notified of the safety implications of any change in operating procedure created by the work.

In addition to test and safety facilities, the field force usually needs a small complement of extra instruments and storage space in which to keep supplies. Major needs and the more elaborate combinations of test gear are either shared with other areas or are maintained as part of the shop facility.

The instrumentation maintenance shop size depends on the magnitude, type, and frequency of work it is expected to handle. Most organizations prefer to have an instrument shop that is self-sufficient. In a typical $100 million plant, a layout similar to that shown in Fig. 2 is used. It is located close to other maintenance facilities and to central materials storage. Heavy items of instrumentation can be unloaded at the entrance. decontaminated, cleaned, and handled by chain hoist to the central work area. Special vises and jigs have been developed for holding control valves, meters, and other heavy units. Cleaning baths use solvent vapor and ultrasonic agitation. Calibration stands for flow meters are equipped for gas and liquid service.

All work stations are designed to suit specific needs. The majority have precision air-supply regulators and calibration gages of multiple range. They each have quick connectors of various sizes and a variety of special connectors and jigs for routine test and adjustment of all common instrument types. Electrical power supply and calibration stands are also installed at numerous locations, although the main electrical repair facility is located in a separate room.

Calibration assemblies for the flow meters used in chemicals service are usually for smaller capacity than those for petroleum refineries and power plants. A manifolded assembly of rotameters of different ranges, dry-gas meters, a wet-test meter, a weigh scale, and a pair of calibration tanks plus one gasometer-type meter prover can be arranged to suit almost any condition likely to be met. Pressure calibration stands using double-well manometers, precision gages, and dead-weight testers can serve the dual purpose of checking static-pressure and differential-type instruments. Temperature calibration usually requires one small muffle furnace, an oil bath, a molten salt bath, and a checking system composed of certified thermometers, resistance bulbs, thermocouples, a precision resistance bridge, and a potentiometer.

The tools in the electrical and electronics repair facility follow fairly standard patterns. Separation from other shop work is recommended in the interests of cleanliness, as well as freedom from noise and vibration. Separation also prevents accidents that might result from the presence of liquids in an area where high voltages may be grounded through a worker's wet feet. Special jigs and racks are usually needed for handling the explosionproof housings used on most chemical-plant instruments of the electrical type. A chemical hood and ventilator are also needed for electronic instruments used for process-stream analysis; these require calibration with standard mixtures of process fluids, some of which may be toxic. The instrument shop is usually responsible for mobile radio equipment and radio telemetering systems. Standard test equipment is supplied for these and for other plant apparatus—such as fire-alarm systems, plant-protection systems, wired television, weather stations, teletype systems, and computers.

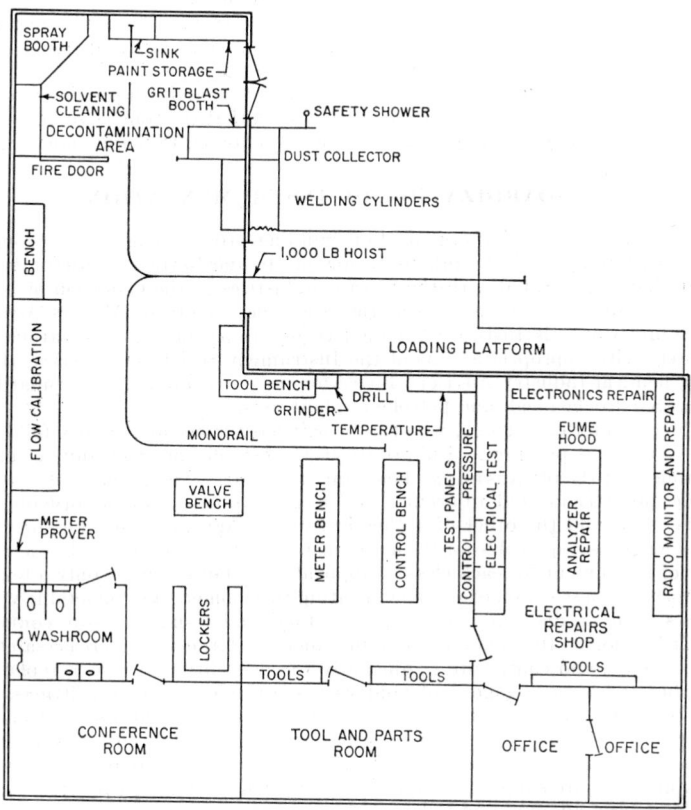

Fig. 2. Typical layout for an instrumentation maintenance shop in a larger processing plant.

Maintenance Records

Despite the difficulty of segregating instrument repair costs from what might be termed "plant continuity insurance," there are numerous advantages besides cost control in keeping good records. To be useful, records must be factual; they must be kept religiously, and their purpose must be clearly in mind in order to ensure that information of value is recorded and subsequently used to improve performance.

Functions other than cost control which a readily accessible system affords include:
1. A central source of original data and specifications.
2. A complete history of alterations, range changes, etc.
3. A complete history of corrosion and wear.

4. A complete history of performance at differing process conditions.
5. A complete history of replacement and repair.
6. A cross comparison between different makes and models of instruments and of similar instruments in different services.
7. An automatic recheck and preventive-maintenance schedule.
8. A quantitative basis for evaluating designs and for supporting the need for redesign or change.
9. A basis for estimating reliability and predicting shutdown schedules.

To best serve these purposes, even in a moderate-sized plant, it is worthwhile to code instrument units on punched cards or other mechanized data forms. A code for critical types of adjustment, replacement, or repair can then be used and a history developed which can be explored by machine search. Patterns and trends are thus discovered which would otherwise remain hidden in a mass of miscellaneous data. Cost information can also be organized on such a system and will provide meaningful guidance for supervision—provided it is remembered that the time devoted to process analysis and control-system adjustment cannot be equated to maintenance costs.

STANDARDIZATION OF INSTRUMENTATION

The larger the contracting organization building process plants, the stronger the tendency to simplify project work by means of standardization. Instrument manufacturers have cooperated with the process industries in the evolution of standard practices. Trade associations such as the Scientific Apparatus Makers Association and the Fluid Controls Institute have active programs for standardization. They work closely with appropriate units of the Instrument Society of America, the API, and the Chemical Industry Advisory Board for the determination of standardization needs peculiar to chemical- and petroleum-plant uses.

It is the user himself, though, who can benefit most from intelligently following an adequate standards program. Uniformity in the establishment of units of measurement throughout the plant is an excellent starting point. Too often one finds a Fahrenheit thermometer located adjacent to a centigrade one, simply because one is part of a power unit and the other of a chemical unit. Agreement on measuring units among all agencies responsible for the use of instruments can be followed by agreement on types of instrument for each class of application. For example, only a few of the choices open in the selection of simple indicating thermometer are bimetallic vs. filled-bulb, glass stem vs. dial, industrial vs. laboratory, and ½-in. vs. ¾-in. connections.

Standardization on the minimum combination of variations can increase safety, save time and money, and contribute significantly to versatility and ease of replacement. This applies to pressure gages, solenoid valves, rotameters, relays, fittings, and all other instrument items used in large quantities for measurement and control.

Connections to piping and vessels, piping and wiring between instrument units, manifolds and mountings, enclosures and winterizing, vents and purges, air supply and power supply—all are subject to improved utility where standard practices and procedures have been adopted and followed.

A uniform basis for conducting tests and a uniform nomenclature and terminology for describing instrument performance are an important result of cooperation between plant instrumentation engineers. This is particularly pertinent with respect to the dynamic characteristics of instruments and to such terms as accuracy, linearity, and resolution.

Standard methods for treating data in the calculation of fluid flow are critical to processing. Different flow-meter manufacturers use different reference pressures and temperatures, and care must be taken in selecting conversion factors for energy and material-balance calculations. Most plants reduce all flows to weight units (e.g., pounds per hour) and adopt a standard reference for base conditions. All orifice-meter installations follow standard patterns, usually employing flange taps, and the orifice diameter is calculated in accordance with one procedure throughout. A standardized program for handling orifice calculations is published under the auspices of AIChE for use on digital computers.

A standard method for calibrating area-type meters and current meters is necessary, along with uniform treatment for factors correcting for viscosity and density. The capacity of automatic control valves is calculated in accordance with whatever standards the manufacturer employs, and it is necessary for the plant to recalculate all of these to a common standard.

Laboratory analysis, so often used to check the accuracy of process-stream analysis, is itself subject to critical examination as a source of discrepancies. Standard practices must be established for the collection of representative samples and for the conduct of each step in the analysis. Here again, care must be taken to ensure that units are reduced correctly to a comparable basis.

The standards used in connection with process-plant instrumentation, if compiled under a single heading, will usually form a fairly large volume.

SPECIALIZED PROCESS-PLANT INSTRUMENTATION

Measurement and control in the process industries are often handicapped by the corrosive or hazardous nature of the materials and the environment encountered. Problems these impose can sometimes be overcome by the use of instrument elements or packages made from stainless steels or other resistant materials. In many cases, however, the fundamental need of the measurement principle dictates otherwise. For example, it may call for a thin-walled elastic member or a material having specific magnetic, electrical, thermal, or other properties which are not compatible with corrosion-resistance or physical-durability requirements. In such cases there are a number of alternate solutions to the problem:

An inert purge can be used, or a sealing fluid. A special sealing bellows can be inserted to isolate the delicate member for direct contact with the offending condition. Also, special designs of instruments for use in such conditions have been added to the standard line offered by most manufacturers.

The schematic diagrams of Fig. 3 give an indication of the varieties of protective techniques used in the process industries. Figure 3a shows methods for liquid-level measurement, using external and internal elements. Figure 3b shows pressure measurement with seals to protect the bourdon tube or other elastic-chamber measuring element. Figure 3c shows three methods for metering the flow of corrosive liquids or slurries.

With the need for resistant materials in instruments have come more suitable materials. Newer alloys, ceramics, and plastics have benefited instrumentation, as they have all other areas of equipment manufacture. One of the greatest boons to instrument versatility is the availability of Teflon, Kel-F, epoxy-impregnated fiberglass, neoprene, and many others for packings, diaphragms, linings, and other parts exposed to corrosive streams.

Many instrument parts are available in glass, Zytel, Lucite, Haveg, Karbate, porcelain, Hastelloy, tantalum, gold, platinum, and the full range of stainless alloys. Furthermore, coating and lining techniques are used extensively for covering critical parts with glass enamel, epoxy resin, neoprene, Teflon and its copolymers, lacquers, and numerous other materials. Complete instruments (such as rotameters, control valves, displacement meters, metering pumps, liquid-level transmitters, and others) can be fabricated from glass, Karbate, Haveg, Teflon, hard rubber, and other special materials.

In some cases, the most common model of a measuring or control device cannot be obtained in a suitable material or be adequately protected without jeopardizing its performance. It is then necessary to turn to ingenious alternatives. Where fluid flow cannot be measured with a head type or an area meter, and no materials are usable for a displacement or current meter, one must resort to completely external means. In some instances the electromagnetic flow meter or the thermal meter can be used; in others, it is necessary to use weigh tanks and scales. The latter alternative is used extensively in the chemical industry, and in numerous instances all the materials used in a chemical formulation are measured in elaborate automatic batch systems.

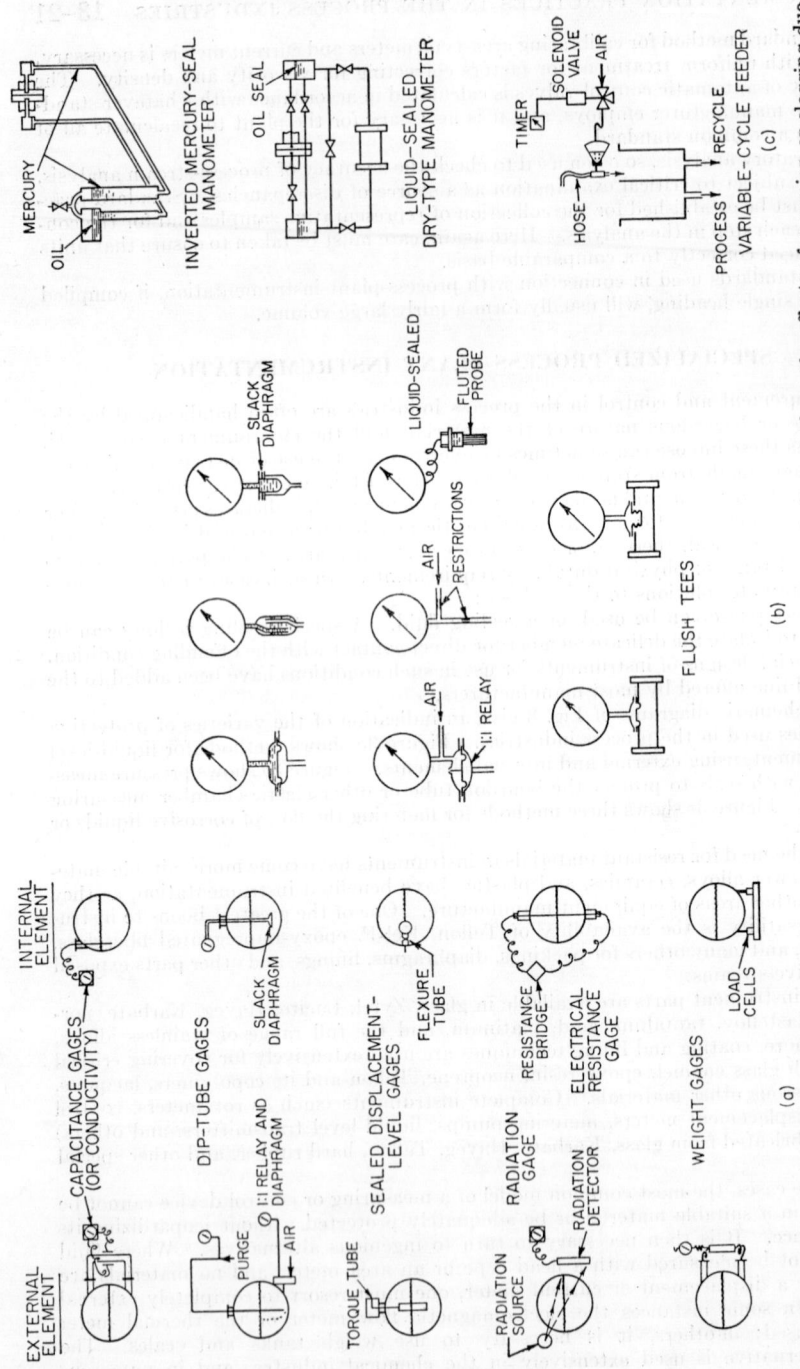

Fig. 3. (a) Samples of techniques in liquid-level measurement for protecting instrument elements from the effects of corrosion or clogging due to solids. (b) Samples of techniques in pressure measurement for protecting instrument elements from the effects of corrosion or clogging due to solids. Note: Diaphragm and bellows gages can also be equipped with steam gage, differential transformer, or other types of displacement detectors instead of with liquid seals. (c) Samples of techniques in flow measurement for protecting instrument elements from the effects of corrosion or clogging due to solids.

The measurement of liquid or solids level in a tank or vessel is often possible only by means of weigh scales, load cells, or strain gages applied to the support structure (see pages 5-55 and 5-68). Radiation-absorption methods offer an alternative technique which avoids contact of the instrument part with the offensive material. A radioactive source and detector combination can be mounted outside the vessel wall in such a geometry as to furnish indications proportional to vessel contents (see page 3-45).

The family of measurements most peculiar to the process industry is the measurement of process-stream chemical composition. Many of the instruments developed for this purpose find commercial application and become part of the standard battery of analytical instruments described in detail in the "Process Instruments and Controls Handbook,* Sec. 6, and briefly reviewed in Sec. 8 of this handbook.

Many analytical instruments are unique to particular processes and never reach a position of commercial availability. These represent ingenious combinations of devices specially developed to measure a property that is characteristic of the material to which they are applied and the equipment with which they are associated. It was from such beginnings that instruments like the infrared analyzer were developed. Others having similar origin are the ultraviolet analyzer, the gas chromatograph, the beta-ray thickness gage, X-ray spectroscopy, the continuous viscosimeter, the colorimeter, the pH meter, and the paramagnetic oxygen analyzer. It will always be typical of the process industries that dramatic advances in instrumentation evolve from development programs of plant organizations.

ECONOMICS OF PROCESS INSTRUMENTATION

Financial justification for an investment in process instrumentation must be calculated on the same basis used for other equipment in a process facility. Instrumentation usually involves a proportionately larger increment of engineering design cost because of the extensive analysis and relatively large amount of detail associated with the intelligent specification of even the lowest-cost instrument item. In addition, cost of installation is high and instrumentation represents an increasingly larger percentage of the equipment investment. All such factors emphasize the need for close scrutiny into the essentiality of every element in the instrumentation system.

Process operations are, in a sense, "blind." Thus, most of the instrumentation investment is justified as necessary for safe, effective operation. Another segment of the investment can be considered an essential part of certain items of operating equipment. The economic justification for these is calculated as a part of the need for the operating unit itself.

The small remainder of the total investment in instrumentation represents choices between automatic vs. nonautomatic operation or choices between approaches to automatic operation. Such choices are a major challenge to the imagination of the engineer confronted with the job of calculating relative returns. There are many cases in chemical work, for example, where a $5,000 quality-control instrument can be clearly credited with reductions in operating costs exceeding $100,000 per year. These are the exception rather than the rule; however, a formal procedure should be set up for evaluating the merit of the more common instrument installations.

Most instrument installations involving economics calculations are justified as tools that reduce direct manufacturing costs. To evaluate such savings, compare the cost without the proposed instrumentation. Consider improvements in operating practice that might be expected solely from the process or equipment changes throughout the period (usually 1 year) covered by the calculation. Anticipated changes in equipment must also be considered for an equitable basis of comparison.

Instrumentation savings are usually calculated on an incremental-cost basis rather than a total-cost basis; i.e., the costs are shown as "incremental operating costs" for the process portion involving the saving, rather than being related to "total operating cost" for the entire process. This reduces the size of the numbers being handled and simplifies the calculation. Generally, each element of the savings calculation is tabulated on an annual-cost basis, using the first year's operation after start-up of the new

* McGraw-Hill Book Company, Inc., New York, 1957.

equipment for the comparison. Typical of the items in such tabulations are the following, which must be shown for both the "present" and "proposed" facilities:

1. Total capacity in pounds.
2. Estimated shipments in pounds.
3. Shipments as per cent of capacity.
4. Unit operating cost per pound.
5. Depreciation rates for:
   a. Direct manufacturing facilities.
   b. Power, general, and service facilities.
   c. Over-all rate.
6. Operating cost:
   a. Labor and other employment costs:
      The savings in operating and maintenance labor are determined by estimating the hours reduction in labor to be obtained by the proposal and then multiplying by the corrected wage rates (dollars per hour = base rate dollar per hour multiplied by the "other employment cost factor"). This takes into account such items as vacation pay, pension, social security, insurance, hospitalization, and shift differential.
   b. Material:
      This includes process material and raw material plus materials for repairs and maintenance. Costs must be predicted, taking into consideration market trends and possible cost advances. Freight and delivery charges are included as are, of course, the cost of processing to the point under consideration.
   c. Depreciation:
      This is calculated on the basis of over-all equipment deterioration, taking into account physical replacement of the entire facility because of wear and/or replacement of the facility as the result of technological obsolescence.
   d. Other operating costs:
      Here are consolidated other costs such as repairs, replacement, power, supervision, overhead, and miscellaneous supplies. *Repairs* includes labor and material. *Replacement* covers items that are replaced periodically because of severe operating conditions or abnormal wear. The term *power* is used to include electricity, steam, water, air, inert gas, vacuum, and fuel. Incremental costs are used unless the change in consumption resulting from the proposal is of such magnitude that powerhouse labor or facilities would be affected. *Overhead* includes such items as salaries and wages of service personnel, cost of operating and maintaining service facilities, plant administration, plant engineering, departmental management, and administration. It is important to investigate the effects of proposed instrumentation on clerical and other indirect work.
7. Total operating cost.
8. Savings in operating cost:
   This is the difference between the operating costs shown for the present and the proposed basis.
9. Less Federal tax on income:
   The rate used in this calculation is applied from accounting records for the period involved. If the facilities are producing material for governmental consumption renegotiation of this expense may be involved.
10. Net savings.
11. Investment:
    a. Project expenditures:
       This is the new money to be spent for permanent facilities. It does not include working capital or any allocated facilities.
    b. Allocated facilities:
       The proposed instrumentation may affect a portion of the available capacity of the existing power or service facilities; thus it is necessary to earn a return only on a fraction of the investment that is consumed as a result of the installation.

c. Working capital:
In addition to the money represented by physical facilities, a certain amount is represented by inventories and cash on hand.   Savings through reductions in working capital are often possible as a result of automatic control.
It should be recognized that a reduction in investment will have a powerful influence on the criterion "per cent return on investment" because it is not subject to the dilution suffered by "operating costs," savings subject to Federal income tax, and other factors.

12. Return on investment:
The percentage represented by the net annual savings to be expected from the proposed investment is the criterion that will be used for evaluation of the proposal.   Different companies have different requirements for the percentage return necessary to obtain authorization for a proposed installation.   This may vary depending on the nature of the savings (whether in labor or materials), the nature of the product and its market, the competitive position of the company, and other factors.
Instrumentation installed on a savings basis will usually affect chemical-plant operation in one of the following ways:
   (1) Reduces equipment investment by the elimination of intermediate storage, excess capacity, standby machinery, unnecessary purification steps, etc.
   (2) Makes possible an operation that would otherwise be inoperable.
   (3) Reduces working capital by speeding production adjustments and by reducing inventory.
   (4) Reduces cost of materials by minimizing waste and increasing yields.
   (5) Maintains product quality at minimum acceptable level.
   (6) Readjusts operating conditions to suit variations in raw-material quality.
   (7) Reduces operating costs by increasing labor efficiency and simplifying operations.
   (8) Increases continuity of operation at lowered maintenance.
   (9) Conserves utilities and supplies consumption.
   (10) Increases production rates per dollar investment in plant equipment.
   (11) Revitalizes obsolescent operation through modernization.

More than half of the instruments used in chemical plants today did not even exist 10 years ago.   In the present picture they are controlling the composition of process streams and the quality of product.   The next step will be to keep the plant at its economic optimum, which is frequently different from the engineering optima.   In the United States currently more than half of instruments bought are for temperature and pressure.   Process-stream analyzers account for less than 10 per cent, although this is rising rapidly.   Flow meters take 10 per cent, weighing and proportioning instruments 20 per cent, with the remainder going for liquid-level and miscellaneous electrical instruments.   Much of the total instrumentation investment in the process industries is for automatic inspectors and special forms of instrumentation developed in the laboratories of the user.   An increasing number of the products developed in these laboratories are licensed for commercial manufacture.   In fact, some users' development programs have been so prolific that they have entered the instrument manufacturing business themselves.

REFERENCES

1. Carroll, Grady C.: "Industrial Instrument Servicing Handbook," McGraw-Hill Book Company, Inc., New York, 1960.
2. Raven, Francis H.: "Automatic Control Engineering," McGraw-Hill Book Company, Inc., New York, 1961.
3. Annett, F. A.: "Practical Industrial Electronics," McGraw-Hill Book Company, Inc., New York, 1952.
4. Miller, Robert E.: "Maintenance Manual of Electronic Control," McGraw-Hill Book Company, Inc., New York, 1949.
5. Perry, R. H., C. H. Chilton, and S. D. Kirkpatrick (eds.): "Chemical Engineers' Handbook," 4th ed., Sec. 22, McGraw-Hill Book Company, Inc., New York, 1963.
6. Considine, D. M.: "Process Instruments and Controls Handbook," McGraw-Hill Book Company, Inc., New York, 1957.

7. Nuneviller, E. D.: Proper Maintenance of Pyrometers Keeps Production Rolling, *Instrumentation*, vol. 1, no. 1, December, 1943.
8. Cusick, Charles F.: The Maintenance of Orifice Type Flow Meters, *Instrumentation*, vol. 1, no. 3, September–October, 1944.
9. Krieg, Alfred: Oil Refinery Instrument Maintenance, *Instrumentation*, vol. 1, no. 6, Part I, October–November, 1945; Part II, vol. 2, no. 1, January–February, 1946.
10. Theissen, W. A: Keeping Them on Stream, *Instrumentation*, vol. 6, no. 5, Second Quarter, 1953.
11. Baker, W. G. Jr.: Making Better Instrument Engineers, *Instrumentation*, vol. 6, no. 6, Third Quarter, 1953.
12. Comstock, C. S.: Instrument Engineering in a Large Chemical Plant, *Instrumentation*, vol. 3, no. 1, Fourth Quarter, 1947.
13. Briggs, Rollin H.: Instrument Department Organization in a Large Chemical Plant, *Instrumentation*, vol. 3, no. 2, First Quarter, 1948.
14. Suber, W. J.: Time-saving Hints for Flow Meter Maintenance, *Instrumentation*, vol. 3, no. 4, Third Quarter, 1948.
15. Staff: Modern Methods for Testing and Inspecting Precision Instruments, *Instrumentation*, vol. 3, no. 5, Fourth Quarter, 1948.
16. Clift, T. L.: Instrument Shop Test Panel, *Instrumentation*, vol. 4, no. 4, First Quarter, 1950.
17. Church, R. J.: Training Tomorrow's Experts, *Instrumentation*, vol. 11, no. 2, March–April, 1958.
18. Staff: How Armco Trains Instrument Men, *Instrumentation*, vol. 9, no. 2, March–April, 1956.
19. Webb, Ralph: Instruments and the Man, *Instrumentation*, vol. 7, no. 3, Second Quarter, 1954.
20. Wright, Joseph F.: Maintenance of Diaphragm Motor Control Valves, *Instrumentation*, vol. 7, no. 6, First Quarter, 1953.
21. DeCourcy, W. D.: A Chemical Plant Program for Instrument Personnel Training, *Inst. & Automation*, vol. 28, no. 8, pp. 1327–1329, August, 1955.

Instrument Society of America-Recommended Practices

Available from the Society, 530 William Penn Place, Pittsburgh 19, Pa.

RP 1.1-7 Thermocouples and Thermocouple Extension Wires, 1959.
RP 2.1   Manometer Tables, rev. 1962.
RP 3.1   Flowmeter Installations Seal and Condensate Chambers, 1960.
RP 3.2   Flange Mounted Sharp Edged Orifice Plates for Flow Measurement, 1960.
RP 4.1   Uniform Face to Face Dimensions for Flanged Control Valve Bodies, 1950.
RP 4.2   Standard Control Valve Manifold Designs, 1956.
RP 5.1   Instrumentation Flow Plan Symbols, 1949.
RP 7.1   Pneumatic Control Circuit Pressure Test, 1956.
RP 7.2   Color Code for Panel Tubing, 1957.
RP 11.1   Mercury Handling, 1956.
RP 12.1   Electrical Instruments in Hazardous Atmospheres, 1960.
RP 12.4   Instrument Purging for Reduction of Hazardous Area Classification, 1960.
RP 16.1,2,3   Terminology, Dimensions and Safety Practices for Indicating Variable Area Meters, 1959.
RP 16.4   Nomenclature and Terminology for Extension Type Variable Area Meters, 1960.
RP 16.5   Installation, Operation, Maintenance Instructions for Glass Tube Variable Area Meters, 1961.
RP 16.6   Methods and Equipment for Calibration of Variable Area Meters, 1961.
RP 20.1   Specifications Forms for Instruments, Gages, Thermocouples, Orifice Plates and Flanges, Control Valves, and Pressure Safety Valves, 1956.
RP 23.1   Miniature Recorder Chart Ranges, 1961.
RP 25.1   Materials for Instruments in Radiation Service, 1957.
RP 26.1   Dynamic Response Testing of Process Control Instrumentation, Part I—General Recommendations, 1957.
RP 26.2   Dynamic Response Testing of Process Control Instrumentation, Part II—Devices with Pneumatic Output Signals, 1960.
RP 26.3   Dynamic Response Testing of Process Control Instrumentation, Part III, Devices with Electric Output Signals, 1960.
RP 26.4   Dynamic Response Testing of Process Control Instrumentation, Part IV, Closed Loop Actuators for Final Control Elements, 1961.
RP 31.1   Terminology and Specifications for Turbine-type Flow Transducers (Volumetric), 1961.

AUTOMATIC CONTROL OF HEAT EXCHANGERS

By Carl W. Sanders*

This subsection covers application engineering information on automatic temperature control of heat exchangers. Particular emphasis is placed on problems of concern when the performance of a heat-exchange system must be improved or control equipment must be selected for a new heat-exchanger installation.

This subject has been treated more quantitatively by others, and reference to these sources[1-4] should be made if a more rigorous treatment of a particular application is required. In Ref. 4, pages 11-4 through 11-16 outline the fundamentals of automatic process control with the heat-exchange operation as an example.

NORMAL CONTROL FOR HEAT EXCHANGERS

The normal method for controlling a heat exchanger is to measure the exit temperature of the fluid being processed and adjust the input of the heating or cooling medium to hold the desired temperature (see Fig. 1). To stabilize this feedback control, in almost all cases, the controller must have a wide proportional band. The band is

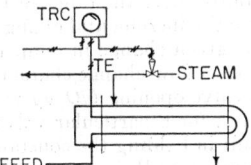

Fig. 1. Conventional heat-exchanger control system.

determined by the gain of other components in the control loop as well as process considerations, and it is an unusual case that the common combination of conventional control elements will permit the use of narrow-band control mechanisms.

Since this type of equipment requires a wide proportional band for stabilization, reset response will normally be required to correct for offset in the controlled variable if load changes are present. This response can be eliminated in cases where disturbances (such as steam header pressure, product flow rate, or inlet temperature changes) have effects which are small in relation to the desired tolerance on the controlled variable.

When the throughput to a heat exchanger is changed rapidly, a short-term error in control temperature will result. The magnitude and/or duration of this error can normally be reduced by a factor of 2 by adding derivative response to the control mechanism and adjusting it properly.[4]

Faulty thermal element or valve specification will materially detract from system performance and will exhibit symptoms that may lead the engineer to make more costly, unnecessary design changes.

* Consulting Engineer in Instrumentation, Engineering Services Division, Engineering Department, E. I. du Pont de Nemours & Company, Inc., Wilmington, Del.

THERMAL-ELEMENT INSTALLATION

Thermal elements in a heat-exchanger installation should be installed as close as possible to the active heat-exchange surface, consistent with the requirements for adequate mixing of the process stream.   As the point of measurement is moved from the active heat-exchange surface (for example, by installing the thermal element several feet downstream in the process pipeline), a time delay is introduced.   This time delay or distance velocity lag has a particularly noticeable effect on control performance since a distance velocity lag rapidly introduces phase shift with little signal attenuation.   A 5-ft length of pipe with a line velocity of 5 fps will produce a distance velocity lag of 1 sec, which can be one of the dominant time constants limiting the dynamic performance of the heat-exchanger control system.

The time response of a thermal element installed with a thermal well is materially slower than when the thermal element is installed bare.   Thermal wells should be omitted where control-system performance is critical and it is also practical to omit them from a maintenance or safety standpoint.   When a thermal well is used, particular care must be taken to assure that the effect of air gap between thermal element and thermal well is minimized by proper installation of conducting sleeves.

CONTROL-VALVE REQUIREMENTS

Valves for the normal heat-exchanger application should be provided with positioners.   Valve friction materially detracts from system performance.   If there is a question on performance of a new installation or if performance requires improvement in an existing installation, it is generally desirable to use or add a valve positioner as a means of eliminating the effects of valve friction.

It is desirable to maintain the same relationship between incremental changes in valve opening and changes in measured temperature for various throughput rates so that the gain of the control system remains reasonably constant.   Satisfying this condition will result in better control over the range of throughput rates.   When the throughput rate is doubled, twice the incremental change in flow of heating or cooling medium is required to produce a given change in temperature.

Equal-percentage valve trim has a flow-change characteristic that is proportional to total flow for a given change in valve opening ($dQ/dy = aQ$, where Q is flow rate, y is valve opening, and a is a constant for a particular valve).   Use of equal-percentage trim will therefore normally result in holding the constant relationship between valve movement and temperature change (see Ref. 4, pages 10-79 to 10-91).

Changes in pressure drop across valves used in steam-heating service must be considered when the over-all valve characteristic is determined since the back pressure from the heat exchanger normally will change over a significant range for various load conditions.   Frequently, a low supply pressure in relation to the maximum operating pressure of the heater must be used as a result of sizing the heat exchanger for minimum heat-exchange area.   When this requirement must be met, difficulty in stabilizing the system at low operating rates may be encountered.   Equal-percentage trim should be used and sized for the maximum allowable pressure drop at full load.

The following example illustrates the importance of sizing and selection to accommodate the wide rangeability that may be encountered:

Heater load at full rate = 1,000 lb/hr of steam
Heater shell pressure at full rate = 95 psig
Heater load at low rate = 200 lb/hr of steam
Heater shell pressure at low rate = 50 psig
Available pressure at the valve at full load = 100 psig

Referring to sizing equations (see Ref. 4, page 10-73), size C_v for 1,250 lb/hr of 100 psig saturated steam at 5 psi drop.

$$C_v = \frac{WK}{3\sqrt{(\Delta P)P_2}}$$

$$C_v = \frac{1,250}{3\sqrt{5(95 + 14.7)}} = 41$$

Compute maximum flow this valve could pass at low-load pressure drop of (100 − 50) = 50 psi

$$W = \frac{3C_v \sqrt{(\Delta P)P_2}}{K}$$

$$W = 3(41) \sqrt{(50)(50 + 14.7)} = 7,000 \text{ lb/hr}$$

The required low-rate flow is 200 lb/hr, which is 100(200/7,000) = 3 per cent of the flow rate which this valve *could* pass at 50 psi drop if wide open.

This 3 per cent would require a 10 per cent opening of an equal-percentage valve having a 2 per cent clearance flow (see Ref. 4, page 10-81, Fig. 7). This compares with 60 per cent opening of the valve if the pressure drop at low load were held to 5 psi. Thus the gain of this valve would depart significantly from that desired. Another unwanted effect occurs because at low flow the valve would be operating near the clearance flow point where the slope of the flow vs. opening changes abruptly in an adverse direction. Best control at the low operating load could be obtained if a separate control valve were used, sized to preserve an effective system gain.

SPECIAL CASES OF HEAT-EXCHANGER CONTROL

Condensate Outlet Control. In some cases, investment in control equipment for steam-heated shell-and-tube exchangers can be reduced by placing the control valve in the condensate line rather than in the steam line (see Fig. 2). This type of control is not so responsive as the systems previously described and, in general, should not be used on critical process applications unless tests can be made to confirm operability of the system. It should be employed only where a special inherent advantage is present. The behavior of this type of control system is quite difficult to predict.

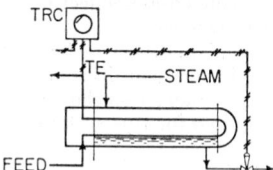

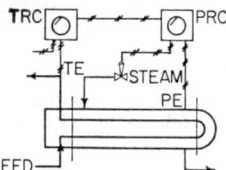

Fig. 2. Condensate-throttling control system.   Fig. 3. Pressure-cascade control system.

Pressure-cascade Control System. This system (see Fig. 3) cascades the output of a standard three-mode temperature controller into the set point of a pressure controller. It achieves a more rapid recovery to load disturbances in a shell-and-tube heat exchanger than can be obtained without the pressure controller. Steam to the heater is regulated by the pressure controller, which is normally provided with proportional and reset responses. A load change is rapidly sensed by a change in shell pressure which is compensated by the pressure controller. The temperature-control system senses the residual error and resets the pressure-control set point.

Bypass Control System. In certain cases, the time-response characteristic of a heat exchanger is too slow to hold temperature deviations resulting from load changes within desired tolerances. In some of these cases, the transient characteristics of the heat exchanger can be circumvented through bypassing the heater with a parallel line and blending cold process fluid with hot fluid from the heater (see Fig. 4).

Care must be taken in sizing the values to obtain the desired flow split with adequate flow vs. travel characteristic. Thermal-element response time is particularly important since this time constant will be a major factor influencing the performance of the system.

Recycled-jacket-water Control System. In certain cases of liquid-to-liquid heat exchangers, particularly of the plate type, where the heavy metal sections increase the thermal capacity on the supply side of the heat exchanger, the heating or cooling fluid is recirculated through the heat exchanger and is externally heated or cooled by a separate system or may be blended from a make-up supply (see Fig. 5).

A change in process-liquid outlet temperature causes a temperature controller to change the set point of a second temperature controller installed in the recirculation system, providing the necessary change in heat-exchange rate to return the process liquid temperature to the desired value. This cascaded control system normally will control exit temperature within closer tolerances than the single control-loop system.

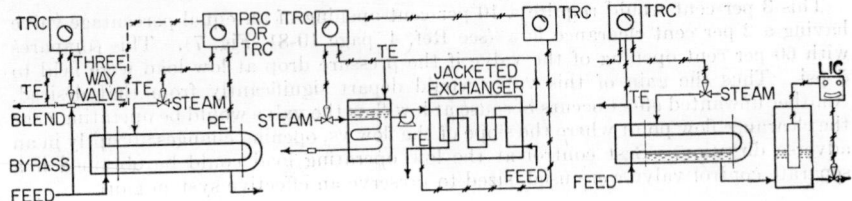

FIG. 4. Bypass control system.   FIG. 5. Recycled-jacket-water system.   FIG. 6. System for raising shell operating pressure.

HEAT-EXCHANGER TRAPPING AND CONDENSATE SYSTEMS*

Care must be taken to assure that condensate-removal systems for steam-heated exchangers are adequately designed for the particular installation. Traps should be selected which will not intermittently introduce disturbances of sufficient magnitude to upset the shell-side pressure as a result of rapid condensate dumping. Disturbances of this nature will cause upsets in the temperature of the controlled process stream.

Steam heat-exchanger traps should discharge into condensate-collecting systems held at pressures below the lowest shell-side operating pressure.

In situations where the minimum condensing pressure might otherwise be lower than the condensate system pressure, the surface of the exchanger can, in some cases, be reduced by holding a condensate level within the heat exchanger as shown in Fig. 6. An external level tank with level control would be required for this system.

REFERENCES

1. Cohen, W. C., and E. F. Johnson: Prediction of the Dynamic Characteristics of Distributed Process Streams; Steam-water Double Pipe Heat Exchanger, *Ind. Eng. Chem.*, vol. 48, pp. 1031–1034, 1956.
2. Mozley, J. M.: Prediction of the Dynamics of a Concentric Pipe Heat Exchanger, *Ind. Eng. Chem.*, vol. 48, pp. 1035–1041, 1956.
3. Paynter, H. M., and Yasundo Takabashi: A New Method of Evaluating Dynamic Response of Counterflow and Parallel-flow Heat Exchangers, *Trans. ASME*, vol. 78, pp. 749–758, 1956.
4. Considine, D. M.: "Process Instruments and Controls Handbook," McGraw-Hill Book Company, Inc., New York, 1957.
5. Morris, H. J.: The Dynamic Response of Shell and Tube Heat Exchanges to Temperature Disturbances, *A.I.Ch.E. Symposium*, St. Paul, Minn., September, 1959.
6. Solheim, O. A.: Guide to Controlling Continuous-flow Chemical Reactors, *Control Engineering*, vol. 7, no. 4, p. 107, April, 1960.
7. Orcutt, J. C., and D. E. Lamb: Stability of a Fixed-bed Catalytic Reactor System with Feed-effluent Heat Exchange, *A.I.Ch.E. Symposium*, St. Paul, Minn., September, 1959.
8. Douglas, J. M., J. C. Orcutt, and P. W. Berthiaums: Design and Control of Feed Effluent Exchanger-Reactor Systems, *A.I.Ch.E. Symposium*, New York University, June, 1962.

* *Acknowledgment:* The assistance of M. J. DePasquale of the M. W. Kellogg Company is gratefully acknowledged.

FILTRATION INSTRUMENTATION

By H. V. Miles*

Basically, the instrumentation of filtering operations is quite simple when compared with several of the other chemical unit operations. Filtration, which may be defined as an operation for the removal of solids from liquids, requires—at a minimum—the maintenance of a pressure differential across the solids-retaining medium. When this basic condition is established, filtration will proceed continuously as long as there is capacity or a discharge means for the solids collected and provided that the driving force (pressure differential) is adequate to assure flow through the unit.

The types of instrumentation problem encountered in connection with filtering operations can best be presented by describing a number of actual installations.

WHITE-WATER FIBER RECOVERY (SAVE-ALL)

In the recovery of fiber from the white water of a Fourdrinier wire paper machine, it is common to employ a rotating drum or disk filter, often referred to as a *save-all*. In

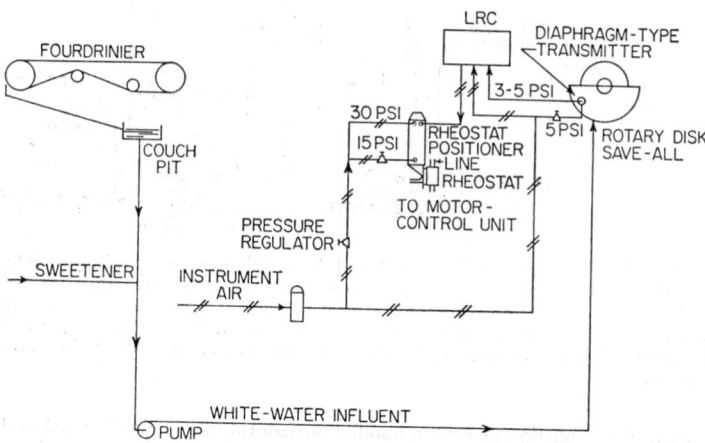

Fig. 1. Disk speed control for save-all unit in white-water recovery.

order to handle the flow from the couch pit and maintain the desired operating level in the filter vat, it is customary to employ a control system that will increase or decrease the disk speed with corresponding changes in the liquid level in the filter tank. A typical arrangement is shown in Fig. 1. Level is sensed by a diaphragm-type transmitter that signals to an indicating proportional-plus-reset controller. This controller actuates a pneumatically operated rheostat in the motor control circuit to provide the desired speed regulation of the disk.

* Manager, Pulp and Paper Development Division, Dorr-Oliver Incorporated, Stamford, Conn.

NITROCELLULOSE DEHYDRATION

In the production of nitrocellulose, the fiber-water slurry is dewatered to obtain the product in dry form. The dewatered fibers must be mildly washed with ethyl alcohol to provide a surface film that will minimize the explosion hazard. This is accomplished on a horizontal vacuum filter by first dewatering the nitrocellulose to a low moisture content (3 per cent) and following with the introduction of a stream of hot air to reduce the moisture content further. Alcohol is then applied in just sufficient quantity to wet the fibers slightly. The nitrocellulose then is discharged by a scroll device which lifts the cake over the outer dam of the filter.

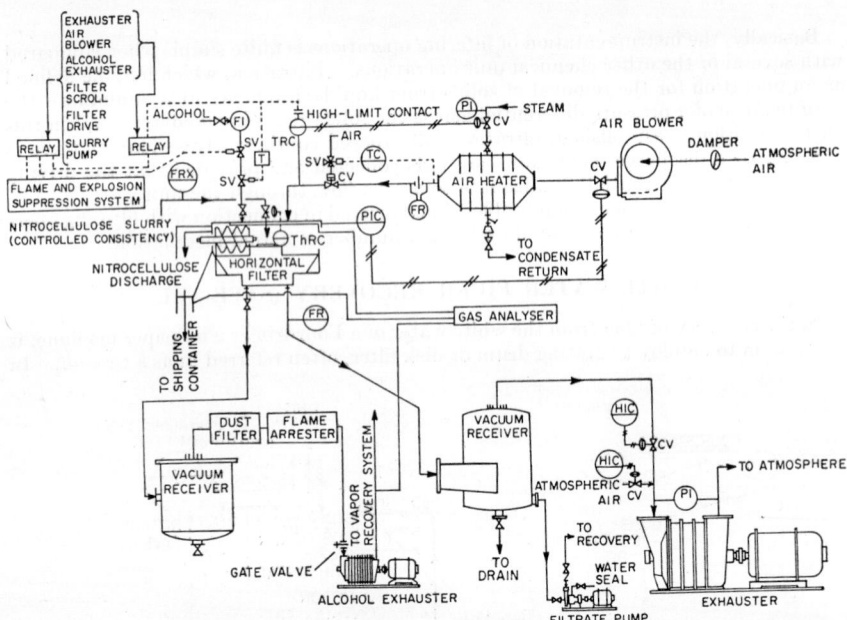

Fig. 2. Filtration instrumentation employed in continuous nitrocellulose dehydration system. (FI) Flow indicator; (FR) flow recorder; (FRX) flow recorder totalizer; (PI) pressure indicator; (PIC) pressure indicator controller; (HIC) remote valve operator; (TR) temperature recorder; (TRC) temperature recorder controller; (ThRC) cake-thickness recorder controller; (TC) temperature controller; (SV) solenoid valve; (CV) control valve.

Because of the explosion hazard, a hooded horizontal filter must be used and instrumented as shown in Fig. 2. As will be noted, the alcohol flow is measured and safety shutoff solenoid valves are provided to operate in connection with the flame- and explosion-suppression system. The hot-air flow for drying is controlled as well as the temperature with a safety shutoff in event of excessive temperature. A cake-thickness recorder regulates the rate of slurry feed from measurements of the thickness of the cake on the horizontal filter. Separate filtrate receivers are provided to collect the alcohol (if present) and water. In the alcohol vapor stream, a gas analyzer will automatically indicate when an explosive mixture is reached. The aim of this instrumentation is to provide for safe operation and essentially automatic operation of the process after start-up.

COAL FILTRATION

In the recovery of coal in a modern coal washery, the fines from various screening operations are concentrated in a thickener of the Dorr type for subsequent dewatering on a rotary-disk filter. A system of this type is shown in Fig. 3.

Unless the per cent solids in the thickener underflow to the filter is maintained at the proper point (usually about 40 per cent by weight), filter performance is adversely affected. When the solids content decreases, poor cake formation, along with loss of vacuum and poor cake discharge, may result. With high feed solids, cake pickup may be difficult, or the thicker cakes may have higher than the desired moisture content.

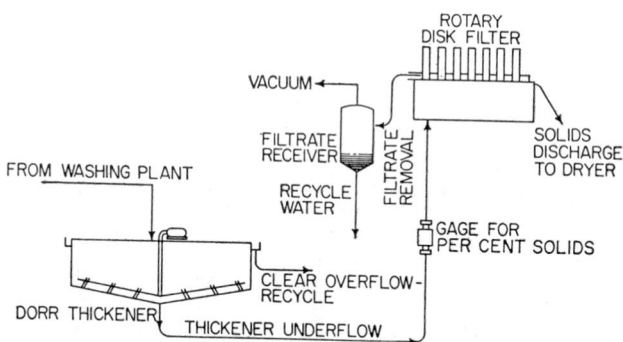

FIG. 3. Instrumentation of rotary-disk filter for coal-fines recovery.

To avoid these difficulties and to maintain a uniform feed to the coal dryer, it is desirable to control feed solids within a narrow range. Use of a density gage of the radioactive type (gamma radiation) to sense changes in slurry specific gravity provides a satisfactory solution to this problem. Based upon this measurement, the operators can maintain a uniform feed to the filters by appropriately governing thickener operation as required. High- and low-limit switches can be installed on the chart recorder to actuate an alarm or other signal device.

BROWNSTOCK WASHING

In the production of pulp by the kraft process, the black liquor must be separated from the pulp following digestion (1) to recover the soda for subsequent recausticizing to avoid chemical loss and (2) to free the pulp from the black liquor so that it will be suitable for subsequent processing (screening, bleaching, and refining). The brownstock-washing operation generally is accomplished on a three-stage rotary vacuum filter system as shown in Fig. 4. This instrumentation is typical for manual operation. There is a trend toward increased instrumentation and consequent automatic control over certain functions, but considerable developments remain to make such a completely automatic system practical and economical.

With reference to Fig. 4, pulp is discharged into the blow tank following digestion. A level recorder (LR) in the blow tank indicates to the digester room when the next cook can be discharged. Some form of consistency instrumentation also is usually employed. The variation in consistency is indicated by measuring the work load on the agitator in the blow tank. Generally, this is done with a thermal converter (CsRC) that measures the current load and, at the same time, compensates for voltage changes. As contrasted with an ammeter, the thermal converter indicates the true motor load.

Stock flow from the blow tank is controlled automatically by an indicating and/or recording flow meter (FRC). Dilution of the stock flow with filtrate from the num-

ber 1 washer is controlled by a similar instrument (FRC). The wash flow for all
filters is recorded (FR's). This is a countercurrent wash applied as follows:

Filter stage	Source of wash
1	No. 2 filtrate
2	No. 3 filtrate
3	Fresh water

The shower flow can be controlled automatically from the liquor level in the service
tanks if desired. A temperature recorder (TR) is used on the fresh-water wash for
the number 3 washer. The other two wash liquors are pumped from the filtrate or
service tanks where the air and foam are separated from the diluted black liquor.

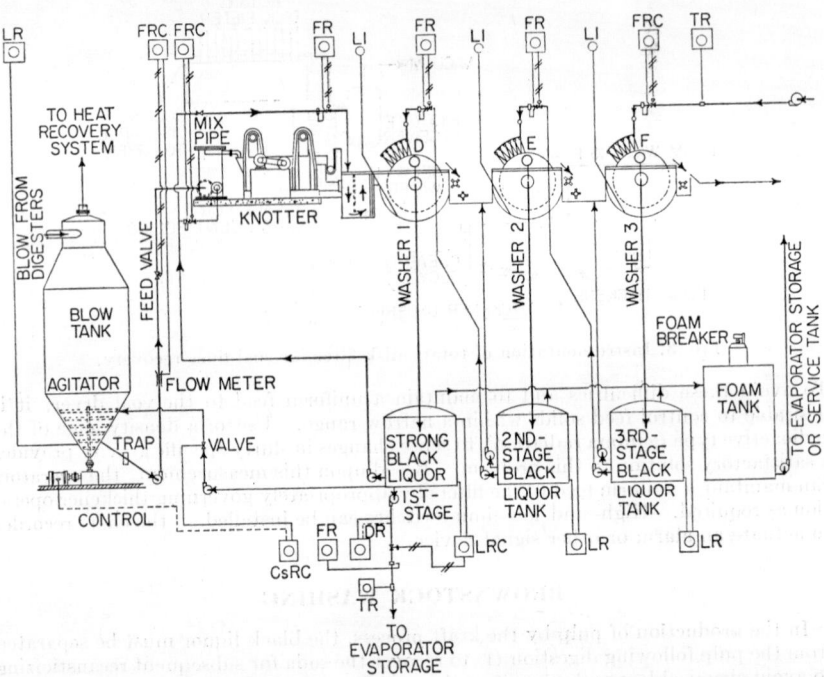

Fig. 4. Brownstock-washing control system. (LR) Level recorder; (FR) flow recorder:
(TR) temperature recorder; (DR) density recorder; (LI) level indicator; (FRC) flow
recorder controller; (LRC) level recorder controller; (CsRC) consistency recorder controller.

Level recorders (LR's) are used on each service tank. The first filtrate represents
the strong liquor that is employed for initial stock dilution and also feed to the
evaporators. A flow recorder (FR) is used on the latter, along with a density recorder
(DR) and a temperature recorder (TR). These measurements provide important
guidance to the evaporator and brownstock-washing operators.

The level in each washer is indicated by level indicators (LI's) because level control
is essential to good washer performance. If not adequately controlled, level variations will affect sheet thickness and therefore washing efficiency.

Dilution of the partially washed, but dewatered stock from the first and second
stages is required to reduce the consistency to the proper level for feed to the subsequent stage. The filtrate from the number 2 and 3 service tanks provides the dilution liquor for the pulp discharge from the number 1 and 2 washers, respectively.
This dilution flow may be manually or automatically controlled in the repulper.

Additional controls sometimes are applied. Automatic control of the wash liquors has been attempted through the use of a conductivity meter on the third-stage filtrate as a control means for regulating the fresh-water flow to the third stage. The first- and second-stage shower flows are controlled from the liquid level in the respective filtrate tanks. Conductivity as a means of control is subject to the variances in the black-liquor composition, and the latter must therefore be taken into account. Successful use of this system requires careful correlation of actual operating results for various conductivity readings so that the optimum conductivity range can be determined.

INSTRUMENTATION OF CYCLE FILTERS

Cycle (or batch) filters are employed for either solution clarification or solids recovery (cake filtration). Clarifying filters often are operated with only a pressure gage to show the pressure increase as solids collect on the filter medium. When the pressure differential reaches a predetermined maximum limit, the flow is shut off and the filter medium is cleaned or replaced.

In cake filtration, a pressure gage may be sufficient to indicate how the filtration cycle is progressing. But the volume of cake held between the leaves or in the filter-press frame also is a limiting factor of operation even if pressure is not. When the cycle is completed, the cake is discharged and a new cycle started.

The removal of cake from a plate-and-frame filter press is a time-consuming task and labor costs therefore are high. One type of modified filter press has been developed to overcome this disadvantage. This device makes use of an alternate set of frames positioned on a common axis with those in service. When the cycle is completed, empty frames swing into place. The entire operation has been automated so that, on a signal from the controller, flows are stopped and the empty frames are moved into place automatically.

With leaf-type filters, the majority of applications require the use of a precoat filter aid such as diatomaceous earth, perlite, or wood fiber. The precoat serves as the filter medium and eventually is discharged with the solids at the end of the cycle. This necessitates a series of steps to ready the filter for the next filtration cycle. From the time the filter is removed from service, the following steps are involved for a dry-cake discharge. Where a wet-cake discharge is satisfactory, the first step indicated below can be omitted and a sluice substituted for step 2. The sluiced cake is discharged as a slurry.

Step 1. Air-blow cake to remove liquid.
Step 2. Discharge cake from leaves by vibrating, scraping, or reverse air blow.
Step 3. Discharge cake from collecting hopper in filter.
Step 4. Fill filter with clear liquid.
Step 5. Prepare filter aid slurry.
Step 6. Precoat leaves.
Step 7. Return filter to service.

Variations are found in the foregoing schedule depending upon the preferred manner of operation. However, a sequence of operations must take place that involves considerable manipulation, including valve positioning. Manual operation of this filter-servicing step is costly when required on short intervals (8 to 12 hr). Consequently, manufacturers have automated this sequence of operations. The necessary instruments (timers, flow indicators, and so on) have been incorporated with the required electric or pneumatic controls to operate all the various components and to position the valves. The end of the normal filtration cycle is signaled by one of several devices. A simple pressure-differential switch may be used, but there is no assurance that the cake may not completely fill the space between the leaves before the preset pressure is reached. To overcome this condition, a cake-thickness detector may be used. The latter may be set for the maximum cake thickness permissible. Accordingly, this device will remove the filter from service for cake discharge and precoating at the proper time. A rate-of-flow controller with a low-flow cutout can also be used to terminate the filter cycle.

Automatic leaf filters are installed in multiples to handle plant flows without interruption. This requires interlocking of controls so that only one filter is removed from service at a time, the others having sufficient capacity to handle the design flow.

Bed-type filters, of either the gravity or pressure type, are used for water filtration and must be backwashed periodically to remove the accumulation of solids. While most installations are operated manually, the trend is toward automatic operation, especially where a large battery of filters is used. The backwashing step is initiated automatically when the loss of head reaches some set point. The necessary valve changes are made and the rate of wash water is flow-controlled. The filter then is returned to service.

REFERENCES

1. Chambers, Ralph, Thomas Kuhn, Ted F. Meinhold, and J. V. Romano: Batch Filtration Goes 100 % Automatic, *Chem. Proc.*, June, 1961, pp. 66, 67, 70, 71.
2. Chalmers, J. M., L. R. Elledge, and H. F. Porter: Filters, *Chem. Eng.*, June, 1955.
3. R. H. Perry, C. H. Chilton, and S. D. Kirkpatrick (eds.): "Chemical Engineers' Handbook," 4th ed., pp. 19-54 to 19-85, McGraw-Hill Book Company, Inc., New York, 1963.
4. Dickey, G. D.: "Filtration," Reinhold Publishing Corporation, New York, 1961.
5. McCabe, W. L., and J. C. Smith: "Unit Operations of Chemical Engineering," pp. 324-353, McGraw-Hill Book Company, Inc., New York, 1956.
6. Foust, A. S., et al.: "Principles of Unit Operations," pp. 484-500, John Wiley & Sons, Inc., New York, 1960.
7. Cole, E. J., and M. Todd: "Pulp and Paper Mill Instrumentation," pp. 54, 61, 63, Lockwood Trade Journal Co., New York, 1957.

CRYSTALLIZER INSTRUMENTATION

By S. A. Laurich* and H. Svanoe†

Crystalline chemicals are produced under widely different operating conditions. The product may be crystallized at temperatures from several hundred degrees centigrade down to considerably below zero degrees. The viscosity of the solution also can vary over a wide range—from 0.5 cp to many hundred times higher than the viscosity of water.

VARIATION IN PRODUCT SPECIFICATIONS

The specifications for the crystalline solds will vary greatly. For food products and materials used in the food industries, purity usually is the main requirement. For fertilizer chemicals, uniform crystal size is important to ensure free-flowing and non-caking materials. For industrial products, the main requirements are high crystal quality, constant bulk density, freedom from dusting, and freedom from caking when stored. In numerous industries, crystalline chemicals often are produced as intermediates. For example, sodium bicarbonate is crystallized for soda-ash production; crystalline raw materials are made for synthetic fiber production.

Although most crystallization operations are on a continuous basis, the nature of the process will, in some cases, dictate batch operation. Equipment and associated instrumentation must be designed accordingly.

CRYSTALLIZATION DRIVING FORCE

To separate chemicals from solutions by crystallization, a driving force in the form of a supersaturated solution must be established. Both the formation of nuclei and the growth of the nuclei to desired dimensions are governed by the degree of supersaturation of the solute. For effective control of the crystallization process, supersaturation of the liquor must be controlled during the entire process. The number of nuclei formed and retained per unit time should be equal to the number of crystals that are withdrawn from the crystallizer as product per unit time.

If supersaturation could be controlled directly by instruments, this would be a simple, straightforward way to control the crystallization process and thereby determine the quality of the finished crystalline material. Unfortunately, less direct instrumental means must be used.

Degree of Supersaturation Required. A brief review of the degree of supersaturation that is used today in some industrial crystallization processes is in order.

Sugar Solutions. Following is a quotation from Webre,[1] who defines the various degrees of supersaturation:

1. "Existing crystals grow in the metastable region, but no new ones form. This zone persists to a supersaturation of about 1.2, this coefficient being the ratio of the actual concentration to the solubility, in terms of units of sugar per unit of water.

2. "Beyond the metastable range is the intermediate, or period of false grain, in which not only existing crystals grow but new ones also form. This zone extends from a supersaturation of about 1.2 to 1.3.

* Manager, Process Plants Division, Struthers Scientific & International Corporation, Warren, Pa.
† Consultant, Struthers Scientific & International Corporation, Warren Pa.

3. "Above this latter value is the labile zone, in which crystals form spontaneously without the presence of others."

Expressing sucrose supersaturation in per cent of solute present above the equilibrium value at the saturation point, the ratio 1.2 is equivalent to 20 per cent, and 1.3 to 30 per cent, respectively.

For sugar solutions, the allowable degree of supersaturation in crystallization practice is about 20 per cent. However, for most inorganic as well as numerous organic solutions, this allowable or workable degree of supersaturation is considerably lower and in the order of 0.5 to 1 per cent. This degree of supersolubility is so small that it cannot be measured directly by instruments normally used in commercial installations.

Even for sugar solutions, where the change in supersaturation during the crystallization process can be detected with instruments, it is only recently that such changes have been used to direct and control the process.

INSTRUMENTAL CONTROL OF SUPERSATURATION

Probably the first scientist to outline the basic principles for application of instruments to control supersaturation was Alfred L. Hoven, and the following quotation is taken from his work:[2] "The degree of coefficient of supersaturation is an important factor in controlling the evaporative processes by which sugar is recovered in crystalline form. There has been no satisfactory means for directly measuring or controlling the degree of supersaturation of boiling sugar solutions. A new method is developed here, based on the hitherto unrecognized fact that supersaturation may be calculated by a mathematical formula derived in these investigations. This formula is based on the discovery that, at all pressures encountered in usual sugar boiling practice, a plot of boiling points of sugar solutions of any degree of supersaturation against the corresponding boiling points of water at these same absolute pressures yields a straight line.

"The development of these graphs and their adaptation to an automatic instrument for continuously recording and controlling the degree of supersaturation is described in detail."

From this study, a sugar supersaturation recorder was developed. The following paragraph is quoted:[3] "The supersaturation recorder is an ElectroniK self-balancing potentiometer which computes supersaturation automatically and continuously from known purity and continuous measurements of the temperature of the boiling massecuite and the temperature of saturated steam under the same pressure."

The following paragraph is quoted:[4] "From the time the vacuum pan is nearly ready for seeding, to the time the grain establishes itself and grows to sufficient size, it is necessary to control the absolute pressure within the vacuum pan, since changes in absolute pressure cause changes in the temperature of the boiling liquor, and consequently, changes in supersaturation."

INDIRECT INSTRUMENTAL MEANS FOR CONTROLLING SUPERSATURATION

In most crystallizer operations, the allowable supersaturation is too small a quantity to be detected by instruments. The degree of supersaturation must therefore be controlled by indirect means. The accompanying diagrams show application of the indirect method.

Temperature Control. In general, the operating temperature of the crystallizer must be kept constant. This usually is done with an absolute-pressure controller which maintains a constant absolute pressure in the vapor section. Assuming that the composition of the solution will remain constant, this will result in a constant operating temperature. Control with an absolute-pressure instrument usually is accomplished by bleeding air into the vaporizer or ejector system to keep the ejectors loaded at a given absolute pressure.

Differential-pressure Measurement. In one type* of crystallizer, the crystals that have been produced are kept in suspension by the hydraulic action of the cir-

* Krystal crystallization system (Struthers Scientific & International Corp.).

culating solution. In most cases, the density of the circulating solution is constant because its temperature and composition are fixed. Therefore, by measuring the differential pressure between two points in the vessel that contains the crystals, the amount of crystals accumulated in that vessel can be measured. Any change in density between these two points is caused directly by the amount of crystal in suspension. This can be calculated by the formula

$$d_{av} = \frac{d_1 d_2}{W_1 d_2 + W_2 d_1}$$

(1)

where W_1 = weight fraction of crystals
W_2 = weight fraction of liquor
d_1 = density of crystals
d_2 = density of mother liquor
d_{av} = density of the suspension of crystals

INSTRUMENTATION OF EVAPORATOR-CRYSTALLIZER

An evaporator-crystallizer utilizing an absolute-pressure controller and a suspension recorder to maintain constant operating conditions is shown in Fig. 1. In addition, a

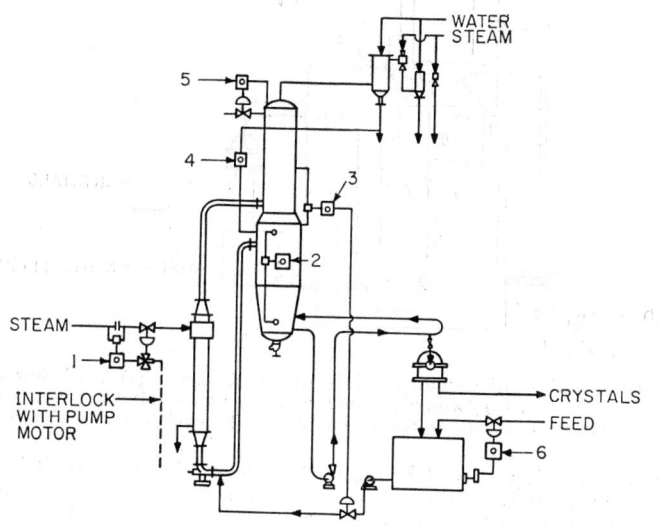

FIG. 1. Instrumentation of evaporator-crystallizer. (1) Steam-flow controller; (2) suspension recorder; (3) level controller, differential-pressure type; (4) temperature recorder; (5) absolute-pressure controller; (6) level controller, differential-pressure type.

differential-pressure-type level controller, with water-purged lines, maintains the proper level in the evaporating section of the crystallizer. Also, a steam-flow controller maintains the rate of steam flow to the heat exchanger. A temperature recorder may or may not be used, depending upon the degree of refinement desired. Since the absolute pressure is constant, the temperature in the crystallizer also will be constant. Therefore, it is only of academic interest to record the actual operating temperature. The second pen of the temperature recorder can be used to measure the temperature of the effluent water from the barometric condenser and to indicate to the operator the amount of water being used by the condenser.

With this flow arrangement, the production rate is a function of the steam flow to the heat exchanger. Once this is set, the production also is fixed, assuming a constant feed composition.

INSTRUMENTATION OF VACUUM-COOLING-TYPE CRYSTALLIZER

The control system for a vacuum-cooling-type crystallizer is shown in Fig. 2.   The only heat added to this type of crystallizer is the sensible heat in the feed and the heat of crystallization.   Here, the production rate is fixed by the operating temperature and the feed flow to the crystallizer.   In order to maintain constant conditions, the feed flow is controlled, and for any given steady feed rate the production will be fixed.

Ratio of Line Product to Mother Liquor.   For a given quantity of feed, a certain amount of crystal and mother liquor is produced.   This ratio of crystalline product to mother liquor usually is of some value other than the ratio of crystals to mother liquor in the slurry being removed from the crystallizer.   If more mother liquor is taken out with the crystals than the amount being produced from the feed,

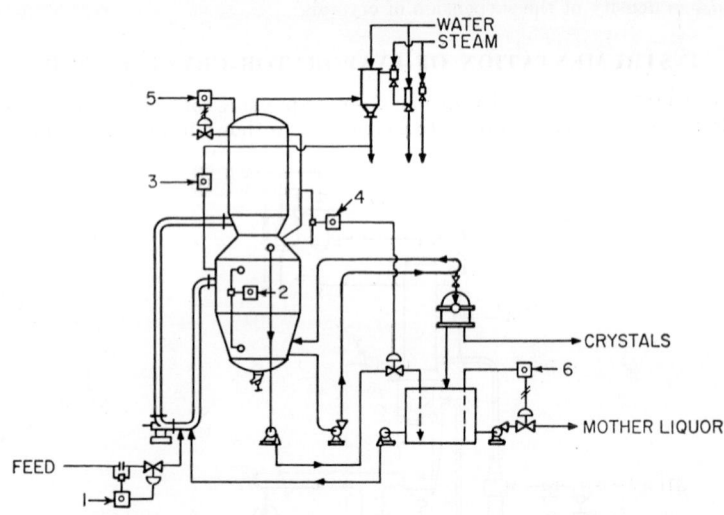

FIG. 2. Instrumentation of vacuum-cooling-type crystallizer.   (1) Feed flow controller; (2) suspension recorder; (3) two-point temperature recorder; (4) level controller, differential-pressure type; (5) absolute-pressure controller; (6) level controller, bubble type.

some of the mother liquor must be returned to the crystallizer after it is separated from the crystals.   It is usual practice to recycle more mother liquor than the calculated amount and allow the excess to overflow from the crystallizer through a control valve operated by the level controller.   This is a simple method for maintaining level in the crystallizer and allows removal of mother liquor during periods of operation when crystals are not being removed from the unit.

Concurrent Chemical Reactions.   It is also possible to carry out reactions in a crystallizer to produce the final product.   The most common example is the production of ammonium sulfate from sulfuric acid and anhydrous ammonia, and the production of diammonium phosphate from phosphoric acid and anhydrous ammonia.

DIAMMONIUM PHOSPHATE PRODUCTION SYSTEM

An arrangement for the production of diammonium phosphate from acid and ammonia is shown in Fig. 3.   Control of pH is very important in an operation of this kind.   Ammonia gas is sparged into a large amount of circulating mother liquor to neutralize phosphoric acid which is added to the system by means of a pH controller. Since the heat of reaction is removed by the evaporation of water which is in excess of the amount being supplied to the crystallizer with the phosphoric acid, additional

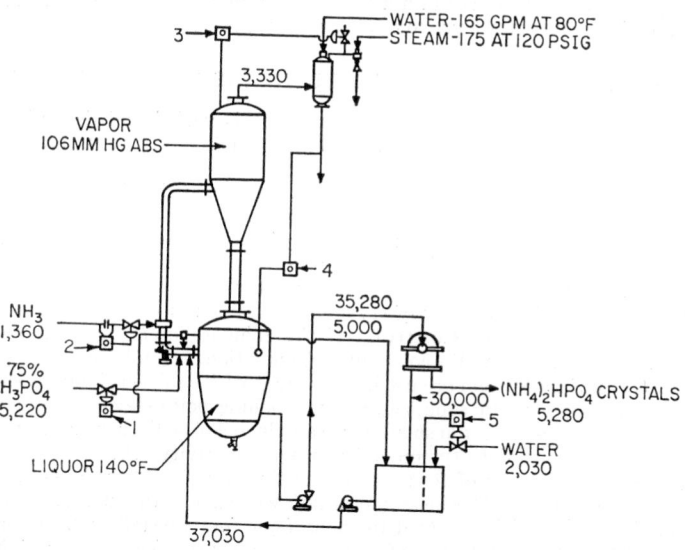

Fig. 3. Diammonium phosphate crystallization system. Quantities given in pounds per hour unless otherwise noted. (1) pH controller; (2) flow controller; (3) absolute-pressure controller; (4) temperature recorder; (5) level controller.

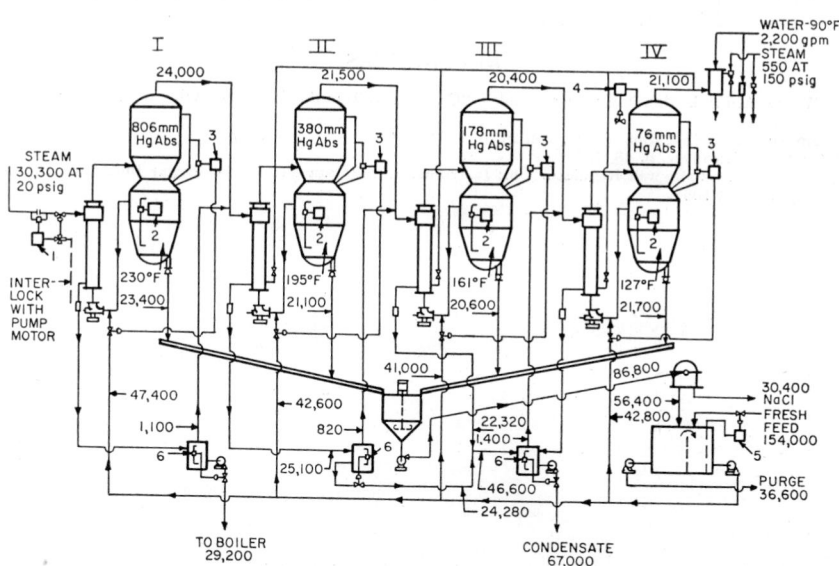

Fig. 4. Instrumentation of multiple-effect system (for production of sodium chloride). Quantities given in pounds per hour unless otherwise noted. (1) Steam-flow controller; (2) suspension recorder; (3) level controller, differential-pressure type; (4) absolute-pressure controller; (5) level controller, bubble type; (6) condensate-level controller.

water must be supplied to maintain a constant volume in the crystallization system. This is done by means of a level controller in the mother-liquor tank.

INSTRUMENTATION OF MULTIPLE-EFFECT SYSTEMS

The instrumentation on multiple-effect systems is varied and depends upon the material being crystallized and the desired end results. Figure 4 shows a general arrangement of this kind.

Level is controlled in each unit by a level controller, usually of the differential-pressure type, with water purge to prevent crystallization in the meter lines. Steam to the first-effect evaporator usually is fixed by means of a steam-flow controller, but this can be changed to vary with the feed depending upon the material entering the multiple-effect system. If crystals are produced in each of the crystallizers, the rate of removal of crystals is governed by density of crystals maintained in the crystallizer.

Boiling-point Elevation Control. If the effluent liquor from the evaporator system must be maintained at a constant concentration, several methods for controlling the flow out of the system are used. Solutions with sufficient boiling-point elevation, as, for example, caustic soda, can utilize a boiling-point elevation controller. If the boiling-point elevation is too small, for example, below 5°F, then some other means must be used, such as a density controller or a conductivity cell.

Feed Control. Supply of feed to the multiple-effect system depends upon the material being handled. The instrumentation will vary with the feed system used. In Fig. 4, the same feed is supplied to each of the four crystallizers. Often feed is supplied to the last effect and passed forward from effect to effect counter to the flow of vapor. Also, it may be supplied to the first effect so that the feed flow is parallel to the vapor flow. Or a combination of arrangements may be used.

Maximum Heat Utilization. A flash system to utilize the maximum amount of heat supplied to the system also is shown in Fig. 4. Hot condensate from the first effect is flashed to a lower pressure in the second effect, and so on—to the final effect. The only control required is a level controller which regulates the amount of condensate removed from the system.

REFERENCES

1. Webre, A. L.: Section in Spencer Meade's "Cane Sugar Handbook," John Wiley & Sons, Inc., New York, 1945.
2. Hoven, A. L.: *Ind. Eng. Chem.*, vol. 34, p. 1234, 1942.
3. Price, J. D.: *Sugar*, p. 48, 1954.
4. Minneapolis-Honeywell Regulator Co.: Instrumentation Data Sheet 3.8-4 (Sugar Industry), Philadelphia, Pa., 1958.

DISTILLATION-COLUMN CONTROL

By L. Bertrand* and J. B. Jones*

The purification of materials by distillation is a prominent operation of many chemical processes. The problems of distillation control relate directly to the *relative volatility* of the materials being separated. For example, water and ethylene glycol are separated easily because of the high volatility of water relative to ethylene glycol. Similarly, water and methanol are easily separated. Closer-boiling mixtures, such as the isomers of xylene, are more difficult to separate by distillation.

Whether a mixture is fundamentally easy or difficult to separate, the distillation equipment used usually is designed for a particular separation—with only modest margins of performance ability provided. Thus it is the responsibility of the distillation control system to safeguard the purity of distilled products. Although references are made in this section to the control of composition, temperature, pressure, and level, all these simply are instrumental methods for maintaining *purity of product*. Although some control systems regulate productivity and effect economies, it is difficult to suggest an example where composition of product is made subordinate to other control objectives.

DISTILLATION-COLUMN DESIGN

The more elusive aspects of distillation control relate to control of product composition and its responses to disturbances, such as feed rate, feed composition, or feed enthalpy—factors that originate external to the distillation column. The McCabe-Thiele † diagram originally was devised as an aid to the design of distillation columns. This device is helpful toward visualizing the effects of disturbances, several of which are discussed later in this section.

SIMPLE BINARY SEPARATION USED AS A BASE

The majority of distillation columns produce two product streams: (1) a stream rich in "low boilers," i.e., *distillate;* and (2) a stream rich in "high boilers," i.e., *bottoms.* In this sense, all distillation columns parallel in their operation that of a simple binary distillation. The concept of *key* component separation represents a means of treating a complex separation in terms of a simple binary separation.

DEGREES OF FREEDOM FOR CONTROL

It is important to recognize the degrees of freedom of choice in specifying the variables to be controlled. The choice open is the natural consequence of the column heat and material balance and is analyzed here.

A typical distillation column is shown in Fig. 1. The over-all material balance can be represented by the following equations:

$$F = W + D \tag{1}$$
$$Fx_F = Wx_W + Dx_D \tag{2}$$

* Engineering Service Division, Engineering Department, E. I. du Pont de Nemours & Company, Inc., Wilmington, Del.

† See R. H. Perry, C. H. Chilton, and S. D. Kirkpatrick (eds.), "Chemical Engineers' Handbook," 4th ed., McGraw-Hill Book Company, Inc., New York, 1963, or one of several good basic texts on the subject of chemical engineering.

where F = feed rate, lb-moles/unit of time
W = bottom product, lb-moles/unit of time
D = distillate, lb-moles/unit of time
x_F = mole fraction of low boiler in the feed
x_D = mole fraction of low boiler in the distillate
x_W = mole fraction of low boiler in the bottom product

The assumption is made that the molar heat capacities and the latent heats of vaporization of all components are the same. Also, it is assumed that (1) heat losses from the column are negligible, and (2) heats of mixing are negligible. In consequence, the upward vapor flow and the downward liquid flow in both the rectifying and stripping sections are invariant within the sections. Also, the accounting for the column heat balance is independent of the compositions of the product streams.

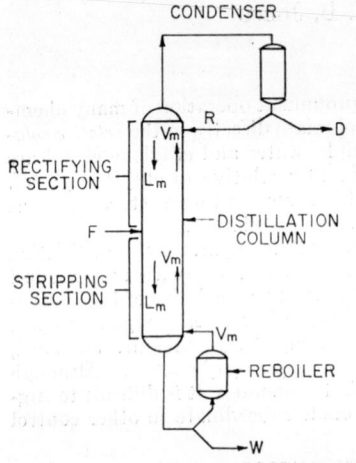

FIG. 1. Typical distillation column.

Internal Material Balance. With these qualifications, the internal material balance and the heat balance can be accounted for by the following equations:

$$L_n = (1 + b)R \tag{3}$$
$$V_n = D + (1 + b)R \tag{4}$$
$$L_m = L_n + qF \tag{5}$$
$$V_m = L_m - W \tag{6}$$

$$x_W = f\left(\frac{L_m}{V_m}\right) \tag{7}$$

$$x_D = g\left(\frac{L_n}{V_n}\right) \tag{8}$$

where V_n = vapor rate in rectifying section, lb-moles/unit of time
V_m = vapor rate in stripping section, lb-moles/unit of time
L_n = liquid rate in rectifying section, lb-moles/unit of time
D = distillate rate, lb-moles/unit of time
R = external reflux, lb-moles/unit of time
b = a numerical factor, depending upon the reflux enthalpy or temperature (*Note*: $b > 0$ whenever the reflux temperature is below that at the top of the column.)
q = a numerical factor, depending on the feed enthalpy whose value satisfies the following constraints
Constraints:
$q > 1$ when the feed temperature is below that of the feed plate
$q = 1$ when the feed temperature and composition are identical to those of the feed plate
$1 > q > 0$ when the feed enters the column partially vaporized
$q = 0$ when the feed is completely vaporized and is at saturated temperature
$q < 0$ when the feed is superheated vapor
f and g = factors which account for functional relationships which are dependent upon column-design criteria, such as (1) number of plates in the column, (2) location of control plates, (3) location of feed plate, and (4) temperature and other criteria specified for the control plates

Fixed Controls. From the standpoint of controls, the foregoing factors are fixed. For example, the automatic controls are unable to shift the injection of feed up or down the column in response to departures from optimum column operation. Nor is the location of the control plates ordinarily regarded as a variable. Even the control-plate temperatures usually are considered as fixed, although it is possible to regard these as variable and arrange feedback controls for their regulation.

The consequence of fixing all factors on which the functionals f and g depend is to fix the functionals in turn. Thus these items are excluded from the list of variables in the analysis which follows.

Free, Controllable, and Dependent Variables. Excluding f and g, Eqs. (1) through (8) include 13 variables. One possible classification of the 13 variables is given here.

Free Variables (usually not controllable). x_F, q, and b.

Controllable Variables (whose values usually are controlled). F and x_D, or (alternately) F and x_W.

Dependent Variables. W, D, V_n, V_m, L_n, L_m, and x_W, or (alternately) x_D.

Thus 5 of the 13 variables, that is, the free and controllable variables, are defined. This leaves eight unknowns in eight independent equations. There exists a unique solution for this set, and thereby the column performance is completely determined.

POSSIBLE MODES OF COLUMN CONTROL

It is necessary to assign values to five of the 13 variables in order to reduce the system to eight equations in eight unknowns. Usually some of the five variables represent conditions imposed by preceding process steps, such as feed composition, feed rate, and feed enthalpy. Usually the overhead or the bottoms composition is specified. It is not permissible to overdefine the system; that is, the sum of the *free* and *controllable* variables must equal five. Conversely, the *dependent* variables must equal eight. Within this framework, there are a number of possible combinations open that characterize different modes of column control. Several of these are shown in Table 1.

Table 1. Possible Modes of Distillation-column Control

Distillate and feed	Distillate and bottoms	Distillate and boil-up	Distillate bottoms and boil-up	Bottoms and boil-up	Bottoms boil-up and reflux	Feed rate distillate and boil-up	Distillate bottoms and boil-up	Distillate bottoms and feed enthalpy
Free Variables								
x_F, q, b	x_F, q, b	x_F, q, b	x_F, b	x_F, q, b	x_F, b	x_F, b	F, x_F, b	x_F, b
Specified Variables								
F, x_D	x_D, x_W	x_D, V_m	x_D, x_W, V_m	x_W, V_m	x_W, V_m, L_n	x_D, F, V_m	x_D, x_W, V_m	x_D, x_W, q
Dependent Variables								
W, D, R $V_m, L_m,$ V_n, L_n, x_W	$W, D, R,$ $V_m, L_m,$ V_n, L_n, F	$W, D, R,$ $L_m, V_n,$ L_n, x_W, F	$W, D, R,$ $L_m, V_n, L_n,$ F, q	$W, D, R,$ $L_m, V_n,$ $L_n, x_D,$ F	$W, D,$ $L_m, V_n,$ $x_D, F,$ q	W, D, R $L_m, V_n, L_n,$ q	$W, D, R,$ $L_m, V_n,$ L_n, q	$W, D, R,$ $L_m, V_m,$ V_n, L_n, F

Composition Measurement

Distillation is invariably concerned with composition control, of either the distillate or the bottom product, or at times both. Because composition bears an unvarying relationship to boiling temperature (assuming fixed pressure), it is natural that product composition is usually controlled via temperature.

Other possible methods of control exist, of course, and may assume more importance in the future. For example, product-stream composition could be measured continuously by other physical measurements, such as vapor-phase chromatography, ultraviolet photometry, or infrared photometry. Also, automatic chemical analyzers are under development that could find applications in distillation-product quality control.

Differential-vapor-pressure Measurement

The actual boiling temperature at the temperature-control point will change with changes in column pressure. This can be troublesome, particularly in low-pressure columns where minor variations in absolute pressure are substantial compared with the operating pressure.

A differential-vapor-pressure-type temperature-measuring system (known as dV_p transmitter) avoids this problem. Figure 2 shows such a system. To one of two opposing bellows is connected a thermometer bulb filled with liquid of the composition desired at that point in the column. When subjected to column temperature, this bulb develops an internal pressure dependent upon the temperature and the composition of liquid in the bulb. Simultaneously the liquid in the column exerts a pressure dependent upon the temperature and the liquid in the column. Since the bulb and

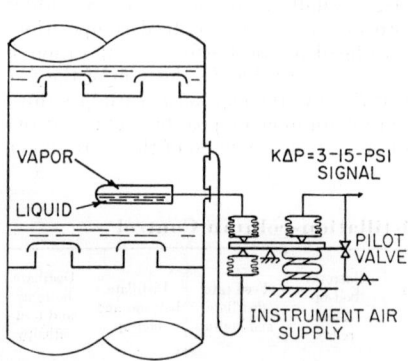

FIG. 2. Differential-vapor-pressure-type temperature-measuring system used in distillation column.

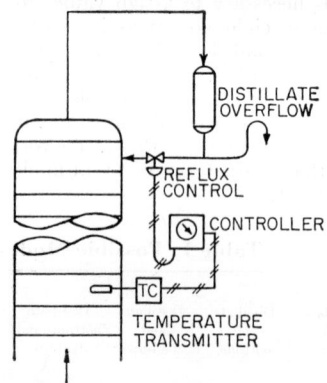

FIG. 3. Distillate-composition control based upon temperature-sensing element located in column.

column contents are at the same temperature, the pressures exerted in the column and in the bulb are each a function of the liquid composition, i.e., that in the bulb and that in the column. Comparing the two pressures in the differential force-balance system shown affords a sensitive index of composition in the column as referred to the fixed composition in the bulb, which is independent of minor variations in absolute operating pressure.

COMPOSITION CONTROL SYSTEMS BASED UPON TEMPERATURE

The following systems are based upon the use of temperature to indicate the composition of material. The techniques described, however, are equally applicable to control loops based upon other composition-measurement methods. Regulation of product composition is normally accomplished by manipulating boil-up and/or reflux rates which change the relative rates of vapor and liquid flow throughout the column.

Distillate- and Bottoms-composition Regulation

The control systems shown in Figs. 3, 4, and 5 affect product composition only. Other aspects of distillation-column control are covered later in this subsection. In Fig. 3, distillate composition, as measured by a temperature element in the column, is controlled by regulation of reflux. In Figs. 4 and 5, bottoms composition is controlled by regulation of either boil-up (Fig. 4) or feed rate (Fig. 5).

Relationship of Reflux and Boil-up

It would appear advantageous to maintain both reflux and boil-up on automatic control. If a simple binary separation is being controlled, this presents no problem, but rarely is this the case. If the feed contains an impurity having a volatility between those of the distillate and bottoms, automatic control on both reflux and boil-up presents problems. The reflux control operates to return the impurity down the column, while the boil-up control operates to send the impurity up the column. The result is that the intermediate boiler accumulates in mid-column and displaces the composition to such an extent that separation fails.

It is possible to devise controls, either continuous or intermittent, to sense mid-column temperature and purge the accumulated impurity (the so-called "pasteurization column"). Unless the accumulation is minor, however, it would be uneconomical to discard the purge without redistillation. An alternative is to allow the intermediate boiler to be discharged with either bottoms or distillate and provide an additional column for its separation if justified. The final choice rests, of course, on a comparison of the economics of the methods.

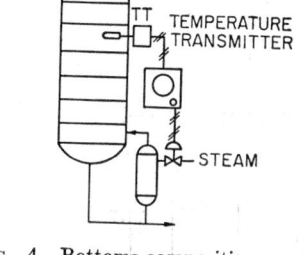
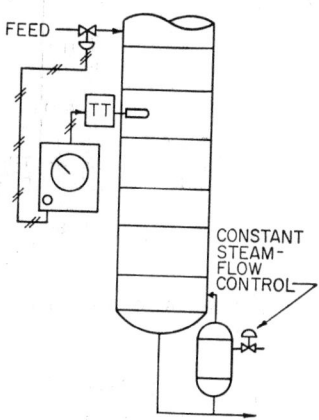

FIG. 4. Bottoms-composition control by regulation of boil-up.

FIG. 5. Bottoms-composition control by regulation of feed rate.

The foregoing discussion was based upon consideration of binary mixtures contaminated with intermediate-boiling impurities. It applies equally to the separation of key components from multicomponent mixtures.

Location of Composition Temperature-sensing Element

Common to both distillate- and bottoms-temperature controllers is the fact that the pertinent temperature measurement is made not on the product directly but in the column a number of plates away. Usually a product stream that is relatively pure is being separated from the binary mixture; so that boiling point or any other quality test of the product is often insensitive to changes in its concentration. Location of the temperature-sensing element some plates away makes it possible to obtain a greater change in temperature for a fixed change in final composition. Fixing the composition at such a point in the column suffices to control the column-product composition within narrow limits, even with wide variations in other factors, such as vapor and liquid flows.

Selecting the optimum location can best be done based on the column temperature profile, which the distillation designer should provide. Figure 6 shows a typical temperature profile in a distillation column. As indicated, the temperature-sensing element should be located where the temperature profile is steep but not too far removed from the end of the column. The rectifying-section temperature measurement must be located above the feed plate; the stripping-section temperature measurement, below the feed plate.

COLUMN-INVENTORY CONTROL SYSTEMS

Distillation columns provide little or no surge capacity so that it is necessary to remove the distillate and bottoms as fast as they accumulate. Controls for product draw-off should be no more complicated than absolutely necessary. In columns operating at atmospheric pressure, loop-sealed overflows may be all that are required. At positive pressures, level-controlled letdown valves are suitable. For vacuum operation, discharge pumps in combination with liquid-level-controlled valves are required.

Atmospheric Pressure. Figures 7 and 8 show systems for columns operating at atmospheric pressure with inventory control by loop-sealed overflows on distillate and bottoms, respectively.

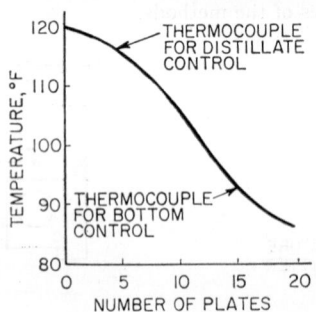

Fig. 6. Typical temperature profile in a distillation column.

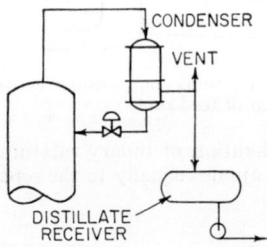

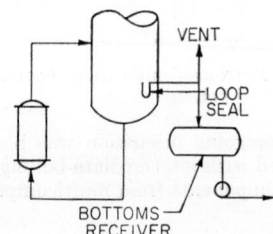

Fig. 7. Distillation-column-inventory control system where column operates at atmospheric pressure, showing regulation of distillate.

Fig. 8. Distillation-column-inventory control system where column is operated at atmospheric pressure, showing loop-sealed overflow on bottoms.

Positive Pressure. Figures 9 and 10 show systems for columns under positive pressure where liquid-level controllers are used.

Vacuum. A typical system for a column under vacuum is shown in Fig. 11.

Low-product Flow. Where the feed composition is very low or very high in low boiler content (i.e., as x_F approaches 1.0 or 0), controlling the flow of the minor product stream can be difficult because of the low rates of flow involved. In such cases, a high-low level control system can be used, allowing the product to accumulate in a receiver until the upper limit of level is reached. Then the system opens a valve fully to discharge until the lower limit of level is reached. For pressure or vacuum operation, Fig. 12 shows a typical control system employing a float switch with magnetic

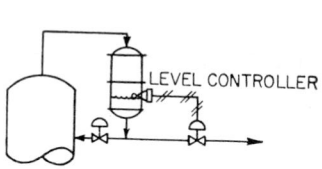

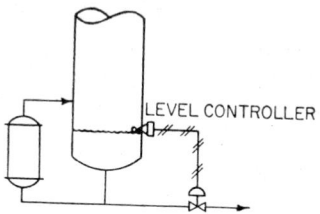

FIG. 9. Distillation-column-inventory control system where column operates under positive pressure, showing level control on condenser.

FIG. 10. Distillation-column-inventory control system where column operates under positive pressure, showing level control of column bottoms.

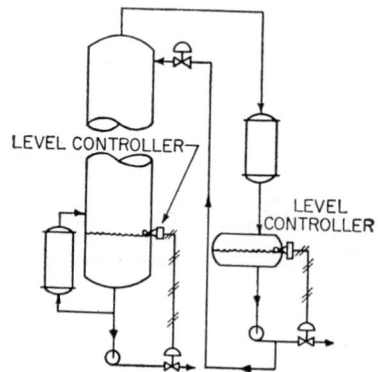

FIG. 11. Distillation-column-inventory control system where column operates under vacuum

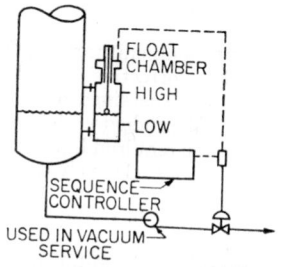

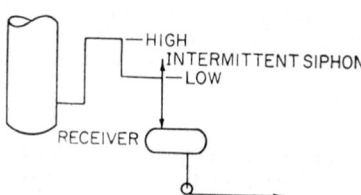

FIG. 12. Distillation-column-inventory control system where control of the flow of the minor product stream is difficult. This system, for pressure or vacuum operation, employs a float switch with magnetic pickup to sense the high- and low-level points.

FIG. 13. Distillation-column-inventory control system where control of the flow of the minor product stream is difficult. This system, for a column operating at atmospheric pressure, utilizes an intermittent siphon to accomplish the high-low level control.

pickup to sense the high- and low-level points and to actuate electrically the discharge valve through a solenoid valve on the air-to-valve line. Note that, in the case of vacuum operation where a discharge pump is used, a sequence controller is employed to start the pump, etc. Figure 13 shows the use of an intermittent siphon to accomplish the high-low level control on a column operating at atmospheric pressure.

PRESSURE CONTROL OF COLUMNS

Distillation systems are invariably designed for uniform-pressure operation, ranging from low vacuum through atmospheric to very high pressures. Atmospheric pressure usually presents no problem to the control-system designer. However, exposure to atmospheric contamination may not be permissible, in which case a modification of vacuum control can be used with a positive bleed-in of a suitable inert gas to keep the control active.

Figure 14 shows a typical pressure control system for a distillation column. Here control is obtained by regulation of the amount of venting on the condenser. Figure 15 shows a vacuum control system with control based on the use of a vacuum-tempering bleed valve ahead of the vacuum source.

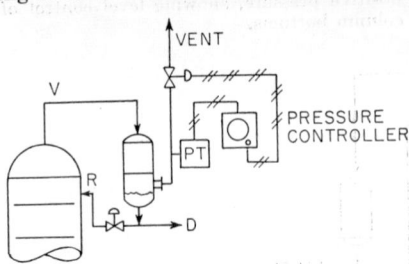
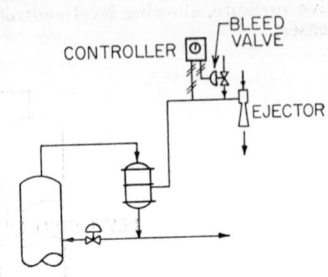

Fig. 14. Typical pressure control system for a distillation column. Control is obtained by regulation of the amount of venting on the condenser.

Fig. 15. Vacuum control system for distillation column. Control is based on the use of a vacuum-tempering valve ahead of vacuum source.

EFFECTS OF COLUMN DYNAMIC BEHAVIOR ON CONTROL

Most of the individual control loops referred to in the previous discussions are of standard design and are covered in standard texts on instrumentation.

In the application of such systems to distillation columns, however, the particular dynamic behavior of such columns often can cause problems. These are discussed briefly in the following paragraphs.

Effect of Sluggish Temperature Response. Column temperatures respond in a relative sluggish manner to operational disturbances and to corrective control action. Improper selection and application of controls can produce sustained oscillation of operating variables. The result can easily be off-standard products. There is the ever-present danger that the controls may fail to provide the required liquid-to-vapor ratios in the column, whereupon operation suffers. Indeed it is relatively easy to lose separation entirely.

Application of Automatic-control Theory. Experience in the application of automatic-control theory to distillation columns is limited. One approach to the problem of designing satisfactory controls, which has proved successful, is to design a reserve of performance ability into the distillation system and to place all but one of controllable variables on fixed-value control. For example, make the column higher (more plates) and larger in diameter (higher allowable reflux and vapor rates), place the feed and boil-up on fixed rate control, and control the reflux by temperature at a rate to guarantee satisfactory rectification. Thus the only disturbances to which the column is subjected are feed composition and feed enthalpy. With adequate reserve in reflux capacity, the feed disturbances would have negligible effect. It must be recognized that there is a capital-cost penalty in the overdesigned column and that the excess reflux rates and boil-up increase heat and coolant consumption.

EFFECT OF DISTURBANCES EXTERNAL TO COLUMN

Within the material-handling capacity range of a given distillation column, i.e., between the excessive loadings at which flooding occurs and the very low loading at which drain-down occurs, the quality of separation depends to an important degree upon the column heat balance. The point at which the heat is applied is also of importance. For example, heat introduced at the base of the column as boil-up operates through a greater number of plates than does heat entering as feed enthalpy. Consequently, as will be shown later, it is possible for separation to become poorer if boil-up heat supply is decreased by controls to compensate for an increase in feed enthalpy due to external conditions.

External disturbances, therefore, are quite significant, to the extent that they upset column heat balance.

The composition controls are designed to mitigate the effects of such disturbances, but because of their customary location in the column, some number of plates removed from either the top or bottom of the column, they are deceived to a minor extent in assessing the purity of product and, in fact, deceived to a major extent in assessing quantitative result of external disturbances.

The effects of several typical disturbances have been analyzed and are discussed here. A 15-plate toluene-benzene distillation was chosen for illustration. Plate-by-plate calculations were made using a digital-computer program. Alternatively, the analysis can be made using the McCabe-Thiele graphical solution. The McCabe-Thiele diagrams of the disturbances analyzed are presented here because they convey a clearer picture of the effects of disturbances than does a tabulation of column operating data.

McCabe-Thiele Diagram

Figure 16 represents a typical McCabe-Thiele diagram of a binary distillation. Reference has been made above to the location of the composition-control temperature measurement in the column. The feedback regulation of the composition on the control plate has the effect of maintaining the composition at that point in the column at a preselected value. Once chosen, this composition is constant. Also fixed are the number of plates above and below the feed plate.

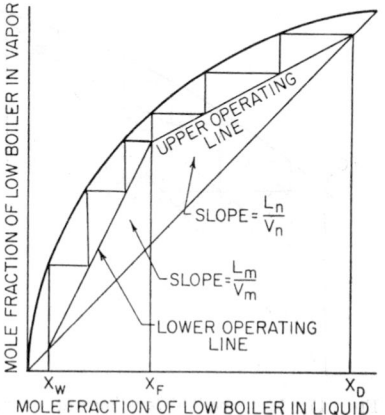

FIG. 16. McCabe-Thiele diagram of simple binary distillation.

As indicated in Fig. 16, the product composition corresponds to the intersection of the operating lines with the 45° diagonal, the upper operating line for distillate composition, and the lower operating line for bottoms composition. The effect on the McCabe-Thiele diagram of external column heat-balance disturbances is to shift the location of the operating lines on the diagram.

The McCabe-Thiele diagram construction implicitly solves the column heat balance; so that this requirement, in analyzing the effect of heat-balance disturbances on composition control, is accounted for. Two other criteria must be met in the graphical solution of the effect of disturbances: (1) as noted above, the composition on the control plate is held invariant by the composition controls; and (2) the number of plates and the locations of the feed and control plates remain unchanged.

Effect of Reduced Reflux. Any response of a distillation column to external disturbances must entail shifts in the location of the column operating lines; in fact, the slopes will change. Moreover, it is clear that the slope changes of the upper and lower operating lines are opposite in sign. The requirement that the control-plate composition remain unchanged and that the number of plates between it and the end of the column remain unchanged requires that the construction of the perturbed operating line be accomplished by displacing it about the control plate in such a way that the number of plates between it and the end of the column remain unchanged. This construction of displacing the operating line is illustrated in Fig. 17, which shows the construction of the upper column diagram for the effect of reduced reflux.

The constraints fulfilled by the construction are: (1) the slope of the operating line declined, and (2) the graphical solution above the control plate corresponded to number of plates in the upper column above the control plate.

Obviously, the decrease in slope of the upper column line can only cause a drop in distillate purity.   A similar analysis holds for the effect of altering the slope of the lower operating line where bottoms-composition control is in operation.

Therefore, to estimate the effect of external disturbances on separation, it remains only to determine the slope change on the operating lines and construct the complete diagram, upper and lower, in such a way as to account for the total number of plates and their deployment.

Figures 18 through 20 represent the analysis of several typical external disturbances. Although the examples shown were precisely calculated by computer, a graphical analysis by means of the McCabe-Thiele construction is just as informative and can be obtained in the absence of a digital computer or a plate-by-plate calculation.

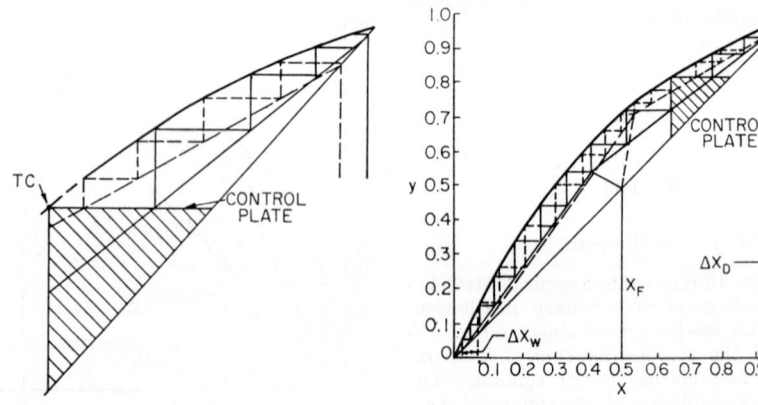

Fig. 17. McCabe-Thiele diagram of the effect of a reflux change under distillate-composition control.

Fig. 18. McCabe-Thiele diagram of the effect of increased feed on a column operating with distillate-composition control of reflux and constant boil-up.

Effect of Increased Feed on Distillate Control.   Figure 18 shows the effect of increased feed on a column operating with distillate-composition control of reflux and constant boil-up.

Since the boil-up remains constant and the feed is increased, the ratio of liquid to vapor flow in the stripping section must increase.   Consequently, the slope of the lower column operating line will increase and the bottoms quality will be degraded. Increasing the feed will increase the bottoms rate substantially and the distillate rate slightly.   The increase in distillate will be at the expense of reflux; so that the upper-column operating-line slope will be decreased.   Consequently, the distillate quality will also be degraded.

It is interesting to note that the column temperature profile below the control plate declines, while that above the control plate increases.

Effect of Decreased Feed Enthalpy on Distillate Control.   Figure 19 shows the effect of decreasing the feed temperature on the toluene-benzene separation operating with constant boil-up and distillate-composition control of reflux.

The effects are similar to those resulting from increased feed.   The ratio of total heat to feed declines.   The slope of the lower-column operating line increases and that of the upper-column operating line decreases, with a consequential degradation of separation.

The column temperature profile shifts downward below the feed plate and increases above it.

Effect of Feed Enthalpy Change on Bottoms-composition Control. Figure 20 shows the effect of feed enthalpy change on the toluene-benzene separation operating with constant reflux and bottoms-composition control of boil-up.

The lower enthalpy feed increases the liquid flow in the stripping section and the boil-up is increased by the control to make up for the loss in feed enthalpy. The total heat supplied is therefore held unchanged or at worst decreased only slightly. Because a substantial fraction of the heat input is shifted from the feed plate to the bottom of the column, the separation actually improves slightly.

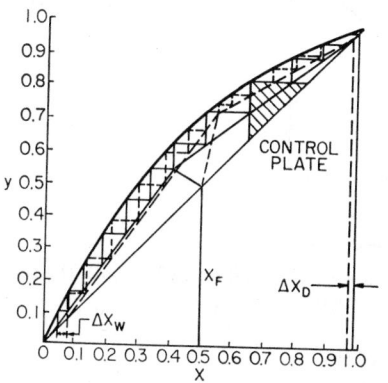

FIG. 19. McCabe-Thiele diagram of the effect of decreased feed temperature on a column operating with constant boil-up and distillate-composition control of reflux.

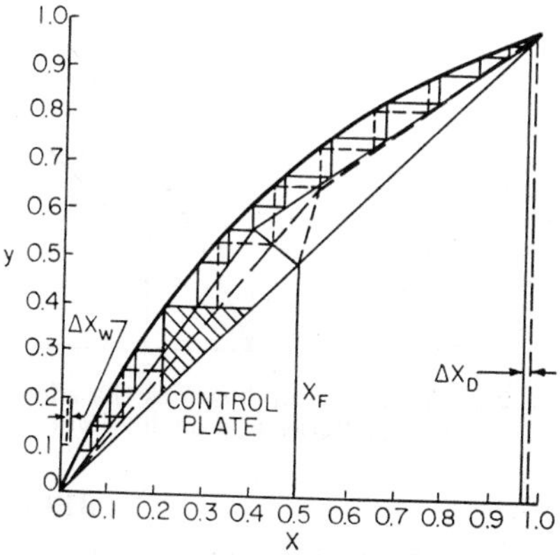

FIG. 20. McCabe-Thiele diagram of the effect of lowered feed enthalpy on a toluene-benzene separation, operating with constant reflux and composition control of boil-up.

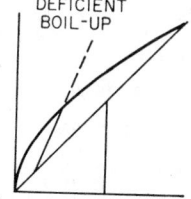

FIG. 21. Diagram that illustrates the cause of flooding when the liquid- and vapor-handling capacity of a distillation column is exceeded. A deficiency of boil-up increases the slope of the lower operating line, calling for an *unobtainable* location of the lower operating line.

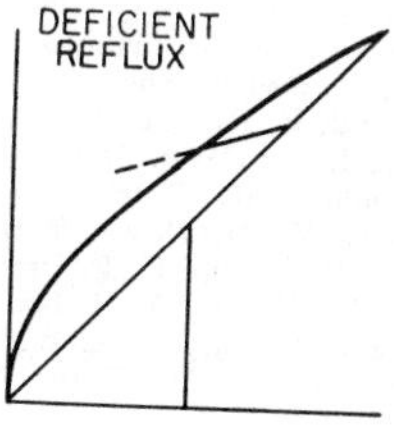

FIG. 22. Diagram that illustrates the effect of a deficiency of reflux which decreases the slope of the upper-column operating line.

The temperature profile above the control plate shifts downward, but below the control plate it shifts to higher temperatures.

Failure of Separation. Any disturbance that increases the demand for reflux or boil-up can cause complete failure of separation if the limits of the column to handle the required liquid or vapor flow are exceeded or if the capacity of the boil-up or reflux facilities is exceeded.

Exceeding the liquid- and vapor-handling capacity causes flooding. Virtually no separation occurs. A deficiency of boil-up increases the slope of the lower operating line, and an unobtainable location of the lower operating line is called for, as illustrated in Fig. 21. A deficiency of reflux decreases the slope of the upper-column operating line, with the similar result that an unobtainable location of the upper operating line is called for in the rectifying section, as shown in Fig. 22. Figure 23 shows the effect that an excessive feed rate can have when both the reflux and boil-up capabilities of a column are overtaxed. It is apparent that the control plate, whether in the top or bottom, would no longer be able to maintain its composition at the prescribed level.

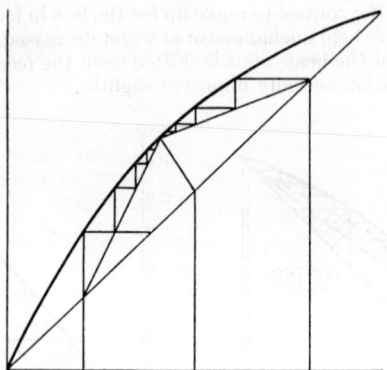

FIG. 23. Diagram that illustrates the effect of an excessive feed rate when both the reflux and boil-up capabilities of a distillation column are overtaxed.

REFERENCES

1. Parkins, R.: Continuous Distillation Plant Controls, *Chem. Eng. Progr.*, vol. 55, no. 7, pp. 60–68, July, 1959.
2. Carney, Thomas P.: "Laboratory Fractional Distillation," The Macmillan Company, New York, 1949.
3. Bogenstatter, G., and K. Hengst: Regelungvon destillationsklonnen, *Chem.-Ing.-Tech.*, vol. 31, no. 7, pp. 425–431, July, 1959.
4. Editorial Staff: Fractionator Control with High-speed Chromatography, *Autom. Control*, vol. 12, no. 5, pp. 23–24, May, 1960.
5. Zuiderweg, R. J.: "Manual of Batch Distillation," Interscience Publishers, Inc., New York, 1957.
6. Rose, A., and E. Rose: "Distillation," Interscience Publishers, Inc., New York, 1951.
7. Gilliland, E. R., and Reed: *Ind. Eng. Chem.*, vol. 34, p. 551, 1942.
8. Perry, J. H.: *Chem. Met. Eng.*, vol. 52, no. 9, p. 108, 1945.
9. Ruhemann, M.: *Trans. Inst. Chem. Engrs. (London)*, vol. 30, p. 25, 1952.
10. Herbert, W. D.: *Petrol. Refiner*, vol. 35, no. 11, p. 151, 1956.
11. Bauer, R. L., and C. P. Orr: *Chem. Eng. Progr.*, vol. 50, no. 6, p. 312, 1954.
12. Balls, B. W., and A. H. Isaac: *Trans. Inst. Chem. Engrs.(London)*, vol. 33, p. 177, 1955.
13. McCabe, W. L., and E. W. Thiele: *Ind. Eng. Chem.*, vol. 17, p. 605, 1925.
14. Rosenbrock, H. H.: *Trans. Inst. Chem. Engrs.(London)*, vol. 35, p. 347, 1957.
15. Armstrong, W. D., and W. L. Wilkinson: *Trans. Inst. Chem. Engrs.(London)*, vol. 35, p. 352, 1957.
16. Rose, A., and T. J. Williams: *Ind. Eng. Chem.*, vol. 47, no. 11, p. 2284, 1955.
17. Boetter: "Plant and Process Dynamic Characteristics," Academic Press Inc., New York, 1957.
18. Robinson, C. S., and E. R. Gilliland: "Elements of Fractional Distillation," 4th ed., McGraw-Hill Book Company, Inc., New York, 1950.
19. Bauer, R. L., and C. P. Orr: *Chem. Eng. Progr.*, vol. 55, no. 7, pp. 60–68, 1959.
20. Perry, R. H., C. H. Chilton, and S. D. Kirkpatrick (eds.): "Chemical Engineers' Handbook," 4th ed., McGraw-Hill Book Company, Inc., New York, 1963.
21. Stanton, B. D., and A. Bremer: Controlling Column Product Composition, *Cont. Eng.*, vol. 9, no. 7, pp. 104–106, July, 1962.
22. Anderson, Z.: Pneumatic Logic Supervises Distillation, *Cont. Eng.*, vol. 10, no. 2, pp. 67–69, February, 1963.
23. Oglesby, M. W., and D. E. Lupfer: Feed Enthalpy Computer Control of Distillation Column, *Cont. Eng.*, vol. 9, no. 2, pp. 87–88, February, 1962.
24. Stanton, B. D., and A. Bremer: Analog Computer Measures BTU (*Distillation Column*) Content, *Cont. Eng.*, vol. 9, no. 12, pp. 97–99, December, 1962.
25. Lupfer, D. E., and J. R. Parsons: A Predictive Control System for Distillation Columns, *Chem. Eng. Progr.*, vol. 58, no. 9, pp. 37–42, September, 1962.
26. Gilliland, E. R., and C. M. Mohr: Transient Behavior in Plate-tower Distillation of a Binary Mixture, *Chem. Eng. Progr.*, vol. 58, no. 9, pp. 59–64, September, 1962.
27. Peiser, A. M., and S. S. Grover: Dynamic Simulation of a Distillation Tower, *Chem. Eng. Progr.*, vol. 58, no. 9, pp. 65–70, September, 1962.
28. Ryle, B. G.: Gamma Density Controls Extraction Column, *Chem. Eng. Progr.*, vol. 53, no. 11, pp. 551–553, November, 1957.

CONTROL OF SOLIDS-DRYING OPERATIONS

By Alfred H. McKinney*

Drying operations have a number of significant variables that are difficult to measure directly. Therefore, general practice is to hold measurable conditions constant so that a uniform product will be made. This practice works out well if there are no load changes.

CONTROL PROBLEMS ARE SIMILAR FOR MANY TYPES OF DRYERS

The basic control problem is essentially the same for rotary, traveling-screen, moving-sheet, tray, and screw-conveyor dryers. The discussion in the first part of this subsection is given in terms of the typical rotary dryer because the discussion also applies to other dryer types—with exception of the spray dryer. Coverage of the spray-dryer control problem is given in a later part of this subsection.

GENERALIZED DRYING SYSTEM

Temperatures and pressures usually are easy to measure in drying systems. These or any other readily applicable measurement can be used for control purposes, since

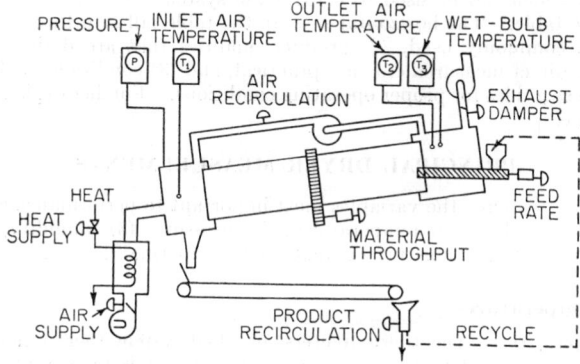

FIG. 1. Generalized drying system (rotary dryer).

system dynamics are favorable in drying systems where there is a reasonably rapid response to change. A generalized drying system, in the form of a rotary dryer, is shown in Fig. 1.

POINTS FOR APPLICATION OF CONTROLS

With reference to Fig. 1, process control for dryers can be supplied at seven points:
Heat Supply. Steam, fuel, or electrical energy.

* Consultant, Instrument Section, Engineering Service Division, E. I. du Pont de Nemours & Company, Inc., Wilmington, Del.

Air Supply. Controlled by damper in the discharge of a fan.

Material Throughput. Movement of the material in a rotary dryer depends on speed, diameter, and slope of the dryer. Belt-dryer throughput rate is controlled easily by regulation of belt speed.

Air-recirculation Rate. In some screen dryers, recirculated air is introduced at one or more points with a separate blower. It is controllable by a damper. Generally, air recirculation is not deliberately introduced or controlled in other types of dryers, but it is present to some extent because of convection currents.

Feed Rate. For material to be dried feed rate is commonly held constant or is determined by production needs. Theoretically, feed rate can be varied in response to a measurement.

Exhaust Damper. Air can be pushed (forced-draft fan) or pulled (induced-draft fan) through a dryer. If both types of fans are used, either but not both can control air flow. The other fan can control pressure at a single point between the two dampers.

Product Recirculation. This is an infrequent practice, used mostly when the material-handling problem becomes troublesome. It is useful in rotary kilns to prevent ring formation of reacting products. Product recirculation takes place in designs with countercurrent air at sufficient velocity to move some of the product to the feed end.

Manual Operation and/or Automatic Control. Probably no single dryer has been built with all seven conditions automatically controlled. However, all these conditions affect dryer performance to some extent. Control of all system variables does not require each controllable condition to be manipulated automatically from a measurement. For example, an operator can adjust a valve in response to a measurement; or the valve can be adjusted automatically.

CONTROL ARRANGEMENTS FOR DRYERS

Possible control arrangements for dryers are given in Table 1. The seven elements just described which can be used to control the system are included in the vertical columns of the table. The horizontal rows in the table fall into three categories: (1) measurements commonly used, (2) product qualities that are desirable to control where suitable direct measurements are practical, and (3) conditions within the dryer that can serve as guides for proper operating conditions. Further explanation accompanies the table.

PRINCIPAL DRYER MEASUREMENTS

In most dryer systems, the variables most important to performance of the control system include (1) dry-bulb temperatures, (2) pressure, (3) humidity, (4) product temperature, (5) product moisture content, and (6) thermal degradation.

Dry-bulb Temperatures

Thermocouples, resistance thermometers, or filled-system thermometers are used according to the range of temperatures encountered. It is important to locate the measuring element at a representative point and to protect it from corrosive and abrasive conditions.

Sensing-element Protection. Signal response time usually is not critical, and heavy-duty protective wells are therefore commonly used. In high-temperature gas streams, the temperature-sensing element must be properly shielded from cold surfaces that would cause radiation errors. Suction pyrometers or very small, shielded, bare thermocouples are necessary to obtain measurements approaching the true gas temperature.

Pressure Measurements

Dryer pressure measurements vary from fractions of an inch of water to much higher values. For dusty conditions, it is good practice to purge the line between the point

Table 1. Possible Control Arrangements for Dryers

Measurement	Heat supply	Air supply	Belt or shell speed	Air-recirculation rate	Feed rate	Exhaust damper	Product recirculation
Common measurements:							
Inlet-air temp T_1	Good	Good	No	No	No	Good	No
Exhaust-air temp T_2	Good	Good	No	Fair	Good	Good	No
Pressure P	...	Good	No	No	No	Good	Fair
Exhaust-air/wet-bulb temp T_3	Good	Good	No	Good	Good	Good	Fair
Product qualities:							
Temperature	Good	Good	Good	Good	Fair	No	Fair
Moisture	Good	Good	Good	Good	Fair	No	Fair
Degradation	Good	Good	Good	Good	Fair	No	Fair
Operating conditions:							
Solids depth	...	No	Good	No	Good*	No	No
Retention time	...	No	Good	No	Good	No	Good
Air velocity	...	Good	No	Good	No	Good	No
Evaporation rate	Good	Good	No	No	No	Good	No
Solids/gas ratio	...	Good	No	Fair	Good	Good	No

* Applies to rotary-type dryers only. Unsatisfactory for screen dryers.

Explanation of Table. A "Good" at the intersection of a row and column means that the measurement in that row can be used to control the condition shown in the given column. In column 1 (Heat supply), for example, "Good" appears opposite the measurements of inlet-air temperature, exhaust-air temperature, exhaust-air wet-bulb temperature, product temperature, product moisture, product degradation, and evaporation rate. This means that all these measurements, actual or potential, would be influenced by a change in the position of the valve in the heat supply. Thus any one of these measurements, if practical, can be used to control the heat supply. However, the moisture content of the final product would be ideal in principle.

A "Fair" in the table means that closing this control loop is not so satisfactory but is possible in most installations. A "No" indicates that control would be unsatisfactory.

Where "Good" is indicated, it is assumed that the equipment is operating with all control elements in a fixed position and that the one element under question is varied.

An underlined "Good" indicates a variable commonly controlled by each measurement. The three such choices indicated on the table are (1) inlet-air temperature T_1 controls the heat supply; (2) pressure P controls the exhaust damper; and (3) exhaust-air wet-bulb temperature T_3 controls the air-recirculation rate. Only the measurements T_1, T_3, and P will be controlled. The others are free to wander within certain limits.

Points of instrumentation T_1, T_2, T_3, and P are shown in Fig. 1.

of measurement and the instrument. This prevents the accumulation of dust in the stagnant pipe run.

Pressure Control. Pressure can be controlled equally well from either of the dampers in the suction fan duct or the forced-draft fan duct. The damper not used to control pressure must be operated by some other measurement within the system. In combination, the two dampers will control the air flow and also the system pressure at the point of measurement. The pressure measurement can be located at any place between the two dampers.

Controlled pressure may be either positive or negative within the range of the two fans. If pressure is to be controlled close to atmospheric at a single point in the system, the pressure will differ from atmospheric at every other point in the system. A chimney effect will occur between different heights if the gas density inside the dryer differs from atmospheric density. There also will be a pressure drop due to air motion inside the dryer.

Humidity

Exhaust-air humidity usually is measured with a wet-bulb psychrometer. Several electrical humidity-sensing elements also are available. See section in this handbook on Humidity and Dew-point Measurements.

Product Moisture Content

This measurement is used extensively for drying continuous sheets. However, moisture content for granular solids and powders is difficult to measure continuously. Dielectric constant has been used as a moisture-content measurement of powders where the final desired moisture content is above 2 to 3 per cent and the handling characteristics are favorable.

Thermal Degradation

This degradation of the product is a serious problem in many drying operations. It has rarely been successfully measured continuously and with sufficient rapidity for good control. Maximum temperature and minimum flow conditions usually are the basis for safe operating limits.

SCREEN-DRYER CONTROLS

The screen dryer shown in Fig. 2 illustrates a single drying section only. The controls on each zone of a multisection dryer would be the same except that pressure drop through the material, measured in only one section, is used to control the belt speed. A belt-speed change may be required to hold constant pressure drop through the material being dried if its consistency or throughput rate changes. The uncontrolled variable is feed rate.

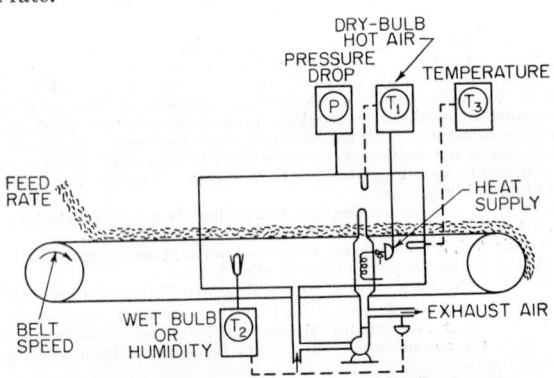

FIG. 2. Traveling screen-dryer control system.

DRUM-DRYER CONTROLS

With the drum dryer shown in Fig. 3, the product moisture content is measured and used to control steam input to the drum. The evaporation from the drum depends upon the external air condition and velocity as well as the drum temperature. The feed rate is controlled to hold the level in the feed hopper. The feed stream will vary with the capacity of the drum at various speeds, web thickness, and temperature conditions.

SPRAY-DRYER CONTROLS

Table 2 shows the control loops possible between the various controllable conditions and the measurements that might be made in spray-dryer operation.

A commonly used system for the control of spray dryers is shown in Fig. 4. Air-inlet temperature controls the heat input. Exhaust-air temperature controls the feed rate. Pressure drop through the atomizing nozzle will also alter the average size and size distribution of the drops of liquid as well as the feed rate. Size distribution may be the most important factor in spray-drying operation because the smaller drops

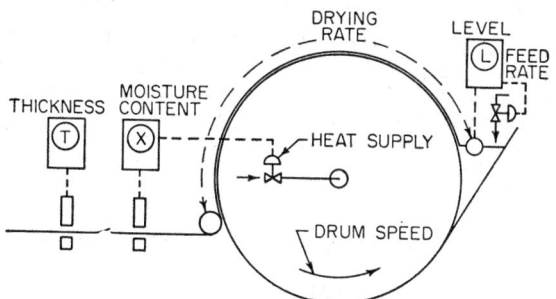

FIG. 3. Drum-dryer control system.

Table 2. Guide to Control of Spray Dryers

Measurement	Fixed or controllable conditions						
	Heat supply	Inlet-air damper	Atomizer speed	Air recirculation	Feed rate	Exhaust damper	Product recirculation
Common measurements:							
Inlet-air temp..............	Good	Good	No	No	No	Good	No
Exhaust-air temp...........	Good	Good	No	Fair	Good	No	Fair
Product qualities:							
Temperature................	Good	No	No	Fair	Good	Good	Fair
Moisture...................	Good	Good	Good	Good	Good	Good	Fair
Thermal degradation........	Good	Good	Fair	Fair	Good	Good	Good
Size.......................	...	No	Good	No	No	No	No
Size distribution..........	...	No	Good	No	No	No	No
Operating conditions:							
Pressure...................	...	Good	No	No	No	Good	No
Exhaust-air humidity.......	Good	Good	No	Fair	No	Good	No
Retention time.............	...	Good	No	No	No	Good	No
Air velocity...............	...	Good	No	Good	No	Good	No
Evaporation rate...........	Good	Good	No	No	Good	Good	No
Solids/gas ratio...........	...	Good	No	Fair	No	Good	No

Explanation of Table. See explanation at bottom of Table 1.

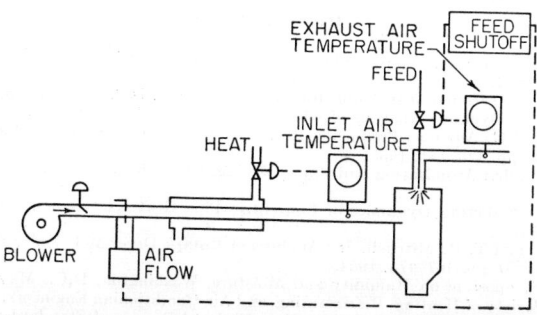

FIG. 4. Spray-dryer control system.

dehydrate more rapidly. Also, the smaller the drops, the more variable is the distribution.

Air Flow–Feed Interlock. If the air flow is interrupted, an interlock is desirable to shut off the feed quickly. This arrangement avoids accumulation of wet material in the bottom of the dryer. Alarms sometimes are provided for high exit-gas temperatures or pluggage of the product-collecting system.

System Response. Dead times for some spray dryers are shorter (response is rapid) than for other types and control may be easier. Faster measuring and control elements may be required. The system usually is sealed and there is no need to balance pressures. The system pressure usually is not controlled. An interlock is usually provided in the spray dryer to shut off the feed if the temperature falls below operating limits.

SYSTEM SAFEGUARDS

Two types of specific safety problems may arise in the use of dryers: (1) flame failure in direct-fired units, and (2) formation of combustible mixtures where dusts or solvent vapors are involved.

Flame Failure

The equipment generally provided to guard against flame failure is substantially more elaborate than the remaining instrumentation. Failure of combustion safeguards because of manual cutout arrangements often is so serious that dual control systems should be used. The first system detects the existence of a flame in the conventional manner. The second system continuously analyzes the combustion products in the dryer for explosibility. Either system can shut down the equipment.

Presence of Combustible Mixtures

The presence of dusts or solvent vapors has caused serious accidents even though no source of ignition presumably existed within the system. Inert atmospheres are used to dry hazardous materials in some cases. Where this is impractical, the solvent concentrations have been controlled by continuous analyzers that sample the atmosphere in the dryer and control the drying rate to keep the mixture below the lower explosive limit.

The sample line leading to the gas analyzer must be kept above the dew point of the solvent (or combustible material) throughout its length to prevent erroneous readings. Both automatic blanketing with inert gases and shutdown of the operation are triggered if the analyzer detects mixtures approaching the explosive limit. Alarm signals and interlock arrangements should be provided. They are actuated whenever unsafe conditions exist or are developing.

Note: The author gratefully acknowledges the assistance of the following persons for descriptions of recommended control systems for dryers: J. R. Boyd, Louisville Dryer Div., General American Transportation Corp.; V. A. Cheney, Link-Belt Co.; and E. H. Rasmussen, Nichols Engineering & Research Corp.

REFERENCES

1. Aikman, A. R.: Frequency-response and Controllability of a Chemical Plant, *Trans. ASME*, vol. 76, pp. 1312–1323, 1954.
2. Hickman, M. J.: Experimental Instrumentation for Dryers, *J. Inst. Fuel*, vol. 28, pp. 113–116, 1955.
3. Holzmann, E. G.: Dynamic Analysis of Chemical Processes, *Trans. ASME*, vol. 78, p. 251, 1956.
4. Kramers, H., and G. Alberda: Frequency Response Analysis of Continuous Flow Systems, *Chem. Eng. Sci.*, vol. 2, pp. 173–181, 1953.
5. Marshall, W. R., Jr.: Atomization and Spray Drying, *Chem. Eng. Progr. Monograph Ser.* 2, p. 50, 1954.
6. Mozley, J. M.: Predicting Dynamics of Concentric Pipe Heat Exchangers, *Ind. Eng. Chem.*, vol. 48, p. 1035, 1956.
7. Saeman, W. C., and T. R. Mitchell, Jr.: Analysis of Rotary Dryer and Cooler Performance, *Chem. Eng. Progr.*, vol. 50, pp. 467–475, 1954.
8. International Symposium on Humidity and Moisture, Washington, D.C., May, 1963. Sponsored by: American Society of Heating, Refrigerating, and Air-Conditioning Engineers; American Meteorological Society; Instrument Society of America, National Bureau of Standards; and U.S, Weather Bureau.

EVAPORATOR CONTROL SYSTEMS

By Henry S. Drinker* and Robert F. Barber†

Several evaporator control systems which are representative of the kinds encountered in the chemical process industries, such as in pulp and paper manufacture, are discussed and illustrated in this subsection.

SINGLE-EFFECT NATURAL-CIRCULATION CALANDRIA

A simple, single-effect, natural-circulation calandria-type evaporator, commonly used for production of high-concentration caustic and the recovery of small amounts of material in by-product operations, is shown in Fig. 1.

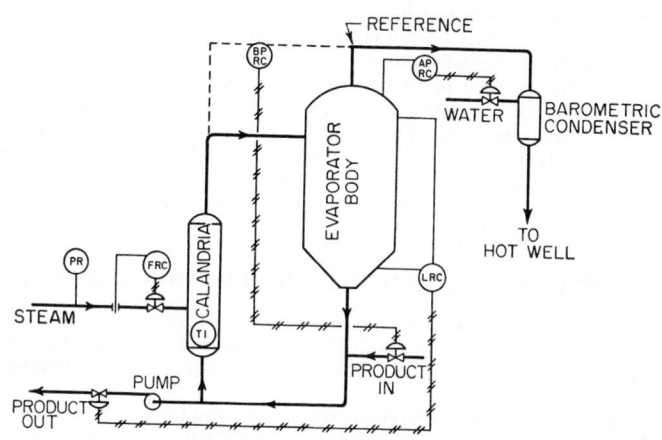

FIG. 1. Instrumentation for single-effect, natural-circulation calandria.

Total Evaporation Rate. The operating rate of the system normally is determined by establishing a fixed flow of steam to the calandria. In this way, the total heat rate is fixed and, consequently, the total evaporation rate is established.

Usually a simple steam-flow controller (FRC) is adequate to assure the required steady heat input. However, with widely varying steam supply pressure or steam superheat, some compensation for these factors may be necessary. At a minimum, the steam supply pressure should be indicated and recorded (PR) for guidance of the operator.

Level Control in Evaporator Body. This variable may be measured by either (1) a differential-pressure transmitter; or (2) a float-type level transmitter. The differential-pressure measurement method, with purged connections, is much more satisfactory on materials which are to be crystallized. In noncrystallization evaporators, either type of level measurement is satisfactory. The level in the evaporator body most often is used as a means of controlling (LRC) the inlet feed to the evaporator.

Boiling-point-rise Measurement. The concentration of material in most evaporators is measured by boiling-point rise, that is, the temperature difference between

* Manager, Systems Development Division, The Foxboro Company, Foxboro, Mass.
† Manager, Pulp and Paper Division, The Foxboro Company, Foxboro, Mass.

the boiling liquid in the evaporator and the boiling steam condensate at the same absolute pressure.

Evaporator Inlet and Discharge Control. Two alternate methods are in use for controlling the inlet and discharge of the evaporator: (1) inlet is controlled by level and discharge is controlled by concentration; or (2) inlet is controlled by concentration and discharge is controlled by level.

Alternate (1) will start automatically, since the level controller establishes the level in the evaporator body and maintains that level until the desired concentration is reached.   At that time, the differential-temperature measurement begins to allow the material to be discharged and an equilibrium is established.

Alternate (2) has been found to hold the end product closer to specification during initial operations, and some operators believe that this advantage overcomes the problem of filling the evaporator body by manually operating the valves.

Absolute-pressure Control. The absolute pressure (compensated for atmospheric-pressure changes) in the evaporator body must be controlled (APRC) for optimum operation of the evaporator and differential-temperature control of the concentration. This can be done either by controlling the amount of water going to the barometric condenser, as shown in Fig. 1, or by bleeding air into the steam-operated ejectors used for air removal.   The choice between these two methods is primarily one of economics. Where water costs are not excessive, the use of an air bleed into the ejector usually is the more economical.   However, where water costs are high, throttling of the water to the barometric condenser may be the better answer.

Control for Wide Range of Operations. If extremely wide ranges of operation are required, it is impractical to depend entirely upon control of water to the barometric condenser—since the barometric condenser will operate satisfactorily only within relatively narrow ranges.   In such cases, it is conventional practice to provide for throttling of the barometric-condenser water supply down to some fixed minimum flow and then to allow an air bleed to the steam ejector to handle the balance over the required range.   It is not practical to throttle the steam to the steam ejectors because of their tendency to "stall" at flow rates quite close to their normal operating levels.

MULTIPLE-EFFECT NATURAL-CIRCULATION CALANDRIA

Extension of the simple control system of Fig. 1 to multiple-effect evaporators of the natural-circulation calandria type requires some consideration of the over-all dynamics of the controlled system.   Figure 2 illustrates a simple control system applied to a triple-effect unit.

Because of the large storage throughout the system and the slowness with which new conditions establish themselves, the control system is designed to maintain steady conditions and to minimize the effects of variations outside the process.

Evaporator Operating Rate. The system shown in Fig. 2 is typical of commonly applied instrumentation to units of this kind.   The evaporator operating rate is established by the pressure in the first vapor-effect calandria—which pneumatically sets the flow controller (FRC) on the steam to this calandria.   No further attempt is made to control the steam rate to the various effects, since these rates already have been established by the design pressure drops in the steam lines and in the second- and third-effect calandrias.

Removal of Concentrated Liquid from Third Liquid Effect. This is controlled by a concentration-measuring instrument based on temperature difference.

Transfer of Liquids between Effects. This is controlled by holding the liquid level (LRC) in each effect to a fixed point.   This control not only takes care of level but also assures proper operation of the natural-circulation effect on the calandria. The feed to the first liquid effect is controlled by level in the first effect.

Thermal-transfer Efficiency. Temperature-difference measurement is made across each calandria as a measure of the thermal-transfer efficiency of the heat-exchange surface.   The absolute pressure (APRC) in the third vapor effect (first liquid effect) is established by an absolute-pressure controller which regulates either the water to the barometric condenser or air bleed to the steam ejector.   Pressures in the balance of the effects are maintained by pressure drop between the effects.

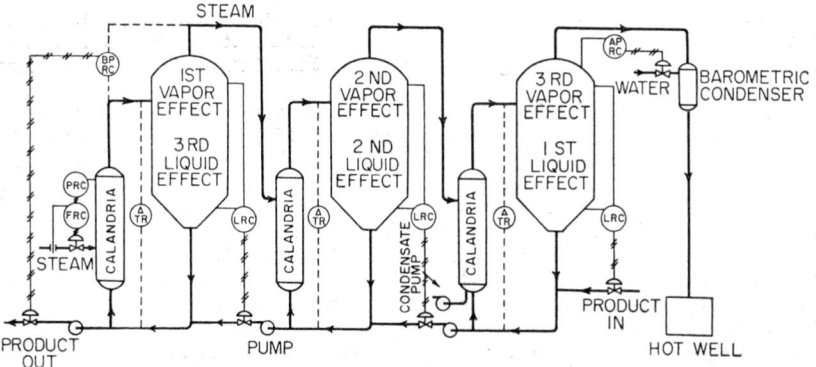

Fɪɢ. 2. Instrumentation for multiple-effect natural-circulation calandria.

BLACK-LIQUOR EVAPORATOR

A multiple-effect evaporator system for black-liquor concentration, as found in the pulp and paper industry alkaline pulping process, is shown in Fig. 3.   The backward-feed system is used, with the weak liquor entering the sixth effect at 15 to 20 per cent solids and leaving the first effect at 50 to 55 per cent solids.   Further concentration in cascade evaporators ahead of the recovery furnace, followed by "burning" in the furnace proper, rids the liquor of all organic materials.   The residue can then be processed by chemical means to recover the soda for recycling to the pulp-cooking operation.

Control of Steam to First Effect.   Steam enters the first effect under pressure control (PRC-1).   The set point of this instrument is dictated by the boiling-point-rise controller (BPRC-2).   The latter controller utilizes two resistance bulbs to provide a reliable, accurate measurement of temperature difference.   The boiling-point-

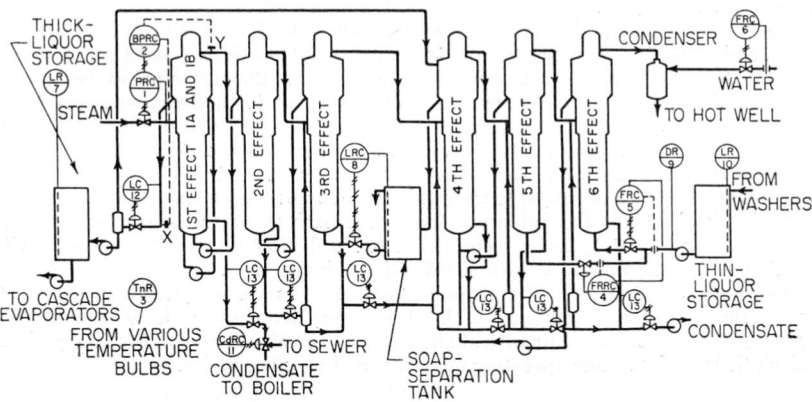

Fɪɢ. 3. Control system for black-liquor evaporator.   (PRC-1) Steam-pressure controller, pneumatic set from BPRC-2; (BPRC-2) boiling-point-rise controller; (TnR-3) electronic multipoint temperature controller; (FRRC-4) thin-liquor-ratio flow controller; (FRC-5) total thin-liquor flow controller; (FRC-6) water flow controller; (LR-7) thick-liquor-level recorder; (LRC-8) soap-tank-level controller; (DR-9) thin-liquor-density recorder; (LR-10) thin-liquor-level recorder; (CdRC-11) condensate-conductivity controller; (LC-12) liquor-level controller; (LC-13) condensate-level controllers.

rise relationship to per cent solids in the liquor leaving the evaporators is established by each mill and has proved to be a useful operating guide. Bulb X as shown in Fig. 3 is assured of continual submergence in the liquor by level controller (LC-12). Bulb Y is installed in a pitot condensing chamber to read the saturated steam temperature, devoid of superheat.

A variation in liquor density to the first effect will cause a pressure change and be partially corrected by the pressure controller as it modifies the steam flow rate. The boiling-point-rise measurement will sense the change in temperature difference and alter the set point of the pressure controller gradually to a new value. This will cause a further change in steam flow rate to complete the compensation and to reestablish the boiling-point rise to its original and desired value.

Condensate Return. Condensate from the first effect is returned to the boiler house through its level controller (LC-13). If contamination with black liquor occurs, conductivity controller CdRC-11 immediately operates the three-way valve to dump the condensate and to sound an alarm.

Feed-liquor Control. Feed liquor to the evaporators is divided equally between the fifth and sixth effects inasmuch as the sixth effect alone could not handle the total feed rate without flooding. Total-flow controller FRC-5 senses the total feed rate at a point ahead of the take-off to the fifth effect. Ratio-flow controller FRRC-4 delivers its desired portion as set by the operator on its ratio dial to the fifth effect. The balance, to give total feed rate as set on FRC-5, is delivered to the sixth effect. Flow measurement using orifice plates and pneumatic differential-pressure transmitters has proved satisfactory.

Liquor-density Measurement. A continuous Baumé recorder (DR-9) operates on the liquor feed. This instrument shows a continuous trend of weak-liquor gravity change to the operator so that he may periodically reset (FRC-5) total volume feed rate to maintain a reasonably constant total solids throughput. A bubbler-type density-sensing device usually is used.

Soap Separation. Liquor leaving the fourth effect is concentrated to the point where the resin soaps formed during the cooking operation no longer will remain dissolved but will tend to separate out. If allowed to continue to the following effects, the soaps would cause excessive foaming, scaling, and reduced capacity. Moreover, the soap, converted to tall oil, represents an appreciable form of revenue. The soap-separation tank therefore skims the soap from the black liquor, and LRC-8 regulates the feed rate on the third effect. Level measurement is by bubble tube or by a special type of level transmitter mounted flush on the tank wall.

Vacuum Control. The LC-13 level controllers hold a condensate seal on each effect and allow flash condensate from the valves to add to the overhead vapor of each stage. The LC-13 for the sixth effect is not used if the condensate pump discharge is submerged to prevent loss of vacuum. The water rate to the barometric condenser for controlling vacuum in the sixth effect is regulated by FRC-6 so that a minimum of excess water will be used. The sixth-effect vacuum also establishes the pressure to the vacuum gradient for all other effects. If water costs to the mill are modest, the water is delivered uncontrolled to the condenser in excess of the maximum rate needed to maintain the highest possible vacuum in the sixth effect—regardless of the maximum feed liquor rate. Level recorders LR-10 and LR-7 provide the operator with continuous information on thin- and thick-liquor inventories.

Temperature Survey. A multipoint temperature recorder (TnR-3) using resistance bulbs provides the operator with all necessary temperature information. The operator may use these readings to aid in efficient operation as well as to make periodic calculations to determine operating efficiency and cleaning cycles.

REFERENCES

1. Johnson, D. E.: Simulation and Analysis Improve Evaporator Control, *ISA J.*, vol. 7, no. 7, pp. 46–47, July, 1960.
2. Mair, James: Evaporator Instrumentation, *Paper Mill News*, Nov. 20, 1954, pp. 19–22.
3. Stephenson, J. N. (ed.): "Preparation of Stock for Paper Making," Chap. 5, McGraw-Hill Book Company, Inc., New York, 1951.

PULP AND PAPER PRODUCTION INSTRUMENTATION

By H. O. Ehrisman*

Many different operations are employed in the manufacture of pulp and paper—batch, semibatch, and continuous—with a strong trend toward the latter. These operations offer an exceptionally wide variety of complex instrument applications, measurements, and control problems. Because of space limitations, only representative control systems are described.

ALKALINE PULPING

Alkaline pulping is the chemical separation of cellulose fiber from prepared wood chips by means of a caustic liquor under controlled temperature and pressure. The caustic dissolves the fiber bonds, releasing individual fibers for subsequent papermaking.

Digester-liquor-measuring System

Preparatory to heating, the pulp digester is charged with wood chips and chemical cooking liquors (white liquor and black liquor). Accurate cooking volumes of both liquors are essential to digest the wood chips and make uniform high-quality pulp.

In a typical liquor-measuring system, the respective liquors are pumped into measuring tanks, from which they are charged into the digester. One common arrangement is shown in Fig. 1.

Liquor levels are measured by recording gages equipped with electrical contacts to actuate the required control valves and pumps. The measurements are translated into appropriate units of liquor volume by the instrument chart calibration. Two completely independent recording-controlling systems are employed—one for white liquor, the other for black liquor.

The instrument control panel is usually located on the digester floor for the convenience of the operator. Electric controls, located on the panel, provide either full automatic or remote manual valve and pump operation.

Operation always starts with full measuring tanks, at which point the instruments read zero. After setting the control index pointers at the desired volumes of liquor draw-down, the operator presses the "Start" push button. As the measuring-tank liquor levels drop, the recording pens move toward their index pointers. When each pen reaches its control index, electric contacts close the liquor-discharge valves and start the pumps to refill the measuring tanks. When each measuring tank has been refilled, its recorder pen again reads zero on the chart and the refilling operation is automatically stopped. The draw-down and refill operations can be stopped or placed on remote manual control at any point in the cycle.

Digester Pulp Cooking Controls

Cooking separates the individual cellulose fibers by dissolving the lignin bonds. Good circulation of liquor is essential for the production of uniform high-quality pulp.

* General Sales Manager, The Foxboro Company, Foxboro, Mass.
Editor's Note. Although some of the diagrams in this section indicate conventional large-case instruments, the small-case strip chart is also in common use. Both types apply equally well to the various pulp- and papermaking processes.

Faulty circulation results in overcooked pulp in some sections and undercooked chips in others.

Indirect-heated digesters are in common use. A continuous flow of cooking liquor is withdrawn from the digester, pumped through a heater, and returned to the digester.

Equally common are direct-heated digesters wherein high-pressure steam is fed directly into the digester through nozzles located in the bottom cone. The entering steam thus comes into actual contact with the wood and cooking liquor. Figure 2 shows such a digester with its control system. Uniform temperature distribution and elimination of noncondensable gases are the primary objectives in good control of cooking.

Temperatures at the top and bottom of the digester, together with digester pressure, are recorded on a single chart in instrument 1, which serves as an essential guide to the proper adjustment of the steaming and relief controls.

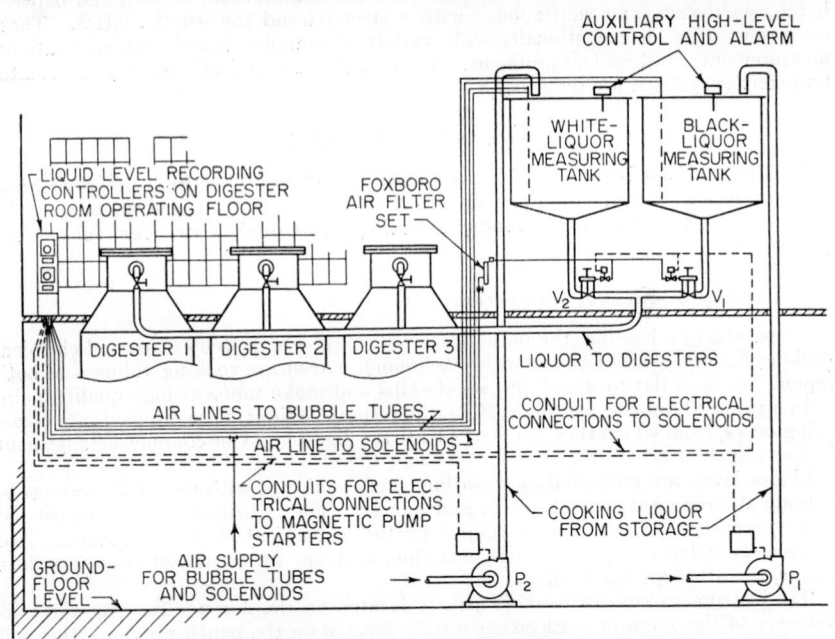

Fig. 1. Digester-liquor-measuring system.

Steaming Controls. At the start of the cook, a steam-flow controller, instrument 2, prevents excessive boiler load by admitting steam at a constant controlled rate. This contributes to establishing good digester circulation without boring or channeling. As the cook progresses, a time-cam pressure controller, instrument 3, takes over to complete the pressure rise on schedule and to hold the pressure at the desired value for the duration of the cook.

Relief Control. Several different types of relief control are in use throughout the industry. Of these, the most modern and successful type is flow control of the actual amount of relief, instrument 4. This method automatically compensates for relief screen plugging and liquor pull-over. At the same time, noncondensables are adequately relieved and good circulation is insured. Automatic blowback is added on digesters utilizing small-area relief screens which tend to plug more readily than the large-area basket type.

Other Digester Instrumentation. Other instrumentation frequently used in the control of a pulp digester may include a steam-flow recorder with totalizer.

On the indirect digester, it is desirable to incorporate a two-pen flow recorder for liquor circulation to permit proper adjustment of the distribution of liquor in the two return lines from the heater to the digester.   Condensate from the heaters of indirect-steamed digesters is normally returned to the boiler house where it is used as feed-water makeup.   Because of the ever-present danger of a liquor leak contaminating the condensate, a recorder measures the condensate conductivity.   It also operates electrical contacts to actuate an alarm and a condensate-to-sewer diversion valve when the condensate becomes contaminated.

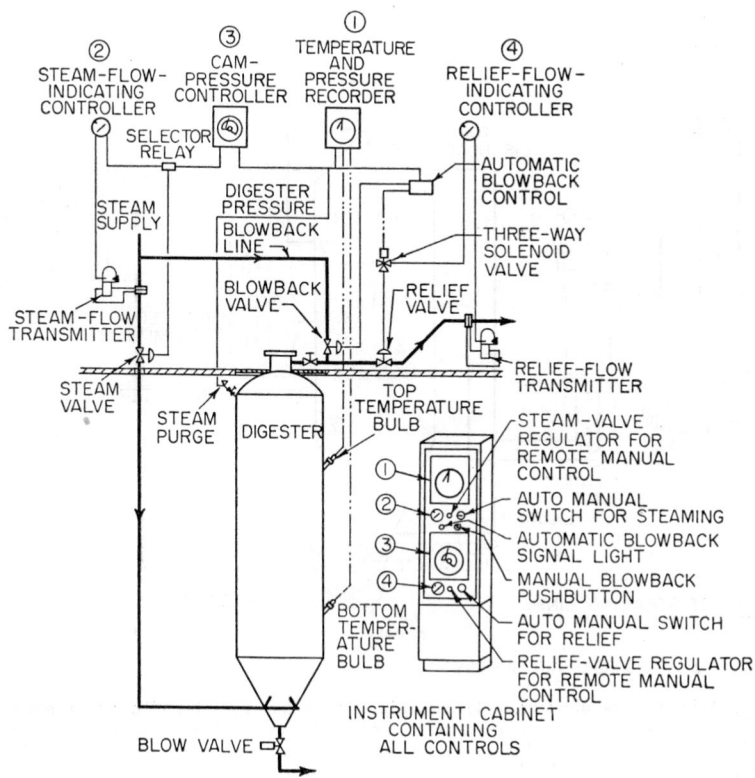

FIG. 2. Direct-steaming alkaline digester.

ACID PULPING

Acid pulping presents the same general problems as described for alkaline pulping. The difference is primarily that the cooking liquor is a solution of calcium bisulfite and sulfurous acid.   Figure 3 shows a digester control system which can be easily converted to either direct or indirect steaming.

Steaming Controls.   In the indirect-heated digester, accurate temperature measurement is possible because liquor flow is external to the cooking process.   Thus instrument 1 receives its temperature signal transmitted from temperature recorder instrument 2, and reproduces the temperature curve for which the cam is cut.

In the direct-heated digester, circulation is established by natural convection and steam velocity.   The temperature at any single location in the digester is not representative of the average digester condition. However, if the heat input is controlled by a predetermined schedule of steaming rate, average digester temperature will follow the prescribed curve. Thus steaming in the direct-heated digester is more accurately controlled from a steam-flow measurement. In this situation instrument 1 is switched to receive its signal from the steam-flow transmitter.

Relief Control. As the digester temperature increases, sulfur dioxide gas is given off which develops a substantial partial pressure. The total digester pressure, made up of SO_2 gas, air, and steam is controlled throughout the cooking cycle. Some mills

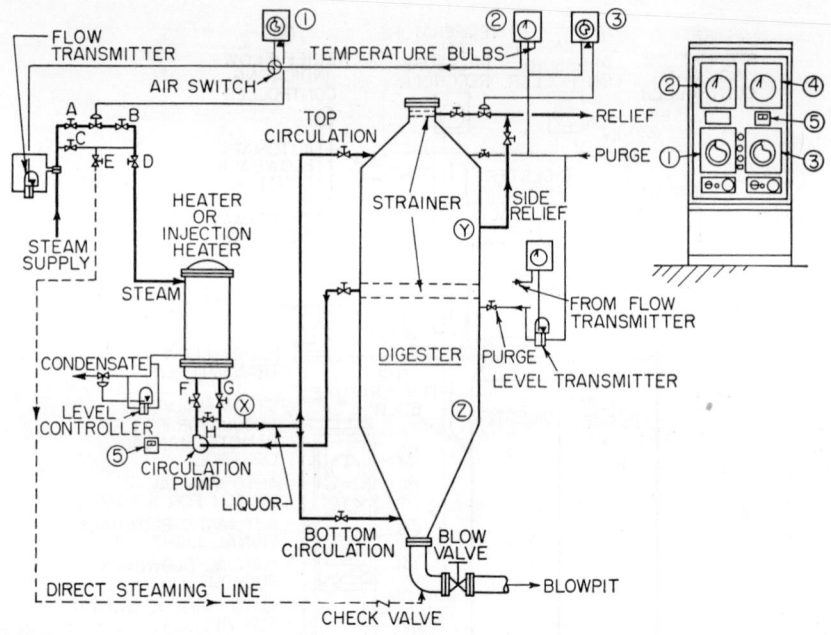

FIG. 3. Acid-cooking digester. (1) Steaming controller regulates either temperature at point X or steam flow; (2) temperature recorder with transmitter and extra pressure pen records temperatures at points X, Y, and Z; (3) relief pressure controller; (4) acid level and steam-flow recorder with integrator; (5) ammeter for circulation pump; (6) condensate floatless level controller. *Alternate systems:* Forced circulation with injection heater—omit 6. Forced circulation with direct steaming, without heater—omit item 6 and control steaming from bulb X. Direct steaming without heater or pump—omit 5 and 6 and control from steam flow. Optional additional equipment may include a 6-in. digester pressure gage, push buttons for circulating pump, ratio flow control for top and bottom circulation, and remote transmission for digester pressure.

cook at a fixed pressure of 75 psi. Others use a time-cam pressure controller, instrument 3, to control pressure to a predetermined schedule.

Other Instruments. Acid-pulping instrumentation frequently includes steam-flow recording and liquor-circulation flow recording and condensate-conductivity control on indirect-heated digester.

OTHER PULPING SYSTEMS

Semichemical pulp treatment is another pulping system. Instrumentation for this method is similar to that discussed in the foregoing material.

Continuous digesters have received increasing acceptance in the industry.   There are quite a few different types, each by a different manufacturer and each having a unique approach.   The instrumentation varies greatly with equipment design and is therefore beyond the scope of this presentation.

ALKALINE PULP WASHING

After the pulp is cooked, it must be washed to remove combustible organic material (principally lignin) and recoverable cooking chemicals.   The lignin and other organic

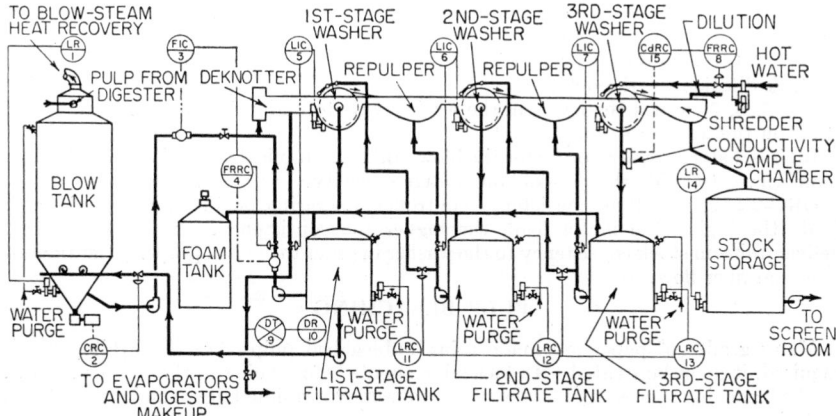

   ⱶ — FLANGE-MOUNTED DIFFERENTIAL-PRESSURE TRANSMITTER
   ⬚ — DIFFERENTIAL-PRESSURE TRANSMITTER
  —□— ORIFICE PLATE
  ⊐◯⊐ — MAGNETIC FLOW METER
   ⋈ — PNEUMATICALLY OPERATED CONTROL VALVE

Fig. 4. Alkaline pulp washing.   (LR-1) Level recorder; (CRC-2) motor load dilution recording controller (proportional plus reset); (FIC-3) magnetic flow meter indicating controller (proportional plus reset) with pneumatic transmitter; (FRRC-4) two-pen recording ratio flow controller (proportional plus reset); (LIC-5) indicating level controller (proportional plus reset); (LIC-6 and 7) indicating level controller (proportional plus reset); (FRRC-8) two-pen recording ratio flow controller (proportional plus reset); (DT-9) Baumé transmitter; (DR-10) Baumé recorder; (LRC-11) recording level controller (proportional); (LRC-12 and 13) recording level controller (proportional); (LR-14) level recorder; (CdRC-15) recording conductivity controller (proportional).

materials constitute a valuable source of heat for the recovery boiler, and the recoverable cooking chemicals are expensive.   Both must be removed from the pulp during this washing process; otherwise they are lost in subsequent operations.

In a typical multistage washer, the flows of pulp and wash water are countercurrent, with fresh water introduced at the last stage.   By the time this water has passed through the first stage, it has picked up sufficient organic and chemicals to be highly discolored; hence it is known as black liquor.

An excessive amount of wash water at the final washing stage dilutes the black liquor unnecessarily, resulting in evaporator and recovery furnace problems.   Insufficient wash water at the final stage results in inadequate, uneconomical recovery of organics and chemicals.

In Fig. 4, instrument CdRC-15 measures the electrical conductivity of the last-stage filtrate as a function of its chemical content.   The more chemical there is in the filtrate, the less is lost to sewer in subsequent screening and washing operations.   A measurement of the chemical content of the filtrate is generally preferred to a measurement of chemicals in the stock for ease of maintenance on the measuring cell.   CdRC-15 automatically adjusts the flow of shower (wash) water to the final stage.   Thus the optimum compromise is established between chemical loss and recovery furnace efficiency.

Any changes in the last-stage shower flow result in equivalent changes in the other shower flows through the action of the filtrate-tank level controllers LRC-12 and 13.

Baumé recorder DR-10 is a final check on the black-liquor density leaving the washer.   This is important in anticipating the evaporator load.

FIC-3 is a flow controller for stock to the washers.   A uniform rate of delivery is essential to good washer operation.   The control is accomplished by automatic valve or variable-speed pump drive.

FRRC-4 is a ratio controller which automatically proportions the required amount of stock dilution for the deknotter.

LIC-5, 6, 7 control the levels in the washer vats to maintain uniform formation of stock on the washer drums.

LR-1 is a stock-level recorder for the blow tank.   A duplicate recorder in the digester room serves to guide the "cooks" on storage capacity.

CRC-2 is a blow-tank consistency controller, operating on the principle of motor load—the higher the motor load, the higher the consistency.   It is important to deliver uniform stock consistency to the washer in order to control properly the amount of actual fiber flow.

PULP BLEACHING

Many grades of paper are made using unbleached pulps, but if a white pulp is required, it must be treated with bleaching agents before going to the paper machine.

The bleaching process further dissolves and removes residual lignin and other foreign materials by chemical treatment.   At the same time, the amount of bleaching chemicals and the reaction time on the fiber must be closely controlled to obtain the desired brightness without loss of fiber strength.

Bleaching is usually a multistage process.   An example of four-stage operation is shown in Fig. 5.   A wide variety of instrumentation is required to control accurately the critical factors of time, temperature, level, flow, consistency, pH, and other variables.

Level recorders LR-1 and LR-44 for the storage chests at both ends of the process guide the operator on stock supply and paper-machine demand.

Level recorders LR-13, LR-24, LR-35 are essential in maintaining constant retention time in each tower.   The chlorination tower is always full; hence no level recorder is required.

Consistency controller CRC-2, stock flow controller FRC-3, and chlorine-gas flow controller FRC-4 are vital to establishing a continuous load to the bleaching system. By proper flow and consistency regulation, the pulp tonnage feed rate is held constant. Chlorine-demand tests made on the pulp determine the chlorine-gas flow-control setting, which is adjusted periodically by the operator to meet changes in bleachability requirements of the pulp.   The result is maximum usage of chlorine to remove the lignin, which in turn permits economical use of the more expensive chemicals in the following stages.

Flow indicators FI-11 and FI-22 plus flow controller FRC-34 measure exact amounts of expensive bleaching chemical into the pulp.   Settings are made by the operator periodically, depending upon the efficiency of the chlorination stage and degree of pulp brightness needed.   In many mills the caustic and hypo meters are automatic flow controllers, to ensure the utmost in precise chemical addition.

Temperature controllers TRC-12, TRC-23, and TRC-33 heat the pulp with steam to temperatures required for optimum action of the chemicals on the impurities.

Included in the temperature controllers are recording temperature pens for showing the approximate interface level between the heavy stock and diluted stock in the bottom of the towers.   Modern bleach plants operate their caustic, hypo, and ClO$_2$ stages at high consistencies.   However, to pump the stock to the washer it must be

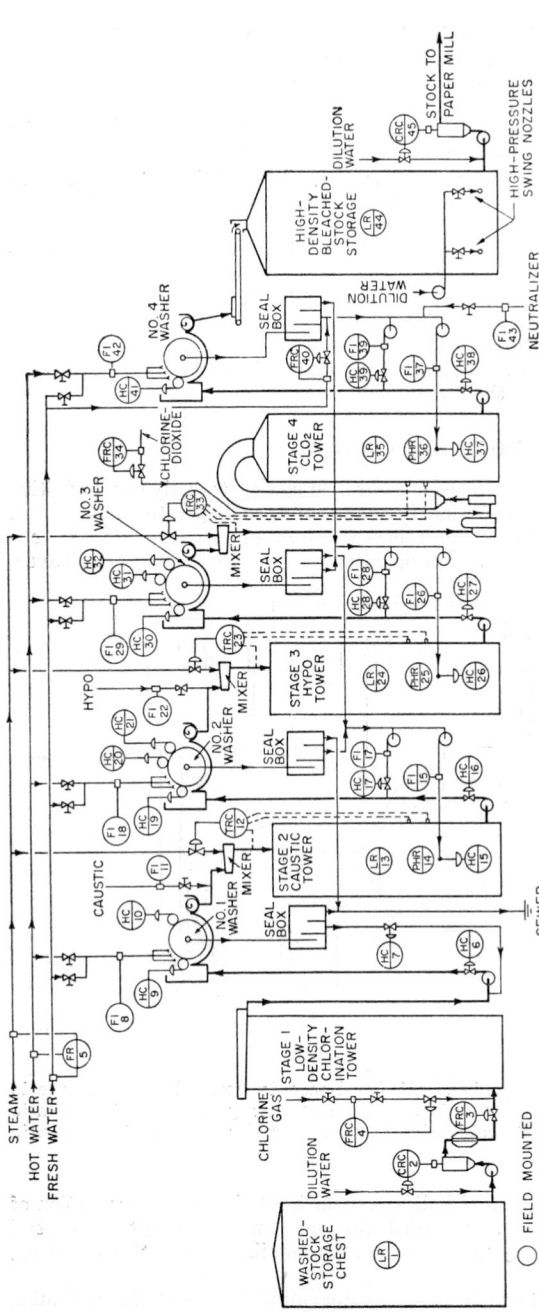

Fig. 5. Bleaching instrumentation. (LR-1) Level recorder; (CRC-2) consistency recorder controller; (FRC-3) flow recorder controller, magnetic meter type; (FRC-4) flow recorder controller; (FR-5) flow recorder with integrators; (HC-6) remote valve control; (HC-7) remote valve control, dilution to No. 1 washer; (FI-8) flow indicator; (HC-9 and 10) remote pneumatic loading control; (FI-11) flow indicator; (TRC-12) temperature recorder controller with additional temperature bulbs; (LR-13) level recorder; (PHR-14) pH recorder; (FI-15) flow indicator, dilution to No. 2 washer; (HC-15) remote valve control, dilution nozzles; (HC-16) remote valve control; (HC-17) remote valve control, dilution to No. 2 washer; (FI-17) flow indicator, dilution to No. 2 washer; (FI-18) flow indicator; (HC-19–21) remote pneumatic loading control; (FI-22) flow indicator; (TRC-23) temperature recorder controller with additional temperature bulbs; (LR-24) level recorder; (PHR-25) pH recorder; (HC-26) remote valve control, dilution nozzles; (FI-26) flow indicator, dilution nozzles; (HC-27) remote valve control; (HC-28) remote valve control, dilution to No. 3 washer; (FI-28) flow indicator, dilution to No. 3 washer; (FI-29) flow indicator; (HC-30–32) remote pneumatic loading control; (TRC-33) temperature recorder controller with additional temperature bulbs; (FRC-34) flow recorder controller; (LR-35) level recorder; (PHR-36) pH recorder; (HC-37) remote valve control, dilution nozzles; (FI-37) flow indicator, dilution nozzles; (HC-38) remote valve control; (HC-39) remote valve operator, dilution to No. 4 washer; (FI-39) flow indicator, dilution to No. 4 washer; (FRC-40) flow recorder controller; (HC-41) remote pneumatic loading control; (FI-42) flow indicator; (FI-43) flow indicator; (LR-44) level recorder; (CRC-45) consistency recorder controller.

diluted to about 3 per cent.   Manual dilution is achieved through remote-operated valves HC-15, HC-26, and HC-37.   Under normal conditions the upper bulb will sense the warm stock and the lower bulb will sense the cooler diluted stock.   Dilution water is added to maintain the interface level between the two bulbs.

FR-5 records steam and water flows to the bleach plant; FRC-40 regulates the amount of fresh water added to the system.

pH recorders 14, 25, and 36 indicate the acid or alkalinity conditions at each stage as these conditions have profound influence on the ability of the bleaching chemicals to remove the impurities without affecting pulp strength.

Successful operation of a multistage bleach plant requires operator ability and techniques of the highest order, yet desired results depend heavily on proper, trouble-free performance of the instrumentation.   Controls selected to meet specific plant requirements include the magnetic flow meter for stock feed and bleaching-chemical flow measurement.   Tower-temperature measurements use special heavy-duty sockets to withstand heavy pulp stock loads.   All instrument materials are selected to meet the severe corrosive services found in the bleach plant.

PULP BLENDING

In a typical modern blending system, each controlled flow of stock, dye, and additive is usually brought to a central panel for maximum flexibility and operator convenience. On the panel also are the chest-level controllers which ensure an adequate and continuous supply to the paper machine.   For operator convenience the panel can have a

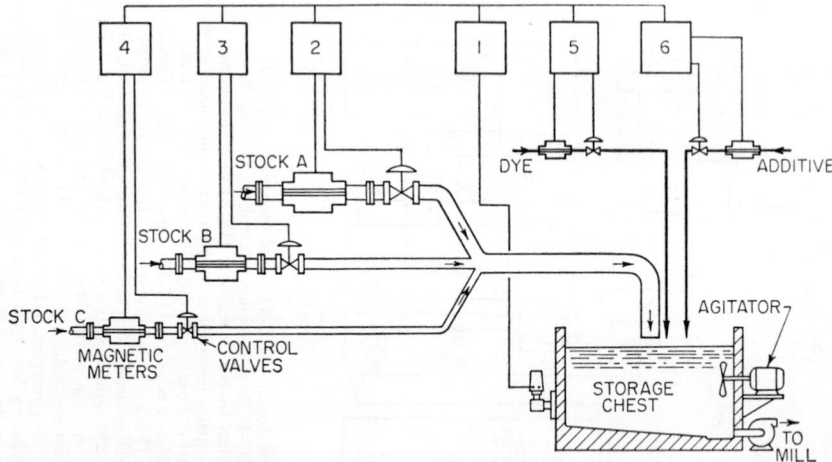

Fig. 6. Stock-dye-additive proportioning.   (1) Level recorder-controller; (2) flow-ratio recorder-controller (stock *A*); (3) flow-ratio recorder-controller (stock *B*); (4) flow-ratio recorder-controller (stock *C*); (5) flow-ratio recorder-controller (dye); (6) flow-ratio recorder-controller (additive).

sloping console area for regulators and indicators which set the proper proportion of each stream to total system flow.   Included also are pumps and agitator control accessories, as well as all necessary interlocks, alarms, totalizers, and other operating aids.

Such a centralized control station gives the operator complete control of the blending operation.   To start up from empty chests, he sets the desired proportion for each flow, then simply pushes a button.   The controls then automatically bring the whole system into operation, delivering properly blended furnish at a rate dictated by machine demand.

Complete automatic blending systems have been made practical by the advent of the magnetic-type flow meter. This meter, with its absence of all flow restrictions, has received wide acceptance in the industry. Small sizes, down to $\frac{1}{10}$ in., are particularly applicable to the control of dyes and additives in the desired proportions.

The flow diagram in Fig. 6 shows a typical blending system. A change in storage chest level, resulting from a change in machine demand, is measured by instrument 1. This instrument automatically varies the amount of stock, dye, and additive flows, in the exact desired proportions to restore the chest level to normal. The proportions of each for a given run are preset by the operator on the control panel. Thus each individual flow is automatically controlled in the desired proportion regardless of variation in machine demand.

PAPER MACHINE

There are two general types of paper machines, the Fourdrinier and the cylinder machine. Among both types there are substantial differences dictated by the grades of paper the machine is designed to make. The discussion here is restricted to the Fourdrinier machine.

Older, smaller Fourdrinier machines run at speeds up to 300 fpm, producing a sheet of perhaps 72-in. width; whereas newer, larger machines may run at speeds of 2,500 fpm and over, making a sheet up to 280-in. width. A modern high-production paper machine may produce as much as 800 tons of paper a day, and represent an investment of 6 to 7 million dollars.

The Fourdrinier papermaking machine is commonly thought of as having a wet end and a dry end. The wet end has two sections. The first includes the stock-regulating box, stock-screen head box, and Fourdrinier wire. The second, or press section, includes two or more pairs of press rolls through which the wet paper is carried between woolen felts. The dry-end section includes the dryer section, calenders, and reel.

Wet-end Instrumentation

In the operation of a typical Fourdrinier machine (Fig. 7) stock is pumped from the machine chest, at 3 per cent consistency, to the stock-regulating box. A continuous record of available stock is necessary operating information. This is provided by a liquid-level recorder, instrument 5, which records machine-chest level and broke-chest stock level. Broke is rejected paper that has to be repulped.

To ensure stock of uniform consistency, the recording consistency controller, instrument 8, automatically controls dilution water added at the suction of the stock pump which serves as a mixer. Instrument 4 controls the consistency of stock from the broke chest.

The stock then passes through a Jordan for refining. Instrument 9 controls the degree of refining by adjusting the Jordan plug position to maintain constant load. The rate of stock delivery may be automatically controlled using a basis-weight (beta-ray) recorder-controller, instrument 11, which resets a magnetic-type flow meter, instrument 10. Screened stock then flows to a head box, through an adjustable orifice running the width of the machine (known as a "slice") and onto the Fourdrinier wire. As the stock flows onto the wire, water drains from it through the mesh.

The head of stock in the head box determines the velocity of stock flow onto the wire. It is essential to good sheet formation that this velocity be the same as the wire speed. For this reason, the range of the head-box level controller (instrument 13) is selected to cover only the necessary operating-level range of the head box to obtain maximum sensitivity.

Many grades of paper are sized for resistance to water penetration by adding rosin to the stock before it goes to the machine test. Alum is generally used to "set" the size on the individual fibers. This is best accomplished at a definite pH. A pH recorder-controller, instrument 12, regulates the addition of alum to the paper stock.

The rate at which water drains from the stock through the Fourdrinier wire has a great effect on the uniformity of sheet formation. This is considerably affected by viscosity which changes markedly with temperature. Especially in northern mills, seasonal variations in water temperature make it necessary to add steam to maintain uniform stock temperature. A recording temperature controller, instrument 14, controls stock temperature, adding steam to the white water as required to ensure a uniform drainage rate.

Excess process water, flowing from the fan pump pit into the white-water tank, contains some usable paper fibers which must be recovered. Therefore, it is pumped to a save-all which filters out the fibers for reuse. Flow rate to the save-all is determined by recording liquid-level controller, instrument 7. The filtering capacity of the

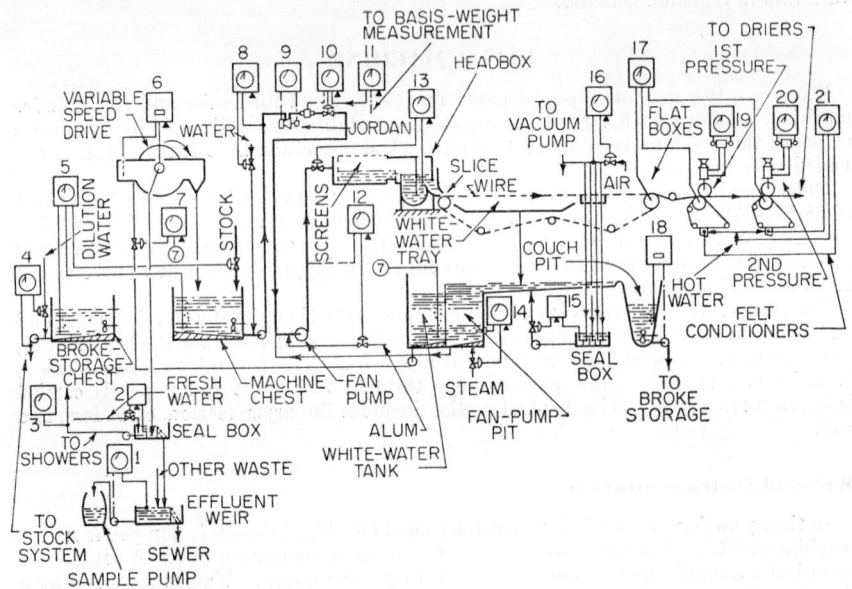

FIG. 7. Fourdrinier paper machine—wet end. (1) Effluent flow recorder with integrator (float and cable type with sampler); (2) save-all seal box level makeup controller; (3) shower water pressure recorder; (4) broke consistency controller; (5) level controller for machine chest with extra pen showing level in broke storage; (6) save-all level controller (varies speed of save-all); (7) white-water-tank level controller; (8) consistency controller; (9) Jordan motor load controller, operates plug-positioning motor; (10) magnetic-meter stock-flow controller (reset by 11); (11) basis weight controller; (12) pH controller; (13) head-box level controller; (14) head-box temperature controller; (15) seal-box level controller; (16) flat-boxes vacuum controller; (17) vacuum recorder couch, first and second pressure; (18) couch-pit level controller; (19) first pressure pneumatic loading control; (20) second pressure pneumatic loading control; (21) felt conditioners vacuum and temperature recorder. (Δ) 20 psi air supply.

save-all increases with its speed of rotation. As more white water flows into the save-all, its vat level tends to increase. The liquid-level controller, instrument 6, maintains vat level by increasing the speed of the save-all. Makeup to the save-all seal box is added as required to maintain constant level by instrument 2.

Shower water is used at various points on the paper machine, and the filtrate from the save-all is generally clean enough for this purpose. It is important to maintain the pressure of this shower water for effective operation. To provide a continuous check on this variable, a pressure recorder, instrument 3, is commonly used. The various showers often require different water pressures. These pressures can also be recorded by adding pressure pens to instrument 3, so that all are recorded on the same chart.

There are a variety of overflow water drains on the paper machine. All are brought together in an effluent weir where it is important to obtain a continuous measurement of sewer losses. A recording meter, instrument 1, records effluent flow.

With regard to the Fourdrinier wire, although initial water drainage is by gravity, as the sheet is formed this is supplemented by the application of a vacuum under the wire through a series of flat boxes. Vacuum is applied to these boxes and water removed (through a barometric loop) by a vacuum pump. The amount of vacuum applied must be closely controlled to ensure a uniform rate of water removal and to avoid excessive wear due to drag of the wire on the vacuum-box covers. The recording vacuum controller, instrument 16, controls the flat-box vacuum by automatically bleeding air into the vacuum header or bypassing the vacuum pump, the selection depending on local conditions.

The vacuum-system seal-box level must be maintained uniform at all times and this control is obtained by a level controller, instrument 15.

The upper portion of this Fourdrinier wire is driven by a couch roll around which it travels continuously. The couch roll and press rolls apply a vacuum under the sheet to remove as much water mechanically as is possible. Mechanical water removal is far more economical than by evaporation in dryers. A uniform rate of water removal is important at each of these points; so a three-pen vacuum recorder, instrument 17, is provided to guide this operation.

Paper which is trimmed off the edge of the sheet is washed into a couch pit by showers. On occasion, the entire sheet may be temporarily discharged into the couch pit when a paper "break" occurs. This accumulation of paper in the couch pit is periodically pumped into the broke storage chest by a level controller, instrument 18, which operates the pump at preset high and low levels.

When the sheet reaches the end of the wire, sufficient water has been removed so that it will support its own weight and make its first jump from the wire to a woolen blanket or press felt. This carries the sheet through several presses to remove more water, just as a wringer removes water from clothes.

To maintain the woolen felts at high water-absorbing efficiency, *felt conditioners* are often used. Hot water is forced through the felt and then removed by a vacuum box. This washes and reconditions the felt. Instrument 21 gives a continuous record of water temperature and vacuum on each conditioner to guide this operation.

Press-roll loading on the various presses is accomplished with pneumatic loading controllers, instruments 19 and 20. These permit independent nip-pressure control of each end of the press. Paper enters the press section with about 80 per cent water and leaves with 25 to 30 per cent. Most of the remaining water is driven from the paper as it passes through the dryer section.

Dryer-section Instrumentation

This section includes the dryers, calenders, and reel as illustrated in Fig. 8. The dryers consist of a series of steam-heated cylinders. The two sides of the sheet are brought alternately into contact with the heated cylinders. Water vaporized from the paper is carried away by exhaust fans through a paper-machine ventilating hood.

Because of the large quantity of steam used in the dryers, it is good practice to account for this costly item with a steam-flow recorder and totalizing integrator, FR-9.

Steam is introduced at the dry end, controlled by the recording pressure controller, PRC-8. To ensure paper of proper moisture content at the dry-end section, recording moisture controller MRC-14 is provided. It measures the moisture content of the paper ahead of the reel and automatically adjusts the control point of PRC-8, as required, to maintain uniform moisture in the sheet.

Low-pressure steam is taken from a turbine exhaust to supply the main dryer section. Since the turbine exhaust does not necessarily meet the dryer demand, PRC-7 is used to control supply pressure. It admits high-pressure steam to the machine header when the demand exceeds the turbine extraction capacity and vents excess pressure when the machine load falls off. Flow recorder FR-5 records the amount of low-pressure steam to the main, middle, and wet-end sections.

Steam to the main section header is supplied by pressure controller PRC-6, ensuring an uninterrupted supply of steam to the main-section dryers. Condensate from the main section is flashed into steam, to supply steam for the middle-section dryers. If flashed steam becomes insufficient for the load demanded, the differential-pressure controller PRC-4 will provide makeup steam from the low-pressure header of the main section. PRC-3 serves the identical purpose between the middle section and wet-end section.

The wet-end (in this diagram the wet-end section is the felt dryer which follows the felt condition previously described) steam is supplied by flashed condensate from the middle section. Should flashed steam pressure fall below that required by pressure

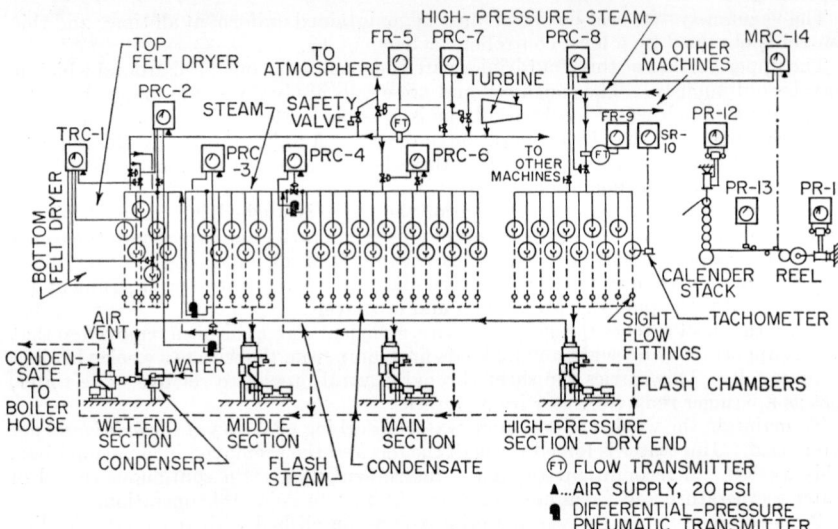

Fig. 8. Paper machine—dry end. (1) Dual temperature controller top and bottom felt dryers; (2) wet-end autoselector controller; (3) middle section differential-pressure controller; (4) main middle section differential-pressure controller; (5) low-pressure steam-flow recorder with integrator; (6) main-section pressure controller; (7) turbine back-pressure controller; (8) high-pressure section pressure controller (reset by moisture); (9) high-pressure steam-flow recorder with integrator; (10) production recorder; (11) reel pneumatic loading controls; (12) calender stack pneumatic loading controls; (13) tension monitor; (14) moisture controller (resets 9).

controller PRC-2, steam supply is automatically shifted to direct exhaust steam until flashed steam can provide sufficient pressure for the required steaming rate.

Generally, several felt dryers are employed, one group to dry the top felt and a second group to dry the bottom. Each group may be run at different temperatures. Control of both groups is accomplished with dual-recording temperature controller TRC-1.

Calender Instrumentation

From the dry-end section, the paper is fed through one or more calender stacks for a final finishing operation. The roll nip pressures should be controlled for uniform finishing and for this purpose pneumatic roll loading controller PR-12 is provided. Graduations of finish are obtained by regulating air pressure applied through air cylinders.

Most paper machines are equipped with a recording tachometer SR-10, calibrated in feet per minute production rate. The speed is measured at the dry end of the machine. Sheet tension is recorded by PR-13.

The nip pressure between the reel and windup drum must be maintained uniform to produce a uniform reel. This pressure is controlled by PR-12.

BLACK-LIQUOR EVAPORATOR (ALKALINE PULPING)

Black liquor is pumped from the pulp washer to thin-liquor storage (see Fig. 9). From here it passes through a multieffect long-tube evaporator where it is concentrated to a density sufficiently high to support combustion in the recovery furnace. The density of thick liquor to the furnace must be closely controlled. Excessive evaporation will plug the evaporator tubes. Insufficient evaporation results in poor furnace control.

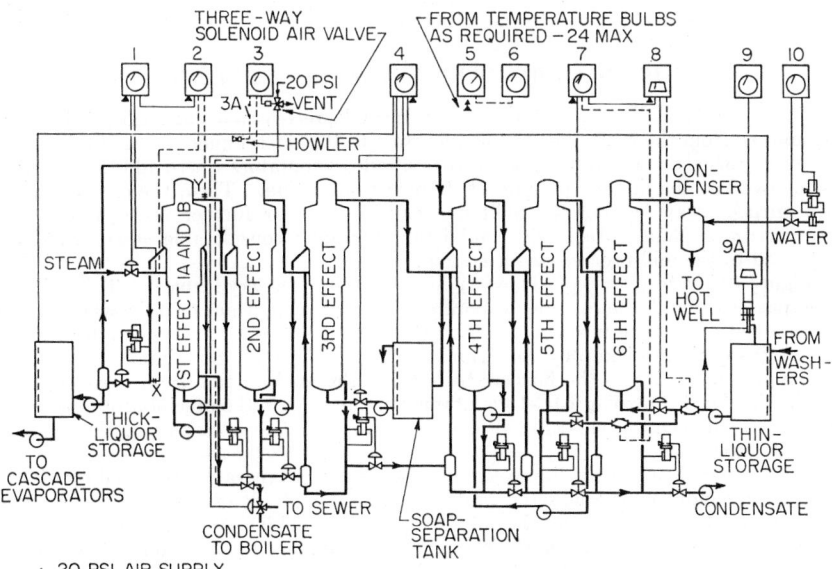

FIG. 9. Black-liquor evaporator. (1) Steam pressure controller, pneumatic set from 2; (2) electronic boiling-point-rise controller; (3) condensate-conductivity controller; (3A) howler switch; (4) soap-tank level controller; (5) auto-manual temperature-scanning unit; (6) electronic temperature recorder; (7) thin-liquor ratio-flow controller; (8) total thin-liquor flow-controller transmitter; (9) thin-liquor density recorder; (9A) thin-liquor density transmitter; (10) water-flow controller.

Instrument 2 measures the boiling-point elevation of the thick liquor. A known relationship exists between boiling-point elevation and density. The instrument measures the difference between the temperature of the concentrated liquor at point X and the boiling temperature of water at the same elevated pressure at point Y. The latter temperature is measured by a bulb in a condensing chamber teed into the vapor line. This condensing chamber serves to desuperheat the vapor by radiation to atmosphere.

Instrument 1 controls steam-chest pressure at the first effect and its set point is automatically adjusted by instrument 2 to maintain uniform boiling-point rise.

Instruments 7 and 8 control the feed of thin liquor to the evaporator. The feed is usually split between the fifth and sixth effects, as shown, and the split flow automatically ratioed in the desired proportions. The magnetic-type flow meter is used because it presents no obstruction to entrained solids.

Levels of thick- and thin-liquor storage are recorded by instrument 4, which also measures and controls the level of the soap-separation tank. An air-bubbler-type level-measuring means is used to avoid problems of clogging or corrosion of other methods.

Condensate removal is controlled individually in each effect to maintain condensate level below the tubes but prevent steam blow-through. Instrument 3 measures the conductivity of the condensate from the first effect with automatic diversion to sewer if contamination occurs.

Instrument 9 records the Baumé of the thin liquor from storage as a guide to adjusting liquor feed rate. Instrument 10 controls water flow to the condenser, in which a constant vacuum is desirable for good evaporator operation.

As a further guide to the operator in periodically checking tube cleanliness and evaporator capacity, a multipoint temperature recorder and a switching unit (instruments 5 and 6) are provided to show the temperature drop across each effect.

CHEMICAL RECOVERY (ALKALINE PULPING)

The heavy black liquor, containing organic solids, recoverable chemicals, and water, is pumped from the multiple-effect evaporators to the thick-liquor storage tank at the recovery furnace. From here, it goes to cascade evaporators where additional water is removed so the liquor can support its own combustion. Then, the organic solids are burned away leaving the molten chemicals free for further processing in the recausticizing plant. Also, the combustion produces considerable steam which is returned to the mill-process steam system.

Figure 10 illustrates a typical modern high-capacity recovery furnace. Thick-liquor feed passes through the cascade evaporators over rotating drums or disks where water is removed by coming into contact with the hot flue gases. Level controller LRC-2 maintains uniform pickup by the drums or disks while DRC-8 maintains a constant liquor density going to the burners by sensing the load on the drum drive motors. Indicating level controller LIC-1 maintains a liquor pool in the precipitator bottom to recover small chemical particles not extracted from the flue gases passing through the cascade evaporators.

Salt-cake makeup to the liquor leaving the cascade evaporators is added by HC-7, which is a remote-set special screw feeder. Reduction of salt cake in the furnace gives sodium sulfide which replaces the sulfide lost in the cooking and washing operations. The liquor is further heated in the primary and secondary heaters, controlled, respectively, by temperature instruments TRC-9 and TRC-10. The liquor flow rate is measured by a magnetic meter and is controlled by FRC-11, which regulates a variable-speed pump. The liquor is then sprayed through burner nozzles into the furnace where it ignites and burns.

Total combustion air to the furnace is supplied by a forced-draft fan. Primary air combines with the sprayed liquor while secondary air enters tangentially above the nozzles to ensure complete combustion before the gases pass the boiler tubes. To ensure proper fuel-air mixture, the total air flow is ratioed to liquor flow and controlled by FRRC-17.

The feedwater-control system for the boiler is monitored by drum-level controller LRC-15. It senses rise or fall of drum level and in turn readjusts the ratio mechanism in the feedwater-flow controller FRRC-14. Thus, under any load conditions, exactly 1 lb of feedwater is delivered for every pound of steam produced.

Oxygen recorder O_2R-12 in the rear pass shows combustion efficiency so that proper fuel-air ratio settings can be maintained. Draft controller FRC-4 regulates the induced-draft fan speed to ensure proper furnace negative pressure. Various temperatures, pressures, and drafts are recorded or indicated depending upon furnace design to give the operator all the necessary information for complete furnace control.

The molten chemicals, called "smelt," pass from the furnace floor into a dissolving tank where they become raw green liquor. Tank level is maintained by LRC-19.

Green-liquor feed of constant gravity is important to uniform operation of the recausticizing plant. Therefore, DRC-18 controls liquor density by regulating weak-green-liquor input to the dissolving tank.

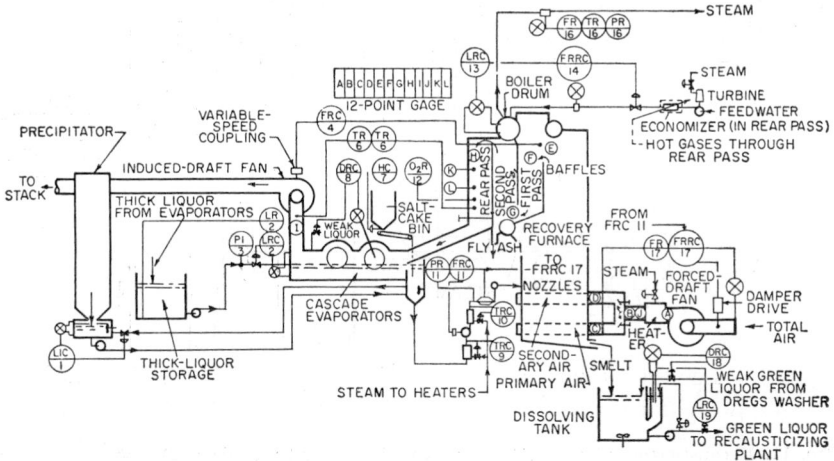

FIG. 10. Kraft recovery furnace. (1) Precipitator level controller; (2) evaporator level recorder-controller with record of thick-liquor storage level; (3) pressure gage, thick liquor to cascade; (4) furnace-draft controller; (5) 12-point indicating gage: (A) air heater inlet pressure; (B) main duct pressure; (C) primary air pressure; (D) secondary air pressure; (E) furnace draft; (F) draft first pass; (G) draft second pass; (H) draft rear pass; (I) draft evaporator outlet; (J) air heater outlet temperature; (K) economizer inlet draft; (L) economizer outlet draft; (6) temperature recorder-evaporator, inlet, outlet; (7) remote-setting salt-cake addition; (8) density controller, cascade outlet; (9) primary heater, temperature controller; (10) secondary heater, temperature controller; (11) heavy-black-liquor flow controller with pressure recorder; (12) oxygen recorder; (13) boiler-drum level controller, resets 14; (14) feedwater-ratio controller; (15) feedwater differential-pressure controller; (16) steam temperature, pressure, and flow recorder; (17) total air-ratio controller with primary air-flow recorder; (18) dissolving-tank density controller; (19) dissolving-tank level controller.

RECAUSTICIZING (ALKALINE PULPING)

Green liquor from the smelt-dissolving tank is pumped to storage in the recausticizing plant. A typical process-control system is illustrated in Fig. 11.

Instrument 1 records raw-green-liquor storage level. From storage the liquor flows to a clarifier. The production rate of the process is maintained by a flow controller, instrument 2. The magnetic-type flow meter is employed for this application because it does not become plugged. The level in the clarified liquor tank is recorded by instrument 3. The liquor is then heated to the optimum causticizing temperature by instrument 4, which includes a second recording pen for causticizer temperature.

Lime is added to the hot liquor at the slakers, and it then passes to the causticizers where the green liquor is converted to white liquor and precipitated lime mud.

Torque recorders 5 and 9 are provided to guide the operator in preventing overload on the rake drives of the white-liquor clarifier and lime-mud thickener. The temperature of wash water to the various units is controlled by instrument 8. Flow controller 10 regulates the flow of hot wash water to the lime-mud thickener. A flow indicator, 7, aids in regulating hot wash water to the dregs washer. Instruments 6 and 11 record the levels in white-liquor storage and lime-mud storage.

The chemical cycle of the cooking liquor is completed with the recausticizing process, and new white liquor is thus made available for digester cooking.

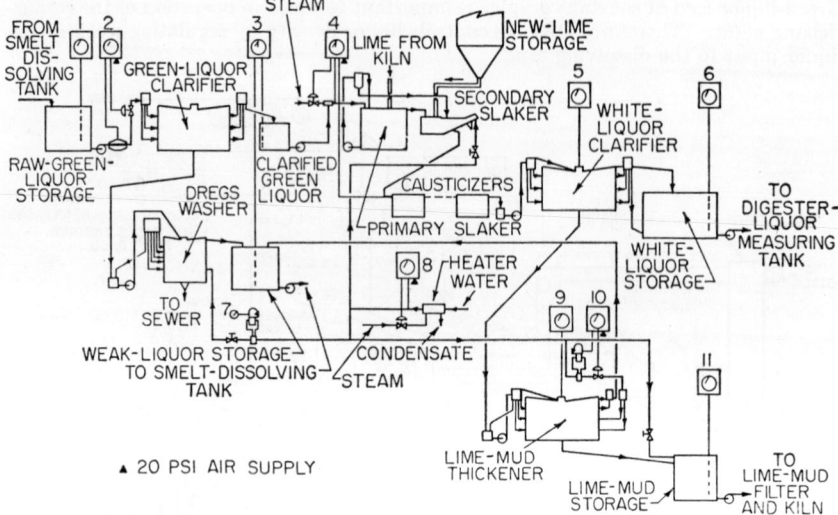

Fig. 11. Recausticizing plant.   (1) Raw-green-liquor level recorder; (2) raw-green-liquor flow controller; (3) clarified-green-liquor and weak-liquor level recorder; (4) clarified-green-liquor temperature controller with after slaker temperature record; (5) torque recorder; (6) white-liquor level recorder; (7) wash-water flow indicator; (8) wash-water temperature controller; (9) torque recorder; (10) wash water to thickener flow controller; (11) lime-mud-storage level recorder.

LIME KILN (ALKALINE PULPING)

The lime kiln converts lime mud (calcium carbonate) from the recausticizing plant back to quicklime (calcium oxide) for reuse at the slakers.   Instrumentation includes controls for the lime-mud filter, where the problem is to maintain a uniform filter-cake thickness and thereby ensure adequate mud washing as well as uniform feed to the kiln.

Controls for the kiln include the regulation of fuel oil and air feeds as well as the control of kiln temperature and draft.   Recorders show temperatures in the kiln and of the exit gas.   Indicating gages normally show draft and kiln speed.

ACID MAKING FOR ACID PULPING

Acid cooking liquor is produced by burning sulfur-forming sulfur dioxide gas and reacting this gas with limestone or milk of lime to produce a cooking acid of the desired composition and strength.

Sulfur Burner

Figure 12 shows typical instrumentation for a rotary sulfur burner making sulfur dioxide.   The controls provide a uniform feed of molten sulfur to the burner and a constant burning area necessary for optimum combustion of the sulfur with minimum losses.

Instrument 2 is an indicating sulfur-level controller, operating from a temperature-bulb system.   The bulb is located at the desired maximum level in the melter.   When the bulb is exposed by a drop in level, it cools to the preset control point and actuates electrical control contacts which start the conveyor to refill the melter.   At the same time, it opens a steam valve to melt the incoming sulfur.   When the temperature bulb is again contacted by the molten sulfur, the bulb temperature rises and shuts off the supply.

Instrument 3 is an indicating temperature controller to maintain the desired uniform melted-sulfur temperature.   Unless the sulfur is kept sufficiently hot, incompletely melted lumps may plug the lines and objectionable moisture in the sulfur will not be driven off.   On the other hand, sulfur when overheated becomes extremely viscous and will not flow readily.

Instrument 4 is a recording liquid-level controller and is the most important instrument on the system.   This instrument controls the area of burning sulfur by controlling the level of sulfur in the burner.   Level is maintained by regulating the flow of sulfur from the melter to the burner through a steam-jacketed valve.   The steam pressure to the sulfur-line jacket is controlled by instrument 1.

Instrument 6 is an SO_2 recorder-controller which governs the strength of the gas as it leaves the combustion chamber.   It does this by regulating the dampers which control input of primary and secondary air.

To guide the operation of the sulfur burner the temperature of the hot gas leaving the system is continuously recorded by instrument 5.

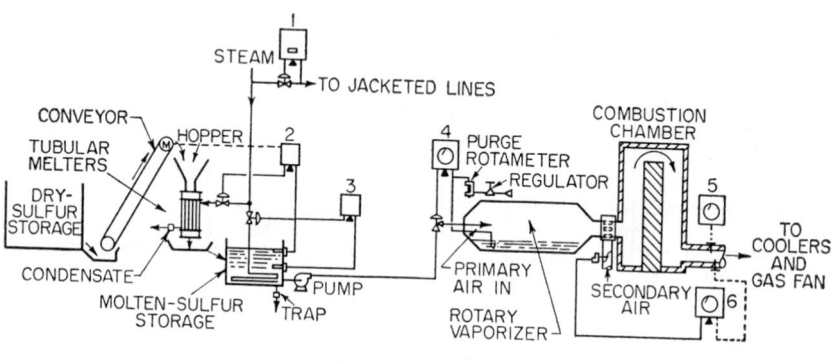

Fig. 12. Rotary sulfur burner.   (1) Steam-pressure controller; (2) molten-sulfur level controller; (3) molten-sulfur temperature controller; (4) vaporizer level controller; (5) combustion-chamber exit-gas temperature recorder; (6) per cent SO_2 controller.

Sulfur Dioxide Gas Cooler

The hot gas enters the coolers and must be cooled quickly, under carefully controlled conditions to prevent chemical losses or the formation of undesirable SO_3.   Figure 13 shows the instrumentation for a typical cooler system.   This integrated control system enables the acid maker to produce more uniform cooking acid at lower sulfur cost per ton of pulp.

Excess sulfuric acid in the effluent from the primary cooler is shown by instrument 1, which records pH of the effluent.   It guides the acid maker in balancing the system variables to prevent such chemical losses.

Flow of spray water through the primary cooler must be sufficient to cool the gas but not excessive to form waste acid which is lost through the effluent.   A weir-meter-type recording flow controller, instrument 3, controls the rate of spray-water flow.

If the normal water supply to the coolers should fail, cooling would not take place and excessive loss of SO_2 would occur in the absorption towers, and serious equipment damage might be caused from the high gas temperature.   A secondary source of emergency water prevents this.   A pressure switch connected to the primary water line instantly detects the loss of water pressure and activates an electrically operated valve to cut in the emergency water supply, and an alarm signal light warns the acid maker.

For balanced operation and most effective cooling in the secondary tower, an indicating liquid-level controller, instrument 4, is used.   This instrument adds just enough makeup water to counteract losses by maintaining the necessary level in the bottom of the secondary cooler.

Temperatures must be known at various critical points to maintain the proper balance throughout the system so that the acid maker can get the desired cooled gas temperature. A multirecord temperature recorder (instrument 5) provides a continuous temperature record from bulbs installed at points A, B, C, D, and E.

To obtain the highest degree of gas cooling, the circulating water must be cooled in a heat exchanger. To stabilize the heat extracted in this exchanger, the flow of cooling water is regulated in accordance with its temperature by a recording temperature controller (instrument 6).

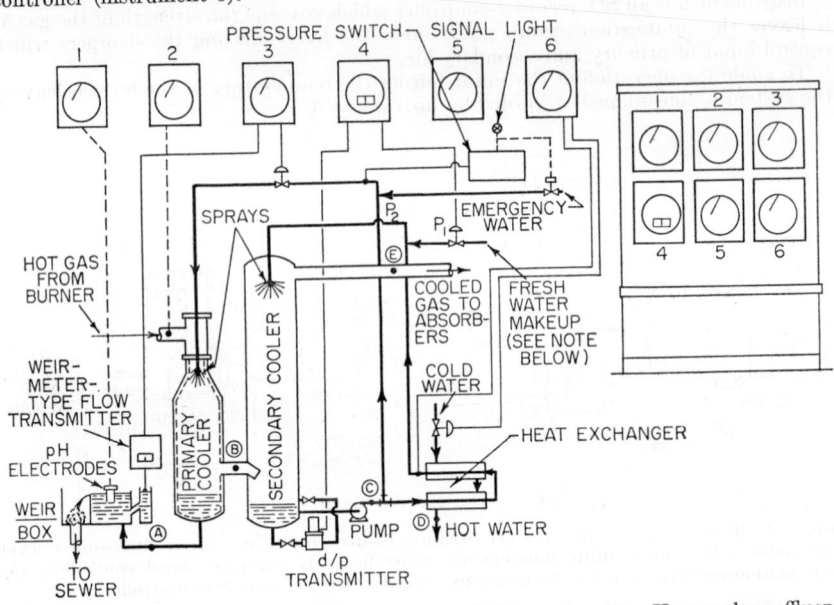

Fig. 13. Sulfur dioxide gas cooler (Chemipulp K-C type). (1) pH recorder, effluent primary cooler; (2) temperature recorder, hot gas from burner; (3) recording flow controller, primary cooler water; (4) indicating level controller, secondary cooler; (5) multirecord dynalog recorder for temperatures at A, B, C, D, and E; (6) recording temperature controller, secondary cooler spray water. *Note:* P_1 must always exceed P_2.

Sulfur Dioxide Gas Absorption

The next step is to allow the cooled SO_2 to react with limestone and water in absorption towers to produce a raw cooking acid. It is very important that this acid be of uniformly high quality because of its bearing on the yield, quality, and cost of the sulfite pulp (acid pulping).

The control system shown in Fig. 14 is typical. SO_2 gas from the coolers flows through the absorption towers packed with limestone, with flow controlled by the gas fan. Water flows countercurrent to gas flow, and it is imperative that the water flow be controlled in proportion. Recording flow controller 5 gives positive control of this important variable regardless of variations in supply pressure.

Temperature of water to the weak tower must be controlled to effect proper combination with the SO_2 gas. It may be necessary to heat the water in winter and cool it in summer. Instrument 4, a temperature recorder or a recording-controller, is used for control of the water temperature.

A pressure tower may be used to build up the "free" SO_2 strength during periods of warm weather. Part of the SO_2 flow is diverted through a compressor to the tower where it flows countercurrent to flow of raw acid.   Since the amount of gas absorbed is proportional to the pressure in the tower, a recording pressure controller (instrument 2) is used to control the tower pressure by bleeding off excess pressure.

Liquor supply must be maintained in the storage and weak absorption towers, the pressure tower, and the settling tank.   Indicating liquid-level controllers, instruments 1, 3, 6, and 7, are used for this purpose.

Instrument 8 gives a continuous record of the specific gravity of the acid leaving the system, an important check on the uniformity of operations.   The density measurement is made in the acid line and the measurement transmitted to the density recorder on the centralized control panel.

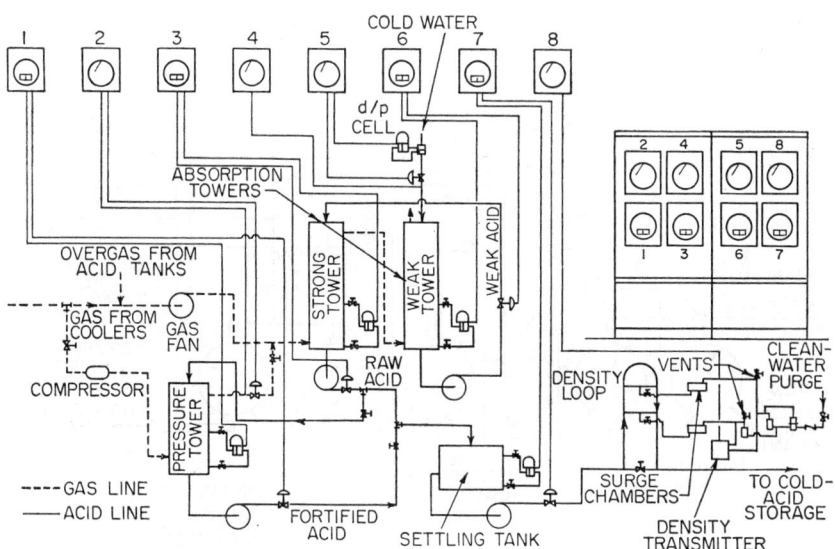

FIG. 14. Sulfur dioxide gas absorption.   (1) Indicating liquid-level controller, pressure tower; (2) recording pressure controller, pressure tower; (3) indicating liquid-level controller, strong tower; (4) temperature recorder, cold water; (5) recording flow controller, cold water to weak tower; (6) indicating level controller, weak tower; (7) indicating liquid-level controller, settling tank; (8) density recorder, cold acid.

Chemipulp K-C* Hot Acid Accumulator

The cold acid coming from the absorption system is fortified before use as cooking chemical in the digesters.   The Chemipulp K-C process,* known as the hot-acid accumulator system, is widely used.   In this system, SO_2 is recovered and the acid is heated by the digester relief gases, producing a uniform hot acid for cooking.

The hot-acid system provides a series of pressure vessels or accumulators to collect gas and entrained liquor relieved from the digester.   The SO_2 in this gas is absorbed by the acid in the accumulators, and its heat elevates the temperature of the acid.   The system is divided into zones so that the gas can be reclaimed at the various pressures encountered in the cooking cycle.   The control instrumentation for this system is shown in Fig. 15.

High-pressure digester relief gas and liquor go through the high-pressure eductor where they mix with the warm acid being pumped to the high-pressure accumulators. The high-pressure accumulators are vented to the low-pressure accumulator under control of the pressure controller, instrument 11.

When the digester pressure falls off, relief gas is diverted to the low-pressure eductor,

* Registered trademark, Chemipulp Process Inc.

mixing with the acid being pumped to the low-pressure accumulators. The low-pressure accumulator is controlled by recording pressure controller 8 by venting excess pressure to the pressure-recovery tower.

Similarly, excess pressure in the pressure tower is vented to the cold acid and from there to the absorption system. Pressures are controlled by recording pressure controllers 4 and 2.

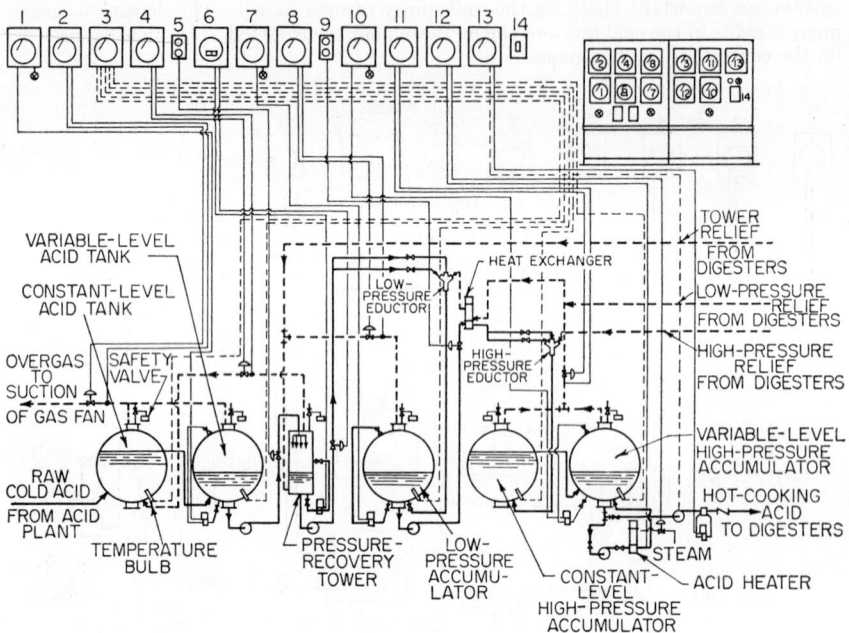

Fig. 15. Chemipulp (K-C type) hot-acid accumulator. (1) Liquid-level recorder for variable-level acid tank (with low-level alarm); (2) recording pressure controller, acid tanks; (3) multirecord dynalog temperature recorder, acid tanks, low- and high-pressure accumulators; (4) recording pressure controller, pressure-recovery tower; (5) remote valve operator for controlling flow of acid to pressure-recovery tower; (6) indicating liquid-level controller, pressure tower; (7) liquid-level recorder, low-pressure accumulator with low-level alarm; (8) recording pressure controller, low-pressure accumulator; (9) remote valve operator for controlling flow of acid from low- to high-pressure accumulators; (10) liquid-level recorder, variable-level high-pressure accumulator with low-level alarm; (11) recording pressure controller, high-pressure accumulators; (12) recording temperature controller, high-pressure accumulator—regulates steam to acid heater; (13) flow recorder with automatic totalizing shutoff for adding controller amount of cooking acid to digesters; (14) acid-batch setting mechanism.

The several supply tanks involved use level recorders with alarm (instruments 1, 7, and 10) or controls (instrument 6). Acid transfer is controlled manually by the operator through observation of instruments 5 and 9.

Multirecord temperature recorder (instrument 3) gives continuous records of the acid in each of the five tanks. Recording temperature controller (instrument 12) controls the final temperature of the acid by regulating steam to the acid heater.

Instrument 13 is a flow recorder with totalizing integrator on hot-acid flow to the digesters to show rate of acid flow, digester charging time, and number of digester charges.

Instrument 14, the digester acid-charge controller, is set for the desired acid charge, and when a "Start" button is pressed it charges this volume to the digester.

REFERENCES

Technical Articles

1. Ehrisman, H. O.: The Instrument Manufacturer's Problems in Meeting the Demands of the Paper Making and Converting Industries, *TAPPI*, vol. 36, no. 9, pp. 138–141, September, 1953.
2. Ehrisman, H. O.: Maintenance of Control Instruments in Pulp, Paper and Paperboard Industry, *Paper Mill News*, June 20, 1942.
3. Yunker, W. S.: Process Instrumentation, *Pulp Paper Mag. Can.*, January (Convention Issue), 1951.
4. Mann, A. A.: Automatic Shower Control in Vacuum Washing, *TAPPI*, vol. 39, no. 11, pp. 790–792, November, 1956.
5. Rich, J. H., and R. J. Becker: The Beta Gage Profiler as a Control Instrument, *Paper Mill News*, pp. 14–15, Nov. 27, 1954.
6. Drinker, H. S.: The Foxboro Moisture Control System, *Pulp Paper Mag. Can.*, Feb. 19, 1956.
7. Drinker, H. S.: The Foxboro Magnetic Flow Meter, *TAPPI*, vol. 39, no. 4, pp. 185A–187A, April, 1958.
8. Ciannamea, N. A.: Caliper Profiler Used as Control Instrument at W. Va. Mill, *Paper Trade J.*, Apr. 25, 1960.
9. Wrase, W. A.: Controlling the Variables, *TAPPI*, vol. 45, no. 7, pp. 189A–194A, July, 1962.
10. Seymour, G. W.: Oxidation Potential Central of Pulp Chlorination, *TAPPI*, vol. 40, no. 6, pp. 426–428, June, 1957.
11. Barber, R. F.: Process Instrumentation for Continuous Refining, *Paper Trade J.*, Oct. 31, 1960.

Textbooks

12. Stephenson, J. N. (ed.): "Pulp and Paper Manufacture," vols. 1, 2, 3, 4, McGraw-Hill Book Company, Inc., New York, 1950, 1951, 1953, 1955.
13. Calkin, J. B.: "Modern Pulp and Paper Making," 3d ed., Reinhold Publishing Corporation, New York, 1957.
14. Sutermeister, E.: "Chemistry of Pulp and Paper Making," 3d ed., John Wiley & Sons, Inc., New York, 1948.
15. Casey, J. P.: "Pulp and Paper," vols. 1, 2, Interscience Publishers, Inc., New York, 1952.
16. Cole, E. J., and M. Todd: "Pulp and Paper Mill Instrumentation," Lockwood Trade Journal Co., Inc., 1957.

FOOD INDUSTRY INSTRUMENTATION

By Robert Barclay Beahm*

The food industry is a complex of many separate industries. Generalities from the standpoint of instrumentation applications are difficult to make and could be more confusing than meaningful. Therefore, the editors have selected a cross section of applications, each highly specific, from five major segments of the food industry, namely, (1) brewing, (2) canning, (3) baking, (4) dairy, and (5) meat packing.

Although there are numerous instrumentation trends going on within the food industry, the applications selected for this section illustrate by far the large majority of in-being installations. Thus most of the applications shown utilize conventional instruments and control consoles. Also, for the temperature applications, filled-system thermometers are shown. In all instances, the reader may logically extend these applications to incorporate miniaturized panel instruments, highly complex centralized and integrated control panels, and electric and electronic measurements and controls instead of the mechanical and pneumatic means shown in the diagrams. The current practice, as covered by this text, has been established over the years as dictated by economic considerations.

INSTRUMENTATION IN THE BREWING INDUSTRY

A panoramic flow chart of the instrumentation utilized in a modern brewery is shown in Fig. 1. Each of the instrumentation applications shown on this chart is described in this section. The more complex applications are diagramed and detailed operating descriptions are provided. To those readers who are seeking solutions to instrumentation problems in related fields but who may not be familiar with the process engineering involved in brewing, reading of the reference material listed at the end of this section is suggested.

Control of Hot-water Tanks

Effective brewery operations require an ample supply of hot water at all times. The uniform temperature at which the water is maintained is a great asset on the many processes where it is used. For example, in sparging, a supply of water at constant temperature assures maximum yield from grains with a minimum amount of water consumption.

A modern instrumentation system will provide (1) regulation of the level in the hot-water tank to assure a constant supply and (2) regulation of water temperature. A system of this kind is shown in Fig. 2. The instrumentation required is as follows:

1. TRC and LRC (combined in one instrument case). Pneumatic control. Temperature range: 0 to 200°F with stainless-steel thermometer bulb. Liquid-level range: 0 to 100 using a bubbler-type measuring system.
2. Diaphragm control valve for water supply; reverse action (air to open).
3. Diaphragm control valve for steam supply; reverse action (air to open).
4. Air-reducing valve to provide 20 psi instrument supply air.
5. Filter for instrument air.
6. Rotameter for bubbler service.

* Systems Engineer, Taylor Instrument Companies, Rochester, N.Y.

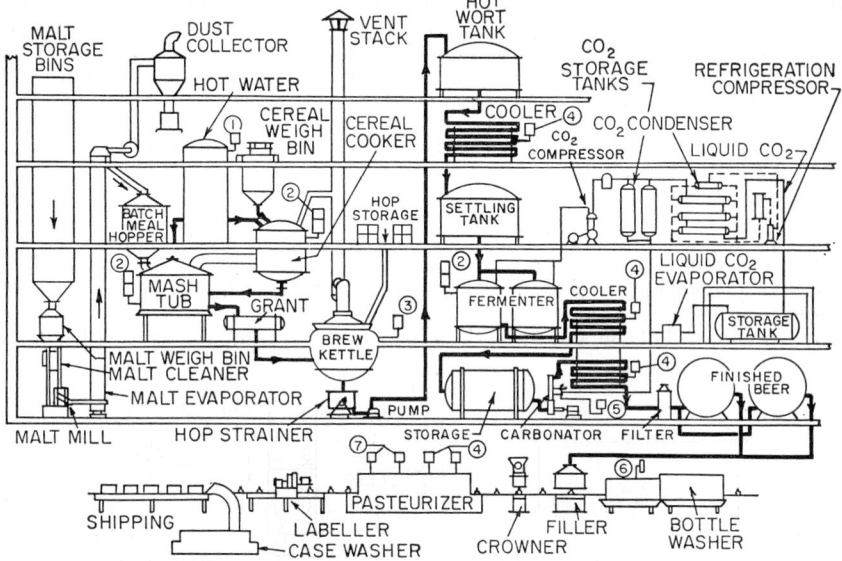

FIG. 1. Panoramic flow sheet showing key points for instrumentation in a brewery. (1) Recording temperature and liquid-level controller; (2) time-schedule controller; (3) recording temperature and pressure controller; (4) recording temperature controller; (5) indicating pressure controller; (6) self-acting temperature controller; (7) indicating temperature controller.

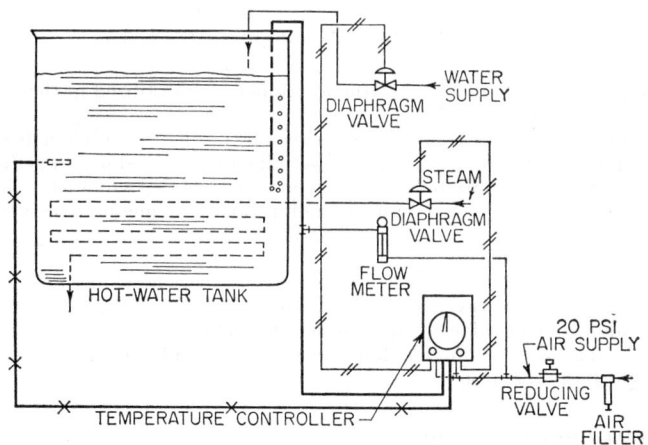

FIG. 2. Hot-water-tank control system.

Control of Cereal Cookers and Mash Tubs

An automatic control system is required to assure maximum starch extraction from each batch of cereal. The ideal time-temperature cooking schedule is determined by the brewmaster from past experience. In a given situation, for example, the brewmaster may have found that best results are obtained (for a specific mixture of malt and cereal) when the mixture is held at a temperature of 113°F for 30 min; then raising the temperature 1° per minute until a temperature of 158°F is obtained; followed by holding the mixture at that temperature for an additional 30 min; and finally bringing the mixture to the boiling point as rapidly as possible and boiling for 1 hr. This complex schedule would be very difficult to control manually and would be subject to numerous possibilities of human error.

The control system shown in Fig. 3 accomplishes the following actions automatically:

1. When the operator presses the start button, the cam in the time-schedule controller commences to turn.
2. A hold period is maintained for complete mixing.
3. Steam is then admitted for the first temperature-rise period.
4. The desired temperature is maintained during the second hold period.
5. Additional steam is admitted for the second rise period.
6. The temperature is controlled for the final hold period by the pressure-control mechanism.
7. The steam shuts off and the cycle is completed and ready for the next batch.

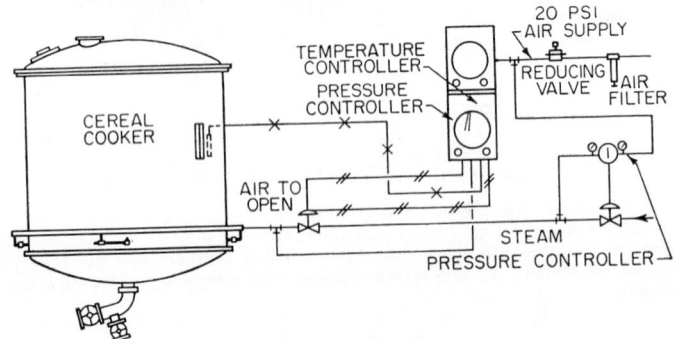

Fɪɢ. 3. Cereal-cooker and mash-tub control system.

Following are the characteristics of the instrumentation utilized in the control system of Fig. 3:

1. TRC equipped with a time-schedule cam for automatically changing the temperature set point in accordance with predetermined time schedule. Pneumatic control. Temperature range: 80 to 240°F with stainless-steel thermometer bulb. Period of cam clock is 2 hr. Instrument panel is equipped with momentary start button and signal lights.
2. PC for steam-supply line.
3. Diaphragm control valve on steam supply line, actuated by PC to maintain constant supply pressure; reverse acting (air to open).
4. Diaphragm control valve on steam supply to cooker; reverse acting (air to open).
5. Air-reducing valve to provide 20 psi instrument supply air.
6. Filter for instrument air.

Sparge-water Control System

After the wort reaches the top of spent grains, sparging with water at temperatures controlled within specific tolerances washes out all the extractives from the grain. The proper water temperature ensures maximum washing of the extractives. A sparge-water control system is shown in Fig. 4. This illustration shows how temperature control is accomplished, especially where an uncontrolled source of hot water is always above the sparging temperature.

The temperature-sensitive element is located on the downstream side of the mixing tee to assure an accurate indication of the mixed streams of water. Hot water comes in on one side of the tee, usually from the storage tanks located at a higher level.

Characteristics of the instrumentation employed in the system of Fig. 4 are as follows:

1. TRC with pneumatic control. Range: 0 to 200°F with stainless-steel thermometer bulb.
2. Diaphragm-control valve on cold-water line.
3. Air-reducing valve to provide 20 psi instrument supply air.
4. Filter for instrument air.

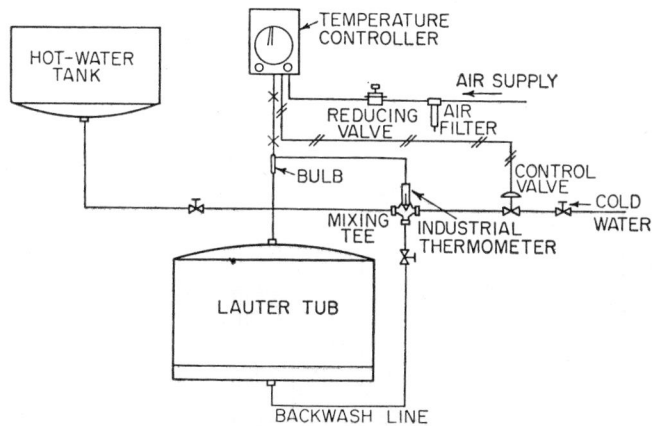

FIG. 4. Sparge-water control system.

Brew-kettle Control System

Beer obtains its stability and character in the brew kettle. A properly engineered automatic control system will yield a more uniform brewing, better wort concentration, and complete enzymic inactivation. The rate of evaporation, temperature, and movement of the wort in the kettle is important to the distinctive flavor of the final brew. A control system makes it easier to calculate hourly evaporation, secure a good "break," and achieve the desired flavor.

The system shown in Fig. 5 is based, first, on controlling the admission of steam to the coil and percolator according to the liquid level in the kettle. Secondly, temperature control during filling prevents burn-on. Finally, pressure control of the steam in the coil and percolator maintains the desired rate of boil or kettle roll.

Instrumentation used in the system of Fig. 5 comprises:

1. TRC and PRC (combined in one instrument case). Pneumatic control. Temperature range: 0 to 250°F using stainless-steel thermometer bulb. Pressure range: 0 to 100 psi.
2. Liquid-level transmitter.
3. Diaphragm-control valves.
4. Booster relay with 3:1 ratio.
5. Pilot valves.
6. Air-reducing valve to provide 20 psi instrument supply air.
7. Filter for instrument air.

Operation of the System. As the kettle is filled to the proper level, the liquid-level transmitter's output pressure goes to the two pilot valves which control the admission of steam to the coil and percolator through diaphragm valves 1 and 2. Then the operator adjusts the temperature controller to the proper simmering temperature on the chart and turns a three-way stopcock to the "on" position. The temperature is controlled at this point during filling or for a hold period, depending upon instructions received from the brewmaster. Next, the brew is brought to the boiling point by changing the set pointer to 212°F on the chart.

As the temperature approaches the boiling point, the operator turns the stopcock to a second position which switches the instrument to pressure control of the steam flowing to the coil and percolator so that it regulates the desired kettle roll.   The pressure controller is adjusted to a value which, from experience, will give the desired results. With a combined temperature-pressure control system of this type, the brewmaster does not have to consider barometric-pressure changes which demand continual manual adjustments.

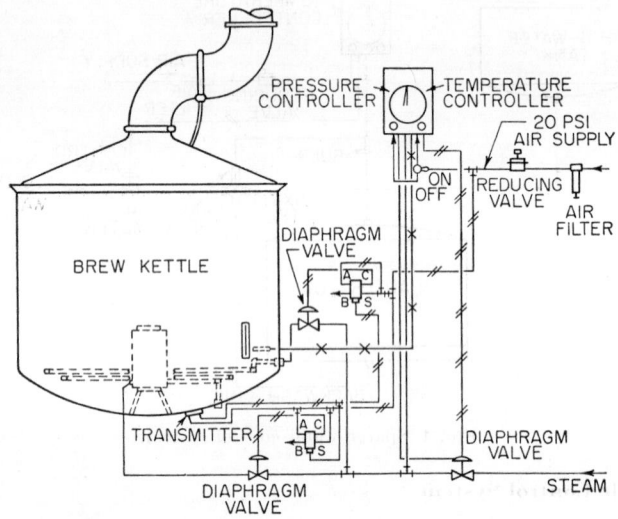

Fɪɢ. 5. Brew-kettle control system.

Wort-cooling Control System

After the spent hops and sludge are filtered off the wort, the wort goes through a cooling process to precipitate proteins, tannin, and hop-resin compounds in the form of coarse and fine flocs.   This "cold break" is of considerable importance to subsequent fermentation and character of the beer.

Types of Coolers.   Three types of coolers are used: (1) open types, such as the Baudelot cooler; (2) closed types of the shell-and-tube construction variety; and (3) plate types.   The control systems described here are applicable to either brine or ammonia coolants.   A control system for a plate-type heat exchanger is shown in Fig. 6; the control system for a shell-and-tube cooler is shown in Fig. 7.

Instrumentation Required.   The basic instrumentation required for the systems of Figs. 6 and 7 is as follows:

1. TRC with pneumatic control.   Temperature range: 0 to 100°F using a copper bulb contained in a stainless-steel well.   The instrument is equipped with electric contacts (low).   A direct-acting controller is used for ammonia systems, a reverse-acting controller for brine systems.

2. Sanitary pressure switch, or magnetic relay.

3. Air-reducing valve to provide 20 psi instrument supply air.

4. Filter for instrument air.

5. Diaphragm-control valve; reverse action (air to open).   An alloy valve is used for brine systems.

6. Three-way solenoid valve—for brine systems only.

7. Back-pressure control valve—for ammonia systems only.

8. Solenoid valve (normally closed)—for ammonia systems only.

Operation of Brine Systems. Wort is circulated at a fixed rate and the TRC regulates the control valve in the brine inlet line. The temperature-sensitive element is located in the wort outlet, and any increase or decrease in temperature is detected and corrected by the controller.

Operation of Ammonia Systems. When ammonia is used as a refrigerant, the same procedure is followed with one exception. A back-pressure control valve is

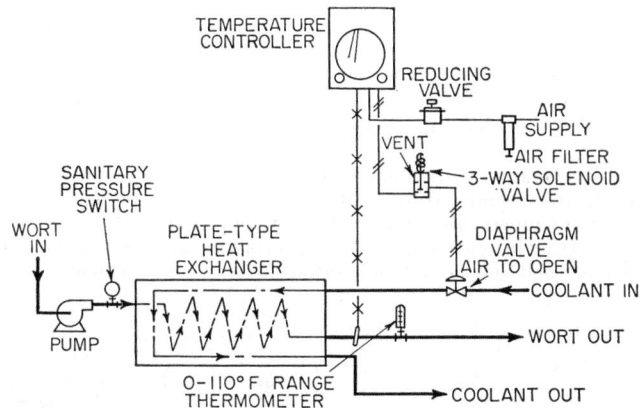

FIG. 6. Wort-cooling control system for a plate-type heat exchanger.

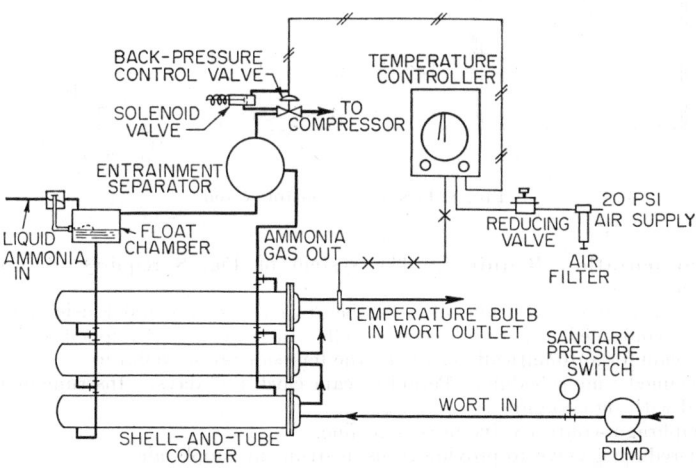

FIG. 7. Wort-cooling control system for a shell-and-tube cooler.

located in the ammonia suction line. To start the system, the operator presses a start button and holds it closed until the beer-pump discharge pressure is sufficient to close the pressure switch and sustain the pump motor circuit. With the temperature-sensitive element located in the wort outlet line, the changes in temperature are detected and the TRC regulates the valve to maintain constant temperature. The control system is designed to provide maximum back pressure in the cooler to prevent freeze-up in the case of beer-flow stoppage or electrical failure.

Fermenter Control System

Temperature affects the progress of fermentation. If the temperature is too low, there may be a slow start; too rapid cooling can cause yeast degeneration and yeast will be found floating on top. The importance of temperature control is obvious from a review of a typical schedule which many breweries follow.

Fermenter Schedule. Wort is pumped into the fermenter at about 44°F and yeast is added. During the first 3 days of fermentation, no coolant is added to the coils and the temperature in the vessel rises slowly to 53°F, this rise being due to heat liberated by the action of the yeast. At this point, introduction of coolant is started in order to keep the temperature from rising above 54.5°F. This temperature is maintained for a specific period. During the remainder of the fermentation period, the temperature of the beer gradually is lowered at a rate not exceeding 5°F over a 24-hr period.

Two methods generally are used in the control of temperature during fermentation: (1) control of the room temperature—not always satisfactory because of the difficulty in maintaining the temperatures of individual fermenters; and (2) control of wort temperature by admitting brine through submerged coils in each fermenter. The latter is a more flexible method and permits separate control of each fermenter.

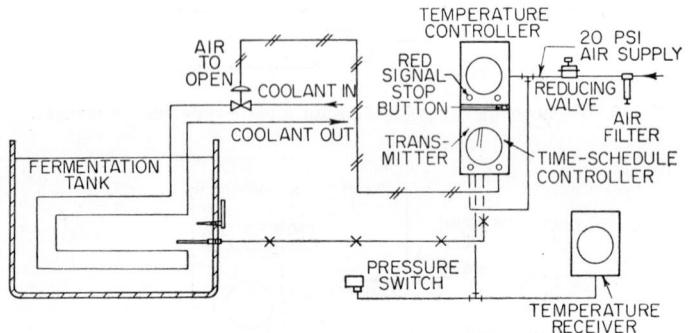

Fig. 8. Fermenter control system.

Instrumentation Required. The system of Fig. 8 requires the following instrumentation:

1. TRC with pneumatic control. Temperature range: 0 to 100°F using a stainless-steel bulb contained in a stainless-steel well. Instrument is equipped with a time-schedule cam for automatically changing the temperature set point in accordance with predetermined time schedule. Period of cam clock is 7 days. Instrument panel is equipped with signal lights.

2. Diaphragm-control valve in coolant line.

3. Air-reducing valve to provide 20 psi instrument supply air.

4. Filter for instrument air.

Operation of the System. The instrument cam is cut to conform with the fermenting schedule. The operation is commenced when the start button is pressed. When the temperature of the wort reaches 53°F it regulates a coolant valve to maintain, for the duration of the fermenting period, any desired temperature as called for by the brewmaster. The schedule may include a hold period for 1 or 2 days at the maximum temperature before the gradual lowering is started. The control system may be shut off at any time by manually cutting off the air to the controller, thus closing the coolant valve.

Bottle-washer Control System

Modern automatic bottle washers contain from one to six solution tanks which require temperature control. An automatic temperature-control system provides more complete sterilization, economically, with less bottle breakage.

Self-acting temperature controllers, as shown in the system of Fig. 9, provide sufficiently accurate control for most installations. Where closer temperature control may be desired, more sophisticated pneumatic or electronic temperature controllers can be used. Where self-acting controllers are used, a temperature recorder usually is incorporated in the system to inform the operator of the temperatures in the various tanks.

With reference to Fig. 9, as the bottles enter the washer, they are submerged in an alkali solution at an elevated temperature. The self-acting controllers maintain a steady temperature automatically and continuously to assure effective sterilization.

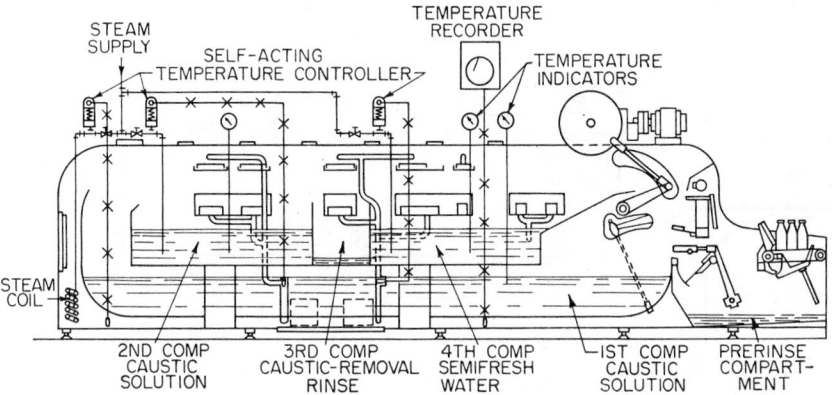

Fig. 9. Bottle-washer control system.

The remaining compartments further cleanse the bottles by hot-water jets and finally hot and cold rinse water prepares the bottles for the beer fillers. Temperature of the rinse water also is controlled by the self-acting controllers.

Control System for Beer Pasteurizers

The biological stability of beer is improved by pasteurization. Beer usually is bottled at about 34°F. As the bottles enter the pasteurizer, the temperature gradually is increased until it reaches 140°F. The temperature is held at this point for 15 to 20 min and then cooled as rapidly as possible. Automatic temperature control assures correct pasteurization and guards against excessive bottle breakage.

With the continued need for higher operating efficiency, individual instruments mounted locally at points of measurement are being superseded by integrated instrumentation systems which form an important part of modern brewery operations.

Operation of Pasteurizer Control System. The basic control is of temperature in five zones of the pasteurizer: (1) preheat, (2) superheat, (3) pasteurizing, (4) precooling, and (5) cool. Pneumatic temperature transmitters are mounted at each zone of the machine and a pneumatic signal proportional to temperature is transmitted to the recording and controlling instruments mounted on the panel. The multipen recorders provide a record of temperatures throughout the machine. The controllers mounted inside the panel modulate the steam and water valves to maintain control of zone temperatures at all times.

A completely interlocked electrical system incorporates an alarm system which guards against over- or underpasteurization.   Green signals at each zone indicate correct temperature and the machine runs automatically.   Should the temperature rise or fall beyond accepted limits, alarms will sound.   In addition to the readings shown by the instruments, colored lights will indicate the extent of off normal.   In event of excessive temperature variations due to equipment malfunction, the machine will be automatically stopped.

Additional instrumentation may include indicators for deck-speed timing, bottling rate, totalizing, and down time.

INSTRUMENTATION IN THE CANNING INDUSTRY

A panoramic flow chart of the instrumentation utilized in a broad cross section of the canning industry is shown in Fig. 10.   Most of the instrumentation applications shown on this chart are described in this section.   The more complex applications are

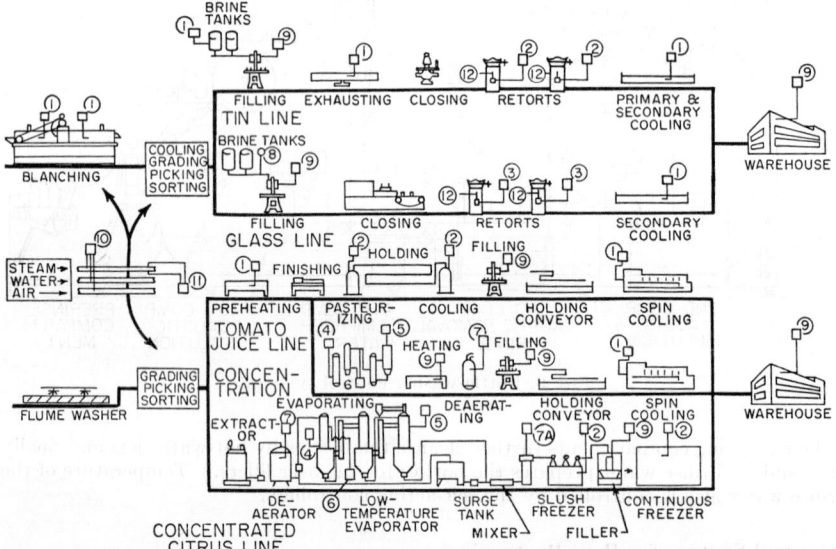

Fig. 10. Panoramic flow sheet showing key points for instrumentation in the canning industry.   (1, 4) Indicating temperature and pressure controller; (2) recording temperature controller; (3) recording temperature and pressure controller   (5) recording absolute pressure controller; (6) recording level controller; (7) ratio temperature controller; (8) dial-type indicating thermometer; (9, 10) temperature and pressure recorder; (11) flow recorder; (12) industrial-type indicating thermometer.

diagramed and detailed operating descriptions are provided.   To those readers who are seeking solutions to instrumentation problems in related fields but who may not be familiar with the process engineering involved in canning operations, reading of the reference material listed at the end of this section is suggested.

Blancher Control System

Correct blanching is the first step in the production of quality canned products. Blanching is the first operation in a series of temperature-processing operations wherein the product is subjected to elevated temperatures for the purpose of creating a chemical change.

Without automatic controls, it would be nearly impossible to maintain correct operating temperatures throughout the entire length of the blancher. A cold product of varying load conditions is introduced into the blancher, causing the inlet end of the blancher to fluctuate considerably in temperature. If an attempt is made to control only the inlet temperatures, the outlet end of the blancher will almost invariably be above the desired temperature, producing a detrimental effect on the product. Therefore, to ensure uniform temperatures under all normal load conditions, the most satisfactory method is to control the temperature automatically at both ends. This is the type of system shown in Fig. 11, as applied to a pipe blancher.

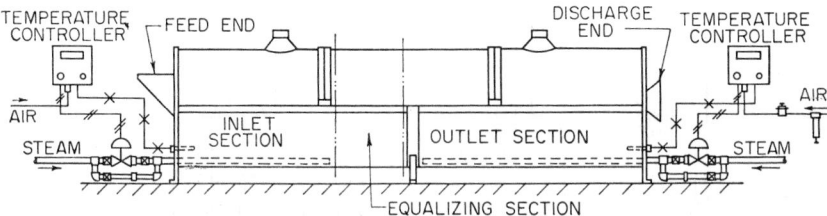

Fig. 11. Control system for blancher employing dual temperature controls.

Operation of the System. As the product enters the blancher, the inlet end cools faster. Then, the inlet controller responds immediately to admit the quantity of heat necessary in proportion to the rate at which the raw product is fed. By the time the product reaches the middle or equalizing section, it is thoroughly heated. Therefore, the controller at the outlet end admits just enough heat for a proper temperature balance.

Retort Control Systems

Three basic cooking methods are encountered: (1) steam cooking, (2) steam cooking with pressure cooling, and (3) water cooking with pressure cooling. A modern, fully automatic retort control system for water cooking with pressure cooling is shown in Fig. 12.

Operation of the System. The operator's responsibilities and the functions of the automatic control system are as follows:

Operator Responsibilities

1. Sets preheat temperature and depresses preheat push button.
2. Loads and locks retort.
3. Sets cook temperature and depresses start push button. Red signal is "on."
4. Unloads retort on completion of cycle. Red signal is "off."

Automatic Control System Functions

1. Controls temperature during the preheat period.
2. Operates agitation air for come-up, cook, and cool periods.
3. Operates steam valve for temperature control.
4. Operates air valve for pressure control.
5. When retort reaches process temperature, timing begins.
6. Times cook period.
7. At end of cook:
   Shuts off steam valve.
   Gradually introduces cooling water.
   Maintains pressure on retort.
8. Controls cooling on a time-temperature basis.
9. End of cool time—pressure control, cooling water, and agitation air "off."
10. Releases pressure on retort.
11. Sequence control resets for next cycle.

Control Systems for Steam Cook—Pressure Cooling.   When No. 2½ cans or larger are processed at temperatures above 212°F, it generally is necessary to carry on a pressure cooling period until the internal can pressure is low enough so that the cans may be handled under atmospheric conditions without deformation.

In these systems, the temperature is automatically maintained during the entire cooking period.   At the end of the cook, the pressure controller is turned on by the operator.   This instrument takes over to maintain the necessary introduction of

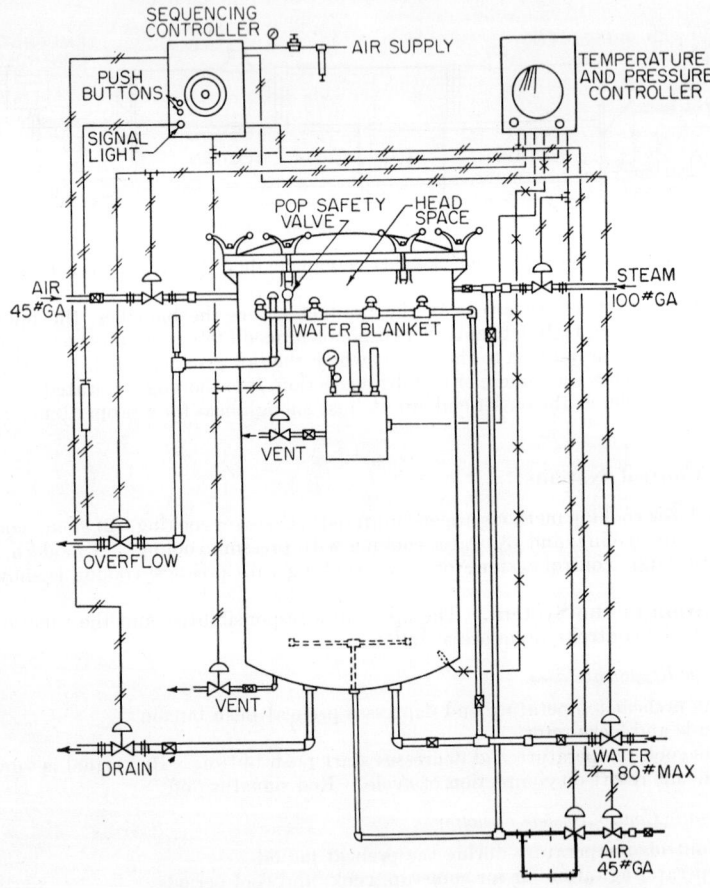

Fig. 12. Automatic retort control system.

compressed air into the retort and to keep that pressure at the proper point during the entire cooling period.

Control System for Water Cook.   Systems of this type are required for processing glass packages with standard caps in water.   These systems provide full advantage of positive water circulation in obtaining uniform cooling at a minimum of air consumption.   A larger quantity of air is supplied during the maximum steam demand and most active water-circulation requirements of the "come-up" period.   A smaller quantity sufficient for good processing is supplied during the "process and cooling" operations.   This is done automatically as the process temperature is reached.   With compressed air supplied continuously to a retort for circulation purposes, such a system reduces the total quantity of air used during a cook to a practical minimum.

The expenditure for air-compressor capacity also is materially reduced as a particular retort uses only air at the higher rate for a relatively small portion of the total processing period. "Come-up" from 150 to 240°F will average about 6 min, as compared with the total made-up processing period of about 40 min or more, depending on product and a cooling period of 15 to 20 min.

The processing of food in glass packages presents problems different from those found with tin containers. One of the problems is primarily due to the type of closure. This is often a cap having a pliable rubber ring held in contact with the "furnish" of the jar by means of higher pressure on the outside than exists on the inside of the container.

The jar is sealed off under vacuum, but during the processing, the internal pressure of the jar may rise to a point as high as 15 to 18 psi. It is necessary, therefore, that the pressure on the outside of the jars (inside the retort) be sufficient to keep the caps in place at all times. Accordingly, processing is done with the jars submerged in water which is under superimposed air pressure maintained throughout the processing cycle. Air space or head space above the water level also is an essential factor in obtaining satisfactory pressure control of the retort. The amount of head space determines the ease with which pressure control can be accomplished.

Horizontal Retorts. The same general types of controls can be applied on horizontal retorts as have been discussed for vertical retorts.

The control instruments for "steam-cook" horizontal retorts are quite similar while the system for "water-cook" horizontal retorts varies slightly because the retort must be empty of water prior to loading and unloading.

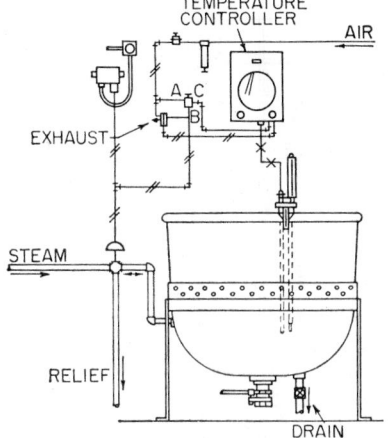

Fig. 13. Jam and jelly cooking-kettle control system.

Jam and Jelly Cooking-kettle Control

To ensure uniform consistency of jelly or jam, the steam supply must be shut off at exactly the correct "end-point" temperature. With the type of system shown in Fig. 13, the desired end-point temperature of the contents will not be exceeded. The system eliminates overcooking or undercooking, and thus each batch is concentrated with maximum yield.

The only manual operation required is that of depressing the starting valve button at the beginning of the cook. The control system automatically modulates the steam supply, shuts the supply off without supervision, and compensates for the effects of atmospheric-pressure changes on the boiling point.

Juice-heater Control System

Properly applied automatic controls for short-time high-temperature juice heaters help to make possible a higher-quality product of consistent purity, maximum flavor, and vitamin retention. The system maintains final juice temperature at the desired point even during complete stoppage of flow. The system also minimizes "burning on."

System for Steam-type Heaters. On this type of heater, the preheating of the product is automatically maintained and the final high temperature secured on the finished juice product assures complete sterilization of the juices.

As shown in Fig. 14, the product to be preheated is introduced into one section of the juice heater and passes directly into the extractor. Juices then are returned to different sections of the heater where the temperature is raised to about 250°F; and then to 255 or 260°F. The product, when heated to above boiling temperature, is constantly circulated and maintained under pressure to preclude boiling. In the holding section, high temperatures are maintained for a period to ensure complete sterilization. Temperatures in the cooling section also are automatically controlled to permit filling without further processing.

System for Water-type Heaters. In the type of system shown in Fig. 15, juice is pumped through the heat exchanger where it is heated to desired pasteurizing temperatures by means of a water-circulating system. Juice enters the regenerative section and is bypassed around this section as necessary in relation to the finished juice temperature.

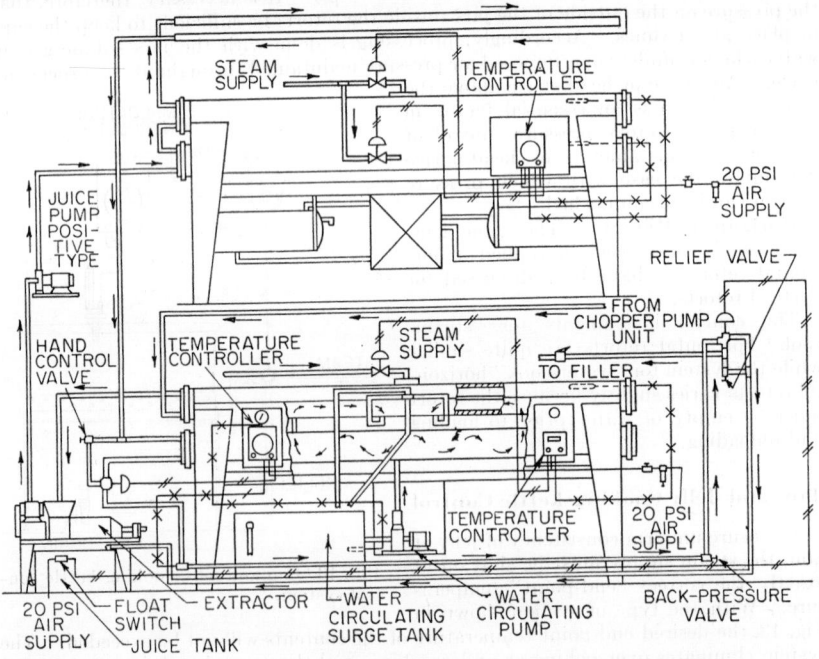

FIG. 14. Juice-heater control system for steam-type heater.

One of the systems of the three-pen instrument shown controls and records the temperature of the juice from the exchanger to the filler by means of the control valve in the raw-juice bypass. The second and third pens of this instrument record the temperature of the juice in the holding tubes, and the raw juice entering the heater. As shown, a second instrument controls the temperature of the circulating water and records the temperature of the water returning to the surge tank.

Zoned System for Steam-type Heaters. In this kind of juice heater, as shown in Fig. 16, four distinct temperatures are maintained as the juice passes successively through (1) the preheater, (2) the pasteurizer, (3) the holder, and (4) the cooler sections. In addition, a steady steam supply pressure is maintained in the preheater and pasteurizer sections.

If the juice is not properly processed as it leaves the cooler section, it is diverted through a sanitary-type stainless-steel three-way valve back to the pasteurizer section to be reprocessed.

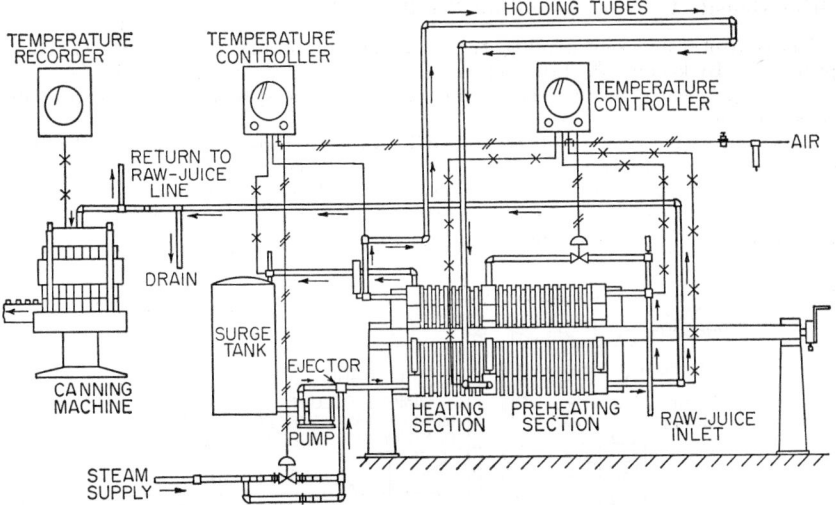

Fig. 15. Juice-heater control system for water-type heater.

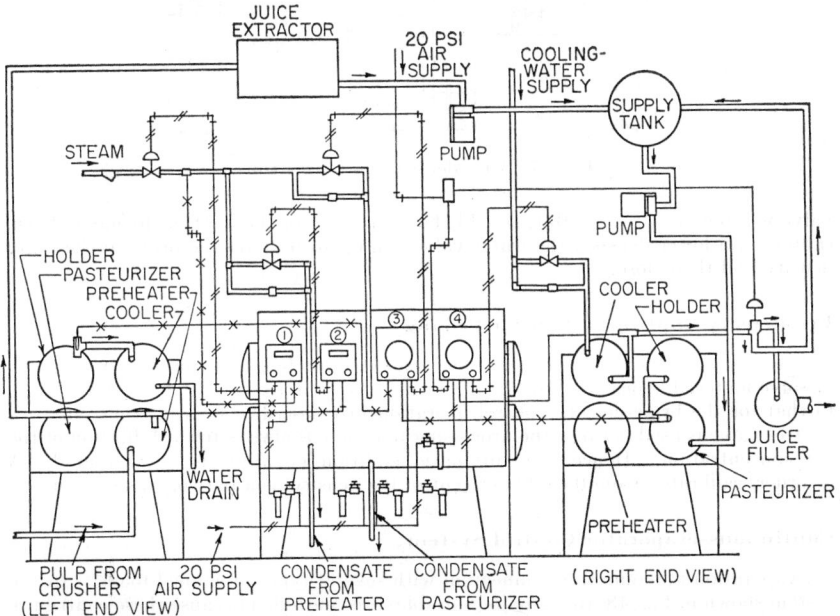

Fig. 16. Zoned juice-heater control system for steam-type heater. (1, 2) Indicating temperature controller; (3, 4) recording temperature controller.

Brine Density Control System for Pea Processing

The control system shown in Fig. 17 is applied to the grading of peas by the flotation principle. By keeping the brine solution within close limits, the top-quality peas are separated efficiently to give the maximum yield. The system also conserves salt consumption.

Brine density is automatically controlled by use of two throttling-type control valves on the brine and water lines. Both these valves are operated by the same controller, but they are staggered in their operation so that only one can be opened at a time. Normally, because of the water entering the grader with the peas, the brine valve will operate to add the required amount of brine to maintain the density; and the water

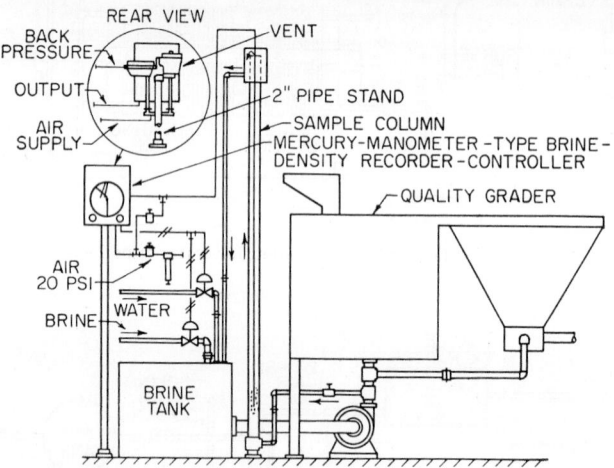

FIG. 17. Brine-density control system.

valve will remain closed. But, should the set pointer of the instrument move down, calling for a lower density, the water valve will open to dilute the brine to the lower density and then close.

Lye-density Control System for Peeling

The control of lye density in fruit and vegetable peeling is important so as not to dissolve material beneath the peel. The control system for this application is similar to that for the brine-density control operation just described—except that only one control valve is used because the processing tendency is always toward dilution of the lye concentration. Hence the automatic addition of water is not required. A recorder, calibrated from 0 to 20 per cent of lye density, is a part of this system.

Continuous-evaporator Control System

Evaporation is required in connection with several types of canned foods. In the system shown in Fig. 18, three major variables are controlled because these contribute largely to burn-on, entrainment, and nonuniform operation. These are (1) absolute pressure, (2) product level, and (3) steam pressure.

The absolute-pressure controller is used to provide a uniform pressure within the second effect by regulating the flow of water to the condenser. This instrument maintains a constant boiling temperature and tends to prevent entrainment by eliminating sudden decreases in pressure.

The pressure controller regulates the steam pressure to the first-effect heater, thus preventing fluctuations in the steam pressure.   By maintaining a constant steam supply, the controller aids in the reduction of burn-on and contributes to uniform operation of the system as a whole.

A separate controller for each effect maintains a uniform level.   The product feed to each effect is regulated to provide the desired level, and thus maximum efficiency of the heating surface is obtained and entrainment and burn-on are reduced.

If final-product-concentration control is desired, the product being discharged from the last effect of the evaporator can be fed to a measuring element which measures a variable related to composition.   Such a variable may be density or refractive index.

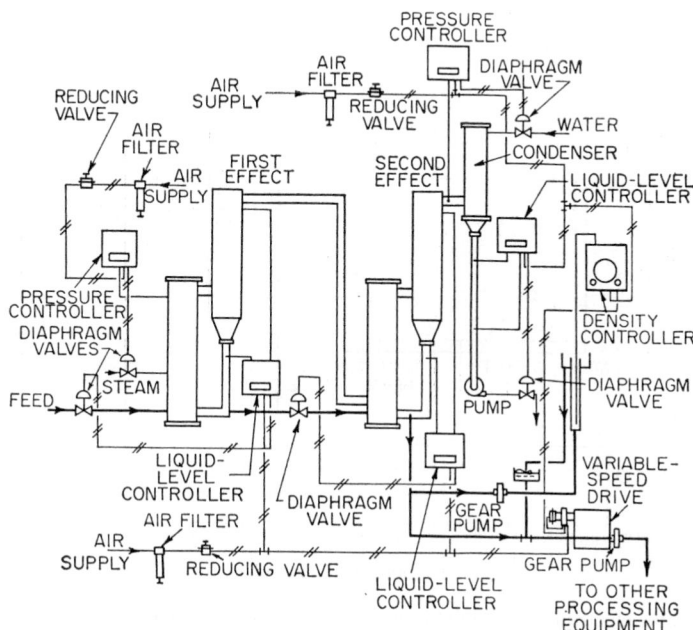

FIG. 18. Continuous-evaporator control system.

A feasible way of controlling composition then is to throttle the evaporator discharge to maintain the desired composition.   As long as the evaporator utilizes level controls in each effect, decreasing the rate of product discharge will allow time for more evaporation with subsequent higher product concentration.   Increasing the rate of product discharge will have the opposite effect.

INSTRUMENTATION IN THE BAKING INDUSTRY

Baking processes are many and varied.   A comprehensive coverage of all instrumentation applications in this industry is beyond the scope of this text.   Several of the major applications are discussed, and these generally illustrate the types of instrumentation problems encountered in the industry.

Dough-mixer Controls

Mixing time and temperature influence the proper development of dough gluten and hence the quality and yield of baked goods.   Excessive temperatures must be retarded during mixing.   In the system shown in Fig. 19, high temperatures are avoided by (1) equipping the mixer with a refrigerated cooling jacket; (2) adding metered, chilled ingredient water to the batch; and (3) employing a temperature-sensitive element in the mixer to measure the dough temperature directly and continuously.

To avoid too little or too much mixing, which will result in irregular gluten development, mixing time must be measured and controlled.   A satisfactory method involves the measurement of dough consistency, the latter being a criterion of the degree of dough development.   Motor load, in turn, is an indirect measure of dough consistency.

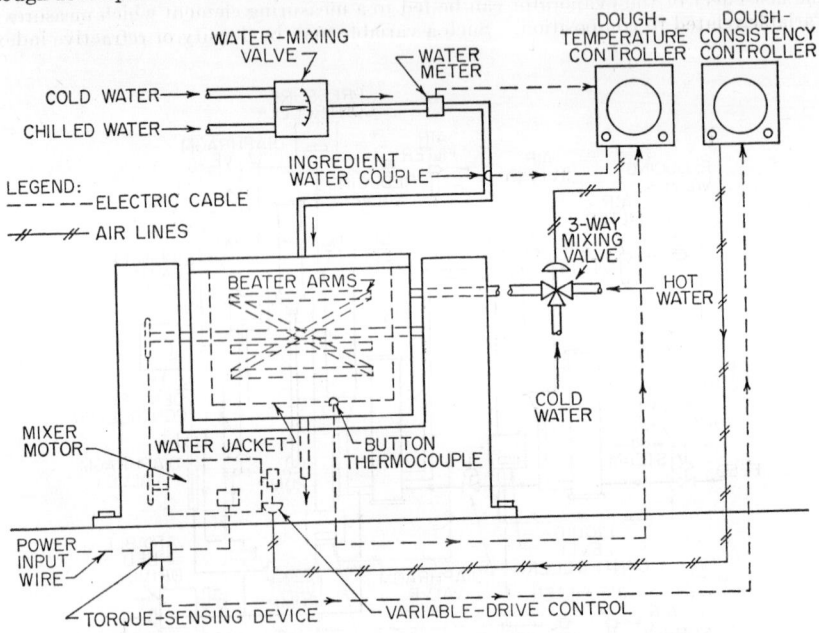

FIG. 19. Dough-mixer control system.

Therefore, an electronic instrument can be tied into motor loading and calibrated in terms of dough consistency.   Consistency measurements also detect formulation errors before the dough is fully mixed, indicate optimum mixing time for different kinds of flour, and provide a record of consistency for later reference and processing quality-control checks.

Fermenting- or Proof-room Control System

The automatic control system shown in Fig. 20 prevents skin formation on dough while it is proofing.   The system maintains a constant desired temperature and humidity on the basis of measuring the wet- and dry-bulb temperatures in the air-supply ducts to the proofing room.

The controller operates a diaphragm valve in the steam supply line to the heater to maintain the correct temperature.   The controller also operates a diaphragm valve in the steam line to the humidifying section to maintain a proper moisture content in the circulated air.   Thus air entering the proof room is held at a constant temperature and relative humidity.

Control System for Continuous Dough Making

The type of automatic control system shown in Fig. 21 makes it possible for two operators to produce 6,000 one-pound loaves of white bread per hour.   The system accomplishes in one continuous operation the functions of three different bakery departments, namely, (1) mixing, (2) fermentation, and (3) makeup (forming and panning the dough loaves).

The brew is formed from dry ingredients, liquid sugar, and water in a blending tank. Mixed for an hour, salt, milk, and a small amount of flour are added to form a liquid sponge. The latter is transferred to a holding tank. Water then is added and the batch is agitated for another hour. The sponge proceeds to a fermentation trough for an hour of mild agitation while undergoing fermentation. The jacket of the trough is temperature-controlled (at approximately 86 to 90°F) for optimum yeast activity.

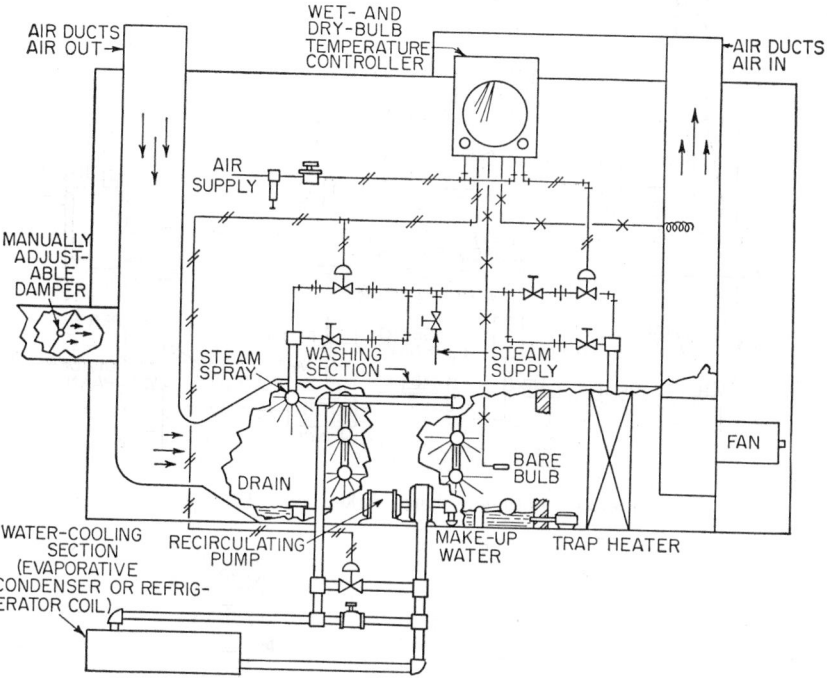

Fig. 20. Control system for proof or fermenting room.

Upon leaving the fermenter, some sponge is fed back to the blending tank for seeding. The remainder of the sponge goes to a constant-level tank. Sponge then proceeds to a heat exchanger to lower the product temperature—and then to an incorporator where shortening, oxidizing agents, and the remaining flours are added. A variable-pitch screw conveyor in the incorporator blends materials into a premix that passes to a developer. In the developer, counterrotating double-arc paddles shear and stretch the dough mass into finished dough. The dough then flows into a divider-panner which deposits required-size pieces of dough into the pans.

Baking-oven Controls

Traveling-tray and tunnel-type ovens require close control to maintain the balanced temperature gradients needed for multiple-zone ovens. The instrumentation utilized should be sufficiently flexible to accommodate for situations where a given oven may be used to bake a variety of products.

With the advent of continuous ovens, modulating control of fuel to burners was required to match oven load at any given time. A "high-low" method of oven firing has been found to be satisfactory.   Often, thermocouple-actuated controllers position a valve motor in the fuel line when the oven temperature rises above or drops below the instrument set point.   When the temperature exceeds the set point, the "high"

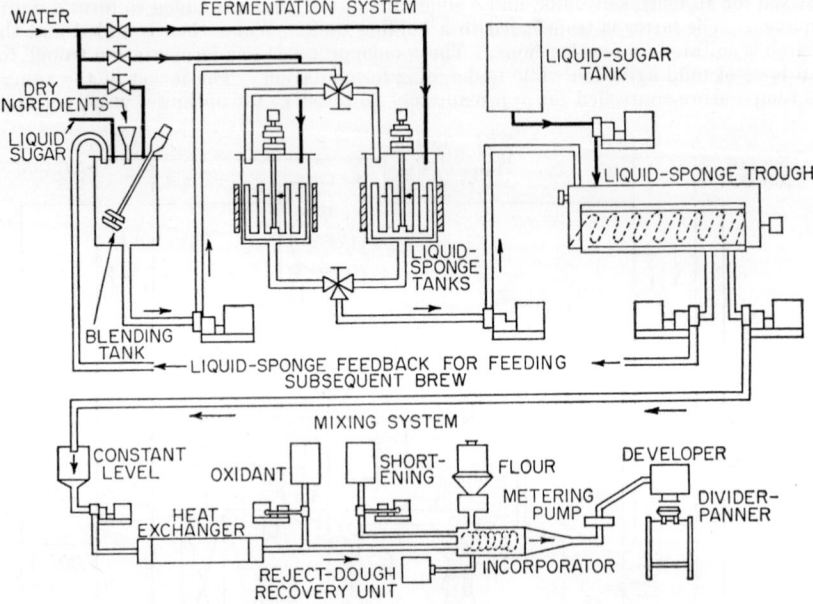

FIG. 21. Continuous dough-making control system.

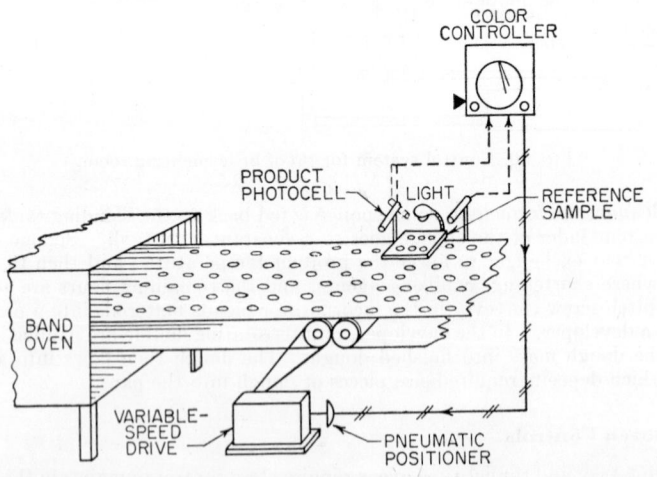

FIG. 22. "On-belt" color-scanning system used in cracker production.

instrument contact drives the control valve to a predetermined low-flame opening. On the other hand, when the temperature falls below the set point, a "low" contact drives the valve back to the full-flame position.

Since baking time is directly influenced by oven conveyor speed, traveling ovens are equipped with indicating tachometers geared to chain drives. An instrument automatically correlates conveyor-drive rpm with the time that the product travels through the oven. Also employed are electronic tachometers with built-in controls for automatically regulating baking time. Oven time is preset on the instrument, which controls baking time by way of adjusting the oven-drive mechanism.

To supervise more easily modern ovens which may be up to 300 ft in length, the automatic controllers are centralized on a single panel. Four-zone traveling-hearth ovens, for example, have indicating air-control thermometers for each zone, an indicating oven-speed tachometer, an indicating zone-temperature pyrometer, and a 12-thermocouple selector switch. Thermocouples are spotted throughout each zone as a check on heat distribution. Through the rotary selector switch, the operator may scan all thermocouples.

One cracker baker has adapted a color-control concept that originally was used by a dry-cereal processor. The system is used to control automatically the color of baked crackers exiting from a continuous oven. The "on-belt" color scanning developed by the cereal manufacturer is shown in Fig. 22. A photometric instrument measures the cracker color, and adjustments are automatically made by varying the speed of the conveyor.

Coating-machine Control System

To produce a uniform coating of chocolate, for example, on a product, a constant temperature must be maintained. If the temperature is allowed to drop, the coating

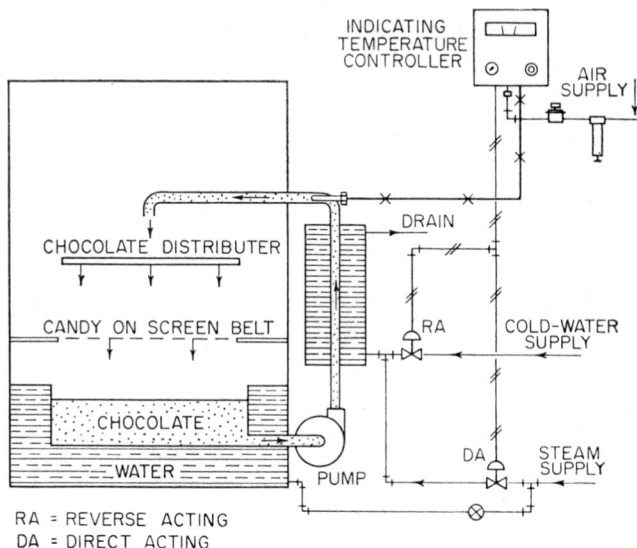

FIG. 23. Coating-machine control system.

thickness increases; if the temperature is permitted to rise, the coating thickness becomes too thin. A coating-machine instrumentation system is shown in Fig. 23.

The temperature-sensitive element is located in the chocolate at the nearest point to the outlet of the chocolate distributor as may be practical. This element senses any variation in temperature of the coat and transmits this detection to the mechanism of an automatic controller. The latter immediately operates the valves on the steam and the cold-water supply lines by output air pressure to maintain the chocolate at the required viscosity.

Potato-chip-cooker Control Systems

Properly applied automatic controls help to maintain uniform potato-chip color and to keep shrinkage and rejects to a minimum.   These controls also contribute to upping the rate of production.   One control system is shown in Fig. 24.

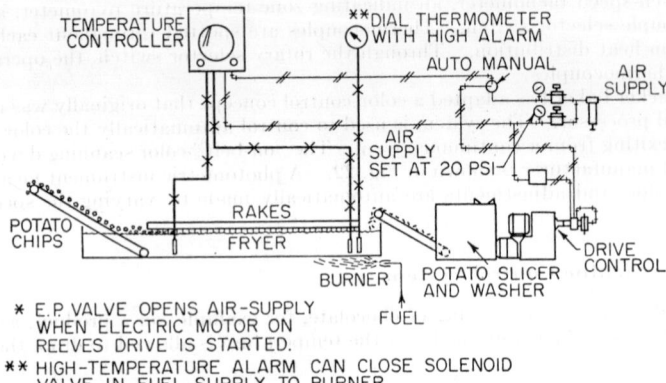

* E.P VALVE OPENS AIR-SUPPLY
  WHEN ELECTRIC MOTOR ON
  REEVES DRIVE IS STARTED.
** HIGH-TEMPERATURE ALARM CAN CLOSE SOLENOID
   VALVE IN FUEL SUPPLY TO BURNER.

FIG. 24. Potato-chip-cooker control system.

In this system, a temperature controller with an extra recording pen controls the process by regulating the speed of a Reeves drive in accordance with the temperature at the hot end of the cooker.   The flame is adjusted for the highest value, and the control system maintains the temperature at the desired point at the inlet end by changing the rate of feed, that is, the amount of potato chips fed into the cooker.

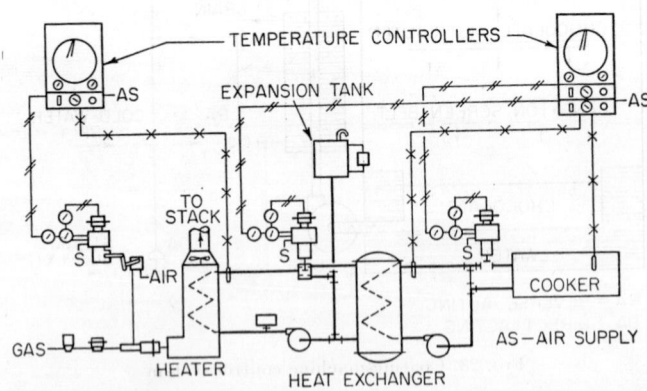

FIG. 25. Control system for indirectly heated potato-chip cooker.

As a further aid to adjusting the rate of chip travel, the instrument has an additional temperature system with its sensitive element located at the discharge end.   This keeps the operator informed of the temperature at this point and enables him to adjust the rake movement manually so that the desired quality and color may be maintained.

Control System for Indirectly Heated Potato-chip Cooker. With direct heating of the cooking fat by a gas-burner flame, high vapor-film temperatures may cause deterioration of the fat with resultant adverse effects on the appearance, flavor, and shelf life of the potato chips.

These difficulties may be alleviated through the use of an indirect-heating method that uses a special heat-transfer fluid.* This fluid boils at 650°F and may be subjected continuously to temperatures of 600°F without decomposition. In the system shown in Fig. 25, the heat-transfer fluid is heated to 480°F in a gas-fired heater and is pumped through the shell compartment of a tube-in-shell heater while the cooking fat is pumped through the tubes. With this arrangement of equipment, the cooking fat may be maintained at the correct processing temperature.

As shown by the diagram, one instrument controls the temperature of the heat-transfer fluid. This instrument regulates the quantity of the gas-air mixture going to the heater burner to maintain the desired heater outlet temperature. The cooking-oil temperature is automatically controlled by a double-duty controller. The left-hand control system regulates the inlet temperature of the cooking kettle by allowing the hot heat-exchange fluid to enter the heat exchanger if the kettle inlet temperature is below 380°F, or to bypass it if the temperature is at or above 380°F. The right-hand control system regulates the kettle outlet temperature by varying the quantity of hot cooking oil flowing through the kettle. An outlet temperature of 350°F is thus maintained. The elevated expansion tank vented to atmosphere maintains a steady pressure head on the heat-exchange medium circulating system.

INSTRUMENTATION IN THE DAIRY INDUSTRY

A panoramic flow chart of the instrumentation utilized in the dairy industry is shown in Fig. 26. Several of the instrumentation applications shown on this chart are described in this section. The more complex applications are diagramed, and detailed operating descriptions are given. To those readers who are seeking solutions to instrumentation problems in related fields but who may not be familiar with the process engineering involved in dairy operations, reading of the reference material listed at the end of this section is suggested.

High-temperature Short-time Pasteurization Control Systems

The instrumentation for HTST operations varies somewhat with the type of processing equipment used. Systems for three kinds of heaters will be described: (1) the water-circulating type, (2) the steam-heated type, and (3) the tubular-heater type.

Controls for Water-circulating Equipment. The left-hand instrument of Fig. 27 houses a safety thermal-limit recorder. Milk leaving the heating section of the pasteurizer passes through the holding tube and is checked by the safety thermal-limit recorder bulb which is located at the end of the holding tube. If the milk is properly pasteurized, the instrument places the flow-diversion valve in the forward flow position, allowing the milk to enter the regenerator and cooler sections of the pasteurizer. A green light informs the operator that proper pasteurization has been accomplished.

Should the milk leaving the holding tube be below the legal temperature, the safety thermal-limit recorder will immediately cause the flow-diversion valve to assume the diverted flow position, returning the milk to the raw-milk tank to be reprocessed. A red light signals the operator when diversion occurs, and a pen at the outer edge of the chart provides a record of the time and duration of the diversion. The flow-diversion valve is designed so that failure of the valve to assume the proper position under sublegal-temperature conditions will result in stoppage of the milk pump.

The control mechanism in the right-hand instrument of Fig. 27 maintains a constant heater temperature by way of a diaphragm control valve in the steam line. This instrument incorporates automatic reset which compensates for fluctuations in steam pressure. With this feature, separate steam-pressure control devices are not required.

* Aroclor, Monsanto Chemical Co.

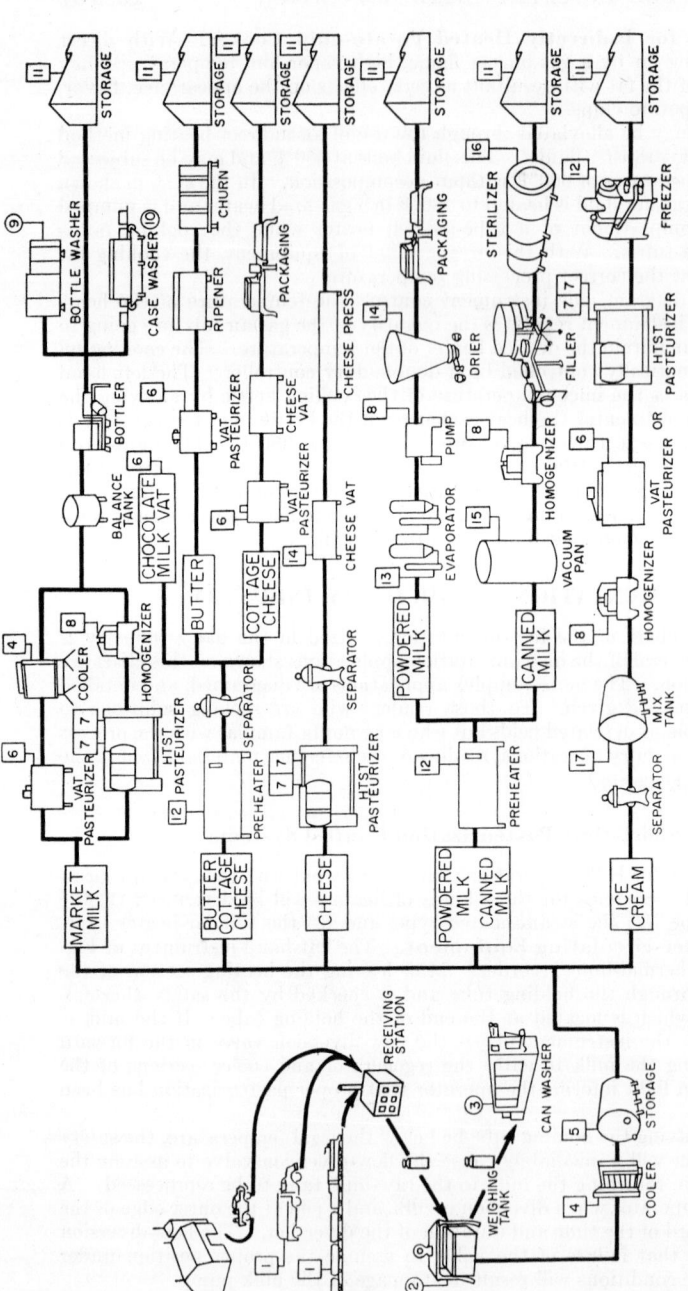

Fig. 26. Panoramic flow sheet showing key points for instrumentation in the dairy industry. (1) *Truck and tank cars*: recording thermometers. (2) *Receiving room*: thermometers for sanitary service tanks and receiving platforms. (3) *Can washers*: dial thermometers and self-acting temperature controllers. (4) *Coolers*: recording and controlling thermometers. (5) *Storage tanks*: dial and recording thermometers; sanitary-type stem thermometers. (6) *Vat pasteurizers*: recording thermometers and self-acting temperature controllers; sanitary-type stem thermometers. (7) *HTST pasteurizers*: HTST controllers; flow-diversion valves; sanitary-type stem thermometers. (8) *Homogenizers*: indicating and recording volumetric pressure gages. (9) *Bottle washers*: dial thermometers and self-acting temperature controllers. (10) *Case washer*: dial thermometers and recording temperature controllers. (11) *Storage rooms*: recording thermometers. (12) *Preheaters*: indicating and recording temperature controllers; self-acting temperature controllers; sanitary-type stem thermometers. (13) *Evaporators*: pressure, liquid-level, and absolute-pressure controllers. (14) *Spray dryers*: temperature and volumetric-pressure controllers. (15) *Vacuum pans*: pressure, absolute-pressure, and liquid-level controllers. (16) *Sterilizers*: recording temperature controllers; industrial thermometers. (17) *Mix tanks*: thermometers; sanitary-type stem

With a system of this type, full boiler pressure is immediately used to establish forward flow in the least possible time. This is of primary importance because it limits overheating at start-up or following diversion. Diversion periods are less likely and will be of shorter duration if they do occur.

A cold-milk recorder also is located in the right-hand instrument. The thermal element is located in the outlet of the cooler, and the instrument records the temperature of the milk as it leaves the cooling section.

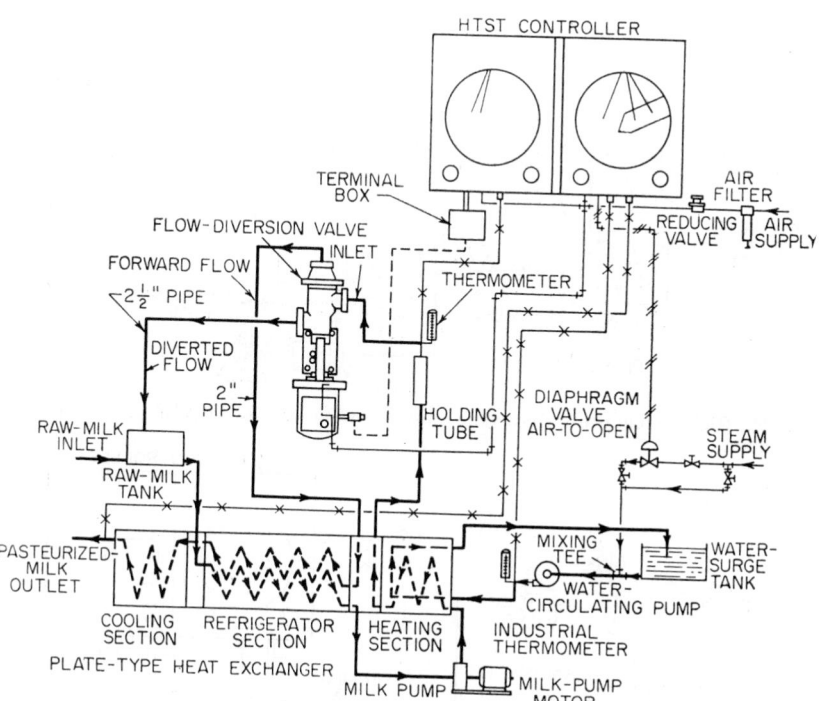

F<small>IG</small>. 27. High-temperature short-time pasteurization control system for water-circulating equipment.

Controls for Steam-heated Equipment. Modification of the foregoing system for application to steam-heated equipment is clearly evident from inspection of Fig. 28.

Controls for Tubular-heater Equipment. Modification of the system for application to tubular-heater equipment is clearly evident from inspection of Fig. 29.

Cleaning-in-place (CIP) Systems. Formerly, the piping and equipment used in dairy processing and instrumentation equipment had to be dismantled for frequent cleaning. Improvements in cleaning solutions and circulation techniques have made in-place cleaning possible. Obvious advantages include the saving of labor required for dismantling, cleaning, and repiping operations, as well as the lengthening of equipment life through elimination of wear and tear caused by frequent dismantling operations.

In-place cleaning involves the use of control systems for ensuring the correct cleaning-solution temperature during all phases of the cleaning cycle. A time-cycle controller regulates the sequence and duration of each cleaning cycle. Sanitary control valves and actuators are operated by temperature and time-cycle controllers to select

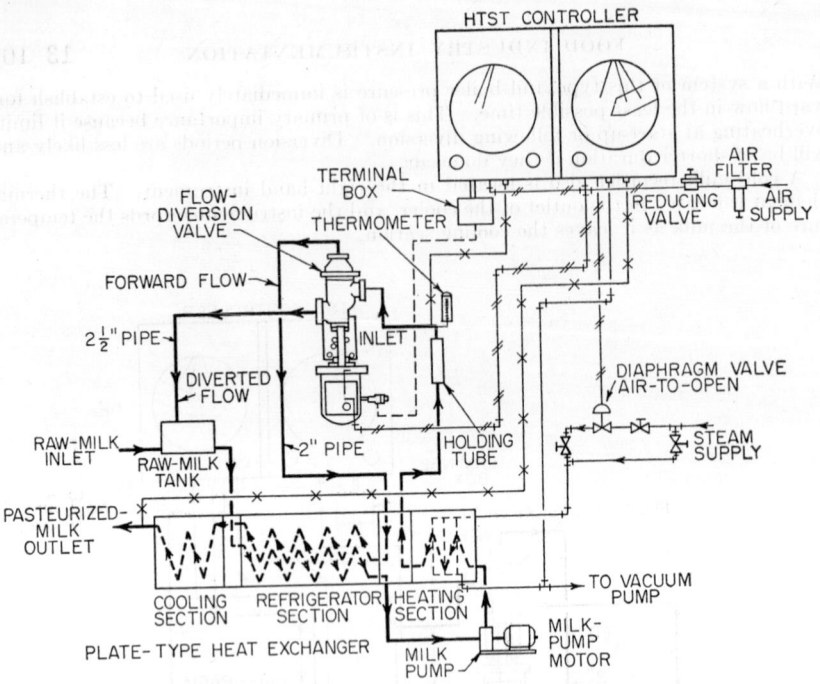

FIG. 28. High-temperature short-time pasteurization control system for steam-heater equipment.

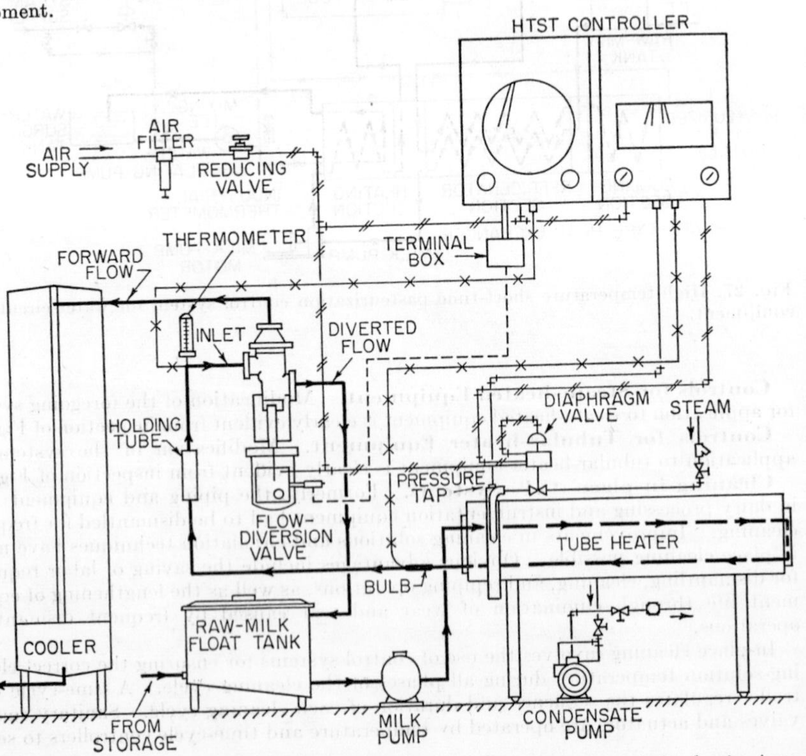

FIG. 29. High-temperature short-time pasteurization control system for tubular-heater-type equipment.

the particular section of processing equipment to be cleaned.   They also regulate the direction and flow of the cleaning solutions.

Vacuum-pan Control System

In the vacuum-pan control system shown in Fig. 30, an absolute-pressure controller maintains a uniform pressure within the pan by regulating flow of water to the condenser.   This assures that only the water required to maintain the desired pressure is

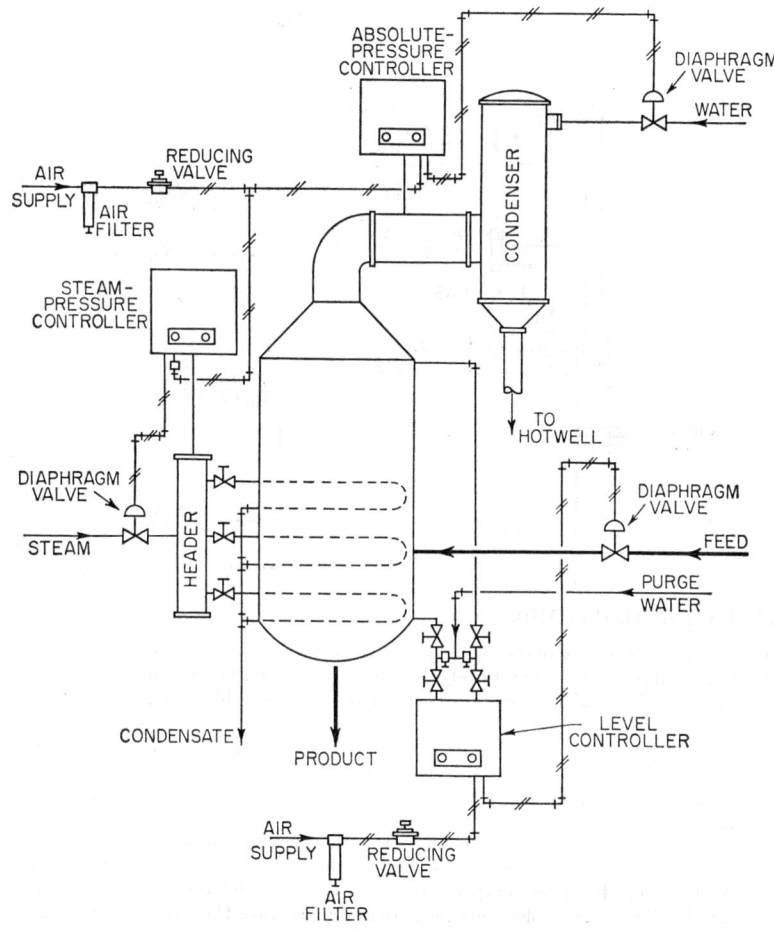

FIG. 30. Vacuum-pan control system.

used.   The instrument maintains a constant boiling temperature and tends to prevent entrainment by eliminating sudden decreases in pressure.

The pressure controller regulates steam pressure to the heating coils in the pan.   By preventing fluctuations in steam pressure, this instrument tends to reduce burn-on and gives more uniform operation.   A level controller regulates the product feed to the pan in order to maintain a proper product level.   This assures a maximum efficiency of the heating surface and helps to reduce entrainment and burn-on.

Spray-dryer Control System

The system shown in Fig. 31 automatically controls the gas and air mixture to the burners by measuring the dryer outlet temperature. In addition, a dial thermometer with rising alarm contact is provided to limit the maximum air inlet temperature to the dryer. The quantity of product sprayed into the dryer is determined by manual adjustment of the high-pressure pump. In normal operation, once this adjustment is made, it remains constant.

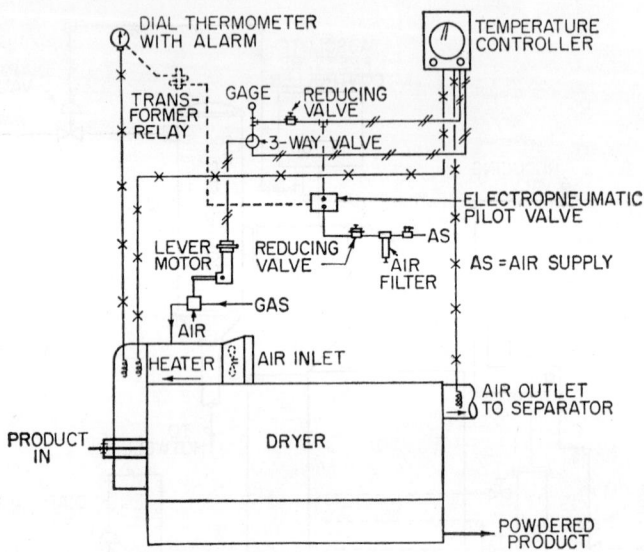

Fɪɢ. 31. Spray-dryer control system.

Deodorizing of Market Milk

Deodorizer units are available with one, two or three bodies. As an example, in a two-body installation the second body is subject to the higher vacuum. This deaerates the milk and deodorization results. Steam may be added to wash the product giving further deodorization.

Controlling product composition is vital. Dilution results if less water is vaporized in the second body than is produced by condensation of steam in the first body. Concentration results if more is vaporized in the second body than is condensed in the first body.

If inlet and outlet temperatures are the same, there will be no change in composition. But should the outlet temperature be higher than the inlet temperature, the end product is diluted. If outlet temperature is lower than inlet temperature, product is concentrated. These variables and these demands dictate the instrumentation.

INSTRUMENTATION IN THE MEAT-PACKING INDUSTRY

A panoramic flow chart of the instrumentation utilized in the meat industry, typified by pork processing, is shown in Fig. 32. Several of the instrument applications shown on this chart are described in this subsection. The more complex applications are diagramed and detailed operating descriptions are given. To those readers who are seeking solutions to instrumentation problems in related fields but who may not be familiar with the process engineering involved in meat-packing operations, reading of the reference material listed at the end of this section is suggested.

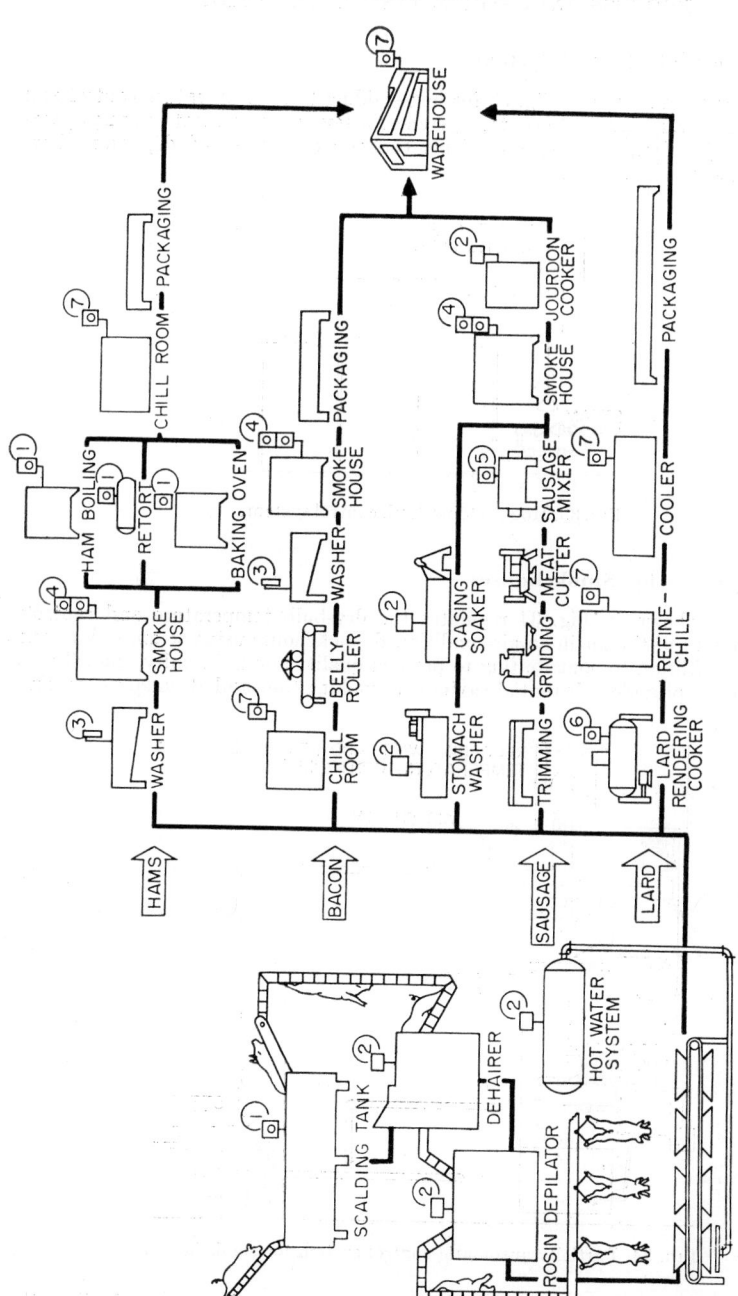

Fig. 32. Panoramic flow chart showing key points for instrumentation in pork processing. (1) Recording temperature controller; (2) indicating temperature controller; (3) self-acting temperature controller and expansion-stem temperature indicator; (4) time-schedule temperature controller; (5) time-temperature and time-pressure controllers; (6) indicating pressure controller; (7) recording thermometers.

Hog-scalding-tank Control System

The basic system shown in Fig. 33 for hog-scalding-tank temperature control and described here is very similar to control systems for several other meat-packing operations such as jourdon, sausage, and wiener cookers; dehairers and depilators; hot-water tanks; and ham boilers.

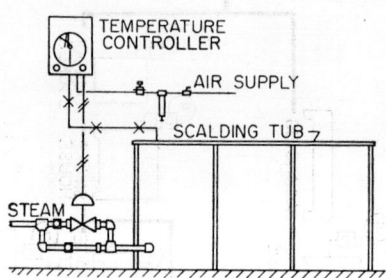

FIG. 33. Hog-scalding-tank control system.

Control Systems for Smokehouses

The system shown in Fig. 34 regulates the dry-bulb temperature and controls moisture content of the air in an air-conditioned smokehouse using steam as a means of heat. In addition to contributing to product quality and uniformity, smokehouse control systems provide substantial savings in smoking time and attendant operating

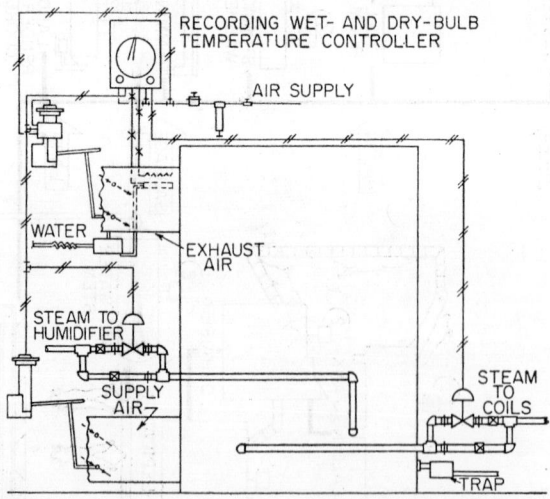

FIG. 34. Partially automatic control system for smokehouse.

costs. Two diaphragm valves are used, one in the steam-heater supply line, the other on the steam spray line. Smoke density is regulated by opening or closing dampers in ducts from the smoke unit to the smokehouse.

It must be stressed that no control system can make up for the deficiencies of a poorly designed, inefficient smokehouse. Where an old smokehouse is a problem,

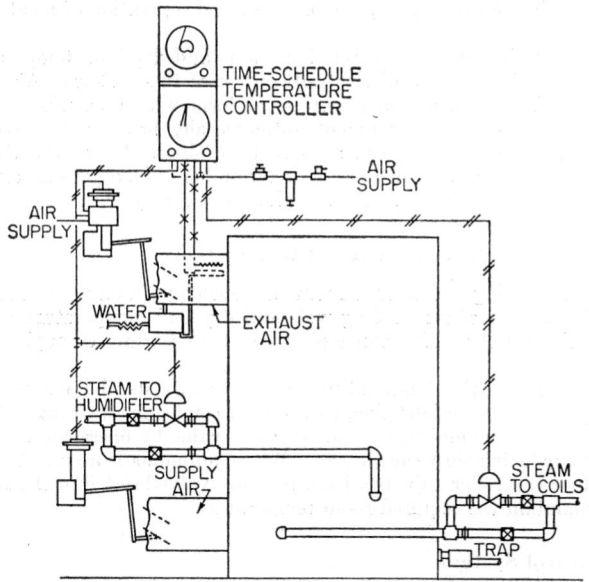

Fig. 35. Fully automatic control system for smokehouse.

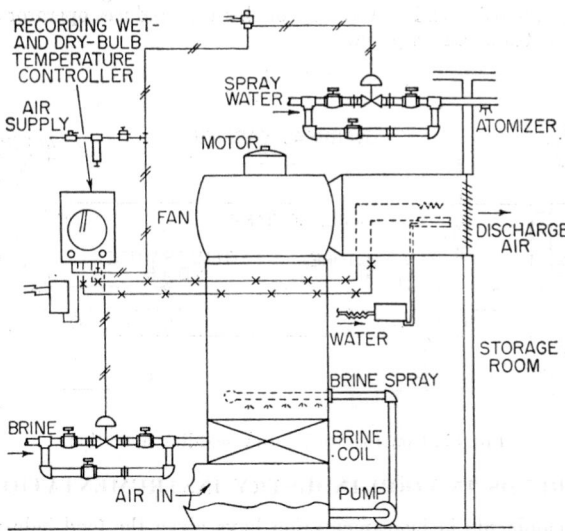

Fig. 36. Meat-packing and storage-room control system.

extensive modernization is indicated, and in some instances, a completely new house may be required if maximum product quality and operating efficiency are targets.

Completely Automatic Smokehouse Control System. With the system shown in Fig. 35, all operations are automatic once the operator depresses the start button. Experience has shown that a system of this type will guard against flavorless hams, tough casings on sausages, green centers, and separation of meat from bone or casing.

A time-schedule instrument controls both wet- and dry-bulb temperatures within the smokehouse. Timing is a distinct advantage of this system. After a predetermined schedule is set up, a cam is cut to conform with that schedule. Once the start button is pushed, the instrument automatically brings the smokehouse up to temperature—gradually or rapidly, as may be desired. Then, the temperature is maintained at a constant value for the desired number of hold periods, after which the system shuts off automatically at the completion of the cycle.

Meat-packing and Storage-room Control System

Most meat-packaging and storage rooms are cooled with either a flooded ammonia or brine system. Use of a recording wet- and dry-bulb temperature controller is a practical solution for the maintenance of proper temperatures throughout the rooms. See Fig. 36.

In a brine system, the temperature controller adjusts the temperature-control valve, located in the brine inlet line, as the load conditions warrant. The wet-bulb controller maintains the moisture at an optimum value by operating the humidifying valve, thereby reducing shrinkage losses. Where ammonia is used as the refrigerant, the temperature controller adjusts a back-pressure valve located in the ammonia suction line to maintain the required room temperature.

Scalding Control System

From a processing standpoint, poultry scalding can be described as delicate. In the system shown in Fig. 37, a recording temperature controller maintains constant temperatures in the first half of each tank and an indicating controller regulates the temperature at the other end. A system of this type reduces operator attention and usually reduces steam consumption.

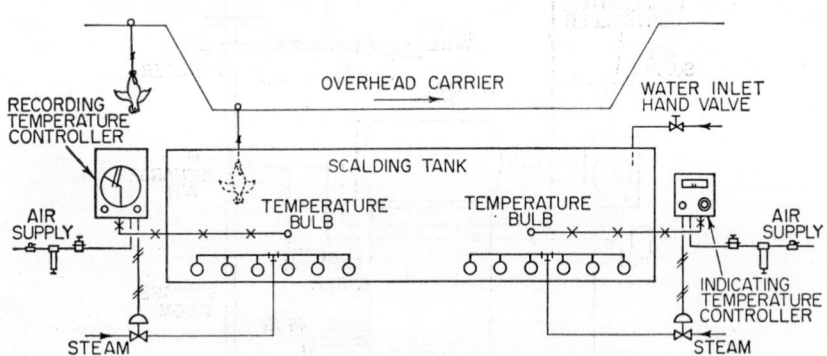

Fig. 37. Controls for poultry-scalding operation.

TRENDS IN FOOD-INDUSTRY INSTRUMENTATION

Recent developments in instrumentation have given the food industry new tools with which to work, and enthusiasm has been shown in the application of these tools to help produce better product, more efficiently and with less labor. Further trends toward continuous-processing operations are requiring the food industry to take advantage of the most up-to-date instrumentation. Major advances which the industry has been able to take advantage of have been in two categories: (1) the development of improved and new sensing elements and (2) the development of improved and new readout and control equipment.

Sensing Elements

Historically, the primary process measurements with which the food industry was conversant were temperature, pressure, and weight. Although these variables are still of extreme importance, the measurement of other process variables and the use of these measurements in coordinated feedback control systems is assuming tremendous importance. Notably, such measurements as flow, density, liquid level, pH, refractive index, conductivity, chromatographic properties, and ultrasonic properties are either practical now or are fast becoming practical as tools of industrial process control and not merely as laboratory measurements.

Improvements in some of the traditional measurements, particularly *weight*, have allowed more complete automation of batching operations, a process of tremendous importance in the food industry.

Specifically, load cells of the strain gage, LVDT (linear variable differential transformer), or "filled-system" type are now being used more and more on weighing applications. These cells are self-contained units which are placed under the supporting members of batching or storage vessels. Their output signals, whether electrical or hydraulic pressure, are linearly proportional to the load on the cell. The signals from the various cells are usually transformed into a usable electrical or pneumatic signal, summed and then compared with a reference "set" value which is the value of weight desired for the ingredient being batched. When the "weight" signal equals the "set" signal the ingredient flow is stopped.

As well as being used for batching purposes, load cells can be used for inventory purposes and less expensive load-cell systems can be used for control where absolute accuracy is not of paramount importance.

Also significant improvements in the standard "lever-type" scale systems have promoted the use of these devices in coordinated, completely automatic batching control systems.

As the need to sense temperatures at many different spots in a process has become a frequent requirement, the use of thermocouple temperature-sensing elements has been dictated. The readout equipment generally associated with temperature "spot checking" is a multipoint indicating or recording device. Thermocouples also lend themselves to those applications where simple temperature computation, such as addition and subtraction, is to be performed. They are particularly useful when their small size allows them to be inserted where other forms of temperature-measuring elements will not fit. However, many modern "filled-system" type temperature transmitters have temperature-sensing bulbs which are roughly the size of a cigarette.

Product *flow* can be one of the most useful measurements in the transformation of a process from batch to continuous. Continuous flow measurement can be used as an aid in coordinating a complex continuous process or as the measurement of primary importance as in the case of a continuous-flow blending system. Flow can be used for batching processes as well as for continuous processes. Because practically all flow-measuring elements measure volume rate of flow, holding density constant or compensating for density changes will provide, in essence, mass flow. Total mass flow or weight is the basic measurement in batching. Flow measurement is particularly useful when it is desired to batch accurately minor ingredients which form only a small fraction of the total batch weight.

Flow-measuring equipment which satisfies the requirement of sanitation and affords greater accuracy than in the past is finding wide application. Such a flow-metering device is the electromagnetic flow meter consisting primarily of an obstructionless section of pipe through which the product flow is passed. As long as the liquid is above a minimum conductivity (most food products satisfy this requirement) the flowing liquid creates a potential which is proportional to the average liquid velocity. This small electrical voltage is sensed by two electrodes mounted flush in the pipe wall. This small voltage is transformed into a usable signal which in turn can be used to drive recording, indicating, controlling, totalizing, and computing devices. The unit poses no obstruction to flow and is sanitary and accurate.

Another flow-measuring device for use where an obstruction can be tolerated but where high accuracy of totalized flow is required is the turbine meter. A rotating turbine in a small section of pipe provides a pulse frequency proportional to flow. Counting the pulses gives total or accumulated flow readout with very high resolution. A converter can convert the pulse rate to an analog signal proportional to flow rate.

Density in many cases is directly related to product concentration and hence can be a valuable measurement in concentrating processes such as evaporation or in the very commonplace blending process. The simplicity of instrumentation is apparent in the bubble tube–sample column methods of measuring specific gravity as is discussed under Lye-density Control System. Other density-measuring devices can be used to sample the entire product stream. Radiation methods of density determination use a radiation source and detector. The amount of radiation received by the detector is inversely proportional to the product density. This type of density-measuring device can readily be incorporated into an integrated control system.

Also *refractive index* as a measure of product concentration (particularly dissolved solids) is being used as a process-control technique as more equipment becomes available for measuring refractive index continuously in a processing operation. (Laboratory equipment for measuring this variable has been available for years.)

Readout and Control Equipment

Miniature instrumentation is finding wider application in the industry because of the inherent small size, serviceability, and the flexibility of transmitter-receiver combinations whether pneumatic or electronic. These features can be used to advantage in the larger food-processing plants which lend themselves to centralized control.

For the same chart width as standard large-case instruments, miniature recorders consume approximately one-quarter the panel face area. This can be an important consideration in the establishment of a control center where fewer operators can observe and control a complex coordinated processing operation. Further, small instrument size allows graphic or semigraphic representation of the process to be displayed on the control panel. In this technique process equipment, color-coded flow lines, and valve symbols graphically represent the process. Miniature instruments can often be located in the symbol of the processing equipment with which it is associated, and valve positions can be shown by pilot lights located in the valve symbols. This form of control-panel layout is particularly useful when there are many switching operations of remotely controlled valves which an operator must perform.

As controls become centralized, the operator is no longer available at the processing equipment to manipulate hand valves during start-up or an emergency. Automatic-to-manual switching units are standard equipment on miniature instrumentation so the control valve can be switched from automatic to manual control at any time.

Serviceability is enhanced by chart and ink supply which need changing only once a month. Since the controller and recorder or indicator slide are "plug-in" components, either may be removed for servicing or replacement even while the process is in operation by switching to manual control.

Although large-case instruments can be used as transmitter-receiver combinations, miniature instrumentation is designed in most cases for use in transmitter-receiver systems. The sensing element, whether it be a thermal system or an orifice plate, is directly connected to a transmitting device which transmits a signal (either pneumatic or electrical) to standard receiving equipment. This allows readout and control equipment to be located at great distances from the transmitters, which are usually located in the immediate vicinity of the process. Where extremely fast transmission and/or where the transmission distance is extremely long (greater than 1,000 ft) electrical transmission is to be preferred.

The flexibility of this system can be realized because all receiving components, whether they be recording, indicating, computing, or for alarm purposes, are designed to operate from the same signal. Hence standard components can be used to form a complex system which can be tailored to the individual control problem.

For instance, in the brewing industry the "fixed-cam" type of time-schedule controller sometimes does not give sufficient flexibility for the cereal- and/or mash-cooking cycles because of the need to change both the rate of temperature rise and the end-point temperatures as the cycle progresses. This can be done easily by combining a series of pneumatic components to make a system where the rate of temperature rise and the end-point temperatures can be changed by a dial setting while the process is in operation.

Some processing equipment in the food industry is reaching a stage of sophistication which requires control equipment with superior control features. Certain special forms of miniature pneumatic instrumentation and electronic instrumentation can provide such features as no overpeaking on process start-up and a wide range of control responses to control even the most tricky and fastest of processes.

REFERENCES

General Instrument Applications

1. Lindsey, E.: Evaporation, *Chem. Eng.*, vol. 60, no. 4, p. 227, April, 1953.
2. Moore, R. R.: Striking Advances in "On Line" Moisture Control, *Food Eng.*, August, 1959, p. 82.
3. Slater, L. E.: Controllers Up Output and Quality, *Food Eng.*, October, 1949, p. 66.
4. Slater, L. E.: New Evaporator Robotized for High Output Efficiency, *Food Eng.*, September, 1950, p. 58.
5. Slater, L. E.: Check pH Right on the Line, *Food Eng.*, September, 1951, p. 98.
6. Ziemba, J. V.: Solving Your Clarifying Problem by Filtration, *Food Eng.*, July, 1960, p. 49.
7. Ziemba, J. V.: Weighing Makes Big Strides, *Food Eng.*, March, 1960, p. 93.
8. Ziemba, J. V.: Better Container Inspection, *Food Eng.*, March, 1960, p. 63.
9. Ziemba, J. V.: Timed-program System "Robots" Processing, *Food Eng.*, December, 1959, p. 62.
10. Ziemba, J. V.: FE Roundup on Radioisotope Devices, *Food Eng.*, May, 1959, p. 88.
11. Considine, D. M.: Industrial Weighing, Chap. 12, "Food Industries," Reinhold Publishing Corporation, New York, 1948.

Brewing

1. Brestedt, R. C.: Introduction to Brewing Instrumentation, *Wallerstein Lab. Commun.*, March, 1952, p. 45.
2. Slater, L. E.: Instruments Take Hold in Brewing, *Food Eng.*, April, 1951, p. 93.
3. Staff: Brewery Sweet Water pH Control, *Application Engineering Data*, no. 208-2, The Foxboro Co., Foxboro, Mass., February, 1950.
4. Ziemba, J. V.: "Eyes" Control Case Count, Cut Handling, Up Output, *Food Ind.*, vol. 22, no. 6, p. 1000, June, 1950.

Canning

1. Desrosier, N. W.: Meter Simplifies Color Grading of Fruits and Vegetables, *Food Eng.*, May 1952, p. 92.
2. Garrison, K. E.: Old-time Quality via New-time Process, *Food Eng.*, May, 1952, p. 94.
3. Geise, C. E.: A Comparison of Three Methods for Determining Moisture in Sweet Corn, *Food Tech.*, vol. 7, no. 6, p. 250, 1951.
4. Gemmill, A. V.: Automatic Controls and Signals Feature Pushbutton Syrup Preparation, *Food Eng.*, May, 1952, p. 82.
5. Gumperts, D. C.: Magnetic Sorting of Unlabeled Food Cans, *Electronics*, vol. 25, no. 9, p. 100, September, 1952.
6. Gehlberg, E. R.: Technics in the Automatic Control of Low-temperature Vacuum Food Concentration, *Food Tech.*, vol. 5, no. 11, p. 68, November, 1951.
7. Shaw, H. L.: Thin Diaphragm Setup Controls Viscous Liquid Density, *Food Eng.*, July, 1952, p. 95.
8. Slater, L. E.: Controls Optimize Canning, *Food Eng.*, February, 1953, p. 85.
9. Staff: Aseptic Canning Process, Instrumentation Data Sheet, no. 3.2-15a, Minneapolis-Honeywell Regulator Co., Industrial Div., Philadelphia, Pa., January, 1952.
10. Staff: Foxboro Juice Weighing System, Instruction Form no. 15–641, pp. 1–6, The Foxboro Co., Foxboro, Mass., February, 1951.

Baking

1. Asproyerakas, M.: Ingenious Bulk-flow Controls Up Handling Efficiency, *Food Eng.*, February, 1959, p. 88.
2. Hlyaka, I.: A Comparative Study of Ten Electrical Meters for Determining Moisture Content of Wheat, *Can. J. Research*, vol. 27, no. 10, p. 382, October, 1949.
3. Slater, L. E.: Robotized Bakery Operations, *Food Eng.*, August, 1951, p. 73.
4. Price, J.: Meter Takes Mixers Pulse, *Food Eng.*, September, 1952, p. 88.
5. Staff: Automatic Control of Macaroni Drying Cycle Increases Production and Improves Quality,

Instrumentation Data Sheet, no. 3.6-6, Minneapolis-Honeywell Regulator Co., Industrial Div., Philadelphia, Pa., May, 1951.

Dairy

1. Barnum, G. F.: Strict Requirements for Dairy Industry Demand Precision Control, *Taylor Technology*, p. 5, autumn, 1950.
2. Slater, L. E.: New Controls Presage Automatic Dairy Processes, *Food Eng.*, October, 1954, p. 78.
3. Welch, E. S.: Teams CIP with Process Control, *Food Eng.*, February, 1959, p. 92.
4. Ziemba, J. V.: Refines Process and CIP Control, *Food Eng.*, July, 1960, p. 44.

Meat Packing

1. Staff: Old Recipes Plus New Tools Equal Top Quality Meat Delicacies, *Food Eng.*, June, 1951, p. 85.
2. Ziemba, J. V.: Swift Engineers Basic Advances, *Food Eng.*, July, 1960, p. 44.

Section 14

ENERGY CONVERSION
PROCESS INSTRUMENTATION

By

DON G. BLODGETT, B.S. (E.F.), *Superintendent of Electrical System Performance, The Detroit Edison Company, Detroit, Mich.; Member, Institute of Electrical and Electronics Engineers; Engineering Society of Detroit.* (*Electric Power Generation and Distribution Control*)

NATHAN COHN, B.S. (E.E.), *Vice President—Technical Affairs, Leeds & Northrup Company, Philadelphia, Pa.; Fellow, Institute of Electrical and Electronics Engineers; Senior Member, and Past President (1962–1963), The Instrument Society of America; Registered Professional Engineer (Ill.).* (*Electric Power Generation and Distribution Control*)

N. S. COURTRIGHT, M.S. (M.E.), *Manager of Education, Bailey Meter Company, Cleveland, Ohio; Member, American Society of Engineering Educators.* (*Steam Power Plant Instrumentation*)

J. H. DENNIS, B.S., *Vice President, Manufacturing, Bailey Meter Company, Cleveland, Ohio; Member, American Society of Mechanical Engineers, Registered Professional Engineer (Ohio).* (*Steam Power Plant Instrumentation*)

A. K. FALK, B.S. (E.F.), *Interconnection Coordinator, The Detroit Edison Company, Detroit, Mich.; Member, Institute of Electrical and Electronics Engineers; Registered Professional Engineer (Mich.).* (*Electric Power Generation and Distribution Control*)

T. W. HISSEY, B.S. (E.E.), *Assistant Sales Manager, Electric Power, Leeds & Northrup Company, North Wales, Pa.; Member, Institute of Electrical and Electronics Engineers.* (*Electric Power Generation and Distribution Control*)

S. B. MOREHOUSE, B.S. (E.E.), *Assistant to the Vice President, Systems Department, Leeds & Northrup Company, North Wales, Pa.; Member, Institute of Electrical and Electronics Engineers, The Franklin Institute, C.I.G.R.E; Registered Professional Engineer (Pa).* (*Electric Power Generation and Distribution Control*)

W. B. SCHULTZ, B.S. (E.E.), *Principal Engineer, Digital Systems, Leeds & Northrup Company, North Wales, Pa.; Member, Operations Research Society of America, Association for Computing Machinery.* (*Electric Power Generation and Distribution Control*).

V. S. UNDERKOFFLER, (M.A., *Physics), Head, Systems Equipment Section, Research and Development Department, Leeds & Northrup Company, North*

14–1

Wales, Pa.; Member, American Nuclear Society, The Franklin Institute. (Nuclear Reactor Instrumentation)

C. W. WORK, B.S. (**M.E.**), *formerly Mechanical Engineer, Hagan Chemicals & Controls, Inc., Pittsburgh, Pa. (deceased); Member, American Society of Mechanical Engineers, U.S. Naval Institute, Carnegie Institute Society, National Industrial Advertisers Association. (Combustion Instrumentation)*

AUTOMATIC COMBUSTION CONTROL SYSTEMS
FOR BOILERS

By Clayton W. Work*

The successful operation of any boiler plant, regardless of size or number of units, requires that the highest possible efficiency in the use of fuel shall be maintained constantly. To effect fuel economy, all necessary adjustments of dampers, valves, and other adjustable equipment must be made when required and in the amount required. The use of automatic equipment to accomplish these purposes assures a constant monitoring of operation and coordination of adjustments. Automatic devices provide a satisfactory solution over the entire range of boiler sizes and for any combinations of fuel-firing and auxiliary equipment.

OBJECTIVES OF AUTOMATIC COMBUSTION CONTROL

A system of automatic combustion control must provide for the accomplishment of three basic functions:

1. Adjust the supply of fuel to secure the heat release necessary for maintaining the *master* condition. This is usually steam pressure but may be steam flow or some other measure of boiler output.

2. Adjust the supply of combustion air in the proper ratio to fuel supply and thereby maintain optimum efficiency in the combustion process.

3. Adjust equipment to maintain the rate of removal of products of combustion in step with the rate at which those products are evolved by the combustion process.

Certain designs of boiler units require automatic control for other functions, such as steam temperature, forced-draft air pressure, induced-draft fan suction, or coal-air mixture temperature. Such factors are separate from but must be coordinated with the control of the three afore-mentioned basic functions of combustion control (see "Steam Power Plant Instrumentation," pp. 14-25 through 14-45).

BASIC COMBUSTION CONTROL SYSTEMS

The method selected for controlling input of fuel and air is based upon consideration of the fuel or fuels to be burned, the physical equipment to be operated, and the engineering practice of the manufacturer of the combustion control equipment. From the standpoint of control of fuel and air, all systems of combustion control may be classed as either (1) *parallel* or (2) *series* systems. The engineering practice of various manufacturers results in many modifications of each of these types. Functional diagrams of the two types are given in Fig. 1a and b.

Series Control Systems

The two most common versions of the *unmodified* series control system are shown as alternates 1 and 2 in Fig. 1a. In each case, steam pressure is considered as the

* Combustion Engineer, Hagan Chemicals & Controls, Inc., Pittsburgh, Pa. (deceased).
Editor's Note. Following the death of the author material for this section was reviewed and updated by C. S. Cotton, Jr., Assistant to Sales Manager, Hagan Controls Corporation, a subsidiary of Westinghouse.

master condition to be satisfied. The indication of any departure of steam pressure from the standard desired value is used to adjust either *air flow*, as in the case of alternate 1; or *fuel supply*, as in the case of alternate 2. A measurement of the controlled variable then is used for adjustment of the second variable—which, in the first alternate example, is fuel supply; in the second, air flow.

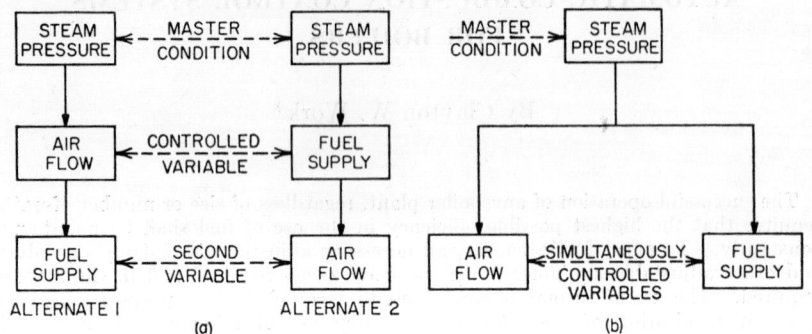

Fig. 1. Basic types of combustion control systems: (*a*) Series-type combustion control systems; (*b*) parallel-type combustion control system.

Several combustion-controls manufacturers employ either one or the other of these unmodified series methods when (1) the fuel fired has a relatively constant calorific value and (2) a satisfactory measurement of fuel-firing rate is available. Small- or medium-sized boilers, up to approximately 100,000 lb/hr capacity, fired with fuel oil, natural gas, or coke-oven gas, are in this category.

Parallel Control Systems

In the parallel system of control, shown in Fig. 1*b*, steam pressure also is used as the master condition, and indications of steam pressure changes are used to make *simultaneous* adjustments of input of both fuel and air.

The principal use of the *unmodified* parallel system is for installations which do not require exactly accurate simultaneous adjustments of fuel supply and air flow. Good examples of such installations are boilers fired with a single retort stoker, a multiple-retort underfeed stoker, or a traveling-chain grate stoker. Such equipment normally operates with a reserve of unburned solid fuel on the grates where the fuel-bed thickness can change materially without appreciable effect on the fuel-air relationship.

The parallel system also is applicable to the control of oil or gas firing, when provision is made for accurate control of fuel supply pressure upstream from the fuel control valve.

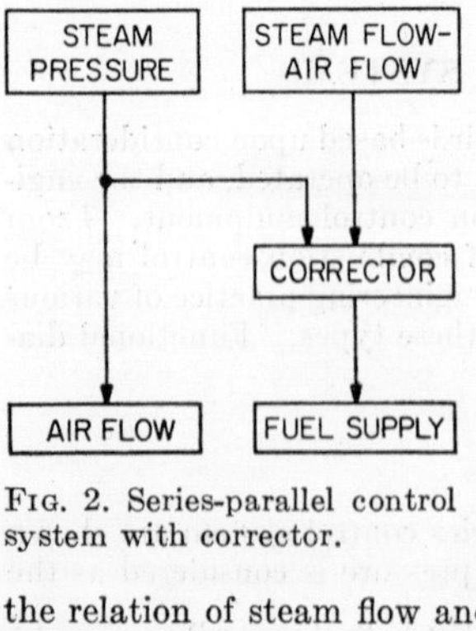

Fig. 2. Series-parallel control system with corrector.

Series-Parallel Systems

A common modification of the series and parallel methods of control involves the application of a correcting factor to the input of fuel or combustion air supply. The correcting factor usually is indicative of the relation of steam flow and air flow, and the correction is on a ratio basis. This is shown in Fig. 2.

The series-parallel control method can be used with virtually every type of fuel and firing equipment.

A second series-parallel method commonly used is shown functionally in Fig. 3. In this arrangement, the measure of the master condition (steam pressure) is used to adjust fuel supply. The rate of steam output is used as a measure of the heat liberation from the fuel supplied. The quantity of heat required per pound of steam delivered is a constant value in the usual installation, and it is established that the combustion-air requirements for each heat unit released by a particular fuel remain constant. Following this reasoning, the measure of steam-flow rate is an accurate means for adjusting the supply of combustion air.

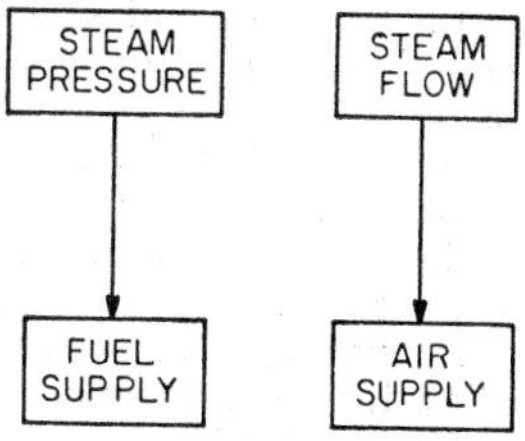

Fig. 3. Series-parallel control system wherein (a) measurement of steam pressure is used to adjust fuel supply, and (b) steam flow rate is used to adjust air supply.

This alternative usually is employed in the firing of fuels which require that the instantaneous values of fuel input and air supply be maintained in the proper ratio, and particularly when the calorific value of the fuel used may vary unpredictably; or where the fuel-feeding mechanisms may be subject to variations in delivery rate. Pulverized coal and oil-refinery waste gas are good examples of fuels of variable heat value.

OPERATING MEDIA

In the choice of the medium for operating the units of a combustion control system, a majority of control manufacturers adhere to a single choice between pneumatic, hydraulic, and electrical means. However, combinations of these are also used.

A common practice is to use pneumatic pressure for both signal and power purposes. Pneumatic signals are also used in combination with hydraulic power. A few systems use pneumatic signals and electric units for power. Hydraulic fluids are used rarely for signal purposes, although very commonly for power purposes. There is also wide acceptance of electronic signals with either electric, pneumatic, or hydraulic power units. Each medium for each service has proponents, and engineering discussions rapidly become complex. However, the chief advantages of each medium are as follows:

1. Pneumatic pressure (usually compressed air) is disposable and does not burn. Elements of a pneumatic control system are easily understood and maintained.

2. A pneumatic system equals any other in reliability, flexibility, and adjustability when transmission of signals does not exceed approximately 1,000 ft.

3. Pneumatically operated units of different manufacturers are compatible.

4. Hydraulic fluid is suited as a power medium for moving heavy loads at relatively slow speed.

5. Electronic systems are definitely indicated for long signal transmission and when data logging and scanning of measurements and outputs are required, and for use with computers for automated plants. Electronic computers are usually the best choice economically where centralization and miniaturization are desired.

Effects of Loss of Power Medium

Reliability is only a relative term, because the supply of any power medium can fail. Precautions, such as the availability of standby power equipment, should be included in the automatic combustion control system to prevent such failure from becoming a dangerous operating emergency.

The effects of power-medium failure in the several types of systems are summarized as follows:

Pneumatic Systems. As the power-medium pressure decreases, any positioning device loses power to move the controlled mechanism. At the same time, signal pressures are lowered, and components of the control system tend to move in the direction corresponding to zero signal.

Locking systems, which trip at a definite minimum air pressure, are available and their operation is reliable. Such locking will maintain boiler operation at the level which existed at the time of tripping.

In the absence of a locking system, dampers or control valves may move open or closed, depending upon the direction of motion corresponding to loss of air pressure; or by unbalanced weights. Problems of this type can be overcome by laying out the system to fail in the direction of safe operation and by using suitable counterweights to cause dampers or valves to drift in the safe direction.

Since a compressed-air system normally has appreciable reservoir capacity, air-pressure reduction to a point dangerous to operation is subject to some delay. This delay usually is sufficiently long to permit the use of manual locking devices and to allow for further manual adjustments during an emergency.

Hydraulic Systems. Upon loss of power, usually caused by pump-motor failure, the power-medium pressure drops to zero almost immediately. This has the effect of locking hydraulic fluid in the system and normally locks all positioning devices in place. Some drift may occur because of the combined effect of unbalanced valves or dampers. This problem can be overcome by using counterweights to eliminate drift or to allow drift in the direction of safe operation.

Electric or Electronic Systems. Power failure interrupts all operation, and electrically operated power units automatically lock in place. Means must be included for emergency manual operation, usually by handwheels.

For any system which uses one medium for signal purposes and another medium for power, the failure of the signal medium only will cause positioning devices to move in the direction corresponding to zero signal. Provision must be made to have this for safe operating conditions. Guarding against loss of power only necessitates provision for any drift to be in the direction of safe operation.

COMBUSTION CONTROL COMPONENTS

All control components* may be divided into the following classifications, based on the service for which they are designed:

1. *High pressure*, as in steam headers, and usually higher than 10 psig.

The pressure-sensitive element may be a flexible diaphragm, a corrugated metal bellows, or a bourdon tube. Deflections usually are transmitted to a beam, with the standard pressure balanced by spring tension or compression, or by weights. Final beam movements may operate a pilot valve for positioning, or a signal-transmitting valve.

2. *Low pressure*, positive or negative, such as forced air pressure or draft within a furnace; or suction in an induced-draft fan system.

3. *Positioning devices*, usually designed to respond to a signal from a measuring unit and capable of exerting sufficient force to move another mechanism, such as a damper or valve, to a position corresponding with the signal received.

Pneumatic positioning units may be double-acting piston type, spring-loaded single diaphragm type, or double opposed diaphragms.

Hydraulic positioning units are usually double-acting piston type.

Electrical positioning units are normally motors, which are driven through reducing gears.

4. *Relays and Controls.* Panel relays for pneumatic and hydraulic systems are usually handwheel-operated and are designed for various functions, such as repeating or modifying signals and permitting transfer between automatic and manual operation. The counterparts of electrical systems are push-button stations or rotary switches.

Controls for temperature and other functions are handled by components of a wide variety of design and are selected for the particular service involved.

COMBUSTION CONTROL SYSTEMS

The combustion control systems which have been selected for description here are of the conventional type usually recommended by controls manufacturers. All these systems are subject to modification.

* For engineering details concerning these components, see D. M. Considine, "Process Instruments and Controls Handbook," Sec. 9, McGraw-Hill Book Company, Inc., New York, 1957.

Criteria for Design

The design of an appropriate combustion control system must be based on the examination of the entire project. The following tabulation is a summary of information categories which must be known prior to the design of the combustion control system best suited for a given installation:

1. Number of boilers.
   - *a.* Controlled individually.
   - *b.* Controlled as a group.
2. Boiler capacity.
   - *a.* Below approximately 10,000 lb of steam/hr.
   - *b.* In the range of 10,000 to 50,000 lb/hr.
   - *c.* In the range of 50,000 to 150,000 lb/hr.
   - *d.* In the range of 150,000 to 500,000 lb/hr.
   - *e.* In the range of 500,000 to 1,000,000 lb/hr.
   - *f.* Greater than 1,000,000 lb/hr.
3. Steam pressure—boiler or superheater outlet.
   - *a.* Below 100 psig.
   - *b.* Between 100 and 300 psig.
   - *c.* Between 300 and 900 psig.
   - *d.* Between 900 and 1,500 psig.
   - *e.* Above 1,500 psig.
4. Type of fuel.
   - *a.* Solid fuel (coal).
   - *b.* Pulverized coal.
   - *c.* Natural gas.
   - *d.* Fuel oil.
   - *e.* Process gases.
   - *f.* Waste fuels (gas, liquid, or solid).
   - *g.* Multiple fuels.
     - (1) Fired separately.
     - (2) Fired simultaneously.
5. Fuel-burning equipment.
   - *a.* Natural draft—chain grate or overfeed stokers.
   - *b.* Forced draft—chain grate or underfeed stokers.
   - *c.* Pulverizers.
   - *d.* Gas burners.
     - (1) Natural draft.
     - (2) Forced draft.
     - (3) Integral fan.
   - *e.* Oil burners.
     - (1) Rotating cup.
     - (2) Mechanical atomizing.
     - (3) Steam, air, or gas atomizing.
     - (4) Return, or wide range type.
   - *f.* Combination burners.
6. Forced-draft equipment.
   - *a.* Number of fans per boiler.
   - *b.* Fan drives—motor or turbine.
   - *c.* Damper locations at each fan.
7. Induced-draft equipment.
   - *a.* Number of fans per boiler.
   - *b.* Fan drives—motor or turbine.
   - *c.* Damper locations at each fan.
8. Associated control functions.
   - *a.* Preheated-air temperature.
   - *b.* Steam-temperature control.
     - (1) Superheat.
     - (2) Reheat.
   - *c.* Coal-air mixture temperature.
   - *d.* Boiler-drum water level.
   - *e.* Means available for measuring air flow, steam flow, fuel input.
   - *f.* Other special conditions.

With the data obtained from such a tabulation, the combustion control engineer is able to recommend a suitable system that will provide optimum fuel saving and the lowest compatible first cost.

COAL FIRING—UNDERFEED AND CHAIN-GRATE STOKERS

For small boilers, up to approximately 20,000 lb of steam/hr, the combustion control system must be simple and of low cost.

Start-Stop Control System

With motor drives for a chain-grate stoker and with a motor-driven forced-draft fan, alternate starting and stopping of the stoker and fan is sometimes used. This system is shown in Fig. 4.

The master unit is a bellows-operated switch which responds to steam-pressure changes and actuates the electric circuit. When the steam pressure drops to a preset value, the stoker and blower motors start and run at constant speed until the steam pressure increases to the higher preset value. The blower and stoker then are stopped until the steam pressure again drops to the "trip-in" value. The pressure proportional band is fixed, but normally there is an appreciable overrun in each direction.

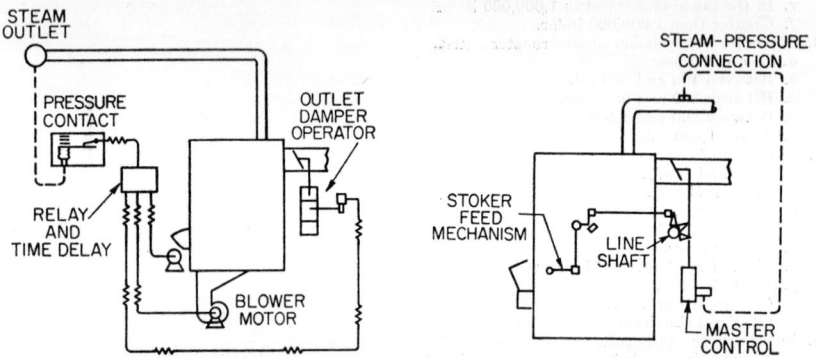

FIG. 4. Start-stop control on coal-fired boiler.

FIG. 5. Modulating control system used on stoker-fired natural-draft boiler.

The boiler outlet damper may be in a fixed position, or it may be operated by a separate power unit. In the latter case, the damper must be opened prior to starting the blower so as to prevent positive pressure in the furnace.

Modulating Control System

The alternate cooling and heating of the boiler with a start-stop control system results in lowered efficiency. A better solution is the use of a modulating combustion control system which will follow the load and steam-pressure changes and which will make *gradual* adjustments of the firing rate. The simplest form of such a modulating system is shown in Fig. 5, for use with natural-draft chain-grate or overfeed-type stokers.

The steam-pressure-responsive unit positions a mechanical linkage system in accordance with changes in steam pressure. The linkage system provides mechanical operation of the stoker-feed mechanism and the boiler outlet damper for each boiler individually. This may be styled a mechanical-type parallel-input system.

Forced Draft—Stoker Firing. When the stoker equipment is of the forced-draft type, as in single- and multiple-retort underfeed stokers, traveling-chain grates, and spreader stokers, the combustion control system must be extended to include means for maintaining a constant value of draft in the furnace above the fuel bed. This provides a substantially constant discharge pressure on the forced-draft supply system and aids materially in the control of combustion-air flow.

Two other common methods of control for forced-draft stokers are shown in Fig. 6a and b. These systems are termed *compensated draft* and *balanced draft*, respectively.

Compensated-draft System. In the compensated-draft system, the master control makes simultaneous adjustments of the outlet damper and the stoker-feed mechanism of all boilers. Each boiler is then provided with a separate component o maintain a constant draft in the boiler furnace, by adjusting the delivery of forced-draft air to the fuel bed.

In theory, the opening of the outlet dampers will result in equality of gas flow, followed by equality of combustion-air supply to all boilers. This would result in even distribution of load among all boilers in service.

In actual practice, any holes or thin spots in the fuel bed will allow combustion air to flow through with little resistance and without coming in contact with a proper quantity of incandescent fuel. This results in furnace-draft changes which call upon the draft-responsive unit to reduce the combustion-air supply. Finally, the boiler output is reduced because the combustion rate is lowered. Successful operation depends largely on keeping fuel beds relatively equal in thickness and free of holes and thin spots. Care must be taken also to have each boiler operate at a rate which will assure a flow of comparatively cool combustion air through the stoker structure at all times. If this flow is reduced too much or for too long, overheating of the stoker parts will occur.

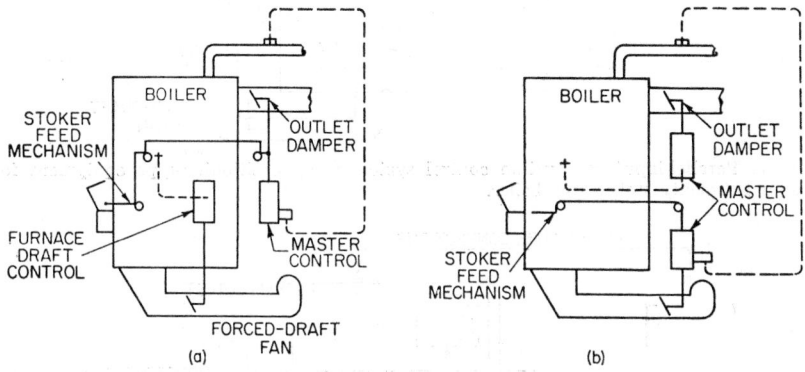

Fig. 6. Combustion control systems for forced-draft stoker-fired boiler: (a) Compensated draft; (b) balanced draft.

Balanced-draft System. The balanced-draft system differs from the compensated-draft system in that the delivery of combustion air is adjusted from the master control, and constant furnace draft is maintained through operation of the outlet damper. This method has the marked advantage of assuring a supply of combustion air to the stoker at all times. Also, it is more dependent on the condition of the fuel bed for proper distribution of air through the fuel.

As regards over-all combustion efficiency, there is little difference between the compensated-draft and the balanced-draft systems, provided that the operators always keep the fuel beds in proper condition. Statistics indicate, however, that the balanced draft apparently is first choice of a majority of stoker manufacturers and plant operators.

Choice of Actuating Equipment

Mechanical linkage systems are recommended by some combustion control manufacturer for units up to about 15,000 lb/hr capacity. As the physical size of a boiler unit increases, it is necessary to consider the mechanical factors of installing the system as well as the ease and convenience of the personnel who operate the unit.

Positioning-type Equipment. In larger installations, these factors indicate definitely that equipment of the signal-transmitting and -receiving type must be used. Such signals may be pneumatic, hydraulic, or electric. The remote character of such systems makes it possible for the operators to make manual adjustments to the control system without leaving their normal stations.

When stoker-fired boilers are in the medium-capacity class (up to about 50,000 lb/hr), it is common practice to use positioning devices only for control of fan delivery and fuel feed. A diagram typical of the positioning-type equipment used for this category of boiler is shown in Fig. 7. The more completely automatic equipment used in the larger installations is shown in Fig. 8.

Correction for Relation of Steam Flow and Combustion-air Supply. While the automatic combustion control systems described here are considered as basic, several

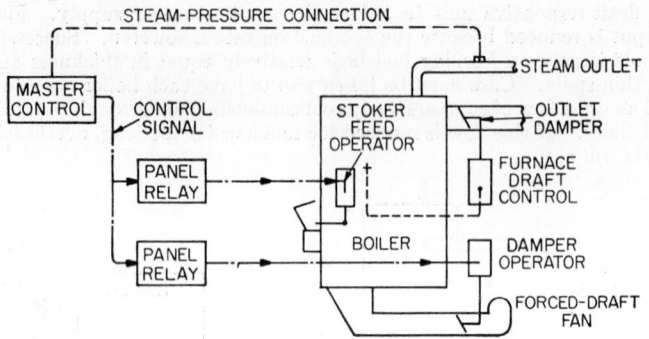

FIG. 7. Parallel-input combustion control system using positioning-type equipment for medium-capacity stoker-fired boiler.

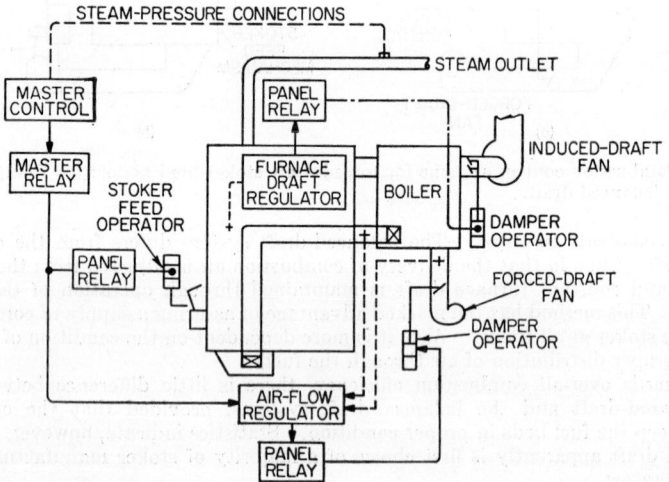

FIG. 8. Parallel-input combustion control system using metering-type equipment on large-capacity stoker-fired boiler.

modifications are made by individual manufacturers of controls. The most common modification is the application, to all sizes of boilers, of a correction in the system based on a measurement of changes in the relation between steam flow and flow of combustion gases. For any installation, a definite relation exists between gas flow through the boiler, measured as a differential between taps in the boiler passes, and combustion-air supply to the stoker. The gas-flow measurement commonly is termed *air flow*. This method is shown functionally in Fig. 9.

In the two alternate examples shown, input of fuel and air are basically parallel, with the correction applied either to the fuel supply or to the combustion-air supply—

on the basis of maintaining a definite rate of air supply for every rate of steam output. This ratio may be a constant value at all ratings or may be adjusted to vary automatically as output rate changes.

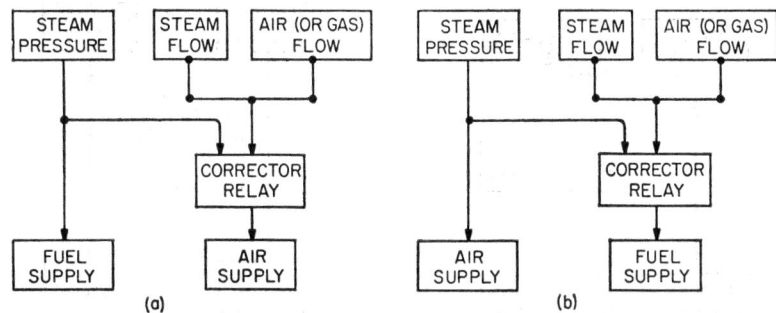

FIG. 9. Parallel-input combustion control system for stoker-fired boiler in which correction is applied to either (a) the air supply, or (b) the fuel supply.

COAL FIRING—SPREADER STOKERS

The spreader-type stoker, with either traveling flat grates or stationary grates arranged for sectionalized dumping, has largely displaced the underfeed-type stoker. This newer design has introduced the additional problem of having a relatively large proportion of the fuel burned while in suspension. The condition approaches that of a pulverized-coal-fired furnace, but there is still a small reserve of fuel on the grates.

To achieve efficient combustion of the fuel in suspension, the relation between the instantaneous values of fuel and air input must be maintained within closer limits than is true for underfeed stokers.

For smaller boiler units, up to approximately 15,000 lb of steam/hr, the simplicity and low first cost of the mechanical-type combustion control, which employs the balanced-draft method shown in Fig. 6b, is often specified by combustion engineers. A majority of such boilers are supplied with a constant-speed motor-driven forced-draft fan. At the same time, the resistance to air flow through the grates and fuel bed is nearly constant. Thus the angular motion of the fan discharge damper is capable of accurate calibration in terms of fan delivery.

As the boiler capacity increases up to about 60,000 lb/hr, the method shown in Fig. 7 is usually considered adequate; while the method of Fig. 8 is often used for larger boiler units.

The modifications shown in Fig. 9 also are applied by combustion engineers for the control of spreader-stoker firing of virtually all sizes of boiler units.

Thus far, the general plan of parallel control of fuel and air input for spreader stokers has been described. The characteristics of the spreader stoker, however, permit considerable freedom in the choice between some forms of parallel and series input.

Series Control Systems. The functional diagram of one series system is shown in Fig. 10. The steam-pressure measurement adjusts, through a suitable signal system, the input of combustion air; and a measurement of the air input is then made the means for controlling fuel input. For boilers of medium size, and with proper calibration, a high level of efficiency can be maintained over a wide range of load conditions with this system.

Series Control with Rate Action. Another modification of the simple series system which uses a measure of steam output from the boiler as an indication of heat input to the furnace is shown in Fig. 11. This system is particularly suited to boiler units which are larger than approximately 60,000 lb/hr capacity; but it is also well suited for boilers in the medium-sized category and often used.

In Fig. 11, the option is indicated of using the master control signal, with rate action adjustable to suit the individual installation, as a temporary correction is most desirable when steam demand is subject to quick and relatively wide changes.

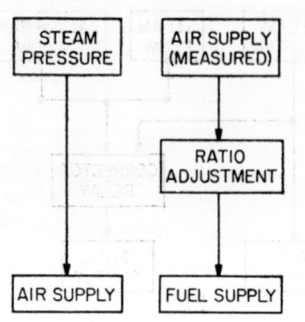

Fig. 10. Series-type combustion control system for spreader-stoker-fired boiler of medium capacity.

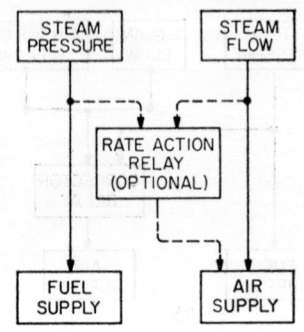

Fig. 11. Series-type combustion control system with optional rate action for spreader stoker or pulverized-coal-fired boiler of high capacity.

PULVERIZED-COAL FIRING

The primary requirement in operating the automatic combustion control system for a pulverized-coal-fired boiler is that the proper quantities of fuel and air shall be supplied to the burners continuously and as nearly simultaneously as is mechanically possible. This imposes full consideration of the boiler design, economizer design, and air preheater design; the characteristics of the control means available at the forced- and induced-draft fans; and the rate of response of the pulverizer equipment. These variables all affect the rate at which the several components of the combustion control system must function. Successful operation requires the closest possible correlation of equipment and time factors.

The time required for various types of pulverizers to respond to indications of a load change is one of the more important considerations in the choice of combustion control method. Some designs are known to have fast response, meaning that the load-change indication secures practically instantaneous change in coal supplied to the burners. Other designs require an appreciable time to accomplish the same cycle. Such equipment requires additions in the combustion control system, such as anticipating, accelerating, or overcorrecting relays to speed up pulverizer response temporarily and therefore nullify the effect of time lags. Unfortunately, these are special factors, for which no fixed rules can be made to apply universally.

Systems for Small Pulverized-coal-fired Boilers

For the smaller boilers, up to about 30,000 lb/hr capacity, and with only a single pulverizer in operation at any given time, it is often feasible to use the unmodified parallel input system with positioning-type components, as indicated in Fig. 7. Successful operation demands very careful calibration and constant monitoring by plant personnel.

For most installations, however, even for the smaller boiler units, it is normally advisable to introduce a corrective factor which is based upon the relation between steam flow and air flow or upon a measurement of heat liberation. In some cases, oxygen and combustible content in flue gases is applied as the correcting factor. All have the advantage of making it easier to maintain test efficiencies, and evaluation of their value in any individual case warrants special study.

The functional diagram of Fig. 11 becomes applicable to these conditions. The further modifications shown in Fig. 12 represent the extension of the use of corrective factors.

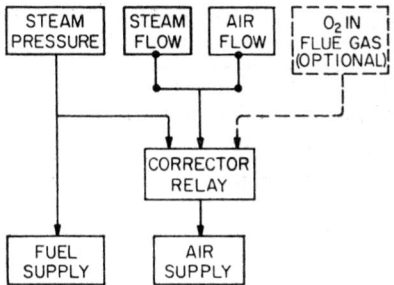

Fɪɢ. 12. Parallel-type combustion control system with optional correctors for pulverized-coal-fired boilers of large capacity.

Systems for Large Pulverized-coal-fired Boilers

As consideration is given to the combustion control systems for pulverized-coal-fired boilers in the range above 30,000 lb/hr and up to approximately 250,000 lb/hr it is rarely that a combustion engineer recommends an unmodified series or parallel system. Series control alone will be subject to the time factors introduced by the physical distance between the points of control of fuel and air and the response characteristics of the pulverizers. The accumulation of these time factors will make it impossible for any combustion control system to maintain the required synchronization of input for fuel and combustion air. Parallel control is subject to errors which are introduced by the pulverizer response rate and delivery of fuel to the burners and other factors which militate against synchronizing coal and air input.

Various combinations of series and parallel systems are used. The most common system is based on Figs. 11 and 12. This particular correction factor recommended by a combustion control supplier is chosen after a study of characteristics of the boiler and auxiliary equipment and a knowledge of the control equipment offered. Very few pulverized-coal installations can be classified on an arbitrary basis, and thus the recommendations of a reputable controls manufacturer are the safest guide available in choosing the best automatic combustion control system for the job.

Each boiler installation at capacities about 500,000 lb/hr is a matter for special consideration. Basically, the control system must be one form of the series-parallel input method previously discussed. The special features of the boiler installation also must be tied in as they may be dependent upon boiler operation. These include proportioning of air flow and fuel supply between two sections of the boiler furnace, operation of bypass dampers or other devices for steam-temperature control, modulated control of boiler-drum water level from steam output rate, and similar special functions which may be incorporated in the boiler-plant design.

Many larger boilers above 1,000,000 lb/hr are designed with two furnaces in which the fuel and air must be properly proportioned in each furnace. The two furnaces may have a common drum and superheater section. When the boiler is fired with a fuel which can be accurately measured, the control system is arranged to proportion fuel and air to each furnace by means of flow measurement.

In cases where fuel cannot be accurately measured, such as pulverized coal, the common system is to regulate air flow to each furnace in accordance with steam flow, while regulating fuel from steam pressure. In order to regulate equal fuel flows to each furnace, the oxygen content of the flue gases is analyzed and compared. Fuel is added to one furnace and subtracted from the other to obtain the proper balance.

SAFETY DEVICES FOR COAL FIRING

Very few safety devices are used in conjunction with combustion control systems for stoker-fired boilers. Even loss of combustion-air supply leads to nothing worse than an excess of coal in the furnace and probable smoking at the stack.

Pulverized-coal firing presents different problems, since the condition of the fuel as fired corresponds to a gas. In certain proportions, a mixture of coal and air is explosive. Temperatures in pulverizers and in the transport system must not exceed well-defined limits. The conditions most likely to lead to hazardous operation are (1) loss of combustion air, (2) loss of draft at the boiler outlet, and (3) excessive temperatures in coal transport or grinding.

Normally, draft-sensitive switches are connected into the rear passes of the boiler or into the inlet ducting of induced-draft fans. Reduction of available draft below the predetermined safe minimum initiates electrical or pneumatic tripping of forced-draft fan drives, tripping of pulverizers, and opening of forced-draft fan dampers to take advantage of available natural draft for purging the setting.

Similar switches are connected in the combustion-air supply system and arranged to trip pulverizers upon reduction of the supply pressure below safe limits.

Temperature-sensitive switches are mounted at strategic points in the coal transport or grinding system and are used for tripping cold air dampers open and hot air dampers closed.

Many combinations are used for automatic tripping and for operating signal lights and alarms. Invariably, the safety system is interlocked to require restarting fans and pulverizers in the correct sequence.

GAS-FIRED SYSTEMS

Natural gas as a boiler fuel is considered as having a substantially constant heat value, and thus measurements of the volume of gas fired may be used by the combustion engineer as a direct measure of heat input to the boiler furnace. Normally, provision is made to assure a constant gas supply pressure. Such factors permit fully adequate combustion control with a minimum of equipment. The selection of the type of combustion control system is governed by the size of the boiler units and the characteristics of the gas burners, the latter being of major importance.

Natural-gas burners are either natural-draft or forced-draft type. The natural-draft type depends upon the negative pressure in the boiler furnace to draw combustion air through the burner registers. Forced-draft burners receive combustion air under positive pressure from a fan or blower. The fan is usually separate from the burner, although one design includes a propeller-type fan in the burner housing. Speed of the fan is made a function of fuel-gas velocity through the burner, and it is claimed that the ratio of fuel to air is maintained accurately over the full load range of the burner.

Many of the gas-fired boiler units of the present era are designed for operation with the furnaces under positive pressures. This is particularly true of the so-called "package" unit, which is readily available in capacities up to approximately 100,000 lb of steam/hr.

Pressurizing the furnace eliminates the use of any controls of furnace draft, and the combustion control system performs only the functions of controlling fuel and combustion-air input.

Gas-fired-package Boiler Control

In the simple form, the combustion control system for a gas-fired-package boiler comprises a master control which operates the gas input and forced-draft damper on either an "on-off" basis or by modulating fuel and air input. The on-off type of operation is best accomplished electrically, whereas modulating control can be handled well pneumatically. The system is parallel input unmodified. Successful operation depends largely upon a constant gas supply pressure and upon reliable safety devices which will prevent the introduction of fuel to the furnace except when the pilot burner is lighted and when combustion-air supply is available. Any delayed ignition of the main burners may result in a furnace explosion of more or less drastic consequences.

On-off control is the logical choice when maximum steam demand is below the minimum steam-producing capacity of a single burner firing continuously. The combination of modulating and on-off control often is used when it is desired to take advantage of the more gradual changes in firing rate inherent in modulating control. Transfer between the two types of operation may be automatic or manual.

The positioning-type system is shown in Fig. 13.

There is no definition available for a choice between the positioning-type control and a metering type, in which gas and air quantities are maintained on a measured basis. The best criterion is a balancing of the initial investment against the expectation of additional fuel saving.

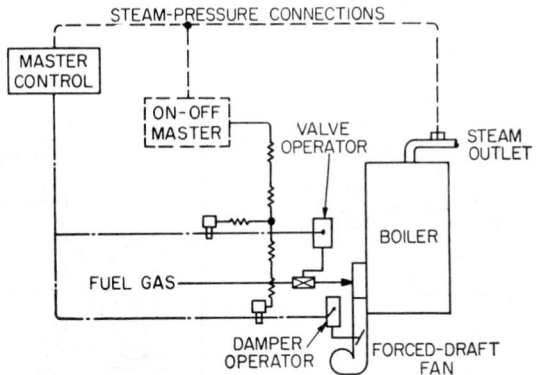

Fig. 13. Positioning-type combustion control system for gas- or oil-fired-package boiler.

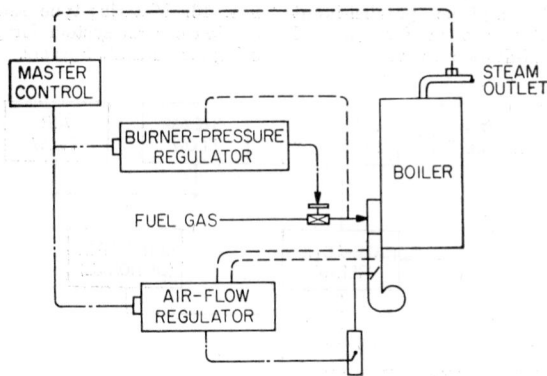

Fig. 14. Metering-type combustion control system for gas- or oil-fired-package boiler.

A typical metering system is shown in Fig. 14. Study of this diagram indicates that it is basically parallel input, with correction available for the relation between steam flow and air flow. While it is possible to operate the metering system for on-off response to steam pressure, it is usually considered advisable to limit on-off operation to a positioning-type system.

Nonpressurized Furnace Operation

When gas-fired boilers are not designed for pressurized furnace operation and when forced draft is supplied to the burners, it becomes essential that the draft in the boiler furnace shall be maintained constant at or below atmospheric pressure. This is usually controlled by operation of the boiler outlet damper, thereby permitting direct comparison of gas and combustion-air flow measurements. It is possible to control furnace draft by operating the forced-draft damper, but this introduces the possibility of having the outlet damper close far enough to call for reduction of forced-draft supply below the safe operating point for the quantity of gas being fired.

Control of Conventional Gas-fired Boilers

For conventional boilers, up to about 15,000 lb/hr capacity, the simple positioning-type parallel-input system of Fig. 15 is applicable. As boiler size increases up to approximately 100,000 lb/hr, it is to be expected that two or more gas burners will

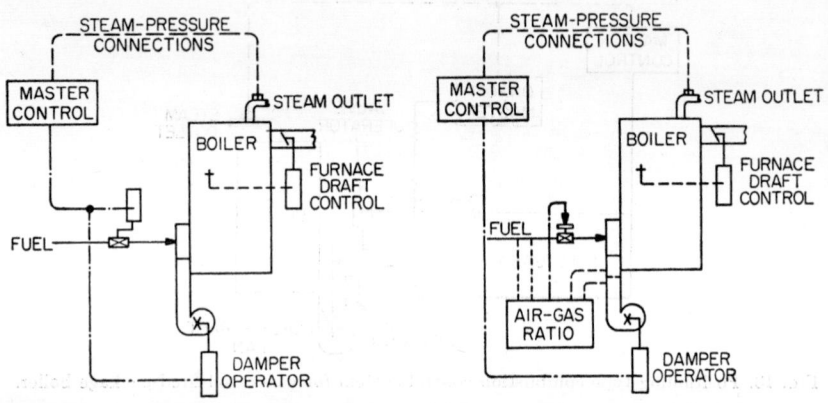

Fig. 15. Positioning-type parallel-input combustion control system for gas-fired boiler using forced-draft burners.

Fig. 16. Metering-type series-input combustion control system for gas-fired boiler using forced-draft burners.

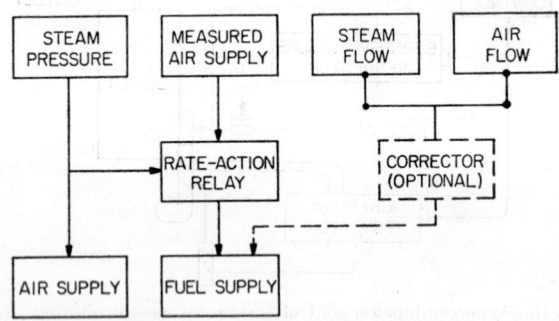

Fig. 17. Series-parallel combustion control system for gas-fired boiler of high capacity.

be operated in parallel. The variations of air and gas distribution through piping and ducts make it advisable to include a means for accurate measurements of gas and combustion-air flows, with the gas-air ratio controlled automatically. It is preferable, from the standpoint of safe operation, to control gas supply from the air-flow measurement. This is not arbitrarily required, provided that suitable safety devices are installed to prevent flow of fuel gas in excess of the combustion-air supply available (see Fig. 16).

The physical size and the resultant response lags in boilers larger than 100,000 lb/hr capacity make it necessary to modify the diagram of Fig. 16. The system shown in Fig. 17 is the series-parallel type, wherein the effect of the master signal on either fuel or combustion-air supply is of a temporary nature; effective during a change in firing rate, but permitting normal functioning of gas-air ratio equipment during periods of relatively steady firing rate.

Gas burners with integral blowers normally are used for smaller boiler units, up to about 15,000 lb of steam/hr capacity, and usually with only a single burner to the boiler. The applicable combustion control system is that shown in Fig. 15, but with omission of the control of the forced-draft damper—thus with dependence only upon the master control to introduce fuel as required to meet the load demand.

Natural-draft burners also are used on relatively small boilers which require the simplest type of combustion control system. A mechanical system, with the master control operating the gas valve and the boiler outlet damper simultaneously through a linkage system, usually is sufficient.

SAFETY DEVICES FOR GAS FIRING

The ideal safety system should include programming equipment which will shut off the fuel supply upon flame failure for any reason, or for a low level of water in the boiler drum. Following such a shutdown, the pilot burner must be lighted and air flow must be established through the burners and setting for a purging period before gas can be admitted to the main burners; also, the water level must be at or above the proper point. These are the primary precautions and must be observed whether the combustion control system is designed for automatically or manually controlled relighting of the burners. The usual safety devices are pressure- or level-operated electric switches which actuate solenoid valves. Pneumatic devices are also available but are rarely used.

OIL-FIRED SYSTEMS

Automatic control of oil firing parallels almost exactly the methods described for gas firing. The principal exceptions are refinements of control brought about by the differences in burner design.

Fuel-oil burners which operate without a supply of forced draft are of the straight mechanical-atomizing or steam-atomizing style. Both styles have a narrow range of adjustment. The mechanical atomizing design depends entirely upon the oil-supply pressure to force oil through a small orifice in the burner tip. Steam-atomizing burners use steam at high velocity to assist the atomizing process. Compressed air or natural gas are used occasionally as the atomizing medium. Air for combustion is drawn through the burner registers because of the negative pressure in the furnace.

Mechanical- and steam-atomizing oil burners are also used when a separate fan supplies combustion air. The forced-draft mechanical-atomizing burner may depend upon oil-supply pressure only for controlling oil firing or may be of the "return" type, which receives oil at a constant pressure, but where part of the oil flows back to the pumping system through a return line from the burner chamber. For this type of burner, control is secured by locating a modulating valve in the return line. Wide-range adjustment of firing rate is a marked characteristic of the return-type burner.

The automatic control system for oil-fired boilers must contain provision for measuring heat input and combustion-air flow. This is facilitated by the usual operating practice of fixing the opening of burner air registers and using a uniform size of burner-tip orifice. Variations in this orifice size are arranged on a definite basis for a particular installation. The necessary readjustments are made easily in the combustion control system. Further, the calorific value and combustion-air requirement of the fuel remain nearly constant on each installation.

The measurement of fuel input may be accomplished by one of three methods. In the first method, a fuel-oil control valve of known flow characteristics is positioned according to requirements. This requires a constant supply pressure at the inlet of the control valve.

In the second method, oil-flow meters which originate electrical or pneumatic signals bearing a definite relation to the oil-flow rate are installed. Such meters may be of the positive-displacement type or of the internal-orifice type, that is, area-type flow meters.

In the third method, advantage is taken of the relation between oil-supply or return-line pressure and the rate of flow through the burners.

Because of the very close similarity between oil-fired and gas-fired boilers, repetitious descriptions are not included here. Reference is made to Fig. 18, which shows a positioning type system, and to Fig. 19, which illustrates a typical metering system

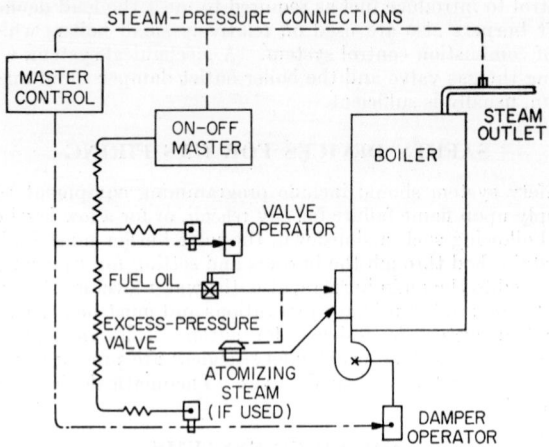

FIG. 18. Positioning-type combustion control system for oil-fired boiler.

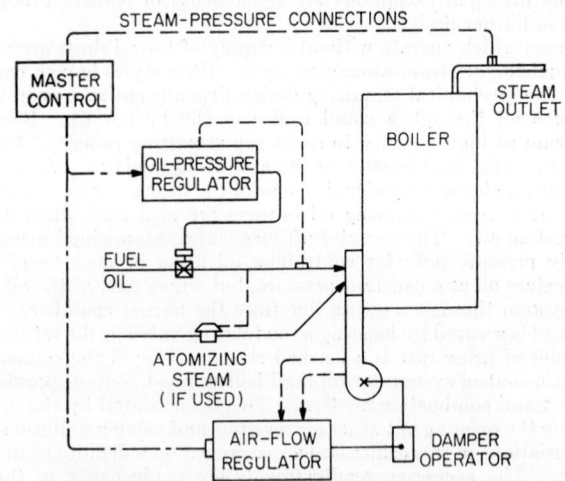

FIG. 19. Metering-type combustion control system for oil-fired boiler.

as applied to a package oil-fired boiler. Combustion control systems for conventional oil-fired boilers having capacities to about 15,000 lb of steam/hr are shown in Fig. 20 (simple positioning-type parallel-input system), Fig. 21 (series system), and Fig. 22 (series-parallel-input system). Figure 23 shows a series-input system for a small boiler in which natural-draft oil burners are used.

Safety devices used in connection with oil-fired equipment are also very similar in principle to those used with gas-fired equipment.

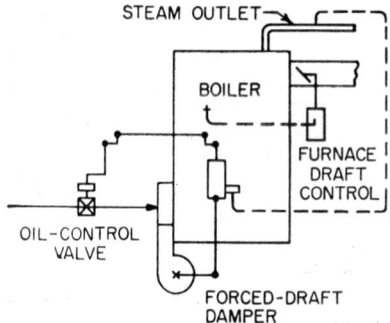

FIG. 20. Positioning-type parallel-input combustion control system for conventional oil-fired boiler of medium capacity.

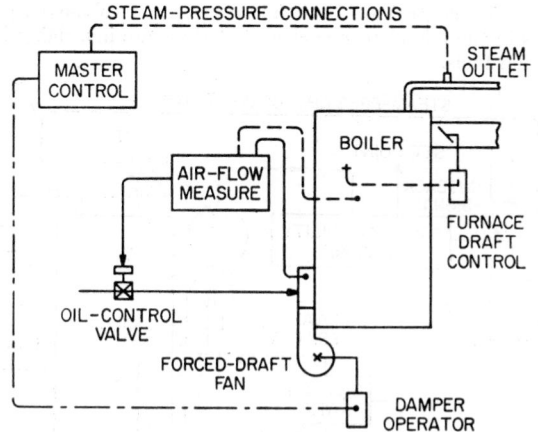

FIG. 21. Series-input combustion control system for conventional oil-fired boiler of medium capacity.

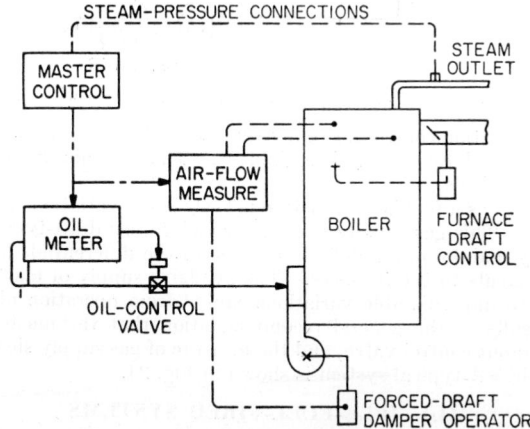

FIG. 22. Series-parallel-input combustion control system for conventional oil-fired boiler of medium capacity.

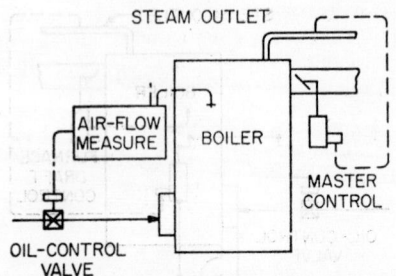

FIG. 23. Series-input combustion control system for boiler of small capacity in which natural-draft oil burners are used.

OTHER FUEL-FIRED SYSTEMS

The more common miscellaneous fuels which are fired singly and made subject to automatic combustion control are blast-furnace gas and coke-oven gas. There are many other fuels in the miscellaneous class, but these are invariably burned in combination with other fuels.

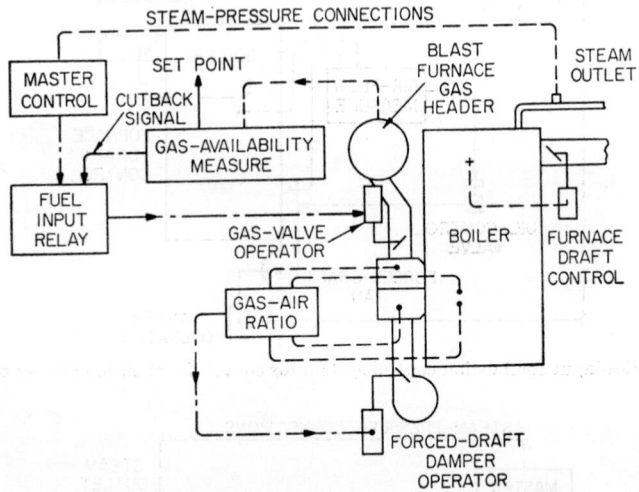

FIG. 24. Series-input combustion control system for boilers fired with blast-furnace gas.

Automatic combustion control for coke-oven gas-fired boilers follows almost exactly the systems described for gas-fired equipment. The calorific value of this gas is uniform at any particular plant.

Blast-furnace gas as a fuel is subject to relatively small variations in calorific value in a given plant. This gas usually is fired through forced-draft-type burners which are controlled by maintaining a definite relation of the differentials between the gas and air compartments to the furnaces. The available supply of blast-furnace gas is usually subject to unpredictable variations caused from operation of the blast furnaces. This results in the general recommendation that the master control shall operate the gas-input control valve, and the measure of gas supply shall control input of combustion air. A typical system is shown in Fig. 24.

MULTIPLE-FUEL-FIRED SYSTEMS

The selection of the method for applying automatic combustion control to the firing of multiple fuels in a particular boiler depends upon several factors. The final system selected usually is composed of the equipment normally recommended in connection with the individual fuels, using as much as possible the particular control elements common to the several types of firing. First, it must be determined what fuels are involved; then, whether two or more fuels are to be fired separately or simultaneously. The types and characteristics of fuel-burning equipment form the next factor. In the case of simultaneous firing of several fuels, the order of firing must be established, since one fuel will be considered as primary, with secondary or supplementary fuels to be fired as needed to meet the steam demand.

A very common example is the use of natural gas as the primary fuel with fuel oil as the supplementary fuel. The use of this combination is more popular with the increasing application of package boilers in sizes up to about 30,000 lb of steam/hr. These fuels usually are fired separately, with a combustion control system actuated

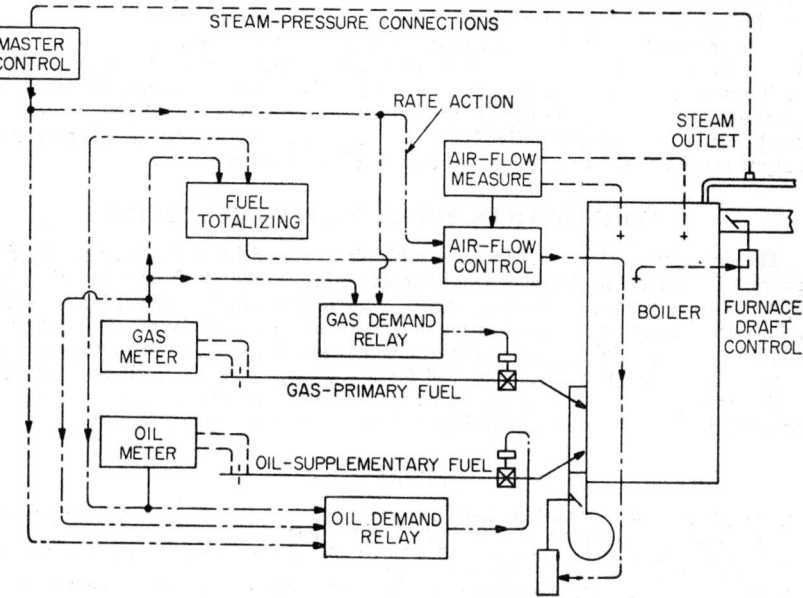

Fig. 25. Combustion-control system for boiler employing simultaneous multiple-fuel firing.

from a common master control, and with provision for separate control of gas and oil firing. Reference is made to Figs. 13, 14, 18, and 19.

Waste gases from various processes usually are treated as the primary fuel, to be burned up to the limit of availability, and with any supplementary fuels fired only as the waste-gas supply is not ample to supply-steam requirements. More common examples are blast-furnace gas, oil-refinery gas, and coke-oven gas. Supplementary fuels are usually fuel oil, pulverized coal, and natural gas.

With the exception of blast-furnace gas, the air requirements for other hydrocarbon gases and for pulverized coal usually are supplied from a common duct system. The control of air for blast-furnace-gas burners should be by means of separate dampers. This is necessary because the air-pressure requirement is unpredictable because of changes in gas availability and pressure and will only occasionally synchronize with air demands of the other fuels involved.

The combustion control system for simultaneous multiple-fuel firing usually is based on the master control signal being the indication of demand for total heat input to the furnace. As shown in Fig. 25, the master signal is connected to regulate input of the primary fuel and is also available for temporary correction of air-supply rate during load changes.

The actual firing rate of the primary fuel is measured, and a signal representing heat units fired is subtracted from the master demand. The remainder then becomes the demand for supplementary fuel.

At the same time, measurement of primary-fuel input, corrected for air requirements of the particular fuel, is connected to air-flow-totalizing equipment.

All fuel-firing rates are measured and utilized in the manner described for the primary fuel. In some cases, certain waste fuels may be supplied on a fixed-flow basis. The proportionate corrections of air flow are required as in automatic control.

The variety of combinations of fuels and the practice of combustion engineers make any arbitrary layouts inadvisable for presentation here, except on a generalized basis. Correction factors may be introduced into the combustion control system which are based on oxygen or combustible content of flue gases. Properly applied and maintained in operation, these can be of such benefit as to warrant thorough consideration. This also applies to air-supply corrections which originate from comparisons of steam output and air flow. If any of the fuels have a variable calorific value, the combustion control system must include means to secure an indication of such changes. Measurement of gas density usually is the most convenient measurement and is accurate for nearly all hydrocarbon gases. When no such measurements can be made, the best alternative is to resort to a series-parallel control, wherein steam output is used as the basic factor for control of combustion air.

SAFETY DEVICES FOR MULTIPLE-FUEL FIRING

The automatic safety system for a multiple-fuel-fired boiler must be arranged to prevent introduction of fuel beyond the available supply of combustion air. The same measurements used for proportioning air to fuel can be utilized for automatic reduction of fuel-firing rate, operating pneumatic or electrical relays in the combustion control system. Emergency conditions which result from outage of fans or malfunctioning of damper-drive units call for immediate closing of fuel control valves.

Restarting after an emergency can be a safe procedure if the safety devices are properly interlocked to require a definite sequence in the starting cycle.

REFERENCES

1. Peth, Herbert W., and John A. Kotsch: Highly Automatic Steam Generation, *Instruments and Automation*, vol. 28, no. 12, pp. 2116–2122, December, 1955.
2. Katzenmeyer, J. P.: Instrumentation of a Modern Steam Generating Facility, *Instruments and Automation*, vol. 28, no. 6, pp. 974–981, June, 1955.
3. Marks, Lionel S. (ed.): Steam Boilers, Engines, and Turbines, Sec. 9, "Mechanical Engineers' Handbook," 6th ed., McGraw-Hill Book Company, Inc., New York, 1958.
4. Feller, Eugene W. F.: "Instrument and Control Manual for Operating Engineers," McGraw-Hill Book Company, Inc., New York, 1947.
5. Gaffert, Gustav A.: "Steam Power Stations," 4th ed., McGraw-Hill Book Company, Inc., New York, 1952.
6. Higgins, Alex: "Boiler Room Questions and Answers," McGraw-Hill Book Company, Inc., New York, 1945.
7. Moyer, J. A.: "Power Plant Testing," 4th ed., McGraw-Hill Book Company, Inc., New York, 1934.
8. Woodruff, E. B., and H. B. Lammers: "Steam Plant Operation," 2d ed., McGraw-Hill Book Company, Inc., New York, 1950.
9. Boho, M. J.: Automatic Combustion Control for Steel Plant Boilers, *Blast Furnace Steel Plant*, April, 1943.
10. Boho, M. J.: Automatic Control for Multiple Fuel Fired Boilers, *Iron Steel Engr.*, February, 1951.
11. Markson, A. A.: Pneumatic Servomechanisms in the Iron and Steel Industry, *Instruments*, June, 1951.
12. Instruments, Controls, and Interlocks, Chap. 27 of "Combustion Engineering," Combustion Engineering, Inc., 1950.
13. "Steam, Its Generation and Use," 37th ed., Babcock & Wilcox Company, New York, 1955.
14. "Power Plant Innovations of the Last Ten Years," Hagan Controls Corporation, *Bull. MSA 225*, Pittsburgh, Pa., 1962.
15. Skrotzki, B. G. A.: Central Control Rooms: Ideas Differ, *Power*, April, 1961.
16. Roth, J. W.: Boiler Combustion Control Systems, *Inst. Control Systems*, vol. 34, no. 12, pp. 2276–2279, December, 1961.

STEAM POWER PLANT INSTRUMENTATION

By J. H. Dennis*
and N. S. Courtright†

Instrumentation and control serve three basic purposes in the steam power plant:

Safety. (1) A record or indication of vital factors such as steam pressure, turbine and tube temperatures, and drum level is essential to the operators. Alarms and signals supplement the instruments. (2) Automatic controls furnish anticipatory action to reduce the magnitude of change compared with hand operation. (3) Interlocks ensure proper sequence during start-up, shutdown, and emergency operations; and offset the danger of confusion.

Automatic Operation. (1) Continuous output at required ratings and prescribed conditions is the day-to-day job of the control system. (2) Centralized instrumentation provides supervisory check readings to reduce operating staff.

Efficiency. (1) Multielement control systems, paralleled by proportioning guides such as a steam flow–air flow recorder, ensure optimum efficiency. (2) Complete instrumentation provides information for either a complete heat balance or efficiency checks on all or part of the system.

SELECTION OF INSTRUMENTATION

Several factors affect the selection of instrumentation and control for the steam power plant:

Boiler and Cycle Design. Size, complexity, and operating conditions.

Fuel. Coal, oil, or gas and resultant type of fuel-burning equipment.

Feedwater System. Drum size, pump characteristics.

Auxiliary Equipment. Fan arrangement, source of fan power, damper arrangement.

Load Characteristics. Anticipated range, rate, and magnitude of change.

Instruments most commonly used are listed in Table 1 with a key to their importance and the desirability of recording or indicating types. The item numbers refer to the typical locations of points of measurement shown in Fig. 1.

Recording Meters. These may be limited to (1) measurements of major operating importance, (2) trend indexes where rate and/or magnitude of change during variations in load or other conditions has significant bearing on effectiveness of control, and (3) need for a history of the measurement for efficiency checks or a picture of the change in condition of the boiler and allied auxiliaries.

Indicators. These are employed for all factors requiring only spot checks or intermittent observation.

Scanners. These are used for checking tube temperatures and values subject to infrequent data keeping. Here a single indicator can be used with a large number of thermocouples on a selective or sequential basis.

* Vice President, Manufacturing, Bailey Meter Company, Cleveland, Ohio.
† Manager of Education, Bailey Meter Company, Cleveland, Ohio.

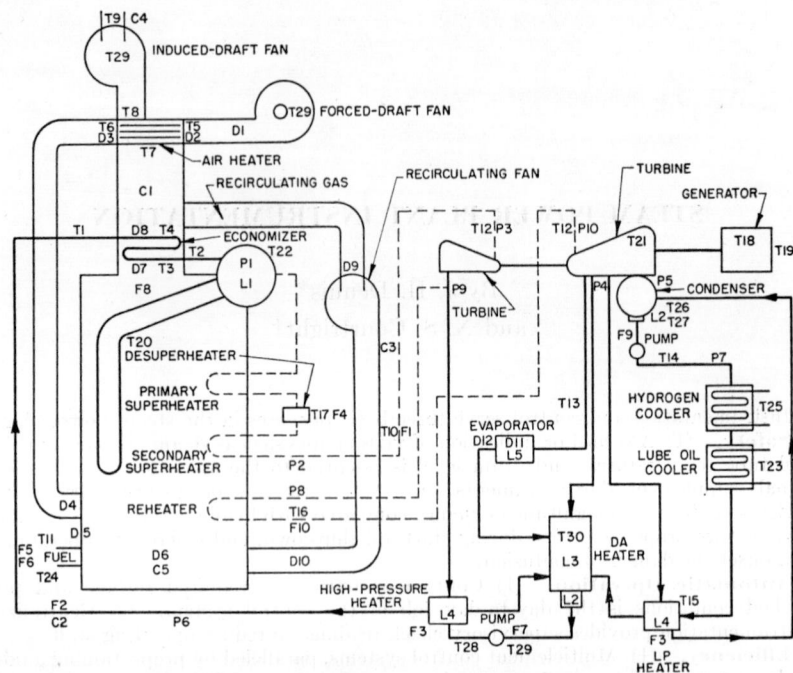

FIG. 1. Location of measurements in typical power plant. Also see Table 1.

PRIMARY POWER-PLANT MEASUREMENTS

Pressures

The important pressure measurements vary from several thousands of pounds per square inch to furnace draft readings over a range of less than 1 in. of water. Only the most vital measurements are discussed here.

Steam Pressure. This is measured at the drum, in the main steam header, and frequently at the turbine throttle. These instruments usually incorporate bourdon-spring measuring elements. Frequently, as with other major measurements, recorders are supplemented by separately connected indicators with large dials. The latter provide for accurate observation even when operators are some distance away.

Feedwater Pressures. These are also in the high-pressure category and incorporate bourdon-spring elements.

Turbine Bleed, Heater Steam, and Deaerator. These pressures may be measured by bourdon-spring elements, but in the lower ranges (50 psi and below), bellows units are frequently employed for recorders. For indicators, the comparatively small motion and friction requirements permit bourdon elements in this range.

Drafts and Very Low Pressures

These are measured by slack-diaphragm units, which are sensitive to changes on the order of hundredths of an inch of water or less.

Furnace Draft. This measurement is vital as a guide to prevent the danger of positive pressure in the furnace and to standardize air infiltration.

Uptake Draft and Forced Draft. Measurement of these factors serves to indicate the proper balance of combustion-air regulating devices, such as fans and dampers.

Pulverizer Differentials. Instruments here provide a measure of primary-air

Table 1. Power-plant Instrumentation Selection

Measurement	Item (see Fig. 1)	Recording (R) Indicating (I) Scanning (S)	Order of preference (see Note)
Pressure			
Boiler drum	P1	I	1
Main steam header	P2	R	1
Turbine throttle	P3	I–R	1–2
Turbine bleeds	P4	I	1
Turbine exhaust	P5	I	1
Feedwater	P6	I–R	1
Condensate	P7	I–R	2
Reheat header	P8	I–R	1
Turbine seal	*	I	1–2
Desuperheater spray water	*	I	1
Pump discharge condensate, main feed, oil, service water	*	I–R	1–2
Lighter oil	*	I	2
Cold reheat	P9	I–R	1–2
Hot reheat	P10	I–R	1–2
Pump suction	*	I	2
Deaerator	*	I	2
Compressed air to soot blowers	*	I	2
Control air header	*	I	2–3
Gas pressure	*	I	1
Low Pressure or Draft	*	I	1
Forced-draft fan outlet	D1	I	2–3
Air heater inlet air and gas side	D2	I	2–3
Air heater outlet air and gas side	D3	I	2–3
Windbox	D4	I	1
Burner differential	D5	I	1
Furnace	D6	I	1
Economizer inlet	D7	I	2–3
Economizer outlet	D8	I	2–3
Gas entering recirculating fan	D9	I	2
Recirculating gas header	D10	I	2
Evaporator steam pressure	D11	I	2
Stoker compartment	*	I	1–2
Pulverizer inlet	*	I	1–2
Pulverizer outlet	*	I	1–2
Temperature			
Feedwater entering economizer	T1	I–R	1–2
Feedwater leaving economizer	T2	I–R	1–2
Flue gas entering economizer	T3	I–R	1–2
Flue gas leaving economizer	T4	I–R	1–2
Air entering preheater	T5	I–R	1–2
Air leaving preheater	T6	I–R	1–2
Flue gas entering preheater	T7	I–R	1–2
Flue gas leaving preheater	T8	I–R	1–2
Flue gas at boiler exit	T9	I–R	2–2
Main steam header	T10	R	1
Coal-air mixture leaving mills	T11	I	1
Steam at turbine	T12	I–R	1–1
Steam at turbine bleeds	T13	I	2
Condensate	T14	S–I	2
Feedwater heaters	T15	S–I	2
Reheater; inlet, outlet	T16	I–R	2
Desuperheater (steam)	*	I–R	2–2
Desuperheater (water)	T17	I–R	2–2
Generator windings	T18	S–I	1
Generator bearings	T19	S–I	1
Tubes	T20	S	2–3
Turbine shell	T21	S	2–3
Drum metal	T22	S	2–3
Lubricating oil	T23	S–I	1–2
Fuel oil	T24	I	2
Hydrogen (gland seal) cooler temperature	T25	S	1
Cooling water into condenser	T26	I–R	1–2
Cooling water out of condenser	T27	I–R	1–2
Feed pump	T28	I–R	1–2
Pump and fan bearing	T29	S–I	2–2

Table 1. Power-plant Instrumentation Selection (*Continued*)

Measurement	Item (see Fig. 1)	Recording (R) Indicating (I) Scanning (S)	Order of preference (see Note)
Deaerator heater...	T 30	I	2
Turbine oil...	*	I	1
Transformer oil...	*	I	1
Flow			
Steam flow...	F 1	R	1
Feedwater..	F 2	R	1
Heater drains disk..	F 3	R	3
Desuperheater water..	F 4	R	2
Coal to mills...	F 5	I–R	1–2
Fuel oil to burners..	F 6	I–R	1
Feedwater pump inlet.......................................	F 7	I–R	1–2
Mass flue gas..	F 8	R	2
Gas flow to burners...	*	R	1
Condensate...	F 9	R	2
Steam to reheat..	F 10	R	2
Heating steam..	*	R	2
Air to soot blower..	*	I–R	3
Evaporator feed..	*	I	2
Level			
Drum water level...	L 1	R	1
Hot well...	L 2	I–R	1–2
Deaerator heater...	L 3	I	2
Low- and high-pressure heaters..............................	L 4	I	2
Evaporator...	L 5	I	2
Condensate storage...	*	R	1–2
Oil-storage tanks...	*	I	2
Condition Analyzers			
Oxygen in flue gas (or CO_2).............................	C 1	R	2
Feedwater and steam:			
Conductivity.......................................	C 2	I–R	1–2
pH...	C 3	I–R	2
Smoke density..	C 4	I–R	1
Fire TV or flame-failure detector............................	C 5	I	3
Dissolved O_2 recorder....................................	*	R	2–3
Auxiliary			
Position indicators (tilting burners, dampers, valves)........	*	I	2
Speed indicators (fans, pumps, turbine)....................	*	I	2
Turbine vibration...	*	I	2
Turbine throttle position....................................	*	I	1–2
Turbine camshaft position...................................	*	I–R	1–2
Eccentricity..	*	I–R	1–2
Motor-switch lights...	*		
Motor ammeters..	*	I	1
Machine-load megawatts....................................	*	I	1
Annunciators and alarms....................................	†	I	1
Flame failure.....................................	...	I	1
Fuel-pressure loss.................................	...	I	1
Control-air loss...................................	...	I	1
Circuit failures...................................	...	I	1
Fan failures......................................	...	I	1
High and low pressures, temperatures, and levels.........	...	I	1

Note. In order of preference the numbers signify:
1. Necessary, in the average installation.
2. Not necessary, but desirable.
3. Desirability varies with specific installation.

* Not shown in Fig. 1.
† Frequently incorporated with other measuring devices.

rate and drop across the pulverizer. They are essential to controlling the transport of coal to the burners and feeding the crushed coal to the mills.

Temperatures

Major temperature measurements are largely those concerned with safety and the maintenance of operating conditions required for maximum efficiency.

Superheated-steam Temperature. This is critical from the standpoint of safety and efficiency. It is commonly the source of regulation for final steam-temperature control to prevent damage to the turbine. Ranging from 900 to 1100°F in modern practice, these temperatures are commonly measured by electrical instruments which employ resistance elements or thermocouples and electronic amplification.

Feedwater and Gas Temperatures. These temperatures are frequently recorded and are measured, in many instances, by thermometers with pressure-filled thermal-expansion elements. Excessive length of connecting capillary can be overcome by use of pneumatic or electric transmission. Resistance- or thermocouple-type elements are also used.

Flows

Flow measurements of major importance are best provided by recording rate meters employing orifices, flow nozzles, or venturi tubes. They generally read directly in terms of flow on uniformly graduated charts.

Superheated-steam Flow. This measurement may be utilized in several ways. The recorder flow rate is an instantaneous measure of output and, through integration, provides vital heat-balance data. Combined with feedwater-flow rate and drum level, the measurement provides for the optimum in feedwater control. It may be employed as an element of steam-temperature control, and when combined with the relative measurement of air flow to the boiler it serves as one of the principal combustion guides.

Boiler-feedwater Flow. Records of this variable provide a comparison with steam flow to determine blowdown, detect leakage, and measure inaccuracies. The measurement also serves as a factor in feedwater control systems (see Three-element Feedwater Control, later in this subsection).

Fuel Flow. In the case of liquid or gaseous fuel, this variable is preferably recorded for heat-input data. Also, it may be combined with relative-air-flow measurement for the control of combustion efficiency.

Air Flow. This variable is rarely recorded on a quantitative basis and frequently employs the effect of the total flow of gases through the boiler. Close proportionality exists between actual air flow and total gas-produce flow. Even in the measurement of combustion air on the "clean-air" side, the relative effect of the flow of required air for optimum combustion conditions compared with the load index (steam flow or fuel flow) is all that is necessary (see Combustion Guides, later in this subsection).

Integration and Compensators. Integration of steam flow, feedwater flow, plant steam takeoff, and fuel oil to burners is useful and desirable, in addition to the coal flow shown in Table 1. This function is incorporated in the recording meters. Integration of coal flow can be provided by either volumetric methods or intermittent-dump weighing counters.

Compensation to correct flow readings for the effect of variation in pressure and temperature is often required in low-pressure steam measurement where considerable variation may be anticipated.

Levels

There are numerous points of storage throughout the feedwater and condensate systems where relatively constant levels should be maintained. Knowledge of the level conditions by record or indication is desirable.

Boiler-drum Water Level. This is one of the most vital measurements in the entire power plant. Safety is the prime consideration, but poor regulation can also affect operating efficiency. Low level endangers circulation and may lead to burned-out generating or superheater tubes, while high level may induce carryover of moisture with resultant damage to the turbine. Thus drum level is commonly indicated and recorded by multiple devices.

Gage Glass. This provides local indication at the drum end, frequently viewed from the firing floor by a system of mirrors.

Remote Indicators. These instruments are often color-coded for quick appraisal and are mounted on the main instrument and control panel. Recorders, with provision for a signal to the automatic feedwater control system, are also mounted on the main panel. *Televised gage-glass* indication is used in many modern installations to transmit an actual picture of the glass to a receiver mounted on the main operating panel. Connections to the drum for each of the measuring devices should be independent of the others to ensure availability of a guide in the event one or more units should become inoperative.

Hot-well Level, Deaerator Level. These and other level measurements in the cycle are frequently served by local indicators or recorders.

Combustion Guides

Efficient combustion within a boiler furnace requires the complete combination of the elements of a fuel with oxygen in the presence of minimum excess air. The theoretical air can be accurately calculated for a given fuel. The type of fuel, fuel-burning equipment, and furnace design determine the additional air required to ensure contact of combustible material and oxygen while such material exists within the combustion zone at or above ignition temperatures. Determination of total air supplied was at one time provided solely by Orsat analysis. However, several means are now available for a recorded determination of combustion conditions.

Fuel Flow–Air Flow

Where liquid and gaseous fuels are concerned, the rate of fuel flow to the furnace is coincident with the rate of combustion. In such cases, the rate of fuel flow to the boiler and the relative rate of air flow can be measured. These records can be combined within a single instrument and calibrated to ensure proper fuel-air ratio. Because the requirements of a boiler, in terms of total air, may vary at different ratings, calibration of the instrument requires tests of the boiler over a range of ratings.

Steam Flow–Air Flow

In the burning of solid fuels (coal is employed for the majority of steam produced) the use of fuel-air ratio is affected by two factors:

1. In an all grate-fired installation, there is no direct relationship between the rate of coal input and the rate of release of heat.

2. There is no satisfactory means of measuring the *rate* of coal input, even in pulverized-coal firing.

The rate of heat release is measured in terms of steam flow from the boiler. The boiler efficiency and the Btu's added per pound of steam remain substantially constant because the outlet pressure and temperature are kept constant.

In either fuel-air, or steam flow-air flow comparisons, the measurement of relative air flow is frequently accomplished by using sections of the flow path of flue gases through the boiler as a primary element. Flue-gas flow is proportional to combustion air. The use of existent pressure drop across tube sections eliminates increased fan power otherwise required to overcome the head loss that would be created by adding a primary element to the system.

Where lack of proportionality or design features of the boiler make it impractical to utilize existing differentials in pressure, venturi sections or orifice segments can be provided in the ductwork. This is done on the clean-air side to eliminate fouling of connections.

Carbon Dioxide Recorders. For a given fuel composition this CO_2 record can be related to total air and an index of combustion conditions is obtained. Where the carbon-hydrogen ratio of a fuel varies or the percentage of fuels in a multifuel installation changes, the required CO_2 at each condition must be supplied from the operator's knowledge.

Automatic analyzers for recording purposes based on absorption, thermal conductivity, and density difference are available.

Oxygen Recorders.   The percentage of oxygen in the flue gases is often employed as a supplementary combustion guide.   This remaining oxygen represents excess air by direct proportion and may be recorded as either excess or total air.   An advantage of oxygen measurement lies in the very slight variance of the relationship between total air and percentage of oxygen in the flue gases over a wide range of fuels.   This is shown in Fig. 2 and is particularly advantageous in mixed-fuel firing.   Oxygen recorders are based on thermal conductivity, catalytic combustion, or paramagnetism principles.

Flame Condition.   To analyze conditions within the furnace further, some installations now employ industrial television.   A camera, directed on the flame path, particularly in pulverized-coal burning, provides a picture on a panel-mounted receiver.

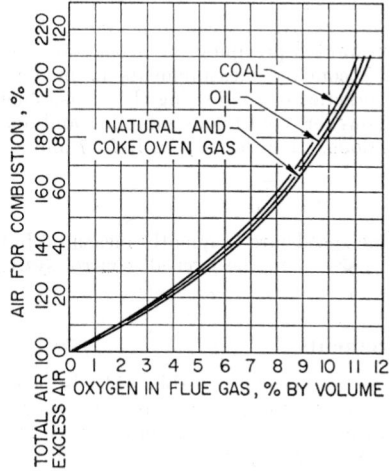

Fig. 2. Oxygen in flue gas vs. total air used for combustion.

This permits the operator to judge turbulence, flame length, and other important factors.

Dependable flame-failure devices are also available and can be used for alarms or automatic fuel shutoff.

SECONDARY POWER-PLANT MEASUREMENTS

The term secondary applied to any measurement in the power plant is relative to (1) system design, (2) plant size, and (3) many associated factors.   For example, a smoke recorder which may be considered only desirable in a plant located in a remote area may be a real necessity in a plant located in a densely populated area regulated by a strict air-pollution ordinance.

In general, a secondary measurement, designated in Table 1 as desirable or optional (2 or 3), may be considered as one which is not directly associated with the automatic control system, or one from which only occasional data taking is required.

Smoke Density

For many years smoke density was determined simply by observing the stack appearance, either directly or through an illuminated duct and mirror arrangement. Recent practice in smoke-density (soot and fly-ash particles) measurement incorporates panel-mounted recorders or indicators of the light-energy-extinction type. These measurements can be used to cut in overfire air blowers automatically to increase turbulence and excess air during temporary smoking conditions. This is accomplished through contacts in the recording instrument.

pH and Electrical Conductivity

Scale formation and corrosion in heat-transfer sections of the boiler, due to high concentration of solids and acidity (low pH), require careful conditioning of feedwater and close observation of steam and water through such measurements as pH and electrical conductivity.

High electrical conductivity of boiler waters is indicative of the presence of considerable solids. These result in foaming in the drums, excessive carryover, and scaling of superheater tubes and turbine blades. Scaling of tubes, in turn, may cause improper circulation, burning of tubes, and at a minimum, excessive maintenance.

Corrosion resulting from low pH will lead to pitting and eventual failure of tubes. A high pH will lead to unwanted precipitation of scale-forming components.

These are considered secondary measurements in moderate-sized power plants because of the common practice of conducting the measurements in the laboratory. In recent large plants, however, continuous sampling and analysis by indicating or recording instruments is done. For details, see Refs. 7 and 8.

Speed

Fan, pump, and turbine speeds are frequently measured but are seldom recorded. Fan speed may be incorporated in the over-all control system as an index of air-delivery rate.

Miscellaneous Measurements

Pressures. These are measured at turbine bleeds, condensate headers, soot blowers, and other auxiliary equipment and are commonly indicated on dial-type bourdon-spring-operated gages. Manometers may be used for turbine-exhaust gland seal and condenser vacuum readings, but here too, sensitive bourdon units may be employed.

Temperature. These measurements of a secondary nature include lubricating and cooling oils, hydrogen coolers, turbine gland seals, and bearings. Usually non-recording devices are used. These vary from mercury-in-glass thermometers to multipoint thermocouple-actuated recorders.

Scanning. Instruments capable of sequencing 60 or more temperatures are often used to measure temperatures of tubes, turbine shells, drum metal, and other points which must be checked, but with insufficient frequency to merit individual instruments. These instruments are usually thermocouple-actuated.

Flow. The flow measurements of secondary importance are measured by indicating rotameters, differential gages, and manometers. In instances where only total flow per unit of time is required, integration alone may suffice.

Several examples could be cited as to the degrees of importance placed on certain types of instrumentation. For example, feed-pump bypass flow may go unmeasured in some systems, while in others it may be an important control factor in safe pump-temperature regulation. In the same manner, heater drains discharge, air to soot blowers, and evaporator feed may be unmeasured in some systems; in some it will simply be indicated; in others it will be recorded.

Signals and Alarms

Audible or visual signals and alarm devices intended to call for action from operating personnel, as distinguished from automatic interlocks, are used on all units regardless of size. Since a multiplicity of audible alarms would lead to confusion, these are usually kept to a minimum, supplementing the sound with annunciators of either the drop type or illuminated-glass-panel design. Contacts located in the indicators and recorders activate the signals in event of emergency conditions, such as high or low drum levels, flame failure, excessive temperatures, or smoking.

Signal lights are used to indicate valve and damper positions. They are also included with nearly all switch gear to show energized motor circuits for fans, pumps, pulverizer mills, feeders, exhausters, and safety trip-outs.

AUTOMATIC CONTROL SYSTEMS

The steam generating electric power plant is one of the most complex processes or combinations of processes. Consequently, automatic regulation of a power plant requires a combination of several closely allied control systems.

Automatic combustion control systems are described in other sections. Other major control problems include:

1. Regulation of flow of feedwater to boiler.
2. Control of reheat and final steam temperature.

Measurements involved in the combustion control system may be employed in one or both of these systems. Thus duplication of recorders or primary elements for such factors as steam flow and air flow need not be provided. The interdependence of systems goes beyond this consideration, because poor regulation of feedwater flow may affect the controllability of each of the others and vice versa. The complexity of any control system is dependent upon the size, rate of change of load, and number of operating units involved in regulation as well as the limits within which the process must be controlled. Thus boiler design, auxiliaries, and load characteristics combine as control-selection determinants.

Feedwater Control

Systems employed for control of feedwater flow to the boiler vary from direct-operating single-element* devices to multielement pneumatic or electronic control systems. Selection of the desired system depends upon:

1. Size of the drum, relative to anticipated rate of change of steam output.
2. Temperature and pressure at which steam is to be produced.
3. Capacity of the boiler.

Single-element* Direct-acting System. A single-element feedwater control depends solely upon variations in drum level as the index for change in feedwater input rate to match a variation in steam flow. A simple method is shown in Fig. 3.

The connecting tube between the drum connections is surrounded, for a considerable distance above and below the normal water level, by a generator tube which is filled with water and connected to an operator on the feedwater valve. Changes in level, within the drum and thus the connecting tube, will vary the heat transfer to the generator. This changes the amount of water which is converted to steam in the closed-pressure system. This, in turn, varies the pressure on the valve operator, and hence the position of the inner valve to increase or decrease feedwater flow rate.

A variation of this type of regulator employs the expansion of a metal tube which is subjected to the change in levels in a similar leg between the drum connections. By linkage, the expansion or contraction of the tube acts upon a lever-operated valve to vary input to the drum.

Single-element Air-operated System. Direct-acting devices, as just described depend upon a change in level to maintain a change in feedwater flow. They are incapable of restoring normal level to a constant value at all ratings. To do so requires proportional-plus-reset control action, and this, in turn, necessitates a more flexible system arrangement.

* The term element is synonymous with measured variable. For example, a single-element system is concerned with the measurement of a single variable; a two-element system is concerned with the measurement of two variables; etc.

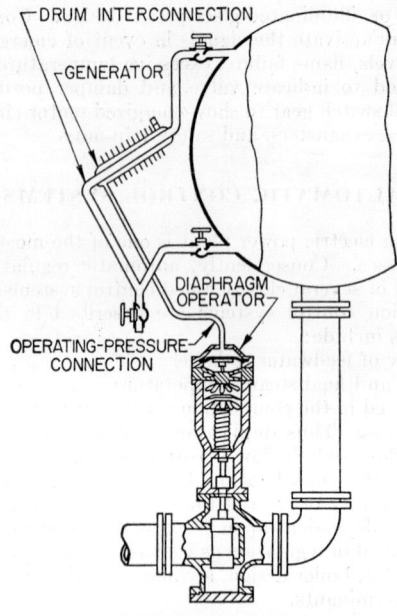

FIG. 3. Direct-acting single-element feedwater control system.

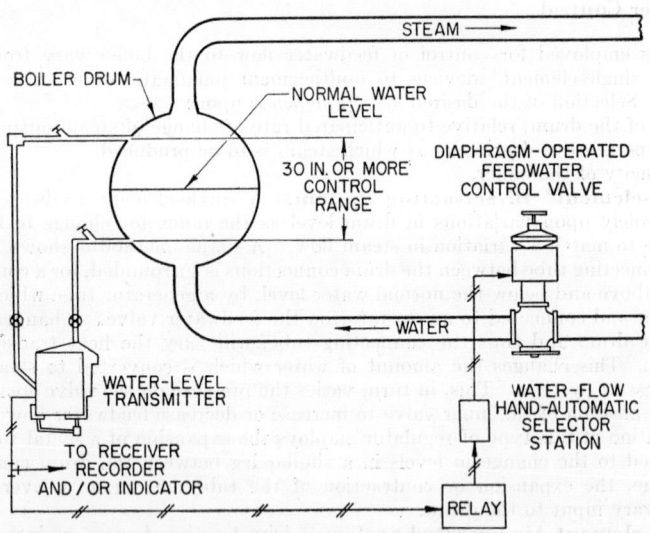

FIG. 4. Layout of typical single-element air-operated feedwater control system.

In the single-element air-operated system shown in Fig. 4, the variation in drum level is measured by a recorder-controller and a control signal is developed which is proportional to deviation. This loading pressure is then modified in a relay which provides for amplification or reduction of the signal to (1) position the diaphragm-operated control valve sufficiently to restore level to the desired constant, and (2) stabilize at a value sufficient to maintain the level at the new rating.

Two-element System. To reduce the magnitude of level variation in event of frequent and considerable load variation, a second element (measured value) of control, steam flow, can be added to drum level. In Fig. 5, the measurement of steam flow provides a control signal which varies the flow of feedwater to offset the change in output, and readjustment from the change in drum level provides for restoration of the normal level. Here the differential across the superheater acts upon a diaphragm operator which positions the feedwater valve with compensating action from the expansion tube–linkage drive actuated from drum level, as previously described in the single-acting direct-operated system.

A similar action can be derived from a pneumatic control system in which air-loading pressures from metering controllers responsive to level and steam flow are

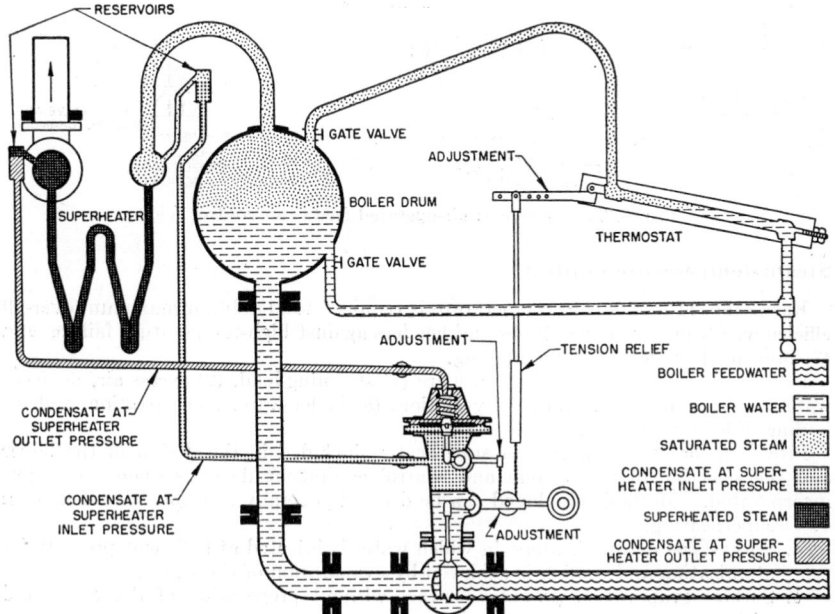

FIG. 5. Two-element feedwater control system.

combined through a relay to position the feedwater valve. This arrangement consists of the steam flow and level elements of the three-element system shown in Fig. 6. The pneumatic systems, in all forms, provide the added advantage or remote manual operation of the feedwater valve by means of panel-mounted selector devices.

Three-element Systems. In boilers where greater capacities are accompanied by high pressures and temperatures—and where the drum and other water-capacity sections of the boiler are relatively small, a change in steam demand will create (1) considerable swell on an increase in load, or (2) shrinkage of the drum level with a decrease in load. Swell, caused by an increase of steam volume within the water, due to an instantaneous reduction of drum pressure, results in an increase in level which causes a drum-level controller to call for a reduction in feedwater valve opening. This, of course, directly opposes the needs of the boiler.

To counteract this effect, three-element systems employ the ratio of steam flow and feedwater flow as the source of principal control action. This is shown in Fig. 6. Readjustment of the final control loading pressure from drum level provides compensation of the steam flow–water impulse, should actual variation in drum level remain.

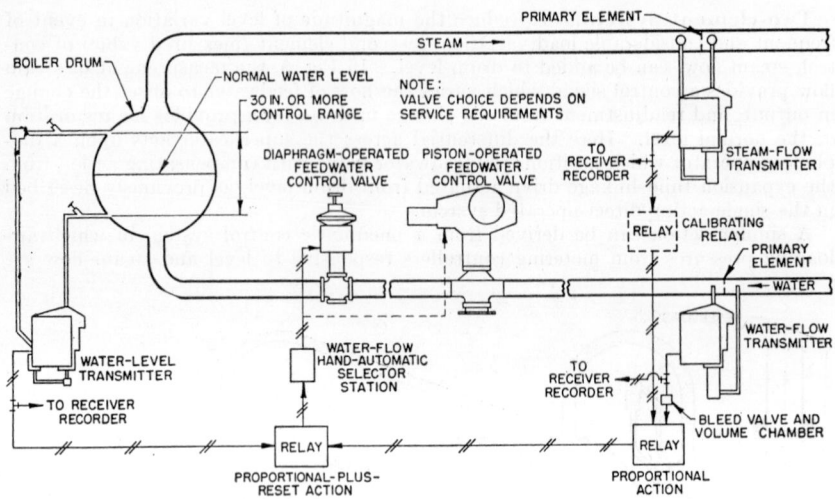

Fɪɢ. 6. Three-element air-operated feedwater control system.

Steam-temperature Control

Regulation of final steam temperature is required to (1) obtain maximum over-all efficiency, (2) protect boiler tubes and headers against high-temperature failure, and (3) prevent damage to turbine blading.

Factors affecting steam temperature are (1) steaming load, (2) excess air, (3) feedwater temperature, (4) drum-level variation, (5) boiler-water concentration, and (6) fouling of heater surfaces.

Provision for the regulation of superheat is included in the design of the boiler and the selection of the system and control equipment depends upon the means incorporated.   Methods employed can be divided generally into four classifications:
1. Desuperheating.
   a. Spray-type desuperheaters, in which water is injected at sufficient pressure for atomization and cooling is achieved by evaporation of the water.
   b. Surface evaporation provides contact without preparation of the water and with slower response to demand change.
   c. Heat exchangers, such as submerged desuperheaters employing boiler water.
2. Flue-gas regulation.
   a. Bypass dampers by which gases are diverted from the superheater, frequently through an open pass of negligible resistance.
   b. Divided-pass boilers in which louver-type dampers provide for adjustment of quantity of gas passing over superheater.
   c. Multiple superheaters provide for regulation of the amounts of steam subjected to varying flue-gas temperatures.
   d. Gas recirculation whereby lessening amounts of flue gas at lower ratings are compensated for by returning a portion of the gas to the boiler at a point below the superheater.   Increased-mass gas flow tends to maintain steam temperature.
3. Firing-control.
   a. Separately fired superheaters may be provided in which firing rate is directly proportioned to steam-temperature requirements.
   b. Tilting burners make it possible to vary the heat available to the superheater by varying the flame path.
4. Combinations of above methods may act in sequence as in gas recirculation for lower operating ranges with spray desuperheat becoming effective at high ratings.
Measurements required for the control system will vary with the number of control

elements involved.   While it is not possible to describe all the available systems here, some typical systems are outlined in the following paragraphs.

Single-element Systems.   In these systems, final steam temperature is the regulating factor.   In Fig. 7, the valve regulating the per cent of steam flowing through the submerged heat exchanger prior to the second-stage superheater is positioned from temperature in the superheater outlet header.   To this may be added a load factor—air flow or steam flow—which can provide for anticipation of change in steam temperature due to load change.   This will improve regulation in critical installations.

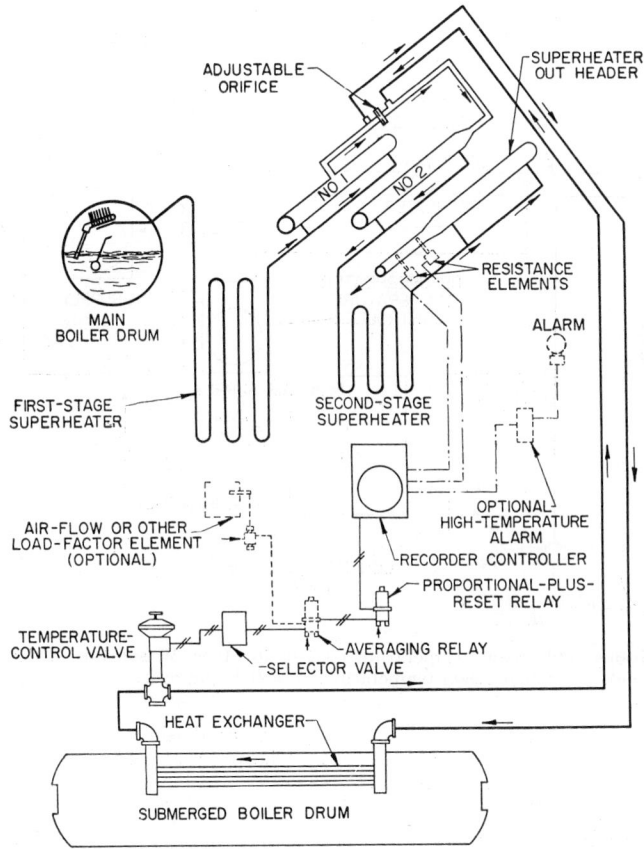

Fig. 7. Superheat control (for Riley Stoker Corp. boiler).

Three-element Systems.   In a three-element system, as shown in Fig. 8, the addition of an intermediate-temperature recorder-controller, which measures the steam temperature between superheater sections but following the spray-type desuperheater, modifies the control action in accordance with the desuperheat effect.   The output pressures of these controllers are combined through averaging relays.   The response rate of the entire system can be modified by adjustment of the reset rate in the relay following the final steam-temperature controller.

Because of the high temperatures involved and the high accuracy and fast response required, platinum resistance and thermocouple-type measuring elements are commonly employed.

Since steam temperature tends to increase with load rate, methods of control based on desuperheating require that steam be at or above design temperature at low ratings, with increasing desuperheating required as the load increases.

Control of the flow of gases, or burner tilt position in boilers fired with fuel in suspension, can provide for substantially constant outlet temperature without desuperheating.

Where controllability does not extend into the highest load range, or where safety provisions require a combination of heat input control, desuperheating can be effected as shown in Fig. 9. The mass of flowing gas is varied by recirculation to maintain constant steam temperature over normal operating ranges. When the load has reached a point where recirculation is no longer necessary, the damper is closed and any further increase in load will call for desuperheat to prevent excessive steam

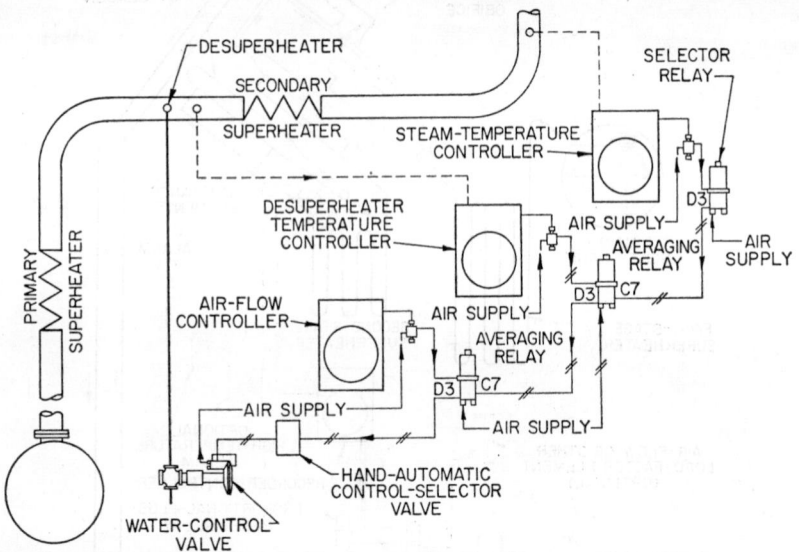

FIG. 8. Air-operated steam temperature control for B & W spray-type desuperheater (single-nozzle). For final steam temperatures of 950°F and over.

temperature. Calibrating relays are employed to establish the point of changeover from reduced recirculation to desuperheat. The per cent load at which this takes place is dependent upon boiler design.

Auxiliary System Controls

In addition to the major control systems previously described, auxiliary portions of the plant circuit may be automatically controlled. On-off or single-element floating controls may be utilized. In large installations, multielement systems are frequently employed. Condenser and deaerator hot-well levels, heater and condensate flow, and lubricating and fuel-oil temperatures are auxiliary operations commonly controlled as independent units.

Where a portion of the steam output is to be used for plant heating or other low-pressure uses, and where the turbine is not designed to provide bleed capacity for such purposes, pressure reducing and desuperheat of the superheated high-pressure steam may be required. Control of these factors may be provided as shown in Fig. 10. A simpler form of pressure reducing may also be effected through use of a direct-acting pressure-reducing valve.

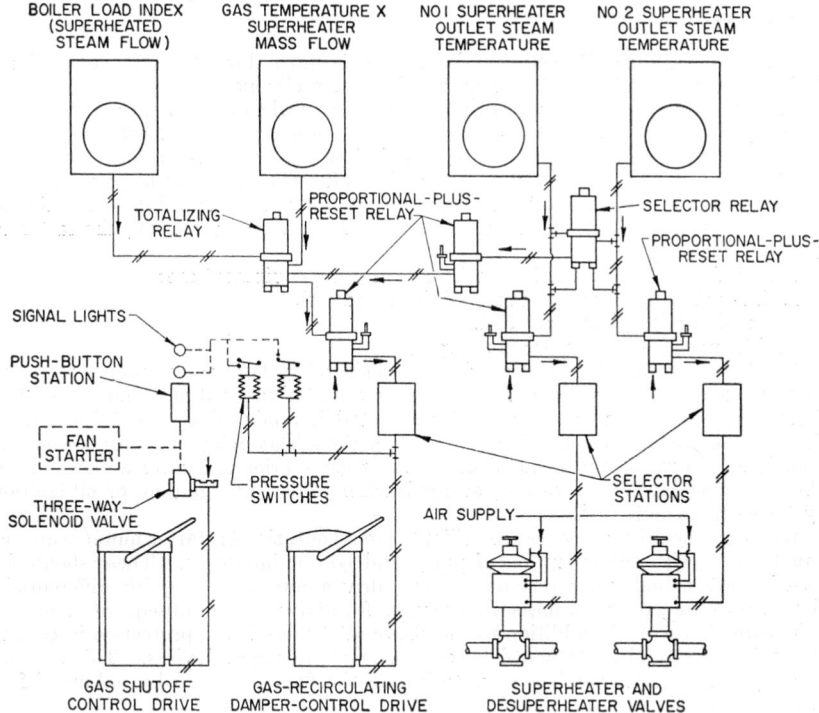

FIG. 9. Basic control arrangement for superheat temperature control using gas-recirculation desuperheater.

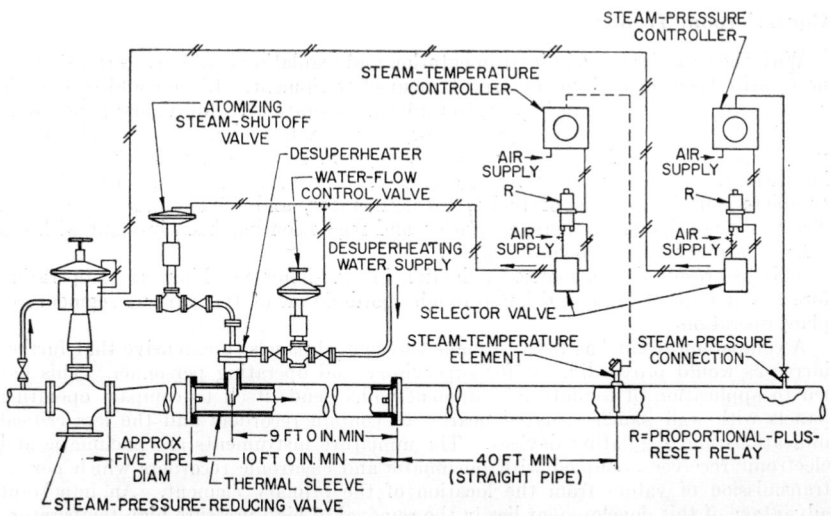

FIG. 10. Pressure-reducing and desuperheat control system.

Interlocks

For safe operation of the compound elements of control throughout the power plant, it is necessary that interlocks be provided. These eliminate the possibility of one or more portions of the system continuing to operate when other vital portions have failed or have been secured out of sequence by the operator. For example, should the supply of combustion air be interrupted, fuel supply in suspension-fired boilers must be stopped so as to prevent filling the boiler with combustible material. If both forced- and induced-draft fans are used, failure of the induced-draft fan should immediately provide for shutoff of the fuel and the forced-draft fan, the latter to prevent a buildup of furnace pressure.

Standard arrangements of interlocks on operating equipment are:
1. Induced-draft failure stops forced draft and fuel.
2. Forced-draft failure stops fuel.
Fuel failure does not require fan stoppage.

Boiler Start-up.   Interlocks and burner light-off systems assure proper sequence in starting up a boiler. The start-up order calls for induced-draft fans and then forced-draft, after the predetermined differential is produced by the induced-draft fan. At this point, a purge interlock with a time-delay relay can provide for the complete removal of any residual combustible gases prior to ignitor action. Then the pulverizer blower, pulverizer, and pulverizer feeder, gas-ignition, or oil-ignition systems can be activated.

Where motor-driven devices are utilized throughout, simple starter motor sequence can be used with the exception of purge and ignitor interlocks. These should be based upon actual measurements. Steam-driven auxiliaries require differential-, draft-, pressure-, or speed-measuring devices for activation of subsequent elements.

Flame Failure.   In addition to the above safety interlocks, protection in case of flame failure should be provided by shutoff or alarm signal devices. This may be accomplished by elements sensitive to loss of the flame, assisted by unburned-fuel detectors (combustible analyzers).

Safety or sequential interlocking devices should be provided with signal lights to afford the operator with the knowledge of units in or out of operation. All safety interlocks should be furnished with audible alarms, in addition to means of identification by either individual signal lights or some form of annunciator.

Control-room Trends

With the great increase in instrumentation and regulating equipment attention has necessarily been focused on the arrangement of equipment. Little could be done in earlier stations, as shown in Fig. 11, to centralize operations. Individual panels were used for each boiler, the turbine, generator, pumps, and condenser. With the reduction in the number of boilers per turbine, improvement in coordination of records, indications, and control panels was obtained. The recent trend has been toward complete units consisting of boiler, turbogenerator, and all auxiliary equipment. The vast majority of all instrumentation and regulation has been brought within a single control room as shown in Fig. 12.

This has resulted in a marked reduction in man-hours per kilowatt in operating forces and improved supervision through coordination of the various segments of plant operation.

A companion result has been to make the control panels so extensive that further increases would prove difficult for supervisory and operating personnel. This has led to application of miniature equipment which lends itself to compact operating panels with wall panels situated nearby to contain recorders and the less critical indications and regulating devices. The miniature instruments are pneumatic and electronic receivers, coupled with pneumatic and electronic recorders which permit transmission of values from the location of the primary element. An important advantage of this development lies in the removal of high-pressure high-temperature steam and water lines from the control-room area.

In order to assist the operators in visualizing the relationship of panel-mounted indicating and regulating devices with the portion of the system or operating equipment affected, graphic representation of the boiler system has been incorporated in some panels. This is of assistance in training new personnel and involves mounting

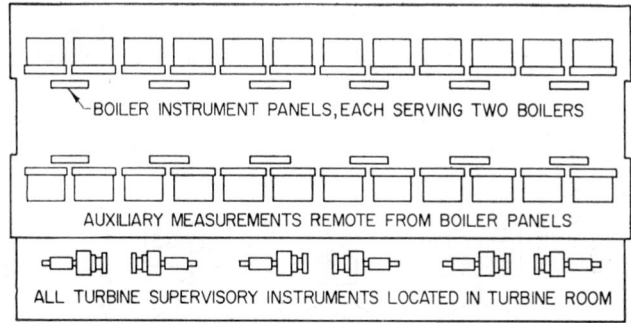

FIG. 11. Plan of boiler, panel, and turbine layout typical of 1930–1940 era.

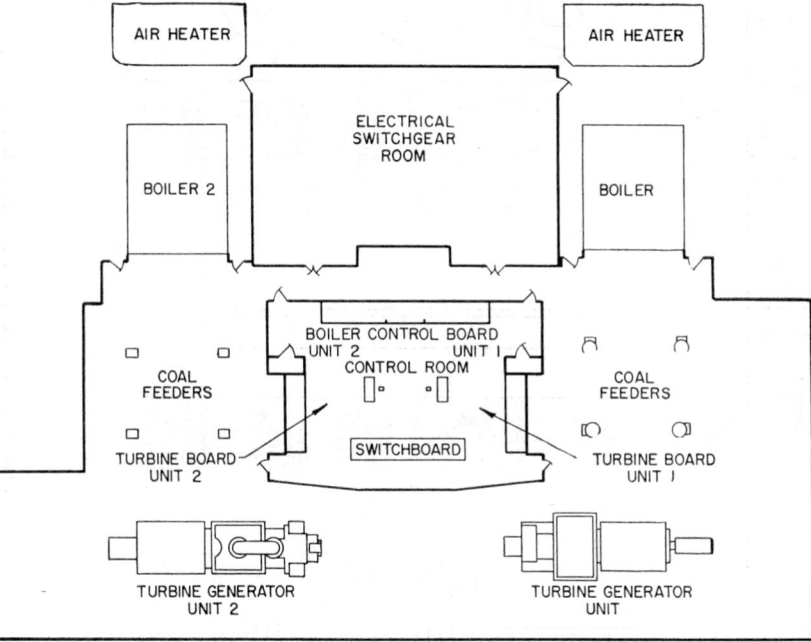

FIG. 12. Plan of modern boiler, panel, and turbine layout.

push-button switches, miniature indicators, and signal lights within the panel diagram at points representative of the respective locations of points of measurement or regulation in the actual boiler. Graphic instrumentation as applied to processes in general is described in Ref. 9.

DATA-LOGGING AND COMPUTING EQUIPMENT

The application of data-logging and calculating systems is causing continued evolution in the methods of presenting measurements and calculated performance values.

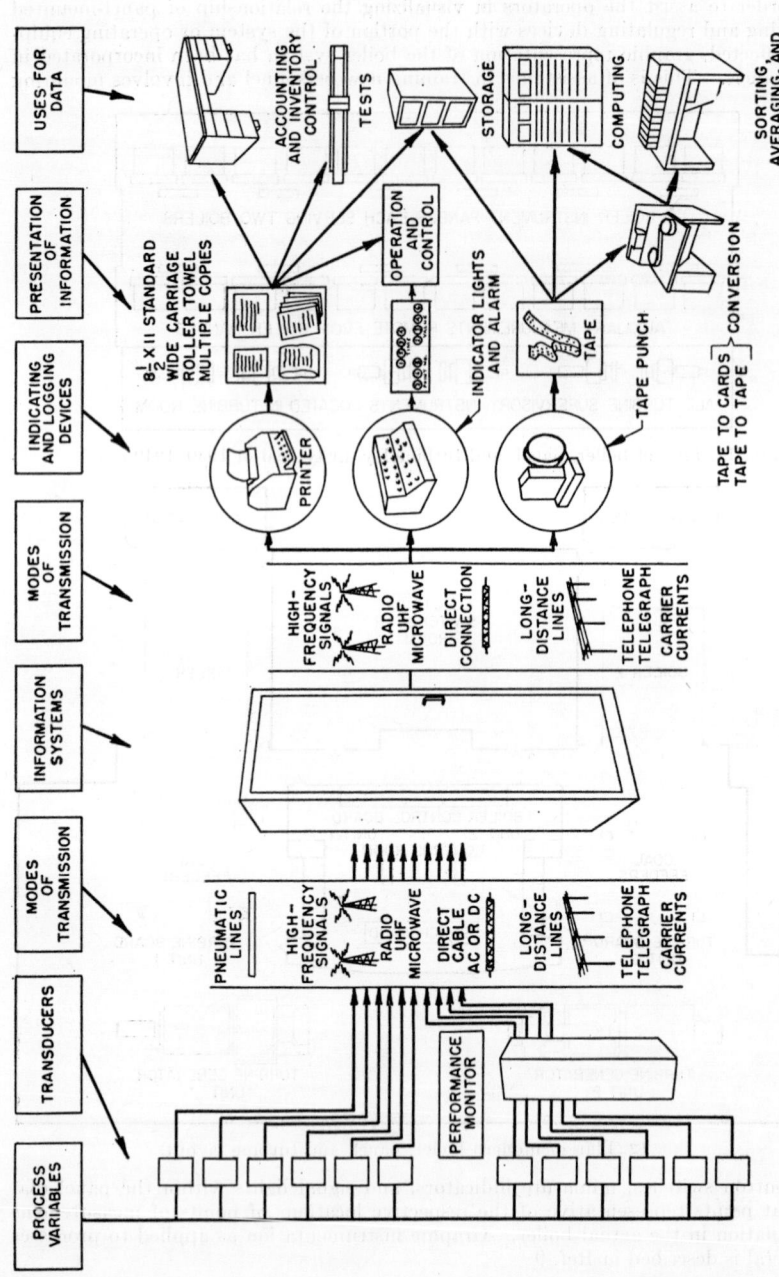

FIG. 13. Information system for data gathering and digital recording.

USES FOR DATA

ACCOUNTING AND INVENTORY CONTROL

TESTS

STORAGE

COMPUTING

SORTING, AVERAGING, AND TOTALIZING

PRESENTATION OF INFORMATION

8½ X 11 STANDARD
WIDE CARRIAGE
ROLLER TOWEL
MULTIPLE COPIES

OPERATION AND CONTROL

INDICATING AND LOGGING DEVICES

INDICATOR LIGHTS AND ALARM

TAPE

TAPE PUNCH

TAPE TO CARDS
TAPE TO TAPE } CONVERSION

PRINTER

MODES OF TRANSMISSION

HIGH-FREQUENCY SIGNALS
RADIO UHF MICROWAVE
DIRECT CONNECTION
LONG-DISTANCE LINES
TELEPHONE TELEGRAPH CARRIER CURRENTS

INFORMATION SYSTEMS

MODES OF TRANSMISSION

PNEUMATIC LINES
HIGH-FREQUENCY SIGNALS
RADIO UHF MICROWAVE
DIRECT CABLE AC OR DC
LONG-DISTANCE LINES
TELEPHONE TELEGRAPH CARRIER CURRENTS

TRANSDUCERS

PROCESS VARIABLES

PERFORMANCE MONITOR

14–42

Reliable high-speed digital systems scan many variables each second, sounding alarms for off-normal conditions and printing compact log sheets automatically.   Figure 13 illustrates the flexibility of digital systems.

The growing acceptance of this equipment suggests the possibility of radical departure from traditional control-room design.   Manual logging, many analog records, and longhand processing of data can be completely eliminated.   Current performance results can be presented to the operators as operating guides.   The calculating equipment can also be called upon to adjust plant control settings automatically.

The availability of accurate sensing devices, reliable pneumatic and electronic control systems, and dependable data-logging and calculating systems makes the automatic steam plant an approaching reality.

Figure 14 has been developed to illustrate the use of new and old instrumentation as essentially independent but interrelated systems, each performing specific operating functions which alone or in combination with others will satisfy certain basic objectives.

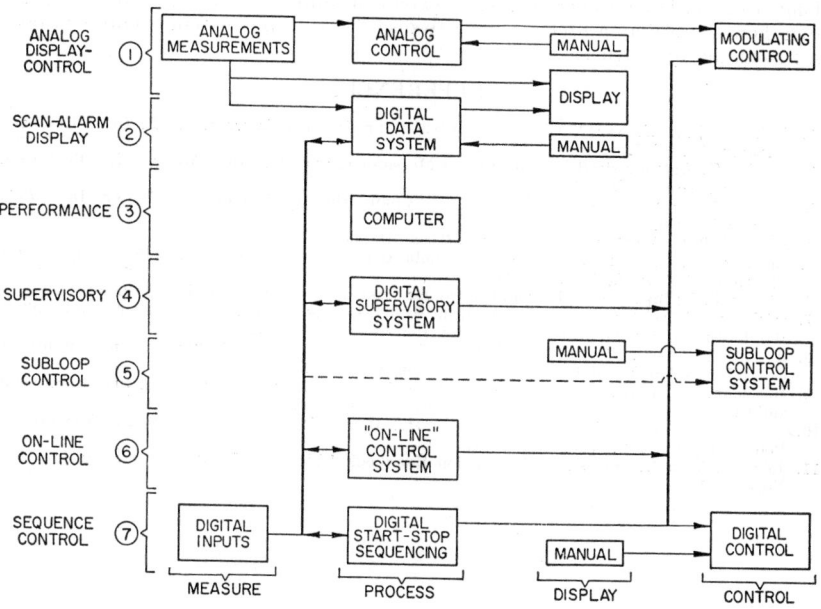

FIG. 14. Steps to the automatic plant.

Section 1 illustrates the commonly used instrument and control concept consisting of analog measurement-sensing devices, standard recording and indicating devices, and analog control systems.

Section 2 illustrates the addition of a digital scan-alarm-printing system to simplify the data presented to the operator.   It minimizes the information under continuous surveillance by directing attention to critical or problem areas.

Section 3 illustrates the addition of a performance computer (analog or digital) to provide the operator with up-to-date performance information so that he may take corrective measures, schedule maintenance, or adjust the control system as required to improve performance.   The computer may also apply automatic corrective actions to the analog control system.

Section 4 illustrates the addition of a digital logic system to extend supervisory control for trip-out, rate-of-change limiting, load reduction, safety interlocking, and disabling controls to prevent the operator from making some of the more common operating errors.

Section 5 illustrates the addition of manual or automatic subloop controls utilizing analog and digital equipment where applicable in order to simplify operation greatly. Automatic burner lighting is an excellent example of a more complete subloop control system.

Section 6 illustrates the addition of "on-line" controls to operate the unit automatically over its full range of operation once it has been placed on the line. This will involve the automatic cut-in and cutout of burners, fuel-handling equipment, fans, pumps, and other auxiliary apparatus but does not involve the complexity and investment required for "automatic" start-up.

Section 7 illustrates the addition of automatic start-stop control to the system to complete the concept of a fully automatic plant.

The present utility power plant is essentially automatic in normal operation between one-third and full load. This is provided by the equipment shown in section 1 of Fig. 14. It can be seen that an extension of automation must come through the acceptance of instrumentation in areas previously considered to be the exclusive domain of the human operator. The extent of automation must therefore depend upon how completely a user is willing to dispense with the human operator in favor of automatic equipment.

REFERENCES

1. Zerban, R., and E. Nye: "Steam Power Plants," pp. 362–406, International Textbook Company, Scranton, Pa., 1952.
2. Rhodes, T. J.: "Industrial Instruments for Measurement and Control," McGraw-Hill Book Company, Inc., New York, 1941.
3. deLorenzi, L.: "Combustion Engineering," Combustion Engineering-Superheater, Inc., *Bull.*, New York, 1950.
4. Skrotzki, B. G. A.: Combustion Control, *Power*, December, 1949.
5. Lammers, H. B., and E. B. Woodruff: Combustion Analysis in the Modern Boiler Unit, *Edison Electric Institute Bull.*, August, 1941.
6. Dickey, P. S.: Instrumentation and Control for Small Power Plants, *Gen. Elec. Rev.*, July, 1951.
7. Stickney, Michael E.: pH and pH Measuring Systems, Sec. 6–7, "Process Instruments and Controls Handbook," McGraw-Hill Book Company, Inc., New York, 1957.
8. Rosenthal, Robert: Electrical Conductivity Measurements, Sec. 6–11, "Process Instruments and Controls Handbook," McGraw-Hill Book Company, Inc., New York, 1957.
9. Hermann, Arthur D., and C. S. Conner: Graphic Instrumentation, Sec. 8, "Process Instruments and Controls Handbook," McGraw-Hill Book Company, Inc., New York, 1957.
10. Feller, Eugene W. F.: "Instrument and Control Manual for Operating Engineers," McGraw-Hill Book Company, Inc., New York, 1947.
11. Johnson, Allen J.: "Fuels and Combustion Handbook," McGraw-Hill Book Company, Inc., New York, 1951.

ELECTRIC POWER GENERATION
AND DISTRIBUTION CONTROL

By D. G. Blodgett,* N. Cohn,† A. K. Falk,‡ T. W. Hissey,§
S. B. Morehouse,‖ and W. B. Schultz.¶   Compiled by J. F. Springman**

Power systems have two fundamental operating objectives.   The first is to maintain a balance between generation and customer load.   The second is to so regulate
generating units that this balance is maintained at lowest cost.   Automatic control
systems have been developed to assist operators in the attainment of these objectives.
In this discussion, we shall explore a definition of the basic regulating problem, review
the classes of control systems which have evolved, and survey the computers—both
analog and digital—which have been developed for automatic economy-dispatch
purposes.

DEFINITION OF THE REGULATING PROBLEM

In this country, interconnected power systems are composed of separate and independent private utilities, public utilities, or large industrial plants which represent a
wide variety of financial interests.   Participation is voluntary and over-all coordination is obtained on a purely cooperative basis.

Figure 1 shows the approximate areas served by major interconnected systems in
this country.   Most of these systems have operational subdivisions called *areas*.   The
economic and technical aspects of these areas have accelerated development of tie-
line interchange and economy-dispatch controls.

General Characteristics of Interconnected Systems

Operating benefits of interconnected-system operation include improved continuity
of service, increased reserve-generating capacity, and better voltage and frequency
regulation.   Most important, excess power can be shipped to the area where it can
most economically be used.   As a given power company approaches the upper limits
of its capacity, added increments of power become increasingly costly (high-cost units
are all that remain).   Under these conditions, the company can frequently find less
expensive power elsewhere on the interconnection and have it delivered under a price
arrangement which is favorable to all concerned.

Obligations include an operating responsibility which is essential to interconnected-
system operation: *each area must absorb its own load changes*.   All regulating requirements are determined, ultimately, by this responsibility.   In turn, these requirements
are the mold in which dispatch-control developments have been made.   Determination of the *area requirement* (difference between generation and load) is therefore an
appropriate starting point.

* Superintendent of Electrical System Performance, The Detroit Edison Company, Detroit, Mich.
† Vice President—Technical Affairs, Leeds & Northrup Company, Philadelphia, Pa.
‡ Interconnection Coordinator, The Detroit Edison Company, Detroit, Mich.
§ Assistant Sales Manager, Electric Power, Leeds & Northrup Company, Philadelphia, Pa.
‖ Assistant to the Vice-President, Systems Department, Leeds & Northrup Company, Philadelphia,
Pa.
¶ Principal Engineer, Digital Systems, Leeds & Northrup Company, Philadelphia, Pa.
** Advertising Specialist, Radio Corporation of America, Cherry Hill, N.J.

The voluntary cooperative nature of these interconnections stems from the reasoning that, when each area was operating isolated, it necessarily kept its generation equal to its load. A changing frequency was the index which showed how well this balance between load and generation was maintained. If an area joins a large interconnection and continues to keep its generation equal to its load, it will not introduce

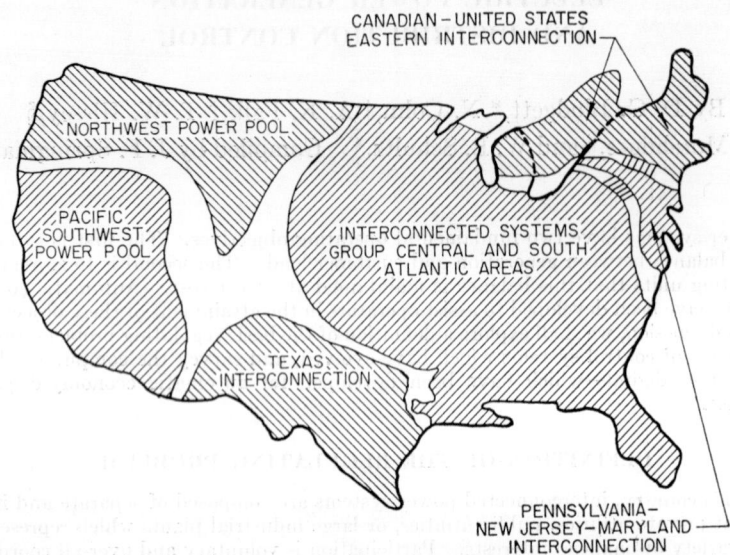

CANADIAN – UNITED STATES
EASTERN INTERCONNECTION

NORTHWEST POWER POOL

PACIFIC
SOUTHWEST
POWER POOL

INTERCONNECTED SYSTEMS
GROUP CENTRAL AND SOUTH
ATLANTIC AREAS

TEXAS
INTERCONNECTION

PENNSYLVANIA–
NEW JERSEY–MARYLAND
INTERCONNECTION

Fig. 1. Major interconnected power systems in the United States.

additional operating problems for the interconnection. If, on the other hand, load changes of an area are corrected by generation changes from another area, unsolicited power flows will occur between areas.

It is necessary, then, that each area measure the interchange of power at all its tie points with the interconnection. Further, it must compare *actual* net power flow with *scheduled* net power flow to determine whether the load changes of the area are being corrected by generation changes made within that area. Naturally, an area may schedule a certain net delivery of power to its neighbors if it has excess capacity at a given time; but this value then becomes a part of the base for determining how closely area generation is following area load changes.

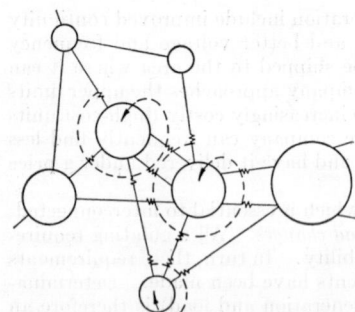

Fig. 2. Principle of area control.

Area-control Principle

Figure 2 illustrates the area-control principle which is the accepted practice on practically all the large interconnected power systems in the United States. Extensive telemetering is used to obtain the net interchange which takes place between an area and the rest of the interconnection. This net interchange value is obtained at a central location—usually in the load dispatcher's office—and used to derive area requirement (Fig. 3) in accordance with the general equation

Area requirement = (scheduled net interchange − actual net interchange)

− bias (scheduled frequency − actual frequency)   (1)

Notice that this equation incorporates a bias factor which recognizes the inherent regulating characteristics of the area as it responds to a load change. This regulating characteristic is expressed in megawatts per 0.1 cycle of frequency change and is experimentally determined for each area. Its magnitude is dependent on such factors as the type of load being serviced, inertia constants, and prime-mover governor action. To date, values for systems in this country have been in the range of ½ to 3 per cent of the spinning capacity on the line in that area. For any given area, this bias characteristic provides an essential part of the reference which determines whether a given load change has occurred locally or in a remote area.

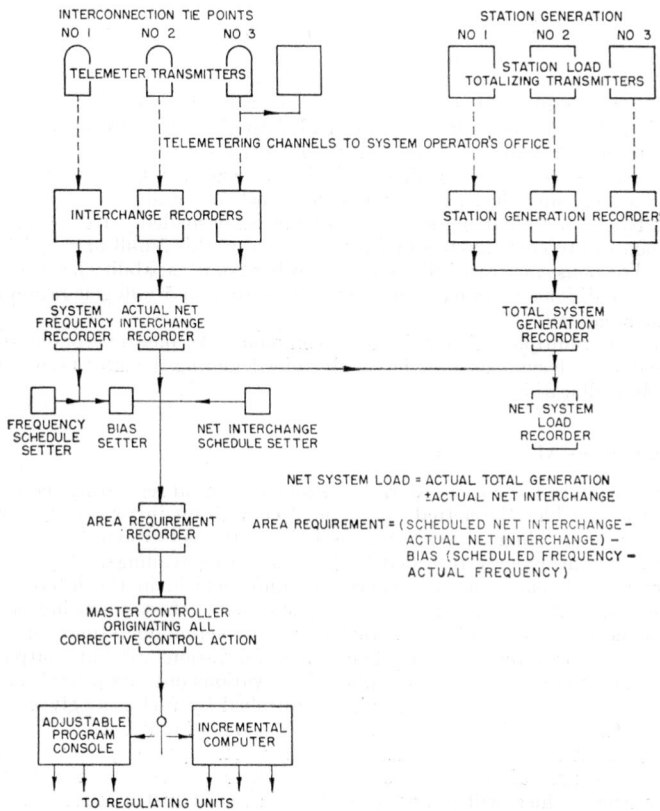

FIG. 3. Schematic diagram of typical centralized control system.

Area requirement is continuously computed and normally recorded for the benefit of the system operator. It shows him the total megawatt change required in generation to balance load changes which have occurred in his area. The area-requirement value, incidentally, is derived and utilized by virtually all automatic economy-dispatch controls.

The next step is to determine how these changes can be most economically allocated among the available generating units. In early control practice, this regulating function was assigned to a single unit. At present, it is allocated to most of the available units in accordance with the need to obtain:

1. Sufficient regulating range.
2. Sufficient regulating rate.

3. An incremental rate which will match the system average over the proposed regulating range.

On some systems which have numerous large hydro units, the principal system WR^2 (rotational energy) is located in these hydro units. Where these units represent a large percentage of the system capacity, a single unit may have ample regulating range over appreciable periods of the operating day. Usually a unit of this size will provide both sufficient regulating range and regulating rate—thereby meeting two of the three essential regulating requirements listed above. If a plant is operating on run-of-the-river with limited storage capacity, the assigned incremental rate of its generation may be satisfactory over the entire regulating range in accordance with the third essential regulating requirement.

In those areas where steam generation predominates, regulating range and rate requirements may be readily met by the newer gas-fired steam units. Usually, however, the incremental rate requirement is such that it is not economical to limit regulation to those units. It is now well established that the most economical system operation is obtained when all units are loaded so that they are operating at the same *incremental rate*, which is expressed in mills per kilowatthour.

For example, as the early-morning load pickup begins, it may be economical to regulate on a large unit, having an incremental cost of 2 mills/kwhr, because other large units having a similar incremental cost are being loaded. However, it would not be economical to continue to provide regulation on this 2-mill unit just to utilize its range and response rate capabilities after all other 2-mill capability has been loaded. Under those conditions, it would be necessary to go to 4- or 5-mill generation in order to carry the area load.

Consequently, multiple-unit regulation is common. Automatic economy-dispatch controls used on such systems must be capable of calculating the most favorable incremental loading allocation among units.

Operating Curves Aid Analysis

In order to analyze further the above-mentioned essential regulating requirements, it is helpful to consider the actual system-load curve, unit incremental curves, and resultant economy-loading curves. Figure 4 shows the net-system-load curve for a typical area as continuously recorded from telemetering readings. Figure 5 shows typical incremental-generating-cost curves for units which, in the interests of simplicity, are assumed to represent single-unit stations. Most operating companies obtain these data for each of their generating units by actual operating tests to determine the Btu per hour fuel input to their boilers for various kilowatt output. The slope of this input-output curve when plotted for various outputs provides an incremental-heat-rate curve. This curve, when multiplied by fuel cost, gives the *incremental cost* of the next unit of generation for each value of unit output.

If the data from Fig. 5 are then plotted against total system generation, the economic-dispatch curves of Fig. 6 are obtained. These curves show the desired unit megawatt outputs which will provide the lowest over-all system generating cost at any given system-load level.

A comparison of these curves (Figs. 4, 5, and 6) yields helpful information. For example, since Fig. 4 shows the minimum system load is above that of both the run-of-the-river hydro plant and station 5, it would appear that both would be full loaded during the entire 24-hr period—and that no automatic regulation could be applied to them.

In Fig. 6 we note that, as load comes on the system above the 180 megawatts minimum, unit 4 will be called upon to increase generation since it has the lowest incremental cost. The actual rate of system load increase, as shown in Fig. 4, may be 10 megawatts/min. If unit 4 should have limitations on rate of load change which prevent it from meeting regulating-rate requirements, it would therefore be unable to satisfy system-load changes.

Power would flow in over ties from adjacent areas in order to supply the load, and this area would have failed to meet its fundamental operating requirement.

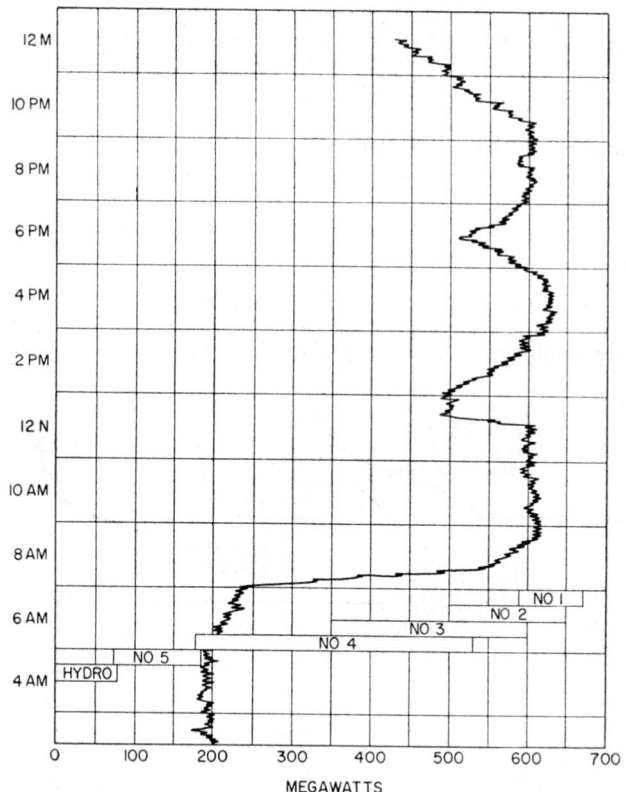

FIG. 4. Typical instantaneous net-system-load curve and incremental-loading program

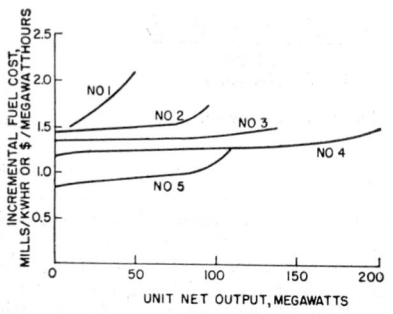

FIG. 5. Incremental-generating-cost curve.

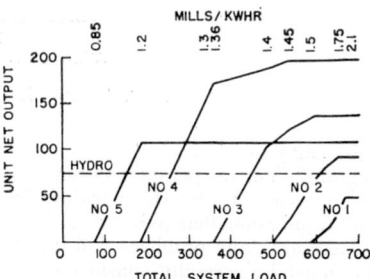

FIG. 6. Economic-dispatch curve.

Under these conditions, then, the system in question does not have sufficient rate to operate in accordance with its strict incremental loading program. Since satisfaction of the area requirement is paramount in power-system operation, an economic loading program cannot always be adhered to throughout the operating day.

In the instance cited, simultaneous responses of additional units are required in order to obtain the required response rate. For the program shown, this means that, instead of fully loading unit 5, or the hydro unit, some regulating range must be left on one or both—and that they should be equipped with automatic control. Also, it might be necessary to bring unit 3 on line earlier than required based on the incremental dispatch curves alone, in order to obtain the additional required megawatts per minute rate most economically.

Versatile Controls Generally Required

As a practical matter, some plants are arbitrarily limited in their response rate because of local plant conditions. Few plants or units are initially designed specifically to handle regulating loads. However, after their base-load period is terminated by the addition of new capacity which has a still lower incremental rate, they will inevitably be used to handle variable loads.

The need for control-system flexibility is further illustrated in Fig. 4. When system generation is within the range of 500 to 525 megawatts, units 2, 3, and 4 should be responding simultaneously. During this period, the obtainable rate of generation

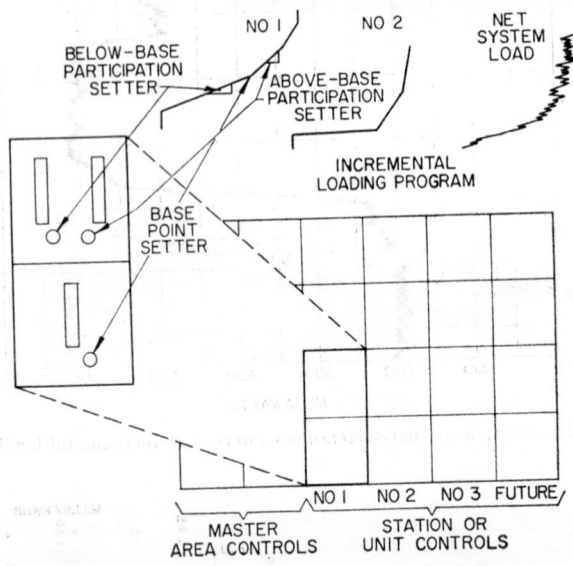

Fig. 7. Adjustable economic-program console.

change may satisfy both the regulating-rate requirements and the incremental-loading program. For the period between 525 and 600 megawatts, units 2 and 3 may provide the necessary rate for the morning pickup but be unable to handle satisfactorily the more rapid noon change between these same values of system generation.

These are daily problems which face the system operator. Naturally, such problems must be carefully considered when control requirements for a given area are being studied. And for this same reason, an adjustable-program type of automatic control has been extensively used in those instances where economic-dispatch schedules —such as those shown in Fig. 6—can be prepared in advance from available data. This control obtains its primary response from area requirement, and its corrective action is selectively routed to regulating units in accordance with an adjustable economy-loading schedule.

The control does not fix the loading schedule, but the system operator can match any given schedule by console setter dials which determine the break point of the curve and slope of the curve above or below that point as shown in Fig. 7. While a complete schedule could be set up in advance, most operating groups prefer to work with loading curves of approximately two or three segments because of the closeness of the match and the simplicity with which new conditions can be met.

Area requirement feeds information into the flexible-program console, which allows a definite destination for each unit and station to be established as each increase or decrease in system load occurs.   Therefore, each unit knows its destination at any instant and can achieve it promptly with a minimum of generation changes on all units.

Another control approach utilizes a fixed economy-loading program which rigidly relates unit loading to station total or a similar reference.   As can be seen from the foregoing discussion, such controls are effective for some fraction of the 24-hr operating day.   Because of this and other disadvantages, the fixed-program control has found only limited application.

CLASSIFICATION OF ECONOMY-DISPATCH CONTROL SYSTEMS

Because of technological advances in energy conversion during recent years, heat rates of generating units on a given system may vary widely.   Also, these units can

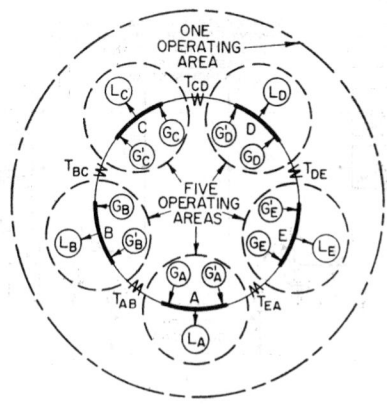

FIG. 8. Simplified representation of an interconnected system.

have divergent fuel costs and may also have their loading influenced by transmission-line loss factors.   Thus it has become possible to obtain significant fuel savings through the use of automatic economic-dispatch control systems.

Most of the major interconnections in the United States operate on a "multiple-area" basis, a term which denotes an interconnected system which contains constituent operating areas, each of which must absorb its own load changes.   With the advent of the economy-dispatch control systems, it has become feasible to transfer low-cost energy from an area which has an excess of power to another area which uses it in place of local high-cost energy.   Such economy power interchange is most commonly achieved automatically in order to realize the potential economics available. Most multiple-area interconnections have an arrangement under which each area regulates its own interchange with the interconnection as a whole, using the area-control principle.

Figure 8 shows a simplified representation of an interconnection having five participating companies, A through E.   Illustrated are tie lines T, loads L, and alternative generating sources G and G', corresponding to all the company's tie lines, loads, and generating sources.   These companies may operate as a single-area system, as five separate areas, or with any intermediate combination of areas which are defined in each instance by the boundary ties that are metered to make up each area's net interchange with the interconnection as a whole.

Three Steps to Generation Control

Figure 9 shows a functional and block diagram of generation controls for a typical area of multiple-area system—as, for example, area A of Fig. 8. It illustrates the three steps by which generation control is achieved:

1. *System Governing.* The match of total generation to total system load is achieved initially by the aggregate action of speed governors on all operating sources of area A acting in concert with all other speed governors of the system, and by the frequency coefficient of all connected system load.

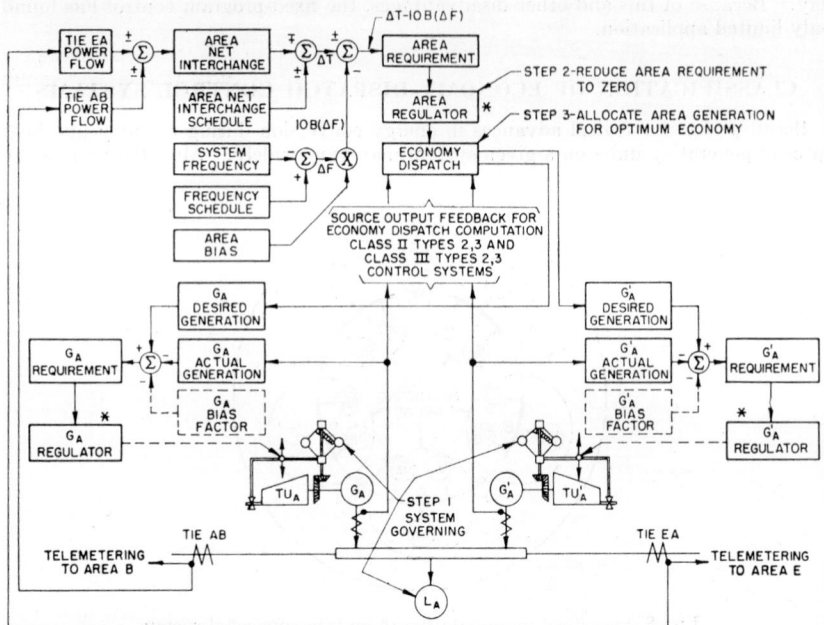

* G_A and G'_A may be regulated by mandatory control signals from their individual regulators, or permissively from control signals derived from the area regulator and routed through the individual regulators.

FIG. 9. Functional and block diagram of an automatic generation and power-flow control system as applied to area A of Fig. 8. Source bias factors, shown in broken lines, are used where control execution is of the mandatory type.

2. *Area Regulation.* The regulating requirement for the area is satisfaction of the area requirement.

3. *Economy Dispatch.* Not only must the control reduce area requirement to zero, but it must also determine the extent to which each generating source will participate in control action so that area requirement is satisfied at lowest cost. Such loading is achieved when sources are loaded to incremental costs of power delivered, as defined by the following general equation:

$$\lambda = \frac{dF_n/dP_n}{1 - \partial P_L/\partial P_n} = \frac{(dH_n/dP_n)f_n}{1 - \partial P_L/\partial P_n} \qquad (2)$$

where λ = area incremental cost of power delivered

$\dfrac{dF_n}{dP_n}$ = incremental cost of power generated at source n

$$\frac{\partial P_L}{\partial P_n} = \text{incremental transmission loss for source } n$$

$$\frac{dH_n}{dP_n} = \text{incremental heat rate for source } n$$

$$f_n = \text{incremental fuel cost for source } n$$

The basic problem of economy-dispatch control is to compute the desired outputs for each of the area generating sources in such a way that (1) all sources will be operating to a common area lambda (λ) and (2) the sum of the desired outputs of all sources will equal the total generation required of the area at that time. Lambda, which is equal to the area incremental cost of power delivered, has a specific value for each prevailing area condition.

In order to establish the correct lambda, therefore, it is necessary to determine the total generation required of the area. In some control executions this determination is inferential; in others it is direct.

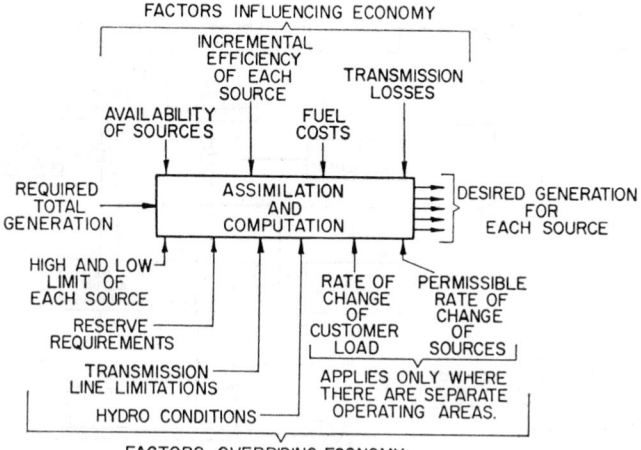

Fig. 10. Factors to be considered in determining how the total generation required of an area is to be allocated among available area sources.

In any practical execution, consideration must be given not only to the parameters that appear in the equation shown above but also to the constraints and overriding factors which appear summarized in Fig. 10.

These factors vary from one system to the next, and as a result, automatic economy-dispatch control executions have taken different forms. Then too, some of these differences reflect progress in the state of the art, while others reflect the control philosophies of the respective control manufacturers.

Three broad classifications of economy-dispatch control systems have been established. They are based on portions of the integrated multiple-area problem which the control system undertakes to solve—and on the nature of the programming technique employed.

Class I Controls—Area Regulation Only

The class I control system provides area regulation only. Economy dispatch within the area is achieved by manual adjustment of generating-unit output. Many of the earliest multiple-area installations were of this type. This class of control has three subdivisions:

Type 1—Single-source Regulation. The control is applied to a single regulating unit at a time.

Type 2—Multiple-source Regulation, Single-source Output Reference.
Control is applied simultaneously to two or more units. Units under control are
automatically loaded with respect to the load level on the one which is designated
as the "master."

Type 3—Multiple-source Regulation, Area-requirement Reference. Two
or more units are automatically loaded with respect to a reference derived from
area requirement. This arrangement is seldom used alone, though it is frequently
employed in combination with economy-dispatch systems to achieve more rapid
area regulation.

Class II Controls—Economy Dispatch, Flexible Programming

In this class, control action is applied to many or all of an area's generators.
Further, it simultaneously achieves the objectives of area regulation and automatic

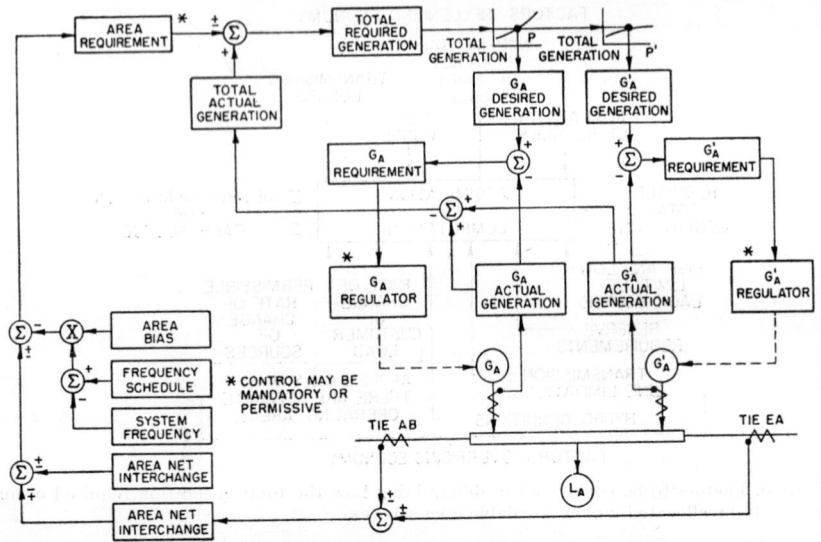

Fig. 11. Class II, type 3 flexible program control. The common reference is a feedback
from source outputs combined with a feed-forward from area requirement.

economy dispatch. A common reference is used for source loading. The distinguishing characteristic of this class is the manual adjustment of loading programs. Programs may be set to provide strict economy dispatch or to recognize the need for
overriding economy dispatch with the factors shown in Fig. 10. Programs must normally be reset for each new combination of participating sources. Subdivisions in
this class are identified by the nature of the common reference which is used for programming source loading.

Type 1—Single-source Output Reference. Here, the reference for source
programming is the output of one of the sources which is designated as "master."

Type 2—Total-area Output Reference. This designation applies where the
reference for source programming is the total prevailing output of participating sources.
Data for the reference are obtained from an intermediate feedback related to the
prevailing output of participating sources. A limitation of this approach is the
manner in which the desired generation from each source is computed: the computation is based on the total generation already *existing* in the area, and not on the total
generation then *required* in the area.

Type 3—Total-area Required Output Reference.   In this instance, the reference for source programming is the total generation which is required of participating sources in order to satisfy area load, and to reduce any prevailing area requirement to zero.   A common reference is derived by using an intermediate feedback relating to prevailing generation of the participating sources, as in class II, type 2, but combining with it a feed-forward which is related to the prevailing area requirement as shown in the block diagram in Fig. 11.   Thus the desired generation computation for each source represents the source output which is to prevail *after* area requirement has been reduced to zero.

An important feature of this control is that, for a given area load, the rate at which each source responds to control action does not influence the desired generation computed for other area sources.   Such controls are now widely applied.   The same basic concept of an intermediate feedback combined with feed-forward is utilized for a unique lambda reference computation in the newest economy-dispatch systems described under class III, type 3.

Class III Controls—Economy Dispatch, Fixed Programming

Here again, automatic control is applied to most or all the generators of the area. Also, area regulation is fulfilled while achieving automatic-economy dispatch, using a common reference for source loading.   The distinguishing characteristic of this class is the use of an incremental-cost value as a common reference.

Control execution for economy dispatch is based on loading participating sources to equal incremental cost.   Programming of a source is based on a preset relationship between incremental cost and output for that source.   In that sense, the economy-dispatch program is fixed and has the advantage that it need not be changed for various combinations of participating sources.   Also, this approach coordinates well with automatic computation and application of transmission-loss factors.

This approach has been used increasingly in recent years.   The common reference may be incremental cost of power delivered, or in instances where transmission losses are not considered significant, a common reference may be incremental cost of power produced.

Grouped in this class, but subdivided as different types, are control systems in which the common reference can either continuously represent incremental cost or can represent incremental cost only during conditions of balance.

Type 1—Lambda Reference Adjustment through Area Control Loop. This is a control in which the common reference is not independently computed as incremental cost but is established by a floating search which is initiated from the prevailing area requirement.   In this method of lambda determination, there is not a direct use of total generation required of the area.   Instead, lambda is continually adjusted until area requirement is reduced to zero.   By that inferential means, required area lambda is related to required area generation.   Control is a series cascade with feedback through the power network.   The boundary tie lines determine when the reference lambda has been properly established.

A simplified block diagram showing the application of this type of control is shown in Fig. 12.   For simplicity, no transmission loss or fuel factors are shown or considered in this schematic (or in Figs. 13 and 14) and it is assumed that the lambda reference may be regarded as incremental heat rate of the sources.   Several installations of this type are in operation at present.

Type 2—Lambda Reference Adjustment Using Area Output Feedback. In this type of control, the common lambda reference is established independently of the area control loop.   Instead, it is based on a comparison between summated desired generations of the participating sources and total prevailing generation.   A basic block diagram is shown in Fig. 13.

Note that this method of lambda determination uses an intermediate feedback from the output of participating sources as illustrated in Fig. 9.   It can be seen, however, that the lambda thus determined corresponds only to the total generation already carried in the area and makes no provision for further area generation changes needed to reduce any prevailing area requirement to zero.

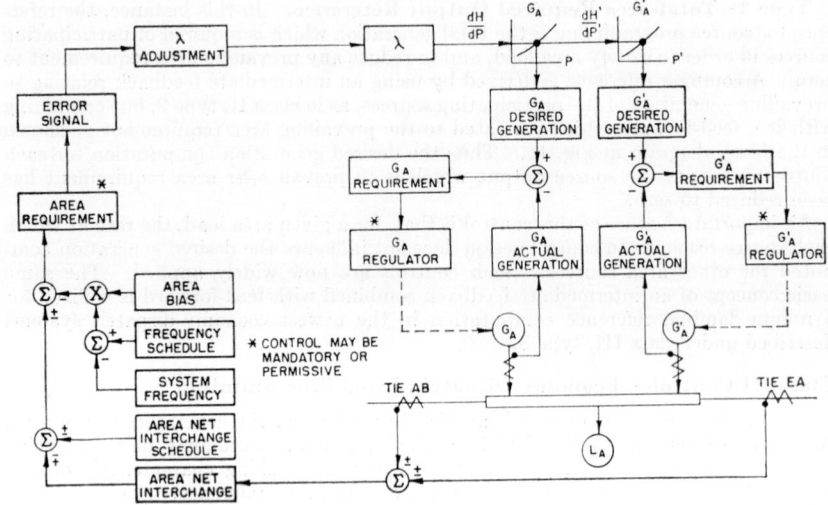

FIG. 12. Class III, type 1 economy dispatch. The common reference lambda adjustment is achieved by feedback through the area control loop.

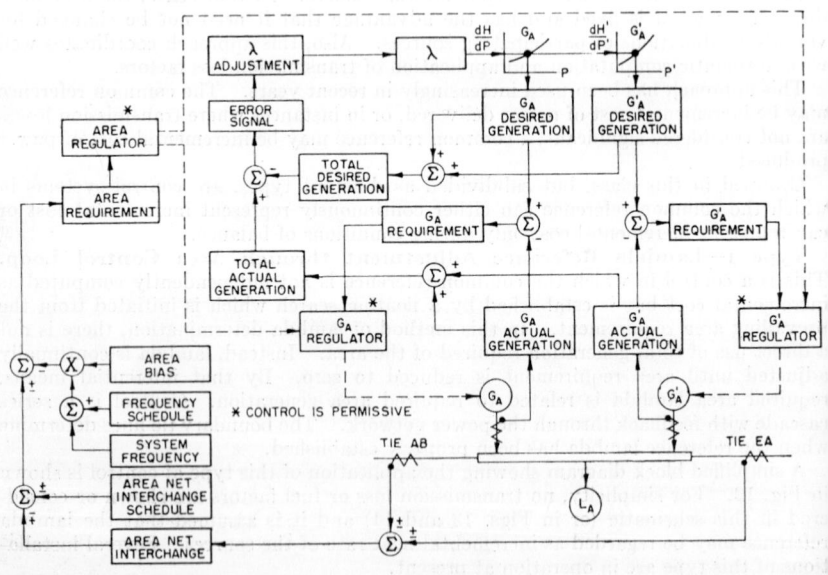

FIG. 13. Class III, type 2 economy dispatch. Lambda adjustment is achieved by intermediate feedback from source outputs.

Type 3—Lambda Reference Adjustment Using Area Output Feedback and Area-requirement Feed-forward. This is the newest of the economy-dispatch techniques. Here, the computation for lambda adjustment is carried a step further than in class III, type 2. In addition to the intermediate feedback from participating sources, the lambda-adjustment computation includes a feed-forward from prevailing area requirement. This is illustrated in the block diagram of Fig. 14.

As shown, the lambda adjustment is based on an error signal which is derived by comparing the computed desired generation for all participating sources, with the total generation required of the area to carry its own load and reduce any prevailing area requirement to zero.   The lambda determination, therefore, corresponds to the total area generation that will prevail *after* area requirement has been reduced to zero.   Similarly, the desired generation thus computed for each of the participating sources represents the generation level which is to prevail for each source after area requirement has been reduced to zero.

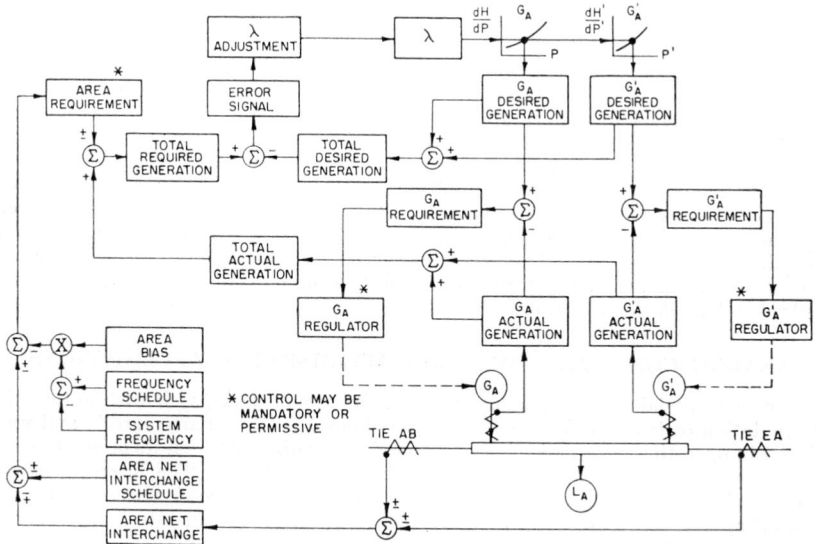

Fig. 14. Class III, type 3 economy dispatch.   Lambda adjustment is achieved by intermediate feedback from source outputs combined with feed-forward from area requirement, and is independent of area control loop.

The equations for lambda adjustment in Fig. 14 are as follows:

$$E + \Sigma G_n = \Sigma D_n \tag{3}$$

or

$$E = \Sigma D_n - \Sigma G_n = \Sigma(D_n - G_n) \tag{4}$$

from which

$$E = \Sigma S_n \tag{5}$$

where E = area requirement
G_n = actual generation of each participating source
D_n = computed desired generation for each participating source
S_n = source requirement for each participating source

The lambda thus established continuously represents the incremental cost which *will* prevail for the total generation then required of the area.   It is based on a direct rather than inferred knowledge of the total generation need of the area.   Control execution is still a cascade, but the common lambda reference is now uniquely established independently of the action or response of source regulators and is independent of feedback through the area control loop and the boundary tie lines.   It is a noninteracting control, and the desired generations computed for participating sources at a prevailing area load are unaffected by the rate at which each participating source responds to control action.

Permissive or Mandatory Control

Figures 8 through 14 show methods of calculating generation-change requirements of sources participating in economy-dispatch control.   These source-requirement blocks are shown feeding "regulator" blocks which, in turn, are shown as operating on the sources to reduce a corresponding area requirement to zero. In so doing, they operate as either "permissive" or mandatory" devices.

With the permissive control execution, all corrective control action originates from area requirement. Source-requirement regulators determine whether or not the prevailing control action from area requirement will or will not be applied to their respective sources. When the area requirement, for example, is to raise, all sources which have a raise requirement are permitted by their respective source regulators to respond to the control signal. Figure 13, for example, would normally involve such permissive control.

In the mandatory control execution, control action on each source is directly from its own source-requirement regulator. Action on each source takes place to restore its own requirement to zero, regardless of the then prevailing nature of area requirement. Figures 9, 11, 12, and 14 could utilize either permissive or mandatory control.

In a mandatory scheme, the source-requirement computation should include a source-frequency bias factor if the supplementary control is to coordinate with source governor responses to remote load changes. If this were not done, mandatory source regulators would oppose the governor response to remote load changes and would introduce corresponding and unnecessary generation swings on the source. Such source bias factors, though omitted for simplicity from most of the figures, are shown as broken-line blocks in Fig. 9.

ANALOG COMPUTER FOR ECONOMY-DISPATCH APPLICATIONS

The first special-purpose analog computer for economy dispatch with automatic computation of transmission loss factors was known as the "Early Bird"[*] and was placed in operation on a major interconnected system in 1954. In its original form, the computer was a complete analog of the entire system. The interconnection transmission network was represented by a B constant (line loss) matrix, and each generating unit was represented by its respective incremental heat-rate characteristic. Two readouts showed the incremental cost of delivered power and the incremental per cent of delivered power for any selected sources.

Desired-generation Computer

Present computers are similar in appearance except that additional readouts have been provided. Also, the readouts today are more accurate dial-indicating servos. This analog assembly, which is referred to as the "desired generation" computer, is used on the majority of economy-dispatch computer-controlled installations in the United States. Its name is derived from the fact that it provides desired generation readings for all units at lowest cost under each prescribed set of system conditions.

In this computation, the computer utilizes the economic-dispatch equation [Eq. (2)], described again in Fig. 15, for coordinating incremental production cost at a particular plant bus with the incremental transmission losses associated with that plant. The incremental production costs are broken down into three parts: incremental heat rule, unit performance, and fuel cost.

The incremental-heat-rate characteristic is determined from data obtained during periodic unit tests and is the derivative of the input-output curve. The unit performance setting enables operating personnel to compensate for the more frequent unit variations such as cooling-water changes, fouled turbine blading, or dirty condensers. Fuel cost is the cost of the variable fuel in cents per million Btu's.

In order to compute the desired generation for any particular unit, it is first necessary to solve the equation for the desired incremental heat rate which appears as a voltage within the computer. The unit incremental heat-rate characteristic (which is set into the computer by means of a function generator) then converts the incremental heat-rate voltage into desired unit generation.

---

[*] So named because of its proponent E. D. Early of Southern Services, Birmingham, Ala. See Refs. 17, 18, and 20.

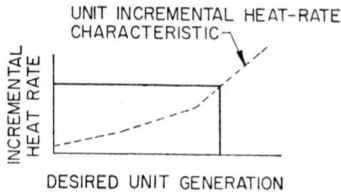

Fig. 15. Economic-dispatch equation.

$$\text{Incremental cost of power delivered to the load from any plant} = \frac{\text{incremental production cost at the bus of that plant}}{1 - \text{incremental transmission loss of power delivered to the load from that plant}}$$

$$\text{System } \lambda = \frac{(\text{incremental heat rate})(\text{unit performance})(\text{fuel cost})}{1 - \partial P_1/\partial P_n}$$

$$\lambda(1 - \partial P_L/\partial P_n) = (\text{incremental heat rate})(\text{unit performance})(\text{fuel cost})$$

$$\frac{\lambda(1 - \partial P_L/\partial P_n)}{(\text{Unit performance})(\text{fuel cost})} = \text{incremental heat rate}$$

Functional Block Diagram of Computer

Referring to the diagram in Fig. 16, all the computing components are shown separately in order to differentiate them from the readouts which have no computing function. These readouts are provided for the convenience of operating personnel during both computer-control and computer-study types of operation.

The B constant matrix which computes the expression $(1 - \partial P_L/\partial P_n)$ for each source is in the upper left-hand corner. The output from the matrix, which we also refer to as the incremental per cent of delivered power, energizes a slidewire on the lambda

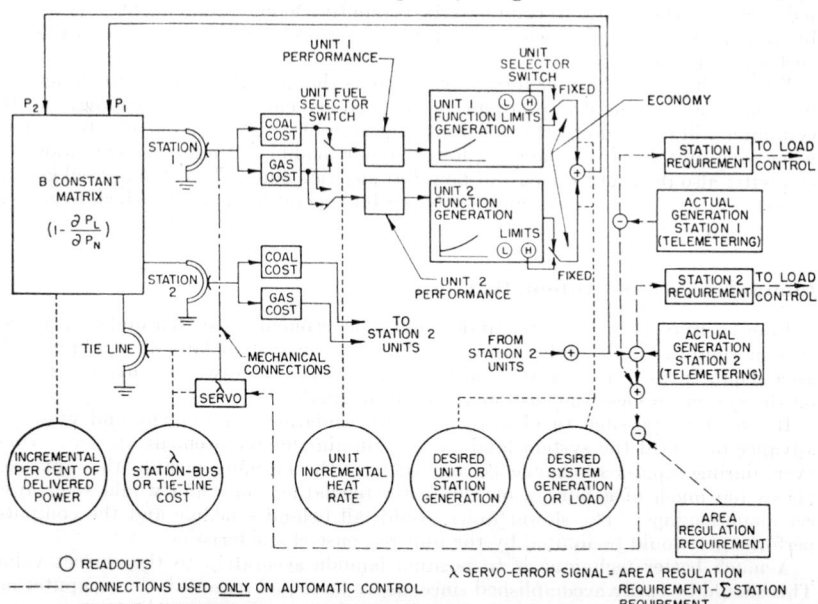

Fig. 16. Functional block diagram of computer.

servo for each generating station and tie line. Let us assume that the computer is isolated from the telemetering and control equipment, and review an example to the unit desired generation readings.

Let us also assume that we have manually positioned the lambda servo to some value, such as 3.0 mills/kwhr. By so doing, we determine the position of a gang-operated group of slidewire contacts on the associated slidewires for all sources. The voltage produced by the matrix is actually $(1 - \partial P_L/\partial P_n)$ whereas the contact position is proportional to lambda. Consequently, the voltage output at each slidewire contact is the numerator of the equation, or $\lambda(1 - \partial P_L/\partial P_n)$. This voltage is equivalent to the incremental bus cost at any station or the incremental tie-line cost at any tie point. Push-button-operated readout servos are provided so that the operator can monitor these readings if desired.

Referring to the equation, the station bus cost must next be divided by fuel cost. The computer is arranged so that any fuel burned at a station can be considered as the incremental fuel. The cost of each fuel in cents per million Btu is set into the computer, and a unit-fuel selector switch is provided to determine which fuel applies to a particular unit. It is possible for one unit at a station to use coal as the incremental fuel, whereas an adjacent unit could be burning gas as its incremental fuel if such a condition should arise. The voltage following the fuel-cost block is equivalent to the desired incremental heat rate which can be read out for each unit if and when desired.

The next adjustment provided for each unit is the performance setting, which is normally calibrated from 90 to 110 per cent to permit operating personnel to compensate for minor variations in the unit.

The output of the function generator is unit desired generation which is monitored by adjustable high- and low-limit setters most often provided on top of the console. The limits are automatic, and indicating lights show the operator when a unit no longer has regulating range.

Once the unit desired-generation reading is made available, the desired-station- and desired-system-generation readings are automatically obtained by simple addition with corresponding readouts provided as shown. The station generations are then fed back into the matrix to complete the computing loop. Consequently, it is possible to read out the desired generation for all units and stations corresponding to this cost level (3 mills).

When using the computer for study purposes the next step is obvious: adjust the lambda reading manually to obtain any desired system generation reading, and the computer will break down this total into the sum of its parts so one can tell exactly what each unit should be carrying for economic loading. Under this condition, the computer also reads out the cost of power at any tie point. Thus the desired generation computer is a convenient and flexible system operating tool with which to analyze any system condition.

Computer Used for Automatic Control

In order to use the computer to determine the economic dispatch under prevailing system conditions, the index used in adjusting the system lambda reading is the area requirement. This reading indicates how many megawatts must be changed on the system to meet the customer load at all times.

It would be possible to observe the area-regulating requirement and gradually advance or retard the system lambda to maintain this requirement at zero. However, during rapid-load-change periods such as the morning pickup, this approach places too much of a burden on the faster regulating units and would soon upset economic loading. The slower units would fall behind schedule and the computer performance would be limited by the unit response characteristics.

A much better technique is to position lambda accurately to the correct value. This should be easily accomplished since, as we have just reasoned, the computer has intelligence to determine the desired loading under any system condition.

In order to analyze how lambda is positioned to the correct value, consider the measurements required to tie the computer to the load-control equipment. In addition to area requirement, we can determine a station-requirement reading which shows exactly how many megawatts must be changed on any particular station to place that station at its economic dispatch. This station-requirement reading is simply the difference between the desired station generation from the computer and the actual measured generation obtained by telemetering.

Assume a set of balanced conditions in which the area requirement and both station requirements are zero when a new customer load comes on the system as a step change causing an area requirement of 50 megawatts. If the computer could position system lambda to that value which would produce a total of 50 megawatts of requirement on the two station-requirement readings and stop at that point, it would restore area requirement to zero with minimum generation changes. This value of system lambda would stay fixed all during the time that regulating stations were picking up their required amounts of generation by means of the automatic control. Each station's control would be independent, and it would make no difference to the system lambda how fast or slowly each station changed generation to reduce its requirement to zero.

With this technique, it would be possible to predict exactly what the value of system lambda should be for any area-requirement deviation. This is exactly what is accomplished in the computer by automatically adjusting lambda until the sum of the station requirements is equal to the area requirement. Hence lambda is a true measurement of that system cost which procures exactly the required desired generation necessary to restore the area-regulating requirement to zero.

DIGITAL COMPUTER FOR ECONOMY-DISPATCH APPLICATIONS

The analog computer's relative simplicity and low cost have gained for it a numerical lead over digital computers on this application. The capabilities of a digital computer, however, make it a superior tool in many instances.

Capabilities of Digital Computers

One important feature of the digital computer for economy-dispatch use is accuracy. Quite literally, digital computers can be built to process data with any desired degree of accuracy. Signals represented in digital form are, in effect, restandardized frequently while they are being processed by the computer. This eliminates the small, but inevitable, deterioration which occurs in analog signals when they are required to take part in repeated calculations.

The second advantage is the digital computer's ability to handle calculations of almost any degree of complexity. A related advantage is the ability to perform logical decisions—a characteristic which leads to new fields of application for the digital computer.

Third, the general-purpose digital computer can more readily adapt itself to changing conditions. The program can be readily modified, for example, so that as operating conditions change—or system operators develop a better insight into power-system operation—the digital computer can, within limits, keep pace.

And, because of its computational speed, the digital computer can supplement its economy-dispatch function with additional, valuable services: it can be programmed to provide interchange billing data, scheduling information, and interchange evaluations.

Interchange billing refers to the cost of power shipped to or from other areas of the interconnection. Such data are of obvious value to the participating companies, for large amounts of power and money are involved. For the same reason, it is helpful to know in advance the probable costs of a proposed interchange. The digital computer provides both sets of data with speed and accuracy.

Also, the digital machine enables the operator to determine when to put units on the line—and take them off—in the most economic pattern. The analog machine makes the most economic use of a given set of generating units; it will not, in most instances, periodically tell the operator which units should be used.

Even with its great speed, of course, the digital computer cannot perform diverse computations simultaneously. Accordingly, various functions are programmed serially, so that a typical digital computer will make an economy-dispatch determination, for example, every three minutes.

Operation of Over-all System

One should not infer that the power system is left at the mercy of random events in the interim; for, as shown in Fig. 17, the over-all system includes an advanced analog control section.

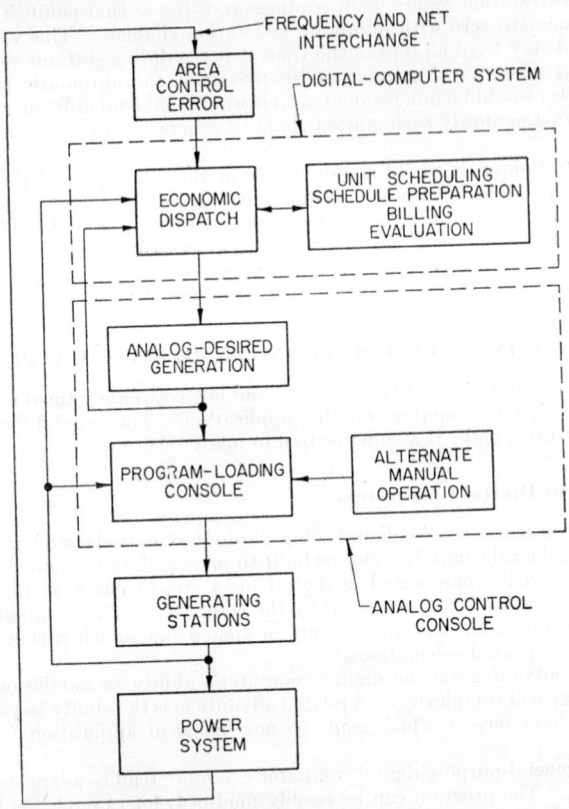

Fig. 17. Block diagram of over-all system.

In this (and most other) digital schemes, area requirement is still the basic parameter. The digital computer, working with stored incremental heat rates and stored unit capabilities, determines the optimum loading of each unit under control. The resulting values are transferred to desired-generation setters in the analog section through the medium of stepping motors. It is these motors which, in this particular approach, provide the digital-to-analog conversion at the output end of the system.

The basic control philosophy of this system is premised on a continual balance of the sum of the dispatch-unit requirements vs. the area requirement. As a load change is required on the system, the analog control portion must automatically establish the megawatt destination for each regulating dispatch unit so that the sum of these is equal to area requirement. In this scheme then, the digital computer's function is to supply vernier adjustments to analog control action.

This approach ensures a noninteracting technique between the respective units because the rate at which any source may respond to the basic control action will not influence the desired megawatt values computed for other dispatch units.

Telemetering is used to feed the computer with various tie-line loadings and source outputs for transmission-loss and security calculations. These security calculations, incidentally, relate to the many constraints required to ensure continuity of service. Finally, telemetering feedback advises the control of the efficiency with which each unit meets its economic loading schedule.

Unit dispatch controls at the various generating stations relay control intelligence to the respective governor motors in order to obtain the required generation changes.

Role of Digital Dispatch Computer

Although one may expect to find numerous digital dispatch computer approaches, the computer system can normally be viewed as the "economy director" whose function is to set targets for the remainder of the analog control system.

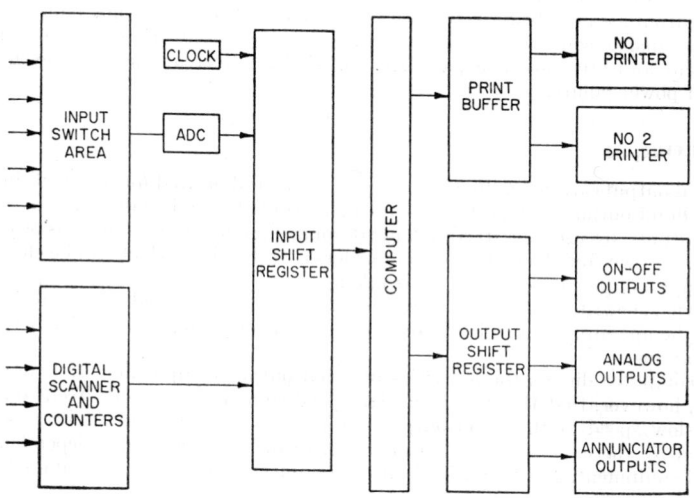

Fig. 18. Computer-system block diagram.

With this in mind, the role of the computer system and its associated equipment can be examined more closely. In Fig. 18, the computer-system portion is expanded to give a clear view of functional components. By comparison with computers used in commercial data-processing systems, the digital economy-dispatch computer is a medium-speed machine. Its basic memory is normally a combination of both core and drum memories.

The computer word may be expected to have at least 22 bits which include sign and parity. The parity bit and an associated parity network provide a check on each word as it is transferred from memory. In the typical instance, the structure of the instruction word is such that all words stored on the magnetic core can be directly addressable from approximately 64 different commands.

Inputs can be divided into three principal areas: one analog and the other two digital. Components concerned with analog inputs are a low-level scanner, an analog signal processor, a high-level scanner, and an analog-to-digital converter. Although a complete description of the analog input equipment will not be given here, its structure can be generally understood through a knowledge of its basic design criterion.

This criterion is accuracy, which in turn demands rejection of noise signals. In order to achieve this aim and still maintain appropriate scanning rates, multiple input channels are normally used. Upon receiving a request from the computer, the low-level scanner connects from 5 to 20 input points to separate assemblies in the analog signal processor. After a delay to allow adequate filtering, the high-level scanner, under the direction of the computer, selects these points for conversion. This delay is sufficient to allow the signal representing each point to reach 99.99 per cent of its actual value, and still maintain the desired system scanning.

Digital Inputs

One of two types of digital inputs is selected by the computer through a digital input scanner. Such data include contact closures, the state of indicating equipment, pulse-count totals, and various switches and dials associated with the computer communications panel. Typewriter and paper-tape reader inputs are the other digital inputs received by the computer.

Another important input has to do with address checking. The low-level scanner, the high-level scanner, and the digital-inputs scanner all return addresses to the computer so that it can be sure it has received the particular input requested.

The output area can be broken down in a similar fashion. Stepping motors provide digital-to-analog conversion. These units convert a series of pulses to rotational movement and are chosen in part because of their inherent memory—even in the event of power failure.

Digital Outputs

Digital outputs are generally of two types: on-off and printed (or punched) information. On-off outputs are used for operating alarm lights and similar devices. When on-off outputs are used for starting and stopping equipment, and safety is of primary importance, two complementary output orders must be given which can be checked by the output equipment before action is taken.

Electric typewriters and paper-tape punches are normally used to present information to the operators and to log information for historical purposes or further off-line computations.

The computer directs the activities of input-output equipment. It is important to note, however, that the relatively high speed computer is never forced to operate at the slow speed of the input-output equipment. This is possible because, after initiation by the computer, certain operations can be carried out independently by external equipment, and because all input and output operations are buffered.

Functions Performed by Computer System

It is the computer program which takes various component parts and molds them into a system with specific operating goals, as shown in the generalized flow diagram of Fig. 19. This diagram cannot be viewed as a representation of the program as finally implemented but rather as an indication of the operations the computer system can be expected to perform.

In an actual economy-dispatch program, one may expect to find the following broad steps:

1. Read all required inputs such as dispatch-unit generation, interconnection and internal tie-line flow, and area requirement.

2. Perform the security checks to ensure proper tie-line flow and area capability.

3. Determine regulating margin restraints.

4. Determine an economic dispatch by computing transmission losses and incremental loading for all dispatch units on the basis of system load, and the incremental heat-rate curve. Take into account any restraints placed upon dispatch for reasons of security or regulating margin.

5. Change the base-point setters (stepping motors) on the basis of the dispatch, and check to see that they have arrived at their proper destinations.

The other major functions can be broken down in a similar fashion. Billing, unit scheduling, and schedule preparations all normally use the basic economic-dispatch routines in order to study hypothetical system conditions.

The analog portion of the system under discussion is a modified class II, type 3 control having a permissive execution. This is a widely used mode of automatic generation control using solid-state components with the modification referred to earlier: automatic setting of the base or desired-generation setters by the digital computer.

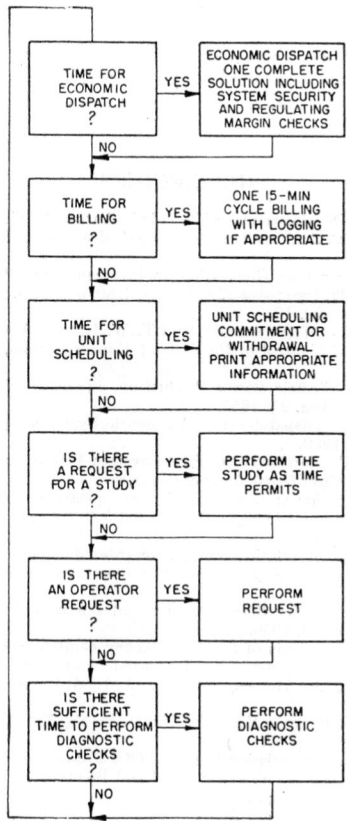

FIG. 19. Block diagram of executive routine.

Again, the heart of this control is area requirement—the number of megawatts required to balance generation and load. This value is used to initiate the basic control action, as well as instructing the dispatch units just how many megawatts each must advance in order to satisfy area requirement and maintain economic loading.

An Emergency-conscious Solution

Most analog and digital economic-dispatch systems will be "emergency-conscious." That is to say, they will automatically shift into supplementary assist types of control when emergencies arise. Various points of deviation of the area requirement from its zero value will cause this automatic change in operation. The two normal degrees of assist action are:

1. Normal assist—certain preselected dispatch units are permitted to deviate a prescribed number of megawatts from their loading schedules.

2. Emergency assist—all dispatch units within regulating range receive control intelligence regardless of economic loading.

A further zone of assist used in some locations will call for continuous control intelligence to be relayed to all units at once, but this is not the normal approach.

Present-day economy-dispatch computer controls, similar to those just reviewed, can do many remarkable things automatically and dependably. They make real contributions to continuity of service and also to minimum cost of production.

These computers—combined with the various types of controls defined in this discussion—allow a utility to take a look at its complete system to provide automatic economic coordination of all operating sources to accomplish the solution to their regulating problem. The combined computer-control systems represent significant advances in the power industry's use of automation as part of its progressing technology.

REFERENCES

1. Morehouse, S. B.: Recent Development and Trends in Automatic Control of Large Interconnected Power Systems, Presented at Conference Internationale des Grands Reseaux Electriques (CIGRE), 1956.
2. Cohn, N.: Methods of Controlling Generation on Interconnected Power Systems, *AIEE Trans.*, vol. 80, part III, pp. 270–282, 1961; also *Elec. Eng.*, vol. 80, no. 3, pp. 202–209, March, 1961. AIEE Paper 60-846, Summer Meeting, 1960.
3. Blodgett, D. G., A. K. Falk, W. B. Schultz, and T. W. Hissey: Application of an On-line Digital Computer for Dispatch and Control of the Detroit Edison System, AIEE Paper 62-247, Winter Meeting, 1962.
4. Cohn, N.: Automation in Electric Power Production, *Electric Light & Power*, Annual Practices Issue, vol. 33, pp. 122–129, Aug. 25, 1955.
5. Cohn, N.: Economic Power Dispatch—The Automatic Systems Approach, *Power Eng.*, vol. 64, no. 1, pp. 70–72, January, 1960.
6. Cohn, N.: Telemetering—Fundamentals for Power System Applications, *Proc. Midwest Power Conf.*, vol. 10, pp. 335–342, 1948; also *Electric Light & Power*, vol. 26, no. 8, p. 72, August, 1948.
7. AIEE-ASME Committee Report: Recommended Specifications for Speed Governing of Steam Turbines Intended to Drive Electric Generators Rated 500 KW and Larger, *AIEE Trans.*, vol. 76, part III, pp. 1404–1411, 1957.
8. Cohn, N.: Power Flow Control—Basic Concepts for Interconnected Systems, *Proc. Midwest Power Conf.*, vol. 12, pp. 159–175, 1950; also *Electric Light & Power*, vol. 28, no. 8, p. 82, August, 1950, no. 9, p. 100, September, 1950.
9. Cohn, N.: Some Aspects of Tie-line Bias Control on Interconnected Power Systems, *AIEE Trans.*, vol. 75, part III, pp. 1415–1428, 1956.
10. Cohn, N.: A Step-by-step Analysis of Load Frequency Control Showing the System Regulating Responses Associated with Frequency Bias, Presented before the 1956 Meeting of the Interconnected Systems Committee, Des Moines, Iowa, March, 1956; Leeds & Northrup Company Reprint 461-5(6).
11. Electric Utility Industry Statistics in the United States for the Year 1958, Edison Electric Institute, New York, *Publ.* 59-122, May, 1959.
12. George, E. E.: Intrasystem Transmission Losses, *AIEE Trans.*, vol. 62, pp. 153–158, 1943.
13. Ward, J. B., J. R. Eaton, and H. W. Hale: Total and Incremental Losses in Power Transmission Networks, *AIEE Trans.*, vol. 69, pp. 626–632, 1950.
14. Kirchmayer, L. K., and G. W. Stagg: Analysis of Total and Incremental Losses in Transmission Systems, *AIEE Trans.*, vol. 70, pp. 1197–1204, 1951.
15. Harder, E. L., R. W. Ferguson, W. E. Jacobs, and D. C. Harker: Loss Evaluation—Part II—Current—Power—Form Loss Formulas, *AIEE Trans.*, vol. 73, pp. 716–731, 1954.
16. Steinberg, M. J., and T. H. Smith: Economy Loading of Power Plants and Electric Systems, John Wiley & Sons, Inc., New York, 1943.
17. Early, E. D.: Central Power Coordination Control for Maximum System Economy, *Electric Light & Power*, vol. 31, pp. 96–98, December, 1953.
18. Early, E. D., R. E. Watson, and G. L. Smith: A General Transmission Loss Equation, *AIEE Trans.*, vol. 74, part III, pp. 510–520, June, 1955.
19. Brownlee, W. R.: Coordination of Incremental Fuel Costs and Incremental Transmission Losses by Functions of Voltage Phase Angles, *AIEE Trans.*, vol. 73, part III, pp. 529–541, June, 1954.
20. Early, E. D., W. E. Phillips, and W. T. Shreve: An Incremental Cost of Power Delivered Computer, *AIEE Trans.*, vol. 74, part III, pp. 529–535, June, 1955.
21. Kaufmann, P. G.: Load Distribution between Interconnected Power Stations, *J. Inst. Elec. Engrs.* (London), vol. 90, part II, no. 14, pp. 119–130, April, 1943.
22. Fereshetian, H., M. D. Liechty, and N. E. Brown: Coordination of Desired Generation Computer with Area Control, *Proc. Am. Power Conf.*, vol. 21, pp. 554–563, 1959.
23. Kirchmayer, L. K., and G. W. Stagg: Evaluation of Methods of Coordinating Incremental Fuel Costs and Incremental Transmission Losses, *AIEE Trans.*, vol. 71, part III, pp. 513–521, 1952.
24. Campbell, W. J.: New System Control Allocates Changing Load among 29 Generating Units, *Elec. World*, vol. 137, no. 1, pp. 32–33, 1952.

25. Cohn, N.: Area-wide Generation Control—A New Method for Interconnected Systems, *Proc. Am. Power Conf.*, vol. 15, pp. 316–344, 1953; also *Electric Light & Power*, vol. 31, no. 7, pp. 167*ff.*, June, 1953; no. 8, pp. 96*ff.*, July, 1953; no. 9, pp. 77*ff.*, August, 1953.
26. Stites, H. E.: Automatic Area Control Meets Pool Operation Requirements, *Electric Light & Power*, vol. 31, pp. 104–108, November, 1953.
27. Bauman, H. A., C. N. Metcalf, J. G. Noest, and J. B. Carolus: Design and Operation of System Wide Automatic Load-frequency Control, *AIEE Trans.*, vol. 73, part III, pp. 1315–1317, 1954.
28. Glass, E. C.: Operating Experience with Area Control, *Proc. Am. Power Conf.*, vol. 19, pp. 502–512, 1957.
29. Radford, R. A.: A New Load-frequency Installation, *Elec. West*, vol. 116, no. 3, pp. 84–86, March, 1956.
30. Cameron, D. H., and E. L. Mueller: A New Type Automatic Dispatching System at Kansas City, *Trans. ASME*, vol. 7, pp. 1663–1668, 1957.
31. Mochon, H. H., Jr.: Operating Experience with an Automatic Dispatching System, *Proc. Am. Power Conf.*, vol. 21, pp. 545–554, 1959.
32. Travers, R. H.: Automatic Economic Dispatching and Load Control—Ohio Edison System, *AIEE Trans.*, vol. 76, part III, pp. 291–297, 1957.
33. Brogdan, W. J., and T. W. Hissey: Closing the Computer Loop on System Regulation, *Proc. Am. Power Conf.*, vol. 22, 1960.
34. Cohn, N.: Common Denominators in the Control of Generation on Interconnected Power Systems, Presented before Systems Operation Committee, Pennsylvania Electric Association, May, 1957; Leeds & Northrup Company Reprint 461-5(8).
35. Preston, E. H.: Desired Generation Computer Aids System Operation and Production, Presented before Missouri Valley Electric Association Engineering Conference, 1960.
36. Cohn, N., H. W. Phillips, W. D. Wilder, A. W. Willennar, and E. K. Corporon: Symposium on Scheduling and Billing of Economy Interchange on Interconnected Power Systems, *Proc. Am. Power Conf.*, vol. 20, pp. 447–489, 1958.
37. Kirchmayer, L. K.: "Economic Operation of Power Systems," John Wiley & Sons, Inc., New York, 1958.
38. Kirchmayer, L. K.: "Economic Control of Interconnected Systems," John Wiley & Sons, Inc., New York, 1959.
39. Kirchmayer, L. K.: Optimalizing Computer Control in the Electric Utility Industry, *Proceedings, of the First Congress of the International Federation of Automatic Control*, Moscow. U.S.S.R., June, 1960.
40. Cohn, N.: Developments in the Control of Generation and Power Flow on Interconnected Power Systems in the United States, *Proceedings of the First Congress of the International Federation of Automatic Control*, Moscow, U.S.S.R., June, 1960.
41. Pender, Harold: "Electrical Engineers' Handbook—Electric Power," John Wiley & Sons. Inc., New York, 1949.
42. Waddicor, H.: "The Principles of Electric Power Transmission." Chapman & Hall, Ltd., London, 1959.
43. Beeman, Donald L. (ed.): "Industrial Power Systems Handbook," McGraw-Hill Book Company, Inc., New York, 1955.
44. McPartland, J. F. (ed.): "Electrical System Design," 2d ed., McGraw-Hill Book Company, Inc., New York, 1960.
45. Lovell, A. H.: "Generating Stations," 4th ed., McGraw-Hill Book Company, Inc., New York, 1951.
46. Long, R. W., R. T. Byerly, and L. J. Rindt: Digital Computer Programs in Electric Utilities, *Elec. Eng.*, pp. 912–916, September, 1959.
47. Hartranft, A. C., and F. H. Light: A Survey of the Application of Automatic Devices for Electric Power Generation, *Trans. IRE*, PGIE-7, p. 55, August, 1958.
48. Mochron, H. H., Jr.: Dispatch Instrumentation Saves $35,000 Annually for CVPE, *ISA J.*, vol. 7, no. 2, p. 45, February, 1960.
49. High-power Instrumentation, *ISA J.*, vol. 7, no. 5, p. 84, May 1960.
50. Robinson, P. B.: Telemetering Channel Sharing for Electrical Utilities, *ISA J.*, vol. 9, no. 9, pp. 85–86, September, 1962.

NUCLEAR REACTOR INSTRUMENTATION

By V. S. Underkoffler*

Instrumentation for nuclear reactors includes a wide spectrum of components and systems. In some respects the problems are similar to those encountered in many conventional applications, whether it be a fossil-fueled power plant or a chemical process. An instrumentation system for a nuclear reactor, however, presents many unique problems that are both interesting and challenging, taxing the skill of the designer, the researcher, and the engineer.

One of the obvious roles for nuclear reactor instrumentation is to provide both information and control functions. The large number of parameters involved in a reactor installation makes the interplay of these parameters extremely important from the information and control viewpoint. The instrumentation system for a reactor is not made up of a simple detector, amplifier, and controller combination but is an aggregate of many subsystems, all performing important functions in the systems complex. The many functions that such instrumentation must perform and the methods for implementing it are of major concern.

The purpose of this subsection is to provide a foundation for an understanding of the needs for instrumentation in nuclear reactor applications. The unique aspects of such instrumentation are discussed. Examples and illustrations are utilized for emphasis.

NUCLEAR REACTOR DYNAMICS

Before considering nuclear reactor instrumentation, let us first review the general nature of the process involved. Obviously great detail is not possible, even for a single reactor type, but the concepts involved can help establish guidelines and a better understanding of the requirements for instrumentation and control.†

In a nuclear reactor, the time-dependent behavior is of extreme importance. It is this behavior which determines the basic control problem and encompasses the safety aspects of reactor operation as well.

The fission process and fission chain reaction are adequately covered in the literature listed. Establishment of a chain reaction, the maintenance of the self-sustained chain reaction, and the control of this reaction are functions of the instrumentation complex. As a minimum for a self-sustained chain reaction, each nucleus capturing a neutron and undergoing fission must produce, on the average, at least one neutron which causes the fission process to take place in another nucleus.

Excess Reactivity (*Delta k*)

This process of maintaining a balance of neutron production or losses (the chain reaction) can be expressed in terms of an *effective multiplication factor*, k_{eff}. The multiplication factor can be defined as the ratio of the number of neutrons produced in any one generation to the number of neutrons produced in the immediately preceding generation.

If $k_{eff} \geq 1$, a chain reaction is self-sustaining. When $k_{eff} < 1$ by even a very

---

* Head of Systems Equipment Section, Research and Development Department, Leeds & Northrup Company, North Wales, Pa.

† See bibliography at the end of this section for greater detail on the physics and dynamics involved.

small amount, a chain reaction cannot be maintained (the reactor is subcritical). When $k_{eff} = 1$, the reactor system is said to be critical; if $k_{eff} > 1$, the reactor is supercritical. The k_{eff} of a reactor system must be capable of exceeding "one" if the neutron flux level or power level is to be increased.

The range of k_{eff} is small for normal reactor operation, and control of k_{eff} is fundamental to the reactor dynamics. Change of k_{eff} can be expressed as follows:

$$k_{eff} - 1 = \Delta k \text{ (can be positive or negative)} \tag{1}$$

The excess multiplication factor, Δk, is often called *excess reactivity* or *delta k*.

Let us examine the effects of a change in Δk in a reactor system operating very close to $k_{eff} = 1$, or critically.

Let n_0 be the number of neutrons per unit volume at time zero before k_{eff} is disturbed.

Let n_1 be the number of neutrons per unit volume or the neutron population in the next generation after k_{eff} is changed.

Then in the first generation:

$$n_1 = n_0(1 + \Delta k) \tag{2}$$

The increase in the number of neutrons per unit volume is $n_0 \Delta k$. If the generation time or neutron lifetime is 1*, then the increase in the number of neutrons in each second is $\dfrac{n_0 \Delta k}{1*}$. A general equation describing the rate of change of n can be written as

$$\frac{dn}{dt} = \frac{n\Delta k}{1*} \tag{3}$$

Integrating:

$$n = n_0 e^{\Delta kt/1*} \tag{4}$$

Reactor Period

The concept of reactor period will assist in understanding the dynamics. If we define T as the time required for the neutron flux to change by a factor of e, Eq. (4) becomes

$$n = n_0 e^{t/T} \tag{5}$$

Therefore, $T = \dfrac{1*}{\Delta k}$ or the "e folding time," since during each T seconds n/n_0 changes by a factor of $e(2.718)$. Typical values of the neutron lifetime range from 10^{-3} sec to 10^{-7} sec.

Assume a 1* of 1×10^{-4} sec and a change in k_{eff} of 0.1 per cent ($\Delta k = 0.001$).

We have

$$T = \frac{1*}{\Delta k} = \frac{1 \times 10^{-4}}{0.001} = 0.1 \text{ sec}$$

Therefore, in one second the neutron density or the power level (since they are proportional) would increase by a factor of e^{10}. Indeed this is a striking increase in power in a very short time.

We have assumed in the above illustration that all neutrons are produced promptly. Fortunately, this is not a valid assumption because a small percentage of the neutrons produced in the fission process are delayed for some time after their precursors were produced. The percentage delayed neutron yield is dependent on the type of fuel utilized in the reactor. These delayed neutrons permit the production of an enormous amount of power with reasonable control schemes.

Effect of Delayed Neutrons

An examination of these delayed neutrons and their effect on the average neutron lifetime will illustrate the important influence of delayed neutrons on nuclear reactor control problems. When U^{235} is utilized as the reactor fuel, approximately 0.075 per cent of the neutrons produced are delayed neutrons. In the above fuel, there are six delayed neutron groups having mean delay times from approximately 0.07 second to 80.3 seconds. If we take the yield for each delay group (β_i) and the average delay time for each delay group (t_i) and sum the product, we have:

$$\sum_{1:1}^{1:6} \beta_i t_i = 0.094 \approx 0.1 \text{ sec} \tag{6}$$

Thus we can now define the average time $\bar{1}$ between generations, accounting for the delayed as well as prompt neutrons.

Then

$$\sum_{1=1}^{1=6} \beta_i t_i + 1^* \tag{7}$$

1^* has the range indicated before; therefore, $\bar{1}$ is approximately 0.1 sec.

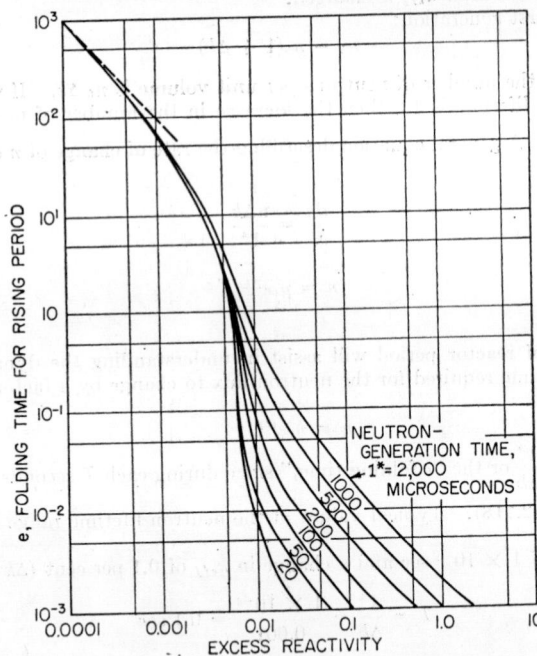

FIG. 1. Effect of neutron-generation time on reactor period.

Using the same values of Δk as in the previous illustration but using $\bar{1}$, we have

$$T = \frac{\bar{1}}{\Delta k}$$

$$= \frac{0.1}{0.001}$$

$$= 100 \text{ sec}$$

and

$$n = n_0 e^{t/T}$$

$$\frac{n}{n_0} = e^{1/100}$$

It is seen that the process is controllable with the presence of delayed neutrons. Figure 1 is a graph of "e folding time" vs. excess reactivity, for various neutron-generation times. The effect of the delayed neutrons is easily seen. As soon as

$\Delta k \approx \beta$, the prompt neutron-generation time is the controlling time. Therefore, k_{eff} must be kept below $k_{eff} \approx 1 + \beta$.

Neutron Density

The equations defining the neutron density which is proportional to thermal power are:

$$\frac{dn}{dt} = \frac{k_{eff} - 1}{1^*} n - \frac{k_{eff}\,\beta n}{1^*} + \sum \lambda_i C_i + S \tag{8}$$

$$\frac{dC_i}{dt} = \lambda_i C_i + \frac{k_{eff}\beta_i n}{1^*} \tag{9}$$

where n = density of neutrons
C_i = concentration of delayed neutrons for a particular group of precursors
β_i = fraction of delayed neutrons for a particular group
1^* = effective mean lifetime
λ_i = disintegration constant of the delayed neutrons from group i
S = artificial or natural source of neutrons

These equations assume only one energy group of neutrons and independency from spatial effects. If one assumes that a reactor has been operating at $k_{eff} = 1$

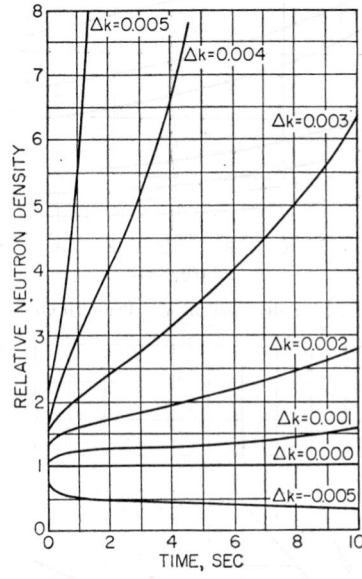

FIG. 2. Ratio of neutron density following a sudden change in k to the equilibrium value prior to the change.

(steady power level) for some time and a sudden change in k_{eff} is made, n will increase, for a positive Δk, very suddenly. Near $t = 0$, the rate will be determined as if all neutrons were prompt (prompt jump). After a short time the effect of delayed neutrons is seen and the rate of rise of power (neutron flux) will become constant on a stable reactor period.

Figure 2 indicates the effects on the relative neutron density for various changes in Δk.

Other Effects of Changing Δk

Figure 3 is an indication of the change in power as a negative Δk is applied suddenly. The influence of delayed neutrons is again illustrated. The prompt negative jump is smaller, but a stable negative period is reached again. After a large negative-reactivity change, there is a rapid decrease in neutron flux, but after a short time neutron flux will decay with a period of about 80.3 seconds, which is the particular delayed-neutron group (using U^{235}) that has the longest mean life.

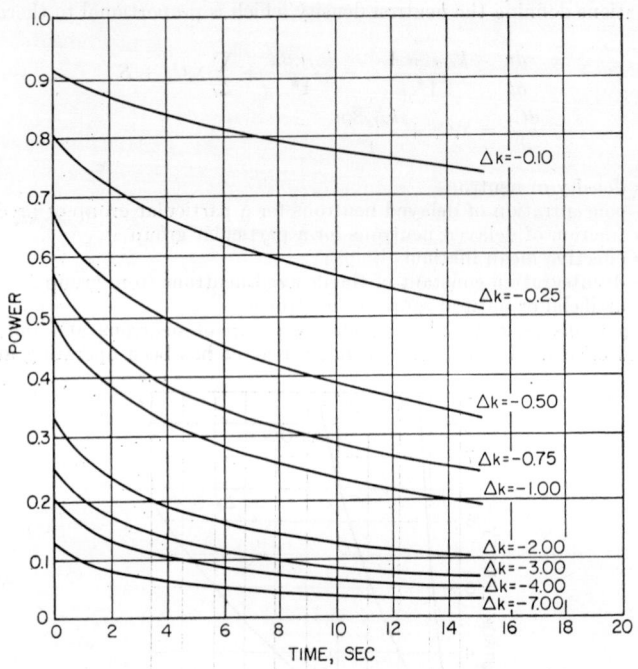

FIG. 3. Power following a sudden decrease in k. Prior to changes, the power was unity.

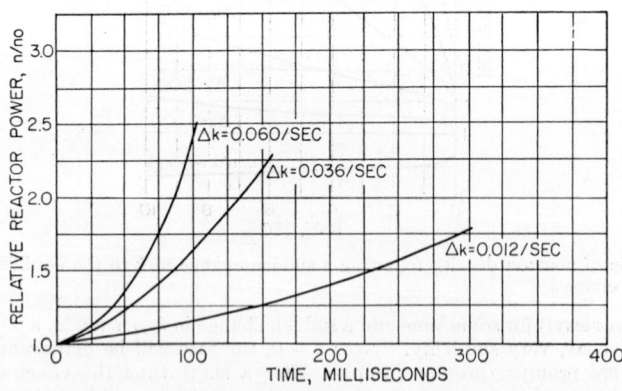

FIG. 4. Relative reactor power for k varying linearly with time. At time zero, the reactor is critical and k commences to rise at the rate shown on each curve.

Figure 4 indicates the effects of changing Δk linearly with time. N_0 is the power at time zero with the reactor critical. The k_{eff} is changed at the various rates shown. Even at small rates of change (i.e., 0.012/sec) the relative increase in power N/N_0 is substantial and takes place in a short time.

Reactivity Determines Dynamics

When the effective multiplication factor, k_{eff}, is exactly unity, the reactor is in a steady-state condition regardless of the power level at which the reactor is operating. Control means and control instrumentation are required to start the reactor, change power level, and shut it down. Recall:

$$\Delta k \text{ or } p = k_{eff} - 1/k_{eff}$$

The numerator is the neutron multiplication in excess of that required to sustain a chain reaction. Therefore, the reactivity determines the dynamics of the reactor. If the reactivity is negative, the flux or power level will decrease, as shown earlier. There are many factors tending to reduce the reactivity. They are:

1. The build-up of fission-product poisons, such as Xe^{135} and samorium^{149}.
2. Fuel depletion due to burn-up.
3. The effects of poisons such as structural materials, coolants, and moderators.
4. The removal of fuel or neutron reflector.
5. The intentional introduction of control poisons.

As shown previously, a nuclear reactor can be controlled by adjustment of k_{eff}. The power level can be changed and the reactor can be shut down. One of the most common methods for adjusting k_{eff} is through the use of poison rods. These rods contain a material which absorb neutrons (high cross section). Boron enriched in the B^{10} isotope is an example. It is often used in the boron carbide form.

Control and Safety Rods

Operating-control and safety-rod positions in the reactor vessel of the Enrico Fermi Power Reactor are shown in Fig. 5. The core is the dotted section surrounded by

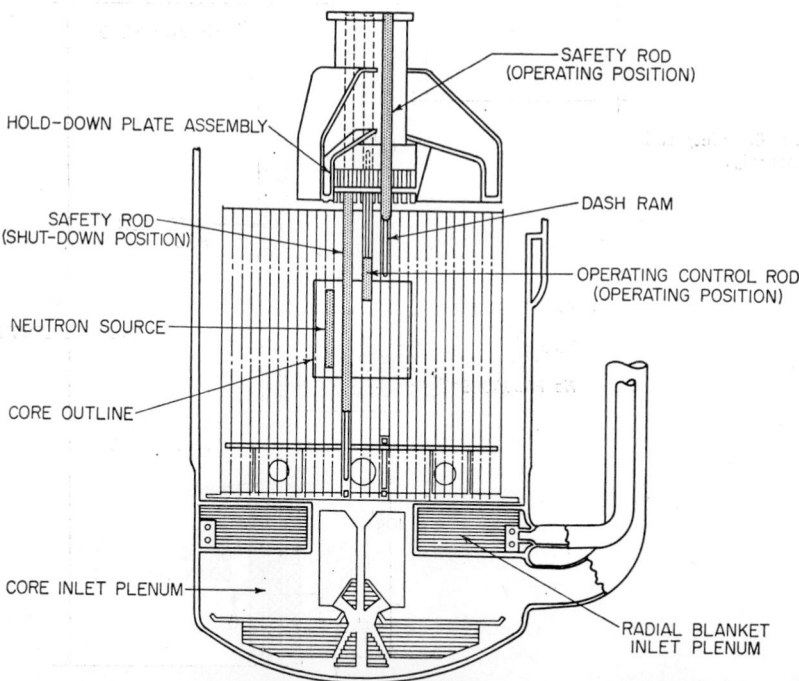

FIG. 5. Operating-control and safety-rod positions in a reactor vessel.

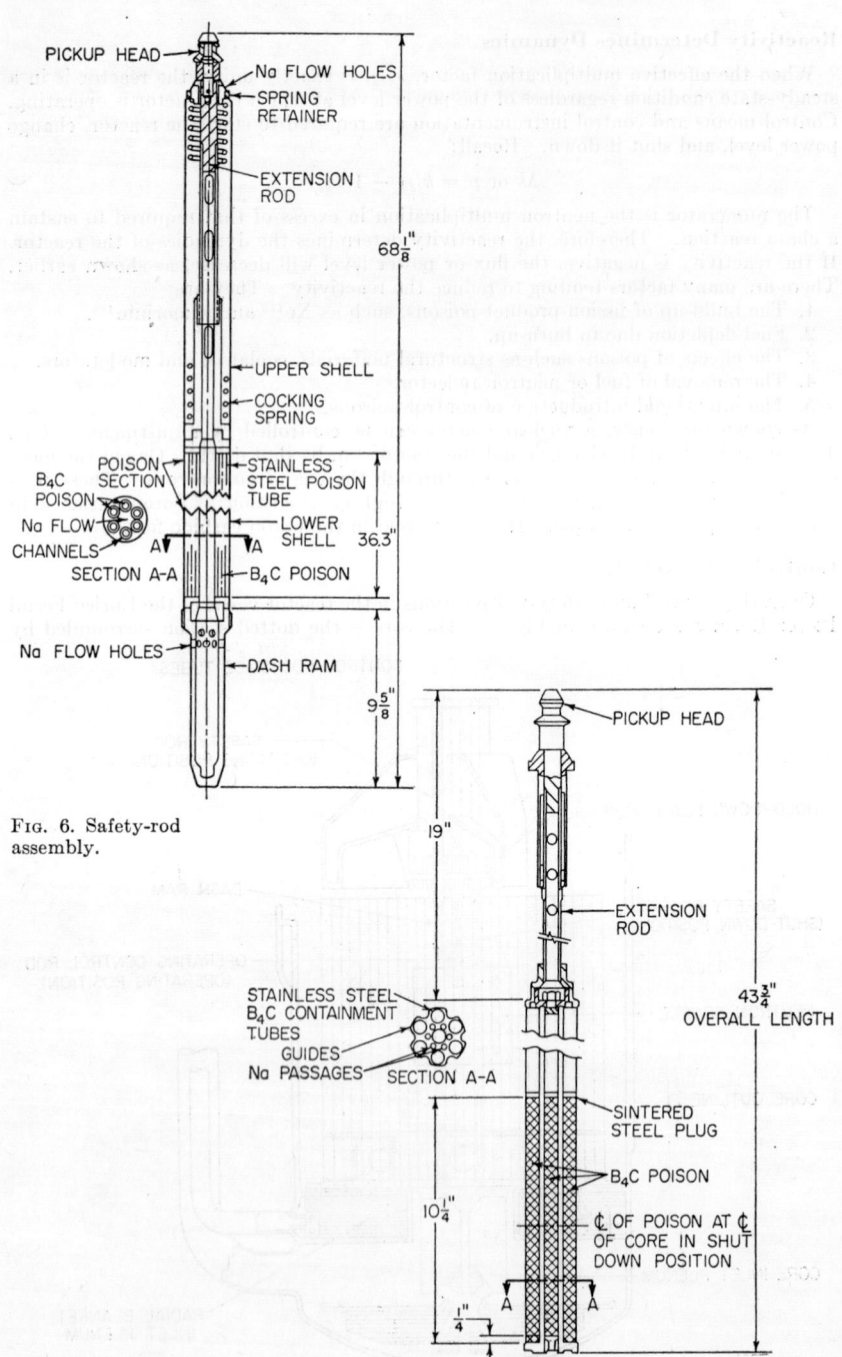

FIG. 6. Safety-rod
assembly.

FIG. 7. Operating-control rod.

the blanket volume. The control rod is shown in the operating position. It is inserted or withdrawn as desired to control k_{eff}. It can be operated manually or automatically. The safety rod is shown in the fully inserted position and in the withdrawn position. These rods are suspended above the core in normal operation and inserted rapidly when called upon by the instrumentation or by the operator.

In Fig. 6, a safety rod is shown with the cocking spring unlatched to provide additional acceleration when shutdown is desired. The control rod is shown in Fig. 7. It is operated by means of a servo system, and movement of this rod is determined by the control system or by the operator.

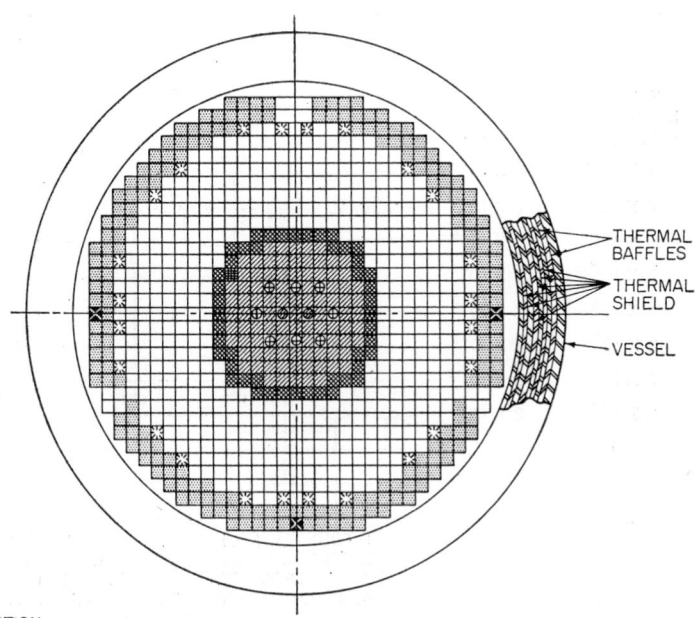

THERMAL BAFFLES

THERMAL SHIELD

VESSEL

DESCRIPTION

◉ OPERATING CONTROL RODS
⊕ SAFETY RODS
▨ CORE SUBASSEMBLIES
▩ INNER RADIAL–BLANKET SUBASSEMBLIES
☐ OUTER RADIAL–BLANKET SUBASSEMBLIES
■ TYPICAL NEUTRON–SOURCE LOCATION

▨ POSSIBLE STORAGE FOR CORE OR INNER RADIAL–BLANKET SUBASSEMBLIES
▨ THERMAL SHIELD IN FORM OF STEEL RODS
⊠ THERMAL SHIELDING IN FORM OF STEEL RODS USED FOR SURVEILLANCE TUBES

FIG. 8. Reactor cross section.

The rods are held by latches or electromagnets. When the current is turned off in the latch or electromagnet, the rod is free to move rapidly into the core, thus inserting an absorber of neutrons to reduce k_{eff} below criticality. In some reactors, removal of fuel or movement of reflectors can cause the reactor to shut down.

Figure 8 is a cross section of the reactor, showing locations of the safety and control rods. Also shown is the neutron source required for start-up. Ra-Be or Po-Be is often used for such sources.

NUCLEAR REACTOR INSTRUMENTATION

A nuclear reactor is unique in its extremely wide range of operation. It is not unusual for a reactor to cover 11 decades or more of power from source level to full-power operation. This wide range places rather difficult requirements on the instrumentation. It is imperative that the operator of a plant have adequate information and control throughout this wide range. Careful design is therefore necessary to insure that this is indeed the case.

To cover adequately the full range of reactor operation from an instrumentation point of view, four classifications of nuclear instrumentation are used, each carrying out its functions in a particular portion of the entire operating range. The classifications are:

1. Pulse-channel instrumentation.
2. Logarithmic (log N) = channel instrumentation.
3. Control instrumentation.
4. Protective system ("safety") instrumentation.

The information from all of the above instrumentation must overlap, so that a continuance of information and control is provided over the entire range of flux or

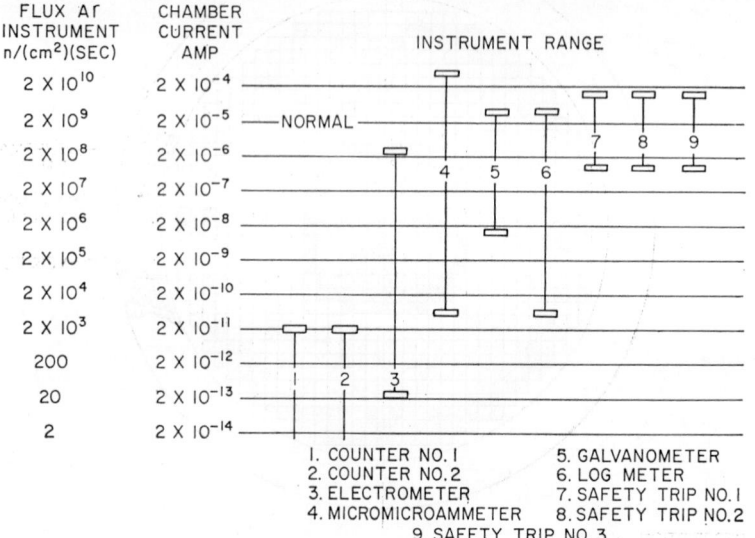

FIG. 9. Ranges of nuclear instrumentation in a typical installation.

power levels. Figure 9 presents graphically a typical installation, with the overlap of ranges of the various classifications of instrumentation.

Pulse-channel Instrumentation

Pulse-channel instrumentation, as its name implies, detects the individual pulses produced by neutrons in a reactor. Because the neutron population at source levels (and for several decades of flux level above) is relatively small, instruments must detect the individual neutrons being produced. Either a BF_3 or a fission counter is used to determine the neutron flux at these lower flux levels. The neutrons produce electrical pulses in such detectors. The resultant pulses are subsequently amplified, shaped, counted, and integrated.

The block diagram of Fig. 10 illustrates typical pulse-channel instrumentation utilized to provide information and control at the lowest levels of flux encountered in normal reactor operation. Ionization produced in the detector generates electrical pulses which are amplified by the linear pulse amplifier. The pulses are then shaped to produce output pulses of a predetermined pulse width and height.

At this point a binary or decade counter can be used to count the pulses. In addition, an integrating circuit is commonly used to produce an output current proportional to the rate at which pulses are being produced. This output current can

)e presented in a linear or logarithmic fashion. The logarithmic presentation is
generally preferred because of the wide range of information desired (usually four
ır five decades).   The number of pulses produced over this range varies from 1 count
ıer second to 100,000 counts per second.

Thus the integrating circuitry, the log count rate meter, produce a varying output
:urrent representing 1 pulse per second to 100,000 pulses per second.

It is desirable to know the rate at which the neutron flux is changing, even at these
ow levels.   Hence a "period amplifier" is used to differentiate the output from the
og count rate meter and produces a signal which represents the reactor period.

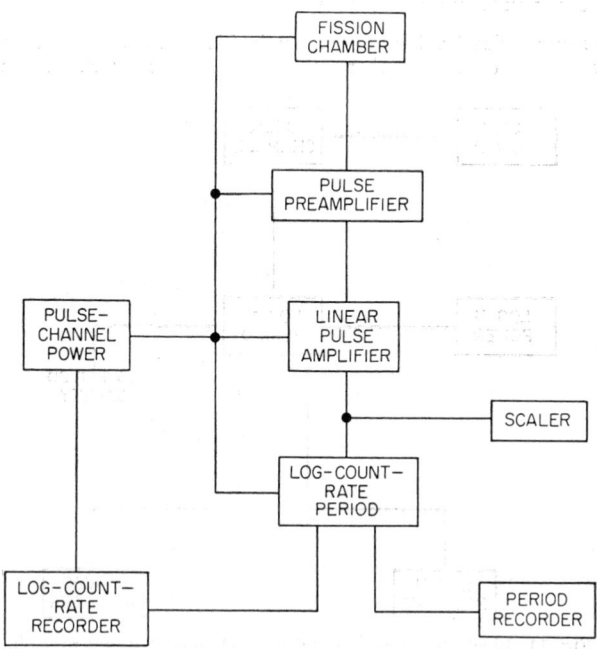

Fɪɢ. 10. Block diagram of pulse-channel instrumentation.

In certain installations and with some detectors, it may be necessary to provide a
ɔulse preamplifier when excessive cable lengths are required.   The preamplifier provides some preamplification of the pulse and is also a means for matching impedances
ɪf the cable and the final amplifier.

Included in the pulse-channel circuitry is a discriminator which is adjustable to
ɔermit (1) the counting of pulses above a certain pulse height and (2) the rejecting of
ıll pulses below this predetermined discrimination level.   Because the detectors
·espond to both neutrons and gamma radiation, the instrumentation must be capable
ɪf selecting those pulses produced by neutrons.   The pulse heights from the detector
ıre different for gamma radiation as compared to those produced by neutrons.   The
ıeutrons produce pulse heights significantly larger than those from gammas.   Thus
:he discriminator setting can be adjusted to reject pulses below a certain pulse height
ɔroduced by gamma radiation.

The following output signals from the pulse-channel instrumentation are used in the
ɔver-all nuclear reactor instrumentation:

1. An output signal proportional to the log of the number of pulses produced per
second.

2. A signal proportional to the reactor period,

3. Output pulses suitable for counting.

 4. An adjustable bistate signal for low counting rate.
 5. An adjustable bistate signal for high counting rate.
 6. An adjustable bistate signal for the reactor period.

Bistate signals are signals which indicate that a certain condition exists or does not exist. For instance, the low counting rate bistate signal indicates the presence of a certain minimum counting rate (i.e., five counts per second), which can be represented by the "1" state. If the counting rate is below this minimum, the signal representing this condition would be in the opposite state, or the "0" state.

Logarithmic-channel Instrumentation

When the neutron flux level is approximately four decades above source level, the neutron flux is large enough so that less sensitive detectors can be employed. A

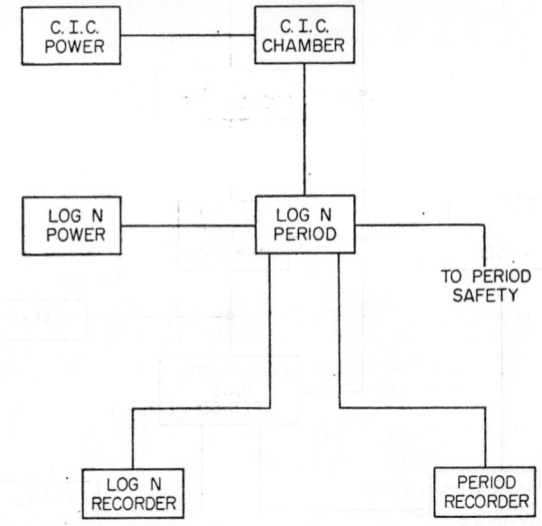

FIG. 11. Block diagram of logarithmic-channel instrumentation.

compensated ion chamber is the most commonly used detector. Such chambers produce a current which is proportional to neutron flux over a wide range. These chambers are designed to eliminate or cancel signals produced by gamma radiation in the presence of neutrons. This compensation is accomplished mechanically or electrically, so that only an output current proportional to the neutron flux is produced.

Logarithmic amplifiers are employed to utilize the wide-range capabilities of compensated ion chambers. The range of flux levels covered by such an amplifier extends from above normal full-power flux levels to six or seven decades below full power. The information and control output signals are in logarithmic form, so that the wide range can be easily accommodated. The input currents to the logarithmic amplifiers are usually in the range of 10^{-10} amp to 10^{-4} or even 10^{-3} amp.

Figure 11 is a block diagram of instrumentation for a typical log N channel. Here again, a period amplifier is used to give the rate of rise of neutron flux information in this range of operation. A compensated ion chamber (C.I.C.) power supply is shown which provides the various voltages necessary for the compensating and accelerating potentials. Typical information and control signals from the log N channel are:

 1. An output signal proportional to the logarithm of neutron flux over six to seven decades.

 2. A signal representing the reactor period. Typical period information and control outputs are -30 seconds to ∞ to $+3$ seconds.

 3. An adjustable bistate output for low end of log N range.

4. An adjustable bistate output for high end of log N range.

5. An adjustable bistate output for the reactor period (more than one such output may be utilized for several period settings; i.e., 25-second period and 5-second period).

Control Instrumentation

The control or linear-flux instrumentation varies quite widely in sophistication and is dependent upon the reactor application. It normally covers approximately two decades of neutron flux. In many installations, particularly power reactors, the

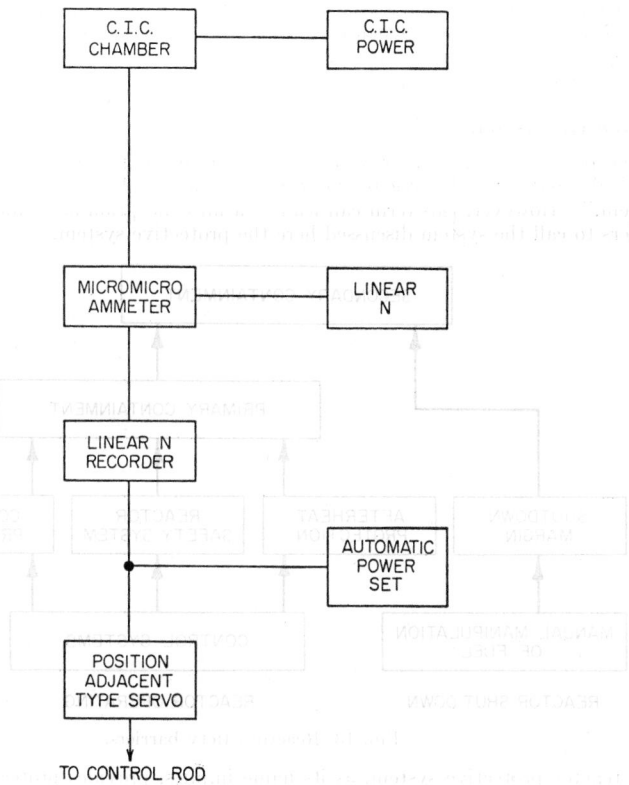

TO CONTROL ROD

FIG. 12. Block diagram of linear-channel instrumentation.

range extends from above full power to flux levels representing 1 per cent of full power. However, it is not unusual to utilize the full neutron sensitivity of ion chambers in order to cover the widest practical neutron flux range. Thus, a wide flux range can be covered in two-decade increments. This type of operation is typical of research and testing reactors where a wide range of fluxes is desired, and is used in power plant start-up operations. Obviously the more sophisticated control schemes are required in the power range (1 per cent to $>$100 per cent of power).

Compensated or uncompensated ion chambers are used as the sensors, and a micro-microammeter as the current amplifier. Manual or automatic switching of ranges is employed to cover the chamber-current ranges desired. The output from such current amplifiers is used in the control scheme. This output is compared with a reference current representing the desired current from the ion chamber (corresponding to the desired neutron flux). The error signal is utilized in a servo system to actuate the final control element—an absorbing rod in many cases. This type of system (Fig. 12) is quite simple and is found in most research-type reactors.

The control schemes for many reactors, particularly power reactors, are much more complex. A combination of neutron flux, period, temperature, and flow signals are used to provide the control action required. Determination of the required control action necessitates extensive simulation of the particular reactor under consideration. Mathematical models of the reactor which take into account spatial and neutron energy considerations are often required. These simulation studies permit the designer to predict the type of control best suited for his particular application.

Many times the simulation program is coupled with reactor-operator training. A mock-up of the final control room is used and the important controls are coupled to at least a portion of an analog computer. Thus plant performance can be realistically displayed and the operators can become skilled in operation. Emergency conditions which can be simulated but not intentionally inserted in the actual plant can provide excellent training for the operator.

Protective System

The role of the protective system instrumentation is extremely important. This particular portion of the reactor instrumentation complex is often called the "safety system." However, this term can lead to a misconception as to its role; the author prefers to call the system discussed here the protective system.

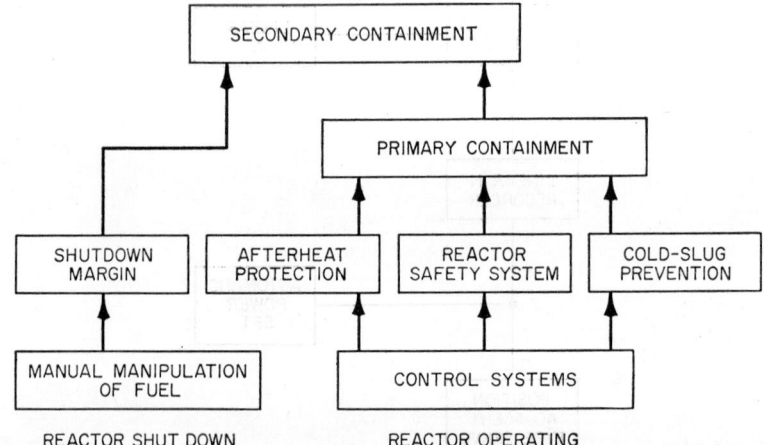

Fig. 13. Reactor safety barriers.

A reactor protective system, as its name implies, provides protection against certain accidents, but it is not the ultimate system which prevents the creation of a public hazard. Protection against accidents which could cause a public hazard is covered in the bibliography and in Federal regulations. The principle of containment and multiple containment is utilized as a barrier against the release of radioactive material. In general, the fuel cladding, the reactor vessel, radiation shielding, primary containment barriers, and the containment vessel itself, all play a part in retaining the radioactive material that may be released accidentally.

Figure 13 indicates the reactor safety barriers which are involved in a reactor installation. All of these barriers would have to be breached before one could violate the ultimate safety of the reactor plant in the operating and the shutdown conditions.[23]

As can be seen, the reactor protective system is in the chain of reactor safety barriers. This instrumentation protects against certain accidents, but not all accidents, which could lead to fuel melting, reactor damage, or exposure of personnel to radiation hazards. It is important to recognize that the protective system can prevent certain accidents which, if allowed to proceed unchecked, would lead to more serious consequences. In Fig. 13, the control system is considered as playing a part

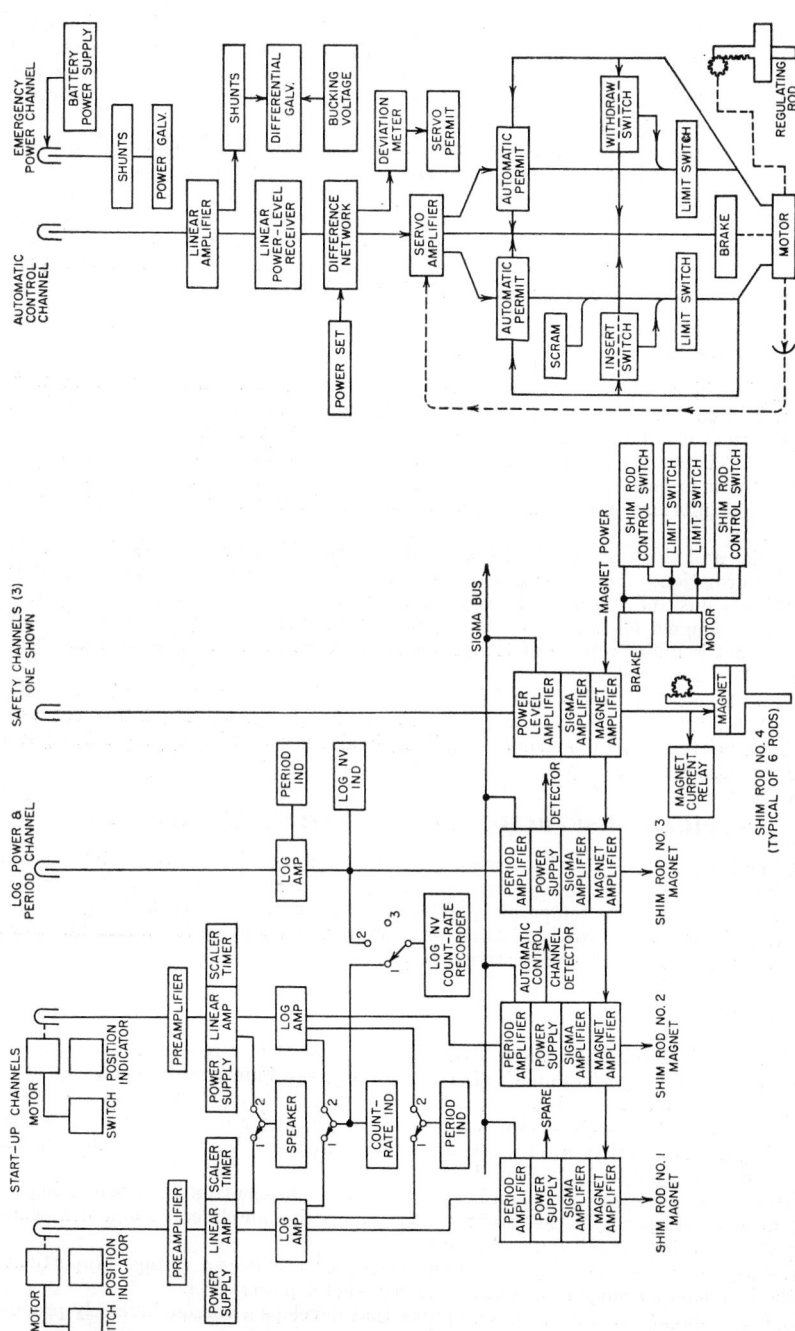

FIG. 14. Nuclear instrumentation for a research reactor.

in reactor safety, but to a lesser degree than the protective system.   Thus the reliability of the control system is not of as great importance as the reliability of the reactor protective system.   The protective system should be considered as a separate reliability problem.   Epler[24] points out that "the coupling together of safety and control can change reliability into danger."

The role of the protective system is somewhat of a paradox.   It must act in those infrequent circumstances when a potentially unsafe condition exists; the remainder of the time it is standing by, performing no active function.   However it is closely coupled to the reactor, and its shutdown mechanisms, and must perform all its functions rapidly and reliably when required to do so.   In those nuclear-reactor operations where continuity of operation is required, such as a power reactor, the protective system design is a challenging reliability problem.

Over-all System

A typical block diagram of the nuclear instrumentation for a research reactor is shown in Fig. 14.   Two start-up channels for pulse instrumentation are utilized. The counters can be moved by motor operators so that an appropriate physical relationship between the reactor core and the detector is obtained.   The channels are identical with appropriate switching for the various functions in each channel.   Note that period information from the log count rate amplifier is utilized in the protective system.

A scaler and timer are used with a speaker to provide an audible signal for additional information.   The log N (log power) and period channels are also shown feeding the "sigma bus" of the protective system.*   Only one of three safety channels is shown.   A magnet for holding the safety (shim) rod is shown.   The current through this magnet is under control of the protective system.   A number of rods are used for the shutdown mode.

The automatic control channel is also shown with the motor operator and the power set for the servo system.   The servo system, with the regulating rod and feedback information to the servo amplifier, is also shown with appropriate signals for automatic control.

TYPICAL RESEARCH REACTOR PROTECTIVE SYSTEMS

It has been shown that the neutron flux and reactor-period instrumentations can give information about the dynamic performance of a reactor system.   Contrasted with such parameters as temperature, where the time constants may be long, neutron flux can change very rapidly.   Therefore, neutron flux and period information are most often used in reactor protective systems.

Composite-amplifier Protective System

Figure 15 is a simplified block diagram of a typical four-rod protective system, using four Oak Ridge National Laboratory Composite Safety Amplifiers.   Electromagnets A, B, C, and D hold poison rods which are suspended above the reactor core. When required, the electromagnets are de-energized, and release the poison rods so that they can be inserted into the reactor core rapidly, thus shutting down the reactor. This is commonly called a "scram."

UIC (uncompensated ion chamber) units detect the neutron flux, producing an output current which is the input to the composite amplifier.   Each composite amplifier contains the following circuits:

1. *Level preamplifier:* A linear d-c amplifier which produces a voltage proportional to the ion-chamber output current and hence reactor power level.

2. *Period amplifier:* A linear d-c amplifier that develops a voltage inversely proportional to the reactor period.

3. *Sigma amplifier:* A nonlinear d-c amplifier driven by either the level or period

---

* The detailed portion of this system is described in a later part of this section.

preamplifier. Composite safety amplifier No. 4 has the period preamplifier driving the sigma amplifier.

When the reactor power level is in the 0–100 per cent region, the sigma amplifier output changes by 0.3 volts per 10 per cent change in power level. For a 10 per cent change in power level above 100 per cent, the output changes by 0.6 volts.

4. *Magnet amplifier:* A linear d-c amplifier which controls the current to an electro-magnet supporting a poison rod. This current varies inversely as the input from the sigma amplifier.

5. *Fast-scram circuit:* A linear d-c amplifier which operates a relay to warn the operator that a scram (fast shutdown) is being approached.

6. *Magnet power supply:* An unregulated d-c supply which provides the magnet current.

7. *Amplifier power supply:* A regulated d-c power supply which furnishes the plate and bias voltages for the amplifiers.

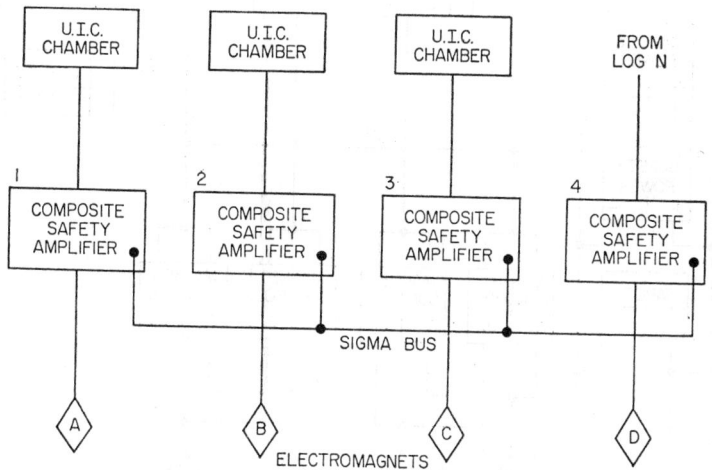

Fig. 15. Four-rod safety system using composite safety amplifiers.

The outputs of the sigma amplifiers are the only interconnection between the composite safety amplifiers. This common connection is called a "sigma bus." The safety amplifier with the highest output has control of the sigma bus.

The one signal on the sigma bus is the input to all four magnet amplifiers. It therefore determines every electromagnet current. Consequently, a reactor scram is possible from any one or all safety amplifiers. This is a "one out of four" or auctioneering type of system. A degree of monitoring is provided in each amplifier by the use of relay and meter-measuring techniques. A number of functions are monitored to warn the operator through an external annunciator when portions of the system have been disabled. Protective systems utilizing such components have been in service for a number of years and provide excellent performance.

In the auctioneering mode, certain electronic failures result in automatic shutdown of the reactor or a disablement of part of the protective system. In addition, maintenance is difficult without a loss of continuity of system operation.

Modular Protective System

Figure 16 is a block diagram of a modular protective system utilizing the basic circuits of the ORNL instrumentation. However, certain advantages have been incorporated into the system so that certain failures will not shut down the reactor, maintenance is easier, and all instrumentation is functioning continuously.

The level-amplifier and period-amplifier modules contain only a single preamplifier, a sigma amplifier, and a fast-scram circuit. Thus, the number of amplifiers is equal to the number of primary input signals whether they are for neutron flux or reactor period. The signal outputs from the amplifiers are again auctioneered and the voltage at the sigma bus is determined by the highest amplifier output, as in the previous system. Plug-in features of the module make it possible to remove any level amplifier without a shutdown.

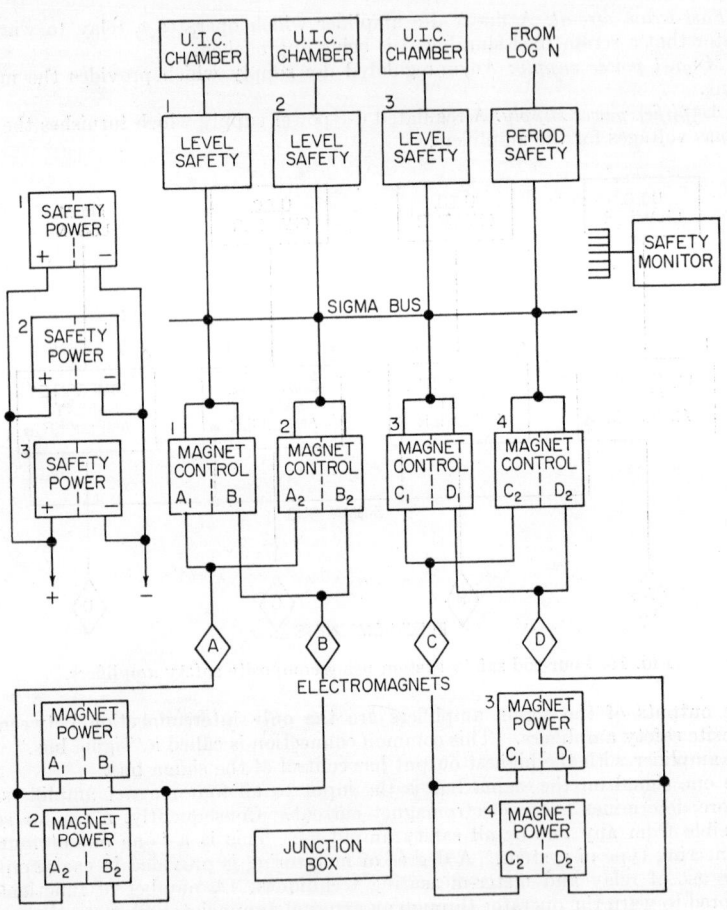

FIG. 16. Four-rod modular safety system.

The magnet-control modules are connected to the sigma bus. Each magnet-control module contains two identical magnet amplifiers whose outputs feed two different electromagnets. The total current in any one electromagnet is the sum of two currents from two magnet amplifiers located on different magnet-control modules. This load-sharing feature makes it possible for magnet-control modules *1* and *3*, *1* and *4*, *2* and *3*, or *2* and *4* to fail or be removed from the system without causing reactor shutdown.

The magnet-power modules are the source of the electromagnet current under the control of the magnet-control module. The modules designated as "safety power" provide the appropriate voltages for the level and period amplifiers as well as for the magnet-control modules. The failure or removal from the system of one safety-power module can be tolerated because only two safety-power supplies are required for the system to operate. The monitoring scheme for this protective system monitors various functions as before. Figure 17 is an artist's conception of a typical reactor installation for research-type reactors.

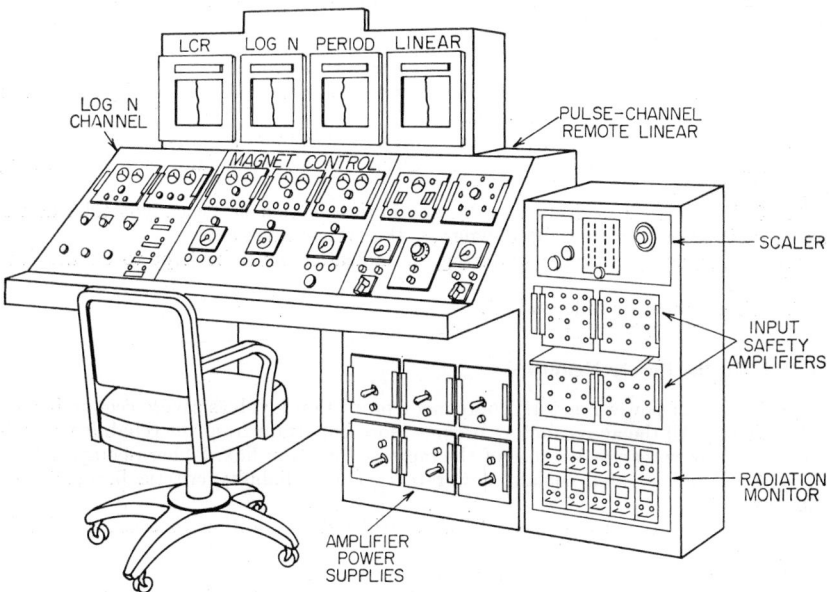

Fig. 17. Typical instrumentation installation for a research-type reactor.

RELIABILITY ASPECTS OF PROTECTIVE SYSTEMS

The fulfillment of the reliability requirements for reactor instrumentation is more than a matter of making reliable components for individual instruments. It is more important to know the system's reliability requirements in general and the specialized needs of the particular installation.

Reliability can be defined as the ability of the system to function properly in its assigned role for a predetermined period of time. This should be accomplished with a minimum of maintenance.

The public hazard aspect of reactor installations has been discussed earlier. Since the inherent stability of a reactor is a basic consideration and is independent of any instrumentation, the basic reactor design must demonstrate this ability through the means of "built-in" factors such as large, prompt, negative temperature coefficient or a combination of factors which leads to inherent stability.

Reliability Requirements

In view of the inherent stability aspects, what is the role of the protective system? *The major role of the protective system is reactor- and plant-component protection.*

Component protection involves the following:

1. Prevention of fuel melting.
2. Reduction of the severity of an abnormal condition.
3. Protection of personnel from exposure to an abnormal radioactive environment.

Reliability as applied to the protective system requires that the system be fast and perform its prime function with an absolute minimum chance of failure. First, let us define "fast." This does not always mean that the faster the system the better it is. However, the system must be fast enough to stop the more probable accidents. With the instrumentation available today, a typical value of over-all actuating time—from the detection of an abnormal condition until the shutdown mechanism is actuated—is 10 milliseconds or less. It is obvious that the instrumentation delay times should be considerably less than the over-all actuating delay (approximately 10 per cent of the total delay time).

The reliability requirements can be stated as follows:

1. Reliable components and individual instruments should be used.

2. Availability of the entire plant must be considered when reliability criteria of the protective system are established.

3. Adequate monitoring of the system performance must be provided.

4. The reactor operator must be made aware of instrument failures, maloperation, and potentially dangerous situations.

5. If the operator neglects to take appropriate action or he cannot take action and the situation is potentially dangerous, the system must act.

6. Ambiguous operation, monitoring, or indication must be kept to an absolute minimum.

7. Failures must be independent.

Reliability in Power Reactors

In the case of power reactors, component protection involves severe economic considerations. All the above factors become more important from the reliability standpoint in a power reactor because of the increased neutron flux, higher operating temperatures, and increased coolant flow rates. Thus reliability criteria become more difficult.

Some of the factors that need to be considered in addition to those previously mentioned are:

1. Effects of thermal shock to reactor components.

2. Effects of the nonnuclear behavior of the power plant on the reactor.

3. Magnitude and number of power excursions which the reactor can sustain without damage to the reactor.

4. Behavior of the reactor and the entire plant under normal and abnormal conditions.

Determination of these factors is a complex matter and involves in many cases the expenditure of a great deal of time and money. However, all this is well spent when one considers the total investment of the reactor power plant. Involved simulation programs, control studies, dynamic studies, and studies of reactor characteristics are beyond the scope of this discussion. Nevertheless these are important aspects of the over-all reactor instrumentation complex.

In power-reactor operation, the necessity for staying "on the line" means that reliability plays an important part in the economic aspect of the plant. False shutdowns must be kept to an absolute minimum. This is more stringent than for most research reactors.

The reliability aspects of protective systems can be stated simply as follows:

The protective system must be available under any abnormal condition and must operate effectively in those situations for which it is designed to act and correct the abnormal conditions without precipitating false action because of the existence of the protective system itself.

In most reactor installations, absorbing rods that are inserted rapidly into the reactor core are used for shutdown. Therefore, we will refer to this operational mode to discuss reliability problems in detail. The rods are usually held outside the reactor and then inserted into the core on the appropriate signals from the protective system. The action can be considered bistate—the rods are either in the "ready" state or in the inserted state. Thus either nonlinear or bistate techniques can be employed. For simplicity, only the fast-shutdown mode will be discussed here. Controlled run back of the rods and other modes are not considered. From an operational viewpoint these modes of operation are important. They should be considered from an over-all criterion that the action taken is commensurate with the severity of the abnormal condition.

System with Redundant Inputs

Consider the simplified system shown in Fig. 18. This is a protective system having redundant inputs. In this system an increase in reliability is being attempted by using a multiplicity of detectors for the measurement of a single parameter (neutron flux). The block labeled *I.U.* is a device for converting the analog neutron flux information to a bistate output. The analog neutron flux information is compared with a desired flux level, and any flux level below this reference point does not cause a change in the output state of the device. However a neutron flux level greater than the reference point (set point) causes the output of *I.U.* to change state.

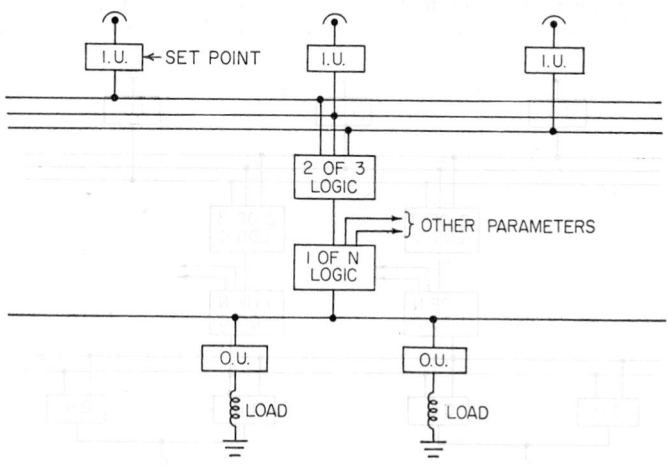

Fɪɢ. 18. Block diagram of a protective system with input redundancy.

For illustrative purposes, output signals below the set point are in the "0" state and signals above the set point are in the "1" state. The input unit acts as a switch, changing its output state when the measured parameter is greater than a desired set point. All subsequent information is handled in the bistate mode (digital signals). The three detectors shown in Fig. 18 are connected in such a way that any two or more signals greater than the desired level will cause the output of the block labeled "two of three logic" to change state. That is, if only one detector indicates high neutron flux level, no change in state of the block labeled "two of three logic" will occur.

This is redundancy in a sense. At least two signals must show an "abnormal" condition before some further action is taken. The output ("0" or "1" state) of the logic unit is combined with a similar logic unit so that signals can be combined from other parameters in what we shall call a "one out of *n*" type logic. The output of this latter device is connected to the output unit which provides the power to the load. In this case the load could be the electromagnet supporting the poison rod.

Thus this system has input redundancy which will only "trigger" the output when two or more input signals are in an abnormal condition. Also, we can combine information from other parameters—such as temperature, flow, pressure, high radiation levels, and the like. This information can be arranged similarly, so that two or more signals from these parameters must be above some desired reference level.

From a reliability standpoint, there is some advantage in such a system. A failure in an input unit or a detector can still provide continuity of operation without loss of protection.   However, if we examine the logic unit or the output unit, we see that a failure can cause a malfunction of the system.   Consider failures of two types:

1. *A failure that prevents the passage of a legitimate signal.*   For example: The "two of three logic" unit could fail to change its output state when two high neutron flux signals are present at its input.

2. *A failure that generates a false signal.*   For example: This same unit could generate a false signal (change its output state to the "1" state) when the necessary conditions for such a change are not present at its input.

Either of these is undesirable.   Therefore continuity of operation is jeopardized and the availability of this system to provide protection is compromised.

System with One Direction of Failure Only

Figure 19 is a block diagram of a system which will tolerate one direction of failure only and is often considered an improvement in reliability over the previous system. However, look at this system in detail.   At the input level we are considering a two

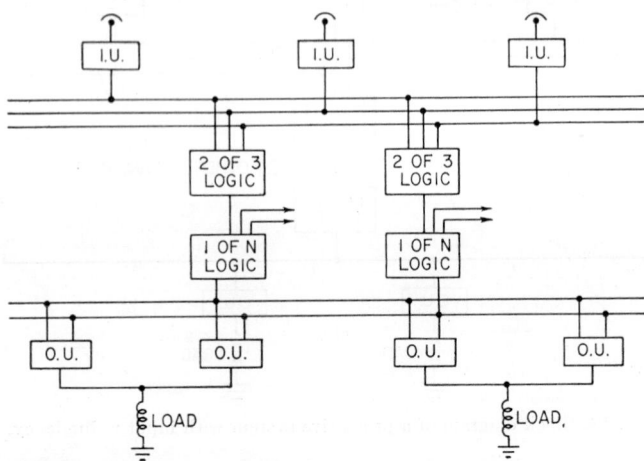

FIG. 19. Block diagram of a protective system with one direction of failure only.

out of three type redundancy as before, but in the secondary portion of this system we have paralleled the equipment.   The functions are as in the previous block diagram.   The inputs of the output units are connected such that any one input going to the abnormal condition will cause the load to drop a "one out of two" logic in these inputs.   The load is shared by the output units.

An examination of this simplified system (Fig. 19) shows that the following single failures can be tolerated without loss of the over-all system function:

1. An input device can fail to indicate an abnormal condition or it can pass false information.

2. Any one logic device can fail to pass legitimate signals because there is an alternate path to carry out the function.   Obviously two logic elements can fail to pass information in a single path without loss of the system.

3. An output device can fail to carry the load, but the parallel device can provide the power to the load.

It is apparent that certain multiple-failure configurations can be tolerated without loss of the over-all system function.   For example, an input device can fail, both logic elements in a single path can fail to pass information, and an output device can fail to provide power, but the system still is available.   As discussed below, detection of the first failure and its repair is of importance in the over-all reliability of the system.

Further examination of the above system will show that the first failure of a logic element which generates a false signal will cause a malfunction of the system. Also, an output device which fails to "turn off" its output to the load when required will prevent the release of the rod. Thus the above system cannot tolerate a single failure in view of the bidirectional failure considerations.

The Fully Redundant System

Figure 20 is a portion of a fully redundant system, which can tolerate both directions of failures. Two out of three redundancy has been carried throughout the system. Only one load is shown, but in a reactor a redundancy of shutdown mechanisms is utilized and the mechanisms can be connected to the "two of three buses" shown as Schedule 2 in Fig. 20. They could have their own redundant output units as well. Close examination will show that both types of failures are possible without losing

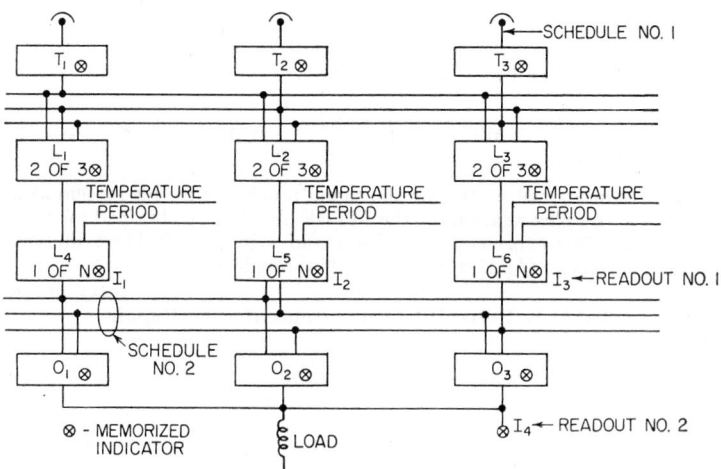

Fig. 20. Block diagram of a fully-redundant system.

the protection provided by the system. It should be carefully noted that only a single failure can be tolerated at any one level in the system, although multiple failures in some cases are possible without negating the functions of the over-all system.

If we examine the output level of the system we find certain requirements of the individual output unit. One out of two logic internal to the unit must be carefully considered. Two independent switches connected in the one of two logic mode must be incorporated in order that the functions of the system can be carried out.

Functions other than neutron level, which is shown here, can be incorporated into the system. Such parameters as period information or temperature can be brought into the appropriate logic level as shown in Fig. 20.

The fully redundant two out of three systems just described will ultimately fail because of a multiplicity of component failures. As indicated before, such systems will only tolerate a single failure at each level of the system. In addition, more components are involved in this system than in a single channel, so that failures may be more frequent in the fully redundant systems because of the component count. What then does the redundant system really give us? *It can provide the means for repair without system shutdown.* In Fig. 20, any one of the blocks shown in the system can be removed without compromising system operation. Therefore, the opportunity to repair presents itself. It has been shown that the mean-time-between-failures (MTBF) of triple redundant systems can be increased many fold, thus affording long-term availability and reliability.[51]

Effects of Repair

The significant factor in obtaining such increases in MTBF is accomplished through the repair activity. This ability to repair in redundant systems, without loss of continuity of operation, is of major importance to reactor operations as well as to other complex systems.

The general approach to repair is illustrated by Fig. 18. Consider the three input units feeding the "two of three logic" units as a simple system in itself. The outputs of the input units (signal processors) are fed to the "two of three logic" (majority selector). At least two or more signals are required at the input of the majority selector to obtain an output.

The signal processors are characterized by a statistical quantity λ, called the "failure rate," which determines their reliability according to the relation

$$R(t) = e^{-\lambda t}$$

where t = time

The MTBF or signal processors are numerically equal to the area under the curve generated by this relation from $t = 0$ to $t = \infty$. Therefore

$$\text{MTBF} = \int_0^\infty e^{-\lambda t} = \frac{1}{\lambda} = M_1$$

When failures occur, we can repair the signal processors in this system at a repair rate μ. A more complex relation for the MTBF, M_2, can be obtained for the scheme utilizing repair. This expression includes μ and λ in M_2. Table 1 illustrates the reliability improvement M_2/M_1 that is possible through repair.

Table 1. Reliability Improvement with Repair

$\dfrac{\mu}{\lambda} = \dfrac{repair\ rate}{failure\ rate}$	$\dfrac{M_2}{M_1} = \dfrac{MTBF\ with\ repair}{MTBF\ without\ repair}$
0	1.3
1	1.5
10	2.7
100	1.9
1,000	169
10,000	1,667

It should be noted that if repair is not included $\left(\dfrac{\mu}{\lambda} = 0\right)$, not much reliability improvement is obtained. On the other hand, large values of μ/λ show marked improvement. For the fully redundant systems, similar improvements in reliability are obtained when μ/λ is large. Repair has the following aspects contained within its concept:

1. The detection of a failure within the system.
2. The location of the failure in the systems complex.
3. The repair of the failure.

Dynamic Monitoring

Because a protective system is normally "standing by," one must be assured that the system will perform the prescribed functions when required. In addition, it must be determined that a failure does or does not exist in the system. This must be determined even though the system can still carry out its function with the failure present. It is important that the first failure be detected and the system restored before a second failure occurs. This is indicative of a fast repair rate.

In order to establish the above, a method of continuously testing or monitoring the system dynamically is required. The purpose of the monitor is to determine system operability. In a redundant system, where failures can occur without loss of system function, the monitor detects failures as they occur and facilitates the location of failed components or units. Replacement of the failed units will reestablish the long-term reliability with no loss in over-all system function.

For a monitor to perform the function of locating a failed component, it is necessary that the operation of each component be tested in accordance with the intention of the system design. The dynamic testing of a system can be accomplished by inserting pulse signals of a short duration into the system at the appropriate points and in such a manner as to completely test the system. It is desirable that the testing include as much of the system as possible from input to output loads. It is necessary that the pulses in the testing scheme be much shorter than the response time of the output load.

With reference to Fig. 20, pulses are inserted at "Schedule No. 1," and the performance of the system up to this point is determined at "Readout No. 1." After the mixing operation, it is necessary to reinsert the pulses at "Schedule No. 2," and performance of the output section is determined at "Readout No. 2." The "memorized indicators" indicate the presence of the pulses at various points in the system and are reset by the monitor as part of its over-all cycle. They also assist in the location of the failed unit.

In a redundant system, such as "two of three system," it is necessary to test all combinations possible in a redundant system. It is not enough to determine that the system functions when two or more inputs are present. In the "two of three system" there are eight combinations of the inputs which must be tested. Table 2 illustrates this fact.

Table 2. Combinations of Inputs to Be Tested

Signal condition	Number of combinations
No inputs	1
One input	3
Two inputs	3
Three inputs	1
Total	8

For a complete test of even this simple system, all combinations must be tested and operability of the system determined. Note that only the two-input and three-input signal conditions are valid for this particular system.

Various levels of sophistication of monitoring can be incorporated, from manual testing to a fully automatic monitoring scheme. Systems with automatic monitoring schemes incorporating many inputs have been designed. Complex systems can be continuously and completely tested for their intended operation every few seconds.

REFERENCES

1. Belchem, R. F. K.: "A Guide to Nuclear Energy," Philosophical Library, Inc., New York, 1959.
2. Taylor, E. Openshaw: "Nuclear Reactors for Power Generation," Philosophical Library, Inc., New York, 1959.
3. Dietrich, J. R., and W. H. Zinn: "Solid Fuel Reactors," Addison-Wesley Publishing Company, Inc., Reading, Mass., 1958.
4. Goldstein, Herbert: "Fundamental Aspects of Reactor Shielding," Addison-Wesley Publishing Company, Inc., Reading, Mass., 1959.
5. Goodman, Clark: "The Science and Engineering of Nuclear Power," Addison-Wesley Publishing Company, Inc., Reading, Mass., 1952.
6. Beck, Clifford K.: "Nuclear Reactors for Research," D. Van Nostrand Company, Inc., Princeton, N.J., 1957.
7. Pickard, James K.: "Nuclear Power Reactors," D. Van Nostrand Company, Inc., Princeton, N.J., 1957.
8. Glasstone, Samuel: "Principles of Nuclear Reactor Engineering," D. Van Nostrand Company, Inc., Princeton, N.J., 1955.
9. Glasstone, Samuel, and Milton C. Edlund: "The Elements of Nuclear Reactor Theory," D. Van Nostrand Company, Inc., Princeton, N.J., 1952.
10. Hoag, J. Barton: "Nuclear Reactor Experiments," D. Van Nostrand, Company, Inc., Princeton, N.J., 1958.
11. United States Atomic Energy Commission: "Research Reactors," McGraw-Hill Book Company, Inc., New York, 1955.

12. American Institute of Chemical Engineers: Nuclear Engineering, Parts I, II, III, and IV, *AIChE Monographs*, New York, 1959.
13. Taylor, E. Openshaw: "Nuclear Power Plants," Philosophical Library, Inc., New York, 1958.
14. National Standards for Nuclear Instruments and Controls, American Standards Association, New York, 1959.
15. Control for a Pressurized Critical Assembly, *Automatic Control*, vol. 12, no. 6, p. 22, June, 1960.
16. Thermocouples for Nuclear Plants, *Automatic Control*, vol. 12, no. 3, p. 50, March, 1960.
17. Hill, H. L., and H. G. Pinder: Self-Testing Safety System Minimizes Power Reactor Downtime, *Cont. Eng.*, vol. 9, no. 5, pp. 89–93, May, 1962.
18. Auricoste, J.: Digital Computers Monitor Nuclear Steam Generator, *Control Eng.*, vol. 8, no. 3, pp. 127–131, March, 1961.
19. Loving, J. J.: Designing an Automatic Rod Control System, *Automatic Control*, vol. 14, no. 4, pp. 51–56, April, 1961.
20. Moss, A. I.: Protection System Experience at Shippingport, *Automatic Control*, vol. 14, no. 12, pp. 53–57, December, 1961.
21. Wakefield, E. H.: "Nuclear Reactors for Industry and Universities," Instruments Publishing Company, Pittsburgh, 1954.
22. Sangren, W. C.: "Digital Computers and Nuclear Reactor Calculations," John Wiley & Sons, Inc., New York, 1960.
23. Epler, E. P.: Reliability of Reactor Systems, *Nuclear Safety*, vol. 4, no. 4, June, 1962.
24. Epler, E. P.: HTRF-3 Excursion, *Nuclear Safety*, vol. 1, no. 2, pp. 57–59, December, 1959.
25. Underkoffler, V. S., and R. G. Olson: Design Consideration in a Nuclear Safety System, *Elec. Eng.*, vol. 77, no. 11, November, 1958.
26. Bush, John H., and Carl F. Leyse: "Materials Testing Reactor Project Handbook," Technical Information Service, Oak Ridge, Tenn., 1951.
27. Stephenson, Richard: "Introduction to Nuclear Engineering," 2d ed., McGraw-Hill Book Company, Inc., New York, 1958.
28. Lapp, Ralph E., and Howard L. Anders: "Nuclear Radiation Physics," Prentice-Hall, Inc., Englewood Cliffs, N.J., 1954.
29. Hughes, D. J.: "Pile Neutron Research," Addison-Wesley Publishing Company, Inc., Reading, Mass., 1953.
30. Murray, Raymond L.: "Introduction to Nuclear Engineering," Prentice-Hall, Inc., Englewood Cliffs, N.J., 1954.
31. ORNL-2695, "Nuclear Process Instrumentation and Controls Conference," Oak Ridge National Laboratories, Oak Ridge, Tenn., 1958.
32. L'Electronique nucleaire, *Proc. Intern. Symp. Nucl. Electron.*, International Atomic Energy Agency, Wien, Austria, 1959.
33. Reactor Safety and Control, *Proc. Second Intern. Conf.*, vol. 11, Geneva, Switzerland, 1958.
34. Lennox, C. G., and A. Pearson: NRU Reactor Neutron Level Control System, *IRE Trans. Nucl. Sci.*, No. 2, August, 1958.
35. United States Atomic Energy Commission: "Reactor Handbook—Engineering," McGraw-Hill Book Company, Inc., New York, 1955.
36. United States Atomic Energy Commission: "Reactor Handbook—Physics," McGraw-Hill Book Company, Inc., New York, 1955.
37. Epstein, B., and J. Hosford: Reliability of Some Two Unit Redundant Systems, *Proc. Sixth Natl. Symp. Reliability Quality Control*, pp. 469–477, January, 1961.
38. Kneale, S. G.: Reliability of Parallel Systems with Repair and Switching *Proc. Seventh Natl. Symp. Reliability Quality Control*, pp. 129–133, January, 1961.
39. Moore, E. F., and C. E. Shannon: Reliable Circuits Using Less Reliable Relays, *J. Franklin Inst.*, vol. 262, pp. 191–208, September, 1956.
40. Von Neumann, J.: "Probabalistic Logics and the Synthesis of Reliable Organisms from Unreliable Components," pp. 43–49, Automata Studies, Annals of Mathematics Studies, no. 34, Princeton University Press, Princeton, N.J., 1956.
41. Enrico Fermi Atomic Power Plant, APDA124, Atomic Power Development Associates, Inc., January, 1959.
42. Schwartzenberg, J. W., J. M. Finan, V. S. Underkoffler, and R. N. Brey, Jr.: Controls for an EBWR Simulator, *Automatic Control*, February, 1960.
43. Cockrell, J. L.: "Modular Concepts in Reactor Control Instrumentation," Nuclear Congress, Chicago, 1958 (published by the Institute of Electrical and Electronics Engineers).
44. United States Atomic Energy Commission: "Research Reactors," McGraw-Hill Book Company, Inc., New York, 1955.
45. Lawrence, George C.: Reactor Safety in Canada, *Nucleonics*, vol. 18, no. 10, 1960.
46. Wilcox, Richard H., and Wm. C. Mann: "Redundancy Techniques for Computing Systems," Spartan Books, Washington, D.C., 1962.
47. *Nuclear Safety*, vol. 4, no. 4, June, 1963.
48. Siddall, E.: Reliable Reactor Protection, *Nucleonics*, June, 1957.
49. Siddall, E.: Statistical Analysis of Reactor Standards, *Nucleonics*, vol. 17, no. 2, p. 64, 1959.
50. Siddall, E.: Reliable Reactor Protection, *Nucleonics*, vol. 15, no. 6, p. 124, 1957.
51. Cockrell, J. L., J. H. Magee, and V. S. Underkoffler: Nuclear Protective System Design, *IEEE Trans. Commun. Electronics*, July, 1963.

Section 15

LABORATORY AND
PILOT-PLANT INSTRUMENTATION

By

CLYDE BERG, Ph. D. (Ch.E.), *Director, Clyde Berg Associates, Long Beach, Calif.; Member, American Institute of Chemical Engineers, American Chemical Society, American Petroleum Institute; Registered Professional Engineer (Calif.). (Pilot-plant Instrumentation)*

HAROLD C. JONES, B.S.E.E., M.S.E.E., *Electrical Engineering Department, University of Maryland; Consulting Engineer (Environmental Testing and Reliability); Registered Professional Engineer (Md.); Member, Institute of Radio Engineers, American Institute of Electrical Engineers; Fellow, Institute of Environmental Sciences. (Environmental Test Instrumentation)*

ALFRED H. McKINNEY, B.S. (Chem.Eng.), *Consultant, Instrument Section, Engineering Service Division, Engineering Department, E. I. du Pont de Nemours & Company, Inc., Wilmington, Del.; Member, American Institute of Chemical Engineers, American Chemical Society, Instrument Society of America; Registered Professional Engineer (Pa.). (Process Laboratory Instrumentation)*

PROCESS LABORATORY INSTRUMENTATION

By Alfred H. McKinney*

The laboratory worker, whether technician or holder of a high degree, is frequently confronted with the need to measure automatically and/or control some physical property. In some instances, this will result in permanent equipment to satisfy continuing needs. In other cases, the setup will be temporary. In either case, the inexperienced worker will probably find it more economical of time and money to purchase available equipment than to develop makeshift arrangements.

Some laboratory directors discourage any but the most conventional arrangements of standard apparatus. Others have found it necessary to develop special equipment and arrangements to solve specific problems. Success in the latter direction requires a good knowledge of physics and instrumentation and a large degree of imagination so as to anticipate the many difficulties which may be encountered. The skepticism with which some look upon any but the most tried and trusted standard equipment may be due to unimaginative attempts to borrow an idea from another field without a careful analysis of the new problems involved.

Judging from the literature, however, the number of laboratory workers who are devising new ways to measure and to control and who use homemade equipment where required is increasing. Some of these setups are described in this subsection. The parenthetical numbers which appear throughout this text pertain to references in the exhaustive list at the end of this subsection. Mention is made of the use of conventional and commercial measuring and controlling equipment, but these are not covered in detail here. Prime emphasis is placed here upon the measuring and controlling systems which are not available as readily purchased assemblies.

Once a concept is obtained for a new product or process, it is often more economical to perform at least some preliminary investigation in the laboratory. It is sometimes possible to simulate a complete process or operation on a laboratory bench and place all the variables under automatic control. This usually has the following advantages:

1. Laboratory arrangements can be made more compactly and simply, with savings effected—in both time and equipment.

2. The quantities of raw materials required are less.

3. Time required to obtain test data is reduced. This is due to better time response; conditions stabilize quickly to the desired operating values.

4. Control is frequently better. With shorter time constants, the variables are easier to control; simpler control functions may be adequate.

CONTINUOUS PROCESS CONTROL IN THE LABORATORY

Laboratory process control on a continuous basis is preliminary to, but not a substitute for semiplant and semiworks evaluation or production. Favorable results which are obtained in the laboratory frequently require confirmation on a larger scale. However, a considerable savings can result if the laboratory tests can demonstrate that a project should be discontinued or requires modifications.

As distinguished from conventional industrial process-control equipment, laboratory controllers for process evaluation may have the following characteristics:

* Consultant, Instrument Section, Engineering Service Division, E. I. du Pont de Nemours & Company, Inc., Wilmington, Del.

1. *Small Size.* In respect to both quantities handled and physical bulk.

2. *Expendable.* Such equipment does not necessarily have to be designed for rugged service.

3. *Conventional Calibration Is Not Necessary.* Resulting process indications can be read on manometers or any arbitrary scale, not necessarily on 3 to 15 psi receivers. Nonindicating self-actuated regulators frequently are satisfactory.

4. *Made from Easily Available Components.*

The principles of automatic control common to all systems apply equally well to laboratory arrangements. Several general discussions of chemical process control are available (11), (25), (39), (43), and (45), as well as more thorough treatments of such processes wherein control engineering theory is used (9) and (16). Roth and others discuss some aspects of control in pilot plants and laboratories (1), (2), (6), (13), (21), (30), (31), and (32).

A pilot plant is a model of a full-scale plant about which information is desired.* To be successful, it must be able to duplicate some of the unknown features of the larger unit of which it is a prototype. This requires the use of dimensional analysis. Scale-up practices are discussed by several authors (4), (22), (28), and Jordan (17)

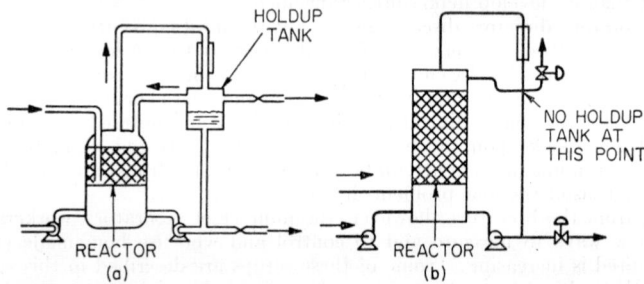

Fig. 1. Example of how system controllability may be changed during scale-up: (*a*) Laboratory arrangement; (*b*) plant arrangement.

carefully points out that, in general, a small-scale model can only duplicate one parameter (e.g., Reynolds number) of the larger prototype. Sometimes more than one feature can be duplicated or compared.

Good experimentation requires that the model duplicate the final design in all respects where doubt exists as to the performance of the full-size plant, and leave as untested only those parameters where basic design data are established. One such parameter is *controllability*, a factor which is usually assumed to be unchanged with the scale of the equipment. Ziegler and Nichols (45) have proposed the ratio C/L as a measure of the controllability of a process, C being the capacity and L being the closed-loop time lag of the process, including the instrumentation. Both these characteristics can be expressed in time units so that the ratio is dimensionless.

Figure 1 shows a pilot-plant model of a reactor which worked well, but where the final full-scale plant, designed from this experience, was difficult to control. One difference to be noted is that the hold-up tank in the smaller-scale experiment, which contributed to process stability and improved control, was eliminated in the final plant. It is usually easier to control a process on a small scale than on a large scale.

A more precise measure of the relative controllability of large and small prototypes would be a comparison of their open-loop frequency responses. If these are identical, then the two will be equally controllable. It will probably be true that any two processing units dissimilar in size, but designed to be dimensionally similar with respect to one dimensionless parameter, will be dissimilar with respect to all other parameters, including controllability.

In simulating the larger plant on a small scale, one can be guided by the published data on the control of certain unit operations and can adopt some of the techniques

* See under Pilot-plant Instrumentation, where the application of commercial instrumentation to pilot plants is described.

that have been reported. Automatic control of laboratory distillation is discussed (7), (8), (14), (20), (33), (40), and (41). Pratt (29) describes a small automatic liquid-liquid separator and Nickels provides a steam trap (27). Standard recorders can be used for many purposes (24); can be converted to controllers (18); and strip-chart controllers can be made into program controllers causing the process variable to follow a mark previously drawn on the chart (5). Shaft seals (38) and sealed, enclosed stirrers (10), (23), (37) are useful in assembling small process models as well as the techniques of coupling glass tubing to other materials (35). In devising transmitters and controllers, the data on flexure pivots (36) and nozzle-baffle systems (19) will be useful.

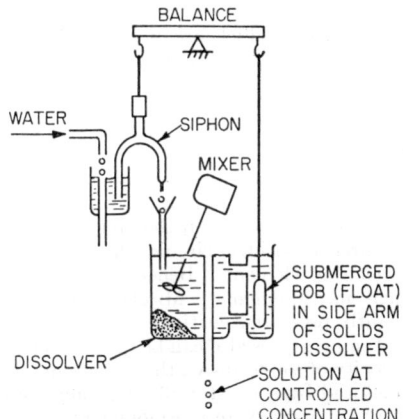

Figure 2 shows a chemical balance which has been modified to control the solution concentration in a dissolver. As the density increases, the submerged bob is raised, thus lowering the siphon and adding more water to reduce the solution density. This example is typical of the arrangements possible in the

Fig. 2. Modification of chemical balance for controlling solution concentration in dissolver.

laboratory and pilot plant to bring various process measurements under continuous and automatic control. The main limitation is simply the lack of imagination and experience on the part of the laboratory worker.

Ability to control a process depends on the physical measurements that can be made on the system and whether or not such intelligence can be used to modify or dominate some source of material or energy that shows an effect on the measurement made. Thus the control loop is closed.

In the following paragraphs, some of the measurements that may be useful are described or references are cited. This list is incomplete but contains many ideas which have been selected for their simplicity, ease of construction, and application to laboratory process control. The techniques are classified in accordance with the process variable which they are controlling.

GENERAL REFERENCES

The texts by Eckman (12) and Holzbock (15) describe many of the conventional process-measuring instruments. Witherspoon (42) gives a general discussion of electrical transducers and Nadir (26) lists in detail the properties of approximately 1,000 transducers which are suited to displaying a wide range of measurements on an oscilloscope.

TEMPERATURE MEASUREMENT

Calibrations for thermocouples made from silver-gold (2-A), silver-palladium (6-A), nickel-molybdenum (50-A), tungsten-molybdenum (51-A), tungsten-carbon (55-A), tungsten-iridium (62-A), and 20 per cent rhodium-platinum (35-A) are reported for high temperatures, and copper-tin-copper alloy (48-A) for low temperatures. Thermocouples made from small wires (9-A), (10-A), (15-A), (21-A), and (59-A) show rapid response to temperature changes. Fuschillo (25-A) discusses errors in thermocouple wires. Burton and Weeks (14-A) review the limitations of thermocouples in the region of 1500 to 1900°F.

The electrical resistance of many materials is a measure of their temperature; carbon for low temperatures (18-A); copper; tungsten (56-A); nickel; platinum (41-A); transistors (65-A); and metal films on glass surfaces (57-A). Postage-stamp-sized units can be applied to many surfaces for temperature measurements.*

Thermistors are commonly used to measure temperatures by their resistance change (1-A), (3-A), (61-A), and (62-A). Measuring circuits can be arranged to provide output readings which are linear with temperature (44-A) and (53-A). Bennett discusses the selection (7-A) and use (8-A) of thermistors. With the large change in resistance per degree, very small temperature changes can be measured (22-A), (28-A), (47-A), (52-A), and (66-A).

Thermistors, in common with all resistance-changing elements, have one drawback; they require the flow of some current to provide a resistance measurement. This current raises the temperature of the measuring element above that of the temperature being measured. In liquids, this is usually negligible because of the high heat-transfer coefficients present; but the precise measurement of temperature, particularly in gases, requires an electric circuit which is designed to liberate a minimum of power in the measuring element. Benedict (11-A) compares thermocouples and thermistors for such measurements. Misener and Thompson (46-A) show that thermistors are slightly pressure-sensitive.

Temperatures also can be measured by bimetal strips, bifibers of plastics (49-A), the velocity of sound (5-A), the density of a liquid (24-A), and the electrical noise which is produced by a resistor (33-A). Joy (36-A) describes the construction and filling of glass vapor-pressure thermometers.

Stillwell (58-A) evaluates the errors of thermocouples which are affixed with adhesive tapes. An alignment chart is given for emergent stem correction for glass thermometers (19-A). Thermocouple errors due to a large number of causes are considered by Guthmann (30-A). Methods of reducing the errors in the measurement of gases at high temperatures (12-A), (16-A), (29-A), (40-A), (42-A), and (64-A); and high velocities (17-A) and (23-A); and the use of suction pyrometers (4-A), (26-A), (34-A), (38-A), and (43-A) are reported. Optical pyrometers to measure gas and flame temperatures are described (20-A), (32-A), (45-A), and (60-A).

Bruce lists suitable melting points for thermometer calibration (13-A).

Shepard and Warshawsky propose an electrical network to compensate for the time lag of thermoelements (54-A).

Temperature Control

On-off temperature control of baths, ovens, and similar equipment in the laboratory is commonly accomplished by using mercury contacts in thermoregulators which operate power relays. If thyratrons are used in place of relays, the current broken by the thermoregulator is greatly reduced, thus making it more sensitive and stable (29-B) and (30-B). Swinehart (34-B) describes a thyratron relay which uses the circuit shown in Fig. 3. This circuit is easily assembled and requires a minimum of parts. The relay† described by Muller (26-B) avoids the mercury contact in the regulator; a metal clip on the outside of a conventional glass thermometer is connected to a capacitance-operated circuit and detects the rise or fall of the mercury column. The height of the mercury also

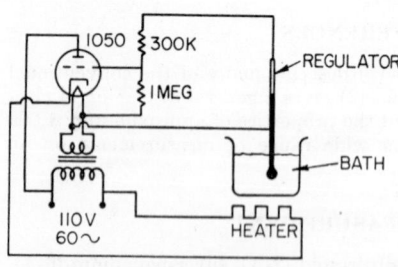

FIG. 3. Laboratory temperature regulator in which relay is replaced with thyratron.

can be detected with inductance coils (2-B) and (15-B). The sensitivity of the mercury regulator can be increased if it confines a more volatile liquid or a gas within the temperature-sensing element (28-B) and (36-B).

* Stikon units, Ruge-DeForest, Inc., Cambridge, Mass.
† Niagara Electron Laboratories, Andover, N.Y.

Other devices can be used as thermostats. Blauvelt (7-B) gives a description of many types. A bimetal thermostat* which is ⅜ in. in diameter and less than 1 in. long is commercially available. Thermistors have been used (13-B) and (27-B); and resistance bulbs (5-B) also have been used. A small contact-making meter† which operates directly from thermocouples is commercially available.

Figure 4 shows the temperature cycle that will result from the use of an on-off regulator. The amplitude of the temperature oscillation will depend on the system

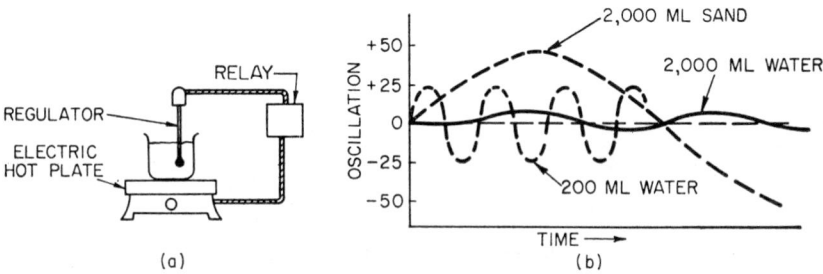

FIG. 4. Common laboratory temperature regulation scheme is shown in (*a*). Oscillation or cycling, with on-off regulation, for various capacities is shown in (*b*).

being controlled. Where reports state that control is achieved to within ±1, or 0.1, or 0.01°C, credit usually should be given to the capacity of the system and to the small time lag involved—and not to the regulator. The amplitude of the cycle that necessarily results from on-off control can be reduced by one or more of the following improvements:

1. Increase in the heat capacity (volume) of the system being controlled.
2. Reduction in the temperature differential (i.e., between on and off) of the regulator.
3. Reduction of the heat insulation between the heater and the process.
4. Relocation of the temperature-measuring element to a point nearer to the heat source.
5. Increase in the agitation.
6. Reduction in the heat capacity of the heater.
7. Provision for supplying some of the heat continuously; and reduction of the amount of heat being controlled.

The foregoing actions will tend to reduce the amplitude of the temperature oscillations, but the cycle cannot be eliminated entirely so long as purely on-off control is used. It is also possible to reduce the amplitude and to increase the frequency of the cycle by using a bridge which contains a thermistor in addition to the regulator. The thermistor is heated when the main source of power is on and provides an *anticipatory* action (14-B) and (33-B). Henschen and others (18-B) describe a control system with constant-speed floating action. Improvement in the cycle amplitude with this system will make recovery from a load change more sluggish.

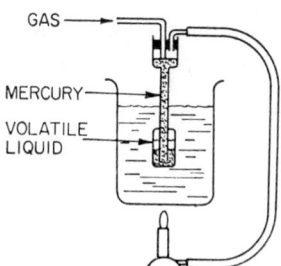

FIG. 5. Arrangement in which flow of gas to burner is controlled by mercury level and thus by temperature to provide proportional control. (*Note:* Gas piping has small hole just above mercury level, as pilot light.)

To eliminate the cycle of the controlled temperature, it is necessary to use an arrangement that modulates the heat supply at intermediate values between fully on and fully off. Several manufacturers supply small bimetal or vapor-filled sensing elements that operate throttling valves. Brown (9-B) describes the regulator shown in Fig. 5.

* Valverde Labs., New York, N.Y.
† Assembly Products, Inc., Chagrin Falls, Ohio.

The flow of gas to the burner is throttled by the mercury level in response to the expansion of the fluid in the bulb. Holms (20-B) builds heater and temperature-measuring element into a single immersible glass assembly. Heat is supplied by current flowing through a salt-glycerol mixture. As the temperature increases, the solution separates from the carbon electrode because of the expansion of the gas volume trapped above the electrolyte. Roberts (12-B) uses the resistance of the platinum heating element as a measure of the furnace temperature. This requires a rather elaborate electronic system. A similar arrangement is reported by the Bureau of Standards (6-B). Conventional recording-controlling potentiometers have been used wherein the pneumatic output positions variable transformers (19-B) or less expensive current-adjusting devices (24-B). Other proportional control arrangements have been described which use thermistors and resistance wires to measure the temperature and where thyratrons or saturable reactors provide proportional control of the heat (10-B), (22-B), (23-B), (25-B), (37-B), and (40-B).

Control of the rate of temperature change also has been reported (8-B) and (35-B).

PRESSURE MEASUREMENT

Eckman (12) gives a good review of conventional pressure-measuring equipment. Deflections of the conventional bourdon tube can be measured with an electrical micrometer (15-C) or with a differential transformer (22-C) to obtain higher sensitivity than that provided by the usual pressure gage. Kerris and Weidemann operate a mirror by a diaphragm and observe small deflections on a photoelectric cell (9-C). Small pressure pickups are described in (16-C) and (20-C).

For pressures below atmospheric, Dushman (5-C), Lawrance (11-C), and Steckelmacher (19-C) provide reviews on measuring methods. Hot-wire vacuum gages are reported as easy to construct (4-C), rugged (2-C), and suitable for high vacuums (1-C), (3-C), (7-C), and (10-C); and for low vacuums (14-C). Von Ubisch reports the response of hot-wire gages for 19 different gases at various pressures (21-C). Havens, Koll, and LaGow (8-C) obtain an a-c signal from a d-c pressure gage by varying the volume at a relatively high mechanical frequency. The amplitude of the a-c signal is proportional to the absolute pressure (8-C).

Manometers

The usual laboratory U-tube manometer has been modified into many forms for various purposes: precision reading (17-D), high pressures (22-D), as a null detector (7-D), and to record on photographic paper (26-D) and (27-D). External inductance pickups are used to sense the change in mercury level (1-D), (12-D), and (28-D). Capacitance between the mercury level and a horizontal plate can be used to detect pressures with an accuracy of $\pm 0.1\ \mu$ (14-D).

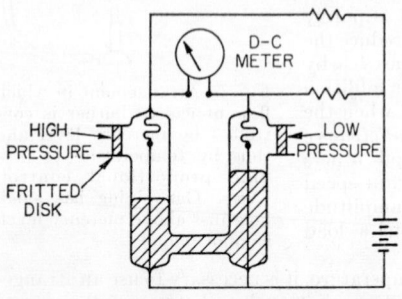

An ordinary manometer can be modified for precise and remote reading as indicated in Fig. 6. A wire through the axis of each arm indicates pressure or differential-pressure changes by measurement of the resistance of the exposed length. Nichrome V wire works well inasmuch as it can be fused to Pyrex and

FIG. 6. Modification of ordinary manometer for precise and remote reading.

is unaffected by mercury. With a spring of heavier Nichrome V at the top to keep the wire at constant tension, the various temperature effects largely balance out, as shown below:

Reading, full scale, %	Temperature error, % per °C
20	+0.02
40	+0.02
60	+0.01
80	0
100	−0.04

Similar arrangements are described for transmitting mercury (20-D) and electrolyte (21-D) levels which are based on resistance changes in submerged wires.   Young and Taylor (29-D) describe the sensitive two-liquid manometer illustrated in Fig. 7. Brow and Schwertz (8-D) report a similar arrangement.

Bean and Morey (6-D) point out that a correction may be necessary for the density of the gas above the liquid level. Simmons (24-D) heats the air in one column to obtain a balance, wherein the temperature is used as a sensitive measure of pressure difference.   MacMillan (15-D) describes a manometer with a short time constant.

Minter (19-D) applies pressure to a bellows which is transferred to a sealed volume of helium and read as a change in thermal conductivity.   Hindley (13-D) uses two inverted bells suspended over oil from a torsion balance to obtain differential pressures

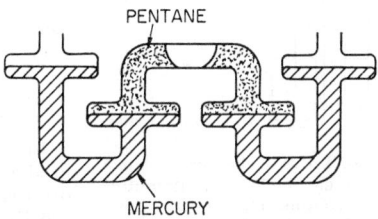

Fig. 7. Arrangement of manometer to effect high amplification in range 0.1 to 0.001 mm Hg.

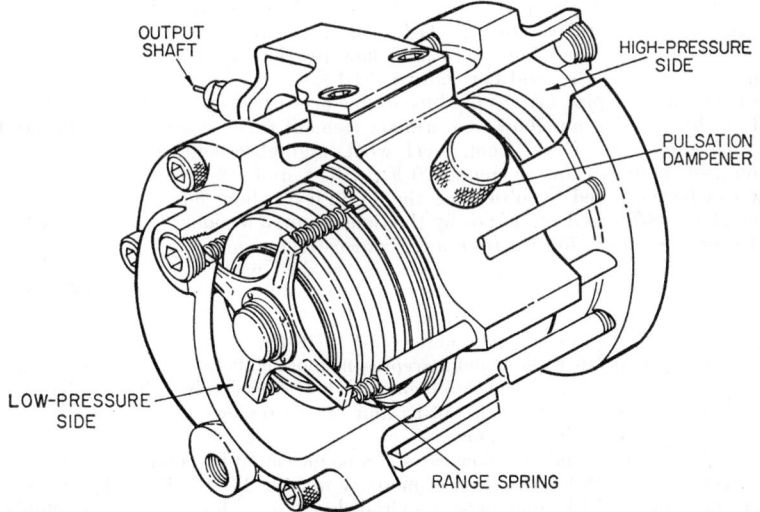

Fig. 8. Small differential-pressure unit.   (*Courtesy of Industrial Instrument Corp.*)

sensitive to 2×10^{-4} mm Hg.   A small differential-pressure unit is now commercially available.*   See Fig. 8.   Units of this type are available for pressures from 500 to 10,000 psi swp and measure some $2\frac{1}{2}$ by $4\frac{1}{2}$ in. long.

* Industrial Instrument Corp., Odessa, Tex.

Pressure Control

On-off regulation of pressure or vacuum can be obtained with an adjustable electric contact which operates through one arm of a U-tube manometer (1-E), (3-E), (12-E), and (8). Melpolder (6-E) describes a contact-making McLeod gage for low pressures. Usually pressure can be controlled so readily that the cycle due to on-off regulation is not bothersome. Proportional control can be obtained with a spring-loaded valve (8-E), a cartesian diver (4-E), or a manometer sealing a fritted glass disk (5-E). Figure 9 illustrates a chemical balance which has been modified to measure near-atmospheric pressures. Nisbet (9-E) controls such pressures by automatically adjusting the speed of an electric fan. Redhead and McNarry (13-E) control very low pressures from an ionization manometer.

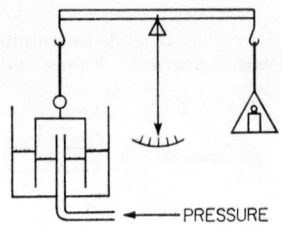

Fig. 9. Use of chemical balance to measure low gas pressures or volume change for flow-meter calibration.

FLOW MEASUREMENT

In the laboratory, flow is most commonly measured with a rotameter or inferred from the pressure drop created by a restriction in the flow line. Head (26-F) provides a correlation for rotameters. In small sizes, the usual orifice calculations (12-F) require special calibrations. Data are given for pipe sizes below 2 in. (50-F); for small orifices (22-F) and (54-F); for small nozzles (7-F); in the region of viscous flow (28-F), (33-F), and (41-F); and for low-pressure gases (10-F). Osburn and Kammermeyer report flow through small holes in thin plastic film (40-F). Critical velocity orifices are made from sapphire jewels (1-F) and (13-F). A length of pipe or tubing can be used as a flow restriction. If the flow is viscous, the pressure drop will respond to viscosity as well as to pressure. Charts and equations are presented for pipes (42-F) and glass capillary tubes (16-F), (25-F), (36-F), (44-F), (55-F), and (58-F). An adjustable glass stopcock can serve as a flow restriction with means to position it accurately for each of several flow ranges (53-F).

In nominal sizes, pipe fittings produce usable pressure drops (21-F) and (39-F); and Stoll (49-F) describes an elbow as a primary element which is sensitive to rate of flow in either direction. Glass wool, steel wool, and absorbent cotton provide linear flow elements (20-F), (24-F), and (48-F). Flow can be measured by observing the liquid level above an orifice (4-F), (8-F), and (27-F); or by the time required for ions (14-F) or ammonia (59-F) to appear downstream from a point of injection; or the emergent velocity of dust particles (9-F); or the saline content downstream after the addition of a known flow of brine (35-F). Ultrasonic (2-F), (3-F), (29-F), (31-F), and (51-F) and electromagnetic (17-F), (19-F), (30-F), (37-F), (45-F), and (46-F) flow meters represent more elaborate arrangements that provide measurement which is completely external to the pipe. Wildhack (56-F) and (57-F) reviews a number of flow-measuring systems.

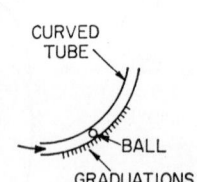

Fig. 10. Gradishar flow meter for corrosive fluids.

In the laboratory, small gas flow rates can be measured by counting bubbles (5-F), (6-F), (11-F), and (38-F), the slugs of mercury in a gas lift (61-F); or by the force of impact on a plate (18-F) and (52-F). Gradishar (23-F) describes a very simple gas flow meter which is comprised of a ball in a curved tube, as shown in Fig. 10. A similar arrangement for liquids is used for measuring flow in either direction (47-F).

Sakai and Sakai measure flow by the apparent change in solution conductivity due to liquid velocity between electrodes (43-F).

Hot-surface Flow Measurements

Fluid flowing past a heated surface results in cooling of this surface in proportion to the flow rate. Usually there are two heated elements in the fluid, where one element is shielded from the actual flow.   The temperature difference provides a flow measurement.   Hastings (4-G) describes an instrument of this type, now commercially available, which is based on heated thermistors.   Henry (5-G) describes thermistors in various circuit arrangements.   Hot-wire probes are described for measurements of low-water velocities (10-G) and compared with other types of thermally sensitive elements (3-G), (6-G), and (12-G).

If small thermistors or single wires are used, the response speed can be increased; even more rapidly by using circuits that keep the sensitive elements at a constant temperature and measure the power required to accomplish this.   Flow changes as fast as 10,000 cps can be followed in this manner.   This speed is required principally in studying turbulence rather than flow (1-G), (2-G), (7-G), (8-G), (9-G), and (11-G).

Flow Regulators

A very simple regulator is available from several suppliers* which consists of a flexible restriction that stretches to produce a smaller opening with increasing pressure drop.   These units are nonadjustable but are available for each of several specific flow

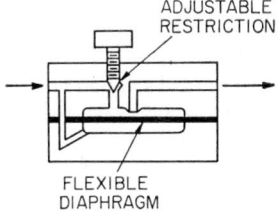

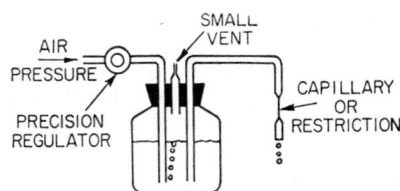

Fig. 11. Constant differential relay.

Fig. 12. Use of storage bottle to provide pressure drop across restriction that is independent of liquid level in container.

rates—from 0.2 to 10 gpm and for pressure drops from 10 to 150 psi.   Conway (2-H) describes the various mechanical arrangements which can be used to hold a constant pressure drop across a restriction.   Figure 11 shows the design of one form of a constant differential regulator.   The device can be purchased in brass or stainless-steel construction or readily made in the laboratory to withstand higher pressures or more corrosive fluids.   A manostat (4-H) or liquid column (1-H), (5-H), (12-H), (13-H) and (14-H) also can be used.

Figure 12 shows a controller for liquid flow rates.   A precision regulator supplies air or other gas at a controlled pressure which produces a constant pressure difference across the fixed-flow restriction.   By using the *Mariotte bottle* principle, the flow rate will be unaffected by changes of liquid level in the storage bottle.   The flow rate can be changed by adjusting the pressure regulator.   Other forms of this arrangement are described (6-H), (7-H), (8-H), (11-H), and (15-H).

Any of the previously described flow-measuring systems can be tied in with the actuation of a valve or pump.   Weiss (16-H) controls liquid flow by lowering a rod into a filled vessel.   Small, controlled flow rates of steam can be modulated electrically (3-H) or by reboiling a controlled volume of condensate (10-H).

Pumps

The laboratory worker who is mechanically inclined will be able to fabricate arrangements for pumping gases or liquids at low flow rates and at low pressures.   Flexible tubing can be squeezed (8-I); an oscillating column of mercury can serve as a piston (2-I), (4-I), (5-I), and (7-I); compressed air can operate a diaphragm (1-I); or a glass-enclosed plunger can be actuated by a solenoid (3-I) and (6-I).   However, except for very special purposes, it usually is more economical to purchase a commercially available pump.

* Bell and Gossett Co., Morton Grove, Ill.   Hays Manufacturing Co., Erie, Pa.

Reciprocating pumps for controlled-volume pumping of liquids consist of a mechanism which periodically changes the volume between the suction and discharge check valves. Such pump applications usually are limited to clean liquids. Reproducibility of flow to better than 1 per cent can be expected if the suction and discharge pressures and fluid properties remain constant. There are three main sources of error which can appreciably affect the volume of fluid pumped if the pressures or fluid properties do change:

1. Metal-to-metal seals leak more than soft-seat valves. The quantity of backflow varies directly with the pressure difference and the duration of the cycle, inversely with the fluid viscosity.

2. The spring rate of the system and the discharge suction-pressure difference determine the required piston or bellows travel between no suction-line flow and no discharge flow. A very small bubble of gas can have a large effect on this spring rate and the sensitivity of the flow rate to pressure change. Soft-seated check valves contribute to this error. After the suction flow stops, the ball or plug continues to squeeze the seal, with a corresponding volume change, until the discharge pressure is reached.

3. At moderate speeds—around 200 cycles per minute—the inertia of an unloaded ball prevents if from reseating until an appreciable interval after the flow reverses. This effect produces a pumping error in direct proportion to speed and to solution viscosity.

Thus, for most precise pumping where variable pressures and solution viscosities are present, the check valves should be spring-loaded and have soft seats with metal-to-metal stops to limit the compression of the seals.

Feeders

Most of the flow-regulating devices previously described could also be listed as liquid feeders. Several arrangements (7-J), (9-J), (11-J), (13-J), and (14-J) produce a constant pressure drop across a fixed-flow restriction. Motor-driven syringes provide very low liquid-feed rates (5-J) and (12-J). Barns (4-J) ratios one liquid stream in proportion to a second flow rate. Ashman (3-J) controls the electrolysis of water to obtain controlled feeds of oxygen and hydrogen. Several arrangements provide for the controlled addition of solids (1-J), (2-J), (6-J), and (8-J).

Valves

On-off laboratory shut-off valves for low pressures have been made of all-glass construction by sealing a magnetic core in glass and operating this with an external solenoid (1-K), (2-K), (9-K), and (12-K). To provide proportional control, valves are required that can throttle the flow in small increments. Control valves in a wide range of metals for corrosion resistance and for low flow ratings are commercially available. Miniature-size control valves of metal construction* and a throttling control valve of all-Teflon® construction† are available. Several throttling-valve designs are described which can be fabricated from glass (4-K), (6-K), (7-K), (11-K), or from metal and plastic materials (3-K), (5-K), and (8-K). Also see (13-K).

LEVEL MEASUREMENT

A sight glass or a manometer provide common methods for measuring liquid level. Illuminated glass beads (6-L) can be used to accentuate the meniscus. Prism-phototube arrangements provide high-, intermediate-, and low-level signals (1-L) and (12-L). Conducting probes can operate an array of neon lights (16-L). A spirit level on a float arm is sensitive to 1μ of level change (8-L). An all-glass hydraulic bridge circuit

---

* Research Controls Co., Tulsa, Okla.
† George W. Dahl Co., Bristol, R.I.

provides a high-level, low-level indication within a laboratory flask (7-L). The temperature of heated wires changes to provide level measurement and control (2-L) and (17-L).

The capacitance of a pair of electrodes provides a sensitive measure of the dielectric constant of the medium between them. This can be used to measure composition if

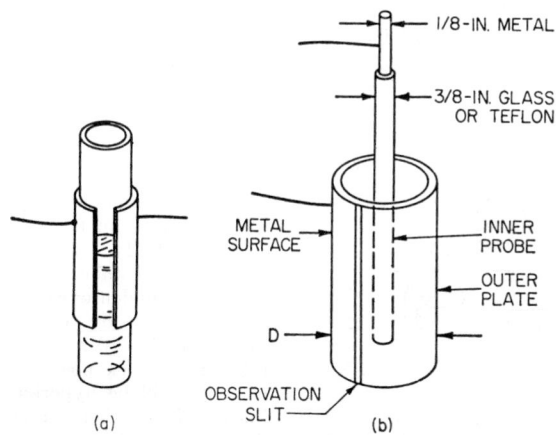

(a)   (b)

Fig. 13. Capacitance-type level detectors: (a) Parallel plate; (b) concentric plate.

the level remains constant or to measure level if the liquid composition does not change appreciably. Figure 13 shows two cell arrangements for level measurement in which essentially parallel plates and concentric plates are used. The concentric-plate arrangement has the advantage that it can be designed for a predictable capacity for given dimensions and a known material, as shown in Fig. 14. Advantages of the dielectric-capacitance type of measurement for laboratory use are: (1) the sensing element can be quite small; and (2) only glass is in contact with the process fluid. Such level measurements can be indicated, recorded, and/or controlled with conventional instruments.

Liquid level can be used to infer other process-variable measurements. Figure 15 shows an all-glass distillation unit with liquid-level sensing elements used to measure feed rate (flow) at A, pressure at B, differential pressure at C, level at D, and top-and-bottom take-off rates at E and F. In this manner, the operation can be completely controlled by using only glass equipment in contact with the process streams.

Broadhurst (3-L) uses an inductance coil around a tube to detect conducting liquids. Marzolf and Gardner (10-L)

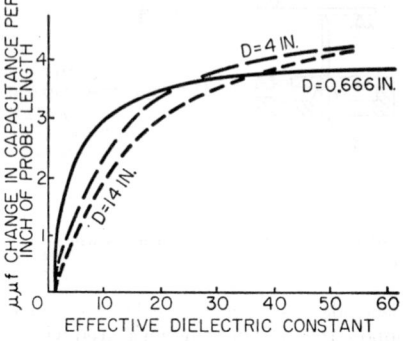

Fig. 14. Design data for determining capacity of concentric plate cells. (Courtesy of Fielden Instruments Div., Robertshaw-Fulton Controls Co.) Note: D is outer plate diameter as indicated in Fig. 13. Inner probe is ⅛-in.-diameter metal contained in a ⅜-in. glass tube.

compare inductance and capacitance measurements for two-liquid systems. Rod (13-L) describes sonic level measurement and Caywood (4-L) changes the volume of air above the stored liquid with a reciprocating diaphragm. The amplitude of the gas-pressure change is a measure of the gas volume and, by difference, a measure of

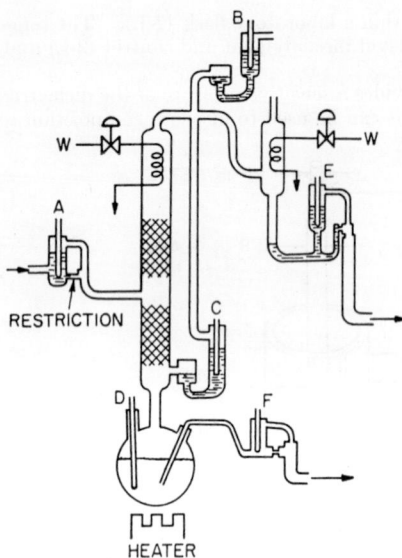

FIG. 15. Application of liquid-level detectors to measure flow, level, pressure, and differential pressure.

the liquid volume. Several constant-level float controllers are described (5-L), (9-L), and (11-L).

Spangler and Cooper (14-L) describe a simple vibrating reed for detecting the presence of liquids or solids. Also see (18-L).

POSITION MEASUREMENT

In the category of devices for detecting the position of rigid mechanical parts while offering negligible restraint to free motion, the differential transformer is probably the smallest and most versatile for motions from 0.01 to 10 in., (2-M), (6-M), (11-M), (14-M), and (15-M). For smaller distances, the motion of one plate of a condenser can provide a highly sensitive measure of position (3-M), (4-M), (7-M), (10-M), and (13-M). The relation between position and measured capacity will be hyperbolic rather than linear. For either method, transducer detecting, indicating, and recording equipment is available. The glow-discharge tube (1-M) and (8-M) has recently been applied to position measurement.

Two small, economical arrangements are available which are based on Wildhack's (16-M) variable-tension spring and the RCA 5734 miniature tube (12-M)—with an

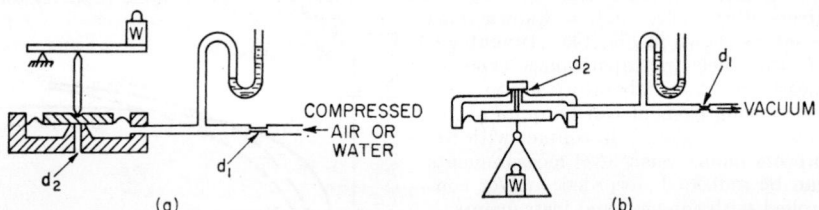

FIG. 16. Force-balance transmitters for measuring force or weight: (a) Unit in which compressed air or water is used as power source; (b) unit in which a vacuum is used as power source.

electrode that can be mechanically positioned externally. Both these transducers oppose the motion being measured with a small spring rate.

FORCE AND WEIGHT MEASUREMENT

Force-balance transmitters to satisfy many laboratory needs can be fabricated easily, as indicated in Fig. 16. Air, water, or vacuum can serve as the power source. The ratio of the restrictions d_2/d_1 should be roughly 2:1. As the diameter d_1 increases, more air is required to operate the system. As d_2 decreases, the response time, particularly for long transmission lines, suffers. For typical laboratory use, with short connections, it is almost impossible to drill the holes too small.

Chemical balances may be modified to provide a record of weight change (2-N), (9-N), and (11-N). Servosystems detect unbalance with a light-beam photocell arrangement; balance is restored with automatically controlled current to the coil of an electromagnet (3-N), (5-N), (6-N), and (10-N). Chemical balances may be modified to cut off flow at a predetermined sample weight (8-N) and (14-N). Pedersen (12-N) controls flow from a storage tank placed on a scale.

A chemical balance may be adopted to measure and control a number of measurements in the laboratory, as indicated in Figs. 2 and 9. This device will respond to any variable pressure, density, viscosity, and level that can be converted into force or weight. The measurement can be controlled directly and can be transmitted by using a position or force-measuring transducer.

DENSITY MEASUREMENT

Gas-density instruments are reviewed by Smith and others (11-P). In the laboratory, changes in gas density usually are measured by the change in buoyancy of a suspended bob. The bob may be supported by a flexible glass tube (13-P) or from a chemical balance (9-P) and (10-P)—and the balance repositioned with an electromagnet (6-P).

Liquid densities also are commonly measured by submerged floats.* The position of the float is a measure of the density and is transmitted by a differential transformer. Similar arrangements are described for a heavy ball (12-P) and for floats which are repositioned by the current in a coil surrounding the float (4-P) and (7-P).

A bubbler system detects the difference in pressure at two levels of a liquid, thus providing a measure of density (8-P). Modifications for zero suppression (2-P) and for temperature compensation (1-P) can be made.

VISCOSITY MEASUREMENT

Several reviews of viscometers for liquids have appeared (2-Q), (6-Q), (9-Q), and (11-Q). Merrington (8-Q) covers viscosity measurements for both liquids and gases. The viscosity of liquids can be measured by the pressure required to produce constant flow through a tube (12-Q) or by the power to rotate a paddle (1-Q), (5-Q), and (7-Q). The dampening of an oscillating cylinder (4-Q) and (10-Q) or sphere (3-Q) can be used to measure viscosity. A thin probe† which oscillates at high frequency also provides continuous viscosity measurement.

INDEX OF REFRACTION AND OTHER MEASUREMENTS

Refractometers suitable for continuous measurements on flowing liquid streams are described (8-R), (10-R), and (11-R). Somewhat higher stability and sensitivity can be obtained from a differential refractometer in which an unknown sample is compared with a reference material. Continuous refractometers for liquids (2-R), (6-R), and (12-R); in explosionproof housings (9-R); and for gases (1-R) and (3-R) are reported.

Colthup and Torley (4-R) measure surface tension by the pressure required to force air through a submerged nozzle, as shown in Fig. 17. Surface tension is also measured by bubble length (5-R) and by the lowest flow rate of a vertical jet producing a changed flow pattern (7-R).

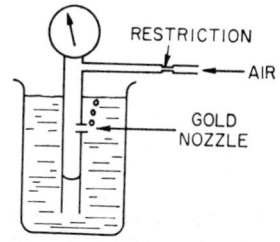

FIG. 17. Apparatus for surface-tension measurement.

Electrodes for pH, made from low-resistance glass, are described (17-R), for pressures up to 75 psi (14-R), and for fused salts (15-R). Arrangements are provided to wipe electrode surfaces continuously to keep them clean (13-R) and (16-R).

---

* Densitrol, Precision Thermometer Co., Philadelphia, Pa.
† Ultra-Viscoson, Bendix Aviation Corp., Cincinnati, Ohio.

CONDUCTIVITY AND DIELECTRIC CONSTANT

Conductivity measurements usually are made with 60 cps or 1,000 cps alternating instead of direct current to avoid polarization difficulties. Figure 18 shows a practical and easily constructed arrangement which operates directly from regular power supply. This system gives high sensitivity and requires a minimum of apparatus. Stainless-steel or copper strips can serve as electrodes provided the solution is not too corrosive. A number of circuits which use simplified or improved arrangements are described (4-S), (16-S), (18-S), (19-S), and (21-S). A thermistor can be used for temperature compensation (1-S).

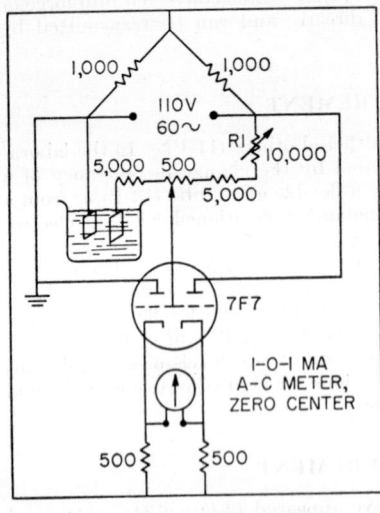

FIG. 18. Circuit for measurement of electrical conductivity at 60 cps.

As the frequency of the power supply is increased, the electrodes can be placed external to the sample container, providing a measure of the apparent dielectric constant of the condenser. It is found that this measurement depends on both the d-c conductivity of the material and the frequency of the power supply. However, for a given frequency, a useful measurement is obtained. The advantage is that relatively simple equipment which remains free from direct contact with the process stream and which is very sensitive to some composition changes can be used.

The philosophy of high-frequency measurements and appropriate circuits is reviewed (3-S), (5-S), (6-S), (7-S), (9-S), (10-S), (11-S), (15-S), (17-S), and (22-S). Circuit designs are considered (2-S) and the application to solids and gases (12-S).

The method has been applied to measure brine (14-S), toluene (20-S), water (24-S), and solids (8-S) and (13-S), and to distinguish between gasoline grades (23-S).

ANALYZERS

Automatic titrators are of two types. The first step in automation is to replace the manual operation of adding the reagent to a definite end point (19-T) and (20-T). Lingane (30-T) describes a multipurpose instrument. Neilands and Cannon (33-T) follow a reaction by titrating at constant pH. The continuous titrator controls the flow of unknown and measures the flow of reacting agent added to hold a constant end point (5-T). This is most easily applied to testing gases and generating the titrating agent electrically (3-T), (16-T), (25-T), (34-T), (35-T), (38-T), and (39-T).

Composition also can be inferred by measuring some physical property. If the measurement is to provide analytical data, care needs to be taken to eliminate or compensate for the effects of materials other than the unknown being measured. If the measurement is to actuate an automatic controller to maintain a physical property at some constant value, however, complete independence from the effects of more than one constituent is unnecessary. For laboratory process control, any of the following techniques can be used in the simplest forms, with no particular effort being required to make the measurement selective for a particular component.

As nonselective measurements, these operations become equivalent to and interchangeable with the previously described process variables.

The laboratory worker is in a particularly fortunate position, in that, to provide automatic control he can use any of the measuring techniques developed for either analysis or plant control. And he requires neither the specificity of the analytical method nor the ruggedness of plant equipment.

The color of a liquid or gas can be measured (4-T), (6-T) and (27-T)—even in a capillary tube (14-T). One device* provides a measure of suspended solids, and fluorescence (31-T) can be applied to a continuously flowing liquid. Conductivity measurements are easy to apply; thermal conductivity for gases (32-T) and (38-T); and electrical conductivity for liquids (22-T). Figure 19*a* shows an enhancement technique in which a soluble portion of the unknown gas stream transfers to a parallel flow of liquid. The electrical conductivity of the resulting liquid becomes a measure of the desired constituent in the gas phase (40-T). Similarly, the thermal conductivity of the gas phase in Fig. 19*b* provides a measure of a volatile component, e.g., ammonia, in the unknown solution. Polarography for continuously flowing liquid samples is discussed in (23-T) and (24-T).

Gases can be analyzed by measuring the velocity of sound (8-T) and (26-T); known (21-T) or critical (41-T) flow rates; or rate of diffusion (15-T). Temperature rise can be used to measure the oxidation of carbon monoxide by Hopcalite (29-T), or absorp-

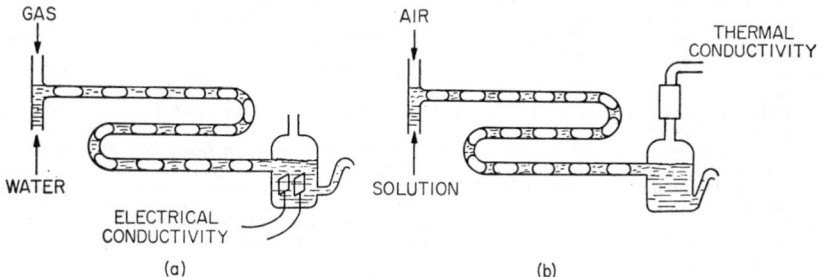

Fig. 19. Phase-transfer and enhancement arrangement for single property measurement: (*a*) Gas analysis; (*b*) liquid analysis.

tion of water vapor in CaH_2 (18-T). The amount of light produced by a continuous flow of gas through an electrical discharge tube measures nitrogen content (28-T) or, at low pressures, water (22-T). Ions produced by radiation vary with gas composition (1-T) and (11-T).

ELECTRIC CIRCUITS

Special-purpose circuits to control voltage (10), (40), and (50); current (80) and (10-U); and to shut off titrations (3-U) and (7-U) using transistors (9-U) are described. Sivertsen (11-U) describes an electrolytic current-integrating arrangement. Relatively simple magnetic amplifier circuits can drive reversible motors in servosystems (6-U) and (12-U).

REFERENCES

This extensive list of references is classified for maximum convenience. Each reference is tied into the foregoing text by the numbers given.

Laboratory Instrumentation—General

1. Anon.: How to Pilot-plant in the Lab, *Chem. Week*, vol. 75, no. 9, pp. 66, 68, 70, Aug. 28, 1954.
2. Berg, Clyde, and others (a panel discussion): Instrumentation for Pilot Plants, *Ind. Eng. Chem.*, vol. 45, no. 9, pp. 1836–1852, September, 1953.
3. Beroza, M.: Small Glass Circulating Evaporator for Laboratory Use, *Anal. Chem.*, vol. 26, no. 7, pp. 1251–1252, July, 1954.
4. Bosworth, R. C. L.: Dimensional Methods in the Design of Industrial Chemical Research, *J. Proc. Roy. Soc. N. S. Wales*, vol. 81, pp. 15–23, 1947.

* Turbidimeter, General Electric Company, Schenectady, N.Y.

5. Brandt, F. A., and Floyd Brown: Simple Device for Converting an Automatic Recorder into a Flexible Program Controller, *Rev. Sci. Instr.*, vol. 26, no. 11, p. 1077, November, 1955.
6. Bukala, V. M., J. Majewski, and A. Semkowicz: Automatization of Laboratory Fractional-distillation Process, *Przemysl Chem.*, vol. 9, pp. 224–228, 283–287, 1953.
7. Cannon, W. A.: An Automatic Mercury Vacuum Still, *J. Chem. Educ.*, vol. 28, no. 6, pp. 272–274, June, 1951.
8. Carney, T. P.: "Laboratory Fractional Distillation," The Macmillan Company, New York, 1949.
9. Ceaglske, N. H.: "Automatic Process Control for Chemical Engineers," John Wiley & Sons, Inc., New York, 1956.
10. Claff, C. Lloyd: Magnetic-flea Agitating Device for Micro-titration, *Science*, vol. 108, p. 67, 1948.
11. Eckman, D. P.: "Principles of Industrial Process Control," John Wiley & Sons, Inc., New York, 1945.
12. Eckman, D. P.: "Industrial Instrumentation," John Wiley & Sons, Inc., New York, 1953.
13. Friden, J. E., and T. G. Stack: A Laboratory for Pilot-scale Distillation, *Chem. Eng. Progr.*, vol. 50, no. 3, p. 151, March, 1954.
14. Highet, H. C.: Automatic Continuous Laboratory Still, *Chem. Ind.* (*London*), Dec. 9, 1950, pp. 783–786, N49.
15. Holzbock, Werner G.: "Instruments for Measurement and Control," Reinhold Publishing Corporation, New York, 1955.
16. Johnson, Ernest F.: The Use of Frequency Response Analysis in Chemical Engineering Process Control, *Chem. Eng. Progr.*, vol. 52, no. 2, p. 64F, February, 1956.
17. Jordan, Donald G.: "Chemical Pilot Plant Practice," Interscience Publishers, Inc., New York, 1955.
18. Kaufman, E. N.: Controlling Potentiometer, *Rev. Sci. Instr.*, vol. 23, no. 7, p. 385, July, 1952.
19. Levinger, J. S.: Nozzle-baffle Force Balance Control System, *U.S. Commerce Dept.*, *Bibliography of Sci. and Ind.*, *Repts*, 2, July 5, 1946, Sept. 27, 1946.
20. Lloyd, L. E., and H. G. Hornbacher: Constant-ratio Still Head, *Anal. Chem.*, vol. 19, pp. 120–123, 1947.
21. McIntosh, C. H.: Specialized Recorders and Controllers on Pilot Units, *Ind. Eng. Chem.*, vol. 45, no. 9, pp. 1849–1852, September, 1953.
22. Michel, Laurent, Robert D. Beattie, and T. H. Goodgame: Census of Equipment Scale-up Practice, *Chem. Eng. Progr.*, vol. 50, pp. 332–335, 1954.
23. Michell, D.: Magnetic Stirrer for Vacuum Work, *J. Appl. Chem.* (*London*), vol 1, p. S8–9, suppl. issue 1, 1951.
24. Muller, R. H.: Instrumentation, *Anal. Chem.*, vol. 27, pp. 49A–52A, November, 1955.
25. McMahon, J. B., and R. A. Ackley: How Processes Effect Control, *Chem. Eng.*, vol. 59, no. 12, p. 191, December, 1952.
26. Nadir, Mark T.: "Transducers, A Compilation of Analog Devices of Oscillography," 2d ed., DuMont Laboratories, Inc., Clifton, N.J., 1953.
27. Nickels, J. E.: Automatic Trap for Laboratory Steam Distillation, *J. Chem. Educ.*, vol. 26, p. 533, 1949.
28. Pierce, J. E., Put Dimensional Analysis to Work, *Chem. Eng.* vol., 61, no. 4, pp. 185–190, April, 1954.
29. Pratt, Ernest F.: Adjustable Liquid Separator, *Anal. Chem.*, vol. 19, p. 288, 1947.
30. Rockley, C. W., and C. W. Shelley: Bench Scale Cracking Unit Reproduces Plant Conditions, *Petrol. Refiner*, vol. 31, no. 6, pp. 89–92, June, 1952.
31. Roth, E. R., and G. S. Masologites: Recorders and Controllers in Pilot Unit Instrumentation, *Ind. Eng. Chem.* vol. 45, no. 9, p. 1836, September, 1953.
32. Roth, E. R.: Automatic Instrumentation for Bench Scale Units, *Ind. Eng. Chem.*, vol. 46, no. 7, p. 1428, July, 1954.
33. Shrader, R. E., and E. J. Wood: Automatic Control of Stills, *Electronics*, vol. 17, no. 9, p. 98, September, 1944.
34. Stenger, F., and R. M. Ancell: Automatic Controls, A Bibliography Compiled, *U.S. Commerce Dept.*, *Bibliography of Tech. Repts.*, 17–18, January–December, 1952.
35. Tait, T. R., and J. T. Mitchell: Coupling Glass Tubing to Other Materials, *Ind. Eng. Chem.*, vol. 45, no. 1, pp. 239–240, January, 1953.
36. Thorp, A. G., II: Flexure Pivots—Design Formulas and Charts, *Prod. Eng.*, vol. 24 pp. 192–200, February, 1953.
37. Tolbert, B. M., W. C. Dauben, and James C. Reid: Laboratory Induction Stirrer for Closed Systems, *Anal. Chem.*, vol. 21, no. 8, pp. 1014–1015, 1949.
38. Trevoy, D. J., and W. A. Torpey: Shaft Seals for Vacuum Apparatus, *Anal. Chem.*, vol. 24, no. 8, p. 1382, 1952.
39. Walter, L.: Controllability and Plant Design, *Chem. Age*, vol. 65, no. 1680, pp. 391–395, Sept. 22, 1951.
40. Wilkinson, W. R., and H. A. Beatty: Proportion-flow Controller for Liquids, *Ind. Eng. Chem.*, *Anal. Ed.*, vol. 18, pp. 725–726, 1946.
41. Willard, Hobart H., Taft Y. Toribara, and Lewis N. Holland: Electronically Controlled Apparatus for Distillation of Fluorine as Hydrofluosilicic Acid, *Anal. Chem.*, vol. 19, no. 5, pp. 343–344, 1947.
42. Witherspoon, J. E.: Electronic-recorder Pickups, *Instruments*, vol. 26, no. 3, pp. 429–431, March, 1953.
43. Young, A. J.: Automatic Control of Chemical Processes, *Instruments*, vol. 26, pp. 254–255, 292–294, 1953, and subsequent issues.
44. Young, A. J.: "An Introduction to Process Control System Design," Instruments Publishing Co., Pittsburgh, Pa., 1955.
45. Ziegler, J. G., and N. B. Nichols: Industrial Process Control, *Chem. Eng. Progr.*, vol. 43, no. 6, pp. 309–314, June, 1947.

46. Hall, George A., Jr.: Fundamentals of Automatic Control, Sec. 11-1 of "Process Instruments and Controls Handbook," McGraw-Hill Book Company, Inc., New York, 1957.
47. Higgins, Stephen P., Jr.: Mathematical Techniques for Solving Automatic Control Problems, Sec. 11-2 of "Process Instruments and Controls Handbook," McGraw-Hill Book Company, Inc., New York, 1957.

Temperature Measurement

1-A. Anon.: Temperature Indicator, *Rev. Sci. Instr.*, vol. 23, no. 2, p. 104, February, 1952.
2-A. Anon.: Silver and Gold Wires: Thermoelectric Power, *Appl. Sci. Res.*, Sec. 4B, no. 5, pp. 388–392, 1955.
3-A. Aoki, Ikno: A Thermometer Using a Thermistor Element, *J. Coll. Arts Sci. Chiba Univ.*, vol. 1, pp. 22–25, 1952.
4-A. Barber, R., R. Jackson, T. Land, and G. G. Thurlow: A Suction Pyrometer for Measuring Gas Exit Temperatures from the Combustion Chambers of Water-tube Boilers, *Appl. Mech. Rev.*, vol. 8, no. 1, p. 314, January, 1955.
5-A. Baruch, J. J., J. C. Livengood, et al.: Ultrasonic Temperature Measurements, *J. Acoust. Soc. Am.*, vol. 26, no. 5, pp. 824–830, September, 1954.
6-A. Barber, C. R., and L. H. Pemberton: Silver-Palladium Thermocouples, *J. Sci. Instr.*, vol. 32, no. 12, p. 487, December, 1955.
7-A. Bennett, F. E.: Which Type Thermistor? *Prod. Eng.*, vol. 25, no. 3, pp. 182–185, March, 1954.
8-A. Bennett, F. E.: How to Use Thermistors, *Prod. Eng.*, vol. 25, no. 4, pp. 200–204, April, 1954.
9-A. Beede, H. M., and C. R. Droms: Simplified Thermocouple for Temperature Measurements in High Velocity Gas Streams, *Instruments*, vol. 24, no. 3, pp. 338–341, March, 1951.
10-A. Bendersky, D.: Special Thermocouple for Measuring Transient Temperatures, *Mech. Eng.*, vol. 75, no. 2, pp. 117–121, February, 1953.
11-A. Benedict, R. P.: Thermistors vs. Thermocouples for Temperature Measurements, *Elec. Mfg.*, vol. 54, no. 2, pp. 120–125, 1954.
12-A. Berry, J. M., and D. L. Martin: Thermocouple Immersion Errors, *ASTM Preprint 95*, pp. 1–8, 1955.
13-A. Bruce, Stanley C.: Thermometer Calibration for Determination of Capillary Melting Points, *Anal. Chem.*, vol. 25, no. 5, p. 825, May, 1953.
14-A. Burton, E. J., and D. J. Weeks: Temperature Measurement Progress Review No. 27, *J. Inst. Fuel*, vol. 26, pp. 260–273, 1953.
15-A. Carbon, M. W., H. J. Kutsch, and G. A. Hawkins: The Response of Thermocouples to Rapid Gas-temperature Changes, *Trans. ASME*, vol. 72, pp. 655–657, 1950.
16-A. Cichelli, Mario T.: Design of Temperature Measuring Elements, *Ind. Eng. Chem.*, vol. 40, pp. 1032–1039, 1948.
17-A. Clark, J. A., and W. M. Rohsenow: New Method for Determining Static Temperature of High-velocity Gas Streams, *Trans. ASME*, vol. 74, no. 2, pp. 219–228, February, 1952.
18-A. Clement, J. R., and E. H. Quinnell: Low Temperature Characteristics of Carbon-composition Thermometers, *Rev. Sci. Instr.*, vol. 23, no. 5, pp. 213–216, May, 1952.
19-A. Cole, P. J.: Nomograph for Thermometer Temperature Correction, *Anal. Chem.*, vol. 22, pp. 946–947, 1950.
20-A. Curcio, J. A., and D. V. Estes: Photoelectric Pyrometer for the Measurement of Luminous Flame Temperatures, *U.S. Commerce Dept. Bibliography of Tech. Repts.*, vol. 19, p. 16, January–June, 1953.
21-A. Daish, C. B., D. H. Fender, and A. J. Woodall: Temperature Measurement, *Phil. Mag.*, vol. 41, pp. 729–730, 1950.
22-A. Doucet, Y.: Thermistors: Use, *J. Phys. Radium*, vol. 12, no. 8, p. 823, October, 1951.
23-A. Fiock, E. F., and A. I. Dahl: The Measurement of Gas Temperatures by Immersion-type Thermometers, *J. Am. Rocket Soc.*, vol. 23, pp. 155–164, 1953.
24-A. Forch, J. H.: A Simple Device for Direct Indication of Small Temperature Differences between Reactor Inlet and Exit, *Chem. Week*, vol. 47, pp. 259–260, 1951.
25-A. Fuschillo, N.: Inhomogeneity Electromotive Forces in Thermoelectric Thermometers, *J. Sci. Instr.*, vol. 31, no. 4, pp. 133–136, April, 1954.
26-A. Godridge, A. M., R. Jackson, and G. G. Thurlow: A Small Pneumatic Pyrometer, *J. Sci. Instr.*, vol. 32, pp. 279–282, July, 1955.
27-A. Goldsmith, A.: Design Study of True Free Air Thermometer, *U.S. Commerce Dept. Bibliography of Tech. Repts.*, 17, 18, January–December, 1952.
28-A. Greenhill, E. G., and J. R. Whitehead: Apparatus for Measuring Small Temperature Changes in Liquids, *J. Sci. Instr.*, vol. 26, pp. 92–95, 1949.
29-A. Guthmann, K.: Temperature Measurement in High Temperature Furnaces, *Chem. Eng. Tech.*, vol. 25, no. 4, pp. 169–176, April, 1953.
30-A. Guthmann, K.: Toleranzen und Fehler bei der Temperatur-messung mit Thermoelementen, *Arch. Eisenhuettenw.*, vol. 25, no. 11/12, pp. 535–561, November–December, 1954.
31-A. Hayes, E. W.: Resistance Thermometry, *Instruments*, vol. 25, no. 5, pp. 629–631, May, 1952.
32-A. Heidmann, M. F., and R. J. Priem: Application of an Electro-optical Two-color Pyrometer to Measurement of Flame Temperature for Liquid Oxygen-hydrocarbon Propellant Combination, *U.S. Commerce Dept. Bibliography of Tech. Repts.*, vol. 21, no. 1, p. 8, January, 1954.
33-A. Hogue, E. W.: Determination of Temperature with Noise Thermometer, *Natl. Bur. Std. (U.S.) Rept. 1683*, April, 1952.
34-A. Jackson, R.: The Measurement of Gas Temperature by Thermocouples, *Bull. Brit. Coal Util. Res. Assoc.*, vol. 14, pp. 33–39, 1950.

35-A. Jewell, R. C., E. G. Knowles, et al.: High Temperature Thermocouple, *Metal Ind. (London)*, vol. 87, no. 11, pp. 217–221, Sept. 11, 1955.

36-A. Joy, A. S.: The Construction and Filling of Glass Vapor-pressure Thermometers, *Chem. Ind. (London)*, no. 36, pp. 874–875, Sept. 6, 1952.

37-A. Kalbfell, D. C.: Rapid-response Thermometer, *U.S. Commerce Dept. Sci. and Ind., Repts.*, vol. 3, p. 400, October–December, 1946.

38-A. Land, T., and R. Barber: The Design of Suction Pyrometers, *Trans. Soc. Instr. Technol. (London)*, vol. 6, no. 3, pp. 112–130, September, 1954.

39-A. LeLacheur, R. M.: Low Range Decade for Resistance Thermometer Bridges, *Rev. Sci. Instr.*, vol. 23, no. 7, p. 383, July, 1952.

40-A. Lieneweg, Fritz: Determination of Temperature Measurement Errors by Means of Thermometer Constants, *Festschr. 100 Jaehr. W. C. Heraeus GmbH Hanau*, 1951, pp. 243–268.

41-A. Lieneweg, Fritz, and Hubert Vanvor: Rapid-response Resistance Thermometers for Wide Temperature Ranges, *Chem. Eng. Tech.*, vol. 27, pp. 309–312, 1955.

42-A. Moffatt, E. Marston: Methods of Minimizing Errors in the Measurement of High Temperature in Gases, *Instruments*, vol. 22, pp. 122–132, 1949.

43-A. Mayorcas, R.: Gas Temperature Measurement above 1,500°, *J. Inst. Fuel*, vol. 22, pp. 251–255, 1949.

44-A. McLean, J. A.: A Method for Constructing Direct Reading Thermistor Thermometers, *J. Sci. Instr.*, vol. 31, p. 455, December, 1954.

45-A. Millar, G. H., et al.: Fast, Electro-optical Hot-gas Pyrometer, *J. Opt. Soc. Am.*, vol. 43, no. 7, pp. 609–617, July, 1953.

46-A. Misener, A. D., and L. G. D. Thompson: The Pressure Coefficient of Resistance of Thermistors, *Can. J. Tech.*, vol. 30, no. 4, pp. 89–94, April, 1952.

47-A. Muller, Ralph H., and Hans J. Stolten: Use of Thermistors in Precise Measurement of Small Temperature Differences. Thermometric Determination of Molecular Weights, *Anal. Chem.*, vol. 25, pp. 1103–1106, 1953.

48-A. Pearson, W. B.: Thermocouples for Use at Low Temperature, *J. Sci. Instr. (London)*, vol. 31, no. 12, p. 444, December, 1954.

49-A. Pohl, Herbert A.: Supersensitive Thermoelements, *Rev. Sci. Instr.*, vol. 22, p. 345, 1951.

50-A. Potter, R. D.: Thermocouples: Calibration, *J. Appl. Phys.*, vol. 25, no. 11, pp. 1383–1384, November, 1954.

51-A. Potter, R. D., and N. J. Grant: Tungsten-molybdenum Thermocouples, *Iron Age*, vol. 163, no. 13, pp. 65–69, 1949.

52-A. Richards, L. A., and R. B. Campbell: Use of Thermistors for Measuring the Freezing Point of Solutions and Soils, *Soil Sci.*, vol. 65, pp. 429–436, 1948.

53-A. Seay, P. A., and W. E. Gordon: Temperature Recording with Thermistors, *U.S. Commerce Dept. Bibliography of Tech. Repts.*, 14, July–December, 1950.

54-A. Shepard, Charles E., and Isidore Warshawsky: Electrical Techniques for Compensation of Thermal Time Lag of Thermocouples and Resistance Thermometer Elements, *U.S. Commerce Dept. Bibliography of Tech. Repts.*, 17, 18, January–December, 1952.

55-A. Shehukin, P. A., and L. V. Pegushina: New Construction of a Tungsten-graphite Thermocouple, *Zavodsk. Lab.*, vol. 14, pp. 632–633, 1948.

56-A. Sias, F. R., J. R. Macintyre, and Albert Hansen, Jr.: A Tungsten Resistance Thermometer, *Elec. Eng.*, vol. 73, p. 442, 1954.

57-A. Simpson, T. B., and C. C. Winding: Properties of Evaporated Metal Films Related to Their Use for Surface Temperature Measurement, *Am. Inst. Chem. Engrs. J.*, vol. 2, no. 1, p. 113, March, 1956.

58-A. Stillwell, G. K.: The Evaluation of the Errors Involved in the Measurement of Surface Temperature with Thermocouples Affixed to the Surface by Adhesive Tape, *Dissertation Abstr.*, vol. 14, no. 6, p. 889, June, 1954.

59-A. Sutter, D. M., and J. E. Brock: Construction of Thermocouples, *Proc. Indiana Acad. Sci.*, vol. 63, pp. 266–268, 1954.

60-A. Strong, H. M., and F. P. Bundy: Flames: Temperature Measurement, *J. Appl. Phys.*, vol. 25, no. 12, pp. 1521–1526, 1527–1530, December, 1954.

61-A. Thomas, W. A., and R. J. Horval: Measurement of Induction Motor Rotor Temperatures, *Instruments*, vol. 24, no. 4, p. 410, April, 1951.

62-A. Troy, Walter C., and Gary Steven: High Temperature Thermocouple, *The Frontier*, vol. 12, no. 4, pp. 6–8, 22–24, 1949.

63-A. Vos, A. S.: Methods of Temperature Measurement, and the Use of Thermistors (Thermally Sensitive Resistors) as a New Possibility for This Purpose, *Chem. Weekblad*, vol. 49, pp. 68–76, 1953.

64-A. West, W. E., Jr., and J. W. Westwater: Radiation-conduction Correction for Temperature Measurement in Hot Gases, *Ind. Eng. Chem.*, vol. 45, pp. 2152–2156, 1953.

65-A. White, A. G.: A Note on the Transistor as a Thermometer, *J. Sci. Instr.*, vol. 32, no. 11, pp. 451–452, November, 1955.

66-A. Zeffert, B. M., and Saul Hormets: Application of Thermistors to Cryoscopy, *Anal. Chem.*, vol. 21, pp. 1420–1422, 1949.

Temperature Control

1-B. Anon.: Electronic Temperature Controller, *J. Elec. Chem. Soc.*, vol. 98, no. 8, p. 118C, August, 1951.

2-B. Anon.: Laboratory Oven Temperature Control, *Electronics*, vol. 17, no. 10, p. 108, 1944.

3-B. Anon.: Precision Thermostat for High Temperature, *Instr. Practice*, vol. 6, no. 10, pp. 701–702, August, 1952.
4-B. Anon.: Temperature Controller, *Chem. Process.*, vol. 15, no. 6, pp. 104–105, June, 1952.
5-B. Benedict, Mason: Use of an Alternating-current Bridge in Laboratory Temperature Control, *Rev. Sci. Instr.*, vol. 8, pp. 252–254, August, 1937.
6-B. Anon.: Temperature Controller Controls within 0.001°C, *Insts. and Automation*, vol. 28, p. 1726, November, 1955.
7-B. Blauvelt, L. C.: Temperature Regulators, *Proc. Eng.*, vol. 24, no. 7, pp. 172–173, July, 1953.
8-B. Brown, B. Floyd: A Single Temperature-controller Adapter for Controlled Heating and Cooling Rates, *N. Carolina State Coll. Eng. School Bull.* 3, pp. 2–4, 1953.
9-B. Brown, John A.: A Thermoregulator for Gas-heated Baths, *J. Chem. Educ.*, vol. 28, no. 9, p. 465, September, 1951.
10-B. Burwell, Robert L., Axel H. Peterson, and George B. Rathman: Temperature Control Device Employing Thermistors and a Saturable Reactor, *Rev. Sci. Instr.*, vol. 19, pp. 608–609, 1948.
11-B. Coops, J., K. VanNes, A. Kentie, and J. W. Dienske: Heat of Combustion V Correction for Heat Exchange between the Calorimeter Vessel and Its Surroundings, *Rec. Trav. Chim.*, vol. 66, pp. 113–130, 1947.
12-B. Crandall, W. B., M. Burzycki, and V. D. Frichette: Improved Roberts Control for Laboratory Furnaces, *Am. Ceram. Soc. Bull.* 29, pp. 87–89, 1950.
13-B. Doucet, Y.: Thermistors: Use, *J. Phys. Radium*, vol. 12, no. 8, p. 823, October, 1951.
14-B. Dymott, E. R., and J. C. Evans: Constant Temperature Water Apparatus, *J. Sci. Instr.*, vol. 28, no. 12, pp. 369–370, December, 1951.
15-B. Firdman, R. A., and V. S. Pellinets: Automatic High-frequency Thermostat, *Zavodsk. Lab.*, vol. 14, pp. 1139–1141, 1948.
16-B. Galloway, K. A.: Laboratory Apparatus: Temperature Control, *Lab. Pract.*, vol. 4, no. 7, pp. 283–286, July, 1955; no. 8, pp. 334–338, August, 1955.
17-B. Gingburg, S. A.: An Automatic Thyratron Temperature Regulator for Electric Laboratory Heaters, *Zavodsk. Lab.*, vol. 15, pp. 369–372, 1949.
18-B. Henschen, Homer, E., James W. Edwards, and Herrick L. Johnston: Temperature Regulation in Induction Heating, *Rev. Sci. Instr.*, vol. 22, no. 12, p. 987, December, 1951.
19-B. Hluchan, S. A.: Process Instrumentation for Crystal Growing Equipment, *Instruments*, vol. 24, no. 5, pp. 585–590, May, 1951.
20-B. Holms, Frederick E.: A Self-regulating Electrolytic Immersion Heater, *J. Chem. Educ.*, vol. 24, pp. 166–168, 1947.
21-B. Horowitz, J.: Temperature Regulation of Electric Ovens, *Glass Ind.*, vol. 36, no. 3, pp. 140–144, March, 1955.
22-B. Lacy, E. D.: Modified Thermostatic Controller, *J. Sci. Instr.*, vol. 29, no. 3, p. 96, 1952.
23-B. Malmberg, Paul R., and Carl C. Matland: Thermistor Temperature Control, *Phys. Rev.*, vol. 93, p. 1249, 1954.
24-B. McKinney, Alfred H.: Using Pneumatic Controllers to Control Electric Heat, *Chem. Eng.*, vol. 57, no. 5, pp. 164–165, 1950.
25-B. Mouzon, J. C.: A Simple Temperature Controller, *Rev. Sci. Instr.*, vol. 19, no. 10, pp. 659–662, 1948.
26-B. Muller, Ralph H.: Instrumentation, *Anal. Chem.*, vol. 19, no. 10, pp. 21A–22A, 1947.
27-B. Noltingk, B. E., and M. A. Snelling: An Electronic Temperature Controller, *J. Sci. Instr.*, vol. 30, no. 10, pp. 349–351, October, 1953.
28-B. Osterlof, J., and S. Olsson: Thermostat, *Acta Chem. Scand.*, vol. 5, no. 4, pp. 626–629, 1951.
29-B. Proctor, Charles M.: An Inexpensive Thyratron Controlled Thermostat Relay, *Rev. Sci. Instr.*, vol. 22, nol 12, pp. 1023–1024, December, 1951.
30-B. Pitha, J. J.: Electronic Relay, *J. Chem. Educ.*, vol. 28, no. 8, p. 429, August, 1951.
31-B. Popjak, G., and M. L. Beeckmans: Water Bath Suitable for the Maintenance of Temperatures to Within 1.5 × 10⁻³ C°, *Biochem. J.*, vol. 46, pp. 558–560, Appendix I, 1950.
32-B. Simon, A., and R. Schrader: Temperature Controller for an Electric Furnace, *Z. anorg. allgem. Chem.*, vol. 266, nos. 4–5, pp. 208–215, November, 1951.
33-B. James, W.: A New Method of Improving Furnace Temperature Control, *J. Sci. Instr.*, vol. 31, no. 1, pp. 23–25, 1954.
34-B. Swinehart, D. F.: Modified Thyratron, *Anal. Chem.*, vol. 21, no. 12, p. 1577, 1949.
35-B. Taxwood, V. C., and C. R. Stock: Apparatus for Automatic Uniform Controlled Rise of Temperature, *ASTM Bull.* 157, pp. 76–77, 1949.
36-B. Tunnicliff, D. D.: Gas Thermometer for Automatic Control of Low Temperatures, *Anal. Chem.*, vol. 20, no. 10, pp. 962–964, 1948.
37-B. Vodden, H. A.: Temperature Control—Laboratory, *J. Soc. Chem. Ind. (London)*, vol. 69, pp. 51–52, 1950.
38-B. Webb, T. L.: An Electronic Temperature Regulator for Muffle Furnaces, *S. African Ind. Chemist*, vol. 3, pp. 187–188, 1949.
39-B. Weil, Louis: A Precision Temperature Regulator, *Compt. Rend.*, vol. 224, pp. 810–812, 1947.
40-B. Wright, P.: Controller—Temperature, *J. Sci. Instr.*, vol. 24, pp. 258–261, 1947.

Pressure Measurement

1-C. Beck, A. H., and A. D. Brisband: Cylindrical Magnetron Ionization Gauge, *Chem. Abstr.*, vol. 47, no. 16, p. 7834, August, 1953 (abstracts from *Vacuum*, vol. 2, pp. 137–146, 1952).
2-C. Bollinger, Loren E.: 3C24 Ionization Gage, *Instr. and Automation*, vol. 28, no. 9, pp. 1507–1509, September, 1955.

3-C. Bradley, R. S.: A Thermistor McLeod Gauge for a Pressure Range $1-10^{-7}$ mm of Mercury, *J. Sci. Instr.*, vol. 31, no. 4, pp. 129–130, 1954.

4-C. Conn, E. K. T., and H. N. Daglish: A Simple Thermionic Vacuum Gauge, *J. Sci. Instr.*, vol. 31, no. 3, pp. 95–96, 1954.

5-C. Dushman, Saul: Manometers for Low Pressures, *Instruments*, vol. 20, pp. 234–239, 1947.

6-C. Garrod, R. L., and K. A. Gross: Combined Thermocouple and Cold-cathode Vacuum Gage, *J. Sci. Instr.*, vol. 25, pp. 378–383, 1948.

7-C. Florescu, N. A.: Shunted Thermocouple Vacuum Gage, *J. Sci. Instr.*, vol. 29, p. 298, 1952.

8-C. Havens, R., R. Koll, and H. LaGow: A New Vacuum Gage, *Rev. Sci. Instr.*, vol. 21, pp. 596–598, 1950.

9-C. Kerris and Weidemann: Development of a Photo-electric Cell Pressure Gage, *U.S. Commerce Dept. Bibliography Tech. Repts.*, vol. 3, p. 562, October–December, 1946.

10-C. Kersten, J. A. H., and H. Brinkman: Construction and Theoretical Analysis of a Direct-reading Hot-wire Vacuum Gauge with Zero Point Control, *Appl. Sci. Res.*, Sec. 1A, pp. 289–305, 1949.

11-C. Lawrance, R. B.: A Survey of Gauges for Measurement of Low Absolute Gas Pressure, *Chem. Eng. Progr.*, vol. 50, no. 3, pp. 155–160, March, 1954.

12-C. LaBlan, Louis: The Precise Measurement of Pressure, *Chim. Ind. (Paris)*, vol. 61, pp. 235–239, 349–354, 1949.

13-C. Nester, R. G.: New Type of Absolute Manometer, *Rev. Sci. Instr.*, vol. 25, no. 11, pp. 1136–1137, 1954. !

14-C. Reynolds, F. H.: An Electrical Manometer for Gas Pressures up to Forty mm Mercury, *J. Sci. Instr.*, vol. 30, no. 3, pp. 92–96, March, 1953.

15-C. Rodgers, Max T., J. G. Malik, and H. B. Thompson: Micrometer Mechanism for Reading a Bourdon Gage, *J. Sci. Instr.*, vol. 26, no. 7, pp. 230–231, July, 1955.

16-C. Rondeau, H. F.: Miniature Force and Pressure Cells, *Gen. Elec. Rev.*, vol. 54, no. 11, pp. 24–26, November, 1951.

17-C. Seddig, M., and G. Hasse: Low Pressures: Measurement, *Z. Angew. Phys.*, vol. 4, no. 3, pp. 105–108, March, 1952.

18-C. Steckelmacher, W.: Knudsen Gauges, *Vacuum*, vol. 1, no. 4, pp. 266–282, October, 1951.

19-C. Steckelmacher, W.: Vacuum Gauges: Review, *J. Sci. Instr.*, vol. 27, pp. 10–19, 1950.

20-C. Verhagen, C. J. D. M., J. P. Palm, et al.: Construction and Properties of Subminiature Pressure Pickups, *Appl. Sci. Res.*, Sec. B3, no. 6, pp. 409–416, 1954.

21-C. Von Ubisch, Hans: Hot-wire Manometers for Chemical Applications, *Anal. Chem.*, vol. 24, no. 6, p. 931-B, June, 1952.

22-C. Wallen, R.: The Recording of Pressures by Means of a Measuring Device Having Slight Displacement, *J. Phys. Radium*, vol. 7, pp. 342–344, 1946.

Manometers

1-D. Anon.: Manometer Height Sensing, *Instruments*, vol. 25, no. 4, p. 490, 1952.

2-D. Anon.: Mercury-column Height: Electronic Sensing, *Natl. Bur. Std. (U.S.) Tech. News Bull.*, vol. 36, no. 2, pp. 26–27, February, 1952.

3-D. Anon.: High-resolution Micromanometer, *Natl. Bur. Std. (U.S.) Tech. News Bull.*, vol. 37, no. 9, pp. 140–142, September, 1953.

4-D. Balson, E. W.: Effusion Manometer Sensitive to 5×10^{-6} mm Hg. Vapor Pressure of DDT and Other Slightly Volatile Substances, *Trans. Faraday Soc.*, vol. 43, pp. 54–60, 1947.

5-D. Baxter, I. G.: Differential Capacitance Manometer, *J. Sci. Instr.*, vol. 30, no. 10, pp. 358–360, October, 1953.

6-D. Bean, Howard S., and Francis C. Morey: Correction Factors for the Balancing Effect of a Gas in One Leg of a Manometer, *Instruments*, vol. 24, no. 5, pp. 528–529, 1951.

7-D. Berry, J. P.: A Sensitive Null-point Manometer, *J. Sci. Instr.*, vol. 33, p. 161, April, 1956.

8-D. Brow, Jeanne E., and F. A. Schwertz: A Simple Micromanometer, *Rev. Sci. Instr.*, vol. 18, pp. 183–186, 1947.

9-D. Brown, O. L. I., and C. M. Delaney: New Type Differential Manometer, *J. Phys. Chem.*, vol. 58, no. 3, pp. 255–258, March, 1954.

10-D. Cook, S. B., and G. J. Sanby: A Simple Diaphragm Micromanometer, *J. Sci. Instr.*, vol. 30, no. 7, pp. 238–240, 1953.

11-D. Gaffee, D. I., and A. G. Monroe: Measurement of Small Differential Pressures at Low Absolute Pressure, *Nature*, vol. 174, p. 756, 1954.

12-D. Galperin, B., J. Saurel, et al.: Hydrostatic Pressure: Level Variations, *J. Phys. Radium*, vol. 16, no. 6, pp. 492–493, June, 1955.

13-D. Hindley, H. R.: A Direct-reading Differential Micromanometer, *J. Sci. Instr.*, vol. 24, pp. 295–297, 1947.

14-D. Los, J. M., and J. A. Morrison: A Sensitive Differential Manometer, *Rev. Sci. Instr.*, vol. 22, no. 11, pp. 805–809, November, 1951.

15-D. MacMillan, F. A.: Liquid Manometers with High Sensitivity and Small Time Lag, *J. Sci. Instr.*, vol. 31, no. 1, pp. 17–20, 1954.

16-D. Malmquist, Lars: A Recording Mercury Manometer, *Tek. Tidskr.*, vol. 84, pp. 955–956, 1954.

17-D. Masloch, G. J.: A Precision Differential Manometer, *Rev. Sci. Instr.*, vol. 23, no. 7, pp. 367–374, July, 1952.

18-D. Matheson, Harry, and Eden Murray: A Highly Sensitive Differential Manometer, *Rev. Sci. Instr.*, vol. 19, no. pp. 502–506, 1948.

19-D. Minter, Clarke C.: A New Type Displacement Transducer, *ISAJ.*, vol. 3, no. 1, pp. 21–22, January, 1956.

20-D. Newell, F. G.: High-pressure Differential Manometers, *Instruments*, vol. 26, pp. 1523, 1553–1554, 1953.
21-D. Pappenheimer, J. R.: Differential Conductance Manometer, *Rev. Sci. Instr.*, vol. 25, pp. 912–917, 1954.
22-D. Peak, Roy D.: O-ring Sandwich Makes Manometer for High Pressures, *Chem. Eng.*, vol. 61, no. 12, pp. 212–214, December, 1954.
23-D. Puddington, I. E.: Sensitive Mercury Manometer, *Rev. Sci. Instr.*, vol. 19, pp. 577–579, 1948.
24-D. Simmons, L. F. G.: A Sensitive Air Manometer, *J. Sci. Instr.*, vol. 31, no. 6, pp. 195–197, June, 1954.
25-D. Stock, J. T., and M. A. Fill: An Electrolytic Remote-indicating Manometer, *Chem. Age (London)*, vol. 69, no. 1784, pp. 599–602, Sept. 19, 1953.
26-D. Tadayon, J.: Pressure Recorder, *Rev. Sci. Instr.*, vol. 22, pp. 534–535, 1951.
27-D. Thomas, George R., and Norman N. Lichtin: An Inexpensive Recording Differential Manometer for Reaction-kinetics Measurements, *Rev. Sci. Instr.*, vol. 24, no. 8 pp. 661–664, 1953.
28-D. Williams, W. E., Jr.: Mercury-level Detecting Unit, *J. Res. Natl. Bur. Std.*, vol. 48, no. 1, pp. 54–58, January, 1952 (also *Res. Paper* 2283).
29-D. Young, W. S., and R. C. Taylor: Vacuum Micromanometer, *Anal. Chem.*, vol. 19, no. 2, pp. 133–135, 1947.

Pressure Regulators

1-E. Booth, H. C., and R. L. Jarry: Device for Control of Still-head Pressures during Iso-thermal Distillation, *Anal. Chem.*, vol. 21, pp. 1416–1417, 1949.
2-E. Digney, P. J., and Stephen Yerazunic: Design and Use of an Electronic Pressure Controller, *Anal. Chem.*, vol. 25, no. 6, p. 921, June, 1953.
3-E. Erner, W. E., and K. N. Campbell: Versatile Electronic Manostat, *Anal. Chem.*, vol. 24, no. 7, pp. 1232–1233, 1952.
4-E. Goodwin, Robert D.: A Simple Manostat of Constant Sensitivity, *J. Chem. Educ.*, vol. 24, pp. 511–512, 1947.
5-E. James, D. H., and C. S. G. Phillips: A Simple Pressure Controller, *J. Sci. Instr.*, vol. 31, no. 5, p. 193, May, 1954.
6-E. Melpolder, F. W.: Vacuum Pressure Regulator, *Anal. Chem.*, vol. 19, p. 617, 1947.
7-E. Merkens, J. C.: Gas-pressure Regulator, *Lab. Prac.*, vol. 4, no. 9, p. 381, September, 1955.
8-E. Mueller, J. H.: A Simple Device for Regulating Pressure or Vacuum, *J. Am. Chem. Soc.*, vol. 71, pp. 1505–1506, 1949.
9-E. Nisbet, John S.: Pressure Controller Sensitive to 10^{-4} mm of Hg, *J. Sci. Instr.*, vol. 26, pp. 271–273, 1949.
10-E. O'Gorman, John M.: An Accurate Pressure-regulating Device, *Anal. Chem.*, vol. 19, p. 506, 1947.
11-E. Prentice, J. H.: A Device for Maintaining a Small Pressure against a Steady Leak, *J. Sci. Instr.*, vol. 31, no. 10, p. 386, October, 1954.
12-E. Ratchford, W. P., and M. L. Fein: Improved Manostat and Manometer, *Anal. Chem.*, vol. 22, pp. 838–839, 1950.
13-E. Redhead, P. A., and L. R. McNarry: An Ionization Manometer and Control Unit for Extremely Low Pressures, *Can. J. Phys.*, vol. 32, pp. 267–274, 1954.
14-E. Todd, Floyd: New Design for Vacuum Pressure Regulator, *Anal. Chem.*, vol. 20, pp. 1248–1249, 1948.

Flow Measurement

1-F. Anderson, J. W., and R. Friedman: An Accurate Gas-metering System for Laminar Flow Studies, *Rev. Sci. Instr.*, vol. 20, pp. 61–66, 1949.
2-F. Anon.: Electronic Flowmeter Is New Method for Measuring Air Currents or Fluid Flow, *Mech. Eng.*, vol. 75, no. 4, p. 313, April, 1953.
3-F. Anon.: Sonic Flow-metering, *Ind. Lab. (USSR) English Trans.*, vol. 4, no. 4, pp. 46–47, April, 1953.
4-F. Austin, George T.: Pot, Pipe, and Hole, *Chem. Eng.*, vol. 57, no. 3, pp. 97–98, 1950.
5-F. Bartlett, W. W.: An Electronic Gas Flowmeter, *Natl. Bur. Std. (U.S.) Rept.* 1693, Fifth Conference on Basic Instrumentation, 1952.
6-F. Bartlett, W. W., and P. Smoot, Electric Gas Flowmeter, *ISAJ.*, vol. 1, no. 10, p. 58, October, 1954.
7-F. Bean, H. S., R. M. Johnson, and T. R. Blakeslee: Small Nozzles and Low Values of Diameter Ratio, *Trans. ASME*, vol. 76, no. 6, p. 863, August, 1954.
8-F. Bergholm, Arne: Indicating Flow Meter for Process Liquids, *Chem. Eng.*, vol. 54, no. 6, p. 119, 1947.
9-F. Blight, D. P.: A Photographic Determination of the Emergent Velocity of Dust Particles, *J. Phot. Sci.*, vol. 3, no. 5, pp. 137–139, September–October, 1955.
10-F. Bohnet, W. J., and L. S. Stinson: Fanning Friction Factors for Air Flow at Low Absolute Pressures in Cylindrical Pipes, *Trans. ASME*, vol. 77, no. 5, p. 683, July, 1955.
11-F. Briggs, W. R. S., and R. L. Werner: A Photoelectric Flow Rate Meter, *J. Sci. Instr.*, vol. 31, no. 7, p. 259, 1954.
12-F. Buckingham, Edgar: Notes on Some Recently Published Experiments on Orifice Meters, *Trans. ASME*, vol. 78, no. 2, p. 379, February, 1956; also *Trans. ASME*, vol. 77, no. 4, p. 352, April, 1955.

13-F. Colcote, H. F.: Accurate Control and Vaporizing Systems for Small Liquid Flows, *Anal. Chem.*, vol. 22, pp. 1058–1060, 1950.

14-F. Cooley, W. C., and H. G. Stever: Determination of Air Velocity by Ion Transit-time Measurements, *Rev. Sci. Instr.*, vol. 23, nol 4, pp. 151, 154, April, 1952.

15-F. Cunningham, R. G.: Orifice Meters with Supercritical Compressible Flow, *Trans. ASME*, vol. 73, no. 5, pp. 625–638, July, 1951.

16-F. Dennis, W. L.: Flowmeter for Measuring Small Liquid Flows, *J. Sci. Instr.*, vol. 25, pp. 317–318, 1948.

17-F. Elrod, H. G., Jr., and R. R. Fouse: Investigation of Electromagnetic Flowmeters, *Trans. ASME*, vol. 74, no. 4, pp. 589–592, May, 1952.

18-F. Estermann, I., and E. D. Kane: A Torsion Balance for Measuring Forces in Low-density Gas Flows, *J. Appl. Phys.*, vol. 20, pp. 608–610, 1949.

19-F. Fisher, J. H.: Flowmeter, *Dissertation Abstr.*, vol. 14, no. 3, p. 506, March, 1954.

20-F. Fleming, F. W., and R. C. Binder: Linear-resistance Flowmeters, *Trans. ASME*, vol. 73, pp. 621–624, 1951.

21-F. Garwin, L., and S. H. Landes: Pipe Tee Fluid Flowmeter, *Ind. Eng. Chem.*, vol. 46, no. 4, pp. 665–669, 1954.

22-F. Grace, H. P., and C. E. Lapple: Small Diameter Orifices and Flow Nozzles, *Trans. ASME*, vol. 73, no. 5, pp. 639–647, July, 1951.

23-F. Gradishar, F. J.: Simplified Visual Flowmeter for Corrosive Gases, *Chem. Eng.*, vol. 57, no. 10, p. 122, 1950.

24-F. Grumer, J., H. Schultz, and M. E. Harris: Calibration of Glass-wool Flowmeters, *Anal. Chem.*, vol. 25, no. 5, p. 840, May, 1953.

25-F. Haines, G. S.: A Laboratory Flowmeter, *Anal. Chem.*, vol. 21, p. 1154, 1949.

26-F. Head, V. P.: Coefficients of Float-type Variable-area Flowmeters, *Trans. ASME*, vol. 76, pp. 851–862, 1954.

27-F. Heiss, J. F., and J. Coull: Nomographs Speed Flow Calculation, *Chem. Eng.*, vol. 56, no. 4, pp. 104–107, 1949.

28-F. Iversen, H. W.: Orifice Coefficients for Reynolds Numbers from 4 to 50,000, *Trans. ASME*, vol. 78, no. 2, p. 359, February, 1956.

29-F. Kalmus, Henry P.: A New Method for Measuring Fluid Flow, *Mod. Refrig.*, vol. 57, pp. 395–396, 1954.

30-F. Kolin, Alexander: Improved Apparatus and Technique for Electromagnetic Determination of Blood Flow, *Rev. Sci. Instr.*, vol. 23, no. 5, pp. 235–242, May, 1952.

31-F. Kritz, Jack: Ultrasonic Flowmeter, *Instr. and Automation*, vol. 28, no. 11, pp. 1912–1913, November, 1955.

32-F. Lavender, J. G., and C. R. Webb: Direct-reading Low-range Liquid Flowmeter, *Engineering*, vol. 171, no. 4452, pp. 619–621, May 25, 1951.

33-F. Linden, H. R., and D. F. Othmer: Air Flow through Small Orifices in the Viscous Region, *Trans. ASME*, vol. 71, pp. 765–772, 1949.

34-F. Lowenstein, Jack: Simple Orifice Construction Based on Pipe Union, *Chem. Eng.*, vol. 60, no. 5, p. 233, May, 1953.

35-F. Mappus, J. H., Jr.: The Saline Dilution Method of Liquid Flow Measurements, *Tappi*, vol. 37, no. 5, pp. 188A–189A, May, 1954.

36-F. Muller, Otto, H.: Selection of Capillaries with Predetermined Characteristics for Dropping Mercury Electrodes, *Anal. Chem.*, vol. 23, no. 8, pp. 1175–1177, 1951.

37-F. Munch, Ralph: Instrumentation, *Ind. Eng. Chem.*, vol. 44, pp. 83A–84A, 1952.

38-F. Newman, A. O., and B. J. Lerner: Electronic Method for Buffle Frequency Measurement, *Anal. Chem.*, vol. 26, no. 2, pp. 417–418, 1954.

39-F. Nord, Melvin: How to Measure Flow by Impact with Ordinary Pipe Fittings, *Chem. Eng.*, vol. 57, no. 7, p. 113, 1950.

40-F. Osburn, J. O., and K. Kammermeyer: Gas Flow through Small Orifices, *Chem. Eng. Progr.*, vol. 50, no. 4, p. 198, 1954.

41-F. Peterson, G. S.: Orifice Discharge Coefficients in the Viscous-flow Range, *Trans. ASME*, vol. 69, pp. 765–768, 1947.

42-F. Rood, Alvin: Calculating Pressure Loss, *Machine Design*, vol. 24, no. 9, pp. 179–182, September, 1952.

43-F. Sakai, I., and N. Sakai: Flowing Water: Electric Conductivity, *Chem. Eng. (Tokyo)*, vol. 17, pp. 136–143, 1953.

44-F. Schumacher, R.: Nomograph for the Determination of Capillary Flow Meters, *Angew. Chem.*, vol. B20, pp. 175–176, 1948.

45-F. Shercliff, J. A.: Experiments on the Dependence of Sensitivity on Velocity Profile in Electromagnetic Flowmeters, *J. Sci. Instr.*, vol. 32, pp. 441–442, November, 1955.

46-F. Shercliff, J. A.: Theory of the D.C. Electromagnetic Flowmeter for Liquid Metals, *U.S. Commerce Dept. Bibliography of Tech. Repts.*, vol. 21, no. 3, p. 70, Mar. 5, 1954.

47-F. Shulman, H. L., F. M. Steiger, and A. R. Leist: Versatile Flow Meter Measures Flow in Two Directions, *Chem. Eng.*, vol. 60, no. 2, pp. 173–174, February, 1953.

48-F. Souers, R. C., and R. C. Binder: Linear-resistance Flowmeters, *Trans. ASME*, vol. 74, no. 5, pp. 837–840, July, 1952.

49-F. Stoll, Henry W.: The Elbow as a Primary Flow Element, *Taylor Tech.*, vol. 7, nos. 1, 2, spring-fall, 1954.

50-F. Stoll, Henry W.: Fluid-flow Measurement in Pipe Sizes Below 2", *Petrol. Refiner*, vol. 27, no. 11, pp. 578–582, 1948.

51-F. Stull, K. S., Jr.: Ultrasonic Phase Meter Measures Water Velocity, *Electronics*, vol. 28, no. 9, pp. 128–131, September, 1955.

52-F. Sugiura, Y.: Experimental Studies on the Force Exerted on a Disk Placed in a Flow of Rarefied Gas, *J. Phys. Soc.*, vol. 9, no. 2, pp. 244–248, March–April, 1954.
53-F. Todhunter, K. H., and B. Wolsten-Home: A Multi-range Laboratory Flowmeter, *J. Sci. Instr.*, vol. 32, no. 1, p. 35, January, 1955.
54-F. Trasher, L. W.: Orifice Meters: Variables, *Dissertation Abstr.*, vol. 14, no. 11, November, 1954.
55-F. Wallick, G. C.: Steady State Gas Flow through Capillary Tubes, *J. Petrol. Technol.*, vol. 5, no. 11, pp. 20–23, November, 1953.
56-F. Wildhack, W. A.: Review of Some Methods of Flow Measurement, *ISAJ.*, vol. 1, no. 12, p. 54, December, 1954.
57-F. Wildhack, W. A.: Review of Some Methods of Flow Measurement, *Science*, vol. 120, no. 3110, pp. 191–197, Aug. 6, 1954.
58-F. Williams, S. F.: Improved Flowmeter for Small Flow Rates, *Chem. Eng.*, vol. 53, no. 10, pp. 123–124, 1946.
59-F. Wright, J. C.: New Displacement Method for Measuring Gas Flow, *Gas*, vol. 26, no. 5, pp. 116, 118, 121, 1950.
60-F. Wunsch, Walter, and Fritz Herning: Fluid-quantity Measurement with Adjustable Segment Diaphragm, *Gas-Wasserfach*, vol. 94, pp. 497–500, 1953.
61-F. Yudowitch, K. L.: Mercury Slug Flowmeter, *Anal. Chem.*, vol. 20, no. 1, p. 86, 1948.

Hot-surface Flow Measurement

1-G. Corrsin, Stanley: Extended Applications of the Hot-wire Anemometer, *U.S. Commerce Dept. Bibliography of Tech. Repts.*, vol. 12, no. 3, p. 79, September, 1949.
2-G. Gould, R. W. F.: Circuit for Compensating Hot Wires Used in the Measurement of Turbulence, *U.S. Commerce Dept. Bibliography of Tech. Repts.*, vol. 10, no. 5, pp. 444, November, 1948.
3-G. Grey, I. R.: Investigation of Possible Dimensional Inadequacies of Hot Wire Anemometer Probes and Their Effect on the Heat Transfer from the Wire, *U.S. Commerce Dept. Bibliography of Tech. Repts.*, vol. 15, no. 6, p. 172, June 15, 1951.
4-G. Hastings, C. E., and C. R. Scislo: Anemometer and Flowmeter, *Elec. Eng.*, vol. 70, no. 7, p. 597, July, 1951.
5-G. Henry, E. E., Jr.: Thermistor Application to Flow Measurements, *Dissertation Abstr.*, vol. 13, no. 5, pp. 743–744, October, 1953.
6-G. Judson, Howard C., Jr.: Calibration of the Hot Wire Anemometer at Low Speeds by Photographic Method, *U.S. Commerce Dept. Bibliography of Tech. Repts.*, vol. 16, no. 1, p. 16, July, 13, 1951.
7-G. Kovasznay, Leslie S. G.: Simple Hot-wire Anemometer, *U.S. Commerce Dept. Bibliography of Tech. Repts.*, vol. 16, no. 1, p. 19, July 13, 1951.
8-G. Laurence, J. C., and L. G. Landes: Application of the Constant Temperature Hot-wire Anemometer to Study of Transient Air Flow Phenomena, *Instrument.*, vol. 26, no. 12, pp. 1809–1804, December, 1953.
9-G. Ling, S. C.: Flow-characteristic: Measurement, *Dissertation Abstr.*, vol. 15, no. 10, pp. 1818–1819, 1955.
10-G. Middlebrook, G. B., and E. L. Piret: Hot Wire Anemometry—Solution of Some Difficulties in Measurement of Low Water Velocities, *Ind. Eng. Chem.*, vol. 42, pp. 1511–1513, 1950.
11-G. Pearson, C. E.: Air Velocity Measurements, *J. Aeron. Sci.*, vol. 19, no. 2. pp. 73–82, February, 1952.
12-G. Syerle, W.: Air-flow Measurements with Hot Wire Electric Probes, *U.S. Commerce Dept. Bibliography of Sci. Ind. Repts.*, vol. 4, no. 8, p. 682, Feb. 21, 1947.
13-G. Van Dilla, M., N. A. Inkpen, and R. D. Donovan: An Improved Hot-wire Anemometer for the Measurement of Wind Velocity, *U.S. Commerce Dept. Bibliography of Tech. Repts.*, vol. 1, no. 13, p. 640, Apr. 5, 1946.

Flow Regulators

1-H. Austin, George T.: Pot, Pipe and Hole, *Chem. Eng.*, vol. 57, no. 3, pp. 97–98, 1950.
2-H. Conway, H. G.: Basic Design of Hydraulic Flow Regulators, *Machine Design*, vol. 24, no. 9, pp. 131–135, 1952.
3-H. Fowler, Frank C.: How to Get Constant Steam Flow for Pilot Plant Use, *Chem. Eng.*, vol. 57, no. 8, p. 124, 1950.
4-H. Gilmont, Roger: Design and Operational Characteristics of Cartesian Manostats, *Anal. Chem.*, vol. 23, no. 1, p. 1571, January, 1951.
5-H. Hall, R. A.: A Laboratory Gas-flow Control Device, *J. Sci. Instr.*, vol. 32, pp. 116–117, 1955.
6-H. Humphlett, W. J.: Automatic Constant-rate Addition Funnel, *Org. Chem. Bull.*, vol. 27, no. 3, p. 6, 1955.
7-H. Lundsted, L. G., A. B. Ash, and N. L. Koslin: Constant-rate Feed Device, *Anal. Chem.*, vol. 22, p. 626, 1950.
8-H. Oliver, Earl D.: Constant-rate Liquid Feeder for Variable Outlet Pressures, *Chem. Eng.*, vol. 57, no. 5, p. 165, May, 1950.
9-H. Richards, A. R.: Apparatus for Continuous Delivery of Liquids at Constant Rate, *Anal. Chem.*, vol. 19, p. 281, 1947.
10-H. Riggio, Vincent A.: Small Steam Flow Control, *Taylor Tech.*, vol. 5, no. 4, p. 25, spring, 1953.
11-H. Schwertz, F. A.: Rate-indicating Mariotte Bottle, *Anal. Chem.*, vol. 22, no. 9, pp. 1214–1216, 1950.

12-H. Serfass, E. J., D. A. Shermer, and Ralph G. Steinhardt, Jr.: Adjustable Constant Flow Regulators for Corrosive Gases, *Anal. Chem.*, vol. 22, no. 4, pp. 618–620, 1950.
13-H. Shaw, T.: Constant Rate Feeder for "Drop-rate" Flow, *Chem. Eng.*, vol. 59, no. 2, p. 160, 1952.
14-H. Tarrant, G. T. P.: Controlled Water Flows, *J. Sci. Instr.*, vol. 32, no. 5, pp. 191–192, May, 1955.
15-H. Taylor, John K., and Enrique Eocridero-Molins: Constant Flow Buret Based on Principle of Mariotte Flask, *Anal. Chem.*, vol. 21, pp. 1576–1577, 1949.
16-H. Weiss, D. E.: Variable Flow Device for Small Flows, *Chem. Ind. (London)*, vol. 21, no. 52, p. 1573, December, 1954.

Pumps

1-I. Bates, H.T .: Inexpensive Diaphragm Pump Operates on Air Pressure, *Chem. Eng.*, vol. 56, no. 4 p. 112, 1949.
2-I. Burke, W. H., Jr., and W. G. Meinschein: Gas Circulating Pump for Closed Systems, *Anal. Chem.*, vol. 26, no. 12, p. 2004, December, 1954.
3-I. Lake, L. E.: A Glass Solenoid Pump, *J. Sci. Instr.*, vol. 30, no. 11, p. 434, 1953.
4-I. Marshall, P. R.: A New Type of Gas-circulating Pump, *J. Sci. Instr.*, vol. 24, p. 192, 1947.
5-I. Nickels, J. E.: Glass Circulating Pump for Gases and Liquids, *Anal. Chem.*, vol. 19, no. 3, p. 216, 1947.
6-I. Rodgers, E.: A Glass Pump for Very Low Rates of Flow, *Ind. Chemist*, vol. 24, pp. 245–246, 1948.
7-I. Stein, K. C., and F. J. Schoeneweis: Mercury Piston Gives Close Control of Liquid Feed Rates, *Chem. Eng.*, vol. 60, no. 8, p. 200, August, 1953.
8-I. Swenson, B.: Laboratory Pump, *Appl. Sci. Res.*, vol. A3, no. 2, pp. 163–164, 1951.

Feeders

1-J. Albright, C. W., J. H. Holden, H. P. Simons, and L. D. Schmidt: Pneumatic Feeder for Finely Divided Solids, *Chem. Eng.*, vol. 56, no. 6, pp. 108–111, 1949.
2-J. Anderson, H. M., and R. P. Zelinski: Device for Introduction of Solid Reagents, *Anal. Chem.*, vol. 22, no. 11, p. 1461, 1950.
3-J. Ashman, A. O.: An Accurate Reagent Feeder for Small Quantities, *Eng. Mining J.*, vol. 151, no. 4, pp. 98–100, 1950.
4-J. Barns, Roy W.: Low-cost Proportioner for Liquids Works Automatically without Attention, *Chem. Eng.*, vol. 55, no. 9, p. 124, 1948.
5-J. Fainman, M. Z.: Automatic Apparatus for Feeding Liquid at Low Rate, *Anal. Chem.*, vol. 21, p. 1438, 1949.
6-J. Karabinos, J. V.: Addition of Solids to a Sealed System, *J. Chem. Educ.*, vol. 31, no. 11, p. 571, November, 1954.
7-J. Lundsted, L. G., A. B. Ash, and N. L. Koslin: Constant-rate Feed Device, *Anal. Chem.*, vol. 22, p. 626, 1950.
8-J. McKinney, Alfred H.: Solids Feeder, *Chem. Eng.*, vol. 54, no. 4, p. 124, April, 1947.
9-J. Oliver, Earl D.: Constant-rate Liquid Feeder for Variable Outlet Pressures, *Chem. Eng.*, vol. 57, no. 5, p. 165, May, 1950.
10-J. Richards, A. R.: Apparatus for Continuous Delivery of Liquids at Constant Rate, *Anal. Chem.*, vol. 19, p. 281, 1947.
11-J. Shaw, T.: Constant Rate Feeder for "Drop-rate" Flow, *Chem. Eng.*, vol. 59, no. 2, p. 160, 1952.
12-J. Smith, I. C. P.: New Automatic Pipet-double-action Precision Instrument, *Chem. Age (London)*, vol. 55, pp. 413–415, 1946.
13-J. Shpolyanskii, M. A., and A. S. Likhacheva: Apparatus for Continuous Delivery of Liquid at a Slow Rate, *Zavodsk. Lab.*, 16, pp. 1500–1502, 1950.
14-J. Weiss, D. E.: Devices for the Continuous Addition of Solids and Liquids on the Laboratory Scale, *Chem. Ind. (London)*, vol. 29, no. 29, pp. 741–742, July 18, 1953.

Valves

1-K. Bartleson, J. D., A. L. Conrad, and P. S. Fay: New Head for Laboratory Fractionating Columns, *Ind. Eng. Chem., Anal. Ed.*, vol. 18, p. 724, 1946.
2-K. Blake, G. G.: An Electromagnetic Liquid Flow-stop Valve, *Chem. Ind. (London)*, 1947, pp. 536–537.
3-K. Boeke, J.: Diaphragm Valve, *Chem. Weekblad*, vol. 43, pp. 503–505, 1947.
4-K. Flinta, J.: A Gas Leak, *J. Sci. Instr.*, vol. 31, no. 10, p. 388, October, 1954.
5-K. Fuller, J.: A Simple Gas Valve, *J. Sci. Instr.*, vol. 33, no. 4, p. 160, April, 1956.
6-K. Harrison, E. R.: Glass Leak and Control Valves, *J. Sci. Instr.*, vol. 30, no. 5, p. 170, May, 1953.
7-K. Kirsls, S. S., and C. Asmanes: Improved Glass Valve, *Rev. Sci. Instr.*, vol. 26, no. 6, p. 615, June, 1955.
8-K. Kunkel, Karl E.: Automatic Valves for Control of Small Flows of Acids or Other Corrosive Materials, *Chem. Eng.*, vol. 54, no. 7, pp. 120–121, 1947.
9-K. McKay, B. P., and C. H. Eades, Jr.: Electromagnetic Laboratory Valve, *Anal. Chem.*, vol. 27, p. 163, 1955.
10-K. Mitchell, John, Jr., and J. W. Henderson: Improved Dropping Funnel, *Anal. Chem.*, vol. 22, no. 2, p. 374, 1950.

11-K. Stern, Joshua: A Vacuum Valve for Glass Systems, *Rev. Sci. Instr.*, vol. 22, no. 9, p. 703, September, 1951.
12-K. Wilcox, A. C., K. E. Coulter, and L. E. Loyd: Laboratory and Continuous Stills—Apparatus and Operation, *Petrol. Refiner*, vol. 31, no. 2, p. 134, February, 1952.
13-K. Brockett, Glenn F.: Control Valve Bodies, Sec. 10-1 of "Process Instruments and Controls Handbook," McGraw-Hill Book Company, Inc., New York, 1957.

Level Measurement

1-L. Anon.: Liquid: Detection in Piping, *Chem. Week*, vol. 70, no. 26, p. 38, 1952.
2-L. Bloch, R., L. Farkas, and G. Stein: Level Control Instrument for Brine Containing Solid Salt, *Chem. Eng.*, vol. 56, no. 4, pp. 113–114, 1949.
3-L. Broadhurst, J. W.: Circuit for Detecting Conducting Liquids, *J. Sci. Instr.*, vol. 21, pp. 108–109, October, 1944.
4-L. Caywood, W.: An Acoustic Volume-measuring Device, *U.S. Commerce Dept. Bibliography of Sci. Ind. Repts.*, vol. 2, no. 3, p. 191, July 19, 1946.
5-L. Couch, Dwight E., and Abner Brenner: Constant Level Device for Liquids, *Anal. Chem.*, vol. 24, no. 5, p. 992, May, 1952.
6-L. Farkas, A.: Photoelectric Control Adapted to Colorless Liquids, *Chem. Eng.*, vol. 55, no. 6, p. 128, 1948.
7-L. Jones, G. O., and C. A. Swenson: A One-point or "Full" Level Indicator for Use with Low-boiling Point Liquids, *J. Sci. Instr.*, vol. 25, pp. 72–73, 1948.
8-L. Kovacic, E.: An Apparatus for Measuring Small Changes in Liquid Level, *J. Sci. Instr.*, vol. 31, no. 6, pp. 205–206, 1954.
9-L. Maguire, R. D.: Constant-level Siphoning Device, *Anal. Chem.*, vol. 20, p. 394, 1948.
10-L. Marzolf, J. M., Jr., and W. K. Gardner: Remote Tank-level Indicators, *U.S. Dept. Commerce Bibliography of Tech. Repts.*, 17–18, January–December, 1952.
11-L. Schechter, Milton S., and H. L. Haller: Constant Level Apparatus, *Anal. Chem.*, vol. 20, p. 596, 1948.
12-L. Phillips, Dell: Liquid Sensing Device with No Moving Parts, *Prod. Eng.*, vol. 23, no. 6, pp. 134–135, 1952.
13-L. Rod, R. L.: Ultrasonic Liquid Level Indicator Systems, *Electronics*, vol. 27, no. 4, pp. 156–161, 1954.
14-L. Spangler, R. D., and E. B. Cooper: Vibrating Reed Makes Level Detector for Dry Powders and Liquids, *Chem. Eng.*, vol. 59, no. 2, p. 204, 1952.
15-L. Terry, P., and R. W. Urie: Fluid Bed Reactors: Level Control, *Australian J. Appl. Sci.*, vol. 1, no. 4, pp. 373–375, December, 1950.
16-L. Thompson, Walter C.: Simple Electrical Arrangement for Liquid Level Determination, *Chem. Eng.*, vol. 56, no. 2, p. 144, 1949.
17-L. Wexler, Aaron, and William S. Corak: Measurement and Control of the Level of Low-boiling Liquids, *Rev. Sci. Instr.*, vol. 22, no. 12, p. 941, December, 1951.
18-L. Elfers, P. A.: Liquid Level Detectors, Sec. 5-1 of "Process Instruments and Controls Handbook," McGraw-Hill Book Company, Inc., New York, 1957.

Position Measurement

1-M. Anon.: New Transducer, *Ind. Lab. (USSR) English Trans.*, vol. 5, no. 9, p. 65, September, 1954.
2-M. Anon.: Electro-mechanical Converter Instrument, *J. Sci. Instr.*, vol. 24, no. 7, p. 238, July, 1952.
3-M. Corner, W. D., and G. H. Hunt: A Direct Reading Instrument for the Measurement of Small Displacements, *J. Sci. Instr.*, vol. 31, no. 12, p. 445, December, 1954.
4-M. Frommer, Joseph C.: Detecting Small Mechanical Movements, *Electronics*, vol. 16, no. 7, p. 104, 1943.
5-M. Jaffe, J. A., and L. H. Cirker: Inexpensive Remote Indication Shows Level of Lift in Gas Holder, *Chem. Eng.*, vol. 53, no. 10, p. 122, 1946.
6-M. Joseph, H. M.: Mutual Inductance Transducer, *Natl. Bur. Std. (U.S.) Rept.* 1753, pp. 43–45, Mar. 31, 1952.
7-M. Lilly, John C., Victor Legallais, and Ruth Cherry: Variable Capacitor for Measurements of Pressure and Mechanical Displacements; A Theoretical Analysis and Its Experimental Evaluation, *J. Appl. Phys.*, vol. 18, pp. 613–628, 1947.
8-M. Lion, K. S.: Displacement Transducer Using Glow-discharge Tube, *Natl. Bur. Std. (U.S.) Rept.* 1693, Fifth Conference on Basic Instrumentation, p. 8, Apr. 29–30, 1952.
9-M. Macgeorge, W. D.: Differential Transformers for Control Indication, *Prod. Eng. Handbook*, Sec. I, pp. 16–21, 1953.
10-M. Maurice, D. M.: A Simple Circuit for Use with Condenser Strain Gauges, *J. Sci. Instr.*, vol. 31, no. 12, pp. 442–443, December, 1954.
11-M. Newton, A. E.: Movable and Fixed Coil Motion Transducers, *Proc. Eng.*, vol. 25, no. 5. pp. 170–173, May, 1954.
12-M. Stovall, J. R.: Movable-electrode Transducer, *Elec. Mfg.*, vol. 48, no. 1, p. 102–106, July, 1951.
13-M. Todd, W. M.: A Capacitance Displacement Gauge, *J. Sci. Instr.*, vol. 31, no. 7, p. 246, 1954.
14-M. Trott, J. J.: An Electronic Device for Measuring Small Transient Displacement, *J. Sci. Instr.*, vol. 29, no. 7, pp. 212–214, July, 1952.
15-M. Walen, R.: The Recording of Pressures by Means of a Measuring Device Having Slight Displacement, *J. Phys. Radium*, vol. 7, pp. 342–344, 1946.

16-M. Wildhack, W. A., and T. A. Peels: Spring Transducer, *Natl. Bur. Std. (U.S.) Rept.*, 1923, pp. 64–66.

Force and Weight Transducers

1-N. Allen, P. W., and R. D. Wright: A Photoelectric Balancing Device for Use with "Chainomatic" Type Balances, *J. Sci. Instr.*, vol. 29, no. 7, p. 235, July, 1952.
2-N. Caule, E. J., and G. McCully: Automatic Recording Analytical Balance, *Can. J. Tech.*, vol. 33, no. 1, pp. 1–11, 1955.
3-N. Clark, John W.: An Electronic Analytical Balance, *Rev. Sci. Instr.*, vol. 18, pp. 915–918, December, 1947.
4-N. Edwards, Frank C., and R. R. Baldwin: Magnetically Controlled Quartz Fiber Microbalance, *Anal. Chem.*, vol. 23, no. 2, p. 357, February, 1951.
5-N. Eisler, Joseph D., George R. Newton, and Willis A. Adcock: An Automatically Recording Magnetic Balance, *Rev. Sci. Instr.*, vol. 23, no. 1, p. 17, January, 1952.
6-N. Gregg, S. J., and M. F. Wintle: An Automatically Recording Electrical Sorption Balance, *J. Sci. Instr.*, vol. 23, pp. 259–264, 1946.
7-N. Hodsman, G. F., and E. R. Brooke: Electromagnetic Weight-loader for Fine Balances, *J. Sci. Instr.*, vol. 28, pp. 348–351, 1951.
8-N. Kellogg, H. W., E. F. Mahlke, and J. T. Jones: Save Time in TEL Analyses, *Petrol. Process.*, vol. 7, no. 10, pp. 1430–1432. October, 1942.
9-N. Lohmann, Idas W.: Electronic Recording Analytical Balance, *Rev. Sci. Instr.*, vol. 21, pp. 999–1002, 1950.
10-N. Mauer, Floyd A.: Recording Rapid Weight Changes, *Instrumentation*, vol. 7, no. 4, p. 36, third quarter, 1954.
11-N. Muller, Ralph H.: Research in Analytical Instrumentation, *Analyst*, vol. 77, pp. 557–563, 1952.
12-N. Pedersen, Svend R.: Liquid Weighing Scale Controls Rate of Flow, *Electronics*, vol. 25, no. 6, pp. 104–106, 1952.
13-N. Rulfs, Charles L.: Photoelectric Balance Indicator, *Anal. Chem.*, vol. 20, pp. 262–264, 1948.
14-N. Simmonds, D. H., and K. I. Wood: Magnetically Operated Balance for Collection of Liquid Fractions of Equal Weight, *Anal. Chem.*, vol. 26, no. 11, pp. 1860–1862, November, 1954.

Density

1-P. Carr, Frederick: Continuous Density Indicator Is Compensated for Temperature, *Chem. Eng.*, vol. 57, no. 6, p. 118, 1950.
2-P. Denyes, R. O., and C. L. Fox, Jr.: Continuous Specific Gravity Measurement, *Chem. Eng.*, vol. 56, no. 12, pp. 92–93, 1949.
3-P. Duff, R. D.: Glass Density Determinations by Sink-float Method, *J. Am. Ceram. Soc.*, vol. 30, no. 1, pp. 12–21, 1947.
4-P. Honik, K. R.: An Electromagnetic Method of Measurement of Density or Specific Gravity of Liquids, *J. Sci. Instr.*, vol. 31, no. 1, pp. 1–2, 1954.
5-P. Krutzsch, Johsef: Measurement of the Density of Solids, *Deut. Apotheker-Ztg.*, vol. 94, pp. 280–282, 1954.
6-P. Johnson, E. W., and L. K. Nash: Magnetically Compensated Vapor-density Balance, *Rev. Sci. Instr.*, vol. 22, no. 4, pp. 240–244, 1951.
7-P. MacInnes, D. A., et al.: Solutions: Determining Densities, *Rev. Sci. Instr.*, vol. 22, no. 8, pp. 642–646, August, 1951.
8-P. Marquard, C. M.: An Accurate Density Meter—How It Works—How It Is Used, *Eng. Mining J.*, vol. 152, no. 1, pp. 78–82, 1951.
9-P. Schumacher, E., H. Mollet, and K. Clusius: Gas Density Measurements with the Suspension Balance, *Helv. Chim. Acta*, vol. 33, pp. 2217–2221, 1950.
10-P. Simons, J. H., W. H. Pearlson, and W. A. Wilson: Fluorocarbon Vapor Balance, *Anal. Chem.*, vol. 20, no. 10, p. 983, 1948.
11-P. Smith, Francis A., John H. Eiseman, and E. C. Creitz: Tests of Instruments for Determination, Indication or Recording of Specific Gravity of Gases, *Natl. Bur. Std. (U.S.) Pub.* M177, p. 143, 1947.
12-P. Tabani, A. Y.: Automatic Detector for Density Changes, *Chem. Eng.*, vol. 55, no. 1, p. 135, 1948.
13-P. Takeda, Eiichi: The Bending Balance for Measuring the Density of Gases, *Proc. Phys. Math. Soc.*, vol. 23, pp. 1020–1031, 1941.

Viscosity

1-Q. Boyle, A. R.: An Electrical Viscometer, *J. Sci. Instr.*, vol. 27, no. 1, pp. 41–43, 1950.
2-Q. Epprecht, A. G.: The Determination of Viscosity in the Laboratory and in the Plant, *Ind. Vernice (Milan)*, vol. 6, pp. 197–201, 1952.
3-Q. Hammond, J. L., Jr.: Electromechanical Device for Measuring Viscosity, *J. Alabama Acad. Sci.*, vol. 26, pp. 56–58, 1954.
4-Q. Ibrahim, Ali A. K., and A. M. Kabril: The Oscillating-cylinder Viscometer, *J. Appl. Phys.*, vol. 23, p. 1190, 1952.
5-Q. Kesler, C. C., and W. G. Bechtel: Recording Viscometer for Starches, *Ind. Eng. Chem.*, vol. 39, pp. 16–21, 1947.

6-Q. Markwood, W. H., Jr.: Report on the Principles Involved in the Determination of Absolute Viscosity, *ASTM Proc.*, vol. 51, pp. 441–474, 1952.
7-Q. Massa, A. P.: End-point Control Via Viscosity, *Chem. Eng.*, vol. 62, no. 11, p. 226, November, 1955.
8-Q. Merrington, A. C.: "Viscometry," Edward Arnold, London, 1949.
9-Q. Miller, J. T.: Viscosity: Its Measurement and Control, part 2, *Instr. Prac.*, vol. 8, no. 9, 1954.
10-Q. Okaye, Tokiharu: An Approximate Formula for the Determination of Viscosity by Torsional Vibration, *Proc. Phys. Math. Soc.*, vol. 23, pp. 1031–1036, 1941.
11-Q. Thomas, B. W.: Instrumentation, *Ind. Eng. Chem.*, vol. 47, no. 4, p. 83A, April, 1955.
12-Q. Trigg, W. M.: Continuous Viscosity Control, *Chem. Eng.*, vol. 57, no. 5, pp. 156–157, 1950.

Miscellaneous Measurements

1-R. Bezinger, Theodor H., and Charlotte Kitzinger: Method for Continuous Recording of Gas Composition by Means of an Interferometer, *U.S. Commerce Dept. Bibliography of Tech. Rept.*, vol. 10, no. 3, p. 279, September, 1948.
2-R. Campbell, D. N., C. G. Fellows, S. B. Spracklen, and C. F. Hwang: Pneumatic Refractometer, *Ind. Eng. Chem.*, vol. 46, no. 7, p. 1409, July, 1954.
3-R. Clamann, H. G.: Gas Analyzer, *Instruments*, vol. 26, no. 5, pp. 740–742, May, 1953.
4-R. Colthup, N. B., and R. E. Torley: A Simple Surface Tensiometer, *Anal. Chem.*, vol. 23, no. 5, pp. 804–805, May, 1951.
5-R. Enverard, Maynard R., and Daniel R. Hurley: Surface Tension Measurement, *Anal. Chem.*, vol. 21, pp. 1178–1780, 1949.
6-R. Fink, Alfred: Recording Refractometer as an Analytical and Control Apparatus, *Oesterr. Chemiker-Ztg.*, vol. 55, pp. 93–102, 1954.
7-R. Ikeda, T.: Liquids: Surface Tension, *Bull. Chem. Soc. Japan*, vol. 26, no. 6, p. 352, August, 1953.
8-R. Jones, H. E., L. E. Ashman, and E. E. Stahly: A Recording Refractometer, *Anal. Chem.*, vol. 21, pp. 1470–1474, 1949.
9-R. Miller, E. C., F. W. Crawford, and B. J. Simons: A Differential Refractometer for Process Control, *Anal. Chem.*, vol. 24, no. 7, pp. 1087–1090, July, 1952.
10-R. Muller, Ralph H.: Instrumentation, *Anal. Chem.*, vol. 25, no. 3, p. 27A, March, 1953.
11-R. Svennson, Harry: Refractometric Analysis of Flowing Solutions, *Anal. Chem.*, vol. 25, no. 6, p. 913, June, 1953.
12-R. Thomas, George R., Chester T. O'Konski, and Charles D. Hurd: Automatically and Continuously Recording Flow Refractometer, *Anal. Chem.*, vol. 22, no. 9, pp. 1221–1223, 1950.
13-R. Anon.: Self-cleaning Antimony Electrode, *Chem. Eng. News*, vol. 31, no. 14, pp. 1475–1476, Apr. 6, 1953.
14-R. Anon.: pH Recorder-controller Operates under Pressure, *Chem. Eng.*, vol. 59, no. 12, p. 228, December, 1952.
15-R. Flood, H., et al.: Oxygen Electrode in Molten Salts, *Acta Chem. Scand.* vol. 6, no. 2, pp. 257–269, 1952.
16-R. Ingold, Werner: Continuous Checking the pH of an Aqueous Solution, Swiss Patent 292,451 (Nov. 2, 1953).
17-R. Yui, N.: Low-resistance Glass Electrodes, *Sci. Repts. Tohoku Univ., First Ser.*, vol. 33, pp. 238–242, 1949.

Conductivity

1-S. Ashma, L. E., R. S. Cohen, and J. A. Glass: Recording Conductometer for Electrolytes, *Instruments*, vol. 24, no. 6, pp. 710–715, 1951.
2-S. Axtman, Robert C.: Dielectric Constant Meter for Kinetic Studies, *Anal. Chem.*, vol. 24, no. 5, pp. 783–785, May, 1952.
3-S. Blake, G. G.: Conductimetric Analysis at Radio-frequency, *Analyst*, vol. 79, no. 935, pp. 108–109, February, 1954.
4-S. Blake, G. G.: A Simplified Circuit and Conductimetric Tube for Chemical Analysis at Low Frequency, *Electron. Eng.*, vol. 26, no. 317, pp. 316–317, July, 1954.
5-S. Brot C.: The Dielectric Behaviour at 9 cm. Wavelength of 6 Normal Primary Alcohols between −60° and +60°, *Comp. Rend.* vol. 239, no. 9, pp. 612–613, Aug. 30, 1954.
6-S. Chien, Jen-Yuan: Dielectric Constant-measurement, *J. Chem. Educ.*, vol. 24, pp. 494–497, 1947.
7-S. Cruse, K.: Die Titration mit Hochfrequenz (High Frequency Titration), *Arch. Eisenhuettenv.*, vol. 25, no. 11/12, pp. 563–568, November–December, 1954.
8-S. Dotson, J. M., et al.: Dielectric Constant, *Chem. Eng.*, vol. 56, no. 10, pp. 128–130, 1949.
9-S. Fischer, Robert B.: Simplified Instrument for Wide-range Dielectric-constant Measurement, *Anal. Chem.*, vol. 19, pp. 835–837, 1947.
10-S. Hall, James L.: Practical High-frequency Titration Apparatus for General Laboratory Use, *Anal. Chem.*, vol. 24, no. 8, pp. 1244–1247, August, 1952.
11-S. Hall, James L., John A. Gibson, Jr., Harold O. Philips, and Frank E. Critchfield: Some Evaluations of High-frequency Titration, *Anal. Chem.*, vol. 26, no. 10, p. 1539, October, 1954.
12-S. Howe, W. H.: Continuous Processes: Dielectric Measurement, *Instruments*, vol. 24, no. 12, pp. 1434–1438, 1484, 1486, 1488, 1490, December, 1951
13-S. Jottrand, R.: Decantation Study by Dielectric Constants, *Chem. Eng. Sci.*, vol. 1, no. 2, pp. 81–85, December, 1951.

14-S. McGourt, C. F.: An Instrument for the Measurement of the Density of Aqueous Ionic Solutions, *J. Sci. Instr.*, vol. 30, no. 7, pp. 241–244, 1953.

15-S. Oehme, F., and S. Wolf: The Significance of Dielectric Constants for the Chemist and Their Measurement with a New Dielectrometer, *Chem. Tech. (Berlin)*, vol. 2, pp. 216–219, 1950.

16-S. Pfundt, O., and G. Jander: Conductivity Measurements, *Chem.-Ztg. (Heidelberg, Germany)*, vol. 77, no. 12, pp. 403–404, June 20, 1953.

17-S. Pulley, Charles N., and W. H. McCurdy, Jr.: Principles of High Frequency Titration, *Anal. Chem.*, vol. 25, no. 1, pp. 86–93, January, 1953.

18-S. Svec, H. J.: A New Bridge Circuit for Conductometric Titrations, *J. Chem. Educ.*, vol. 31, no. 4, pp. 193–194, April, 1954.

19-S. Thiessen, G. W., and J. Wertz: Direct Reading, Inexpensive, Conductivity Bridge, *Chemist-Analyst*, vol. 42, no. 4, pp. 91–92, December, 1953.

20-S. Thomas, B. W., F. J. Faegin, and G. W. Wilson: Dielectric Constant Measurement for Continuous Determination of Toluene, *Anal. Chem.*, vol. 23, no. 12, pp. 1750–1754, 1951.

21-S. Thomas, E. B., and R. J. Nook: A Simple Bridge Balance Indicator for Conductance Measurements, *J. Chem. Educ.*, vol. 27, no. 1, pp. 25–26, 1950.

22-S. Thomas, B. W.: Instrumentation, *Ind. Eng. Chem.*, vol. 45, no. 10, pp. 87A–90A, October, 1953.

23-S. Thomas, B. W.: Instrumentation, *Ind. Eng. Chem.*, vol. 46, no. 12, pp. 71–72A, 74A, December, 1954.

24-S. West, Philip W., Senis Paschool, and T. S. Burkhalter: Determination of Water in Alcohols by Means of High Frequency Oscillators, *Anal. Chem.*, vol. 24, no. 8, pp. 1250–1252, August, 1952

Analyzers

1-T. Anon.: Hydrogen: Quantitative Measurement Apparatus, *Chem. Week*, vol. 73, no. 20, pp. 47–48, Nov. 14, 1953.

2-T. Anon.: Water Purity Tester, *Electronics*, vol. 16, no. 12, p. 162, 1943.

3-T. Austin, R. A., G. K. Turner, and L. E. Percy: Applications of Feed-back Electronic Control to Automatic Continuous Titration, *Instruments*, vol. 22, pp. 588–589, 1949.

4-T. Bertein, F., C. Cherrier, L. Verot, and R. Wagner: Photoelectric Analyzers Determination of Colored Gases, *Comp. Rend.*, vol. 230, pp. 1866–1867, 1950.

5-T. Bett, N., W. Nock, et al.: Colometric Titrimeter, *Analyst*, vol. 79, no. 943, pp. 607–616, October, 1954.

6-T. Bishop, John F., and Ralph S. White: Beckman Flow Colorimeter, *Ind. Eng. Chem.*, vol. 46, no. 7, p. 1432, July, 1954.

7-T. Cook, Warren A.: Review of Automatic Indicating and Recording Instruments for Determination of Industrial Atmospheric Contaminants, *Am. Ind. Hyg. Assoc. Quart.*, vol. 8, pp. 42–48, 1947.

8-T. Crouthamel, Carl E., and Harvey Diehl: Gas Analysis Apparatus Employing the Velocity of Sound, *Anal. Chem.*, vol. 20, pp. 515–520, 1948.

9-T. Crumpler, Thomas B., William H. Dyre, and Aldenlee Spell: Simple Photoelectric Polarimeter, *Anal. Chem.*, vol. 20, no. 10, pp. 1645–1648, October, 1955.

10-T. Cutting, C. L., A. C. Jason, and J. L. Wood: A Capacitance-resistance Hygrometer, *J. Sci. Instr.*, vol. 32, pp. 425–431, November, 1955.

11-T. Deisler, Paul F., Jr., Keith W. McHenry, Jr., and Richard H. Wilhelm: Rapid Gas Analyzer Using Ionization by Alpha Particles, *Anal. Chem.*, vol. 27, no. 9, pp. 1366–1374, September, 1955.

12-T. Ewing, Galen W.: "Instrumental Methods of Chemical Analysis," McGraw-Hill Book Company, Inc., New York, 1954 (especially see p. 425).

13-T. Furman, N. Howell: Electrical Methods of Chemical Analysis, *Record Chem. Progr. Kresge-Hooker Sci. Lib.*, vol. 11, pp. 33–45, 1950.

14-T. Gorbach, G.: Capillary Photometer, *Mikrochim. Acta*, no. 4, pp. 879–881, 1955.

15-T. Greinacher, H.: The Diffusion Hygrometer, *Helv. Phys. (Swiss)*, vol. 17, pp. 437–454, 1944.

16-T. Hallikainen, K. E., and D. J. Pompeo: Continuous Recording Electrometric Titrometer, *Instruments*, vol. 25, no. 4, pp. 335–338, 1952.

17-T. Harley, John H., and Stephen E. Wiberley: "Instrumental Analysis," John Wiley & Sons, Inc., New York, 1954 (especially see p. 400).

18-T. Harris, Frank E., and Leonard K. Nash: Determination of Traces of Water Vapor in Gases, *Anal. Chem.*, vol. 23, no. 5, p. 737, May, 1951.

19-T. Harwell, K. E.: Precise Electronic Titration Instrument, *Anal. Chem.*, vol. 26, no. 3, pp. 616–619, 1954.

20-T. Haslam, J., and D. C. M. Squirrell: An Automatic Titrimeter, *Analyst*, vol. 79, no. 944, pp. 689–696, November, 1954.

21-T. Hine, T. B.: Theory of the Flow-meter Type Continuous Gas Analysis Apparatus, *U.S. Commerce Department Bibliography of Tech. Repts.*, vol. 1, no. 24, p. 1484, June 21, 1946.

22-T. Hinzpeter, A., and W. Meyer: Moisture Determination by Glow Potential, *Z. Angew. Phys.*, vol. 3, no. 6, pp. 216–218, June, 1951.

23-T. Ippen, A. T., and C. E. Carver, Jr.: Dissolved-oxygen Measurement, *Instr. and Automation*, vol. 27, no. 1, p. 128, January, 1955.

24-T. Jura, W. H.: Shielded Dropping Mercury Electrode for Polarographic Analysis of Flowing Solutions, *Anal Chem.*, vol. 26, no. 7, p. 1121, 1954.

25-T. Landsberg, Henry, and Edward E. Escher: Potentiometric Instrument for Sulfur Determination, *Ind. Eng. Chem.*, vol. 46, no. 7, p. 1422, July, 1954.

26-T. Lawley, L. E.: Gas Analysis by Acoustic Methods, *Chem. Ind. (London)*, no. 8, pp. 200–203, Feb. 20, 1954.

27-T. Liebhafsky, H. A., and E. H. Winslow: Photoelectric Colorimetry Is Inexpensive Equipment, *J. Chem. Educ.*, vol. 27, no. 1, pp. 61–62, 1950.
28-T. Lilly, John C., and Thomas F. Anderson: The Nitrogen Meter: An Instrument for Continuously Recording the Concentration of Nitrogen in Gas Mixtures, *U.S. Commerce Dept. Bibliography of Tech. Repts.*, vol. 11, p. 130, February, 1949.
29-T. Lindsley, Charles H., and John H. Yoe: Simple Thermometric Apparatus for the Estimation of Carbon Monoxide in Air, *Anal. Chim. Acta*, vol. 2, pp. 127–132, 1948.
30-T. Lingane, James J.: Multipurpose Electroanalytical Servo Instrument, *Anal. Chem.*, vol. 21, pp. 497–499, 1949.
31-T. Lynch, Frank J., and James B. Baumgardner: New Fluorescence Photometer, *Rev. Sci. Instr.*, vol. 26, no. 5, pp. 435–440, May, 1955.
32-T. Minter, Clarke C., and Lyle M. J. Burdy: Thermal Conductivity Bridge for Gases, *Anal. Chem.*, vol. 23, no. 1, pp. 143–147, 1951.
33-T. Neilands, J. B., and M. D. Cannon: Automatic Recording pH Instrumentation, *Anal. Chem.*, vol. 27, no. 1, p. 29, January, 1955.
34-T. Northrop, John H.: Apparatus for Automatic Detection of H_2 or Other Gases Which React with Bromine by Means of the Bromine Electrode to Sept. 8, 1944, *U.S. Commerce Dept. Bibliography of Sci. Ind. Repts.* vol. 1, no. 11, p. 513, Mar. 22, 1946.
35-T. Northrop, John H., and John N. Gettemans: Portable Apparatus for Rapid Estimation of H_2 or Other Gases Which React with Bromine to Dec. 20, 1944, *U.S. Commerce Dept. Bibliography of Sci. Ind. Repts.*, vol. 1, no. 12, p. 578, Mar. 29, 1946.
36-T. Patterson, Gordon D., Jr.: Automatic Operations in Analytical Chemistry, *Anal. Chem.*, vol. 27, no. 4, p. 574, April, 1955.
37-T. Perley, G. A., and E. L. Eckfeldt: Automatic Potentiometric Recording Equipment for Determination of Chemical Warfare Agents, to March 20, 1944, *U.S. Commerce Dept. Bibliography of Sci. Ind. Repts.*, vol. 1, no. 9, p. 383, Mar. 8, 1946.
38-T. Pritchard, F. W.: Gas Analyzer, *J. Sci. Instr.*, vol. 29, no. 4, pp. 116–117, April, 1952.
39-T. Shaffer, Philip A., Jr., Anthony Briglio, Jr. and John A. Brockman, Jr.: Instrument for Automatic Continuous Titration, *Anal. Chem.*, vol. 20, no. 11, pp. 1008–1014, 1948.
40-T. Thomas, Moyer D., James O. Ivie, and T. Cleon Fitt: Automatic Apparatus for Determination of Small Concentrations of Sulfur Dioxide in Air, *Anal. Chem.*, vol. 18, no. 6, pp. 383–387, 1946.
41-T. Wildhack, W. A., and T. A. Perls: Pneumatic Instruments Based on Critical Flow, *Natl. Bur. Std. (U.S.) Rept.* 1753, pp. 56–57, Mar. 31, 1952.
42-T. Willard, H. H., L. L. Merritt, and J. A. Dean: "Instrumental Methods of Analysis," D. Van Nostrand Company, Inc., Princeton, N.J., 1951 (especially see p. 340).

Electrical Measurements

1-U. Benson, F. A.: Glow-discharge Voltage-regulator Tubes, *J. Sci. Instr.*, vol. 28, no. 11, pp. 339–341, November, 1951.
2-U. Dauphinee, T. M., and S. B. Woods: Low Level Thermocouple Amplifier and a Temperature Regulation System, *J. Sci. Instr.*, vol. 26, no. 7, pp. 693–695, July, 1955.
3-U. Frediani, H. A.: Automatic Karl Fischer Titration Apparatus Using Dead-stop Principle, *Anal. Chem.*, vol. 24, no. 7, pp. 1126–1128, July, 1952.
4-U. Kiriloff, A. A.: Reference-voltage Circuits for Automatic Controls, *Elec. Mfg.*, vol. 53, no. 1, pp. 97–103, 296, 298, January, 1954.
5-U. Lanphere, R. W., and C. B. Rodgers: Instrument for Controlled-potential Electrolysis, *Anal. Chem.*, vol. 22, pp. 463–468, 1950.
6-U. Lufey, Carooll W., A. C. Schmid, and P. W. Barnhart: Half-wave Magnetic Servo Amplifier, *Electronics*, vol. 25, no. 8, pp. 124–125, August, 1952.
7-U. Muller, Ralph H., and James J. Lingane: Electronic Trigger Circuit for Automatic Potentiometric and Photometric Titrations, *Anal. Chem.*, vol. 20, no. 9, pp. 795–797, 1948.
8-U. Patton, H. W.: A Constant Current Regulator, *Anal. Chem.*, vol. 23, no. 2, pp. 393–394, 1951.
9-U. Phillips, J. P.: A Transistor Amplifier for the Dead-stop End Point, *Chemist-Analyst*, vol. 44, no. 3, pp. 80–81, September, 1955.
10-U. Reilley, C. H., Ralph N. Adams, and N. H. Furnier: Improved Constant-current Supply for Coulombic Analysis, *Anal. Chem.*, vol. 24, no. 6, pp. 1044–1045, June, 1952.
11-U. Sivertsen, Jens: Electrolytic Integrating Tube, *Electronics*, vol. 20, no. 6, pp. 92–95, 1947.
12-U. Suozzi, Joseph J.: Magnetic-amplifier Two-speed Servo System, *Electronics*, vol. 29, no. 2, p. 141, February, 1956.

PILOT-PLANT INSTRUMENTATION

By Dr. Clyde Berg*

Continuous processing and handling over recent years have brought increased economy and efficiency to many of the major industries. On the other hand, the staggering investment required to design, construct, and put the average continuous process on stream has accentuated the need to *prove* a process on a comparatively small scale. A grave financial risk is involved in extrapolating laboratory test data, accumulated in terms of cubic centimeters or ounces per hour, into plant-design data where volumes of hundreds of thousands of barrels per hour may be involved. Hence the basic need for intermediate stages between laboratory and process—bench-scale units, pilot plants, and semiworks. It is in these midway units that the researcher can seek and try and even make mistakes on a small scale in an effort to achieve perfection and profit on a large scale.

The fundamental objective of a pilot plant is to prove the correctness of deductions arrived at from prior laboratory and literature researches—to prove not only that a process or reaction will work beyond the test-tube stage, but also that it can be carried forward on a profitable basis. Pilot-plant operations by and large are data-gathering operations, and any tools available to expedite and refine the collection and collation of data, such as modern instruments, are a must.

NEED FOR THOROUGH INSTRUMENTATION

Although it is impractical to list all the reasons which justify the thorough instrumentation of a pilot plant, some of the highlights are:

1. *Simulation of Conditions Expected in the Large-scale Unit to Be Constructed.* Obviously, if a pilot plant is considered as a small-scale version of a forthcoming commercial unit, it follows that all variable conditions—such as critical processing temperatures, pressures, velocities, reflux ratios, pH values, movement of fluids and solids, levels, quality of product—must be controlled and maintained in a manner representative of the commercial process in the course of creation. It is folly to relegate critical pilot-plant variables to manual or crude automatic controls when much more sophisticated controls will be used in a larger manufacturing unit. It is unfair to the process under development not to extend to it the refinements of control which can mean much to its success. In many cases instrumentation in excess of that required by large units is required by the pilot plant to simulate large-unit operating conditions.

2. *Confirmation of Controlled Conditions.* Only graphic records can reliably yield the confirming evidence that the desired uniform process conditions have or have not been maintained. They also make possible determination of the required accuracy of control. Even more important, graphic records give rapid and direct evidence as to causes and effects of upsets in unexpected phenomena which occur so frequently in pilot-plant development studies.

3. *Determination of Inherent Control Problems.* Certain processes as conceived may incorporate inherent control problems of substantial time lags which can, in the extreme, actually render the process virtually inoperable. Such control problems must be tackled in the pilot-plant stage and specialized instrumentation developed if

* Director, Clyde Berg Associates, Long Beach, Calif.

necessary, or critical process variables substituted by others more indicative of trends in the processing conditions.   Pilot-plant instrumentation permits the determination of the proper process variables to be controlled.

4. *Conservation of Scientific Man-hours.*   Even an experienced Ph.D. cannot manually log data with the accuracy and rapidity now afforded by modern instruments. Instruments not only do a better, quicker job but also, at the same time, accomplish these tasks at lower cost.   Regardless of cost, however, management today can ill afford to expend valuable scientific man-hours on any task which can be done equally well or better automatically.   Around-the-clock operation also is much more easily attained with automatic control.

5. *Testing of Indicating and Recording Instruments.*   Testing of the suitability of various indicating and control instruments can most readily be done in the pilot-plant stage.

FUNDAMENTAL REQUIREMENTS OF INSTRUMENTS

Pilot-plant instrumentation has objectives similar to those of large-scale installations, but it must contend with a number of additional complicating factors, including:

1. *Ability to Handle a Wide Range of Operating Conditions.*   In the crystallization of a new process, a range of processing conditions must be explored to establish optimum conditions in the new process.   Once these optimum conditions are established, it follows that the instruments ordered for the large-scale process can be specified to operate within fairly narrow parameters.   This is not true in the pilot plant. Some factors which should be considered when specifying instruments for pilot plants include:

 a. *Ease of changing from one range to another,* as, for example, from a temperature range of 0 to 200°F to a range of 100 to 400°F, or from an average flow of 20 gpm to one of 2 gpm or lower.

 b. *Ease of changing from one variable to another,* as, for example, actuation from a thermocouple to actuation from a pH electrode assembly, or the activation of a controller by either a flow, level, pressure, or temperature indicator.

The above factors not only are important in connection with the flexibility required for a given pilot process but also are steps toward economy in the operation of a pilot plant or development division.   Obviously, fewer instruments are required where those which are available possess maximum flexibility.

2. *Ability for Installation in Small Space.*   Since pilot plants operate at reduced capacities, available dimensions for installation of measuring elements and final control elements frequently limit or prevent the use of conventional commercial equipment. Special designs and precautions are often required in pilot-plant instrumentation to assure the accurate measurement of the process condition being measured.   This need, however, has been realized during recent years, and before undertaking costly instrument-development work, it is wise to check all possible sources of commercially available equipment to handle the job.

3. *Ability to Cope with Magnified Fouling Effects.*   The surface-to-volume ratio employed in pilot-plant operations is considerably higher than that of large-scale commercial units—yet commercial construction materials are generally employed.   What may be considered a negligible amount of corrosion and sealing in commercial operations oftentimes subjects pilot-plant valves and measuring elements to a much magnified and possibly serious fouling effect.   This fouling effect is normally accelerated during shutdown periods, which are usually quite frequent in pilot-plant work.   Since the apertures of control valves, orifices, and piping for pilot plants are usually relatively small, they are readily affected by scale particles.   Ease of disassembly, because of this condition, is something to look for in control equipment.

4. *Ability to Measure without Materially Affecting the Process.*   Because of the small quantities handled in pilot plants, precautions must be taken to assure that the process conditions are not altered by the measuring element.

ADAPTABILITY OF ELECTRICAL INSTRUMENTS

Electrical indicating and recording instruments are extremely versatile, and it is important that the pilot-plant instrument engineer know of the many possible uses of these instruments. By changing ranges, it is possible to use these instruments over a virtually unlimited number of applications. Or, by changing primary sensing elements, the pilot-plant engineer can use the same instrument to detect a large number of different variables.

Some of the methods of range changing are:

Manual Span Switch. With this device, two or more slidewire voltage ranges are supplied in one instrument, with two or more sets on markings on the scale. Toggle or rotary switches are used to change range.

Manually Adjustable Slidewire Span. This provides a continuously adjustable range to compensate for variations in calibration constants of primary elements, which are used with recorders calibrated directly in units of temperature or some other variable. A split potentiometer circuit with continuous manual span control is available for these applications. Maximum range can be as much as twenty times minimum. Adjustable zero suppression may also be included in these instruments.

Manual Zero Suppression or Elevation. In these instruments, slidewire span is limited to a portion of the voltage from the sensing element. Steps of zero suppression are precisely adjusted for minimum error. Two or more steps of suppression are selected by a toggle or rotary switch.

Table 1. Measurement of Typical Pilot-plant Variables by Means of D-C Voltages

Primary sensing element	Quantity measured	Electrical span	Typical external circuit resistance	Added features
Thermocouple...	Temperature	Up to 100 mv	Up to several hundred ohms	Automatic reference junction compensation
Thermocouple or thermopile	Temperature difference	100 μv to 10 mv	Up to several hundred ohms	Stabilized d-c microvolt amplifier available for shorter spans
Radiation detector	Temperature	100 μv to 30 mv	Under 50 ohms	Range adjustable to match detector calibration and permit emissivity correction
Glass electrode..	pH	200–900 mv	To 2,000 megohms	High impedance measuring circuit. Compensation for electrode temperature
Strain gage......	Stress, load, weight, pressure	To 30 mv	100 to 3,000 ohms	Circuit includes zero and range adjustments
Thermal-conductivity cell	Gas analysis	5–15 mv	To 50 ohms	Zero and range rheostats
Infrared-absorption cell	Gas analysis	100 μv to 1 mv	100 ohms	Adjustable span
Oxidation-reduction electrode	Millivolt potential	As specified	To 2,000 megohms or more	High input impedance to eliminate electrode polarization
Tachometer.....	Speed	To 1,250 mv	To 100 ohms	Filter eliminates commutator ripple and eccentricity effects
Direct measurement	D-C volts	100 μv to 1,000 mv	To several megohms	Detector amplifier selected accordingly
Thermal converter	A-C volts, amps, watts	To 700 mv	To 3,000 ohms	
Thermocouple pressure gage	Absolute pressure	15–50 mv	To 100 ohms	
Photomultiplier tube	Light	200 mv	Over 500 megohms	Measuring circuit includes multirange shunt

Adjustable Zero, Adjustable Range. A number of specialized recorders are available with built-in adjustable zero and range. Calibrated knobs are used to change settings.

Automatic Selector Switch. Such a switch in a multiple-point instrument can be used to measure different ranges from a number of primary sensing elements with a synchronously driven selector switch. This switch automatically connects the primary elements in sequence and simultaneously chooses the range suited to the connected element.

Manual Range Change. A kit of parts can be obtained from the instrument manufacturer to permit range changes by the user or by a company serviceman.

In pilot-plant work it is frequently necessary to convert an electrical measuring instrument from one primary sensing element to another. For this reason, Table 1 is included here, listing the most common sensing elements, the quantities they measure, the electrical span, and typical external circuit resistance. While this table is based upon information furnished by Leeds & Northrup Co., approximately the same information would apply to similar instruments manufactured by other suppliers.

PECULIAR PROBLEMS OF MEASUREMENT

Following are discussions of the major process variables, the measurement of which often poses particular problems in pilot-plant operation. Under each variable, special equipment designed to solve such problems is described.

Temperature

Because multiple recording of temperature is so important for reasons of coordination and economy in pilot plants, electrical methods such as thermocouples, radiation

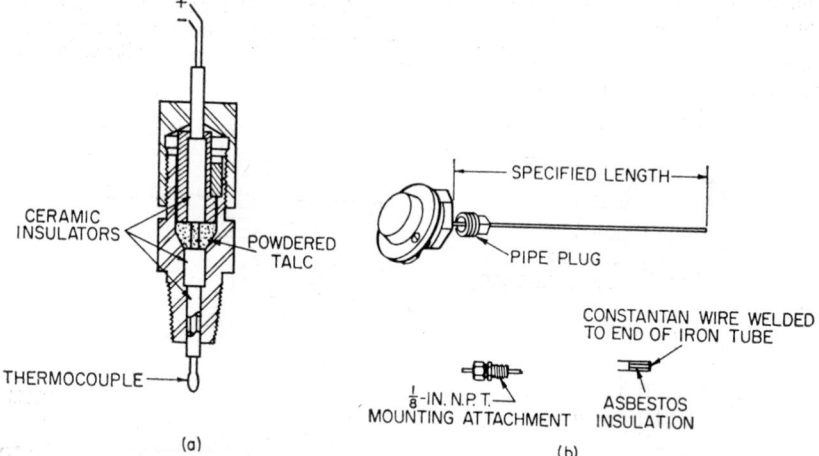

FIG. 1. Thermocouples for pilot-plant measurements: (a) bare-wire-type thermocouple (*Conax Sales Company, Inc.*); (b) pencil-type thermocouple (*Minneapolis-Honeywell Regulator Co.*).

pyrometers, and resistance thermometers are usually preferred over filled-system thermometers. The ease of installing and repairing extension wires, as compared with capillary tubing, is also a factor in favor of the electrical systems. Electrical instruments also have the advantage of ease of change of range, and changeover to the measurement of another variable, such as pressure from a strain element, pH from an electrode assembly, and others. Particular attention must be paid to possible errors in pilot-plant work.

A bare thermocouple with high-pressure insulating seal (Fig. 1a) offers both small size and high sensitivity for pilot-plant measurements. It avoids use of a well with its added lag in measurement, except of course where a well is required for protection against corrosive fluids. Figure 1*b* shows another highly responsive and small-sized thermocouple, called a "pencil" type, wherein one of the thermocouple elements is in the form of a tube and the other element is a wire inside the tube.

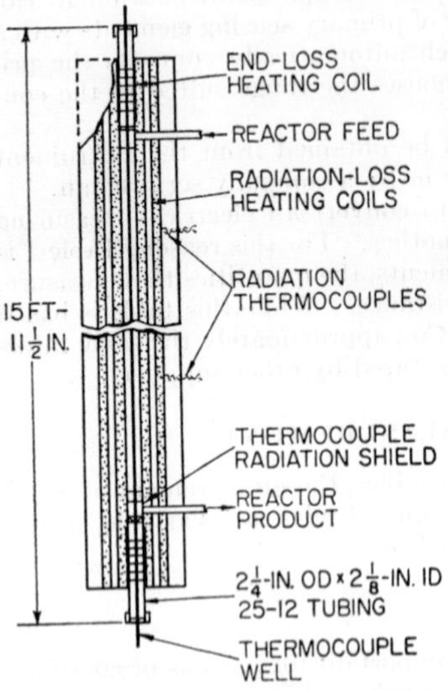

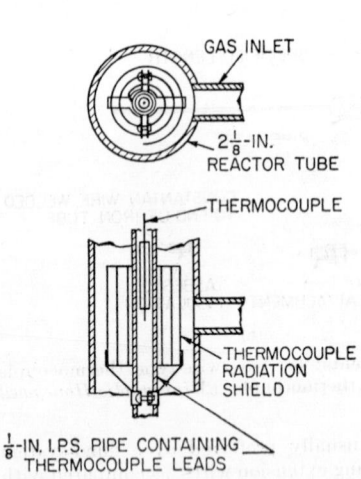

FIG. 2. Pilot-plant reactor thermo-couple installation.

Where a thermocouple is used with a well, careful attention should be given to possible errors due to heat conduction along the metal wall of the well. Because available distances for use of full-length standard wells are limited, errors of this type can be a problem. Figure 2 shows the cross section of a pilot-plant reactor in which the problem of heat conduction along a thermocouple well was solved by use of a longitudinal tube running the full height of the reactor; inside the tube, multiple thermocouples of fine-gage wire covered with braided-glass insulation were introduced. This construction effectively eliminated heat conduction that would be present if the thermocouples were installed through the side of the reactor.

Errors in thermocouple readings caused by radiation effects are accentuated when high-temperature gas streams are measured in pilot plants. Often gas velocities are quite low, so that radiation to low-temperature surfaces can be serious. Figure 3 shows a thermocouple construction which has concentric shields to minimize such errors in spite of low gas velocities at the entrance of a pilot-plant reactor.

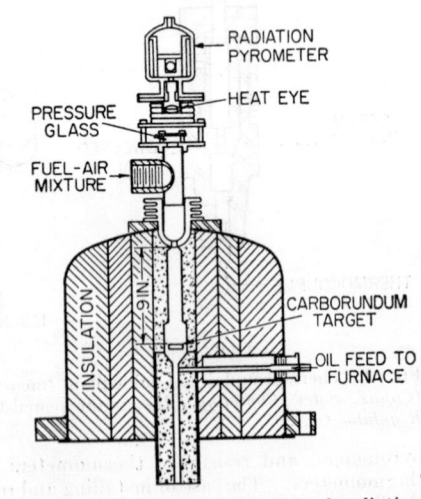

FIG. 3. Radiation shield in pilot-plant thermocouples.

FIG. 4. Pilot-plant installation of radiation pyrometer.

Resistance thermometers provide a very sensitive element for temperature measurement where high accuracy over a rather narrow range of temperature is desired. For higher temperatures, particularly where an element cannot contact the measured medium, the radiation pyrometer may provide the ideal measuring means.   Figure 4 shows a typical installation of a radiation-detecting unit in a pilot-plant application.

Flow

Pilot-plant flow rates must be measured with high precision because they are the basis of determining yields and material balances.   In many cases, the various commercially available metering methods can be readily adapted for such measurements, but more care must be taken in applications where pipe sizes of less than 2 in. are

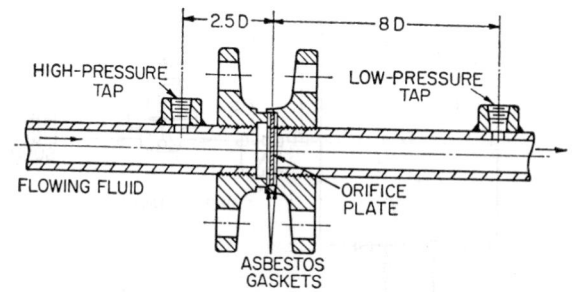

FIG. 5. Pilot-plant orifice installation.

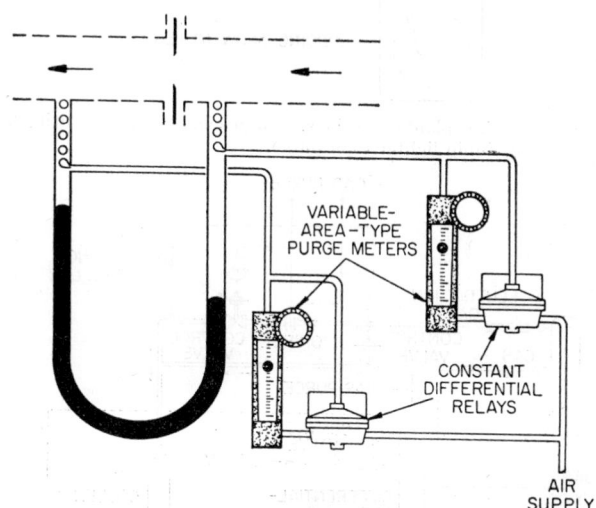

FIG. 6. Purge system in pilot-plant flow measurement—purge in pressure taps.

involved.   In these cases, dimensional tolerances and ratios of pipe surfaces to volumes become increasingly important as pipe size decreases.   Viscosity, kinetic-energy state, and turbulence of the fluid all become more serious factors affecting the different metering methods to varying degrees.[10]

Variable-head meters employing orifices are frequently used in pilot plants.   Figure 5 shows a typical orifice installation with pipeline taps for connection to the differential-pressure-measuring device.   Where flow quantities are very small, a capillary tube can be substituted for the orifice; in this case, the pressure-differential response characteristic becomes linear, rather than square root.   For small flow rates, differential-pressure transmitters of the diaphragm type are offered with a small orifice integrally mounted as part of the assembly.

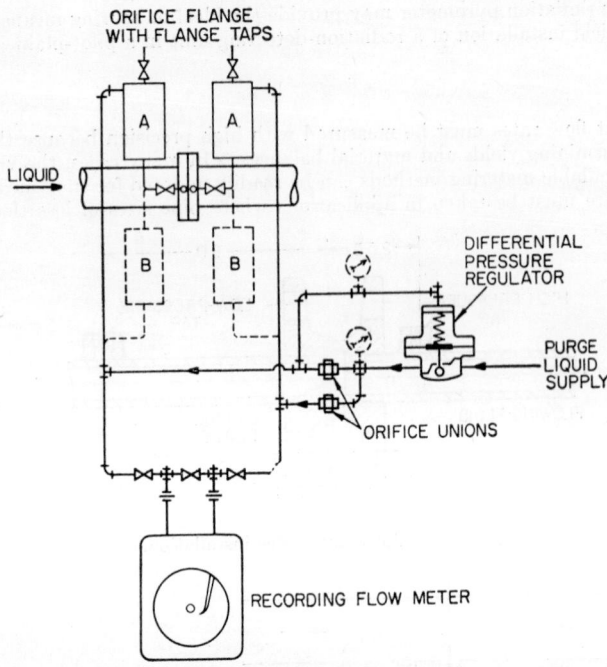

Fig. 7. Purge system for pilot-plant operations. Surge chambers: (A) flowing liquid heavier than purge, (B) flowing liquid lighter than purge.

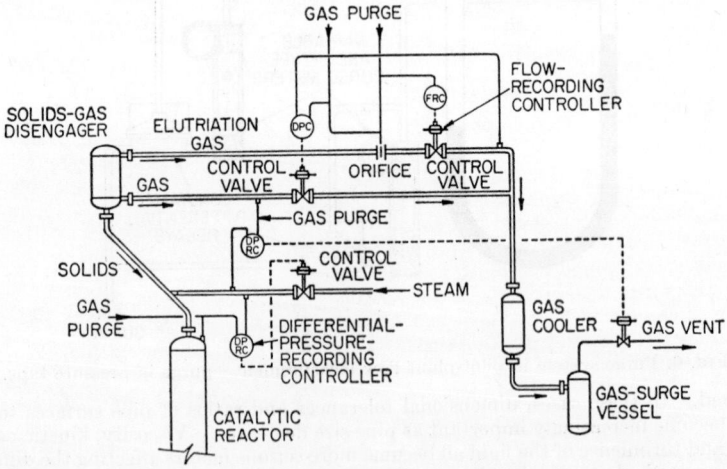

Fig. 8. Gas purges applied with seal pot in typical pilot-plant operation.

The variable-area meter (rotameter) is also widely used for flow measurements in pilot plants, particularly where line sizes are 2 in. and smaller. This type of meter can measure flows accurately (± 1 per cent of maximum flow) at rates down to 1 cc of water/min. Simple indicators provide an inexpensive means of directly indicating flow rates on a linear scale with a 12½:1 usable span. Models are available to provide either electric or pneumatic transmission to remote panel-mounted instruments,

Other methods of flow measurement are also used, such as time required to fill or empty a tank of calibrated volume. Methods such as this can be used, however, only in cases where a change in inventory within the calibrated tank will have no effect on the process.

Purging of pressure taps with a liquid or gas may be required where condensable vapors or entrained solids are carried by the measured fluid. Figure 6 shows one type of purge system, using indicating rotameters to meter the purge flow rate found to be needed and constant-differential-pressure relays to maintain the desired flow rate in spite of variations in system pressure. For higher pressures, surge chambers with capillary metering orifices and differential-pressure controllers become preferable (Fig. 7). Shown in Fig. 8 is a method of employing gas purges with a seal pot in differential-pressure measurements.

Pressure

Pressure measurement for pilot plants has been traditionally handled by dial-indicating-type bourdon-tube gages for static pressures from 5 to 50,000 psi. Below 5 psi, diaphragm-type gages, manometers, and inclined-tube draft gages are used. However, there is a current trend in larger pilot plants and semiworks installations away from manual control with handwritten log sheet records. There has also been an increase in the use of conventional instrumentation identical to the type that will be used with the full-scale operation. This has the marked advantage of indicating the controllability and control problems of the full-scale unit.

Pneumatic transmitter units are commonly used because their small size allows placing them near the point of measurement. Their small total fluid volume and displacement volume are valuable on units where total flow is small, to avoid an increase in the fluid inventory by unnecessary piping. Flexibility of the transmitter-receiver arrangement is also valuable, because at the end of the development work the instruments can be either transferred to the full-scale operation or reused in other pilot-plant activities. The potential flexibility and versatility of all-electric control systems seem to indicate that, as these systems come into greater use, they will find wide acceptance in pilot-plant installations.[13]

Standard industrial instruments are generally applicable to pilot-plant pressure measurement without much modification. However, for highly critical measurements and high precision, individually calibrated indicating gages are used. Consolidated Engineering offers a force-balance electronic pressure indicator for which they claim an accuracy of better than one-tenth of 1 per cent. At the other extreme, for measurements of pressure of from 50,000 psi and up, the manganin-type gage has been used.

A number of pressure-measurement problems in pilot-plant operation are completely neglected in some operations. In particular, there is a strong tendency in pilot-plant design toward the use of small positive-displacement pumps. The intermittent discharge of these pumps, combined with the limited volumetric capacity of most pilot-plant structures, tends to cause a considerable pulsation in pressure.

In most cases, this fluctuation is handled either by allowing the indicator to fluctuate and taking a visual average, by introducing a choke to damp the fluctuation, or by a combination of the two, using a choke to reduce the fluctuation to a small value and then visually averaging. In many cases, pressure fluctuation is a major factor in the functioning of the process. Visual averages or choked readings are decidedly questionable. Occasionally, major difficulties in full-scale operation have occurred because of neglect of these factors in the pilot plant.

The investigation of these dynamic data should be considered, because they represent a major factor in the over-all performance of the plant. Electrical transmitting units, originated primarily in the aircraft testing fields, can be used with excellent results. A true indication can be obtained by photographing an oscilloscopic presentation of the fluctuating pressure, or by a high-speed pen and ink recording on a Brush, Sanborn, or equivalent recorder. A storage or memory-type oscilloscope (Hughes Aircraft Co.) also can be used to freeze the trace for a short or long period.

Level of Liquids and Solids

Commercially available units are normally used in measuring the level of liquids and solids.[12]   In pilot plants, changes in liquid inventory appreciably affect yield, and care must be taken to eliminate the possibility of major inventory changes. Displacement-type level controllers are quite frequently used because of their high sensitivity.   Differential-pressure-type units have the advantage of less liquid contained but usually do not have as high sensitivity as is desired for level control. Electrical conductive-type level controls are used but they usually produce an on-and-off type of flow which is unsuitable for many applications.

Grid-type solid-level indicators are sometimes applied along with electrical conductive- and capacitance-type level indicators for solids.

pH and Redox

Measurement of pH in a pilot plant is used, initially, to determine the optimum pH level for proper reaction, maximum yield, and least hazard.   It is also used to control the pH at a static level to permit the proper evaluation of other variables in the system, such as reaction time, temperature, and concentration.   The pilot-plant installation can also provide valuable control-loop information, such as time constants and dead times of pH equipment and accessories.   If systems contain solids or corrosives, a study can also be made to determine what difficulties might be encountered with respect to electrode fouling and deterioration.   Remedies can usually be developed, and the pilot plant offers the surest means of establishing the proper procedure to avoid electrode coating, abrasion, or fouling.

The corrosion potential of the pilot plant lines will determine the construction of the pH flow assembly.   One type of electrode assembly has individual glands to hold each electrode.   These units can be located directly in pipelines, sides of tanks, or in special flow or immersion assemblies.   Such gland assemblies are desirable in pilot plants, where equipment size is such a critical factor.

In addition to standard industrial equipment a line-operated laboratory meter offers additional versatility and convenience.   This laboratory instrument can be used for indirect standardization of the continuous meter, for grab sampling of the less critical flows in the system, and if necessary, can be operated as a continuous pH meter that in turn can drive a recorder for auxiliary measurement.

The major problem in applying pH equipment to the pilot plant is the scaling down of the reaction vessels, the introduction of dead times corresponding to those which will be encountered in the process, and procuring adequately sized control elements.

All the equipment described above for pH measurement can be used in redox measurement as well.   The only substitution will involve interchanging the glass electrode for a metallic one, and changing the pH scale for a millivolt scale.

Other Variables

Commercially available instruments are used to measure other variables such as viscosity, gas analysis, and density.   In addition, pilot plants are used to develop or adapt instruments for measuring the quality or true composition of the product to be produced.[15]

SELECTION OF CONTROLLERS FOR PILOT PLANTS

In selecting controllers for pilot-plant operations, one must keep in mind the accuracy and sensitivity and mode required to maintain the variables within the desired limits.   Also controllers and indicators must be selected which will, over-all, perform in a manner similar to the large-scale unit.

1. Because of the low process lag in most pilot-plant applications, controllers and indicators with high response rates are required.

2. Testing prior to installation in commercial plants can best be done in pilot plants, which in some cases, include identical applications.

The pilot plant provides an unsurpassed opportunity for determination of process-control characteristics by experimental methods. Both the semiempirical method of Ziegler-Nichols[1] and the frequency-response technique[2] can be used for determination of the nature and character of control problems. The Ziegler-Nichols method will be found adequate for most pilot-plant instrumentation problems. However, the frequency-response technique is the more refined of the experimental methods. This technique requires a special variable-frequency generator in the control circuit but provides considerable additional detail in the analysis of response characteristics (see Sec. 10 also and Ref. 9, 14, and 16).

FINAL CONTROL ELEMENTS

In the pilot-plant control operations, the control valve may be subject to wide extremes of temperature, pressure, and fouling. Selection of the proper-sized valves for pilot-plant operations is particularly difficult because of the unusual capacity range in which they frequently operate and the lack of knowledge and true capacity of the flow characteristics of the valve actually installed.[11] Incorrect sizing of pilot-plant valves perhaps more than any other single factor contributes to poor performance of experimental automatic control systems. The control valve is an integral part of the over-all instrument control loop.

Proper operating characteristics of the control valve are as important as any part of the control system. Since pilot-plant studies cover a range of flow quantities and pressure drops far greater in fluctuating nature than those normally encountered in commercial operations, there is frequent need for alternation of valve-port sizes to maintain operating characteristics consistent with altered process conditions. The more frequent error in pilot-plant valve selection is that of oversizing. When this condition exists, controller sensitivity is penalized. This may lead to wide deviations in the control point when minor changes cause a disturbance and, of course, unsatis-factory control of the variable.

Control valves used in pilot plants must of course be specially designed for these applications. Commercial valves in general have excessively large valve-port leakage or valve-port clearing, and in many cases the valve must ride on its seat to control the flow, which results in poor operation. For small flows found in pilot plants, valves should be the single-seated type, actuated by a pneumatic diaphragm or piston and incorporating a valve-positioning mechanism.

For satisfactory control, a pilot-plant valve should be operated near the mid-point of its maximum travel. For this reason, it is important that actual flow-capacity characteristics of the pilot-plant valve be readily and accurately evaluated and that means be available to change the internal port construction readily when processing conditions are altered. Percentage-type valve characteristics are to be preferred, permitting flow control over a wide range of conditions.

REFERENCES

1. Ziegler, J. G., and N. B. Nichols: Optimum Settings for Automatic Controllers, *Trans. ASME*, vol. 64, pp. 759–768, 1942.
2. Caldwell, W. I.: Frequency Response Analysis, *Taylor Tech.*, vol. 4, no. 2, pp. 13–19, fall, 1951.
3. Pollard, E. F., R. M. Persell, H. J. Malaison, and E. A. Castrock: Instrumentation for Pilot Plants —Operations and Applications Fundamentals, *Ind. Eng. Chem.*, vol. 42, no. 4, pp. 748–752, 1950.
4. Anon.: Instruments Accelerate Research, *Instrumentation*, vol. 4, no. 5, pp. 17–24, 1950.
5. Goldstein, W. A.: Instrumentation for Pilot and Experimental Scale Chemical Plant, *Trans. Soc. Instr. Technol.*, vol. 5, no. 3, pp. 126–137, September, 1953.
6. Wedner, B. M., W. A. Horne, and T. P. Joyce: Pilot Plant Design, *Petrol. Eng.*, vol. 23, no. 13, pp. 11–14, 16, December, 1951.
7. Dowding, C. W., and F. R. Russell: Process Evaluation in Computer-run Microplant, *Chem. Eng.*, Oct. 30, 1961, pp. 97–103.
8. Munson, J. K., and W. F. Wagner: Computer Simulation of Pilot Plant Reactors, *Proc. Fourth Natl. ISA Chem. Petrol. Instrumentation Symposium* on Instrumentation Scale-up from Laboratory to Pilot Plant to Production, sponsored by the Wilmington Section of ISA, Apr. 9–10, 1962, pp. 35–39.
9. Bronson, S. O.: Design and Control of Small Pilot Plants, *ibid.*, pp. 43–48.
10. Stoll, H. W.: Low Flow Rate Measurement, *ibid.*, pp. 49–55.

11. Coe, B. F.: Low Flow Control Valves, *ibid.*, pp. 65–73.
12. Herbster, E. J.: Liquid Level Measurement and Control in Pilot Plant Operation, *ibid.*, pp. 75–79.
13. Clarridge, R. E.: Small Pressure Transducers, *ibid.*, pp. 81–99.
14. Read, J. E., and J. T. Ward: How to Instrument a Pilot Plant, *ibid.*, pp. 101–107.
15. Williams, J. W.: Adapting Analyzers to Process Scale-up, *ibid.*, pp. 109–113.
16. Keeler, R. M., and R. D. McCoy: How Much Instrumentation, *ibid.*, pp. 115–123.
17. Garrett, D. G.: Bench-scale Pilot Plants, *Tech. Paper No.* 39, A.I.Ch.E. Symposium, New York, March 17, 1959.
18. Kern, D. Q.: Scale-up Aspects of the Pilot Plant, *Tech. Paper No.* 65, A.I.Ch.E. Symposium, New York, March 17, 1959.
19. Lindstedt, P. M., C. T. Winchester, K. G. Roquemore, and E. W. Campbell: The Design and Layout of a Pilot Plant for Elastomers, *Tech. Paper No.* 68, A.I.Ch.E. Symposium, New York, March 17, 1959.
20. Boycks, E. C., W. Priestley, Jr., and W. F. Taylor: The Performance of a Computer Controlled Pilot Plant, *Tech. Paper No.* 67, A.I.Ch.E. Symposium, New York, March 17, 1959.
21. Fanning, R. J., and J. D. Bryan: What About Pilot Plant Instrumentation?, *Automatic Control*, vol. 7, no. 7, p. 85, July, 1960.
22. Hamilton, L. W., W. F. Johnston, and R. D. Petersen: Pilot Plants and Technical Considerations, *Chem. Eng. Progr.*, vol. 58, no. 2, pp. 51–54, February, 1962.
23. Novack, Joseph, Robert O. Lynn, and Edwin C. Harrington: Process Scale-up by Sequential Experimentation and Mathematical Optimization, *Chem. Eng. Progr.*, vol. 58, no. 2, pp. 55–59, February, 1962.
24. Davis, Robert S.: Statistically Designed Pilot Plant Experiments, *Chem. Eng. Progr.*, vol. 58, no. 2, pp. 60–63, February, 1962.
25. Karasek, F. W.: Stream Analysis and Data Reduction in Pilot Plants, *Control Eng.*, vol. 8, no. 12, pp. 93–95, December, 1961.
26. Crandall, W. A.: A Pilot Plant for Evaluating Analyzers, *ISA J.*, vol. 8, no. 8, pp. 83–87, August, 1961.

ENVIRONMENTAL TEST INSTRUMENTATION

By Harold C. Jones*

Environmental test instrumentation could be construed to include any instrument used for measuring the surroundings of an object during test. This broad definition would include devices such as those used on satellites for measurement of cosmic radiation, meteorite density, and similar instrumentation. This definition also would include the entire gamut of meteorological instrumentation as used in connection with weather balloons. On the other hand, a rather large body of information and scientific discipline has come into being in the last several decades which embraces the field of *environmental simulation*. It is this field of environmental simulation that is covered in this subsection.

SCOPE OF INSTRUMENTATION

Environmental simulation pertains to the simulation on earth of the environment in which a person or an object may be placed in any of its travels and uses throughout the surface of the earth and, more recently, travel into space.

Environmental simulation requires instrumentation for the measurement of (1) vibration, (2) shock, (3) temperature, (4) altitude, (5) relative humidity, and (6) acceleration. There are additional, more specialized environmental variables which must be measured from time to time, but these shall not be covered in detail here. The instruments used may be direct-reading or recording.

VIBRATION INSTRUMENTATION

An extensive line of instrumentation has been developed for the measurement of vibration.

Paste-on Triangle

The simplest vibration measurement device is the paste-on triangle shown in Fig. 1a. This triangle is observed during the test period of vibration. By noting where the figure "becomes double," as shown in Fig. 1b, a rough determination of vibration can be made. Although it is a rather crude type of instrumental technique, the paste-on triangle has the peculiar property that it can be used where the addition of weight (thus changing the characteristics of the specimen) cannot be tolerated.

Criteria for Vibration Measurements

This peculiar property of the paste-on triangle emphasizes the first criterion for the measurement of vibration—*the instrument used must not appreciably change either the amplitude or frequency of the vibrating member to which it is attached.*

The second criterion is common to practically all forms of instrumentation—*the instrument shall reproduce faithfully the parameter which it is measuring.* A vibration-

* Electrical Engineering Department, University of Maryland; Consulting Engineer (Environmental Testing and Reliability).

measuring device measures the *amplitude, frequency,* and *waveshape* of a vibration. The vibration per se may or may not be sinusoidal, and it is important to know this fact.

Basic Types of Vibration-measuring Instruments

With these basic criteria in mind, the instruments used for vibration measurement can be categorized roughly as (1) vibrometers, (2) velocity pickups, and (3) accelerometers. These three classes of instruments correspond to the three categories of measurements which they make.

A vibrometer measures the amplitude x of the vibration. The velocity pickup measures the velocity dx/dt of the motion. The accelerometer measures the acceleration d^2x/dt^2 of the motion. If the motion is known to be sinusoidal, it is a simple

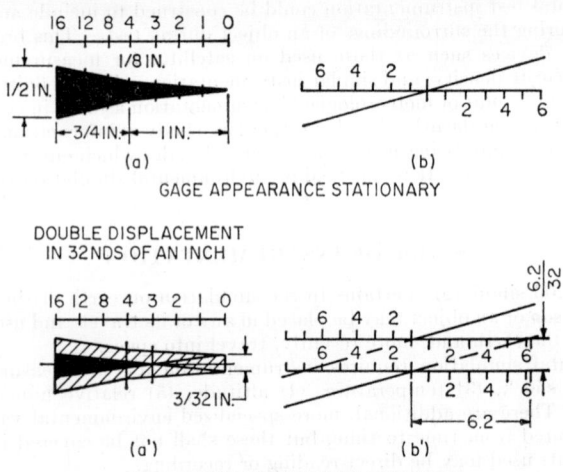

FIG. 1. Paste-on type vibration-amplitude indicator.

matter to insert the time dependence of the displacement x into the original formula and by differentiating to find dx/dt and d^2x/dt^2. In the case of the accelerometer, the value of the acceleration expressed in units of gravity (g's) is

$$a = 0.0511 \, Df^2 \tag{1}$$

where D = peak-to-peak displacement of motion, in.
$\quad f$ = frequency of the motion, cps

Vibrometers

As shown in Fig. 2, one type of vibrometer simply is a dial indicator which is held by hand and has associated with it quite a heavy block (mass). The actuating tip of the dial indicator is placed against the vibrating member, and the indicator is read by noting the extremities of motion of the indicating pointer. The useful range of this instrument is from approximately 10 to 100 cps. The device will measure amplitudes from approximately 0.001 in. to slightly greater than 0.100 in. The device is limited to from 2 to 3 g's.

The basic principle of this device has been modified in several ways. One commercially available instrument* incorporates a mirror and light source. The prod actuates the mirror. In another instrument,† the amplitude changes a capacitor and

* General Electric.
† General Radio.

this, in turn, is amplified.   Still another modification uses a moving element suspended in a seismic mass and actually scribes a plate so that a record of the vibration amplitude can be made.   A further modification uses a plate that moves at a known rate, thereby indicating the frequency and waveshape of the vibration.

The general use for this kind of device lies in probing machinery for undesirable vibrations.   Generally, it is not used for laboratory analysis of equipment that is mounted on a vibration exciter.

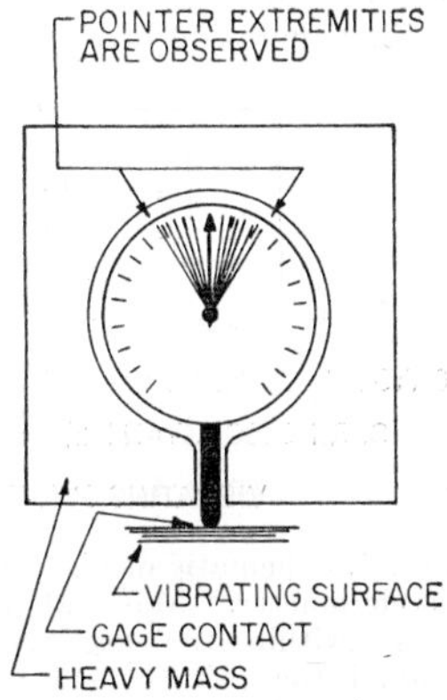

Vibration-velocity Pickups

Of large value for low-frequency work in laboratory vibration analysis is the properly applied velocity pickup. As shown in Fig. 3, the basic construction utilizes a seismic mass with a coil, such that the coil can move through a magnetic structure.   Thus, as the case of the instrument is vibrated, there is a motion of the magnetic field about a coil which will induce in that coil a voltage proportional to the rate of change of the flux linkages—which, in turn, is proportional to the velocity of the relative motion.

POINTER EXTREMITIES ARE OBSERVED

VIBRATING SURFACE
GAGE CONTACT
HEAVY MASS

FIG. 2. Vibrometer.

Several precautions must be observed in the use of this type of instrument.   Where a seismic mass is involved, the *natural frequency* of such a system must be considered. Commercially available pickups have a natural frequency in the vicinity of 5 cps and may be obtained either damped to about 0.65 of critical damping or undamped.   The lower frequency limit for use of this instrument, therefore, should be considered in the vicinity of 10 cps—since even the undamped instrument will have a 30 per cent error,

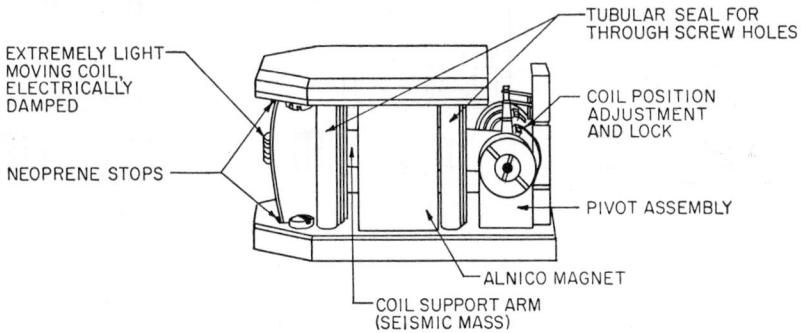

EXTREMELY LIGHT MOVING COIL, ELECTRICALLY DAMPED

NEOPRENE STOPS

TUBULAR SEAL FOR THROUGH SCREW HOLES

COIL POSITION ADJUSTMENT AND LOCK

PIVOT ASSEMBLY

ALNICO MAGNET
COIL SUPPORT ARM (SEISMIC MASS)

FIG. 3. Velocity pickup (type 124) manufactured by MB Electronics, a Division of Textron Electronics, Inc.

approximately, at this frequency.   A unit with a damping factor of approximately 0.6 of critical can be used as low as 7 to 8 cps.*

Another precaution is concerned with the existence of any magnetic field in the vicinity of the vibration exciter.   Since most vibration exciters use a very large magnetic field, even the use of a degaussing coil may allow a fairly high field to exist in the vicinity where the measurements are made.   This field can drastically affect the calibration of the velocity pickup.   If it is essential to use this type of pickup under such conditions, magnetic shielding is the best approach for protection—but a calibration must be made with the instrument mounted in exactly the position in which it will be used during the test period.

---

* For further details concerning the relationship between damping and error, refer to Cyril M. Harris (ed.), "Handbook of Noise Control," McGraw-Hill Book Company, Inc., New York, 1957.

Piezoelectric Pickups

This widely used device depends for its output upon the principle of piezoelectricity —the generation of a voltage by a crystal when it is deformed in a certain direction.

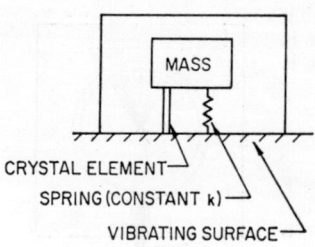

CRYSTAL ELEMENT—
SPRING (CONSTANT k)—
VIBRATING SURFACE—

FIG. 4. Schematic of piezoelectric accelerometer. Mass M is the mass within the accelerometer and k is the spring constant of the crystal and any other associated spring elements.

Generally, barium titanate or a similar crystal material is used. Deformation of the crystal is provided by a mass which, through its inertia, acts on the crystal element and thereby deforms it. Essentially, this can be considered as a single-degree-of-freedom system without damping, as shown in Fig. 4.

The voltage is proportional to the compression of the crystal, or the x direction in the illustration. This compression, in turn, is proportional to the force exerted on it by the mass. The force exerted by the mass is, in turn, proportional to the acceleration which is applied to the case. Thus these pickups are termed *accelerometers*, since they directly measure the acceleration of the members to which they are attached.

As shown in Fig. 5, several methods can be used to produce this action.

Basic Compression Design (Fig. 5a). The mass is torqued down against the crystal. The spring elements are the crystal and the case wall.

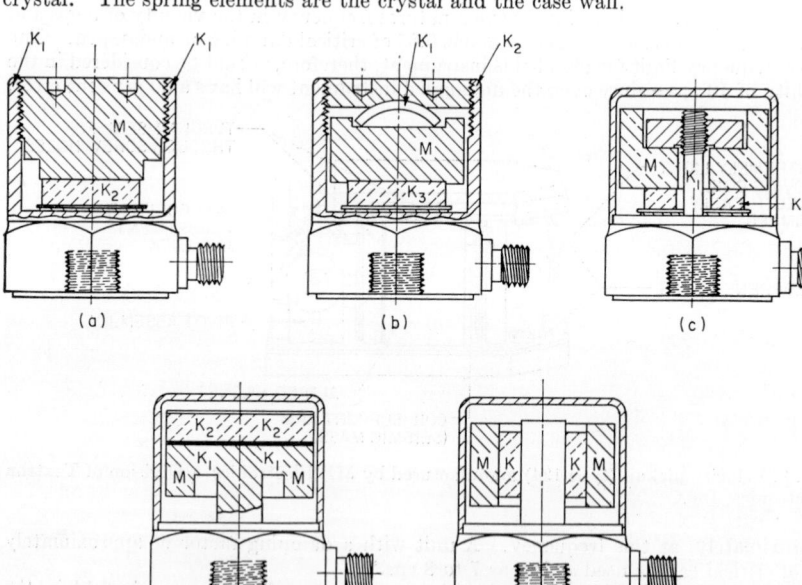

FIG. 5. Types of vibration-measuring accelerometers: (a) Basic compression design; (b) isolated compression; (c) single-ended compression; (d) bender design; and (e) shear design. (*Courtesy of Endevco Corp.*)

Isolated Compression Design (Fig. 5b). An internal spring in part isolates the system from case-wall phenomena. The spring members are the crystal, internal spring, and case wall.

Single-ended Compression Design (Fig. 5c). In this design, there is no direct mechanical connection from the internal elements to the case wall or lid. The mass is compressed against the crystal by using a nut which is torqued down on a threaded center post. The spring members are the crystal and the center post.

Bender Design (Fig. 5d). The mass consists of an element with uniform or non-uniform cross section (nonuniform illustrated). The spring members are the stiffnesses of the crystal and mass assemblies in bending.

Shear Design (Fig. 5e). The annular crystal is bonded to a center post, and an annular mass is bonded to the outside of the crystal.

The basic compression design is the most sensitive to case effects such as (1) outputs due to transient temperature; (2) outputs due to varying cable tension as a result of cable whip; or (3) in some cases, changes in sensitivity due to variations in mounting torque.

The single-ended compression, shear, and bender designs all feature no contact of the internal elements to the case wall or lid. The single-ended compression and shear designs are more rugged in that there are no stress concentrations, such as are present with the bender design. The shear design is particularly appropriate for small transducers; the single-ended compression designs are appropriate for high output.

The primary differences of the methods shown in Fig. 5 lie in the sensitivity, total range, and errors. In the bender design, the outer portion of the mass tends to remain still as the center portion of the mass moves, thus providing a bending action of the crystal. If the vibratory motion is sinusoidal, there also will be sinusoidal velocity and sinusoidal acceleration—all related by different constants. Since this device uses the acceleration to produce a voltage, it must be stressed that, if a constant displacement amplitude is maintained and the frequency is decreased, the force decreases as the frequency decreases. This is true because there is a frequency-squared term in the expression for the force in g's. This means that, at lower frequencies, even relatively large amplitudes will produce a fairly low force and, therefore, a fairly low output.

In application, the light weight of the accelerometer makes it a very versatile tool for use in the study of vibration where the g levels are appreciable. Care must be exercised in the handling of the output of the accelerometer. As can be noted from Fig. 6, the accelerometer can be represented as a charge generator in parallel with a capacitor, the value of which is determined by the type of crystal material used, the construction, and the external cable capacitance. The voltage output, therefore, will be equal to the charge divided by this external capacitor—the charge being a direct function of the deformation of the crystal. Thus the voltage output of the accelerometer will be a function of the acceleration that is applied to the instrument.

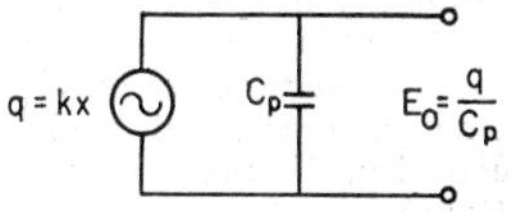

Fig. 6. Electrical equivalent circuit of a piezoelectric accelerometer. The charge generator produces a charge proportional to the deformation.

The fact that there is a capacitance in the denominator of the equation means that care must be taken in the choice of loads into which the charge generator can be operated. If the capacitance is in the order of magnitude of 500 $\mu\mu f$, the input impedance of the electronic system for measurement should be at least 100 megohms. Some accelerometers have been made with a considerably larger capacitance in the order of magnitude of 7,000 $\mu\mu f$. These can be used with a vacuum-tube voltmeter or other device with an input impedance in the range of 2 to 10 megohms.

Care must be taken concerning the cable leading from the accelerometer to the input of the amplifier, or cathode follower, which usually is used. With such a high input impedance, pickup in the cable is a strong possibility. Thus only the highest-quality shielded cable should be used for this connection.

Care must also be taken as regards the design of the cathode follower in order to maintain this high input impedance. It has been the author's experience that money is saved in the long run by purchasing commercially built and marketed amplifiers and cathode followers which are matched to the system and which, over many years of development, have been perfected to the point where the impedance is correct and very little noise is introduced by the system.

The frequency response of the accelerometer on the high end of the spectrum can be taken directly from an undamped system, since the damping of the crystal is very low.   Therefore, the relative frequency response can be stated as

$$R = \frac{1}{1 - B^2} \qquad (2)$$

where B = ratio of operating frequency to natural frequency of the crystal, mass combination

Thus an accelerometer when used at approximately one-fifth of the natural frequency will have a response which is only 4 per cent above the input amplitude.   For most vibration measurements, this is quite adequate, and 20 per cent is a widely used upper range.

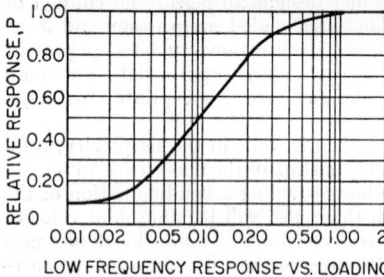

LOW FREQUENCY RESPONSE VS. LOADING

Fig. 7.   Crystal frequency response. Actual response at any frequency can be measured from this graph, where f is the frequency in cps; r is the input impedance in ohms of the matching amplifier; c is the total capacitance in farads of the accelerometer, plus additional applied shunt capacity, if any.   (*Courtesy of Endevco Corp.*)

The low-frequency response of the crystal is shown in Fig. 7, where the relative response is on the ordinate and the product of the frequency times the amplifier input impedance times the total capacitance of the accelerometer, shunt cable, and calibrating capacitors (if used) is plotted on the abscissa.   Therefore, given any lower frequency and the physical characteristics of the system, the response can be calculated.

Auxiliary Instruments

In addition to the instrumentation used for directly measuring the vibration amplitude and frequency, it is also necessary in many instances to use visual observation— in order to establish the mode of vibration and also to indicate possible methods of correction of high resonance.   A stroboscope is a very useful instrument for this purpose.   An instrument* of recent design is very convenient for this use—since it provides a high level of illumination and still is reasonably lightweight for moving around and holding in the position to give maximum information.

In using the stroboscope, the frequency of the flashes must be kept slightly above or slightly below the frequency of the actual vibration—so that a very slow relative motion is seen to exist.   A precaution—it is very common to interpret an increasing frequency as an increasing amplitude when observing a resonance under a strobe light. Therefore, it is important to keep approximately the same slip between the vibration frequency and the strobe-lamp frequency.   The operator should track the changing frequency as resonance is approached, peaked, and passed.

While this is not too difficult to do manually, an instrument† is commercially available which will do this automatically.   This is a very useful piece of equipment to have in the well-equipped vibration laboratory.

Observation under a strobe lamp is useful only when the amplitude to be observed is reasonably high.   Thus this method is more useful in the low-frequency vibration ranges than as higher frequencies are approached.   In the high-frequency areas, the audible effects of resonances often are noted.   The use of a "well-educated finger" often is quite good for determining individual amplifications when the amplitude is too small to be observed with a strobe lamp.   For quantitative information, the lighter accelerometers would have to be used.

* General Radio.
† Hellmuth Company, Slip-Sync.

General Arrangements and Accuracy Requirements

Figure 8 represents, in block-diagram form, a typical setup for the performance of a vibration test. The command for either displacement or acceleration is set into the control console. This, in turn, is amplified by the amplifier and drives the moving element of the vibration exciter. Either from the built-in pickup coil in the exciter or from an external accelerometer, a return signal is presented to the control console. When the level of this return signal matches that of the command signal (as in any servosystem), an equilibrium is reached and that level is maintained.

If a displacement is being commanded, two integrations of the acceleration signal are required in the electric circuitry of the control console. If a signal coil is used, the return signal must be integrated once. It is desirable for the operator to have the alternative available within the control console of either (1) controlling by the motion of the vibration exciter coil or (2) controlling by an accelerometer (or velocity pickup)

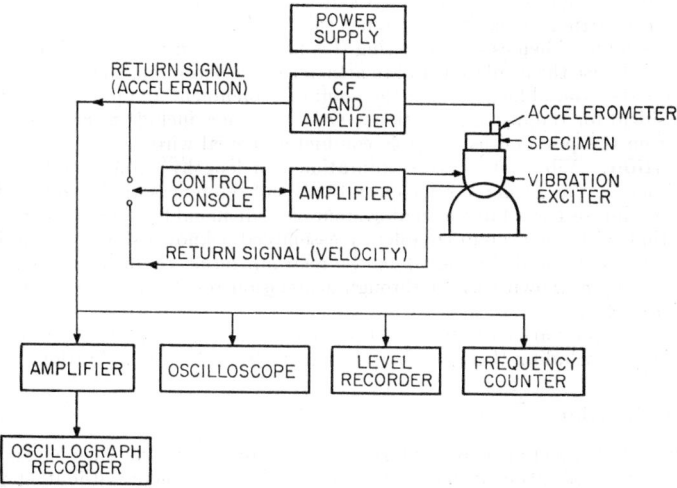

FIG. 8. Typical vibration-test setup.

mounted on the specimen itself. In many instances, it is desired to maintain the vibration at a given level at a given spot within the specimen. This type of flexible setup makes this possible.

A modification of the same basic approach, when fixture problems arise, involves mounting the accelerometer or velocity pickup adjacent to the mounting point of the specimen on the vibration fixture. Thus, if the fixture goes into resonance, the amplitude of the vibration exciter automatically will be decreased to maintain the proper input at the mounting point of the specimen.

The remainder of the instrumentation shown in Fig. 8 is concerned with the formation of a recorded set of data which may be retained and analyzed. The accelerometer signal passes through a cathode follower and amplifier and then is put through an amplifier which, in turn, will drive the low-impedance galvanometers on an oscillographic recorder. It is always desirable to view the signal on an oscilloscope, to detect its rms amplitude by a level recorder, and to make a record of this amplitude. An oscillographic recorder and a level recorder are both desirable because, in many instances, the level recorder will be of the rms type and will give a distorted answer as to the peak. This can occur if the waveshape is not truly a sine wave. On the other hand, reading information from an oscillographic record is quite time-consuming and also adds to the cost of the setup.

Lastly, in Fig. 8, a frequency counter is included so that at any given time the frequency of the vibration can be determined. The setup of Fig. 8 is simple. In most instances, several accelerometers will be recorded.

Accuracy requirements for a vibration test must be tailored to the needs for each specific test. Thus generalizations on this subject are difficult to provide. General military requirements for the accuracy of measurement of the vibration amplitude are plus or minus 10 per cent. The allowable tolerance on the frequency is plus or minus 2 per cent.

Some precautions are in order concerning the proper use of accelerometers. Since the signal being generated by the accelerometer is at a very low level, the creation of a "ground loop" can destroy the accuracy of measurement. In any type of accelerometer, where one side of the signal is tied to the case, an incipient ground loop always is present. If the accelerometer is screwed down to the test specimen, which in turn is tied to the vibration exciter, there will be a complete ground loop from the console to the shaker to the specimen to the case of the accelerometer to the shield of the cable—and back to the console. Such a loop can carry relatively high currents and thus induce a fairly strong signal into the inner conductor.

There are many schemes for correcting this problem. In the author's opinion, one of the best is to use the insulated stud to attach the accelerometer to the test specimen. This breaks the ground loop right at the point of attachment and provides an effective Faraday shield for the instrumentation. Other schemes include isolating the pickup element from the case and using triple-conductor coaxial wire.

Calibration. Vibration pickup calibration is a lengthy subject and will not be discussed here. The reader is referred to the list of references at the end of this subsection. Of interest is a fairly recent system of built-in calibration in the accelerometer—so that with the system completely assembled a known signal can be inserted in the accelerometer and this signal can be read out on all the instruments. This is done by passing a known current through a precision resistor mounted in the base of the accelerometer.

Table 1 lists the various types of accelerometers and pickups which are available commercially—including their peculiar characteristics and self-calibration.

Random Vibration

Although the subject of random vibration is too complex for detailed coverage here, no discussion of vibration instrumentation would be complete without at least an introduction to this field. Historically, the concept of random vibration is the out-

Table 1a. Characteristics of Various Accelerometers and Vibration Pickups

	Manufacturer*					
	MB	MB	MB	MB	Cons.	Cons.
Type.........	Moving coil not damped	Moving coil electrically damped	Moving coil electrically damped	Moving coil not damped	Moving coil fluid damped	Moving coil electrically damped
Velocity sensitivity, volts/cm/sec	25.6×10^{-3}	25.6×10^{-3}	25.6×10^{-3}	25.6×10^{-3}	30.0×10^{-3}	28.24×10^{-3}
Weight, oz.....	12	11.2	9.7	9.7	10	1.25
Max acceleration G.......	500	100	100	100	35	
Frequency range, cps....	3–2,000	2–1,500	2–1,500	7–2,500	2–700	70–500
Max temp., °F.	250	500	250	250	150	500

* MB = MB Electronics, Division of Textron Electronics, Inc. Cons. = Consolidated Electrodynamics, Subsidiary of Bell & Howell.

Table 1b. Characteristics of Three Accelerometer Types

Manufacturer and type No.[a]	Sensitivity, pk. mv/pk. g	Frequency response, 100-megohm load	Weight, oz.	Remarks
E-2206	12	±5% 7.5–2,000 cps	1.6	Liquid-cooled 2000°F at base, 500°F at cable
E-2211	25	±5% 7–8,000 cps	0.83	For shock use
E-2213	45	±5% 6–6,000 cps	1.0	General purpose
E-2213M5	43	±5% 6–6,000 cps	1.0	Hermetic seal
E-2215	8	±5% 2–6,000 cps	1.0	High cap. to 2,000 pk. g sine
E-2217	72	±5% 20–6,000 cps	1.1	High sensitivity Hermetic seal
E-2219	350	±5% 2.5–2,500 cps	2.75	Very high sensitivity
E-2221C	13	±5% 7.5–6,000 cps	0.39	350°F ring-shaped
E2223C	10	±5% 5.5–4,000 cps	1.4	Triaxial 350°F
E-2224C	9.5	±5% 5.5–5,000 cps	0.56	Top connector 350°F
E-2225	0.6	±10% 5–15,000 cps	0.46	Top connector for shock use
E-2226	5.0	±5% 10–5,000 cps	0.1	Microminiature adhesion mounting
E-2227	5.0	±5% 10–5,000 cps	0.11	Microminiature stud mounting
E-2232	8	±5% 5–8,000 cps	0.70	Hermetic seal −65°F to +350°F
E-2235	25	±5% 5–8,000 cps	1.09	Hermetic seal −65°F to +350°F
E-2242	10	±5% 25–7,000 cps	1.0	−320°F to +500°F
E-2242M5	12	±5% 5–4,000 cps[b]	1.4	−65°F to +500°F
C-300	25	±5% 1–10,000 cps[c]	0.8	Calibration insertion Freq. standard and shock use
C-301	20	±5% 1–10,000 cps	0.55	Calibration insertion Freq. standard and shock use
C-302	50	±5% 1–6,000 cps[c]	0.80	General purpose
C-303	40	±5% 1–6,000 cps[c]	0.55	General purpose
C-200	25	±5% 2–10,000 cps[c]	0.19	Miniature
C-201	15	±5% 2–10,000 cps[c]	0.10	Miniature
C-606	2.5	±5% 5–10,000 cps[c]	0.035	Subminiature
C-607	20.0	±5% 5–10,000 cps[c]	0.018	Subminiature
C-408-TX	40	±5% 1–2,500 cps[c]	2.1	Triaxial
C-200HT	20	±5% 2–10,000 cps[c]	0.19	500°F
C-300HT	20	±5% 1–10,000 cps[d]	0.81	500°F
C-302HT	45	±5% 1–6,000 cps[d]	0.81	500°F
C-402HT	40	±5% 1–10,000 cps[d]	1.41	500°F
C-606HT	2	±5% 5–10,000 cps[c]	0.035	500°F
C-408TXHT	35	±5% 1–2,500 cps[d]	2.1	500°F Triaxial
C-508TXHT	20	±5% 2–6,000 cps[c]	0.66	500°F Triaxial
G-A314	1.25[e]	3–9,000 cps[f]	0.11	−65 to +250°F
G-A314TM	1.25[e]	3–9,000 cps[f]	0.20	−65 to +250°F
G-A315	1.25[e]	3–9,000 cps[f]	0.07	−65 to +250°F
G-A315TM	1.25[e]	3–9,000 cps[f]	0.12	−65 to +250°F
G-A316T	1.25[e]	3–9,000 cps[f]	0.12	−65 to +250°F
G-A320	7[e]	3–3,000 cps[f]	0.18	−65 to +250°F
G-A320TM	7[e]	3–3,000 cps[f]	0.29	−65 to +250°F
G-A321	7[e]	3–3,000 cps[f]	0.14	−65 to +250°F
G-A321TM	7[e]	3–3,000 cps[f]	0.19	−65 to +250°F
G-A323T	7[e]	3–3,000 cps[f]	0.16	−65 to +250°F
G-A371TM	2.5[e]	3–5,000 cps[f]	0.22	−65 to +250°F
G-A380M	35[e]	3–2,500	0.64	−65 to +250°F
G-A380TM	35[e]	3–2,500	0.81	−65 to +250°F
G-A381M	35[e]	3–2,500	0.49	−65 to +250°F
G-A381TM	35[e]	3–2,500	0.63	−65 to +250°F
G-A391TM	0.75[e]	3–18,000	0.14	−65 to +250°F
G-A395M	8.5[e]	3–8,500	0.49	−65 to +250°F
G-A395TM	8.5[e]	3–8,500	0.63	−65 to +250°F
G-A396M	8.5[e]	3–8,500	0.42	−65 to +250°F
G-A396TM	8.5[e]	3–8,500	0.56	−65 to +250°F
G-A3540MU	2.5[e]	3–2,500	0.99	−65 to +250°F
G-AC475TK	7.5[e]	3–7,500	0.78	−65 to +250°F
G-A30	15[e]	3–15,000	1.23	−65 to +250°F Triaxial
G-A31	10[e]	3–10,000	1.23	−65 to +250°F Triaxial
G-A40	3[e]	3–3,000	1.23	−65 to +250°F Triaxial
G-AHT315TM	1[e]	3–8,000	0.12	−100°F to +350°F
G-AHT321TM	5[e]	3–3,000	0.29	−100°F to +350°F

Table 1b. Characteristics of Three Accelerometer Types (*Continued*)

Manufacturer and type No.[a]	Sensitivity, pk. mv/pk. g	Frequency response, 100-megohm load	Weight, oz.	Remarks
G-AHT381TM...	30	3–2,500	0.71	−100°F to +350°F
G-AHT396TM...	7.5	3–8,500	0.56	−100°F to +350°F
G-AHTC475TK..	20	3–7,500	0.77	−100°F to +350°F
G-AHT31T......	1	3–9,000	1.83	−100 to +350°F Triaxial
G-AHT40T......	5	3–3,000	1.83	−100 to +350°F Triaxial
G-A3100T.......	4.5	10–2,000	0.35	Ring-shaped

   [a] E indicates Endevco Corporation.
   C indicates Columbia Research Laboratory.
   G indicates Gulton Industries.
   [b] Working into a 1,000-megohm load.
   [c] Manufacturer gives response working into his cathode follower with input of about 5,000 megohms.
   [d] 1,600-megohm load.
   [e] All Gulton sensitivities are minimum.   Other listed sensitivities are nominal.
   [f] Upper limit is 30% of lowest resonant frequency.

growth of the analysis of recordings of vibrations on aircraft and missiles.   In analyzing these recordings, it was found that the recordings contained a background of random vibration with certain predominant frequencies superimposed on it.   It was proposed, therefore, that this type of vibration should be imposed on specimens which were to be used in this type of environment.

To do this, the basic system presented earlier in this section must be modified—first by the introduction of a random noise generator—and also by the inclusion of many filters and other measuring devices.

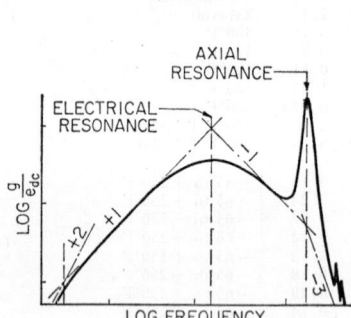

Fɪɢ. 9. Exciter frequency response. (*Courtesy of MB Electronics, a Division of Textron Electronics, Inc.*)

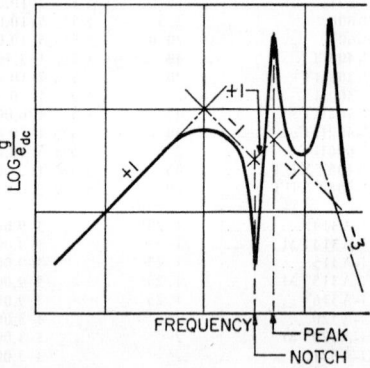

Fɪɢ. 10. Frequency response of mechanical system with one load resonance. (*Courtesy of MB Electronics, a Division of Textron Electronics, Inc.*)

Compensating Vibration-table Frequency Response.   The first filter system which must be installed is for the purpose of correcting the response of the exciter itself.   If a random-noise signal is to be reproduced faithfully by the vibration table, then the vibration table must have a frequency response which is flat, that is, with no variation over the range of frequencies desired.   Unfortunately, the existing vibration exciters do not have such a response.   Instead, their response is of the nature shown in Fig. 9.   Since the response of the system cannot be changed, the system must be electrically compensated for this response by preemphasis and deemphasis in an equalization network.   Once the unloaded table has been equalized, a load can be put on the table, and if the load has a resonance, the uncompensated system will have been changed to one as shown in Fig. 10.

Compensating for Load Resonance. To compensate for this additional notch and peak, a "peak-notch" filter is installed which will compensate for the effects of the load resonance. This is *not* to say that the load will not go into resonance. This merely keeps the input at the mounting points of the equipment constant as a function of frequency, thereby maintaining the true random-noise input to the specimen.

Principle of Basic Random-vibration System. To facilitate the performance of the equalization function, much work has been done by the major vibration-equipment manufacturers. Recently, automatically compensated systems, or systems in which compensation may be rapidly obtained, have become commercially available. It will serve to better illustrate what is occurring, however, if one of the older random-vibration systems is examined. The basic block diagram of such a system is presented in Fig. 11 and operates as follows: In the input section are four different possible inputs. The first is the noise generator, which generates a random noise over a wide spectrum. In addition to this, a continuous-wave signal generator is mounted in the input section and can be patched into the shaker and mixer which is fed from the bandpass filter—through which the random-noise-generator output is passed. Use of the bandpass filter limits the noise spectrum—usually between 20 and 2,000 cycles. The continuous-wave signal generator adds any discrete frequencies which may be required to simulate the environment properly, such as a known burning-reasonance frequency of a rocket engine.

The wideband oscillator and frequency cycler are included so that the system can be used for purely sinusoidal operation, as discussed previously. The tape playback allows the system to be used for vibration of a specimen in the same manner as it will be subjected to vibration in its environment. Thus, if tape data from the vehicle are available, this information may be played directly into the vibration exciter—and the unit then subjected to this vibration. Then, analysis and reconstruction of the characteristics of the signal are not necessary.

In the equalization section, the necessary peak-notch filters and exciter equalizers are included. The displacement limiter also is necessary, since by its very nature, random noise contains some extremely high displacements—occurring at very wide intervals. Such displacements, if allowed to actuate the table of the system, will trip out the limit switches and put the entire system out of operation. Thus they should be removed before they reach the power amplifier.

In the output section are the various recording and monitoring systems which have been found to be useful for this type of vibration work. It will be noted that the signal monitors feed back to a crossover control and servo control in the same manner as previously discussed for sinusoidal vibration. In use, the signal generator is swept over the entire system —with the signal going directly to the exciter equalizer, bypassing the peak-notch filters. This is done with no load on the table. Using the panoramic analyzer, the exciter equalizer is adjusted to obtain a flat frequency response over the desired frequency range.

Then, the specimen is placed on the table and the sinusoidal signal is sent into the peak-notch filter—and then on to the exciter equalizer. The peak-notch filters then are used to take out the peak and notch characteristics which are caused by the resonance in the specimen. This, of course, is all done at a low level of amplitude so that the specimen is not damaged during this initial setup period of testing. The continuous-wave signal generator then is taken out of the circuit, the noise generator is introduced, and the proper level is set for the random-noise test. While this appears straightforward, it should be understood that this process of equalization and compensation is quite long and contains a great deal of "cutting and trying." The automatic, or at least improved manual compensation systems do essentially the same thing but do it in a far more expeditious manner than the system just discussed.

Modern Random-vibration System. A more up-to-date system for random vibration is shown in Fig. 12. In this type of system, the main difference lies in the use of a *filter analyzer* and *filter equalizer* to perform the equalization. The peak-notch filter is included only for use in case a peak or notch exceeds the range of the equalizer. The filter analyzer consists of a group of bandpass filters with bandwidths of from 14 cps on the low end to 100 cps on the high end. The filter bands are in juxtaposition

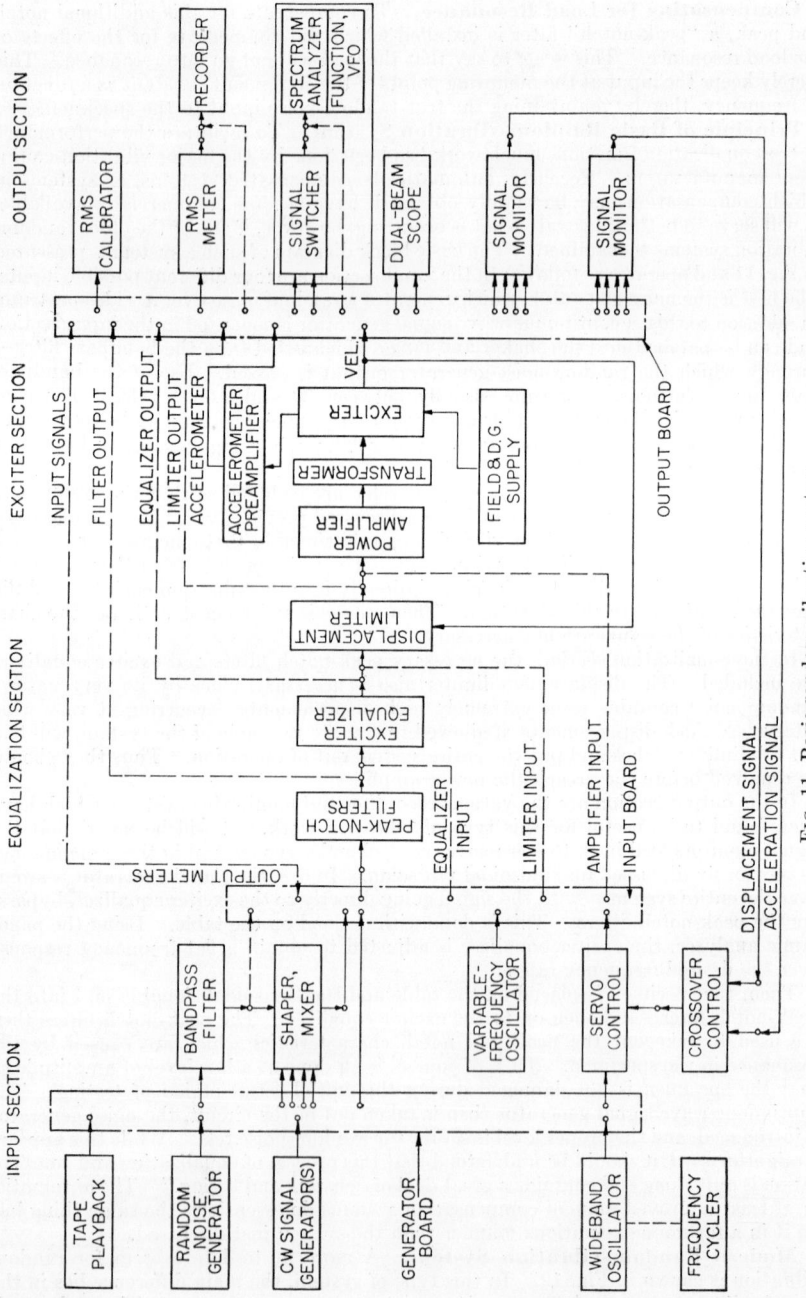

Fig. 11. Basic random-vibration system.

in the frequency spectrum. Thus each filter samples a section of the spectrum of the accelerometer signal and the output of the filter is fed to a meter which records the power level of the sample. Thus, by using 26 filters to cover the band from 10 to 2,000 cps, a reading of the power spectral density in each band can be combined to give the power spectral density distribution as a function of frequency. The readout meters are arranged in a line so that a flat distribution will appear as a straight line. The filter equalizer is composed of a similar series of filters, the output of each of which is individually controlled by a vertically moving lever which controls a potentiometer.

In use, a low level of random excitation is applied and the filter equalizer is adjusted to provide the desired spectral-density shape as read on the analyzer. The level of the random excitation then is increased to the desired level, final adjustment is made

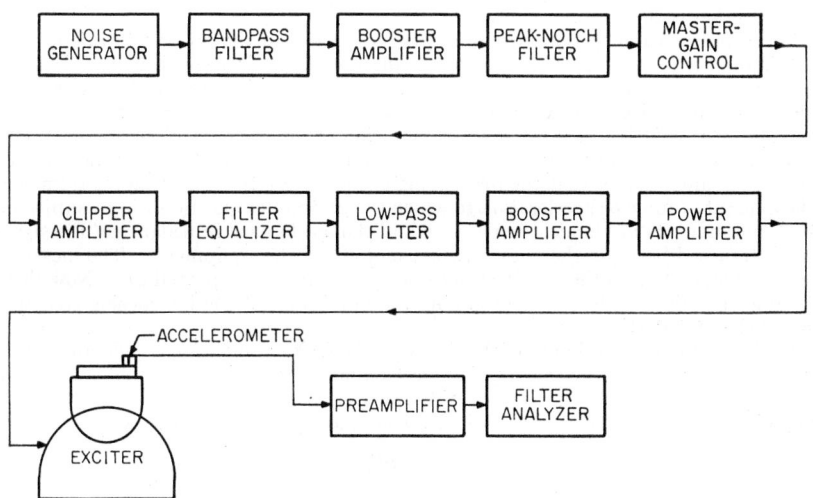

Fig. 12. Modern random-vibration system. (*Courtesy of Endevco Corp.*)

on the equalizer, and the equalization is complete. This method of equalization is much faster than the system previously described and also gives a more accurate equalization.

Force and Mechanical-impedance Testing

One of the newer techniques in the field of vibration testing involves the use of force as a specified parameter as opposed to the usual displacement or velocity specification. The logic for this derives from the fact that a jet engine or rocket engine, which is generating a vibration, actually is generating a vibratory force. When this force is transmitted into a piece of equipment, the equipment sees a driving force—not a driving acceleration.

Some investigation has been made into the feasibility of using force as a controlling parameter on vibration exciters, as opposed to the usual acceleration. There are many approaches to the problem of measuring this force and care must be exercised in selection of method. A method which the author has found satisfactory under some conditions involves the use of a piezoelectric-crystal type of force transducer. With this system, the force transducer is placed between the vibration-exciter head and the mounting fixture for the specimen. The servosystem of the vibration exciter then is controlled from the output of the force gage, rather than from the output from an accelerometer. In this way, a constant force may be programmed into the vibration machine and testing with a force as a parameter can be achieved.

Another system for achieving the same end would involve the use of a ring on which bonded strain gages (SR-4 type) are placed.   The output from these strain gages then is amplified and used as the force signal.   In either the piezoelectric or strain-gage system, the transducer element (crystal or strain gage) actually is measuring the deformation of the spring.   In the case of the strain-gage system, this would be the ring on which the strain gages are mounted.   In the case of the piezoelectric transducer, this is a combination of both the crystal per se and the outside stainless-steel shell of the gage.

Mechanical Impedance.   Another development which holds great promise in the field of vibration analysis is that of mechanical impedance.   The concept of impedance has been used by the electrical engineering field for a great many years, and although known in the mechanical engineering field, little use has been made of it in connection with vibration analysis.

Recently, instrumentation developments have renewed the interest in this type of testing, and it is quite possible that much analysis in the future will be based on the mechanical impedance of a given system.   The underlying reason for desiring to know the mechanical impedance of a piece of equipment lies in the difficult area of deciding just what is a realistic level of vibration to impose on a piece of equipment.

Assume that there is a component mounted on a piece of structure within a missile body.   Assume further that the force being generated by the rocket motor is known. The amount of force or acceleration to which the unit under question will be subjected is a function of its mechanical impedance and the mechanical impedance of the structure interposed between the missile-rocket motor and the equipment.   The measurement of the force from the rocket motor is a commonly known procedure.   Now there remains only the problem of determining the impedance of the interposing structure and of the unit itself.

To determine the mechanical impedance of the unit, a small force generator and a system composed of a force gage and an accelerometer can be used to excite the equipment.   Thus a measurement is made of the acceleration and of the force at the mounting point of the equipment in question.   The following formula applies:

$$Z = \frac{\omega F}{x} \tag{3}$$

where Z = mechanical impedance
ω = radial frequency
F = driving sinusoidal force
x = sinusoidal acceleration

The instruments for accomplishing this can be either a force gage and an accelerometer, as discussed previously, or a mechanical-impedance head in which the force-measuring element and the accelerometer are compactly packaged together with two outputs from the over-all unit.   The output from this unit is fed into the conventional cathode followers and the output from the cathode followers then either may be read directly or a simple analog circuit can be prepared to divide the force by the acceleration and multiply by the frequency—thus giving the answer for the mechanical impedance directly.   The frequency then can be swept over the band of interest and a direct recording made of the mechanical impedance.

SHOCK INSTRUMENTATION

Instrumentation for shock testing is very similar to that for vibration testing. Most techniques used in vibration testing are directly applicable to shock testing. There are a few additional problems and precautions which must be taken.

In the observation and measurement of a shock pulse, it may be necessary to eliminate some of the high-frequency hash, or ringing, which occurs in the system lest no readable data be obtained.   To do this, a bandpass filter usually is inserted in the electric circuit ahead of the recorder or oscilloscope on which the pulse is to be observed.

These high frequencies come from two sources: (1) there is the ringing of the structure itself after the shock pulse; and (2) there is the resonance in the accelerometer or associated instruments. If an instrumentation system is used which has no resonance below approximately 50,000 cps and if the specimen which is being measured is a reasonably dead structure (has very few high-frequency resonances in it), it is then advantageous either to eliminate the bandpass filter or to observe simultaneously the unfiltered pulse from the shock. Ideally, the system of measurement should reproduce the frequency band of from 2 cps to well above 10,000 cps with complete fidelity. This means that the resonant frequency of the sensing element and instrumentation should be above 100,000 cps ideally. If the Q of the resonance is quite high, 50,000 cps is fairly acceptable. Figure 13 shows a typical setup for the measurement of the shock impulse. The purpose of the miniature switch is to trigger the trace on the oscilloscope prior to the actual start of the pulse—so that the actual impulse which is to be observed is moved to the right on the face of the oscilloscope, thereby allowing the entire pulse to be seen.

An impact recorder also proves useful in certain shock-testing applications. This device uses a pendulum weight which scribes a moving waxed tape to indicate the amplitude of the shock. Units of this type* are capable of recording up to 35 g and

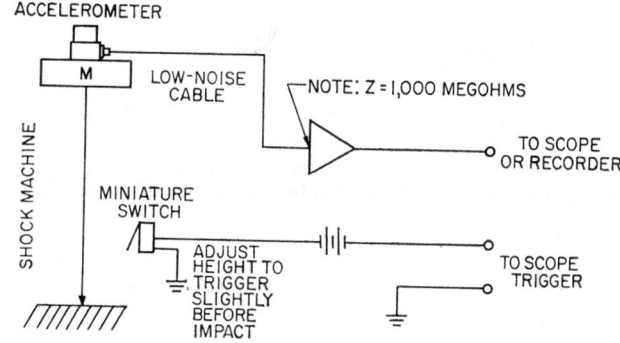

Fig. 13. Typical shock-measurement setup. (*Courtesy of Endevco Corp.*)

show the magnitude of the shock pulse in all three planes. These instruments are particularly useful when it is desired to know the shipping shocks a unit is being subjected to, since they are small enough to be packed right with the unit. The tape is transported by a clock mechanism and can run as long as 28 days.

ACCELERATION MEASUREMENT

Although not performed as frequently as shock and vibration testing, a third type of dynamic test performed in the environmental laboratory is the *constant-acceleration test.* This usually is done on a centrifuge where the specimen is rotated on an arm to provide the desired acceleration. The purpose of the test is to simulate either launch, reentry, or retrieval of a missile or aircraft where fairly long periods of sustained acceleration are encountered, thus differentiating it from a shock pulse. Therefore, any instrument used to measure this force must be capable of reproducing the steady-state condition, meaning that its frequency response must be as low as zero cps. An instrument† of this type is shown in Fig. 14.

An unbonded strain-gage wire is wrapped around sapphire pins and a weight is connected to the movable portion of the system. Thus, when it is subjected to acceleration, the weight tends to move out and is restrained by the spring action of the instrument. The distance of motion will be indicated as a strain of the strain-gage wires, and this may be detected by the usual bridge circuit for measurement of resist-

---

* Impact-O-Graph.
† Statham Instruments, Inc.

ance. The natural frequency of such an instrument is between 20 and 500 cps. Also see Sec. 5 under Acceleration Measurement.

It is important to note in this type of testing that a constant acceleration cannot be applied to all portions of a specimen whose dimensions are an appreciable fraction of

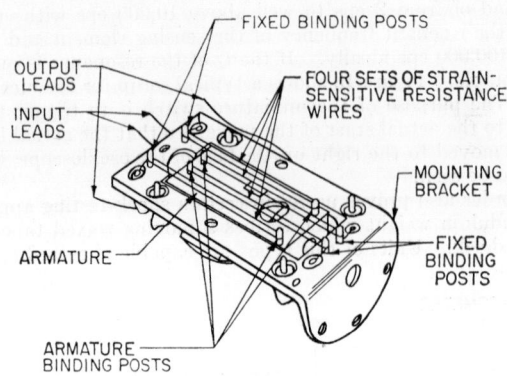

Fig. 14. Unbonded strain-gage accelerometer. (*Courtesy of Statham Instruments, Inc.*)

the total length of the distance from the center of rotation to the center of mounting of the specimen. Therefore, it is important to define the point where a given acceleration is to be experienced by the specimen.

The accelerometer should be mounted at this point. In most tests of this type, the reaction of structural members of the specimen is of prime importance.

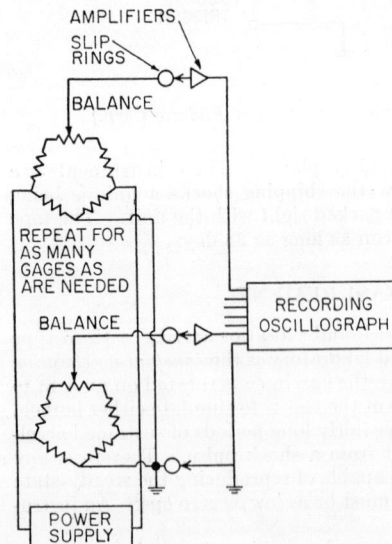

Fig. 15. Strain-gage recording setup for centrifuge testing.

Centrifuge Test Setups. The only strain-gage techniques peculiar to environmental instrumentation arise from the necessity for transmitting the signal from a rotating body. To do this, several systems may be used. First, and probably most unsatisfactory, is to mount the measuring instrument, which essentially is a Wheatstone bridge, outside the centrifuge and pass the wires from the strain-gage element through slip rings to the instrument. The very minute change in resistance of the strain gage (to indicate the strain on the member to which it is cemented) is *swamped out* completely by the variation in the slip-ring resistance as a function of angular position. This system, therefore, is difficult to apply successfully.

A far better system is to mount a strain-gage measuring instrument within the rotating member in as close a proximity to the center of rotation as possibly can be achieved and bring out only the leads to a recording system. In this type of system, as shown in Fig. 15, the bridge is balanced under static conditions and the output from the bridge is fed through an amplifier into a galvanometer. When the centrifuge is rotated and a strain occurs, the bridge will be unbalanced and the amount of unbalance will be indicated by the galvanometer. In practice, a recording oscillograph will be used so that many strain gages can be recorded simultaneously as a function of time. This type of system, where the amplifier has a high input impedance, does not introduce any appreciable error as a function of the rotational variations in the resistance of the slip rings.

Still another system which is useful for this type of test involves a carrier system in which the change in resistance modulates a carrier. Thus a frequency-modulated signal is generated—with the frequency change proportional to strain. Any change in resistance of the slip rings only results in a change in amplitude, and this is not sensed by the instrument. Further details of this type of system can be found in texts on telemetry and strain-gage measurements.

When more than one strain gage is used, an automatic stepping switch mounted within the rotating system usually is included. Care must be taken to be certain that only the highest-quality switch is used—so that there will be no appreciable change in resistance due to change in contact resistance of the switch. Telephone-type switches with gold-plated contacts have been used satisfactorily. Rotary switches with solenoid drives, using the multileaf wiper and stud principle for their operation, have also been used. The simple wafer switch, as generally found in most radio circuits, should be avoided, since its resistance is not sufficiently constant.

Use of Television. Closed-circuit television also is used for centrifuge tests. Whenever there is anything on the centrifuge which can be visibly observed to indicate the action of the centrifugal force on the specimen, it is highly useful to have a television camera mounted on the rotating element—transmitting to a remote monitor at the operator's position. In the performance of centrifuge tests on human beings, this is essential. In many types of equipment, visual observation greatly enhances the tests. The television camera should be mounted as close to the axis of rotation as possible—so that it sees only a minimum amount of force on its circuitry.

ALTITUDE MEASUREMENT

In the simulation of flight conditions, a chamber is used to simulate the altitude under which the equipment will be required to operate. Therefore, the simulated altitude must be measured inside the chamber. In considering the altitudes normally encountered by manned flight in aerodynamic aircraft, altitudes ranging up to 100,000 ft are involved. Since 100,000 ft is equivalent to approximately 0.159 psia, in accordance with the standard NACA atmosphere chart, it can be seen that this resolves itself into the measurement of a not-too-high vacuum.

Most of the specifications to which the environmental engineer will be working require that this altitude be measured with an accuracy of plus or minus 5 per cent. At first glance, this seems quite reasonable. However, in applying this 5 per cent figure, which of course refers to the altitude in feet, to the pressure, it can be seen that, at an altitude of 100,000 ft, 5 per cent represents plus or minus 5,000 ft. This represents a pressure change of only 0.201 to 0.125 psia. If one instrument were required to read from sea level (15 psia) to 100,000 ft, the plus or minus 5 per cent altitude tolerance would impose a pressure tolerance of 0.076 psia—which is an accuracy of 0.506 per cent of full scale.

This accuracy can be achieved in several ways. First, by splitting the range, that is, by having a bellows-type recording controller which covers from sea level to 40,000 ft, from 40,000 to 80,000 ft, and from 80,000 to 150,000 ft. The instruments used are standard pressure-measuring instruments commercially available. These instruments are of the bellows-and-spring type.

To the environmental test engineer, the measurement is far less difficult than obtaining a controlling system with accuracy and stability. To do this, it is almost mandatory to use a full proportioning system where both the vacuum valve and the bleed valve are proportioned from signals out of the control instruments. One system which works very well is to use three pneumatic-type recorders with the ranges as previously given, and with a three-position valve which will automatically switch to the recorder which is controlling. The control air pressure then operates the vacuum and bleed valves through proportioning devices. The bleed is set to open on rising pressures above the mid-pressure and the vacuum valve is set to open on falling pressures below the mid-pressure.

Through experimentation with such a system, it has been found that an actual overlap at the mid-pressure, allowing both the vacuum and bleed valves to be slightly open, provides the best control. The use of a purely on-off system, either of the vacuum pump or of the vacuum valve, usually results in inadequate control.

Another system utilizes a sensing element that is basically a bellows which actuates strain gages. The strain gages are then measured with a conventional recording-controlling instrument. In order to achieve the required accuracy up to 100,000 ft, two ranges must be used.

When the simulation enters the field of satellites, rockets, and other spacecraft, an entirely new set of problems is encountered. Under these conditions, the term *altitude* no longer is applicable. Absolute pressure, expressed in microns, usually is used just as it is employed in other high-vacuum work. The measurement techniques involved are approximately the same as have been developed for high-vacuum measurements.*

TEMPERATURE MEASUREMENT

In the performance of an environmental test, it is necessary to measure the temperatures of many different parts of the equipment, and also the ambient atmospheric conditions within the test chamber. Three general classes of sensing elements are used: (1) thermocouples, (2) resistance-thermometer bulbs, and (3) filled-system thermometers.† Since thermocouples are commonly used and since these devices pose certain special problems in their application, a few basic principles are reviewed here.

Thermocouple Use. In standard commercially available thermocouples (or thermocouples made from commercially available thermocouple wire), lack of homogeneity in the wires can cause deviation in temperature readings of as great as plus or minus 1°C. Errors also will be introduced by the readout instruments. Therefore, thermocouple measurements under such conditions cannot be classified as precision measurements.

In general, if the thermocouple is calibrated and properly used, however, the results will be within the required accuracy of the test. The specifications on accuracy which are most usually used and which are quoted by many government test specifications are plus or minus 2°C.

The following general rules must be observed in order to obtain good thermocouple measurements:

1. The thermocouple must be in intimate thermal contact with the point on the equipment which is to be measured. This can be accomplished by way of clamping or gluing. An epoxy resin is widely used.

2. The thermocouple wires must be small enough so that the heat flow which occurs along them will not appreciably affect the temperature of the specimen being measured. If the thermocouple is very small or made of low-specific-heat material, the wires must be kept very small. On the other hand, they must be as large as practical, keeping the heat-flow problem in mind, so as to prevent breakage and other application problems which attend very small thermocouples.

3. The thermocouple must be small enough so that the junction will accurately follow any variations in temperature which the specimen may undergo. The largest wire (up to approximately No. 20 wire gage) which meets these conditions will make the best thermocouple for the purpose.

4. The thermocouple junction must be insulated from the ambient air surrounding the specimen so that it reads the true temperature of the specimen and does not assume some interim temperature between that of the specimen and that of the surrounding atmosphere.

* See Richard B. Lawrance, High-vacuum Measurement, pp. 3-64 to 3-89, "Process Instruments and Controls Handbook," McGraw-Hill Book Company, Inc., New York, 1957.
† See Sec. 2, p. 2-13, also Sec. 2, pp. 2-5 to 2-20, pp. 2-47 to 2-59, and pp. 2-68 to 2-84 of "Process Instruments and Controls Handbook," McGraw-Hill Book Company, Inc., New York, 1957.

Multiple-temperature Recorders. Of much use in the performance of an environmental test in which temperatures must be measured is an automatic multipoint thermocouple recorder. In many instances, a six- or twelve-point recorder is used for most important temperatures; while secondary temperatures are measured by a switch and potentiometric bridge. In general, the potentiometric bridges which have built-in reference junctions are useful for temperatures above 0°C. If it is anticipated that many measurements are to be made in the low-temperature range, it is useful to provide a 0°C melting-ice-bath reference junction. If this in not available, the reference-junction compensator must be set to zero, the temperature at the binding posts of the potentiometer measured, the thermocouple tables consulted to obtain the number of millivolts which correspond to this junction temperature, and this number of millivolts added algebraically to the actual reading of the potentiometric bridge. Temperatures below zero are defined as giving a negative voltage.

Although resistance-thermometer bulbs are excellent as regards their sensitivity and linearity when used in a Wheatstone-bridge circuit, most designs do not provide satisfactory results where rapid temperature changes are to be recorded. The relatively large thermal mass of the sensing element itself tends to integrate any rapid fluctuations. Thus these fluctuations will not be seen in the final recording.

Another device which, in many instances, is of use is heat-sensitive enamel* which changes color at a given temperature. This device proves useful for the indication of temperatures on parts to which a thermocouple cannot be affixed, or where thermocouple leads cannot be brought out to an instrument. This paint is applied on the spot where the temperature is desired to be known. At the conclusion of the test, the color is compared with a standard color chart from which the temperature can be determined. Although not a very accurate measurement, this method can be used to determine temperatures within plus or minus 1 per cent.

A similar device† changes to jet black when exposed to the stated temperature on the plate. These devices are adhesive-backed and can be obtained for temperatures from 100 to 500°F in 50°F increments.

RELATIVE HUMIDITY

In the field of environmental simulation, more controversy has centered around the measurement and control of relative humidity than has occurred for almost any other environment which is treated. The reason for this can be seen by examining the basic specifications which are used by the military for the performance of these tests. The allowable tolerance on test-condition *measurements* is required to be, in some instances, plus or minus 5 per cent; in other instances, plus 5 per cent and minus 0 per cent. The latter tolerance band is the more recent requirement.

Since tests frequently are operated in a cycle in which the temperature rises from approximately 102 to 122°F and where it is required to maintain a relative humidity in excess of 95 per cent, it can be seen from the psychrometric chart that the wet-bulb depression varies between 1.5 and 1.7°F. One measurement system which is used quite widely consists of a wet-bulb recording-controlling instrument and a dry-bulb recording-controlling instrument. These instruments are guaranteed to have an accuracy of plus or minus 1 per cent of total span and usually cover the span of from 0 to 200°F.

Assume that an instrument is indicating 120°F and exactly 95 per cent relative humidity (which is 1.5°F depression of the wet bulb). This means an indicated 120°F dry bulb and a 118.5°F wet bulb. If the dry-bulb instrument indicates on the low end of its tolerance band and if the wet-bulb instrument indicates on the high end of its tolerance band, there actually will be dry bulb of 122°F and a wet bulb of 116.5°F, giving a relative humidity of 84 per cent. This type of system, therefore, clearly does not meet the requirements of the military specifications. The solution to this problem lies in a different type of instrumentation.

---

* Tempelac.
† Temp-Plates (manufactured by Pyrodyne).

For high-humidity measurements, differential thermocouples can be used, that is, one thermocouple which measures the dry-bulb and another thermocouple which measures the wet-bulb temperature. The two thermocouples are connected in series opposition and feed the basic instrument. This system tends to be reasonably *noisy* in that there are very low signal levels with such a system.

A superior system, if the thermal lag of the sensing element can be tolerated, is that of using a Wheatstone-bridge type of circuit with the wet-bulb temperature in one leg of the bridge and the dry-bulb sensing element in the other leg of the bridge. A circuit is shown in Fig. 16.

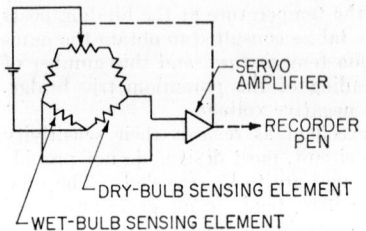

FIG. 16. Wet-bulb depression recorder-controller.

This system can be checked rapidly by removing the wet sock from the wet-bulb sensor and by varying the temperature over its required range and ascertaining that a zero degree differential is maintained. With this system, it is not uncommon to have the full range of the instrument calibrated to 5°F difference between wet bulb and dry bulb. Once this type of measurement has been attained, however, it usually is found that it is very difficult to maintain the actual test chamber within the required limits. Thus care should be exercised in the application of such a sensing system— unless the entire humidity chamber has been designed to use this very fine type of control.

In the measurement of lower humidities, the Dunmore cell, as originally developed by the National Bureau of Standards, can be useful. The principle involved is that of the variation in resistance between the two wires of a bifilar winding as humidity is absorbed in a surface coating which exists between the two wires. Then the only item that is basically needed is a high-resistance ohmmeter to measure the resistance change. This type of instrument is portable. Since the sensing elements are relatively inexpensive, they have been included on the inside of cocoons which house the guns of the mothball fleet. The leads from the sensing element are brought out to a plug on the outside of the cocoon.

The measuring instrument is hooked up at this point and a relative-humidity measurement on the inside can be made. If the humidity is too high, the drying agent is replaced.

This type of system does not solve the problem of measuring high humidities, since the curves for the instrument become quite close together as the higher humidities are approached, and hence accuracy is lost. One approach to this problem is to heat the air through a known increase in temperature, measure the humidity with a Dunmore cell, and knowing the temperature at which the measurement is made, enter the steam tables and refer this measurement back to what the condition must have been in the incoming air before it was heated. On the surface, this appears to be a relatively satisfactory measurement, but with existing steam tables and psychrometric charts, much of the accuracy is lost in reconversion. Thus the need is for a measurement of high humidity with some type of a differential hookup between wet bulb and dry bulb.

Of large importance in any system using a wet-bulb thermometer is the proper care of the wet sock. First, the velocity of the air over the sock should be between 800 and 1,000 fpm. Second, the water for wetting the sock should be distilled and must be at the chamber temperature. Third, the sock must cover all of the wet-bulb sensing element. Proper maintenance and frequent replacement of the wet sock are absolutely essential if the accuracy of the system is to be maintained.

For other discussions of humidity measurement, see Sec. 7, page 7-11 of this handbook; and W. F. Hickes, Psychrometers, Hygrometers, and Dew-point Meters, pp. 7-60 to 7-72, "Process Instruments and Controls Handbook," McGraw-Hill Book Company, Inc., New York, 1957.

ACOUSTICAL ENVIRONMENTS*

Acoustical noise is encountered as high-intensity noise and as low-intensity noise. High-intensity noise is generated by rockets and jets and is of concern because of the damage it causes to both personnel and equipment. Low-intensity noise as generated by shipboard equipment may permit detection in the case of submarines. In either case, the noise may be airborne or structure-borne.

The measurement of low-intensity noise differs from vibration measurement in that the item to be measured creates the motion and the amplitudes usually are quite low. A second complication results from the frequency range of from 25 to 8,000 cps required by some specifications. In some instances, measurements to 15 kc have been specified. A third difference results from the complex signal which is generated. This corresponds to random vibration except that it is of a lower level. The last major difference is the airborne noise where the measurements are made of the vibration of the air.

The instrumentation for measuring low-intensity airborne and structure-borne noise over measurable levels and frequencies required by many noise specifications is the same except for the transducer and the type of analysis. High-level noise recording utilizes the same basic instrumentation except that a high-level microphone† must be used. The transducer for airborne noise will be a microphone. This may be a capacitor microphone, a dynamic microphone, or a crystal microphone. The transducer for structure-borne noise will be a high-sensitivity crystal accelerometer. The frequency range prohibits the use of other transducers, and the low-level signal requires extremely high sensitivities.

The cathode follower or preamplifier must have a very low internal noise level. Older specifications permit correction for noisy instrumentation, but the newer specifications do not. The amplifier should be equipped with a weighting network and a meter. This then will serve as a sound-level meter for broadband noise measurements.

Present specifications require a one-third octave analysis of both airborne and structure-borne noise. Older specifications require the use of an octave-band analysis. Conflict can be avoided here by obtaining a filter which has both octave band and one-third octave bands. The Amercian Standards Association is rewriting ASA Z24.10-1953 to include a one-third octave–octave band filter.

The presentation of the data remains a very real problem, especially if production quantities are involved. The level recorder provides a tape which can be preprinted with frequency and level. This can be used with a transparent overlay to permit acceptance or rejection immediately.

A narrow-band analyzer also is required. This is used to determine the discrete frequencies which are causing trouble and must be corrected.

For further information concerning acoustical measurements, see Sec. 3, page 3-83 of this handbook.

OTHER ENVIRONMENTS

The other environments, such as sand and dust, salt spray, fungi, sunshine, and explosive atmosphere, require little other than the normal type of instrumentation.

Explosion Testing. One instrument is useful in the conduction of an explosion test. This is a small sampling chamber‡ which is used to test the explosiveness of a mixture in the explosion chamber without, at the same time, exploding the entire chamber. The device is built on the principle of a miner's lamp in that a double-screened enclosure is made. Within this enclosure, a spark plug is mounted to ignite the mixture. To determine whether or not an explosion has taken place, a thermo-

---

* The information in this section was provided by Joseph Hobbins, Litton Systems, Inc., Maryland Div.

† Gulton MA299501.

‡ To the best of the author's knowledge, this device is not presently marketed commercially, but its construction is relatively simple. The author first encountered this device at Westinghouse where the device was conceived by W. W. Hill.

couple is mounted within the enclosure.   Just before the spark plug is fired, the bridge-reading thermocouple is balanced so that the galvanometer is on zero.   The spark plug then is fired.   If the mixture within the sampling chamber ingites, the galvanometer travels upscale, thus verifying the explosiveness of the mixture—as required by military specifications.

REFERENCES

1. Roeser, W. F., and S. T. Lonberger: Methods of Testing Thermocouples and Thermocouple Materials, *Natl. Bur. Std. (U.S.) Circ.* 590.
2. Roeser, W. F.: Thermocouple Electricity, *J. Appl. Phys.*, vol. 2, no. 6, pp. 388–497, June, 1940.
3. Swindell: Calibration of Liquid-in-glass Thermometers, *Natl. Bur. Std. (U.S.) Circ.* 600.
4. Mueller, E. F.: Precision Resistance Thermometry, in "Temperature, Its Measurement and Control in Science and Industry," Reinhold Publishing Corporation, New York, 1941.
5. Thomas, J. L.: Precision Resistors and Their Measurement, *Natl. Bur. Std. (U.S.) Circ.* 470.
6. Roeser, W. F., and E. F. Mueller: Measurement of Surface Temperatures, *J. Res. Natl. Bur. Std.*, vol. 5, p. 793, *Res. Paper* 231.
7. Wexler, Arnold, and W. G. Brombacher: Methods of Measuring Humidity and Testing Hygrometers, *Natl. Bur. Std. (U.S.) Circ.* 512.
8. Wexler, Arnold: Electric Hygrometers, *Natl. Bur. Std. (U.S.) Circ.* 586.
9. Den Hartog, J. P.: "Mechanical Vibrations," 4th ed., McGraw-Hill Book Company, Inc., New York, 1956.
10. Weiss, D. E.: Design and Application of Accelerometers, *Proc. SESA*, vol. 4, no. 2, p. 89, 1947.
11. Welch, W. P.: A Proposed New Shock Measuring Instrument, *Proc. SESA*, vol. 5, no. 1, p. 39, 1947.
12. Wildhack, W. A., and R. O. Smith: A Basic Method of Determining the Dynamic Characteristics of Accelerometers by Rotation, *Proc. ISA*, 9th Annual Meeting, Instrument Society of America, Pittsburgh, Pa.
13. Harris, Cyril M.: "Handbook of Noise Control," McGraw-Hill Book Company, Inc., New York, 1957.

ELECTRONIC LABORATORY
AND RESEARCH INSTRUMENTATION*

The following text discusses some of the various types of instrumentation used to make many of the electrical measurements required in an electronics research and development laboratory. The text is divided into five sections: (1) voltage and current; (2) power; (3) frequency; (4) resistance, inductance, capacitance; and (5) impedance.

Space limitations preclude a comprehensive treatment of measurements, or even a complete listing of all available types of instruments and methods of measurement. Emphasis has been placed on instruments and measuring systems used routinely rather than on laboratory-standards instruments. Several more comprehensive treatments are listed in the references.

VOLTAGE AND CURRENT

Perhaps the key of any dynamic measuring system is a sensitive ammeter in which the movement of the indicating element varies in accordance with the current through an associated control apparatus or circuit. Many different methods are used for relating the amount of meter movement to accurate determination of the desired electrical quantity.

In recent years digital measuring techniques have been developed in which the result is displayed in digital form rather than as a meter deflection. Digital techniques are widely used for measuring frequency, time interval, and voltage. They are also applicable, and used to a lesser extent, for making measurements of other quantities such as resistance and capacitance.

Oscilloscopes are also used for making certain types of voltage and current measurements. Presentation of the actual waveform on a calibrated graticule is particularly useful for measurement of complex waveforms and high-speed transients.

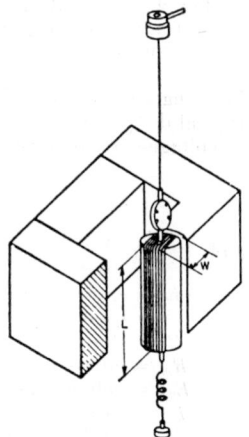

D-C Measuring Instruments and Methods

Most d-c and voltage measurements are made with small portable instruments which incorporate moving-coil meters. Inexpensive single- or general-purpose (voltage-current-resistance) meters are available to measure a wide range of values. Measurement accuracy will depend on the quality

FIG. 1. d'Arsonval galvanometer.

of the meter movement and on the type of circuitry associated with the meter.

The Indicating Meter. In the basic moving-coil meter (d'Arsonval galvanometer) as shown in Fig. 1, current is indicated as a function of coil motion or rotation within a magnetic field. A pointer is attached to the current-carrying coil, which moves freely and has an angular motion of

$$\theta = \frac{BNAI}{K} \qquad (1)$$

* By Engineering Staff of the Hewlett-Packard Company, Palo Alto, Calif.

where θ = angular rotation of the coil, radians
B = flux density of the magnetic field, webers/sq m
N = number of turns of wire in the coil
A = cross-sectional area of the coil, sq m
I = current in the coil, amp
K = spring constant of the system

The moving-coil meter is basically a *current*-measuring device. Shunt resistors of various values are used to control the amount of current which causes full-scale deflection for a given voltage. *Voltage* can be measured with the same meter by

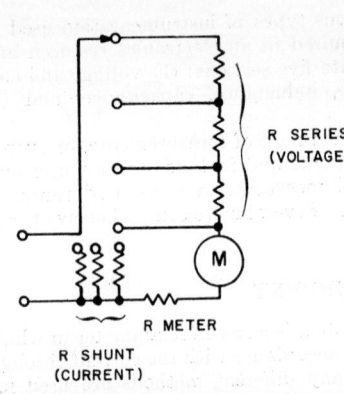

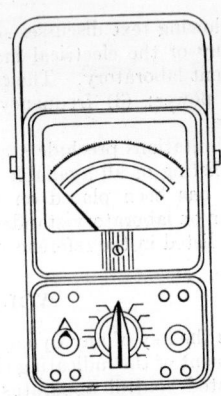

FIG. 2. Series and shunt resistors for measuring voltage and current with a moving-coil meter.

FIG. 3. Typical volt-ohm-ammeter.

inserting series resistors in the current path. Series-shunt resistor arrangement for a typical d-c volt-ammeter is shown in Fig. 2.

Voltage-current-resistance relationships are

$$i_m = \frac{R_s i}{R_s + R_m} \tag{2}$$

and

$$E_{fs} = I_{fs} R_t \tag{3}$$

where i_m = current through the meter movement, amp
i = total current into the metering circuit, amp
R_s = shunt resistance across the meter, ohms
R_m = internal resistance of the meter movement, ohms
E_{fs} = voltage applied for full-scale meter deflection, volts
I_{fs} = current through the meter movement for full-scale deflection, amp
R_t = total resistance of the metering circuit as seen from the input terminals, ohms

An important characteristic of a d-c volt-ammeter is its sensitivity.

$$\frac{R_t}{E_{fs}} = \frac{1}{I_{fs}} \quad \text{ohms/volt} \tag{4}$$

Typical full-scale currents for moving-coil meters are from 1 ma to 50 μa; typical sensitivities are 1,000 to 50,000 ohms/volt. Measurement accuracy is determined by the characteristics of the moving-coil meter and by the extent that the circuit being measured is loaded by connection of the metering circuit. Accuracies of ± 1 to ± 2 per cent are typically obtained. Figure 3 shows a typical volt-ohm-ammeter—a combination triple-purpose laboratory instrument.

Null-detection Potentiometers. A calibrated precision potentiometer in conjunction with a standard cell can be used for extremely accurate measurements of *voltage* or *current*. Potentiometric instruments are often called null detectors since the measuring process involves adjusting the potentiometer for a null (center position) on a zero-center meter.

A typical potentiometer circuit is shown in Fig. 4. The potentiometer is first calibrated by placing the switch in position *A*, setting the precision potentiometer to some convenient position, and adjusting the calibrating rheostat for a null on the meter. The voltage drop from 1 to 2 then equals the voltage of the standard cell.

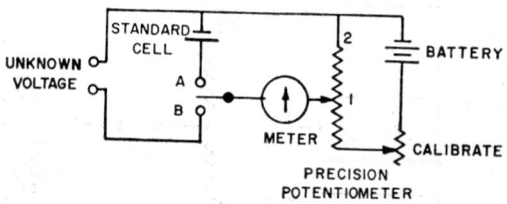

FIG. 4. Potentiometric null detector for measuring voltage and current.

Then, with the switch set to position *B* and the unknown voltage applied, the potentiometer is adjusted until the meter again indicates a null. The value of the unknown voltage is determined from the position of the potentiometer.

The potentiometric method can be used to measure *currents* by measuring the voltage which the unknown current develops across a standard resistor.

D-C Vacuum-tube Voltmeter. Since vacuum-tube circuits can easily achieve input impedances measured in megohms/volt, they provide a number of important advantages over the simple resistance input circuits of moving-coil indicating meters. The high sensitivity and input impedance mean negligible loading of the circuits under test as well as uniform accuracy over a wider range of measurement.

In its simplest form, a d-c vacuum-tube voltmeter consists of an input voltage divider, a d-c amplifier, and a metering circuit to measure the plate current of the amplifier tube (see Fig. 5).

Accuracy of a d-c VTVM is determined by the characteristics of the amplifier, in terms of both linearity and drift, and by the quality of the meter movement used. With appropriate feedback circuitry, and with a chopper-stabilized amplifier, accuracies of ±1 per cent of full scale may be obtained.

Any d-c voltmeter can be adapted for *resistance measurements* by providing a d-c test potential and a calibrated resistance scale on the meter face. Measurements of *a-c voltage* can be made with a d-c VTVM by using a special probe which includes a peak-detecting diode to obtain wide frequency coverage. The

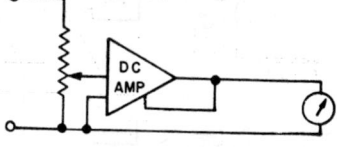

FIG. 5. Simple d-c vacuum-tube voltmeter.

d-c output from the probe is then read from suitably calibrated meter scales. Figure 6 shows a block diagram of a typical d-c VTVM with accessory circuitry for measuring *resistance* and *a-c voltage*.

Drift in d-c amplifiers can be made insignificant by using chopper stabilization. The chopper converts the incoming d-c signal to a-c, thus permitting amplification by a highly stable a-c amplifier. The amplified a-c voltage is then demodulated to d-c to drive the metering circuits. Electromechanical choppers (essentially a vibrating relay) have been used extensively for achieving stable d-c amplification.

Recently developed photoelectric chopper systems provide a more economical method of reducing noise and thus permit greater sensitivity in low-cost measuring instruments. A block diagram of a multipurpose d-c VTVM with photoelectric chopper is shown in Fig. 7. A typical instrument is shown in Fig. 8. This instrument provides input resistance of 10 to 200 megohms (depending on range) and measurement accuracies of ±1 per cent on voltages as low as 1 mv full scale, or ±2 per cent on currents as low as 1 μa full scale. Chopper techniques are used in other commercially available instruments which provide measurement accuracy of ±3 per cent for full-scale voltages as low as 10 μv and full-scale currents as low as 10 ma.

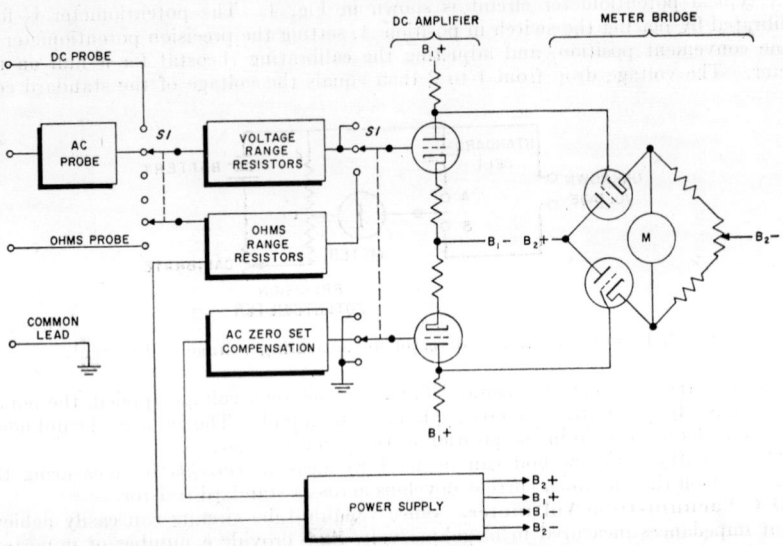

FIG. 6. Multipurpose VTVM for measuring d-c voltage, a-c voltage, and resistance.

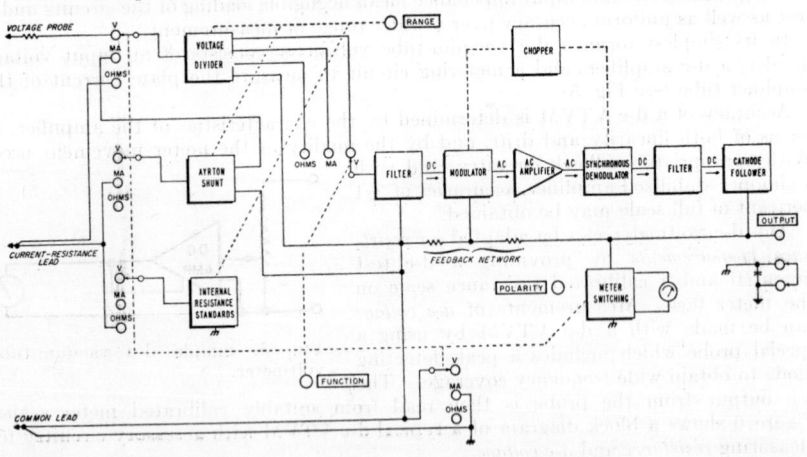

FIG. 7. A d-c VTVM with chopper stabilization.

Current measurements can be made with a d-c VTVM by measuring the voltage drop across a precision resistor. This method is commonly used in general-purpose laboratory instruments. A more convenient means of measuring direct current is the clip-on probe which permits current to be measured without breaking the circuit. A typical clip-on current probe and meter are shown in Fig. 9. The probe and meter are designed as a single instrument and must be used together. The probe operates by sensing the magnetic field of the current being measured. An a-c output from the probe is applied to the meter for direct reading of the d-c value. The clip-on probe provides measurement accuracy of ±3 per cent for currents as low as 3 ma full scale.

The name "electrometer" is sometimes given to a d-c VTVM with very high input resistance. Strictly speaking, however, an electrometer is a device which measures current by determining the rate of change of voltage across a small capacitor being charged by the current being measured. The same effect can be achieved by utilizing the grid-to-cathode capacitance of a vacuum tube with a very high input resistance. These tubes are called electrometer tubes.

Commercial vacuum-tube electrometers capable of measurement accuracies of ±2 to ±4 per cent are available for measuring currents as low as 10^{-13} amp.

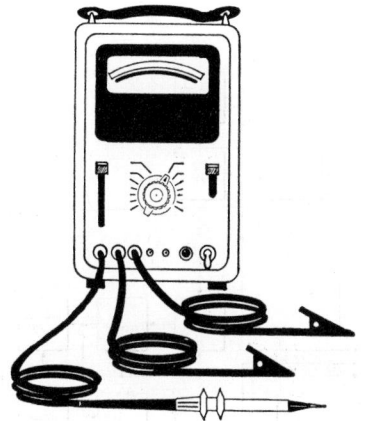

FIG. 8. A precision d-c voltmeter-ohmmeter-ammeter.

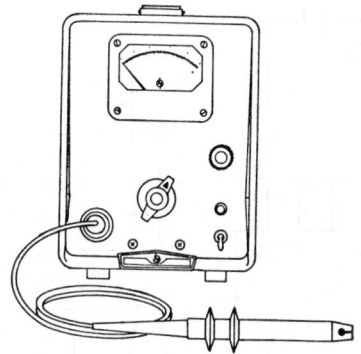

FIG. 9. Clip-on probe for measuring direct current.

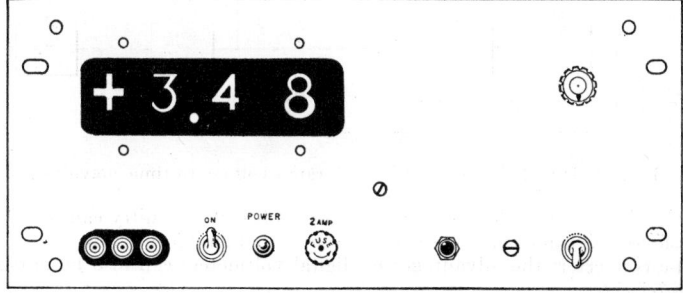

FIG. 10. D-C digital voltmeter.

Digital Voltmeters. Accuracies better than ±1 per cent are difficult to achieve with any d-c voltmeter incorporating a moving-coil meter movement. The digital voltmeter, however, by utilizing a completely different principle of voltage measurement, can be constructed to give measurement accuracy in the order of ±0.1 per cent, or even ±0.01 per cent with sufficiently linear circuit elements. Even where the greater accuracy is not required, digital voltmeters provide the advantages of easier operation and more readily interpreted information presentation. A particularly important feature is the high resolution possible with digital techniques. This permits accurate measurement of small incremental changes.

A typical general-purpose d-c digital voltmeter is shown in Fig. 10. This instrument provides accuracy of ±0.2 per cent, in-line digital readout, automatic indication of polarity, and automatic decimal-point positioning. In addition, output terminals are provided so that the measured data can be permanently printed by means of a digital recorder.

Among the measurement techniques commonly used in digital voltmeters is that of *time measurement*. The circuits are arranged so that a measured time interval corresponds with a particular voltage level. Figure 11 shows a block diagram of a typical instrument which operates as a ramp-voltage-to-time converter. When a d-c voltage is applied, a ramp voltage is generated, and its value is compared with the input. During the time required for the ramp to reach the input-voltage level, pulses from a stable oscillator are counted by three-decade counting units. Sampling rate can be as fast as several times per second for accurate observation of varying voltages.

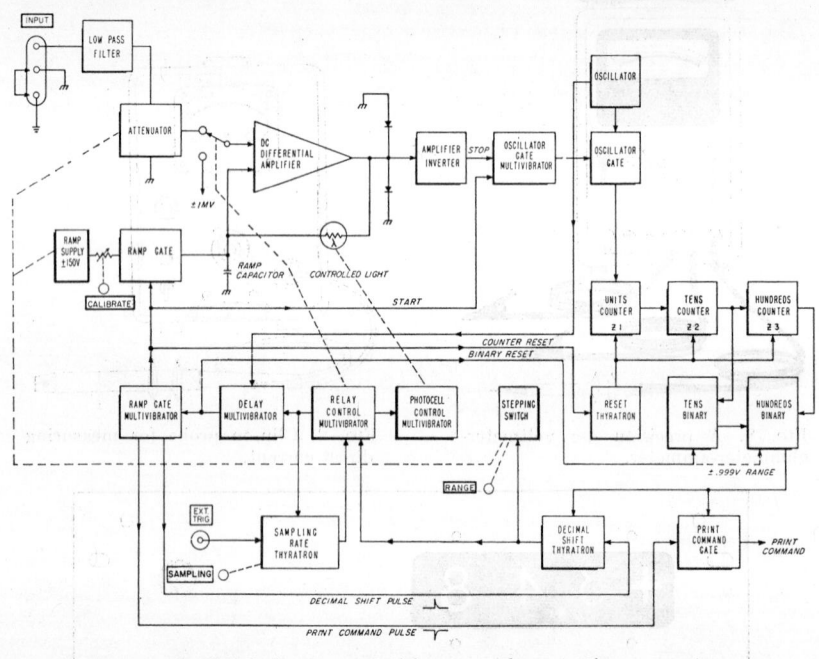

FIG. 11. D-C digital voltmeter with ramp-voltage-to-time converter.

As with other types of d-c voltmeters, digital voltmeter circuitry can be adapted so that *resistance* also can be measured and presented in digital form. By using a suitable a-c to d-c converter, the advantages of digital voltmeters can also be obtained for *a-c measurements*.

A-C Measuring Instruments and Methods

Measurements of time-varying voltages and currents must take into account the nature of the waveform and the response of the measuring circuits to the variations in the input. For convenience and uniformity, the scales of most instruments are calibrated to read the rms value of an assumed sine-wave input. The actual measurement may be made with circuitry which responds to the average, effective, or peak value of the input.

Dynamometer-type A-C Voltmeters. The a-c equivalent of the d-c moving-coil meter is the dynamometer illustrated in Fig. 12. If an a-c voltage were impressed across a moving-coil meter, the indicator would vibrate around zero since the torque would be reversed every half cycle. In the dynamometer, the magnet of the moving-coil meter is replaced by a fixed coil in series with the moving coil. Thus the direction of the torque remains constant as the magnetic field is reversed. The torque developed in a dynamometer-type meter is

$$T = I_m I_s \frac{dM}{d\theta}$$ (5)

where T = torque, newton-meters
I_m and I_s = currents through the movable and stationary coils, amp
$\frac{dM}{d\theta}$ = rate of change of the mutual inductance with respect to change in angular position

The dynamometer is a true rms reading instrument with maximum sensitivity of about 100 ohms/volt. It can provide measurement accuracy of the order of $\pm \frac{1}{2}$ per cent at frequencies up to a few hundred cycles. Standard a-c voltmeters and ammeters are constructed with dynamometers in the same manner that similar d-c instruments are constructed with moving-coil meters.

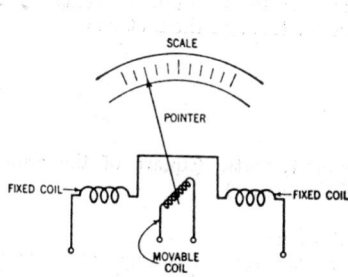

Fig. 12. Dynamometer-type a-c meter movement.

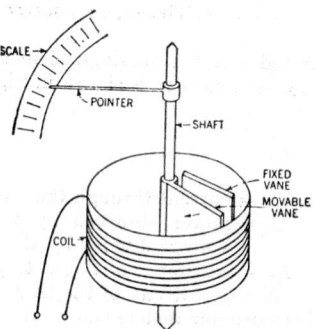

Fig. 13. Iron-vane-type a-c meter movement.

Iron-vane Meters. Less accurate than dynamometer-type movements, but essentially more rugged, the moving-iron or iron-vane-type meter movement is used extensively in the electrical power field. As shown in Fig. 13, this type of meter consists of two vanes or iron pieces placed within a wire coil. One of the vanes is stationary. The other is movable and mounted on a shaft with attached pointer. When an alternating current passes through the coil, the vanes become magnetized, and a force or torque is created which repels the vanes. This causes the movable vane to rotate and move the pointer shaft. The torque developed is

$$T = \frac{1}{2} I^2 \frac{dL}{d\theta}$$ (6)

where T = torque, newton-meters
I = current through the coil, amp
$\frac{dL}{d\theta}$ = time rate of change of the inductance with respect to angular position

Series and shunt resistors may be added to an iron-vane-type meter movement for measuring various ranges of currents and voltages. Instruments of this type are generally used for 60-cycle measurements with sensitivity in the order of 100 ohms/volt and accuracy of measurement of about ± 1 per cent.

Rectifier-type A-C Voltmeters. A-C meter movements of the dynamometer or iron-vane types have limited frequency range and relatively low sensitivity. One of the more common methods of eliminating these disadvantages is to use a d-c moving-coil meter with a half- or full-wave rectifier to convert the a-c input into a direct current for measurement. Figure 14 shows a full-wave bridge rectifier circuit.

A typical commercial instrument will provide sensitivity of about 2,000 ohms/volt with measurement accuracy of ±5 per cent at frequencies up to about 10,000 cycles.

Thermocouple A-C Meters. Power applied to a thermocouple meter heats a short section of resistance wire which is thermally associated with a thermocouple element consisting of two lengths of dissimilar wire connected together. Heat applied to the thermocouple junction causes a voltage to be generated. The thermocouple is

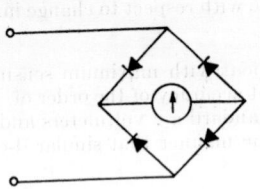

FIG. 14. Rectifier-type a-c meter.

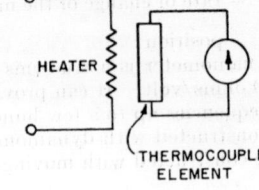

FIG. 15. Thermocouple-type a-c meter.

connected across a moving-coil meter which measures the resultant current. In an arrangement such as is shown in Fig. 15, the current through the meter is

$$I_m = \frac{KP}{R_m} \tag{7}$$

where I_m = current through the meter, amp

P = power absorbed by the heater element, watts (square of the applied voltage divided by the resistance of the heater)

R_m = resistance of the meter, ohms

K = sensitivity of the thermocouple, volts/degree

Thermocouple meters can be calibrated for various ranges of voltages and currents in the same manner as other meter types. They are particularly useful for measuring

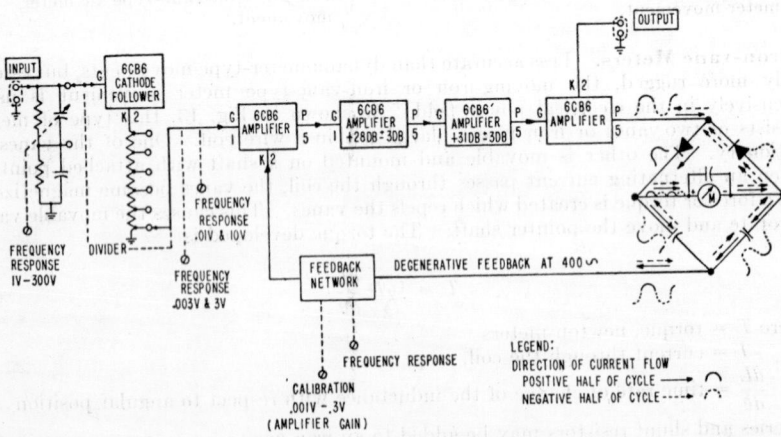

FIG. 16. A-C vacuum-tube voltmeter.

alternating currents at frequencies up to 100 mc with accuracies of the order of ±1 to ±2 per cent. When used as voltmeters, sensitivities of about 500 ohms/volt are common. Thermocouple meters can be damaged by relatively low overloads as compared with other types of a-c meters.

A-C Vacuum-tube Voltmeters. Vacuum-tube voltmeters specifically designed for making measurements of alternating currents and voltages are similar in operation to rectifier-type a-c meters. By the use of vacuum tubes they provide greater input sensitivity over a wide range of current and voltage measurements. Figure 16 shows a typical circuit arrangement for an a-c VTVM. As in the d-c VTVM, the instrument includes an input voltage divider, various stages of amplification, and a feedback network. The a-c VTVM also has rectifying circuits to provide a direct current to the moving-coil indicating meter.

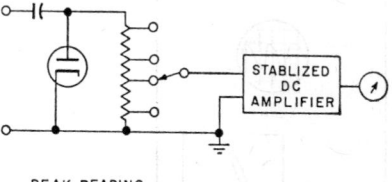

AVERAGE READING

PEAK READING

Two types of a-c VTVM's are commonly used—those which respond to the average current, and those which respond to the peak current. Meter scales of both types are usually calibrated to read rms values. In the first type, the meter deflection is proportional to the average value of a rectified cycle of the applied waveform. A-C VTVM's of this type typically measure a-c voltages at frequencies up to 4 mc with accuracies ranging from ±1 to ±5 per cent. Input impedances of the order of 10 megohms are typical.

In the second type of a-c VTVM, the deflection is proportional to the positive peak value. With specially designed

FIG. 17. Average- and peak-reading a-c vacuum-tube voltmeters.

probes, this type of meter is suitable for measurements at much higher frequencies. Figure 18 shows a probe designed for measuring a-c voltages at frequencies up to 700 mc with mid-range accuracy of the order of ±3 per cent.

Equivalent circuits for both types are shown in Fig. 17. Note that the meter shown in Fig. 16 is of the average-reading type. In general, average-reading meters are less sensitive to waveform effects; peak-reading types are applicable to higher frequencies. Many a-c VTVM's also include output terminals which can be connected to an oscilloscope so the unknown waveform can be observed while it is being measured.

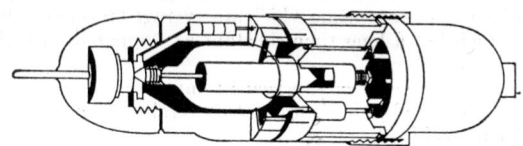

FIG. 18. A-C VTVM probe for measurements to 700 mc.

Oscilloscopes. Complex waveforms and transients are often measured on cathode-ray oscilloscopes. Because the waveform can be seen, measurement errors caused by nonsinusoidal signals are minimized, and system parameters can be adjusted to give optimum waveshape and amplitude while the waveform is displayed.

Many laboratory oscilloscopes include calibrated internal amplifiers and precision-etched graticules in front of the cathode-ray tube. Amplifier calibration is usually in terms of volts or millivolts per centimeter. Peak instantaneous voltage of any waveform can be read with oscilloscopes at frequencies of up to 1,000 mc with accuracies of the order of ±3 to ±5 per cent. A typical measurement arrangement is shown in Fig. 19.

The upper limit of frequency which can be measured on an oscilloscope is usually limited by the frequency response of the vertical amplifier in the instrument. Inexpensive general-purpose oscilloscopes can normally be used up to several hundred kilocycles. Laboratory-type oscilloscopes provide satisfactory performance up to 80 mc. Specially designed oscilloscopes using sampling or traveling-wave techniques are used at higher frequencies.

In many oscilloscopes external signals can also be coupled directly to the vertical deflection plates. Accurate measurements can then be made by calibrating the oscilloscope presentation with an external voltage source. By this method both frequency response can be extended well beyond the normal range.

Current waveforms can be measured by displaying the voltage developed across a known resistor. Many oscilloscopes provide two vertical amplifiers so that both current and voltage waveforms can be displayed simultaneously.

Electrostatic Voltmeters. In contrast to all other indicating meters, which are current-actuated, the electrostatic voltmeter is a voltage-operated device. The deviation of the indicating pointer is determined by the applied voltage directly rather than by the current flowing through a coil.

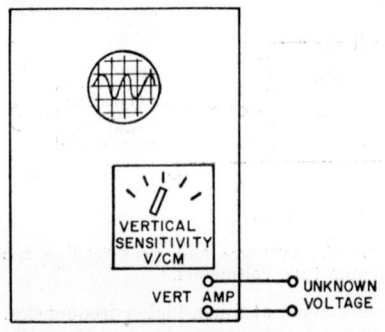

FIG. 19. Voltage measurement with cathode-ray oscilloscope.

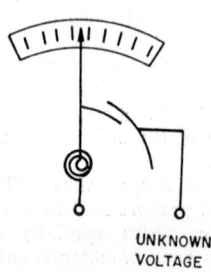

FIG. 20. Electrostatic voltmeter.

An electrostatic meter movement consists essentially of two plates forming a capacitor. One plate is fixed, the other movable as shown in Fig. 20. The movable plate is attached to a shaft or pointer which is free to rotate. When a voltage is applied to the input terminals, a torque is generated to move the shaft and indicating pointer.

Most electrostatic measurements are true rms, regardless of the applied waveform. A-C voltages produce surges of torque which are averaged mechanically by the instrument movement which has a long time constant as compared with the period of the applied voltage.

Electrostatic voltmeters are characterized by input resistance in the order of 10^{12} to 10^{15} ohms. They are used to measure up to several thousand volts with accuracies in the order of ± 1 per cent.

POWER

Three methods are commonly used in the electronics research and development laboratory for measuring power. Voltmeter-ammeter arrangements give excellent results at low frequencies where average power can easily be computed from the measured values of current and voltage. At microwave frequencies, where power is constant while voltage and current values vary with position, power is usually measured directly. Bolometers with associated bridge measuring circuitry are used for measuring low power levels. Calorimetric methods are normally used for power levels above about 10 mw.

Instantaneous and Average Power

The instantaneous power supplied to a cricuit from a source is

$$p = ei \qquad (8)$$

where p = instantaneous power, watts
$ e$ = instantaneous voltage, volts
$ i$ = instantaneous current, amp
For a sinusoidal voltage

$$e = \sqrt{2}V \sin(\omega t + \alpha) \tag{9}$$

and $ i = \sqrt{2}I \sin(\omega t + \beta) \tag{10}$

where V = effective (rms) voltage, volts
$ I$ = effective (rms) voltage, amp
$ \omega$ = angular frequency, radians/sec
$ \alpha, \beta$ = initial angles of V and I, respectively
Then the equation for power is

$$p = 2VI \sin(\omega t + \alpha) \sin(\omega t + \beta) \tag{11}$$

which can be reduced trigonometrically to

$$p = VI \cos(\alpha - \beta) - VI \cos(2\omega t + \alpha + \beta) \tag{12}$$

Equation (12) shows that the instantaneous power consists of a constant component and a superimposed double-frequency alternating component. The constant component is the average power and can be represented by

$$P = VI \cos\theta \tag{13}$$

where P = average power, watts
$ \theta = \alpha - \beta$ or $\beta - \alpha$, deg
Thus average power is the product of the effective voltage, effective current, and the cosine of the angle between them.

Consideration of impedance angle can be eliminated when measuring actual power delivered to a load by measuring the resistive component of the load and calculating power from

$$P = I^2 R = \frac{V^2}{R} \tag{14}$$

where R = load resistance, ohms

Volt-Ammeter System

The greatest accuracy in measuring power at low frequencies can be obtained by connecting an ammeter and a voltmeter as shown in Fig. 21. A switch alternately connects the two meters to the circuit. By following this procedure, neither the current through the voltmeter nor the voltage drop across the ammeter will be measured. The power may then be computed from Eq. (13) or (14).

By eliminating the loading effects of the meters, the measurement accuracy is limited only by the basic accuracies of the two meters.

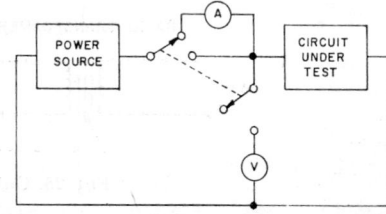

Electrodynamic Wattmeter

The electrodynamic wattmeter, as shown in Fig. 22, consists of two separate coil systems. A fixed set of coils is connected to measure current through the load, and a

Fig. 21. Voltmeter-ammeter methods of power measurement.

movable coil is connected to respond to the voltage across the load. Since the deflection of the meter is the result of the two magnetic fields set up by the coils, the pointer will indicate the product of the voltage and current, which is a measure of power.

Since the current and voltage leads are brought out separately, the wattmeter may be connected in either of two ways, as shown in Fig. 23. In the first case, the current I_f through the stationary (current) coil is equal to the sum of the current I_m in the moving (voltage) coil and the load z. Thus the meter indicates the power drawn by

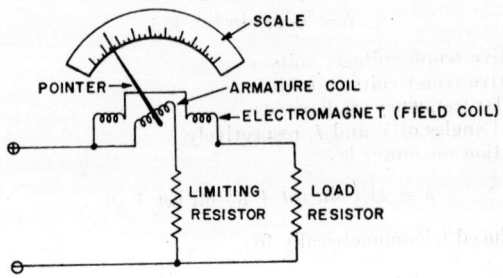

Fig. 22. Wattmeter connected to measure power consumed by load resistor.

the meter itself as well as that drawn by the load. With the other method of connection, the voltage drop across the fixed coil is included in the measurement. This voltage drop is small for most current coils so that the measurement approximates the power dissipated by the load within approximately 1 per cent.

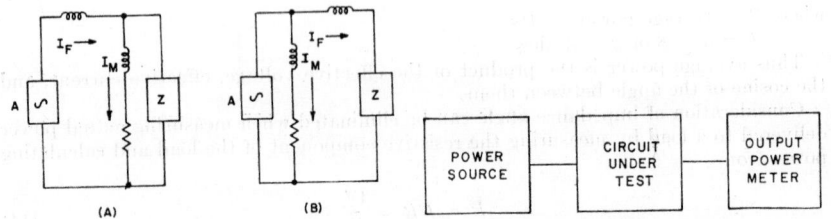

Fig. 23. Two methods of connecting a two-coil wattmeter.

Fig. 24. Direct-reading output power meter for four-terminal devices.

Output Power Meters

One general type of output power meter consists of an adjustable load impedance and a voltmeter calibrated directly in watts dissipated by the load. The meter measures the output power of four terminal devices as shown in Fig. 24.

To measure the power output, the impedance of the meter is adjusted to match that of the circuit under test. The meter scale should be set for the most sensitive

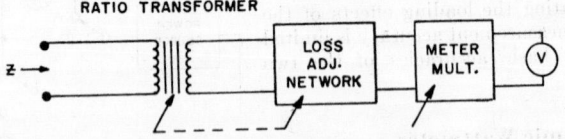

Fig. 25. Output power meter.

reading. Figure 25 shows a block diagram of an output power meter with its impedance-matching network.

Power can generally be measured to an accuracy of between $\frac{1}{4}$ and $1\frac{1}{2}$ db, although errors in impedance matching can cause variations in measurement of from 2 to 10 per cent. If the meter used is calibrated to read the rms value of a true sine wave, nonsinusoidal signals will cause measurement errors. The significance of the error depends on the magnitude and phase relationship of the harmonics present in the signal.

Bolometer Mounts

Microwave power in the range of 0.1 to 10 mw is customarily measured by using a bolometer element and a bridge-type meter. The unknown power is absorbed in a specially constructed bolometer of resistive material. The change in bolometer resistance is then measured by the bridge, which is calibrated to read power directly.

Two types of bolometers are used, *barretters* and *thermistors*. Barretters have a positive temperature coefficient; thermistors have a negative temperature coefficient. In general, barretters are delicate and will burn out with small power overloads. They are most suitable as microwave signal detectors. Thermistors, on the other hand, are more rugged devices. Their actual burnout value is several hundred milliwatts. Thermistors are generally preferred for measuring the average value of pulsed power because of their longer time constant. Table 1 compares the characteristics of these bolometers as power-measuring devices.

Table 1. Comparison of Thermistor and Barretter Characteristics

Characteristic	Thermistors	Barretters	Application notes
Time constant.............	1 sec	350 μsec	Long time constant to measure pulsed power
Max average power.........	>25 mw	<10 mw	Thermistor good for pulsed power and sudden overloads
Peak overload power........	400 mw	<25 mw	Same as above
Power sensitivity...........	35 ohms/mw	5 ohms/mw	Thermistor good for low-duty-cycle pulsed power and low-level signals
Drift (200 ohms)...........	1.8 %/°C	0.15 %/°C	Barretter has accuracy and low drift
Temperature coefficient.......	Negative	Positive	Negative coefficient less sensitive to burnout

The bolometer element itself must be mounted and well matched to the transmission system and the power meter. The power measured by the bolometer mount will depend on the relationship between the load and the source impedance. In order to receive maximum available power, the load should present a conjugate match to the source impedance. This can be achieved by properly adjusting a double-stub tuner, a stub-line stretcher, an E-H tuner, or a slide-screw tuner. These tuners transform the magnitude and phase of the source impedance to match it to the load impedance.

Microwave Power Meter

A microwave power meter will give direct instantaneous reading of microwave power when connected to a suitable bolometer mount. The bias current necessary to bring the bolometer to the correct operating resistance is supplied in many cases by the power meter itself. A typical microwave power meter as shown in Fig. 26 consists of a self-balancing bridge and an audio voltmeter which is used to indicate the bridge amplifier output.

The self-balancing bridge uses an external bolometer element in one of the bridge arms. The other bridge resistors are selected and arranged so that the bridge will be nearly balanced when the bolometer is brought to a predetermined operating resistance. Note that actual balance never occurs in a self-balancing bridge since a small amount of unbalance is needed to provide oscillator feedback.

The bolometer receives audio-frequency power from the bridge oscillator and d-c power from the bias supply. These two power sources heat the bolometer until it reaches its operating resistance. At this predetermined value the bridge circuit is balanced. The resistance of the bolometer is sustained at this value by the applied a-f and d-c power.

When the bolometer is placed in an r-f field, it absorbs energy which heats the element, thereby changing its resistance and unbalancing the bridge. This action causes the oscillator output to decrease, which in turn brings the bolometer temperature, hence its resistance, back to the original operating value, once again balancing the bridge.

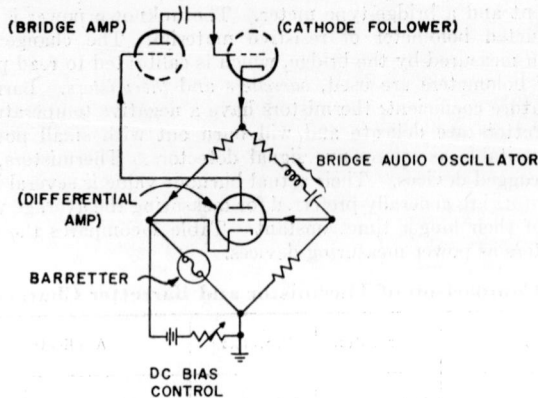

FIG. 26. Simplified schematic of typical microwave power meter.

Operation of the bridge is represented by

$$P_{bal} = P_{dc} + P_{af} + P_{rf} \tag{15}$$

where P_{bal} = power needed to establish the resistance of the bolometer at the balanced value

P_{dc} = d-c power supplied by the bias supply

P_{af} = a-f power supplied by the bridge oscillator

P_{rf} = r-f power absorbed by the bolometer element when placed in an r-f field

Since the d-c biasing power is constant for a particular bolometer, the a-f power must decrease as r-f energy is absorbed by the bolometer. The voltmeter circuit in the power meter is connected to measure the decrease in the a-f power supplied by the bridge oscillator. If the voltmeter is zero-set after the bolometer has reached its operating resistance, and is not in the presence of an r-f field, it can be calibrated to read the amount of r-f power which is presented to the power meter, if the balance resistance is known and constant.

Microwave Power Meter System

An equipment arrangement for measuring microwave power with a bolometer mount and a microwave power meter is shown in Fig. 27. In a waveguide system a slide-screw tuner or E-H tuner would be used to match the bolometer mount to the source impedance. A double-stub tuner or stub-line stretcher would be used in a coaxial system. Figure 26 shows how the bolometer element is connected in the bridge circuit.

When properly adjusted, a microwave-power-meter measuring system will provide measurement accuracy within about ±5 per cent. Two major sources of meter inaccuracy to be considered are envelope-tracking errors and beat-frequency errors. An additional error which should be recognized is the bolometer mount efficiency, which typically varies between 95 and 97 per cent.

At low modulation frequencies, bolometer response attempts to follow the envelope of a sine- or square-wave modulated signal. This causes the bridge oscillator also to attempt to follow the envelope by adjusting its power output to follow the changes in bolometer resistance.. The modulation frequency is then impressed on the bridge oscillator's audio signal. The panel meter, which is an average responding device, will then indicate the average of the modulation envelope.

The fast response of a barretter with its short time constant is particularly susceptible to this action. The meter indication will remain steady until the modulation frequency is reduced to a critical value, at which point the meter indication will increase. The critical frequency for a barretter is about 200 to 300 cycles, for a thermistor about 100 cycles. At frequencies above the critical value, power readings for sine- or square-wave-modulated signals are accurate.

When pulse-modulated signals are measured, repetition rates which are submultiples of the bridge oscillator audio-frequency cause a beating effect in the bridge circuit which is reflected in the panel meter indications. This effect is particularly noticeable when using a barretter. At certain low power levels a strong beating may occur followed by the meter dropping to a low value as the bridge oscillator locks in. A

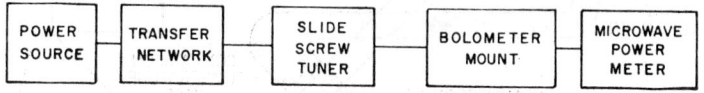

Fig. 27. Power measurement with a bolometer and power meter.

slight variation of the pulse-repetition rate will eliminate this beating effect. When using thermistors, the beats are smaller, the oscillator does not lock in, and the instrument readings are not affected.

Calorimetric Power Meter

Calorimetric methods are normally used to measure power above the range which can be accommodated with bolometer mounts. There are several types of calorimetric meters which measure power above 5 to 10 watts. One type measures average power automatically over full-scale ranges from 10 mw (upper limit of bolometers) to 10 watts over a frequency range of direct current to 12.4 kmc. Higher powers can also be measured by using suitable input attenuators and directional couplers. Nominal accuracy of this instrument is ±5 per cent of full scale.

As shown in Fig. 28, the calorimetric power meter consists of (1) a self-balancing bridge with identical temperature-sensitive resistors (gages) in two arms, (2) a d-c meter calibrated to read watts, and (3) two load resistors, one for the comparison power and one for the unknown input power. The input-load resistor and one gage are in close thermal proximity so the heat generated in the input-load resistor heats the gage and unbalances the bridge circuit.

The unbalance signal is amplified and applied to the comparison-load resistor, which is in close thermal proximity to the other gage. Heat generated in the comparison-load resistor is transferred to its gage which tends to rebalance the bridge. The meter is calibrated to read the d-c power supplied to the comparison load. Since the gage characteristics are the same, and the heat-transfer properties of each load are the same, the meter can be calibrated to read input power directly.

All the components of the calorimetric power meter are immersed in an oil stream to obtain efficient and uniform heat transfer from the resistive loads to the gages. The oil-flow rate through each of the heads is equal and, to ensure constant temperature, both oil streams are passed through a parallel-flow heat exchanger just prior to entering the measuring heads. Identical flow rates are obtained by placing all elements of the oil system in series.

Higher accuracies than the nominal ±5 per cent can be obtained by minimizing frequency and impedance mismatch effects. For example, accuracy can be improved by applying a correction to compensate for the internal power loss in the r-f termination. Accuracy can also be improved by accurately matching the calorimetric power meter to the circuit under test. Tuning is easier to achieve in a waveguide system than in a coaxial system because a waveguide slide-screw tuner may be used in front of the waveguide-to-coaxial adaptor. Such waveguide tuners have less loss than coaxial tuners. With good measurement techniques, accuracy of up to ±½ per cent at

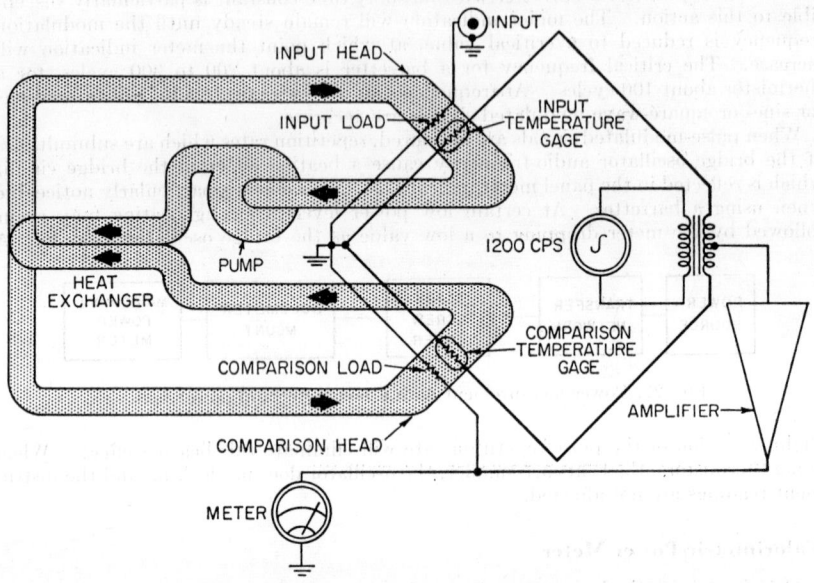

FIG. 28. Calorimetric power meter.

direct current can be obtained on the 3.0- and 10.0-watt scales, with the limit of error approaching ±5 per cent at 12.4 kmc on the 10- and 30-mw ranges.

FREQUENCY

Several methods are available for measuring frequency. The choice of method depends on the accuracy desired and the frequency range concerned. In the a-f and r-f regions, one of several comparison methods may be used in which the unknown frequency is compared with a standard frequency. The comparison can be made on an oscilloscope or in a circuit which detects the difference between the two frequencies. In the uhf and microwave regions, tuned wavemeters calibrated against accurate frequency standards are very common. Slotted-line techniques based on wavelength measurement are also used. Electronic counters are employed for very accurate determination of frequency from fractions of a cycle up to as high as 40 kmc. There are also electronic frequency meters for operation up to about 100 kc in which the periodic input is translated into a quantity which can be presented on an indicating meter.

Frequency-Time Relationship

Since frequency is defined as the number of periodic phenomena occurring in a unit time, many methods of measuring frequency require precise measurement of a specific period of time. Frequency and time are related by

$$F = \frac{1}{t} \tag{16}$$

where f = frequency, cps
t = time, sec

The most widely accepted standard of frequency is the period of rotation of the earth. Because this rotation is not absolutely constant, very precise methods have been devised to establish correction factors which are then applied to establish what is called the "mean solar day." A basic unit of time is then defined to be the second, which is equal to 1/86,400 of a mean solar day.

Accurate time signals for calibrating frequency- and time-measuring instruments are broadcast by the National Bureau of Standards through HF stations at Beltsville, Md., and Maui, Hawaii. The Bureau also has LF and VLF stations near Boulder, Colo. Complete information concerning these broadcasts may be obtained from the National Bureau of Standards, Boulder Laboratories, Boulder, Colo.

Oscilloscope Frequency Comparison

Perhaps the most common method of oscilloscopic frequency comparison is the generation of Lissajous figures on the face of a cathode-ray tube. A source of known

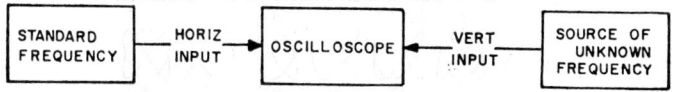

Fig. 29. Equipment arrangement for observation of Lissajous figures.

frequencies is connected to one set of deflection plates, and the unknown frequency is connected to the other set. If one frequency is a harmonic or integral multiple of the other, a stable and recognizable pattern will be displayed.

Measuring Procedure. Figure 29 shows the usual arrangement for Lissajous-figure observation with the unknown frequency connected to the vertical input of the oscilloscope and the known frequency to the horizontal input. Using this arrangement the horizontal time-base generator can be switched in to observe the waveform of the unknown signal. The unknown frequency can then be estimated to within about ±5 per cent by setting the horizontal input to the internal sweep position and measuring the period of the unknown waveform.

The known frequency is then applied to the horizontal input and the sensitivity and gain controls for both the vertical and horizontal amplifiers are adjusted to obtain equal vertical and horizontal deflections.

By adjusting the two applied frequencies, various Lissajous patterns will be formed to indicate the ratio between them. Figure 30 illustrates the most often used frequency ratios. Note that, as the phase between the two applied signals is varied, the Lissajous pattern shifts. A specific unknown frequency can be measured by the Lissajous method if an interpolation oscillator is used with a frequency standard to provide continuous adjustment of the standard frequency.

Lissajous figures are extremely valuable in calibrating oscillators and signal generators. The method provides an immediate analog representation of the frequency and simplifies the adjustment of the frequency output, yet provides extreme accuracy. Some of the sources of error which may be encountered are instability in the applied frequency signals which will result in phase changes with resultant pattern rotation, jitter in the input signals, and observation errors in determining the more complex ratio patterns.

Electronic Frequency Meters

Several frequency-measurement devices are available which display the measured frequency on an indicating meter. A typical circuit is shown in Fig. 31. The input signal is amplified and converted to a square wave which charges a capacitor through a diode. A second diode discharges the capacitor through the meter circuit. The current through the meter is proportional to the rate of the charging pulses; hence it is proportional to the frequency of the input signal.

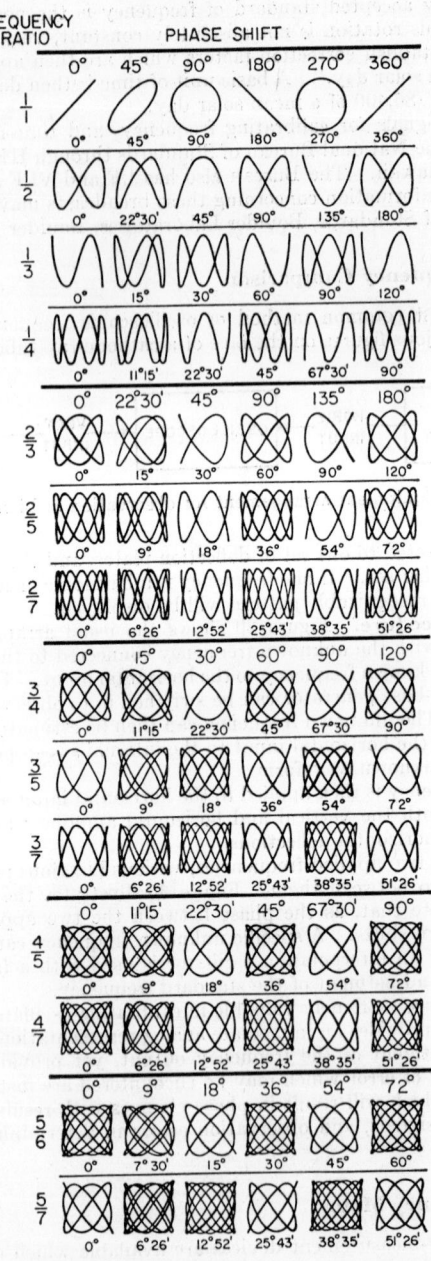

FIG. 30. Lissajous figures for the most often used frequency ratios.

Another type of electronic frequency meter is illustrated in Fig. 32.   In this instrument the input waveform drives a trigger circuit which generates a series of negative trigger pulses.   These actuate a constant current source and a phantastron (linear-timing) circuit.   The phantastron output turns off the current source, thus determin-

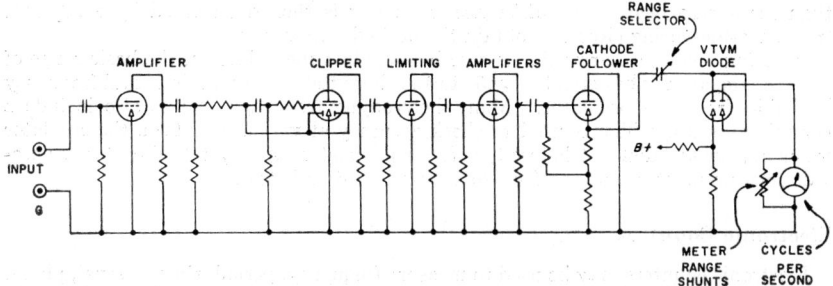

FIG. 31. Electronic frequency meter.

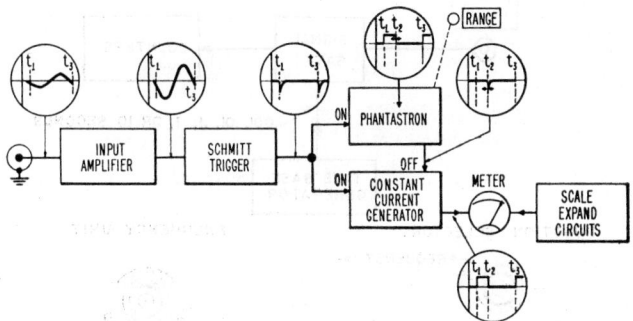

FIG. 32. Electronic frequency meter.

ing the width of the current pulse delivered to the meter circuit (and establishing the meter "range").   One such stable pulse is passed through the meter circuit for each input cycle.   The meter averages the pulses it receives and presents an indication proportional to the average frequency.

Heterodyne Frequency Meter

The heterodyne frequency meter consists essentially of a frequency-calibrated stable oscillator, a mixer, and an indicator.   An amplifier for driving the beat indicator is also usually included, as shown in Fig. 33.   The output signal may be used to drive external indicators such as headphones, oscilloscopes, or recorders.

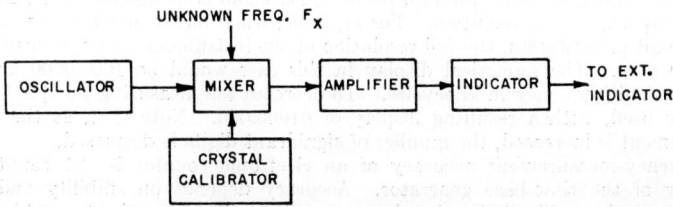

FIG. 33. Heterodyne frequency meter.

With the unknown frequency applied to the mixer, the oscillator is usually tuned for a zero beat, which is indicated by a minimum or zero output from the mixer. The oscillator frequency will then be the same as, or a subharmonic of, the frequency of the applied signal. Heterodyne techniques are also used in some applications where the oscillator is tuned for a difference frequency in a specified range of frequencies rather than for a zero beat. The difference frequency is then determined by a separate frequency-measuring circuit to obtain the desired information.

By using oscillator harmonics to zero beat against the unknown, the basic range of the instrument may be extended by 50 to 100 times the highest fundamental frequency available from the oscillator. Many heterodyne frequency meters also include a crystal calibrator which is used to check accuracy at various points on the oscillator frequency-control dial. A heterodyne frequency meter with crystal calibrator can provide measurement accuracy of up to 1 part in 10^6 or better.

Electronic Counters

Electronic counters may be used to measure frequency, period, time interval, phase, ratio, or simply to totalize periodic or random events over a given time period. As

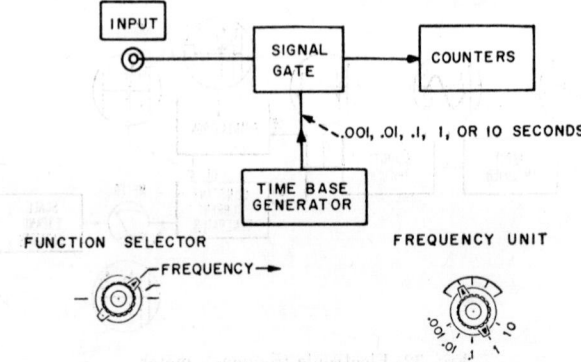

FIG. 34. Basic electronic counter.

shown in Fig. 34 an electronic counter consists of a time-base generator, a signal gate, and decade-counting units. Frequency is measured by counting the number of input cycles over a precisely controlled period of time.

The time-base generator develops control signals which are applied to the signal gate. When the first or "start" signal is received, the signal gate opens to pass input pulses from the unknown frequency source to the decade-counting units. When the second or "stop" signal is received, the signal gate closes to prevent further input pulses from reaching the decade-counting units. Totalization of input pulses by the decade-counting units during the interval when the gate was open is then a measure of the input-signal frequency.

The standard gate time (interval between start and stop signals) may be adjusted for various degrees of resolution. For example, when measuring 1 mc on a counter with 8-digit presentation, the full resolution of the instrument would be utilized with a 10-sec gate. The numerical display in this case would be 1000.0000 kc. Most counters read frequency in kilocycles. To decrease the measuring time, a 1-sec gate could be used, with a resulting display of 01000.000. Note that, as the speed of measurement is increased, the number of significant digits is decreased.

Frequency-measurement accuracy of an electronic counter is ±1 count ± the accuracy of the time-base generator. Accuracy depends on stability and on the technique used to calibrate the time-base generator. The error of ±1 count is inherent in the gating methods used.

Consider the measurement of a frequency of 100.1115 kc using a commercially available electronic counter with time-base stability of ±5 parts in 10^8 per week. Assume that the time-base generator is accurately calibrated. The error contributed to the measurement by the oscillator drift can be neglected since it is only a small fraction of the error introduced by the ±1 count factor. As frequency increases the ±1 count error becomes insignificant and the time base becomes the limiting factor. If a 10-sec gate time is used the reading would be 0100.1115 kc ±1 count in the last digit. This would mean the frequency is between 100.1114 and 100.1116 lc. The error would be about 1 part in 10^6. If a gate time of 1 sec were used the reading would be 00100.111 kc ±1 count. This would indicate a frequency between 100.110 and 100.112 kc. The error would be about 10 parts in 10^6. Figure 35 illustrates the

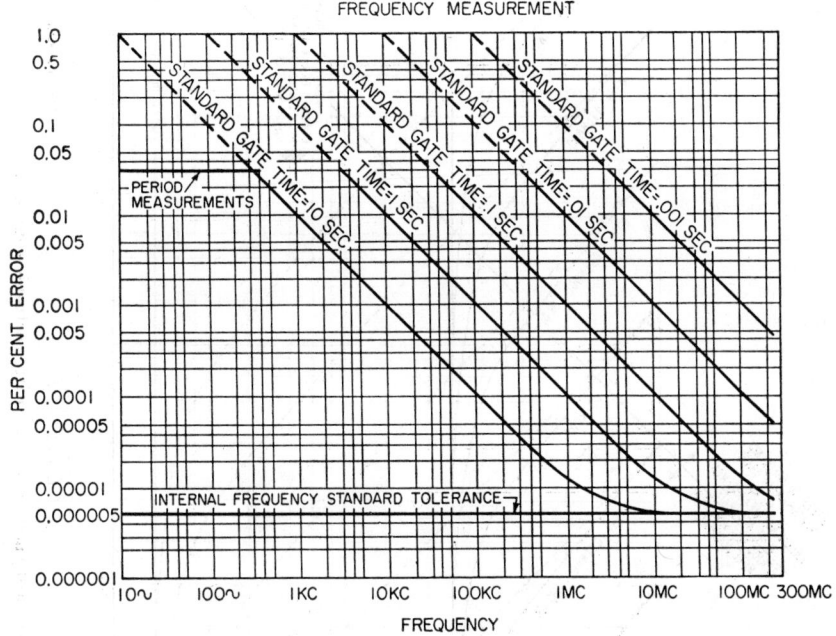

Fig. 35. Frequency-measurement accuracy with electronic counters.

accuracy of frequency measurements with various gate times for an electronic counter with a stability of ±5 parts in 10^8 per week.

In addition to measuring frequency, the electronic counter can measure the period of the input signal. This is particularly valuable when measuring low frequencies. For most electronic counters, the accuracy for single-period measurements is ±0.3 per cent ±1 count, and the accuracy for 10- period average measurements is ±0.03 per cent ±1 count. Figure 36 shows the accuracy which can be obtained in period measurement when counting the various standard frequencies normally available in an electronic counter. Note that greater accuracy can be achieved for a square-wave input than for a sine wave and that noise superimposed on an input sine wave introduces additional measurement error.

Figure 37 illustrates the operation of an electronic counter when making period measurements. For 10-period average measurements the input signal passes through a 10:1 decade-divider unit. When measuring period the input signal is converted to a series of trigger pulses which open and close the signal gate while the decade-counting units count a standard frequency supplied by the time-base generator.

Consider the case of a 400-cycle input signal. During a single-period measurement, the signal gate receives the start trigger pulse and, 2.5 msec later, the stop trigger pulse. With a standard frequency of 100 kc, 250 pulses from the time-base generator would be counted during the 2.5-msec interval. The period of the unknown frequency would be displayed as 000002.50 msec. If a 10-period average measurement were made, the

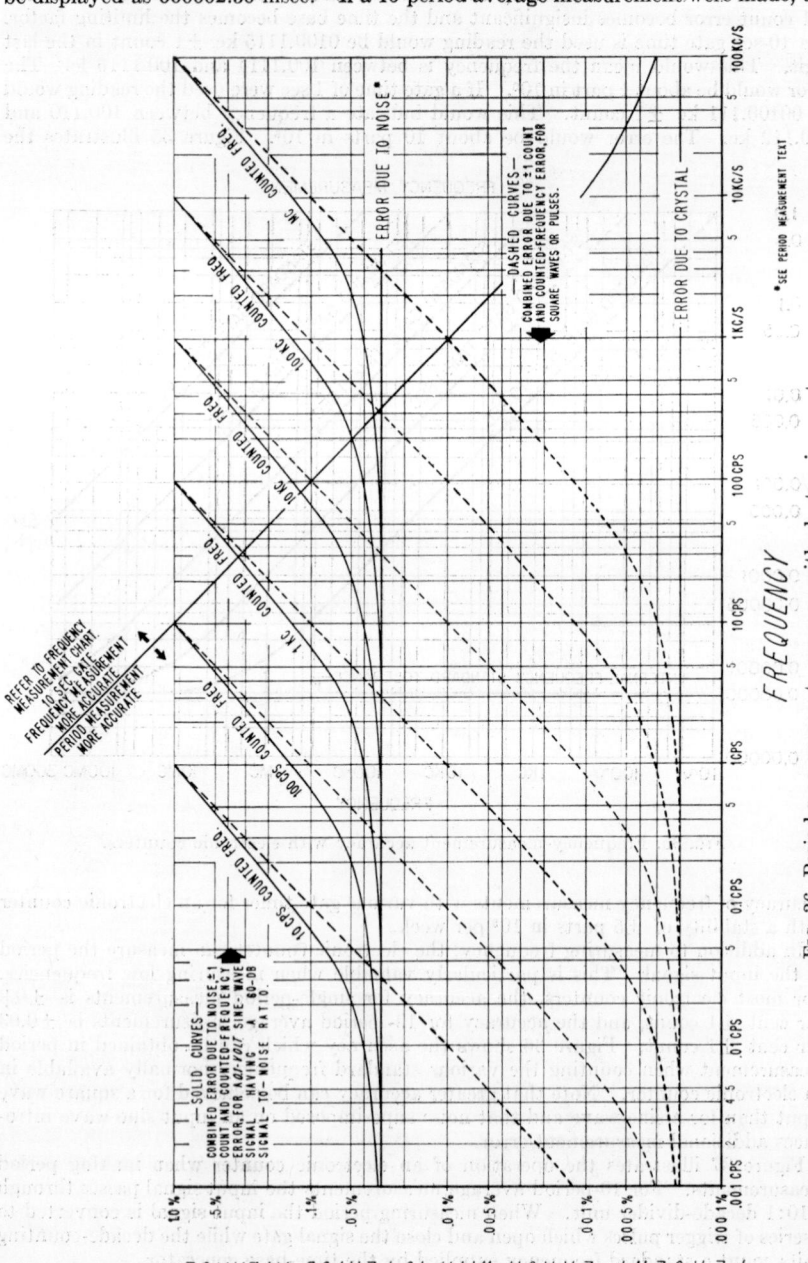

Fig. 36. Period-measurement accuracy with electronic counters.

input signal would be applied to a decade divider which would produce 1 output pulse for every 10 input pulses. The signal gate would be open for 25 msec, and 2,500 pulses from the time-base generator would pass through to the decade-counting units. The decimal point would be automatically shifted to provide a display of 00002.500 msec.

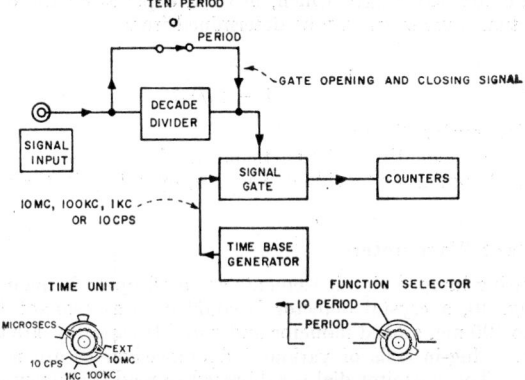

Fig. 37. Period measurement with electronic counter.

By using accessory heterodyne frequency converters and transfer oscillators, the basic frequency range of an electronic counter may be extended. In one standard arrangement, frequencies up to 500 mc or more are measured with a 10-mc counter and a frequency converter. With a transfer oscillator or other harmonic generator-mixer arrangements, the range is further extended to at least 40 kmc. At lower frequencies the accuracy of measurement is the basic accuracy of the electronic counter. Inherent characteristics of transfer oscillators limit the measurement accuracy at higher frequencies to about 1 part in 10^7.

Slotted-line Measurement of Frequency

An electronic counter and transfer oscillator can be used for extremely accurate measurements of frequency in the uhf and microwave bands. For some measurements, where an accuracy of the order of ± 0.5 per cent is sufficient, the frequency may

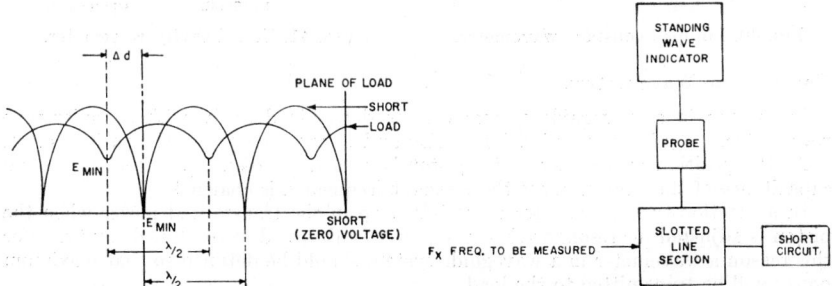

Fig. 38. Standing-wave patterns.   Fig. 39. Slotted-line measurement.

be determined with a slotted line or slotted section by observing the standing-wave pattern. Figure 38 shows the standing-wave pattern obtained with a measuring arrangement as shown in Fig. 39. Adjacent minima are one-half wavelength apart. In free space, where the velocity of propagation of electromagnetic waves is approximately 3×10^{10} cm/sec,

$$\lambda = \frac{3 \times 10^{10}}{f} \qquad (17)$$

where λ = wavelength, cm
f = frequency, cps

Equation (17) may be used directly with reasonable accuracy in an air-dielectric coaxial slotted-line measuring system where velocity of propagation approximates that of free space. In a waveguide system, where velocity of propagation is somewhat less than in free space, the measured distance between standing-wave minima corresponds to the guide wavelength which, in turn, depends on the dimensions of the waveguide. Actual wavelength can be determined from

$$\lambda_g = \frac{\lambda}{\sqrt{1 - (\lambda/\lambda_c)^2}} \tag{18}$$

where λ_g = guide wavelength, cm

λ_c = cutoff wavelength for the particular mode

In an air-filled rectangular waveguide of width a, operating in the usual TE_{10} mode, $\lambda_c = 2a$.

Lumped-constant Wavemeters

Wavemeters using lumped circuit elements are used up to frequencies of 1,200 mc. As shown in Fig. 40, a crystal detector is coupled to a resonant LC circuit. For frequencies up to 100 mc, a fixed inductor and variable capacitor are used to form the resonant circuit. Plug-in coils of various inductance values are used to cover the frequency range. The capacitor dial is calibrated to read frequency directly. When the wavemeter coil is in the presence of the field of the unknown signal, the circuit is tuned to resonate at the same frequency as the unknown signal. The output of the crystal detector is indicated by the meter circuit. In wavemeters operating above 100 mc, a butterfly resonant circuit is used. This element varies both the inductance and capacitance simultaneously. The accuracy of a lumped-constant wavemeter depends on the Q of the resonant circuit. In general, accuracies between $\frac{1}{4}$ and 1 per cent may be obtained.

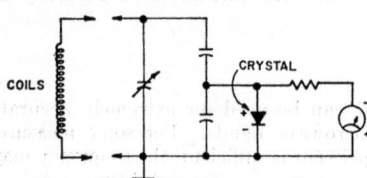

FIG. 40. Lumped-constant wavemeter.

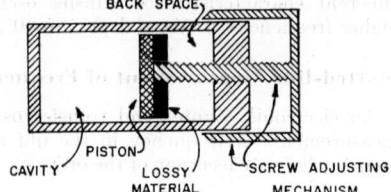

FIG. 41. Tuned-cavity wavemeter.

Cavity-type Wavemeters

Frequency in a waveguide system can be determined easily with a cavity-type wavemeter. Typically a cylindrical cavity and piston are used as shown in Fig. 41.

As the position of the wavemeter piston is varied, the distributed inductance and capacitance of the cavity, hence the resonant frequency, is changed.

In a reaction-type wavemeter there is a drop in the transmitted power when the piston is adjusted for resonance because power is absorbed in the tuned cavity. For this reason a wavemeter in a waveguide system should be detuned so that maximum power will be transmitted to the load.

Accuracy of a cavity-type wavemeter depends on the selectivity, or Q, of the resonant cavity. Commercial wavemeters are available with accuracies of 0.1 to 0.01 per cent.

RESISTANCE, INDUCTANCE, CAPACITANCE

Resistance

Resistance is defined by Ohm's law as

$$R = \frac{E}{I} \tag{19}$$

where R = resistance, ohms
  E = voltage or potential across the resistor, volts
  I = current through the resistor, amp
  All methods of resistance measurement make use of the basic relationship of Eq. (19). Depending on the resistance value and the accuracy required, an unknown resistance may be measured directly with an indicating meter, or it may be measured by comparison with a standard resistance in a bridge circuit.

Voltmeter-Ammeter Method. A simple and accurate method of measuring resistance with a voltmeter and an ammeter is shown in Fig. 42. Resistance is then determined from Eq. (19). By alternately connecting the two meters to the circuit, neither the current through the voltmeter not the voltage drop across the ammeter will be measured. The accuracy of measurement depends on the accuracy of the meters used.

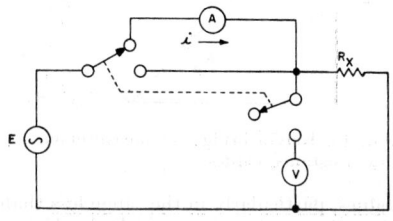

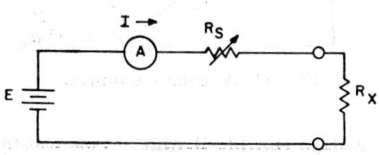

FIG. 42. Voltmeter-ammeter method for measuring resistance.

FIG. 43. Simple ohmmeter.

Ohmmeter. The standard ohmmeter consists essentially of a standard or known voltage source, an internal resistance, and an indicating meter as shown in Fig. 43. The unknown resistance is

$$R_x = \frac{E - IR_s}{I} \tag{20}$$

where E = voltage applied to the circuit, volts
  I = current through the meter, amp
  R_s = internal resistance, ohms
  R_x = unknown resistance, ohms
  The internal resistance is used to zero-set the instrument and to correct for changes in the voltage source. Accuracies of ± 5 to ± 10 per cent are easily obtained. Somewhat greater accuracy can be achieved by an ohmmeter arrangement using a VTVM with a regulated supply as the voltage source.

Megohmmeter. The megohmmeter, a modification of the standard ohmmeter, is used for measuring very high values of resistance and for insulation testing. The range of resistance covered is generally from about 50 megohms to 5 megamegohms (5×10^{12}). To measure these high resistance values, the applied voltage may be as high as 500 to 1,000 volts. The necessary test potential is usually supplied by a highly regulated or stabilized power supply. In a standard ohmmeter, by contrast, the measuring voltage may be supplied by a flashlight-type battery. A d-c VTVM circuit is used in most megohmmeters to obtain the required high input impedance. Megohmmeters provide measurement accuracies in the order of ± 5 to ± 10 per cent.

Wheatstone Bridge. Highly precise resistance measurements are made with bridge circuits in which a known resistance is matched with the unknown to obtain bridge balance. Perhaps the most common type of resistance-measuring bridge is the Wheatstone bridge illustrated in Fig. 44.

In the Wheatstone bridge a d-c potential is applied across the bridge and a sensitive meter is used as a null detector to indicate the balanced condition. No current flows through the meter when the bridge is balanced and the potential difference across it is zero. In this condition

$$R_x = \frac{R_a}{R_b} R_s \tag{21}$$

where R_x = unknown resistance, ohms
R_a and R_b = resistances of the bridge ratio arms, ohms
R_s = variable resistance, ohms

Accuracy of a Wheatstone-bridge measurement depends upon the precision of the resistors used in the bridge and the sensitivity of the null-detecting meter. Accuracies of the order of ±0.1 per cent are easily achieved for measurements from 1 ohm to 1 megohm.

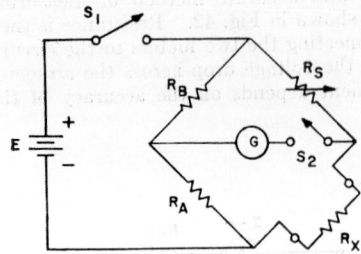

FIG. 44. Wheatstone bridge.

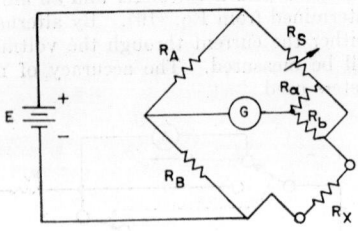

FIG. 45. Kelvin bridge for measurement of low resistance values.

Kelvin Double Bridge. Low resistance values, particularly in the range less than 1 ohm, may be measured with accuracy of less than ±2 per cent with the Kelvin double bridge illustrated in Fig. 45. The unknown resistor is connected to the standard resistor by a very low resistance conductor with the bridge resistors selected so

$$\frac{R_a}{R_b} = \frac{R_A}{R_B} \tag{22}$$

Then, when the bridge is balanced,

$$R_x = \frac{R_B}{R_A} R_s \tag{23}$$

Oscillator-VTVM Method. A resistance-measuring system with accuracy intermediate between ohmmeter and bridge methods uses an audio oscillator and a VTVM as shown in Fig. 46. The 1-megohm resistor in the circuit causes the audio oscillator to appear as a constant current source.

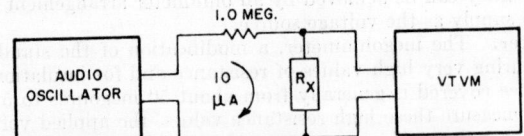

FIG. 46. Measuring resistance with an oscillator and a VTVM.

The measurement system is first adjusted by placing a precision 100-ohm resistor in place of the unknown and adjusting the oscillator output for a reading of 1 mv on the VTVM. Then, with the unknown resistor in the circuit, its value in ohms can be read directly as the VTVM indication times 10^6. Accuracy of the measurement depends on the accuracy of the VTVM and the precision of the 100-ohm resistor.

Inductance

Inductance is usually measured with bridge techniques which indicate Q as well as inductance value. Several well-known bridge configurations are used to meet various measurement requirements.

Maxwell Bridge. Figure 47 illustrates the Maxwell bridge, a variation of the basic Wheatstone circuit. Equations at balance are

$$L_x = R_s R_A C_q \tag{24}$$

$$Q = \frac{\omega L_x}{R_x} = \omega R_q C_q \tag{25}$$

$$R_x = \frac{R_A}{R_q} R_s \tag{26}$$

where L_x = unknown inductance, henrys
R_x = resistive losses in L_x, ohms
R_A, R_s, R_q = bridge resistors, ohms
ω = frequency, radians/sec ($2\pi f$ where f is frequency, cps)
Q = measure of quality of the coil

In the Maxwell bridge, the measured inductance value is independent of both frequency and the resistive losses in the inductor. The variable resistance standard

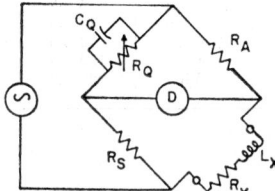

FIG. 47. Maxwell bridge.

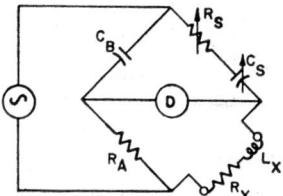

FIG. 48. Owen bridge.

R_s is calibrated to read inductance directly, and when operated at a specific frequency, resistance R_q may be calibrated to read Q directly. If the Q is very large, the value of R_q becomes too high for practical purposes.

Owen Bridge. The Owen-bridge circuit is often used to measure incremental inductance in iron-core coils with both alternating and direct current flowing. A fixed capacitance standard is used, and the bridge is balanced with a variable resistor and an air-dielectric capacitor. The Owen-bridge circuit is shown in Fig. 48. Balance equations are

$$L_x = R_a R_s C_b \tag{27}$$

$$Q = \omega C_s R_s \tag{28}$$

$$R_x = \frac{R_a C_b}{C_s} \tag{29}$$

Oscillator-VTVM Method. An oscillator and VTVM can be used to measure inductance by establishing the oscillator as a constant current source and measuring

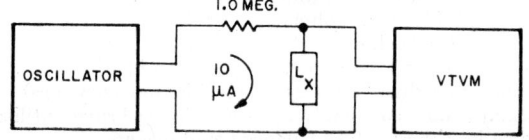

FIG. 49. Measuring inductance with an oscillator and a VTVM.

the voltage across the unknown inductor with the VTVM. Since the impedance of the inductor is a function of frequency, the oscillator may be set to specific frequencies which will provide direct reading of inductance on the VTVM. Figure 49 shows the measuring circuit.

The system is calibrated by replacing the unknown inductor with a precision 100-ohm resistor and adjusting the oscillator output for a VTVM reading of 1 mv. This establishes the oscillator output at 10 μa. The value of the unknown inductance is then directly related to the VTVM reading. Table 2 shows the value of the unknown inductance as a function of VTVM reading for various frequencies and meter ranges.

Table 2. Values of Inductances as Function of VTVM Reading for Various Frequencies and Meter Ranges

Voltmeter range	Inductance (full-scale value)				
	Frequencies, cps				
	24	48	160	1,600	16,000
0.3 mv	.....	100 mh	30 mh	3 mh	300 μh
1.0 mv	.....	333 mh	100 mh	10 mh	1 mh
3.0 mv	.....	1 h	300 mh	30 mh	3 mh
10 mv	.....	3.33 h	1 h	100 mh	10 mh
30 mv	.....	10 h	3 h	300 mh	30 mh
0.1 volt	.....	33.3 h	10 h	1 h	100 mh
0.3 volt	200 h	100 h	30 h	3 h	300 mh

Capacitance

Capacitance is measured by essentially the same methods as are used for measuring inductance.

Wien Bridge. The Wien-bridge circuit shown in Fig. 50 is often used to measure capacitance to a high degree of accuracy. Balance equations are

$$\frac{C_x}{C_s} = \frac{R_B}{R_A} - \frac{R_s}{R_x} \tag{30}$$

and

$$C_s C_x = \frac{1}{R_s R_x \omega^2} \tag{31}$$

From these

$$C_x = \frac{R_B C_s}{R_A(1 + R_s^2 C_s^2 \omega^2)} \tag{32}$$

$$R_x = \frac{R_A(1 + R_s^2 C_s^2 \omega^2)}{R_B R_s^2 C_s^2 \omega^2} \tag{33}$$

where
C_x = unknown capacitance, farads
C_s = standard capacitance, farads
R_B and R_A = resistive bridge arms, ohms
R_s = resistive standard, ohms
R_x = resistive loss in the unknown capacitor, ohms
ω = frequency, radians/sec ($2\pi f$, where f is frequency, cps)

Schering Bridge. Many commonly used capacitance bridges utilize the Schering-bridge circuit shown in Fig. 51. The balance equations are

$$C_x = \frac{C_s R_B}{R_A} \tag{34}$$

$$\text{D.F.} = \omega C_x R_x = \omega C_B R_B \tag{35}$$

From Eqs. (34) and (35),

$$R_x = \frac{C_B R_A}{C_s} \tag{36}$$

where C_x = unknown capacitance, farads
        C_s = capacitance standard of the bridge, farads
        R_x = resistive losses in C_x, ohms
R_A and R_B = resistive bridge arms, ohms
        C_B = capacitive bridge arm, farads
        ω = frequency, radians/sec ($2\pi f$ where f is frequency, cps)
        D.F. = dissipation factor of the unknown capacitance
Oscillator-VTVM Method. The value of an unknown capacitance can be determined by using an oscillator and a VTVM in the same manner as inductance is measured with these instruments. Measurement range of 800 pf to 800 μf is practical with the measuring circuit shown in Fig. 49.

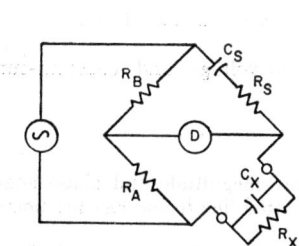

FIG. 50. Wien bridge.

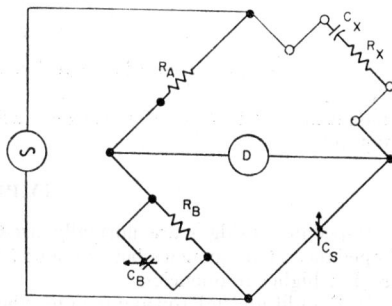

FIG. 51. Schering bridge.

The oscillator output is first adjusted for a reading of 1 mv with a precision 100-ohm resistor in place of the unknown capacitor. Oscillator output is then 10 μa, and the unknown capacitance value will be indicated directly on the VTVM when the unknown is connected into the circuit. Table 3 shows the value of the unknown capacitor as a function of frequency and VTVM range.

Table 3. Values of Capacitances as Function of VTVM Reading for Various Frequencies and Meter Ranges

Volt-meter-range	Frequencies, cps			
	20	200	2,000	20,000
Constant-voltage method (7.96 volts)				
0.3 mv	0.003 μf	300 pf	30 pf	3 pf
1.0 mv	0.01 μf	0.001 μf	100 pf	10 pf
3.0 mv	0.03 μf	0.003 μf	300 pf	30 pf
10 mv	0.1 μf	0.01 μf	0.001 μf	100 pf
30 mv	0.3 μf	0.03 μf	0.003 μf	300 pf
0.1 volt	1 μf	0.1 μf	0.01 μf	0.001 μf
0.3 volt	3 μf	0.3 μf	0.03 μf	0.003 μf
Constant-current method (10 μa)				
0.3 mv	800–250 μf	80–25 μf	8–2.5 μf	0.8–0.25 μf
1.0 mv	250–80 μf	25–8 μf	2.5–0.8 μf	0.25–0.08 μf
3.0 mv	80–25 μf	8–2.5 μf	0.8–0.25 μf	0.08–0.025 μf
10 mv	25–8 μf	2.5–0.8 μf	0.25–0.08 μf	0.025–0.008 μf
30 mv	8–2.5 μf	0.8–0.25 μf	0.08–0.025 μf	0.008–0.0025 μf
0.1 volt	2.5–0.8 μf	0.25–0.08 μf	0.025–0.008 μf	0.0025 μf–800 pf
0.3 volt	0.8–0.25 μf	0.08–0.025 μf	0.008–0.0025 μf	800–250 pf

Another method of measuring capacitance with an oscillator and VTVM is shown in Fig. 52. The oscillator output is connected directly to the VTVM before the unknown capacitor and the precision 100-ohm resistor are placed in the circuit and adjusted for an output of 7.96 volts. Then, with the unknown capacitance and the precision resistor connected, the capacitance value can be read on the VTVM. Table 3 correlates the value of the unknown capacitor and the VTVM reading for various

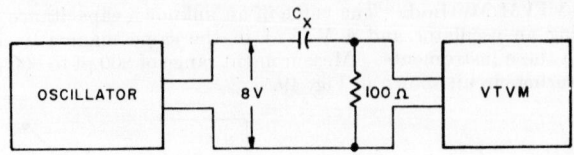

Fig. 52. Measuring capacitance with an oscillator and a VTVM.

frequencies and voltmeter ranges, using constant-voltage and constant-current methods.

IMPEDANCE

Impedance bridges are normally used to measure magnitude and phase angle of impedance at frequencies below about 500 mc. Slotted-line techniques are normally used at higher frequencies.

In the uhf and microwave regions where slotted-line techniques are used, the R, L, and C parameters are distributed and the impedance varies with position of measurement. For this reason impedance measurements must be referred to some reference position. There are well-established methods for obtaining desired impedance information from slotted-line measuring systems.

Since impedance determines energy reflected from the load, information concerning the load may be obtained by determining the reflection coefficient ρ_v.

$$\rho_v = \frac{Z_L - Z_0}{Z_L + Z_0} \tag{37}$$

where Z_L = load impedance
Z_0 = characteristic impedance of the line

Magnitude of the reflection coefficient may be determined by measuring the standing-wave ratio (SWR) with a slotted line or slotted section.

$$\rho_v = \frac{\sigma - 1}{\sigma + 1} \tag{38}$$

where σ = standing-wave ratio

Equating Eqs. (37) and (38),

$$Z_{L\ max} = Z_0 \sigma \tag{39}$$

$$Z_{L\ min} = \frac{Z_0}{\sigma} \tag{40}$$

The reflection coefficient may also be measured directly with a reflectometer system, a group of instruments which samples the incident and reflected waves, computes their ratio, and indicates reflection coefficient as a percentage. The reflectometer system is based on

$$\rho_v = \frac{E_{r0}}{E_{i0}} \tag{41}$$

where E_{r0} = magnitude of the reflected wave at the load
E_{i0} = magnitude of the incident wave at the load

Reflectometer systems are most useful for fast swept-frequency production measurements. Such a system provides information on the magnitude of the reflection coefficient by measuring reflected power with a directional coupler but does not indicate phase angle. An oscilloscope with a long-persistence cathode-ray tube may be used with a reflectometer system to provide a graphic indication of reflection coefficient for go/no go production testing. A recorder may be added to the system to provide a permanent record of the frequency response/reflection coefficient characteristic of the device under test. Basic accuracy of a reflectometer system is sufficient for most production-test requirements.

Impedance Bridge System

Several types of bridges are available to measure impedance at frequencies in the 50- to 500-mc range within ±2 to ±5 per cent. One of the most reliable of these is the Byrne bridge illustrated in Fig. 53. The bridge has a magnetic probe which

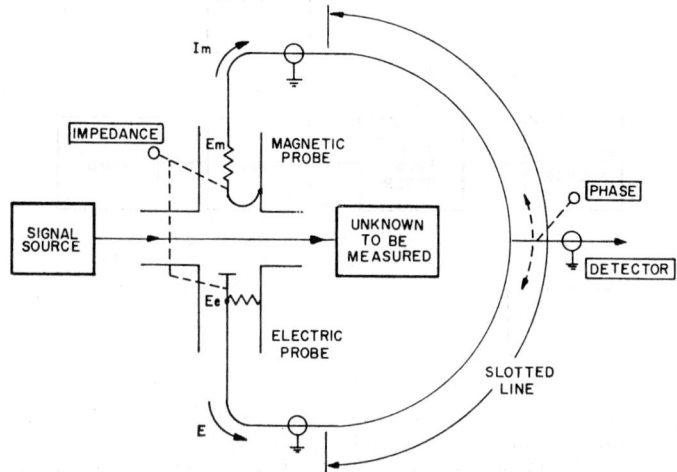

Fig. 53. Byrne bridge.

samples the field proportional to current, and an electric probe which samples the field proportional to voltage. The ratio of these fields is then mechanically derived to indicate impedance. Phase information is determined by comparing the relative positions of the current and voltage waves in a circular slotted line.

Unlike many bridges which read impedance in rectangular coordinates (resistance + reactance, or conductance + susceptance) the Byrne bridge measures impedance in polar coordinates to give direct indication of magnitude and phase angle. Impedance transformations and graphic solutions of transmission-line problems are then done on a Z-θ chart.

Measuring Procedure. The Byrne bridge is used in a measuring system as shown in Fig. 54. If a sensitive detector is used (5-μv sensitivity or greater), the connection between bridge output and detector should be a solid coaxial conductor.

With the load connected to the "unknown" terminals, the bridge is balanced to a null by adjusting the phase and magnitude controls for a reduction in the detected signal level. The region of the null is most easily established with a low signal level, or low sensitivity. Once the null region has been established either the signal level or the detector sensitivity is increased, and the bridge controls are further adjusted for more accurate localization of the null.

Phase and magnitude of the impedance are read on the bridge dials. The magnitude of the impedance is direct-reading and is independent of frequency. The phase angle, however, is dependent on the frequency of the input signal and is referenced to a convenient frequency (such as 100 mc).

Measuring Effects.   At high frequencies, several nulls can be obtained on the phase dial of a Byrne bridge.   The significant null is the one closest to 0°.

In many cases the impedance of an antenna or other load must be measured under conditions where appreciable radiated energy is present.   Precautions must be taken to prevent erroneous readings since even small amounts of leakage into the detector will affect the measurement.   Ordinary r-f double-braid shielded cable is often inadequate, and copper-clad cable should be used, with both cable ends carefully grounded to the connectors.   Inadequate shielding becomes evident from inconsistent repeat measurements or from a shifting of the null when the detector or bridge is touched or grounded.

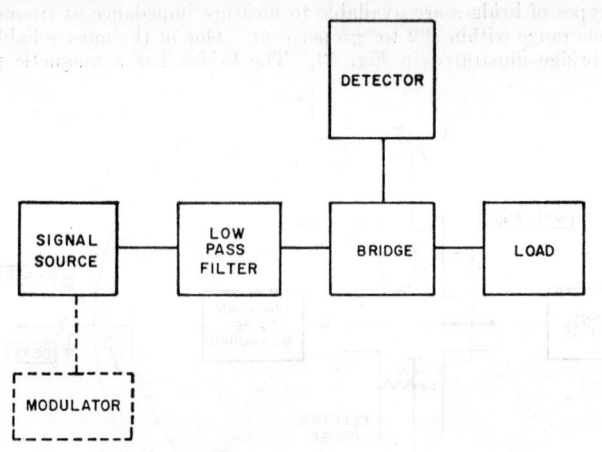

FIG. 54. Impedance bridge measuring system.

For measurement accuracy the source should be capable of delivering an a-m signal with a power output of at least 1 mw to the bridge.   The signal should be relatively free from harmonics.   If necessary, a low-pass filter can be used to assure freedom from harmonics.

Slotted-line System

Equipment Required:

Signal source: signal generator + low-pass filter
    or   oscillator + modulator + low-pass filter
Transmission line: slotted section + probe carriage
    or   slotted line and probe
Detector: broadband or untuned probe, crystal detector, etc. + standing-wave indicator or tuned voltmeter
Coaxial or waveguide short circuit

Measuring Procedure.   The equipment is connected as shown in Fig. 55.   Then the probe depth is adjusted, the gain setting of the SWR indicator is established, and the frequency of the modulating signal is set to peak the response of the detected signal.   With the load connected to the transmission line, the SWR is measured, and the position of the minima in the standing-wave pattern is noted.

Then the load is replaced with a short, and the new position of the standing-wave minimum is noted.   Typical standing-wave patterns are shown in Fig. 56.   Since voltage minima occur each half wavelength, the maximum shift necessary to consider is a quarter wavelength each side of the reference point.   Interpretation of the shift is illustrated in Fig. 58.

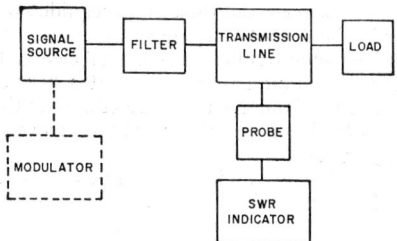

Fig. 55. Slotted-line impedance-measuring system.

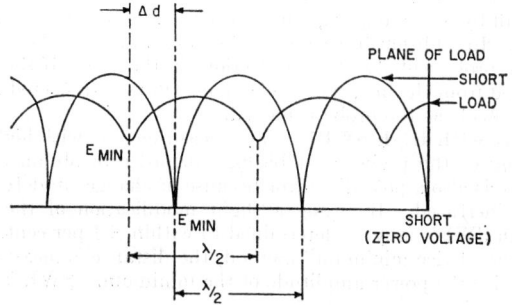

Fig. 56. Standing-wave patterns with load and short.

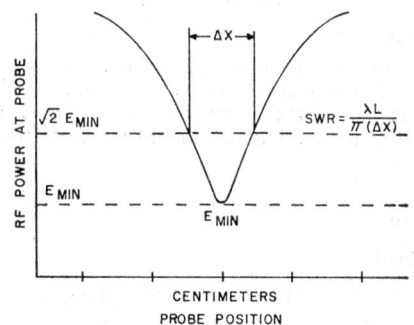

Fig. 57. Double-minimum method for computing SWR.

The normalized impedance may be computed from

$$\frac{Z_{\text{load}}}{Z_0} = \frac{1 - j(\text{SWR})\tan X}{(\text{SWR}) - j\tan X} \tag{42}$$

where

$$X = \frac{180° (\pm \Delta d)}{\lambda/2} \tag{43}$$

$\pm \Delta d$ = shift in centimeters of the minimum point when the short is applied. d is positive (+) when the minimum shifts toward the load, negative (−) when the minimum shifts toward the generator

$\lambda/2$ = one-half the guide wavelength, cm (distance measured between two adjacent minima)

Measuring Effects. Signal sources can introduce r-f harmonics, f-m and spurious signals, any of which will affect slotted-line measurements.

Harmonics have more effect in coaxial than in waveguide systems because excessive harmonics often appear in the output from coaxial sources. Also, broadband coaxial pickups will often pass harmonic frequencies more efficiently than the fundamental. The over-all effect is that the standing-wave ratio at a harmonic frequency may be considerably higher than at the fundamental.   To eliminate the effect of harmonics a low-pass filter with about 50 db of rejection to the harmonic frequencies is inserted into coaxial slotted-line systems.

F-M or other spurious signals are undesirable in any system because they obscure the minimum points and increase the difficulties of accurately measuring the voltage minima.

Since the sampling probe must extract power from the transmission line, it will have an effect on the fields within the line. Probe penetration should therefore normally be reduced to a minimum.   In making slotted-line measurements the shunt admittance is kept small by loose coupling (small penetration) and by using a signal power of at least 1 mw.   In addition to its power absorption and its effect on the standing-wave pattern, the probe will also cause reflections in the line.   If the generator is not matched or padded from the line, these reflections will be re-reflected toward the load, causing variable effects as the probe is moved.

Measurements with High SWR's.   For measurements with high SWR's (above 10:1) the coupling of the probe must be high in order to obtain a reading at the minimum.   There is also a potential error because of changes in detector characteristics as power is increased.   By using a slight modification of the standard SWR measurement high SWR's may be determined to within ± 1 per cent.   In this procedure, known as the "twice-minimum" method, the distance is measured between the points that are twice the power amplitude of the minimum.   SWR is then calculated from

$$\text{SWR} = \frac{\lambda L}{\pi \, \Delta X} \tag{44}$$

where λL = guide wavelength, cm
  ΔX = distance between twice-minimum points, cm

Note that the points referred to are the twice *power* points.   Thus, if a linear voltage indicator is used with a square-law detector, the voltage indication of the twice power point will be twice that of the minimum.   Figure 57 illustrates the use of the twice-minimum method.

Reflectometer System

Equipment Required

Signal source: Sweep oscillator + variable attenuator
                  + modulator or signal generator

Multihole directional couplers

Detector: Crystal or barretter mounts

Ratio meter
Short circuit

A typical reflectometer system is shown in Fig. 59.   Either a crystal or a barretter detector can be used to detect the signals from the directional couplers.   The barretter, however, offers distinct advantages over the crystal detector.   Among these advantages are precisely known response characteristics, relative ruggedness of construction, comparable dynamic range, and superior stability.   The time constant of the barretter must be small compared with the period of the modulation envelope.

Since barretters require biasing power for operation, some ratio meters provide a d-c output for this purpose.   In general, the barretter must be connected to the ratio meter through an impedance-matching transformer.   Figure 60 illustrates the necessary cable connections for a reflectometer system with barretter detectors.

Measuring Procedure. After connecting the equipment as shown in Fig. 59, adjust the frequency of the modulating signal to correspond with that of the ratio meter. Then replace the load with a short at the load end of the system. With the output shorted, all of the transmitted signal will be reflected. Thus the reflection coefficient will be 100 per cent.

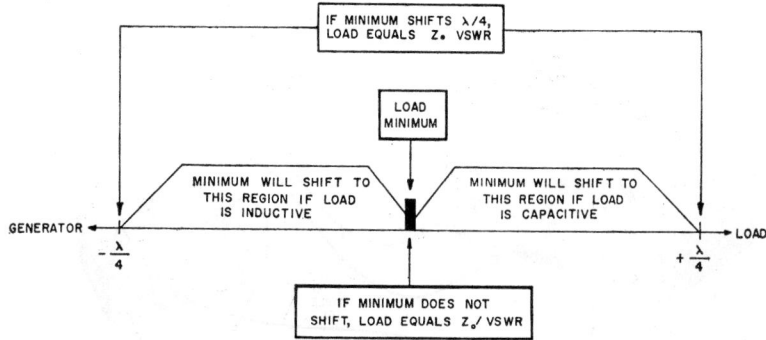

FIG. 58. Summary of rules for impedance measurement.

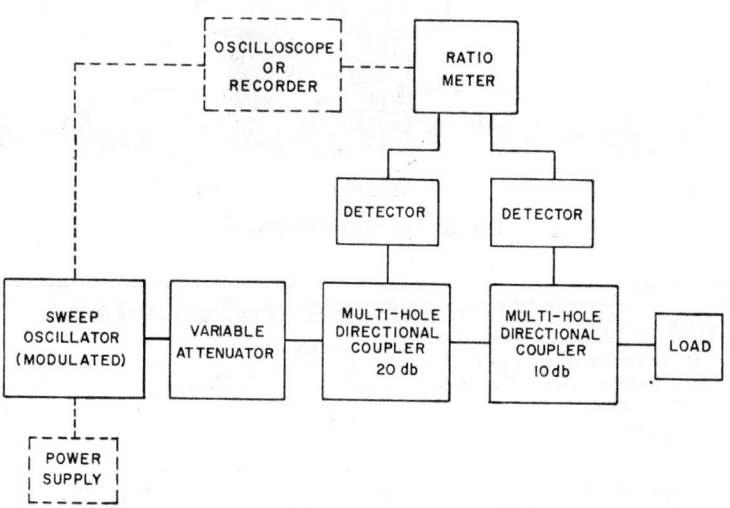

FIG. 59. Reflectometer system.

The ratio meter, oscilloscope, or recorder should be calibrated for 100 per cent reflection coefficient. Then the ratio meter will be direct-reading for any specific frequency or will present the dynamic characteristics of the reflection coefficient over the sweep range of the oscillator. The SWR of the system may be calculated from

$$\text{SWR} = \frac{1 + \rho}{1 - \rho} \tag{45}$$

or computed by using a reflectometer calculator shown in Fig. 61. This calculator is available from the Hewlett-Packard Company, Palo Alto, Calif. In addition to computing the reflection coefficient, the calculator determines return loss, mismatch loss, and voltage, current, or power ratio expressed in decibels.

Measuring Effects.   There are two general sources of inaccuracies in a reflectometer system.   Those which result from sensitivity variations are scalar quantities. Those which result from spurious signals are vector quantities.

Sensitivity variations can be traced to sources such as variation with frequency of the coupling coefficients of the directional couplers, and variation with frequency and

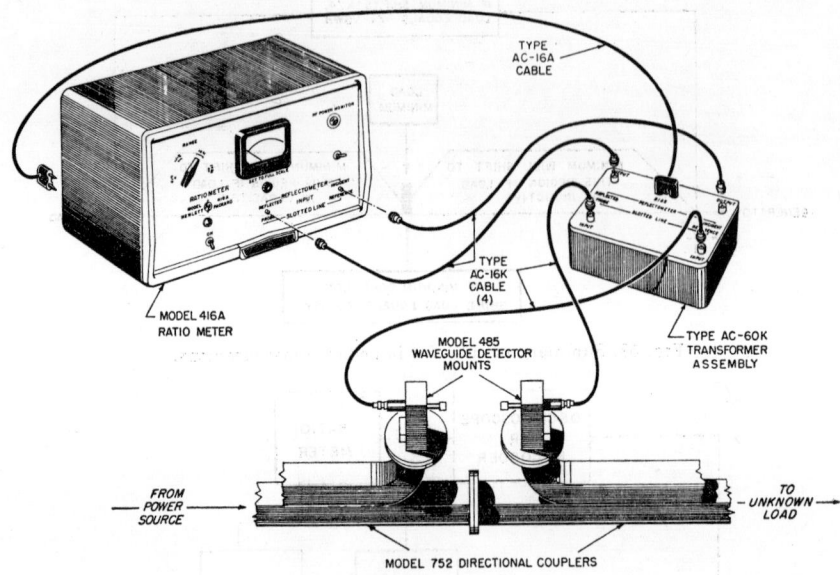

FIG. 60. Reflectometer system.

FIG. 61. Reflectometer calculator.

amplitude of the detection efficiency of the detectors.   If high measurement accuracy is needed these errors can be determined and compensation can be made.

Where inaccuracies are due to spurious signals which appear at the reverse detector the magnitude can be determined but the phase angle cannot be.   The effect of these spurious signals can therefore only be estimated.

A thorough discussion of the errors in a reflectometer system, and the means of compensating for them, will be found in published literature such as the Application Note "Applications of the Ratio Meter" published by the Hewlett-Packard Company.

REFERENCES

1. Guillemin, E. A.: "Introductory Circuit Theory," pp. 343–354, John Wiley & Sons, Inc., New York, 1953.
2. Hand, B. P.: An Automatic DC to X-band Power Meter, *Hewlett-Packard J.*, vol. 9, no. 12, August, 1958.
3. Strata, R.: Microwave Power Detectors, *Electronics*, July 17, 1959.
4. Wind, M.: "Handbook of Electronic Measurements," Polytechnic Institute of Brooklyn, Microwave Research Institute, 1956.
5. Terman, F. E., and J. M. Pettit: "Electronic Measurements," 2d ed., McGraw-Hill Book Company, Inc., New York, 1952.
6. Considine, D. M.: "Process Instruments and Controls Handbook," McGraw-Hill Book Company, Inc., New York, 1957.
7. Lepage, W. R.: "Analysis of Alternating-current Circuits," McGraw-Hill Book Company, Inc., New York, 1952.
8. Michels, W. C.: "Electrical Measurements and Their Applications," D. Van Nostrand Company, Inc., Princeton, N.J., 1957.
9. Wind, Moe, and Harold Rapaport: "Handbook of Microwave Measurements," Interscience Publishers, New York, 1955.
10. Czech, J.: "The Cathode Ray Oscilloscope," Interscience Publishers, New York, 1957.
11. Rider, J. F., and S. D. Uslan: "Encyclopedia on Cathode-ray Oscilloscopes and Their Uses," Chapman & Hall, Ltd., London, 1959.
12. Banner, E. H. W.: "Electronic Measuring Instruments," Chapman & Hall, Ltd., London, 1958.
13. Wilson, W.: "The Cathode Ray Oscillograph in Industry," Chapman & Hall, Ltd., London, 1953.
14. Elliott, A., and J. Home Dickson: "Laboratory Instruments," Chapman & Hall, Ltd., London, 1959.
15. Ginzton, Edward L.: "Microwave Measurements," McGraw-Hill Book Company, Inc., New York, 1957.
16. Montgomery, Carol G.: "Technique of Microwave Measurements," McGraw-Hill Book Company, Inc., New York, 1947.
17. Hund, August: "High Frequency Measurements," McGraw-Hill Book Company, Inc., New York, 1951.
18. Saul, Robert, and Elaine Luloff: Designing a Spectrum Analyzer, *Electronic Industries*, pp. 66–71, April, 1959.
19. Carroll, John M.: "Electron Devices and Circuits," McGraw-Hill Book Company, Inc., New York, 1962.
20. Soisson, Harold E.: "Electronic Measuring Instruments," McGraw-Hill Book Company, Inc., New York, 1961.
21. Griffiths, V. S., and W. H. Lee: "The Electronics of Laboratory and Process Instruments," John Wiley & Sons, Inc., New York, 1962.

Section 16

AEROSPACE INSTRUMENTATION

By

PAUL F. BECHBERGER, B.E.E., M.Sc., *Manager, Advanced Systems and Development Department, Eclipse-Pioneer Division, The Bendix Corporation, Teterboro, N.J.; Member, Institute of Aeronautical Sciences. (Aircraft and Aerospace Vehicle Instrumentation)*

RICHARD C. DEHMEL, Ph.D., *Vice President—Engineering, Curtiss-Wright Corporation, Wood-Ridge, N.J.; Member, Institute of the Aeronautical Sciences. (Aircraft Flight-simulation Instrumentation)*

AIRCRAFT FLIGHT-SIMULATION INSTRU-MENTATION

AIRCRAFT AND AEROSPACE VEHICLE INSTRUMENTATION

By P. F. Bechberger*

There are few, if any, applications where instruments and controls are so concentrated as aboard the modern aircraft. Although much of the instrumentation used in aircraft and spacecraft closely parallels that used in process-control work, many instruments have been developed strictly for this specific use. Because each device must operate in any attitude and over a wide range of pressures, temperatures, and accelerations, careful design is required to avoid errors due to these causes. Most aircraft instrumentation involves some form of remote indication, usually electrical and often synchro. Peculiar to these applications are the gyroscopes which give space reference not required in any other type of transportation.

Although aerospace instruments and controls can be classified in several ways, for the purposes of this handbook, the following grouping will be used:
1. Flight and attitude instruments and controls.
2. Navigational devices.
3. Engine instruments and controls.
4. Miscellaneous instruments and controls.
5. Satellite and space-vehicle instrumentation.

FLIGHT AND ATTITUDE INSTRUMENTS AND CONTROLS

Air-data Instruments

Pitot Tube. Most air-data flight instruments operate from "static" or "pitot" pressure or both. These pressures are commonly obtained from a pitot-static tube (see Fig. 1), which heads into the airstream. On the smooth side of the pitot tube, the pressure differs little from that of undisturbed air, and slots placed there provide a source of static pressure. Most tubes are electrically heated to melt off any ice that might form, which could partly or completely seal off the openings and give erroneous and possibly disastrous readings in airspeed, altitude, and rate of climb.

Altimeter. Most aircraft altimeters operate in a manner similar to the aneroid barometer, except they are calibrated in height instead of pressure. The relationship between altitude and static pressure is given in Table 1. Static pressure surrounding one or more evacuated-diaphragm capsules operates a pointer through a system of linkages and gears.

Corrections must be made for changes of barometric pressure. (One inch of mercury-pressure change corresponds to roughly 1,000 ft of altitude at sea level.)

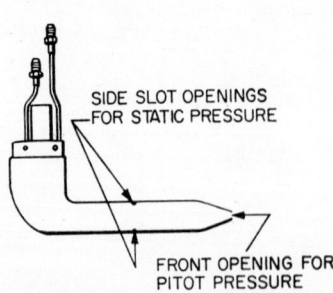

SIDE SLOT OPENINGS FOR STATIC PRESSURE

FRONT OPENING FOR PITOT PRESSURE

FIG. 1. Pitot tube. (*Source: Kollsman Instrument Corporation.*)

* Manager, Advanced Systems and Development Department, Eclipse-Pioneer Division, The Bendix Corporation, Teterboro, N.J.

**Table 1. U.S. Standard Atmospheric Altitude Versus
Static Pressure and Temperature***

Altitude, geometric, ft above sea level	Altitude in geopotential measure, ft above sea level	Temp., °R	Pressure, in. Hg
−1,500	−1,500	524.04	31.579
0	0	518.69	29.921
+1,500	+1,500	513.34	28.335
2,500	2,500	509.77	27.315
5,000	4,999	500.86	24.897
10,000	9,995	483.04	20.581
15,000	14,989	465.23	16.893
20,000	19,981	447.43	13.761
25,000	24,970	429.64	11.118
30,000	29,957	411.86	8.902
35,000	34,941	394.08	7.060
40,000	39,923	389.99	5.558
45,000	44,903	389.99	4.375
50,000	49,880	389.99	3.444
60,000	59,828	389.99	2.135
70,000	69,766	389.99	1.324
80,000	79,694	389.99	0.822
100,000	99,523	418.79	0.326
200,000	198,100	449.00	0.006
300,000	295,746	299.20	0.022×10^{-3}
500,000	488,292	1,933.00	1.465×10^{-7}

* ARDC Model atmosphere, 1959, AFCRC-TR-59-267.

Sensitive altimeters will show changes in altitude of 10 ft or less (see Fig. 2a). Although atmospheric pressure reduces in half for approximately each 18,000-ft increase in altitude, nonlinear diaphragm-linkage systems are used which give linear altitude vs. pointer deflection over the complete altitude range.

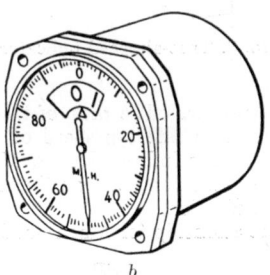

a   *b*

FIG. 2. (*a*) Sensitive altimeter. (*Source: Kollsman Instrument Corporation.*)   (*b*) Air-speed indicator. (*Source: The Bendix Corporation, Eclipse-Pioneer Div.*)

Airspeed Indicator.   The difference between the "pitot" pressure and "static" pressure is a measure of indicated airspeed.   Airspeed indicators are calibrated in knots (nautical mph) under standard pressure (29.92 in. Hg) and temperature (59°F). See Fig. 2*b*.   Table 2 gives the relationship between pressure difference and airspeed in knots.

The indicated airspeed at which a given aircraft with a given load stalls is a constant over wide ranges of pressure and temperature.   A knowledge of this is of the utmost importance during takeoffs and landings, as well as during flight.

True-airspeed Indicator.   True airspeed is primarily useful in navigation.   It is a function not only of difference in pitot and static pressure, but also of temperature.

A true-airspeed indicator (Fig. 3) is in effect a computer operating on a rather complex relationship of two pressures and air temperature to give the correct indication. This relationship is expressed by the following:

$$V = 38.94 \frac{M \sqrt{T_1}}{\sqrt{1 + 0.2KM^2}} \tag{1}$$

where V = true airspeed, knots

T_1 = indicated air temperature, °K

K = recovery factor of temperature probe (usually 1.00 or slightly less than 1.00)

M = Mach number [described under Machmeter, Eqs. (2) and (3) following]

Another approach is to use the speed of an air turbine as a measure of true airspeed, the intake pressure to the turbine being static pressure and the output being balanced to pitot pressure.

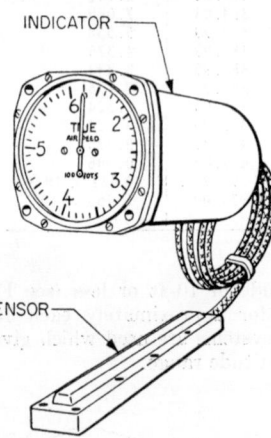

Fig. 3. True-airspeed indicator.

Fig. 4. Mach indicator. (*Source: The Bendix Corporation, Eclipse-Pioneer Div.*)

Machmeter. The ratio of true airspeed to the velocity of sound is called Mach number. The safe top speed of fast planes is usually expressed in terms of Mach number and is indicated on a Machmeter (see Fig. 4). This is also a mechanical computer which operates on the relationship of static and pitot pressures to give the proper indication.

The relationship for Mach number is as follows:

$$\frac{dP}{Ps} = (1 + 0.2M)^{23.5} - 1 \qquad \text{for } M < 1 \tag{2}$$

and

$$\frac{dP}{Ps} = \frac{166.9215M^7}{(7M^2 - 1)^{2.5}} - 1 \qquad \text{for } M > 1 \tag{3}$$

where M = Mach number

dP = differential (pitot less static) pressure

Ps = static pressure

A combined true airspeed and Mach number computer indicator is shown in Fig. 5.

Rate-of-climb Indicator. This instrument (Fig. 6) comprises an enclosed volume of air connected to atmospheric (static) pressure through a constriction. As the altitude changes the enclosed pressure lags that outside, and the pressure difference is measured in terms of rate of change of altitude, or rate of climb. Because any change of temperature causes a proportional change in pressure of an enclosed volume of air, such as used in the rate-of-climb indicator, the container must be a good thermal insulator to prevent all but very gradual temperature changes of the enclosed air, or proper correction must be made.   Other corrections must be introduced for both static pressure and temperature to provide correct rate-of-climb indication.

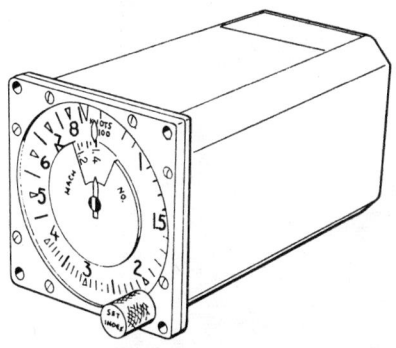

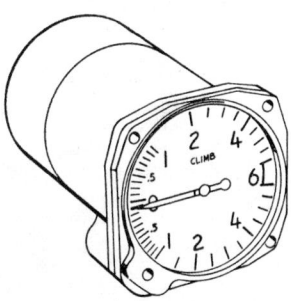

Fig. 5. True airspeed–Mach number indicator. (*Source: The Bendix Corporation, Central Div.*)

Fig. 6. Rate-of-climb indicator.   (*Source: The Bendix Corporation, Eclipse-Pioneer Div.*)

Air-data Computers.   Since a number of useful functions may be derived from the inputs of static and pitot pressure and air temperature, it is sometimes convenient to combine these measurements and computations in one package.   Such units are illustrated in Fig. 7*a* and *b* and are called air-data computers.

Air-data computers have been built in many shapes, sizes, and ranges and with a wide variety of outputs.   These computers sense pitot and static pneumatic pressure inputs by means of elastic sensors which are servo followed or balanced so as to provide

Table 2. Airspeed vs. Differential Pressure at Standard Temperature and Pressure*

Indicated airspeed, knots	Differential pressure, in. Hg	Indicated airspeed, knots	Differential pressure, in. Hg
0	0.000000	400	8.3973
10	0.004795	450	10.8837
20	0.019185	500	13.7967
30	0.043179	550	17.1859
40	0.076793	600	21.1088
50	0.1200	650	25.6316
60	0.1729	700	30.8159
70	0.2355	750	36.6276
80	0.3079	800	43.0092
90	0.3901	850	49.9241
100	0.4821	900	57.3481
125	0.7557	950	65.2644
150	1.0924	1,000	73.6613
175	1.4938	1,050	82.5302
200	1.9616	1,100	91.8649
225	2.4977	1,150	101.6606
250	3.1045	1,200	111.9136
275	3.7845	1,250	122.6210
300	4.5407	1,300	133.7805
350	6.2949	1,320	138.3706

* WADC 1952 model atmosphere, prepared by Batelle Memorial Institute, 1953.

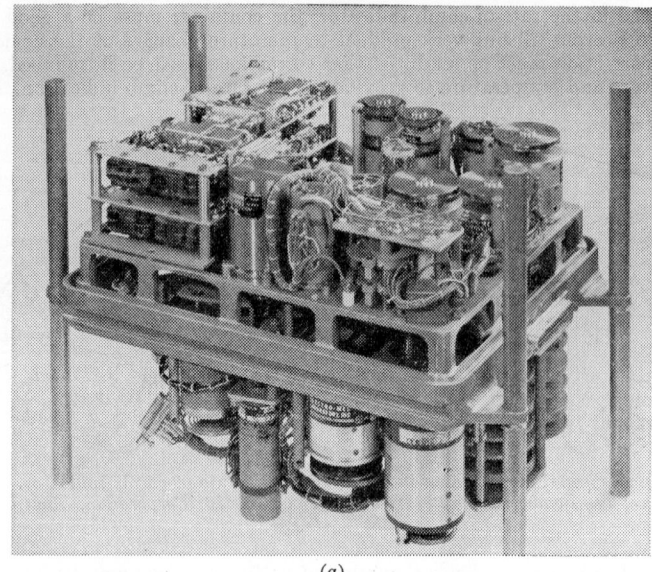

(a)

(b)

Fig. 7. Air-data computers. (a) Upper view: central unit. (*Courtesy of The Bendix Corp., Eclipse-Pioneer Div.*) (b) Lower view: indicating unit. (*Courtesy of The Garrett Corp., AiResearch Manufacturing Div.*).

a powered displacement output for the computer. Temperature is also servoed but usually from electrical inputs of variable resistance or thermocouple voltage. These input quantities are then processed by either mechanical or electrical computing devices to produce the desired output functions. These output functions can then be transmitted to other systems such as the ballistics, navigation, engine-control, and flight-control systems as well as the vertical scale or rotational display of altitude, airspeed, Mach number, and other required cockpit information.

Gyroscopic Instruments

Accelerations, both translational and due to turns, combine with gravity to give what is known as dynamic vertical. A human being can fairly well sense dynamic vertical but is poor at sensing rate of turn or at remembering true vertical. Pilots become hopelessly confused when they lose visual contact with some reference to true vertical. Thus some instrumentation to assist the aircraft pilot in the problem of attitude is essential.

The gyroscope provides a means of getting large values of effective inertia in compact form. The basic formula for all gyro instruments is

$$T = IV\omega \tag{4}$$

where T = precession torque, dyne-cm
I = moment of inertia of the gyro wheel about its spin axis, g cm^2
V = spin velocity, radians/sec
ω = precession velocity, radians/sec

The spin axis, the precessional velocity axis, and the precession torque axis are mutually perpendicular.

Turn-and-Bank Indicator. The simple instrument used for many years is the turn-and-bank (or turn-and-slip) indicator (Fig. 8a). It contains a rate gyro mounted with its spin axis athwartship and its spring-restrained gimbal axis aligned fore and aft as shown in Fig. 8b. Deflection of this axis from its neutral position is translated to motion of pointer A (Fig. 8a), calibrated in rate of turn. Below the pointer is a ball

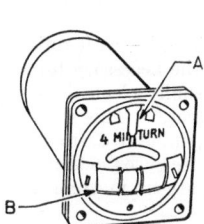

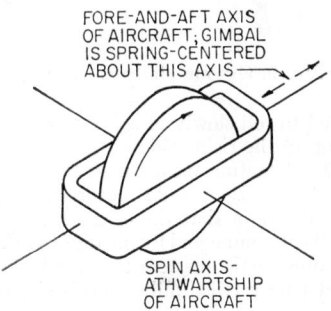

FORE-AND-AFT AXIS OF AIRCRAFT; GIMBAL IS SPRING-CENTERED ABOUT THIS AXIS

SPIN AXIS- ATHWARTSHIP OF AIRCRAFT

FIG. 8. (a) Turn-and-bank indicator. (*Source: The Bendix Corporation, Eclipse-Pioneer Div.*)

FIG. 8. (b) Turn gyro. (*Source: The Bendix Corporation, Eclipse-Pioneer Div.*)

bank indicator B, which acts as a damped pendulum showing the dynamic vertical component lying in the plane perpendicular to the fore-and-aft axis. If both pointer and ball are centered, the pilot knows he is flying straight and his wings are horizontal. The turn-and-bank indicator is also used in making desired rates of turn with proper bank. Both air-driven and electric (a-c or d-c) gyros are in common use. In a properly designed turn-and-bank indicator, the direction of rotation, moment of inertia, and speed of the gyro wheel are so matched to the centering spring that the gyro spin axis remains nearly horizontal as the plane makes a turn with proper bank at its normal cruising speed. Instead of a rotating gyro wheel, other sensing elements have been used, as a vibrating reed or a stream of liquid.

Gyros of the type used in turn-and-bank indicators are often called rate gyros and, when their speed is accurately controlled, can serve as accurate measures of rate of turn. They find many uses, as in gunsights, bombsights, autopilots, and navigation devices. Instead of driving a pointer, they may mechanically perform some function in a computer or may drive a potentiometer, synchro, switch, or similar device giving an electrical function of rate of turn.

Artificial Horizons or Gyro Verticals. If a gyro is arranged with gimbals as shown in Fig. 9, and there is some method of detecting instantaneous or dynamic vertical by which a torque may be applied to the gyro with its downward component being directed to the rising side of the gyro wheel (see Fig. 10), the spin axis will move toward the dynamic vertical. As the rate of this movement toward vertical, called "erection rate," is relatively slow, of the order of a few degrees a minute, the spin axis indicates a long-time average of dynamic vertical which usually is very close to true vertical.

If the gyro is air-driven, the detection of vertical may be pendulums controlling air jets whose reaction causes the gyro to erect. In the "ball-erection system," the detection means is combined with the method of applying the erecting torque in a ball

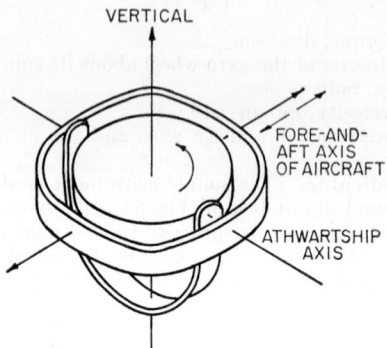

Fig. 9. Gyro in gimbals.   (*Source: The Bendix Corporation, Eclipse-Pioneer Div.*)

forced to roll slowly in a track perpendicular to the spin axis—as it rolls more slowly going up the rising side of the gyro, its weight applies the necessary force.

The detecting means may be pendulous switches, usually electrolytic or mercury, which energize torque motors causing the gyro to "erect." This system has many advantages in that the erection rate is externally adjustable if desired, and erection may be disconnected by means of switches actuated by rate gyros so that only when the dynamic vertical is near true vertical is it averaged by the gyro mechanism and a more exact true vertical indication is obtained.

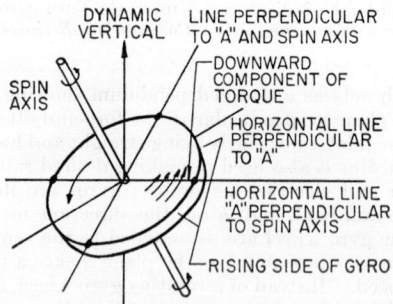

Fig. 10. Gyro vertical.   (*Source: The Bendix Corporation, Eclipse-Pioneer Div.*)

Many ways of presenting to the pilot the information as to direction of true vertical have been used, but one that is now in common use is shown in Fig. 11.

Artificial horizons usually have a caging device which allows the pilot to move the spin axis perpendicular to the fore-aft and athwartship axes, thus avoiding a long erecting time at takeoff and also to protect the gyro during shipment and in nonoperating periods. Gyro verticals may have signal takeoffs from each axis, and this information may be used for remote indication as in integrated cockpit display (Fig. 14b), for automatic pilots, for radar-antenna stabilization, and for many other applications.

Directional Gyros. If the gyro spin axis is maintained horizontal instead of vertical, it will tend to keep its spin axis pointed in one direction. By putting a degree calibration on the gyro, as shown in Fig. 12a, it forms a stable heading reference. A knob is usually provided to set in desired heading. Methods of drive, erection, and signal takeoff are quite similar to those used on gyro verticals. Conventional directional gyros (two gimbals) exhibit errors during turns. Therefore, roll stabilization illustrated in Fig. 12b is added to eliminate gyro turning errors. Its outstanding feature is that it provides continuously accurate heading information during bank and high-roll-rate maneuvers. Directional gyros of both types are often used in conjunction with magnetic-compass elements as described under All-latitude Compass Systems.

Schuler Tuning. Horizontal accelerations cause errors in vertical gyros of the type described previously. These errors can be canceled by precessing the gyro at a rate V_h/R, in which V_h is the horizontal velocity and R the radius of the earth. V_h is usually obtained by integration from an accelerometer, mounted on the gyro gimbal to maintain its sensitive axis horizontal. A signal proportional to V_h then torques the gyro. Figure 12c shows a single-axis erection loop. Any disturbance of the vertical gyro causes the accelerometer to sense a component of gravity. Since the accelerometer cannot distinguish between gravity and acceleration, it generates a signal

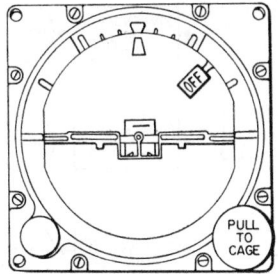

Fig. 11. Artificial horizon. (*Source: The Bendix Corporation, Eclipse-Pioneer Div.*)

that torques the gyro back toward vertical; however, when the gyro overshoots, it oscillates about the vertical with a period of 84 min near the earth's surface.

It is interesting that the 84-min period is also characteristic of a simple pendulum having its weight at the center of the earth and its pivot at the surface. A vertical gyro having this characteristic is said to be "Schuler-tuned."

Stable Platforms. A three-axis attitude reference can be obtained by mounting a directional gyro on the inner gimbal of a vertical gyro, or on gimbals slaved to those of a vertical gyro. However, a stable platform provides advantages over such a combination because servomotors cancel disturbing torques that would otherwise be transmitted through the gimbals to the gyros.

The platform is gimbaled to allow rotation in any direction and carries gyros to sense disturbances about three orthogonal axes, two of which are usually aligned horizontally. Either three single-degree-of-freedom gyros, as shown in Fig. 12d, or two two-degree-of-freedom gyros will suffice. The latter arrangement provides one extra sensitive axis. The gyro(s) sensing horizontal disturbances and associated accelerometers form parts of the two-axis Schuler-tuned erection system. V_h, determined by Doppler radar and properly inserted into the Schuler loop, gives improved results.

If the platform employs two-degree-of-freedom gyros, the redundant sensitive axis is often aligned vertically so that platform drift about this axis is the average of two gyro drifts. The azimuth gyro(s) may operate unrestrained as a directional gyro, it may be slaved to an external directional reference, or it may be made intrinsically north-seeking by application of a principle similar to that long used in the marine gyrocompass.

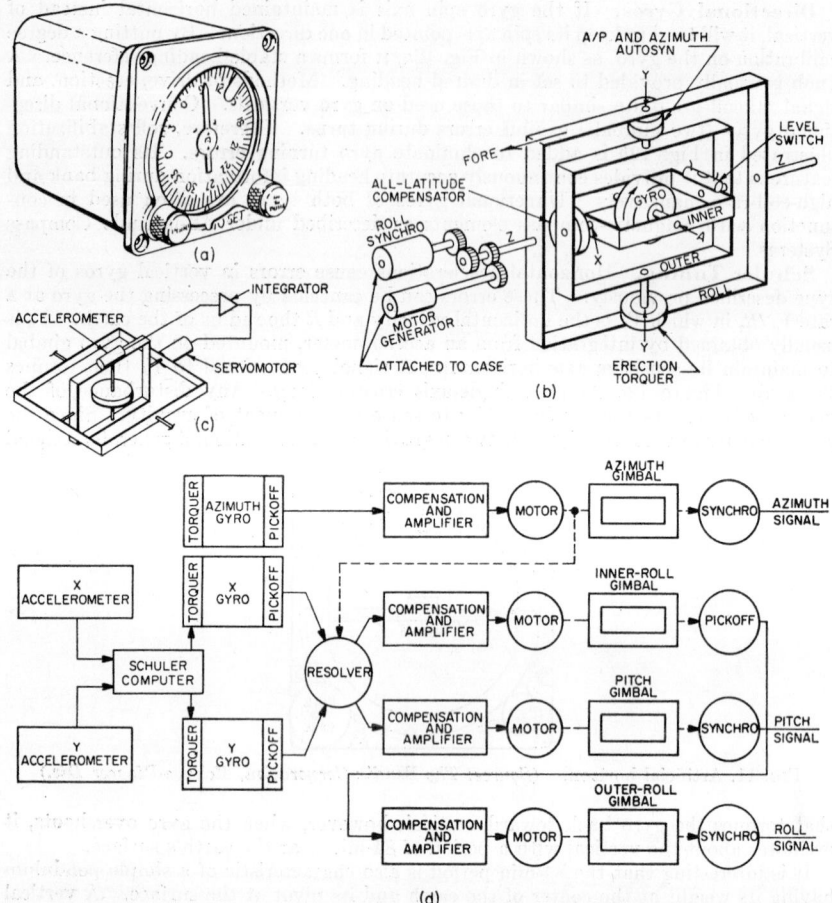

Fig. 12. (a) Directional gyro. (*Source: Engineering Research Associates.*) (b) Roll-stabilized gyro. (*Source: The Bendix Corporation, Eclipse-Pioneer Div.*) (c) Schuler-tuning single-axis erection system. (*Source: The Bendix Corporation, Eclipse-Pioneer Div.*) (d) Block diagram of typical three-gyro platform. (*Source: The Bendix Corporation, Eclipse-Pioneer Div.*)

Automatic Pilots

By combining signals from a gyro vertical as a vertical reference source, from **rate** gyros as needed to give stability on any or all of the aircraft's three axes, and from a directional gyro or stabilized compass for heading reference, actuators can be made to operate on the aircraft's control surfaces (ailerons, rudder, and elevators) and provide automatic flight. Altitude may be controlled by signals from an altimeter and maneuvers are readily accomplished by the human pilot using simple controls conveniently arranged. In addition, signals from existing ground radio navigational facilities are used to control the aircraft automatically along the airways and to the runway. Block diagrams of two types of autopilots used extensively on commercial aircraft are shown in Figs. 13 and 14a.

For aircraft operating over a wide range of speeds and altitudes, such as supersonic jets, it is necessary to change some of the control parameters as a function of flight

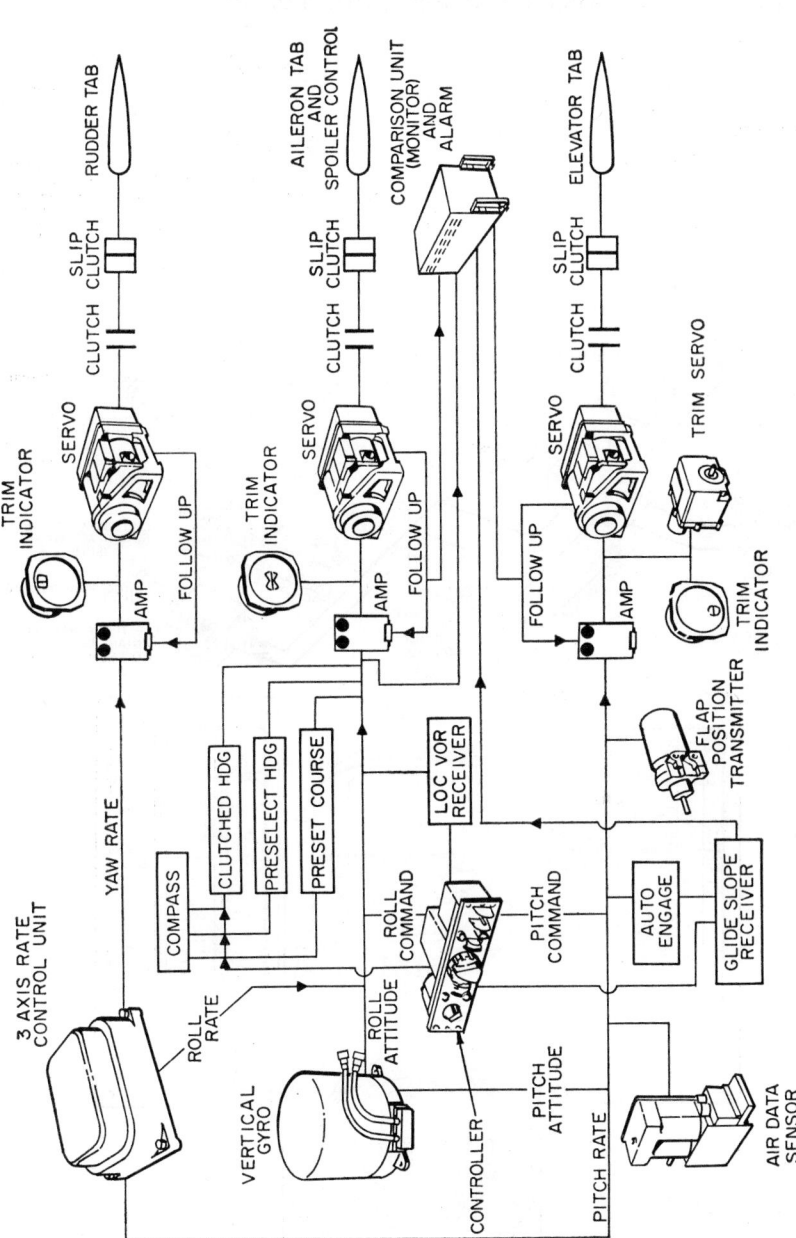

FIG. 13. PB-20 automatic flight-control system.   (*Source: The Bendix Corporation, Eclipse-Pioneer Div.*)

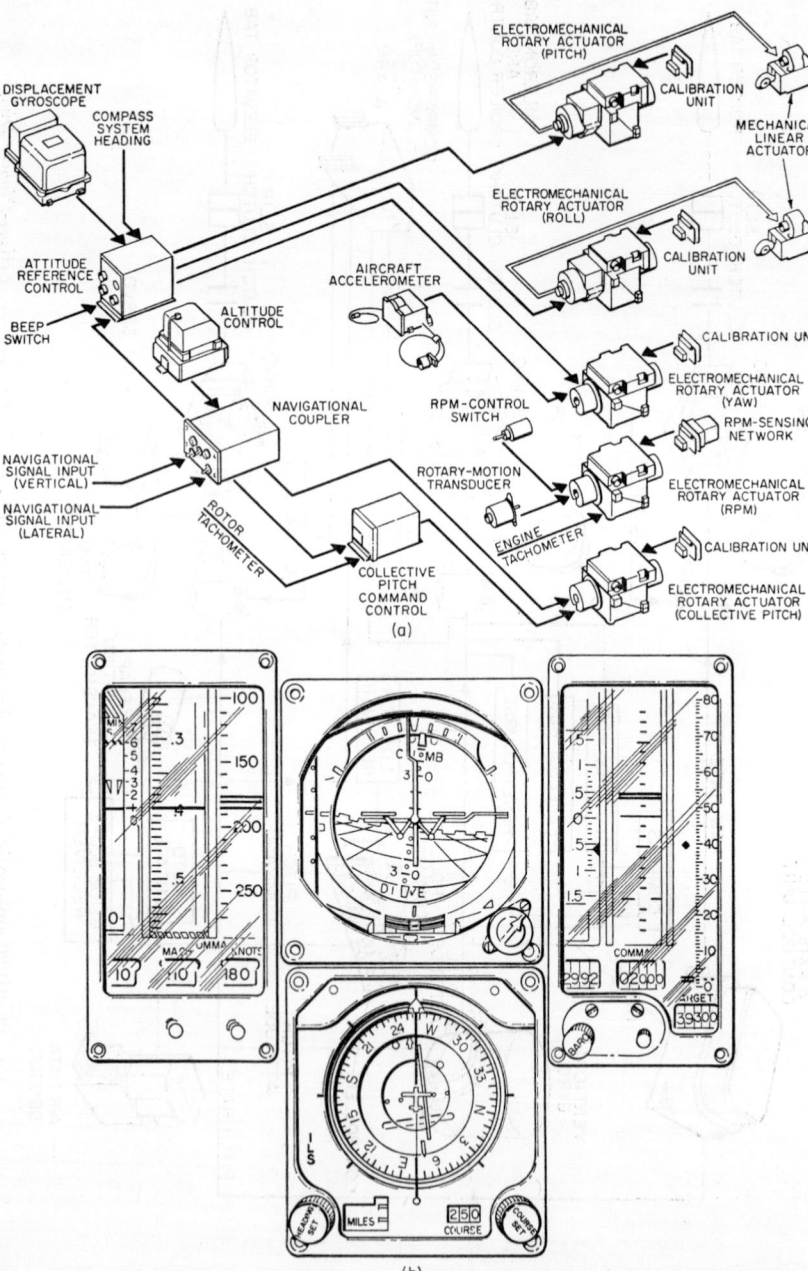

FIG. 14. (*a*) PB-20 automatic-pilot system. Heavy black arrows indicate electrical relationship. Light double-line arrows indicate flexible shaft. (*Source: Sperry Phoenix Company.*) (*b*) Integrated cockpit display. (*Source: U.S. Air Force.*)

environment. This may be done by using inputs from total and static pressure sensors on an open-loop basis, or by sensing performance response of the aircraft and adjusting the control parameters to give the desired response by "self-adaptive" means.

Integrated Flight Instruments

Various instruments are sometimes grouped together into integrated displays with improved readability and understandability. Figure 14*b* shows an integrated flight instrument panel giving air data, attitude, and navigational information as used in certain military aircraft. Of particular interest are the vertical scale indicators, the right and left instruments. Aircraft performance is read from vertically moving tapes against horizontal reference lines. Command information, set in locally by switches on the instrument front, or by data link from the ground, is displayed by markers read against the tape so that the position of the markers relative to the horizontal reference presents the pilot with immediate quantitive information as to his desired performance. The direction of tape motion, to display increasing or decreasing values of speed and altitude, is tied in with control stick and throttle motion, which is an important consideration from a human-factors aspect, and facilitates pilot familiarization. The Mach-airspeed-*g* indicator is shown to the left, and the rate-of-climb and altitude indicator is on the right. The latter contains both course and sensitive altitude indications, as well as target data. Most of the inputs to the vertical-scale indicators are received from a central air-data computer. The upper center instrument presents attitude (vertical situation) and the bottom instrument presents navigational information (horizontal situation).

NAVIGATIONAL DEVICES

Magnetic Compasses

Float-type Magnetic Compasses (Direct-reading). If a two-pole magnet is pivoted about a vertical axis with its poles in the horizontal plane, the magnet will align itself with the horizontal component of the earth's magnetic field and indicate magnetic north, which differs from true north by the variation angle, which varies from place to place on the earth's surface. In aircraft compasses (Fig. 15), the magnet is

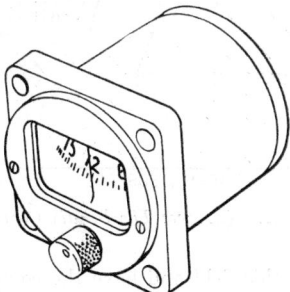

Fig. 15. Float-type magnetic compass. (*Source: The Bendix Corporation, Eclipse-Pioneer Div.*)

attached to a float in an approximately spherical chamber filled with liquid, thus removing most of the weight of the magnet from its pivot to reduce pivot wear and friction, the float providing a reference to dynamic vertical, and the liquid providing damping. Usually there is considerable magnetic material in the aircraft acting as a disturbing influence on the compass; so a compensator of small adjustable magnets is usually provided to balance out this effect.

Float-type Magnetic Compasses (Remote-reading). Because of strong and varying disturbances near the pilot (heavy electric currents affect compasses), it is desirable to place the compass as far away from as many disturbances as possible, as in a wing tip or in the tail of the plane, and by some means of takeoff remotely indicate the reading to the pilot. A system of this type is shown in Fig. 16.

Stabilized Magnetic Compasses. Satisfactory compass readings are obtained in aircraft in straight flight without accelerations; however, with any acceleration, the dynamic vertical differs from true vertical and the float-type compass no longer looks at only the horizontal component of the earth's field. In fact, in a turn in one direction in high latitudes, it is possible to have the dynamic vertical at a greater angle to the

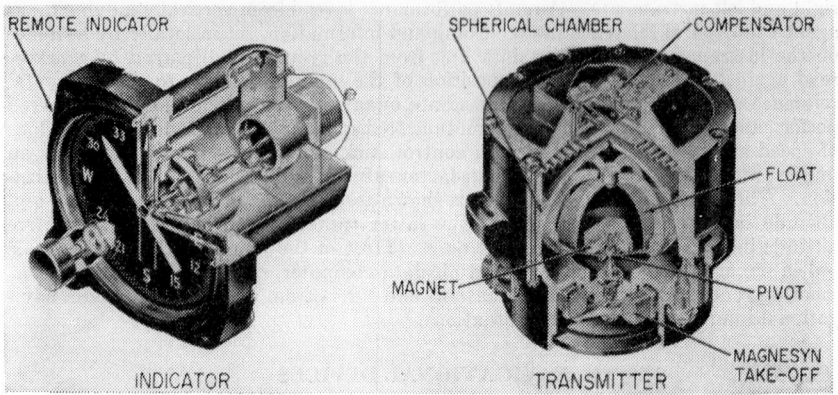

Fig. 16. Remote-reading Magnesyn compass. (*Source: The Bendix Corporation, Eclipse-Pioneer Div.*)

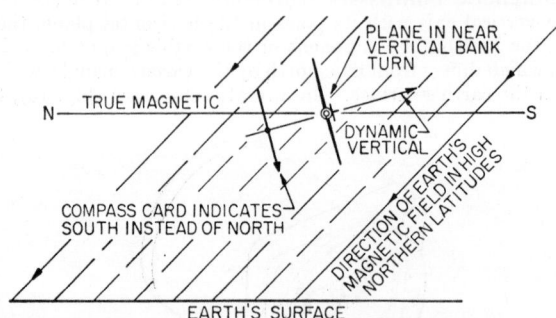

Fig. 17. Northerly turning error. (*Source: The Bendix Corporation, Eclipse-Pioneer Div.*)

true vertical than is the earth's field and have the compass point south instead of north. This effect, shown in Fig. 17, is called northerly turning error.

Flux Gate Compass. If a magnetic element is mounted on a gyro vertical, it can be quite accurately aligned with true vertical, and instabilities that the ordinary compass suffers from accelerations can be eliminated. Such a compass with amplifier and indicator is shown in Fig. 18.

Gyrosyn Compass. If a signal can be taken from a float or pendulous type compass and used as a slow correction on a directional gyro, such a long time average will be very nearly correct. Such a device is shown in Fig. 19.

All-latitude Compass. Directional gyros are in use with drift rates of 1 to 4° an hour. However, correction must be applied for rotation of the earth and aircraft

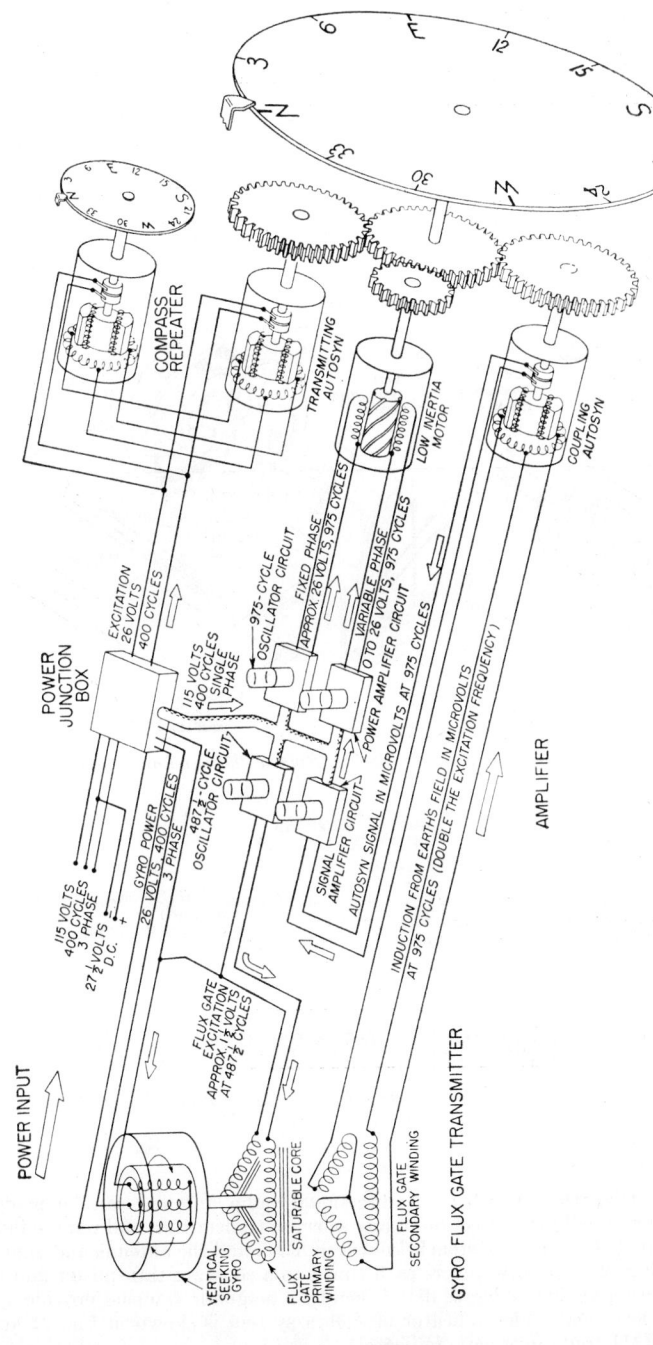

MASTER DIRECTION INDICATOR

POWER INPUT

POWER JUNCTION BOX

COMPASS REPEATER

TRANSMITTING AUTOSYN

LOW INERTIA MOTOR

COUPLING AUTOSYN

AMPLIFIER

EXCITATION 26 VOLTS 400 CYCLES

115 VOLTS 400 CYCLES SINGLE PHASE

975-CYCLE OSCILLATOR CIRCUIT

FIXED PHASE APPROX 26 VOLTS, 975 CYCLES

VARIABLE PHASE 0 TO 26 VOLTS, 975 CYCLES

POWER AMPLIFIER CIRCUIT

115 VOLTS 400 CYCLES 3 PHASE

272 VOLTS D.C.

GYRO POWER 26 VOLTS, 400 CYCLES 3 PHASE

487½-CYCLE OSCILLATOR CIRCUIT

SIGNAL AMPLIFIER CIRCUIT

AUTOSYN SIGNAL IN MICROVOLTS AT 975 CYCLES

INDUCTION FROM EARTH'S FIELD IN MICROVOLTS AT 975 CYCLES (DOUBLE THE EXCITATION FREQUENCY)

FLUX GATE EXCITATION APPROX 6 VOLTS AT 487½ CYCLES

VERTICAL SEEKING GYRO

FLUX GATE PRIMARY WINDING

SATURABLE CORE

FLUX GATE SECONDARY WINDING

GYRO FLUX GATE TRANSMITTER

Fig. 18. Flux gate compass and indicator. (*Source: The Bendix Corporation, Eclipse-Pioneer Div.*)

16–17

velocity. Such corrections properly applied give a satisfactory heading reference in very high latitudes where magnetic headings are valueless. Such a system, with magnetic correction available when usable, is shown in Fig. 20 and is known as the all-latitude "polar-path" compass.

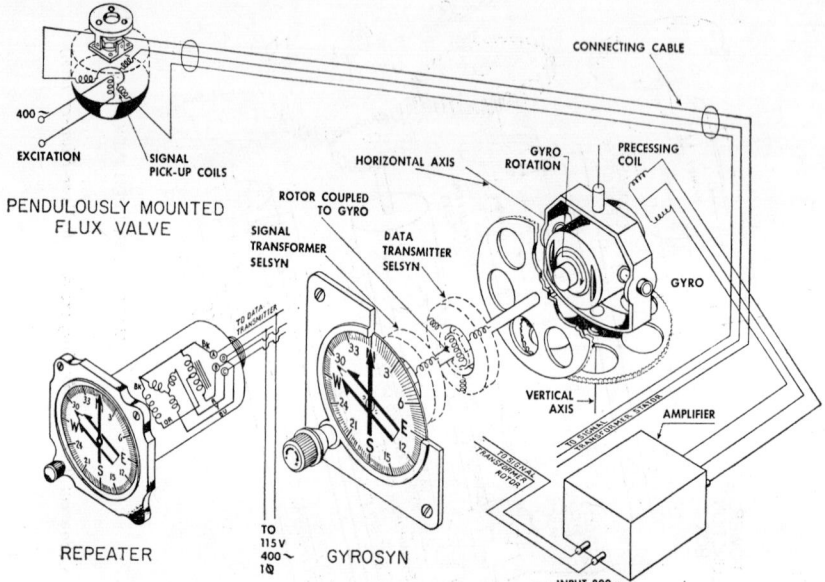

FIG. 19. Gyrosyn compass. *(Source: Sperry Phoenix Company.)*

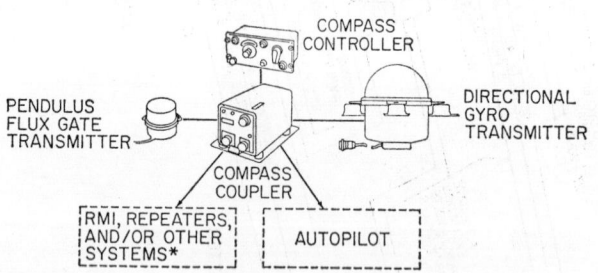

FIG. 20. Polar-path compass. *(Source: The Bendix Corporation, Eclipse-Pioneer Div.)*

Radio Navigation Aids

Automatic Direction Finder. A directional radio antenna can be made to determine automatically the direction of any selected transmitter, and remote indication can be provided. Such a system is known as ADF, and the antenna and indicator are shown in Fig. 21. Two receivers each tuned to a separate transmitter and each operating a pointer against a degree dial driven by a magnetic compass provide useful navigation information. The indicator of such a system is shown in Fig. 22 and is known as an RMI (radio magnetic indicator).

Instrument Landing System. Near the end of an airport runway, transmitters are set up to emit two narrow beams of signals which differ slightly in angular elevation.

If an aircraft flies the glide slope established by equal intensity between the two beams, it will make a proper letdown toward the runway. Two more beams displaced in azimuth provide similar alignment to the runway. This is known as ILS and is shown in Fig. 23.

Visual Omnirange. A ground transmitter emits two signals, both modulated by the same low frequency. The one signal has the same phase in all directions while the other has a phase depending on direction, rotating once per low-frequency cycle. If the two signals are in phase at north, then the phase difference between the two

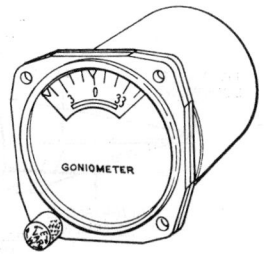

FIG. 21. ADF antenna (left) and goniometer (right). (*Source: The Bendix Corporation Radio Div.*)

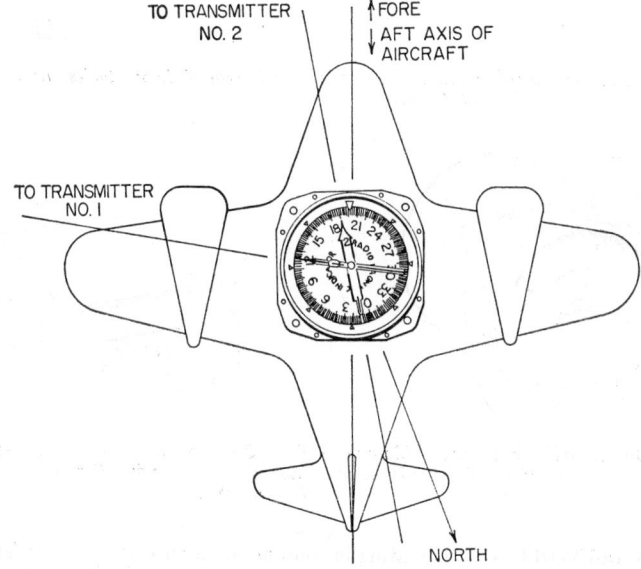

FIG. 22. Radio magnetic indicator. (*Source: The Bendix Corporation, Eclipse-Pioneer Div.*)

indicates the angle between north, the transmitter, and the receiver. Such a system is called VOR (visual omnirange) and gives azimuth indication on an OBI (omnibearing indicator), Fig. 24, containing a differential synchro which, when connected to a magnetic-heading synchro gives the same type of information as does an ADF and may be used to operate a pointer of an RMI.

Distance-measuring Equipment. A transmitter in the aircraft sends a coded pulse that is received by a ground receiver which immediately responds by transmitting back a signal (this ground station is called a transponder), and the signal is again received by the aircraft. The time required for the signal to make this round trip is an exact measure of distance. This system is called DME, and the distance indicator is shown in Fig. 25. (A warning bar appears across the numerals when the received signal is weak or undependable.)

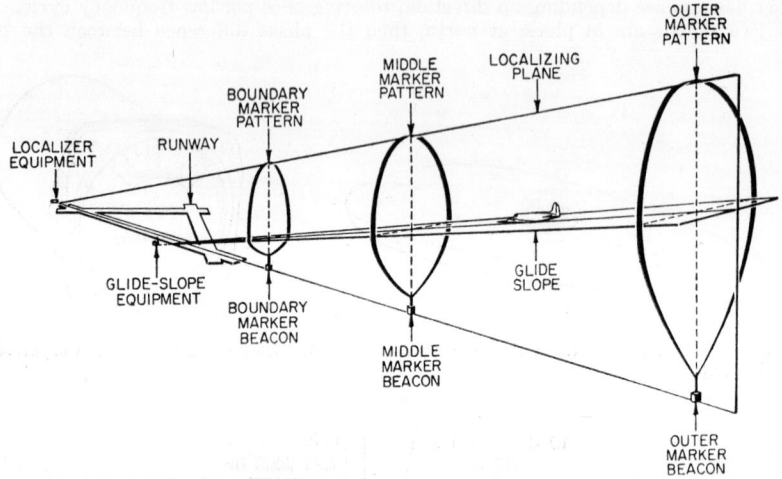

Fig. 23. Instrument landing system. (*Source: ITT Federal Laboratories, Div. of International Telephone and Telegraph Corp.*)

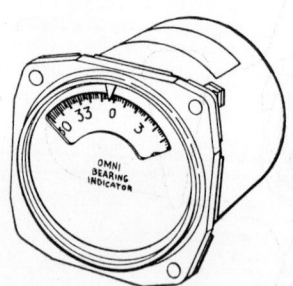

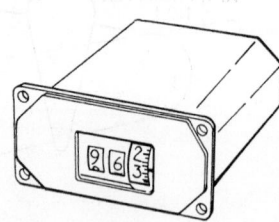

Fig. 24. Omnibearing indicator. (*Source: The Bendix Corporation, Eclipse-Pioneer Div.*)

Fig. 25. Distance-measuring equipment. (*Source: The Bendix Corporation, Eclipse-Pioneer Div.*)

TACAN and VORTAC. Operating in conjunction with DME, a directional signal similar to VOR but in a different frequency band is combined with a signal shifting phase at nine times the azimuth angle, giving nine times the azimuth accuracy obtainable by VOR. This combination is called TACAN (tactical air navigation) and is used primarily by the military. The combination of VOR and the DME portion only of TACAN is called VORTAC and is used by airlines and others.

Radar

A radar transmits sharp bursts of energy and measures the times for reflections of each to return. By moving the antenna back and forth, the reflections can be made to sweep out on a cathode-ray tube a picture or map of reflecting objects.   Such an airborne radar is extremely useful in making foul-weather landings, avoiding intense storm areas, and preventing collision with other aircraft and with mountains (Fig. 26). One version of radar looking fore, aft, and athwartship provides an accurate measure of ground velocity and drift and is called Doppler radar.

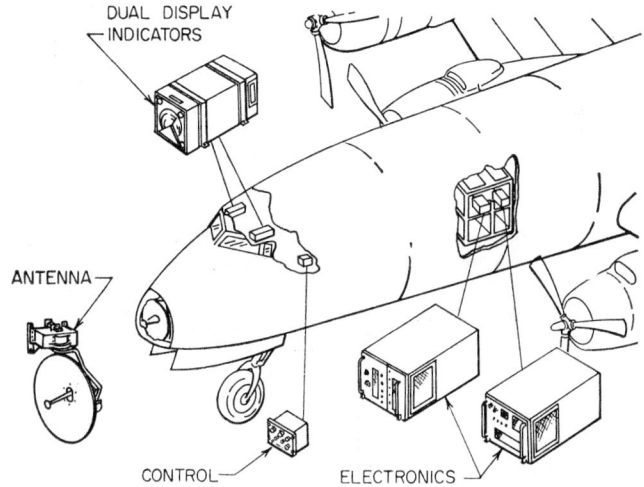

FIG. 26. Airborne radar.   (*Source: The Bendix Corporation, Eclipse-Pioneer Div.*)

Optical Instruments

Octants and Sextants.   Modernized, lightweight octants and sextants (Fig. 27*a*) are used to establish one's location quite accurately by determining the angle between the horizon and the sun or some other heavenly body.   A small bubble made visible in the instrument's optical system is used as an indication of vertical if no horizon is visible, and in many provision is made to average the reading over a period of a minute or two; this is particularly useful in "rough air."   A gyro vertical has also been used as a horizon reference.

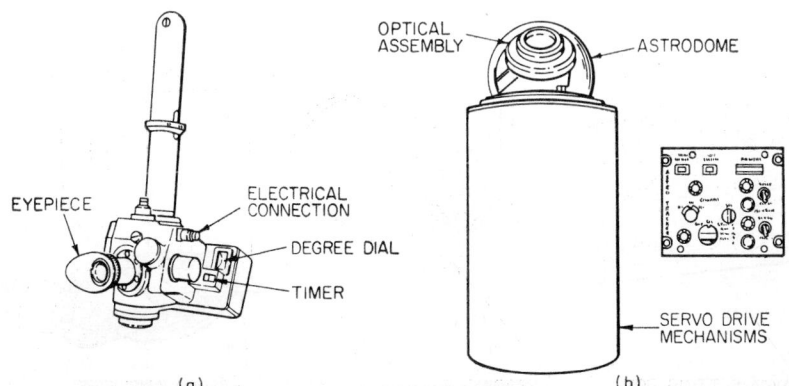

FIG. 27. (*a*) Periscopic sextant; (*b*) photoelectric sextant tracker and control panel. (*Source: Kollsman Instrument Corporation.*)

Star Trackers. At high altitudes, clouds obscure the skies a smaller portion of the time, and stars also may be visible during daylight hours, making them more useful for navigation. An optical system is used to collect and focus the light from a star onto sensitive photocells, which, through amplifiers and servos, keep the star tracker aimed at the selected star. Accurate readouts provide angular information with respect to vertical or aircraft coordinates. One such unit is shown in Fig. 27b.

Driftmeters. Cross winds, if not properly compensated, can drive a plane far off course. True course compared with plane heading can be determined by finding the drift angle. This may be done by watching the angle of the track of objects on the ground. Parallel lines on an optical system make it easy to determine when the objects travel parallel to the lines, and stabilizing the optical system by a vertical gyro makes the drift sight usable in rough weather. Figure 28 shows such a gyro driftmeter.

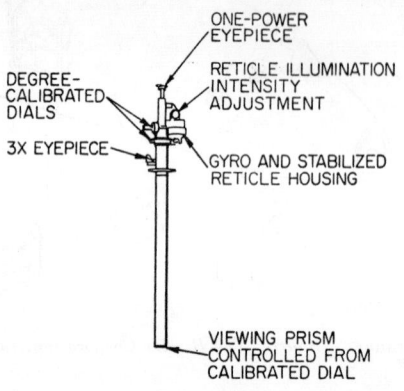

FIG. 28. Lightweight B3 driftmeter.   (*Source: The Bendix Corporation, Eclipse-Pioneer Div.*)

Navigational Computers

In fast-flying single-place or minimum-crew aircraft, the pilot or his assistants, if any, seldom have time to navigate their craft properly. Many types of computers have been developed to simplify the problem of navigation. A typical computer is shown in Fig. 29. It is a dead-reckoning computer (ASN/19) and accepts compass heading and true airspeed. Latitude and longitude at start is set into the counters,

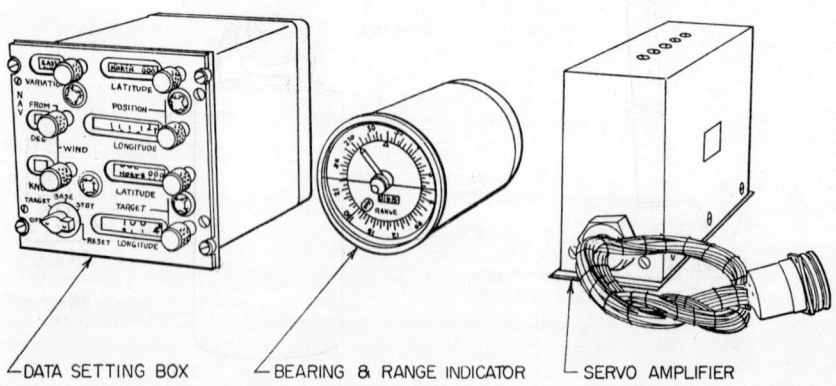

FIG. 29. Dead-reckoning computer, ASN/19. (*Source: The Bendix Corporation, Eclipse-Pioneer Div.*)

which thereafter update the reading to present position. The destination latitude and longitude are set into their counters, and the computer solves for the great-circle distance and heading, presenting them on the indicator, along with compass heading. Magnetic variation, wind heading, and velocity are set in manually. Other versions accept Doppler information and, in turn, give information to operate a wind-memory computer. This provides latest wind information in the event of Doppler failure.

ENGINE INSTRUMENTS AND CONTROLS

Pressure Instruments

The use of direct indication of engine pressures is seldom required except on small single-engine private planes. The two major reasons for this are that it is highly desirable to keep fuel and oil under pressure out of the cockpit, and most lubricating oils congeal sufficiently at cold temperatures to prevent pressure readings at the end of a long tubing.

The number of synchros in aircraft probably outnumber those in all other uses. The electrical power is 400 cps, 26 volts, and although 115 volts is sometimes used, the lower voltage is considered somewhat more reliable.

Synchro Indicators. A synchro, with its shaft carrying a pointer which reads against a dial enclosed in a case with a cover glass in front and a standard electrical

Fig. 30. Single Autosyn indicators. (*Source: The Bendix Corporation, Montrose Div.*)

Fig. 31. Dual Autosyn indicator. (*Source: The Bendix Corporation, Montrose Div.*)

connector in back, constitutes the typical synchro indicator. Three common sizes are shown in Fig. 30. Two synchros with concentric pointers provide "dual indication" of the same function for two engines, or fast-slow, minute-hour-hand type of presentation (Fig. 31). Many indicators are hermetically sealed with the edge of the cover glass metallized and soldered to the metal case and the electrical connections being taken out through glass or ceramic-insulated terminal or connector pins. Such indicators are filled with a dry, inert atmosphere, usually partly helium, to permit detection of leaks with a sensitive mass spectrometer. Accuracies and readings of closer than ±1° are difficult when using torque receivers. These errors can be reduced to about one-fourth by using a servoed indicator, the synchro being used as a control transformer driving a small motor through a self-contained amplifier (Fig. 32). With such a system sufficient torque is available to drive a fast pointer, and the added accuracy thus becomes readable.

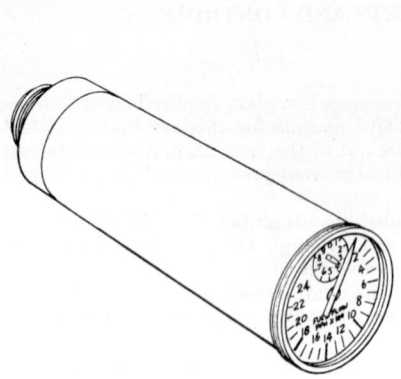

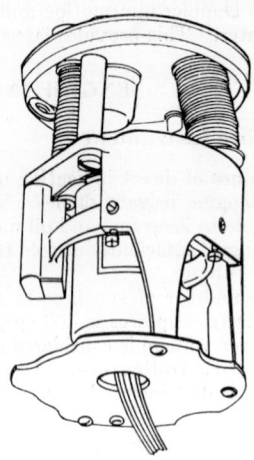

Fig. 32. Servoed Autosyn indicator. (*Source: The Bendix Corporation, Eclipse-Pioneer Div.*)

Fig. 33. Bourdon-tube transmitter (case removed). (*Source: U.S. Gauge Div., Ametek, Inc.*)

Bourdon-tube Transmitters. In oil and hydraulic pressure transmitters in ranges 0 to 1,000 psi to 0 to 5,000 psi a single bourdon tube is usually used (Fig. 33). If the maximum pressure exceeds that which the sealed case can withstand if there is an internal leak, a blowout or overpressure relief should be provided, perferably by means of a vent connection through which leakage can be disposed of safely. With most types of hermetically sealed transmitters, provision is made to adjust zero from the outside of the case.

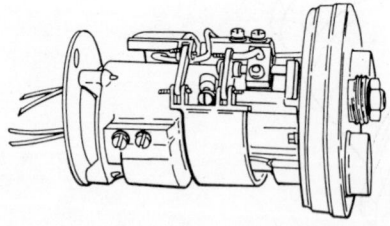

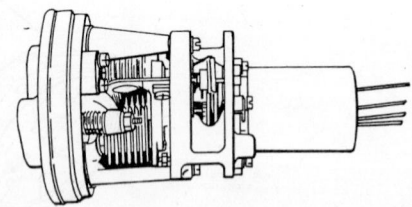

Fig. 34. Dual bourdon-tube transmitter (case removed).

Fig. 35. Fuel-pressure meter (case removed). (*Source: The Bendix Corporation, Montrose Div.*)

If differential-pressure measurement is required, or if the range is below 1,000 psi and the case is sealed (pressure differences between inside and outside the case may vary enough because of altitude and temperature to put the error of a single-tube unit out of tolerance), then a second bourdon tube is used, connected to external or sump pressure. Figure 34 shows such a differential-pressure transmitter.

Two common methods of minimizing temperature errors are to use bourdon tubes of low temperature coefficient or modulus of elasticity or to use a bimetal that changes linkage length to compensate for temperature effects.

In common with most synchro pressure transmitters, the synchro of a bourdon-tube-type transmitter usually carries a hairspring applying sufficient torque to remove backlash from gear trains and pivots.

In most engines driving propellers (both piston and jet), the gear reduction to the propeller is a planetary system with the ring gear held from rotating by hydraulic pressure in several cylinders spaced around the circumference of the gear. The pressure required to prevent rotation is an accurate measure of the torque transmitted to

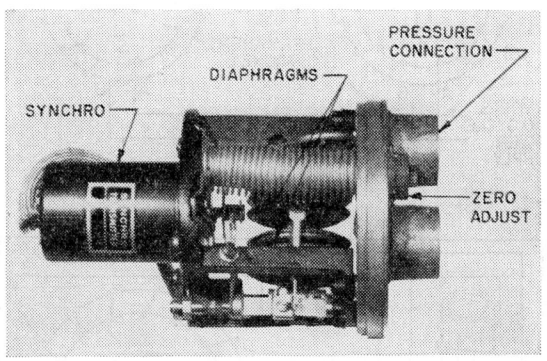

Fig. 36. Manifold-pressure synchro transmitter. (*Source: U.S. Gauge Div., American Machine and Metals, Inc.*)

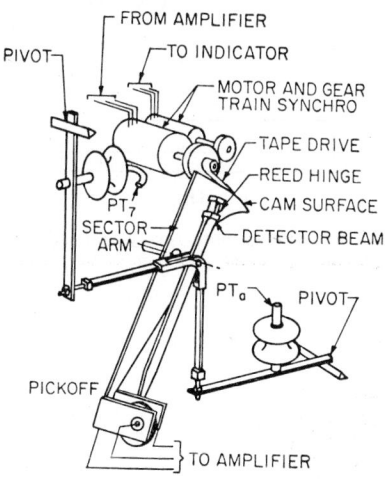

Fig. 37. Pressure-ratio-thrust indicator system. (*Source: Minneapolis-Honeywell Regulator Co.*)

the propeller. Engine-mounted transmitters are coming into use for this application as well as others as the problems of operation under high vibration are solved.

Bellows and Diaphragm Transmitters. For low ranges to pressures up to several hundred psi, bellows or diaphragms may be used. Oil-pressure transmitters for reciprocating engines must be designed to withstand pressures as high as 600 psi which may occur on cold start. Fuel-pressure transmitters are normally 0 to 25 to 0 to 50 psi, one type being shown in Fig. 35, with two balanced bellows and a cantilever spring which is the principal spring member.

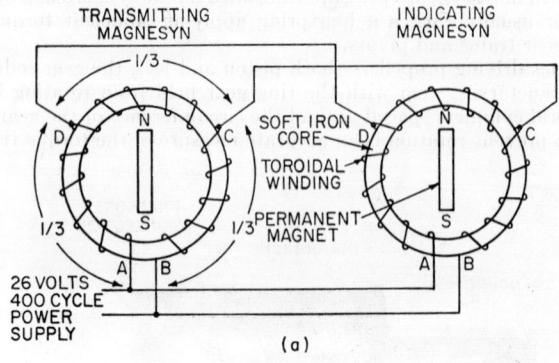

(a)

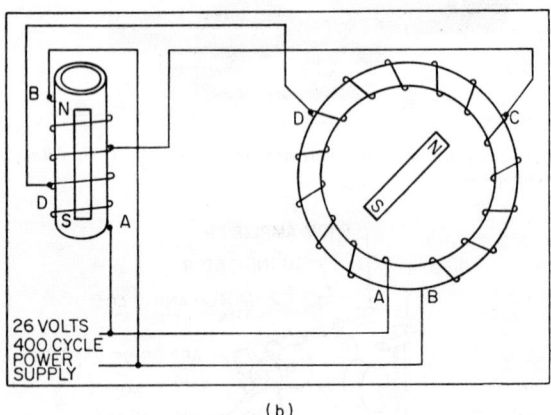

(b)

FIG. 38. (a) Schematic of Magnesyn system, rotary transmitter and receiver; (b) linear Magnesyn transmitting system. (*Source: The Bendix Corporation, Eclipse-Pioneer Div.*)

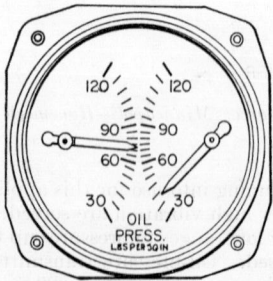

FIG. 39. Magnesyn indicator, dual. (*Source: The Bendix Corporation, Eclipse-Pioneer Div.*)

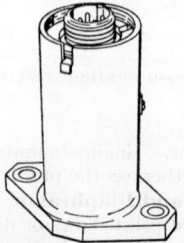

FIG. 40. Translation-type transmitter. (*Source: The Bendix Corporation, Eclipse-Pioneer Div.*)

Manifold-pressure transmitters have ranges of 10 to 60 to 10 to 100 in. of Hg absolute.   Such a transmitter is shown in Fig. 36.

Thrust meters show engine performance in terms of pressure ratios.   These, however, are more complicated and are a form of computer (Fig. 37).

Magnesyn Remote Indication.   Magnesyns serve the same purpose as torque synchros, being excited from alternating current but they have no moving windings, slip rings, or brushes.   The moving parts are permanent magnets, and the effective signal currents are even harmonics of the excitation frequency.   Receivers and some transmitters are rotary, while other transmitters are translational, with a bar magnet acting as a piston in a cylinder.   Schematics are shown in Fig. 38a and b; illustration

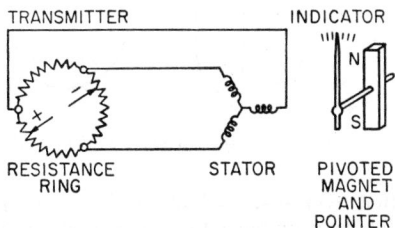

FIG. 41. D-C synchro system.   (*Source: The Bendix Corporation, Eclipse-Pioneer Div.*)

of a dual indicator is shown in Fig. 39.   The translational-type transmitter withstands heavy vibration and forms the basis for engine-mounted oil, fuel, and torque transmitters (Fig. 40).

D-C Synchro and Potentiometer Remote Indication.   Remote indication is sometimes accomplished by means of variable resistance or potentiometer devices. One such system has a transmitter consisting of a wound resistance ring with three equally spaced taps and two sliding contacts supplying d-c power to opposite sides of the ring.   The three taps are connected to the stator of the indicator which has a permanent-magnet rotor (Fig. 41).

Thermometers

Resistance Thermometers.   Engine oil and air temperatures are often indicated by a system using a resistance bulb containing a resistance varying with temperature in accordance with Table 3.   The indicator may be either a meter movement or a servoed potentiometer, and both types of indicators are shown in Fig. 42.

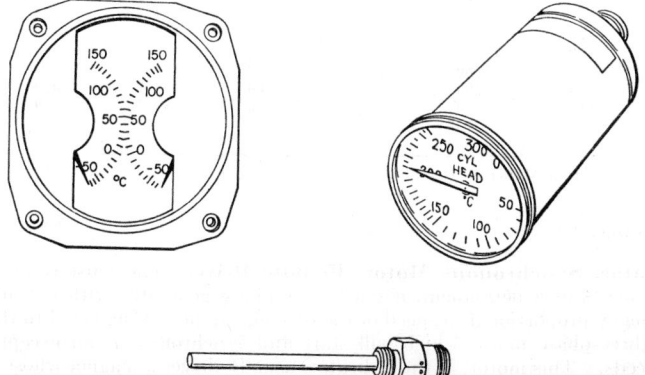

FIG. 42. Resistance-thermometer (bulb and meter).   (*Source: Thomas A. Edison, Inc.*)

Thermocouple Indicators. Chromel-Alumel couples are used to measure the higher temperatures such as exhaust or tailpipe. Eight-ohm circuits are customary. Indicators are normally d'Arsonval meter movements, though servoed indicators are sometimes used (see Fig. 43).

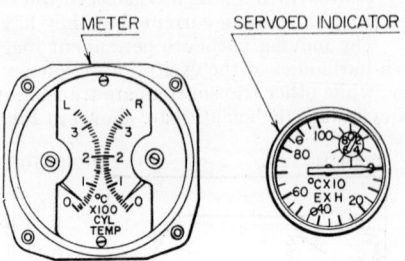

Fɪɢ. 43. Thermocouple indicator.   (*Source: Thomas A. Edison, Inc.*)

Vapor-pressure-filled Systems. A bulb filled with a liquid having measurable vapor pressure in the temperature range to be measured is connected by a small tube to a bourdon-tube-type transmitter with the bourdon tube curbed by a number of adjustable screws to provide a linear calibration from the very nonlinear temperature–vapor pressure curve.

Table 3. Temperature-Resistance Characteristics for Cylinder-head Bulb*

Temp., °C	Resistance, ohms	Tolerance, ± ohms	Temp., °C	Resistance, ohms	Tolerance, ± ohms
− 50	37.95	0.6	130	89.20	0.6
− 40	40.15	0.6	140	92.80	0.6
− 30	42.45	0.6	150	96.50	0.7
− 20	44.85	0.6	160	100.30	0.7
− 10	47.35	0.6	170	104.20	0.7
0	50.00	0.5	180	108.20	0.7
10	52.60	0.5	190	112.30	0.7
20	55.25	0.5	200	116.50	0.8
30	57.95	0.5	210	120.80	0.8
40	60.70	0.5	220	125.20	0.8
50	63.55	0.6	230	129.70	0.8
60	66.50	0.6	240	134.30	0.8
70	69.55	0.6	250	139.00	0.9
80	72.65	0.6	260	143.80	0.9
90	75.80	0.6	270	148.75	0.9
100	79.00	0.6	280	153.85	0.9
110	82.30	0.6	290	159.10	0.9
120	85.70	0.6	300	164.50	1.0

* From Specification MIL-B-5491 bulb temperature, cylinder head.

Tachometers

Generator, Synchronous Motor, Remote Drive. The most commonly used system consists of a permanent-magnet three-phase generator with output voltages and frequency proportional to speed of the driving engine. Connected to this generator is a three-phase motor which will start and synchronize at all except the very lowest speeds. This motor, in the indicator, usually drives a magnet whose field pulls an alloy rotor against a hairspring. Such a deflection is accurately linear, and two-pointer (fast and slow) presentation is common (Fig. 44). The alloy rotor is made of a material having low change of conductivity with temperature.   Such compensation as required is accomplished by shunting the magnet with a magnetic alloy having its Curie points in the operating-temperature range.

Instead of magnetic-drag indication, generator–synchronous motor–remote drive has been used with centrifugal or other forms of speed indication.

Flexible-shaft Drive.   Flexible shafts have been used to drive tachometer indicators up to a maximum distance of about 12 ft, beyond which whipping and poor performance in cold are experienced.   Magnetic drag, chronometric, and centrifugal are the most common types of shaft-driven indicators.

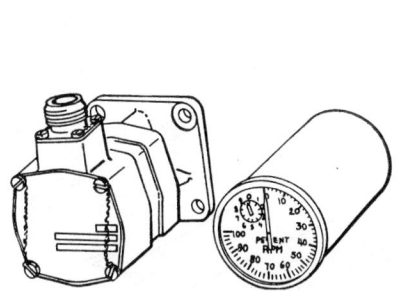

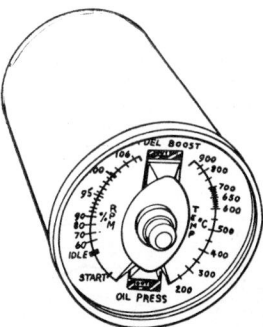

Fig. 44. Tachometer generator and indicator.   (*Source: General Electric Co.*)

Fig.   45.   Engine-performance   indicator. (*Source: The Bendix Corporation, Eclipse-Pioneer Div.*)

Combined Indication

To save instrument-panel space and to simplify the pilot's task, indications of several functions may be combined into one unit.   Figure 45 shows an engine-performance indicator giving engine speed on the left with expanded scale in the top operating range.   Exhaust temperature is shown on the right with warning flags at top and bottom to indicate anything amiss with oil or fuel pressures.   The scales are not calibrated in rpm or degrees but in operating levels.

Engine and Ignition Analyzers

Signals from vibration pickups on the engine or from an electrical link to the ignition system are viewed on a cathode-ray oscilloscope.   A synchronizing breaker or generator provides starting reference for the oscilloscope sweep.   Means are provided to identify each signal with the cylinder producing it.   Many types of engine malfunctions have characteristic signal shapes.   Normal firing and three types of ignition malfunctions are shown in Fig. 46.   Analyzers provide a rapid, convenient method of diagnosis

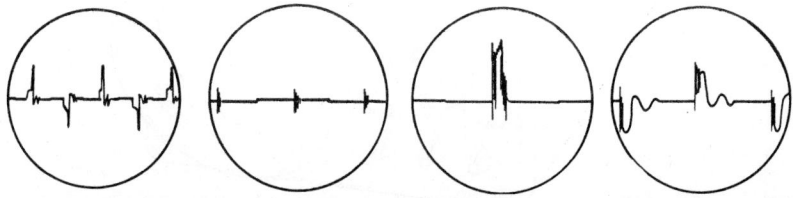

Fig. 46. Typical ignition-analyzer waveforms.   (*Source: The Bendix Corporation, Scintilla Div.*)

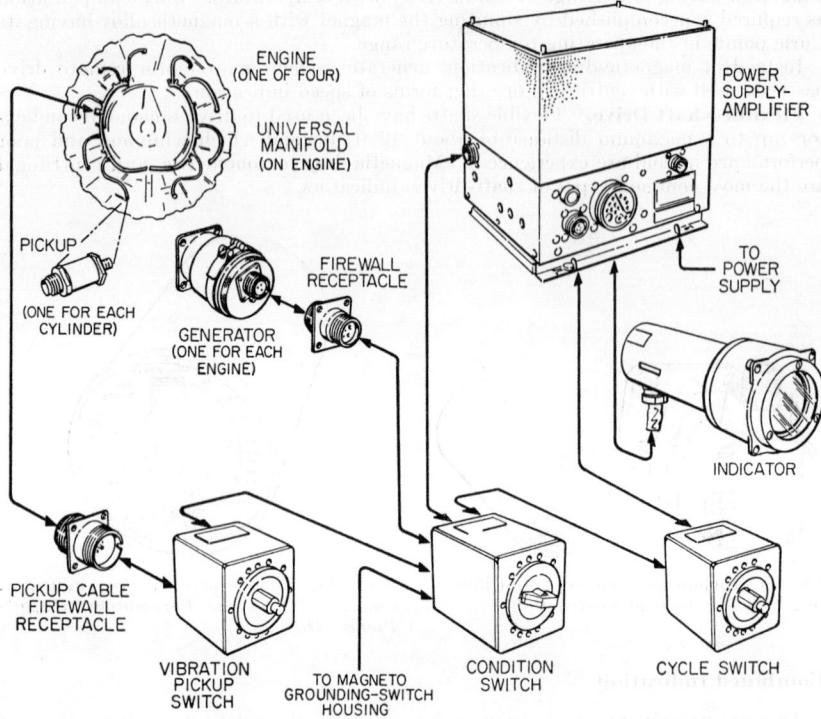

FIG. 47. Engine analyzer.   (*Source: Sperry Phoenix Company.*)

Within the image:
ENGINE
(ONE OF FOUR)

UNIVERSAL
MANIFOLD
(ON ENGINE)

PICKUP

(ONE FOR EACH
CYLINDER)

GENERATOR
(ONE FOR EACH
ENGINE)

FIREWALL
RECEPTACLE

POWER
SUPPLY-
AMPLIFIER

TO
POWER
SUPPLY

INDICATOR

PICKUP CABLE
FIREWALL
RECEPTACLE

VIBRATION
PICKUP
SWITCH

TO MAGNETO
GROUNDING–SWITCH
HOUSING

CONDITION
SWITCH

CYCLE SWITCH

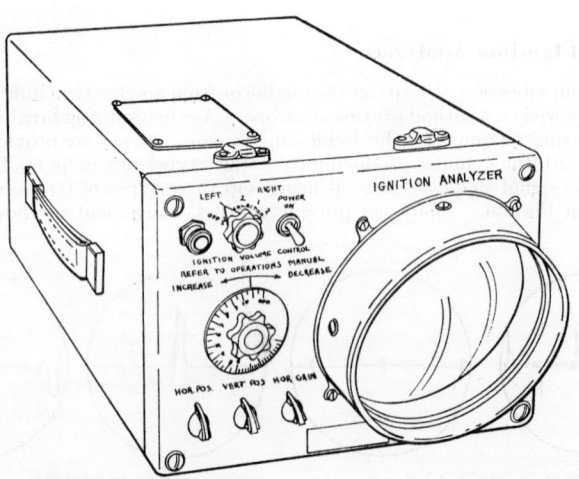

FIG. 48. Ignition analyzer.   (*Source: The Bendix Corporation, Scintilla Div.*)

Within the image:
IGNITION ANALYZER

LEFT   RIGHT   POWER
ON

IGNITION VOLUME CONTROL
REFER TO OPERATIONS MANUAL
INCREASE   DECREASE

HOR. POS. VERT. POS. HOR. GAIN

indicating the bad cylinder and the kind of trouble.   Two types of analyzers are shown in Figs. 47 and 48.

Automatic Engine Controls

Automatic engine controls for reciprocating engines have been built but have not received wide acceptance by the industry.

Both turbo-propeller and turbojet engines require automatic controls for proper engine operation.   Figure 49 shows a turbo-propeller engine control which emphasizes the control of part-power levels at fixed engine speed, providing for accurate and complex acceleration and part-throttle schedules through use of a three-dimensional cam-type hydromechanical computer.   Variations in fuels, engines, and hydromechanical

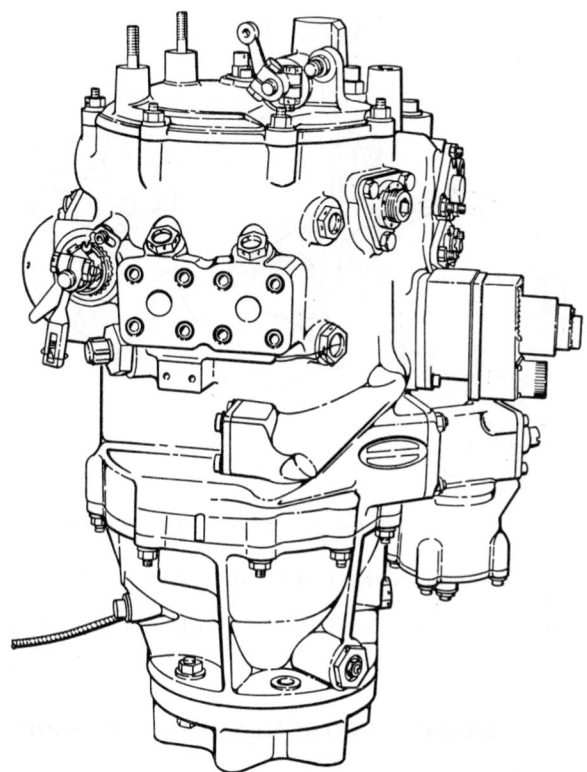

Fɪɢ. 49. Fuel control for turboprop   (*Source: The Bendix Corporation, Bendix Products Div.*)

controls are eliminated during power operation by trimming fuel flows to hold a pilot-scheduled turbine in temperature.   Not shown are an electronic control and a temperature datum valve which takes the turbine inlet temperature signal, compares it with request, and trims to precise requirements.

Precise control is assured using throttle power lever angle, turbine inlet temperature, compressor inlet pressure, and engine speeds as parameters.

Overtemperature limiting, deceleration schedule, overspeed protection, and reduced-speed idle are provided.

Figure 50 shows a turbojet fuel control, a hydromechanical unit which incorporates both primary and manual emergency fuel-metering systems.

All critical factors influencing the metering of fuel for engine performance under all variable conditions are sensed and interpreted by the computing and metering sections of the control.

The fuel control senses the temperature and pressure of the air entering the engine, the pressure of the air in the engine burner section and the speed of the engine, to regulate and maintain the fuel flow at the required value.

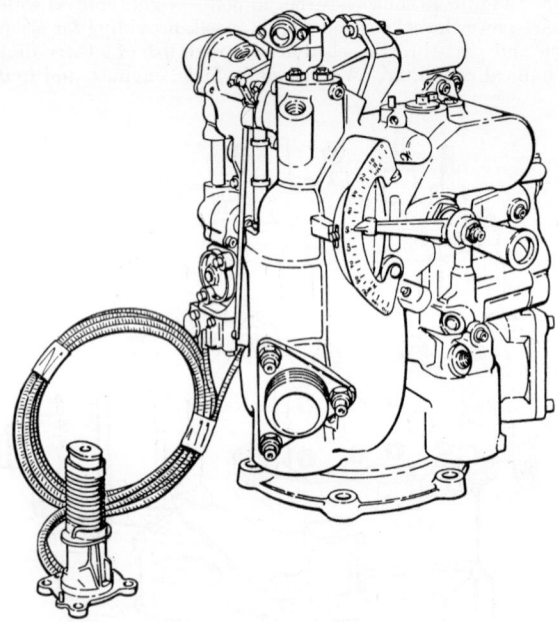

FIG. 50. Turbojet fuel control.   (*Source: Holley Carburetor.*)

The emergency fuel-control system may be selected by the pilot any time during operation.

When the power-lever setting is changed and while accommodating to a new steady-state fuel rate, the fuel control varies the fuel flow between the limiting values established.

MISCELLANEOUS INSTRUMENTS AND CONTROLS

Fuel-flow Indication

Rate of Fuel Flow.   This quantity not only is used to indicate rate of fuel consumption but also serves as a sensitive indication of engine performance.   In the flow meter shown in Fig. 51, the fuel forces its way between the end of a vane and the shaped walls of a housing.   The higher the flow, the further the vane is forced against its spring and the wider the opening at the end of the vane.   On the outside of the fuel chamber is a synchro driven through a magnetic coupling by the vane.   Indication is either by torque synchro indicator or by sensitive servoed synchro indicator.   Such a system reads in pounds per hour and is almost independent of fuel density.   This is important because of the large variation in density between different fuels and even the same fuel over normal temperature ranges (see Table 4).

Another type of flow meter operates by imparting a constant angular velocity to the

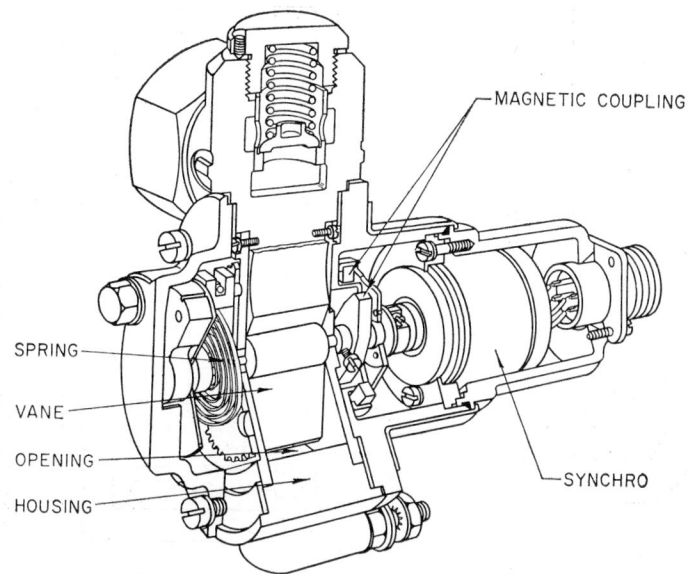

FIG. 51. Fuel-flow meter.   (*Source: The Bendix Corporation, Eclipse-Pioneer Div.*)

Table 4. Density of Fuels, Pounds per Gallon

Temp., °F	Jet fuel	High-octane aviation gas
−65	6.78	6.37
−30	6.65	6.24
0	6.54	6.12
+30	6.43	6.00
+60	6.33	5.89
+90	6.22	5.77
+125	6.09	5.64

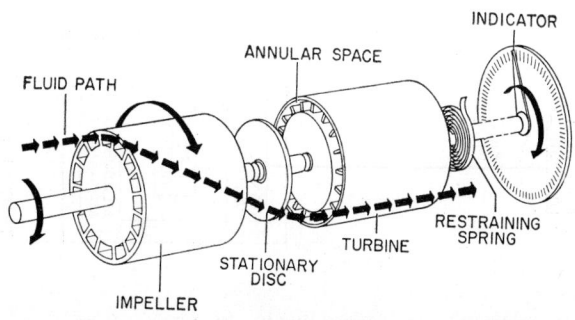

FIG. 52. Mass-flow meter.

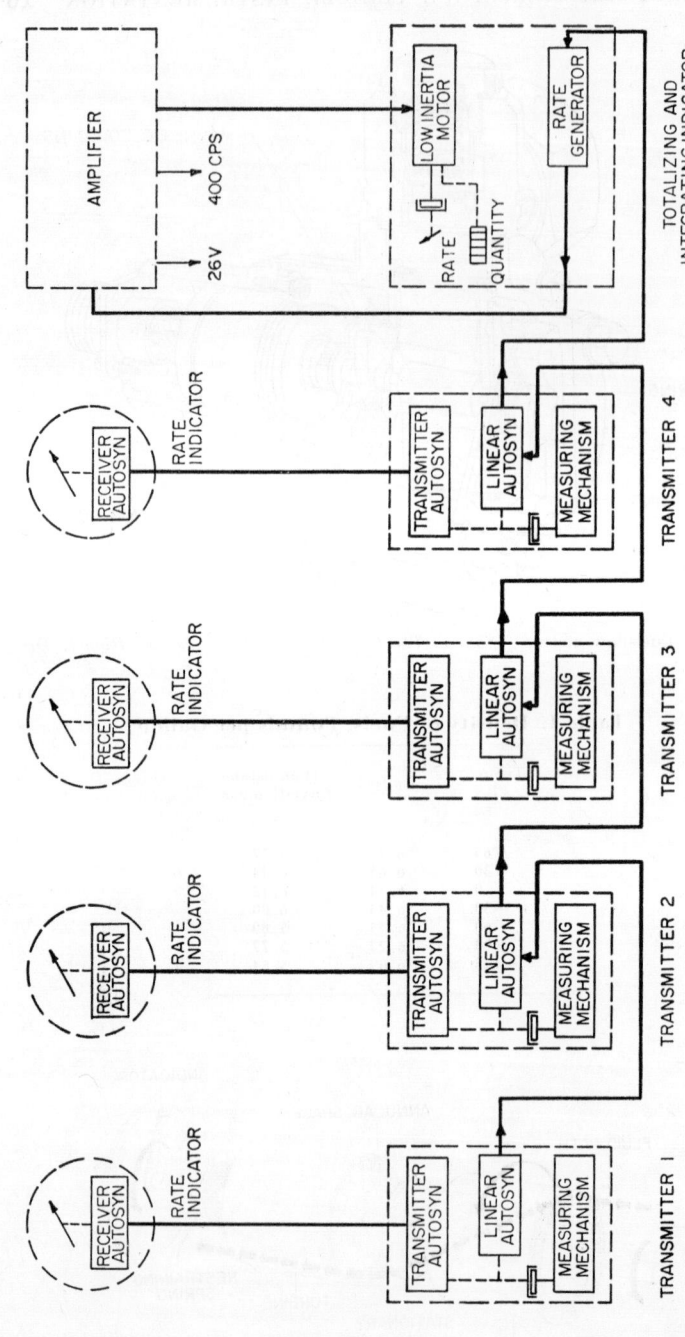

Fig. 53. Schematic of fuel-flow totalizer-integrator system.   (Source: The Bendix Corporation, Eclipse-Pioneer Div.)

16–34

liquid flowing through a cylinder, regardless of its rate of flow or axial velocity. The liquid then flows through a spring-restrained turbine which is deflected by the liquid's angular momentum; because its angular velocity is constant, the turbine deflection depends only on the mass flow. A schematic of this flow meter is given in Fig. 52.

Other types of flow meters, such as those measuring the pressure drop across a venturi, and volumetric meters, require correction for fuel density if good accuracy is to be obtained.

Fuel-flow Totalizer and Integrator. By using a synchro-like device giving a voltage proportional to angle on each flow meter, flow from any number of meters can be added to give total flow. This total indication can be integrated with respect to time to give total fuel consumed. The schematic for a totalizer-integrator system is given in Fig. 53.

Densitometer. When it is desirable to make density corrections to fuel-flow measurements, a densitometer is used. It measures the density by sensing the buoyancy of a body submerged in a sample of the liquid, and provides a signal to give the proper correction to the flow meter.

Fuel- and Oil-quantity Gages

Float Type. A float, supported on the end of an arm pivoted at about half the tank height or guided by a vertical spiral through the tank, can drive through a magnetic coupling to a synchro or other remote-transmitting device on the outside of the tank. Although simple, this type suffers from bad position errors and indicates volume and not mass.

Capacity. Closely spaced conductor plates placed in the tank with fuel or liquid serving as dielectric can be used as a condenser whose capacity is an indication of the mass of liquid present. With proper placing of the plates, indication can be made relatively independent of aircraft attitude (see Fig. 54).

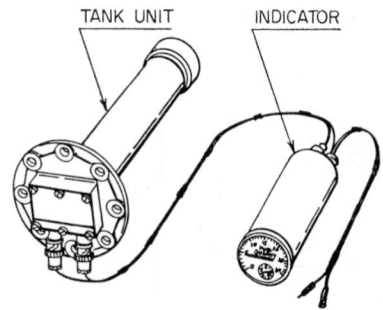

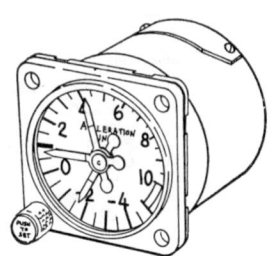

FIG. 54. Capacity-type fuel-quantity gage. (*Source: Avien-Knickerbocker, Inc.*)

FIG. 55. Accelerometer. (*Source: The Bendix Corporation, Eclipse-Pioneer Div.*)

Other Types. A sensitive pressure transmitter placed in a sump at the bottom of a tank can be calibrated in terms of head of fuel or oil and is termed a hydrostatic quantity gage. Strain-gage tank supports have been used to give fuel weight, and attenuation of gamma radiation between top and bottom of tank has been used as an indication of the mass remaining.

Accelerometers

In fast-maneuvering aircraft or in "rough air," accelerations of several g's or even higher are experienced perpendicular to the fore-aft and athwartship axes. It is desirable to indicate the amount of this acceleration and show the most positive and most negative values attained. A damped weight sliding on vertical guides drives a set of three pointers to give the required indications. A knob is provided to reset the maximum pointers (see Fig. 55).

Smoke and Fire Detection

Smoke warning to cargo holds or other places inaccessible in flight is usually accomplished by means of lights and photocells. Fire detectors may operate by sensing either elevated temperature or rate of temperature rise—usually by thermocouples or coaxial cable wherein the insulator is a semiconductor which becomes conductive at elevated temperature. A simplified system is shown in Fig. 56.

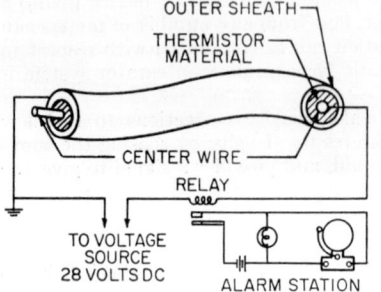

FIG. 56. Fire-detection system. (*Source: Thomas A. Edison, Inc.*)

Deicing Indicators and Controls

Numerous type of ice detectors or rate-of-icing indicators have been built. One measures the time for ice to seal shut an orifice, after which it is thermally deiced and a new cycle begun. Another measures the thickness of ice built up on disk rotating slowly at constant speed. After the ice passes under a measuring feeler, it is scraped off and again builds up until it again passes under the feeler. Rate of icing may range in excess of 1 in./min. While ice can overload a plane, change its aerodynamics, and increase its drag to the point where it can no longer maintain altitude, there are certain critical areas where icing effects are most severe.

Carburetor jets readily ice under certain conditions of temperature and humidity. Carburetor heat is an effective remedy.

Intake manifolds of both jet and reciprocating engines ice up, closing off the engines' air supply. Heating of the affected areas is common practice.

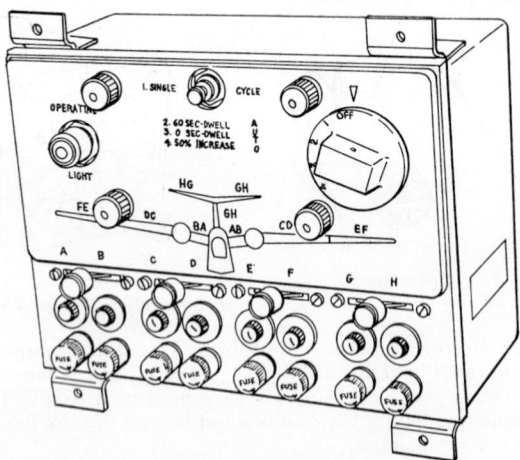

FIG. 57. Electronic deicer timer.

Leading edges of wings, elevator, and rudder rapidly collect ice. Rubber "boots" may be used containing tubes which are alternately inflated and deflated in a slow cyclic manner to crack off any layer of ice that forms. An automatic timer for controlling this cycling is shown in Fig. 57. Some planes use heat for deicing these areas, usually obtained from hot exhaust gases or combustion heaters.

Propellers may be deiced electrically on their leading edges or by a flow of anti-icing fluid onto the blades.

Windshields may be deiced by electrically heating a transparent conducting coating on the glass. Power as high as 1 kw/sq ft may be required. Temperature can be indicated and controlled by embedded resistance-thermometer elements.

Cabin Pressure and Temperature Control

All high-flying commercial aircraft, as well as many military aircraft, have pressurized cabins. Pressure is not always maintained equivalent to sea level but is usually decreased to a much less extent than external pressure.

Automatic controls on some superjet commercial aircraft maintain a pressure equivalent to sea level at altitudes up to 20,000 ft and then raise the internal altitude as high as 8,000 ft when flying at 45,000 ft. Pressure rates of change are kept low to reduce passenger discomfort. Pressurized air is obtained from engine-driven superchargers controlled to give the desired schedule. As the adiabatic compression of air

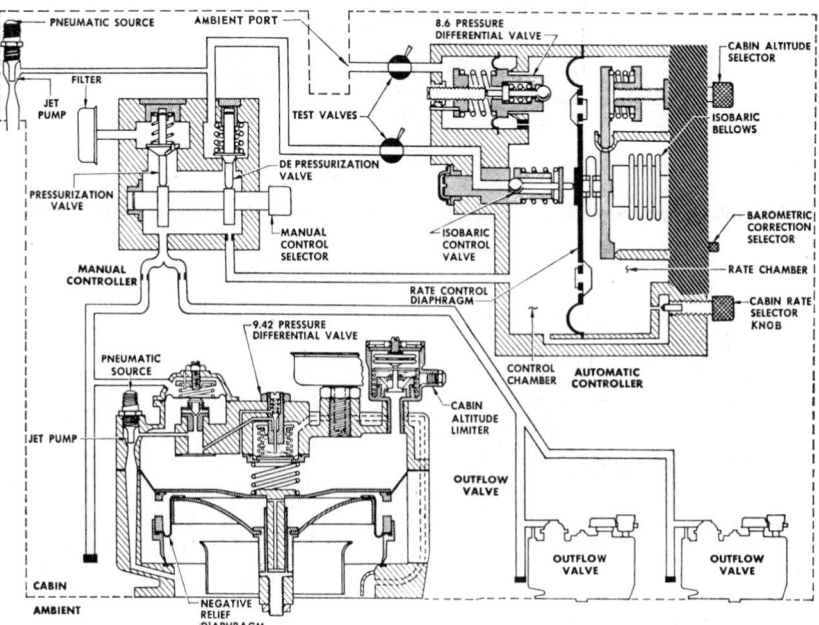

Fig. 58. Pressurization-control indicating system.

by the superchargers raises its temperature, it may be unbearably hot, especially in the summer. The pressurized air is cooled by passing through an aftercooler with the heat being carried away by ram air. In hot weather, it may be necessary to cool the air even further for comfort. This may be accomplished by further compression of the air with resultant temperature rise, cooling it in a heat interchanger with ram air and then expanding back to the desired pressure with resultant adiabatic cooling. The power obtained from the expansion turbine may be used to help drive the compressor, thereby reducing the power requirements of such a cooling system. In cold weather, heat rather than cooling may be required and is obtained from fuel-burning heaters. The cooled or heated air is mixed with recirculated air from the cabin to keep cabin temperatures at desired levels. The schematic of a typical pressurization-control indicating system is shown in Fig. 58.

Oxygen Systems

In order to maintain consciousness or even to survive, human beings must get more oxygen than the atmosphere provides at high altitude. Oxygen is carried as gas under pressure (normally 2,000 psi) in tanks or as a liquid under pressure and at reduced temperatures. The oxygen regulator receives this high-pressure oxygen and controls its flow at low pressures to the oxygen mask of the flier. Starting at about 10,000 ft, oxygen will be diluted with air. The percentage of oxygen increases with altitude until pure oxygen flows into the mask at about 30,000 ft. The demand regulator (Fig. 59) supplies oxygen to meet the flier's needs, delivering only on inhalation. At

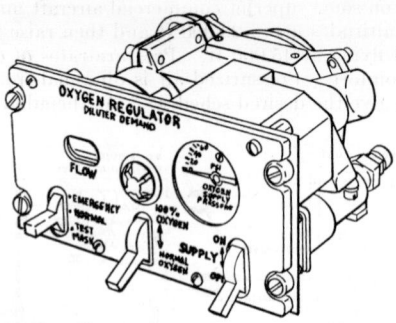

FIG. 59. Oxygen regulator.   (*Source: The Bendix Corporation, Eclipse-Pioneer Div.*)

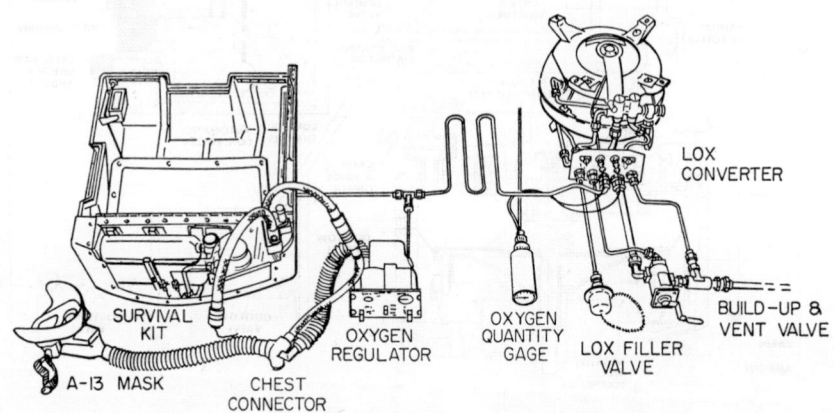

FIG. 60. Oxygen system.   (*Source: The Bendix Corporation, Eclipse-Pioneer Div.*)

altitudes over 35,000 ft, pressure above atmospheric, increasing with altitude, is required in the lungs to force oxygen into the blood stream. The complete oxygen system for a single-seat high-performance military plane is shown in Fig. 60.

Oxygen is kept instantly available to pilots of most high-flying pressurized commercial airliners to be used in case of accidental loss of cabin pressure. In high-flying commercial jet aircraft, oxygen masks are also instantly available to the passengers.

Propeller Controls

It is the propeller that converts the torque of reciprocating or turboprop engines into the thrust required to fly the airplane. Variable-pitch propellers are used on all but the smallest engines and must be controlled to meet varied requirements.   Controls must be of highest mechanical and electrical integrity with all possible fail-safe provisions because malfunction could be disastrous.

For both types of engines, propellers are controlled to provide low loading (flat pitch) for starting.   During most operating conditions, engine speed is controlled by adjusting the pitch of the propeller to provide the load required to keep the engine at the desired speed.   In automatic engine controls, this function is coupled with control of fuel.   In case of an engine failure, the controls should feather the propeller so as to

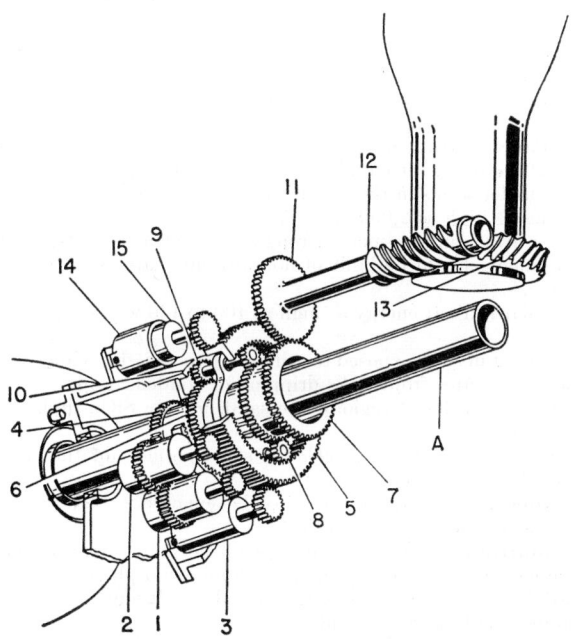

1. Increase-pitch clutch.	9. Reaction planet gear.
2. Decrease-pitch clutch.	10. Fixed-ring gear.
3. Fixed-pitch brake.	11. Pitch-change pinion gear.
4. Propeller-shaft gear.	12. Blade wormgear.
5. Movable-ring gear.	13. Blade gear.
6. Propeller-shaft sun gear.	14. Electric motor (feathering).
7. Pitch-change sun gear.	15. Electric motor clutch.
8. Pitch-change planet gear.	A. Propeller shaft.

Fig. 61. Schematic of pitch-changing mechanism for turboelectric propeller.

minimize its drag but this should be done at the proper rate with turboprop engines to avoid damage from too sudden engine deceleration.   During approach, low thrust is required, but engine speed must be maintained high, ready for fast response to a possible sudden demand for high thrust.   On landing, the length of runway to decelerate can be materially reduced by applying power to propellers in reverse pitch.

Synchronization is accomplished by comparing the frequency of a master generator with the speed of the slave engines.   This master generator may be driven by one of the engines or may be driven by any independently controlled speed source, which serves as well to control engine speed.   The cutaway sketch of a turboprop propeller control is shown in Fig. 61.

Environmental Conditions

In aircraft, instruments and controls must operate under many abnormal conditions. Although flight decks are normally heated and pressurized, instruments must continue to operate even though cabin heat or pressurization should fail.   Some exposures may be more severe than the conditions listed as follows, particularly as to high temperature and vibration near engines, and in high-performance military aircraft where altitude and temperature conditions may vary extensively simultaneously.

Temperature.   Ranges from -54 to $+125°C$ can be expected in flight, with rates of change as high as $1°/\text{sec}$.   In storage, ranges of -62 to $+71°C$ may be encountered.

Humidity.   Ranges from very dry to 100 per cent relative, including condensation in the form of water or frost.

Altitude.   Ranges from sea level to 80,000 ft or higher in some military craft with rates of change up to 0.5 in. Hg/sec.   High altitudes are usually accompanied by cold temperatures.

Vibration.   Panel instruments usually are required to withstand vibrations no greater than 0.020-in.-diameter circular vibration from 5 to 50 cps, while equipment mounted at various locations in the aircraft may be subjected to vibration as high as 20 g to frequencies as high as 2,000 cps white noise.

Fungus.   In warm humid climates, fungus may attack anything that acts as a nutrient.   Equipment should be built of nonnutrient materials or be treated with an effective fungicide.

Sunshine.   Solar radiant energy as high as 100 to 140 watts/sq ft may be encountered.

Rain.   Equipment may be exposed to rainfall up to 4 in./hr, while even "protected" units may experience water from leaks dripping or condensations.

Sand and Dust.   In desert regions fine sand, mostly SiO_2, finds its way into all but perfectly sealed areas.

Acceleration.   Rates as high as 30 g for short periods (of the order to 0.01 sec) may be occasionally experienced.

Salt Atmosphere.   The equipment may be exposed to salt-sea atmosphere in both operating and nonoperating condition.

Explosive Conditions.   The equipment shall not cause ignition of an explosive gaseous mixture with air when operating is such an atmosphere.

Space Conditions.   In space, environmental conditions are even more severe. There is no cooling due to gaseous conduction or convection, but only due to conduction through solids and by radiation.   This means small exposed objects get quite warm in the sun and quite cold in the shade.   Optical surfaces eventually get "sandblasted" by meteoric dust.   Any lubrication—that thin layer which allows low friction slip between metal surfaces—soon evaporates in space vacuum leaving raw, open pores of metal rubbing on a similar surface.   When two pieces of metal are pressed together under these conditions, they often "cold-weld."   This seriously affects such things as bearings, gears, slip rings, and brushes.   Likewise, most organic things such as elastomers and Teflon and even some metals evaporate.   At certain altitudes high-energy charged particles of the Van Allen belts will cause radiation damage to materials.

SATELLITE AND SPACE-VEHICLE INSTRUMENTATION

Propulsion Controls

In space, where there is no atmosphere to react against, change of the velocity vector can be accomplished only by ejection of mass at considerable velocity, as by rocket motors, plasma jets, etc.   Liquid propellants can, in general, be controlled over a much greater thrust ratio than solid propellants.   Direction of thrust control requires movable nozzles or deflecting vanes.

Stabilization

Disturbing torques come from a number of sources and must be evaluated before proper controls can be designed. They are:

1. Angular momentum imparted before or at separation from the final stage, the latter being due to unbalanced forces from explosive rivets or spring-loaded release mechanisms; angular rates depend on many factors, such as size, but are of the order of 1 rpm.

2. Gravity gradient, which tends to align the axis of minor moment of inertia along the gravity-gradient vector according to formulas.

$$T_X = \frac{3\omega^2}{2} (I_Y - I_Z) \sin 2\theta \tag{5}$$

T_X represents the gravitational torque in dyne-centimeters on a body of revolution, typical of the general shape of spacecraft or satellites, for which two of the three principal moments of inertia are essentially equal. In this case $I_X = I_Y$, the moments of inertia in g-cm^2 about the transverse axes; and I_Z is the moment of inertia about the longitudinal axis. θ represents the angle between the longitudinal axis and the vertical; and ω is the orbital angular velocity in radians per second. This expression is valid for circular orbits.

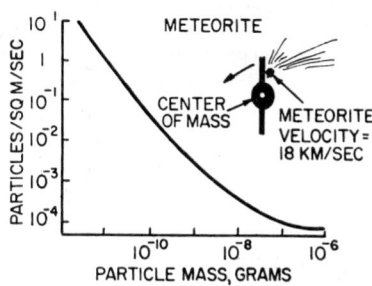

Fig. 62. Effect of meteorites on body rotation. (*Source: The Bendix Corporation, Eclipse-Pioneer Div.*)

3. Solar-radiation pressure, though amounting only 0.94×10^{-7} psf at our distance from the sun, which can cause small but continuous torque if the center of pressure does not coincide with the center of mass.

4. Atmospheric dynamic pressure at altitudes up to several hundred miles (velocities of 18,000 mph or more required for orbit or escape); the atmosphere exerts a small but appreciable dynamic pressure. If the center of this pressure differs from the center of mass, we have a disturbing torque.

5. Magnetic fields coupling to the earth's field, major offenders being:

   a. D-C loops in the satellite wiring, as leads to solar-cell arrays.

   b. Permanent magnets aboard, perhaps in electronic gear, or magnetic material used in the structure having appreciable coercive force.

   c. Fields from electric equipment, such as motors or solenoids.

   d. Soft magnetic material, as might be used in the structure.

6. Meteorites traveling at high velocities are present in appreciable quantities. There are many particles of dust size, but fewer and fewer of the larger and larger sizes. Impact of a meteorite any place but in line with the center of mass causes body rotation as illustrated in Fig. 62.* Direction, velocity, and quantity vs. size information is required to estimate control requirements.

* Legow, H. E., and W. M. Alexander, *Proc. First International Space Science Symposium*, Nice, France, January, 1960, p. 1033.

7. Motion or rotation of anything (including personnel of manned vehicles) in or on the craft such as changes in speed of a motor or gyro, or rotation of solar arrays must be accounted for. Wherever possible devices should be designed so that motions or rotations are accomplished without momentum change as referred to the vehicle.

Stabilization Sensors

Some of the sensors required for stabilization are as follows:

Gyros. Three rate gyros arranged mutually perpendicular are sometimes used to measure rate of rotation, especially immediately after separation. These are very similar to those used in autopilots for conventional aircraft, except that reliability requirements are more severe. After the initial phase of stabilization is accomplished, they may be turned off and only used again in case of vehicle disorientation due to a heavy meteorite strike or temporary malfunction of controls. The change of momentum resulting from starting and stopping of these gyros must be accommodated by the control system.

Inertial platforms are required at launch and perhaps again at reentry, also to measure velocity changes during thrust periods of orbit change, but are not normally

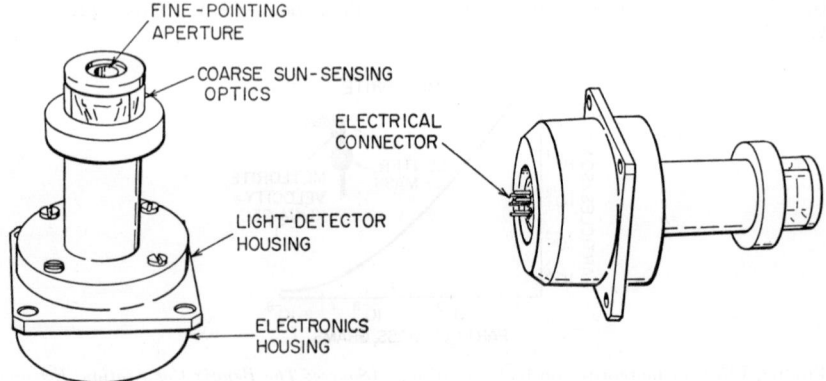

FINE-POINTING
APERTURE

COARSE SUN-SENSING
OPTICS

ELECTRICAL
CONNECTOR

LIGHT-DETECTOR
HOUSING

ELECTRONICS
HOUSING

FIG. 63. Sun sensor. (*Source: The Bendix Corporation, Eclipse-Pioneer Div.*)

required for purposes of stabilization. Gyro horizons or gyro verticals, as used in conventional aircraft, do not operate in orbit because the pull of gravity is balanced by centrifugal force and objects are "weightless."

Sun Sensors. Sun sensors are used in determining the direction to the sun and to provide the signals necessary to point either the vehicle or a seeker toward the sun. The sensor shown in Fig. 63 can sense over a hemisphere, and by itself it contains no moving parts.

Horizon Scanners. Horizon scanners provide an approximate vertical to the earth.

They operate in the infrared region, rather than on visible light. They sense the difference between the temperature of outer space and that of the earth at the horizon, be that day or night, polar or tropic. They should be designed to contain an absolute minimum number of moving parts (zero, if possible). They are used primarily on vehicles in the vicinity of the earth, though they will be useful in approaching the moon or planets. Figure 64 illustrates a wide-angle horizon sensor.

Star Trackers. While sun sensors and horizon scanners can provide stabilization information to an accuracy of 1° or so, star trackers such as the one shown in Fig. 65 can be used to a fraction of a minute. Their field of view is limited normally to a few degrees. They may be operated to exclude all stars but those of a given magnitude. They must be gimbaled to allow pointing to cover, if possible, a hemisphere and usually

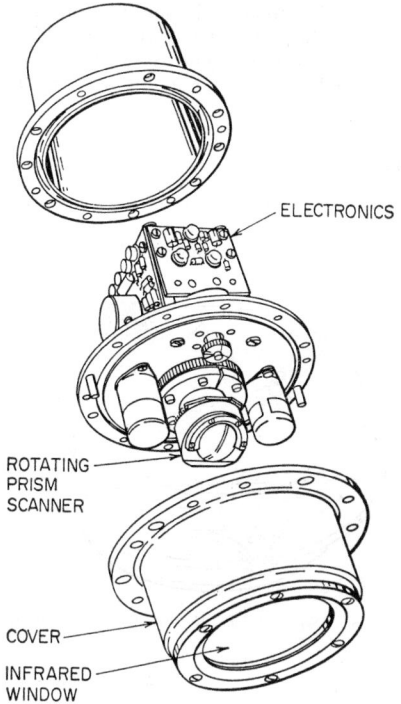

FIG. 64. Wide-angle horizon sensor

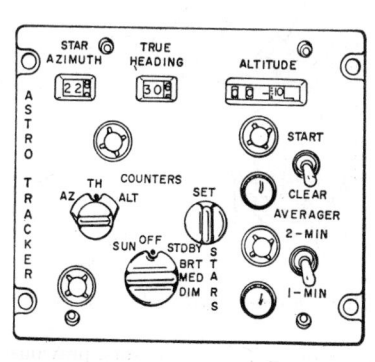

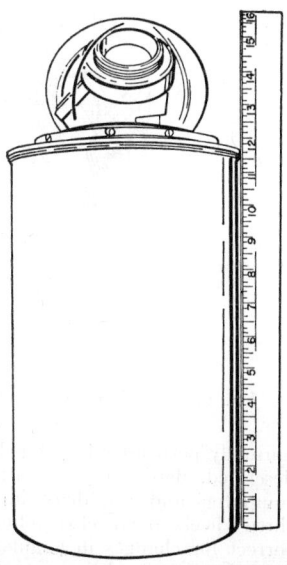

FIG. 65. Star tracker.   (*Source: Kollsman Instrument Corporation.*)

contain a solar shutter mechanism to keep the sun from inadvertently focusing on and damaging the sensitive phototube. Angular readouts must be made from each of the two gimbals, either digital or two-speed synchros with an accuracy of a few seconds of angle. Similar to those used on aircraft, a motor and gear train are needed to servo each axis.

Stabilization Controls

Some of the controls required for stabilization are as follows:

Reaction Wheels. When a disturbance applies a torque to the vehicle, the resulting momentum can be absorbed by the change in rotation of three flywheels, arranged

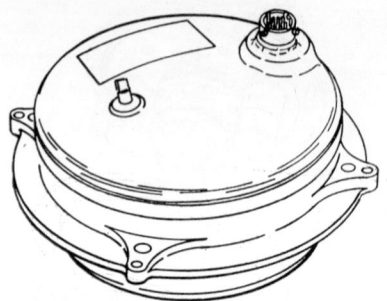

FIG. 66. Reaction wheel. (*Source: The Bendix Corporation, Eclipse-Pioneer Div.*)

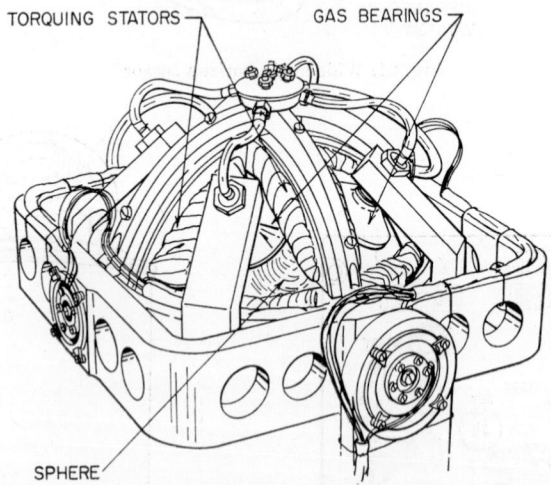

TORQUING STATORS   GAS BEARINGS

SPHERE

FIG. 67. Reaction sphere. (*Source: The Bendix Corporation, Eclipse-Pioneer Div.*)

mutually perpendicular, thus keeping the vehicle from rotation. Sensors, previously described, detect any deviation from desired orientation and, through associated computers and amplifiers, impart the proper combination of speed changes in the three wheels to cancel the effect of the disturbance. If any of the wheels is asked to correct for changes in angular momentum beyond its design capabilities, its speed increases to a fixed limit at which reaction jets are operated to "unload" the wheel. Figure 66 shows a reaction wheel for the Nimbus weather satellite.

Reaction Spheres.   A sphere, suspended by some low-friction support, can do the work of three reaction wheels and part of their associated computer.   Figure 67 shows an experimental gas-bearing model, and electrostatically supported spheres appear to be practical.

Reaction Gyro Pairs.   Three pairs of identical gyros, with each pair arranged to have their spin axes movable relative to each other, as shown in Fig. 68, are sometimes used in place of three reaction wheels.

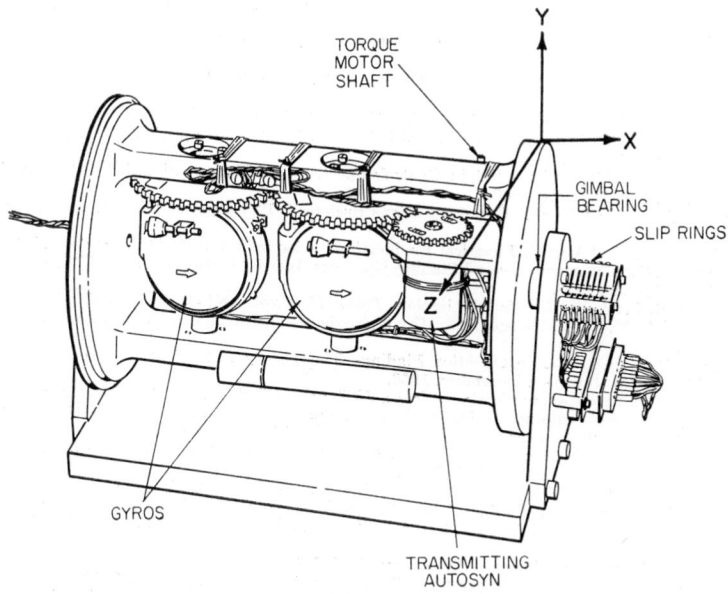

FIG. 68. A pair of reaction gyros for X axis.   (*Source: The Bendix Corporation, Eclipse-Pioneer Div.*)

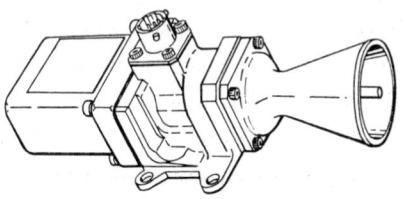

FIG. 69. Reaction jet.   (*Source: The Bendix Corporation, Eclipse-Pioneer Div.*)

Reaction Jets.   Highly compressed gas, escaping through a nozzle at high velocity, imparts a high force compared with the mass of the escaping gas, much higher than obtainable for a solid or liquid.   In space we have no means of replacing ejected mass; so we must use it sparingly and efficiently.   Reaction jets are usually used to stop initial spins occurring at launch and separation, and to "unload" momentum storage wheels or spheres.   Figure 69 shows a reaction jet used on the Discoverer series of satellites.

Magnetic Torquing.   If the vehicle is in an appreciable magnetic field (as a first approximation, the earth's field varies as the inverse cube of the radius from the earth's center), it can be used to produce a torque on the vehicle except around the magnetic vector. Controlled currents in three mutually perpendicular coils or a gimbaled permanent magnet which turns freely except when torque is required are two useful methods.

REFERENCES

1. Nikolsky, Alexander A.: "Helicopter Analysis," John Wiley & Sons, Inc., New York, 1951.
2. Perkins, Courtland D., and Robert E. Hage: "Airplane Performance, Stability, and Control," John Wiley & Sons, Inc., New York, 1949.
3. Moe, Harris G.: "Flight Engineering and Cruise Control," John Wiley & Sons, Inc., New York, 1947.
4. Kaufmann, R. H., and H. J. Finison: "D. C. Power Systems for Aircraft," John Wiley & Sons, Inc., New York, 1952.
5. Scarborough, J. B.: "The Gyroscope—Theory and Applications," Interscience Publishers, Inc., New York, 1957.
6. Jaques, C. N.: "Instrumentation in Testing Aircraft," Chapman & Hall, Ltd., London, 1957.
7. Stotz, John K., Jr.: Instruments Prove Out Safety of Aircraft Pilot Ejection, *ISA J.*, vol. 6, no. 12, p. 54, December, 1959.
8. Staff: Controllability and the 'Flight Controls Spectrum for Manned Aircraft,' *Automatic Control*, vol. 13, no. 1, p. 26, June, 1960.
9. Grover, J. H. H.: "Aircraft Communications Systems," Philosophical Library, Inc., New York, 1959.
10. Wakefield, G. G.: "Aircraft Electrical Engineering," Chapman & Hall, Ltd., London, 1959.
11. Mitsutomi, T.: Characteristics and Stabilization of an Inertial Platform, *Trans. IRE*, vol. ANE-5, no. 2, p. 95, June, 1958.
12. Horsfall, R. B.: Stellar Inertial Navigation, *Trans. IRE*, vol. ANE-5, no. 2, p. 106, June, 1958.
13. Majendie, A. M. A.: The Display and Use of Navigational Intelligence, *Trans. IRE*, vol. ANE-5, no. 3, p. 142, September, 1958.
14. Ancker, C. J. Jr.: Airborne Direction Finding—The Theory of Navigation Errors, *Trans. IRE*, vol. ANE-5, No. 4, p. 199, December, 1958.
15. Swerling, P.: Space Communications, *Trans. IRE*, vol. MIL-2, no. 1, p. 20, December, 1958.
16. Draper, C. S.: Self-contained Guidance Systems, *Trans. IRE*, vol. MIL-2, p. 25, December, 1958.
17. Dow, Richard B.: "Fundamentals of Advanced Missiles," John Wiley & Sons, Inc., New York, 1958.
18. Sutton, George P.: "Rocket Propulsion Elements, An Introduction to the Engineering of Rockets," John Wiley & Sons, Inc., New York, 1956.
19. Zucrow, M. J.: "Aircraft and Missile Propulsion, Volume I: Thermodynamics of Fluid Flow and Application to Propulsion Engines," John Wiley & Sons, Inc., New York, 1958.
20. Barker, A., T. R. F. Nonweiler, and R. Smelt: "Jets and Rockets," Chapman & Hall, Ltd., London, 1959.
21. Burgess, Eric: "Guided Weapons," Chapman & Hall, Ltd., London, 1957.
22. Burgess, Eric: "Satellites and Spaceflight," Chapman & Hall, Ltd., London, 1957.
23. Ley, Willy: "Rockets, Missiles and Space Travel," Chapman & Hall, Ltd., London, 1957.
24. Adams, Carsbie C.: "Space Flight," McGraw-Hill Book Company, Inc., New York, 1958.
25. Department of the Air Force: "Guided Missiles," McGraw-Hill Book Company, Inc., New York, 1958.
26. Puckett, Allen E., and Simon Ramo: "Guided Missile Engineering," McGraw-Hill Book Company, Inc., New York, 1958.
27. Tanzman, H. D.: High Accuracy Time Interval Measurements, *Electronic Industries*, pp. 62–67, January, 1959.
28. Staff: Cold Testing Techniques, *Automatic Control*, vol. 12, no. 5, p. 31, May, 1960.
29. Staff: Space Vehicle Controller, *Automatic Control*, vol. 12, no. 3, p. 53, March, 1960.
30. Oppenheim, J., and B. Sachs: Data File 35—Short Guide to Military Synchro Specs, *Control Eng.*, vol. 7, no. 3, p. 137, March, 1960.
31. Meyer, Jerry: How to Use Strain Gages for Missile Data Acquisition, *ISA J.*, vol. 7, no. 2, p. 58, February, 1960.
32. Staff: Component Development for Atlas Inertial System, *Automatic Control*, vol. 13, no. 1, p. 23, June, 1960.
33. Staff: Advanced Data Acquisition System Aids Polaris Missile Development, *Automatic Control*, vol. 13, no 1, p. 49, June, 1960.
34. Staff: X-15 Altitude Flight Data System, *Automatic Control*, vol. 13, no. 1, p. 55, June, 1960.
35. Reed, D. L.: Computers and Programmers in Hybrid Checkout Systems, *Control Eng.*, vol. 10, no. 4, p. 79, April, 1963.
36. Bradshaw, C. E.: Structural Testing with Hybrid Checkout Systems, *Control Eng.*, vol. 10, no. 4, p. 83, April, 1963.
37. Hellings, G.: All-weather TV Display Adds Realism to Simulated Landings, *Control Eng.*, vol. 10, no. 2, p. 107, February, 1963.
38. Hardy, H. S.: What You Can Learn from Regulus Control System, *Control Eng.*, vol. 8, no. 3, pp. 143–144, March, 1961.
39. Sinnott, F.: Designing Communications Links for Space Vehicles, *Control Eng.*, vol. 8, no. 3, pp. 155–159, March, 1961.
40. Fischel, E. M.: Two vs. Three-gyro Guidance Platforms, *Control Eng.*, vol. 8, no. 4, pp. 122–126, April, 1961.

41. Slater, J. M.: Autocompensation of Errors in Gyros and Accelerometers, *Control Eng.*, vol. 8, no. 5, pp. 121–122, May, 1961.
42. Forman, R. V.: The DC-8 that Didn't Fly, *Control Eng.*, vol. 8, no. 6, pp. 111–113, June, 1961.
43. Traynelis, K. A., and D. L. Ryan: Hot Gas Control Systems—Using Reaction to Control Vehicle Attitude, *Control Eng.*, vol. 8, no. 7, pp. 109–114, July, 1961.
44. Meyer, L. L.: Controlling a Missile by Swiveling the Rocket Motor, *Control Eng.*, vol. 8, no. 10, pp. 71–74, October, 1961.
45. Pallme, E. H., and V. B. Corey: Steadying a Stabilized Platform, *Control Eng.*, vol. 8, no. 11, pp. 85–88, November, 1961.
46. Fischel, E. M.: The Case for Dead Reckoning, *Control Eng.*, vol. 9, no. 10, pp. 71–72, October, 1962.
47. Holzbock, W. G.: Adjusting Pump RPM for Flow Regulation, *Control Eng.*, vol. 9, no. 10, pp. 87–89, October, 1962.
48. Lindahl, J. H., and W. M. McGuire: Adaptive Control Flies the X-15, *Control Eng.*, vol. 9, no. 10, pp. 93–97, October, 1962.
49. Slater, J. M.: Exotic Gyros—What They Offer, Where They Stand, *Control Eng.*, vol. 9, no. 11, pp. 92–97, November, 1962.
50. Fischel, E. M.: Schuler Detuning Cuts Verticality Errors, *Control Eng.*, vol. 9, no. 7, pp. 95–96, July, 1962.
51. Buscher, R. G.: Self-adaptive vs. Conventional Flight Control, *Control Eng.*, vol. 10, no. 1, pp. 102–103, January, 1963.
52. Raytheon staff: Guidance and Control for Manned Lunar Flight, *Control Eng.*, vol. 9, no. 6, pp. 87–102, June, 1962.
53. Lee, S.: Automatic Gyro Trimming during Platform Alignment, *Control Eng.*, vol. 9, no. 5, pp. 85–88, May, 1962.
54. Whitford, R. K.: Attitude Control of Earth Satellites, *Control Eng.*, vol. 9, no. 2, pp. 93–97, February, 1962.
55. Norton, H. N.: Formulating Aero-Space Transducer Standards, *ISA J.*, vol. 8, no. 9, pp. 44–47, September, 1961.
56. Blitzer, F., G. Bonelle, and B. Kriegsman: Orbital Rendezvous Control, *ISA J.*, vol. 9, no. 8, pp. 33–38, August, 1962.
57. Staff: Digital Techniques in Airborne Data Acquisition, *Automatic Control*, vol. 14, no. 2, pp. 37–41, February, 1961.
58. Staff: Digital Techniques in Airborne Data Acquisition, *Automatic Control*, vol. 14, no. 3, pp. 41–51, March, 1961.
59. Arck, M. H.: Simulator Proves Operation of Horizon Sensors, *Automatic Control*, vol. 14, no. 9, pp. 21–26, September, 1961.

AIRCRAFT FLIGHT-SIMULATION INSTRUMENTATION

By Richard C. Dehmel*

Introduction and widespread use by both military and commercial operators of high-performance aircraft have not been without complications. Modern aircraft operate close to their design limits and the pilots must be completely familiar with the aircraft capabilities to avoid exceeding structural limits in flight. Pressure, temperature, density, and several other variables all have a direct effect on the performance capabilities of turbine-powered aircraft.

All-weather flight capability requires that these aircraft be flown primarily by reference to instruments. These factors, together with the need to handle new tactical-weapons systems and radar and radio navigational aids, combine to create a difficult training problem. Although training requirements are well understood, they have become critical with the increasing congestion of airspace and the sharp rise in aircraft operating costs.

One type of flight simulator,† described here, exemplifies the solution to this problem. Simulators of this type are used world-wide by commercial and military aircraft operators, as well as by aircraft manufacturers. Use of simulators has included virtually every production-model aircraft built since 1948, from eight-jet bombers to four-engine transports and many single-jet fighters.

BASIC DESCRIPTION OF A FLIGHT SIMULATOR

Basically, a flight simulator is a grounded aircraft in which engines, wings, control surfaces, radio, radar, and tactical-weapons systems have been replaced by electronic computers. These actuate the cockpit instruments, pressurize the controls, and create radio signals, radar displays, etc., exactly as would occur in the actual aircraft with similar manipulation of all controls. Environmental realism is provided by using actual aircraft parts for the cockpits, crew seats, instrument panels, control consoles, and other equipment surrounding the crew. A "trouble panel" is provided so that engine failure, fires, system malfunctions, and other emergency conditions can be introduced at will by the instructor to enhance training.

The full regimes of flight and aircraft ground operations are encompassed. To obtain this high degree of simulation, approximately fifteen computer cabinets are necessary. These computers may be of either the analog or digital type.‡

Analog computers use physical measurements, such as shaft rotation and voltages to represent problem information. The solution to the problem is computed continuously and, therefore, is correct at every instant of time. For example, the flight computer continuously solves, for any change in configuration, power, or other effects of the controls, equations which represent flight path, acceleration, and attitude. Those variables which are presented on instruments in the actual aircraft are identically displayed in the simulator. Provision is made for adapting the simulator to new facilities as they become operational.

* Vice President—Engineering, Curtiss-Wright Corporation, Wood-Ridge, N.J.
† Curtiss-Wright Dehmel flight simulator.

‡ Currently, extensive research and development (Curtiss-Wright Electronics Division) is underway to adapt digital computers to flight simulation. The largest digital training device yet built has been delivered to the U.S. Navy for use in its fleet ballistic-missile submarine program.

Solution of Aerodynamic Equations

Judicious handling of the aerodynamic equations and careful selection of the axes systems used is essential to avoid excessive complications in computer design.   For a typical supersonic jet fighter, which may be used as a general illustration, aerodynamic equations representing six degrees of freedom are utilized.   Three of these are translational equations and three are rotational equations.   Three force equations may be

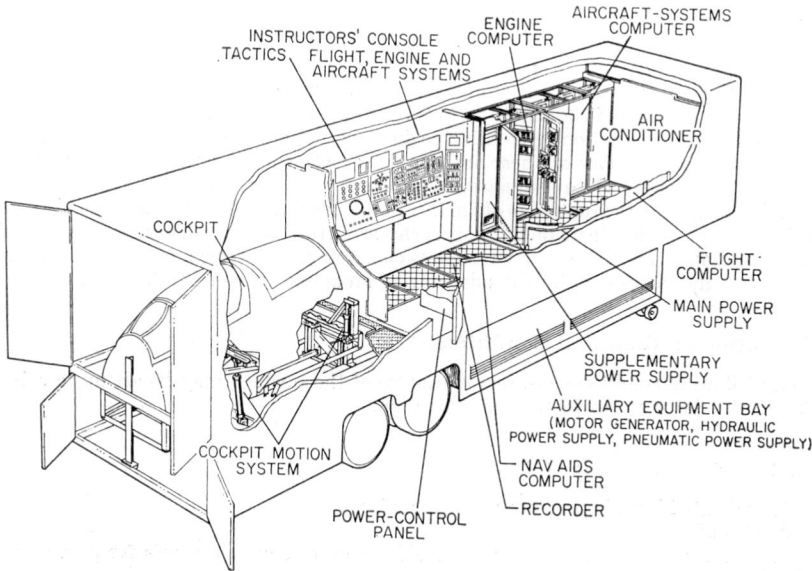

FIG. 1. Flight and weapons-system trainer mounted in transportable van.

presented with respect to the wind axes, while the three moment equations may be presented with respect to airplane body axes.   The equations involve complicated functions of many variables, such as:

Angle of attack.
Angle of sideslip.
Rate of change of angle of attack.
Rate of change of angle of sideslip.
Rate of change of angle of flight path.
Tail angle of attack.
Rates of roll, pitch, and turn.
Wing-flap deflection.
Speed-brake extension.
Spoiler extension.
Primary-control-surface deflection.
Landing-gear position.
Lift coefficient.
Dynamic pressure.
Mach number.
Aeroelastic effects.
Gross weight.
Center of gravity.
Distance to ground plane.

In addition, in the solution of the equations, the basic aerodynamic coefficients are computed, that is:

Lift coefficient.

Drag coefficient.

Crosswind-force coefficient.

Pitching-moment coefficient.

Rolling-moment coefficient.

Yawing-moment coefficient.

The necessary transformations and the resolution of the airplane weight are performed by generating certain trigonometric relations as functions of angles of attack, sideslip, roll, pitch, and flight path.

Realistic cockpit motion is being utilized on an optional basis. Experience has revealed that instrument flight training is greatly enhanced when the pilot receives cues from the motion of the aircraft which supplement his reactions to the instrument indications. The equations governing cockpit motion represent an appropriate combination of the sensations of linear and angular acceleration of actual aircraft cockpit.

Through the use of visual simulation equipment, the utility of the flight simulator is extended to include the full range of flight training from takeoff under visual conditions, to instrument flight, instrument approach, breakthrough and visual contact, and visual landing. The visual presentation is completely nonprogrammed and faithfully reproduces the views seen by the pilot during an actual takeoff and landing.

Simulation of Abnormal Conditions

The following conditions are representative of those which may be set up in a jet-aircraft simulator at the instructor's direction.

Jet-engine power-plant Troubles

1. Indicated low air pressure on each engine.
2. Cut off each engine separately.
3. Install compressor pulsation, or malfunction of engine controls for preventing pulsations.
4. Install combustion flameout.
5. Fail the spike position system.
6. Cut off each afterburner separately.
7. Indicate engine fire.
8. Indicate engine icing.
9. Establish incorrect nozzle.
10. The panel indicates a bad start automatically in accordance with the use of improper starting procedures.

Fuel-system Troubles

1. Failure of airplane transfer and boost pumps individually.
2. Fail engine fuel pump.
3. Fail automatic fuel-regulation device.
4. Fail fuel valve.
5. Failure of high-level cutoff switch to operate while refueling.
6. Variable fuel-flow indications as controlled by the operator.

Flight Controls and Autopilot Troubles

1. Failure of either primary or utility hydraulic power system.
2. Failure of manual trim on control stick.
3. Elevator and aileron ratio chargers being driven from existing positions to either extreme positions.

4. Each of autopilot servos being driven from existing positions to either extreme positions at maximum rate.

5. Each of the damper servos being driven from existing positions to either extreme positions at maximum rate.

6. The automatic trim actuator being driven from existing position to either extreme position at maximum rate.

7. Failure of autopilot servo to disengage.

8. Failure of constant Mach functions.

9. Failure of Mach altitude functions.

10. Failure of omnilocalizer and glide slope functions.

11. Failure of resolutions surface.

12. Failure of autopilot to energize.

13. Failure of autopilot elevator or aileron synchronizer.

14. Failure of heading signal from bomb-navigation system.

15. Failure of wing-flap control or torque tube.

16. Blocking of pitot-static system.

17. Failure of gyro instrument.

18. Runaway trim control.

Electrical Troubles

1. Alternator operation.
   De-excite, due to:
   Increased voltage.
   Differential current between phases.
2. Alternator off frequency.
3. Alternator overload.
4. Interruption of power supply to any of the power circuits at the pilot's station, as by power loss due to blown fuses.

Hydraulic-system Troubles

1. Pressure loss on each system separately.
2. Loss of any one of hydraulic pumps.
3. Loss of normal brake systems.
4. Loss of nose wheel steering.

Landing-gear Troubles

1. Main gear fails to retract.
2. Nose gear fails to retract.
3. Main gear fails to extend.
4. Nose gear fails to extend.

Cabin-conditioning Troubles

1. Left-hand or right-hand conditioning systems failure.
2. Electronic cooling-system failure.

Indication of Unsafe Canopy

Icing Conditions

Controls for simulating wing, prop spinner, and engine icing conditions.

Radio Troubles

1. Command set transmitter failure.
2. Receiver failure of either command or navigational sets.

Separate Generator-system Troubles

1. Variable torque-meter indications as controlled by the operator.
2. Variable turbine-inlet-temperature indications, as controlled by the operator.
3. Variable engine-pressure-ratio indications, as controlled by the operator.
4. Variable per cent nozzle position indications, as controlled by the operator.
5. Variable per cent rpm indications, as controlled by the operator.
6. Variable oil-pressure indications, as controlled by the operator.

Trouble-indicator Lights

Lights are installed on the operator's panel for all "trouble" switches. Corrective action will automatically turn off these indicator lights.

Other Functions

Controls are located at the operator's station for operation of the following:
1. Gross weight.
2. Center of gravity.
3. Rough air.
4. Fuel filling and depletion.
5. Lightning.
6. Clouds (shadow effect only).
7. Sound control.
8. Crash override.
9. Ground power.
10. Ground air conditioning.
11. Brake chute and oxygen.
12. Ratio-changer reset (required after student uses manual ratio-changer controls).
13. Simulated door warning lights.

ADVANTAGES OF INSTRUMENTED FLIGHT SIMULATION

Flight crews require more hours of flight instructions if the instruction periods are given sporadically, rather than in a continuous flow. While the flight simulator is ready regardless of weather or time of day, training in actual aircraft suffers from such drawbacks as:
1. Unavailability of aircraft due to other commitments.
2. Adverse weather, either canceling or handicapping training periods.
3. Traffic delays at busy airports.
4. Mechanical malfunctions and maintenance delays which can seriously upset training schedules.
5. Dependence upon such outside radio aids as low-frequency and omniranges, beacons and ILS installations—which may be unavailable when desired.
6. It is not prudent to conduct much of the necessary aircraft flight training at night. Emergency procedures require the utmost of the instructor and he should not be burdened with the additional load of operating in darkness. Traffic presents a more difficult problem at night than in daytime.

Cost Reductions. Substantial reductions in crew training costs may be realized through use of flight simulators. Generally, the cost of operating transport-type simulators is less than one-tenth of the cost of operating the actual airplane. For military aircraft, the relationship is even more favorable to the simulator. For example, a medium-range bomber costs approximately $2,000 per hour to operate, as compared with about $83 per hour for the simulator. When these savings are multiplied by the large number of pilots who require training, flight simulation costs savings are large.

A typical commercial-airline flight training program using aircraft requires only about 14 hr for transition or initial-checkout training.   Through use of a combination aircraft and simulator program, the aircraft time can be reduced to 6 hr.   Eighteen hours of simulator time will be required, but the latter at a much lower cost than the aircraft time.

Military Training.   These requirements are not directly comparable with commercial needs because the military pilot has additional requirements, such as aerial refueling, weapons and tactics systems operation, and various other tactical considerations.   The flight simulator is considered essential to the military aviation training program.

Simulation of Difficult Conditions.   In addition to minimizing checkout time, the flight simulator can provide emergency procedure training not easy to duplicate in any other fashion.   For example, inadvertent propeller reversal could not be practiced on aircraft training flights because it is hazardous and extremely difficult to simulate such conditions in the airplane even under carefully controlled test conditions. In the simulator, such difficult and hazardous conditions, including flameout in jets, can be practiced at will.   This information enables a flight crew to avert potential fatal accidents with a minimum of danger during training.

Weapons-system Trainer.   Most military aircraft carry complex weapons systems.   These introduce additional training requirements.   Where the aircraft contain separate stations for a navigator, bombardier, radar fire control or antisubmarine-warfare system operator, these must also be provided in the simulator.   The end objective of military combat crew training is to graduate a proficient and skilled team, fully capable of completing any assigned mission under any given condition. Although a combat crew is composed of highly qualified specialists they must function and operate together as a single unit, rather than as individuals.   These considerations have led to the development of a complete weapons-system trainer,* in which the simulation is extended to include appropriate weapons, fire control, countermeasures, radar, and tactical systems.   Air intercept, low-level bombing, and antisubmarine-warfare missions are among programs that can be scheduled on the simulator.

Need for Realism.   The philosophy of simulation requires the achievement of realism in the training environment, with accuracy and versatility in the simulation of training problems and tactics equipment.   This necessitates faithful duplication of the actual tactical situation in every possible respect.   As in the flight simulator, all controls, indicators, panels, and other devices associated with the simulated tactical system will be the same equipment which the trainees operate in actual tactical situations.

The internal modifications of the equipment necessary for adaptation to the operation of the simulated system will in no way detract from the realism of the training environment.   For example, in addition to simulating the radar-directed armament control system, the simulation equipment must provide the trainees with synthetic radar presentations so that tactical air intercepts and special weapons situations may be accurately and representatively developed.

The requisite degree of fidelity of simulation in each of the many simulated areas depends largely on its influence in the transfer of learning from the device to the actual weapons system.   The design of the training aid must be such as to always add to the skills.   This is positive transfer of learning.   There must be no negative transfer, because wrong learning can make the trainer dangerous to the mission.   The most efficient training device is the one which produces the greatest transfer of learning at minimum financial cost in all areas where additional skills are needed.

Although it is possible, at least in principle, to construct an analog of any physical system to any desired degree of accuracy and completeness, the equipment might be impractically large and costly.   Therefore, the criterion employed in each instance is the relative importance of the amount of positive transfer of training gained by the inclusion of additional components.

Often it is desirable to measure proficiency by means of scoring devices, records of flight path, and response to emergencies.   Under these conditions, simulator variations

* Curtiss-Wright Dehmel weapons-system trainer.

which disturb the momentary performance level of the crew members cannot be tolerated. It is evident that accepted human engineering principles must be applied throughout the design of the weapons-system tactics trainer as well as the flight simulator.

Instrumentation. Just as in the aircraft, inputs to the crew are largely derived from the instrumentation of the simulator. Hence, instrument construction and performance is of the utmost importance. Any noticeable discrepancy in appearance, static accuracy, small-signal, or dynamic response detracts from training value.

Most of the instruments encountered in aircraft may be categorized as follows:

1. Pneumatic (airspeed, altitude, rate of climb).
2. Voltmeters or ammeters (electric load meters, trim indicators, omnirange, localizer, and glide-path indicators).
3. Synchro (flaps, oil pressure, fuel pressure).
4. Gyroscope (heading, turn rate, attitude).
5. Servo (vertical-tape indicators, E.P.R.—engine pressure ratio, E.G.T.—engine gas temp., B.D.H.—bearing direction, heading).
6. Inertial (slip, accelerometer).
7. Cathode-ray tube (ignition analyzer, fire-control display).

Because simulator computer outputs are readily available as shaft positions or voltages, it is most convenient to use instrumentation that can be either shaft-(via synchro) or voltage-actuated. The latter includes servo types. Fortunately, the large majority of modern aircraft instruments are either synchro or voltage controlled. Pneumatic, gyroscope, or inertial units must, however, also be simulated, and this is done by replacing the actuating elements with synchros, servo drives, or voltmeter movements, as appropriate. In all instances, the outward appearance of the instruments is carefully retained and their response to computed signals is designed to match that in the aircraft.

Other types, such as thermometers, tachometers, and magnetic (compass), are normally simulated by use of synchro, servo, or voltmeter drives. Cathode-ray tubes are used unmodified.

Total instrumentation in a multiengine-aircraft simulator comprises several hundred units.

REFERENCES

Editor's Note: Several of the articles listed after the preceding subsection on "Aircraft and Aerospace Vehicle Instrumentation" contain material on flight simulation.

Section 17

INSTRUMENTATION ENGINEERING PRACTICE

By

VIGGO O. ANDERSEN, B.S.(E.E.), *Staff Engineer, Non-Linear Systems, Inc., Del Mar, Calif.; Member, Institute of Radio Engineers. (Basic Electricity and Electronics for Instrumentation Engineers)*

W. T. DUMSER, B.S. (E.E.), *Supervisory Engineer, Republic Flow Meters Company, a division of Rockwell Manufacturing Company, Chicago, Ill.; Member, American Institute of Electrical Engineers, Instrument Society of America; Registered Professional Engineer (Ill.). (Instrument Panelboard Design and Construction)*

GERALD L. EBERLY, B.S. (M.E.), *Technical Staff, Harris D. McKinney, Inc., Philadelphia, Pa.; Member, Instrument Society of America. (Conditioning the Instrument Air Supply)*

J. W. FRIEDMAN, B.S. (E.E.), *Supervisory Engineer, Panel Division, Minneapolis-Honeywell Regulator Co., Philadelphia, Pa. (Preparation of Wiring Diagrams for Instrument Applications)*

JEROME KOHL, B.S.(Ch.E.), *Coordinator of Special Products, General Atomic Division, General Dynamics Corporation, San Diego, Calif.; Member, American Institute of Chemical Engineers, American Nuclear Society; Registered Professional Engineer (Ch.E.) (Calif.). (Basic Electricity and Electronics for Instrumentation Engineers)*

HENRY REINECKE, JR., M.S.(E.E.), *Systems Manager, Non-Linear Systems, Inc., Del Mar, Calif.; Member, Institute of Radio Engineers. (Basic Electricity and Electronics for Instrumentation Engineers)*

BASIC ELECTRICITY AND ELECTRONICS
FOR INSTRUMENTATION ENGINEERS

By J. Kohl,* H. Reinecke, Jr.,† and V. O. Andersen‡

Editor's Note. Although it is not the general policy of this handbook to include fundamental information which is well covered in other handbooks and texts, it was agreed by the Editorial Board of Advisors to include the following material on basic electricity and electronics. This exception was made because of the tremendous importance which a fundamental electrical and electronics knowledge holds for the effective application of modern industrial instruments and because only limited attempts have been made previously to interpret these large subjects in terms particularly appropriate to the instrumentation field.

ELECTRIC-CIRCUIT THEORY

Basic electrical quantities and units are given in Table 1.

Table 1. Basic Electrical Quantities and Units

Symbol	Quantity	Equation	Practical unit
I, i	Current	$I = Q/t \quad I = E/R$ or E/Z	Ampere
Q, q	Quantity	$Q = It$ or 6.24×10^{18} electrons	Coulomb
E, e	Electromotive force	$E = W/Q$ or IR	Volt
R, r	Resistance	$R = P/I^2 = E/I$	Ohm
p	Resistivity	$p = RA/L$	Ohms per circular mil-ft, ohms per cm cube
λ	Conductivity	$\lambda = 1/p$	Mhos per unit volume
C	Capacitance	$C = Q/E$	Farad
L	Inductance	$L = n\phi 10^{-8}/I$	Henry
T	Time constant	L/R, CR	Second
T	Period or cycle	$T = 1/f$	Second
f	Frequency	$f = 1/T$	Cycles per second (cps)
w	Angular velocity	$w = 2\pi f$	Radians per second
X_L	Inductive reactance	$X_L = 2\pi fL$	Ohm
X_C	Capacitive reactance	$X_C = 1/2\pi fC$	Ohm
X, x	Reactance	$X = X_L - X_C$	Ohm
Z, z	Impedance	$Z = \sqrt{R^2 + X^2}$	Ohm
G, g	Conductance	$G = R/Z^2$	Mho
B, b	Susceptance	$B = X/Z^2$	Mho
Y, y	Admittance	$Y = \sqrt{G^2 + B^2}$ $= 1/Z$	Mho
P	Electric power	$P = EI = I^2R$ $= EI \cos\theta$	Watt
W	Electric energy	$W = Pt$	Joule, watthour, kilowatthour
	Power factor (p.f.)	$\dfrac{EI \cos\theta}{EI} = \dfrac{\text{real } P}{\text{apparent } P}$	
	Reactive factor	$\dfrac{EI \sin\theta}{EI} = \dfrac{\text{reactive } P}{\text{apparent } P}$	

n = number of turns; t = time in seconds; f = frequency; ϕ = flux.

* Coordinator of Special Products, General Atomic Division, General Dynamics Corporation, San Diego, Calif.
† Systems Manager, Non-Linear Systems, Inc., Del Mar, Calif.
‡ Staff Engineer, Non-Linear Systems, Inc., Del Mar, Calif.

D-C Circuits

Following are the basic d-c circuit relationships:
Ohm's Law

$$E = IR \tag{1}$$

where a voltage E (volts) across a resistance R (ohms) causes a current I (amperes).
Power

$$W = EI = I^2R = \frac{E^2}{R} \tag{2}$$

where W = power (watts).   746 watts = 1 horsepower.

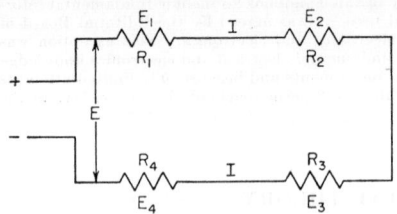

Fig. 1. Series circuit.

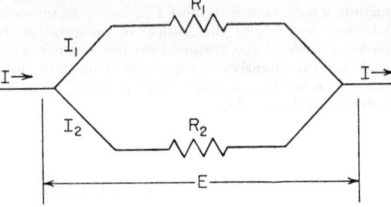

Fig. 2. Parallel circuit.

Series Circuits.   The following basic relationships apply:
1. Current flow is the same at every point in the circuit.
2. The sum of the IR drops around the circuit is equal to the applied voltage.
3. Total resistance for the circuit is the sum of the component resistance (see Fig. 1).

$$E = E_1 + E_2 + E_3 + E_4 \tag{3}$$
$$R = R_1 + R_2 + R_3 + R_4 \tag{4}$$
$$E_1 = IR_1 \qquad E_2 = IR_2 \qquad \text{etc.} \tag{5}$$

Parallel Circuits.   The following basic relationships apply:
1. Voltage drop is the same for each of the parallel paths.
2. Current entering the junction point divides between the branches in inverse proportion to their resistances, and the total current is the sum of the currents in the branches (see Fig. 2).

$$E = E_1 = I_1R_1 = E_2 = I_2R_2 \tag{6}$$
$$I = I_1 + I_2 \tag{7}$$
$$\frac{1}{R} = \frac{1}{R_1} + \frac{1}{R_2} = \frac{R_2 + R_1}{R_1R_2} \tag{8}$$

or

$$R = \frac{R_1R_2}{R_1 + R_2} \tag{9}$$

where R is the equivalent single resistance which would replace R_1 and R_2 in parallel without a change in current.
3. Conductance is defined as

$$G = \frac{1}{R} \tag{10}$$

4. The equivalent resistance of any number of parallel resistances may be found by adding the sum of the conductances of each branch and taking the reciprocal of the result.
Series-Parallel Circuits.   Each group of resistances in series is combined by addition; the sum of each group is converted into a conductance and added to other parallel conductances.

Kirchhoff's Laws

1. The algebraic sum of currents at any junction of conductors is zero, or the total current flowing toward a junction must equal the total current flowing away from the junction.

2. The algebraic sum of the voltages around any closed path in a network is zero.

A-C Circuits

In general, the laws that apply to d-c circuits also apply to a-c circuits. However, there are several important additional considerations in a-c circuits. If a changing voltage is applied to a reactance, the corresponding change in current through the reactance will be displaced in time from the applied voltage. Following are the basic relations and equations concerning this phenomenon. See Ref. 1 for more details.

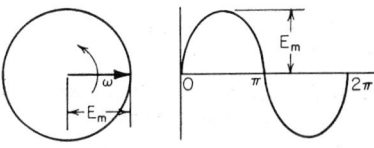

FIG. 3. Sinusoidal varying voltage.

Vectors. An a-c generator or alternator produces a sinusoidal voltage. The frequency of this voltage is related to the speed of rotation of the alternator, and in a simple two-pole alternator one revolution will equal 1 cycle or 360 electrical degrees. The frequency may be represented by

$$f = \frac{\omega}{2\pi} \tag{11}$$

where f = cycles per second
ω = angular velocity, radians/sec

The period is the reciprocal of the frequency, or the time required for 1 cycle.

The instantaneous voltage produced is

$$e = E_m \sin \omega t = E_m \sin 2\pi ft \tag{12}$$

where e = instantaneous voltage
E_m = maximum or peak voltage
t = time, sec

If this voltage is applied to a resistance, the resulting current will be

$$i = \frac{e}{R} = I_m \sin \omega t = I_m \sin 2\pi ft \tag{13}$$

where i = instantaneous current, amp
I_m = maximum or peak current, amp

The effective or root-mean-square (rms) value of alternating current is the equivalent value of direct current necessary to give the same power dissipation in a resistance, and

$$I_{\text{rms}} = \frac{I_m}{\sqrt{2}} \tag{14}$$

Since the effective value I or E, which is $1/\sqrt{2}$ times the peak value I_m or E_m, is the current or voltage most a-c meters are calibrated to read, this is the value usually referred to unless otherwise specified.

The equation $e = E_m \sin \omega t$ describes the length of the projection on the vertical axis of a vector of length E_m rotating counterclockwise around the origin with an angular velocity ω starting from the positive segment of the horizontal axis when t equals zero (Fig. 3). If the rotating vector is displaced from the horizontal axis by θ radians when t equals zero, the voltage and current equations are

$$i = I_m \sin (\omega t + \theta) \tag{15}$$

$$e = E_m \sin (\omega t + \theta) \tag{16}$$

Instantaneous values of current or voltage can be added algebraically. Peak, rms, or average values of current or voltage must be added vectorially unless the currents or voltages are in phase ($\theta_1 = \theta_2$) and of identical frequency ($\omega_1 = \omega_2$).

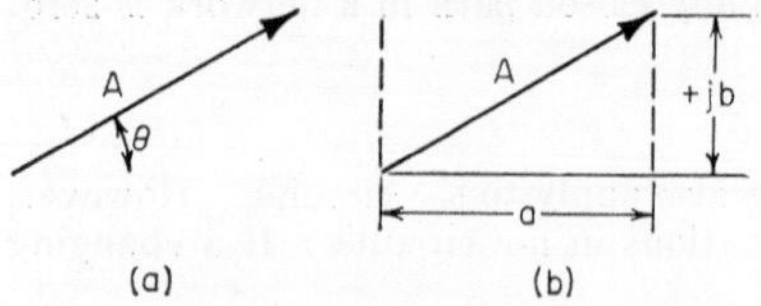

Using the notations **I** and **E** to designate vector quantities, the equations for combining currents or voltages are written simply as

$$\mathbf{I}_{1+2} = \mathbf{I}_1 + \mathbf{I}_2 \tag{17}$$
$$\mathbf{E}_{1+2} = \mathbf{E}_1 + \mathbf{E}_2 \tag{18}$$

Vectors may be expressed in any of several forms (see Fig. 4). The most common are

(a)   (b)

FIG. 4. Vector notation: (a) Polar form; (b) rectangular form.

1. Polar form: $\mathbf{A} = \underline{A}|\theta$ (19)

where **A** = vector quantity of magnitude A at an angle θ with the positive segment of the reference axis

2. Rectangular form: $\mathbf{A} = a + jb$ (20)

where a = projection on the reference axis (axis of reals) of a vector of length A
b = projection of this vector on an axis (axis of imaginaries) at 90° to the reference axis
$+j$ = projection on the segment of the axis of imaginaries 90° counterclockwise from the positive segment of the axis of reals

Resistance, Capacitance, and Inductance.   Ohm's law applies to a-c circuits as well as to d-c circuits.   The relationship $E = IR$ is valid for a-c circuits when E and I are in phase, as is the case in a pure resistive circuit.   When capacitance or inductance is present in the circuit, however, Ohm's law must be considered in the form

$$E = IZ \tag{21}$$

In a circuit containing only capacitance c the relationship between instantaneous current and voltage is given by

$$i = c\,\frac{de}{dt} \tag{22}$$

Since the instantaneous voltage $e = E_m \sin \omega t$,

$$i = \frac{E_m}{X_c} \sin(\omega t + 90°) \tag{22a}$$

where capacitive reactance

$$X_c = \frac{1}{\omega C} = \frac{1}{2\pi f C} \tag{23}$$

Thus the current and voltage are not in phase, the current leading the voltage by 90°.

In a circuit containing only inductance the relationship between instantaneous current and induced voltage is

$$e_1 = -L\,\frac{di}{dt} \tag{24}$$

Since the voltage applied to the inductance must be equal to and opposing the induced voltage

$$e = -L\,\frac{di}{dt} = X_L I_m \sin(\omega t + 90°) \tag{25}$$

where the inductive reactance $X_L = 2\pi f L$.   Again the current and voltage are 90° out of phase but with the voltage leading the current.

In a series combination of resistance R, capacitance C, and inductance L, the same current passes through the resistance, capacitance, and inductance, and

$$I = I_R = I_C = I_L \tag{26}$$

Since the voltage across the resistor is in phase with the current, the voltage across the capacitance lags 90° behind the current, and the voltage across the inductance leads the current by 90°. Thus the total voltage

$$E = E_R - jE_c + jE_L \tag{27}$$

The impedance of the circuit is given by

$$Z = R - jX_c + jX_L \tag{28}$$

When resistance, capacitance, and inductance occur in parallel branches of a circuit, the voltage appearing across the parallel branches is the same in both magnitude and phase. The currents flowing in the branches may differ in phase as well as in magnitude. The circuit laws used in solving parallel d-c circuits are also applicable to a-c circuits when proper recognition is given to the phase difference between current and voltage.

Kirchhoff's current law for vector quantities is: The vector sum of all the vector currents flowing toward a point is equal to the vector sum of all the vector currents flowing away from the point.

The relationships between resistances R, capacitances C, and inductances L in series and in parallel are summarized in Table 2.

Table 2. Relationships between Resistances, Capacitances, and Inductances in Series and Parallel Circuits

Combination	Series	Parallel
R_1, R_2, R_3	$R_1 + R_2 + R_3$	$\dfrac{1}{1/R_1 + 1/R_2 + 1/R_3}$
C_1, C_2, C_3	$\dfrac{1}{1/C_1 + 1/C_2 + 1/C_3}$	$C_1 + C_2 + C_3$
L_1, L_2, L_3	$L_1 + L_2 + L_3$	$\dfrac{1}{1/L_1 + 1/L_2 + 1/L_3}$

Circuits involving impedances in parallel and in series can normally be solved by determining the series circuit equivalent to the parallel circuits and then solving the resulting series circuit. When several parallel branches are involved, the admittance method is frequently employed. The admittance of a circuit Y is the reciprocal of the impedance Z and is expressed in mhos. In a circuit having both resistance and reactance, the admittance

$$Y = \frac{1}{Z} = \frac{1}{R + jX} = \frac{R}{Z^2} - \frac{jX}{Z^2} = G - jB \tag{29}$$

where G is the conductance and B the susceptance of the circuit. Admittances, being equal to the current divided by the voltage, can, when they appear in parallel, be added vectorially to give the admittance of the equivalent series circuit.

The equivalent series admittance may then be converted to the corresponding impedance and added vectorially to the other series impedances to determine the impedance of the combination series and parallel circuit.

Resonance. When the quadrature component of current (i.e., reactive current) in a series or parallel circuit containing capacitance and inductance is zero, the circuit is said to be resonant.

In a series circuit the conditions for resonance indicate that

$$X_L = X_C = 2\pi f L = \frac{1}{2\pi f C} \tag{30}$$

Thus resonance in a circuit containing R, L, and C, or L and C alone (this condition cannot be reached in practice) can be obtained by varying the inductance, capacitance, or frequency.

At resonance:

1. Impedance of the series circuit is minimum, being equal to the resistance of the circuit.

2. Current and voltage are in phase.

3. Resonant frequency is

$$f_r = \frac{1}{2\pi \sqrt{LC}} \quad \text{cps} \tag{30a}$$

when L and C are expressed in henrys and farads, respectively.

A parallel circuit is said to be in parallel resonance or antiresonance when the total current through the parallel branches is in phase with the applied voltage. If the resistances in the parallel branches are very small compared with the reactance, antiresonance occurs when $X_L = X_C$ and the antiresonant frequency

$$f_{\text{ar}} = \frac{1}{2\pi \sqrt{LC}} \tag{30b}$$

which is the same as that for a series circuit.

For fixed values of resistance, inductance, and capacitance, the impedance of a circuit is dependent upon frequency. A series circuit containing resistance, inductance, and capacitance appears to be capacitative at frequencies below resonance and inductive at frequencies greater than the resonant frequency. When the frequency is increased or decreased from the resonant value, there are two frequencies f_1 and f_2 at which the reactance is equal to the resistance. At these points, $Z = \sqrt{2}\,R$ and the magnitude of the current flowing is 0.707 times the current at resonance. The sharpness of the resonant peak depends primarily upon the ratio of the inductive reactance of the circuit to the resistance. This ratio is usually designated by the letter Q. The relationship between the sharpness of the resonant peak and the Q of the circuit is

$$\frac{f_2 - f_1}{f_r} = \frac{R}{2\pi f_r L} = \frac{1}{Q} \tag{31}$$

At resonance, the current through the circuit is E_0/R and the voltage across the inductance and capacitance is $E_0 Q$.

Parallel circuits containing inductance and capacitance in the two branches appear to be inductive at frequencies below resonance and capacitive above resonance. The impedance of the parallel circuit is maximum at resonant frequency and decreases at frequencies both above and below the resonant value. The larger the resistance in a parallel circuit, the less sharp will be the resonant peak. Unlike a series resonant circuit, however, the larger the impedance of the signal source, the sharper will be the resonant peak.

Power. The instantaneous power absorbed in an a-c circuit is equal to the product of the instantaneous values of current and voltage and is expressed in watts when the current and voltage are in amperes and volts, respectively. In a pure resistive circuit where the current and voltage are in phase, the real, average, or measured power P is equal to the product of the effective values of current and voltage. When the current and voltage are not in phase, the real power consumed by the circuit is $EI \cos \theta$, where θ is the phase angle between voltage and current and $\cos \theta$ is equal to R/Z.

The equation $EI \cos \theta$ gives the product of the in-phase components of current and voltage, which is also equal to the product of the square of the measured current and the resistive component of the impedance. The product of the square of the measured current and the reactive component of the impedance is known as the reactive power or reactive volt-amperes (vars). Reactive power is also given by the equation $EI \sin \theta$.

The power factor (p.f.) of a circuit is equal to $\cos \theta$ or to the reading obtained from a wattmeter placed in the circuit divided by the product of simultaneous voltage and current values. The product of current and voltage is termed the apparent power and is usually expressed in volt-amperes.

$$\text{p.f.} = \frac{\text{real power}}{\text{apparent power}} \tag{32}$$

The continuous output limit of rotating equipment or transformers is generally determined by the temperature rise in the windings. Thus they are rated in volt-amperes rather than in watts. The maximum real power obtainable from rotating equipment or from transformers depends upon the power factor of the load; the larger the power factor, the greater the ratio of real to apparent power.

The reactive factor (r.f.) of a circuit is equal to the reactive volt-amperes divided by the apparent volt-amperes.

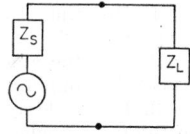

FIG. 5. Power transfer.

Unlike power circuits, where efficiency is a prime consideration, it is often desirable to obtain the maximum power transfer from one part of an electronic circuit to another. Consider a circuit such as Fig. 5, where a power source is represented by an ideal generator in series with output impedance Z_s of the power source delivering power to load Z_L.

If Z_s and Z_L are pure resistances, maximum power will be transferred when $Z_L(R_L) = Z_s(R_s)$.

If Z_s and Z_L contain both resistance and reactance and if the reactance of Z_L or Z_s is adjustable, maximum power will be transferred to Z_L when $Z_s = r \pm jX$ and $Z_L = r \mp jX$ (i.e., Z_L is the conjugate of Z_s).

If only the magnitude and not the phase angle of the impedances is adjustable, then maximum power transfer will take place when the absolute values of Z_L and Z_s are equal.

Nonsinusoidal Waves. According to Fourier, a continuous, periodic, single-valued, finite function can be expressed as the sum of an average component and a series of sine waves of various amplitudes, frequencies, and phases. The frequencies of the terms of the series will be exact multiples of the lowest or fundamental frequency. The higher-frequency components are called harmonics and are multiples of the fundamental; thus the first harmonic is twice the fundamental frequency.

A wave having "half-wave" symmetry (any ordinate will be followed 180° along the time axis by an ordinate of equal magnitude but opposite sign) is commonly called a symmetrical wave. If the d-c value of a wave is equal to zero, the addition of odd harmonics to the fundamental always results in a symmetrical wave regardless of the phase of the components. Even harmonics cannot be present in a symmetrical wave. The effective or rms value of any wave subject to Fourier analysis is

$$E = \sqrt{B_0{}^2 + E_1{}^2 + E_2{}^2 + \cdots + E_n{}^2} \tag{33}$$

where B_0 = average component
$E \ldots E_n$ = effective values of sinusoidal components

Each individual harmonic contributes its part of the total power produced by the wave independent of the existence of the other harmonics. The true power of such waves is automatically indicated by ordinary wattmeters.

The form factor for a symmetrical wave is the ratio of its effective value to its average value over one-half period. For a sine wave the form factor is 1.11; in general, it is greater for a peaked wave and smaller for a flat-topped wave, but never less than 1.

The equivalent sine wave of a nonsinusoidal wave is a sine wave of the fundamental frequency having an effective value equal to the effective value of the nonsinusoidal wave.

Pulses

A transient or a pulse is a momentary voltage and current excursion in a circuit due to a sudden change in either the circuit impedance or the impressed voltage. For example, the momentary closure of a telegraph key that is in series with a battery and a resistance will produce a rectangular voltage pulse across the resistor.

The evaluation of pulse circuits is most conveniently accomplished in quite a different manner from conventional a-c circuits. Although all pulses contain a-c components, it is much more to the point to consider rise time, duration, and duty cycle rather than frequency or bandwidth.[2]

Pulse Characteristics

1. Rise time, the time the leading edge takes to go from 10 to 90 per cent of full amplitude.

2. Sag, the amount of droop in the top of the pulse. Excessive sag is usually caused by coupling time constants that are too short.

3. Fall or decay time, the elapsed time between the 90 and the 10 per cent points of the trailing edge.

4. Amplitude of a pulse refers to its peak value unless otherwise specified.

5. Overshoot, a spike on the leading or the trailing edge of a pulse. A sharp spike on the leading edge is usually caused by inductance in the circuit. In the case of a very fast rising pulse, the inductance of long leads may be sufficient to cause overshoot. This can be overcome by using shorter leads or coaxial cables for the signal leads. Sag in a rectangular pulse will cause overshoot on the trailing edge. Longer coupling time constants will reduce the effect.

6. Ringing, a damped oscillation following a step change in voltage or current. This is a special case of overshoot and is caused by the presence of both inductance and capacitance in a circuit. Both the inductance and capacitance may be due to long signal leads.

7. Duration, for a rectangular pulse, the elapsed time between the 50 per cent points of the leading and the trailing edges. For other pulse shapes the definition is arbitrary, but the 10 per cent points are often used.

8. Base line, the normal value of circuit voltage between pulses. Base-line shift is the change of base-line voltage following a pulse. This change will decay exponentially back to the normal value. However, any pulse occurring while the base line is shifted will be shifted also.

9. Duty cycle, the percentage of time that the pulse is "on." Duty cycle is important in capacity coupled circuits since pulses with d-c components tend to charge up circuit capacity at high repetition rates and cause base-line shift. This shift may ultimately block the circuit. In this respect, bandwidth ratings for pulse circuits are often misleading since they give only an indication of rise-time capability and not maximum repetition rate.

Duty cycle is also an important factor in calculating power dissipation in pulse circuit components.

10. Pulse pair resolution, the minimum time interval between two pulses which will allow the second pulse to be undistorted. This is important in the consideration of random pulse circuits.

Circuit Design.

In designing a circuit to have as fast a rise time as possible, it should be borne in mind that, to obtain a given rise time, there must be driving current available to charge any shunt capacity fast enough, as well as current to the load. An approximation of the peak current necessary to charge the shunt capacity can be calculated by

$$I = \frac{VC}{T} \times 1,000 \tag{34}$$

where I = milliamperes of charging current
V = volts, pulse amplitude
C = micromicrofarads of shunt capacity
T = microseconds, pulse rise time

Magnetic Circuits

Induced Voltage and Inductance. Faraday's law states, "In any coil of conductors where the magnetic flux linking the conductors is changing, an emf is induced in the coil which is proportional to the rate of change of the flux linkages." The emf of self-induction is

$$e_i = -L \frac{di}{dt} \tag{35}$$

where e_i = induced emf, volts
L = inductance, henrys
$\dfrac{di}{dt}$ = rate of change of current, amp/sec

If a change of current of 1 amp/sec in a circuit induces an emf of 1 volt, the inductance is 1 henry.

If the magnetic flux produced by current flowing in one coil links a second coil, the voltage induced in the second coil by a change in current in the first coil is

$$e_{21} = -M_{21} \frac{di_1}{dt} \tag{36}$$

where e_{21} = voltage induced in coil 2 by change in current in coil 1
M_{21} = mutual inductance of coil 2 with respect to coil 1
$\dfrac{di}{dt}$ = rate of change of current in coil 1

Likewise

$$e_{12} = -M_{12} \frac{di_2}{dt} \tag{37}$$

and if the magnetic path has constant permeability, the common mutual inductance of the two coils

$$M = M_{12} = M_{21} \tag{38}$$

Formulas for the calculation of self-inductance and mutual inductance for several conditions are included in Sec. 2-74 to 2-95 of Ref. 1.

Magnetic Quantities and Units. A summary of some commonly used magnetic quantities and units is given in Table 3.

Table 3. Basic Magnetic Quantities and Units

Quantity	Cgs unit	Mks unit	Engineering unit
Magnetomotive force $\mathfrak{F}$.........	Gilbert	Pra-gilbert (0.1 gilbert)	Ampere-turn (gilberts/0.4π)
Field intensity H..............	Oersted	Pra-oersted (10^{-3} oersted)	Ampere-turns/in.
Flux Φ......................	Maxwell	Weber (10^8 maxwells)	Line (1 maxwell)
Flux density β................	Gauss (1 maxwell/sq cm)	Weber/m²	Lines/sq in.
Permeability μ of vacuum.......	1	10^{-7}	

Magnetomotive Force $\mathfrak{F}$. The total influence causing a magnetic field. This force is analogous to the electromotive force in an electric circuit. When an electric current of I amp flow through N turns, the magnetomotive force (mmf) is

$$\mathfrak{F} = 0.4\pi NI \quad \text{gilberts} \tag{39}$$

Field Intensity H.   The magnetomotive force per unit length of magnetic path.

Flux Φ.   Analogous to current in an electric circuit.

$$\Phi = \frac{\mathfrak{F}}{R} = PHI = \mu HA \tag{40}$$

where R = reluctance of the magnetic circuit, equal to $v\,1/A$

v = reluctivity of the material of the magnetic path

I and A = length and cross-sectional area of the path

P = permeance of the magnetic circuit, equal to $\mu\,A/1$

μ = permeability of the material of the magnetic path

Flux Density β.   Flux per unit area.

$$\beta = \frac{\Phi}{A} = \mu H \tag{41}$$

where A = cross-sectional area of the magnetic path normal to the flux

Table 4. Ferromagnetic Materials

	μ	β		μ	β
Cobalt..........	170	3,000	Perminvar.......	2,000	4
Nickel..........	1,000	3,000	Steel...........	1,500	7,000
Permalloy.......	100,000	40,000	Iron...........	7,000	7,000

Magnetization Curves.   Unlike nonmagnetic materials where permeability is unity regardless of flux density, the permeability of magnetic materials is considerably greater than unity and depends upon flux density.   Since the relationship between

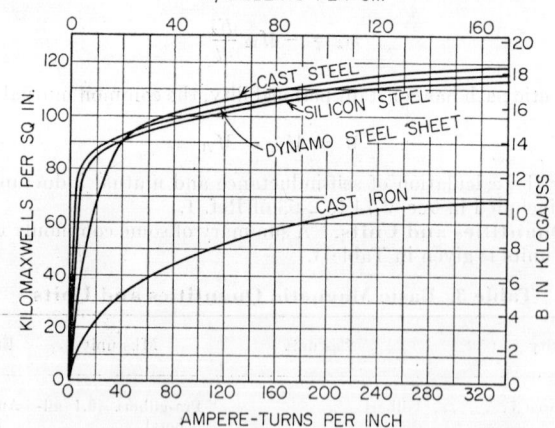

FIG. 6. Typical magnetization curves.

permeability and flux density is not readily expressed by a mathematical equation, graphs and solutions by successive approximations must be used.   The manufacturers of core materials supply permeability and/or magnetization curves on request.   These curves are normally expressed as permeability vs. flux density or as flux density vs. field intensity.   A typical magnetization curve for several materials is shown in Fig. 6.

Saturation.   The shape of the curve in Fig. 6 indicates that for large values of H (amp-turns/in.) the value of $\Delta\beta/\Delta H$ is much lower than at small values of H.   This saturation of the core material results in distortion of the flux waveform.

Core Losses (Iron Losses). Core losses are made up of eddy-current loss and hysteresis loss. The former results from eddy currents induced in the core material by a varying flux, and the latter results from the tendency of magnetic materials to retain a certain amount of residual flux after the magnetizing force has been removed. These losses are given by Eqs. (42) and (43).

$$W_e = e(F_f f \beta_{max} t)^2 \tag{42}$$

where W_e = eddy-current loss, ergs/cm^3/sec
$\quad e$ = eddy-current coefficient
$\quad F_f$ = form factor of the flux wave
$\quad f$ = frequency, cps
$\quad \beta_{max}$ = maximum flux density, gauss
$\quad t$ = thickness of sheets or laminations, cm

$$W_h = Nf \beta^k_{max} \tag{43}$$

where W_h = hysteresis loss, ergs/cm^3/sec
$\quad N$ = hysteresis coefficient (0.0001 to 0.01 for various materials)
$\quad f$ = frequency, cps
$\quad \beta_{max}$ = maximum flux density, gauss
$\quad k$ = empirical exponent ~ 1.6

A typical hysteresis loop is shown in Fig. 7.

Leakage Flux. This is that portion of the total flux established by an mmf which does not follow the desired path in the magnetic circuit and thus fails to contribute to the action of the flux which follows the desired path.

Energy Stored in a Magnetic Field. The energy stored in a magnetic field is

$$\text{Energy} = \tfrac{1}{2}LI^2 \tag{44}$$

where energy = joules (or watt-seconds)
$\quad L$ = henrys
$\quad I$ = amperes

Skin Effect. The magnetic field produced by alternating current flowing in a conductor tends to force the current to flow on the surface of the conductor rather than to be uniformly distributed over its cross sectional area.

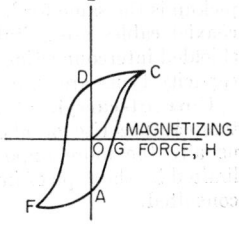

FIG. 7. Hysteresis loop.

This is known as skin effect, and it results in a higher resistance in a particular conductor for alternating current than for direct current. The current distribution is further influenced by the cross-sectional shape of the conductor and by any parallel current-carrying conductor.[1,3] High-powered r-f circuits often use tubing as a conductor since its a-c and d-c resistance differ less than for a solid conductor. The inductance of a straight wire is an important factor in high-frequency a-c and pulse circuits. It can be responsible for such unwelcome effects as spurious oscillations, ringing, overshoot, or cross coupling to adjacent wires. As a result, it is always good practice to keep signal leads short and route them away from other leads.

CIRCUIT COMPONENTS

Only the most common components used in electronic circuits are considered here. No attempt has been made to identify specific components because of the wide variety available and the high rate at which new types appear.* However, some of the factors to be considered in component selection are as follows:

Conductors

Most instruments are wired internally with plastic-insulated stranded copper wire 20 to 24 gage for currents in the milliampere range, and heavier for larger currents.

* Listing of many components will be found in Ref. 8.

Stranded wire is used because it is less likely to fail than solid wire as a result of repeated bending or vibration. Solid wire is particularly susceptible to failure if nicked in stripping insulation. However, solid wire is excellent for short jumpers or heavy busses where wires must be rigid mechanically.

Insulation. The most popular plastic insulations are cheap and easy to strip. However, such insulation is easily damaged by a soldering iron or an unusually hot component. An entire cable of such wires will sometimes be fused together by the heat from a short circuit. Also, plastic insulation will cold flow if a sharp point or edge is pressed against the wire and may result in an eventual short circuit. Teflon insulation overcomes all the above-mentioned shortcomings, tolerating even the recklessly used soldering iron. However, Teflon is expensive both to buy and to install, as it is hard to form and difficult to strip.

Between the two extremes of plastic and Teflon are many other insulations such as rubber and cotton, all of which have their places depending on circuit environment. Considerations may include puncture voltage, insulation resistance, dielectric constant, presence of moisture, operating temperature, abrasion, and vibration.

Flexible Coaxial Cables. These cables are designed for carrying high-frequency signals or pulses. They consist of a center conductor (solid or stranded), an insulating sheath (most commonly polyethylene), a metal braid encasing the insulating sheath, and a plastic outer jacket. Coaxial cables are characterized by relatively high voltage ratings and low attenuation of high-frequency components, even over long lines. Voltage ratings are usually for rms radio frequencies. When coaxial cables are used for direct current, much higher voltages are often permissible before the corona point is reached. Other factors being equal, the cable with the thickest inner insulation and the largest-diameter center conductor will have the highest corona point. Inductive pickup is the same for both inner conductor and the shield and has little net effect; so coaxial cables are suitable for carrying low-level signals. They are convenient as shielded interconnecting cables and can be terminated to eliminate reflections. Low-capacity types are available.

Current-carrying Capacity. Current-carrying capacity is set by either allowable temperature rise or allowable voltage drop. In instrument work voltage drop is usually the more important consideration. Allowable temperature rise is usually limited by the type of insulation used, and the manufacturer's specifications should be consulted.

Resistors

Fixed Resistors. Fixed resistors can be divided into three general categories: composition, deposited film, and wirewound. Composition resistors find the most widespread usage as they are inexpensive, small in size, reliable, and have good frequency characteristics. However, they are temperature-sensitive to various degrees (depending on the manufacturer), voltage-sensitive, and relatively inaccurate. The long-term stability may be affected by humidity and operating temperature. In very sensitive circuits composition resistors may contribute objectional noise.

Deposited-film resistors are constantly finding more widespread usage. Various types are available which, compared with composition resistors, are more accurate initially and have lower temperature coefficients, higher voltage ratings, and better combinations of accuracy and frequency. Cost of deposited-film resistors is becoming more competitive with composition resistors, but reliability is the major drawback so far.

Wirewound resistors are often used where the power dissipation is high or where extreme accuracy or stability is necessary. Power types usually have the wire sealed in ceramic and will tolerate high temperatures without damage. Contrary to composition and deposited-film types (which should not dissipate more than about half their rated wattage), wirewound resistors can safely dissipate full ratings. However, care should be taken that adjacent components are not damaged by the heat. Precision wirewound resistors are best for initial accuracy, long-term stability, temperature coefficient, and reliability. However, they are expensive in higher values and have both high inductance and high capacity. Special winding helps to reduce inductance and capacity but does not make them suitable for radio frequencies.

Adjustable Resistors and Potentiometers. Considerable confusion exists in the terms "adjustable resistor," "rheostat," and "potentiometer." A potentiometer can mean either a voltage-measuring device or a resistor with an adjustable center tap. In this instance it refers to the resistor. Likewise, adjustable resistors are often confused with rheostats and potentiometers. An adjustable resistor uses only two terminals while a rheostat or potentiometer uses three.

As the name implies, an adjustable resistor is one whose resistance can be changed by either a contact or clamp which selects a portion of its total length.

Rheostats and potentiometers both have sliding contacts that can be moved to any point between the end terminals and tap off a fraction of whatever voltage may be across the end terminals. A rheostat usually is considered to be a wirewound high-current device; while a potentiometer (which may be either composition or wirewound) is a low-current device.

Capacitors

Definitions and Basic Relationships. A capacitor consists of two conductors separated by an insulator. Capacitance is the ratio of charge to the potential difference between the conductors ($C = Q/V$). For a charge of 1 coulomb (the quantity of charge which in 1 sec crosses a section of a conductor in which there is a constant current of 1 ampere) and a potential of 1 volt, the capacitance is 1 farad. This is a virtually unattainable value, and measurements of capacitance are therefore usually on the order of *microfarads*.

The dielectric constant of a material is the ratio of the number of lines of electric force in the material to the number in air. Owing to the effect of temperature and frequency on the dielectric, capacitance varies with temperature and frequency.

Capacitor Rating. Capacitors usually are rated according to capacitance and operating voltage. In selecting a capacitor, some factors to be considered are working voltage, operating temperature, frequency, power factor, humidity, and soakage. Soakage may be thought of as the hysteresis of the capacitor.

Capacitor Types. Capacitor types vary widely in both physical makeup and electrical characteristics. Some of the more common capacitors are the following:

Ceramic. Ceramic capacitors generally consist of two conducting plates with a ceramic dielectric and a ceramic outer coating. They are commonly available in sizes from 10 $\mu\mu$f to 0.1 μf with 500-volt ratings, and a smaller size range is available with ratings to 30 kv. These capacitors find common usage in coupling and bypass applications where tolerance is not critical. They are excellent at high frequencies and very compact. However, with the exception of a few special types, they are very temperature-sensitive and vary widely in initial value. Capacity ratings are often given as guaranteed minimum values, and the actual capacity may be almost double this. In general, ceramic capacitors should not be used in applications where capacity tolerances are critical unless the capacitor is one of the special temperature-compensated types. Several different types of ceramic trimmers are available with various degrees of temperature compensation.

Mica. Mica capacitors use a mica dielectric and commonly range in values from 1 to 1,000 $\mu\mu$f with voltage ratings from 500 to 100,000 volts direct current. They are available with very close tolerances and have excellent frequency characteristics. However, they tend to be both bulky and expensive in the higher voltage and capacity types.

Paper. Paper capacitors comprise several layers of mineral-oil-impregnated paper separated by a thin sheet of aluminum foil (about 0.025 in.). Ratings vary from 0.001 to 50 μf for d-c voltages up to 100,000 and a-c voltages at 60 cps up to about 15,000. Paper capacitors are popular in the smaller sizes since they are relatively inexpensive and not so temperature-sensitive as ceramic capacitors. They have a certain amount of inductance that limits their usefulness to the lower radio frequencies.

Air and Gas. Usually, these have values less than 0.01 μf, temperature sensitivity of about 0.002 per cent per °F, and voltage ratings up to 50 kv. Most variable capacitors are of the air-dielectric type, ranging in capacitance from 1 $\mu\mu$f to 0.001 μf. Oil is often used as a dielectric for variable capacitors, providing up to five times the capacitance of similar air types and increasing the voltage rating by as much as ten times.

Electrolytic. Electrolytic capacitors utilize a metal-foil anode (usually aluminum or tantalum) and an electrolyte cathode separated by the anode oxide, which acts as a thin dielectric. This type of capacitor is suitable for use only in d-c circuits. For a-c service, two anode foils are used with an electrolyte common to both. Ratings cover the range up to several thousand microfarads at voltages of not more than about 700 direct current. Electrolytic capacitors provide more capacity for less price and bulk than any other type. However, most electrolytic capacitors have a considerable amount of leakage and tend to lose capacity as they age. A number of miniature types are available for use in printed circuits, and special stable types (usually tantalum) are suitable for moderately accurate timing circuits.

Vacuum. These are limited to values below 0.001 μf with voltage ratings up to 50 kv. Leakage currents are low and life is infinite unless a leak develops in the envelope.

Special Types. Metallized paper and metallized Mylar capacitors have relatively large capacity values for their physical size. However, they not infrequently develop internal shorts which can be burned out if enough current is available in the circuit. Mylar capacitors are becoming steadily more popular because they offer better temperature and stability characteristics and are smaller than equivalent paper capacitors.

Polystyrene capacitors have very high insulation resistance and low soakage. They are often used in circuits where high stability and low loss are necessary.

Inductors

A reactor, choke, or simply a coil consists of one or more conducting coils wound on either magnetic or nonmagnetic cores. Reactors store energy in electromagnetic form during periods of increasing current and deliver this energy to the circuit during periods of decreasing current. They prevent sudden fluctuations in current by extracting energy during sudden increases in current and delivering energy during sudden decreases. The two principal types of reactors are (1) iron-core reactors and (2) air-core reactors.

The self-inductance of a reactor is

$$L = \frac{4\pi N^2 A \mu}{l} 10^{-9} \tag{45}$$

where L = inductance, henrys
N = number of turns
A = effective area of core, in.^{2}
l = effective length of path in core, in.
μ = effective permeability

Since μ is not constant with current in iron-core reactors, the effective permeability depends upon the operating characteristics of the reactor.

The basic formula relating the voltage across a coil with the flux density in its core is (also see Ref. 1)

$$E = 4 \times 10^{-8} F_f N f A \beta_{max} \tag{46}$$

where E = voltage across coil, volts
F_f = form factor of voltage wave = 1.11 for sine wave
N = number of turns in coil
f = frequency, cps
A = effective cross-sectional area of core, sq in.
β_{max} = maximum flux density in core, lines/sq in.

Iron-core Reactors. *Power-supply Filter Chokes.* Filter chokes are specified on the basis of inductance, allowable direct current, and d-c resistance. These reactors are available in sizes ranging from fractions of 1 henry to more than 50 henrys and, on special order, can be obtained in any reasonable size. The current ratings and d-c resistance values of standard filter chokes vary over wide ranges.

In power-supply filter sections, the actual inductance value is not usually important so long as it exceeds a certain value. The allowable current is limited chiefly by saturation of the magnetic core and by the IR drop in the conductor.

The d-c resistance of the choke appears across the filter in series with the load and, unless the power supply is regulated between the reactor and the load, the variation in voltage drop across the choke with changes in load may be objectionable.

Swinging Chokes. When the load current varies over a wide range, better regulation can be obtained with a filter reactor whose inductance is high when the load current is low, but lower at higher currents. Such a reactor is called a swinging choke. It uses the decrease in permeability of magnetic-core materials at high flux densities. Typically, swinging chokes have an inductance decrease of 5:1 for a load-current change from minimum to maximum.

Air-core Reactors. Because air has a much lower permeability than iron, the flux and thus the inductance of an air-core reactor is considerably less than the flux and inductance of an iron-core reactor of the same number of turns and the same core size. The air-core reactor is not subject to variations in permeability with flux density; so distortion from magnetic saturation is not present.

Air-core reactors find their greatest application in tuned circuits, lumped constant delay lines, peaking circuits, and r-f filters. In all cases their reactance to high frequencies combined with relatively small resistance to direct current allows signals of high frequency to be developed across them without causing an appreciable voltage drop or power loss from the direct current.

Air-core reactors are normally specified on the basis of their inductance, allowable direct current, and ratio of inductive reactance to resistance at a specified frequency Q.

Since the inductance of a reactor is a function of the permeability of the core material, it is possible to increase the inductance by inserting a core (or slug) of high-permeability material into the winding. By varying the projection of the slug into the winding, the inductance of the reactor can be varied.

Shielding. Since the magnetic flux cannot be completely confined to the desired path, it frequently is necessary to shield reactors to prevent stray flux from linking conductors external to the reactor.

Certain types of windings minimize the amount of shielding required to protect external conductors. Frequently the metal cans of reactors do not present sufficient shielding; so that magnetic shields of Mu-metal or other high-permeability metals are required. The disturbances from stray magnetic fields can be reduced sometimes by proper physical arrangement of components and conductors.

Transformers

The basic formula relating the voltage across a coil to the flux density in its core is

$$E = 4 \times 10^{-8} f N F_f A \beta_{max} \tag{47}$$

where E = volts
f = frequency, cps
N = number of turns in coil
F_f = form factor of voltage wave (equals 1.11 per sine wave)
A = effective cross-sectional area of core, sq in.
β_{max} = maximum flux density in core, lines/sq in.

The Ideal Transformer. If n is equal to the number of turns in the primary winding divided by the number of turns in the secondary winding, and assuming no losses, the transformer output voltage is equal in magnitude to the primary voltage divided by n.

The current in the secondary is equal in magnitude to the primary current times n. The impedance of the transformer load appears across the primary terminals as $n^2 \times Z_L$ where Z_L is the load impedance.

Deviations of real transformers from ideal transformers are caused by:
1. Copper losses in both primary and secondary windings.
2. Losses in the core material.
3. Leakage fluxes.
4. Reluctance of the magnetic circuit.
5. Capacitance between turns or windings.

Coefficient of Coupling. The ratio of the actual mutual inductance between two coils to maximum mutual inductance is the coefficient of coupling between the coils and is designated by the letter k. Maximum mutual inductance is obtained where the flux set up by one coil cuts all the turns of the other coil. Coils approaching the maximum possible inductance are said to be closely or tightly coupled coils, as opposed to loosely coupled coils in which the actual mutual inductance is considerably less than the maximum value.

Power and similar transformers employing high-permeability cores generally approach unity coupling. The coupling coefficient for air-core coils usually is less than 0.6. The coupling coefficient depends on the physical spacing of the coils and their mutual orientation.

Numerically, the coefficient of coupling is

$$k = \frac{M}{L_1 L_2} \tag{48}$$

where L_1 and L_2 = self-inductance of the two coils
M = mutual inductance of the two coils

Distortion. Distortion of a sinusoidal waveform results from variation in permeability of the core material with flux density and also from hysteresis of the core material. This type of distortion is not great when the core material is operated well below saturation. Presence of direct current in the windings tends to saturate the core material.

When the signal contains harmonics, the fundamental and the various harmonics can be passed through the transformer with different phase shifts and with frequency-dependent distortion.

Since a transformer presents different reactances to signals of different frequency, the voltage appearing across the load for constant signal amplitude is a function of signal frequency. While this effect is unimportant in power transformers where the applied voltage rarely contains appreciable harmonics, it is extremely important in transformers which are required to pass a band of frequencies.

Power Transformers. Power transformers are normally wound on laminated iron cores and have one or several secondary windings. Power transformers are normally specified by frequency, primary voltage, secondary voltages, and secondary currents. Large power transformers are usually specified by volt-ampere output rather than by secondary current.

Other factors which influence the selection of power transformers include direct current in any of the windings (i.e., secondary windings of transformers used with half-wave rectifiers), special insulation requirements for secondary windings, amount and type of shielding required, and of course, weight and size of the transformer.

Most transformers designed for use at one frequency can be operated successfully at frequencies considerably above the design frequency, and also below the design frequency at reduced power. The use of power transformers at frequencies above design frequency, however, is wasteful of space and weight. Transformer operation at frequencies much below design frequency results in overheating the transformer from excessive copper loss, and the maximum frequency of operation may be limited by core losses.

Most transformers operate satisfactorily at voltages ranging from considerably below to slightly above the design voltage. Any transformer winding can be used as the primary winding if it has enough inductance to oppose the applied voltage without requiring excessive current, and if the remaining windings are sufficiently well insulated to withstand the voltages induced in them.

Transformer operation at less than rated secondary current results in no damage to the transformer, although the efficiency is less than that obtained at rated load. The efficiency of most power transformers ranges from 70 to 90 per cent.

Excessive direct current in any of the windings may result in overheating or in distortion due to saturation of the core material.

Either electrostatic or electromagnetic shielding, or both, can be incorporated into transformers. Electrostatic shields of copper foil or similar conducting materials are inserted between or around windings to prevent capacitive coupling between windings, or between windings and external fields. These shields are usually connected to ground. Electromagnetic shielding is sometimes necessary to prevent stray flux from linking conductors outside the transformer. The use of iron cases to house transformers results in slight electromagnetic shielding. Electromagnetic shields consisting of several alternate layers of Mu-metal or other high-permeability material and copper, completely enclosing the transformer, result in considerable decrease in stray flux. When this type of shielding is combined with astatic coil construction, problems arising from stray flux can be minimized.

Since the windings have resistance and reactance, the voltage appearing at the load decreases as the load current increases. The voltage regulation of a transformer is a measure of its ability to maintain its output voltage as the load changes. The regulation is defined as the change of terminal voltage when the current is reduced from full load to zero divided by the rated output voltage.

The power rating of a power transformer is usually limited by the allowable temperature rise of the winding insulation. Operation at increased temperature shortens the life of the transformer. Most manufacturers supply information on the allowable temperature rise of their transformers. Since most small transformers depend largely on free convection for cooling, the temperature rise at constant load increases with altitude.

Low-frequency Broadband Transformers. Low-frequency broadband transformers differ from power transformers principally because they must "pass" a band of frequencies rather than the single frequency or narrow band of frequencies required of a power transformer. Audio-frequency interstage coupling and output transformers are of this type.

Since the driving source for a broadband transformer is usually of considerably higher impedance than power sources, the primary impedance of a broadband transformer has an increased effect on the voltage output. It is desirable to keep the primary inductance high in order to develop a large signal across it. The low-frequency response of a broadband transformer is normally limited by the decrease in reactance of the primary at low frequency, although saturation of the core may be the limiting factor in some cases. For a constant transformation ratio, increasing the number of turns in the primary in order to increase the primary inductance requires a corresponding increase in the number of turns in the secondary winding. This increases the secondary and leakage inductance and may increase the distributed capacitance of the secondary. Since the secondary leakage inductance and shunt capacitance largely control the high-frequency response of these transformers, extension of either the high- or low-frequency response is accomplished at the expense of the other.

The shape of the frequency-response curve at the high end of the "passband" is influenced by resonance of the effective output series inductance and the secondary distributed capacitance. A typical response curve for a low-frequency broadband transformer is shown in Fig. 8.

Selection or design of a broadband transformer is frequently influenced by the requirement of passing direct current several times greater than that of the signal through the windings.

Intermediate-frequency Transformers. Untuned i-f transformers (not resonated at a particular frequency which it is desired to pass) are similar to low-frequency broadband transformers. Since these transformers operate at higher frequencies than low-frequency broadband transformers, the problems of secondary inductance and shunt capacitance are more pronounced. Untuned i-f transformers have either powdered-iron or air cores.

The primary, secondary, or both windings of tuned i-f transformers are resonated at a particular frequency to provide a larger output signal for a given input at that frequency. These transformers have either air or powdered-iron cores. Transformers can be tuned by adjusting a variable capacitor across the windings. Powdered-iron-core transformers can also be tuned by placing a fixed capacitor across the windings and varying the position of the powdered-iron core within the windings. Such tuning is called permeability tuning.

Air capacitors are normally recommended for good stability when variable-capacitance tuning is employed. High-stability fixed-mica condensers are suitable for tuned i-f transformers using permeability tuning.

High-frequency Transformers. Transformers operating at frequencies above 100 mc are usually air-core transformers. Distributed capacitance in the winding and between windings is of great importance. Temperature rise is important only with respect to copper losses, and considerable amounts of direct current may flow through the windings without damage, provided the copper losses do not become excessive. Skin effect may become of considerable importance, and at very high frequencies the inductance of the transformer leads may influence the circuit.

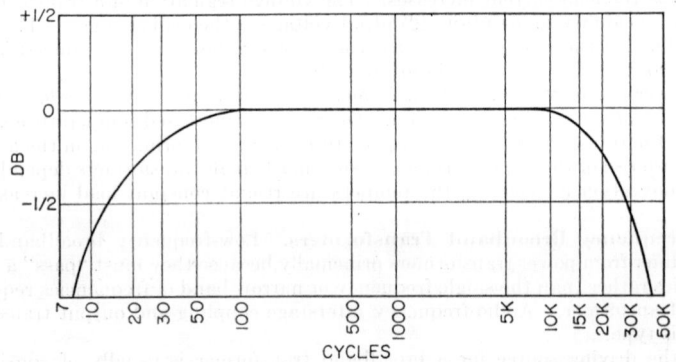

FIG. 8. Typical response curve for low-frequency broadband transformer.

High-frequency transformers can be tuned or untuned. Since magnetic or high-permeability cores are not usable because of their high core losses, capacitive tuning must be used. At very high frequencies the distributed capacity of the windings may be adjusted to give resonance at the desired frequency by varying the space between adjacent turns.

Electronic Tubes

Electronic tubes comprise two or more electrodes (usually called elements) enclosed in an envelope of glass or metal, which is evacuated or filled with an inert or conducting gas at low pressure.[3,4] Their operation utilizes the phenomenon of electron flow between a cathode emitter and a positively charged anode receiver. In some types of tubes (directly heated) the cathode is a filament which is heated by its own resistance when a current is passed through it. In other tubes (indirectly heated) a separate coil heats the cathode. This heating coil is made of metal with a high melting point, such as tungsten, since it is usually operated at temperatures approximating 1800°F. The number preceding the first letter of many tubes indicates the approximate filament voltage to be used. For example, a 6V6 tube requires a filament voltage of 6.3 volts.

Indirectly heated cathodes are made of metals such as tungsten and are coated with compounds such as thorium or oxides of barium and strontium to increase their ability to emit electrons. Filamentary (direct-heated) cathodes heat more quickly than indirectly heated cathodes but have only relatively small emitting surfaces.

Most thermionic-emission tubes are evacuated.   However, thyratrons are the exception.   They are gas-filled and contain a grid which will "turn on" the tube if it is
made positive.   However, once the tube is in conduction the grid has no further
control.

Several special types of tubes do not use heated cathodes and fall in the category of
"cold-cathode" types.   Phototubes utilize light to liberate electrons from the cathode.
Glow tubes are gas-filled and are actuated by the potential gradient between the
cathode and anode.   Mercury-pool tubes (ignitrons) release electrons by means of an
arc drawn between the surface of the mercury and the anode, which is started by a
special electrode called the ignitor.

The anode or plate of a tube attracts or repels (depending on its polarity with
respect to the cathode) the electrons that are boiled off the cathode.   The current that
flows from the cathode to the anode is frequently termed plate current.   When positive
voltage is applied to the anode, it attracts electrons; when negative, electrons are
repelled.   Anodes are made of tungsten, molybdenum, Nichrome, graphite, nickel, or
tantalum.

The grid is the controlling electrode of the tube and is located between the cathode
and anode in the path of the electrons.   Some tubes have several grids to permit
greater range of control over the plate or anode current.   In
a thyratron the grid controls the point in the cycle at which
current starts to flow.   In high-vacuum tubes the current flow
is usually proportional to the voltage applied to the grid.   The
power required in the grid is small; thus a relatively small
amount of power in the grid of an electronic tube can precisely
control a much larger current flow in the plate circuit.

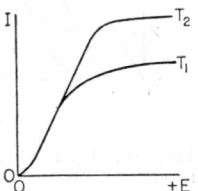

FIG. 9. Diode char-
acteristics for two
temperatures.
$T_2 > T_1$.

Vacuum Diodes.   A tube with an electron source and
collector is called a diode.   When the number of electrons
emitted by the cathode limits the current, the diode is said to
be operating in the emission-limited current region.   In this
region, the current-voltage relationship is shown in Fig. 9,
where T_2 is a higher cathode temperature than T_1.   In most
diodes, current limits are established by the space charge, which is created by the electron cloud surrounding the cathode.

Because current flows in only one direction through a diode, the diode is often used
as a rectifier.   Unidirectional output occurs for an a-c input, thereby providing
pulsating direct current from an a-c signal.

Triodes.   A tube with a third electrode or grid which controls the plate current is
called a triode.   This control grid is placed close to, and sometimes surrounding, the
cathode; it must be open in construction to permit free passage of electrons.   It is
frequently made of a spiral winding of thin wire.

Two sets of static characteristic curves describe the behavior of a triode.   Plate
current as a function of plate voltage at constant grid voltage is termed the plate
characteristic; and plate current as a function of grid voltage at constant plate voltage
is called the transfer characteristic or gain.   Figures 10 and 11 give these curves for a
6J5 triode.

Plate resistance or r_p is the ratio of a change in plate voltage e_b to the corresponding
change in plate current i_b, at constant grid voltage e_c ($r_p = \Delta e_b / \Delta i_b$ with e_c constant).

Transconductance or g_m is the ratio $\Delta i_b / \Delta e_c$ at constant e_b.   It has units of mhos
(reciprocal ohms) or more frequently micromhos and is a measure of how effectively
the grid controls the plate current.

Amplification factor, or μ, is the ratio $\Delta e_b / \Delta e_c$ at constant i_b.   It is the maximum
possible gain of the tube.   The foregoing three factors are related by

$$\mu = g_m r_p \tag{49}$$

A triode has two disadvantages: (1) high interelectrode capacitance and (2) low
amplification.   Capacitance between electrodes of a triode is a few $\mu\mu f$ and is called
interelectrode capacitance; it couples high-frequency a-c signals from plate to grid,

grid to cathode, and plate to cathode, and limits the use of triodes as amplifiers at high frequencies. Also, an increase in plate current in a triode causes a decrease in plate potential, which tends to counteract the grid control and to limit the amplification.

Tetrodes. Grid-plate capacitance of a tube and the effect of the plate on the grid can be reduced by adding a fourth element, called a screen grid, between the control grid and plate. The screen grid is held at a fixed potential determined by the tube

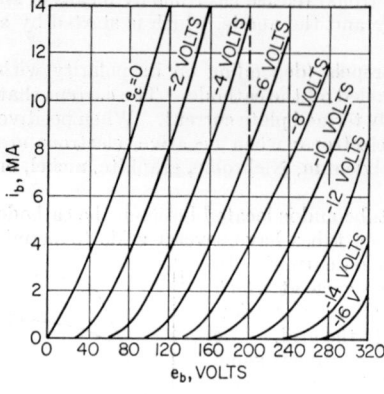

FIG. 10. Plate characteristics, type 6J5 triode.

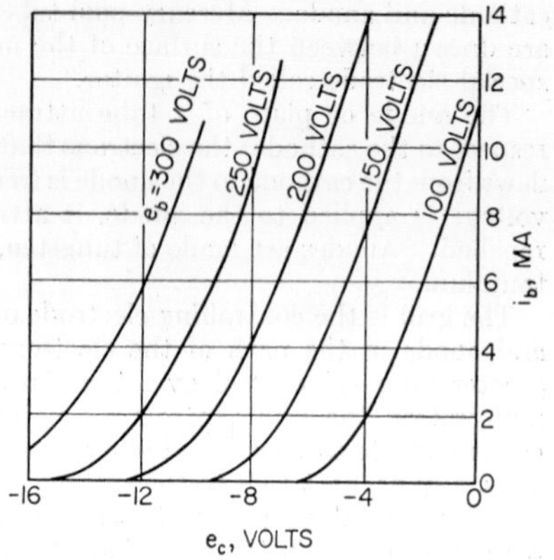

FIG. 11. Transfer characteristics, type 6J5 triode.

construction. The screen is a spiral winding of wire. A typical family of characteristic curves for a tetrode (four-element tube) is shown in Fig. 12.

In the region where the characteristic curves have a negative slope, the output signal is distorted and the tube can oscillate. The negative slope is the result of secondary emission from the plate. To overcome this effect, a fifth element can be added.

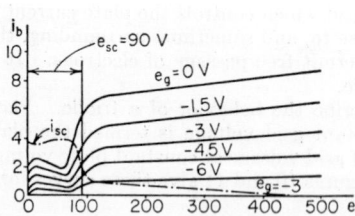

FIG. 12. Characteristic curves of a type 24A tetrode with screen grid at 90 volts.

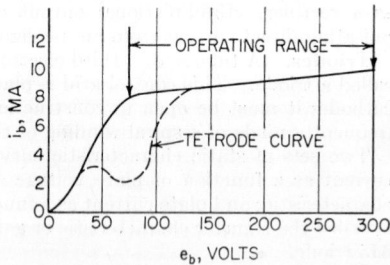

FIG. 13. One characteristic curve of a pentode, showing how the suppressor grid overcomes the negative resistance of a tetrode.

Pentodes. In a pentode, a fifth element, called the suppressor grid, is placed between the screen and the plate, and is normally at or near the cathode potential. The suppressor does not prevent secondary emission from the plate; it prevents the effects of secondary emission by driving the electrons back to the plate after they have been emitted. Figure 13 shows one pentode characteristic. The linear range of operation is greater than for a tetrode. The amplification of pentodes and tetrodes is higher than that of triodes because the opposing action of the plate is shielded from the grid.

Beam-power Tubes. A beam-power tube is a four- or five-element tube in which the cathode deflecting plates help form the electron stream into narrow, well-defined beams. The concentration of electrons between the screen and plate acts like a suppressor grid but intercepts no electrons between the screen and plate. This combination of beam formation, small screen current, and effective suppressor action gives a plate characteristic that is reasonably linear over a wide range of plate current. Beam tubes such as 6L6 and 6V6 are useful as power amplifiers.

Cathode-ray Tubes. In a cathode-ray tube, electrons from a cathode are made to move at high speed in a narrow beam and then strike a fluorescent screen or a

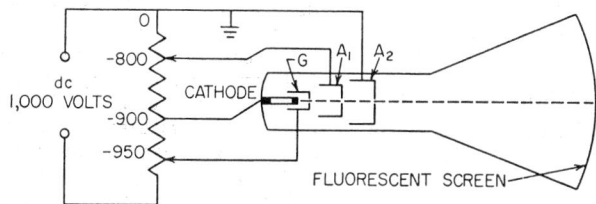

Fig. 14. Elements of a typical cathode-ray tube, showing the various d-c potentials applied to the electron gun. These potentials are measured with respect to the second anode, which usually is grounded.

photographic film. The cathode-ray tube presents electrical data visually by means of the lateral deflection of the beam by electric or magnetic fields. The small mass of the electron permits rapid beam deflection. The source that furnishes the electron beam is often called the electron gun. Figure 14 shows the components of a representative cathode-ray tube.

The potential to be observed deflects the beam after it passes the shaping and accelerating anodes, but before it reaches the screen. The deflection can be by electrostatic or electromagnetic means. Electrostatic tubes demand less energy from the signal source and respond more rapidly than electromagnetic tubes, which are shorter, more rugged, and can function with a low-voltage power supply.

An intensifying ring is sometimes used near the screen to provide additional acceleration of the electrons after they have left the deflection plates. A graphite coating is usually applied on the glass in the screen to prevent the accumulation of charge and to complete the circuit back to the power supply.

Electrometer Tubes. Special tubes which have unusually small grid current are called electrometer tubes, since they resemble the ideal electrostatic measuring device that requires no continuous signal current. The range of grid current is from 10^{-12} to 10^{-16} amp. The low grid currents are achieved by use of high vacuum; operation at relatively low voltage; grid supports of high-resistance material; and, usually, direct connection of the leads to minimize surface leakage.

Gas-filled Tubes. Gas-filled tubes are used as switches, rectifiers, voltage regulators, and to carry currents which are too large for similar-sized high-vacuum tubes. Measurement of the current in a gas-filled diode as a function of plate voltage results in Fig. 15. At point a at which the current increases rapidly, the tube is said to "fire."

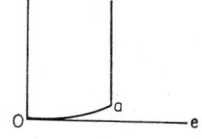

FIG. 15. Simplified volt-ampere curve of a thermionic gas diode.

Voltage Regulators. The firing potential of a cold-cathode diode is a function of the fill gas and its pressure. The voltage required to fire the tube is approximately 30 per cent above the operating voltage. After the tube has fired, the voltage drop across it is almost constant for a considerable range of current. This type of diode, known as a voltage regulator or VR tube, can maintain a nearly constant voltage (± 1 per cent) across a load in parallel with it in spite of fluctuation in the source voltage or load resistance (see Fig. 15).

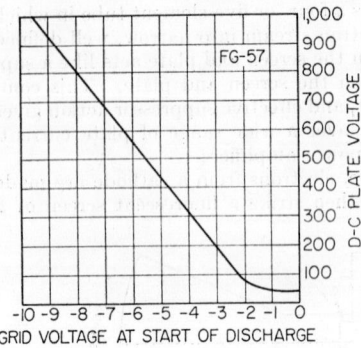

Fig. 16. Grid-control characteristic of a thyratron.

Thyratrons. A grid between the cathode and plate of a gas-filled tube can control the breakdown process and thus the start of the plate current. This type of tube is called a thyratron and is widely used as a switch for currents too large for ordinary vacuum tubes. Some thyratrons will handle several amperes of current, and a typical forward voltage drop is around 15 volts. Once the tube is conducting the grid has no further control, and only lowering the plate voltage will extinguish it. For this reason an a-c supply is often convenient if the load will operate on pulsating direct current.

Thyratrons can be used to operate d-c relays or motors directly from an a-c source, and thyratron control can be used to govern motor speed. Thyratrons also find use in some types of pulse circuits and as controlled rectifiers. A typical grid-control characteristic is shown in Fig. 16. The grid is usually biased well below the firing potential and is triggered by a fast-rising voltage pulse.

Semiconductors

Semiconductor devices such as diodes and transistors are being used in an ever-increasing number of applications such as rectifiers, amplifiers, shapers, and switches. They are advantageous in size, reliability, and power consumption as compared with vacuum tubes.[5,6]

Semiconductor materials lie between the metals and the insulators in their ability to conduct electricity. Germanium or silicon crystals may have impurities added which contribute either an excess or a deficiency of electrons in the crystal. If the impurity contributes an extra electron (arsenic, for example), it is called a donor. The extra electrons from arsenic atoms are easily freed and can move in the crystal. The donor atom stays fixed but assumes a positive charge. A semiconductor which containes donor atoms is called an "n-type" semiconductor.

If the impurity lacks valence electrons sufficient to form all the covalent bonds with the crystal (four bonds for germanium or silicon), missing electrons may be acquired from neighboring atoms. This action produces free "holes" in the crystal. Impurities (such as aluminum) which produce holes are called acceptor atoms, and a semiconductor which conducts by virtue of free holes is called a "p-type" semiconductor. Free holes or electrons are known as carriers.

Vibration of the atoms caused by heating will also cause an occasional valence electron to break away, thus forming a hole-electron pair. This is known as thermal generation.

Diffusion and Drift. Carriers can move through a semiconductor by two different mechanisms: (1) diffusion, or (2) drift. *Diffusion* occurs whenever there is a difference in the concentration of the carriers in any adjacent regions of the crystal. The carriers have a random motion owing to the temperature of the crystal; so that carriers will move in a random fashion from one region to another. However, more carriers will move from the region of higher concentration to the region of lower concentration than will move in the opposite direction. *Drift* of carriers occurs whenever there is a difference in voltage between one region of the semiconductor and another. The voltage difference produces a force on the carriers, causing the holes to move toward the more negative voltage and the electrons to move toward the more positive voltage.

The mechanism of drift is shown in Fig. 17 for both n-type and p-type semiconductors. For the n-type material, the electrons enter the semiconductor at the lower electrode, move upward through the semiconductor, and leave through the upper electrode, passing then through the wire to the positive terminal of the battery. Note that, in accordance with the principle of space-charge neutrality, the total number of electrons in the semiconductor is determined by the total number of acceptor atoms in the crystal.   For the case of the p-type semiconductor, hole-electron pairs are generated at the upper terminal.   The electrons flow through the wire to the positive terminal of the battery and the holes move downward through the semiconductor and recombine with electrons at the lower terminal.

Diodes.   If a p-type region and an n-type region are formed in the same crystal structure, we have a device known as a rectifier or diode.   The boundary between the two regions is called a *junction,* the terminal connected to the p region is called the *anode,* and the terminal connected to the n region is called the *cathode.*

A rectifier is shown in Fig. 18 for two conditions of applied voltage.   In Fig. 18*a,* the anode is at a negative voltage with respect to the cathode and the rectifier is said

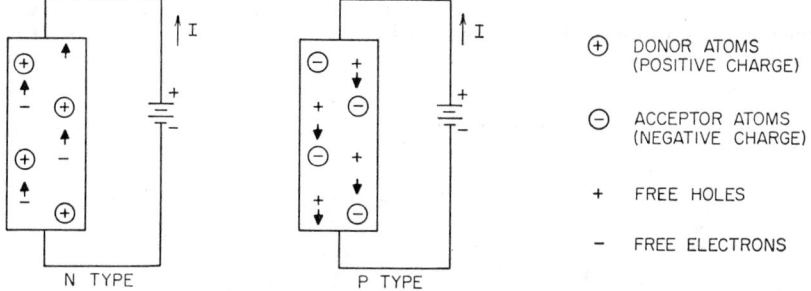

Fig. 17. Conduction in n-type and p-type semiconductors.

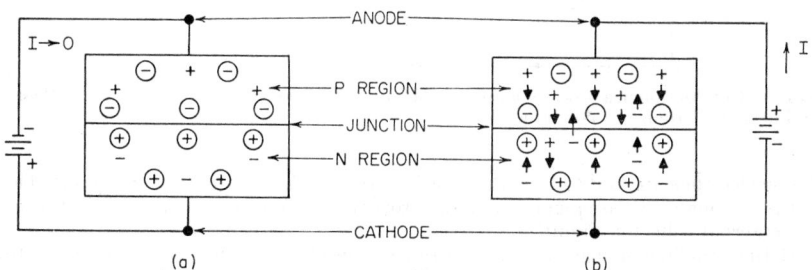

Fig. 18. Conduction in a p-n junction rectifier: (*a*) Reverse bias; (*b*) forward bias.

to be *reverse-biased.*   The holes in the p region are attracted toward the anode terminal (away from the junction), and the electrons in the n region are attracted toward the cathode terminal (away from the junction).   Consequently, no carriers can flow across the junction and no current will flow through the rectifier.

Actually, a small *leakage* current will flow because of the few hole-electron pairs which are thermally generated in the vicinity of the junction.   Note that there is a region near the junction where there are no carriers (*depletion layer*).   The charges of the donor and acceptor atoms in the depletion layer generate a voltage which is equal and opposite to the voltage which is applied between the anode and cathode terminals. As the applied voltage is increased, a point will be reached where the electrons crossing the junction (leakage current) can acquire enough energy to produce additional hole-electron pairs on collision with the semiconductor atoms (*avalanche multiplication*). The voltage at which this occurs is called the *avalanche voltage* or *breakdown voltage* of the junction.   If the voltage is increased above the breakdown voltage, large currents can flow through the junction, and unless limited by the external circuitry, this current can result in destruction of the rectifier.

In Fig. 18*b*, the anode of the rectifier is at a positive voltage with respect to the cathode and the rectifier is said to be *forward-biased*. In this case, the holes in the p region will flow across the junction and recombine with electrons in the n region. Similarly, the electrons in the n region will flow across the junction and recombine with the holes in the p region. The net result will be a large current flow through the rectifier for only a small applied voltage.

Transistors.   An n-p-n transistor is formed by a thin p region between two n regions. The center p region is called the *base* and in practical transistors is *generally* less than 0.001 in. wide. One junction is called the *emitter* junction and the other junction is called the *collector* junction. In most applications, the transistor is used in the common emitter configuration as shown in Fig. 19, where the current through the output or load R_L flows between the emitter and collector and the control of input signal V_{BE} is applied between the emitter and base. In the normal mode of operation, the collector junction is reverse-biased by the supply voltage V_{CC} and the emitter junction is forward-biased by the applied base voltage V_{BE}. As in the case of the rectifier, electrons flow across the forward-biased emitter junction into the base region.

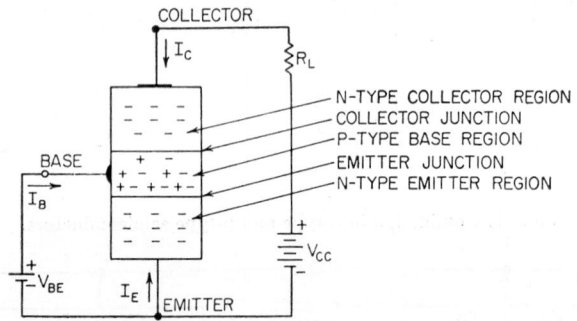

Fɪɢ. 19. Conduction in a n-p-n junction transistor (common emitter configuration).   (Donor and acceptor atoms are not shown.)

These electrons are said to be emitted or injected by the emitter into the base. They diffuse through the base region and flow across the collector junction and then through the external collector circuit.

If the principle of space-charge neutrality is used in the analysis of the transistor, it is evident that the collector current is controlled by means of the positive charge (hole concentration) in the base region. As the base voltage V_{BE} is increased the positive charge in the base region will be increased, which in turn will permit an equivalent increase in the number of electrons flowing between the emitter and collector across the base region. In an ideal transistor it would be necessary to allow base current to flow for only a short time to establish the desired positive charge. The base circuit could then be opened and the desired collector current would flow indefinitely. The collector current could be stopped by applying a negative voltage to the base and allowing the positive charge to flow out of the base region. In actual transistors, however, this cannot be done because of several basic limitations. Some of the holes in the base region will flow across the emitter junction and some will combine with the electrons in the base region. For this reason, it is necessary to supply a current to the base to make up for these losses. The ratio of the collector current to the base current is known as the current gain of the transistor $h_{FE} = I_C/I_B$. For a-c signals, the current gain is $\beta = h_{fe} = i_c/i_b$. The ratio of the a-c collector current to a-c emitter current is designated by $a = h_{fb} = i_c/i_e$.

When a transistor is used at higher frequencies, the fundamental limitation is the time it takes for carriers to diffuse across the base region from the emitter to the collector. Obviously, the time can be reduced by decreasing the width of the base region.

The frequency capabilities of the transistor are usually expressed in terms of the *alpha-cutoff frequency* f_{ab}. This is defined as the frequency at which α decreases to 0.707 of its low-frequency value. The alpha-cutoff frequency may be related to the base-charge characteristics and the base width by the equations

$$T_E = \frac{Q_B}{I_E} = \frac{W^2}{2D} = \frac{0.19}{f_{ab}} \tag{50}$$

where T_E = the emitter time constant
Q_B = base charge required for an emitter current I_E
W = base width
D = diffusion constant which depends on the semiconductor material in the base region

As evident from Fig. 19, the n-p-n transistor has some similarity to the vacuum-tube triode. Positive voltage is applied to the collector of the transistor which corresponds

Circuit configuration	Characteristics*
Common emitter (CE)	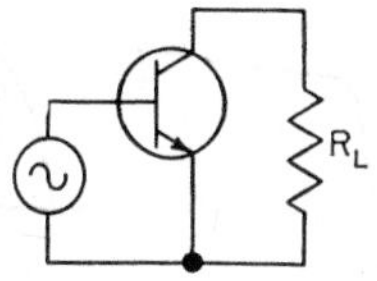 Moderate input impedance (1.3 K) Moderate output impedance (50 K) High current gain (35) High voltage gain (−270) Highest power gain (40 db)
Common base (CB)	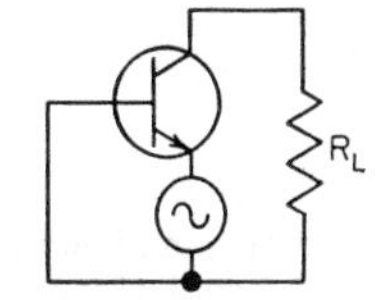 Lowest input impedance (35 ohms) Highest output impedance (1 M) Low current gain (−0.98) High voltage gain (380) Moderate power gain (26 db)
Common collector (CC) (emitter follower)	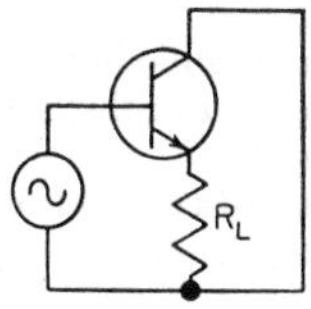 Highest input impedance (350 K) Lowest output impedance (500 ohms) High current gain (−36) Unity voltage gain (1.00) Lowest power gain (15 db)

* Numerical values are typical for the 2N525 at audio frequencies with a bias of 5 volts and 1 ma, a load resistance of 10K, and a source (generator) resistance of 1K.

Fig. 20. Transistor circuit configurations.

to the plate of the tube, electrons are "emitted" by the cathode and are "collected" by the plate of the tube, and the control signal is applied to the base of the transistor which corresponds to the grid of the tube. One important difference between transistors and tubes is that the input impedance of the transistor is generally much lower than that of a tube. It is for this reason that transistors usually are considered as current-controlled devices and tubes usually are considered as voltage-controlled devices. Another important difference between transistors and tubes is the existence of *complimentary* transistors. That is, a p-n-p transistor will have characteristics similar to an n-p-n transistor except that in normal operation the polarities of all the voltages and currents will be reversed. This permits many circuits which would not be possible with tubes (since no tube can operate with negative plate voltage). Examples of complementary circuits are frequently found.

The operation of the transistor has been described in terms of the common emitter configuration. The term grounded emitter is used frequently instead of common emitter, but both terms mean only that the emitter is common to both the input circuit and output circuit. It is possible and often advantageous to use transistors in the common-base or common-collector configuration. The different configurations are shown in Fig. 20 together with their comparative characteristics in class A amplifiers.

Tunnel Diodes.   A tunnel diode is a two-terminal device consisting of a single p-n junction. The essential difference between a tunnel diode and a conventional diode is due to the fact that the conductivity of the p and n material used in the fabrication of a tunnel diode is more than 1,000 times as high as the conductivity of the material used in the fabrication of conventional diodes. This higher conductivity is obtained by increasing the concentration of acceptor and donor impurities in the semiconductor material when it is formed.

Owing to the very high conductivity of the p and n materials used in tunnel diodes, the width of the junction (the depletion layer) is very small, of the order of 10^{-6} in. Because of the extremely narrow junction it is possible for electrons to tunnel through the junction even though they do not have enough energy to surmount the potential

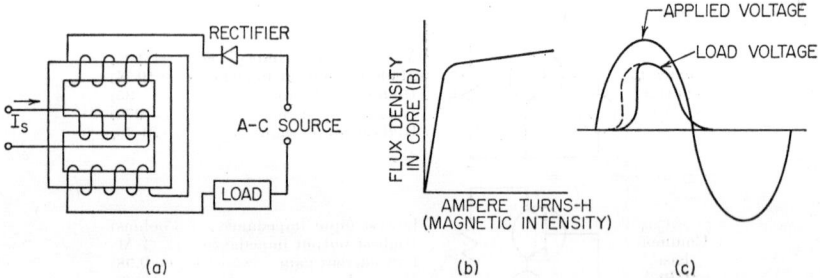

FIG. 21. Circuit and characteristic curves of a magnetic amplifier: (*a*) Half-wave magnetic amplifier; (*b*) *B-H* characteristics of core material; (*c*) control characteristics.

barrier of the junction. Although tunneling is impossible in terms of classical physics, it can be explained in terms of quantum mechanics. For this reason the mechanism is commonly called quantum-mechanical tunneling.

Amplifying action is possible because the forward-conduction characteristic has a negative slope for a narrow range of applied voltages. Thus, in this range, increasing the voltage across the diode actually decreases the current through it.

Tunnel diodes have excellent switching characteristics at very high frequencies and can, within narrow voltage limits, be used as linear amplifiers. Their maximum frequency limitation is due primarily to lead inductance and in a transistor package is in the order of 1,000 mc. Chief advantages of tunnel diodes are high speed, circuit simplicity, and relative indifference to radiation.

Magnetic Amplifiers

A magnetic amplifier consists of a saturable reactor with a control or bias winding and a load winding (see Fig. 21). The reactance of the load winding is determined by the extent to which the reactor core is saturated by current in the control winding.

Push-pull amplifiers are used in most applications in order to have a symmetrical output. Although magnetic amplifiers are bulky and have a very limited frequency range, they offer important advantages in circuit simplicity and indifference to temperature, radiation, and shock, and they are extremely reliable. Magnetic amplifiers find many applications in servo and control systems. For further information, see Ref. 7.

Dielectric Amplifiers

Dielectric amplifiers make use of the fact that, in some dielectric materials such as barium titrate, strontium titrate, and lead zirconate, the electrostatic-field strength produced is not a linear function of charge (applied voltage). A signal applied across a capacitor with a nonlinear dielectric can control the current flow to a load if an a-c power source, the capacitor, and the load are all placed in series. In practice, two nonlinear capacitors are usually used as two arms of a bridge circuit so that the a-c power source does not reflect back on the signal (see Fig. 22).

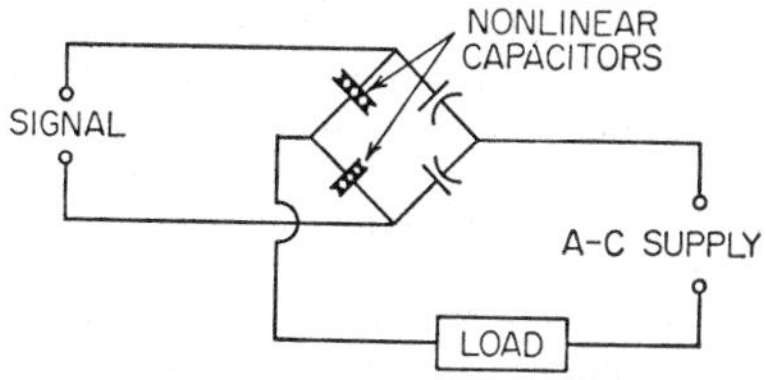

Dielectric amplifiers offer advantages in ruggedness, efficiency, and low cost.

FIG. 22. Simplified balanced dielectric amplifier circuit.

However, they have temperature sensitivity and frequency limitations.

D-C POWER SUPPLIES

Direct current for instrumentation electronics can be obtained from rotating d-c generating equipment, from batteries, or from rectification of alternating current. Where suitable, rectifier power supplies are usually smaller and cheaper than rotating equipment or batteries.

D-C Generators (Rotating)

D-C generators are available in sizes ranging from fractions of watts to numbers of kilowatts.

D-C generators may be shunt, series, or compound wound and self or externally excited. Generators can be obtained with or without output voltage regulators. Shunt or compound wound, self-excited generators with regulators are commonly used for d-c power required for instrumentation. When power from rotating d-c equipment is to operate sensitive electronic equipment, the generator output is usually filtered to eliminate commutator ripple and noise.

The life of the brushes of rotating equipment is considerably shortened by operation at high altitude.

Converters

When either a-c or d-c power, differing from the required voltage, is available, one of several types of converters may be used.

Dynamotors are sometimes employed where moderate amounts of power are required at voltages under approximately 1,500 volts. Essentially, a dynamotor is a d-c motor and d-c generator that utilizes a common field winding. Dynamotors can have separate primary and secondary windings, or a single tapped winding. In either case, the ratio of output to input voltage depends on the relative number of turns in the secondary and primary windings.

A rotary inverter which consists of a d-c motor driving an a-c generator can, with an associated rectifier, be used to supply the necessary d-c voltage and power from d-c power of other than the desired voltage. Regulation of frequency, as well as of voltage, is of importance in this type of system.

Electronic inverters are sometimes used in instrument applications. Either tubes or transistors are used to convert direct current into pulsating current which is stepped up or down by a transformer and then rectified and filtered. Usually a push-pull arrangement of the tubes or transistors (which are driven into saturation) works into a center-tapped transformer primary.   The driving signal is supplied from an external source.

Vibrators, which consist of interrupting contacts, can be used to convert direct current into unidirectional pulsating current or into square-wave alternating current. When used with appropriate transformers, rectifiers, and filters, satisfactory d-c outputs can be obtained.   Vibrators can be of either the synchronous or nonsynchronous types.   In addition to the normal interrupting contacts present in both types of vibrators, the synchronous type contains an additional set of contacts which are used to rectify the secondary output of a transformer whose primary is connected to the interrupting contacts.   Thus rectification is achieved without employing a separate rectifier.

The voltage waveform of the vibrator, which roughly approximates a square wave, may require a considerable amount of filtering if the power supply is to be used with sensitive electronic circuits.

Rectification and Rectifiers

Rectification is the process of converting alternating current or voltage into unidirectional pulsating current or voltage.   A number of devices possess the property of presenting a small resistance to the passage of current in one direction, and a greater resistance to the passage of current in the opposite direction.   Included among these devices are vacuum diodes, gas-filled diodes, and semiconductors.

The rectifiers of greatest importance in the instrumentation field may be divided into three classes:

1. High-vacuum diodes and some types of semiconductor rectifiers whose resistance in the conducting direction is approximately constant.

2. Gas-filled diodes and semiconductors (silicon, for example) which present a nearly constant voltage drop in the conducting direction.

3. Multielement gas-filled tubes and silicon-controlled rectifiers which do not start to conduct until a small voltage is applied to the control element.

With the continuing development of semiconductor diodes, these are rapidly replacing high-vacuum and gas-filled diodes because of their compactness, greater reliability, and lower resistance in the conducting direction.   Since no filament current is necessary, greater simplicity and compactness of design may be realized.

Rectifier Circuits.   In the selection and design of rectifier circuits the circuit considerations should include load voltage and current, peak inverse voltage, maximum surge current, and ripple.   In some high-frequency applications the capacity of the diode and its recovery time (restoration of reverse resistance after carrying a forward current) are important factors.   When semiconductors are used as rectifying elements, front-to-back resistance ratios and ambient temperatures may be important considerations.   The circuits described are considered to have sine-wave inputs and ideal rectifiers with no forward voltage drop.[3-5]

Half-wave Rectifier.   Half-wave rectifiers are commonly used in applications where the current demand is modest, such as high-voltage cathode-ray-tube power supplies and inexpensive transformerless supplies.   Large capacitor input filters are usually used.   A typical circuit and output waveform are shown in Fig. 23.   In designing a filter, only the lowest frequency present in the output need be considered, and for a half-wave rectifier this is the frequency of the applied alternating current.

If a capacitor input filter is used, at no load the input capacitor will charge up to the peak value of the applied voltage.   Then at the negative half of the cycle the inverse voltage across the rectifier will be twice the peak value.   Consequently the inverse-voltage rating of the rectifier should be at least twice the peak value of the applied voltage, or $2\sqrt{2}\,E_{rms}$.   Output voltage of the rectifier will depend on the amount of load current drawn and at maximum load will be only about 0.45 times the rms applied voltage.

Full-wave Rectifier.   A full-wave rectifier is essentially two half-wave rectifiers back to back.   It is a convenient circuit for vacuum diodes, since the cathodes of both diodes have a common connection (see Fig. 24) and the filaments can be supplied from a common winding. A center-tapped transformer is necessary, and maximum peak inverse voltage with no load will be twice the peak voltage from center tap to one end of the transformer secondary. Substantially less filtering is necessary than with a half-wave rectifier since the ripple frequency is twice that of the applied voltage. Maximum load current is twice that of a corresponding half-wave rectifier, and output

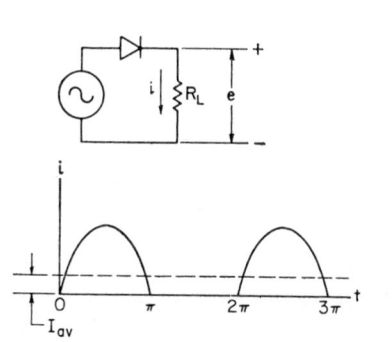

Fig. 23. Half-wave rectifier.

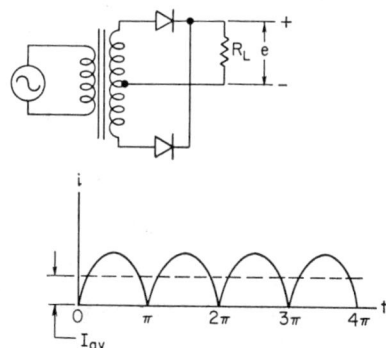

Fig. 24. Full-wave rectifier.

voltage at maximum load will be about 0.9 times the rms voltage from center tap to one end of the transformer secondary. Transformer losses are greatly reduced since load current is drawn on both halves of the cycle.

Full-wave Bridge Rectifier. The full-wave bridge rectifier is commonly used with selenium, copper oxide, or silicon diodes (see Fig. 25). In this arrangement, two rectifiers operate in series on each half of the cycle, one rectifier being in the lead to the load and the other in the return. Though the circuit contains four rectifiers, peak inverse voltage is only equal to the peak of the applied voltage. This is convenient for semiconductors, since they do not generally have as high inverse voltage ratings as thermionic diodes. The circuit is seldom used with thermionic diodes, since the cathodes are all at different potentials, and a separate filament winding is necessary for each diode. Output voltage is twice that of a standard full-wave rectifier, or about

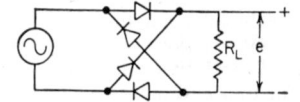

Fig. 25. Full-wave bridge rectifier.

0.9 times the rms end-to-end voltage of the transformer with maximum load current. Ripple frequency and transformer losses are the same as for a standard full-wave rectifier. Note that there is no common lead between the rectifier input and output.

Voltage Doublers. Circuits for half-wave and full-wave voltage doublers are shown in Fig. 26. These circuits are used where an output voltage greater than the input voltage is needed and the current demand is very moderate. Both circuits involve the principle of charging condensers in parallel from the input and adding them in series for the output, the switching being accomplished by the rectifying diodes. In the half-wave circuit (Fig. 26a) on one half of the cycle the condenser C_1 is charged through CR_1; this voltage is then added in series on the next half cycle to the voltage of the condenser C_2 charged through CR_2. A voltage of approximately twice the peak input will appear across the load, depending on the rectifier type, load resistance, and values of C_1 and C_2. Ripple frequency will be the same as the supply frequency. C_1 must be rated for the peak value of the applied voltage, and C_2 for twice this value.

In the full-wave doubler (Fig. 26b) C_1 and C_2 are charged on alternate half cycles, approximately twice the peak applied voltage appearing across the two in series. The ripple frequency will be twice the supply frequency, and the condenser ratings should be equal to the peak supply voltage.

In both circuits the larger the capacitors the nearer the output voltage will be to twice the peak input, and the better the voltage regulation. However, care must be taken not to exceed the peak current rating of the rectifiers.

Special-purpose Rectifiers. In addition to the more common voltage-doubler rectifiers, voltage triplers and quadruplers can be used to boost the output voltage. For more information on these and other special supplies, see Ref. 9.

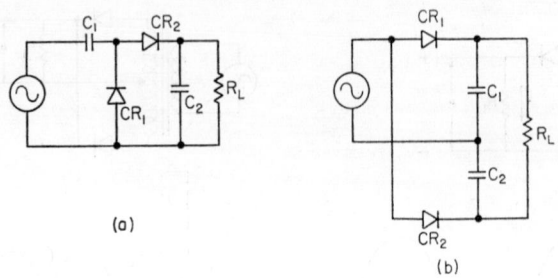

(a)

(b)

Fig. 26. (a) Half-wave voltage doubler; (b) full-wave voltage doubler.

In applications where very heavy currents are required various types of gas-filled diodes and ignition-type rectifiers are used. The ignition types have the property that they do not start to conduct until triggered by an external signal, and this property may be used to control the output voltage.[1] Among the family of ignition or controlled rectifiers are thyratrons, ignitrons, and silicon-controlled rectifiers.

Filters

In most cases the pulsating output of a rectifier does not satisfy the requirements for a d-c supply. Smoothing filters are usually required to reduce the alternating components of the rectified waveform.[3,4,7]

The following description gives an approximate analysis for a better understanding of filter action.

The relative values of the direct, fundamental, and harmonic alternating components for full- and half-wave rectifiers are summarized in Table 5.

Table 5. Relative Values of Direct, Fundamental, and Harmonic Alternating Components for Full- and Half-wave Rectifiers

	Direct component	Fundamental	Second harmonic	Third harmonic	Fourth harmonic
Half-wave............	$0.318I_m$	$0.5I_m$	$0.212I_m$	0	$0.042I_m$
Full-wave............	$0.636I_m$	0	$0.424I_m$	0	$0.085I_m$

Capacitor Filters. A capacitor in parallel with the rectifier load, as shown in Fig. 27a, acts as a smoothing filter. When the rectifier(s) conduct, the capacitor receives energy through the rectifying element(s). The capacitor delivers energy to the load when the rectifying elements are not conducting. This results in output waveforms of the type shown in Fig. 27b.

Larger products of capacitance and load resistance result in more nearly constant voltages across the load. As the load current increases, the amount of ripple increases and the d-c component of the output voltage decreases. The maximum d-c output voltage approaches the peak value of the rectified wave for small values of load current.

The portion of the positive half cycle, during which the rectifier(s) must supply sufficient energy to recharge the capacitor, decreases as the capacitance is increased.

Thus, for constant-load current, the peak current drawn through the rectifier(s) and transformer secondary increases as the capacitance increases. This, in turn, gives a higher I^2R loss in the transformer and the rectifier. Also, there is a practical limit to the value of the capacitor. When a large capacitor filter is used it is advisable to have a small resistance between the rectifier and the capacitor to limit the initial charging current. This high charging current could otherwise damage the rectifier or the capacitor or blow the line fuse of an instrument.

Though simple capacitor filters alone do not usually provide sufficient filtering except for very modest load currents, they are often used in conjunction with electronic regulators. The electronic regulator then provides the remainder of the necessary filtering. Such a system can have excellent transient response since it contains no inductance.

Inductor Filters. An inductor in series with the load, as shown in Fig. 27c, results in output waveforms of the type shown in Fig. 27d. The inductance presents a greater impedance to the alternating component of the rectifier current than it does to the direct component. Thus, considering the series combination of the inductor and load resistor, the alternating component of the voltage across the load resistor will be attenuated to a greater degree than will the direct component.

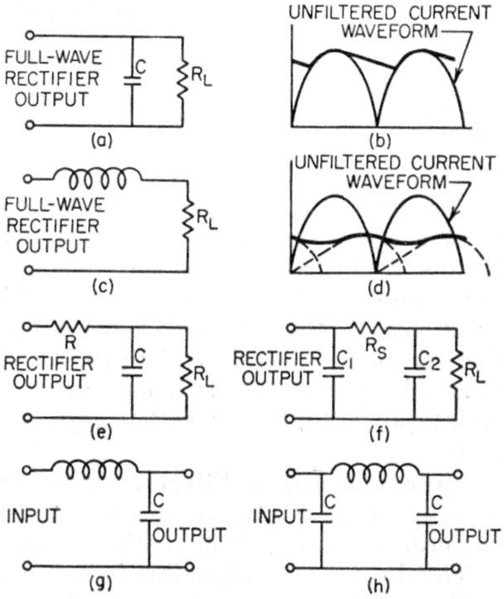

The amount of filtering and the direct component of voltage across the load are substantially independent of the load current. The maximum d-c output voltage approaches the average value of the rectified wave as the d-c resistance of the series inductor approaches zero.

Inductor filters are not suitable for circuits which present suddenly varying loads, since the inducter will resist any sudden change in load current.

Combination RC Filters. The circuit shown in Fig. 27e is sometimes employed when the load current is small. The series resistance is large compared with the impedance of the capacitor to ripple frequencies. Thus only a small fraction of the alternating components of the input waveform appear across the capacitor and the load resistance. Since the capacitor does not pass the direct component of the signal current, this component must pass through the load resistance.

The circuit shown in Fig. 27f finds application where medium-load currents (10 to 100 ma, for example) are required. For the same degree of filtering to the load resistor, the value of C_1 can be considerably smaller than when only a smoothing capacitor is employed.

The tolerable ripple across C_1 is equal to the tolerable ripple across R_L times the ratio of R_s to the impedance of C_2 at the ripple frequency. The direct component of voltage across R_L is equal to the direct component of voltage across C_1 less the voltage drop across R_s.

If, for example, $R_L = 10,000$ ohms, $X_{C_2} = 300$ ohms (approximately 8 μf at 60 cps), and $R_s = 3,000$ ohms, the ripple is attenuated by a factor of 10 while the direct component is decreased by only 23 per cent of the value appearing across C_1.

Since the allowable percentage ripple across C_1 is 10 times the permissible value when only a smoothing capacitor is used, the capacitance C_1 can be reduced by a factor somewhat less than 10. Reduction in the capacitance of C_1 permits conduction of the rectifier(s) over a greater portion of the positive half cycle, and thus for constant

FIG. 27. Rectifier filters: (a) Parallel filtering capacitor; (b) effect of filtering capacitor; (c) series filtering choke; (d) effect of filtering choke; (e) RC filter; (f) CRC filter; (g) LC filter (choke input); and (h) LC filter (capacitor input).

load the necessary charge can be supplied at smaller peak values of rectifier current. This reduces the IR drop in the rectifier and the transformer.

Combination LC Filters. The combination of an inductor and a capacitor as shown in Fig. 27g is called an LC filter section, or sometimes an L section. The reduction of ripple between input and output terminals is given by the relationship

$$\frac{\text{Output ripple}}{\text{Input ripple}} = \frac{1}{\omega^2 LC + 1} \tag{51}$$

Choke Input Filters. When the output of the rectifier is connected directly across the input terminals of an LC filter section, the filter is said to be "choke input." The addition of filter sections in series results in increased filtering action.

The d-c voltage obtainable under load from a choke input filter cannot approach the peak value of the rectifier wave because of the voltage drop across the series choke. The output ripple is relatively independent of load current. The regulation is generally considerably better than that obtainable with capacitor input filters. For the same peak current through the rectifier, a higher load current can usually be drawn from a choke input filter than from a capacitor input filter. Choke input filters are particularly well suited for use with variable load currents of relatively large magnitudes, provided, however, that any change in load current occurs at a relatively slow rate.

Capacitor Input Filters. If the LC section is preceded by a filter capacitor, as shown in Fig. 27h, the resulting filter is called a capacitor input filter. This circuit is sometimes called the conventional capacitor input filter or the π section filter.

The d-c voltage obtainable from a capacitor input filter approaches the peak value of the rectified wave. As the load current increases, the ripple increases and the d-c output voltage decreases; thus the regulation is relatively poor.

In practice, condenser input filters are sometimes preceded by a small resistor (approximately 100 ohms) to limit the initial condenser charging current drawn through the rectifier. This resistor or resistors should withstand the peak voltage of the rectified wave.

When electrolytic capacitors are to be used as filters, it is often desirable to use capacitors of two to four times the calculated required capacitance to compensate for their gradual deterioration.

Regulators for Rectifier Power Supplies

Most electronic circuits are sensitive, in varying degree, to variations in the voltage supplied by the source of d-c power. Three fundamental types of regulation are required of most d-c power supplies:

1. Regulation against changes in line voltage.
2. Regulation against changes in load.
3. Regulation against long-term drifts in the rectifier power supply and regulator. The long-term drift, or lack of stability of the system, is of practical importance only when extremely good regulation is required.

Regulators are classified as shunt or series regulators. Shunt regulation is accomplished by varying the amount of current flow through a fixed-series dropping element. Series regulation is accomplished by varying the effective resistance of the series element.

Glow-tube Regulator. The simplest type of voltage regulator utilizes the substantially constant voltage characteristic of a glow tube (or VR tube, as this type of tube is frequently called). The volt-ampere characteristic of a typical glow tube is shown in Fig. 28b. A glow-tube regulator circuit is shown in Fig. 28a. The voltage drop across R_s is equal to R_s times the sum of the load current and the current through the regulator tube. Any attempted change in load voltage is reflected in a change in current in the regulator tube to restore the desired load voltage by increasing or decreasing the voltage drop across R_s. Sufficient current must be drawn through the regulator tube to permit the current to increase and decrease the necessary amount without exceeding the minimum or maximum current ratings of the tube. However, large overloads can be tolerated for short periods such as initial warmup.

Several glow tubes can be connected in series if the desired voltage is higher than that of a single tube.

This type of regulator is satisfactory where:

1. One or a series combination of tubes deliver the required voltage.
2. Load current is not more than about 100 ma.
3. Regulation of roughly 1 per cent is satisfactory.

The voltage required to fire most glow tubes is roughly 30 per cent higher than the operating voltage. The series dropping resistor must be small enough to permit the minimum required current to flow through the regulator tube and large enough to prevent the maximum current rating of the tube being exceeded under no-load conditions.

The glow-tube regulator (in common with all shunt-type regulators) is relatively inefficient, but this is normally not a problem in low-current supplies.

Zener Diode. Operation of a zener diode in a circuit is similar to a glow-tube regulator since it presents a constant voltage drop that is independent of current within certain limits. This type of diode makes use of the breakdown properties of a p-n junction. If the reverse voltage applied to a p-n junction is progressively increased, a value will be reached at which the current will increase greatly from its normal cutoff value. This value is called the breakdown or zener voltage. Zener diodes are often

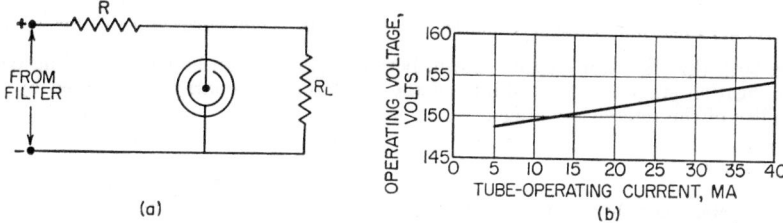

(a)   (b)

Fig. 28. Glow-tube regulator: (a) VR tube circuit; (b) typical regulation characteristics.

used as precision voltage reference elements, since they can be made to have a very constant voltage drop over the operating current range, provided, however, that temperature of the diode is maintained constant. Current in excess of the rated value will destroy the diode, but unlike a glow tube, it does not require a higher voltage to start conduction or a minimum current to prevent oscillation. Zener diodes, while limited to modest currents, may be obtained with wattage ratings up to 10 watts and zener voltages from a few volts to several hundred volts.

Electronic Voltage Regulator. Electronic voltage regulation can be used to give short- or long-term voltage stability. A voltage regulator can also be used to reduce ripple and to maintain a constant voltage with varying line voltage or varying load current. The effective internal impedance of a power supply can be reduced from several hundred ohms to an ohm or less.

Electronic regulators fall into the two general categories of (1) shunt, and (2) series types. A shunt regulator has a fixed resistance in series with the supply. A vacuum tube or a transistor shunted beyond the resistance acts as an artificial load and regulates the output by varying the current and thus the voltage drop across the series resistor. A series regulator uses a vacuum tube or transistor as a variable resistance in series with the supply and controls the output by varying the apparent resistance of the tube or transistor. Though either type may achieve a high degree of regulation, the shunt regulator is suitable only for very modest currents.

In order to maintain a constant output voltage, a regulator must contain some source of reference or comparison voltage. This may be a VR tube, a zener diode, or a standard cell and servoamplifier. The circuit of a basic vacuum-tube regulator is shown in Fig. 29. The cathode of voltage amplifier V_1 is maintained at a constant potential by the reference element (commonly a VR tube). If the output voltage increases, a portion of this increase is applied to the grid of V_1. The error signal which is applied to V_1 is amplified and inverted and applied to the grid of V_2. The lower grid voltage of V_2 causes it to conduct less and restores the output voltage to its original value.

A simple transistorized regulator is shown in Fig. 30. The regulated output voltage will be determined by the zener voltage of the reference diode. The series transistor acts as an emitter follower, so that the voltage on the emitter will be almost identical with the base voltage even though the supply voltage or the load current changes. The value of R_1 must be selected so that, when the load is drawing maximum current, current through R_1 will be greater than I_L/B, where I_L = load current and B = the d-c current gain of the transistor. At no load most of the current through R_1 will flow through the reference zener, whose maximum wattage rating must not be exceeded.

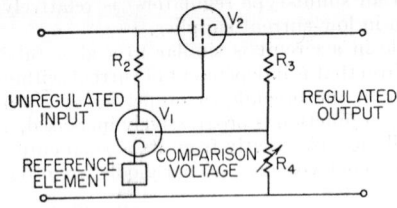

FIG. 29. Electronic voltage regulator.

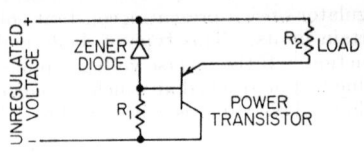

FIG. 30. Transistorized regulator circuit.

In both the regulator circuits shown, better regulation can be had by adding more amplification stages. The stability of the regulator is primarily determined by the stability of the reference voltage.

Batteries*

An electric battery is a device for the direct transformation of chemical energy into electrical energy.

Primary Cells. In a "primary" battery the parts which react chemically require renewal. Primary cells may be divided into wet and dry cells.

Wet Cells. There are several types, as follows:

Gravity Cell. This cell consists of a copper positive electrode in $CuSO_4$ and a zinc negative electrode in a $ZnSO_4$ solution. The cell provides 1.07 volts.

Leclanché Cell. This cell consists of a positive electrode of carbon with a depolarizer of MnO_2, a negative electrode of zinc, and an electrolyte of NH_4Cl. This cell was used for bell ringing but has been largely replaced by the dry cell.

CuO–Zn in Caustic Soda. This is the commonest wet primary cell now in use. It is used in railway signal service.

Standard Cells. The Weston normal cell uses a cadmium amalgam negative pole, a mercury positive pole, and has an emf of 1.01823 volts at 20°C. The Clark standard cell uses zinc amalgam and has an output of 1.4328 volts at 15°C. Temperature equations are

Clark cell $\quad E_t = 1.4328[1 - 0.00119(t - 15) - 0.000007(t - 15)^2] \quad$ volts $\quad$ (52)

Weston cell $E_t = 1.0183[1 - 0.0000406(t - 20) - 9.5 \times 10^{-7}(t - 20)^2$
$$+ (1 \times 10^{-8})(t - 20)^3] \quad \text{volts} \quad (53)$$

Dry Cells. There are several types, as follows:

Leclanché Cell. The most common dry cell is a modification of the Leclanché cell in which the NH_4Cl is held by capillary action in a porous medium separating the zinc from the carbon electrode. These cells are made round (flashlight cells) or flat so that they can be stacked to provide B, C, and other purpose batteries. Each cell provides ~1.6 volts. Shelf life of dry cells is usually under 1 year and is improved by low-

---

* Reprinted with permission from Kohl, Zentner, and Lukens, "Radioisotope Applications Engineering," pp. 450, 451, 452, D. Van Nostrand Company, Inc.

temperature storage.   The ampere-hour output of dry cells is increased by providing adequate recuperation time between uses and by minimizing the discharge current. Dry cells can be recharged under certain controlled conditions.   In an improvement to this cell, magnesium has been added to the MnO_2 to enhance shelf life and high drain performance.

Mercury Cell.   The Ruben dry cell developed during World War II utilizes the system $Zn/Zn(ON)_2$ (solid) KOH aqueous HgO (solid) Hg.   This is a small cell with a uniform discharge voltage of approximately 1.25 volts, suitable for hearing aids and portable equipment.   The mercury cells, when compared with "dry" cells, offer the following advantages:

1. High ratio of energy to volume and weight—three to four times the energy-volume ratio of other batteries.

2. Long shelf life, at 70°F service-life reduction of only 7 per cent in 18 months obtained with some mercury cells.

3. Leakproof, dimensionally stable.

4. Uniform discharge voltage, for Mallory RM-4 A battery, 40 ma drain at 1.25 volts and 95°F—12 hours on, 12 off, in 72 hr of service—voltage range from 1.3 to 1.1 volts; for light (under 1 ma) or no load, voltage regulation approximately 1 per cent for 1,000 hr of operation.

5. Ability to operate over wide temperature ranges (to 250°F).

6. Unaffected by pressure (to over 1,000 psia) to impact or acceleration (100 *g*).

The indium cell developed for wrist watch and instrument applications uses a mercuric oxide cathode and an indium anode.   It provides essentially constant discharge voltage through its service life.

Magnesium–Cuprous Chloride Cell.   The magnesium–cuprous chloride cell was developed during World War II to operate under certain extreme conditions where normal dry batteries do not function efficiently.   The battery is shipped dry in moistureproof container and water (fresh or salt) is added, allowed to remain in the battery for a short period, and the excess removed.   The battery must be "activated" at temperatures above freezing but can be used in the range from +140 to −60°F.   If the battery is kept free from moisture, it can be stored almost indefinitely.   Power outputs in the range from 1 ma to 100 amp are available with a single cell voltage of 1.0 to 1.6 volts.   Units have been built to deliver up to 30 watts/cu in. or 360 watts/lb.

Solid Electrolyte Batteries.   These "solid-state" batteries are dry cells which operate by diffusion of ions through a solid, rather than by movement of particles through a liquid or gas.   The National Carbon W-618 uses vanadium pentoxide as the cathode with a silver anode and silver iodide electrolyte.   The P. R. Mallory Solidon has a conductor of tin sulfate and a depolarizer cathode that is a mixture of barium permanganate and lead peroxide.   The battery can be completely encapsulated in epoxy resin.

Storage Batteries.   Cells which are reversible to a high degree, or in which the chemical and physical state of the electrodes after discharge can be brought back to the original condition by causing current to pass in the opposite direction, are called storage cells.   There are two major types:

Pb–H_2SO_4 Cell.   This is the most common type of storage cell used in automobiles, aircraft, and industry.   It contains PbO_2 on the positive plate, a sponge, lead on the negative plate, and an electrolyte of H_2SO_4 and H_2O.   The voltage per cell is approximately 2 volts and depends on the specific gravity of the electrolyte.

Alkaline Storage Cell—Nickel-iron or Edison Cell.   This cell comprises NiO in the positive plate, iron in the negative plate, and an electrolyte of a 21 per cent solution of potassium and lithium hydroxides.   This cell differs from the lead H_2SO_4 cell in that it is not damaged by short circuit or complete discharge.   Edison cells have a life of approximately 1,500 cycles (5 to 25 years).   Sealed rechargeable small cells using silver-zinc, nickel-cadmium, and silver-cadmium are finding wide use in hand tools, portable shavers, and radio and TV sets.   The silver-zinc batteries are available as high-rate units for applications where the total energy is delivered in a matter of minutes and as low-rate units which deliver their power over long periods of time. The silver-cadmium batteries are only 40 per cent of the size of the nickel-cadmium batteries of the same power range.

Nuclear Batteries. Four principal types of nuclear batteries, all based on beta-emitting radioisotopes, exist, and a variety of other types have been considered. In the first type, the β-current nuclear battery, beta rays emitted by a radioisotope source coated on a central source electrode are collected on a surrounding conducting collector electrode which is insulated from the source electrode. Classically, the two electrodes are separated by an evacuated and therefore insulating space, but lately it has been found that synthetic organic plastic separators, such as polystyrene, may be used, provided they are not thick enough to absorb all the beta radiation. In this way, voltages as high as 7,000 volts, with currents on the order of 40 $\mu\mu a$, may be attained. Such batteries are characterized by ruggedness because of their simple construction and by long life.

The second type of nuclear battery is the junction-type battery, in which an avalanche of 200,000 electrons is produced for every emitted beta particle. The battery consists of a wafer of silicon coated on one surface with a beta-emitting isotope such as Sr^{90}, and having electrical leads connected to the silicon and to an antimony inset on the wafer. These cells develop only about 0.2 volt and, because of radiation damage to the silicon, have a life of only a few weeks.

The contact-potential-difference cell is a third type under commercial development. In 1924, it was demonstrated that a continuous current could be generated by the effect of radiation on cells which consist of two electrochemically dissimilar electrodes separated by a filling gas. When the gas is exposed to ionizing radiation, orbital electrons are knocked out of position, and both free electrons and positive ions are formed between the electrodes. Since the electrodes are electrochemically dissimilar and therefore have different work functions, a bias field exists between them. Under the influence of this field, the positive ions are attracted to the more electronegative electrode, where they acquire an electron and return to electrical neutrality.

If there is an external load between the electrodes, the free electrons which migrate to the electropositive electrode flow through the external circuit to the other electrode to replace the electrons which have been lost by neutralizing the positive ions. Thus a continuous flow of current, which is a function of the intensity of the incident radiation, can be generated. This principle is effectively employed in various types of radiation-measurement equipment.

The intervening gas in such cells can be a radioactive one, such as tritium mixed with heavy gases such as argon or krypton. In this way cells having voltages on the order of 1 to 2 volts can be prepared.

A fourth type of nuclear battery, in which beta particles excite a phosphor and the resulting light is converted to electrical energy by a photocell, has been reported. In such cells, an isotope such as Pm^{147} is embedded in a cadmium sulfide phosphor, and the phosphor is sandwiched between two photocells. By using 4.5 curies of the promethium, a cell delivering 20 μa at about 1 volt can be obtained. Such cells are constant-current sources which, like other nuclear batteries, have outputs which decay as the radioisotope decays.

Summary of Battery Characteristics. The principal characteristics of common battery types are given in Table 6. The properties of miniature batteries are summarized in Table 7.

ELECTRONIC CIRCUITS

Amplifiers

Amplifiers are devices which increase the power level in an electrical signal. The amplifiers described in this section are considered to be linear; i.e., the output signal is proportional to the input signal.[2,3,4,5,6,9]

Voltage Amplifiers. Voltage amplifiers give primarily a voltage rather than a current increase to the signal. They fall into the three general categories of narrow-band, wideband, and d-c amplifiers.

Narrow-band amplifiers pass only a small band of signal frequencies as compared with the mean frequency. (Almost all r-f amplifiers fall in this category.) They are usually transformer or inductance-capacitance coupled, and at radio frequencies the

Table 6. Cell Characteristics of Common Batteries

Battery type	Composition charged state			Cell potential, volts		Time to discharge			Shelf life if discharged (wet)	Shelf life in charged condition				Life in operation	
										Without maintenance		With maintenance			
	Positive	Negative	Electrolyte	Open circuit	Discharging	Fastest, min	Avg, hr	Slowest, days		Charge loss, %	Shelf life	If charged each	Shelf life is	Cycles	Float
Lead-acid............	PbO_2	Pb	H_2SO_4	2.14	2.1–1.46	3–5	8	> 3	Not permitted	High rate: 50% in 10 days Low rate: 15–20% per year	Days Months	30–45 days	Years	10–400	Up to 14 years
Nickel-iron...........	NiO_2	Fe	KOH	1.34	1.3–0.75	10	5	> 3	Decades	15–25% per month	Weeks	30–45 days	Years	100–3,000	8–20 years
Nickel-cadmium........	NiO_2	Cd	KOH	1.34	1.3–0.75	5	5	> 3	Years	Pocket: 20–40% per year Sintered: 10–15% per month	Months Weeks	30–45 days	Years	100–2,000 25–1,000	8–14 years 4–8 years
Silver-zinc..........	AgO	Zn	KOH	1.86	1.55–1.1	0.5	5	>90	Years	15–20% per year	3–12 months	6 months	1–2 years	100–300	1–2 years
Silver-cadmium.........	AgO	Cd	KOH	1.34	1.34–0.8	5	5	>90	Years	50% in 2 years	1–2 years	6 months	2–3 years	500–3,000	2–3 years

SOURCE: *Product Engineering.*

Table 7. Properties of Miniature Batteries

Battery type and commercial designation	Voltage	Continuous current drain, per sq in. of cell area	Available charge or capacity	Flash or peak discharge current, per sq in. of cell area	Shelf life, 70°F	Weight, oz	Dimensions, in. Diam	Dimensions, in. Length or thickness	Operating temp., °F
Flat cells:									
Mercuric oxide (Mallory RM 400)....	1.4	150 ma, 100 hr	80 ma-hr	250 ma	3-5 years	0.04	0.45	1/8	25-125
Manganese dioxide (National Carbon —W307)	1.5 nom.	60 μa, 1,000 hr	60 ma-hr, to 1.3 volts	200 ma	>1 year	0.05	1/2	1/8	0-120
Indium anode (Elgin—microcell)	1.15-1.37	100 ma, 100 hr	150 ma-hr	1,100 ma	1-3 years	0.13	3/4	0.2	-20-190
Solid state and solid electrolyte:									
Tellurium (Patterson, Moos—Dynox 95)	95	1 μa; to 85 volts 100 hr	0.33 ma-hr, to 85 volts	200 μa	>20 years	0.395	1 1/4	3/8	-65-130
Tin sulfate (Mallory—Solidion ELS-2)	100	1 μa.	...	100 μa	>20 years	0.6	1	...	-100-170
Vanadium pentoxide (National Carbon—W618)	95	0.007 μa; to 70 volts, 100 hr	0.03 ma-hr to 70 volts	2 μa	>20 years	0.25	...	0.33	-70-170
Nuclear batteries:									
Strontium (Patterson, Moos)..........	5,000-20,000	10-200 μμa	...	...	25 years	0.5	0.6	1/2	-175-175
Krypton (Patterson, Moos)..........	3,000	40-700 μμa	...	...	10 years	0.6	1 7/16	7/16	-200-200
Fully rechargeable batteries:									
Lead-silver oxide (Nav. Ord. Lab. XA-10E)	0.6 nom.	0.1-30 μa	1,500 ma-hr, to 0.77 volts	>3 ma	>3 years	1.5	1 3/16	0.586	-65-165
Nickel-cadmium (Sonotone)..........	1.3	1.35 ma, 100 hr	60 ma-hr	200-300 ma	>10 years	0.2	3/4	1/4	-65-160
Silver-zinc (Yardney—HK01)..........	1.5 nom.	0.5 amp	200 ma-hr, to 1 volt	3 amp	1 year	0.1	...	...	0-160
Mercury (Mallory #4)..........	1.35	125 ma	2,000 ma-hr, to 0.9 volt	1.2 amp	>3 years	1.54	1 3/16	1/2	25-125

SOURCE: *Product Engineering.*

coupling circuits are generally designed to resonate at the mean frequency. Common examples are the tuned r-f and i-f stages in a radio, which may have a bandpass of about 2 per cent of the mean frequency.

Wideband amplifiers pass a large variance of signal frequencies as compared with the mean frequency. Audio amplifiers, for example, may have a bandwidth extending from 20 to 20,000 cycles, while the bandwidth of a pulse amplifier may extend to several megacycles. If a rectangular pulse is impressed on a wideband amplifier, the rise time of the signal out will be related to the high-frequency response of the amplifier. The maximum pulse duration the amplifier can accommodate without introducing appreciable sag will be related to its low-frequency response.

There are numerous types of d-c amplifiers with upper frequency limits ranging from a few cycles to several megacycles. A problem common to any d-c coupled amplifier is drift (d-c stability). This is compensated for in some circuits by using balanced or differential amplifier pairs so that the drift of one is offset by the drift of the other. This system, however, still requires zero adjustment from time to time. Chopper-stabilized d-c amplifiers circumvent drift problems by converting the direct current to alternating current with a chopping relay and amplifying the chopped signal with an a-c amplifier. The amplified alternating current is rectified and filtered to give a d-c output proportional to the d-c input.

Power Amplifiers. Power amplifiers are essentially dynamic impedance-matching devices. A higher-impedance source is matched to a lower-impedance load by increasing the current and thus the power in the signal. The general categories of narrowband, wideband, and d-c amplifiers apply to power as well as to voltage amplifiers.

The Decibel. The impression of sound on the human ear is roughly proportional to the logarithm of its magnitude. This has led to the use of the decibel, which is a logarithmic power ratio referring to the power increase effected by an amplifier. It is defined by the equation

$$db = 10 \log \frac{P_2}{P_1} \qquad (54)$$

where db = decibels
P_1 = power in
P_2 = power out

If the input and output impedances of an amplifier are the same, the decibel change can be calculated by

$$db = 20 \log \frac{V_2}{V_1} \text{ or } 20 \log \frac{I_2}{I_1} \qquad (55)$$

where V_1 and I_1 = voltage and current in
V_2 and I_2 = voltage and current out

Gains or losses in power are expressed by using either a plus or a minus sign as a prefix, as +6 db or −3db.

Since the decibel is a ratio, it has no absolute value unless related to a reference level. Unfortunately there is no universal reference level, though in telephone work 0.006 watt into 500 ohms is generally used. The decibel is convenient in audio work, but in industrial amplifiers it is usually more informative to state gain in terms of voltage increase ($K = V_2/V_1$) and state input and output impedances.

Distortion. Distortion is any change in waveform produced in an amplifier. The most common types are nonlinear, frequency, and phase-shift. •

Nonlinear distortion occurs when the amplifying element does not have a linear relationship between the input and the output, so that the signal is not amplified the same amount at each instantaneous amplitude. This injects harmonics and sum and difference frequencies in the output that were not present in the input.

Frequency distortion occurs when the gain of a stage is not constant with respect to frequency; hence some signal frequencies are amplified more than others.

Phase-shift (delay distortion) occurs when the circuit offers different amounts of reactance to different signal frequencies, and the delay through the amplifier varies with the applied frequency.

Feedback. When a portion of the output signal of an amplifier is returned to its input, it is termed feedback. Feedback can be used to change the performance characteristics of an amplifier favorably. A block diagram of an amplifier with feedback is shown in Fig. 31. The effective gain is given by

$$K_e = \frac{K}{1 - \beta K} \tag{56}$$

where K_e = effective gain
  K = open-loop gain (no feedback)
  β = the fraction of the output voltage fed back—carries the same sign as the feedback (+ for positive and − for negative feedback)
  βK = feedback factor

Negative feedback can be used to stabilize the gain, lower the output impedance, improve the linearity, reduce distortion, and in some cases reduce the noise in an amplifier. All this is at the expense of gain, however. The expression for K_e shows that the larger the value of open-loop gain K, the less effect a change in K will have on the effective gain. However, there is always a phase lag in the passage of a signal through an amplifier; and thus, as the signal frequency increases, some point will be

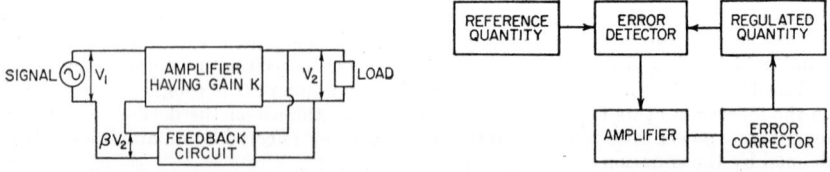

FIG. 31. Feedback in amplifiers.   FIG. 32. Servoamplifier loop.

reached at which the feedback adds to rather than subtracts from the input. This can cause high-frequency oscillation unless the bandwidth of the amplifier is restricted to frequencies having a phase shift less than 180°.[3,4,10]

A special type of negative-feedback amplifier is the servoamplifier. A servosystem controls a regulated quantity by means of an error-detecting device which detects any difference between the regulated quantity and a reference, an amplifier that amplifies any error signal, and an error-correcting device which changes the regulated quantity to match the reference quantity. The essential components are shown in Fig. 32.

Positive feedback tends to increase the gain and decrease the range of uniform response of an amplifier. Therefore, positive feedback can be used to increase gain and selectivity. However, if the feedback factor βK becomes as large as 1, the amplifier will become unstable.

Sinusoidal Oscillators. Sinusoidal oscillators in the audio- and radio-frequency range are usually positive-feedback amplifiers which have in-phase signal returned from the output to the input to sustain oscillation. According to the Barkhausen criterion, the voltage multiplication in the amplifier must exactly counterbalance the voltage-dividing ratio of the feedback circuit.

Representative feedback oscillators are the Colpitts, Hartley, Meissner, tuned-plate, tuned-grid, and Wien bridge. Most vhf oscillators are Colpitts or Hartley, with the Colpitts predominating. In several of these circuits, a fixed-frequency element can be substituted for the frequency-determining tuned circuit to give very stable operation on a single frequency. These fixed-frequency elements include quartz crystals in the r-f range and tuning forks and magnetostrictive devices in the low-frequency range.

In the microwave region, ordinary oscillators do not produce sufficient power, and transit-time oscillators of the klystron or magnetron type are used. Either may be able to produce enough power so that no further amplification is needed, even in high-powered transmitters. The klystron is more readily tunable than the magnetron and is more generally used in local oscillator applications.[3,4,9,10,11]

Modulation and Demodulation

Modulation is the process of impressing lower-frequency information on a higher-frequency carrier. Usually the carrier wave is sinusoidal, though this is not necessarily so. The three basic types of modulation are amplitude (a-m), frequency (f-m), and phase modulation (p-m). Amplitude modulation consists of varying the amplitude of the carrier-wave envelope in proportion to the modulating signal. In frequency modulation, the amplitude of the carrier wave remains constant and the frequency of the carrier is varied in proportion to the modulating signal. Phase modulation is similar to frequency modulation with the exception that the phase of the carrier wave is varied.[3,4,9,10,12]

Other types of modulation are basically variations of a-m, f-m, or p-m. Single-sideband and pulse modulation, for example, are essentially variations of amplitude modulation.

Demodulation is the process by which the carrier frequency is removed and the modulating signal retained. A demodulating circuit is usually called a detector. A great many circuits have been devised for demodulating an amplitude-modulated wave, but all operate by rectifying the modulated wave and then filtering out the carrier frequency. The diode detector is very simple. It consists (in its simplest form) of a rectifying diode, a filter which blocks out the carrier but not the signal frequency, and a load resistor.

There are two general methods for demodulating a frequency-modulated wave. The first is to subject the modulated wave to a circuit which has a nonlinear frequency response. The frequency modulation is thus converted to amplitude modulation which can be demodulated by a conventional a-m detector. This method does not give particularly good linearity.

Though it is not strictly correct, the second method might be thought of as beating the frequency-modulated wave against its center frequency to give the modulating frequency as the difference.

The ratio detector, which detects the phase deviation of the modulated wave, falls into the second category. It is a popular f-m detector because of its relative simplicity and high quality. It also has an advantage over other f-m detectors in being insensitive to amplitude variations in the modulated signal.

A phase-modulated wave can be demodulated by the same types of detectors as a frequency-modulated wave.

Pulse and Switching Circuits

This category of circuits includes configurations which exhibit behavior typified by either short duration, discontinuous signal conditions, or signal excursions driving the active device into nonlinear regions of its characteristics. In particular, the operating characteristics of any active device may be divided into three general areas: (1) the nonconducting or cutoff portion, (2) the linear region, and (3) the saturation or high-conducting portion. The application of these criteria will be demonstrated in the following discussions of a few representative circuits.[2,10,13]

Nonsinusoidal oscillators exist in several forms: relaxation oscillators, free-running multivibrators, and blocking oscillators. Relaxation oscillators generally depend on the nonlinear characteristics of a gas tube to control the circuit operation and are frequently used for sawtooth sweep generators. The circuit is simple but restricted to frequencies below 25 kc and applications where time stability is not of paramount importance. With the use of vacuum tubes, the frequency of operation may be extended to 100 kc. Some form of synchronization is usually desirable for timing purposes.

Free-running multivibrators are astable and depend on the RC time constants of the cross-coupling networks for the frequency of their operation. This type of circuit is frequently encountered in digital-computer systems where it provides the timing at frequencies up to approximately 10 mc.

Blocking oscillators require an external feedback network (consisting of a trans-former) to provide the required phase shift and the resulting regeneration. The circuit may be used free-running or triggered. Durations as short as 0.1 μsec are possible, and the output impedance is characteristically very low.

A second class of circuits includes the group normally referred to as trigger circuits. The operation is typically initiated by an input pulse which causes a change of state of the circuit. The circuit may either reset itself automatically or await a second pulse to return it to its original state.

"One-shot" blocking oscillators and multivibrators and bistable multivibrators are switching circuits of the triggered type. A monostable (one-shot) multivibrator is used for obtaining a fixed or variable delay or for pulse shaping. Phantastron circuits are capable of generating pulses with durations of many seconds with a repeatability of 1 per cent.

Bistable multivibrators or "flip-flops" are used for both pulse shaping and digital logic circuits. The Eccles-Jordan circuit can be considered to be composed of two inverters in cascade, with the input tied to the output. Many variations of this circuit exist, but in all cases it is a device capable of exhibiting two stable states.

Logic circuits are a very important segment of the switching category and find extensive application in both computers and data-processing systems. In addition to the multivibrators which are used to establish timing and to provide storage, diode networks are employed to perform the needed gating and logical operations. Basic and- and or-gates consist of several diodes connected to a common point, and the circuit function is defined by its transmission. As an example, the and-circuit produces an output only when all inputs are the same, and the converse applies to or-gates. In specific instances resistors may be substituted for the diodes and the circuit will function on a majority-input basis. Gate circuits which utilize an inverter to restore the logic levels are called nor- and nand-gates.

Practical considerations to be considered in the design of gate circuits include the attenuation of logic levels, fan-in and fan-out ratios, number of cascaded gates, and the speed of the nets in special cases. A serious problem in many large systems is the cable capacitance, which often turns out to be the limiting factor in speed of response.

The determination of logical transmission functions advantageously uses Boolean algebra to optimize the design. The principles of Boolean algebra are applicable to any system of variables exhibiting two unique stable states. In addition, a large number of techniques are available for the minimization of switching functions and the design of logical equations using digital computers.[13]

Data Storage and Recorders

A necessary part of any computer or automatic data system is the recorder, printer, or punch used to store the outputs. Analog recorders include strip-chart recorders, magnetic tape, oscillographs, and recording galvanometers. Digital storage includes magnetic-core memories, magnetic tape and drums, electrostatic cathode-ray-tube storage, punched cards and paper tape, photographic records, and printed copy. The applicable method is primarily a function of speed, utility for reprocessing, cost, size, or space required per unit of storage and permanency of record. In addition, the data record is frequently the end product and the only tangible item describing the conclusions of the system and must therefore perform with a very high degree of reliability.

REFERENCES

1. Knowlton, Archer E.: "Standard Handbook for Electrical Engineers," 9th ed., McGraw-Hill Book Company, Inc., New York, 1957.
2. Moskowitz and Racker: "Pulse Techniques," Prentice-Hall, Inc., Englewood Cliffs, N.J., 1958.
3. Craft Laboratory: "Electronic Circuits and Tubes," McGraw-Hill Book Company, Inc., New York, 1947.
4. Terman, F. E.: "Radio Engineers' Handbook," McGraw-Hill Book Company, Inc., New York, 1943.
5. Hunter, L. P.: "Handbook of Semiconductor Electronics," 2d ed., McGraw-Hill Book Company, Inc., New York, 1962.

6. Riddle and Ristenbatt: "Transistor Physics and Circuits," Prentice-Hall, Inc., Englewood Cliffs, N.J., 1957.
7. Darling, Horace E.: Magnetic Amplifiers for Instrumentation, *ISA J.*, vol. 7, no. 1, pp. 59–72, January, 1960.
8. *The Radio-Electronic Master*, United Catalog Publishers, Inc., New York (published periodically).
9. Langford and Smith: "Radiotron Designer's Handbook," distributed by Radio Corp. of America, Camden, N.J.
10. Landee, Davis, and Albrecht: "Electronic Designers' Handbook," McGraw-Hill Book Company, Inc., New York, 1957.
11. Martin, T. L., Jr.: "Ultrahigh Frequency Engineering," Prentice-Hall, Inc., Englewood Cliffs, N.J., 1956.
12. Hund, August: "Frequency Modulation," McGraw-Hill Book Company, Inc., New York, 1942.
13. Caldwell: "Switching Circuits and Logical Design," John Wiley & Sons, Inc., New York, 1952.
14. Kohl, Zenter, and Lukens: "Radioisotope Applications Engineering," D. Van Nostrand Company, Inc., Princeton, N.J., 1953.
15. Massachusetts Institute of Technology, Electrical Engineering Staff: "Electric Circuits," John Wiley & Sons, Inc., New York, 1940.
16. Skilling, Hugh Hildreth: "Electrical Engineering Circuits," John Wiley & Sons, Inc., New York, 1957.
17. Middendorf, William H.: "Analysis of Electric Circuits," John Wiley & Sons, Inc., New York, 1956.
18. Timbie, William H., and Alexander Kusko: "Elements of Electricity," John Wiley & Sons, Inc., New York, 1953.
19. Timbie, William H.: "Industrial Electricity, Volume I: Direct-current Practice," John Wiley & Sons, Inc., New York, 1939.
20. Timbie, William H., and Frank G. Willson: "Industrial Electricity, Volume II: Alternating Currents," John Wiley & Sons, Inc., New York, 1949.
21. Andres, Paul G.: "Survey of Modern Electronics," John Wiley & Sons, Inc., New York, 1950.
22. Brown, Thomas Benjamin: "A Textbook for Students in Science and Engineering," John Wiley & Sons, Inc., New York, 1954.
23. Corcoran, George F., and Henry W. Price: "Electronics," John Wiley & Sons, Inc., New York, 1954.
24. Dow, William G.: "Fundamentals of Engineering Electronics," John Wiley & Sons, Inc., New York, 1952.
25. Westinghouse Electric Corporation: "Industrial Electronics Reference Book," John Wiley & Sons, Inc., New York, 1948.
26. Harris, Forest K.: "Electrical Measurements," John Wiley & Sons, Inc., New York, 1952.
27. Dunlap, W. Crawford, Jr.: "An Introduction to Semiconductors," John Wiley & Sons, Inc., New York, 1958.
28. Buckingham, H., and E. M. Price: "Principles of Electrical Measurements," Philosophical Library, Inc., New York, 1959.
29. Kloeffler, Royce G., and Karl L. Sitz: "Basic Theory in Electrical Engineering," The Macmillan Company, New York, 1955.
30. Albert, Arthur L.: "Electronics and Electron Devices," The Macmillan Company, New York, 1956.
31. Frost-Smith, E. H.: "The Theory and Design of Magnetic Amplifiers," Chapman & Hall, Ltd., London, 1958.
32. Meares, J. W., and R. E. Neale: "Electrical Engineering Practice," Chapman & Hall, Ltd., London, 1958.
33. Perrigo, A. E. B.: "Electricity Supply Meters," Chapman & Hall, Ltd., London, 1947.
34. Stubbings, G. W.: "Commercial A.C. Measurements," Chapman & Hall, Ltd., London, 1952.
35. Stubbings, G. W.: "Electricity Meters and Meter Testing," Chapman & Hall, Ltd., London, 1947.
36. Krugman, L. M.: "Fundamentals of Transistors," Chapman & Hall, Ltd., London, 1959.
37. Lovell, Bernard: "Electronics and Their Application in Industry and Research," Chapman & Hall, Ltd., London, 1950.
38. Marrows, H. E.: "Transistor Engineering Reference Handbook," Chapman & Hall, Ltd., London, 1957.
39. Windred, G.: "Elements of Electronics," Chapman & Hall, Ltd., London, 1949.
40. Dawes, Chester L.: "A Course in Electrical Engineering," McGraw-Hill Book Company, New York, 1952.
41. Gray, Alexander and G. A. Wallace: "Principles and Practice of Electrical Engineering," 8th ed., McGraw-Hill Book Company, New York, 1962.
42. Angelo, E. J., Jr.: "Electronic Circuits," McGraw-Hill Book Company, New York, 1958.
43. Hunter, Lloyd P.: "Handbook of Semiconductor Electronics," 2d ed., McGraw-Hill Book Company, New York, 1962.
44. Markus, John, and Vin Zeluff: "Handbook of Industrial Electronic Circuits," McGraw-Hill Book Company, New York, 1948.
45. Markus, John, and Vin Zeluff: "Electronics for Engineers," McGraw-Hill Book Company, New York, 1945.
46. Miller, Robert E.: "Maintenance Manual of Electronic Control," McGraw-Hill Book Company, New York, 1949.
47. Walker, J. R.: Applying Magnetic Amplifiers, Part I, Controlling Positioning Motors, *Automation*, vol. 3, no. 11, pp. 71–72, November, 1956.
48. Walker, J. R.: Applying Magnetic Amplifiers, Part II, Motor Speed Control, *Automation*, vol. 3, no. 12, pp. 72–73, December, 1956.
49. Walker, J. R.: Applying Magnetic Amplifiers, Part III, Switching Functions, *Automation*, vol. 4, no. 1, pp. 65–67, January, 1957.

50. Walker, J. R.: Applying Magnetic Amplifiers, Part IV, Timing Functions, *Automation*, vol. 4, no. 2, pp. 87–88, February, 1957.
51. Walker, J. R.: Applying Magnetic Amplifiers, Part V, Process Control, *Automation*, vol. 4, no. 3, pp. 73–74, March, 1957.
52. Walker, J. R.: Applying Magnetic Amplifiers, Part VI, Voltage Regulation, *Automation*, vol. 4, no. 4, pp. 73–74, April, 1957.
53. Bloom, L.: Guide to Semiconductor Voltage Regulators, *Control Eng.*, vol. 8, no. 1, pp. 83–84, January, 1961.
54. Bowen, T. J., and R. M. Walp: Transistorized Circuitry for Road Machinery Control, *Control Eng.*, vol. 8, no. 2, pp. 99–104, February, 1961.
55. Sturman, J. C.: Transistor Switches for Industrial Service, *Control Eng.*, vol. 8, no. 2, pp. 103–126, March, 1961.
56. Stelzried, C. T.: Loaded Parallel-T RC Filters, *Control Eng.*, vol. 8, no. 5, pp. 113–114, May, 1961.
57. Cook, C. R., Jr., and J. Luecke: High-speed Switching with the Tetrode Transistor, *Control Eng.*, vol. 8, no. 6, pp. 118–121, June, 1961.
58. DeSautels, A. N.: A Proportional Transistor Switch, *Control Eng.*, vol. 8, no. 7, pp. 87–88, July, 1961.
59. Walker, C. S., and A. M. Roberts: Constant Watts Protection Keeps Power Transistors Cool, *Control Eng.*, vol. 8, no. 8, pp 67–71, August, 1961.
60. Van Praag, V. A.: Magnetic Core Converts Voltage to Pulse Duration, *Control Eng.*, vol. 8, no. 8, pp. 87–88, August, 1961.
61. Stuart-Williams, R.: Magnetic Cores, Characteristics, and Applications, *Autom. Control*, vol. 14, no. 4, pp. 44–47, April, 1961.
62. Stuart-Williams, R.: Magnetic Cores, Characteristics, and Applications, *Autom. Control*, vol. 14, no. 5, pp. 56–61, May, 1961.
63. Motto, J. W., Jr.: Using the Hall Generator: A New Control and Instrumentation Component, *Autom. Control*, vol. 14, no. 6, pp. 48–53, June, 1961.
64. Stuart-Williams, R.: Magnetic Cores, Characteristics, and Applications, *Autom. Control*, vol. 14, no. 6, pp. 54–57, June, 1961.
65. Staff: Transistorized Relay Servo Amplifier, *Autom. Control*, vol. 14, no. 6, pp. 58–59, June, 1961.
66. Motto, J. W., Jr.: Using the Hall Generator: A New Control and Instrumentation Component, *Autom. Cont.*, vol. 14, no. 7, pp. 24–29, July, 1961.
67. Stuart-Williams, R.: Magnetic Cores, Characteristics, and Applications, *Autom. Control*, vol. 14, no. 7, pp. 37–43, July, 1961.
68. Lade, R. W.: Theory and Applications of Avalanche Diodes, *Autom. Control*, vol. 14, no. 10, pp. 31–32, October, 1961.
69. Young, L. H.: What Solid State Means, *Control Eng.*, vol. 8, no. 9, pp. 120–121, September, 1961.
70. Milnes, A. G.: Explaining Solid State Phenomena, *Control Eng.*, vol. 8, no. 9, pp. 122–133, September, 1961.
71. Wood, C.: Hall Effect Transducers; *Control Eng.*, vol. 8, no. 9, pp. 138–141, September, 1961.
72. Eckard, R. D.: How to Build a Constant Current Power Supply, *Control Eng.*, vol. 8, no. 10, pp. 101–104, October, 1961.
73. Perry, J. A.: How Diodes Simplify Circuits, *ISA J.*, vol. 9, no. 1, pp. 42–43, January, 1962.
74. Chass, J.: The Differential Transformer—Its Main Characteristics, *ISA J.*, vol. 9, no. 5, pp. 48–50, May, 1962.
75. Pohl, P.: Signal Conditioning for Semiconductors, *ISA J.*, vol. 9, no. 6, pp. 33–34, June, 1962.
76. Chass, J.: The Differential Transformer—Its Applications and Circuits, *ISA J.*, vol. 9, no. 6, pp. 37–39, June, 1962.
77. Pisarcik, D. A.: Applying a 50-ampere Controlled Rectifier, *Autom. Control*, vol. 13, no. 5, pp. 23–26, November, 1960.
78. Sommer, B.: Chopper Noise Sources and Measurement Techniques, *Autom. Control*, vol. 14, no. 11, pp. 39–42, November, 1961.
79. Bickley, H. D., D. R. Brandt, J. B. Cage, and J. J. Skiles: Digital Techniques for Voltage Regulation Studies, *Autom. Control*, vol. 14, no. 12, pp. 26–30, December 1961.
80. Sussman, N.: Magnetic Modulators, Characteristics, and Applications, *Autom. Control*, vol. 17, no. 3, pp. 16–19, October, 1962.
81. Faulkner, A. H., F. Gurzi, and E. L. Hughes: MAGIC, An Advanced Computer for Spaceborne Guidance Systems, *Autom. Control*, vol. 17, no. 5, pp. 13–20, December, 1962.
82. Aronson, M. H.: "Electronic Circuits," Instruments Publishing Co., Pittsburgh, Pa., 1960.
83. Aronson, M. H.: "Electronic Circuitry for Instruments and Equipment," Instruments Publishing Co., Pittsburgh, Pa., 1957.
84. Lytel, Allan: "Printed Circuitry," Instruments Publishing Co., Pittsburgh, Pa., 1957.
85. Carpenter, J.: Basic Sequencing Circuits, *Automation*, vol. 8, no. 6, pp. 71–76, June, 1961.

PREPARATION OF WIRING DIAGRAMS
FOR INSTRUMENT APPLICATIONS

By J. W. Friedman*

Two types of electrical diagrams generally are required in the application of instruments and control systems: (1) the schematic or elementary diagram; and (2) the connection diagram. There is no clear distinction between the terms schematic and elementary; often the terms are used interchangeably. The term schematic shall be used throughout this section.

OBJECTIVES OF SCHEMATIC WIRING DIAGRAMS

The schematic diagram is one of the most valuable tools for designing, analyzing, and checking electric control systems. In the process industries, for example, where such systems are often complex, it is essential that a schematic diagram be made before attempting to produce a connection diagram or even to establish final equipment specifications. Even with simple control systems, it is much safer and more positive to prepare a schematic diagram before taking other detailed engineering steps. The schematic diagram is the only certain way to reveal the following details:

1. Contact action and number of control contacts required in controlling instruments.
2. Number of auxiliary contacts required in addition to control contacts.
3. Need for separate leads as opposed to common connections, for example, the use of two SPST switches as contrasted with one SPDT switch.
4. Number and type of relays necessary to accomplished desired control.
5. Number and type of auxiliary devices, such as push-button switches, selector switches, and signal lights.
6. Types of timers and their contact action where timed control functions are involved.
7. Feedback or "sneak" circuits. It is unlikely that such circuits would be recognized without a schematic diagram.
8. Electrical conditions that may cause the system to operate incorrectly.

With these details all neatly assembled in the form of a schematic diagram, it is much easier to visualize how the system will perform, and the schematic diagram makes an excellent basis for preparation of the connection diagram, the latter serving as the specific guide to the installation of wires between control devices and components.

PREPARATION OF SCHEMATIC DIAGRAMS

The schematic wiring diagram may be defined as one which shows, in straight-line form without regard for physical relationship, all control circuits, device components, and associated equipment of a control system.

Basic requirements for the preparation of a schematic wiring diagram include:

1. A clear understanding of how the control system is to operate. It is usually helpful to write a short description of the control system, including sequence of opera-

* Supervisory Engineer, Panel Division, Minneapolis-Honeywell Regulator Co., Philadelphia, Pa.

tion.   The diagram should be started by beginning with the first control operation and proceeding, step by step, through the complete cycle.

2. A thorough understanding of the function, operation, and contact action of control instruments and auxiliary devices that have been selected to accomplish the desired results.   In the form of questions, the following are some of the important points which must be covered:

    a. Do instruments have maintained, intermittent, or momentary contact action?

    b. Do instrument contacts have sufficient capacity to handle the load which they control?   Are external relays required?

    c. What is the contact action of control switches when instrument pen is above or below control index?   How are these switches wired?

    d. If a timer is required, how does it operate?   What contact action is necessary before timing, during timing, and after timing?   Should timer reset on power failure?

    e. If relays are required, should normally open (NO), normally closed (NC), or double-throw contacts or combination of these actions be specified?

3. Knowledge that all control components are commercially available with the operating characteristics necessary to satisfy the demands of the schematic diagram.

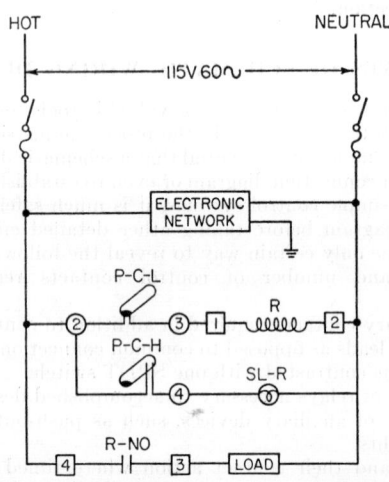

Fig. 1. Schematic wiring diagram for comparatively simple control system.   Connection wiring diagram for this system is given in Fig. 2.   *P* = potentiometer.   *R* = relay.

For instance, timer action or number of contacts may not be commercially available, thus requiring revision of the schematic.   Can one relay be obtained with the number and type of contacts?   Or will two or more be required?

4. Summary of all points that involve safety of operation.

5. Check on features of design from a cost standpoint.   A small change may greatly reduce the complexity of a system and consequently its cost.

The schematic wiring diagram of Fig. 1 represents the comparatively simple control system of Fig. 2.   Following are some of the characteristics of the schematic diagram:

1. Each circuit is drawn in straight-line form between vertical lines from one polarity of its source of power to the other.   The "hot" line is at the left; the neutral line is at the right.

2. Physical relationship is disregarded.   Relay coil and contact and instrument switches are disassociated.   Their position in the schematic diagram is arranged for convenience and simplicity and does not indicate physical position of the equipment.

3. Approximate physical relation of device terminals and connection to their ele-

ments are shown in the pictorial or connection diagram of Fig. 2.   In the schematic diagram of Fig. 1, the connections to the elements of a device are indicated by cross reference between terminal numbers and/or element symbol which appear on the connection diagram.   For example, *P-C-L* in Fig. 1 carries the same terminal and element designation as it does in Fig. 2.

4. All "loads" are arranged in vertical line on the right-hand side, while all contacts are arranged on the left-hand side.   Exception may be made for the sake of simplicity.

5. All elements of the diagram are identified and correspond to the legend description.

The wiring-connection diagram for a fairly complex program control system is shown in Fig. 3*a*.   The schematic wiring diagram for this system is shown in Fig. 3*b*.

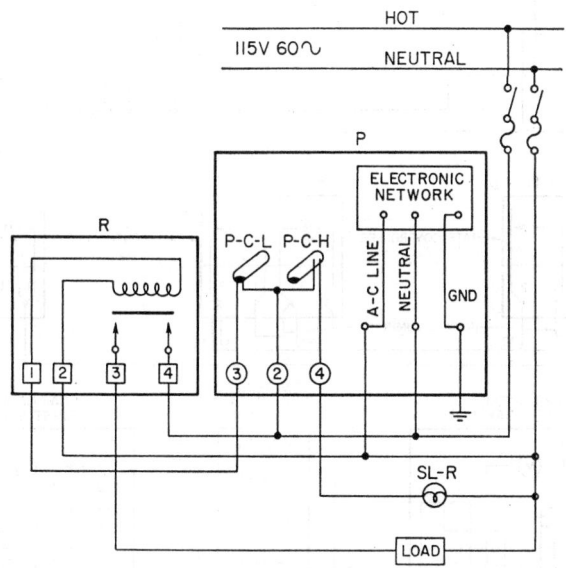

FIG. 2. Connection wiring diagram for comparatively simple control system.   Schematic wiring diagram for this system is given in Fig. 1.   *P* = potentiometer.   *R* = relay.

The time-temperature relationship chart for the system is shown in Fig. 3*c*.   Operation of the system is as follows:

1. Set the following:
a. Potentiometer right-hand index-operated switch to first soaking temperature.
b. Potentiometer left-hand index-operated switch to second soaking temperature.
c. Potentiometer auxiliary switch to point where program is to end.
d. Rate controllers to desired heating and cooling rates.
e. Timers for desired soaking period.
2. Close line switch to energize system.
3. Turn "program on-off" switch to "on" and press "start" button.   Potentiometer upscale index drive is energized and index and temperature will be driven upscale at a controlled rate until index reaches first soaking temperature.   At this point, upscale drive is deenergized, and when instrument pen reaches soaking temperature, timer will be energized, thus starting first timed soaking period.
4. At end of soaking period, timer times out, energizing downscale index drive.   Index and temperature will be driven downscale at a controlled rate until second soaking temperature is reached.   At this point, downscale index drive is deenergized and second soak timer is energized, thus starting second timed soaking period.
5. At end of soaking period, downscale index drive is again energized and index and temperature will be driven downscale at a different controlled rate until index reaches bottom of scale.   At this point, downscale index drive is deenergized, and when pen reaches end of program, the heating load is deenergized.
6. To operate without program control, turn "program on-off" switch to "off" and press "start" button.   Temperature will then be controlled about manually positioned index.

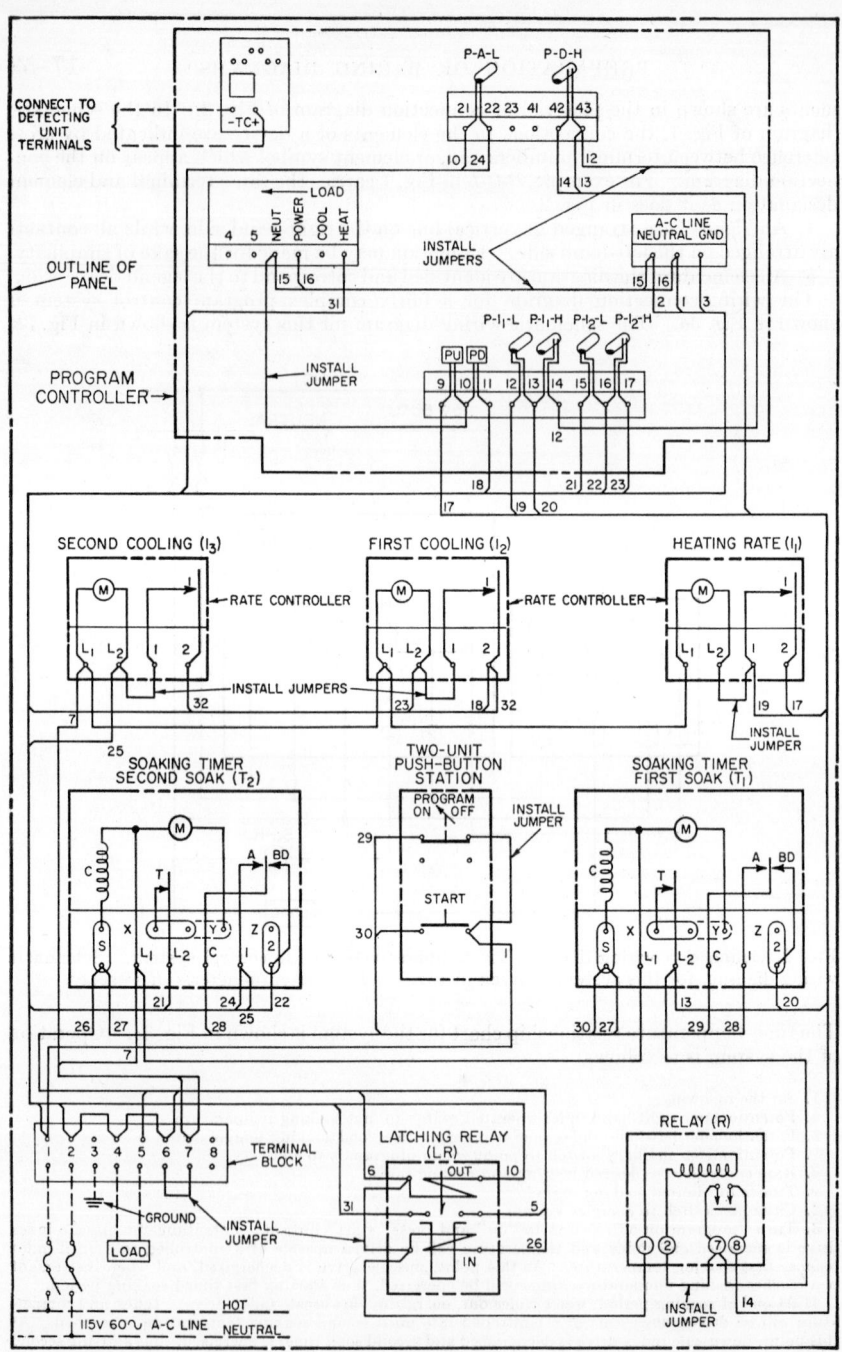

Fig. 3. (a). Program control system—connection wiring diagram.

$T\text{-}C\text{-}D$ = timer clutch, deenergized during timing period
$T\text{-}M$ = timer motor
$T\text{-}T$ = timer timing contact
$T\text{-}BD$ = timer load contact, closed before and during timing
$T\text{-}A$ = timer load contact, closed after timing
$I\text{-}M$ = rate-controller motor
$I\text{-}I$ = rate controller, interrupting contact
R = relay coil

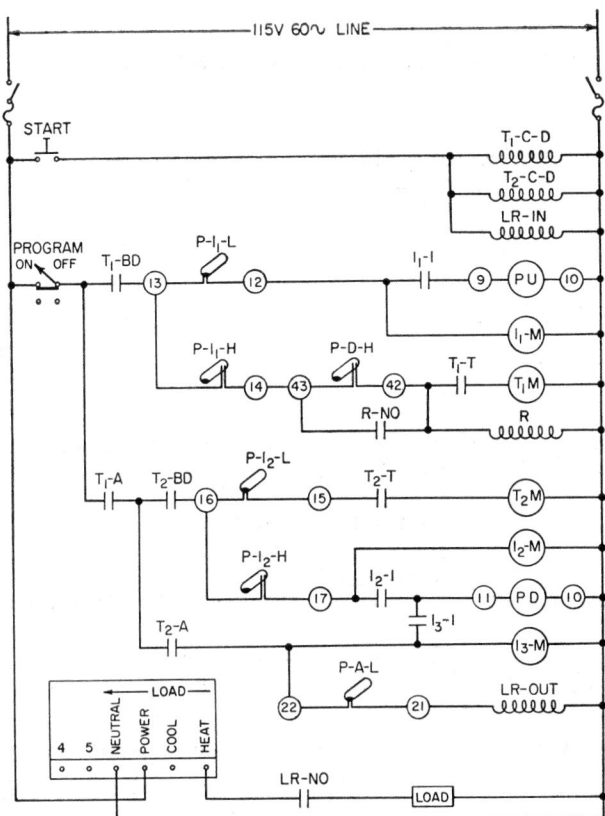

FIG. 3. (b) Program control system—connection wiring diagram.

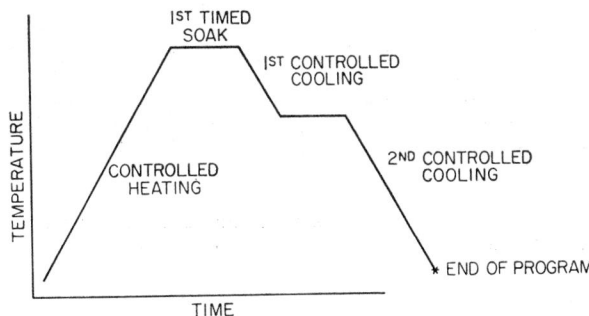

FIG. 3. (c) Program control system—time-temperature relationship.

R-NO = relay normally open contact
LR-IN = mechanically latched relay "in" coil—armature pulls in and latches when energized
LR-OUT = mechanically latched relay "out" coil—armature unlatches when energized
LR-NO = mechanically latched relay contact, open when unlatched
P-I_1-L = potentiometer right-hand index-operated switch closed below switch setting
P-I_1-H = potentiometer right-hand index-operated switch closed above switch setting
P-I_2-L = potentiometer left-hand index-operated switch closed below switch setting
P-I_2-H = potentiometer left-hand index-operated switch closed above switch setting
P-D-H = potentiometer differential operated auxiliary switch, closed when pen is above index
P-A-L = potentiometer auxiliary switch, closed when pen is below switch setting
PU = potentiometer upscale index-drive motor
PD = potentiometer downscale index-drive motor
O = potentiometer terminals

17–55

SYMBOLS

Familiarity with the symbols used is necessary to a clear understanding and interpretation of the schematic diagram.   The fundamental symbols are, in many instances, the same as those used on other types of diagrams.   However, because of the nature of control systems, there may be symbols and designations that are peculiar to these systems or that represent new or less frequently used devices.   Because the schematic diagram is concerned with the elements of devices and their electrical interconnection, the symbols used represent device elements rather than complete devices.   The diagram always must contain a complete legend of symbols used.

The order, from left to right, in which letters and numerical symbols appear in the complete symbol or identification for an element in the schematic diagram is very important.   The system described here has been used successfully for the preparation of literally thousands of schematic wiring diagrams.

As indicated by the example which follows, there are three left-to-right positions, each being separated by a dash.

Example

Position 1 - Position 2 - Position 3

P_1 - C - L

Normally written: P_1-C-L

where P_1 indicates (*first potentiometer*), C indicates (*control switch*), and L indicates (*contact closed when pen is below index*).

Example

Position 1 - Position 2 - Position 3

T_3 - C - E

Normally written: T_3-C-E

where T_3 indicates (*third timer*), C indicates (*clutch*), and E indicates (*energized during timing*).

Through consistent habits of use, the symbol C has come to have two designations—indicating a control switch when associated with a contact-making function and a clutch when associated with a timer.   However, because of this consistent association with other symbols, the dual usage of the C does not cause confusion.

Definitions of Positions.   The function of each position is defined as follows:

Position 1.   Identification of device.   The letter or letters serve to indicate the type of device, and the subscript number further identifies more than one device of the same type.

Position 2.   Identification of component.   The letter or letters identify components of the devices which appear in the schematic diagram.   Subscript numbers may be used where more than one of the same type of component is in the same device.

Position 3.   Identification of operation or function.   The letter or letters designate the conditions of operation or the function of the component.

Symbols and designations which find most common use in schematic wiring diagrams are given in Table 1.

Table 1. Designations and Symbols Commonly used in Schematic Wiring Diagrams

Designation	Symbol	Description
I_1		Rate controller
I_1M	—(I₁M)—	Rate-controller motor
I_1-I	—⊣⊢—	Rate controller, interrupting contact
T_1		Timer
T_1M	—(T₁M)—	Timer motor
T-C-D	—⌐₀₀₀₀₀₀₀—	Timer clutch, deenergized during timing period
T-C-E	—⌐₀₀₀₀₀₀₀—	Timer clutch, energized during timing period
T_1-T	—⊣⊢—	Timer timing contact
T_1-A	—⊣⊢—	Timer load contact, closed after timing
T_1-BD	—⊣⊢—	Timer load contact, closed before and during timing
T_1-BA	—⊣⊢—	Timer load contact, closed before and after timing
T_1-D	—⊣⊢—	Timer load contact, closed during timing
R_1	—₀₀₀₀₀₀₀—	Relay coil
LR_1-IN	—₀₀₀₀₀₀₀—	Mechanically latched relay "in" coil—armature pills in and latches when energized
LR_1-OUT	—₀₀₀₀₀₀₀—	Mechanically latched relay "out" coil—armature unlatches when energized
TDR_1-E	—⌐₀₀₀₀₀₀₀—	Time-delay relay coil—time delay starts when coil is energized
TDR_1-D	—₀₀₀₀₀₀₀⌐—	Time-delay relay coil—time delay starts when coil is deenergized
R_1-NO	—⊣⊢—	Relay normally open contact, i.e., when deenergized
R_1-NC	—⊬⊢—	Relay normally closed contact, i.e., when deenergized
LR_1-NO	—⊣⊢—	Mechanically latched relay contact—open when unlatched
LR_1-NC	—⊬⊢—	Mechanically latched relay contact—closed when unlatched
TDR_1-NO	—⊣⊢—	Time-delay relay normally open contact
TDR_1-NC	—⊣⊢—	Time-delay relay normally closed contact
R_1-1-NO	—⊣⊢—	No. 1 contact of relay 1—normally open
R_1-1-NC	—⊬⊢—	No. 1 contact of relay 1—normally closed

Table 1. Designations and Symbols Commonly used in Schematic Wiring Diagrams (*Continued*)

Designation	Symbol	Description
MV_1-L	—⊣⊢— OR	Millivoltmeter control contact—closed below control index (or closed below neutral zone)
MV_1-N	—⊣⊢— OR	Millivoltmeter control contact—closed within neutral zone
MV_1-H	—⊣⊢— OR	Millivoltmeter control contact—closed above control index (or closed above neutral zone)
MV_1-NH	—⊣⊢— OR	Millivoltmeter control contact—closed in neutral and high zones
MV_1-LN	—⊣⊢— OR	Millivoltmeter control contact—closed in low and neutral zone
TC_1-L	—⊣⊢— OR	Temperature-controller contact—closed on temperature drop (or closed below index)
TC_1-H	—⊣⊢— OR	Temperature-controller contact—closed on temperature rise (or closed above index)
PC_1-L	—⊣⊢— OR	Pressure-controller contact—closed on pressure drop (or closed below index)
PC_1-H	—⊣⊢— OR	Pressure-controller contact—closed on pressure rise (or closed above index)
P_1-I_1-L		Potentiometer right-hand index-operated switch—closed below switch setting*
P_1-I_1-H		Potentiometer right-hand index-operated switch—closed above switch setting*
P_1-I_2-L		Potentiometer center index-operated switch—closed below switch setting*
P_1-I_2-H		Potentiometer center index-operated switch—closed above switch setting*
P_1-I_3-L		Potentiometer left-hand index-operated switch—closed below switch setting*
P_1-I_3-H		Potentiometer left-hand index-operated switch—closed above switch setting*
P_1-D-L		Potentiometer differential-operated auxiliary switch—closed when pen is below index
P_1-D-H		Potentiometer differential-operated auxiliary switch—closed when pen is above index
P_1-A-L		Potentiometer auxiliary switch—closed when pen is below switch setting
P_1-A-H		Potentiometer auxiliary switch—closed when pen is above switch setting

Table 1. Designations and Symbols Commonly used in Schematic Wiring Diagrams (*Continued*)

Designation	Symbol	Description
P_1-C-L		Potentiometer control switch—closed when pen is below index (or below neutral zone)
P_1-C-N		Potentiometer control switch—closed when pen is within neutral zone
P_1-C-H		Potentiometer control switch—closed when pen is above index (or above neutral zone)
P_1-C-LN		Potentiometer control switch—closed when pen is in low and neutral zones
P_1-C-NH		Potentiometer control switch—closed when pen is in neutral and high zones

The designations for control switches given above are only for single-point controllers. If necessary subnumbers may be added to the No. 3 position to indicate the various records on multipoint controllers

Designation	Symbol	Description
P_1U		Potentiometer upscale index-drive motor
P_1D		Potentiometer downscale index-drive motor
S		Switch
S		Switch, fused
Signal	SIGNAL	Signal—this symbol and designation used when exact type is not known
SG_1 or SB_1		Signal gong or signal buzzer
SH_1		Signal horn
SL_1-W	OR (W)	Signal light, white
SL_1-R	OR (R)	Signal light, red
SL_1-B	OR (B)	Signal light, blue
SL_1-G	OR (G)	Signal light, green
SL_1-A	OR (A)	Signal light, amber
SV_1		Solenoid valve
Load	FURNACE	Load

* These descriptions pertain to one specific brand (Brown) instruments. Index-operated switches of other instruments should be described in a manner which permits easy identification of their contact action and location.

INSTRUMENT PANELBOARD DESIGN
AND CONSTRUCTION

By Wesley T. Dumser*

With the growing importance and complexity of instrumentation has developed the practical need of centralizing the instrumentation for a piece of equipment, a process, or an entire plant onto one or more panels. It has become desirable frequently to locate these panels in a central control room convenient to the individual responsible for the operation of the process. The instrumentation on the control panel must be so designed that it can be used as an integral part of the over-all control system and so that the information portrayed on the instrument panel can be coordinated with information available from data-logging and computer-control equipment. The average investment in instrumentation per plant today is of a magnitude which requires that a maximum of convenience in the use of instruments be paramount.

PANELBOARD FUNCTIONS AND TYPES

The primary purposes of a panelboard are to support and protect instruments and, at the same time, to group them in a single location. Thus two purposes are served: (1) the instruments become more convenient to use and (2) the interrelation of process variables and controlling forces becomes more apparent to the operator.

Three definitions are in order:

1. *Unit Panel.* A simple vertical panel on which instruments are mounted for support and protection.

2. *Control Board.* A panel which groups together the related instruments and controls required for the intelligent operation of a process step or steps.

3. *Central Control Room.* A room for housing the control board for large and complex processes. Its fundamental purpose is to bring before the operators all the interrelated factors which must be correlated in order to keep the process operating satisfactorily. The term control center is also used to describe a central control room.

In general, the more information that can be put before an operator, the more intelligently he can operate a complex process. Of course, there is a limit to the total volume of instrumentation which one man can absorb, particularly during times of process disturbances or upset. This limit is often reached in modern control rooms and is a subject worthy of much further study.

Where the amount of data to be brought to the control center is great, the size of the control room tends to become large. This has stimulated the use of miniature instruments in lieu of the large case instruments commonly used until recent years. The development of the reduced-size instruments has made it possible to group large numbers of instruments in a small space, thus extending the scope of the control room beyond that obtainable prior to miniaturization. Control rooms for process instrumentation have begun to take on more of the characteristics of electrical control boards, which have, for many years, used compact indicating and controlling instruments.

Miniature instrumentation has made it possible to group so many instruments before an operator that again his limit of capacity to absorb and correlate information is

* Supervisory Engineer, Republic Flow Meters Company, A Division of Rockwell Manufacturing Company, Chicago, Illinois.

reached. As a result, it is often necessary to employ visual aids and other devices—such as graphic display, alarms, annunciators, and safety devices—to extend the operator's capacity to absorb information. This has led to the development of the semigraphic and graphic panel which, by combining visual-aid techniques with miniature instrumentation, has made it possible to put the maximum amount of instrumentation and control equipment on a panelboard.*

PANELBOARD DESIGN

The design of instrument panelboards must be approached from both the point of view of good initial design and the point of view of potential process modifications. It is important in designing control panels for a new plant to take into account ultimate expansions of this plant as well as minor modifications which can be expected to occur. Here one must reach a compromise between initial cost and providing capacity for future expansions balanced against the possible obsolescence of an instrument installation if adequate flexibility is not provided for. New panelboards should be designed with sufficient flexibility to make future additions as easy to accomplish as possible.

The material presented below applies principally to design for new construction. However, the same general practice can be followed when making revisions to existing instrumentation.

Basic Arrangements of Instrumentation

Two possible approaches to the arrangement of instrumentation are shown in Figs. 1 and 2. In Fig. 1, three identical four-step processes have been laid out, with the raw materials entering from area A and the finished products leaving at area B. Unit panels are shown installed at each step of each process. So long as each step in each process has one operator, this arrangement may work well. But, should one operator attempt to run all four steps of one process, difficulties are encountered. The operator may need to know a rate of flow at step 3, while making adjustments on step 4. Because this information is not available, step 4 can go badly out of control later while he is checking conditions at step 1. If an alarm sounds at step 4, he must run the length of the process line in order to make the necessary correction. An installation of this type usually results from considering instrumentation as an afterthought.

The same three identical four-step processes can also be laid out as shown in Fig. 2. In this case, indicating transmitters are used for many of the readings and information from the transmitters can be recorded on multipoint instruments. Here the operator is able to view the process in its entirety and is able to give quick attention to out-of-control steps. It may also be noted that both the operator and the instruments are isolated from high-pressure high-temperature areas, through use of the principle of retransmission. Laying out the process as shown in Fig. 2 takes full advantage of instrumentation and makes much more efficient use of the operator.

Operator Limitations. The operator is the key to the problem of designing a control room of the type in Fig. 2. He can remain in effective control of the process only when (1) he is receiving and comprehending all the necessary information and (2) he is able to take appropriate corrective action in time.

In most early control rooms, the instruments were installed on a panelboard built along one straight wall. As processes became more complex, the length of the panel began to exceed the operator's range of vision from any one position in the control room. The length of panel that can be controlled by a single operator varies with (1) the density of instrumentation on the panel, (2) the number of variables which affect the stability of the process and which the operator may possibly have to adjust, and (3) the form in which the instrumentation is presented to him.

When a panel exceeds 30 to 40 ft in a straight line, it is almost impossible for one operator to see and visualize what is taking place, let alone control it. The time required in moving from one panel section to another becomes a handicap in time of

* Ref. 1, Sec. 8, pp. 8-48 to 8-56, Graphic Instrumentation.

trouble, but a more serious objection lies in the fact that the operator tends to lose the correlation between various parts of the process when moving from point to point along the panel.

Rectangular Control Room. The better approach is that shown in Fig. 2 in which the control room is arranged in an approximately rectangular shape. With this arrangement, up to 150 ft of control board can be placed within the view of a single operator placed in the center of the room. However, in times of emergency, startup, or shutdown, it does not necessarily follow that one operator can handle all the required controls.

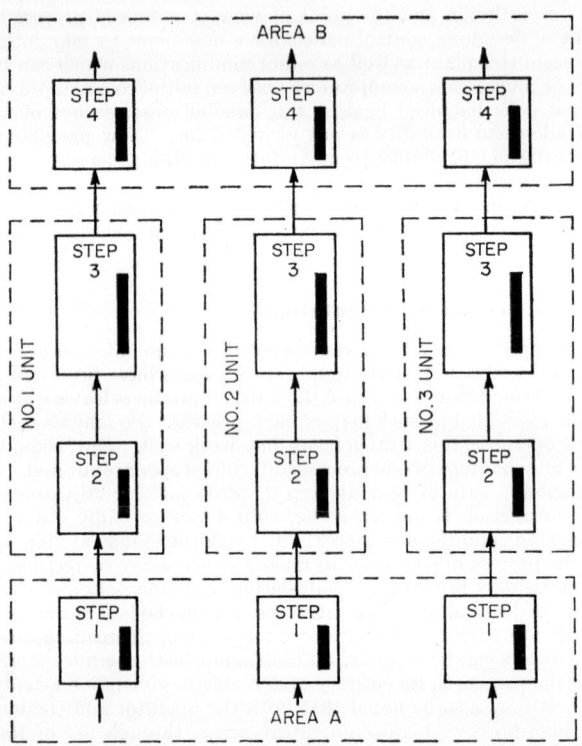

Fig. 1. Arrangement of three four-step processes with separate instrument panelboards for each processing step, giving a total of 12 separate panelboards.

Control rooms of this type can be quite successful when a single operator is employed even where large instruments requiring long panelboards are used. This is particularly true when the panel is equipped with ample signals and safety devices. If a control room this large is equipped with miniature or graphic instrumentation, it is likely to be so condensed and to control so much more process equipment that two or more operators required depends not so much on the size of the control room as on the amount and arrangement of instrumentation which it contains.

Control Rooms for Small Panels. A four-sided control room is not practical when the panelboard is less than 70 ft in length. For shorter boards a U-shaped arrangement is more practical, with windows in the fourth side to give the operator some feeling of contact with the rest of the plant.

There is a real danger, if the control room is made small and completely isolated from the rest of the plant, that the operator may not have enough work and activity to keep him alert. Operators in small and isolated control rooms have found it almost impossible to stay awake during the night shifts. If the control of the process does not keep him busy, he should have other duties that keep him moving about or talking to other attendants.

Panel Dimensions

The over-all length of a panel should be kept as short as practical without undue crowding. Instruments should be grouped to follow the logical sequence of the

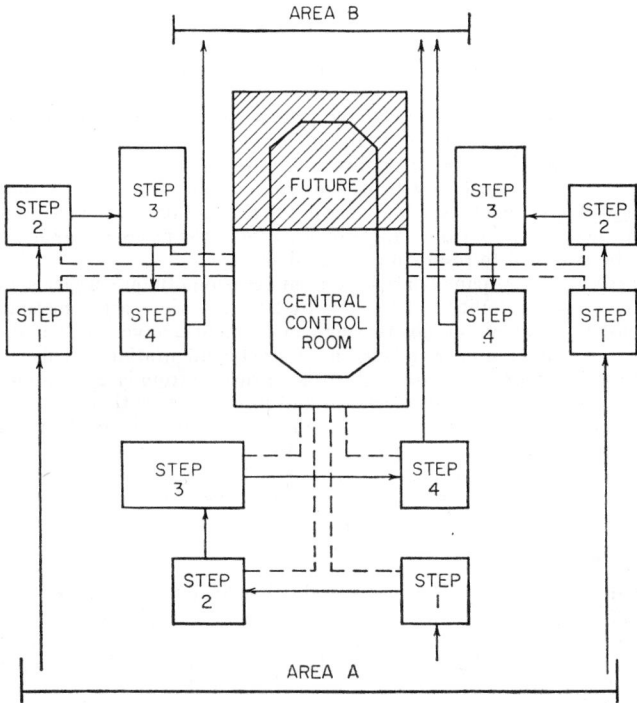

Fig. 2. Arrangement of three four-step processes in which all instrumentation is centered in a central control room. From one to three panelboards in the control room replace the 12 separate boards of Fig. 1. Dashed line indicated instrument piping or wiring from process to central control room.

process. In addition, they should be arranged on the panel with the most important recorders and controllers located at eye level. Indicators, alarms, and push buttons can be located above and below this level. Since usable height is restricted by this arrangement, the length of the panel will be a compromise between the shortest possible length and the best instrument grouping.

The length of individual sections will depend on the total length as well as on the character of the board.

Short panelboards (up to 10 ft in length) are most economically built in one or two sections. This permits the use of the rear structure of the panel for the mounting of appurtenances. Many users prefer to make panel sections in 2- or 4-ft modules. This permits replacement of a panel section for change in the field but complicates the interconnecting of individual components in a complex system.

Long panel sections should be split in equal lengths of optimum size. Optimum size usually relates to the configuration of instruments for each particular process section.

As the width of individual panel sections is decreased, the material-handling problems increase.

Preferred Panel Sizes. Though panel section widths vary greatly in practice, certain sizes appear to be most commonly preferred. For simple arrangements where large-case conventional instrumentation is used, 24-in.-wide panels are quite common since they accommodate one vertical row of almost any standard instrument.

In the more complex panels, 44- and 48-in. widths are the most frequently used. The 44-in. width approaches the narrowest panel which will accommodate two conventional instruments of almost any make, side by side, and still allow room for piping and wiring. With the growing use of miniature instrumentation combined with visual representation of the process, the grouping of process units tends to become more important in the determination of panel sizes. The trend is to select panel sizes so that the instrumentation for the available process unit can be grouped on one panel section with the desired graphic or semigraphic presentation.

The height of a panelboard is often dictated by the standard size of metal or plastic panel available. Thus most panelboards are made from 8-ft stock. Generally, they are about 7½ ft high when finished because of the amount of metal which must be bent back to form the top and bottom flanges or because of the necessary amount of plastic which must be trimmed to result in a squared sheet.

Sectional (or modular) panelboards are used occasionally on simpler arrangements to provide ultimate flexibility. Thus, 24- by 24-in. sections are sometimes selected simply because they can accommodate single instruments conveniently and are racked together on pipe or angle-iron stands in any desired configuration. This method generally requires more room than control boards fabricated from larger sections and does not lend itself so well to complex electrical and piping installations.

One factor which cannot be overlooked with regard to panelboard length and height is the maximum dimensions which can be practically and economically transported by public carriers during shipment of the equipment. Also, size is sometimes dictated by the size of available openings to an existing control room.

Panel Materials

Metal Panels. These are almost invariably constructed of mild steel. Panel sections up to 3 ft wide are frequently made of ⅛-in. cold-rolled steel which has a smooth finish but which is hard to fabricate and tends to buckle when cut by a torch. It is a good material for standardized panel production and for small units which are not likely to be altered after installation.

Larger panel sections are most frequently made from ¼-in. hot-rolled steel, although 3⁄16-in. steel is sometimes preferred. This thicker material is easier to weld and burn with a torch and does not warp so readily when welded or torch-cut. It is much sturdier and requires much less bracing than lighter-gage material. It is also possible to blind-tap ¼-in. steel for mounting equipment on the rear of the panel. Because of its rougher surface, ¼-in. steel is more costly to finish than cold-rolled steel. However, careful finishing methods can result in a good-appearing finished panel. Stainless steel has been used in the food industries for sanitary considerations. It is difficult to work and, of course, quite expensive.

Nonmetal Panels. Nonmetal panels and metal panels with plastic finishes are becoming more popular where corrosion problems make metal panels difficult to maintain. Although marble and slate panels were among the earliest materials used, they are now practically obsolete because of the difficulty in fabricating finished control boards from them.

Ebony Asbestos. Ebony asbestos, ½ to 1 in. thick, can be used to build small panels. This material withstands corrosion well, has a good appearance, and requires no finish. It is not easy to cut, however, and requires rigid supports by a metal framework structure.

Die-Stock Masonite.   Die-Stock Masonite, $\frac{1}{2}$ to $\frac{3}{4}$ in. thick, sometimes is used for sectional panels.   While this is an easy material to work, it requires "bolted-through" construction, since blind-tapped holes are not sufficiently strong.   Also, the corners and edges are fragile and tend to chip.

Fiberglass Panels.   Fiberglass panels, $\frac{1}{4}$ in. thick, are coming into use for extremely corrosive atmospheres.   These panels are available up to 2 ft wide and finished in a variety of colors.   This material, as is the case with other nonmetallic materials, is difficult to patch if instrument cutouts are changed.   As this is a molded material, the face of the panel must have a slight curvature to allow withdrawal from the mold.

Laminated Plastic.   Melamine plastic sheets $\frac{1}{4}$ to $\frac{1}{2}$ in. thick rigidly bolted or riveted to a metal framework have been used successfully for panelboard materials.   These materials have durable finishes and resist scratching and discoloration.   They are available in a wide variety of colors.   This material has also been laminated in $\frac{1}{16}$-in. sheets directly onto an aluminum backing $\frac{3}{16}$ or $\frac{1}{4}$ in. thick.   This combines the finished characteristics of the plastic with the strength of metal.

Plastic materials are generally too brittle to use in thin sheets without considerable framework supports; so the alternate approach of bonding the plastic to the aluminum has been found to be a good solution.   This is a particularly successful approach for graphic or semigraphic panels because nameplates and graphic symbols and lines can be glued directly to the face of the panel without damaging the finish.   The above construction is somewhat more expensive than painted steel.   Panelboards are also now being constructed with $\frac{1}{64}$-in.-thick sheets of vinyl material bonded directly to both sides of the metal backing.   The bonding is done on *both* sides so that the differences in the coefficients of expansion of vinyl material and the metal backing cancel out and eliminate warpage which would otherwise result.

To summarize, almost any structural material can be used for panelboards, but steel seems to have the widest application, particularly for the more complex panelboard structures.   Steel is certainly the most economical material to use for this purpose.

Location of Panels

The ultimate location of a panelboard has much bearing upon its design.   Important factors are (1) corrosive atmosphere, (2) indoor vs. outdoor installation, (3) effects of weather and climate, (4) explosion or fire hazards, (5) cleanliness, and (6) vibration.

Corrosive Atmospheres.   This is an especially important factor where unit panels are mounted close to a process.   Moisture and corrosive fumes, if present, usually attack the instruments or connections.   In extreme cases, corrosion is even a problem in connection with the panelboard material itself.   Location of a panel beneath piping which may drip acid would be such a case, although that would indicate poor initial planning.   Where corrosion of the panel itself is a consideration, ebony asbestos or slate may be used, but a more practical solution may be the use of corrosion-resistant coatings on the panel steel, with plastic covering on the connecting piping.

Where corrosive conditions are such that there is concern with the panelboard material, great consideration should be given to the components where the effects will be even more serious.   Instrument housings can be purged with an inert gas and electrical connections can be enclosed in vaporproof housings.

Humidity.   Humidity may be a factor in corrosion, but principally it is a problem in connection with chart papers.   Expansion or shrinkage of charts can cause recording and reading errors.   In excessively humid locations, special chart materials can be used to reduce these errors.   Thin plastic nameplates have been known to buckle because of humidity and temperature changes; in such climates it is wise to use heavier nameplate materials.

Outdoor Locations.   Panelboards should be protected by rain or weather hoods, which have the additional function of protecting the operator while he changes charts and makes adjustments.   Outdoor-unit panels (located near the equipment which they monitor) are chiefly in connection with start-ups and often are needed even though a central control room is used.

Panels in outdoor locations usually are equipped with the larger conventional instruments because (1) only a few instruments are usually needed on the panel and grouping does not present a serious problem and (2) the larger instruments usually are more rugged and withstand the weather better. Miniature instruments are improving in the latter respect, however, even though there is little reason for using them in such local panels.

Explosion Hazards. These are defined in the National Electrical Code. Requirements are stated in detail in Article 500 of the code. The American Petroleum Institute also has defined requirements in this area.

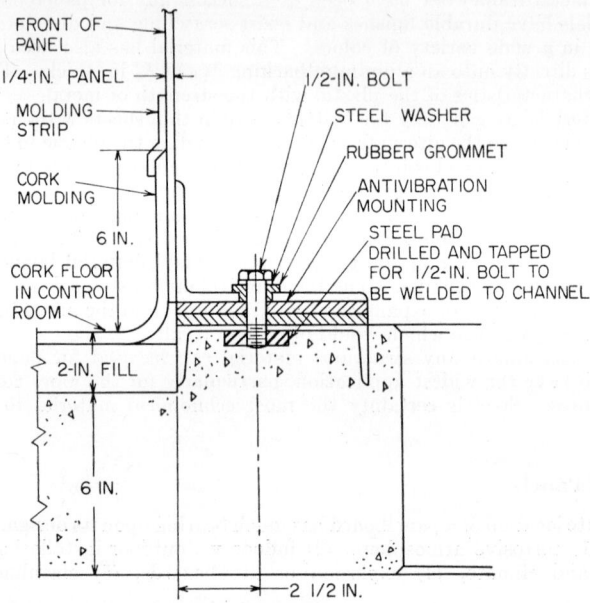

FIG. 3. Antivibration pad used in panelboard mounting.

Class I, *Div.* 1.   These locations are those in which an explosion could occur if any flame or electric arc were present to ignite explosive fumes or dusts. Wiring in such areas must be completely explosionproof.

Class I, *Div.* 2.   These locations* are kept from becoming hazardous by positive ventilation and are safe unless the ventilation breaks down. Equipment used in a Div. 2 location should be safe and unlikely to cause sparks, but it does not have to be completely explosionproof, in accordance with the requirements of the National Electrical Code.

In addition to their greater cost, explosionproof wiring and equipment have the disadvantage of being very bulky. This is a serious limitation in panelboard use, especially in graphic panels where many alarms and electrical devices must be grouped together in a small space.

Dust.   Much harm has been done to instruments during construction periods because of inadequate protection against the inevitable dirt, dust, and mistreatment. Control rooms designed for air conditioning, like others, are usually very dirty before a start-up, and equipment must be installed with this in mind.

---

* In England, the codes recognize certain "intrinsically safe" electrical devices for use even in hazardous areas. The ISA is presently studying the possibility of designing equipment which is inherently safe, even when not enclosed in explosionproof housings.

For installations such as cement plants and dusty boiler rooms, it is best to use pressurized cubicles. A positive blower, drawing air in through an adequate glass-wool filter, can easily keep a slight positive pressure in the cubicles and instrument cabinets and protect them from damage even in very dusty plants. However, such cubicles must be kept under surveillance because, strange as it may seem, operating personnel often are negligent in keeping access doors closed, thus defeating the purpose of pressurizing.

Cleanliness and Safety. These factors become more important as a control board becomes more complex. Service and calibration costs are reduced where clean locations are selected. An operator will also attain a greater efficiency in a clean, safe location.

Vibration. This can cause serious damage to instruments when a panelboard is located near compressors or other heavy reciprocating equipment. Two general approaches are possible: (1) use of anti-vibration mountings and (2) isolation of control-room area.

Even large panelboards can be protected from moderate vibration by some form of shock mounting. In Fig. 3, the panelboard is shown mounted on a pad of resilient material. Figure 4 shows a method in which rubber suspensions are used to reduce vibration in more extreme cases. Control rooms can be isolated from vibration by supporting the rooms on a separate foundation and joining the floor to the surrounding area with a mastic material. This approach is similar to that commonly practiced in the mounting of turbines and has been used satisfactorily for some control rooms installed in the center of a compressor house.

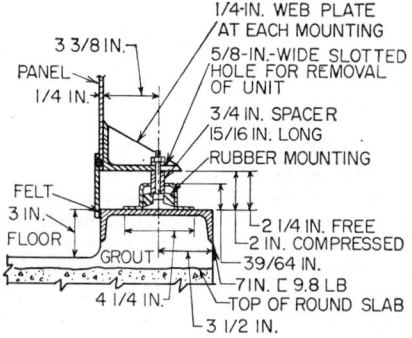

FIG. 4. Antivibration suspension for mounting panelboard.

Where vibration dampeners are used, it is necessary to know the total weight and the approximate center of gravity of the panel so that the proper number and location of dampeners can be applied. The benefit of the dampeners will be nullified if heavy piping or conduit is fastened rigidly to the panelboard.

For very delicate instruments, such as those incorporating galvanometers, individual suspensions must be designed. However, instruments for panelboard mounting are generally designed to withstand a reasonable amount of vibration.

CONTROL-ROOM DESIGN

The layout of the control room will be governed in part by its location, but principally by the arrangement of the instruments, the piping and wiring connections, and the general style and design of panelboard used.

Two considerations are important to control-room layout: (1) provisions for servicing the instruments from the rear of the panel and (2) handling of incoming piping and wiring.

Sufficient space between the rear of the panel and the wall behind it must be allowed so that a maintenance man can conveniently bring in his tools and equipment and work without undue crowding. This usually requires a minimum of 5 ft. Where miniature instruments are used in a graphic panel, the panelboard proper is frequently 2 ft deep. In such cases, it is best to allow at least 6 ft between the face of the panel and the wall of the room.

Piping and Wiring

Centralization of instruments has created new problems in bringing piping and wiring to the panelboard A variety of schemes have been evolved to simplify this problem.   The modern control room is likely to have about ten incoming tubes per foot of panel length.   Where miniature instrumentation is placed in a compact arrangement, there may be as many as twenty incoming tubes per foot.   Thus, in some panels, the designer may have to cope with hundreds of incoming tubes and similar numbers of electrical wires.   These must be arranged so as to reach their destined instruments throughout the panel in a logical manner and one which will facilitate servicing and making later changes.

Separation of Piping and Wiring.   It is good practice to separate piping from wiring at the outset by bringing all piping to the top of the panel and all wiring to the bottom.   Not only does this simplify incoming connections, but it also often reduces the apparent complexity behind the panel, because push buttons and electrical manual controls are usually located near the bottom of the panel.   In instances where many alarms and electrical devices are located along the top of the panel, the procedure can be reversed.

Piping connections to the process can be simplified by bringing all control-room piping out of the room through the wall or ceiling and passing the parallel pipes across the roof of the room.   Pipes coming from load centers throughout the plant are then brought across the roof in a similar manner at right angles to the panel piping.   Not only does this make connection a simple matter (viz., proper crossing pipes are cut and a joint is sweated), but individual pipes and connections are easily found and identified. in times of trouble and change.   Changes in the system can also be made without disturbing the panel connections.

Panel-wiring difficulties usually arise because of the inflexibility of incoming wiring. Individual conduits brought in directly from the plant probably pose the largest problem.   An excellent solution, especially for nonhazardous conditions, is that of installing a common cross duct near the bottom of the panel.   This may be located in the floor or next to the curb.   It serves to make cross connections between panels and also allows control wiring to be interchanged after it has been installed.   This arrangement usually is more economical than attempts to design the board to accommodate incoming leads perfectly at the outset.

When incoming wiring is brought to the top of the panel, this duct can be moved to a more convenient location around the upper periphery of the room.   The electric code specifies that separate ducts must be used for signal wiring, power wiring, and wiring carrying over 600 volts.   It does not require separate ducts for alternating and direct current, or for different power voltages under 600 volts, but many electrical engineers prefer to have separate ducts for these uses.*

Piping and wiring to the panel are no problem when the concentration of instruments is low but become significant when the panel is condensed and, particularly, when a large amount of electrical control wiring is brought to the panel.

Wiring for Outdoor Plants

In most outdoor plants, such as oil refineries, it is customary to locate the control room on the ground floor.   In many large chemical plants, however, it has been found advantageous to locate the control room on the second floor because much of the equipment is located above ground level and the operator may have better access to the important levels of operation.   This has several advantages from an instrumentation viewpoint.   Piping can be brought in overhead, as in the ground-floor control room, but wiring can be brought in from below, thus separating them without the necessity of underground troughs.

Of course, the above considerations apply to panelboards where there is a substantial amount of electrical work on the panels.   In refinery practice the operation is so continuous that, except where electronic instrumentation is employed, electrical wiring on the panelboard is usually limited to (1) chart drives, (2) alarm systems, and (3) thermocouples.   In chemical plants, however, it is usually necessary to bring much of the pump control wiring to the board; this complicates matters considerably.

* Voltages over 125 volts should be avoided on complex instrumentation panels whenever possible, because instrument servicemen are not accustomed to working with higher voltages.

Control-room Lighting

The chief illumination concern is that of minimizing glare from glass enclosures, pilot lights, and alarms. However, this glare usually is a problem only in connection with the upper rows of instruments where the angle is such that light is reflected into the operator's eyes. Indirect, nonglare lighting, which can be obtained with "egg-crate" diffusers and indirect fixtures, is excellent.

Control rooms do not require a high intensity of illumination. Dials and charts on the panel usually stand out with good contrast and are easily visible under normal lighting conditions. High-intensity lighting can be a disadvantage where there are many alarms and bull's-eyes on the panel, since it reduces the contrast, making illuminated alarms less visible. This is particularly true when dim/bright illuminated alarms are employed. Glare on an alarms device also can give the false impression that it is signaling. Consequently, in many modern control rooms, the operators have purposely reduced the over-all room illumination.

Canopy Lighting

Where cubicles are located in isolated sections of the plant, illumination can be obtained by mounting the lights in a canopy above the panel. This type of lighting is convenient to install but requires careful design for the desired results. If the canopy is too close to the top row of instruments, there will be direct glare from the gage glasses and other light-colored materials mounted along the top row. This can be prevented by raising the canopy a few feet above the panel. Another disadvantage of canopy lighting is that it illuminates only the top half of the panel. The ratio of illumination between the top and bottom of the panel is usually on the order of 4:1.

Individual illumination of instrument cases is practical in the larger instruments and in many backlighted pressure gages. However, it is rather difficult to obtain in miniature instruments. Generally, it is a more costly form of illumination, but where it can be applied practically, it may have considerable utility. The main difficulty lies in the fact that not all instruments are available with such lighting. Therefore, other forms of lighting are needed in the control room, and the requirements between the lighting needed for individually illuminated instruments and for the nonilluminated instruments may conflict. Consequently, it is often best to depend on a uniform type of lighting throughout the control room.

Color in Control Rooms

Light colors are being used to a greater extent for finishing modern panels because it is possible to obtain a flat or matte finish with metallic-base lacquers in these colors. Black was almost a standard color for earlier panels, but it has the disadvantage of becoming shiny with continued use and cleaning. A shiny background increases the glare and reduces the contrast between instruments and panel.

Lighter colors also seem to have a better psychological effect on the operator and tend to keep him more alert.

With the advent of graphic panels, the question of color has become more important. Any light color, such as green, gray, blue, or tan, can be used as a satisfactory background against which the symbols and lines of the graphic panel show up distinctly.

Curbs and Panel Supports

Most panels are mounted on a curb which serves two functions: (1) comprises a flat support against which the panel can rest and (2) raises the panelboard above the floor line, tending to keep it more free from dust and dampness which may result from sweeping and washing the floors.

A commonly used curb construction is shown in Fig. 5. Note that the panel is fastened to the curb by means of a removable bolt. Bolts which are cast into the floor require that the panel sections be lifted and lowered into place. This causes installation difficulties and often results in the edges of the panels being damaged. With the removable-bolt construction, as shown in Fig. 5, it is possible to slide the sections into place, thereby reducing the possibility of damage. If it becomes necessary to move the board, it is easy to drill and tap additional holes for support.

The Unistrut channel and curb construction shown in Fig. 6 represents a practical method of clamping panels to a curb. The Unistrut channel is fastened to the floor, making it possible to slide the panel sections along the channel in either direction. This method is flexible and facilitates changing and moving the panels.

For individual panels or short panels, it may not be necessary to have a curb. In such cases, it is customary to lay a standard steel channel on the floor and bolt the panel to it, as shown in Fig. 7.

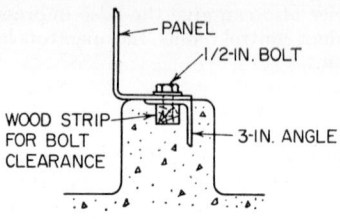 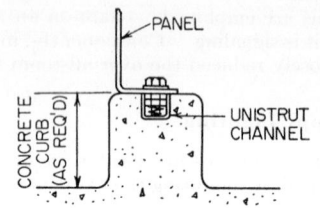

FIG. 5. Commonly used curb construction for panelboards.

FIG. 6. Unistrut curb construction for panelboards.

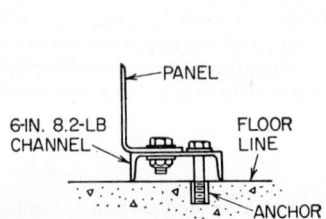 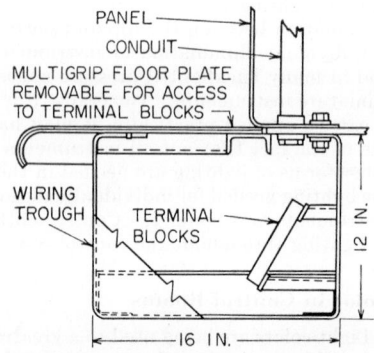

FIG. 7. Standard channel mounting for panelboards.

FIG. 8. Construction where panelboard is mounted on steel curb.

Mounting of a panel on a steel curb is shown in Fig. 8. This curb also serves as a wiring duct and is a convenient place to locate terminal blocks. Where controllers and recorders are arranged three high on a panel, the operator can sit on this curb to work on instruments at the lowest level and can use it as a step for reaching the instruments near the top of the panel.

Long rows of panel sections should be supported by tying them to a longitudinal angle iron running along the top of the board. This, in turn, can be supported from the back wall of the room or ceiling. Wing-back and cubicle panels do not require this type of support.

PANELBOARD COSTS

Many attempts have been made to establish rules of thumb for estimating panelboard costs. The advent of miniature instruments has made such a marked change in the concentration of instruments per foot of panelboard that most cost studies have been limited to special types of boards. A few simple rules may be helpful but care must be used in their application.

Steel Panels. With cutouts for instruments, these will vary in cost from 30 to 40 cents per pound, unpainted; and about 55 cents per pound, painted.

Piped and Wired Panels. With large-sized instruments, which are usually arranged in rows three high, these panels will cost approximately $175 per foot. The same type of panelboard using miniature instrumentation will vary from about $250 to $450 per foot and even higher, depending upon the density of the instrumentation used. Even more importantly, the cost will be dependent upon accessory equipment, the number of controllers used, and the number of pressure switches and solenoid valves that are mounted on the back of the panelboard.

Cubicles and Consoles. These and other special types of panelboards may add $100 to $300 per foot, again depending upon their complexity and refinements of construction, and the piping and wiring problems involved.

Graphic Panels. These may vary in cost from $325 to $650 per foot for standard types, although if much electrical equipment is added the cost may increase considerably above these figures.

The above estimates are for panels with instruments on one side of the cubicle only. Two-sided cubicles or tunnel-type panels may actually double these estimates. Very complex two-sided panelboards, such as used in large power stations, may cost between $1,000 and $2,000 per foot for each side of the panelboard in the case of very crowded boards. These costs result from the complexity of electrical and mechanical interlocking relay equipment and protective equipment mounted within the panelboards.

In all cases, these estimates do not include the cost of instrumentation but include only the panelboard, engineering, piping, wiring, assembly, testing, and the piping and wiring materials. These estimates are based on typical panelboards, both fabricated in the field and prefabricated, of the type commonly used today. Large design variations may cause discrepancies of as much as 100 per cent in any of these estimates.

PANELBOARD PIPING

Pneumatic transmission is used widely in the process fields and consequently becomes an important consideration in panelboard design.

Process Piping

Direct connection of process piping to instrument panelboards is generally restricted to the smaller panels which are installed near to the equipment or in the case of central control boards, it should be confined to safe fluids such as water and air.

Pneumatic Piping

This is encountered in connection with control equipment designed for low pressures (20 to 30 psi or less) with slightly higher pressures for the compressed-air supply. Even though low pressures are involved, care is required to make the systems pressure-tight; very small leaks can cause serious inaccuracies of measurement. Most panelboard piping is effected with tubing rather than pipe because of the relatively large number of bends and turns involved and the consequent reduction in the number of threaded fittings required. Tubing also provides greater compactness. See standards established by ISA on this subject.*

Soft Copper Tubing. This is the most commonly used material and lends itself well to careful and neat configurations behind the panel. It is probably the most economical of all materials available. Ordinarily, it is procured in rolls of dead-soft tubing which must be straightened and stretched before used. Stretching the length between 1 and 2 per cent will usually bring the tubing to a satisfactory hardness (about one-quarter hard) so that it will flare, make up well, and hold its shape.

* ISA Recommended Practice RP7.2, published by The Instrument Society of America, Pittsburgh, Pa.

Hard Copper Tubing. This is obtained in straight lengths of about 20 ft and is used in the refrigeration field. If made up carefully, this material is satisfactory but is much harder to flare and tends to crack at the flares. In general, it is more costly and offers no practical advantages over the soft copper type.

Aluminum Tubing. This can be worked in almost the same manner as copper tubing, but its use is limited by the type of atmosphere to which the panel will be exposed. Caustic atmospheres and welding sparks are very likely to cause damage. If aluminum tubing is used in contact with other metals in a moist atmosphere, electrolysis may take place. Consequently, in outdoor locations it must be carefully protected.

Plastic and Plastic-covered Tubing. In very corrosive atmospheres, these materials offer practical solutions. They can be used with ordinary fittings which can be covered with plastic tape to protect them against the atmosphere. Flexible plastic tubing can be conveniently utilized on control boards with quick-break couplings. The latter are readily disconnected for servicing and interchanging instruments.

Fittings. Almost any type of standard tubing fitting will satisfy the requirements of the ASA Piping Code for the low pressures used with pneumatic piping. The choice of fittings often is made to conform with the standards established throughout the plant.

Valves and Test Connections. It is good practice, on the more complex panelboards, to provide sufficient valves and test connections so that calibration and checking can be effected without removing the instruments. A test tee and valve should be included in all receiver connections if continued accuracy of the pneumatic instruments is essential.

PANELBOARD WIRING

The increased use of electrical and electronic instruments for the control of hazardous equipment has been greatly expedited as a result of the recognition by the National Electrical Code that a properly protected area in a hazardous plant can be made safe enough so that explosionproof wiring is not needed. The widespread use of remote transmission of measurements, either electrical or pneumatic, has made it possible to keep all hazardous fluids out of the control room. This practice, coupled with the use of forced ventilation, if the control room is close to hazardous equipment, has resulted in most central rooms being classified as either nonhazardous or, at most, semihazardous (class I, Div. 2 locations). In either case there is a tremendous saving in both first cost and operating cost as compared with explosionproof installations.

While it is true that electronic equipment can be protected by explosionproof housings, such housings are very costly and bulky and do not lend themselves readily to good panelboard arrangements. But even more significant is the fact that complex analysis instrumentation needs regular attention, and to be safe, all power must be disconnected from the instrument each time the case is opened. This is a serious limitation of their general utility. It is far better to locate the equipment in a safe control room where it can be used and serviced safely and conveniently.

Therefore, class I, Div. 2 wiring is becoming specified most frequently in central control rooms in hazardous plants. This type of wiring generally can be applied to panelboards quite economically and is similar to general-purpose wiring, with the following over-all exceptions:

1. All make-and-break switches and relay contacts must be hermetically sealed or enclosed in class I, group D housings.

2. Lamps, resistors, and the like can be used in that they are not likely to reach a temperature which is within 80 per cent of the safe ignition temperature.

3. Slidewire and adjustable rheostats are considered arcing devices and must be protected. Potentiometers for thermocouple actuation, however, are approved by the 1959 Revision of the Code, and electronic potentiometers therefore are approved by the Code for class I, Div. 2 locations—if they do not contain make-and-break switches or resistors which are likely to overheat. (See latest revision of the National Electrical Code for details.)

Purging of Electrical Devices

Purging of cabinets housing electronic devices is often used in hazardous and corrosive atmospheres. Continuous studies are being made by the Recommended Practices Committee of ISA to establish safe methods. In general, the following precautions should be observed when purging electric equipment in a hazardous area:

1. A bubbler or flow indicator should be provided in the flow line so that there will be a definite indication to the operator that the cabinet is being purged.

2. A pressure switch or other means should be provided which disconnects the power in event the cabinet is opened or the source of pressure fails.

Electric Power Supplies

Any standard method of connecting the power supplies to electronic instruments and electric clocks can be used in panelboard work. A very convenient method of connection is that of using a three-wire cable with a safety-type twist-lock plug and socket. Instruments equipped in this manner can be removed easily for servicing without the necessity of removing any conduits and wiring.

Location of Terminals

When laying out a panelboard, it is advisable to bring in the electrical connections at the opposite point of the panel from which the piping connections are fed into it. This provides a much more serviceable arrangement than where both the piping and wiring are brought in at either the top or bottom of the panel.

Incoming conduits very often are brought into a control room directly behind the individual panels. A practical alternative is that of providing ductwork either at the top of the panelboard or at the floor line. In this manner, incoming wires can be brought in at any convenient place and rerouted behind the panelboard to the specific instrument location. This permits a greater ease of installation and greater flexibility, particularly where revisions may be required.

BASIC PANELBOARD CONSTRUCTION

Actual panelboard structures available for consideration of the designer vary widely, but basically they are divided into general types, each of which is described below.

Subpanel and Frame

Figure 9 shows a typical unit panel arrangement where the panelboard is divided into small panel sections (20 by 24 in. or any standard size). Each of the subpanels is used to support an individual standard instrument or group of small instruments. The entire structure is held together by an angle-iron framework, which is so arranged that the panels can be shifted about easily.

This type of board is quite flexible and is used widely in many of the older process plants. Some of the flexibility is lost, however, when modern instrumentation, with its necessary complex interconnections and wiring on the rear of the panel, is used. In such cases, interchanging of instruments is probably as complicated with this type of unit as with a standard full-sized panel. A further disadvantage is that the board requires from twice as much to four times as much space as a panelboard designed for the purpose at hand. In the large majority of modern plants, where it is desirable to bring a large amount of instrumentation information before one operator, this type of board generally has been replaced by a full-sized steel board.

Full-sized Panels

Figure 10 is a representative design drawing for a typical full-sized steel panel such as is becoming standard for modern control rooms. The two types shown are merely alternate arrangements of bolting and flange construction.

Figure 11 shows an alternate construction for steel panels where flat steel plate is welded to an angle-iron or Unistrut framework. This method of construction is frequently used where there is a large amount of equipment to be mounted on the rear of the panel. The framework is utilized for this purpose.

The $\frac{1}{8}$-in. panels generally are used in widths up to 24 in., while the heavier thicknesses are used in the larger panel widths between 4 and 6 ft. Where a large number of instruments is mounted on the panel and consequently a large number of cutouts is needed, the $\frac{1}{4}$-in. panel offers some advantages. For example, it can be cut by a

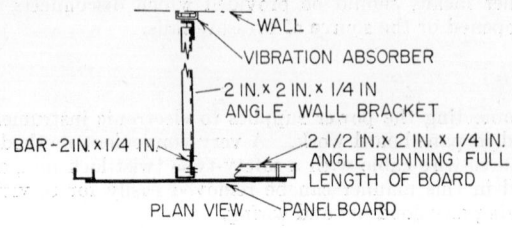

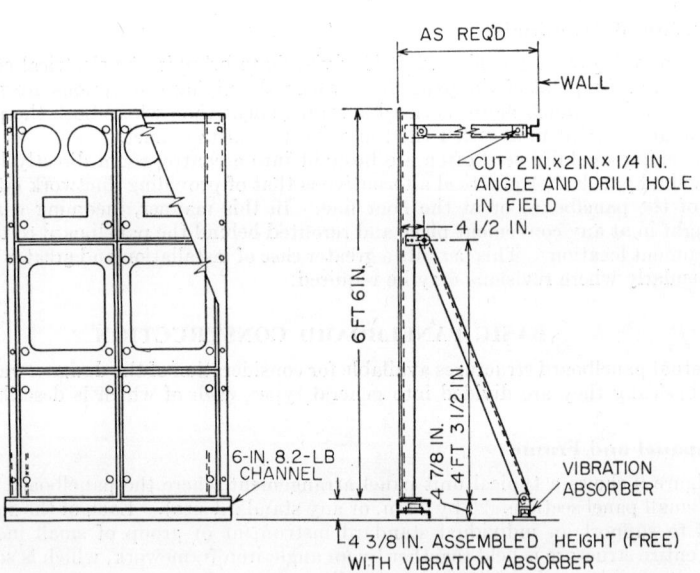

Fig. 9. Representative construction drawing showing a typical unit panel arrangement where panelboard is divided into small panel sections.

welding torch without undue distortion, even in the field. The $\frac{1}{8}$-in. and even the $\frac{3}{16}$-in. panels in the narrower sizes have the advantage that they can be fabricated of cold-rolled steel and lend themselves to a less costly painting and finishing process. Mild steel ($\frac{1}{4}$ in.) usually is hot-rolled and requires much more filling and a much better undercoating to obtain a good finish.

Wingback Panels

As the amount of material to be mounted on the rear of the panel increases, it becomes quite customary to extend the flanges of the panel back 18 to 24 in. or to provide side walls in order to give additional space for mounting accessory equipment. A further advantage is that the panelboard becomes self-supporting and thus requires no bracing at the top.   In general, wingback panels of this type are used in the wider panels (4 to 6 ft wide).

Cubicles and Cabinets

When panelboards are located in an open area, the rear of the panel usually needs some protection.   In such cases, cubicles and cabinets are used to enclose completely

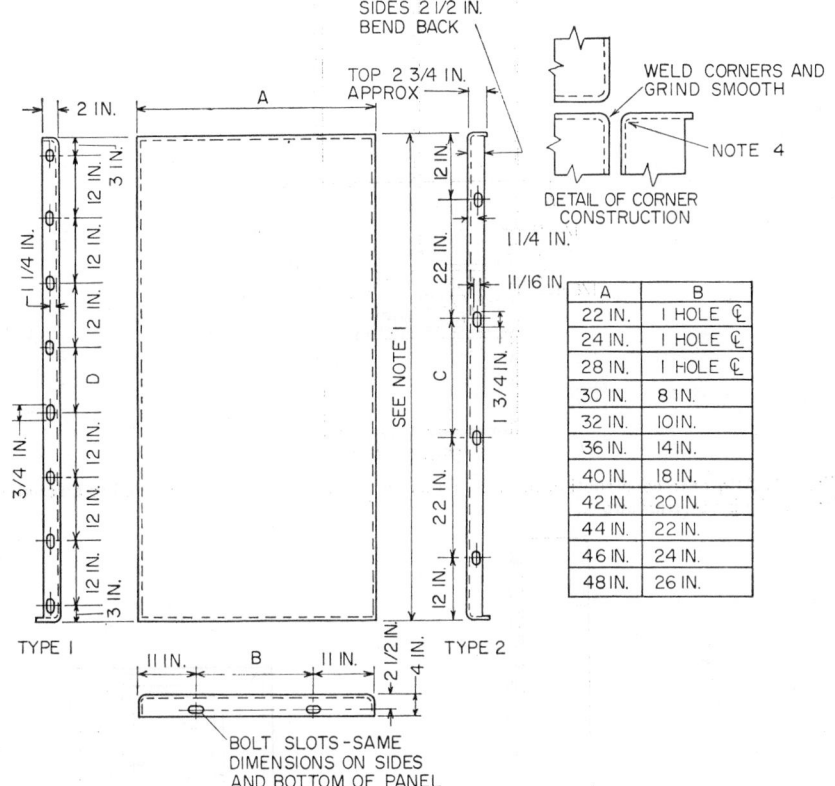

A	B
22 IN.	I HOLE ₵
24 IN.	I HOLE ₵
28 IN.	I HOLE ₵
30 IN.	8 IN.
32 IN.	10 IN.
36 IN.	14 IN.
40 IN.	18 IN.
42 IN.	20 IN.
44 IN.	22 IN.
46 IN.	24 IN.
48 IN.	26 IN.

FIG. 10. Representative construction drawing for a typical full-sized steel panelboard.
1. Panels to be 7 ft 6 in. high or as specified.   Vary C or D dimension only to match height.
2. Over-all dimensions $^{+0}_{-\,1/16}$ tolerance.
3. Panels must be flat and free from warps, surface flaws, and blemishes.
4. Radius R = panel thickness.

the instrumentation and incidental piping and wiring.   Figures 12 and 13 show a standard type of cubicle construction.   Figure 12 shows an 18-in.-deep cubicle of a lighter construction (usually made from $3/8$- to $3/16$-in. steel).   Figure 13 shows a combination attached-bench and walk-through-type cubicle whose depth may vary from 5 to 6 ft.   The cubicle shown in Fig. 13 is sufficiently deep so that instrumentation could be placed on both the front and rear panels.

One factor which must be considered is that it is generally less expensive to build the rear housing of ordinary masonry in the field than it is to supply a prefabricated cubicle.   This is particularly true of large panels.   A long row of panelboards can be made into a cubicle construction by lining them up 4 to 5 ft away from the wall and placing a door at each end. Such construction is commonly used and can approach the effectiveness of a cubicle structure and at much less cost.

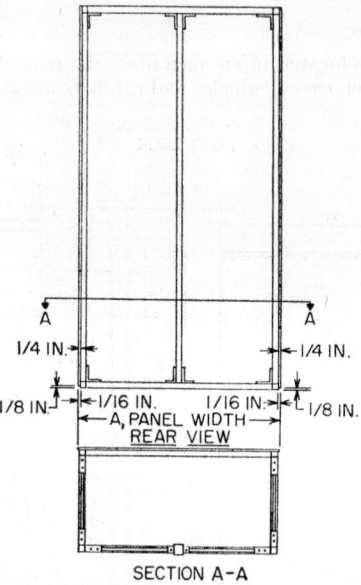

FIG. 11. Alternate construction for steel panels where flat steel plate is welded to an angle-iron or Unistrut framework.

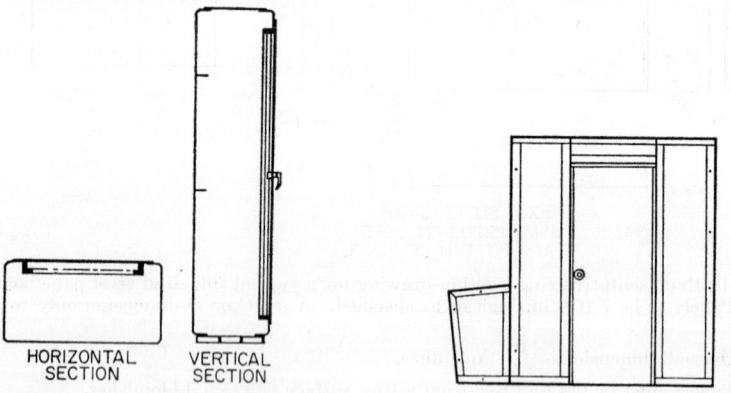

FIG. 12. Typical cubicle construction (light-weight).

FIG. 13. Typical cubicle construction (heavy).

Consoles and Operators' Desks

As the role of the operator has become more important and better understood in connection with the central control room, a number of new panel arrangements have been devised to increase the scope of his activities. The console type of panelboard or benchboard has become increasingly popular and is available in a large variety of arrangements, all of which serve to increase the number of control switches and devices that are within the operator's reach when he is sitting down. The general principles for design of these boards are similar in practice to what has long been "benchboard" practice in central-station power plants.

A slightly different arrangement that has become quite popular is a combination of an "operator's desk" with a temperature-monitoring system.   An ordinary steel type of office desk is equipped with an indicating pyrometer conveniently located at the rear edge along with a multiplicity of switches so that the operator can, at will, read any temperature throughout the plant and if necessary log the temperature in his record book.   In some installations the switches are replaced with a dialing system similar to that on a standard telephone handset.   The amplifiers and terminal plugs are usually mounted in the space behind the drawers in the desk in a convenient manner so that a very functional arrangement ensues.

Relay Racks

With the increased use of electronic equipment, there are many cases where a considerable number of amplifiers and other electrical accessories are part of the installation.   A very practical way of housing these accessories is to mount them in relay racks of the type used in radio and transmission gear.   To facilitate this, many companies are designing their accessory equipment for relay-rack mounting, which offers a very economical panelboard construction for this part of the equipment.

PANELBOARD LAYOUT

From the standpoint of successfully operating a process, the layout of the instrumentation on the front of the panelboard is the most important single element in panelboard design.   As the complexity of processes increases, more and more responsibility will be placed on instrumentation.   Automatic controllers will perform more of the mechanical and repetitive operations, but at the same time, the operator will be faced with increasing responsibility and burden unless the panelboard is so designed that he is relieved from all unnecessary operations.   The purpose of instrumentation in a control room is to extend the eyes and arms of the operator to such a degree that he is free to utilize the unique advantages of his brain for those tasks which require judgment, memory, and experience.

Functional panelboard design therefore dictates the grouping of instruments in such a manner as to make the task of the operator as natural and intuitive as possible.   At the same time, the panelboard must be economical, safe, utilitarian, and readily serviceable.   And it must be acceptable to the surroundings and requirements of a particular plant or location.

Important Layout Factors

The following factors are important in panelboard design:

Safety Considerations.   Designing a panel wherein one operator is expected to control more equipment than he can effectively handle during periods of disturbances, shutdowns, or emergency is unsafe practice.   The practical scope of the operator has been extended many times during the last decade through development of safety devices, alarm systems, shutdown systems, and as a result of the increase in reliability of control instruments.   Nevertheless, in panelboard design, the operator's abilities during emergency periods rather than normal operating periods should be the key criterion.

Operator Experience and Ability.   The more experienced the operator, the more he can control with the least help from instrumentation.   However, experienced operators are in short supply.   New processes are being developed as fast as operators can be trained.   Shorter working hours are making the shift problem more difficult.   Therefore, most new plants must expect to operate with relatively inexperienced people, and such men require more carefully presented information and much more assistance from automatic controls and alarm systems.   Thus it is well to design all panelboards with a below-average rather than the best operator in mind.

Characteristics of the Process.   Some processes are quite stable and safe in their operation.   Such processes can be made practically automatic, and the panelboard serves largely as a means for gathering historical data and as a nerve center during start-ups, shutdowns, and emergency periods. In many cases, processes of this type can be made to shut down automatically. Many of the new processes require the handling of dangerous and corrosive materials, often at high temperatures and pressures. Often, they are not particularly stable and require careful watching. Almost daily, contingencies arise such as sticking valves and vapor-locked lines—matters which require the operator's concentrated attention for varying periods of time. It is not uncommon for one or more of the automatic controllers to be put on manual control for fairly long periods because of some process difficulty. Therefore, in designing for a troublesome and complex process, the entire question of ready access to manual-control switches and switchover devices must receive particular attention.

Servicing Requirements. When panelboards become difficult to service because of overcrowding, this condition outweighs many of the other advantages which they may possess. Miniature instrumentation poses greater problems in this respect than where large instruments are used. In general, it is not good practice to crowd miniature instruments into smaller spacings than those indicated in Fig. 14c and e. Further crowding would result in an unserviceable rear-of-panel arrangement because of crowded piping and wiring.

Handling of Secondary Instruments. Many instruments, such as level controllers in heat exchangers and flow meters for utilities (steam, water, and air), serve secondary functions. While they are important to the over-all picture of the process, they usually are not of immediate concern to the operator. Consequently, they should be (1) omitted from the panel and mounted elsewhere or (2) mounted at the ends of the panel or on wings. They should not be located among the operating instruments because they detract from the operator's attention and also reduce the compactness of the operating section of the board.

Location of Accessories. There are some differences in opinion as regards the acceptable approach to location of such accessory equipment as pump and motor controls and pneumatic switches. In general, it is good practice to mount these on a separate panel or section, unless they are actually needed by the operator in his normal operating routines. Push buttons for start-ups tend to clutter a panel. However, if these controls are needed for normal operation, as in many batch processes, then they should be located together with the operating instruments.

ECONOMICS OF PANELBOARD DESIGN

A balance must be reached between justifiable first cost and the later costs of operation and servicing. Usually it is a simple matter to compare various types of panelboards from the standpoint of first costs, but evaluation of them in terms of possible long-term savings in operating efficiency and throughput is a more complex problem.

Shutdowns can be very costly and may justify large expenditures in the form of warning systems, monitoring devices, and graphic visual aids. Items like these must be viewed not only from their initial costs but also from the savings which might accrue from their value in the case of even one shutdown or emergency.

Case Study of Panel Arrangements

Figure 14a to e shows five groupings of instruments. They are all designed for the same process and utilize the same number of controllers and other instruments. While additional arrangements are possible, these five have been selected because they are typical of real installations.

Conventional* Instrumentation. Figure 14a and b shows a typical panel utilizing rectangular-case recorder-controllers for all key process variables. For levels

---

* Large, generally rectangular-shaped instruments were used almost exclusively until recent years. Their size obviously limited their use within the diagram on a graphic panel. With the introduction of miniature instruments, the term *conventional* became the common method for expressing the former large type of instrumentation. Actually, this is an unfortunate term in that both the larger and miniature instruments undoubtedly will be used separately and often together for years to come. Thus no one type is more conventional than the other.

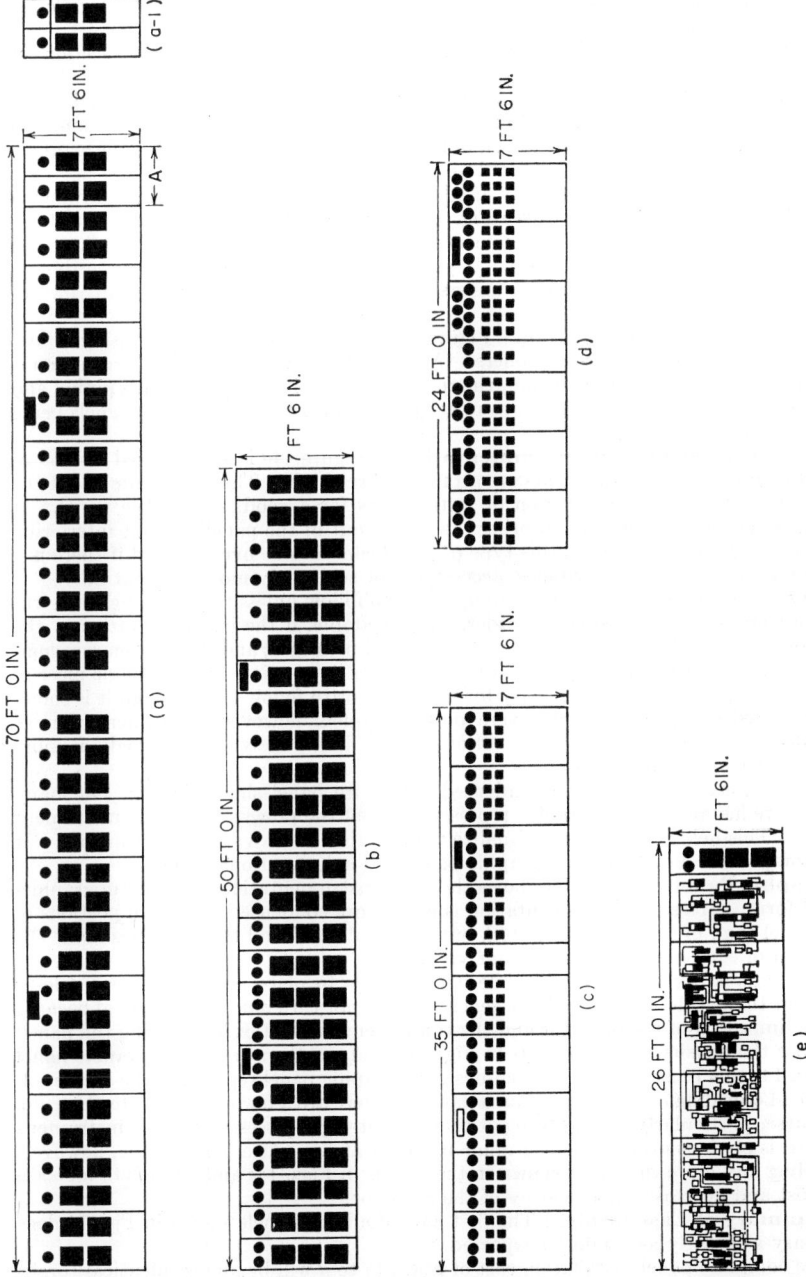

Fig. 14. Various panelboard arrangements for the instrumentation of an identical process. This illustrates how the panelboard length varies with the type of instrumentation and arrangement used.

and pressures, which need not be recorded, 6-in. gages are used.   Multiple alarm units are used for signaling.

Figure 14a is made up of eighteen 44-in. sections, plus one master panel section for the multipoint pyrometers and oxygen analyzers.   This is a type of panelboard very commonly found in process plants where conventional instrumentation is used. This panel could be divided into thirty-six 22-in. sections as shown by section A in Fig. 14a.   Another variation is shown in Fig. 14a-1 to illustrate how the same arrangement can be used in the sectionalized unit type of panel.

Another arrangement frequently used is that of grouping the main instruments three high instead of two high (see Fig. 14b).   This raises the top instrument above eye level and brings the lower instruments down rather low but does reduce the panel length by approximately one-third.   A panel of this type can be used with curb construction shown in Fig. 8, which assists in servicing the top and bottom rows of instruments.

Miniature Instrumentation.   Figure 14c is a typical arrangement using miniature recorders and recorder-controllers.   Although there appears to be a large amount of blank panel space at the bottom, the rear of the panel is well filled with piping and wiring.   This is a serviceable and compact panel and is typical of many panels being built today.

Condensed Miniature Instrumentation.   Figure 14d is a condensed version of Fig. 14c in which the instruments are arranged three high.   This represents about the maximum practical concentration of instruments for a good, serviceable panel.   It requires careful arrangement to bring the piping out of the panel without interfering with the instruments.   Even this type of panel can become overcrowded if there is a large number of pressure switches, accessories, or separately mounted controllers.

This panel would be difficult for an operator to follow unless the grouping of instruments were in a very systematic order.   For example, if the instrumentation in the top row comprised flow, the middle row comprised pressures, and the bottom row comprised temperatures and if the vertical rows were each associated with a single piece of process equipment, it would be easy for the operator to follow even though it is somewhat crowded.   If the process were quite complex, however, instrumentation as crowded as this would require the aid of graphic lines and symbols to make it more comprehensible to the operator.

To facilitate the use of the condensed miniature panel shown in Fig. 14d many people are installing a semigraphic representation of the process above the instrument panel.   This helps the operator visualize the function of the various instruments.   It is a compromise between a full graphic panel and a conventional panel.

Graphic Panel.   Figure 14e shows the same instrumentation arranged in graphic-panel form.   Here the instrumentation is even more condensed in certain portions of the panel than in the case of Fig. 14d, but the symbolism and the flow lines make the function of each instrument stand out vividly.

It is interesting to note that, while it is possible to place instruments even closer together than as done in Fig. 14d, it generally does not turn out to be practical because of the limitation of the operator and the consequent difficulties in servicing the panel. Because the density of instruments in Fig. 14d is about the same in the graphic panel of Fig. 14e this illustrates that a well-designed graphic panel represents the maximum practical crowding of instruments on a panel.   There are a few exceptions to this rule, of course, particularly in highly repetitive unit step processes where the instrument pattern is obvious even without graphic representation.   As a general rule, however, crowding any more than that shown in the panels in Figs. 14c and d tends to increase the cost and decrease the efficiency of the operator.

Summary of Case Study.   The first cost of all the panels shown in Fig. 14 does not vary as greatly as would be expected.

The complete panel-installation cost of Fig. 14a to d will be almost identical (probably will vary less than 5 per cent).   This is because the reduction in cost due to the saving in steel in the shorter panels is approximately offset by the slight increase in the cost of piping and wiring as the density of instrumentation increases.

The cost of panels in Fig. 14d and e can be reduced somewhat (probably by 15 to 20 per cent) where plug-in-type miniature controllers are used.   In the past, one of the big elements of cost in miniature instrumentation resulted from the need for separate mounting of the controllers.   It is evident that a panel like Fig. 14d would be very complex if there were 12 stack controllers mounted on the rear.

The graphic panel (Fig. 14e) will cost from 15 to 20 per cent more than the panel in Fig. 14d, mainly because of the added layout work and the cost of symbols and flow lines.

In summary:

1. The cost of the panel depends more upon the total volume of instrumentation than on the panel length or specific design.

2. The graphic panel roughly represents the maximum concentration of instruments which is practical to place before an operator.

3. Miniature instrumentation and graphic panels can reduce the size of the control room to one-third or one-half that required with conventional instrumentation.

REFERENCES

1. Considine, D. M.: "Process Instruments and Controls Handbook," McGraw-Hill Book Company, Inc., New York, 1957.
2. Dorscheimer, W. T.: A Critical Look at Graphic Panels, *Chem. Eng. Rept.*, 1952.
3. Stachner, J.: Spotlighting Process Control, *Instrumentation*, vol. 5, no. 3, 1951.
4. Howard, G. E.: Graphic Panels Help Solve Processing Industry Problems, *Petrol. Refiner*, March, 1950, pp. 107–109.
5. Boyd, D. M., Jr.: "Process Control by Graphic Panel," no. 49-4-1, Conference of the Instrument Society of America, 1949.
6. Novak, A. V.: Graphic Panels, *Instruments*, vol. 23, pp. 1128–1129, November, 1950.
7. Bowers, W. S.: Graphic Instrumentation: A New Tool for Industry, *Instruments*, vol. 23, pp. 1188–1191, November, 1950.
8. Uhl, W. C.: Automatic Refineries in a Few Years, *Petrol. Processing*, October, 1950, pp. 1064–1067.
9. Pond, R. N.: Graphic Panels: The New Advancement in Continuous Process Control, *Taylor Technol.*, vol. 3, no. 1, pp. 11–13, 1950.
10. Uhl, W. C.: Graphic Panels: The New Look in Instrumentation, *Petrol. Process.*, April, 1950, pp. 361–368.
11. Frost, H. C.: Utilization of Instrumentation in the Wet Milling Industry, *Food Technol.*, vol. 6, no. 2, pp. 50–53, 1952.
12. Weingartner, J.: Centralized Graphic Panels—Unlimited, *Instrumentation*, vol. 6, no. 1, pp. 17–21, 1952.
13. Hurley, R. L., Jr.: Panel Design, *Instruments*, December, 1951, pp. 1425–1429.
14. Jensen, W. R.: Graphic Panels in Operation, *Petrol. Refiner*, September, 1950.
15. Gess, L.: The Graphic Panel in Automatic Process Control, *Petrol. Refiner*, March, 1950.
16. Gess, L.: "Evolution of Graphic Panels," Conference of the Instrument Society of America, 1950.
17. Hermann, A. D., and C. S. Conner: Graphic Instrumentation, Sec. 8-3, "Process Instruments and Controls Handbook," McGraw-Hill Book Company, Inc., New York, 1957.
18. King, R. S.: Applications of Newer Process Control Concepts to a Steam Generating Plant, *ISA J.*, vol. 1, no. 10, pp. 31–36, October, 1954.
19. Madden, J.: A Graphic-panel Centralized-control Waste-treatment Plant, *Instruments and Automation*, vol. 28, no. 1, pp. 102–105, January, 1955.
20. Schwartz, I. R.: Centralized Control Stimulates Electric Transmission, *Automation*, vol. 4, pp. 61–62, March, 1957.
21. Walter, L.: Two Aspects of Automation in Refining, *Petroleum*, vol. 20, pp. 369–371, October, 1957.
22. Automation, *World Petrol.*, vol. 26, pp. 55–74, 160, July 15, 1955.
23. Swamenthan, V. S.: First All-electronic Control Panel, *Petrol. Engr.*, vol. 27, pp. C-13–16, April, 1955.
24. Gallagher, G. G., and R. A. Robinson: Future Trends in Automation, *ASME Paper* 55-PET-5, Sept. 25, 1955.
25. Banner, E. H.: Instrumentation and Automatic Control in Oil Refineries, *Petroleum*, vol. 18, pp. 6–9, 26, January, 1955; pp. 55–60, February, 1955; pp. 91–98, March, 1959; pp. 121–125, April, 1955; pp. 170–175, May, 1955.
26. Boyd, D. M.: Why Electronic Process Control, *ISA J.*, vol. 2, pp. 16–19, November, 1954.
27. Gallager, G. G., and R. A. Robinson: Significant Trends in Control, *Petrol. Process.*, vol. 10, pp. 1740–1743, November, 1955.
28. Yanak, J. D.: Instrument Panels Made for Easier Operation, *Petrol. Refiner*, vol. 37, pp. 224–228, March, 1958.
29. Waldron, A. J.: Take a New Look at the Process Control Center, *Control Eng.*, vol. 7, no. 3, pp. 132–136, March, 1960.
30. Crawford, W. A.: Economic Comparison of Alternate Instrument Systems, *ISA J.*, vol. 7, no. 5, pp. 72–75, May, 1960.

CONDITIONING THE INSTRUMENT AIR SUPPLY

By G. L. Eberly*

A successful installation of pneumatic instruments requires a supply of reasonably clean air. Unreliable performance and high maintenance costs result from excessive contamination of instrument air.

The principal contaminants are moisture, oil, and solids, such as dust, dirt and scale. These substances can be found separately or in any combination. One of the worst possible combinations is an emulsion formed of water and oil, mixed with tar and carbon products from the compressor, together with the scale and rust normally found in corroded air lines. This combination forms a sludge which gums up orifices and restrictions to such an extent that, unless removed, it can shut down an air-controlled installation.

SOURCES OF CONTAMINATION

The primary sources of air contamination are (1) conditions at the atmospheric intake; (2) neglectful operation or inadequate maintenance of the compressor and auxiliary equipment; and (3) corrosion of piping and the presence of foreign material in the pipes.

The compressor intake air always contains some moisture. As intake air is compressed, its capacity to hold this moisture decreases. If this air is then subjected to lower temperatures, as it flows through transmission lines, vaporized moisture will condense out unless precautionary measures are taken.

Every compressor, in addition to supplying compressed air, produces contamination of some kind. The lubricated compressor can add water, oil, and carbon and tar products. Oil-free compressed air can be produced by a carbon-ring compressor, but then carbon dust can become a contaminant. Oily contamination also can be eliminated by use of a water-seal-type compressor, but the most serious contaminant, water, remains a problem.

No single solution to all compressed-air problems can be prescribed for all applications. One plant may be troubled with moisture, another with dirt or oil, and in varying degrees. There is usually a good way to eliminate each of these difficulties, and it is not to be construed that each instrument air system will be complex or costly.

TYPES OF AIR-DISTRIBUTION SYSTEMS

The basic function of the air-supply system is to compress air to a high pressure, store it in a receiver, and deliver it, clean, dry, and at a prescribed pressure, at the point of use. Three of the most common methods of compressing and distributing air for instrument use are:†

Large Installation Using Atmosphere Intake. This system (Fig. 1) is recommended for installations requiring 5 scfm or more where a plant supply of air in such amounts does not exist. To prevent formation of rust particles, piping after the receiver should be only of brass, copper, galvanized iron, stainless steel, aluminum, or plastic.

* Technical Staff, Harris D. McKinney, Inc., Philadelphia, Pa.
† From "Instrument Air Supply Systems," *Application Bulletin* 91-53P-01, Fischer & Porter Company, 1960.

Large Installation Using Plant Air. This system (Fig. 2) uses plant air but employs a back-pressure regulator on the plant air header to assure that the instrument supply has all available air in the event of unusual demands for plant air. It is

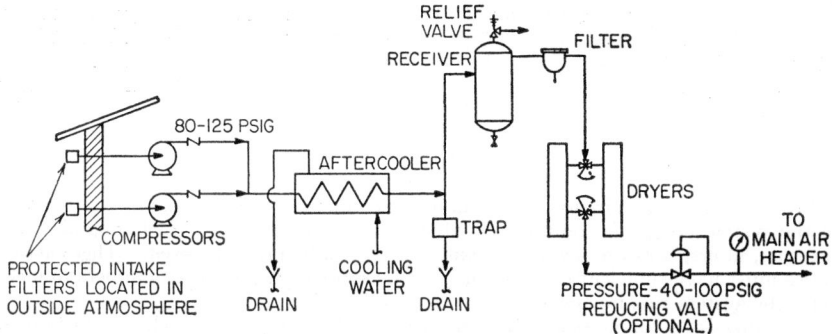

Fig. 1. Large air supply installation using atmospheric intake.

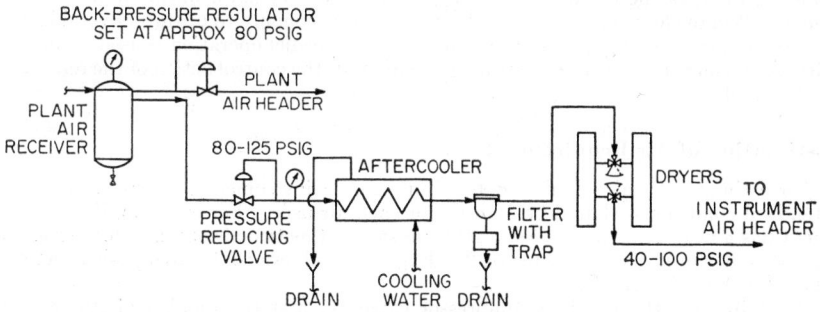

Fig. 2. Large installation using plant air.

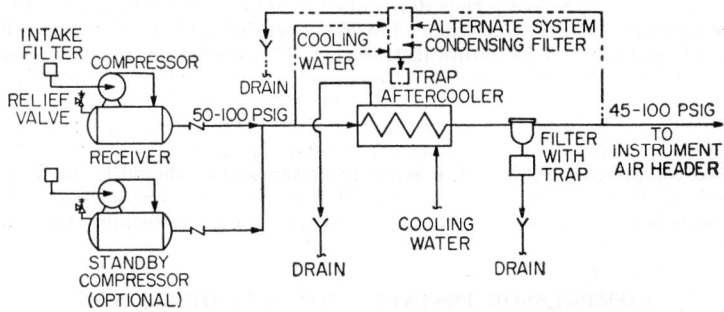

Fig. 3. Small installation, inside building.

not recommended if there is any possibility of its failure to supply air continuously and at sufficient pressure for instrument operation.

Small Installation—Inside Building. This system (Fig. 3), for smaller needs inside a heated building, does not employ dryers. Small-capacity compressors are furnished mounted on a receiver tank. If any of the instrument air lines are run outdoors, a scaled-down version of Fig. 1 should be used.

INSTRUMENT AIR REQUIREMENTS

Pressure

Practically all instruments use air at 20 psig. Since pressure-reducing valves should have a minimum pressure drop of 5 psig, the supply to the valve should be at least 25 psig, preferably 35 psig. A further 10 psig allowance should be made for the pressure drop which occurs through drying apparatus and distribution lines. Thus at least 45 psig should be maintained at the compressor.

Quantity of Air

When designing an air-supply system for instrument use, the mean air requirements of the system must be carefully estimated so that the accessories, including air compressors, filters, pressure regulators, and dryers, can be properly sized. These accessories must have sufficient capacity or must be supplied in sufficient numbers to meet fully the system requirements.

To undersize a compressor is to shorten its life because, under such circumstances, the compressor will operate almost constantly. Inadequately sized filters and dryers can damage instrument components by allowing oil and water carryover. Good practice dictates the use of a properly sized pressure regulator ahead of each instrument. Where close control is important, it is poor economy to operate several instruments in parallel with a single regulator. Where parallel operation is used, large load changes on one or more of the instruments will upset the control action of the remaining instruments.

Estimation of Air Requirements

To estimate the air requirements of pneumatic instruments, $\frac{1}{2}$ cu ft of air (at standard conditions) should be allowed for each air user. For example, if an instrument has a nozzle and a pilot valve, this must count as two air users. Hence such an instrument would have a factor of 2. Experience shows that valve positioners also should carry a factor of 2.

To determine the required compressor capacity, first the number of instruments must be determined and then the factors, as just described, are applied. This gives the equivalent number of users. The number of cubic feet of air (at standard conditions) per minute required is then determined by multiplying the number of equivalent users by the factor of 0.5. Then 15 to 20 per cent of this total should be added to allow for blowdown and for future instruments. Two compressors (one a spare) of this capacity then should be selected. Compressors with the splash system of lubrication should not be specified.

Compressors are available in many standard sizes, with capacities as low as 1.4 cfm. This size is sufficient for one or two instruments. To assure continuous service, at least two compressors, each with a separate power source, should be used for each installation. Compressors often are connected so that only one operates, until its output falls below a specified amount, at which time the second compressor cuts in to service.

COMPRESSOR INSTALLATION AND OPERATION

Although much contamination arises from atmospheric conditions which cannot be modified, several precautions can be taken which result in cleaner air and improved compressor efficiency.

Location of Air Intake

Generally, the intake should be located outside and on the coolest side of the building. Intake air at this lower temperature increases compressor efficiency. At constant pressure, the density of air increases with decreasing temperature. Consequently, a lower inlet temperature increases the quantity of air that is delivered. For example, if the outside air temperature averages 40°F and the indoor temperature is 70°F, only 943 cu ft of outdoor air will be required to deliver 1,000 cu ft of air at the indoor temperature. This represents a direct savings of 5.7 per cent.

Wherever possible, the actual point of intake should be as remote from fine industrial dusts or pigments, steam or air exhaust vents, or other sources of contaminants as is practical. Care must be taken to avoid sulfurous or chlorine-contaminated air because these substances, when mixed with condensate, produce weak acid solutions which eventually corrode screens and piping.

Intake Filters

Air intake filters are essential to remove grit and dust that are present in practically all industrial areas. Removal of these substances is important not only to the desired final air condition, but also to protect the compressor, since any solids in the air will cause wear on pistons, cylinders, valves, and other compressor parts. For best performance, these filters must be cleaned and inspected regularly.

Compressor Operation

Careless operation and maintenance of the compressor can cause dirt and oil to enter the compressed-air lines. Compressor load is another important factor, since compressors usually pump oil when overloaded. Worn pistons or valves and excessive temperatures also increase the amount of oil vapor that is carried over with the air.

THE PROBLEM OF MOISTURE REMOVAL

Water is probably the most serious contaminant in compressed-air systems. Water causes rust and corrosion of piping and connections and also carries dust, dirt, and sediment that can plug tubing and restrictions inside the instrument. Water collects in low spots, causing "water hammer" which, in turn, jars loose any scale, rust, or dirt that may be present in the lines. This action puts excessive loads on the filters. Moisture also reduces pipe capacity and, when present under freezing conditions, may plug or even burst the piping.

Occurrence of Moisture

All atmospheric air contains some water vapor. When the air is compressed, both the temperature and the pressure increase. As the pressure increases, the maximum amount of water vapor that can be contained by the air is reduced. But, at the same time, the temperature increase allows the air to contain more water vapor. The net result is compressed air with a high temperature and a low relative humidity. This is what occurs during adiabatic compression, that is, where all the generated heat is retained and the temperature of the air remains high. In a pneumatic control system, however, all the generated heat is not retained. Radiation through the cylinder walls and the air lines to the receiver accounts for heat loss. More is lost through the receiver itself, and if an aftercooler is used, still more heat is lost.

As the air temperature decreases, its ability to hold water vapor also decreases and the saturation point is soon reached. As this air leaves the receiver, more heat is lost, and if the water is not removed in some manner, it will condense in the lines. For example, if air at 90°F and 20 per cent relative humidity is compressed to 100 psi, the relative humidity will be 60 per cent at 125°F. If this air then is cooled to 90°F, the relative humidity will be 152 per cent; that is, water will be precipitated.

Dew Point

Dew point may be defined as the temperature below which, if air is cooled, water will condense out of the air. If air is cooled below its dew point, more water will condense, and the dew point is reduced to the lowest temperature that has been reached. If the temperature subsequently is increased, the dew point will stay at the lowest temperature to which the air was previously subjected, provided that the condensed water has been trapped out. Otherwise, it will vaporize again and raise the dew point.

The only way to prevent condensation is to reduce the dew point of the instrument air, before distribution, below any temperature that it will encounter in any part of the system. In this way, further condensation will be impossible and the lines will remain dry. Tests made in connection with "mothballing" naval vessels showed that, if air is kept below 25 per cent relative humidity, no corrosion will take place.

Dew point should be reduced *before* pressure reduction. If this is not done, water usually will be generated at the reducing valve. When saturated air passes through a reducing valve, both the temperature and pressure are reduced. The pressure reduction lowers the dew point, but the temperature reduction brings the air closer to the dew point. While these changes are occurring, the saturation point may be momentarily exceeded. Since it is easier to condense a vapor than it is to evaporate the same weight of liquid, the condensed water remains as a liquid and is carried through the lines along with the air. For this reason, the dew point and relative humidity must be reduced before the pressure is reduced.

DEHYDRATORS

The three basic types of dehydrating devices are (1) mechanical type, (2) cooler type, and (3) desiccant type. The mechanical type is least expensive; the desiccant type is most expensive. Selection depends upon the results desired and upon the ambient conditions which prevail.

Traps

The trap is the most commonly used mechanical type of dehydrating device. Traps, however, do not reduce dew point or relatively humidity but simply remove water that has already condensed out. The air downstream of a trap remains saturated with water and oil vapors if such were present upstream of the trap. For maximum effectiveness, the traps must be located as close as possible to the point of air use, and even when this practice is followed, any drop in temperature after the air leaves the trap will cause condensation.

Separators

The separator is another form of mechanical dehydrator. Two types of separators are described here: (1) expansion type and (2) centrifugal or cyclone type.

In the expansion-type separator, the air is allowed to expand through an orifice, which reduces the temperature so that some water condenses out. Expansion depends on the pressure differential between the upstream and the downstream sides of the separator. Therefore, expansion often is retarded by back pressure. A separator of this type reduces temperature a certain amount, but since no cooling agent is used, additional temperature reduction usually is required to yield air that is sufficiently dry for instrument use.

The cyclone-type dehydrator operates by centrifugal action. Air enters a chamber where it is whirled around, dispelling moisture by centrifugal force. The faster the air moves, the more effective is the separation. Since compressed air usually moves through the distribution lines rather slowly, this type of separator, like the expansion type, does not always take out sufficient moisture for instrument use.

With each of these mechanical devices, the temperature of the air seldom is reduced far below ambient. Consequently, the air leaves the device saturated with vapor, at a relatively high temperature. If this air is used immediately after leaving the mechanical separator, it will perform satisfactorily, but if any further temperature drops occur, moisture will condense in the lines.

Cooling Devices

If the air is cooled to a temperature lower (about 30 to 40°F for safety) than any temperature that it will encounter in the distribution system, there will be no condensation in the lines or instruments. Partial cooling is effected in the receiver. An aftercooler is used to condense most of the water vapor. The aftercooler is a heat exchanger located next to the compressor, on the high-pressure side. For this reason, the aftercooler inlet temperature is relatively high; its outlet temperature usually is about 15°F above that of the cooling water used. If this is sufficiently cool, no additional drying is required; but this is not often the case.

Secondary Coolers

More often, either a refrigerated aftercooler or a secondary cooler is required. The secondary cooler should be located downstream from the aftercooler and receiver,

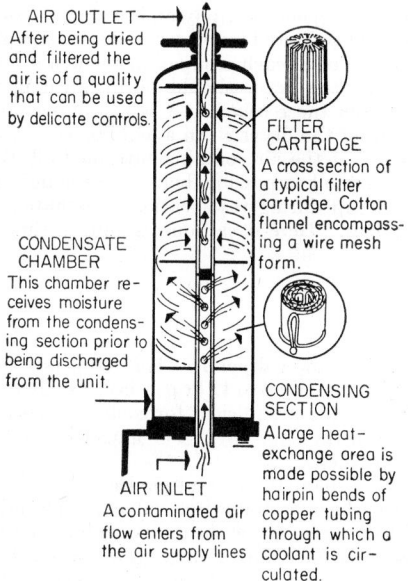

AIR OUTLET——▶
After being dried and filtered the air is of a quality that can be used by delicate controls.

FILTER CARTRIDGE
A cross section of a typical filter cartridge. Cotton flannel encompassing a wire mesh form.

CONDENSATE CHAMBER
This chamber receives moisture from the condensing section prior to being discharged from the unit.

CONDENSING SECTION
A large heat-exchange area is made possible by hairpin bends of copper tubing through which a coolant is circulated.

AIR INLET
A contaminated air flow enters from the air supply lines

FIG. 4. Condensing-type filter (secondary cooler) used to remove moisture and solid particles from instrument air. (*Hankison Corp.*)

where it will reduce the air temperature far below ambient, yielding a dew point only slightly above the cooling-water temperature.

A secondary cooler is shown in Fig. 4. Its effectiveness depends principally on the temperature of the available cooling water. When planning an installation of this type, it must be stressed that the air should be cooled at the highest possible pressure, because as the pressure is reduced later, the dew point and relative humidity also drop.

Desiccant Dehydrators

When air lines are subjected to subfreezing temperatures, or where there is no cooling water, desiccant dryers should be used. The desiccant dehydrator is a device which is filled with a chemical of hygroscopic quality, that is, a chemical material that adsorbs water vapor on its pore surfaces until it becomes saturated. These vapors then are held until the desiccant is heated to a point sufficiently high to drive off the moisture.

Commonly used desiccants include silica gel,* Mobilbead,† activated alumina,‡ and molecular sieves.§ These desiccants can be used either singly or in combinations. Desiccant dehydrators usually comprise two towers. The air is passed on alternate cycles of from 3 to 8 hr, through one tower and then the other. While one tower is used for dehydration, the other is heated to a temperature sufficiently high to reactivate the desiccant. Moisture that is boiled out of the desiccant is either vented to the atmosphere or condensed out. The cycling is arranged so that the instrument air always passes through a tower containing active material.

Dryer Location

It would appear to be less costly to dry the air entering the compressor at atmospheric pressure than to dry it at the higher pressure on the outlet side. However, drying at the lower pressure requires a greater amount of desiccant and cuts down on the effectiveness so much that it is not practical.

The proper dryer location depends entirely on the plant layout. If there are only a few groups of instruments at scattered locations, and the main plant air system is used, small dryers should be installed at each group of instruments. If a special compressor for instrument air is used, the dryer should be installed in the compressor room, preferably downstream from the receiver. A self-draining prefilter and a separator should be used just ahead of the dryer. An adequate aftercooler also is needed.

If there is any chance that the air contains entrained oil, the separator should be designed to remove as much oil as possible because, although desiccants retain oil on their surfaces and give it up during reactivation, a minute amount of the oil does remain on the pore surfaces. Here it carbonizes and, in time, will clog the pores of the desiccant.

Types of Dryers

There are single- and dual-tower dryers. Choice depends upon the schedule of air usage. If dry air is needed for only 8 to 12 hr per day, a single-tower dryer can be used. The 12 to 16 off-duty hours will be sufficient for cooling the tower and for reactivating the desiccant for service on the next day. For longer or continuous service, a dual-tower dryer should be specified.

Dryers are available for manual, semiautomatic, and fully automatic operation. Manual units require that the operator switch valves at the end of each cycle and that he manually operate the regeneration heaters. Semiautomatic units require the attention of an operator only once during each cycle—when the switching valve must be reversed and the timer set to control the period of regeneration. Automatic units require no manual attention after installation (see Fig. 5). The most common drying cycles are 4, 6, 8, and 12 hr. The units with shorter cycles usually are automatic. Units with a longer cycle usually are semiautomatic or manual in operation. The short-cycle fully automatic dryer is generally the most economical because it is of smaller size, requires less desiccant, and has a lower heating load for activation.

* Silica gel (Davison Chemical) has high efficiency, pound for pound, where "solid" water carryover has been eliminated.
† Mobilbead (Socony Mobil) is also a silica product, formed into beads, and very closely approaching the efficiency of silica gel. This is a denser material, weighing approximately 50 lb/cu ft, as contrasted with 45 lb/cu ft for silica gel. As with silica gel, Mobilbead should not be immersed in liquid water. Under such conditions, it will break down because of the high internal stresses which are set up by the heat of adsorption.
‡ Activated alumina (Alcoa) does not tend to break down or explode when immersed in water. For these reasons, it is used often where there is a heavy carryover of water, such as at the bottom of a tower which is filled with silica gel or Mobilbead. Activated alumina usually is activated at a temperature of 400°F.
§ Molecular sieve (Linde) is a desiccant which is best activated at temperatures of 600°F. It has a pore structure which is smaller than the other desiccants described and is reputed to maintain its maximum efficiency even when drying air of low relative humidity.

Combination Systems

It is seldom that only one type of moisture-removal equipment will provide the best solution to drying problems. Usually it is best to use a combination cooler and desiccant dehydrator. By locating a cooler ahead of the desiccant dehydrator, the size of the dehydrator can be reduced and the life of the desiccant can be prolonged— largely as the result of removing oil vapors from the air. In some cases, it is good

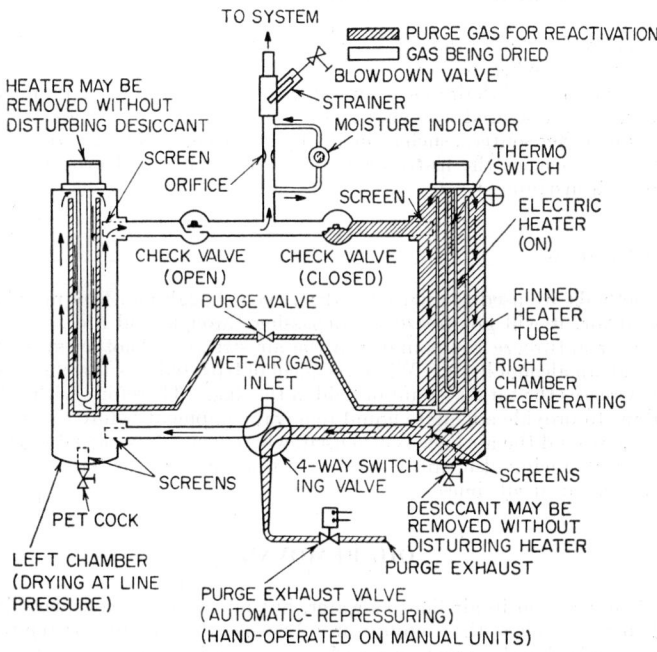

FIG. 5. Desiccant-type dehydrator showing air cycle. Wet gas enters through four-way plug valve and passes into left desiccant chamber. Dry gas leaves chamber top, passes through check valve, and enters distribution system. Small amount of dry gas passes through purge line into heater tube of reactivating chamber (at right). Chamber is heated and dry gas picks up vaporized moisture from desiccant bed. Moisture is carried through drain. Periodically, a timer reverses the chambers. (*Industrol Corp.*)

practice to use a refrigerated cooler to reduce the air temperature to the recommended refrigeration temperature, usually 40°F.

SUMMARY OF GOOD AIR-DRYING PRACTICE

1. The dew point should be reduced to from 35 to 40°F below the lowest temperature through which the air will pass. The relative humidity should be kept below 25 per cent at this lowest temperature.

2. Duplicate equipment should be installed wherever possible to avoid instrument shutdown.

3. When oil-lubricated compressors are used, the oil must be removed from the air stream before the air enters a desiccant dehydrator.

4. When requesting quotations on drying equipment, all local details should be described, including the available activation medium (steam, gas, or electric power), so that the supplier can recommend the best system.

5. Both summer and winter conditions should be considered so that the drying system will be adequate to protect instruments at all times, without being oversized and excessively costly.

6. Bids from desiccant-dryer manufacturers should be requested in two ways: (*a*) with the air cooled by available cooling water and (*b*) with the air precooled to the recommended refrigeration temperature.

OTHER MOISTURE-REMOVAL METHODS

In addition to drying equipment, there are several other ways to minimize the effects of moisture. When piping passes through a cold space, it should have a pitch and should be provided with drains, even though moisture-removal equipment is used. A check valve should be used wherever there is any chance that liquids will back into the system from other sources, such as in air-purge systems. Finally, to make certain that no moisture reaches the instruments, a separate filter and drip well should be installed at each instrument.

Antifreeze Systems

Several methods are used to guarantee that, even though moisture may be present in instrument air, it will not freeze when passing through cold areas. One system feeds alcohol or antifreeze compound into the air stream. Another system bubbles the air through an alcohol bath. While the alcohol bath reduces the chances of freezing, it also increases the total amount of fluid in the line. Therefore, with this system, it is important to provide adequate liquid-removal equipment before the instruments. Steam tracing around the air lines is also used, wherein steam is passed around the air lines. Electrical tracing also can be used. In this way, the air temperature is kept above its dew point at all times

OIL REMOVAL

Oil is not so common in air lines as water, but oil can cause considerable trouble. The usual source of oil is the compressor. Because the compressor temperature is high, oil is vaporized and carried over with the air. Old, worn, overheated, or overloaded compressors permit excessive oil carryover. This oil must be removed because it clogs filters and restrictions and has a tendency to cause deterioration of valve diaphragms.

The best precaution is that of operating the compressor properly and maintaining it well. Most oil is condensed in the aftercooler and the secondary cooler, in the same way that water vapor is condensed. Desiccant dehydrators remove some oil, but in time this clogs the pores of the material.

Where oil conditions are so severe that the foregoing methods are not fully effective, the air can be bubbled through a solvent. Special filters, or precontactors, which contain fuller's earth or low-grade desiccants, also can be used as an auxiliary measure. Activated alumina that has become spent as to the adsorption of water may be set aside for use in these precontactors, since the capacity of the material for adsorbing oil still will be moderately high.

SOLIDS REMOVAL

Solids constitute the most serious form of contamination, although not the most prevalent. Good low-cost solids-removal equipment is available. Solids in air include the dirt and dust carried by the atmosphere, as well as pipe scale, rust, pipe dope, and in some plants, lint. Even very tiny particles can cause instrument failure as the result of clogging transmission lines, instrument nozzles, and restrictions.

Most instruments have integral filters.   However, it is also good practice to install a separate filter ahead of each instrument, as shown in Fig. 6.   Rust is eliminated with assurance only by using nonferrous piping or tubing.

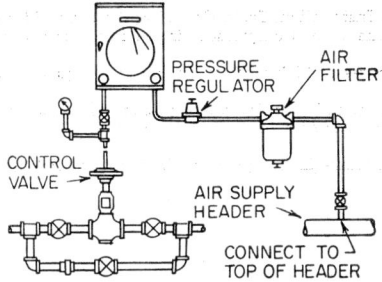

FIG. 6. Piping for pneumatic controller, showing separate filter and regulator for compressed-air supply.

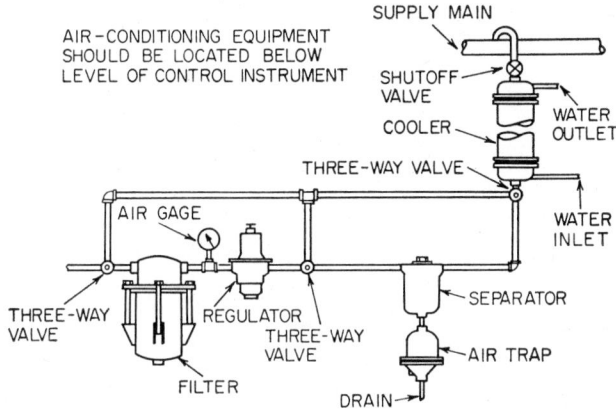

FIG. 7. Piping for air-supply-conditioning system recommended where a complete dehumidification system is not used.

In large plants, where great amounts of air are used, other methods of solids removal sometimes used include (1) gravity separators, (2) spray towers or scrubbers, and (3) centrifugal and cyclone separators.

REFERENCES

1. Anders, E. W.: Discussion of Design and Application of Instrument Air Dryers, *ISA Paper*, Instrument Society of America, Pittsburgh, Pa., 1956.
2. Feeley, E. R.: Dryers Stop Air Line Condensation and Freezing without Tracers, *Power Eng.*, December, 1955.
3. Hankison, P. M.: Instrument-quality Air, *Instruments and Automation*, vol. 28, no. 4, pp. 626–627, April, 1955.
4. Hankinson, P. M.: Theory and Filtering Techniques for Compressed Air, *Instruments*, vol. 26, no. 11, November, 1953.
5. Murphy, L. J.: Engineering Compressed Air Systems, *Petrol. Refiner*, vol. 31, no. 12, pp. 139–143, December, 1952.
6. Parker, W. A.: How to Maintain Your Air Lines, *Mill and Factory*, vol. 48, no. 2, pp. 93–95, February, 1951.
7. Pearse, D. J.: Clean Constant Air for Instruments, *Mill and Factory*, vol. 53, no. 3, pp. **127–128**, September, 1953.

8. Simpson, G. L.: Instrument-air Dryers, *Instruments and Automation*, vol. 28, no. 5, pp. 800–802, May, 1955.
9. Porter, R. W.: Conditioning of Instrument Air, *Instrumentation*, vol. 1, no. 2, pp. 24–25, May–June, 1944.
10. Anon.: Pure Air for Pneumatically-operated Instruments, *Mech. World*, vol. 120, no. 3345, pp. 167–168, Feb. 23, 1951.
11. Benner, D. W.: Air-line Installation, *Instr. Control Systems*, vol. 36, no. 3, p. 129, March, 1963.
12. Benner, D. W.: Maintenance of Compressed-air Devices, *Instr. Control Systems*, vol. 36, no. 2, p. 119, February, 1963.
13. Benner, D. W.: Air Compressor Maintenance, *Instr. Control Systems*, vol. 36, no. 1, p. 115, January, 1963.
14. Benner, D. W.: Cost of Air and Leaks, *Instr. Control Systems*, vol. 35, no. 12, p. 113, December, 1962.
15. Benner, D. W.: Selecting Your Air Compressor, *Instr. Control Systems*, vol. 35, no. 11, pp. 111–112, November, 1962.
16. Benner, D. W.: Contaminants in Compressed Air, *Instr. Control Systems*, vol. 35, no. 10, p. 108, October, 1962.

INDEX

Note: See special classified indexes immediately following this index.

1

Alternating-current bearingless generator, **5**-35
  bridges, **1**-37 to **1**-41, **15**-93 to **15**-98
  distribution of, **14**-45 to **14**-66
  frequency measurements, **9**-9, **15**-52 to **15**-90
  generation of, **14**-45 to **14**-66
  power measurements, **9**-5 to **9**-9, **15**-76 to **15**-82
  reactive voltamperes, **9**-6, **15**-76
  tachometer, **5**-3
  voltage measurements, **9**-2 to **9**-5, **15**-67 to **15**-76
  waveforms, **9**-3, **15**-75, **15**-89
Alternator, aircraft, **16**-51
Altimeter, **16**-4, **16**-15
Altitude, automatic-pilot control of, **16**-12
  gravity change with, **5**-42
  measurement of, **15**-62, **16**-4, **16**-15
  pressure change with, **4**-10, **4**-13
  range of, aircraft, **16**-40
  rate-of-change of, **16**-6
Alum addition, regulation of, **13**-73
Alumel thermocouples (*see* Chromel-Alumel thermocouples)
Aluminum, beta-ray absorption of, **3**-42
  thermal expansion of, **5**-65
Aluminum mirrors, **3**-62
Aluminum tubing, panelboard use of, **17**-72
Ambient influences on measurements (*see* Classified Index—"Ambient Influences and Accuracy Deterrents")
Ambiguous numbers, **1**-26
American Gage Manufacturers, **6**-6
American Gas Association, measurement standards of, **1**-32
American Institute of Chemical Engineers, **13**-20
American Petroleum Institute, **7**-4
American Society of Mechanical Engineers, fluid meters report of, **5**-11
American Society for Testing Materials, calorimetry standards of, **2**-23
American Standards Association, **6**-6
Ammeters, accuracy of, **1**-13, **15**-67
  alternating-current, **9**-2, **15**-72
  direct-current, **9**-2, **15**-67
Ammonia, determination of, **3**-56, **8**-13
Ammonia-cooling system, **13**-91
Ammonia production, digital computer for, **11**-39
Ammonium sulfate, production controls for, **13**-40
Amperometry, **8**-6
Amplifiers, chopper type, **11**-71, **15**-69
  cooling of, **6**-36
  dielectric, **17**-33
  for direct-current signals, **11**-70, **15**-69
  drift of, **15**-69
  electric, for analog computers, **11**-7
  errors in, **1**-14 to **1**-15, **11**-66, **11**-84
  feedback, **11**-68
  Fitgo, **11**-70
  gains of, **1**-14, **1**-15
    (*See also* various types of amplifiers)
  hydraulic, **6**-37
  input, **11**-70
  magnetic, **6**-36, **10**-48, **17**-32
  noise in, **11**-66
  piezoelectric crystal, **15**-51
  pneumatic, for analog computers, **11**-7
  positioning-control, **6**-27
  potentiometric, **11**-69, **15**-69
  power, **6**-37
  scanner switch for, **11**-83
  tachometer, **5**-38
  transducer voltage, **6**-35
  transistor, **5**-38, **6**-36
  transistor-magnetic, **6**-36
  vacuum-tube, **6**-35, **15**-69
  voltage, **17**-42
Amplitude, background noise, **1**-14, **11**-66

Amyl acetate quench, in Geiger counter, **3**-24
Analog computers, applications of, **11**-6 to **11**-25
  (*See also* Computers, analog)
Analog-to-digital converters, **11**-36, **11**-72
Analog signals, **1**-9
  switching of, **11**-85 to **11**-91
Analog systems, in aircraft flight simulator, **16**-48
  in power generation and distribution, **14**-58 to **14**-61
Analysis, of errors, **1**-21 to **1**-27
  of system response, **1**-14, **1**-15
Analytical instruments, absorption spectrometry, **3**-61, **3**-74 to **3**-77
  acoustic, **8**-6, **8**-13
  for air, **8**-12 to **8**-14
  alpha ray detector, beryllium, **3**-53
  amperometry, **8**-6
  calorimetric, for gases, **2**-23
  for carbon dioxide, **8**-13, **8**-14, **14**-30
  characteristics of, **8**-2 to **8**-11
  chromatography, **8**-6, **8**-14, **10**-21, **13**-12
  colorimetry, **3**-65, **3**-73, **3**-76 to **3**-80, **3**-82, **8**-4, **8**-12
  conductivity, electrical, **8**-6, **13**-17, **14**-32
    gaseous, **8**-6
  control system use of, **10**-21
  coulometry, **8**-5
  dead-stop methods, **8**-6
  density methods, **8**-6
  dielectric constant, **8**-6, **8**-12
  distillation column use of, **13**-45
  dryer use of, **13**-60
  electrical conductivity, **8**-6
  electron paramagnetic resonance, **8**-6
  emission spectroscopy, **3**-61, **3**-62, **8**-4, **8**-12, **13**-12
  explosimeter, **13**-13
  flame photometry, **3**-76, **8**-4
  fluorescence, **3**-62, **3**-77, **8**-4
  fractionation methods, **8**-6
  gaseous conduction, **8**-6
  hydrogen ion, **8**-5, **8**-13, **10**-34, **13**-12 to **13**-17, **13**-74, **13**-82, **14**-32, **15**-18, **15**-42
  induced radioactivity, **8**-4
  infrared, **3**-62, **3**-76, **3**-77, **8**-4, **8**-12, **10**-21, **13**-12
  ion chamber, **8**-13
  mass spectrometry, **8**-5, **8**-12, **10**-12, **13**-12
  measurement lags in, **1**-30 to **1**-32, **8**-8
  microwave spectroscopy, **8**-4
  nephelometer, **3**-65
  nonradioactive gas, **3**-56
  nuclear magnetic resonance, **8**-6, **8**-13
  nuclear quadrupole moment, **8**-4
  on-line (*see* on-stream, *below*)
  on-stream, **8**-7, **10**-21
  open hearth use of, **12**-23
  Orsats, **8**-5, **13**-13, **14**-30
  oscillometry, **8**-6
  oxidation-reduction methods, **13**-12
  oxygen, **14**-31
  paramagnetism, **8**-6
  pH (*see* hydrogen ion, *above*)
  polarimetry, **3**-61, **8**-4
  polarography, **8**-5
  radiochemical exchange methods, **3**-56
  radiological, gas, **8**-13
  Raman spectrophotometry, **8**-4
  redox methods, **8**-5, **13**-12
  refractive index, **3**-64, **8**-4, **8**-13, **10**-22, **13**-12, **13**-101, **13**-118
  sound velocity, **8**-6, **8**-13
  for sulfur dioxide, **13**-82
  thermal conductivity, **8**-6, **8**-13, **13**-12, **13**-17
  titrators, **3**-79, **8**-5
  training course in, **13**-17

Gold, freezing point of, 2-6, 2-10
  instrument parts of, 13-21
  light reflection of, 3-64
  radiation detector use of, 3-31
  weighing of, 5-44
Goniophotometric instrument, definition of, 3-76
Goodrich tire experiments, 3-48
Grab samples, 8-2
  (See also Sample preparation)
Graduated-trough consistency measuring device, 7-50
Grain, moisture measurement of, 7-33
  processing of, brewery, 13-88
Grain size, crystalline, 13-38
Granular solids, drying of, 13-58
  moisture measurement of, 7-31
Graphic panels, 12-18, 17-71, 17-80
Gratings, optical, 3-68, 3-75, 6-32
Gravity, center of, 4-4, 5-61, 16-49
  measurement standard of, 4-2, 5-41
  specific (see Specific gravity)
  weight relationship of, 1-5
  (See also Weight)
Gravity-balance pressure-measuring elements, 4-18, 4-22
Gravity cells, electrical, 17-40
Gravity gradient, satellite stabilization effects of, 16-41
Gravity-type humidity measurements, 7-14
Gravity-type pressure gages, 4-18
Greases, calorimetric testing of, 2-23
Green liquor, density measurement of, 7-7
  controls for, 13-79
Grid-type solids-level detectors, 5-80, 15-42
Grinding machine, controls for, 6-40
Grinding mill, solids-level control of, 5-83
Gross weight, definition of, 5-43
Grounding of data processing systems, 11-63
  (See also Electric-circuit theory)
Guidance, missile, 11-28
Gum, rheogram of, 7-41
Gunsights, use of gyros in, 16-9
Gyros, rate, 16-9
  satellite stabilization sensor use of, 16-45
Gyroscope compass, 16-16
Gyroscopic instrumentation, aerospace, 16-9 to 16-22
  air-driven gyros, 16-10
  all-latitude compass, 16-16
  artificial horizons, 16-9
  automatic pilots, 16-12
  directional tyros, 16-11
  driftmeters, 16-22
  flux-gate compass, 16-16
  gyro verticals, 16-9
  gyrosyn compass, 16-16
  stable platform, 16-11
  turn-and-bank indicator, 16-9

Hagan-Poiseuille law, of viscosity, 7-43
Hair hygrometer, 7-13, 7-25
Half-life, radioisotope, 3-40, 3-55
Half-wave rectifier, 15-74, 17-34
Halogen quenched Geiger counter, 3-23
Ham boilers, controls for, 13-114
Hand hydrometer, 7-6
Harbor protection, instruments for, 3-52
Hard rubber, instrument use of, 13-21
Hardness, measurement of, 1-7
Harmonic circuits, 6-42, 15-99, 17-13
Harmonics of stroboscopic tachometer, 5-38
Hartley oscillator, 17-46
Hastelloy, instrument use of, 13-21
Haveg, instrument use of, 13-21

Hazards, radiation, 3-41, 3-18, 13-18
  (See also Combustible mixtures; Explosion protection)
Haze, measurement of, 3-12, 3-76
Head, constant, 5-55
  hydrostatic, 4-11, 5-60
Head box, controls for, 13-73
Head flow meters, 5-8, 5-14
Heading reference, aircraft, 16-11
Hearing, human, noise damage to, 3-88
  sensitivity of, 3-8, 3-83
Heat, of combustion, measurement of, 2-19 to 2-27
  flow of, effect on response time, 1-14, 1-15
  latent, in calorimetry, 2-23
  of mixing, 13-44
  specific, 1-6
  transducers for measurement of, 1-10
  (See also Temperature)
  utilization of, in crystallizers, 13-42
  of vaporization, latent, 13-44
Heat-of-absorption humidity detectors, 7-14
Heat-addition type flow meter, 5-20
Heat balance, of distillation columns, 13-43 to 13-54
  of dryers, 13-56 to 13-60
  of evaporators, 13-61 to 13-64
  of pulp digester, 13-65
Heat-balance computer, 11-19
Heat capacity, molar, 13-44
Heat exchanger, analog simulation of, 11-15
  control systems for, 10-4, 13-27 to 13-30, 13-90, 13-107 to 13-111
  control valves for, 13-28, 13-107
  crystallizer, 13-37 to 13-42
  flow measurements for, 3-56
  measurement lags in, 1-30, 1-31
  (See also Control systems)
Heat losses, distillation column, 13-44
Heat-sensitive enamels, 15-63
Heat-supply control, dryer, 13-55
Heat treating, batch, 12-41
  of glass, 12-61
  of steel sheet, 12-38
Heaters, juice, 13-98
  oil, for blast furnaces, 12-21
Heating, batch, analog computer simulation of, 11-15
  electromagnetic, 3-8
Heating cycle, blast furnace, 12-14
Heating rate, open hearth, 12-30
Heating value (see Calorific value)
Heating zone, furnace, control of, 12-35
Height, fluid, control of, 5-56
  tank, versus capacity, 5-59
Helium, detection of, 8-14
  liquid, bubble-chamber use of, 3-17
  synchro indicator test use of, 16-23
Helium atom, 3-39
Helix, filled-system thermometer use of, 2-16
Helmert's equation of gravity, 5-41
Henry, definition of, 17-7
Hercules moisture detector, 7-34
Herschel, early flow studies of, 5-5
Heterodyne frequency meter, 15-85
High boilers, 13-46
High-frequency oscillation tachometers, 5-38
High-temperature industries, 2-4
High-temperature short-time pasteurization, controls for, 13-107
High vacuum, measurement of, 4-20 to 4-22
Hinged-magnet voltmeter, 9-10
History, of calorific value measurements, 2-19
  of flow measurements, 5-5 to 5-10
  of humidity measurements, 7-13

Polymerization processes, digital computer control of, 11-28
Polymers, analysis of, 8-14
Polystyrene, insulation use of, 3-19
Polystyrene scintillator, 3-26
Porcelain, instrument parts of, 13-21
Porcelain enamel, refractive index of, 3-64
Portable electrical measuring instruments, 9-2 to 9-10, 13-12
Portable test set-up, 13-12
Position, of a body, measurement of, 1-7, 6-27 to 6-44
  counter used for measuring, 11-103
  definition of, 1-7, 6-27
  laboratory measurements of, 15-16, 15-29
  transducers for measurement of, 1-7, 6-27 to 6-39
Positioner, in combustion control system, 14-11
  pneumatic, 10-53
  valve, definition of, 10-5
Positioning-non-balance transmitter, 1-42
Positive-displacement flow meters, 5-11, 5-12
  open-hearth use of, 12-20
Potassium iodide scintillator, 3-25
Potato chip cooker, controls for, 13-106
Potential, a-c, measurement of (see Voltage measuring systems)
Potential energy, flow meter measurement of, 5-11
Potentiometer, accuracy checking of, 1-18
  capacitance-type, 15-94
  deflection-type, 1-37 to 1-41
  electrical variable measurement with, 9-2, 15-69, 15-91
  null-balance, 1-37 to 1-41, 9-2, 15-69
  position-control use of, 6-27
  self-balancing, 1-37 to 1-41
  wirewound, 17-19
Potentiometer-type remote indicators, 16-27
Potentiometric accelerometers, 5-30
Potentiometric amplifiers, 11-69
Potentiometric scales, 5-47
Potentiometric tachometers, 5-37
Poultry mixes, uniformity control of, 3-49
Pound, definition of, 5-41
Poundal, definition of, 4-2, 5-41
Pouring temperature, steel, control of, 12-29
  refractive index of, 3-64
Powder mixers, uniformity measurement of, 3-49
Powders, moisture measurement of, 7-35
Power, of aircraft, testing of, 4-4
  definition of, 3-12, 15-76, 17-8, 17-12, 17-13
  electrical, measurement of, a-c current and voltage, 9-2 to 9-10, 15-76, 17-8, 17-12
    barretters, 15-79
    bolometers, 15-79
    calorimetric-type meters, 15-81
    d-c current and voltage, 9-2 to 9-10, 17-8
    electrodynamic wattmeter, 15-77
    frequency, 9-9, 15-82 to 15-90
    instantaneous average power, 15-76
    microwave power, 15-79 to 15-81
    output power meter, 15-78
    power factor, 9-7, 17-13
    reactive voltamperes, 9-6
    synchronism, 9-9
    thermal converter, 9-6, 15-81
    thermistors, 15-79
  of engines, testing of, 4-7
  resolving, 1-16
Power amplifiers, 6-37
Power distribution, controls for, 14-45 to 14-67
Power generation, digital computer controls for, 11-28, 11-48
  measurements for, 1-28, 9-2, 14-58

Power-line noise, 11-82, 14-63
Power reactors, nuclear, 14-68 to 14-92
Power supplies, d-c, 17-33
  panelboard, 17-73
  rectifier, 17-38
Preact control, definition of, 10-7
Preamplifiers, data-processing systems, 11-68 to 11-72
  (See also Amplifiers)
Precession of gyros (see Gyroscopic instrumentation)
Precision (see Special Classified Index)
Predetermined-count meter, 11-116
Preheaters, blast furnace, controls for, 12-14
  open hearth, 12-25
Preparation of analytical samples, 2-22, 7-31, 8-7, 12-23, 12-56
Presentation of data, 11-58, 11-91 to 11-97
  (See also Recorders)
Press, cut-off, position controls for, 6-28
  hydraulic, pressure-measurement of, 4-12
  pellet, use in calorimetry, 2-22
  photometric safeguards for, 10-45
  platen, controls for, 10-32
Press roll, paper machine, controls for, 13-75
Pressure, absolute, definition of, 4-10
  (See also Absolute pressure)
  ambient effects of (see Classified Index—"Ambient Influences and Accuracy Deterrents")
  atmospheric, definition of, 4-10
  calorimetric standard of, 2-23
  change with altitude, 4-13, 15-61, 16-5
  conversion factors for, 4-12
  differential, definition of, 4-10
  gage, definition of, 4-10
  hydrostatic, definition of, 4-11
  impact, definition of, 4-11
  line, definition of, 4-11
  of solar radiation, 16-41
  sound, 3-84
  standards of, 4-14
  static, definition of, 4-11
  tons-on-ram, 4-12
  total, definition of, 4-11
  units of, 4-12
  vacuum, definition of, 4-11
  velocity, definition of, 4-11
  viscosity changes caused by, 7-44
Pressure-control systems, aircraft, 16-24, 16-37
  brewery, 13-86 to 13-94
  canning retort, 13-95
  cereal cooker, 13-87
  Chemipulp K-C hot-acid accumulator, 13-83
  combustion systems, 14-5 to 14-24
  distillation column, 13-45 to 13-54
  dryers, 13-56, 13-75
  evaporators, 13-62, 13-77, 13-101
  filters, 13-31 to 13-36
  furnaces, 12-24, 12-38, 12-50
  (See also Combustion control systems)
  heat-exchangers, 13-29
  hydraulic ram, 4-12
  laboratory applications, 15-10 to 15-12, 15-23 to 15-25
  paper calender, 13-76
  pasteurizer, 13-107
  pilot-plant applications, 15-41
  process safety, 4-13
  pulp digester, 13-66
  retort, 13-95
  self-operated, 10-16, 10-46
  soaking pit, 12-34
  vacuum pan, 13-111
Pressure drop, control-valve, 13-28

Classified Index

AT-A-GLANCE GUIDE TO INSTRUMENT
SELECTION INFORMATION

Accuracy – Sensitivity – Range

EQUATIONS AND FORMULAS

Classified Index

AMBIENT INFLUENCES AND ACCURACY DETERRENTS